Physics of
Magnetic Flux Ropes

Geophysical Monograph Series

Including

IUGG Volumes
Maurice Ewing Volumes
Mineral Physics Volumes

GEOPHYSICAL MONOGRAPH SERIES

Geophysical Monograph Volumes

1 **Antarctica in the International Geophysical Year** *A. P. Crary, L. M. Gould, E. O. Hulburt, Hugh Odishaw, and Waldo E. Smith (Eds.)*
2 **Geophysics and the IGY** *Hugh Odishaw and Stanley Ruttenberg (Eds.)*
3 **Atmospheric Chemistry of Chlorine and Sulfur Compounds** *James P. Lodge, Jr. (Ed.)*
4 **Contemporary Geodesy** *Charles A. Whitten and Kenneth H. Drummond (Eds.)*
5 **Physics of Precipitation** *Helmut Weickmann (Ed.)*
6 **The Crust of the Pacific Basin** *Gordon A. Macdonald and Hisahi Kuno (Eds.)*
7 **Antarctica Research: The Matthew Fontaine Maury Memorial Symposium** *H. Wexler, M. J. Rubin, and J. E. Caskey, Jr. (Eds.)*
8 **Terrestrial Heat Flow** *William H. K. Lee (Ed.)*
9 **Gravity Anomalies: Unsurveyed Areas** *Hyman Orlin (Ed.)*
10 **The Earth Beneath the Continents: A Volume of Geophysical Studies in Honor of Merle A. Tuve** *John S. Steinhart and T. Jefferson Smith (Eds.)*
11 **Isotope Techniques in the Hydrologic Cycle** *Glenn E. Stout (Ed.)*
12 **The Crust and Upper Mantle of the Pacific Area** *Leon Knopoff, Charles L. Drake, and Pembroke J. Hart (Eds.)*
13 **The Earth's Crust and Upper Mantle** *Pembroke J. Hart (Ed.)*
14 **The Structure and Physical Properties of the Earth's Crust** *John G. Heacock (Ed.)*
15 **The Use of Artificial Satellites for Geodesy** *Soren W. Henricksen, Armando Mancini, and Bernard H. Chovitz (Eds.)*
16 **Flow and Fracture of Rocks** *H. C. Heard, I. Y. Borg, N. L. Carter, and C. B. Raleigh (Eds.)*
17 **Man-Made Lakes: Their Problems and Environmental Effects** *William C. Ackermann, Gilbert F. White, and E. B. Worthington (Eds.)*
18 **The Upper Atmosphere in Motion: A Selection of Papers With Annotation** *C. O. Hines and Colleagues*
19 **The Geophysics of the Pacific Ocean Basin and Its Margin: A Volume in Honor of George P. Woollard** *George H. Sutton, Murli H. Manghnani, and Ralph Moberly (Eds.)*
20 **The Earth's Crust: Its Nature and Physical Properties** *John G. Heacock (Ed.)*
21 **Quantitative Modeling of Magnetospheric Processes** *W. P. Olson (Ed.)*
22 **Derivation, Meaning, and Use of Geomagnetic Indices** *P. N. Mayaud*
23 **The Tectonic and Geologic Evolution of Southeast Asian Seas and Islands** *Dennis E. Hayes (Ed.)*
24 **Mechanical Behavior of Crustal Rocks: The Handin Volume** *N. L. Carter, M. Friedman, J. M. Logan, and D. W. Stearns (Eds.)*
25 **Physics of Auroral Arc Formation** *S.-I. Akasofu and J. R. Kan (Eds.)*
26 **Heterogeneous Atmospheric Chemistry** *David R. Schryer (Ed.)*
27 **The Tectonic and Geologic Evolution of Southeast Asian Seas and Islands: Part 2** *Dennis E. Hayes (Ed.)*
28 **Magnetospheric Currents** *Thomas A. Potemra (Ed.)*
29 **Climate Processes and Climate Sensitivity (Maurice Ewing Volume 5)** *James E. Hansen and Taro Takahashi (Eds.)*
30 **Magnetic Reconnection in Space and Laboratory Plasmas** *Edward W. Hones, Jr. (Ed.)*
31 **Point Defects in Minerals (Mineral Physics Volume 1)** *Robert N. Schock (Ed.)*
32 **The Carbon Cycle and Atmospheric CO_2: Natural Variations Archean to Present** *E. T. Sundquist and W. S. Broecker (Eds.)*
33 **Greenland Ice Core: Geophysics, Geochemistry, and the Environment** *C. C. Langway, Jr., H. Oeschger, and W. Dansgaard (Eds.).*
34 **Collisionless Shocks in the Heliosphere: A Tutorial Review** *Robert G. Stone and Bruce T. Tsurutani (Eds.)*
35 **Collisionless Shocks in the Heliosphere: Reviews of Current Research** *Bruce T. Tsurutani and Robert G. Stone (Eds.)*
36 **Mineral and Rock Deformation: Laboratory Studies—The Paterson Volume** *B. E. Hobbs and H. C. Heard (Eds.)*
37 **Earthquake Source Mechanics (Maurice Ewing Volume 6)** *Shamita Das, John Boatwright, and Christopher H. Scholz (Eds.)*
38 **Ion Acceleration in the Magnetosphere and Ionosphere** *Tom Chang (Ed.)*
39 **High Pressure Research in Mineral Physics (Mineral Physics Volume 2)** *Murli H. Manghnani and Yasuhiko Syono (Eds.)*

40 **Gondwana Six: Structure, Tectonics, and Geophysics** *Garry D. McKenzie (Ed.)*

41 **Gondwana Six: Stratigraphy, Sedimentology, and Paleontology** *Garry D. McKenzie (Ed.)*

42 **Flow and Transport Through Unsaturated Fractured Rock** *Daniel D. Evans and Thomas J. Nicholson (Eds.)*

43 **Seamounts, Islands, and Atolls** *Barbara H. Keating, Patricia Fryer, Rodey Batiza, and George W. Boehlert (Eds.)*

44 **Modeling Magnetospheric Plasma** *T. E. Moore and J. H. Waite, Jr. (Eds.)*

45 **Perovskite: A Structure of Great Interest to Geophysics and Materials Science** *Alexandra Navrotsky and Donald J. Weidner (Eds.)*

46 **Structure and Dynamics of Earth's Deep Interior (IUGG Volume 1)** *D. E. Smylie and Raymond Hide (Eds.)*

47 **Hydrological Regimes and Their Subsurface Thermal Effects (IUGG Volume 2)** *Alan E. Beck, Grant Garven, and Lajos Stegena (Eds.)*

48 **Origin and Evolution of Sedimentary Basins and Their Energy and Mineral Resources (IUGG Volume 3)** *Raymond A. Price (Ed.)*

49 **Slow Deformation and Transmission of Stress in the Earth (IUGG Volume 4)** *Steven C. Cohen and Petr Vaníček (Eds.)*

50 **Deep Structure and Past Kinematics of Accreted Terranes (IUGG Volume 5)** *John W. Hillhouse (Ed.)*

51 **Properties and Processes of Earth's Lower Crust (IUGG Volume 6)** *Robert F. Mereu, Stephan Mueller, and David M. Fountain (Eds.)*

52 **Understanding Climate Change (IUGG Volume 7)** *Andre L. Berger, Robert E. Dickinson, and J. Kidson (Eds.)*

53 **Plasma Waves and Istabilities at Comets and in Magnetospheres** *Bruce T. Tsurutani and Hiroshi Oya (Eds.)*

54 **Solar System Plasma Physics** *J. H. Waite, Jr., J. L. Burch, and R. L. Moore (Eds.)*

55 **Aspects of Climate Variability in the Pacific and Western Americas** *David H. Peterson (Ed.)*

56 **The Brittle-Ductile Transition in Rocks** *A. G. Duba, W. B. Durham, J. W. Handin, and H. F. Wang (Eds).*

Maurice Ewing Volumes

1 **Island Arcs, Deep Sea Trenches, and Back-Arc Basins** *Manik Talwani and Walter C. Pitman III (Eds.)*

2 **Deep Drilling Results in the Atlantic Ocean: Ocean Crust** *Manik Talwani, Christopher G. Harrison, and Dennis E. Hayes (Eds.)*

3 **Deep Drilling Results in the Atlantic Ocean: Continental Margins and Paleoenvironment** *Manik Talwani, William Hay, and William B. F. Ryan (Eds.)*

4 **Earthquake Prediction—An International Review** *David W. Simpson and Paul G. Richards (Eds.)*

5 **Climate Processes and Climate Sensitivity** *James E. Hansen and Taro Takahashi (Eds.)*

6 **Earthquake Source Mechanics** *Shamita Das, John Boatwright, and Christopher H. Scholz (Eds.)*

IUGG Volumes

1 **Structure and Dynamics of Earth's Deep Interior** *D. E. Smylie and Raymond Hide (Eds.)*

2 **Hydrological Regimes and Their Subsurface Thermal Effects** *Alan E. Beck, Grant Garven, and Lajos Stegena (Eds.)*

3 **Origin and Evolution of Sedimentary Basins and Their Energy and Mineral Resources** *Raymond A. Price (Ed.)*

4 **Slow Deformation and Transmission of Stress in the Earth** *Steven C. Cohen and Petr Vaníček (Eds.)*

5 **Deep Structure and Past Kinematics of Accreted Terranes** *John W. Hillhouse (Ed.)*

6 **Properties and Processes of Earth's Lower Crust** *Robert F. Mereu, Stephan Mueller, and David M. Fountain (Eds.)*

7 **Understanding Climate Change** *Andre L. Berger, Robert E. Dickinson, and J. Kidson (Eds.)*

8 **Evolution of Mid Ocean Ridges** *John M. Sinton (Ed.)*

Mineral Physics Volumes

1 **Point Defects in Minerals** *Robert N. Schock (Ed.)*

2 **High Pressure Research in Mineral Physics** *Murli H. Manghnani and Yasuhiko Syono (Eds.)*

Geophysical Monograph 58

Physics of Magnetic Flux Ropes

C. T. Russell
E. R. Priest
L. C. Lee
Editors

American Geophysical Union

Published under the aegis of the AGU Geophysical Monograph Board.

Library of Congress Cataloging-in-Publication Data

Physics of magnetic flux.

(Geophysical monography ; no. 58)
Based on papers presented at the American Geophysical Union Chapman conference on the Physics of Magnetic Flux Ropes was held in Hamilton, Bermuda on Mar. 27-31, 1989.
1. Solar photosphere. 2. Magnetic flux. 3. Astrophysics.
I. Russell, C. T. II. American Geophysical Union. III. American Geophysical Union Chapman Conference on the Physics of Magnetic Flux Ropes (1989 : Hamilton, Bermuda) IV. Series.
QB528.P48 1990 523.7'4 90-745
ISBN 0-87590-026-7

Printed in the United States of America.

CONTENTS

PART 1: Structure, Waves and Instabilities

REVIEW PAPER

RESEARCH REPORTS

PART 3: Structure and Heating of Coronal Loops

TUTORIAL PAPER

REVIEW PAPERS

RESEARCH REPORTS

PART 7: The Magnetopause

TUTORIAL PAPER

REVIEW PAPERS

RESEARCH REPORTS

PART 8: Magnetospheric Field Aligned Currents and Flux Tubes

TUTORIAL PAPER

REVIEW PAPER

RESEARCH REPORTS

PART 9: The Magnetotail

REVIEW PAPER

RESEARCH REPORTS

PREFACE

The American Geophysical Union Chapman Conference on the Physics of Magnetic Flux Ropes was held at the Hamilton Princess Hotel, Hamilton, Bermuda on March 27-31, 1989. Topics discussed ranged from solar flux ropes, such as photospheric flux tubes, coronal loops and prominences, to flux ropes in the solar wind, in planetary ionospheres, at the Earth's magnetopause, in the geomagnetic tail and deep in the Earth's magnetosphere. Papers presented at that conference form the nucleus of this book, but the book is more than just a proceedings of the conference. We have solicited articles from all interested in this topic. Thus, there is some material in the book not discussed at the conference. Even in the case of papers presented at the conference, there is generally a much more detailed and rigorous presentation than was possible in the time allowed by the oral and poster presentations.

The conference consisted of tutorial presentations, review papers, research reports and discussion. The research reports were presented as either oral or poster presentations. We have attempted to preserve this structure in the book which contains the above three different types of paper, together with many of the questions, answers and comments that took place during the discussion periods following the oral presentations. To capture the discussion, questioners were asked to write out their questions and the presenter was then given the opportunity to prepare a written answer.

The book has been divided into parts. In Part 1, we present the papers treating the basic processes occurring in magnetic flux ropes such as: the equilibrium of magnetic flux ropes by E. R. Priest, MHD waves on flux ropes by J. V. Hollweg, ideal instabilities by G. Einaudi, resistive instabilities by K. Schindler and A. Otto and reconnection by B. U. Ö. Sonnerup and colleagues. A paper by H. Zirin on solar flares has been included here because of the basic role flares play in the solar system.

Part 2, treats the properties and structure of photospheric flux tubes. The thin tube approximation is discussed by B. Roberts as is the fine structure of the photospheric field by J. H. Thomas. Dynamical effects, energy transport, current generation, and waves are all examined.

Part 3, examines the structure and heating of coronal loops. R. Rosner provides a tutorial overview of coronal loops. E. N. Parker reviews the spontaneous appearance of discontinuities and the resulting dissipation while S. K. Antiochos reviews the structure and flow in the corona. Numerous research reports on the physics of these structures are also reported, including reconnection, quasi-static evolution, and heating.

Part 4, opens with a tutorial by T. G. Forbes on the properties of solar prominences. This paper is followed by reviews of the structure and stability of prominences by U. Anzer and the cooling and condensation of filaments by G. Van Hoven. Research reports cover fibril structure, equilibria, eruptions and helical flux ropes.

Part 5, changes scales to examine those larger structures which link the Sun and the Earth's magnetosphere: coronal mass ejections and magnetic clouds. J. T. Gosling presents the energetic particle evidence pertaining to the topological structure of coronal mass ejections while N. U. Crooker and L. F. Burlaga and colleagues discuss magnetic constraints on these structures. Also treated in this chapter are MHD simulations of CME's, energetic particles in CME's and slow shocks at CME's.

Part 6, in contrast, treats the smallest scale flux ropes: those in the Venus ionosphere which are only a few kilometers across. J. G. Luhmann provides an introduction to the solar wind interaction with unmagnetized planets while C. T. Russell reviews the observed properties of ionospheric flux ropes. J. G. Luhmann adds an analysis of the properties of these ropes and E. Belova and L. Zelenyi provide a model for their formation.

Part 7, takes us to the Earth's magnetopause where reconnection leads to what appears to be rope-like structures. The chapter opens with a brief review of the magnetopause followed by reviews by R. C. Elphic of the phenomenon known as the flux transfer event and of their theory by M. M. Kuznetsova and L. M. Zelenyi. The various reports in this chapter concern the structure of these flux tubes and possible mechanisms for their generation.

Part 8, moves us into the magnetosphere to examine field-aligned current systems with a tutorial paper by G. Haerendel and a review of satellite observations of the fine-scale structure of auroral field-aligned current systems by E. M. Dubinin. The following reports treat ionospheric observations of possible flux tubes and the response of the magnetosphere-ionosphere system to solar wind dynamic pressure fluctuations which might mimic the expected signature of magnetic flux tubes being dragged through the ionosphere.

The book closes with a discussion of the magnetotail which itself could be considered to be two giant flux ropes anchored in the earth. W. Baumjohann and G. Haerendel discuss the substorm phenomenon and the formation of flux rope structures. Research reports cover the evidence for flux ropes and the structure of plasmoids which are created during substorms.

The organizers wish to thank the many people who helped make the conference and this book possible. We are especially appreciative of the referees who spent much time poring over these papers and at times caused the authors to re-evaluate their hypotheses. These referees included: T. Amari, R. R. Anderson, S. K. Antiochos, U. Anzer, D. N. Baker, M. A. Berger, J. Birn, J. Brackbill, P. K. Browning, L. F. Burlaga, C. R. Clauer, S. W. H. Cowley, T. E. Cravens, N. U. Crooker, P. Demoulin, D. Q. Ding, J. Drake, T. E. Eastman, P. Edwin, G. Einaudi, R. C. Elphic, D. H. Fairfield, J. A. Fedder, T. G. Forbes, G. E. Francis, V. Gaizauskas, C. K. Goertz, D. J. Gorney, J. T. Gosling, R. A. Greenwald, R. Harrison, J. V. Hollweg, A. Hood, W. J. Hughes, A. J. Hundhausen, S. W. Kahler, J. R. Kan, M. G. Kivelson, L. C. Lee, M. Lockwood, B. C. Low, J. G. Luhmann, R. L. Lysak, W. Matthaeus, M. McHenry, K. L. Miller, E. Neilson, A. Nishida, N. Omidi, E. N. Parker, V. Pizzo, E. R. Priest, P. Pritchett, P. H. Reiff, B. Roberts, R. Rosner, M. P. Ryutova, M. Schussler, Y. Shi, D. G. Sibeck, S. K. Solanki, Y. Song, B. U. Ö. Sonnerup, D. J. Southwood, H. Spence, R. S. Steinolfson, D. W. Swift, M. Temerin, J. H. Thomas, B. T. Tsurutani, A. A. Ballegooijen, G. Van Hoven, W. J. Wagner, R. J. Walker, D. F. Webb, Q. C. Wei, D. R. Weimer and R. Wolfson. We would also like to thank the AGU meetings staff, Brenda Weaver and Patrice Dickerson, for their able assistance at the conference, and the AGU publications staff, Patricia Rayner and Lathifah Jocum, for their timely handling of the publication of the book. Finally we are extremely grateful for the assistance of Sarah Suk at UCLA who handled all the correspondence with the attendees at the conference, the UCLA correspondence with AGU and authors, and any necessary retyping of manuscripts. M. Ishiwata is gratefully acknowledged for her redrafting of many figures to improve their readability. All this assistance made our job much easier.

C. T. Russell
Institute of Geophysics
and Planetary Physics
University of California
Los Angeles, CA 90024

E. R. Priest
Applied Mathematics Division
University of St. Andrews
North Haugh
St. Andrews KY16 9SS
Fife, Scotland

L. C. Lee
Geophysical Institute
University of Alaska
Fairbanks, AK 99775-0800

March 1990

Physics of
Magnetic Flux Ropes

THE EQUILIBRIUM OF MAGNETIC FLUX ROPES (TUTORIAL LECTURE)

E. R. Priest
Mathematical Sciences Department, The University, St. Andrews KY 16 9SS, Scotland

Abstract The building blocks of magnetic configurations are magnetic flux tubes and, as an introduction to the theme of this conference, I shall review the theories for their equilibrium. Straight tubes are described by the variation with radius of pressure, axial magnetic field and azimuthal magnetic field or twist, two of which are determined by the radial force balance equation and the defining equation for twist, but a wide variety of models arise from prescribing the other two variables. Linear force-free tubes are of particular interest. The effects of expanding or compressing a tube are discussed, and tubes with varying cross-section are described in some detail. For thick tubes confined by an external pressure, one has a complicated nonlinear integro-partial differential equation to solve with a free boundary. Progress has been made analytically by expanding about a thin tube or twisting a straight tube slightly, the results showing the creation of an intense core with a strong axial field. Numerical solutions of the nonlinear equation show also that the boundary expands to give a shell of weak field and high twist. Curved tubes may be modelled by means of large-aspect-ratio expansions. In the solar photosphere the effect of gravity is to make horizontal tubes rise by magnetic buoyancy and vertical tubes to spread out with height. In the corona there are many types of coronal loop, and the field around a prominence is probably a large twisted flux rope.

1. Introduction

Throughout the universe magnetic fields are interacting with plasma in a variety of subtle and complex ways. Indeed, much of the plasma structure we see is created directly by the magnetic field in the form of flux tubes. In the solar system we are able to observe them over a wide range of parameter regimes, and the localised in-situ magnetospheric observations from spacecraft complement the global remote-sensing observations of the solar atmosphere. So I am delighted that our two communities have come together at this meeting to learn from each other.

Discrete flux tubes may be present in flux transfer events, the Venus ionosphere, the geomagnetic tail and magnetic clouds. They are also found in many different forms on the Sun. Sunspots are present where a large 3000 Gauss flux tube breaks through the surface, typically in pairs with the tube coming up through one spot and going back down through the other. Also a close-up of a spot (of typical diameter 20 Mm) is suggestive that it may itself consist of many smaller tubes (Figure 1). In the chromosphere above a sunspot pair one sees fibril structures joining one spot to the other and presumably outlining the magnetic field. In photospheric magnetic field maps one finds that, outside the active regions surrounding sunspot groups, the solar surface is covered with a fragmentary network structure consisting of many tiny flux tubes at the boundaries of large convection cells (the supergranulation). These tubes are unresolved but are believed to be a few hundred kilometres across and to have field strengths of about 1500 Gauss.

The solar corona is seen in soft X-rays (Figure 2) to consist of myriads of flux loops, both outside active regions, joining active regions and also within an active region (Figure 3). Hot flux loops (~ 10^7K) may also be created by large solar flares which subsequently cool to give an arcade of cool loops joining the two ribbons which make up the flare in its main phase in the chromosphere. Solar prominences are huge vertical sheets of plasma up in the corona but with a density a factor of a hundred higher and a temperature a factor of a hundred lower than the surrounding coronal plasma.

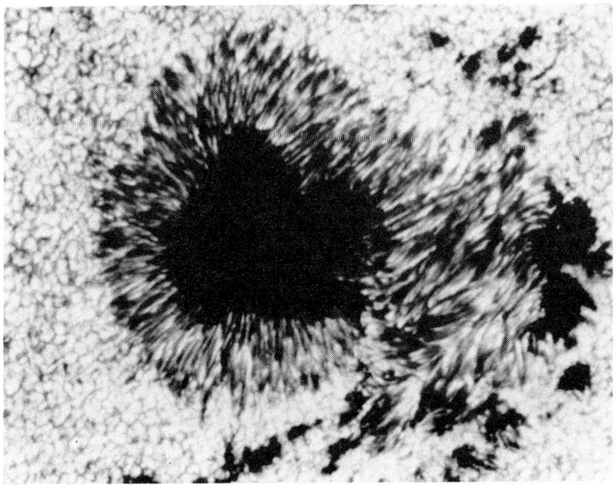

Fig. 1. A close-up of a sunspot (R. Muller)

Geophysical Monograph 58

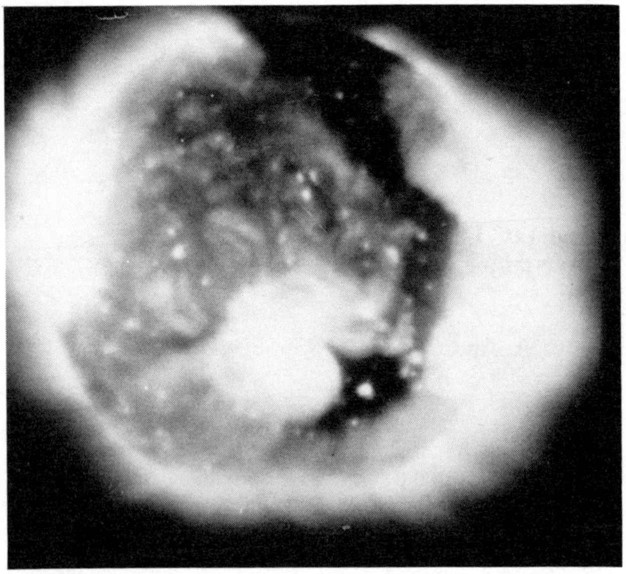

Fig.2. The solar corona in soft X-rays (American Science & Engineering)

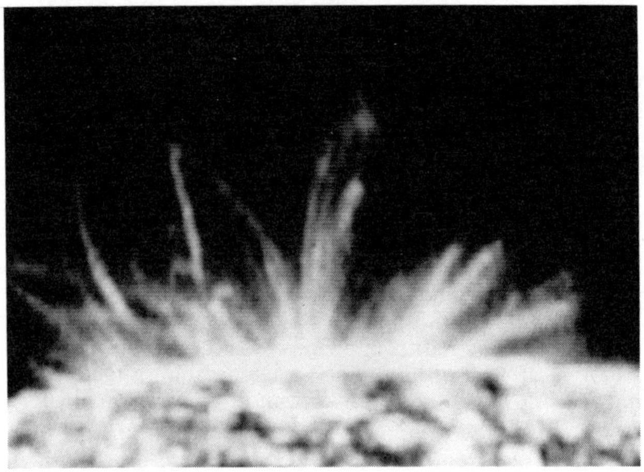

Fig. 3. Loops above an active region

Amazingly, prominences remain in stable equilibrium for months and reach down to the surface in great tree-trunk like structures (Figure 4). Occasionally, they lose equilibrium and erupt outwards when they undergo a metamorphis and take on the appearance of a large twisted flux tube (Figure 5). Ahead of the prominence one finds an expanding bubble known as a coronal mass ejection (Figure 6).

Having seen how beautiful the appearance of flux tubes may be, how do we model them mathematically? Let us start with a few definitions. (Further details of many topics in this tutorial can be found in Parker (1979) and Priest (1982).) A *magnetic field line* is everywhere parallel to the magnetic field $\underline{B}$ and is a solution of

$$\frac{dx}{B_x} = \frac{dy}{B_y} = \frac{dz}{B_z} .$$

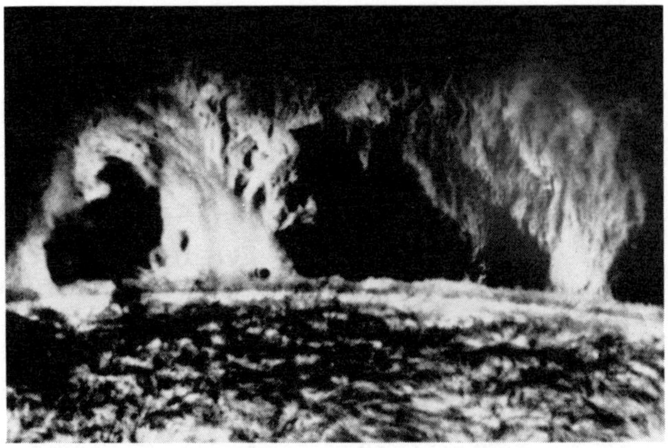

Fig. 4. A solar prominence (Big Bear Solar Observatory)

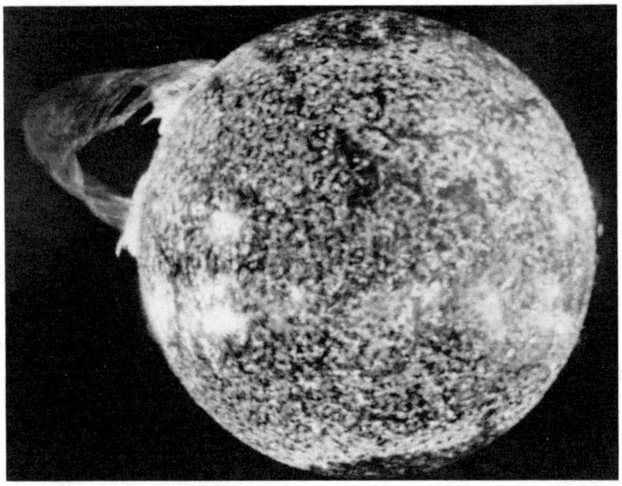

Fig. 5. An erupting prominence (US Naval Research Laboratory)

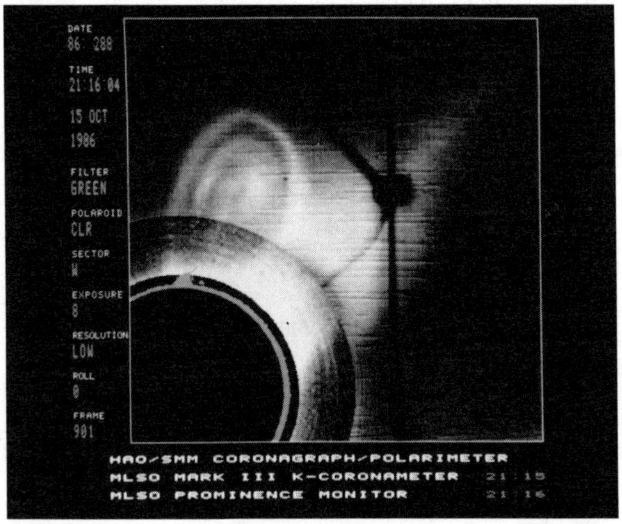

Fig. 6. A coronal mass ejection (High Altitude Observatory)

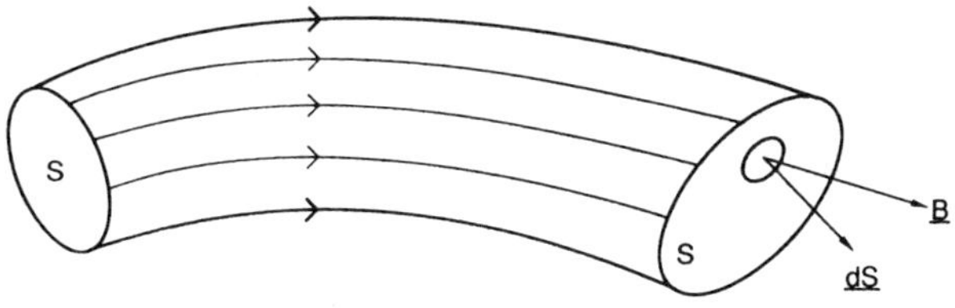

Fig. 7. A curved flux tube

A *magnetic flux tube* (Fig 7) is then the volume enclosed by the set of field lines which intersect a simple closed curve. (Some define it as the curved surface produced by the field lines.) A *flux rope* is a twisted flux tube.

The *strength* F of a flux tube is the flux crossing a section S, namely

$$F = \int_S \underline{B}.d\underline{S} = \int B_n dS = \bar{B}_n A, \qquad (1)$$

where $A = \int dS$ is the cross-sectional area and $B_n = \int B_n dS/A$ is the mean axial field. $d\underline{S}$ is measured in the same sense as $\underline{B}$ and so an integration of the equation div $\underline{B} = 0$ over the volume of a flux tube bounded by two sections S_1 and S_2 gives (by Gauss's divergence theorem)

$$O = \int \text{div}\, \underline{B}\, dV = -\int_{S_1} \underline{B}.d\underline{S} + \int_{S_2} \underline{B}.d\underline{S},$$

since the contribution from the curved surface vanishes. In other words the strength remains constant along the tube. This is a natural consequence of div $\underline{B} = 0$ which implies that there are no magnetic monopoles (nosources or sinks of magnetic flux) and so the flux entering through one end of the tube must be equal to the flux leaving through the other. Thus from (1), B_n is inversely proportional to A so that if a flux tube narrows the mean field B_n must increase and vice versa.

Flux tubes have several physical effects. They store magnetic energy ($\int B^2/(2\mu)dV$) and they may act as a channel for a flow of fast particles, heat and plasma. They are often treated as isolated structures in a passive medium, but this can be highly misleading, since they can interact with their surroundings, both through surface pressure forces and also through reconnections which may exchange mass, momentum, energy and a topological quantity known as *magnetic helicity*

$$H = \int_V \underline{A}.\underline{B}\, dV, \qquad (2)$$

where

$$\underline{B} = \underline{\nabla} \times \underline{A},$$

in terms of the vector potential $\underline{A}$. H is gauge-invariant if the volume V is bounded by a simple connected magnetic surface S, since the addition to $\underline{A}$ of a function $\underline{\nabla} g$ which would not change $\underline{B}$ simply increases H by an amount

$$\int_V \underline{\nabla} g.\underline{B}\, dV = \int_V \underline{\nabla}.(g\underline{B}) = \int_S \underline{n}.g\underline{B}\, dS.$$

Here we have used the equation $\underline{\nabla}.\underline{B} = 0$ and Gauss's divergence theorem, and the latter term vanishes since $\underline{n}.\underline{B} = 0$ on S. If the boundary is not a magnetic surface a related gauge-invariant quantity, called relative magnetic helicity, may be used instead (Berger and Field, 1984). Magnetic helicity is a measure of the twist and linkage of magnetic field lines. For example, for two linked flux ropes having twists Φ_1 and Φ_2 and magnetic fluxes F_1 and F_2, it may be written

$$H = \Phi_1 F_1^2 + \Phi_2 F_2^2 + 2LF_1F_2,$$

where L is the linking number, which is a measure of the number of links between the two tubes (Figure 8). Magnetic reconnection can convert linkage helicity to twist helicity, but it preserves the total helicity.

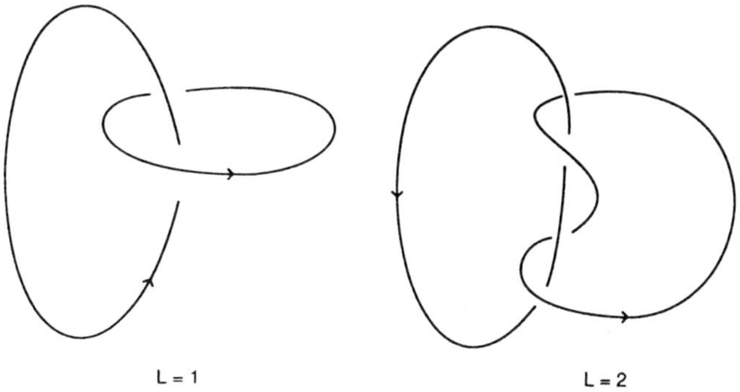

Fig. 8. Two linked flux ropes

The MHD equation for the motion of plasma under the action of a plasma pressure gradient, a magnetic force ($\underline{j} \times \underline{B}$) and gravity is

$$\rho \frac{D\underline{v}}{Dt} = -\underline{\nabla} p + \underline{j} \times \underline{B} + \rho \underline{g}, \qquad (3)$$

where

$$\underline{j} = \underline{\nabla} \times \underline{B}/\mu. \qquad (4)$$

In terms of a typical plasma speed (v_0), density (ρ_0), pressure (p_0), field strength (B_0) and length-scale (L) for plasma and magnetic variations, the orders of magnitude of the terms in (3) are

$$\frac{\rho_0 v_0^2}{L},\ \frac{p_0}{L},\ \frac{B_0^2}{\mu L},\ \rho_0 g,$$

respectively. Thus in a situation where the other terms do not dominate the magnetic force, the first term is negligible if

$$v_0^2 << \frac{B_0^2}{\mu\rho} \equiv v_A^2.$$

In other words, we have a magnetohydrostatic force balance if the flow is much slower than the Alfvén speed (v_A). If in addition

$$L << \frac{B_0^2}{\mu \rho_0 g} \equiv \frac{2H}{\beta},$$

where $H = p_0/(\rho_0 g)$ is the *pressure scale height* and $\beta = 2\mu p_0/B_0^2$ is the *plasma beta* (the ratio of plasma pressure (p_0) to magnetic pressure ($B_0^2/(2\mu)$)), then the gravity term is much smaller than the magnetic term and we are left with a magnetostatic balance

$$\underline{0} = -\underline{\nabla} p + \underline{j} \times \underline{B}. \qquad (5)$$

Finally, the pressure gradient term is negligible if $2\beta << 1$ and the equation reduces to

$$\underline{0} = \underline{j} \times \underline{B} \qquad (6)$$

for *force-free* equilibrium. By using Ampère's law (4) the magnetic force may be written

$$\underline{j} \times \underline{B} = (\underline{B}.\underline{\nabla})\frac{\underline{B}}{\mu} - \underline{\nabla}\left(\frac{B^2}{2\mu}\right),$$

in which the second term is the *magnetic pressure gradient* (which acts from regions of high to low magnetic pressure) and the first term is the *magnetic tension force* (which acts along the inwards normal $\underline{n}$ when a field line is curved): it may be rewritten $B^2/(\mu \mathcal{R})\underline{n}$ in terms of the radius of curvature of the field line. It is a common fault in physical descriptions to overlook the effect of magnetic pressure while including that of the magnetic tension.

2. Straight Tubes

2.1 Basic Solution

Adopting cylindrical polar coordinates (r,θ,z) for a straight cylindrically symmetric tube of length L, the magnetic field and electric current components are (Fig 9)

$$(B_r, B_\theta, B_z) = (0, B_\theta(r), B_z(r)),$$

and from $\underline{j} = \text{curl}\, \underline{B}/\mu$

$$(j_r, j_\theta, j_z) = \left(0, -\frac{1}{\mu}\frac{dB_z}{dr}, \frac{1}{\mu r}\frac{d}{dr}(rB_\theta)\right).$$

Thus the magnetic field lines lie on cylindrical surfaces and the angle Φ through which a field line is twisted about the tube axis in going from one end to the other is

$$\Phi = \frac{LB_\theta}{rB_z}. \qquad (7)$$

This expression for the *twist* Φ may be derived by imagining the cylindrical surface of radius r is flattened out to give a series of straight parallel field lines which travel a horizontal distance Φr as they go a distance L along the tube, so that the equation

$$\frac{\Phi r}{L} = \frac{B_\theta}{B_z}$$

gives two equal expressions for the arctangent of the angle of inclination (Θ) to the axis of the field lines. Θ is known as the *pitch angle* such that

$$\tan \Theta = \frac{B_\theta}{B_z},$$

and by comparison the *pitch* is the distance along the tube for one revolution, namely

$$\frac{2\pi}{\Phi} L = 2\pi r \frac{B_z}{B_\theta}.$$

In general the twist Φ varies with r and so field lines on different flux surfaces have different inclinations (Figure 10).

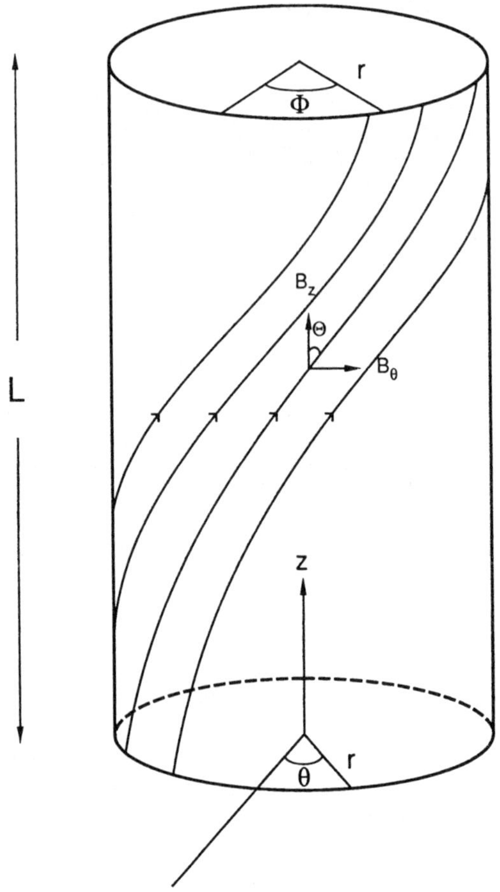

Fig. 9. Notation for a straight tube

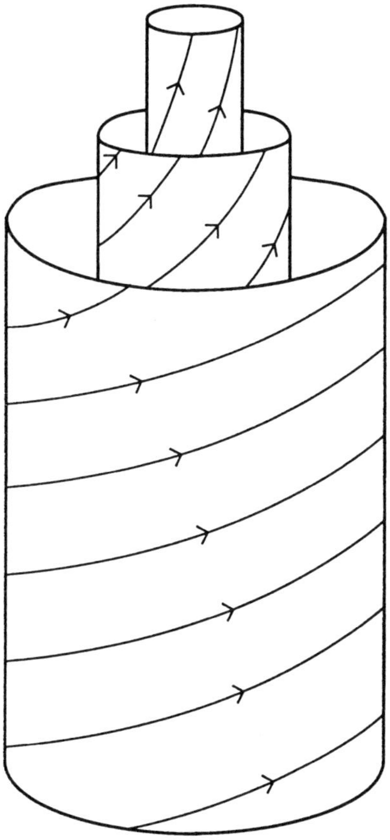

Fig. 10. Field lines on different flux surfaces

In this cylindrically symmetric geometry the force balance equation (5) becomes

$$\frac{dp}{dr} + \frac{d}{dr}\left(\frac{B_\theta^2 + B_z^2}{2\mu}\right) + \frac{B_\theta^2}{\mu r} = 0, \qquad (8)$$

with a balance of three terms representing a plasma pressure gradient, a magnetic pressure gradient and a magnetic tension force, respectively. The first two are directed outwards when the plasma pressure and magnetic pressure decrease with r and the third term is always directed inwards. In general, you may prescribe the functional forms for any two of p(r), $B_\theta(r)$, $B_z(r)$, $\Phi(r)$ and can then use (7) and (8) to deduce the other two, but which two you prescribe depends crucially on the physics of the problem you are modelling. Some examples are as follows.

For a purely axial field (known as a θ-pinch in the laboratory) B_θ and Φ vanish, so that (8) may be integrated to give

$$p = \text{constant} - \frac{B_z^2}{2\mu},$$

and so the plasma pressure is enhanced in regions where the field is reduced. Another example is that of a purely azimuthal field B_θ (known as a linear or z-pinch) for which (4) implies

$$\frac{1}{\mu r}\frac{d}{dr}(rB_\theta) = j_z. \qquad (9)$$

Thus, for instance, a cylinder of uniform current of radius a (Figure 11) has

$$j_z = \begin{cases} j_0, & r < a, \\ 0, & r > a, \end{cases}$$

and so (9) may be integrated to give

$$B_\theta = \begin{cases} \frac{1}{2}\mu j_0 r, & r < a, \\ \dfrac{\frac{1}{2}\mu a^2 j_0}{r}, & r > a, \end{cases} \qquad (10)$$

where continuity of B_θ has been assumed at r=a.

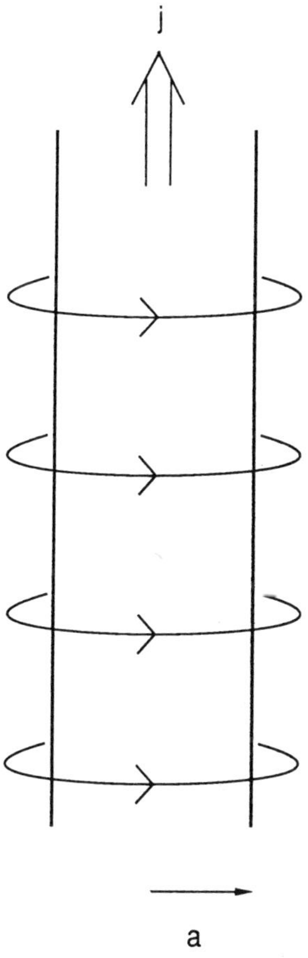

Fig. 11. A z-pinch

Furthermore, (8) may be integrated with $B_z = 0$ to give the pressure (also assumed continuous at r=a) as

$$p = \begin{cases} p_0 + \frac{1}{4}\mu j_0^2(a^2-r^2), & r < a, \\ p_0, & r > a. \end{cases}$$

Thus the magnetic field increases linearly from zero and then falls off as r^{-1}, while the plasma pressure decreases quadratically and then remains constant. Inside the cylinder of radius r=a, both the magnetic pressure and tension forces act inwards and so are balanced by an outwards plasma pressure gradient, whereas outside r=a the magnetic pressure and tension forces are oppositely directed and balance one another while the plasma pressure gradient vanishes. The total current through the column is

$$I = \int_0^a j_z 2\pi r dr = \frac{2\pi a}{\mu} B_\theta(a).$$

Thus, if B_θ vanishes at the surface, the total current vanishes. If the tube is confined by an external plasma pressure, there is in general a surface current with both axial and azimuthal components such that the surface axial current cancels out the axial current in the bulk of the tube. In the particular case when $B_\theta(a) = 0$, the surface current has only an azimuthal component.

An integration of (8) when $p_0 = 0$ and T = constant gives an expression

$$I^2 = \frac{8\pi}{\mu} k_B TN$$

known as *Bennet's relation*, where N is the number of particles per unit length.

If the axial field is uniform (B_z = 1, say) and the twist profile falls off from a maximum value of Φ_0 on the axis like

$$\Phi = \frac{\Phi_0}{1 + r^2},$$

then (7) and (8) give

$$B_\theta = \frac{r\Phi_0}{L(1 + r^2)}, \tag{11}$$

$$p = p_0 + \frac{\Phi_0^2}{2\mu L^2(1+r^2)^2}.$$

Thus B_θ increases with radius to a maximum at r=1 and thereafter declines, while the pressure falls off with radius so as to create the outwards pressure gradient required to balance the inwards tension force due to B_θ. As the twist Φ_0 increases, so the maxima in B_θ and p increase (Figure 12).

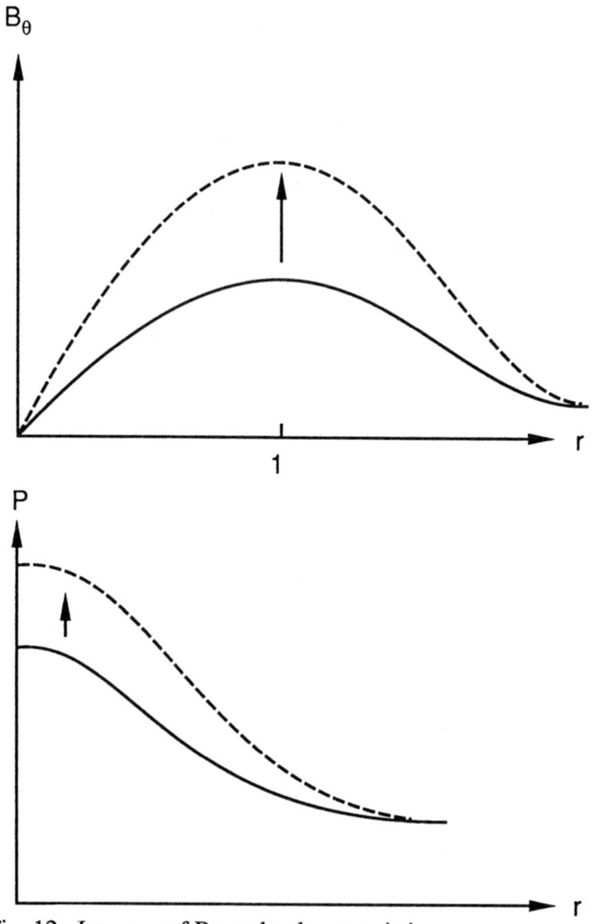

Fig. 12. Increase of B_θ and p due to twisting

If p is assumed to be uniform then we obtain a force- free field, and, if in particular $\Phi(r) = \Phi_0$ is independent of radius, we find from (7) and (8) the so-called *uniform-twist field* (Fig 13) with

$$B_\theta = \frac{B_0 R}{1 + R^2}, \quad B_z = \frac{B_0}{1 + R^2}, \tag{12}$$

where

$$R = \frac{\Phi_0 r}{2L},$$

in which the whole tube is twisted like a rigid body.

2.2 Linear Force-Free Ropes

For a force-free field we have

$$(\underline{\nabla} \times \underline{B}) \times \underline{B} = \underline{0},$$

which implies that $\underline{\nabla} \times \underline{B}$ is parallel to $\underline{B}$ so that

$$\underline{\nabla} \times \underline{B} = \alpha \underline{B}, \tag{13}$$

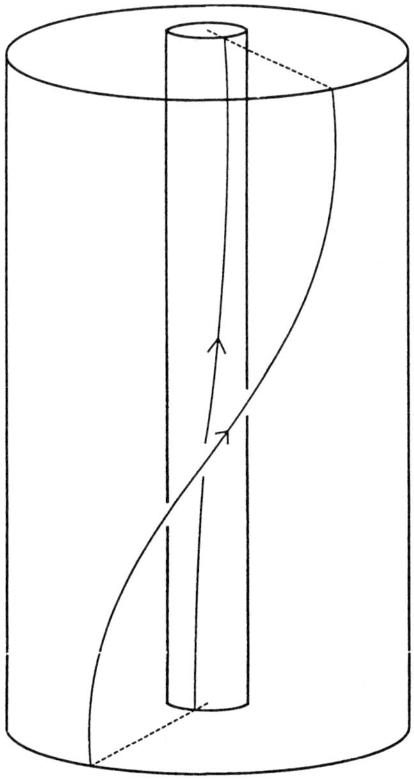

Fig. 13. A uniform twist field

where α is a scalar function of position. Taking the divergence of (13) and using $\underline{\nabla}.\underline{B} = 0$ gives

$$\underline{B}.\underline{\nabla}\alpha = 0,$$

so that α is constant along field lines. If α is uniform, i.e. the same constant on all field lines, we have a linear or constant-α force-free field, and the curl of (13) yields

$$(\nabla^2 + \alpha^2)\underline{B} = \underline{0}.$$

This gives the equation which each of the components of $\underline{B}$ satisfy, but they are in turn related by the components of (13). A particularly useful solution is (Fig 14)

$$B_z = B_0 J_0(\alpha r), \quad B_\theta = B_0 J_1(\alpha r), \qquad (14)$$

the so-called Bessel-function field due to Lundquist (1950). The axial field (B_z) starts from B_0 at r=0 and decreases to zero at $\alpha r = 2.4$, after which it reverses sign, while the azimuthal field (B_θ) increases from zero at the origin to a maximum and then declines. At larger radii both components oscillate indefinitely while their amplitudes decrease. (In practice one often cuts the field off at some radius a and so such a field will contain a reversal when $\alpha a > 2.4$.)

The physical relevance of the Bessel-function field was stressed by Taylor (1974) in a laboratory context extending earlier astrophysical work by Woltjer (1958). *Taylor's hypothesis* suggests that, in a weakly resistive plasma where small-scale reconnections can take place, of all the possible magnetic configurations the one with the minimum magnetic energy is a constant-α force-free field, with the value of α determined by the global magnetic helicity. The hypothesis gives a good agreement with laboratory toroidal configurations known as reversed-field pinches, which are highly twisted and may easily reconnect when not in a minimum energy state, but it does not apply to tokamaks, which are weakly twisted and much more stable. In a solar context, the hypothesis has been extended by Heyvaerts and Priest (1984) to allow a normal component of magnetic field on the boundary S (the photosphere), in which case motions on the boundary may inject or extract magnetic helicity and change it at a rate

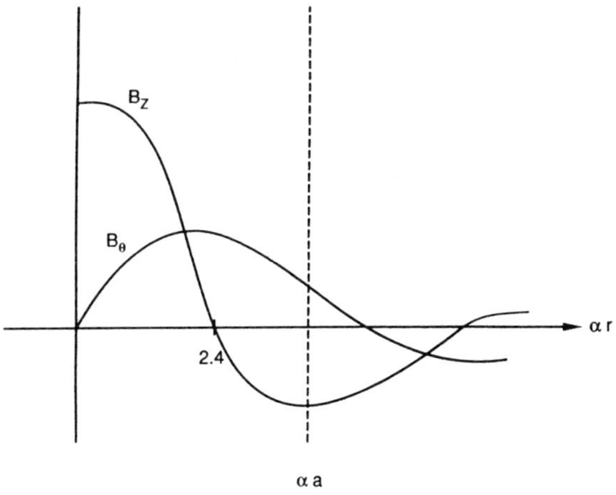

Fig. 14. The Bessel-function field

$$\frac{dH}{dt} = \int_S (\underline{A}.\underline{v})\underline{B}.d\underline{S}. \qquad (15)$$

Furthermore, Dixon, Berger and Priest (1989) have extended it further to allow a free boundary, such as the boundary of an active region.

Now, for the field (14) one may calculate the magnetic flux (F), magnetic energy (W) and magnetic helicity (H) in a cylinder of radius a and unit length (with $\underline{A} = \alpha^{-1}(\underline{B} - B_0 J_0(\alpha a)\underline{z})$ so that $\underline{A}$ vanishes on r=a) to give

$$F = \int_0^a B_z 2\pi r dr = 2\pi a^2 B_0 \frac{J_1(\alpha a)}{\alpha a}$$

$$W = \int_0^a B^2/(2\mu) 2\pi r dr$$
$$= \frac{\pi a^2 B_0^2}{\mu}\left[J_0^2(\alpha a) - \frac{J_0(\alpha a)J_1(\alpha a)}{\alpha a} + J_1^2(\alpha a)\right]$$

$$H = \frac{2\pi a^2 B_0^2}{\alpha}\left[J_0^2(\alpha a) - \frac{2}{\alpha a} J_0(\alpha a)J_1(\alpha a) + \alpha a J_1^2(\alpha a)\right].$$

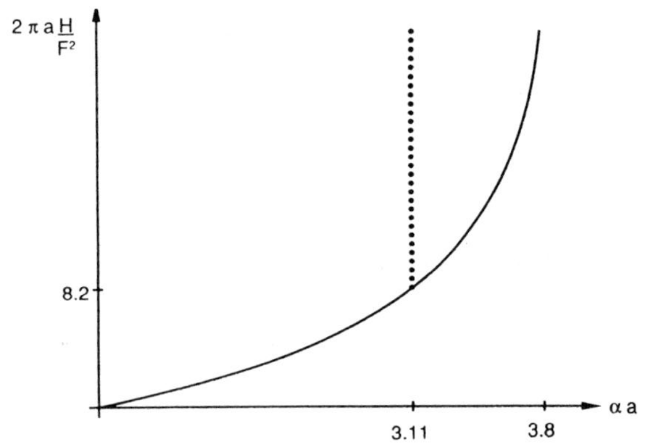

Fig. 15. Magnetic helicity as a function of α for the Bessel-function field

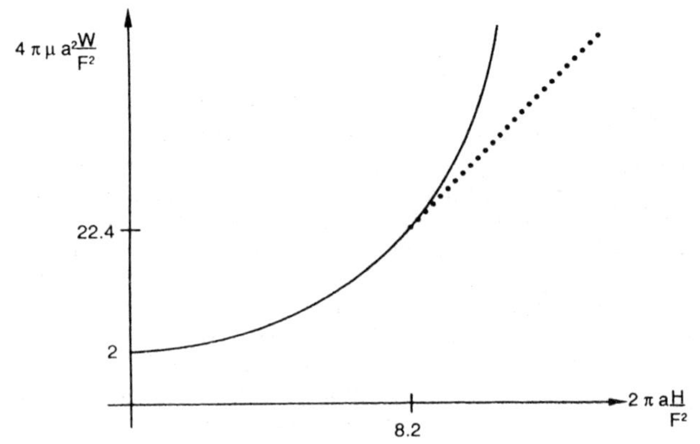

Fig. 16. Energy against magnetic helicity

The continuous curves in Figures 15 and 16 show how the values of α and magnetic energy increase monotonically with magnetic helicity. H has been non- dimensionalised in terms of F^2/a, W in terms of F^2/a^2, and α in terms of a^{-1} so as to eliminate the field magnitude B_0.

For the above cylindrical fields αa is determined by the value of aH/F^2, but there are also helical solutions with αa kept equal to 3.11 and aH/F^2 determining the amplitude of the helix. The helical solutions are

$$B_\theta = B_0 J_1(\alpha r) - \frac{B_1}{\ell}\left[\alpha J_1'(\ell r) + \frac{k}{\ell r} J_1(\ell r)\right]\cos(\theta + kz),$$

$$B_z = B_0 J_0(\alpha r) + B_1 J_1(\ell r)\cos(\theta + kz), \qquad (16)$$

$$B_r = -\frac{B_1}{\ell}\left[k J_1'(\ell r) + \frac{\alpha}{\ell r} J_1(\ell r)\right]\sin(\theta + kz),$$

where $\ell^2 = \alpha^2 - k^2$ and the first terms in the expressions for B_θ and B_z are the usual cylindrical solution of magnitude B_0, while the remaining terms represent an additional helical variation of magnitude B_1. Now the condition that B_r vanish at r=a so as to give a straight cylindrical surface determines a discrete value of αa as a function of ka, and the values which give the minimum energy are

$$\alpha a = 3.11, \qquad ka = 1.23.$$

The expressions for flux, energy and helicity become

$$F = 0.19\,\pi\, B_0 a^2,$$

$$W = \frac{\pi a^2}{\mu}(0.20\, B_0^2 + 0.054\, B_1^2),$$

$$H = \pi a^3 (0.15\, B_0^2 + 0.035\, B_1^2),$$

so that

$$\frac{2\pi a H}{F^2} = 8.2 + 1.9\frac{B_1^2}{B_0^2},$$

which implies that as $2\pi aH/F^2$ increases above a minimum value of 8.2, so B_1 increases, as shown by the dotted line in Figure 16. At the same time the magnetic energy increases (linearly) with helicity, as shown dotted in Figure 17.

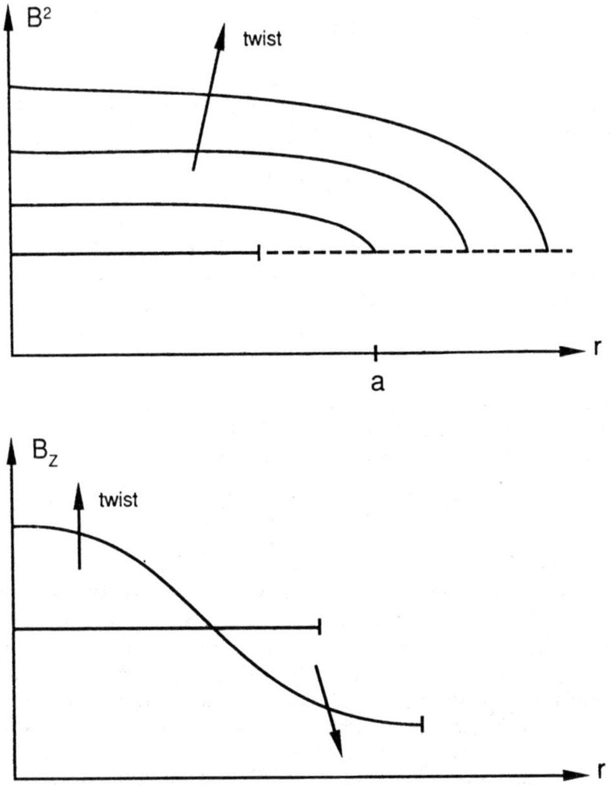

Fig. 17. Change of form of $B^2(r)$ and $B_z(r)$ with twist

However, it can be seen that the energy of the helical state is less than that of the cylindrical one, which implies that as a tube is slowly twisted up so it passes through a series of cylindrical states until $2\pi aH/F^2$ reaches 8.2, beyond which it just passes through a series of helical states. Interestingly, the threshold for the onset of helical equilibria also corresponds to the threshold for linear resistive m=1 kink instability of the cylindrical equilibria.

2.3 Nonlinear Force-Free Fields

For force-free fields (8) reduces to the balance

$$\frac{d}{dr}\left(\frac{B_\theta^2 + B_z^2}{2\mu}\right) + \frac{B_\theta^2}{\mu r} = 0 \qquad (17)$$

between magnetic pressure and tension forces, and so one may impose either B_θ or B_z and deduce the other. A common alternative is to impose the functional form of

$$B^2 = f(r),$$

known as the generating function, and then (17) implies

$$B_\theta^2 = -\frac{r}{2}\frac{df}{dr}, \qquad B_z^2 = f + \frac{r}{2}\frac{df}{dr}. \qquad (18)$$

There are some restrictions on f. For example, $B_\theta^2 \geq 0$ implies that $df/dr \leq 0$, and the condition $B_z^2 \geq 0$ implies that f approaches zero no faster than r^{-2}.

As a tube is twisted, its radius expands because of the increase in magnetic pressure caused by the additional B_θ component. Parker (1979), however, proved an interesting theorem, namely that the mean square B_z of a force-free flux tube confined by a fixed external pressure (p_e) is not affected by twisting. By definition

$$\langle B_z^2 \rangle = \frac{1}{\pi a^2}\int_0^a B_z^2 \, 2\pi r dr,$$

or, after using (18) to substitute for B_z^2,

$$\langle B_z^2 \rangle = \frac{2}{a^2}\int_0^a \left(rf + \frac{r^2}{2}\frac{df}{dr}\right)dr$$

$$= \frac{1}{a^2}\int_0^a \frac{d}{dr}(r^2 f)dr$$

$$= f(a).$$

But $f(a) = B^2(a)$, and pressure balance on the boundary (for a negligible magnetic field outside and a negligible plasma pressure inside) implies that $B^2(a) = 2\mu p_e$, which is being held constant. The general effect on $B^2(r)$ and $B_z(r)$ of twisting is shown schematically in Figure 17, with B^2 and the radius (a) tending to increase. Thus, since the flux

$$F = \pi a^2 \langle B_z \rangle$$

is constant, the mean value of B_z decreases. The value $B_z(0)$ on the axis increases because an outwards magnetic pressure force (namely $-d/dr(B_z^2/2\mu)$) is needed to balance the inwards magnetic tension force. At the same time the value $B_z(a)$ at the edge of the tube decreases because $B_z^2 + B_\theta^2 = B^2$ is held fixed to balance $2\mu p_e$ and twisting makes B_θ increase.

2.4 Effect of Compressing or Expanding a Tube

2.4.1 Untwisted Tube

Consider a uniform untwisted tube of field strength B_0 and plasma density ρ_0 and suppose its length (L_0) and radius (a_0) are changed by factors λ^* and λ, respectively (Fig 18). Then conservation of matter gives

$$\rho\pi(\lambda a_0)^2(\lambda^* L_0) = \rho_0 \pi a_0^2 L_0,$$

so that the new density is

$$\rho = \frac{\rho_0}{\lambda^2 \lambda^*}. \qquad (19)$$

Conservation of magnetic flux implies

$$B\,\pi(\lambda a_0)^2 = B_0 \pi a_0^2,$$

so that the new field strength is

$$B = \frac{B_0}{\lambda^2}. \qquad (20)$$

Thus, what happens to a compressing or expanding tube depends on the physics of the situation. For example, if the length is held fixed ($\lambda^* = 1$), (19) and (20) imply that

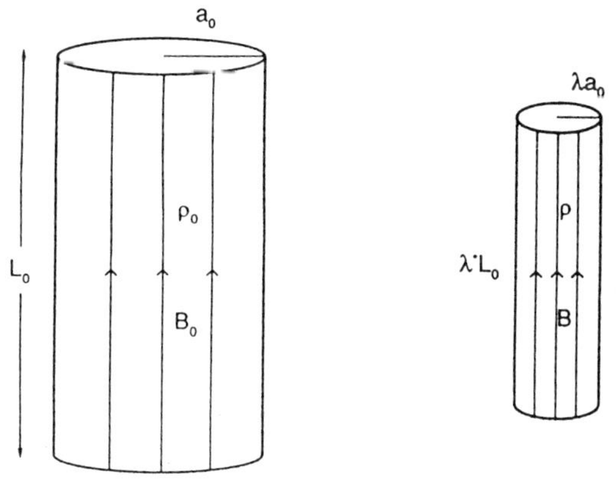

Fig. 18. Compression of an untwisted tube

$$\frac{B}{\rho} = \text{constant},$$

so that a transverse compression ($\lambda < 1$) increases B and ρ by the same proportion, while an expansion ($\lambda > 1$) decreases them. (However, with free flow along a tube, matter is generally not conserved.) Instead, suppose there is no plasma compression ($\lambda^2\lambda^* = 1$), then an extension ($\lambda^* > 1$) of a tube makes it narrow and the field strength rise, while a shortening of the tube does the opposite.

2.4.2 Twisted Tube

Consider next a force-free tube with components $B_\theta(r)$, $B_z(r)$ and generating function whose confining pressure falls so that the tube expands, with the radii r and r+dr becoming $\bar{r}$ and $\bar{r}+d\bar{r}$ and the field components becoming $\bar{B}_\theta(\bar{r})$ and $\bar{B}_z(\bar{r})$ (Fig 19). Parker (1979) noted that the axial and azimuthal fluxes through the annulus between the radii r and r+dr are conserved, so that

$$B_z r\, dr = \bar{B}_z \bar{r}\, d\bar{r}$$

$$B_\theta\, dr = \bar{B}_\theta\, d\bar{r},$$

where the components of $\underline{B}$ and $\underline{\bar{B}}$ can be written in terms of f and f, respectively. Eliminating f between these two equations therefore determines the mapping $r(\bar{r})$ when the initial generating function is prescribed. For example, a flux tube with $f = 1-r^2$ and $a^2 = 1/2$ has field components

$$B_\theta = r, \quad B_z = (1 - 2r^2)^{1/2}$$

and leads to the mapping equation

$$\frac{d^2u}{d\bar{u}^2}(1 - 2u + \bar{u}) + \frac{du}{d\bar{u}} - \left(\frac{du}{d\bar{u}}\right)^2 = 0,$$

where $u = r^2$ and $\bar{u} = \bar{r}^2$. For a large expansion ($a^2 >> 1/2$) the solution is

$$r^2 = \frac{\log(1 + \bar{r}^2)}{4 \log \bar{a}}$$

and the new field components are

$$\bar{B}_\theta = \bar{r}\,\bar{B}_z, \quad \bar{B}_z = \frac{1}{4(1 + \bar{r}^2)\log \bar{a}}.$$

Thus expanding the tube makes it more twisted and the field becomes mainly azimuthal ($\bar{B}_\theta >> \bar{B}_z$) over most of the radius. Similarly, the tube also has its twist increased by an axial compression. However, fields that are initially purely axial, purely azimuthal or have uniform twist are invariant in form with respect to expansion.

3. Tubes with Varying Cross-Section

3.1 Expansion of a Section

Parker (1979) has also considered the expansion of a twisted tube from one straight section with generating function f(r) to another with generating function $\bar{f}$, as shown in Figure 20. The axial flux is again conserved so that

$$B_z\, r\, dr = \bar{B}_z\, \bar{r}\, d\bar{r},$$

but the azimuthal flux is no longer constant. Instead the conservation of the torque of the azimuthal Maxwell stress ($B_\theta B_z/\mu$) gives

$$r\, B_\theta B_z\, r dr = \bar{r}\, \bar{B}_\theta \bar{B}_z\, \bar{r} d\bar{r}.$$

Again, these two equations may be written in terms of f and $\bar{f}$

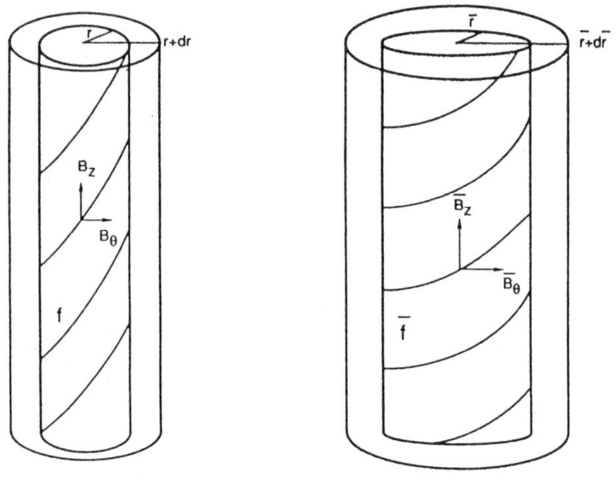

Fig. 19. Expansion of a twisted tube

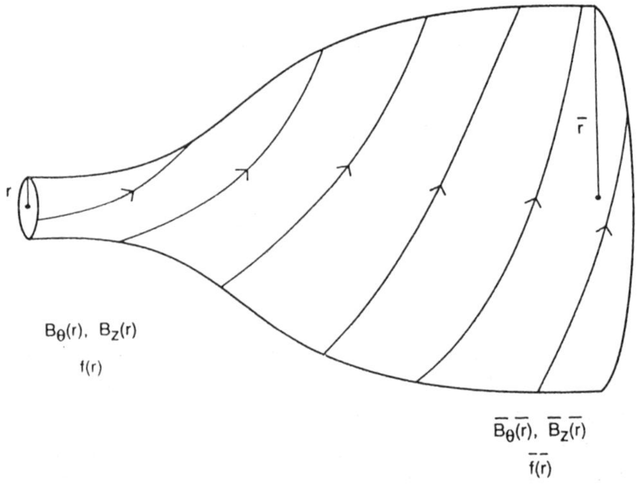

Fig. 20. Expansion of part of a flux rope

and so $\bar{f}$ may be eliminated to give an equation for the mapping r(r). For example, an initial generating function $f = 1-r^2$ gives when $\bar{a}^2 >> 1/2$.

$$r = \bar{r}\, J_1\, (2\bar{r})/\text{constant}.$$

Parker finds that the field becomes mainly azimuthal, so that coils are transferred to the expanded section. It may be noted that dividing one conservation equation by the other gives

$$r\, B_\theta = \bar{r}\, \bar{B}_\theta$$

so that rB_θ is constant along a field line, a result that we shall meet in §3.3.

3.2 Thin Confined Untwisted Tube

One way to make progress with analysing the structure of an expanding tube is to assume that it is so thin that the variations across its section are negligible. Suppose that its plasma pressure $p_i(s)$ and field strength B(s) are functions of distance along the tube. Flux conservation may then be written

$$\pi a^2(s)\, B(s) = F, \qquad (21)$$

which determines the tube radius (a(s)), while surface equilibrium gives

$$p_i + \frac{B^2}{2\mu} = p_e \qquad (22)$$

when the tube is confined by an external plasma pressure p_e. As the pressure difference p_e-p_i falls, so the field strength decreases and the tube radius increases. In the limit as p_e-p_i approaches zero, so B approaches zero and a becomes infinite, but by then the thin-tube approximation has failed.

In order to derive the force balance equation normal to the tube when it is confined by a static external plasma in hydrostatic equilibrium with pressure p_e and density ρ_e, Spruit (1981) considered the general force balance equation

$$-\underline{\nabla}\left(p + \frac{B^2}{2\mu}\right) + (\underline{B}\,.\,\underline{\nabla})\,\frac{\underline{B}}{\mu} + \rho\underline{g} = \underline{O} \qquad (23)$$

with a tube magnetic field

$$\underline{B} = B(r,\theta,s)\hat{\underline{s}}$$

in terms of coordinates r,θ,s (Figure 21), which holds when the tube radius (a) is much smaller than the scale height (H) and the length-scale for variations along the tube. Now the component of (23) along the tube is simply hydrostatic equilibrium for the internal plasma, namely

$$-\frac{\partial p_i}{\partial s} + \rho_i g_{||} = 0$$

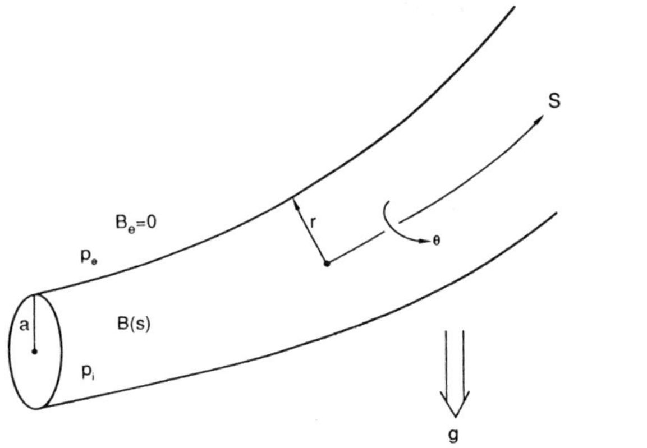

Fig. 21. Notation for a slender tube

where $g_{||}$ is the component of $\underline{g}$ along the tube. The component of (23) normal to the tube is

$$-\underline{\nabla}_\perp\left(p_i + \frac{B^2}{2\mu}\right) + \frac{B^2}{\mu}\frac{\hat{\underline{n}}}{\mathcal{R}} + \rho_i g_\perp = \underline{O},$$

but (22) implies that

$$\underline{\nabla}\left(p_i + \frac{B^2}{2\mu}\right) = \underline{\nabla} p_e,$$

which in turn equals $\rho_e\underline{g}$ from external hydrostatic equilibrium and so this reduces to

$$\frac{B^2}{\mu}\frac{\hat{\underline{n}}}{\mathcal{R}} + (\rho_i - \rho_e)\underline{g}_\perp = \underline{O}, \qquad (24)$$

where the first term represents the magnetic tension force and the second term a *magnetic buoyancy* force acting perpendicular to the tube due to the difference in internal and external densities.

If the tube is locally inclined at Θ to the horizontal so that $g_\perp = g \cos \Theta$ and

$$\frac{\hat{\underline{n}}}{\mathcal{R}} = \frac{\partial\hat{\underline{s}}}{\partial s} = \sin\Theta\,\frac{d\Theta}{dz}\,\hat{\underline{n}},$$

(24) may be rewritten

$$\frac{B^2}{\mu}\frac{d\Theta}{dz} + (\rho_i - \rho_e) g \cot\Theta = 0, \qquad (25)$$

which determines the shape of the tube, i.e. the variation of Θ with z. The above analysis may be extended to include flows, twist and waves.

3.3 Thick Confined Force-Free Tubes
(a) Untwisted Axisymmetric Tube

Consider an axisymmetric tube with field components (Fig 22)

$$(B_r(r,z),\ 0,\ B_z(r,z))$$

in cylindrical polar coordinates, (Browning and Priest, 1982,1983), and with electric current

$$j_\theta = \frac{1}{\mu}\left(\frac{\partial B_r}{\partial z} - \frac{\partial B_z}{\partial r}\right).$$

The equation

$$\underline{\nabla} \cdot \underline{B} = \frac{1}{r}(rB_r) + \frac{\partial B_z}{\partial z} = 0$$

may be satisfied identically by writing

$$B_r = -\frac{1}{r}\frac{\partial \psi}{\partial z}, \quad B_z = \frac{1}{r}\frac{\partial \psi}{\partial r} \tag{26}$$

in terms of a *flux function* ψ. ψ is constant on magnetic field lines and $2\pi\psi$ is the flux through a circle of radius r.

Since the electric current is here normal to the magnetic field, the force-free condition ($\underline{j} \times \underline{B} = \underline{O}$) reduces to $\underline{j} = \underline{O}$, or in terms of ψ

$$\frac{\partial^2 \psi}{\partial r^2} - \frac{1}{r}\frac{\partial \psi}{\partial r} + \frac{\partial^2 \psi}{\partial z^2} = 0. \tag{27}$$

The aim is therefore to solve (27) subject to the boundary conditions that

$$\psi = f(r) \tag{28}$$

is a given function on the ends z = 0 and z = L of the tube and

$$\psi = \psi_s \qquad \text{on } r = a(z), \tag{29}$$

$$B^2 = B_s^2(z) \equiv 2\mu p_e(z) \qquad \text{on } r = a(z), \tag{30}$$

where a(z) is the tube radius and $B_s(z)$ is a prescribed function, namely, the value of the field at the surface needed for pressure balance there.

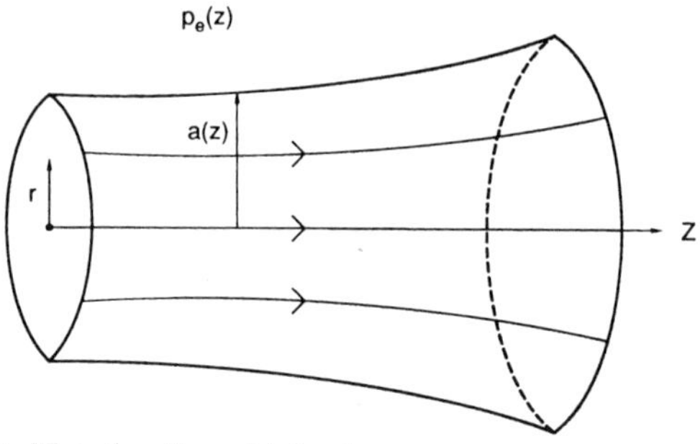

Fig. 22. Notation for a thick tube

Thus, since a(z) is unknown, we have a free boundary-value problem, which is notoriously difficult to solve, and questions of existence and uniqueness of the general solutions are non-trivial. As a preliminary there are several ways of making progress. For example, conservation of flux F gives

$$\pi a^2 \overline{B}_z = F, \tag{31}$$

where B_z is the mean axial field. For a thin tube

$$\psi = \frac{1}{2} r^2 B_s(z)$$

and $B_z = B_s$, so that (31) determines the radius to be

$$a_0^2 = \frac{F}{\pi B_s(z)}. \tag{32}$$

For a correction to this thin-tube result one may expand in powers of $\varepsilon = a_0/L << 1$, where L is the length-scale for variations along the tube. The flux function is now

$$\psi = r^2 f_0 - \frac{1}{8} r^4 f_0'',$$

with

$$B_r = -rf_0', \quad B_z = 2f_0 - r^2 f_0'',$$

and the tube radius becomes

$$a^2 = a_0^2\left\{1 + \frac{a_0^2}{16B_s^2}(B_s'^2 - B_s B_s'') + 0(\varepsilon^4)\right\}. \tag{33}$$

In this correction to (32) the contribution of $B_r^2(a)$ to B_s^2 makes the radius a increase above a_0, whereas the fact that $B_z(a)$ differs from B_z can make a either increase or decrease: for example, if $B_s'' < 0$ then $B_z(a) > B_z$ and a becomes larger than a_0.

A second approach is to consider a particular analytical solution such as

$$\psi = \psi_L\, r\, J_1(kr)\cosh kz, \tag{34}$$

where

$$\psi_L^{-1} = a_L J_1(ka_L)\cosh \frac{1}{2} kL,$$

so that $\psi = 1$ at the ends $z = \pm 1/2\, L$ at the radius $r = a_L$. This represents an expanding and symmetric flux tube, as sketched in Figure 23, with a surface given by $\psi = 1$, so that the radius ($r = a_0$) at the widest part (z=0) is given from (34) by

$$1 = \psi_L a_0 J_1(ka_0).$$

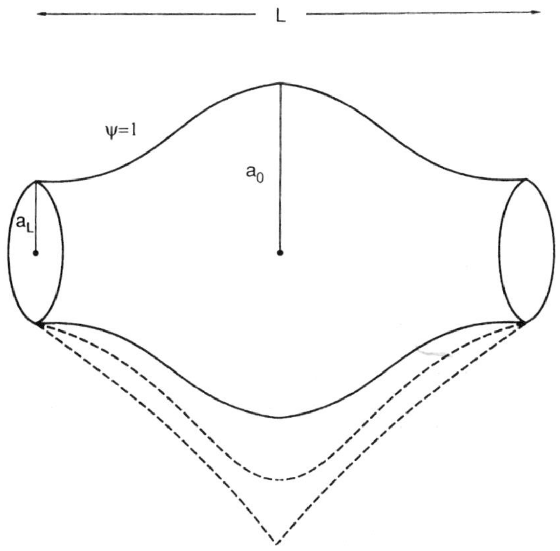

Fig. 23. An expanding flux tube

The way that a_0 varies with the parameter k from this equation is sketched in Figure 24. As k increases so the tube expands, as indicated by the dashed curve. When the radius a_0 reaches a critical value such that $d/da_0(a_0J_1(ka_0)) = 0$ or $J_0(ka_0) = 0$, there is no neighbouring equilibrium with a larger radius. The surface has reached a cusp shape and B_z has fallen to zero at the edge since the external pressure p_e has become equal to the internal one p_i, but unlike the thin tube result the tube has a finite radius at this point rather than becoming infinite. The nonequilibrium consequences when $p_e < p_i$ are not clear but perhaps the tube bursts. The critical radius (a_0*) for bursting depends on the dimensions of the tube, with a_0*/L varying between about 2 and 6 as L/a_L increases from 4 to 20.

A third way of solving (27) is to impose the shape of the flux tube as, say,

$$a = 1 + c\left[\frac{1}{2}(1+z)^2 - \frac{1}{3}(1+z)^3\right]$$

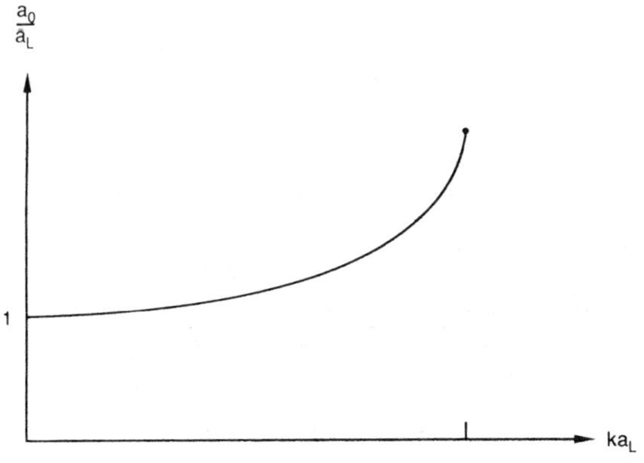

Fig. 24. Variation of maximum radius a_0

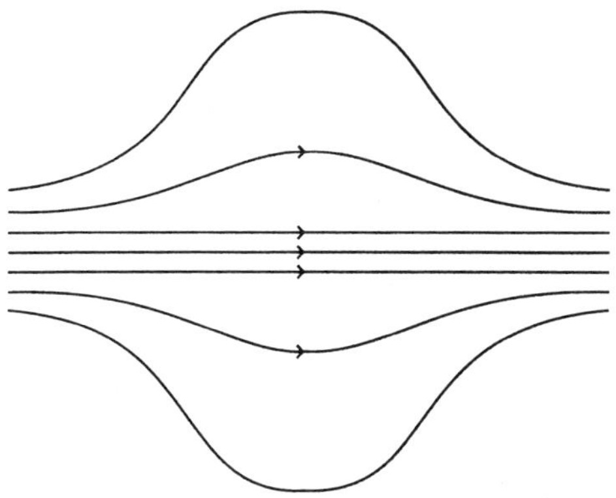

Fig. 25. Sketch of field lines in an expanded tube

and to solve (27) numerically. The result is a shape looking like Figure 25, in which the flux tube possesses a strong core with the magnetic flux concentrated near the axis. This in turn makes the field strength B(a) at the surface lower than the mean value B_z and so the tube expands less than one would expect from the thin-tube result for a given external pressure.

The strong core discovered by Browning and Priest is also a feature of more recent calculations, as we shall see below.

(b) Twisted Axisymmetric Tube

The effect of twisting a tube is to add an extra magnetic component $B_\theta(r,z)$, with the other two components still given in terms of the flux function by (26). The electric current components become

$$(j_r, j_\theta, j_z) = \frac{1}{\mu}\left(-\frac{\partial B_\theta}{\partial z}, \frac{\partial B_r}{\partial z} - \frac{\partial B_z}{\partial r}, \frac{1}{r}\frac{\partial}{\partial r}(rB_\theta)\right),$$

and so the θ-component of $\underset{\sim}{j} \times \underline{B} = \underline{0}$, namely $j_zB_r - j_rB_z = 0$, becomes

$$\frac{\partial}{\partial r}(rB_\theta)\frac{\partial\psi}{\partial z} = \frac{\partial}{\partial z}(rB_\theta)\frac{\partial\psi}{\partial r},$$

which implies that rB_θ is a function of ψ alone

$$rB_\theta = f(\psi),$$

so that rB_θ is constant along field lines, which is the same as Parker's result in §3.1. Equation (27) is then generalised

$$\frac{\partial^2\psi}{\partial r^2} - \frac{1}{r}\frac{\partial\psi}{\partial r} + \frac{\partial^2\psi}{\partial z^2} = -f\frac{df}{d\psi}. \qquad (35)$$

Again we have a free boundary-value problem, which has been tackled approximately by several approaches, but in general one would like to specify not f(ψ) but the twist

$$\Phi(\psi) = \int_0^L \left(\frac{B_\theta}{rB_z}\right)_{\psi=\text{const}} dz. \qquad (36)$$

First of all, the thin tube result (33) is affected at order ε^4 by twisting. The introduction of the component B_θ tends to increase the surface field B_s and so make the tube expand, whereas the resulting reduction of $B_z(a)$ tends to produce a contraction. The extra term of order ε^4 in (33) is found to be

$$\frac{a_0^4}{384}\left[\frac{4B_s^{1v}}{B_s} + 15\frac{B_s'''B_s'}{B_s^2} + 36\left(\frac{B_s'}{B_s}\right)^4 - 90\frac{B_s''(B_s')^2}{B_s^3} - 3\left(\frac{B_s''}{B_s}\right)^2 + 6b^2\frac{B_s''}{B_s} - 9b^2\frac{B_s'^2}{B_s^2} + 5b^4 + 12bf''(0)B_s\right].$$

The positive terms represent an increase in radius due to twisting, as for Parker's result.

When $f = \alpha\psi$, there is a simple analytical solution to (35), namely

$$\psi = \psi_L rJ_1(kr)\cosh mz, \qquad (37)$$

where $m^2 = k^2 - \alpha^2$. The resulting twist is concentrated in the expanded part, and again there are no solutions when p_e is too small (Browning & Priest, 1983).

When the twist is small, progress has been made by Lothian and Hood (1989), (based on earlier work by Zweibel and Boozer (1985)). They twist up a uniform field

$$\psi_0 = r^2$$

by imposing a field

$$B_\theta = \begin{cases} \varepsilon r(1-r^2), & r < 1, \\ 0, & r > 1, \end{cases}$$

at the ends of the tube. This corresponds to imposing the function

$$f(\psi) = \begin{cases} \varepsilon\psi(1-\psi), & \psi < 1, \\ 0, & \psi > 1, \end{cases}$$

or the twist

$$\Phi = \begin{cases} \varepsilon L(1-r^2), & r < 1, \\ 0, & r > 1, \end{cases}$$

at the ends of the flux tube. The linearised form of (35) with

$$\psi = \psi_0 + \varepsilon^2\psi_1$$

is then

$$\frac{\partial^2\psi_1}{\partial r^2} - \frac{1}{r}\frac{\partial\psi_1}{\partial r} + \frac{\partial^2\psi_1}{\partial z^2} = -\left(f\frac{df}{d\psi}\right)_{\psi_0}, \qquad (38)$$

with boundary conditions

$$\psi_1 = 0 \text{ at } z = 0 \text{ and } z = L, \text{ and } \psi_1 = 0 \text{ at } r = a,$$

so that the perturbed flux function vanishes both at the ends and at the surface of the flux tube. The resulting solution (Lothian and Hood, 1989) is

$$\psi_1 = r\sum_1^\infty c_n J_1\left(v_n\frac{r}{a}\right)\left[\frac{\cosh v_n(z - \frac{1}{2}L)/a}{\cosh v_n \frac{1}{2}L/a} - 1\right], \qquad (39)$$

where v_n is the nth zero of the Bessel function J_1 and the constants c_n are given by

$$c_n = -\frac{4}{J_0^2(v_n)v_n^2}\int_0^a r^2(1-r^2)(1-2r^2)J_1\left(v_n\frac{r}{a}\right)dr.$$

The axial parts of the field lines (in an r-z plane) are sketched in Figure 26 and show several interesting features. For a large aspect ratio ($L/a >> 1$) most of the flux tube is straight, with the axial field bowing out only in thin boundary layers of thickness a near the ends of the tube. This occurs because the expression in square brackets in (39) differs substantially from -1 only near the ends: for example, when $\frac{1}{2}L < z < L$, the quotient of cosh functions becomes approximately

$$\frac{\exp v_n(z - \frac{1}{2}L)/a}{\exp v_n \frac{1}{2}L/a} = \exp v_n(z-L)/a,$$

which is much smaller than unity except when z is close to L. Other features of the solution (39) are that the twist is concentrated near the axis of the tube and the outermost

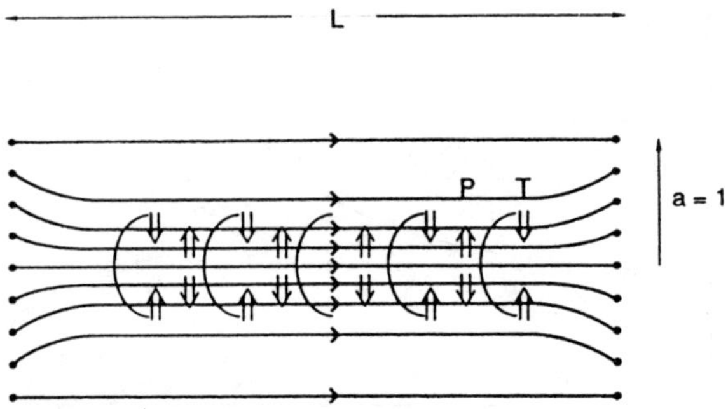

Fig. 26. Sketch of axial field lines for small twist

fieldline remains straight, unaffected in its position by the twisting.

There are clear differences with Parker's results due to the different assumptions. Parker finds an isolated tube that expands has most of the twist in the expanded part. Here the footpoints are fixed with the twist imposed at the ends and there is an external potential field. Also the structure is highly nonuniform across the tube even though the edge remains straight. The work of Lothian and Hood has recently been extended to the nonlinear regime by Browning and Hood (1989), who do find that the tube expands. In their nonlinear solution of (35) they impose

$$f(\psi) = \begin{cases} \psi(1-\psi), & \psi < 1 \\ 0, & \psi > 1 \end{cases}$$

and find again that the core contracts to a region of strong axial field with a straight central region and boundary layers near the ends (Figure 27). Twisting up a loop therefore tends to create filamentary structures in the axial magnetic flux. The outer part of the tube expands because twisting increases B_θ, which lowers the value of B_z at the surface of the tube (for a constant confining pressure $2\mu B_s{}^2$), and this in turn tends to lower the mean value B_z of the axial field, thereby increasing the radius. The resulting twist function is sketched in Figure 28, where a strong peak in the twist can be seen at a location near the edge of the tube. When B_z actually falls to zero there and the twist becomes infinite, there is a loss of equilibrium. This nonequilibrium feature is similar to the one found by Browning and Priest (1982,1983) above and can easily be demonstrated for a one-dimensional equilibrium that is typical of most of the flux tube away from the footpoints. Thus, if we take for simplicity

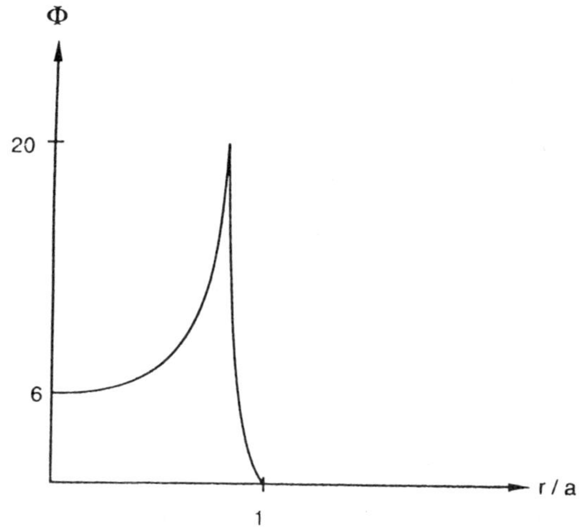

Fig. 28. Twist as a function of radius

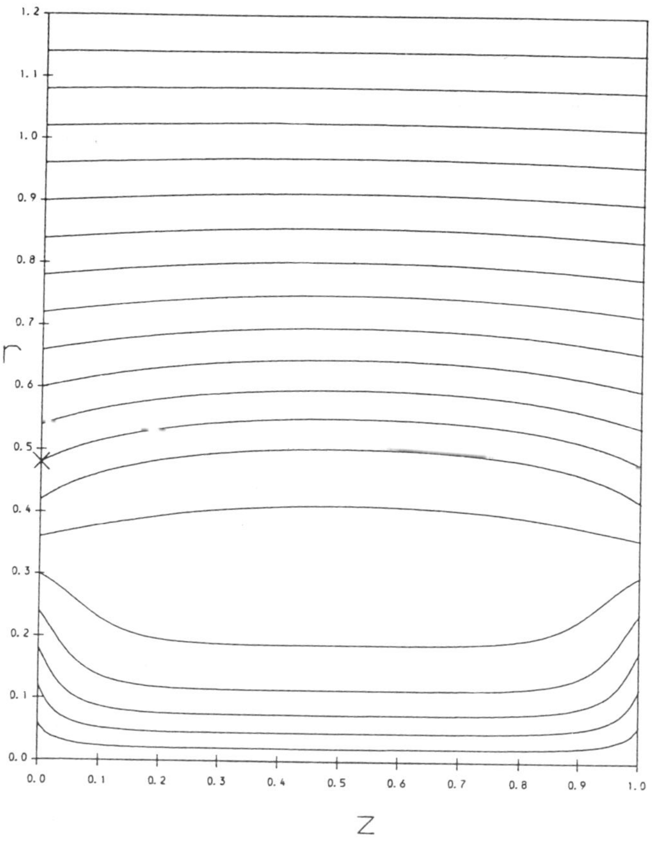

Fig. 27. Axial field lines when twist is nonlinear

$$f(\psi) = \begin{cases} \lambda\psi, & r \le a, \\ 0, & r > a, \end{cases}$$

then the one-dimensional solution of (35) is

$$\psi = \frac{r\, J_1(\lambda r)}{a\, J_1(\lambda a)}$$

for $r \le a$, so that $\psi = 1$ at the edge $r = a$. The radius (a) is then given by the condition

$$B^2(a) = B_s^2,$$

and, as λ increases, so the tube radius expands But the axial field is

$$B_z = \frac{1}{r}\frac{\partial\psi}{\partial r} = \frac{\lambda\, J_0(\lambda r)}{a\, J_1(\lambda a)}$$

for $r \le a$ and this vanishes at the edge when λa has increased up to the first zero of J_0 (i.e. approximately 2.4). For larger values of λ there is no equilibrium.

In an interesting numerical experiment Steinolfson and Tajima (1989) have examined the effect of slowly twisting up the ends of an initially straight flux tube at a speed of about 0.01 v_A. They find that for the first 150 Alfvén times the flux

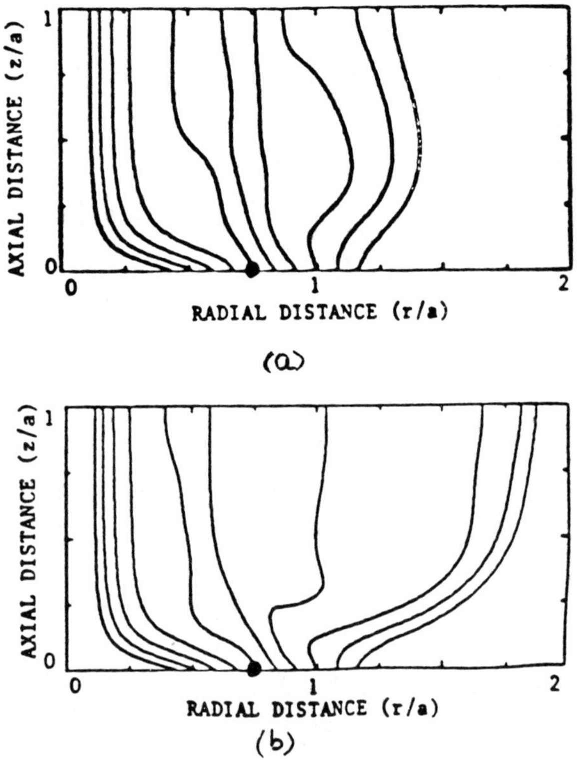

Fig. 29. Field lines in (a) quasi-equilibrium and (b) dynamic phases (Steinolfson and Tajima).

tube evolves through a series of equilibria with a straight core and boundary layers near the ends (Figure 29a) as in the above analytical models. Then just after that time there is a dynamic stage with field lines bursting outwards with rapid twisting and a rapid increase of kinetic energy (Figure 29b).

4. Curved Tubes

Toroidal effects in flux tubes have long been considered in laboratory devices (e.g. Freidberg, 1987; Wesson, 1988). For a general magnetostatic equilibrium satisfying

$$\underline{j} \times \underline{B} - \underline{\nabla} p = \underline{0}, \qquad (40)$$

one can see by taking scalar products with $\underline{B}$ and $\underline{j}$ that

$$\underline{B} \,.\, \underline{\nabla} p = \underline{j} \,.\, \underline{\nabla} p = 0,$$

so that magnetic field lines and electric current lines lie on surfaces of constant pressure. Such flux surfaces are in general nested (Figure 30) but for a toroid there are three types of field line trajectories, namely rational (which close on themselves after a finite number of curcuits of the torus), ergodic (which do not close but cover a whole surface of constant pressure) and stochastic (which fill a whole volume).

In general one may work for convenience either in cylindrical polar coordinates (R,Z,ϕ) or in toroidal coordinates (r, θ, ϕ) such that

$$R = R_0 + r \cos \theta, \quad Z = r \sin \theta, \qquad (41)$$

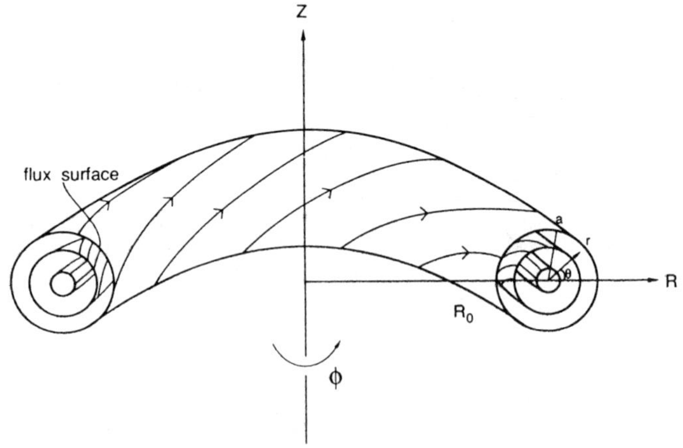

Fig. 30. Notation for a toroidal flux rope

as shown in Figure 30. It can be shown that a purely poloidal field (Figure 31a) is not in equilibrium by itself. The magnetic pressure due to a magnetic field of strength B_1 on the inside and B_2 on the outside of the torus produces an outwards hoop force of

$$\frac{B_1^2 S_1 - B_2^2 S_2}{2\mu},$$

which is positive because $B_1{}^2 > B_2{}^2$ and S_2 exceeds S_1 by a smaller amount. If the flux tube contains a plasma of pressure p, it also exerts an outwards tyre-tube force of magnitude

$$pS_2 - pS_1.$$

These outward forces can be balanced in the laboratory by the effect of a conducting shell or by a vertical field B_0 which gives a force

$$2\pi R_0 I_0 B_0,$$

but the resulting equilibrium is unstable.

Furthermore, a purely toroidal field is not in equilibrium either (Figure 31b). Suppose a current flows in the plasma surface. Then Maxwell's equations imply that outside the torus the field behaves like $B_\phi = B_0 R_0/R$, where $B_0 = \mu I_c/(2\pi R_0)$ and I_c is the current in toroidal field coils. Thus $B_1{}^2 > B_2{}^2$, where B_1 is the field on the inner side of the torus and B_2 on the outer. There is again an outwards force

$$2\pi^2 a^2 \left(p + \frac{B_0^2}{2\mu} \right),$$

but this time neither a conducting shell nor a vertical field helps.

An axisymetric field may be written

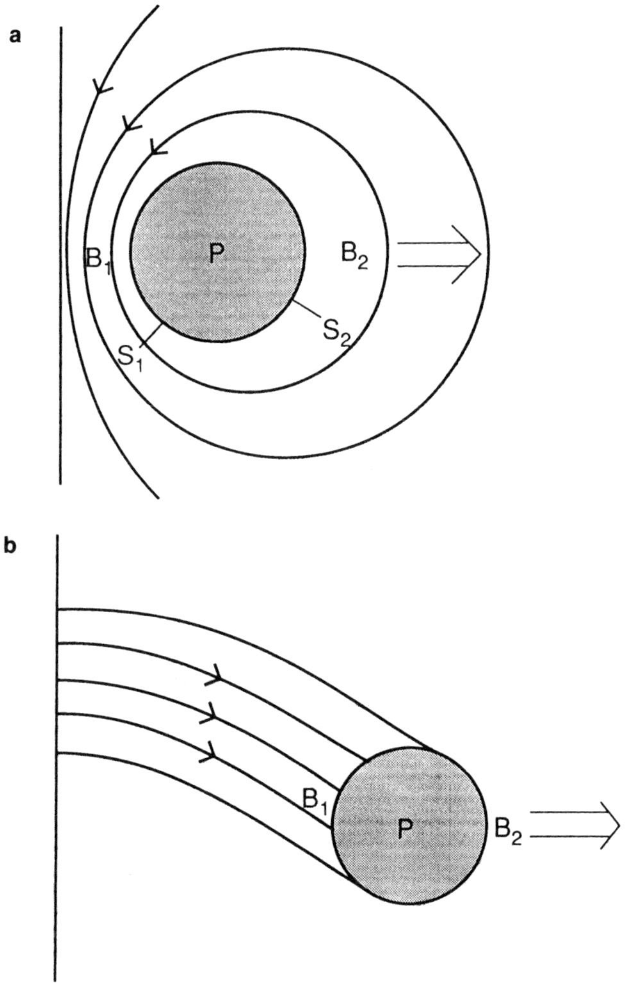

Fig. 31. Field that is purely (a) poloidal or (b) toroidal

$$\underline{B}(R,Z) = \underline{B}_p + \underline{B}_\phi$$

in terms of poloidal and toroidal components, respectively, where the poloidal components are

$$(B_R, B_Z) = \left(-\frac{1}{R}\frac{\partial\psi}{\partial Z}, \frac{1}{R}\frac{\partial\psi}{\partial R}\right)$$

and ψ = constant represents a flux surface. The magnetostatic equation implies that

$$p = p(\psi), \quad RB_\phi = f(\psi),$$

where ψ satisfies a grad-Shafranov equation

$$R\frac{\partial}{\partial R}\left(\frac{1}{R}\frac{\partial\psi}{\partial R}\right) + \frac{\partial^2\psi}{\partial Z^2} = -\mu R^2\frac{dp}{d\psi} - f\frac{df}{d\psi}. \qquad (42)$$

A standard method of seeking solutions is to make a so-called *Reversed Field Pinch Expansion* by expanding in powers of the inverse aspect-ratio ($\varepsilon = a/R_0 << 1$), where a and R_0 are the minor and major radii, such that

$$\psi(r,\theta) = \psi_0(r) + \varepsilon\psi_1(r)\cos\theta + \dots . \qquad (43)$$

The ordering

$$\frac{B_p}{B_\phi} \sim 1, \quad \beta_p \equiv \frac{2\mu p}{B_p^2} \sim 1$$

is made and so the zeroth-order state is just a cylindrical one with

$$B_{0r} = 0, \quad \frac{d}{dr}\left(p_0 + \frac{B_0^2}{2\mu}\right) + \frac{B_{0\theta}^2}{\mu r} = 0.$$

At the next order one finds

$$\left[r\,B_{0\theta}^2\left(\frac{\psi_1}{B_{0\theta}}\right)'\right]' = r\,B_{0\theta}^2 - 2\mu r^2 p_0', \qquad (44)$$

which determines ψ_1. The resulting flux surfaces are displaced outwards from R_0 by a distance

$$\Delta = -\frac{\psi_1(a)}{\psi_0'(a)} \qquad (45)$$

known as the Shafranov shift (Fig 32), which makes the poloidal field decrease on the inside of the torus and increase on the outside. This is derived by noting that the boundary of the flux surface at $r = a + \varepsilon\Delta\cos\theta$ is from (43) the constant

$$\psi_0(a) = \psi_0 + \varepsilon\psi_1\cos\theta$$
$$= \psi_0(a) + \varepsilon(\Delta\cos\theta\,\psi_0'(a) + \psi_1\cos\theta),$$

to lowest order.

A more relevant approach for weakly twisted coronal loops is to adopt a *Tokamak Expansion* by expanding the flux

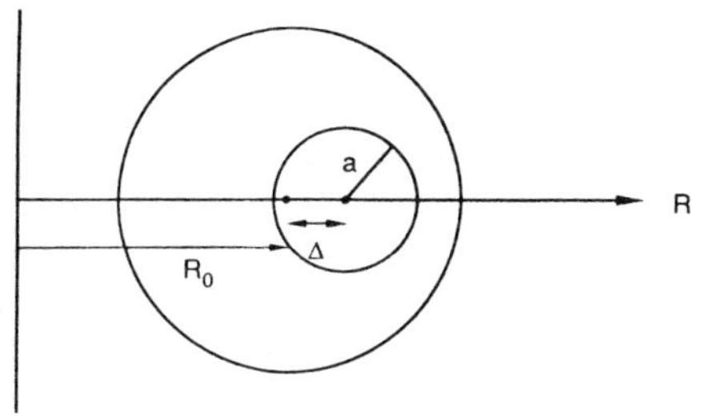

Fig. 32. The Shafranov shift of flux surfaces

function in powers of ε, as in (43), but to adopt the ordering

$$\frac{B_p}{B_\phi} \sim \varepsilon, \quad \beta_t \equiv \frac{2\mu p}{B_\phi^2} \sim \varepsilon^2,$$

so that

$$p = \varepsilon^2 p_2(\psi) + \dots ,$$
$$B_\phi = B_0 + \varepsilon^2 B_{2\phi}(\psi),$$

where B_0 is the field at $R = R_0$. To lowest order, one finds the cylindrical result

$$\frac{d}{dr}\left(p_2 + \frac{B_0 B_{2\phi}}{\mu}\right) + \frac{B_{1\theta}}{\mu r}\frac{d}{dr}(r B_{1\theta}) = 0,$$

and at the next order

$$[r B_{1\theta}^2 (\psi_1/B_{1\theta})']' = r B_{1\theta}^2 - 2\mu r^2 p_2',$$

which determines ψ_1. A vertical field $B_v \sim \varepsilon^2 B_0$ may be added to give an equilibrium, which implies adding the term $\psi_v = R_0 B_v r \cos\theta$ to ψ. The resulting toroidal shift becomes

$$\Delta = -\frac{\psi_1(a)}{\psi_0'(a)} - \frac{R_0 B_v}{B_{1\theta}(R_0)},$$

so that B_v can indeed be chosen to make Δ vanish. In particular, one finds that

$$B_v = \frac{\mu I_0}{4\pi R_0}\left(\beta_p + \frac{\ell_i - 3}{2} + \log_e \frac{8R_0}{a}\right),$$

where

$$\ell_i = \frac{2}{a^2 B_\theta^2(a)} \int_0^a B_\theta^2 \, r \, dr,$$

which is greater than or of order 1/2. Alternatively, the vertical field may be written

$$B_v = \frac{1}{4\pi R_0 I_0}\left[I_0^2 \frac{\partial}{\partial R_0}(L_e + L_i) + 8\pi^2 \int_0^a \left(p - \frac{B_0 B_{2\phi}}{\mu}\right) r dr \right],$$

where

$$L_e = \mu R_0 (\log 8R_0/a - 2),$$
$$L_i = (2/I_0^2) \int B_\theta^2 / 2\mu \, dV,$$

and the terms in the expression for B_v represent the hoop force, the tyre-tube force and the R^{-1} force due to the toroidal field, respectively.

5. Effect of Gravity - the Solar Photosphere

The effects of gravity in the solar atmosphere are most important in the photosphere where the scale-height $H = RT/g$ is lowest. For example, Parker (1955) suggested that an isolated horizontal flux tube in the solar interior would tend to rise by so-called *magnetic buoyancy*. The argument is very simple. if a tube is in lateral equilibrium with its field-free surroundings having a plasma pressure p_e, then its internal pressure (p_i) and magnetic field (B_i) satisfy

$$p_e = p_i + \frac{B_i^2}{2\mu},$$

or, if the temperature (T) is uniform,

$$RT\rho_e = RT\rho_i + \frac{B_i^2}{2\mu}. \tag{46}$$

Thus

$$\rho_e > \rho_i,$$

and the plasma in the tube experiences a buoyancy force, which exceeds the magnetic tension if

$$(\rho_e - \rho_i) g > \frac{B_i^2}{\mu L},$$

where L is the length of tube which is curved upwards (Figure 33). After substituting for $(\rho_e - \rho_i)$ from (46) this condition becomes

$$L > 2H.$$

If such a large flux tube in the interior rises and breaks through the surface it will form a pair of sunspots. In practice, the unbalanced force would make flux tubes rise much faster

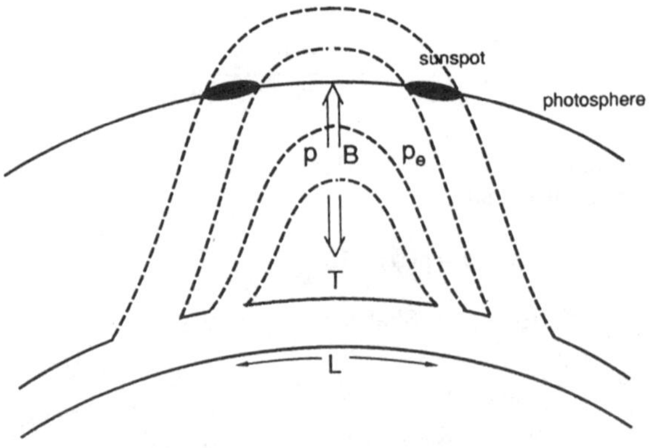

Fig. 33. Rise of flux tube by magnetic buoyancy

than a solar cycle period and so it is thought that the flux tubes are created by dynamo action not throughout the convection zone but only at its base.

Most of the flux tubes which penetrate the solar surface are thought to be almost vertical due to magnetic buoyancy. A whole hierarchy of such tubes exist from the tiniest, only one or two hundred kilometres across, to enormous sunspots with a diameter of thirty megametres (Zwaan, 1978; Roberts, 1989). A basic problem, therefore, is to determine the structure of such a tube (its pressure p_i, density ρ_i, field B_i and radius a) as a function of height (z). As the external pressure (p_e) and density (ρ_e) fall off in value with height, so the internal field strength tends to decrease and the tube spreads out (as the radius a increases). One needs to solve the equilibrium equation

$$-\underline{\nabla}\left(p_i + \frac{B_i^2}{2\mu}\right) + (\underline{B}_i \cdot \underline{\nabla})\,\frac{\underline{B}_i}{\mu} + \rho_i \underline{g} = \underline{0}$$

inside the tube together with the hydrostatic equilibrium equation

$$\frac{dp_e}{dz} = \rho_e g$$

outside and a pressure matching condition

$$p_i + \frac{B_i^2}{2\mu} = p_e$$

on the suface (S) of the tube. For a thin tube with

$$p_i \sim p_e \sim e^{-z/H}$$

pressure balance gives

$$B^2 \sim e^{-z/H}$$

and so the tube radius expands exponentially like

$$a \sim e^{z/(4H)}.$$

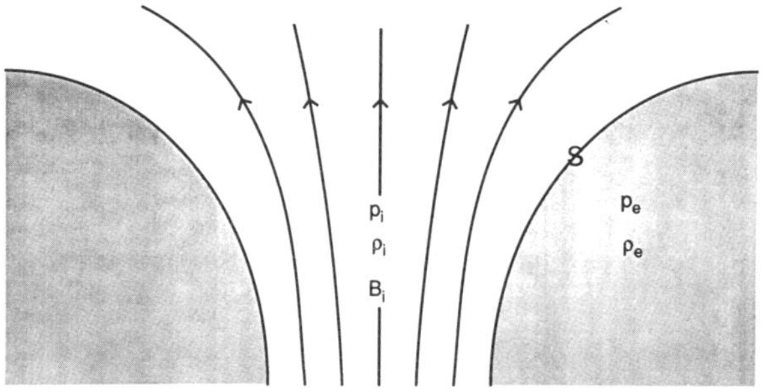

Fig. 34. A spreading vertical flux tube

Numerical solutions of great interest for a thick tube representing a sunspot have been presented by Pizzo (1986). A basic question about the nature of a sunspot is whether it is essentially a single monolithic flux tube or whether it consists of many flux tubes (Parker, 1979). Arguments in favour of the latter include the presence of bright umbral dots, the role of interchange instability in fragmenting a spot and the fact that the temperature of a sunspot tends to be independent of its radius.

6. Solar Coronal Flux Tubes

In the solar corona there are many different types of coronal loop (Priest, 1978, 1981, 1982) which have often for simplicity been modelled as current-free or force-free structures. As well as understanding their plasma and magnetic structure, an important problem is to determine how they are heated. One class of mechanisms involves magnetic waves in response to rapid photospheric motions of the footpoints of coronal field lines. A second class arises from the response to slow footpoint motions, when the coronal field tries to evolve slowly through a sequence of smooth equilibria. But it is possible that smooth three-dimensional equilibria do not general exist, and instead current sheets may form (Parker, 1972, 1989; Berger 1989; Wolfson 1989; Aly and Amari, 1989; Amari and Aly, 1989). Such regions of high magnetic gradients may develop either about X lines that exist between twisted flux tubes, or they may appear due to random braiding of the footpoints. Either way, the current sheets would not persist but would instead give rise to magnetic reconnection (Sonnerup, 1989) and intense local heating.

An interesting three-dimensional ideal numerical experiment has recently been set up by Mikic et al (1988). The plasma is assumed incompressible with a negligible plasma pressure and flow speeds much smaller than the Alfvén speed. A $64 \times 64 \times 64$ mesh is used and a set of initially straight and uniform magnetic field lines is braided by a series of random footpoint motions, at peak speeds of 0.03 v_A (Figure 35). A series of smooth equilibria is produced, but there is a transfer to small scales creating filamentary currents that grow exponentially in time with a growth-time of a hundred Alfvén travel times. An example of one of the current filaments is shown in Figure 36.

One of the most mysterious structures in the solar corona is a prominence, which is a vertical sheet of dense cool plasma supported against gravity by the magnetic field (Figure 4). A new *Flux Tube Model* for prominences has recently been proposed by Priest, Hood and Anzer (1989). They suggest that the basic geometry of the magnetic field around a prominence is a large-scale curved flux tube whose ends are slowly twisted up by Coriolis forces. The initial untwisted field lines are concave downwards (Figure 37a) and are not condusive to prominence formation by injection or radiative condensation, since any cool plasma at the flux tube summit will tend to drain down the legs before it accumulates. The extra component introduced by twist, however, is concave upward and so, when the twist is large enough its curvature will dominate that of the axial component and will produce a dip in the field line with upward curvature at the flux tube summit. Then the prominence can begin to form by injection or condensation in the magnetic dip. As the twist continues, so the region of upward curvature expands along the flux tube and the prominence grows in length (Figure 37c). Eventually, when the twist is too large the prominence may loose

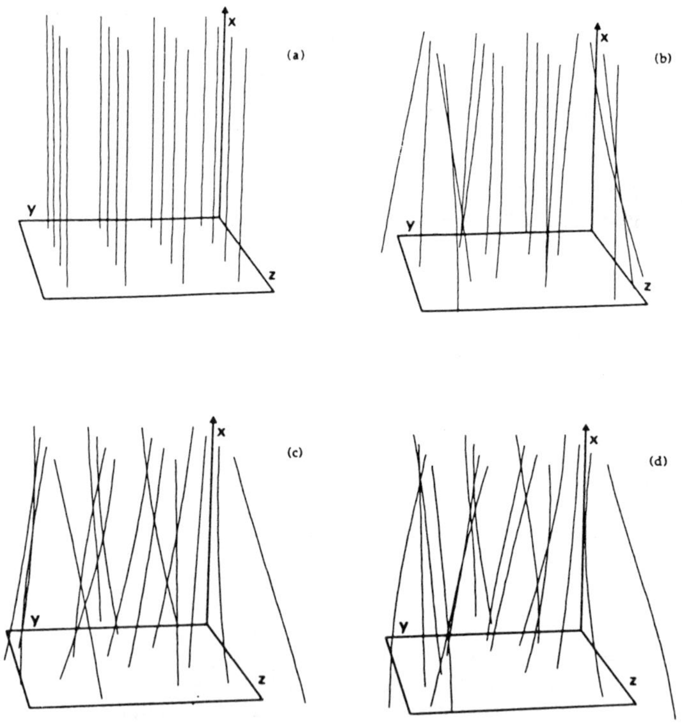

Fig. 35 Braiding of magnetic field lines (Mikic et al)

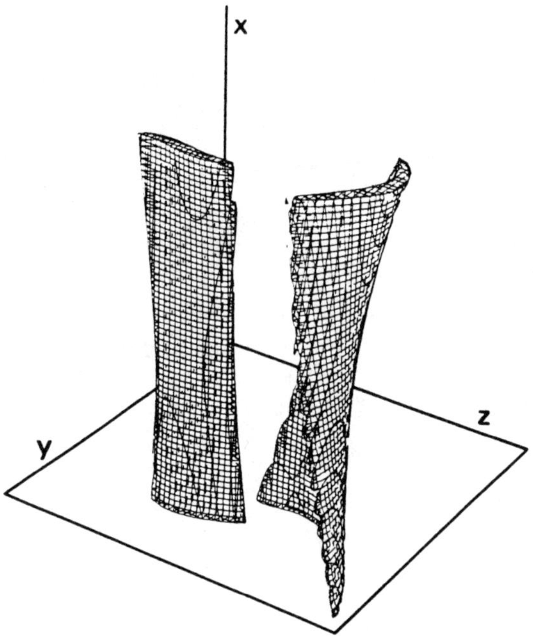

Fig. 36. A current filament

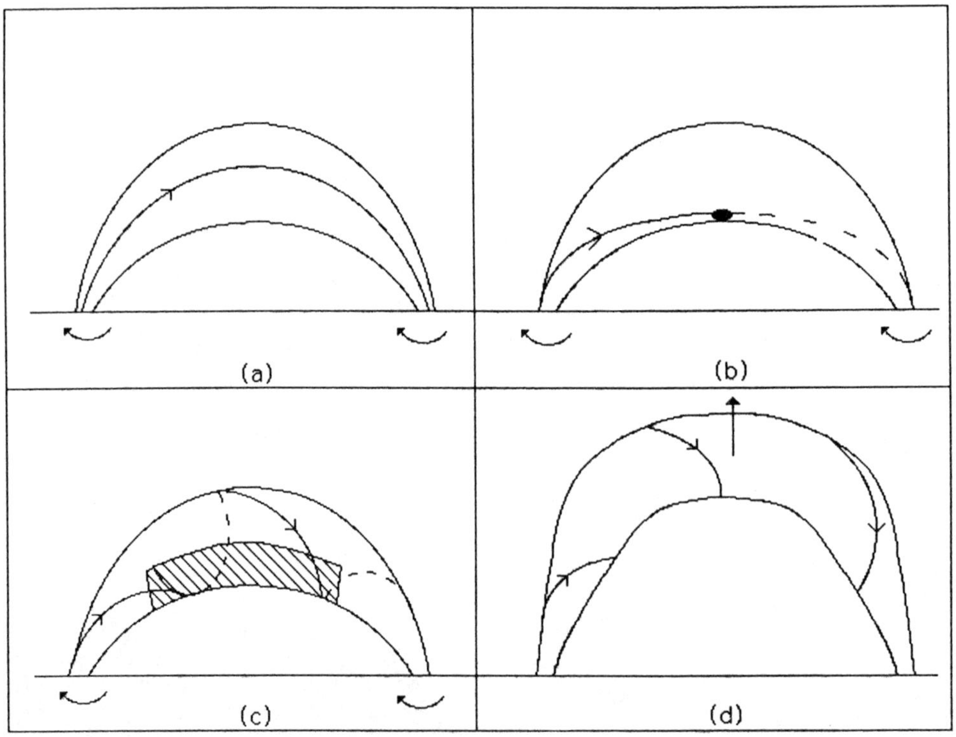

Fig. 37 The Flux Tube Model for Prominences

equilibrium or go unstable and erupt, undergoing a metamorphosis and revealing the flux tube structure for the first time. The force-free equilibrium structure locally around the prominence sheet before the eruption may be modelled by neglecting the large-scale curvature and writing the field components as

$$(B_r, B_\theta, B_z) = \left(\frac{1}{r}\frac{\partial A}{\partial \theta}, -\frac{\partial A}{\partial r}, B_z(A)\right)$$

in terms of a flux function (A) which satisfies

$$\nabla^2 A + \frac{d}{dA}\left(\frac{1}{2}B_z^2\right) = 0.$$

Solutions to this equation may then be found which has the required topology shown in Figure 38.

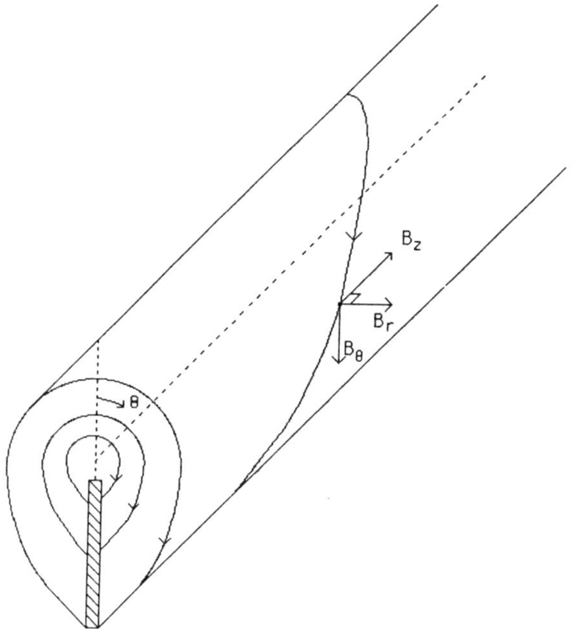

Fig. 38. Support of a Prominence Sheet in a Flux Tube.

7. Conclusion

Even though magnetic flux ropes have at first sight a simple structure, we have seen that many effects need to be incorporated to build a realistic model, such as variations in external pressure, twist, transverse structure and curvature. These have been included in an approximate manner, but the full nonlinear free boundary-value problem that arises when the twist and confining pressure are imposed is only just being tackled. However, in the remainder of these proceedings, having set up a basic equilibrium state, we shall: see how to perturb the equilibrium and study the resulting waves and instabilities; study the interaction of tubes by magnetic reconnection; and compare the behaviour of flux ropes in different parts of the solar system.

Questions and Answers

Mikic: Under what circumstances would you expect flux tubes to balloon out radially (in the middle), as you showed?

Priest: Bulging of flux tubes can be caused either by a decrease of the external confining pressure at the bulge or as increase of the twist (and therefore the internal magnetic pressure).

Akasofu: You showed Hα photographs of loop-like structures of the Sun. I believe that they are embedded in magnetic field lines which have similar (invisible) loop-like structures. The loop-like structure is visible because of the presence of cool and dense gases. How can you justify your treatment of singling out one loop-like structure from the configuration?

Priest: I agree that the whole of the corona is filled with magnetic field lines and that the field may well be slowly varying there. But, the photospheric magnetic field of high strength (≧ 1 kG) is certainly restricted to discrete flux tubes. In soft x-rays the loops that show up have a higher density and temperature than normal, but in a chromospheric, transition region or coronal line, the visible loops will be the ones that have the appropriate temperature. As I stressed, the interaction between loops is most important, but since flux tubes are the building blocks of coronal configurations, it makes sense from a mathematical modeling philosophy to consider separate tubes before studying their interactions. Moreover, one can imagine a situation where one part (i.e., flux tube) of a configuration is subjected preferentially to an enhanced change of pressure or twist or footpoint motion.

Parker: The dramatic coronal loop structure shown in your early slides represents discrete tubes of enhanced plasma density, presumed to be along the lines of force of the magnetic field (the field is invisible, of course). Since the magnetic pressure is generally 10 - 10^3 times larger than the plasma pressure, is there any reason to think that the discrete tubes of plasma imply anything more than a continuous magnetic field - a magnetic field composed of close packed flux tubes, with some tubes containing a tenuous plasma, and most of the tubes relatively empty and hence invisible?

Priest: Certainly I expect plasma structures to lie mainly along magnetic field lines, although it is possible for standing compressions and expansions to produce plasma structures inclined to the field (Priest, App. J., 1988). I also agree that the coronal magnetic field is likely to be mostly smoothly varying, although the flux tube structure I have been describing gives a mechanism for producing filamentary currents with flux tubes having intense cores in the corona. One effect of the low plasma beta in the corona is that small magnetic changes can produce large plasma changes. However, it is not clear to me how much of the plasma tubes are contributing to the observed emission and so, how many of the flux tubes are invisible. Soft x-rays are rather insensitive, summing over a wide range of temperatures, but transition region and coronal lines can pick out loops over a small temperature range and appear to show many more loop structures.

Wolfson: You found a sequence of equilibrium solutions in which a cusp eventually formed and no further equilibrium was possible, but this required an infinite twist. Doesn't this imply that nonequilibrium will never really be achieved in this geometry?

Priest: It is important fully to understand the simpler case of imposing B_θ as a preparation for study of imposing the twist. However, by analogy with the case of an arcade in cylindrical geometry (Zwingmann, 1987) I expected that imposing the twist for a force-free flux tube should always produce an equilibrium, whereas imposing the pressure in a magnetostatic tube is more likely to produce nonequilibrium. It was, therefore, a surprise to me that Steinolfson's force-free tube appears to show nonequilibrium when twisted too much and I do not yet understand why.

Haerendel: Is not the large-scale curvature of the field downwards in a prominence?

Priest: The large-scale structure of a prominence between two feet is curved downwards, but this cannot represent the magnetic field, since gravity would make the plasma flow downwards towards the feet rather than remaining supported as observed. Instead, the local magnetic field lines cannot follow the plasma structure but must be curved upwards so as to provide the support against gravity.

Martens: Does not Leroy find most prominences to have inverse polarity?

Priest: The observations of prominence magnetic field polarity by Leroy refer only to tall enough prominences and so many low-lying active region prominences are exlcuded. Indeed, the new flux tube model of Priest, Hood and Anzer suggests that, as a large flux tube is slowly twisted up and migrates away from an active region, so it may evolve from normal polarity to inverse polarity.

Acknowledgements

The author is most grateful to his colleagues in St Andrews for stimulating comments and to the UK Science and Engineering Research Council, European Economic Community and American Geophysical Union for financial support.

References

Aly J J and Amari, T, Current sheets in 2D potential magnetic fields I, *Astron. Astrophys.*, submitted, 1989.

Amari T and Aly J J, Current sheets in 2D potential magnetic fields II, *Astron. Astrophys.*, submitted, 1989.

Berger M, Braided flux ropes and coronal heating, This Monograph, 1989.

Berger M and Field G, The topological properties of magnetic helicity, *J. Fluid Mech.* 147, 133, 1984.

Browning P K and Hood A W, The shape of twisted, line-tied coronal loops, *Solar Phys.*, submitted, 1989.

Browning P K and Priest E R, The structure of untwisted magnetic flux tubes, *Geophys. Astroph Fluid Dyn.* 21, 237, 1982.

Browning P K and Priest E R, The structure of untwisted magnetic flux tubes, *Astrophys. J.* 266, 848, 1983.

Dixon A, Berger M, Browning P K and Priest E R, A generalisation of the Woltjer minimum energy theorem, *Astron. Astrophys.*, in press, 1989.

Freidberg J P, Ideal Magnetohydrodynamics, Plenum Press, London, 1987.

Heyvaerts J and Priest E R, Coronal heating by reconnection in DC current systems, a theory based on Taylor's hypothesis, *Astron. Astroph.* 137, 63, 1984.

Lothian R M and Hood A W, Twisted magnetic flux tubes: effect of small twist, *Solar Phys.*, in press, 1989.

Lundqvist S, Magnetohydrostatic fields, *Arkiv fur Fysik* 2, 30, 1950.

Parker E N, Formation of sunspots from the solar toroidal field, *Astrophys.J.* 121, 491, 1955.

Parker E N, Topological dissipation and the small-scale fields in turbulent gases, *Astrophys. J.* 174, 499, 1972.

Parker E N, Cosmical Magnetic Fields, Oxford University Press, 1979.

Parker E N, Formal mathematical solutions of the force-free equations and dissipation in large-scale magnetic fields, This Monograph, 1989.

Pizzo V, Numerical solution of the magnetostatic equations for thick flux tubes, with application to sunspots, pores and related structures, *Astrophys. J.*, 302, 785; 1986.

Priest E R, The structure of coronal loops, *Solar Phys.* 58, 57, 1978.

Priest E R, Theory of loop flows and instability, Chap 9 of *Solar Active Regions* (ed F Q Orrall) Colo. Ass. Univ. Press, Boulder, USA, 1981.

Priest E R, Solar Magnetohydrodynamics, D Reidel, 1982.

Priest E R, Hood A W and Anzer U, A twisted flux tube model for solar prominences. I general properties, *Astrophys.J.*, in press, 1989.

Roberts B, Properties and models of photospheric flux tubes, This Monograph, 1989.

Sonnerup B U O, Ip J and Phan T D, Magnetic field reconnection, This Monograph, 1989.

Spruit H C, Motion of magnetic flux tubes in the solar convection zone and chromosphere, *Astron. Astrophys.* 98, 155, 1981.

Steinolfson R S and Tajima T, Energy buildup in coronal magnetic flux tubes, *Astrophys.J.* 322, 503, 1989.

Taylor J B, Relaxation of toroidal plasma and generation of reverse magnetic fields, *Phys. Rev. Lett.* 33, 1139, 1974.

Wesson J, Tokamaks, Oxford University Press, 1988.

Wolfson R, Current sheet formation in sheared force-free fields, This Monograph, 1989.

Wolfjer L, A theorem on force-free fields, *Proc. Nat. Acad. Sci. USA*, 44, 489-91, 1958.

Zwaan C, On the appearance of magnetic flux in the solar photosphere, *Solar Phys.* 60, 213, 1978.

Zweibel E G and Boozer A H, Evolution of twisted magnetic fields, *Astrophys.J.* 295, 642, 1985.

MHD WAVES ON SOLAR MAGNETIC FLUX TUBES
TUTORIAL REVIEW

Joseph V. Hollweg

Physics Department and Institute for the Study of Earth, Oceans and Space
University of New Hampshire

Abstract. Solar magnetic flux tubes are intense (1000-2000G) fields at photospheric levels, concentrated into small (diameter ≈ few x 10^2 km) bundles. They are confined by the external gas pressure, but since this pressure decreases rapidly with height, the flux tubes fan out and lose their individual identities above some 500 km or so. This rapid variation with height makes realistic analysis of wave propagation very difficult, but in this tutorial we will review from a physical point of view some of the highly simplified models which have been studied. The torsional Alfvén mode is one case which can be investigated in detail, and we will discuss its propagation, reflection, and non-WKB properties. The thin flux tube approximation has been used at low heights, and we will discuss the sausage and kink modes described by it. In the corona one sees x-ray emitting loops, which are really plasma tubes in a roughly uniform magnetic field. Here much emphasis has been placed on surface waves and resonance absorption, which we will discuss in simple physical terms, with emphasis on its viability as a coronal heating mechanism. We will also discuss some recent work on resonant instabilities which can occur when bulk flows are present. Finally, we will discuss some nonlinear effects such as solitons, shock formation, and spicules.

Introduction

The study of waves in the solar atmosphere is an old subject, motivated originally by attempts to explain coronal heating by the steepening of sound waves into shocks. That idea has now been discarded, because the solar magnetic field is observed to be associated with regions of strong heating, and because sound waves cannot carry adequate energy fluxes, given the observational constraints on motions that might be associated with waves. However, there is much interest in MHD wave propagation in the solar atmosphere. The fast and Alfvén MHD modes, and MHD surface waves, may be able to heat the atmosphere, and much effort has gone into investigating that possibility. The interaction of the solar p-modes (essentially sound waves trapped in the solar convection zone) with sunspots and other magnetic features may eventually lead to an understanding of how the magnetic field is structured below the visible surface. The propagation of waves in the presence of a magnetic field, gravity, and strong inhomogeneity is a challenging and intriguing problem in its own right. And the strong potential for nonlinear effects in the solar atmosphere promises to yield new insights into the burgeoning field of nonlinear mechanics. Finally, the direct observation of Alfvén waves in the solar wind strongly suggests that the sun does produce MHD waves, and this provides a strong motivation for studies of the generation and propagation of MHD waves in other parts of the solar atmosphere.

In this tutorial discussion we will summarize some current ideas concerning waves in the sun's magnetic field. Our emphasis will be on the basic physical principles, with the intent of providing a guide to the more detailed discussions appearing elsewhere in this volume. We will not give complete references, but we will try to cite papers which provide a fairly complete survey of the literature. The review by Roberts (1985) is particularly recommended, as is the short contribution of Spruit(1981); some of the ideas discussed in this paper have also been reviewed by Hollweg (1985).

The Solar Magnetic Field

In the photosphere and chromosphere, the sun's magnetic field is highly structured. At photospheric levels the field appears to be clumped into intense (1-2 kilogauss) bundles with diameters of a few hundred kilometers. The gas pressure inside these 'flux tubes' is lower than the outside pressure, which provides the confining force. The gas pressure declines with increasing height, and the flux tubes necessarily expand and lose their individual identities above some 500 km or so. The expansion is so rapid that field lines at the edges of the flux tubes are nearly horizontal, and some workers envision a 'magnetic canopy' at a height of about 500 km. When the flux tubes fan out and merge the average field strength is 5-10 Gauss in coronal holes, and some 50-100 Gauss in the active regions. See the tutorial review by Priest (this volume) for a detailed discussion of magnetic field structures.

The Alfvén Mode

The Alfvén mode on magnetic flux tubes has received considerable attention because it involves equations which are easily solved analytically (Hollweg, 1984) and numerically (An et al., 1989). It is an idealization, but it

Geophysical Monograph 58

can be regarded as generally representative of the behavior of more complicated wave motions.

Consider an axisymmetric flux tube, with a vertical symmetry axis. In cylindrical coordinates (R, ϕ, z) the background magnetic field is taken to have only R and z components, which can be arbitrary functions of position consistent with $\nabla \cdot \mathbf{B} = 0$. In linear theory the Alfvén mode consists of axisymmetric twists in the ϕ direction. The only restoring force is the magnetic tension; the mode has no pressure or density fluctuations, and there is no coupling to gravity or radiation. Energy propagates along the individual field lines, and the problem becomes one-dimensional, with distance along a given field line, s, being the only spatial variable.

The linearized equations are:

$$\rho_o \frac{\partial}{\partial t}(R\delta V) = \frac{B_o}{4\pi} \frac{\partial}{\partial s} (R\delta B) \quad (1)$$

$$\frac{\partial \delta B}{\partial t} = R\, B_o \frac{\partial}{\partial s}(\delta V/R) \quad (2)$$

$$\frac{\partial^2 x}{\partial t^2} = \frac{B_o}{4\pi\rho_o R^2} \frac{\partial}{\partial s} (R^2 B_o \frac{\partial x}{\partial s}) \quad (3)$$

where

$$x = \delta V/R$$

and R(s) is the distance from the symmetry axis to the field line. Here ρ is density, B is magnetic field, V is velocity, the subscript 'o' denotes the background and the prefix 'δ' denotes the fluctuation in the ϕ direction. Near the symmetry axis one expects $R^2 B_o$ = constant and then

$$\frac{\partial^2 x}{\partial t^2} = v_A^2 \frac{\partial^2 x}{\partial s^2} \quad (4)$$

where $v_A(s)$ is the Alfvén speed. Equation (4) is the one usually studied.

The Alfvén speed increases rapidly with height, from 7-8 km s^{-1} in the photosphere to 1000-2000 km s^{-1} in the corona, which begins only 2200 km above the photosphere. This large Alfvén speed gradient causes wave reflections and other non-WKB effects. A fundamental problem is how much wave energy can actually be transmitted into the chromosphere and corona. If we take $v_A \propto \exp(s/2h)$ in the lower solar atmosphere, and $v_{A,corona}$ = constant, the energy transmission coefficient is $4\pi h\omega/v_{A,corona}$, where ω is angular frequency and we have taken $2h\omega/v_{A,corona} << 1$. If the magnetic field were not clumped into flux tubes, the transmission would be very small. But the clumping increases the energy which can be transmitted into the corona by increasing h in two ways: the larger Alfvén speeds in the photosphere increase the transmission by a factor of 2, and the longer field lines near the edges of the expanding flux tubes can increase the transmission by a factor of 2-5. There is still some controversy, but it appears to us that interesting energy fluxes can be carried from the photosphere into the chromosphere and corona by Alfvén waves.

There are other interesting non-WKB effects. One does not have local equipartition between kinetic and magnetic energy, and the WKB expressions $\delta V \propto \rho_o^{-1/4}$ and $\delta B \propto \rho_o^{1/4}$ no longer hold. The average Poynting flux is not $\rho_o \langle \delta \mathbf{V}^2 \rangle v_A$ and care must be taken in using observations of nonthermal motions to deduce an energy flux.

As an extreme example, consider waves in a coronal hole and take $\omega \approx 0$ (Hollweg and Lee, 1989). Close to the sun the time-averaged (indicated by the angle brackets) Poynting flux turns out to be

$$P \approx \rho_o \langle \delta \mathbf{V}^2 \rangle v_A^2/v_{A,crit} \quad (5)$$

where $v_{A,crit}$ is v_A at the *Alfvénic* critical point in the solar wind; in deriving (5) we have used $R^2 B_o$ = constant near the symmetry axis. We now find from energy flux conservation that $\delta V \propto B_o^{-1/2}$ and $\delta B \propto B_o^{1/2}$, in contrast to the WKB results. In the photosphere and chromosphere $v_A < v_{A,crit}$, and the Poynting flux is less than the WKB result; this is a consequence of the reflections. At the base of the corona, however, $v_A > v_{A,crit}$ and the energy flux is greater than the value that would be deduced from measurements of nonthermal coronal motions (Hassler et al., 1988) using the usual WKB results. The reason is that more energy resides in the magnetic fluctuations than in the velocity fluctuations; we have

$$\frac{\langle \delta \mathbf{B}^2 \rangle}{8\pi} = \frac{1}{2} \rho \langle \delta \mathbf{V}^2 \rangle \frac{v_A^2}{v_{A,crit}^2} \quad (6)$$

for this case. For example, at the base of a coronal hole we take B_o = 6 Gauss and $\rho_o = 3.5 \times 10^{-16}$ g cm^{-3}, giving v_A = 900 km s^{-1}. In a high-speed solar wind stream we might have $v_{A,crit}$ = 600 km s^{-1}, and the low-frequency Poynting flux is enhanced by 50 percent. Typical nonthermal velocities in the corona are 20-30 km s^{-1} (along the line of sight), and if these motions are waves the energy flux is sufficient to drive high-speed solar wind streams. In reality, neither the WKB nor $\omega \approx 0$ limits apply exclusively, and the actual energy flux probably lies between the two extremes.

If we take $\delta V \propto B_o^{-1/2}$, then a 25 km s^{-1} velocity in a 6 Gauss coronal field corresponds to a 1.6 km s^{-1} velocity in a 1500 Gauss photospheric field. The latter value is comparable to solar granular velocities.

Another interesting feature of Alfvén waves is their ability to excite resonances on coronal field lines which return to the solar surface rather than extending out into the solar wind. The coronal part of the field line can act as a resonant cavity because waves can be strongly reflected from the dense chromosphere and photosphere at each end of the coronal 'loop'. If the wave source is in the photosphere at one end of the field line, it turns out that the transmission into the corona is greatly enhanced at those frequencies for which an integral number of half wavelengths fits into the coronal part of the loop, i.e.

$$\omega_{res} = n\pi v_{A,corona}/L$$

where L is the length of the coronal part of the field line and n = 1,2,... At these resonant frequencies a large-amplitude standing wave builds up in the corona. The transmission is enhanced because some of this wave leaks out of the corona, and destructively interferes with the downgoing wave which originated at the source but was reflected on its way to the corona. The net result can be nearly complete cancellation of the downgoing wave below the corona, resulting in almost no reflected wave and a large transmission coefficient. The effect is analogous to anti-reflection coatings on camera lenses.

Solar spicules may also act as resonant cavities, but in this case the resonance occurs when an odd number of quarter wavelengths fits into the spicule, which is bounded above by the rarefied corona and below by the dense photosphere and low chromosphere.

An interesting feature of the loop resonances is the following: Consider a broadband wave source, and calculate the net coronal heating by doing an appropriate integration under the resonance curve, which has a frequency width determined in part by the coronal wave dissipation rate, γ, and in part by the rate of leakage out of the coronal cavity. If γ is small, then the net coronal heating is proportional to γ. But if γ becomes large enough, then the net coronal heating will be independent of γ, and one can calculate the coronal heating rate without having to worry about the dissipation mechanism. This behavior results when increasing γ leads to a decreased energy density in the cavity, but also to a compensating increase in the width of the resonance curve. This behavior was postulated by Ionson (1982), and studied further by Hollweg (1984, 1985).

The Sausage Mode

In addition to the Alfvén mode, there are sausage and kink modes (Edwin and Roberts, 1983; Roberts, 1985). The sausage mode represents a propagating change in the cross-sectional area, while the kink is a sideways displacement. Consider first the case where the flux tube is infinitessimally thin.

The sausage mode is easily understood physically if we ignore gravity and the fanning out of the flux tubes with height. If we move along with the wave, we see a steady background flow at the phase speed (which is not yet determined) moving through a 'pipe' with a cross-sectional area 'A' which varies spatially due to the wave; see Figure 1. We use the usual one-dimensional channel flow equations

$$\rho VA = \text{constant}$$

$$BA = \text{constant}$$

$$\rho VdV + dp = 0$$

$$p/\rho^{\gamma} = \text{constant}$$

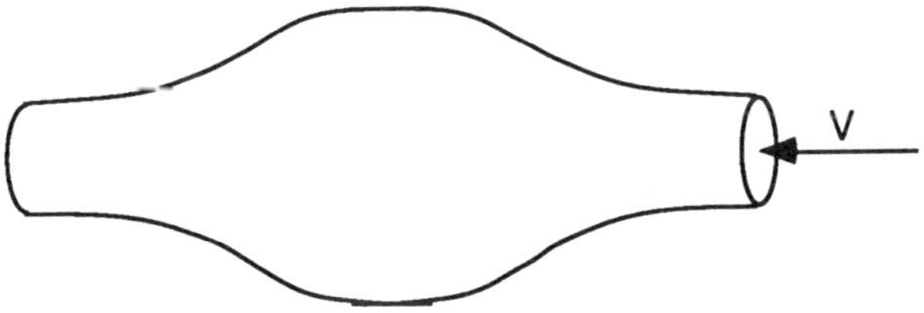

Fig. 1. Sketch of a sausage mode as viewed in the wave frame.

We obtain the following expression for the total pressure (i.e. the sum of the fluid and magnetic pressures) variation:

$$dp_{tot} = B\left(\frac{\rho V^2/B^2}{M^2-1} + \frac{1}{4\pi}\right) dB \tag{7}$$

where M is the Mach number. Since the tube is thin, it hardly disturbs its surroundings, so dp_{tot} must be nearly zero. This requires

$$V^2 = \frac{v_A^2\, v_s^2}{v_A^2 + v_s^2} = (\text{tube speed})^2 \tag{8}$$

where v_s is the sound speed. In linear theory the value of V given by (8) is the background flow speed, i.e. the phase speed. The phase speed is subsonic and sub-Alfvénic, so the sausage mode is basically a slow mode and it does not carry a substantial energy flux. Increases in the magnetic field strength are accompanied by decreases in gas pressure and density, and by a velocity perturbation in the direction opposite to the propagation direction. Note that this derivation is independent of the cross-sectional shape. And since the surroundings are not perturbed, the result is unchanged if there is an external magnetic field or flow (Ryutova, 1988).

Gravity introduces a cutoff frequency and low frequencies don't propagate. The cutoff depends on v_A/v_s and on the fanning out of the flux tube with height. [See Roberts (1985).] If the background temperature is constant, if the tube does not fan out, and if $v_A{}^2 >> v_s{}^2$, then the cutoff is the usual acoustic cutoff frequency

$$\omega_{ac} = \gamma g/2v_s$$

which corresponds to a period of about 225s in the chromosphere if $\gamma = 5/3$.

The presence of a cutoff manifests itself when one considers the impulse response of the system, which is often described by a Klein-Gordon equation (Rae and Roberts, 1982). One obtains a front propagating at the tube speed, followed by a wake which oscillates at approximately the cutoff frequency. Hollweg (1982) has shown that the wake nonlinearly evolves into a train of 'rebound shocks' in the chromosphere. When the shocks hit the chromosphere-corona transition region (TR) from below, they thrust the TR and underlying chromosphere upwards. This has been proposed as a mechanism for generating the solar spicules.

If the finite diameter of the flux tube is taken into account, then the sausage mode perturbs its surroundings and $dp_{tot} \neq 0$. This turns out to make the wave dispersive. It also means that the pulsating tube can radiate sound into the field-free exterior region, and this in effect damps the sausage mode (Spruit, 1982). If we again imagine that we are moving along with the wave, then sound waves will be produced if the flow in the exterior region is supersonic in that frame. The supersonic flow over the wavy surface of the flux tube produces sound waves propagating along downstream-pointing characteristics, and a drag force is exerted on the wavy surface, damping the wave. In the photosphere, this requires a velocity shear between the internal and external regions.

The Kink Mode

We again consider the thin flux tube limit. We include gravity, and take the unperturbed flux tube to be vertical. Then the kink mode is governed by two restoring forces: magnetic tension and buoyancy. The linearized equation for the kink mode is (Spruit, 1981):

$$(\rho_o + \rho_e)\frac{\partial^2 V_\perp}{\partial t^2} = g(\rho_o - \rho_e)\frac{\partial V_\perp}{\partial z} + \frac{B_o^2}{4\pi}\frac{\partial^2 V_\perp}{\partial z^2} \tag{9a}$$

where z is the vertical direction, the subscript 'e' denotes the exterior region, g is the gravitational acceleration, and the subscript '⊥' denotes the direction transverse to the flux tube. The presence of ρ_e on the left-hand side represents the extra inertia associated with displacing the exterior region as the flux tube moves through it.

Compared to the sausage mode, the kink mode has a faster group velocity and a lower cutoff frequency. Both of these effects enhance the energy flux which can be carried by this mode, and it is a candidate for coronal and chromospheric heating. But it should be remembered that the thin flux tube approximation breaks down badly above about 500 km. It should also be kept in mind that the kink mode suffers from the same reflection and transmission problems as were discussed in connection with the Alfvén mode.

If we neglect gravity but allow for an external field B_e along the axis of a thin cylindrical plasma column, the dispersion relation is

$$\frac{\omega^2}{k_z^2} = \frac{B_o^2 + B_e^2}{4\pi(\rho_o + \rho_e)} \qquad (9b)$$

This relation is surprisingly robust (Goossens and Hollweg, 1989). If the motions vary with ϕ as $\exp(im\phi)$, it turns out that the above dispersion relation is valid for any integer $m \neq 0$; the term 'kink mode' is reserved for $|m| = 1$. In particular, waves on active region loops obey this dispersion relation.

Edwin and Roberts (1983) have also pointed out the existence, under photospheric conditions, of a 'slow kink' mode with dispersion relation given by equation (8). We do not understand the physical properties of the slow kink.

Nonlinear Effects

We have already discussed how the sausage mode can lead to a shock train, with possibly important implications for spicules. Other modes can lead to shocks too. Even the Alfvén mode, which is linearly noncompressive, can form fast shocks in the chromosphere (Hollweg et al., 1982). These shocks interact with the chromosphere and TR and produce motions of those regions. Perhaps more importantly, the shocks enter the corona and contribute to the coronal heating, but shocks formed in this way do not dissipate very efficiently and they may only be effective in heating the coronal holes.

Another very interesting nonlinear structure is the soliton. We have seen that the sausage and kink modes can be dispersive. The wave packet spreading caused by dispersion can balance the nonlinear steepening, and result in a soliton. B. Roberts and his colleagues have discussed the very intriguing possibility that this can occur for waves on solar magnetic flux tubes. If it does occur, a possibly important consequence would be that shocks do not develop, so that the implied entropy jumps across them, and plasma heating, do not occur. See the article by Roberts (this volume).

Coronal Loops and Surface Waves

In coronal x-ray pictures one sees an abundance of bright loop structures which apparently outline the magnetic field. They are associated with the solar 'active regions', and they are called 'active region loops'. These structures are not magnetic flux tubes. They are tubes of enhanced plasma density immersed in the relatively uniform coronal magnetic field. They are bright because the emission is basically proportional to the square of the electron number density.

A kink wave on a loop can to a large extent be thought of as a surface wave. To get an idea of how surface waves work, we consider an incompressible fluid and ignore gravity. We again work in the frame moving along the surface with the wave. In that frame there is a background flow at the phase speed, which is not yet determined; see Figure 2. The flow does not cross the surface, and neither does the magnetic field (the surface is thus a planar tangential discontinuity). The flow accelerates and decelerates as it goes over the hills and valleys of the wavy surface. Similarly the magnetic field lines are squeezed together or expanded apart. But this must all happen in such a way that the total pressure is everywhere continuous *across* the surface, even though it varies *along* the surface.

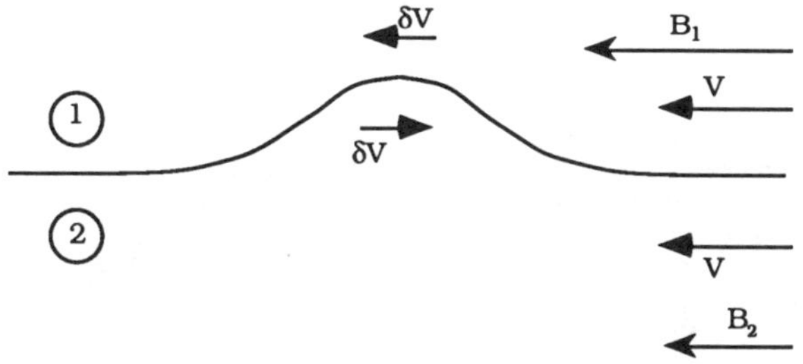

Fig. 2. Sketch of a surface wave as viewed in the wave frame.

Consider for the moment a wave propagating along $\mathbf{B}_o$, which has the same direction on both sides of the surface. Bernoulli's equation can be written

$$p_{tot} + \rho^* V^2/2 = \text{const. on streamlines} \qquad (10)$$

where

$$\rho^* = \rho - B^2/4\pi V^2 = \text{const. on streamlines} \qquad (11)$$

Now in linearized theory an acceleration on one side of the surface is accompanied by an equal deceleration on the other side of the surface. So in order to have continuity of p_{tot} across the surface we require

$$\rho^*_1 = -\rho^*_2 \qquad (12)$$

where the subscripts '1' and '2' refer to the two sides of the surface. This gives the phase speed:

$$V^2_{phase} = \frac{B^2_{o1} + B^2_{o2}}{4\pi(\rho_1 + \rho_2)}$$

Now we can add an arbitrary component of the magnetic field perpendicular to the direction of propagation. That component is carried along with the fluid with no change in its magnitude, because the fluid is incompressible, and it thus does not affect the total pressure balance. So we can write

$$V^2_{phase} = \frac{B^2_{ok1} + B^2_{ok2}}{4\pi(\rho_1 + \rho_2)} \qquad (13)$$

where the subscript 'k' means the component in the propagation direction. Surprisingly, equation (13) also holds when the plasma is cold (the low-β limit) and the propagation direction is nearly perpendicular to $\mathbf{B}_{o1}$ and $\mathbf{B}_{o2}$. Equation (13) is the same as equation (9b), which applies to a thin cylindrical plasma column.

Thus the phase and group velocities are of the order of the Alfvén speed, and these waves can carry sufficient energies to heat the chromosphere and corona, if the observed nonthermal motions are indeed waves.

Resonance Absorption

A nice feature about surface waves is that they can dissipate efficiently by a process called 'resonance absorption' (Ionson 1978). This occurs when the 'surface' has finite thickness, rather than being a true discontinuity.

Usually two cases are considered. The first is the surface wave, which decays due to resonance absorption. Lee and Roberts (1986) studied an initial value problem for a surface wave in an incompressible fluid with no dissipation. They found that the energy which was originally in the surface wave (with a spatial scale of order k^{-1} perpendicular to the surface, where 'k' is the wavenumber) ends up in a thin layer with a thickness of order $ka^2 << k^{-1}$, where 'a' is the surface thickness, which was assumed small so that $ka << 1$. This energy-containing layer surrounds the field line where $\rho^* = 0$ (see below). After the surface wave has decayed away, the fluctuations on neighboring field lines within the energy-containing layer behave independently, and get more and more out of phase with one another. This process is called 'phase mixing', and it leads to large gradients in the direction normal to the surface. If some dissipative mechanism, such as viscosity or electrical resistivity is present, then the motions in the energy-containing layer will eventually be damped and converted into heat. Lee and Roberts also found that the surface wave decay rate scales linearly with surface thickness, and vanishes when $a = 0$.

The second situation involving resonance absorption is when a propagating wave impinges on a surface. If the surface is a true discontinuity, then one has a standard reflection/transmission problem, and the reflected and transmitted energies balance the incident energy. If the 'surface' has a finite thickness, then resonance absorption occurs, energy builds up in the surface and the reflected and transmitted energies are less than the incident energy. This process may be relevant to the interaction of solar p-modes with sunspots. Braun et al. (1987) have investigated the behavior of the p-modes outside of sunspots, and they found that the outgoing p-modes carry substantially less energy than the incoming waves. Hollweg (1988) suggested resonance absorption as a possible mechanism to explain the energy deficit. He used a simple analysis for a *thin* sunspot boundary, and slab geometry, to investigate the requirements for substantial p-mode absorption. He concluded that the requirements were probably too stringent to explain the large observed energy deficit. However, resonance absorption should occur for some of the incident p-mode waves, and it was suggested that those absorbed waves may run up the edge of the umbra and become the 'running penumbral waves'. More recently Lou (1989) has re-examined the problem using cylindrical geometry and a numerical integration of the relevant equations. In agreement with Hollweg, he finds that the energy absorption coefficient is small when the sunspot boundary is thin. But if the plasma and magnetic field vary smoothly over the entire sunspot radius (in a sense this is a *thick* boundary) he finds that the absorption coefficient can be large, as observed.

Hollweg (1987) and Hollweg and Yang (1988) have offered a simple physical picture of resonance absorption. Here we consider again the case of a surface wave in an incompressible fluid. Take $\mathbf{B}_o$ to have the same direction on both sides of the surface, and let the wave propagate along $\mathbf{B}_o$; the components of $\mathbf{B}_{o1}$ and $\mathbf{B}_{o2}$ transverse to the propagation direction again don't change anything. However, we now let the 'surface' have a small thickness, and we assume that all quantities vary smoothly across the surface. We will again work in the frame moving with the wave; see Figure 3. Since the 'surface' is arbitrarily thin, we can make the approximation that p_{tot} is nearly constant *across* it, even though p_{tot} varies *along* the surface; this approximation is equivalent to neglecting the inertia in the 'surface'. The linear dispersion relation for the wave is still approximately given by equation (12), i.e. $\rho^*_1 = -\rho^*_2$. Thus there must be at least one field line within the surface along which $\rho^* = 0$. But Bernoulli's equation (10) then gives the result that p_{tot} is strictly constant along that field line. Thus the requirement that the variations of p_{tot} along the surface be continuous across it can not be satisfied on the $\rho^* = 0$ field line. Something has gone wrong. The error is our implicit assumption that we can find a wave frame in which $\partial/\partial t = 0$. In fact there is no such frame in this case. This is another way of saying there is no normal mode, and a proper analysis requires solution of an initial value problem (Lee and Roberts, 1986).

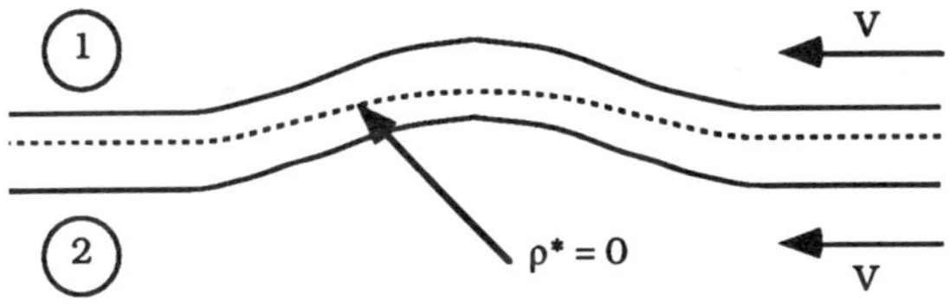

Fig. 3. Sketch of a 'surface' wave on a 'surface' of finite thickness. The field line having $\rho^* = 0$ is in resonance with the surface wave, and field lines in its neighborhood absorb energy from the wave. This leads to wave damping via 'resonance absorption'.

Since p_{tot} is evidently important, we will regard it as known, and write the other variables in terms of it. (In fact, if the 'surface' is thin, then p_{tot} in the surface will be nearly the same as p_{tot} on a true discontinuity, which is easily calculated.) Again linearizing, we obtain

$$\left[\frac{\partial^2}{\partial t^2} - (\mathbf{k}\cdot\mathbf{v}_{Ao})^2\right]\delta V_x = -\frac{1}{\rho}\frac{\partial^2 \delta p_{tot}}{\partial x\,\partial t} \qquad (14)$$

where we have taken the x-direction to be the propagation direction, we have gone back into the frame in which $\mathbf{V}_o = 0$ everywhere, and we have allowed $\mathbf{B}_o$ to have components transverse to $\mathbf{k}$; in equation (14) ρ and $\mathbf{v}_{Ao}$ vary across the surface. (If we take the y-direction to be perpendicular to the surface, we can show that $\delta V_z = 0$.) Equation (14) is the equation for a driven harmonic oscillator. If we use this equation inside the 'surface', then the driver is the total pressure fluctuation applied to the surface by the external surface wave. On field lines satisfying $v_{Ax}^2(y) = \omega^2/k_x^2$, where ω is the angular frequency of the driver, the oscillator is driven at resonance. This resonance condition is precisely equivalent to the condition $\rho^* = 0$ if we replace V^2 with ω^2/k_x^2 in equation (11). The motions driven at resonance are slow mode waves in the incompressible fluid. The resonance leads to secular growth of δV_x (and δB_x). The magnetic and kinetic energies grow on field lines near $\rho^* = 0$, and since energy is conserved in this ideal system the external surface wave must decay. Thus resonance absorption of the surface wave occurs because the total pressure perturbations applied to the surface resonantly drive motions on field lines near $\rho^* = 0$.

This harmonic oscillator model can be used in an initial value problem and the results of Lee and Roberts can be recovered with very little effort (Hollweg, 1987). Their result that the thickness of the energy-containing layer is of the order of ka^2 is easily understood. Strictly speaking, only the single field line on which $\rho^* = 0$ is driven at resonance. But a neighboring field line with natural frequency $|k_x v_{Ax}|$ is effectively in resonance with the driver at frequency $|\omega|$ as long as $|k_x v_{Ax}|t$ and $|\omega|t$ are approximately in phase up to $t = t_{decay}$, where t_{decay} is the decay time of the surface wave. At later times the surface wave has decayed away, and it does not matter if things get out of phase. This requires

$$|k_x v_{Ax} - \omega| t_{decay} < \pi/2 \qquad (15)$$

The criterion $\pi/2$ is somewhat arbitrary, but it happens to yield the exact results obtained by a more formal analysis. Equation (15) contains the essential physics underlying the result that the thickness of the energy-containing layer is of order ka^2. The harmonic oscillator model also reveals why the decay rate is proportional to surface thickness. If the surface is thicker, then $|dv_{Ax}/dy|$ near the resonant field line is smaller and more field lines are effectively in resonance with the driver, leading to a more rapid transfer of energy into the 'resonant' field lines.

It is important to note that energy builds up on field lines which stay nearly *in* phase with the driver. On the other hand, phase-mixing occurs after the energy has resonantly built up, and the driver has decayed away. It is thus wrong to think, as many workers do, that phase-mixing causes resonance absorption. Just the opposite is true.

Hollweg and Yang (1988) have generalized the harmonic oscillator model to a compressible plasma. In that case there are two types of resonances. The 'Alfvén resonance' occurs on field lines satisfying

$$v_{Ax}^2 = \omega^2/k_x^2 \qquad (16)$$

In this case the components of $\delta\mathbf{V}$ and $\delta\mathbf{B}$ perpendicular to $\mathbf{B}_o$, but lying in the plane of the surface, are resonantly driven by δp_{tot}. The motions near the resonant field line closely resemble Alfvén waves, and the surface wave is essentially converted into an Alfvén wave. The other resonance is called the 'cusp resonance'. It occurs on field lines satisfying

$$v_A^2 + v_s^2 - k_x^2 v_{Ax}^2 v_s^2/\omega^2 = 0 \qquad (17)$$

In this case the surface wave is converted into a slow mode near the resonant field line, and the components of $\delta\mathbf{V}$ and $\delta\mathbf{B}$ parallel to $\mathbf{B}_o$ are resonantly driven. Note that the cusp resonance vanishes in a cold plasma. It is usually not of interest in the corona or upper chromosphere, where $v_s^2 << v_A^2$, since the resonance can only occur for slowly propagating surface waves which carry rather little energy. However, the cusp resonance may play a role in dissipating the sausage and kink modes on the photospheric flux tubes, where $v_A \approx v_s$.

Resonance absorption can proceed sufficiently rapidly to heat the corona, if the energy in the energy-containing layer can be converted into heat. As a simple numerical example, we consider a cold plasma (the low-β limit) and $\mathbf{B}_o$ = constant. We take ρ_o to vary linearly across the 'surface'. And we take $\mathbf{k}$ approximately perpendicular to $\mathbf{B}_o$. [This roughly corresponds to the case of a kink mode on a coronal active region loop. If we consider the m = 1 mode (m is azimuthal order), then $k_\perp \approx 2/D$, where D is the loop diameter. If the loop is excited at its fundamental resonance (see the section on Alfvén waves above), then $k_{||} = \pi/L$, where L is the loop length. Thus $k_\perp/k_{||} \approx (2/\pi)(L/D)$. Since $L/D \approx 10$ for active region loops, we have $\tan^{-1}(k_\perp/k_{||}) \approx 80$ degrees. See Goossens and Hollweg (1989), Grossmann and Smith (1988) and Poedts et al. (this volume) for proper discussions of cylindrical geometry.] The surface wave decay time is then given by

$$\omega t_{decay} \approx \frac{8(\rho_{o1} + \rho_{o2})}{\pi k_x a |\rho_{o1} - \rho_{o2}|} \qquad (18)$$

and the thickness of the energy-containing layer is approximately $\pi^2 k_x a^2/8$. If we take 10^4 km for the loop diameter, $\rho_{o2}/\rho_{o1} = 3$, and a = 2000 km, we obtain $\omega t_{decay} \approx 13$. The heating rate, E_H, will be

$$E_H = 2\rho\langle\delta\mathbf{V}^2\rangle/t_{decay} \qquad (19)$$

where we have allowed for equal amounts of fluctuating kinetic and magnetic energy, and the factor 2 comes about because equation (19) deals with energy while t_{decay} is the amplitude decay time. We take $\rho = 5 \times 10^{-15}$ g cm^{-3} and $\langle\delta\mathbf{V}^2\rangle = 2 \cdot (30 \text{ km s}^{-1})^2$ (the factor 2 allows for two degrees of freedom). We then obtain $E_H = 9 \times 10^{-4}$ erg cm^{-3} s^{-1} if $\omega = 2\pi/(100$ sec). This value of E_H is what is required to heat the typical active region loops which we have been considering. The most uncertain numbers are the surface thickness, a, and the wave period; we have chosen a period of 100 sec because it corresponds to the fundamental resonance on a loop with $L = 10^5$ km and $v_A = 2000$ km s^{-1}.

We still have to find a way for the energy which is deposited in the energy-containing layer to be converted into heat. Consider first viscosity or electrical resistivity (Poedts et al., this volume), which we shall simply refer to as η. If the system is continuously driven, then in the steady-state the energy which is pumped into the surface gets dissipated in a layer with thickness proportional to $\eta^{1/3}$, and the maximum velocity and magnetic field fluctuations are proportional to $\eta^{-1/3}$ (Davila, 1987; Hollweg and Yang, 1988). If η is small, so that the background quantities are nearly constant across the heating layer, it turns out that the net heating is independent of η, and equal to the rate at which energy is pumped into the energy-containing layer according to the harmonic oscillator model presented above, which took $\eta = 0$. The result that the net heating can be independent of

η was proposed by Ionson (1978). Unfortunately, the shear viscosity and electrical resistivity are very small in the corona, and the heating layer turns out to be exceedingly thin. This then poses another problem: how does the heat get out of the thin layer and fill up the loop? However, it also turns out that the small coronal dissipation leads to very large velocity and magnetic field fluctuations in the heating layer, and for realistic numbers the assumed linearity fails. So it seems likely that nonlinear processes will occur and distribute the heat and limit the maximum fluctuation amplitudes. Strauss (1989) has suggested that the heated layer will be subject to ballooning instabilities. Hollweg and Yang (1988) suggested that the large velocity shear across the resonant field line will be Kelvin-Helmholz unstable, and that this could initiate a turbulent cascade and ultimately dissipation into heat. Rough estimates of the consequences of these nonlinear processes yield promising numbers: heating layer thicknesses of a few thousand kilometers and maximum velocities of about 40 km s^{-1}. But these ideas should really be tested by numerical simulations.

The postulated nonlinear effects pose another problem. The analysis of resonance absorption assumes well-ordered surfaces or regions which satisfy a simple resonance condition. But nonlinear effects such as turbulence presumably tangle up the field and plasma and destroy the assumed order. It is not known whether resonance absorption can proceed under these circumstances, and numerical simulations will probably be required to settle the issue. See, however, Similon and Sudan (1989) for a discussion of issues related to wave propagation in a disordered environment.

Resonance Absorption and Velocity Shear

We turn now to a fairly new topic. Until recently, resonance absorption has been studied in plasmas having no velocity shear, in which case it is possible to find a frame in which $\mathbf{V}_o = 0$. Introducing velocity shear can make a qualitative difference, however: the resonance can lead to an instability instead of dissipation.

Consider again an incompressible ideal fluid with a thin planar 'surface' separating two uniform regions. The surface is parallel to the x-z plane, the background varies only in the y-direction, and $B_{oy} = V_{oy} = 0$. All fluctuating quantitites are assumed to vary as $\exp(ik_x x - i\omega t)$. The y-displacement of the plasma, ξ, obeys the linearized equation:

$$\frac{d}{dy}\left[\rho\varepsilon\frac{d\xi}{dy}\right] - k_x^2\rho\varepsilon\xi = 0 \qquad (20)$$

where

$$\varepsilon = (\omega - k_x V_{ox})^2 - k_x^2 v_{Ax}^2 \qquad (21)$$

The quantity in square brackets is proportional to δp_{tot}. If we again make the thin surface approximation, δp_{tot} = constant, we can write the following expression for ξ in the surface:

$$\xi = c_3 + c_4\int_{-a}^{y}\frac{dy}{\rho\varepsilon} \qquad (22)$$

where c_3 and c_4 are constants, and the thin surface is taken to be in $-a < y < a$. In the uniform regions 1 ($y > a$) and 2 ($y < a$) we have

$$\xi_1 = c_1 \exp\left[-|k_x|(y-a)\right] \qquad (23)$$

$$\xi_2 = c_2 \exp\left[+|k_x|(y+a)\right] \qquad (24)$$

with c_1 and c_2 constants. At $y = \pm a$ we match ξ and $\rho\varepsilon d\xi/dy$. This then yields the 'dispersion relation'

$$\rho_1\varepsilon_1 + \rho_2\varepsilon_2 + \rho_1\rho_2\varepsilon_1\varepsilon_2|k_x|\int_{-a}^{a}\frac{dy}{\rho\varepsilon} = 0 \qquad (25)$$

If the surface is a true discontinuity ($a = 0$) the dispersion relation is the usual one for a surface wave, i.e. $\rho_1\varepsilon_1 = -\rho_2\varepsilon_2$, and this contains the usual Kelvin-Helmholz instability for an incompressible fluid. If 'a' is small, this is still the approximate dispersion relation. However, the integral in equation (25) contains ε in the denominator, which can be zero. The integral through this pole is handled in the same manner as in the usual derivation of Landau damping, and this introduces a small imaginary part to ω:

$$\omega_i \approx \frac{\pi\,(\rho_1\varepsilon_1/2k_x)^2/(\Sigma\rho)\,(\rho\,|V'_\pm|)_{res}}{\left(\frac{\omega}{k_x}\right)_{c.m.}\left(V_{ox,res} - \frac{\omega}{k_x}\right)} \qquad (26)$$

Here $\Sigma\rho = \rho_1 + \rho_2$, $(\omega/k_x)_{c.m.}$ is the phase velocity in the center-of-mass frame, the subscript 'res' indicates conditions at the location where $\varepsilon = 0$, the prime indicates the y-derivative, $V_\pm$ is

$$V_\pm = V_{ox} \pm v_{Ax}$$

and the sign is chosen according to whether $\omega/k_x = (V_+$ or $V_-)$ at the resonance. If there is more than one resonance in the surface, then the respective ω_i's are added together.

If there is no flow, ω_i is negative definite, and equation (26) yields precisely the decay rate of the surface wave obtained by the simple harmonic oscillator model discussed above. But if there is flow ω_i can be positive giving instability. This instability is resonant, it does not occur if there is a pure discontinuity, and it is thus distinct from the Kelvin-Helmholz instability. Hollweg et al. (1989) have investigated the instability criterion for linear profiles of ρ, V_{ox}, and B_{ox} through the surface. In some cases the instability can occur for velocity shears significantly below the Kelvin-Helmholz threshold. It may play a role in the development of turbulence in solar wind streams (Goldstein et al., 1987). It could perhaps occur at the earth's magnetopause, in association with the Evershed flow in sunspot penumbras, and possibly in association with flows in the photospheric flux tubes (Ryutova, 1988).

Even if the velocity shear is too small to drive the instability, Hollweg et al. found that even small shear can significantly increase or decrease the resonance absorption rate. This may be important for the interaction of p-modes with sunspots.

The nonlinear evolution of the instability is not known. This is a good candidate for a numerical simulation study.

Ryutova (1988) has suggested that 'negative energy waves' can be thought of as being responsible for the instability. This is a very interesting idea which deserves detailed study. But Ryutova's paper contains a puzzle. She uses the concept of negative energy waves to derive an instability criterion for the kink mode on a magnetic flux tube. But she also derives an explicit expression for ω_I for the case where $\rho\varepsilon$ is linear across the flux tube 'surface'. The expression for ω_I can also be used to derive an instability criterion. The two instability criteria do not agree, so the negative energy wave picture is apparently not the whole story. [In fact, Ryutova's expression for ω_I (the second equation following her equation 12) gives an instability even when there is no flow, but we believe that this is due to a sign error. Yang (personal communication, 1989) has shown that there is no instability when $\rho\varepsilon$ is linear across a *planar* surface.] Further studies of negative energy waves in this context will undoubtedly yield some interesting surprises.

Questions and Answers

Hundhausen: In your estimate of an Alfvén wave energy flux at a base of a coronal hole, you used a magnetic field of 15 gauss. Is that field not rather high compared to what you would estimate from the value of the radial field measured in the solar wind and the fraction of the base of the corona covered by coronal holes? Is your conclusion that the Alfvén wave flux is large enough to be interesting not based upon this very optimistic choice of the field?

Hollweg: I agree it's marginal no matter what I do, but Poynting (coronal base) $\propto V_\phi^2 B^2_{corona}$. *My numbers took* $B_{corona} = 15$, $V_\phi = 10$ *km* s^{-1}. B_{corona} *may be overestimated. But,* V_ϕ *may be underestimated since line-broadening data allow "turbulent" rms velocities of 20-30 km/s along the line of sight. As I recall, Dave Webb's favorite number for* B_{corona} *in a hole is 6 Gauss. So, I can still get the flux quoted in my talk (*$3 \times 10_5$ *erg* cm^{-2} s^{-1}*) by increasing* V_ϕ *to 25 km* s^{-1}. *This seems to be allowed, but I agree one has to stretch one's belief. But the Sun doesn't care about what you or I believe.*

Antiochos: Isn't the resonance absorption heating process incompatible with the density profile required to produce it, since wherever the heating is maximum the density will become (on the sound travel time) a local maximum?

Hollweg: A good question which touches on a problem faced by all coronal heating models. I envision the following: suppose one starts out with a weak density gradient or magnetic field gradient. If waves are present one will get heating by resonance absorption in the gradient. The heating will produce a local density maximum as you suggest, with gradients on both sides. There will be more heating there, more mass flows producing new gradients, and so on. The heating will eat its way across the field. Near any local maximum in density, the heating will stop, the density will fall, and that will produce new gradients and renewed heating. So, the whole process is probably not steady but dynamic. The turbulence associated with the heating will tend to smooth things out somewhat, however, and that will probably tend to reduce the on-again, off-again aspect mentioned above. But, a key question is what stops the heating from eating its way completely through the field, i.e., why are only some coronal field lines "lit up" while its neighbors undergo less heating. I don't know the answer to that one. But, that problem is faced by every coronal heating scenario (even yours) and no one has proposed an answer. It must have something to do with what is happening at the magnetic footpoints. I suspect that only higher-revolution photospheric observations will provide the answer.

Acknowledgements. We are grateful for discussions with M. Goossens, M.A. Lee, M. Ryutova, and G.Yang. We also wish to thank E.R. Priest and C.T. Russell for their encouragement. This work is supported in part by the NASA Solar-Terrestrial Theory Program under Grant NAGW-76 and in part by NASA Grant NSG-7411.

References

An, C.-H., Z.E. Musielak, R.L. Moore, and S.T. Suess, Reflection and trapping of transient Alfvén waves propagating in an isothermal atmosphere with constant gravity and uniform magnetic field, Astrophys. J., in press, 1989.

Braun, D.C., T.L. Duvall, Jr., and B.J. Labonte, Acoustic absorption by sunspots, Astrophys. J., 319, L27, 1987.

Davila, J. M., Heating of the solar corona by the resonant absorption of Alfvén waves, Astrophys. J., 317, 514, 1987.

Edwin, P.M., and B. Roberts, Wave propagation in a magnetic cylinder, Solar Phys., 88, 179, 1983.

Goldstein, M.L., D.A. Roberts, S. Ghosh, and W.H. Matthaeus, Numerical simulation of solar wind and magnetospheric phenomena, in Small-Scale Plasma Processes, edited by B. Battrick and E.J. Rolfe, ESA-SP-275, p. 115, 1987.

Goossens, M., and J.V. Hollweg, Work in progress, 1989.

Grossmann, W., and R.A. Smith, Heating of solar coronal loops by resonant absorption of Alfvén waves, Astrophys. J., 332, 476, 1988.

Hassler, D.M., G. J. Rottman, and T.E. Holzer, Line broadening of transition region and coronal emission lines observed above the solar limb (abstract), Bull. Amer. Astron. Soc., 20, 1008, 1988.

Hollweg, J.V., On the origin of solar spicules, Astrophys. J., 257, 345, 1982.

Hollweg, J.V., Resonances of coronal loops, Astrophys. J., 277, 392, 1984.

Hollweg, J.V., Energy and momentum transport by waves in the solar atmosphere, in Advances in Space Plasma Physics, edited by B. Buti, World Scientific, Singapore, p. 77, 1985.

Hollweg, J.V., Resonance absorption of MHD surface waves: physical discussion, Astrophys. J., 312, 880, 1987.

Hollweg, J.V., Resonance absorption of solar p-modes by sunspots, Astrophys. J., 335, 1005, 1988.

Hollweg, J.V., S. Jackson, and D. Galloway, Alfvén waves in the solar atmosphere. III. nonlinear waves on open flux tubes, Solar Phys., 75, 35, 1982.

Hollweg, J.V., and G. Yang, Resonance absorption of compressible MHD waves at thin 'surfaces', J. Geophys. Res., 93, 5423, 1988.

Hollweg, J.V., and M.A. Lee, Slow twists of solar magnetic flux tubes and the polar magnetic field of the sun, Geophysics. Res. Lett., in press, 1989.

Hollweg, J.V., G. Yang, V. Cadez, and B. Gakovic, Surface waves in an incompressible fluid: resonant instability due to velocity shear, Astrophys. J., in press, 1989.

Ionson, J.A., Resonant absorption of Alfvénic surface waves and the heating of solar coronal loops, Astrophys. J., 226, 650, 1978.

Ionson, J.A., Resonant electrodynamic heating of stellar coronal loops: an LRC circuit analog, Astrophys. J., 254, 318, 1982.

Lee, M.A., and B. Roberts, On the behavior of hydromagnetic surface waves, Astrophys. J., 301, 430, 1986.

Lou, Y.-Q., Viscous MHD modes and p-mode absorption by sunspots, NCAR preprint, 1989.

Poedts, S., M. Goossens, and W. Kerner, Coronal loop heating by resonant absorption, this volume, 1989.

Priest, E.R., Equilibrium of flux ropes, this volume, 1989.

Rae, I.C., and B. Roberts, Pulse propagation in a magnetic flux tube, Astrophys. J., 256, 761, 1982.

Roberts, B., Magnetohydrodynamic waves, in Solar System Magnetic Fields, edited by E.R. Priest, D. Reidel, Dordrecht, Holland, p. 37, 1985.

Roberts, B., Properties and models of photospheric flux tubes, this volume, 1989.

Ryutova, M., Negative energy waves in magnetic flux tubes, J.E.T.P., 94, 138, 1988.

Similon, P.L., and R.N. Sudan, Energy dissipation of Alfvén wave packets deformed by irregular magnetic fields in solar coronal arches, Astrophys. J., 336, 442, 1989.

Spruit, H.C., Magnetohydrodynamics of thin flux tubes, in Solar Phenomena in Stars and Stellar Systems, edited by R.M. Bonnet and A.K. Dupree, D.Reidel, Dordrecht, Holland, p. 289, 1981.

Spruit, H.C., Propagation speeds and acoustic damping of waves in magnetic flux tubes, Solar Phys., 75, 3, 1982.

Strauss, H.R., The effect of ballooning modes on thermal transport and magnetic field diffusion in the solar corona, Geophys. Res. Lett., 16, 219, 1989.

SOLAR FLARES

Harold Zirin
Big Bear Solar Observatory,
California Institute of Technology, Pasadena, California 91125

Abstract. We discuss the general characteristics of solar flares. They are a spectacular energy release arising from development of magnetic strain and shear resulting from eruption of new magnetic flux near old. The release of magnetic energy is channeled into hard electrons and possibly nucleons which produce a series of other effects. We point out that the ultimate source of the magnetic complexity that produces flares is deep in the Sun. A number of examples are given, and the difference between various types of flares is discussed.

Introduction

Solar flares are an abrupt release of energy in the solar atmosphere associated with energy stored in magnetic fields. They release up to 10^{33} ergs in the form of energetic particles (electrons and protons) and mass motion. Modern observation techniques have provided insight into the conditions under which flares occur, but the origin of those conditions still remains invisible deep within the Sun. On the other hand it has been possible to observe the relation of the flares to the magnetic fields that beget them, as well as magnetic changes associated with some flares. On a much smaller scale, we have observed the disappearance of magnetic fields from the solar surface. Some of this is simply subduction of fields, but some cases may be small- scale examples of the flare energy release.

Flares are distributed so that roughly equal amounts of energy released are released in each magnitude range. This refers to the Hα area classification: class one, two, three, four. The energy released in a class 4 flare, the largest, is roughly ten times than of a class three, and there are roughly one-tenth as many of them. Another scheme is based on the 1–8 Å SXR flux monitored by the GOES spacecraft. The flares are designated by Cn, Mn or Xn, where the first letter is determined by whether the flux is greater than 10^{-6}, 10^{-5}, or 10^{-4} w/m^2 respectively and the integer n gives the flux for each power of 10. Thus M3 means a flux of 3×10^{-5}w/m^2 at the Earth. Typically M1 is a class 1, X1 a class 2, and X5 or higher class 3 or 4 flares. Each of these uses peak values and ignores duration, so the total energy is not counted, a serious shortcoming. But the scheme has biased toward slow or long-lived flares, because the Hα brightness and the hardness of the particles accelerated, both tied to impulsive events, are not accounted for. Since the original energy release appears (but is not proven!) to be in non-thermal electrons, a better scheme would probably take more account of the microwave intensity or the hard X-ray (HXR) flux, a better measure of the total number of high-energy particles.

There is a clear dichotomy [Kahler et al. 1984] between the class of high-energy impulsive flares which produce strong nuclear gamma rays and electrons up to the highest energy, and the long duration flares with multiple peaks which produce large quantities of interplanetary particles. While the very largest events produce both, most others tend to fall in one class. Presumably this depends on the magnetic configuration.

How do flares occur?

Figure 1 shows a typical sizable solar active region. The preceding (p) spots at left, lead as the spot moves westward across the solar disk with the 27-day solar rotation, and the following (f) spots have the opposite magnetic polarity. The sign of p and f polarities are opposite in the two hemispheres and reverse in each successive eleven-year cycle. The large spots in this region are well separated and one may view this arrangement as a simple loop of magnetic flux coming up through the surface in the one polarity and returning in the other. The magnetic field is in large measure current-free, the lowest possible energy state. There is, however, a dark filament (FI) below the f spots, and we can see that the fibril structure runs along it. It separates p and f polarity, but the lines of force connecting the opposite fields run along

Geophysical Monograph 58

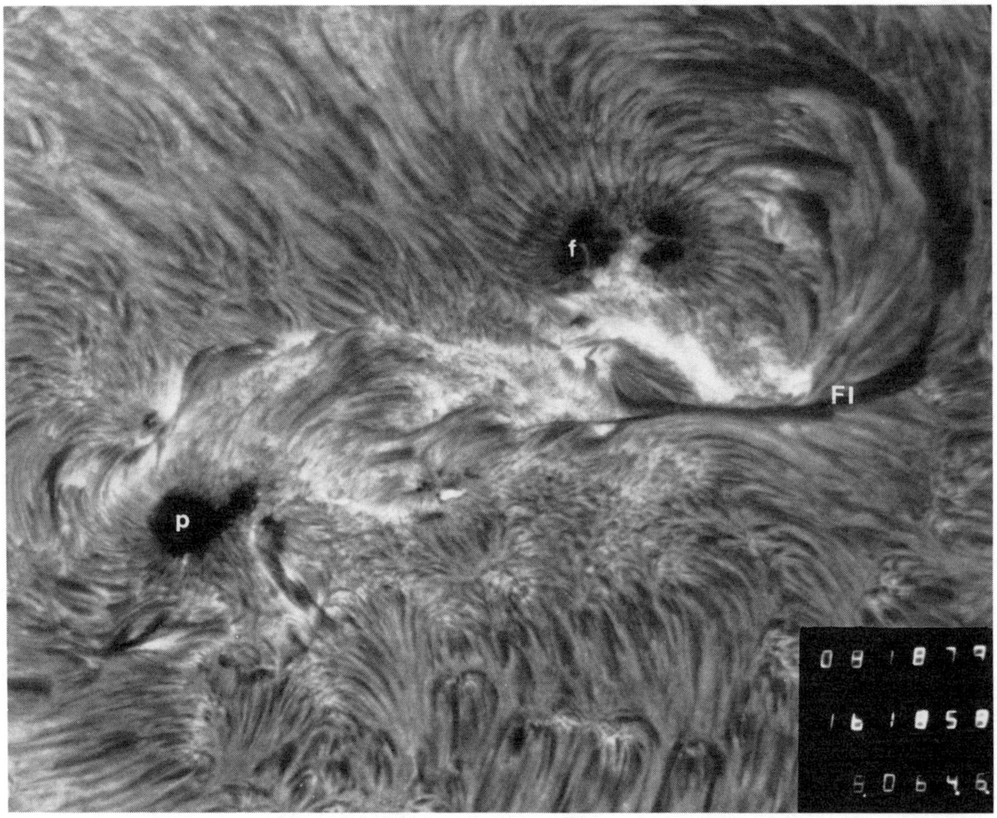

Fig. 1. A typical solar active region photographed in Hα. The spots are widely separated, and the plages (bright areas) are not especially bright. The only possible area for flares is the dark filament (FI) at the right, which bounds a bright plage. While this region had some flares, it was not especially active. W left, S top.

it. This region had some flares, as almost all regions of this size, but was not exciting.

Figure 2 shows, by contrast, the great sunspot group of March 1989 which produced some of the largest flares in history. In this case the magnetic fields are pushed close together in a much higher energy configuration. Penumbral structure (the region around the dark cores or *umbrae* of the sunspots, is spiral rather than radial. This is characteristic of the spot groups producing the largest solar flares. Regions like this are rare, although they may produce as many high energy particles as all the others of that year put together. Most active regions are like Figure 1, but most flares occur in the complex active regions. If we look at those regions we will see that the flares occur where longitudinal magnetic fields of opposite sign are pushed close to one another, where the spots are long and thin instead of round, and where the penumbral structure, instead of diverging outward, is twisted round tangent to the edge of the spot. In that case a current-free magnetic field is not possible and the system has surplus energy to release to a lower energy state. Also one finds the field lines run parallel to the neutral line. But the relaxation of the magnetic field does not lead to an obvious jump in field directions; rather both spot sizes and field gradients decrease.

The flare is an atmospheric phenomenon. Only the largest flares are visible in integrated light, and the typical flare must be seen by observing the atmosphere of the Sun above the photosphere. This is typically done in hydrogen alpha light, in which most of the images here were taken. Figure 3 shows stages in a flare near the limb, mostly imaged in the wing of Hα, which tends to show the lower parts of the flare. In the Hα wing we see mostly the most energetic kernels. The flare is occuring in the complex δ spot (see below) close to us. The initial energy inputs create a hot (40 million deg) coronal plasma, from which the loop prominences flow to the main magnetic poles. In the beginning these loops are dense and appear in emission ($N_e > 10^{13}$). At the end, higher, less dense material rains down and is seen in absorption. The loops terminate in both of the two round spots at left, which are opposite in polarity to the big spot. The impulsive phase ends around the first frame, the rest of the sequence corresponding to the SXR event. This lasts so long that there must be

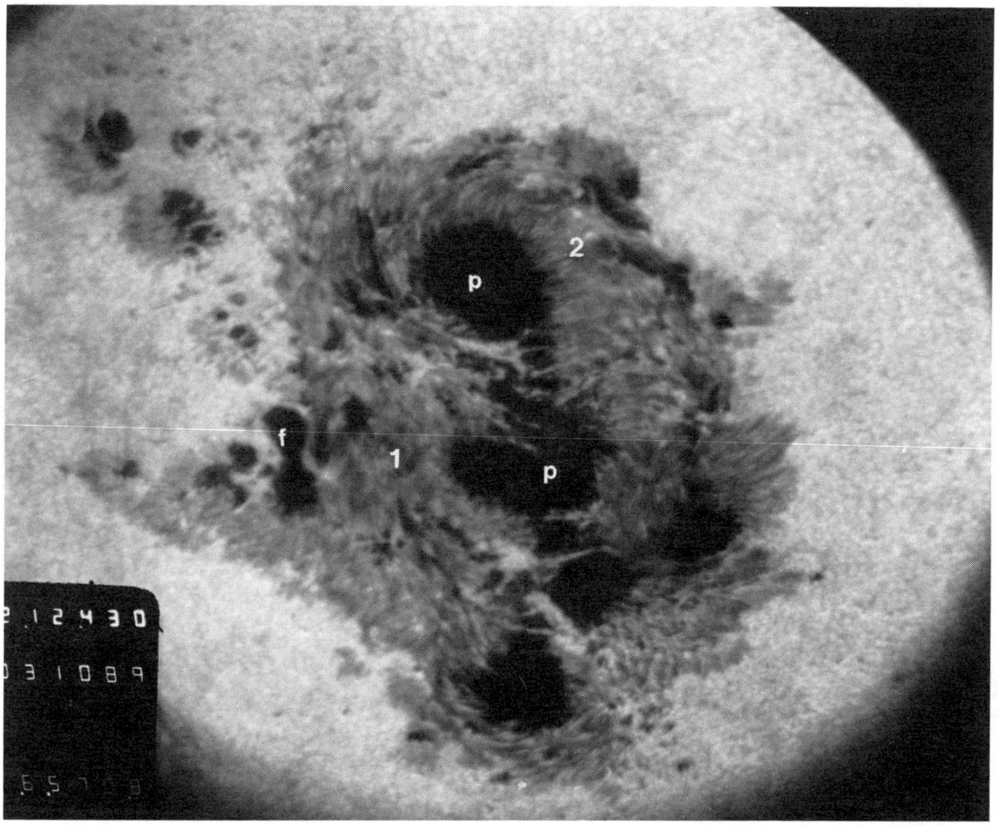

Fig. 2. The great sunspot group of March 1989 photographed in white light. The big spots in the center were *p* polarity, while the spots at right and left are *f*. Since they are all in a single penumbra, they form a δ configuration, the kind responsible for most big flares. The main flares occurred near points 1 and 2, where smaller spots of *f* polarity most closely approached the big ones. The dumbbell shaped spot near 1 changed markedly during the Mar 10 flare.

additional energy release. The bright points at the edge of the penumbra are the poorly understood Ellerman bombs.

Observations in the light of helium D3 (Fig. 4) also emphasize the more energetic kernels which appear in emission sources or ejecta which appear in absorption against the surface. Emission in helium D3, like Hα, occurs when the density in the flare exceeds 10^{12}– 10^{13} particles per cubic centimeter at temperatures above ten thousand degrees. The solar surface temperature is six thousand degrees. The temperature in the flare itself is much higher; initially it is a power law, and as the energy degrades, becomes thermal with temperatures of the order of thirty million degrees indicated by soft X-ray line observations. What we see in the visible reflects heating of the the higher density layers to temperatures above the normal 6000° by the flare.

There is no doubt that flares are a magnetic phenomenon. All flares, without exception, occur on or near magnetic inversion lines separating regions of opposite longitudinal polarity. These are rarely places where the field lines loop nicely from one region to the other, but those places where opposite polarities are squeezed together and the path of the field lines is highly sheared. How do they get that way? In cases such as Figure 2 an extremely complex sunspot in a high energy state appears from below the surface and we have no idea how it got that way. A normal sunspot group pushes up through the surface as a pair or several pairs of poles which gradually separate to reach the form of Figure 1. Shear can then occur when additional magnetic dipoles appear quite close to the first ones. There is a remarkable tendency for the new leader spot to move forward in the direction of solar rotation, while the follower spot remains more or less fixed. If there is a sunspot in the way, it pushes into it. If the new *p* spot has the same polarity as the old they merge and nothing happens. But if they (and only sunspots show this behavior) have the opposite polarity,

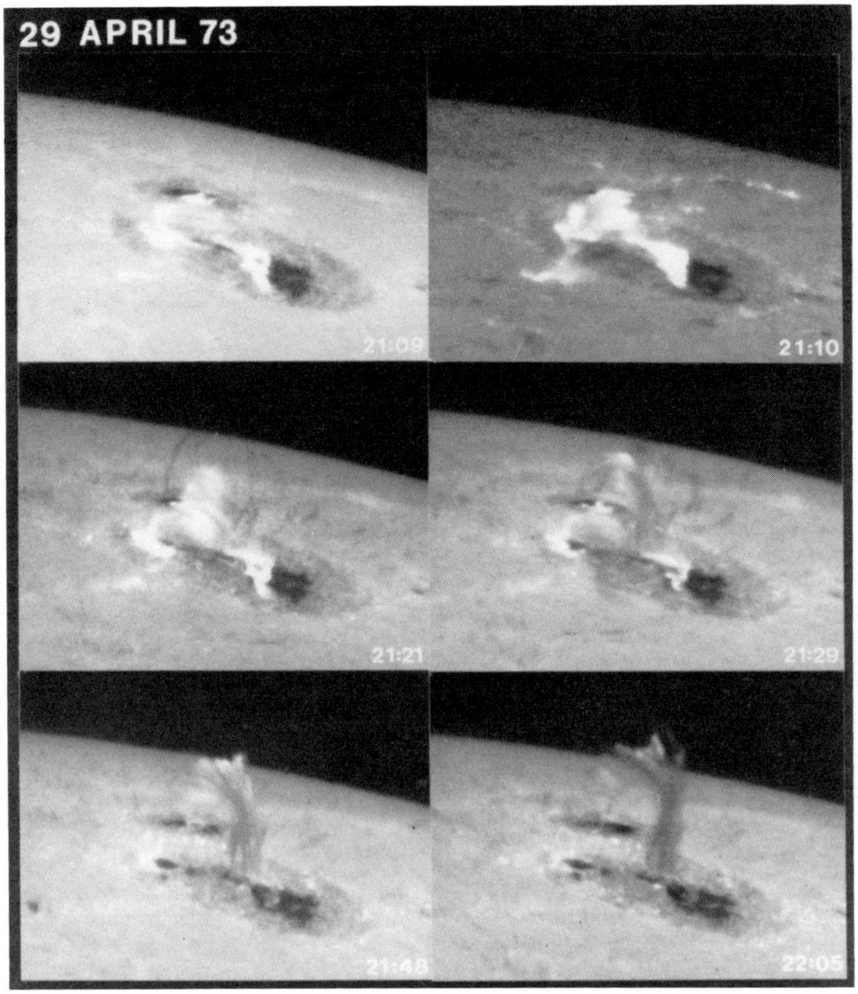

Fig. 3. Stages in a flare near the limb, mostly imaged in the wing of Hα. We see the transition from emission to absorption in the loops.

a δ spot forms with a sheared boundary between the new and old. How the shear boundary forms is unclear, for first the new dipole must form a connection with the older spot [Zirin 1983], so there is a connection to shear. What may happen is simply that intervening fields are squeezed between approaching spots until some connection is made. While a normal sunspot is a round umbra surrounded by a relatively round penumbra of outward-pointing fibrils, in this case the penumbra becomes (as in Fig. 2) a series of fibrils parallel to the boundary. This is called a "delta" configuration and was first recognized by Künzel [1968] as characteristic of sunspots producing the largest flares. The most active δ spot groups seem to rise to the surface already twisted and complex, but others result from the emergence of new spots near existing ones. In either case, the subsequent evolution is in the direction of a simplified lower energy configuration which occurs cataclysmically by means of flares.

Since the flare energy in the atmosphere is liberated in magnetic flux loops, it flows down to the surface at the two points at which the loop intersects it. These footpoints are heated and emit in the optical and UV spectrum. The atmospheric region where the energy release appears to take place is opaque in Hα and microwaves, and has also been imaged in the extreme ultraviolet. However none of the images or movies presently available sheds much light on this complex process.

Lin and Hudson [1971] were the first to show that the major energy input to solar flares is in the form of non-thermal electrons. This model still appears to fit the observations best, although it requires a remarkably efficient acceleration process. Thus the optical observations, which are the only ones with adequate space or time resolution, only show the secondary effects of electron heating of the atmosphere. One reason we feel certain the agent is electrons is the occurrence

6 JUNE 1982

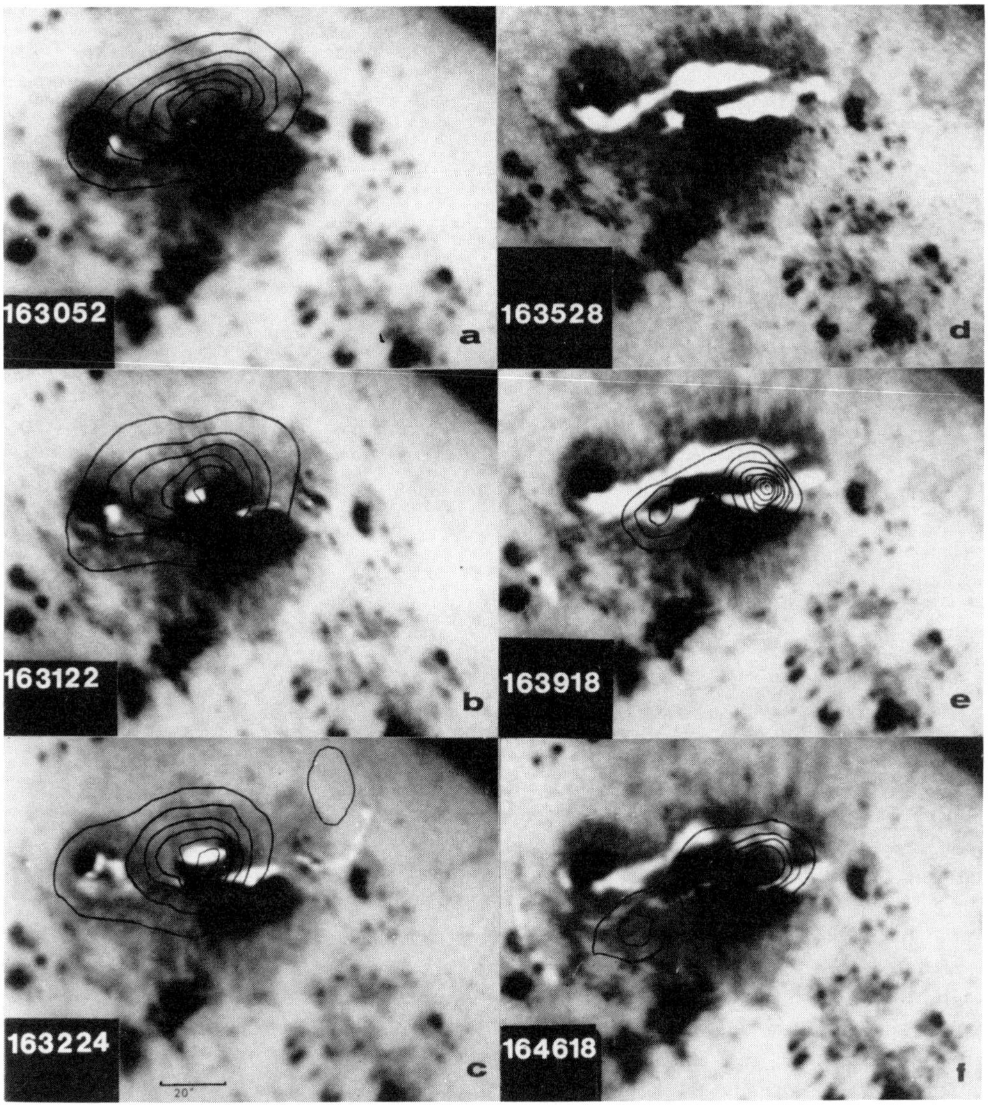

Fig. 4. Images of a great flare in D3. The contours are the HXR flux measured by Hinotori. The contours are missing the the third frame because of saturation [Tanaka and Zirin 1985].

of simultaneous brightenings in distant areas, showing that the exciting agent travels so fast, at least 50 000 km/sec, that it must be electrons. The X-ray observations [Tsuneta et al. 1983], rare and of low resolution, show sources above the neutral line and at footpoints. The microwave data [Dulk et al. 1986; Marsh and Hurford 1980; Shevgaonkar and Kundu 1985] show sources above the neutral line. But in each case only a blob is seen, no structural detail. The Hα centerline observations (Fig. 5), which might reveal the atmospheric structure, are too complex and opaque. Obviously magnetic energy can only be released in regions where non-potential fields occur, and this must only be in these areas of steep magnetic gradients where magnetic poles of opposite sign occur anomalously close to one another.

Beside the regions of sheared penumbra, flares will occur in other odd configurations, such as elongated sunspots or spots without penumbra at all. But the general principle is the same: steep gradients and sheared fields. It is interesting that all published sunspot models refer to nice round sunspots of little interest, and there has been no attempt to understand the really interesting spot phenomena: elongated thin spots, penumbra with no umbra, etc. In Hα the pre-flare state is often marked by

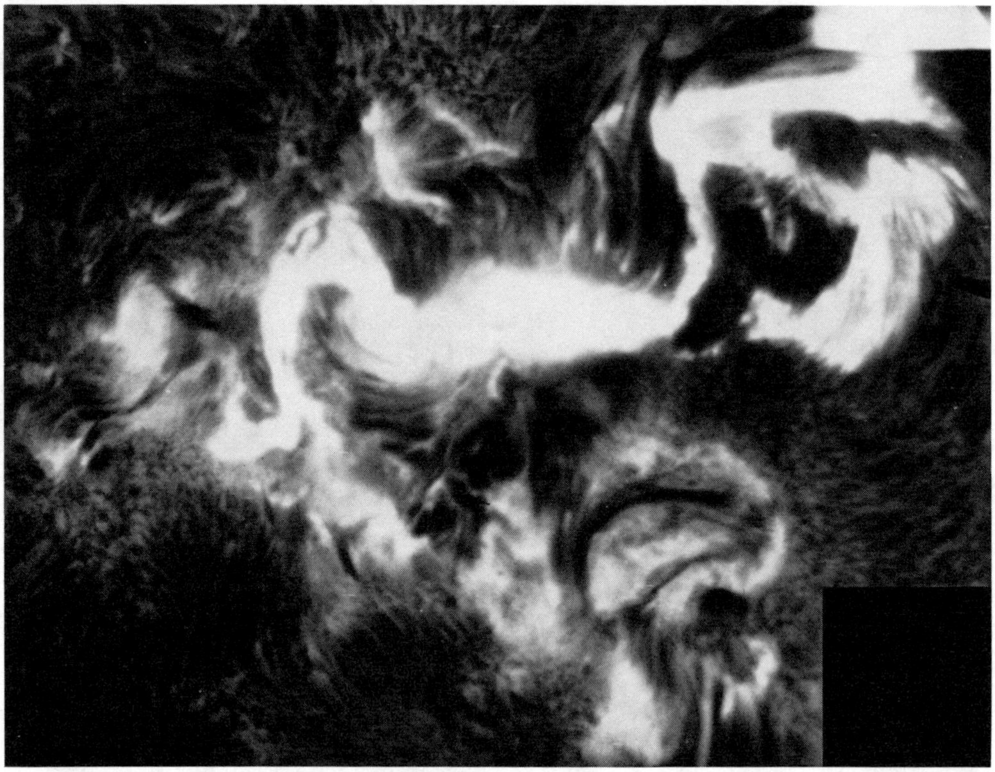

Fig. 5. The same flare in Hα centerline at maximum.
The emission covers all spots and saturates everything.

bright emission associated with flux emergence. Why are these regions bright? We assume (but have no proof) that energy is released by reconnection with existing surface fields. Another flare predictor is the covering of spot umbrae by bright Hα, as Figure 5. The umbra (except for the long skinny spots) is normally a region of vertical field which cannot support overlying dense material, and bright Hα emitting gas has a density $> 10^{13}$ atoms/cm^3. Such dense material can only be supported by horizontal magnetic fields, so there must be a sharp turn in the magnetic field, and of course if we take the curl of a sharp turn we get a strong current, probably in a force free condition.

In the past year we have begun to use these empirical-logical effects to issue BEARALERTS when we feel there is a high probability of flares. These are sent free (which accounts for their popularity) via E-mail to those interested. The success record of these is excellent, and we thus feel we know the circumstances under which flares will occur. But how they got that way we cannot tell.

The fact that big active regions like Figure 2 may erupt without warning in an apparent high energy state raises the important questions: Why didn't it simplify below the surface? Is it possible that all active regions are formed complex, and most are simplified before we see them? Or, conversely, are they formed simple and then twisted up in the convective zone? The fact that they commonly occur away from the sunspot number maximum and that they are typically quite large suggests they were produced in the complex form. Parker [1987] suggests that the suppression of convective transport by a sunspot enhances the possibility of field generation. No matter how the formation, the Hale–Nicholson force pulls the p spots forward and, if the polarity is reversed, increases the shear.

Kinds of flares

The fact that we use this term for any intense and rapid energy release means that the flare occurs in varied forms. To be sure, all flares occur on neutral lines, are bright in Hα and (with maybe a few exceptions) are associated with filaments. But in recent years some patterns have emerged.

Except for the very largest events there is little coincidence between the greatest X-ray or γ-ray line events and the greatest producers of solar energetic particles (SEP) in the interplanetary medium. Kahler et al. [1984] showed that almost all SEP events were associated with coronal mass ejections (CME). The charge on the SEP corresponds [Luhn et al. 1984] to ions common at 1–2 million deg, such as FeXIV. This means that the ions are accelerated in the corona. Mullan and Waldron [1986] show that the ionization is not uniform, but it certainly is not very high. Thus there is a real possibility that the SEP are accelerated in the CME – Moreton wave – type II burst, and second-stage acceleration may actually occur in these cases. Since there are nuclear gamma rays produced simultaneous with flare onset, that means there are at least two quite different nucleon acceleration processes, one in the flash phase and one as the shock wave moves through the corona. Obviously it would be hard to bring the CME particles back down to the surface to produce γ-ray lines.

Cane et al. 1986 selected flares lasting more than one hour at above 10% of their maximum 1–8 Å flux and showed that these were far more likely to produce SEP events than impulsive events. These "long-duration" events are typically associated with filament eruptions and CME.

My uncalibrated experience is that the long duration events are typically marked by development of two-ribbon Hα structure and great area. But one must be quite careful about comparisons. Because indices of soft X-ray or microwave flux are keyed to the peak values, we end up comparing flares of the same peak intensity. Thus the Cane et al. comparison may have simply established that events of the greatest total input flux *times* duration produced the greatest interplanetary effects.

Flares of all sizes show different time and energy profiles. Tanaka [1987] defines three characteristic classes:

A. *Hot thermal*, with $T = 3 - 4 \times 10^7$ deg, but limited and soft HXR emission from a compact source and little radio emission.

B. *Impulsive*, with spiky HXR and microwave emission from footpoints and low corona.

C. *Gradual-hard*, a long-enduring (> 30 min), large event with gradual peaks, a hard spectrum and a strong X-ray-microwave source high in the corona.

Tanaka points out that type B are frequently associated with a filament in a highly sheared neutral line. He shows that one well-observed case of type C shows slow optical evolution beginning with widely spaced flare ribbons, implying reconnection above the surface. Energetic particles reaching the Earth are preferentially produced by flares accompanied by CME and Moreton waves; most gamma-ray line flares, on the other hand, direct their particles inward and do not produce proton storms at the Earth. But the largest flares produce almost all these effects.

Small flares, on the other hand, are mainly impulsive. Zirin and Tang [1989] found 46% of X-ray bursts to occur in less than 20 sec. Possibly we don't see the longer, lower intensity part of such event. In any event, the single impulsive spike is what we usually see.

The size of flares is related in some degree to the size of the sunspots involved. A small satellite spot, even near a big spot, will produce only modest flares, while shear between large spots (Fig. 5) can produce big flares. This is necessary but not sufficient; big spots with no magnetic anomalies will not produce flares.

An exception to the relation between spot and flare size is the spotless flare, the eruption of a filament far from any spot. If the erupting filament is associated with plage from an old active region, an optical flare will occur. This is usually a big two-ribbon event with thermal emission and a CME. Typically the chromosphere underlying the flare will show a fibril structure indicating fields parallel to the filament. If the eruption is far from the active latitudes, only a slight Hα emission will be seen, but a CME may occur. In both cases proton events may be observed, the particles apparently accelerated in the outward moving magnetic fields. Examples are given by Zirin [1988 - chap. 9].

Morphology of the energy release

Flares are one of the few places in nature where we can closely observe how the magnetic energy release process proceeds. It is presently best done with the largest flares where the scale is adequate for us to distinguish the morphology. In all the flares for which we have adequate data – perhaps ten or twenty – the initial flare foot points are marked by a series of bright points along the shear boundary. This is illustrated in Figure 5 [Tanaka and Zirin 1985] which shows a great flare in He D3. The initial points appear in the first frames, and the two ribbons after 1632 UT. Presumably the points correspond to successive loops of flux in which energetic particles are accelerated and subsequently deposited. The HXR source as measured by Hinotori is shown by contours in Figure 5 and came from between the two ribbons even before optical emission appeared. The ribbons are connected by post-flare loops cooling from the hot coronal cloud.

In great flares the strands rapidly elongate (≈100 km/sec) on either side of the neutral line and separate at 5–20 km/sec while the loop prominences connecting them rise higher and higher in the corona and a long soft X-ray (SXR) event takes place. If one ribbon is near a sunspot, it will be small and bright, because many flux lines converge there; the ribbons will not cross the spot since the other side involves flux lines connected away from the flare. In the late stages the strands evolve into two thin lines formed by the intersection of a thin shell of hot coronal material with the surface. It is interesting that the X-ray flux from these great events is unknown; no selection board will pick an insensitive detector, so all those flown have saturated in big flares.

The role of filaments in flares is important and not understood. Besides the spotless flares (which show that a filament is enough to produce a flare all by itself), most of the active region flares occur along filaments or are associated with filament eruptions. Often the filament rises tens of minutes before the flare; it may get exceptionally dark, blue-shifted or broadened in Hα. Then the flare breaks out with brilliant Hα emission and the filament blows away. HXR emission begins only with the Hα emission. The blue-shifted filament is one of the few infallible flare precursors. Why? Probably because the filament only occurs along a sheared neutral line, with horizontal field lines that can be exchanged to a lower energy state.

From the progression from bright points along the neutral line to the arcade of arches of the two-ribbon event, we can gain the picture of triggering of the initial flare in the sheared region followed by extensive magnetic reconnection along the neutral line where the energy injected in each flux eventually reaches the surface in the two ribbons.

While this picture seems reasonable and we can follow the general progression from complex sheared magnetic field to a peaceful low-gradient active region, there are few examples of actual change in magnetic field before and after the flare. Further, when these are detectable, the amount of change does not appear to correspond to what we think the total flare energy is. However, a few examples have been reported. During the flare one gets a false field reversl because absorption lines go into emission. Because traditional magnetographs only measure the longitudinal field, and we don't see striking changes, it has generally been felt that the change is in the transverse field connecting the line of sight poles that we see. This appears reasonable because the highly sheared fields indicate that the transverse fields are indeed very strong, and the shear disappears as the activity dies down. At present, such changes can only be deduced from the Hα structure which marks the connection between the different poles and from the fairly crude transverse magnetograms we can obtain.

In the large, however, there is no doubt that the flares are connected with field changes. Active regions show considerable simplification after big flares, and Wang and Zirin [1988] were able to observe the disappearance of opposite magnetic polarities just after flares.

Under good conditions, continuum emission may be detected from almost all flares class 2 or brighter. And when we go to super-high-resolution techniques, we find that every flare is a white-light flare (WLF). Because the continuum is enhanced in the blue and the background depressed by the many lines, monitoring is best done below 4000 Å. The continuum morphology reflects the fact that we see the photosphere, perhaps 500-1000 km deeper than the D3 source. Only the most intense flare components will penetrate that deeply. But in fact, current calculations show that nothing, neither protons nor electrons, will penetrate to the depth where the continuum is formed. Yet they occur.

The WLF flashes are similar to those in D3, only fainter [Zirin and Tanaka 1973]. The main phase ribbons are also isomorphic, but fade more quickly than the D3 emission, presumably because the photosphere gets rid of the energy more rapidly than the less dense higher layers. An even brighter third point may appear in the thermal phase. In most cases the WLF emission disappears with the end of the impulsive phase. In addition a slowly moving bright wave [Machado and Rust 1974, Zirin and Tanaka 1973] coincident with the brightest part of the spreading Hα double ribbons may be seen. This must result from the thermal conduction that excites the ribbons, and their existence shows that conducted energy can reach the surface if there is enough time and energy.

Understanding the spectrum of the WLF is a real challenge. In the visible the continuum is flat down to 4600 Å, where it begins to rise [Machado and Rust 1974, Zirin and Neidig 1981] to a strong Balmer jump. The blue continuum might be explained by a hot black body source, but the total emission is too small. So the WLF, which is excited mysteriously to a mysterious color, is a mystery indeed.

Future directions of flare research

While the illustrations presented here are pretty, one can see that they are very far from revealing the flare phenomenon. One must at first realize that although there is a great deal of lip service to flare studies, practically no one observes flares from the ground, and with the forthcoming death of SMM, from space either. As a result the data we have leaves much to be desired. Some years ago, NASA even made a decision (perhaps feeling that we already knew too much about flares) to transfer effort to theoretical studies. While lack of knowledge makes modeling easier, it is more productive to know what you're modeling. Really only Ottawa and Big Bear carry out longer time optical studies with reasonable resolution. One must be able to observe with many wavelengths and high resolution for a long time. The Orbiting Solar Laboratory will have all these good properties: long observing runs, many wavelengths, excellent resolution. That will revolutionize the field.

While high resolution will help answer the question of how flares take place, it will not explain how the complex flare-producing magnetic configurations come to exist. That seems to be decided below the surface (Fig. 6), and that is presently one of the main challenges

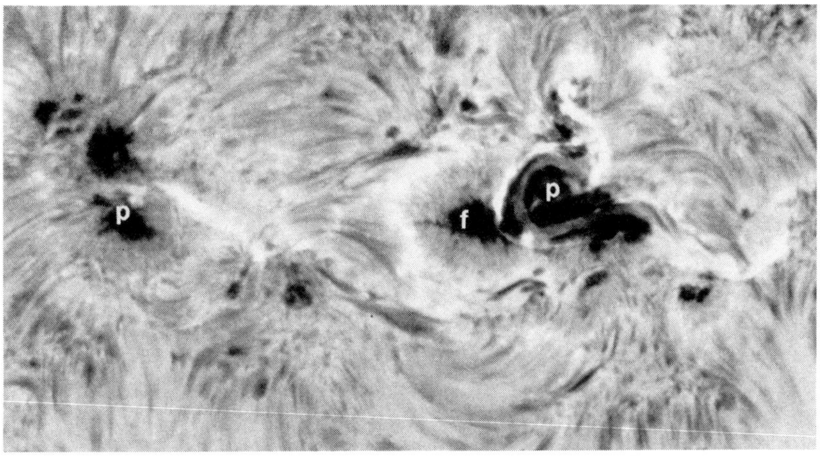

Fig. 6. This active region emerged in 1974 with the p sunspots at right trying to push their way through the spot f. This activity produced a bright rim and several arc-like umbrae along the neutral line, which gave rise to many flares.

in solar physics. If active regions are produced far below the surface, interaction with the convective zone should somehow affect their structure. One would expect this interaction to increase the twists and shears, bu most active regions show little shear. Big spot groups are generally more complex than small ones. This would fit a scenario where there are many twisted ropes deep within the Sun, but only a few survive the rise through the convective zone. Unfortunately one could also guess that all the regions start out as simple flux lines, and a twisting process increases their size and complexity. δ configurations often are created on the solar surface by the successive eruption of new flux in the same place. But this is not a random process; there is a high probability for sunspots to emerge where others already exist [Liggett and Zirin 1985]. Thus the eventual complexity is predetermined by sub-surface structure. There may be additional folds in the loops, or else a path by which flux reaches the surface. It is to be hoped that helioseismology, which gives us some picture of what exists below the surface, will shed some light on this.

We have always hoped that mapping the X-ray and microwave emission from flares would lead to better understanding of the energy release process. And so it has. Microwave imaging established the importance of the loop tops, and X-ray images from Hinotori showed the importance of both atmospheric and footpoint sources. But each observation has shortcomings. X-rays are only produced where the particles lose their energy, while microwaves easily reach self-absorption (which does not limit imaging, only spectroscopy). The two must go together. The Japanese Solar A spacecraft will provide X-ray mapping capability at the 5 arc second (35 hundred kilometer) level. Such data is useless if we don't have the optical observations which tell us in what magnetic configuration the footpoints and sources are located. Similarly, microwave mapping, which might reveal the source variations and sizes of the accelerating region, can only be done at one or two sites with adequate hardware by a small number of solar radio scientists. The theoretical effort in flare research is probably three or four times greater than the observational effort – maybe because anybody with a computer can carry it forward while extensive capital investment is required for actual observation. Mapping of the locus of the energy release relative to three dimensional vector field maps is within the realm of current possibility. Even this leaves the energy release and acceleration problem unsolved.

The search for flux disappearance or field changes associated with the flare depends on continuous observations. We now have extensive observations of the disappearance or cancellation of small magnetic poles in the quiet sun, which may be a model of what happens in the larger regions. A few cases of flux disappearance connected with flares have also been recorded; this appear to occur after, rather than during, the flare. Alas, all that we see is a close approach of two flux elements of opposite sign which come together and disappear. The exact nature of what makes this disappearance occur, reconnection above or below the surface or submergence of flux loops, is not yet clear. However, that it does occur is undoubted and clearly established – and that is a step forward at least. Another intriguing observation is the presence of velocity anomalies in magnetic inversion lines where flares occur. Data is still undigested, but we hope the Doppler observations will give us clues.

As usual, more clever observations and better theories, in particular closer application of theory to observation, is required.

Acknowledgments. Our work at Big Bear has been principally supported by the NSF under grant ATM-8513577 and NASA under grant NGL 05-002-034.

References

Cane, H. V., R. E. McGuire, and T. T. Rosenvinge, 1986. *Ap. J.* **301**, 448.

Dulk, G. A., T. S. Bastian, and S. R. Kane 1986. *Ap. J.* **300**, 43.

Kahler, S. W. *et al.* 1984. *J. G. R.* **89**, 9683.

Künzel, H. 1960. *Astron. Nachr.* **285**, 271.

Liggett, M. A. and H. Zirin 1985. *Solar Phys.* **97**, 51.

Lin, R. P. and H. S. Hudson 1971. *Solar Phys.* **17**, 412.

Machado, M. E. and D. M. Rust 1974. *Solar Phys.* **38**, 399.

Marsh, K. A., and G. J. Hurford 1980. *Ap. J.* **240**, L111.

Parker, E. N. 1987. *Ap. J.* **312**, 868.

Shevgaonkar, R. K., and M. R. Kundu 1985. *Ap. J.* **292**, 733.

Tanaka, K. 1987. *Publ. Astron. Soc. Japan* **39**, 1.

Tanaka, K. and H. Zirin 1985. *Ap. J.* **299**, 1036.

Tsuneta, S. *et al.* 1983. *Solar Phys.* **86**, 313.

Wang, H. and H. Zirin 1989 *to appear in Solar Physics.*

Zirin, H. 1983. *Ap. J.* **274**, 900.

Zirin, H. and D. F. Neidig 1981. *Ap.J.* **248**, L45.

Zirin, H. and F. Tang 1989 *to appear in Ap. J. Suppl.*

IDEAL INSTABILITIES IN A MAGNETIC FLUX TUBE

Giorgio Einaudi

Dipartimento di Astronomia e Scienza dello Spazio
Universita' di Firenze, 50125 Firenze, Italy

Abstract. In this paper we will review some theoretical aspects of ideal instabilities which arise in a axially symmetric magnetic flux tube. The interest in the stability of these structures derives from the fact that their observed lifetime in astrophysical plasmas is often much longer than the relevant hydromagnetic timescale and that they exhibit a dynamical active behavior. Therefore these structures must be globally stable with respect to fast destructive instabilities and, at the same time, must allow some local, non-disruptive process to take place. We will first of all classify the ideal instabilities in terms of their axial and azimuthal wave numbers and describe how the energy principle and normal mode approaches can lead to information on the stability properties of the modes. We will then outline the importance of the axial boundary conditions and show that, when these conditions are taken into account, the nature of the instability strongly depends on the topology of the equilibrium magnetic field. Finally we will discuss the effects of the unstable modes on the magnetic tube.

Introduction.

The stability of the magnetic flux tubes observed in many astrophysical objects has received a great deal of attention in the last decade, since observations have clearly shown the high level of activity present in these structures.

The fact that the magnetic tubes are not in a stationary state is not surprising, since it is well known the tendency of a magnetized plasma to be unstable. The nature of the instability depends on the features of the equilibrium configuration and on the boundary conditions the perturbations must satisfy. The challenge in the last years has been to understand precisely the link between the nature, and therefore the effects and the timescales, of the instability and the physical conditions of the structure in which the instability evolves.

In this paper we will concentrate on phenomena governed by the fluid equations, assuming that the collisional magnetohydrodynamic theory is applicable, that the plasma is adiabatic and that the only non-ideal effects which are important are due to finite constant and isotropic resistivity. The dynamics of such a plasma is then described by the following set of equations:

$$\partial\rho/\partial t + \mathrm{div}\,(\rho \mathbf{v}) = 0 \qquad (1.1)$$

$$\rho\,(\partial/\partial t + \mathbf{v}.\ \mathrm{grad})\,\mathbf{v} = -\mathrm{grad}\ p + (\mathrm{curl}\mathbf{B})\mathbf{x}\mathbf{B}/4\pi \qquad (1.2)$$

$$\partial\mathbf{B}/\partial t = \mathrm{curl}\,(\mathbf{v}\mathbf{x}\mathbf{B}) - (c^2/4\pi)\eta\ \mathrm{div\ grad}\ \mathbf{B} \qquad (1.3)$$

$$(\partial/\partial t + \mathbf{v}.\ \mathrm{grad})(p/\rho^\gamma) = 0 \qquad (1.4)$$

where ρ is the plasma density, p the pressure, **v** and **B** the velocity and magnetic field respectively, η the resistivity and γ the specific heats ratio.

The magnetic term in Eq 2) can drive an ideal magnetic instability due to the curvature of the magnetic lines, whereas pressure gradients can have both stabilizing and destabilizing effects, depending on their sign. The resistive term in Eq. 3) is generally very small with respect to the ideal convective term and therefore negligible everywhere except in the regions close to the zeroes of curl (**vxB**). When these zeroes are present in the structure, the resistive term can either destabilize the system, if ideally stable, or change the nature of the ideal instability, when it evolves on time scales longer than the Alfven time. In both cases reconnection of the magnetic lines occurs with a consequent conversion of the magnetic energy into other forms of energy. It is therefore of crucial importance, when discussing ideal instabilities, to determine under what conditions the small plasma resistivity may be active.

We will be concerned with those instabilities arising in an axially symmetric cylindrical magnetic tube, where currents flow and radial pressure gradients can be present. The equilibrium configurations adopted and the methods to study their stability are introduced in Section II. In Section III we present the main results obtained in the last years on the stability properties of such configurations, examining the possible resistive nature of the instability. The last Section is devoted to a discussion of the effects of the instabilities on the equilibrium configurations depending on the importance of the resistivity.

Equilibrium and linearized equations.

We model the magnetic tubes as axially symmetric plasma columns of finite length 2L. Introducing cylindrical coordinates and unit vectors $\mathbf{e}_r,\mathbf{e}_\theta,\mathbf{e}_z$, the magnetic field,

Geophysical Monograph 58

which may be written as

$$\mathbf{B} = B_\theta (r)\, \mathbf{e}_\theta + B_z (r)\, \mathbf{e}_z, \tag{2.1}$$

must satisfy the static MHD equilibrium condition

$$\partial/\partial r\, (p + B^2/8\pi) = - B_\theta^2/4\pi r \tag{2.2}$$

It is convenient to introduce the angle through which a field line is turned in going from one end of the tube to the other, namely the magnetic "twist" $\phi = 2L\, B_\theta/r\, B_z$, and the ratio between the magnetic field and the parallel current density $\alpha = 4\pi \mathbf{j}.\mathbf{B}/c|\mathbf{B}|^2$. The topology of the magnetic field is related to the functions $\phi(r)$ and $\alpha(r)$, which, along with the pressure profile, completely determine B_θ (r) and B_z (r).

In the following we will consider magnetic fields which can either be force-free or admit pressure gradients. In particular we will discuss the stability properties of the so-called Gold and Hoyle [1960] constant-twist force-free field and of the current confined fields [Chiuderi and Einaudi, 1981] in which currents flow only in the innermost region of the tube (r<a) imbedded in a much more extended potential field.

The Gold and Hoyle field has the simple analytical form

$$B_\theta (r) = B_o(r/b)/(1+r^2b^2), \qquad B_z (r) = B_o/(1+r^2b^2), \tag{2.3}$$

where b is a constant. In this configuration $\phi = 2L/b$, $\alpha = (2/b)/(1+r^2b^2)$ and gradp=0.

The Chiuderi and Einaudi field is characterized by

$$\begin{aligned} \alpha &= \alpha_o && \text{for } 0 \le r \le r_o \\ \alpha &= \alpha_o\, (1 + \cos \pi(r-r_o)/\delta)/2 && \text{for } r_o \le r \le r_o+\delta = a \\ \alpha &= 0 && \text{for } r \ge r_o+\delta. \end{aligned} \tag{2.4}$$

The pressure variation is expressed by a simple truncated Fourier expansion of the type [Chiuderi *et al*., 1977]

$$p(r)/p_c = p_1 - p_2 \cos (\pi r / r_o) - p_3 \cos (2\pi r / r_o). \tag{2.5}$$

With different choices of the p_i parameters it is possible to represent force-free fields ($p_2 = p_3 = 0$, $p_1 = 1$), monotonically increasing pressure profiles (e.g., $p_1 = 0.55$, $p_2 = 0.45$, $p_3 = 0$, model b) and profiles with an internal pressure maximum (e.g., $p_1 = 1.82$, $p_2 = 0.35$, $p_3 = 1.17$, model c). In this configuration the form of $B_\theta(r)$ and $B_z(r)$ is obtained numerically and two limiting examples are shown in Figure 1 in the force-free case. In equilibrium a) there is no inversion in the axial component of the magnetic field, which becomes a very small constant outside the current channel. By increasing the radius at which α begins to vary, a continuous sequence of equilibria moving towards case b) is obtained.

The ideal stability properties of this kind of equilibria can be obtained by linearizing Eqs. (1.1)-(1.4). This set of equations can be reduced to

$$\rho_o \partial^2 \xi/\partial t^2 = \mathbf{F}\, (\xi) \tag{2.6}$$

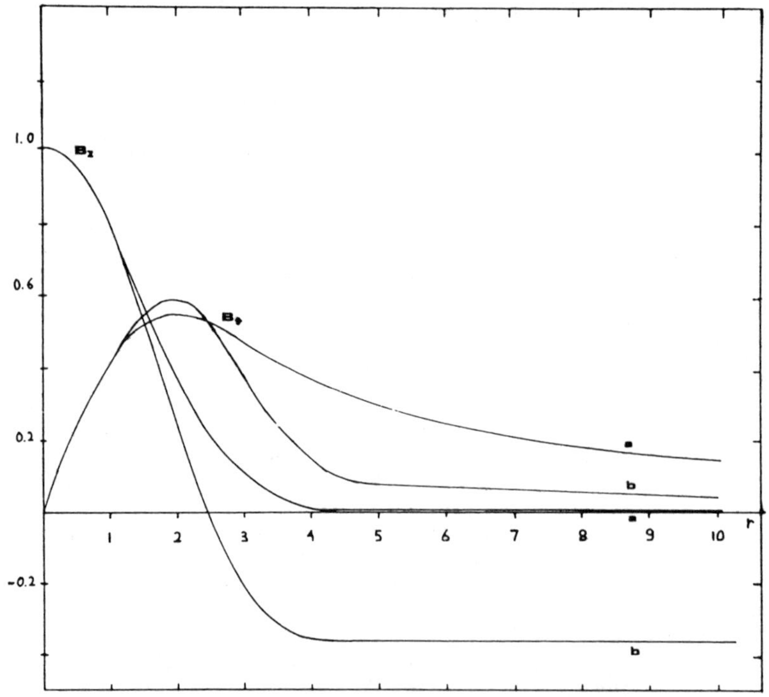

Fig. 1. Profiles of the Chiuderi and Einaudi equilibrium magnetic fields: (a) $\alpha_o r_o = 0.1$, $\alpha_o\delta = 4.9$; (b) $\alpha_o r_o = 2.0$, $\alpha_o\delta = 3.0$. [from Velli *et al*., 1989b]

where ξ is the Lagrangian displacement of a fluid element from its equilibrium position. The force operator **F**, as derived by Bernstein *et al.* [1958], has the form

$$\mathbf{F}(\xi)=\gamma \,\mathrm{grad}p_o \mathrm{div}\xi+\mathrm{grad}(\xi.\mathrm{grad})p_o+((\mathrm{curl}\mathbf{Q})\mathbf{x}\mathbf{B}_o+(\mathrm{curl}\mathbf{B}_o)\mathbf{x}\mathbf{Q})/4\pi,$$

where ρ_o,p_o,and $\mathbf{B}_o$, respectively, are the equilibrium density, pressure and magnetic field and $\mathbf{Q}$=curl($\xi \mathbf{x} \mathbf{B}_o$) is the first order perturbation in the magnetic field. By solving Eq (2.6) with the proper boundary and initial conditions, it is possible, in principle, to follow in time any small motion about the equilibrium state. This is a very difficult task and there are two simpler ways to study the stability of a given equilibrium configuration.

The first is to look for normal mode solutions of Eq (2.6), that is for solutions of the form $\xi(\mathbf{r},t) = \xi(\mathbf{r}) \exp(\gamma t)$. Then Eq (2.6) becomes

$$-\rho_o\gamma^2\xi = \mathbf{F}(\xi), \tag{2.7}$$

which is an eigenvalue equation for γ, since $\xi(\mathbf{r})$ must satisfy the proper boundary conditions. When $\gamma>0$ the system is unstable and the inspection of the corresponding eigenfunction ξ gives information on the effects of the instability on the equilibrium configuration.

One can determine whether a system is stable or not in an even easier way by using the energy principle, if **F** is a self-adjoint operator. To do so, one uses the variation in kinetic energy δK (a volume integral of $\rho_o|\partial\xi/\partial t|^2$) and the change in the potential energy, which is a quadratic form

$$\delta W = -1/2 \int_v dv\, \xi.\mathbf{F}(\xi). \tag{2.8}$$

From Eqs (2.6) and (2.7) it follows that $\partial/\partial t\, (\delta W + \delta K) = 0$ if the surface contribution to δW is zero, which implies that the boundary conditions ξ has to satisfy are such that the operator **F** is self-adjoint. Then if the sign of δW (with respect to every perturbation which can arise in the system) is positive, there is stability, while if there exists a perturbation which makes δW negative, the system is unstable. The use of the energy principle reduces the stability problem to the minimization of the functional ∂W and makes possible to determine stability by considering only all possible initial perturbations, without following their time development.

Newcomb [1960] applied the energy principle method to derive necessary and sufficient conditions for the stability of a diffuse linear pinch, stating a number of theorems which make possible to handle the second order variation of potential energy ∂W in a general axially symmetric magnetic field. The link between the normal modes and the energy principle approaches was clarified by Goedbloed [1970] who showed that for the eigenvalue equation (2.7) a Sturm theorem holds and that the study of the marginal equation (γ=0) produces stability criteria without computing the actual growth rate.

Both methods have been widely used for determining the stability of both laboratory and astrophysical plasma configurations, the difference being the boundary conditions that the perturbations must satisfy. For laboratory applications the toroidicity of the devices implies the periodicity of the perturbations in the axial and azimuthal directions, that may therefore be Fourier analysed in the following way , neglecting the toroidal curvature:

$$\xi(\mathbf{r}) = \Sigma_n\Sigma_m\, \xi(r,m,n) \exp(im\theta + ik_nz), \tag{2.9}$$

where $k_n=n\pi/L$ and n and m = 0,±1,±2,.... The use of this expansion has the effect of reducing the dimensionality of the problem to one and the stability can be tested as function of m and n. The unstable magnetically driven modes in a axially symmetric tube are classified in terms of the corresponding value of m: the sausage mode corresponds to m = 0, the kink mode to m=1, the flute modes to higher values of m. Newcomb [1960] showed that the most dangerous modes are the sausage and the kink in the small wavenumbers limit. The modes driven by localized, unfavourable, pressure gradients are known as interchange (low m) or ballooning (high m) modes.

In astrophysical applications it is unlikely to match the axial periodicity requirement leading to the expansion (2.9), and the problem becomes at least two-dimensional. However, one can insist in Fourier analysing the perturbations also in the z-direction, limiting the analysis to a local one in which $L \rightarrow \infty$ and therefore the axial wavenumbers k_n form a continuum. In order to perform a global stability analysis it is necessary to be able to impose the proper axial boundary conditions, excluding a simple harmonic z-dependence of the perturbations.

Stability properties.

A lot of work has been done to study the stability of tokamak and reversed field pinches configurations against the modes introduced in the previous Section with the conclusion that all the cylindrical pinches have the tendency to disrupt on the very short Alfven time scale [Bateman, 1978]. The toroidal curvature has stabilizing effects that are second order in the inverse tube local aspect ratio (that is the ratio between the minor and local major radius of the tube) [Bussac *et al.*, 1975]. In the laboratory, stabilization is achieved by a combination of strong axial applied fields and close-fitting concentric conductors. Such conditions, as well as the periodic axial boundary conditions leading to the expansion (2.9), are unrealistic in modelling astrophysical plasmas, and therefore the use of laboratory results for astrophysical applications is, generally speaking, unsafe. In this Section we review the stability properties of the equilibrium configurations introduced before, discussing the influence of the axial boundary conditions and outlining the importance of performing global stability analysis.

Kruskal *et al.* [1958] and Shafranov [1957] independently found that an infinitely long tube is unstable to kink perturbations when $B_\theta/r+kB_z \geq 0$, which result, in terms of the twist ϕ, states that for a given wavenumber k, instability occurs when the twist is given by $\phi \geq -2Lk$. In the particular case of a toroidal pinch of length 2L, the above criterion gives instability when $\phi = 2\pi$ at some radius. The Kruskal-Shafranov criterion was obtained for particularly simple equilibrium configurations, but it holds for the constant twist field of Gold and Hoyle [1960], which therefore is unstable against kink perturbations for $0 \geq k \geq -1/b$. The effects of localized pressure gradients has been considered by Suydam [1958], who found that a cylindrical tube is necessarily unstable against interchange perturbations if $8\ dp/dr+rB_z^2(d\phi/dr)^2/\ \phi^2<0$. This result implies that a negative pressure gradient has a destabilizing influence which can be balanced only if $d\phi/dr\neq 0$, but it does not state anything about the possible stabilizing effect of a positive pressure gradient. Giachetti *et al.* [1977] considered the stability of an equilibrium showing a strong axial presure deficit, embedded in an external force-free field. The adopted field structure represents a particular case of the Chiuderi and Einaudi field described in the previous Section ($\delta \rightarrow \infty$). By comparing the range of unstable wavenumbers k's in this case to the range found by Voslamber and Callebaut [1962] min the Lundquist [1951] case ($\delta \rightarrow \infty$, force-free field), they showed that positive pressure gradients improve stability with respect to the force-free case. The stability properties of the Chiuderi and Einaudi field with a monotonically increasing pressure profile for different values of the parameters r_0 and δ are summarized in Figure 2, where $\lambda=\alpha_0\delta$ and $\beta_0=8\pi p(r_0)/B(r_0)^2$ is a measure of the ratio between kinetic and magnetic pressure. The unstable regions are those limited by the corresponding marginal curve and the $\alpha_0 r_0$-axis. The cases shown refer to equilibrium configurations having one reversal in B_z. When the axial field never changes sign the configuration turns out to be always unstable to kink perturbations having $k<0$ and satisfying the Kruskal-Shafranov criterion. For small values of β_0 long wavelength perturbations are unstable in agreement with the force-free results of Chiuderi *et al.* [1980]. The simultaneous inclusion of positive pressure gradients and of an external potential blanket makes the range of unstable k's smaller and produces a complete stabilization for values of β_0 of the order of unity or larger when the magnetic twist is small enough at the border of the current channel.

The results obtained using periodic axial boundary conditions (that is performing a local stability analysis in astrophysical applications) lead to the conclusion that: a) the majority of the equilibria are unstable to long wavelength perturbations; b) favourable pressure gradients and the confinement of the currents reduces the size of the unstable wavelengths range when the axial field exhibits one reversal and the average magnetic twist is small; c) the fastest growth time τ of the instability, as computed by Goedbloed and Hagebeuk [1972] for the Lundquist and the constant twist fields, is $\tau \sim 100\ \tau_a\ a/L$, where τ_a is the axial Alfven time and L/a the aspect ratio of the tube.

As a consequence of these facts a global stability analysis is necessary because the axial boundary conditions can greatly affect the behavior of long axial wavelength perturbations. It is clear that, even in the axially periodic case, for a given radial twist profile, stability can be achieved if the length of the tube does not exceed the value $2\pi/|k_n|_M$, where $|k_n|_M$ is the maximum unstable wavenumber in absolute value. For example for the Gold and Hoyle field a sufficient condition for stability is $2L<2\pi b$, or $\phi<2\pi$, whereas the Chiuderi and Einaudi field with an inversion in B_z would never be stable. When the axial boundary requirement is not periodicity, a more sophysticated analysis is necessary. This kind of analysis has been done in the solar case, where the axial boundary conditions must mimic the influence of the much denser photosphere on the perturbations arising in corona. The coronal magnetic field is embedded in the high-inertia photospheric plasma, which effectively anchors the magnetic lines, producing the so-called line tying. The relatively long photospheric Alfven time is taken to mean that the photospheric plasma cannot move on the relevant coronal time scale. This circumstance, in turn, ties the magnetic field in the photosphere against perpendicular motions and maintains the continuity of the component of the magnetic field perpendicular to the boundary at the ends [Van Hoven *et al.*, 1981, Velli *et al.*, 1989a]. Raadu [1972] was the first to suggest the importance of the line-tying and to investigate its effect.

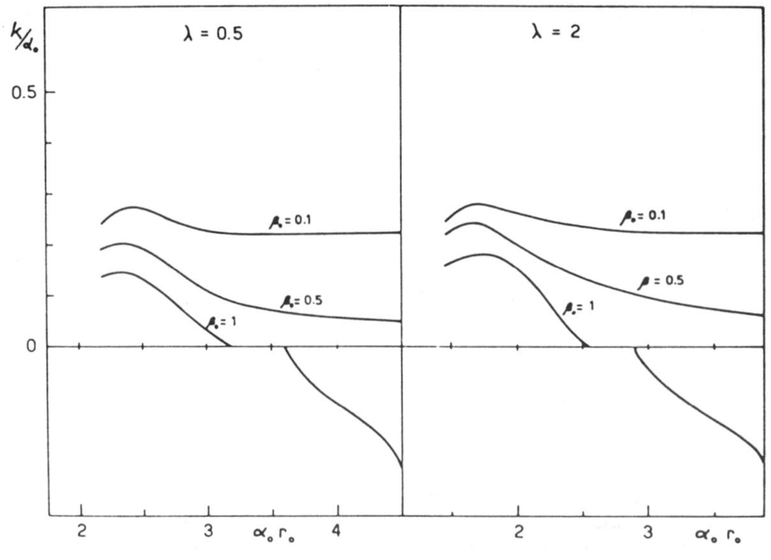

Fig. 2. Curves of marginal stability for non-line-tied Chiuderi and Einaudi fields in the $(k/\alpha_0, \alpha_0 r_0)$ plane. The unstable regions are those limited by the curves and the $\alpha_0 r_0$-axis. [from Chiuderi and Einaudi , 1981]

However, he only derived an upper bound on the stability of the Gold and Hoyle field, since he used trial functions in the energy principle, without examining the sign of ∂W with respect to the most general perturbation satisfying the imposed axial boundary conditions. He found that the tube may still become unstable when it is twisted enough. Hood and Priest [1979,1981] improved the Raadu results and, using a two-dimensional numerical analysis, showed that the tube goes unstable once the twist exceeds the critical value

$$\phi_{crit} = 2.49\pi. \tag{3.1}$$

Einaudi and Van Hoven [1981, 1983] obtained necessary-and-sufficient conditions for the stability of the Chiuderi and Einaudi field, using an energy principle. They made a general Fourier-series expansion along the tube axis and minimized the energy integral subject to the line-tying conditions. They then truncated the resulting infinite set of coupled ordinary equations to a finite number of harmonics and showed that the stability criterion converges rapidly with increasing number of harmonics. The critical line-tied stability limits are presented in Figure 3 for a range of force free and variable-pressure models. The stability depends effectively on three main features of the equilibrium model. The first is the aspect ratio of the tube, the second the orientation angle B_θ/B_z of the field, computed at the border of the current channel, the third the pressure profile. It is interesting to notice that the orientation angle of the field indicates the total current I flowing in the tube ($I=2\pi a B_\theta(a)$) and whether or not there is a reversal in B_z. In terms of the magnetic twist, defining an average twist as

$$\langle\phi\rangle = {}_s\!\int 2LB_\theta \, dr \, / \, {}_s\!\int rB_z \, dr \, , \quad s=[0,a] \tag{3.2}$$

The tube goes unstable once the average twist exceeds the critical value

$$\begin{aligned} \phi_{crit} &= 2.49\pi \quad \text{for equil.(a)} \\ \phi_{crit} &= 8.98\pi \quad \text{for equil.(b)} \end{aligned} \tag{3.3}$$

The first value is the same as that of eq. (3.1) found for the Gold and Hoyle field and this is due to the strong similarity in the equilibrium profiles of the two fields, provided there is no inversion in B_z. When this inversion is present, a much greater twist is necessary for an ideal instability to arise. This is a first important difference in the stability properties of configurations with and without an axial field inversion.

Recently Velli *et al.* [1989b] have realized the importance of studying the growth rate and the spatial profile of the growing perturbations in order to determine the capacity of the small plasma resistivity to enhance instability producing reconnection of magnetic lines. As discussed above, resistive field reconnection involves a localized break-down of the frozen-in-field constraint of infinite conductivity MHD. In the infinite length or periodic cases this break-down occurs in narrow layers around the surfaces at $r=r_s$ where $mB_\theta/r+kB_z=0$ [Coppi *et al.*, 1966], where a θ and z dependence $\sim\exp(i(m\theta+kz))$ of the perturbed quantities has been assumed. Line-tying however excludes a simple harmonic dependence in the axial direction. It has been suggested [Mok and Van Hoven, 1982] that in this case only θ-independent (m=0) perturbations are resistively unstable in configurations where the field component B_z vanishes at some point. Mok and Van Hoven [1982] and Velli and Hood [1988] have found that the instability is localized in vicinity of such points and that line-tying does not influence its properties with respect to the infinite or periodic case. However also θ-dependent perturbations can be resistively unstable, as Velli *et al.* [1989b] have shown for the resistive kink by studying the behavior of the operator **B**.grad on the ideal eigenfunction at marginal stability. They found that the derivative along the magnetic field vanishes on the same magnetic surface ($r=r_s$) as in the infinite case but only at the central height (z=0) of the cylinder and when an inversion of the axial component of the field is present, as shown in Figure 4. It follows that one may expect resistive effects to be important in the regions around the "resonant" point $r=r_s, z=0$ and the instability which arises once the critical average twist is excedeed to have a resistive nature and to produce reconnection of magnetic lines. This feature does not occur when there is no inversion of the axial field, since no resonant points are present in this case. The fact that in one case the instability is ideal and in the other one is resistive represents the second

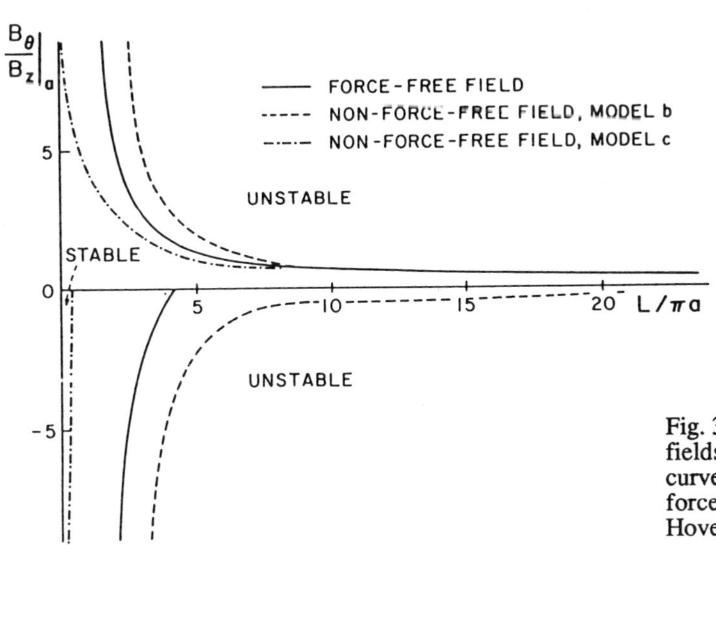

Fig. 3. Stability properties of line-tied Chiuderi and Einaudi fields. Stability exists on the origin side of the limiting curves for a force-free field and for models b) and c) of non force-free fields. Here $\beta_o=0.1$. [from Einaudi and Van Hoven, 1983]

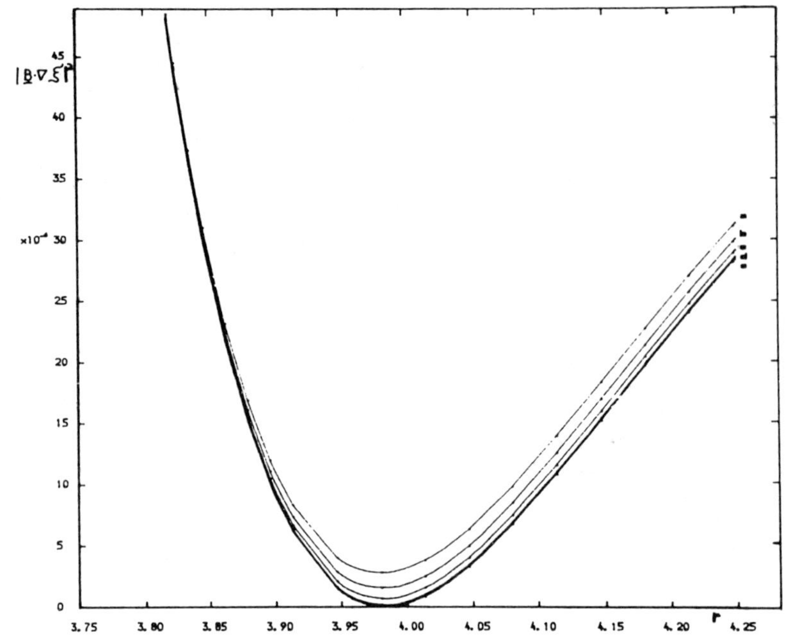

Fig. 4. Graph of $|(\mathbf{B}.\text{grad})\xi|^2$ as function of r for various values of z for the Chiuderi and Einaudi field, equilibrium b). Curve (a) z=o.8L, (b) z=0.6L, (c) z-0.4L, (d) z=0.2L, (e) z=0.0. [from Velli *et al.*, 1989b]

important difference in the behavior of configurations with and without an axial field inversion.

Discussion.

The knowledge of ideal line-tied instabilities in cylindrically symmetric magnetic tubes is fairly complete, from the localized interchange modes to the global current driven kink modes. We may summarize the results presented in the previous Section by saying that all kinds of equilibria become unstable once a critical magnetic twist is excedeed. The critical twist strongly depends on the topology of the magnetic field, on the axial boundary conditions the perturbation must satisfy and on the form of the radial pressure profile. Once the boundary conditions and the pressure profile are given, the critical twist is much smaller when there is no inversion in the axial component of the magnetic field than when an inversion is present. Moreover, in the first case the instability is ideal, whereas in the second case it has a resitive nature, since at marginality there is a resonant point around which the resistivity is important. It is known that when this is the case, the interplay of ideal and resistive effects markedly modifies the plasma behavior [Batistoni *et al.*, 1985] producing an instability which grows on a faster time scale than the ideal one, the so-called internal resistive kink mode [Ara *et al.*, 1978].

The different nature of the instability depending on the topology of the magnetic field may have important consequences as far as the effects of the unstable modes on the magnetic tube are concerned. First of all we want to point out that a magnetic configuration without inversion in B_z can be easily obtained from a potential one just twisting the axial ends of the tube. A configuration in which not all field lines are connected to the axial ends is necessarily the result of a more complicate process, like, for example, the gradual approach of two different magnetic tubes in which currents are flowing in opposite directions. With this point in mind, an interesting conjecture on the possible effects of the instability on axially symmetric magnetic tubes is the following. When no inversion in B_z is present, once the critical twist is exceeded, the initially axially symmetric equilibrium tends towards an helical equilibrium, energetically close to it, $(\mathbf{B}_0(r)\rightarrow\mathbf{B}(r,z)\cos\theta)$, with a smaller content of magnetic energy. The magnetic energy excess is tranformed into kinetic energy during the evolution of the instability and eventually into heat during the non-linear phase of the instability, due to the effects of dissipative processes as resistivity or viscosity. When an instability arises in a configuration with an inversion in B_z, reconnection of magnetic lines takes place in the region around the "resonant" point, with a consequent change in the topology of the magnetic field. It follows that the non-linear evolution of the instability in this case may have much more important consequences from an energetical point of view, since a new equilibrium, close to the initial one, does not exist.

We believe that future work on the stability properties of magnetic flux tubes should attempt to test the validity of the above conjecture, in order to better understand which physical features are important in determining the dynamical behavior of the tube.

References

Ara, G., Basu, B., Coppi, B., Laval, G., Rosenbluth, M.N., Waddell, B.V., Ann. Phys., 112, 443, 1978.

Bateman, G., MHD Instabilities, M.I.T., Cambridge Mass., 1978.

Batistoni, P., Einaudi, G., Chiuderi, C., Solar Phys., 97, 309, 1985.

Bernstein, I.B., Frieman, E.A., Kruskal, M.D., Kulsrud, R.M., Proc. Roy. Soc. London, A244, 17, 1958.

Bussac, M.N., Pellat, R., Edery, D., Soule, J.L., Phys. Rev. Lett., 35, 1638, 1975.

Chiuderi, C., Giachetti, R., Van Hoven, G., Solar Phys., 54, 107, 1977.

Chiuderi, C., Einaudi, G., Ma, S.S, Van Hoven, G., J. Plasma Phys., 24, 39, 1980.

Chiuderi, C., Einaudi, G., Solar Phys., 73, 89, 1981.

Coppi, B., Greene, J.M., Johnson, J.L., Nucl. Fusion, 6, 101, 1966.

Einaudi, G., Van Hoven, G., Phys. Fluids, 24, 1092, 1981.

Einaudi, G., Van Hoven, G., Solar Phys., 88, 163, 1983.

Giachetti, R., Van Hoven, G., Chiuderi, C., Solar Phys., 55, 371, 1977.

Goedbloed, J.P., Physica, 53, 501, 1970.

Goedbloed, J.P., Hagebeuk, H.J.L., Phys. Fluids, 15, 1090, 1972.

Gold, T., Hoyle, F., Mon. Not. Roy. Astron. Soc., 120, 89, 1960.

Hood, A.W., Priest, E.R., Solar Phys., 64, 303, 1979.

Hood, A.W., Priest, E.R., Geophys. Astrophys. Fluid Dynamics, 17, 297, 1981.

Kruskal,M.D.,Johnson,J.L.,Gottlieb,M.B.,Goldman,L.M. , Phys Fluids, 1, 421, 1958.

Lundquist, S., Phys. Rev., 83, 307, 1951.

Mok, Y. and Van Hoven, G., Phys. Fluids, 25, 636, 1982.

Newcomb, W.A., Ann. Phys., 10, 232, 1960.

Raadu, M.A., Solar Phys., 22, 425, 1972.

Shafranov, V.D., J. Nucl. Energy II, 5, 86, 1957.

Suydam, B.R., International Conference on the Paceful Uses of Atomic Energy, 31, 157, 1958.

Van Hoven,G. , Ma, S.S., Einaudi, G., Astron. Astr., 97, 232, 1981.

Velli, M, Hood, A.W., Solar Phys., 106, 354, 1986.

Velli,M., Einaudi,G., Hood,A.W., Ap.J., in press, 1989a ; 1989b.

Voslamber, D.,Callebaut, D.K., Phys. Rev., 128, 2016, 1962.

RESISTIVE INSTABILITY

Karl Schindler and Antonius Otto
Ruhr–Universität Bochum, Bochum, Federal Republic of Germany

Abstract. The concept and the relevance of resistive instabilities in cosmic plasmas is discussed from a general point of view. The discussion emphasizes qualitative changes that occur by proceeding from simple cases where symmetry allows ignoring one or two space coordinates to the general three-dimensional process. The prototype of resistive instability, the tearing mode of a plane sheet, already exhibits important aspects. Particularly, the instability occurs on a time scale that can be much faster than magnetic diffusion for large values of the Lundquist number. This speed-up is due to a strong localization of the resistive effects.

The step from one- to two- or three-dimensional systems brings about large scale consequences of the instability. Such resistive instability processes are found to be powerful mechanisms for releasing previously stored magnetic energy in cosmic plasma configurations and for initiating the magnetic opening of surfaces separating regions of different topological structure. A further important aspect is the occurrence of structures resembling magnetic flux ropes.

The step from two to three dimensions bears on the concept of magnetic reconnection. An analysis of resistive instability processes has demonstrated the necessity for a revision of that concept. It is shown that the recently suggested notion of general magnetic reconnection is adequate to describe three-dimensional resistive instability processes. This includes cases where the magnetic field has no singularity such as magnetic nulls.

A three-dimensional process that illustrates these properties is a model of flux transfer events at the earth's magnetopause. By a 3D-MHD simulation this model analyses the dynamic consequences of a sudden appearance of a localized spot of resistivity at the magnetopause. The results are discussed in the light of general magnetic reconnection.

Geophysical Monograph 58

Introduction

Resistive instabilities like many other plasma instabilities require that a non-equilibrium feature surpasses a critical strength. In particular, we are dealing with electric currents and pressure gradients. The instability in its nonlinear evolution then leads to a reduction of these quantities such that the system develops into states closer to thermodynamic equilibrium. As a consequence, the magnetic field relaxes into configurations approaching the potential field. Plasma transport perpendicular to the original magnetic field takes place faster than by stable diffusion. This can occur by a reconfiguration of the magnetic field such that the plasma flow that reduces pressure gradients can take place largely parallel to **B** (Fig. 1). There is an obvious relationship with magnetic reconnection.

Potential applications of resistive instabilities have a wide range. In the field of magnetospheric plasma transport and activity the prominent candidates for resistive instabilities are flux transfer events at the magnetopause [Russel and Elphic, 1978; Paschmann et al., 1982; Lee and Fu, 1985; Scholer, 1988] and magnetotail dynamics [Hones, 1979; Birn, 1980; Birn and Hones, 1981]. In a certain sense (see section 4) the processes occurring in the auroral acceleration region might also be considered as belonging to a generalized class of resistive instabilities.

Eruptive phenomena in the solar atmosphere such as solar flares, eruptive prominences and coronal transients are widely believed to involve resistive instabilities [Van Hoven et al., 1984; Robertson and Priest, 1987] also. Further it is conceivable that the magnetic structure of magnetic fields draped around non magnetized objects such as the planet Venus [Elphic and Russel, 1983] or comets is influenced by the occurrence of resistive instabilities. Generally, the regions where magnetized plasmas of different origins interact seem to be potential sites of resistive instabilities, examples being the interaction region of interplanetary and interstellar fields [Fahr et al., 1986], of the recently postulated galactic magnetosphere with interstellar fields, and of galactic and intergalactic fields, particularly in connection with galactic jets [Ferrari and Pacholczyk, 1983].

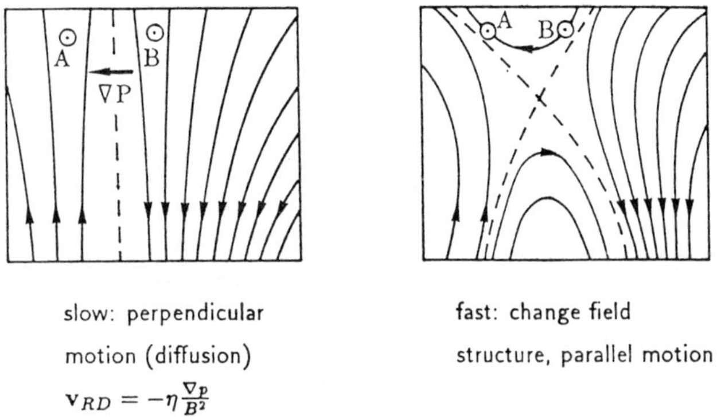

Fig. 1. Qualitative comparison of a slow diffusion and a fast resistive (reconnection) process

The theory of resistive instabilities is based on resistive magnetohydrodynamics. A relatively simple version is given by

$$\frac{\partial \varrho}{\partial t} + \nabla \cdot (\varrho \mathbf{v}) = 0$$

$$\varrho \frac{\partial \mathbf{v}}{\partial t} + \varrho \mathbf{v} \cdot \nabla \mathbf{v} = -\nabla p + \frac{1}{\mu_0} (\nabla \times \mathbf{B}) \times \mathbf{B}$$

$$\frac{\partial}{\partial t}\left(\frac{p}{\varrho^\gamma}\right) + \mathbf{v} \cdot \nabla \left(\frac{p}{\varrho^\gamma}\right) = \frac{\gamma - 1}{\varrho^\gamma} \frac{\eta}{\mu_0^2} (\nabla \times \mathbf{B})^2$$

$$\mathbf{E} + \mathbf{v} \times \mathbf{B} = \frac{\eta}{\mu_0} \nabla \times \mathbf{B}$$

$$\nabla \times \mathbf{E} = -\frac{\partial \mathbf{B}}{\partial t}$$

consisting of the usual ideal MHD equation modified by introducing resistive terms in Ohm's law and in the balance for p/ρ^γ which is a measure of entropy. The notion of instability requires that boundary conditions provide energetic closure or energy decrease.

The origin of the resistivity η in cosmic plasmas is not always obvious, in particular in cases where binary collisions become too rare. In the earth's magnetosphere η can only stem from collective fluctuations which could, for instance, be caused by microinstabilities. As discussed in section 6 particle inertia effects can have similar effects.

Resistive instabilities of any physical relevance must develop faster than resistive diffusion, which typically occurs on the time scale

$$\tau_{RD} \approx \frac{S}{\beta_\Delta} \tau_A$$

where τ_A is the Alfvén time scale and S the Lundquist number

$$S = \frac{\mu_0 L v_A}{\eta}$$

and β_Δ is based on the plasma pressure variation Δp

$$\beta_\Delta = \frac{2\mu_0 \Delta p}{B^2}$$

Through S the time constant τ_{RD} depends on a typical length scale L. The speed-up that occurs in resistive instabilities is based on the ability of the unstable mode to introduce a new length scale that is smaller then L, which leads to a shorter time scale. Quantitative estimates require a more thorough analysis and depend on the details of the configuration. The simplest example is the resistive tearing mode (see section 2).

Further interesting aspects of resistive instabilities appear when discussed from the point of view of nonlinear dynamics. Obviously, particular circumstances must prevail, if a system assumes an unstable equilibrium state. A quasistatic evolution reaching a marginal state might just continue on a new stable branch and there is no fast dynamical evolution at all. Scenarios that can lead to strong dynamic effects are depicted in Figure 2. Here λ denotes the control parameter which could, for instance, be a characteristic pressure or a measure of magnetic shear.

For the purposes of this talk the left diagram of Figure 2 is the most relevant. Suppose the system, when passing the bifurcation point, behaves essentially ideal because resistivity is not available. If the resistively unstable branch (broken line) is stable within ideal MHD the quasistatic evolution can proceed until resistivity becomes available. For instance, the electric current density surpasses a threshold for microinstability. Then, the resistive instability becomes effective deep inside the unstable regime. It is also of interest to note that there exist branches that are stable against small amplitude perturbations but unstable against

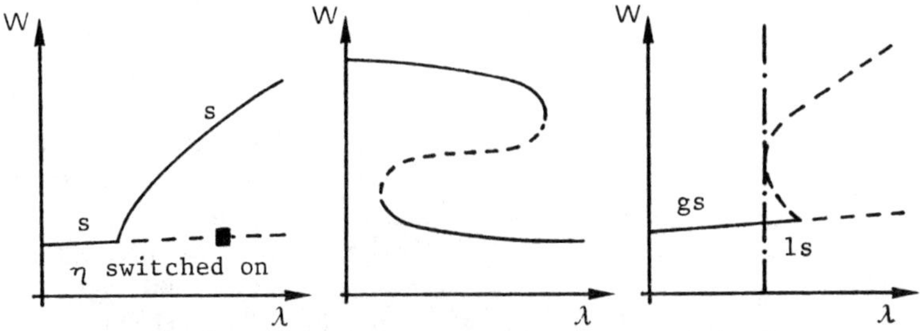

Fig. 2. Bifurcation diagrams relevant to resistive instability processes. W and λ denote a typical quantity such as energy and the control parameter, respectively. Local "ls" and global "gs" stability are indicated.

larger amplitudes (local stability "ls" in the right diagram of Figure 2; "gs" stands for global stability).

A particularly striking property of resistive instabilities is their strong dependence on the number of spatial dimensions considered, where the reduction in the numbers of dimensions is due to ignorable coordinates in the presence of symmetries. Naturally, much of the earlier work was on reduced dimensions, e.g. two-dimensional perturbation of a one-dimensional equilibrium. Starting out from such a case and going to more and more realistic cases, eventually deal-

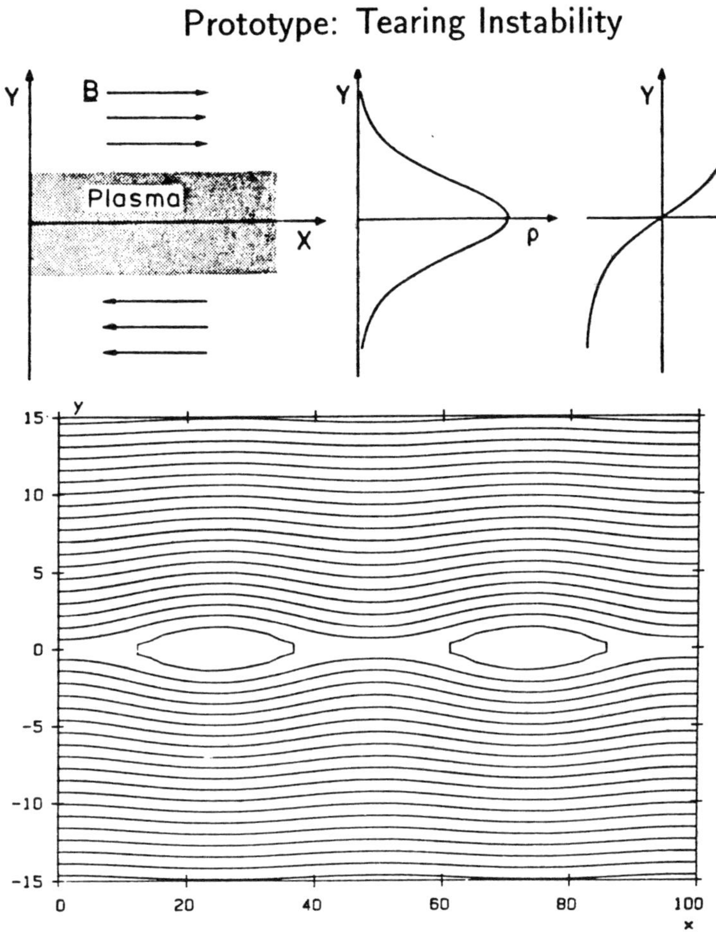

$$A(x,y,t) = A_0 + \hat{A}(y)e^{i\alpha x+qt}$$

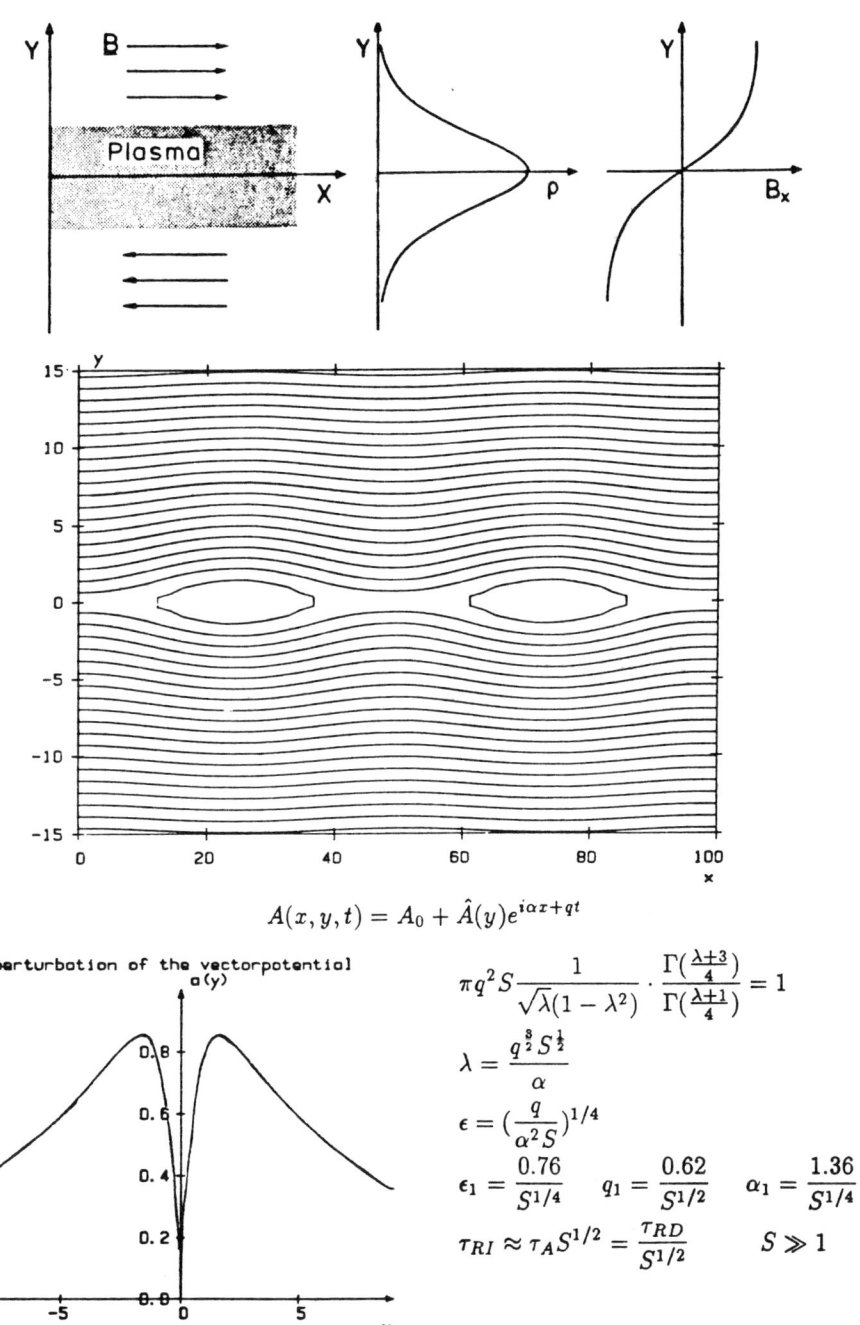

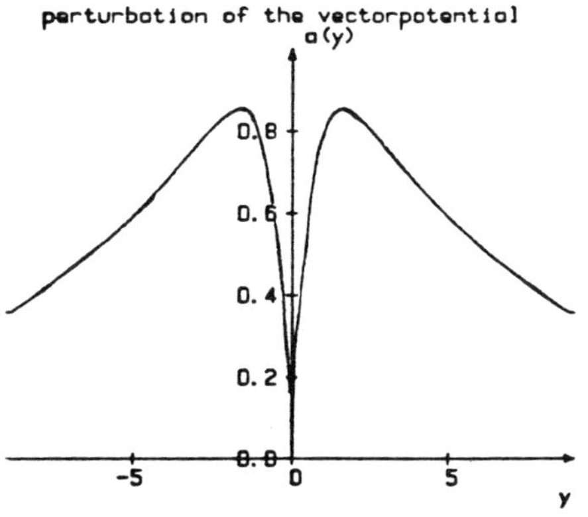

$$\pi q^2 S \frac{1}{\sqrt{\lambda}(1-\lambda^2)} \cdot \frac{\Gamma(\frac{\lambda+3}{4})}{\Gamma(\frac{\lambda+1}{4})} = 1$$

$$\lambda = \frac{q^{\frac{3}{2}} S^{\frac{1}{2}}}{\alpha}$$

$$\epsilon = (\frac{q}{\alpha^2 S})^{1/4}$$

$$\epsilon_1 = \frac{0.76}{S^{1/4}} \qquad q_1 = \frac{0.62}{S^{1/2}} \qquad \alpha_1 = \frac{1.36}{S^{1/4}}$$

$$\tau_{RI} \approx \tau_A S^{1/2} = \frac{\tau_{RD}}{S^{1/2}} \qquad S \gg 1$$

Fig. 3. Illustration of equilibrium, perturbation and dispersion relation of the resistive tearing instability of a Harris sheet.

ing with fully three-dimensional systems, one encounters a number of surprising features appearing with increasing number of dimensions.

In the following sections of this paper we concentrate on the latter effects, beginning with the tearing mode as a simple prototype of a resistive instability. In going to more complicated structures we concentrate on magnetospheric applications. This means that the resistivity will be assumed to be based on collective fluctuations and that pressure forces have to be taken into account. Nevertheless, we will not ignore other cases completely.

Prototype: Tearing Instability

Figure 3 illustrates the simplest case of a resistive instability, the classical resistive tearing instability [Furth et al., 1963] of a Harris-sheet. The mode causes islands that form spontaneously as soon as resistivity becomes available, the sheet is stable within ideal MHD. Thus, we are dealing with a situation as shown in the left diagram of Figure 2.

The dispersion relation, which is also shown in Figure 3, implies that the minimum time scale τ_{RI} is shorter than resistive diffusion by a factor of $S^{1/2}$. This can make a substantial difference in cosmic plasmas, where S is usually rather large (S is several hundreds for magnetospheric applications, and much higher in the solar atmosphere).

It is also clear from the Figure that the mode introduces its own length scale. The amplitude $a(y)$ of the magnetic flux function has a rather large curvature at the origin ($y = 0$). The corresponding scale (not resolved in the plot) is of order $\epsilon_1 L$ with $\epsilon_1 = 0.76/S^{1/4}$. We are dealing with a boundary layer, where a small dissipative term has a large effect.

The mode causes magnetic reconnection manifested by the occurrence of field lines that close on themselves, thus being topologically different from the field lines of the original sheet. Following individual plasma elements with time, one finds that the property of magnetic line conservation [Newcomb, 1958], which is present in ideal MHD, is violated.

The non-linear development of the tearing mode is not very dramatic. A numerical study with $S = 500$ starting with a velocity perturbation of $10^{-2}\, v_A$, grows considerably slower than exponential already after 3 growth times ($\approx$ 300 Alfvén times in that case).

Before drawing a conclusion on the macroscopic significance of tearing we consider this process in a configuration that is more realistic than a Harris-sheet in many cases. Adding a constant y-component of the magnetic field to the Harris-sheet field shown in the upper part of Fig. 3, gives a configuration such as shown in Fig. 4. This configuration has been studied by several authors, among them Janicke [1980] and Nishikawa and Sakai [1982]. Recently Paris [1987] concluded that the growth rate is reduced as the normal component increases. Thus one might expect that the instability is not important in that case either.

$$\mathbf{B} = B_0 \tanh y \mathbf{e}_x + B_n \mathbf{e}_y$$

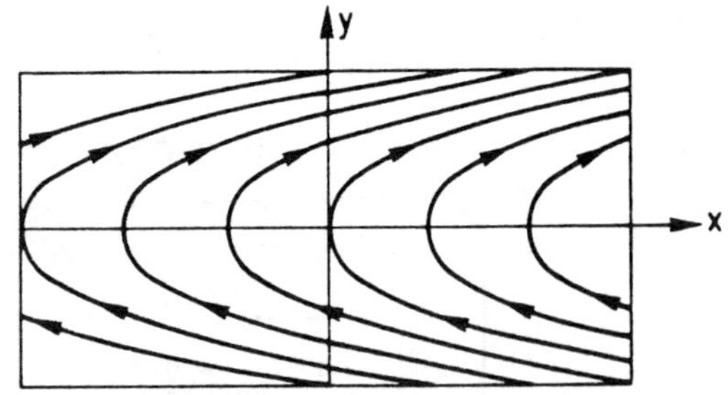

Fig. 4. Magnetic field lines of a Harris sheet with B_y superimposed [Janicke, 1980; Nishikawa and Sakai, 1988].

Therefore it might seem surprising that - as discussed in the following section - numerical simulation finds a strong effect with pronounced macroscopic consequences. This apparent contradiction is related to the fact that the configuration shown in Fig. 4 is not strictly in equilibrium. The x- component of the $\mathbf{j} \times \mathbf{B}$ force resulting from B_y and j_z is not balanced by the pressure gradient force which has a y-component only. Although this seems to be acceptable for WKB-type studies this imbalance has a cumulative effect for the large scale and long time behaviour. To correct for this deficiency, one must introduce a pressure variation in the x-direction also. This however destroys the translational invariance with respect to x, so that the equilibrium configuration becomes genuinely two-dimensional.

Resistive Instability of Two-Dimensional Equilibria

Fig. 5 shows the time-development of a two-dimensional configuration which is in equilibrium initially [Otto, 1989]. The upper panel gives the magnetic field lines of the equilibrium. In the closed flux regions to the left of the (distant) neutral line, the pressure gradient on the axis points to the left such that it cancels the horizontal component of the $\mathbf{j} \times \mathbf{B}$-force. At time zero a resistivity becomes available. A plasmoid forms and it is accelerated in the direction of decreasing pressure. The lower part of Fig. 5 shows the magnetic force and the pressure force acting on the plasmoid. It is obvious that the pressure force dominates initially. Such a force would be absent in a system, where the initial state is translationally invariant with respect to x. We remark that the plasmoid which is defined in terms of a magnetic separatrix, first appears at the time when the growing perturbation has become sufficiently strong to balance the equilibrium y-component of $\mathbf{B}$ at some point. There, a field cusp forms, out of which the plasmoid grows. The plasmoid birth time, corresponds to the instant where the forces start to grow from zero in the lower panel of Fig. 5. ($\Delta t = 0$ is the time when the plasmoid detaches from the

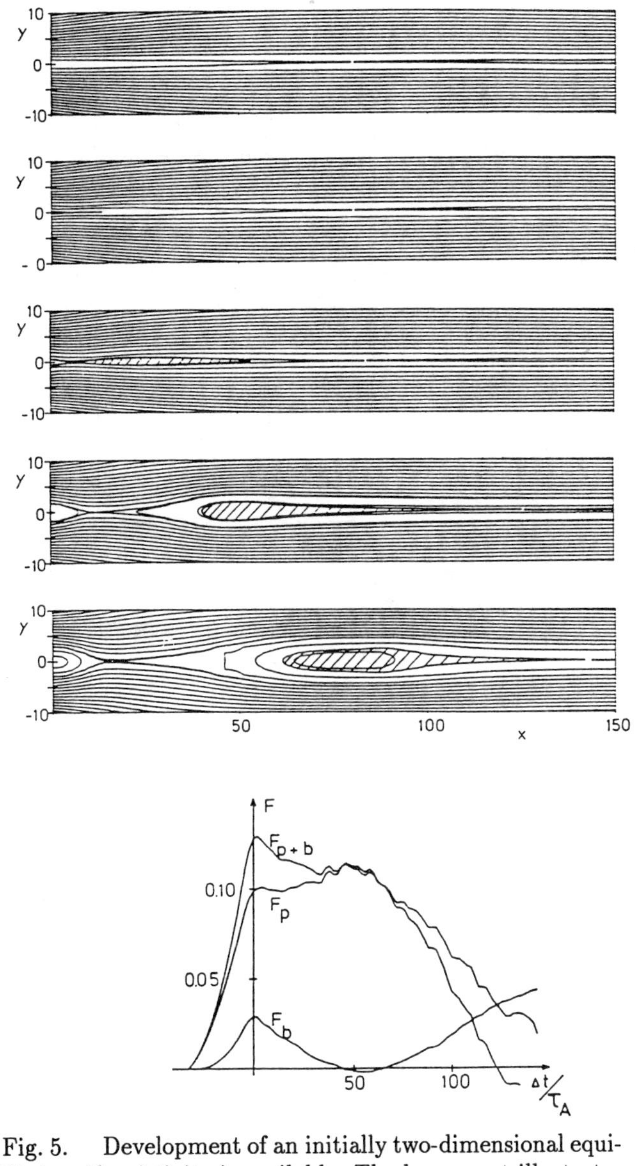

Fig. 5. Development of an initially two-dimensional equilibrium if resistivity is available. The lower part illustrates pressure and magnetic forces acting on the plasmoid.

near-earth X-line.) Obviously, the motion of the plasmoid is closely related to the fact that reconnection at the X-type neutral point is not saturated. In fact, reconnection continues and the plasmoid moves to the right in Fig. 5. The transport by formation, acceleration and eventually ejection of plasmoids is now widely understood as a major loss mechanism, controlling the energy content in highly loaded closed magnetic field regions. In the tail of the magnetosphere plasmoids are believed to form and travel tailward in connection with geomagnetic activity [Baker et al., 1984].

Similar conclusions were reached for arcade-like configurations in the solar atmosphere. Such emission of plasmoids was recently found by Mikic et al. [1988] and by Biskamp and Welter [1988]. Although details depend on the boundary conditions, it seems of interest that plasmoid formation similar to that of Fig. 5 can be obtained in magnetic arcades.

Thus the macroscopic significance of resistive instability in loaded closed field regions is dynamic unloading according to the process sketched in Fig. 6. The lower panel of that figure is an attempt to indicate that the plasmoid moves out to large distances such that the configuration close to the surface of the magnetized body returns to an unloaded state. In view of the main topic of this conference we add that in the presence of a magnetic field component perpendicular to the drawing in Figure 6 the closed field line region would go over into a structure resembling magnetic flux ropes. The dynamics and stability properties would remain roughly unaltered. For a polytropic plasma with $\gamma = 2$ in static equilibrium this equivalence holds even quantitatively.

Further results in the two-dimensional resistive instability context is the formation of shockwaves [Forbes et al., 1988; Forbes and Malherbe, 1986] or the resistive coalescence [Pritchett and Wu, 1979] and the resistive ballooning mode, which is strongly localized across field lines.

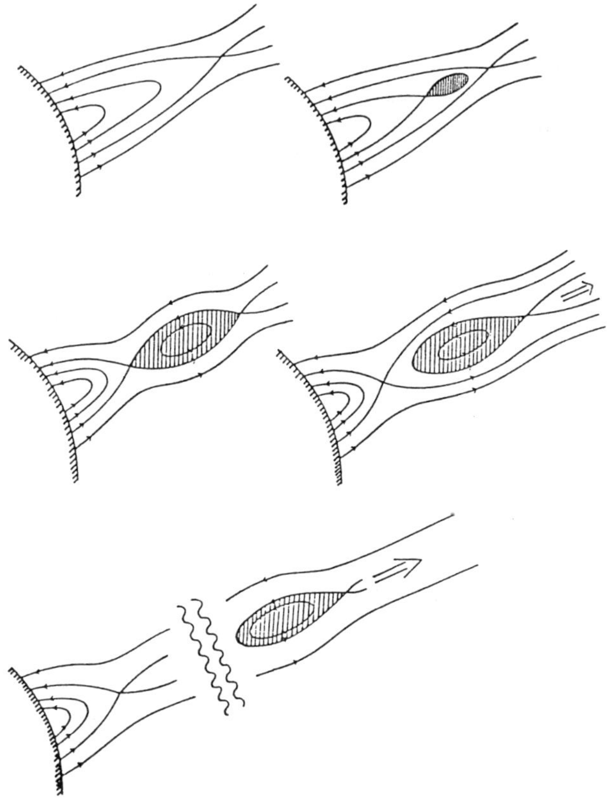

Fig. 6. Sketch of the plasma transport mechanism in a stretched closed magnetic flux region by a reconfiguration due to resistive instability.

In view of the relevance of bifurcation theory for resistive instabilities, as discussed in section 1, we mention criteria that the equilibrium magnetic field must satisfy at a bifurcation point. In strictly one-dimensional fields of the type $\mathbf{B} = B_x(y)\mathbf{e}_x$ the bifurcation point field has point $y = y_0$ where B_x vanishes. Visualizing this system imbedded in 3D space, this means the presence of a neutral sheet.

In the case of 2D fields ($\mathbf{B} = B_x(x,y)\mathbf{e}_x + B_y(x,y)\mathbf{e}_y$) both components must change their signs at the boundary. If one cartesian component vanishes identically it is sufficient that the other component changes its sign. Figure 7 shows three configurations, where this criterion is not satisfied (left column) and three cases where it is satisfied (right column). The upper two cases on the left are on the stable side of a bifurcation point, i.e. closer to a potential field configuration. The lower configuration is considerably more structured and it would be on the resistively unstable side; a boundary located sufficiently farther inside would make the system marginally stable, i.e. satisfying the present necessary bifurcation point criterion. We add that these results are based on a model where boundaries are open to a plasma flow along $\mathbf{B}$. For an outline of the underlying model see the Appendix.

Three-Dimensional Systems

We now generalize the process discussed in the previous section (Fig. 5) to a genuinely three-dimensional case. To that end we add a z-component of the magnetic field and consider a case where the plasmoid has a finite length along the z-direction. As shown in Fig. 8 the topological differences of field lines that close on themselves and field lines

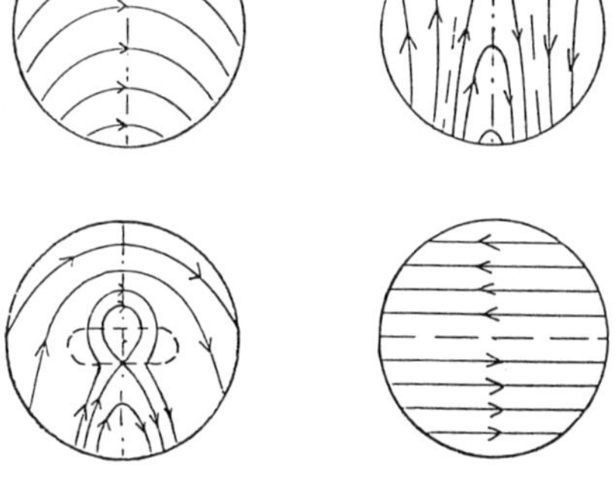

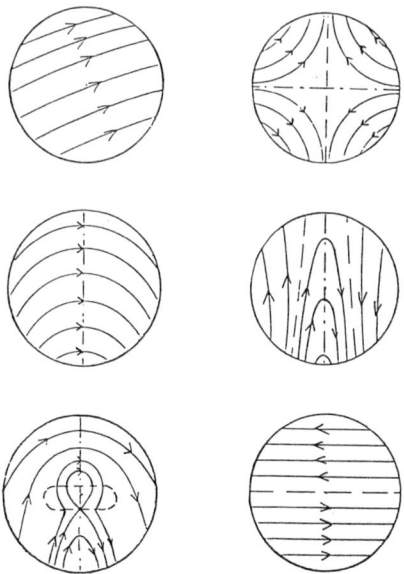

Fig. 7. Sketch of configurations that satisfy the bifurcation criterion (on the right) and that do not (on the left).

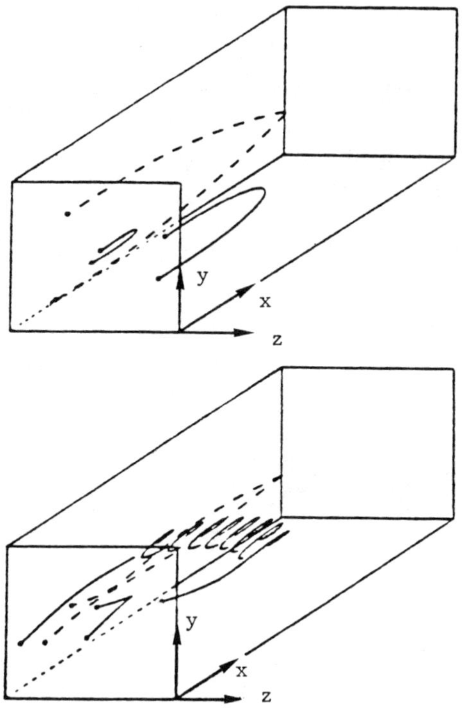

Fig. 8. A 3D generalization of the situation shown in Figure 5: Unperturbed (upper part) and plasmoid (lower part) field lines.

that do not disappears: In the early stage of the development of a three-dimensional plasmoid all field lines have the same topology [Hughes and Sibeck, 1987; Schindler et al., 1988; Hesse and Schindler, 1988]. This is due to the fact that the early plasmoid is exclusively surrounded by field lines with both ends on the earth such that a field line that leaves or enters the plasmoid must join flux that is connected to the earth. (During later stages of plasmoid development other topologies of plasmoid field lines occur, see the discussion by Birn, this volume.)

This has an important consequence for the notion of magnetic reconnection. If one ties reconnection to the topological structure of the two-dimensional case one finds that such a notion is structurally unstable, because it ceases to be applicable in the 3D case. Therefore, a revision of the concept of magnetic reconnection seemed necessary. It is important that reconnection as a basic space plasma physics process is properly defined. A recent suggestion ('general magnetic reconnection' or 'GMR') is based on the violation of the line-conservation property due to a localized non-idealness in Ohm's law [Schindler et al., 1988; Hesse and Schindler, 1988]. In this case the electric field parallel to $\mathbf{B}$ is the central quantity, particularly in cases where magnetic nulls are absent, as in the case of Fig. 8.

It is of interest to return our discussion to the effects associated with points where the magnetic field vanishes (($B = 0$)-singularities). Since neutral sheets and neutral lines are structurally unstable the only remaining candidates for stable ($B = 0$)-singularities are magnetic nulls. The role of magnetic nulls for reconnection is not yet clear. The concept of GMR covers cases both, with and without magnetic nulls. The example of the following section is one without nulls. Recently, Greene [1988] emphasized magnetic nulls as the central element of magnetic reconnection. One of the ingredients of this picture is the fact that magnetic dipole with a homogeneous magnetic field superimposed must contain magnetic nulls. The relevance of such nulls for magnetic reconnection and, in particular, for resistive instabilities has not yet been demonstrated in dynamical models.

A further new aspect of GMR is the occurrence of stochastic magnetic connection [see Birn, this volume]. This is different from stochastic fields occurring in toroidal laboratory devices such as Tokamaks, where the stochasticity is based on multiple recurrence of plasma elements due to parallel motion. We add that the concept of GMR is qualitatively quite different from diffusion. Not only is the resistivity highly localized, but also in the case of 'global effects', where $\int \mathbf{E} \cdot ds \neq 0$ on a suitable set of field lines passing through the non-ideal region, the change of magnetic connection involves plasma elements that never entered the non-ideal zone at all.

Flux Transfer Events

In this section we briefly describe a three-dimensional resistive instability process that provides a model for flux transfer events (FTE's) at the earth's magnetopause. As a typical signature of flux transfer events the observation of energetic particles within the magnetosheath together with a bipolar signature of the magnetic field component normal to the magnetopause was first reported by Russel and Elphic [1978]. They concluded that as a consequence of reconnection magnetic flux of the magnetosphere must have been connected to the interplanetary magnetic field, thus allowing energetic magnetospheric particles to enter the magnetosheath. Estimates of the amount of connected magnetic flux [Rijnbeek et al., 1984] together with the occurrence rate of FTE's indicate that these phenomena might contribute a dominant part to the plasma and energy transport from the solar wind into the magnetosphere. This is in good agreement with the fact that FTE's are strongly correlated to a southward component of the interplanetary magnetic field [Berchem and Russel, 1984] which again is correlated to geomagnetic activity [Baker et al., 1981]. For details of the observations of FTE's we refer to the papers by Russel and by Elphic [this volume].

A number of theoretical models [e.g. Lee and Fu, 1985; Sonnerup, 1987; Scholer, 1988] have been worked out describing the formation and/or the interior structure of flux transfer events. Lee and Fu [e.g. 1985] suggested the development of multiple X-lines in a current sheet leading to the bipolar B_N signature, whereas a model by Scholer [1988] assumes a localized resistivity with the formation of a single X-line. Most of the existing models are two-dimensional, where it should be mentioned that the model of Lee and Fu has been expanded to three dimensions recently [Fu, 1989]. Another three-dimensional model has been presented by Ogino et al. [1989] who carried out a global simulation. However, especially resistive instabilities involve rather small length scales which might not be resolved sufficiently in such a global model.

It is well known that FTE's are strongly localized in at least two dimensions. Further the varying orientation of these structures with respect to the ambient magnetic field strongly indicates that FTE's are intrinsically three-dimensional structures in many cases, which requires 3D modelling with sufficient spatial resolution. Thus we suggest that flux transfer events might be the result of a general magnetic reconnection process due to a spot of three-dimensionally localized nonideality at the magnetopause.

Figure 9 shows the initial configuration of a 3D resistive MHD computation applying to this model. This initial state consists basically of a plane current sheet (the magnetopause) separating regions of 150° sheared magnetic fields with different amplitudes. At time $t = 0$ a localized resistivity, which may be caused by enhanced microturbulence as a consequence of fluctuations in the solar wind, becomes available (sketched in Fig. 9). In the following evolution the localized resistivity leads to an increase of the parallel electric field (Fig. 10) in a narrow layer in the magnetopause within the domain of the localized resistivity. We should emphasize that due to the magnetic shear there are

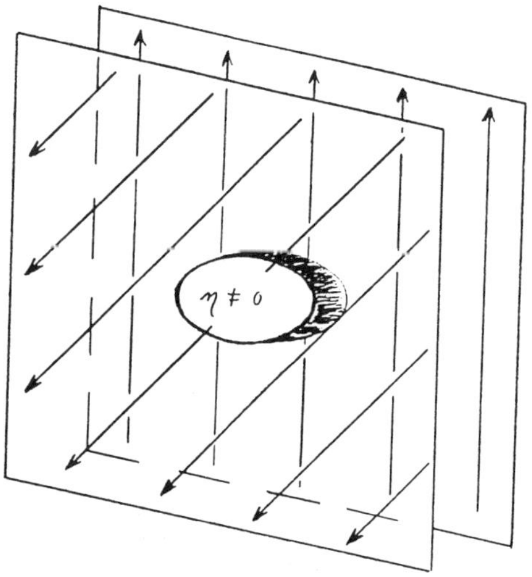

Fig. 9. Sketch of the initial equilibrium, which is assumed for the three-dimensional MHD computations applying to the formation of FTE's.

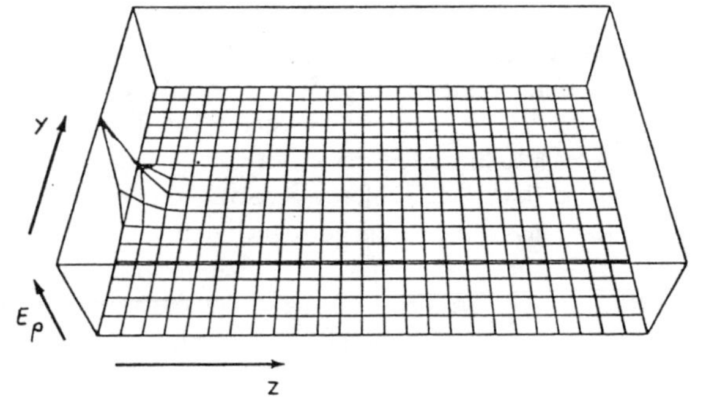

Fig. 10. Parallel electric field in the y, z-plane at $x = 0$ which corresponds to the center of the magnetopause.

no ($B = 0$) singularities such as nulls involved in this evolution, such that the process is typical for finite-B GMR (see section 4).

The characteristic bipolar signature of the normal magnetic field component which develops as a consequence of the reconnection process is obvious in the lower part of Fig. 11 which represents perspective views of the magnetic field lines in the 3D system. The magnetopause layer is defined by the y, z-plane. All field lines in these plots start in the magnetosheath ($x > 0$) go through points with $x = 2.7$, $y = 13.5$ and different values of z (measured in width of the magnetopause) and, depending on whether they turn into the positive z-direction, might enter the magnetosphere. Note that plasma flow is basically in the z-direction such that a satellite would first observe the increase of B_N followed by a decrease and a recovery to $B_N \approx 0$. As an example, Fig. 12 shows the time evolution of magnetic field components and pressure which a simulated satellite, that is located at the position indicated by a cross in Figure 11, has observed (time is measured in Alfvén times with one Alfvén time corresponding to one or a few seconds roughly). Normalization to magnetopause plasma parameters yields realistic values for the length and time scales. Other typical features of FTE's like the increase of the total pressure were recovered in this simulation. Details will be given elsewhere.

We emphasize that a comparison of the upper part of Figure 11, which shows the instant $t = 160$, with the time development of B_N illustrated in Fig. 12 reveals that the bipolar B_N signature and the region of open field lines pass the satellite at the same time.

So far we have illustrated that 3D magnetic reconnection due to localized resistivity is a possible explanation for the occurrence of FTE's. Note that multiple X-line reconnection can occur if several nonideal spot are present, but for the observed FTE signatures multiple X-lines are not necessary to occur.

At last we ought to demonstrate that the process leading to the interconnection of magnetospheric and magnetosheath flux indeed involves a resistive instability. Figure 13 shows the total kinetic energy as a function of time for the above presented simulation and a case of initially parallel magnetic fields of different amplitude separated by the magnetopause. In the latter case the total amount of kinetic energy is by a factor of nearly 10^{-5} smaller than in the 150° shear case at the end of the computation. Thus it can be concluded that for parallel magnetic fields a resistivity produces a slow diffusion of plasma whereas a southward component of the interplanetary magnetic field indeed leads to the development of a resistive instability a soon as resistivity becomes available.

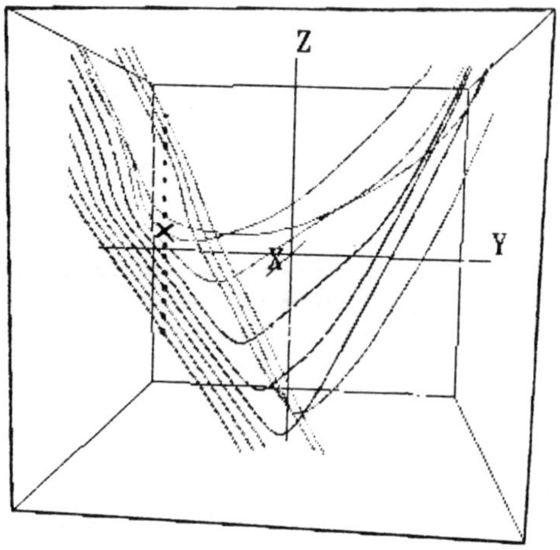

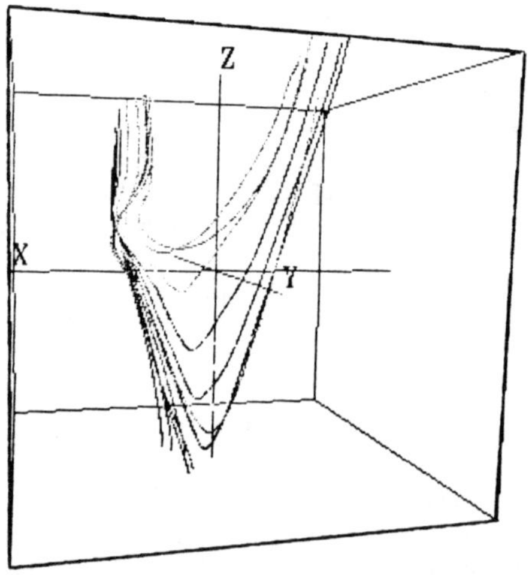

Fig. 11. Perspective view of magnetic field lines in the 3D-system. The y, z-plane corresponds to the magnetopause layer. The upper part shows a view from the interplanetary medium onto the magnetopause, the lower part presents a view approximately parallel to the magnetopause.

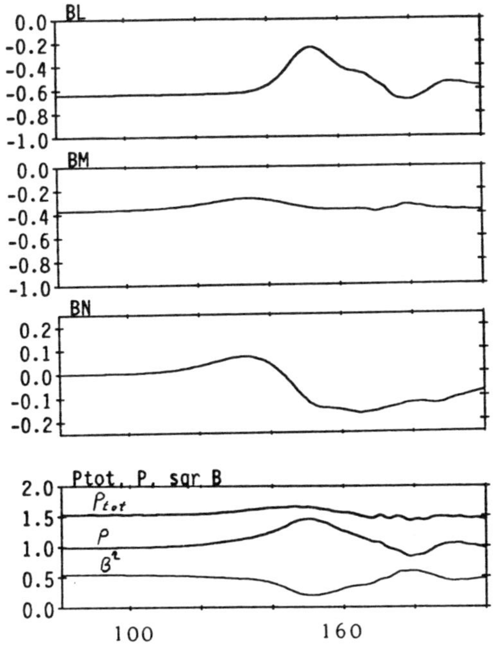

Fig. 12. Magnetic field, kinetic and magnetic pressures and their sum as measured by a hypothetical satellite. All quantities are normalized to typical values and the time is measured in Alfvén times.

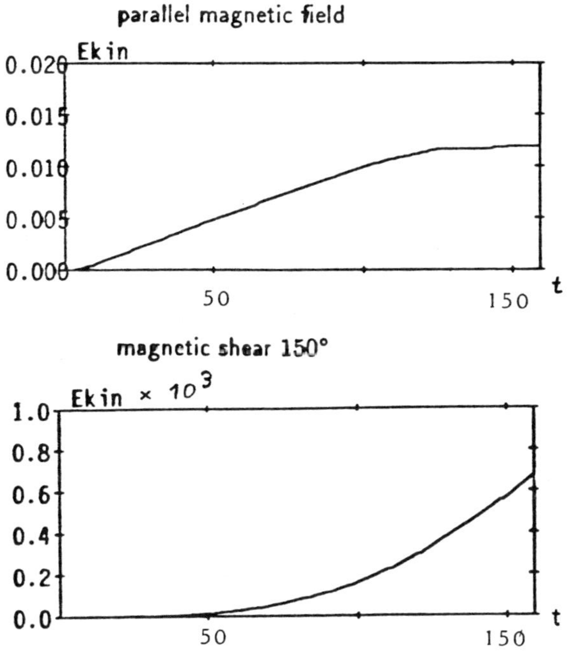

Fig. 13. Kinetic energy as a function of time for the case of parallel magnetic fields (upper plot) and for 150° magnetic shear.

Summary

Important properties of resistive instabilities may be summarized as follows.

Resistive instabilities

- are relevant for almost ideal $S \gg 1$ plasmas
- are faster than magnetic diffusion by localization of resistive effects
- involve relaxation of stressed magnetic fields and plasma transport by magnetic field reconfiguration
- are sensitive to the number of dimensions
- can be understood in terms of general magnetic reconnection
- play a central role in the theories of magnetospheric and stellar activity
- often lead to rope-like structures and may even represent a basic mechanism for the formation of magnetic flux ropes.

In conclusion we remark, that resistive instabilities may be considered to be representative for a wider class of phenomena based on nonideal plasma effects. The process presented in Figure 5 has recently been shown [Zwingmann, 1989] to occur in the same way for a collisionless plasma where the nonidealness is based on particle inertia alone. This process can of course be accelerated by additional turbulence effects.

Appendix

A necessary condition for the first bifurcation of a Grad-Shafranov equilibrium sequence may be obtained as follows. At a bifurcation point, besides the Grad-Shafranov equation with control parameter λ

$$-\Delta A = \lambda G(A) \qquad (A1)$$

the associated eigenvalue problem

$$-\Delta a - \lambda G'(A) = \nu a \qquad (A2)$$

must have a nontrivial solution with eigenvalue $\nu = 0$, with a vanishing on the boundary. Taking the gradient of (A1) one easily finds using (A2) that

$$\oint \frac{\partial a}{\partial n} \nabla A ds = 0 \qquad (A3)$$

where the integral is expended over the boundary. At the first bifurcation, limiting the stable regime which begins at $\lambda = 0$, the eigenstate $\nu = 0$ corresponds to the ground state of the Schrödinger equation (A2) such that, generically, a does not process zeros. This means that (A3) implies that both magnetic field components $B_x = \frac{\partial A}{\partial y}$, $B_y = -\frac{\partial A}{\partial x}$ must assume both signs on the boundary. This is the desired criterion. (This result was first presented to the CECAM-

Workshop, Paris, 1984. We reproduce it here because it is not readily available in the literature.)

Questions and Answers

Atkinson: Did the viewgraph showing colored magnetic field lines represent merging at a three-dimensional blob of finite resistivity? That is a limited region in all three dimensions?

Schindler: Yes. The resistive region is limited in all three directions. The resulting field is also three-dimensional. It has no (B = 0) singularities.

Luhmann: You have "introduced" another method of flux rope formation in your talk, where flux ropes can be formed as tearing islands in current sheets where the fields on either side are not strictly antiparallel. This is in contrast to the picture of flux rope formation by moving foot points. Could you suggest some observational differences that one might look for to distinguish the one type of flux rope from the other? (I note that this second mechanism requires a current sheet.)

Schindler: There are many ways in which a given configuration can be changed so that it becomes resistively unstable. Motion of foot points is just one possibility. A well developed tearing island (or its three-dimensional generalization) with a significant magnetic field component along the direction of the current that forms the island, has the essential properties of flux ropes. Therefore it will probably be difficult to identify its origin observationally. I agree that current sheets do play an essential role in resistive instabilities, where the current can flow partly or entirely (force-free case) along the magnetic field. But the current sheet disappears as the current concentrates in rope like structures due to the instability.

Birn: I would like to comment on Janet Luhmann's question. It just occurred to me that a distinction between the two cases of flux rope formations, which holds in typical cases, might be based on a comparison of the magnetic fields inside and outside. In typical reconnection cases the magnetic field inside the flux rope is smaller than on the outside, while twisting of a flux tube should lead to increased field strength inside.

Schindler: I would expect that this is true for cases where the plasma "beta" is not too small. However, in a force-free field, the magnetic field inside may well exceed the external field strength, even in the case of resistive instability.

Lui: In 2-D resistive tearing with inclusion of ∇p along the x-axis, the growth rate becomes high with the imbalance of ∇p and $\mathbf{j} \times \mathbf{B}$ forces. Are the forces evaluated at a particular point or integrated over the magnetic island structure? Why is there an imbalance? Also, what is the magnetic Reynolds' number for that simulation? How is it compared with the value for the real magnetotail case?

Schindler: The ∇p and $\mathbf{j} \times \mathbf{B}$ forces balance each other in equilibrium. In the early stages of unstable growth the magnetic stresses are reduced such that the plasma accelerates in the direction of the pressure force -∇p, i.e., tailward. It is only at later stages that the magnetic field stresses dominate the pressure force in (tailward) acceleration. The forces are integrated over the plasmoid cross section defined in terms of the magnetic separatrix. (For the actual plasmoid motion the momentum added by the reconnection process itself is also important.) The resistive Reynolds numbers that yield observed time scales lie in the range of several hundreds. The value used in the simulation was 100.

Antiochos: What produces and maintains the small scale of the resistive region in your 3-D model with no *B* reversals?

Schindler: The model simply assumes that a fixed region exists where the resistivity is different from zero. The physical concept behind this assumption is that the magnetopause may become locally sufficiently changed so that the threshold for a microinstability is surpassed. The dynamic modification of the resistivity profile, which would exist in the real case, is ignored in the model.

Acknowledgments. This work was supported by the Deutsche Forschungsgemeinschaft through the Sonderforschungsbereich "Plasmaphysik Bochum / Jülich" and the Schwerpunktsprogramm "Theorie kosmischer Plasmen".

References

Baker, D.N., E.W. Hones, Jr., J.B. Payne, and W.C. Feldman, A High Time Resolution Study of Interplanetary Parameter Correlations with AE, *Geophys. Res. Lett.*, **8**, 179, 1981

Baker, D.N., S.J. Bame, R.D. Belian, W.C. Feldman, J.T. Gosling, P.R. Higbie, E.W. Hones, Jr., D.J. McComas, and R.D. Zwickel, Correlated Dynamical Changes in the Near-Earth and Distant Magnetotail Regions: ISEE 3, *J. Geophys. Res.*, **89**, 3855, 1984

Berchem, J., and C.T. Russel, Flux Transfer Events on the Magnetopause: Spatial Distributions and Controlling Factors, *J. Geophys. Res.*, **89**, 6689, 1984

Birn, J., Computer Studies of the Dynamic Evolution of the Geomagnetic tail, *J. Geophys. Res.*, **85**, 1214, 1980

Birn, J., and E.W. Hones, Jr, Three-Dimensional Computer Modeling of Dynamic Reconnection in the Geomagnetic Tail, *J. Geophys. Res.*, **86**, 6804, 1981

Biskamp D., and H. Welter, Magnetic Arcade Evolution and Instability, submitted to *Solar Phys.*, preprint, 1988

Elphic, R.C., and C.T. Russel, Magnetic Flux Ropes in the Venus Ionosphere: Observations and Models, *J. Geophys. Res.*, **88**, 58, 1983

Fahr, H.J., W. Neutsch, S. Gredzielski, W. Macek, and

R. Ratkiewicz-Landowska, Plasma Transport across the Heliopause, *Space Sci. Rew.*, **43**, 329, 1986

Ferrari, A., and A.G. Pacholczyk, (Eds.) *Astrophysical Jets*, D. Reidel Publ. Co., Dordrecht-Holland, 1983

Forbes, T.G., and J.M. Malherbe, A shock Condensation Mechanism for Loop Prominences, *Ap. J.*, **302**, L67, 1986

Forbes, T.G., J.M. Malherbe, and E.R. Priest, The Formation of Flare Loops by Magnetic Reconnection and Chromospheric Ablation, *Solar Phys.*, **120**, 285, 1989

Fu, Z.F., 2D and 3D Simulation Study of Multiple X Line Reconnection, in *Reconnection in Space Plasmas*, Vol II, 275, Conference Proceedings, Potsdam, ESA, ESTEC, Noordwijk, Netherlands, 1989

Furth, H.P., J. Killeen, and M.N. Rosenbluth, Finite-Resistivity Instabilities of a Sheet Pinch, *Phys. Fluids*, **6**, 459, 1963

Greene J. M., Geometrical Properties of Three-Dimensional Reconnecting Magnetic Fields with Nulls, *J. Geophys. Res.*, preprint, 1988

Hesse, M., and K. Schindler, A Theoretical Foundation of General Magnetic Reconnection, *J. Geophys. Res.*, **93**, 5559, 1988

Hones, E.W., Jr., Plasma Flow in the Magnetotail and its Implication for Substorm, in *Dynamics of the magnetosphere*, ed. by S.I. Akasofu, D. Reidel, Dordrecht-Holland, 545, 1979

Van Hoven, G., T. Tachi, and R.S. Steinolfson, Radiative and Reconnection Instabilities: Filaments and Flare, *Astrophys. J.*, **280**, 391, 1984

Hughes, W.J., and D.G. Sibeck, On the 3-Dimensional Structure of Plasmoids, *Geophys. Res. Lett.*, **92**, 636, 1987

Janicke, L., Resistive Tearing Mode in Weakly Two-Dimensional Neutral Sheet, *Phys. Fluids*, **23**, 1843, 1980

Lee, L.C., and Z.F. Fu, A Theory of Magnetic Flux Transfer at the Earth's magnetopause, *Geophys. Res. Lett.*, **12**, 105, 1985

Mikic, Z., D.C. Barnes, and D.D. Schnack, Dynamical Evolution of a solar Corona Magnetic Field Arcade, *Ap. J.*, **328**, 830, 1988

Newcomb, W.A., Motion of Magnetic Lines Force, *Annals. Phys.*, **3**, 374, 1958

Nishikawa, K.-I., and J. Sakai, Stabilizing Effect of a Normal Magnetic Field on the Collisional Tearing Mode, *Phys. Fluids*, **25**, 1384, 1982

Ogino, T., R.J. Walker, and M. Ashour-Abdalla, A Magnetohydrodynamic Simulation of the Formation of Magnetic Flux Tubes at the Earth's Dayside Magnetopause, *Geophys. Res. Lett.*, **16**, 155, 1989

Otto, A., The role of Magnetic Reconnection in Magnetotail Plasmoid Dynamics, in *Reconnection in Space Plasmas*, Vol II, 223, Conference Proceedings, Potsdam, ESA, ESTEC, Noordwijk, Netherlands, 1989

Paris, R.B., The Resistive Tearing Mode in a Weakly Two-Dimensional Sheet Pinch, *Phys. Fluids*, **30**, 102, 1987

Paschmann, G., G. Haerendel, I.Papamastorakis, N. Sckopke, S.J. Bame, J.T. Gosling, and C.T. Russel, Plasma and Magnetic Field Characteristics of Magnetic Flux Transfer Events, *J. Geophys. Res.*, **87**, 2159, 1982

Pritchett, P.L., and C.C. Wu, Coalescence of Magnetic Islands, *Phys. Fluids*, **22**, 2140, 1979

Rijnbeek, R.P., S.W.H. Cowley, D.J. Southwood, and C.T. Russel, A Survey of Dayside Flux Transfer Events Observed by ISEE 1 and ISSE 2 Magnetometers, *J. Geophys. Res.*, **89**, 768, 1984

Robertson, J.A., and E.R. Priest, Line Field Magnetic Reconnection, *Solar Physics*, **114**, 311, 1987

Scholer, M., Magnetic Flux Transfer at the Magnetopause Based on Single X Line Bursty Reconnection, *Geophys. Res. Lett.*, **15**, 291, 1988

Schindler, K., J. Birn, and L. Janicke, Stability of Two-Dimensional Pre-Flare Structures, *Solar Physics*, **87**, 103, 1983

Schindler, K. , M. Hesse, and J. Birn, General Magnetic Reconnection, Parallel Electric Fields and Helicity, *J. Geophys. Res.*, **93**, 5547, 1988

Sonnerup, B.U.O., On the Stress Balance in Flux Transfer Events, *J. Geophys. Res.*, **92**, 8613, 1987

Russel, C.T., and R.C. Elphic, Initial ISEE Magnetometer Results: Magnetopause Observations, *Space Sci. Rev.*, **22**, 681, 1978

Zwingmann, W., Particle Simulation of Magnetic Reconnection in Collision-Free Plasma, in *Reconnection in Space Plasmas*, Vol II, 67, Conference Proceedings, Potsdam, ESA, ESTEC, Noordwijk, Netherlands, 1989

STEADY MAGNETIC FIELD RECONNECTION

B. U. Ö. Sonnerup, J. Ip, and T.-D. Phan

Thayer School of Engineering, Dartmouth College, Hanover, New Hampshire

Abstract. A brief overview is presented of steady two dimensional magnetic field reconnection, with emphasis on recent developments. First, comments are made on difficulties in arriving at a satisfactory rigorous definition of the reconnection phenomenon which retains the traditional topological aspects of the magnetic field as an integral part. Analytical models of magnetic field annihilation and magnetic field reconnection are then reviewed with emphasis on the description of flux pile up as well as on the structure of exit jets and of vortex layers forming at magnetic separatrices. In particular, a new exact analytic incompressible MHD solution, illustrating flux pile up in the inflow, is presented. A brief discussion is then given of numerical simulation results. Finally, comments are made on the importance of two-fluid effects in the immediate vicinity of the reconnection site, in the so-called diffusion region.

1. Introduction

The classical concept of magnetic field reconnection in a highly conducting plasma is illustrated in Figure 1. Plasma elements in magnetic cell #1, linked together by a strong magnetic field, B_1, move slowly to the right with speed $v_1 = E_0/B_1$. At the same time, plasma elements in magnetic cell #2, linked together by a strong but oppositely directed field, B_1, move slowly to the left, again with speed $v_1 = E_0/B_1$. The motion occurs as a result of the reconnection electric field, E_0, directed out of the plane of the paper, and the speed of motion is small because B_1 is large (in a 2D steady state, E_0 is the same everywhere, as a consequence of Faraday's law). An encounter and subsequent relinking of pairs of oppositely directed magnetic field lines and the plasma located on them occurs at the X-type magnetic null point at the center of the figure. The relinking is made possible by finite electrical resistivity, which serves to break the frozen magnetic-field condition in a small volume, called the diffusion region, around the null point. The result of the relinking is that reconnected magnetic field lines and plasma elements of different origin (cells #1 and 2) leave together at a high speed $v_2 = E_0/B_2$ in magnetic cells #3 and 4. The speed v_2 is large because the magnetic field B_2 is weak. In this sense, reconnection may be thought of as a process that converts, on a continuous or intermittent basis, the high magnetic energy and low plasma kinetic energy in cells #1 and 2 to a state of low magnetic energy and high plasma kinetic energy in cells #3 and 4. In addition, realistic treatments of the process indicate that the thermal energy of the plasma is also increased as it enters the exit flow cells #3 and 4. The pairwise symmetry of cells #1 and 2 and #3 and 4 in the figure is not required and the process may operate also in the presence of a "guide" magnetic field, $B_{\|}$, perpendicular to the plane of the figure.

As indicated in Figure 1, an analogy exists between reconnection and some of the interactions between solar and magnetospheric physicists sought at the present meeting. The reconnection electric field that drives the two research communities together at Bermuda is provided by conference organizers E.R. Priest and C.T. Russell in the solar and the magnetospheric cells, #1 and 2 respectively, while the reconnection process itself is brought about by certain activities of the third conference organizer, L.C. Lee, in the diffusion region (Hamilton). We know that merely pushing the two research communities together does not guarantee that strong interaction and relinking will occur. The same holds true for differently magnetized plasma regions: magnetic reconnection between them often occurs only reluctantly and at low rates although in special circumstances, poorly understood at present, the process is thought to proceed rapidly, even explosively.

Magnetic reconnection is expected to play an important role in flux rope physics. As illustrated in Figure 2, it can be responsible for: (a) the generation of flux ropes via the tearing instability in the presence of a guide magnetic field, $B_{\|}$, along the current; (b) the coalescence of flux ropes via the coalescence instability; (c) the relinking and

Geophysical Monograph 58

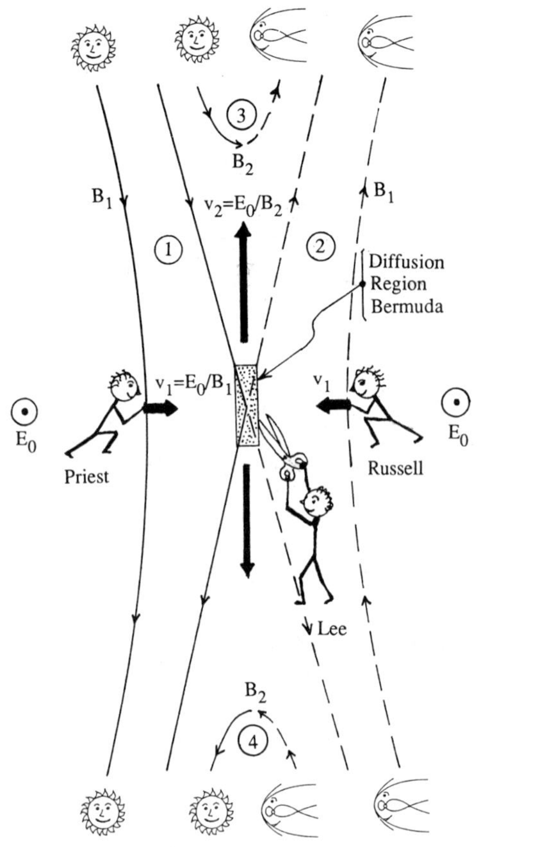

Fig. 1. Schematic drawing showing basic aspects of the magnetic field reconnection process.

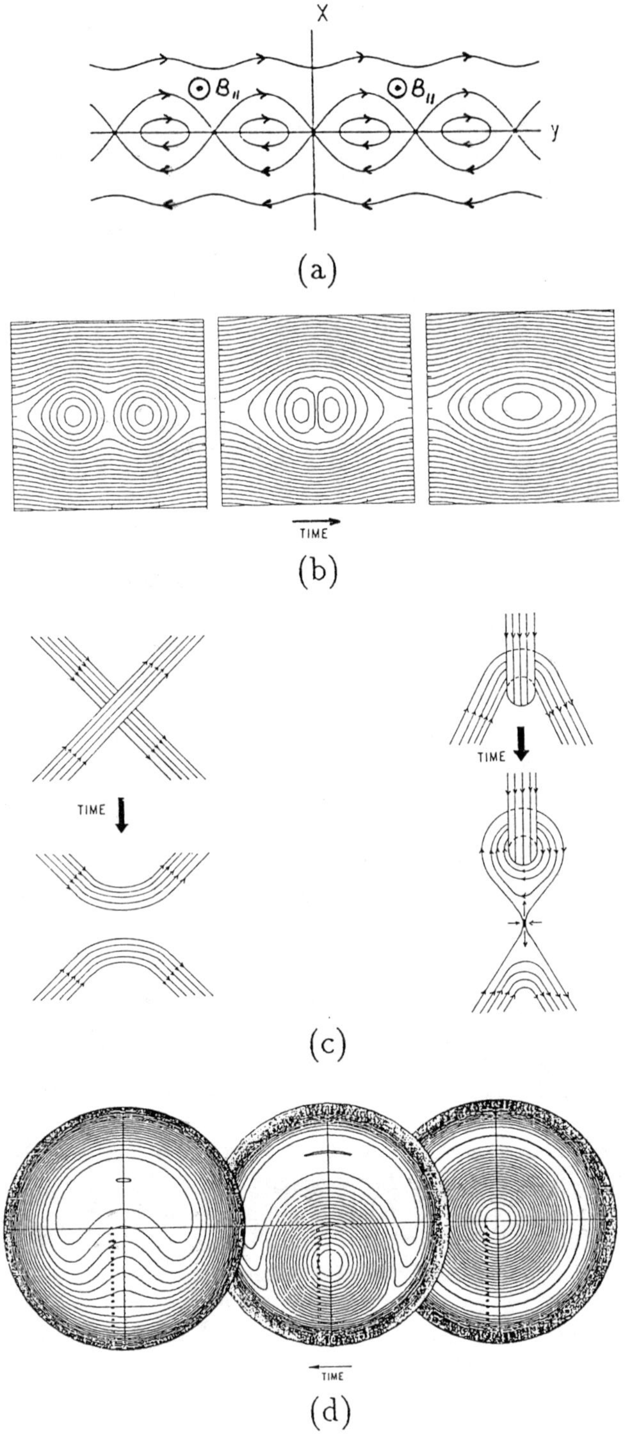

Fig. 2. Role of reconnection in flux rope physics.(a) Flux ropes generated by tearing instability in the presence of a guide field, $B_{\|}$; (b) coalescence of flux ropes [after Pritchett and Wu, 1979]; (c) relinking and unlinking of flux ropes; (d) rearrangement of internal flux rope structure by the internal kink instability [from Park et al., 1984].

unlinking of flux ropes; and (d) the restructuring of a flux rope via the internal kink mode.

In this paper, a brief discussion is presented of certain aspects of steady or quasi-steady reconnection; dynamic aspects are dealt with in the review by Schindler and Otto [1989; this conference]; turbulent aspects are discussed by Lysak and Song [1989; this conference]. The paper is organized as follows. We first comment briefly on the definition of reconnection, an area where certain important developments have occurred recently. We then summarize the main properties of existing 2D analytical models of reconnection, including a recent unification of those models provided by Priest and Forbes [1986]. Numerical simulations are then examined which have brought to light new and, in part, unexpected features of the reconnection process, among them the formation of vortex layers at magnetic separatrices. Finally, we comment briefly on the physical processes in the diffusion region which may allow reconnection to occur even in a collisionless plasma. This outline of the paper has substantial similarities to that of a recent review article on steady state reconnection [Sonnerup, 1988; hereafter referred to as Paper I]. For that reason, the present account is brief and from time to time the reader will be referred to the earlier article for details.

2. Definition

One of the first formal definitions of reconnection was provided by Vasyliunas [1975] who stated that "magnetic field line merging, or reconnection, is the process whereby plasma flows across a surface that separates regions containing topologically different magnetic field lines. The magnitude of the plasma flow is a measure of the merging rate." The separating surface referred to in the definition is called a separatrix: the field lines forming the X in Figure 1 delineate two such surfaces orthogonal to the plane of the figure. Later, the working group on reconnection at the Coolfont Workshop [Butler and Papadopoulos, 1984] agreed on a slightly different, but for practical purposes equivalent definition: "magnetic field reconnection occurs in a plasma whenever an electric field, $E_{\|}$ ($\equiv E_0$ in Figure 1), is present along a magnetic separator, i.e., along a line of intersection of two separatrix surfaces which divide space into different magnetic cells...". The separator in Figure 1 is a straight line through the center point of the X and at right angles to the plane of the figure; it is often referred to as the X line.

Both of the above definitions are based on topological properties of the magnetic field, namely, the separatrices and the separator. Except in special cases, such as the purely 2D case with $B_{\|} \equiv 0$, these properties cannot be ascertained locally but require reference to null points in the magnetic field, points that may be located at large distances from the place where reconnection is being observed or examined. The importance of these nulls has been discussed recently by Greene [1988]. An illustration of magnetic cells, separatrix surfaces, separators and null points is shown in Figure 3 where a cross section is depicted of the field configuration obtained by superposition of a dipole field and a uniform field at right angles to the dipole moment vector. Two X type magnetic null points (see Dungey [1963] for a precise definition) are located at X_1 and X_2 in the plane of the figure. They are connected by a circular [Yeh, 1976] field line that lies in a plane perpendicular to that of the figure. Note that this special field line is located at the intersection of separatrix surfaces that separate three distinct magnetic cells. Thus it comprises the separator in this configuration. At an arbitrarily chosen point on the separator, there is a nonvanishing value, $B_{\|}$, of the parallel magnetic field in general; only at X_1 and X_2 is that field absent. At such an arbitrary location, it does not seem possible to uniquely identify a field line as being the separator without following it in both directions to assure that one ultimately reaches points X_1 and X_2. In the same way, identification

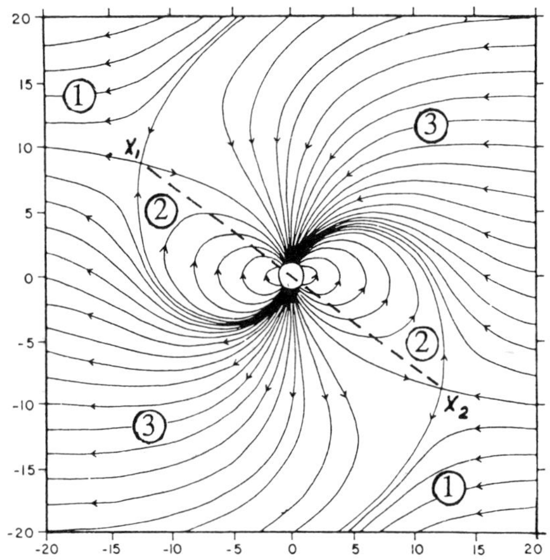

Fig. 3. Magnetic field lines in the meridional plane containing the null points X_1 and X_2 for a uniform horizontal field superimposed on a vertical 3D dipole. Dashed line is the separator circle, viewed edge on [after Cowley, 1973].

of a separatrix surface cannot be done locally but requires an excursion along two neighboring field lines located in the presumed separatrix surface until those two lines meet at X_1 or at X_2. Two dimensional configurations such as the one shown in Figure 1, but with an added transverse constant field component, $B_{\|}$, or toroidal configurations such as tokamaks, have the further difficulty that no null points exist which can form the basis for a unique definition of separator and separatrices. Similar difficulties may arise in many space applications of reconnection, e.g., in plasmoid formation in the geomagnetic tail.

A local definition of reconnection, i.e., a definition that does not depend on finding, possibly distant, null points in the **B** field would be desirable since it seems somewhat questionable whether the local dynamics associated with the reconnection process could be controlled by these distant nulls. An attempt to provide such a local definition that retains the topological aspects of the field consists of examining the local magnetic field topology in a plane perpendicular to a chosen field line, as discussed in some detail in Paper I (see also Podgorny [1986] and Priest and Forbes [1989]). If the field topology in that plane is hyperbolic, then the field line is a potential separator line. However, it turns out that this procedure does not provide a unique separator; typically, entire regions in space will contain potential separator field lines (e.g., in the field $B_x = ay$; $B_y = bx$; $B_z = B_0$, all field lines display the required hyperbolic topology). In Figure 1, say, reconnection then ceases to be localized at the origin when $B_{\|} \neq 0$

but occurs in the entire diffusion region, i.e., wherever the electric field has a component along $\mathbf{B}$.

The difficulties described above can be circumvented by adopting a more general definition of reconnection that does not involve the topology of the magnetic field at all. Thus Axford [1984] described reconnection as "a localized breakdown of the requirement for 'connection' of elements of fluid at one time on a common magnetic field line." Such breakdown by necessity involves the presence, in a localized region, of an electric field component along $\mathbf{B}$. More recently, this idea has been pursued in detail by Schindler et al. [1988] and Hesse and Schindler [1988] who define what they call 'general magnetic reconnection' (GMR) as "the breakdown of magnetic connection due to a localized nonidealness." They have established that such breakdown occurs if and only if the quantity

$$\mathbf{U} \equiv \mathbf{B} \times [\nabla \times (\mathbf{E} + \mathbf{v} \times \mathbf{B})] \neq 0 \; . \tag{1}$$

This broader definition includes all cases, such as the ones depicted in Figures 1 and 2, which we have traditionally associated with the term reconnection, but also cases that are rather different. An example of GMR that does not involve the usual hyperbolic field topology is shown in Figure 4 where a flux rope configuration for which

$$\left.\begin{aligned} \mathbf{B} &= \underline{\hat{\phi}}\mu_0\sigma_2\omega B_0 R^3/8h + \hat{\mathbf{z}} B_0 \\ p &\simeq p_0 - \mu_0(\sigma_2\omega B_0)^2 R^6/96h^2 \end{aligned}\right\} \tag{2}$$

is depicted. Here B_0 and p_0 are constants representing the uniform axial magnetic field and the plasma pressure on the flux tube axis ($R = 0$), respectively. Pressure variations due to fluid motion are neglected. In region 1, i.e., for $z < 0$, the electrical conductivity is assumed infinite and there is no plasma motion so that $\mathbf{v}_1 = \mathbf{E}_1 = 0$. In region 3, i.e., for $z > h$, the conductivity is again infinite

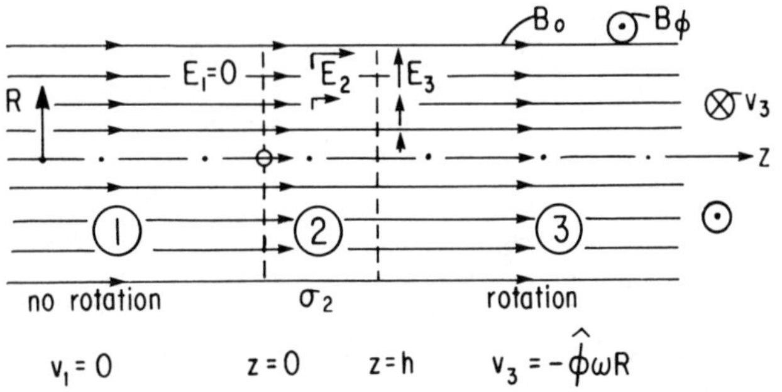

Fig. 4. Flux rope having a twisted magnetic field, (B_ϕ, B_z), given by equation (2). The plasma in region 1 is infinitely conducting and stationary; the plasma in region 3 is also infinitely conducting but is in solid body rotation about the z axis. General reconnection occurs in region 2 where the conductivity is finite.

but this portion of the flux rope is in solid-body rotation at angular rate ω about its axis so that

$$\left.\begin{aligned} \mathbf{v}_3 &= -\omega R\underline{\hat{\phi}} \\ \mathbf{E}_3 &= \omega R B_0 \hat{\mathbf{R}} \end{aligned}\right\} \tag{3}$$

In the intervening region 2, $h > z > 0$, the conductivity, σ_2, is large but finite and the plasma velocity and the electric field are

$$\left.\begin{aligned} \mathbf{v}_2 &= -(z/h)\omega R\underline{\hat{\phi}} \\ \mathbf{E}_2 &= (z/h)\omega R B_0 \hat{\mathbf{R}} + (\omega R^2/2h) B_0 \hat{\mathbf{z}} \end{aligned}\right\} \tag{4}$$

The electric field is curl free everywhere, as required, but there is a positive volume charge density in region 2 as well as positive and negative surface charges at $z = 0$ and at $z = h$, respectively, so that this region has the appearance of an electrical double layer plus a net charge. There is an electric field component along $\mathbf{B}$ in region 2 and the quantity $\mathbf{U}_2 = (\omega R/h)B_0^2\hat{\mathbf{R}} \neq 0$ so that, according to the Schindler et al. definition, reconnection does indeed take place throughout region 2 (except at $R = 0$) as a result of the localized resistive nature of the plasma in that region.

The type of behavior illustrated in Figure 4 is particularly relevant to this conference because it would allow a segment of a strongly twisted flux rope to untwist itself, thereby releasing magnetic energy, a mechanism that has been discussed by P. Carlqvist [1969] in connection with solar flares. But at the same time, we point out that the nature and dynamics of this type of reconnection appears rather different from those of the configuration in Figure 1. Thus the wisdom of using the same name for both cases may be questioned. At present, we tend to prefer the more traditional, topologically based definitions even though they have certain shortcomings. To quote Supreme Court Justice Potter Stewart (in Jacobellis v. Ohio, 1964): "I shall not today further attempt to define the kinds of material I understand to be embraced within that shorthand description; and perhaps I could never succeed in intelligibly doing so. But I know it when I see it, ...".

3. Magnetic Field Annihilation Models

In the usual reconnection configuration, magnetic flux is transported from two magnetic cells (cells #1 and 2 in Figure 1) into two other cells (#3 and 4 in Figure 1) in which the magnetic field lines are differently connected. In the special case of field annihilation the angle occupied by the two exit cells (#3 and 4), which is also the angle between the separatrices, has been collapsed to zero which means that antiparallel magnetic flux is carried from cells #1 and 2 towards the current sheet (the yz plane) separating them where the flux is resistively annihilated. The

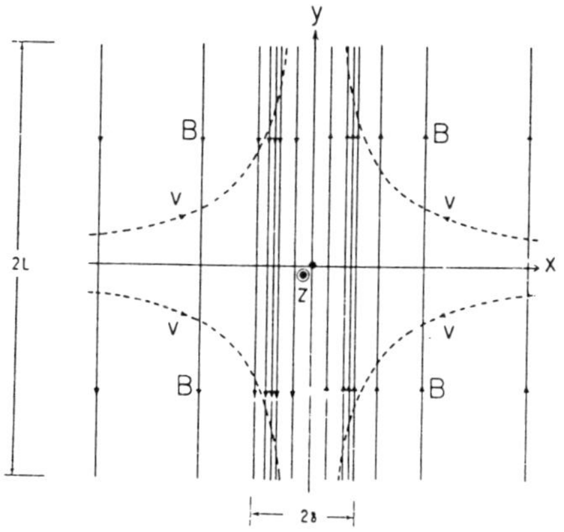

(a)

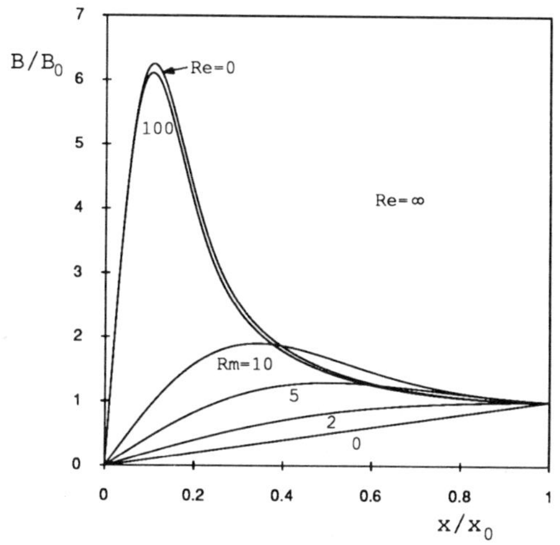

(b)

Fig. 5. Magnetic field annihilation (Sweet-Parker like) models. (*a*) Stagnation point flow [after Sonnerup and Priest, 1975]; (*b*) magnetic field profiles from equation (11) for flow between two fluid emitting walls at $x = \pm x_0$ where $\mathbf{v} = \mp\hat{\mathbf{x}}U_0$.

resulting field annihilation configuration is shown in Figure 5a. In this figure, plasma inflow occurs from the left and right in broad regions (length $= 2L$) and with a small velocity $v_1 = E_0/B_1$ that exactly equals the diffusion velocity $1/\mu_0\sigma\delta$, based on the electrical conductivity, σ, in the current sheet and on the half width, δ, of the sheet. Outflow in the $\pm y$ direction occurs in the narrow region (thickness $= 2\delta$) occupied by the current sheet. The outflow along the y axis is accelerated from $v_y = 0$ at the stagnation point (at $x = y = 0$) to $v_y = v_2 = B_1/\sqrt{\mu_0\rho} \equiv v_{A1}$ at the "exit" sections, $y = \pm L$, where the plasma pressure is assumed to be the same as in the inflow, i.e., $p_2 = p_1$. The acceleration along the current sheet is effected by the pressure gradient, $\partial p/\partial y$, associated with the excess pressure, $p_0 = p_1 + B_1^2/2\mu_0$, at the stagnation point. The key formula describing this model, due to Sweet [1958] and Parker [1957; 1963], is

$$M_{A1} \equiv v_1/v_{A1} = S_1^{-1/2} \tag{5}$$

where $v_{A1} = B_1/\sqrt{\mu_0\rho_1}$ is the Alfvén speed in the inflow and $S_1 \equiv \mu_0\sigma v_{A1}L$ is the Lundquist number based on the length L of the current sheet. This formula is easily obtained from mass conservation and Bernoulli's equation along with Ampère's and Ohm's laws, as reviewed in Paper I. The Alfvén Mach number M_{A1} is a commonly used nondimensional measure of the reconnection rate. Since the Lundquist number is very large in most cosmic applications, mainly on account of assumed large values of L, the predicted reconnection rate is very small. However, larger rates can be achieved if the plasma can find a way to generate an effective scale much smaller than L, the overall size of the reconnection configuration in the y direction or to generate a smaller value of v_{A1}. This is exactly what is assumed to happen in the Petschek model and other reconnection models, to be discussed presently. Note however that a decrease in v_{A1}, while increasing M_{A1}, in fact leads to a decrease in the actual inflow speed, v_1, as well as in the reconnection electric field, E_0.

The configuration discussed above incorporates the basic assumption of uniform flow and field conditions in the upstream regions. This is not necessarily what will occur: the velocity v_∞ and magnetic field B_∞ far upstream may differ from those (v_1 and B_1) immediately adjacent to the current sheet, as long as $E_0 = -v_1B_1 = -v_\infty B_\infty$ remains constant. For incompressible flow we may thus rewrite equation (5) in the form

$$M_{A\infty} = \frac{v_\infty}{v_{A\infty}} = \frac{B_1}{B_\infty}\frac{1}{\sqrt{\mu_0\sigma v_{A\infty}L}} \tag{6}$$

If we think of $v_{A\infty}$ and L as fixed, then an increase in v_∞ (or equivalently E_0) means an increase in $M_{A\infty}$: this can only be accomplished by increasing B_1/B_∞, an effect referred to as flux pile up. An important consequence of the above is that the upstream measure of the reconnection rate, $M_{A\infty}$, is different from the reconnection rate, M_{A1}, evaluated adjacent to the diffusion region, namely

$$M_{A1} = M_{A\infty}(B_\infty/B_1)^{3/2} \tag{7}$$

Thus, flux pile up corresponds to $M_{A1} < M_{A\infty}$. Furthermore, for fixed L, an increase in $M_{A\infty}$ by a factor k, say, simply leads to an increase in B_1/B_∞ by the same factor and an associated decrease of M_{A1} by a factor $k^{1/2}$.

The flux pile up effect is a prominent feature of the 2D MHD stagnation point flow depicted in Figure 5a and studied by Parker [1973] as well as by Sonnerup and Priest [1975]. The latter authors demonstrated that exact solutions of the incompressible MHD equations could be obtained for the type of straight field-line configuration shown in Figure 5a when the flow field is of the form $v_x = -k_1 x$; $v_y = k_2 y$; $v_z = k_3 z$ with $(-k_1 + k_2 + k_3) = 0$. These solutions are reviewed in Paper I. Here we discuss instead a new exact 2D field annihilation solution where a somewhat different flow configuration is used, namely unidirectional uniform inflow $\mathbf{v} = \mp U_0 \hat{\mathbf{x}}$ at two upstream boundaries $x = \pm x_0$ (this set of boundary conditions was brought to our attention by M. F. Heyn in a private communication). This solution, along with the stagnation point flows mentioned above, are members of a general class of 2D and 3D field annihilation solutions in which the magnetic field lines remain straight and parallel to the current sheet. The viscous momentum equation governing the velocity field (here assumed two dimensional for simplicity) is given by

$$R_e^{-1} f''' = f'^2 - f f'' + \kappa \tag{8}$$

where $v_x = -U_0 f(s)$, $v_y = U_0 (y/x_0) f'(s)$ and the constant $\kappa \equiv (x_0^2/\rho U_0^2 y)(\partial p/\partial y)$. Also, $R_e \equiv U_0 x_0/\nu$ is the viscous Reynolds number, based on the ordinary newtonian kinematic viscosity, ν, and a prime denotes differentiation with respect to the variable $s \equiv x/x_0$. Equation (8), with the boundary conditions $f = f'' = 0$ at $s = 0$ and $f = 1$, $f' = 0$ at $s = \pm 1$, has simple solutions both for $R_e \to \infty$ and for $R_e \to 0$:

$$\left.\begin{aligned} f &= \sin \pi s/2 \\ \kappa &= -\pi^2/4 \end{aligned}\right\} \quad R_e \to \infty \tag{9}$$

$$\left.\begin{aligned} f &= (3s - s^3)/2 \\ \kappa &= -3/R_e \end{aligned}\right\} \quad R_e \to 0 \tag{10}$$

The corresponding magnetic field profiles, obtained from Ohm's law, are given by

$$B_y = (E_0 R_m/U_0) e^{-R_m I(s)} \int_0^s e^{R_m I(\tilde{s})} d\tilde{s} \tag{11}$$

where $R_m \equiv \mu_0 \sigma U_0 x_0$ is the magnetic Reynolds number and

$$I(s) \equiv \int_1^s f(\tilde{s}) d\tilde{s}$$

The relation between the reconnection electric field, E_0, and the magnetic field, B_0, at $x = \pm x_0$ is

$$B_0 = (E_0 R_m/U_0) \int_0^1 e^{R_m I(\tilde{s})} d\tilde{s} \ .$$

The pressure distribution is given by a Bernoulli-type formula:

$$p = p_0 - \frac{\rho}{2}(v_x^2 + v_y^2) - B_y^2/2\mu_0 \tag{12}$$

A plot of the magnetic field, B_y, given by equation (11) is shown in Figure 5b for several different values of the magnetic Reynolds number but with $R_e = \infty$. The flux pile-up effect is clearly evident in this diagram for $R_m =$ 5, 10, and 100. In fact, the pile up becomes infinite as $R_m \to \infty$.

A second point to note is that the influence of finite R_e on these results is minimal, as shown by comparison of the curve for $R_m = 100$ labelled $R_e = 0$ with the curve for $R_m = 100$ having $R_e = \infty$. The reason for this small difference is that the two velocity distributions $v_x = -U_0 f(s)$ given by equations (9) and (10) for $R_e = \infty$ and $R_e = 0$, respectively, differ by only 2% in the region $0 \leq |s| \leq 1$. Since it is v_x that enters the term $\mathbf{v} \times \mathbf{B}$ in Ohm's law, the influence of the viscous Reynolds number on the magnetic field is correspondingly weak. Thus it is evident that the viscous forces (which vanish identically in the stagnation point flows discussed by Sonnerup and Priest [1975]) play no important role in these symmetric MHD flows toward a current sheet other than that of changing the pressure gradient κ. This latter effect would modify the Sweet-Parker formula, equation (5): on the basis of approximate calculations, Park et al. [1984] conclude that viscous effects reduce the value of M_{A1} by a factor $(1 + \mu_0 \sigma \nu)^{-1/4}$. Note, however, that this result cannot be general since it does not apply to the stagnation point flows mentioned above.

It should be added that the exact solutions discussed above are valid only for incompressible flow. In that limit, time-dependent versions can also be generated [e.g., Gratton et al., 1988]. For the compressible case, no exact solutions have been found: it appears that the field lines cannot remain straight and parallel to the yz plane in that case. On the other hand, the case of self-similar decay of a one-dimensional current sheet in a compressible medium has been treated [Kirkland and Sonnerup, 1979].

4. Magnetic Field Reconnection Models

As mentioned already, the principal difficulty in applying the Sweet-Parker formula to cosmic reconnection problems lies in the small reconnection rates obtained for L values comparable to the overall size of a cosmic region in which the reconnection occurs. Perhaps the only way to avoid this difficulty is to assume that the field and flow geometries will adjust themselves automatically in such a fashion that a much smaller effective length, y^*, say, is established and replaces L in equation (5). This is the idea developed by Petschek [1964] who proposed the reconnection configuration shown in Figure 6a. In this geometry, the Sweet-Parker current sheet, or some refinement of it to allow a magnetic field component $B_x(y)$ to develop gradually as $|y|$ increases, still applies near the origin, in a diffusion region of size $2y^* \times 2x^*$. (Priest and Cowley [1975] have shown that, assuming analytic behavior near $x = y = 0$, $B_x \sim y^3$ in incompressible resistive flow.) In the outflow wedges, the flow in Petschek's model has speed $v_2 \simeq B_1/\sqrt{\mu_0 \rho}$ regardless of the reconnection rate; the acceleration of most of the plasma to this velocity cannot be achieved by a pressure gradient $\partial p/\partial y$ (only the plasma flowing through the diffusion region is accelerated in this fashion). Rather Maxwell stresses, concentrated in pairs of standing slow shocks (near the switch-off condition) are responsible for the acceleration.

The qualitative behavior of Petschek's reconnection configuration, as the reconnection rate, $M_{A\infty}$, increases is as follows.

- The outflow speed remains approximately

$$v_2 = B_{1\infty}(\mu_0 \rho)^{-1/2}$$

  but the angle θ_2 between the slow shocks increases.
- At the same time, the length, $2y^*$, of the diffusion region decreases. The width, $2x^*$, of that region also decreases but rather less rapidly so that the diffusion region becomes smaller and more square in shape as the reconnection rate increases.

The magnetic field and the flow speed, measured at points on the x axis outside the diffusion region, are nearly independent of x for small reconnection rates. However, as $M_{A\infty}$ increases, the field weakens and the flow speed toward the diffusion region increases near $|x| = x^*$. As pointed out by Vasyliunas [1975], this behavior is ascribable to fast-mode expansion in the inflow; it is the opposite of the flux pile up described in the previous section and it leads to $M_{A1} > M_{A\infty}$ where M_{A1} is the Alfvén Mach number measured at $|x| = x^*$. According to Petschek [1964], the field weakening at $|x| = x^*$ limits the reconnection rate to a maximum value given by the formula (as corrected by Vasyliunas [1975])

$$M_{A\infty}^{max} \simeq \frac{\pi}{8} \{\ln(2M_{A\infty}^{max})S_\infty\}^{-1} \qquad (13)$$

At this maximum rate, M_{A1} is expected to be of order unity.

An important point to note is that Petschek's maximum reconnection rate depends on the upstream Lundquist number, S_∞, only logarithmically. Thus the formula (13) predicts maximum rates that far exceed those obtained from the Sweet-Parker formula (5). It is noted

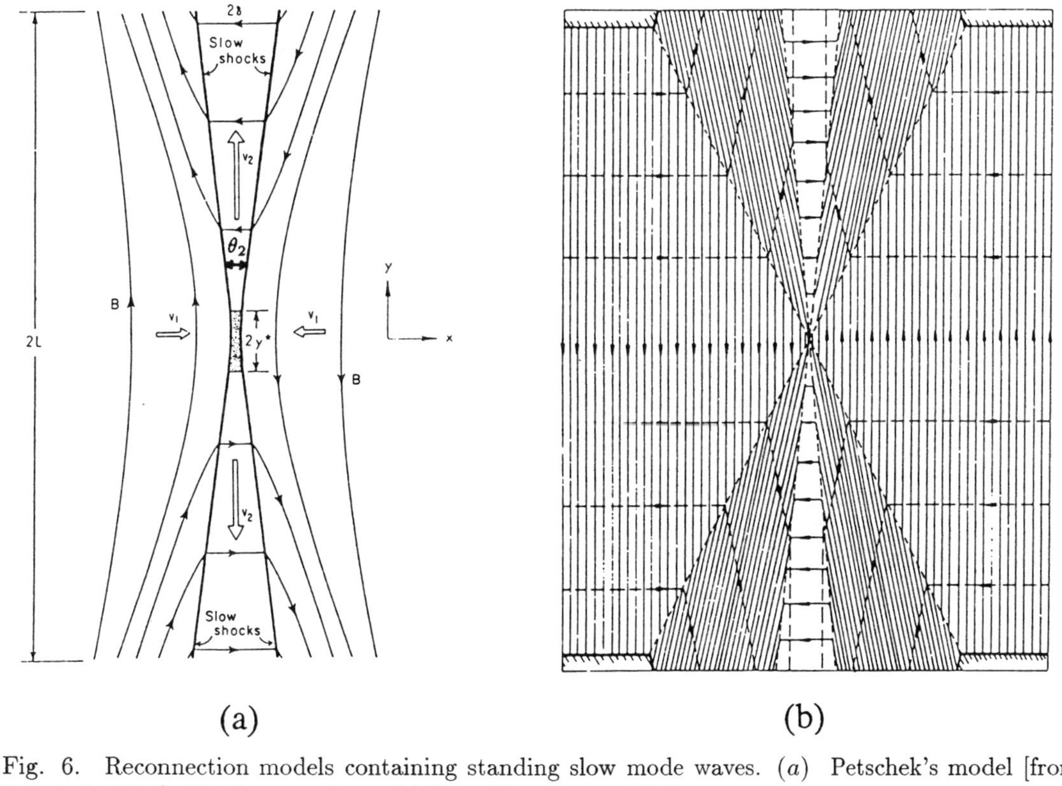

Fig. 6. Reconnection models containing standing slow mode waves. (*a*) Petschek's model [from Petschek, 1964]; (*b*) Sonnerup's model [from Vasyliunas, 1975].

that Petschek employed small perturbation analysis in arriving at equation (13) even though the deviations from the unperturbed state, which consists of oppositely directed uniform magnetic fields in the regions $x > 0$ and $x < 0$, are not small. Thus the formula for the upper limit $M_{A\infty}^{max}$ must be considered approximate.

Petschek's [1964] discussion of the above reconnection configuration also contained the observation that, in incompressible flow, the addition of a constant magnetic field, $B_z = B_{\|}$, is allowed without any important change in the analysis or the behavior of the system. In the same paper, Petschek also dealt with the compressible case but only for antiparallel fields; the addition of a B_z component is now not trivial since B_z by necessity is a function of x and y in compressible flow. Nevertheless, such a component is not expected to have a critical influence on the reconnection process, at least not in the MHD limit.

A special case of asymmetric reconnection, applicable to the earth's magnetopause, was discussed by Levy et al. [1964]; more realistic versions of this case have been produced recently [Heyn et al., 1985; Biernat et al., 1989], including a time-dependent model that may be relevant to the formation of flux transfer events and associated flux ropes on the dayside magnetopause [Biernat et al., 1987].

There exists another type of solution to the incompressible reconnection problem that consists entirely of a set of standing slow-mode waves separating wedges of uniform flow and field [Sonnerup, 1970]. This solution is shown in Figure 6b. It is exact exterior to the diffusion region; in the latter, only approximate analysis is available. A basic property of this model is the appearance of slow-mode rather than fast-mode expansion in the inflow. However, the slow-mode expansion is unrealistically confined to a set of standing waves upstream of Petschek's slow shocks and no expansion is present on the x axis so that $M_{A1} = M_{A\infty}$. The maximum reconnection rate in this case is found to be $M_{A1} = (1+\sqrt{2})$. At the maximum rate, conditions are the same in the inflow and outflow wedges so that no energy conversion takes place. More generally, it is believed (but has never been rigorously proved) that values of the Alfvén number,M_{A1}, in the inflow adjacent to the diffusion region, of order unity represent the maximum reconnection rate possible in any reconnection configuration of the general type described in Figure 6.

Approximate compressible, symmetric as well as asymmetric versions of this model have been given by Yang and Sonnerup [1976;1977] who also pointed out the possibility of fast-mode termination shocks in the exit flow wedges. An exact solution in which all slow-mode waves were replaced by rotational discontinuities has been discussed by Hameiri [1978]. In the latter solution, the magnetic field magnitude is constant everywhere and no conversion of energy from magnetic field to plasma occurs, except in the diffusion region.

Recently, Priest and Forbes [1986] have produced a unified linear perturbation theory of steady 2D reconnection which provides insight into the interrelationship between the various reconnection models discussed above. These authors started with an unperturbed stationary state in which the magnetic field is uniform and directed to the right in the upper half of a square box and uniform but directed to the left in the lower half of that box. A concentrated current sheet separates the two halves. A small electric field, E_0, is then introduced to drive the plasmas in the two halves of the box toward each other; the perturbation procedure used can be viewed as an expansion in powers of this field or, equivalently, in powers of the reconnection rate. The method of expansion is equivalent to the perturbation analysis performed by Petschek [1964]. To lowest order, the expansion leads to a Poisson equation for the perturbation magnetic vector potential in which the nonhomogeneous term, the perturbation current, is a function only of the coordinate perpendicular to the current sheet. Therefore this current in fact can be specified as a boundary condition. The equation is solved by separation of variables and a large variety of solutions, the nature of which depends on the boundary conditions, are possible. The most interesting aspect of the analysis is that it allows, for the first time, a systematic insight into the role played by the boundary conditions in determining the nature of the plasma inflow toward the current sheet. The diffusion region as well as the outflow wedges, the latter being extremely narrow on account of the assumed small reconnection rate, are not treated in detail but have the same character as in the original Petschek calculation. Behavior in the inflow ranging from slow-mode compression to fast-mode and slow-mode expansion and, in the extreme, to flux pile up can be readily generated. Petschek's configuration is included among the solutions as is the case of pure slow-mode expansion, although this expansion occurs in a distributed manner rather than being concentrated into the standing waves shown in Figure 6b. Even the Sweet-Parker solution and the case of stagnation point flow, with its associated extreme flux pile up, can be identified. Illustrations of the type of field and flow maps obtained from the analysis are shown in Figure 7.

The structure of the outflow regions associated with reconnection cannot be obtained from linear analysis. Recently, this problem has been studied in detail by use of the boundary layer approximation [Sonnerup and Wang, 1987]. In this manner a family of self similar incompressible solutions describing the flow and field in such reconnection layers was obtained. The layers, which are nondissipative, are bounded by two slow-mode standing waves in which all the dissipation is concentrated. One member of the family of solutions has precisely the same structure as that assumed in Petschek's analysis; other members cor-

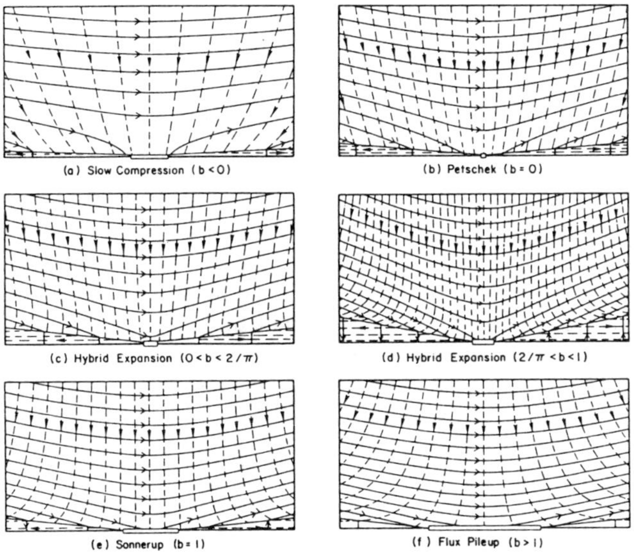

Fig. 7. Magnetic field lines (solid lines) and streamlines (dashed lines) resulting from small perturbation analysis by Priest and Forbes for magnetic Reynolds number = 500, based on inflow conditions and box size. Rectangle at bottom of each frame shows size of diffusion region. The quantity b measures the current density in the inflow [from Priest and Forbes, 1986].

respond to different overall reconnection geometries, e.g., reconnection at a Y type magnetic null. In this latter case, it appears that the reconnection layer, rather than being wedge shaped, can have a constant thickness. An example of this type of layer is shown in Figure 8. An

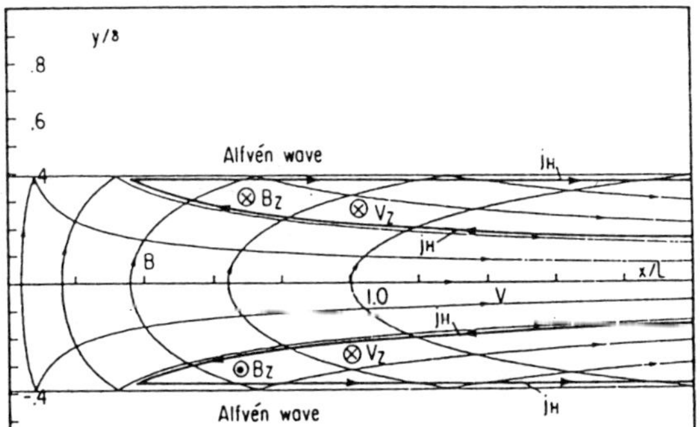

Fig. 8. Magnetic field and streamlines in a self similar incompressible reconnection layer having width independent of x. By interchange of field lines and streamlines, a vortex layer surrounding a magnetic separatrix is obtained. Also shown are Hall current loops, $\mathbf{j}_H$, in the xy plane and associated transverse components, B_z and v_z, of field and flow [from Sonnerup and Wang, 1987].

interesting feature of the reconnection layer solutions is that, in the nondissipative equations describing them, the magnetic field and the velocity field appear in an essentially symmetrical fashion. Therefore these two fields can be interchanged so that the magnetic field lines in Figure 8 become the streamlines and vice versa. If this is done, one finds that plasma flow across a magnetic separatrix appears to be associated with sharp bends in the streamlines, i.e., with the presence of a layer of strong vorticity, centered at the location of the separatrix. The role played by these magnetic separatrix layers and by the reconnection layer is shown schematically in Figure 9 for the case of reconnection at a Y type magnetic null point. It is expected that separatrix layers could occur also in other reconnection configurations, including the usual Petschek geometry. However, the linear analysis performed by Petschek [1964] and by Priest and Forbes [1986] does not permit of such layers. Thus it is not clear that the results of these linear analyses are meaningful, except for very small reconnection rates. In particular, it is not clear that linear analysis can be used to properly predict the maximum reconnection rate, as was done in obtaining equation (13).

The physical reason for the appearance of separatrix layers has been given, e.g., by Soward and Priest [1986] and Schindler and Birn [1987]. As reviewed in detail in Paper I, these layers occur because of the large increase in flux-tube cross section that occurs near a magnetic null point as a tube is convected toward the null. Plasma must flow along the magnetic field toward the weak field region in order to fill the void and, at least for incompressible flow, this effect will be strong as long as the diffusion region (where the frozen-field condition becomes invalid) is small.

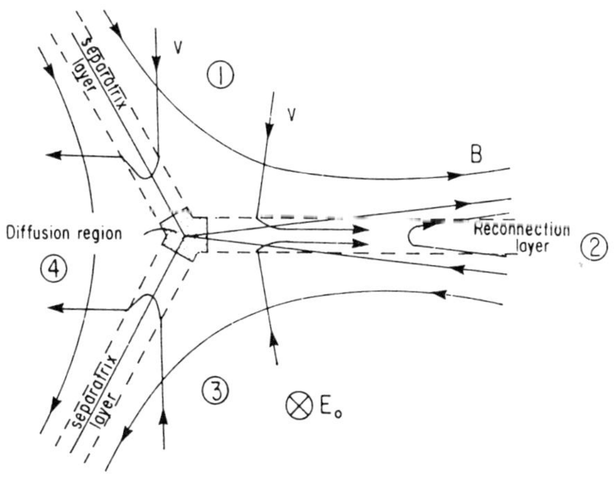

Fig. 9. Schematic drawing of reconnection configuration at a Y-type magnetic null showing how the reconnection layer/separatrix layer solution in Figure 8 would arise [from Sonnerup and Wang, 1987].

5. Numerical Simulations

A fairly large number of numerical simulations of 2D reconnection have been performed to date. These have been reviewed in detail by Forbes and Priest [1987]. Here we mention only three important features of these simulations.

The first point to be made is that it appears fairly difficult to set up boundary conditions such that the precise Petschek configuration emerges, including fast-mode expansion in the inflow. The main feature of Petschek's model, namely the appearance of Alfvénic exit flow jets wedged between slow shocks is seen in a number of driven as well as spontaneous numerical reconnection experiments [e.g., Sato, 1979; Ugai, 1988] but the inflow often has features associated with slow-mode rather than fast-mode expansion. Fast mode termination shocks in the exit flow have been seen in several simulations [e.g., Ugai, 1988].

The second item to be mentioned is that for certain classes of boundary conditions, e.g., those used by Biskamp [1986] and Lee and Fu [1986a,b] the reconnection configuration depends on the reconnection rate in a manner that is exactly opposite from the Petschek model: both the length $2y^*$ and the width $2x^*$ of the diffusion region increase with increasing reconnection rate in these simulations. Increasing reconnection rate also leads to an increasing amount of flux pile up in the inflow. Although it is clear from the work of Priest and Forbes [1986] that this behavior is somehow associated with the particular boundary conditions used in these simulations, a convincing physical explanation for the scaling ($x^* \sim M_{A\infty}$; $y^* \sim M_{A\infty}^4$) observed by Biskamp is not available. And the question of whether real reconnection events occurring on the sun or in the earth's magnetosphere have boundary conditions that will lead to Petschek-like or current-sheet like (Sweet-Parker like plus flux pile up) configurations remains unanswered. However, we note that in a number of geometries where the exit flow is impeded, long diffusion regions, i.e., Sweet-Parker-like behavior seems to be preferred. This appears to be the case in tokamak simulations [Park et al., 1984], in magnetic island coalescence simulations [Pritchett and Wu, 1979], and in simulations of decaying 2D MHD turbulence [Biskamp and Welter, 1989]. The tearing mode also becomes active in such long layers, for $y^*/x^* \geq 10$, say, [Lee and Fu, 1986b; Biskamp, 1986; Biskamp and Welter, 1989].

The third point to be made is that the (incompressible) simulations of driven reconnection by Biskamp, and to some extent those by Lee and Fu, show strong vortex layers at the magnetic separatrices, much as discussed in the previous section. The outflow wedges between these separatrices are much wider than in the Petschek model and the flow speed there is only a fraction of the Alfvén speed, v_{A1}.

Field and flow maps, taken from the work of Lee and Fu [1986b] and illustrating the behavior described above, are shown in Figure 10.

6. Two Fluid Effects

The MHD description of the reconnection process used up to this point in the paper is expected to be adequate, even in the case of a collisionless plasma, provided the characteristic scale sizes for change in plasma flow and field are much larger than the relevant inner plasma scales, viz., the ion and electron inertial lengths, λ_i and λ_e, and gyroradii, R_{Li} and R_{Le}. If that condition is satisfied, the half width, x^*, of the diffusion region will be of the order of the resistive length $\lambda_r = (\mu_0 \sigma v_1)^{-1}$, as discussed in section 3. The above condition for use of the MHD description in the diffusion region is then equivalent to the statement that the resistive length must be substantially larger than the inner plasma scales. Note that the resistive length can be nonzero even in a collisionless plasma if microinstabilities are present in the diffusion region to generate a finite effective electrical conductivity. If the resulting value of λ_r is comparable to λ_i and/or R_{Li} but remains much larger than λ_e and R_{Le}, then the MHD description should be improved by inclusion of the Hall

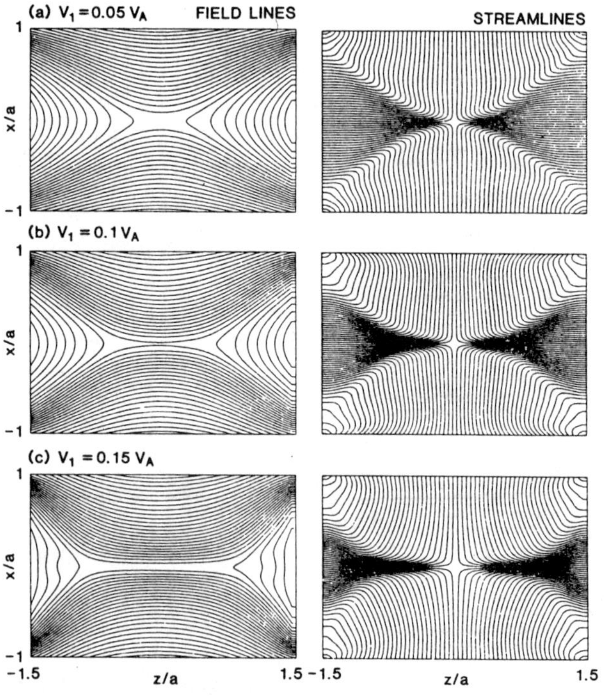

Fig. 10. Formation of a long current sheet as the reconnection rate V_1/V_{A1} (evaluated at $x/a = \pm 1$) increases in incompressible MHD simulation by Lee and Fu. Note also vortex layers at the magnetic separatrices [from Lee and Fu, 1986*b*].

current term, $\mathbf{j} \times \mathbf{B}/ne$, in Ohm's law so that it has the form

$$(\mathbf{E} + \mathbf{v} \times \mathbf{B}) = \mathbf{j}/\sigma + \mathbf{j} \times \mathbf{B}/ne - \frac{1}{ne} \nabla p_e \qquad (14)$$

as well as by inclusion of an improved version of the ion pressure tensor in the momentum equation to incorporate the effects of finite ion gyroradii (if R_{Li} is comparable to λ_r, and λ_i). Neither of these effects has been dealt with in much quantitative detail to date but the role of the Hall term may be understood qualitatively as follows (see also Sonnerup [1979]). If $\nabla \times (\mathbf{j} \times \mathbf{B}/ne)$, with $\mathbf{j} = \hat{\mathbf{z}}j(x,y)$ and $\mathbf{B}$ taken from the MHD solution, vanishes identically, as it does for the field annihilation solutions discussed in section 3, then Hall currents in the xy plane need not flow. The Hall term can then be exactly cancelled by a Hall electric field, provided the boundary conditions permit of the presence of the corresponding Hall potential. If not, Hall currents will flow. A quantitative example of this situation, referring, not to the diffusion region but to a reconnection layer of width comparable to λ_i, is shown in Figure 8. In this particular MHD configuration $\nabla \times (\mathbf{j} \times \mathbf{B}/ne) = 0$ but the condition of constant potential along the slow shocks that form the boundaries of the layer leads to the two Hall current loops in the xy plane shown in the figure. These currents, $\mathbf{j}_H$, are seen to flow along the streamlines but in the opposite sense to the bulk plasma flow. They close by flowing outward along the slow shocks (labelled "Alfvén waves" in the figure). We expect that a set of four such Hall current loops (two in each outflow region) would also be present in the diffusion region but no rigorous analysis producing such loops has been performed. The point we want to emphasize is that such loops will lead to a nonuniform magnetic field component, $B_z(x,y)$, as well as flow, $v_z(x,y)$, in the z direction. There will also be associated electric field components E_x and E_y which were not present in the MHD model. Thus the actual electromagnetic field configuration in the diffusion region and outflow regions would be far more complicated than that indicated in Figure 1.

If the resistive length becomes comparable to, or smaller than λ_e and/or R_{Le}, additional terms, namely the electron inertial terms and the full electron stress tensor, $\mathbf{P}_e$, come into play in Ohm's law which now becomes [e.g., Rossi and Olbert, 1970]

$$\mathbf{E} + \mathbf{v}_e \times \mathbf{B} = \mathbf{j}/\sigma - (1/ne)\nabla \cdot \mathbf{P}_e + (m_e/ne^2)[\partial \mathbf{j}/\partial t + \nabla \cdot (\mathbf{vj} + \mathbf{jv})] \qquad (15)$$

In this version of Ohm's law, the terms $\mathbf{v} \times \mathbf{B}$ and $\mathbf{j} \times \mathbf{B}/ne$ which appear separately in (14) have been combined to form $\mathbf{v}_e \times \mathbf{B}$, where $\mathbf{v}_e$ is the electron fluid velocity. It is then seen that unfreezing of the electron fluid from the magnetic field must be accomplished by one or more terms on the right-hand side of (15). If the first of these terms is unable to do this, or is unable to do it alone, then either the off-diagonal electron stress tensor terms or the electron inertial terms must come into play. It can be shown [Sonnerup, 1979] that the latter terms are unable to provide the unfreezing in 2D steady-state cases (their role in 3D stagnation point flows with magnetic field annihilation has been examined by Sonnerup [1979]). That leaves the off-diagonal (gyroviscous) terms in the electron stress tensor. The importance of these terms to collisionless reconnection was first discussed by Vasyliunas [1975].

If terms such as electron gyroviscosity or electron inertia are required to unfreeze the magnetic field from the electrons, then the diffusion region must contain a substructure at the separator of scale size λ_e and/or R_{Le}. This would add further complexity to the electromagnetic structure of the diffusion region. More detailed discussion of two-fluid phenomena in the diffusion region is provided in Paper I. Here we add one more comment: because of an ongoing controversy in magnetospheric physics concerning antiparallel ($B_\parallel = 0$) merging versus so-called component ($B_\parallel \neq 0$) merging, it would be particularly important to evaluate how the unfreezing provided by the three terms on the right-hand side of equation (15) is influenced by the presence of a guide magnetic field, $B_\parallel$, in the z direction.

Questions and Answers

Zelenyi: What can be the maximum potential reconnection rate (and what are the corresponding boundary conditions) for the two counterstreaming plasma flows with antiparallel magnetic fields with a given Mach number and a given resistivity?

Sonnerup: In my view, the maximum reconnection rate, M_{A1}, evaluated in the inflow region immediately outside the diffusion region, is of order unity. This result is given for the unified theory provided by Priest and Forbes (1986) and also in the self-similar MHD reconnection geometry (Sonnerup, 1970) where the value $M_{A1} = 1 + \sqrt{2}$ was obtained.

Atkinson: In a paper to be given on Thursday, I will be presenting a model which proposes that the Alfvén waves you discussed propagate to the ionosphere and become auroral arcs. I will also present experimental observations which strongly support the idea that there is a one-to-one relationship between arcs and X-lines.

Sonnerup: I shall be very interested in hearing about these new results.

Schindler: I should like to explain why there is a close relationship between the configuration that you referred to as a "double-layer" and more conventional reconnection fields. Consider a situation where the electric field is parallel to the magnetic field, varying in the perpendicular direction such that $\nabla \times \mathbf{E} \neq 0$.

Then there is a time-dependent magnetic field component which has the effect of magnetic shear. The whole process may be described as relaxation of the shear, similar to classical resistive instabilities that are conventionally seen as reconnection processes.

Sonnerup: The electric field that was shown in my example is curl-free, so that I do not see that this comment is applicable.

Greene: Why should a separator be a local quantity? One can imagine that behavior is forced to be two-dimensional by Alfvén waves traveling along the separator. Thus, the two-dimensionality might depend on the nonlocal behavior of Alfvén waves.

Sonnerup: The nonlocal effects you describe could be operative, in particular, in a low-β and low Mach number plasma. But when one considers the actual behavior of the magnetic field in the magnetopause, one finds it to be extremely complicated and often turbulent. In such circumstances, it is hard for me to believe that null points located on the flanks of the magnetosphere could dictate the behavior of the reconnection process near the subsolar point. In my view, such null points would be simply blown downstream by the magnetosheath flow. On the other hand, a magnetic null point trapped in the cusp can probably influence reconnection occurring along a reconnection line originating in the cusp and extending downstream in the manner observed in the 3-D numerical simulations performed by Fedder and Lyon (this conference). In that case, the steady-state induction equation,

$$\mathbf{v} \bullet \nabla\left(\frac{\mathbf{B}}{\rho}\right) = \frac{\mathbf{B}}{\rho} \bullet \nabla\, \mathbf{v}$$

leads to the generation of a true magnetic null line originating in the cusp as proposed a number of years ago by Nancy Crooker (1979). It should be added that magnetic null points do not occur in the tearing mode when a magnetic field component along the current is present. Thus, a separator line cannot be identified by first finding the magnetic nulls. Yet, we would like to consider tearing of this type as a form of reconnection.

Forbes: Can the vorticity layers along the separatrices be considered as discontinuities? If so, what determines their thickness, and what type of discontinuities are they?

***Sonnerup:** Yes, I think they could be considered as discontinuities, but they are of a new type different from either shocks or rotational and tangential discontinuities. The vortex layers are two-dimensional and, consequently, they are not included in the usual set of one-dimensional discontinuities. I believe their strength and thickness are determined non-locally, by the field geometry and the effectiveness of field-line diffusion near the magnetic null point.*

Acknowledgments. The research was supported by the National Science Foundation, Atmospheric Sciences Division, under grant ATM-8807645, and by the Air Force Geophysics Laboratory, under contract F19628-87-K0026.

References

Axford, W. I., Magnetic reconnection, in *Magnetic Reconnection in Space and Laboratory Plasmas, Geophys. Monogr. Ser.*, vol. 30, edited by E. W. Hones, Jr., pp. 1-8, AGU, Washington, D.C., 1984.

Biernat, H. K., M. F. Heyn, and V. S. Semenov, Unsteady Petschek reconnection, *J. Geophys. Res.*, *92*, 3392-3396, 1987.

Biernat, H. K., M. F. Heyn, R. P. Rijnbeek, V. S. Semenov, and C. Farrugia, The structure of reconnection layers: Application to the earth's magnetopause, *J. Geophys. Res.*, *94*, 287-298, 1989.

Biskamp, D., Magnetic reconnection via current sheets, *Phys. Fluids*, *29*, 1520-1531, 1986.

Biskamp, D., and H. Welter, *Dynamics of Decaying Two-Dimensional Magnetohydrodynamic Turbulence*, Max-Planck-Inst. f. Plasmaphysik, Report IPP 6/279, to appear, *Phys. Fluids B, 1*, 1989.

Butler, D. M., and K. Papadopoulos, eds., *Solar Terrestrial Physics: Present and Future*, NASA Ref. Publ. 1120, 1984.

Carlqvist, P., Current limitation and solar flares, *Solar Phys.*, *7*, 377-392, 1969.

Cowley, S. W. H., A qualitative study of the reconnection between the earth's magnetic field and an interplanetary magnetic field of arbitrary orientation, *Radio Sci.*, *8*, 903-913, 1973.

Dungey, J. W., The structure of the exosphere or adventures in velocity space, in *Geophysics: The Earth's Environment*, edited by C. De Witt, J. Hieblot, and L. Le Beau, Gordon and Breach, New York, p. 503, 1963.

Forbes, T. G., and E. R. Priest, A comparison of analytical and numerical models for steadily driven magnetic reconnection, *Revs. Geophys.*, *25*, 1583-1607, 1987.

Gratton, F. T., M. F. Heyn, H. K. Biernat, R. P. Rijnbeek, and G. Gnavi, MHD stagnation point flows in the presence of resistivity and viscosity, *J. Geophys. Res.*, *93*, 7318-7324, 1988.

Greene, J. M., Geometrical properties of three-dimensional reconnecting magnetic fields with nulls, *J. Geophys. Res.*, *93*, 8583-8590, 1988.

Hameiri, E., Compressible magnetic field reconnection, *J. Plasma Phys.*, *22*, 245-256, 1979.

Hesse, M., and K. Schindler, A theoretical foundation of general magnetic reconnection, *J. Geophys. Res.*, *93*, 5559-5567, 1988.

Heyn, M. F., H. K. Biernat, V. S. Semenov, and I. V. Kubyshkin, Dayside magnetopause reconnection, *J. Geophys. Res.*, *90*, 1781-1785, 1985.

Kirkland, K. B., and B. U. Ö. Sonnerup, Self-similar resistive decay of a current sheet in a compressible plasma, *J. Plasma Phys.*, *22*, 289-302, 1979.

Lee, L. C., and Z. F. Fu, A simulation study of magnetic reconnection: Transition from a fast-mode to a slow-mode expansion, *J. Geophys. Res.*, *91*, 4551-4556, 1986*a*.

Lee, L. C. and Z. F. Fu, Multiple x-line reconnection, 1, A criterion for the transition from a single x-line to a multiple x-line reconnection, *J. Geophys. Res.*, *91*, 6807-6815, 1986*b*.

Levy, R. H., H. E. Petschek, and G. L. Siscoe, Aerodynamic aspects of the magnetospheric flow, *AIAA J.*, *2*, 2065-2076, 1964.

Lysak, R. L., and Y. Song, Formation of flux ropes by turbulent reconnection, *This Conference*, 1989.

Park, W., D. A. Monticello, and R. B. White, Reconnection rates of magnetic fields including the effects of viscosity, *Phys. Fluids*, *27*, 137-149, 1984.

Parker, E. N., Sweet's mechanism for merging magnetic fields in conducting fluids, *J. Geophys. Res.*, *62*, 509-520, 1957.

Parker, E. N., The solar flare phenomenon and the theory of reconnection and annihilation of magnetic fields, *Astrophys. J., Suppl. Ser.*, *8*, 177-212, 1963.

Parker, E. N., Comments on the reconnection rate of magnetic fields, *J. Plasma Phys.*, *9*, 49-63, 1973.

Petschek, H. E., Magnetic field annihilation, *NASA Spec. Publ., NASA-SP-50*, pp. 425-439, 1964.

Podgorny, A. I., *A Longitudinal Magnetic Field in the Vicinity of a Singular Line*, Lebedev Phys. Inst., Preprint 358, Moscow, 1986.

Priest, E. R., and S. W. H. Cowley, Some comments on magnetic-field reconnection, *J. Plasma Phys.*, *14*, 271-282, 1975.

Priest, E. R., and T. G. Forbes, New models for fast steady state magnetic reconnection, *J. Geophys. Res.*, *91*, 5579-5588, 1986.

Priest, E. R., and T. G. Forbes, Steady magnetic reconnection in three dimensions, *Solar Physics*, *119*, 211-214, 1989.

Pritchett, P. L., and C. C. Wu, Coalescence of magnetic islands, *Phys. Fluids*, *22*, 2140-2146, 1979.

Rossi, B., and S. Olbert, *Introduction to the Physics of Space*, McGraw-Hill, New York, p. 348, 1970.

Sato, T., Strong plasma acceleration by slow shocks resulting from magnetic reconnection, *J. Geophys. Res.*, *84*, 7177-7189, 1979.

Schindler, K., and J. Birn, On the generation of field-aligned plasma at the boundary of the plasma sheet, *J. Geophys. Res.*, *92*, 95-108, 1987.

Schindler, K., M. Hesse, and J. Birn, General magnetic reconnection, parallel electric fields, and helicity, *J. Geophys. Res.*, *93*, 5547-5557, 1988.

Schindler, K., and A. Otto, Resistive instability, *This Conference*, 1989.

Sonnerup, B. U. Ö., Magnetic field re-connexion in a highly conducting incompressible fluid, *J. Plasma Phys.*, *4*, 161-174, 1970.

Sonnerup, B. U. Ö., Magnetic field reconnection, in *Solar System Plasma Physics*, vol. III, edited by L. T. Lanzerotti, C. F. Kennel, and E. N. Parker, pp. 46-108, North-Holland, Amsterdam, 1979.

Sonnerup, B. U. Ö., On the theory of steady state reconnection, *Computer Phys. Comm.*, *49*, 143-159, 1988.

Sonnerup, B. U. Ö., and E. R. Priest, Resistive MHD stagnation-point flows at a current sheet, *J. Plasma Phys.*, *14*, 283-294, 1975.

Sonnerup, B. U. Ö., and D.-J. Wang, Structure of reconnection layers in incompressible MHD, *J. Geophys. Res.*, *92*, 8621-8633, 1987.

Soward, A. M., and E. R. Priest, Magnetic field-line reconnection with jets, *J. Plasma Phys.*, *35*, 333-350, 1986.

Sweet, P. A., The neutral point theory of solar flares, in *Electromagnetic Phenomena in Cosmical Physics*, edited by B. Lehnert, p. 123, Cambridge University Press, New York, 1958.

Ugai, M., MHD simulations of fast reconnection spontaneously developing in a current sheet, *Computer Phys. Comm.*, *49*, 185-192, 1988.

Vasyliunas, V. M., Theoretical models of magnetic field line merging, *Rev. Geophys.*, *13*, 303-336, 1975.

Yang, C.-K., and B. U. Ö. Sonnerup, Compressible magnetic field reconnection: a slow wave model, *Astrophys. J.*, *206*, 570-582, 1976.

Yang, C.-K., and B. U. Ö. Sonnerup, Compressible magnetopause reconnection, *J. Geophys. Res.*, *82*, 699-703, 1977.

Yeh, T., Day-side reconnection between a dipolar geomagnetic field and a uniform interplanetary field, *J. Geophys. Res.*, *81*, 2140-2144, 1976.

MAGNETIC FLUX TUBES: THEIR ORIGIN AND APPEARANCE

V. D. Kuznetsov

Izmiran, 142092 Moscow region, Troitsk, USSR

Abstract In the chain of physical processes leading to the origin and appearance of magnetic flux tubes in the Sun the following problems are considered: (1) Magnetic fragmentation of flux tubes in subphotospheric layers of the Sun; (2) Global instability of subphotospheric magnetic field and the origin of coronal transients; (3) Force-free twisted flux tubes in the corona. In the framework of magnetohydrodynamics with turbulent viscisity the problem of magnetic fragmentation of flux tubes reduces to the analysis of a dispersion equation. The characteristic scales of the flux tubes in the linear stage of the kink instability are evaluated for an isothermal plane-parallel atmosphere with constant Alfvenic velocity. For such an atmosphere in cylindrical geometry the differential equation describing the behaviour of the small perturbations of the subphotospheric toroidal magnetic field is obtained, the unstable character of which is connected with the origin of emerging magnetic flux tubes and coronal transients. For a description of the force-free magnetic field of twisted loops in the corona and interplanetary space we have utilized a quasi-cylindrical orthogonal coordinate system with an axis which models the curvature and twisting of the loop axis. For a toroidal configuration with an axis of constant curvature and zero twisting an equation is deduced which allows the solution for a force-free field at any order of the aspect ratio.

1. Introduction

Solar magnetic fields and their associated activity are manifested in the form of distinct flux tubes in the photosphere and corona. The tendency of magnetic fields to form discrete structures is common for any cosmic objects and is a consequence of a fundamental mechanism of magnetic field fragmentation under cosmic conditions [Parker, 1979]. The presence of gravity and small-scale turbulence are usual for the convective envelopes of the Sun and stars, interstellar clouds and nebulas supported by magnetic field lines above the galatic plane.

The theoretical explanation of the fine filamentary structure of solar magnetic fields and their associated activity is connected with fragmentation of the subphotospheric magnetic field due to the instability and its subsequent emergence through the surface of the Sun by magnetic buoyancy.

Geophysical Monograph 58

1.1 Magnetic field fragmentation of flux tubes in subphotospheric layers of the Sun

In a gravity field a plasma with horizontal magnetic field lines is subject to the kink instability (magnetic buoyancy instability). The characteristic scales of the emerging magnetic tubes are determined by the scales of perturbations with maximal growth rate of MHD instability. In the framework of ideal MHD zero values of transverse scales ($k_x \rightarrow \infty$, see Fig. 1a) correspond to the maximal growth-rate of kink instability, whereas for a purely longitudial perturbation ($k_x = 0$) the scales are of finite wavelength (see Fig. 1b). The characteristic scales of magnetic tubes originating in the unstable atmosphere may be evaluated by considering the linear stage of the kink instability in the framework of MHD equations with dissipative terms.

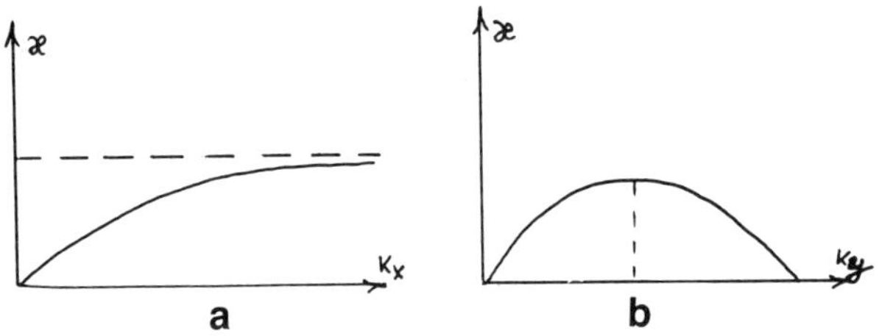

Fig.1. The dependence of the growth rate of instability without dissipation:a) for transverse perturbations, b) for longitudinal perturbations.

The basic equations are

$$\frac{\partial \vec{B}}{\partial t} = \mathrm{rot}\left[\vec{V} \times \vec{B} - \nu_m^1 \,\mathrm{cot}\vec{B}\right] \quad (1)$$

$$\rho \frac{d\vec{V}}{dt} = -\nabla p + \frac{1}{4\pi}\left[\mathrm{rot}\vec{B} \times \vec{B}\right] + \rho \vec{g} + \sum_{k=1}^{3} \frac{\partial}{\partial x_k}\left[\eta_t\left(\frac{\partial V_i}{\partial x_k} + \frac{\partial V_k}{\partial x_i} - \frac{2}{3} S_{ik}\,\mathrm{div}\,\vec{V}\right)\right] \quad (2)$$

$$\frac{d}{dt}\left(\frac{p}{\rho\gamma}\right) = 0, \; \nu_m^1 = \frac{c^2}{4\pi\sigma} \; ; \; \eta_t = \rho\nu_t \qquad (3)$$

$$\frac{\partial\rho}{\partial t} + \operatorname{div}\rho\,\vec{V} = 0 \qquad (4)$$

1.2 Equilibrium state

$$\frac{d}{dZ}\left(p_o + \frac{B_o^2}{8\pi}\right) = -\rho_o g; \quad \vec{B}_o = B_o(z)\vec{e}_y \qquad (5)$$

$$p_o = \rho_o u^2; \quad u^2 = \frac{KT}{mg} = \text{const},$$

$$\alpha = \frac{B_o^2}{8\pi\rho_o} = \text{const} \qquad (6)$$

$$\vec{g} = -g\vec{e}_z, \; g = \text{const} \qquad (7)$$

The simplifications of the equilibrium state made above determine the exponential distribution of the equilibrium parameters for an isothermal atmosphere

$$\frac{p_o(z)}{p_o(0)} = \frac{\rho_o(z)}{\rho_o(0)} = \frac{B_o^2(z)}{B_o^2(0)} = \exp\left(-\frac{z}{H}\right);$$

$$H = \frac{u^2}{g}(1 + d) \qquad (8)$$

and impose strong restrictions on the criteria of instability rise. In particular, they require low values of the adiabatic index γ. However, these simplifications facilitate the solving of the problem because they allow us to reduce the investigation of the small perturbation behaviour in an inhomogeneous system to the analysis of a dispersion equation. The stationarity of the initial distribution of the magnetic field (8) for nonzero magnetic viscosity (equation (1)) is supported by external sources (dynamo action).

1.3 Instability of small perturbations and characteristic equation.

Linearization of the equations (1) - (4) about the equilibrium (8) gives for the small perturbations of the vector-potential A_x' and A_z' a system of two equations. After substitution of all perturbations in the form

$$f(z)\exp(i\omega t + ik_x x + ik_y y) \qquad (9)$$

they become

$$\begin{cases} \hat{P}(A_x') = \hat{G}(A_z') & (10) \\ \hat{L}(A_x') = \hat{M}(A_z') & (11) \end{cases}$$

here $\hat{G}$, $\hat{P}$, $\hat{M}$ and $\hat{L}$ are differential operators (see [Kuznetsov, 1988]), depending on dimensionless parameters

$$\varepsilon_x = k_x H; \quad \varepsilon_y = k_y H; \quad \Omega = \frac{\omega H}{u} = \frac{H}{u}(\omega_o - i\omega_i) = \text{in}$$

$$\nu = \frac{\nu\tau}{uH}; \quad \nu_m = \frac{\nu_m^1}{uH} \qquad (12)$$

The solution for the function f(z) is in the form

$$f(z) \sim \exp\left(q\frac{z}{H}\right), \qquad (13)$$

which in (10) and (11) gives the characteristic equation

$$\phi(q) = L(q)G(q) - P(q)M(q) = \sum_{k=0}^{N=10} a_u q^k = 0 \qquad (14)$$

The general solution for small perturbations is

$$\begin{Bmatrix} A_x'(\xi) \\ A_z'(\xi) \end{Bmatrix} = \sum_{k=1}^{10} \begin{Bmatrix} \tau_k \\ d_k \end{Bmatrix} e^{q_k\xi}; \quad \xi = \frac{Z}{H}, \qquad (15)$$

where q_k are the roots of the equation (14):$q_k = q_k(\nu, \nu_m, \varepsilon_x, \varepsilon_y, n)$, r_k and d_k-constants.

1.4 Boundary conditions

On the lower high conductivity boundary (z = 0) we have: the conditions of sticking and impenetrability of the mass

$$V_x'(0) = V_y'(0) = V_z'(0) = 0, \qquad (16)$$

the nonbending of the boundary field lines

$$B_x'(0) = B_y'(0) = B_z'(0) = 0, \qquad (17)$$

the vanishing of the tangential component of the electric field

$$E_x'(0) = 0. \qquad (18)$$

The boundedness of the perturbations at $z = \infty$ is

$$A_{x,z}'(\infty) < \infty \text{ or } \operatorname{Re} q_k \le 0 \qquad (19)$$

Perturbations leading to instability must not propagate from the region $z = +\infty$ (for the case of oscillatory instability)

$$\operatorname{Re}\omega/\operatorname{Im} q_k \le 0 \qquad (20)$$

1.5 Dispersion equation

In the general solution (15), of the ten roots q_k, three roots must be discarded according to condition (19). The boundary conditions (16) - (18) after taking into account the relation between $\vec{V}'$, B'_x, E'_x with A'_x and A'_z give the following dispersion equation of the investigated system as the condition for a nontrivial solution [Kuznetsov, 1988]:

$$F(n, \nu, \nu_m, \varepsilon_x, \varepsilon_y) =$$

$$\begin{vmatrix} 1 & 1 & 1 & 1 & 1 & 1 & 1 \\ \mathcal{P}(q_1) & \cdot & \cdot & \cdot & \cdot & \cdot & \mathcal{P}(q_7) \\ q_1 & \cdot & \cdot & \cdot & \cdot & \cdot & q_7 \\ q_1\mathcal{P}(q_1) & \cdot & \cdot & \cdot & \cdot & \cdot & q_7\mathcal{P}(q_7) \\ q_1^2 & \cdot & \cdot & \cdot & \cdot & \cdot & q_7^2 \\ q_1^2\mathcal{P}(q_1) & \cdot & \cdot & \cdot & \cdot & \cdot & q_7^2\mathcal{P}(q_7) \\ q_1^3 Q(q_1) & \cdot & \cdot & \cdot & \cdot & \cdot & q_7^3 Q(q_7) \end{vmatrix} = 0 \quad (21)$$

where

$$Q(q_k) = -n\nu q_k \mathcal{P}(q_k) + i\varepsilon_x\left(j + \frac{\nu n}{3}\right)$$

$$\mathcal{P}(q_k) = \frac{L(q_k)}{M(q_k)} = \frac{P(q_k)}{G(q_k)}$$

Here and in (14) L(q), M(q), P(q) and G(q) are polynomials with respect to q corresponding to the differential operators of the system (10) - (11) after substituting the function (13) (see [Kuznetsov, 1988]). Equation (21) defines the growth rate of instability as a function of the wave numbers k_x and k_y. The maximum increment of instability and characteristic scales of the flux tubes corresponding to it along and transverse to the magnetic field are given by the conditions

$$\frac{\partial\omega_i}{\partial\varepsilon_x^2} = 0, \quad \frac{\partial\omega_i}{\partial\varepsilon_y^2} = 0 \quad (22)$$

Thus, the problem of magnetic fragmentation of flux tubes in an unstable atmosphere is reduced to the analysis of the dispersion equation (21) and to the finding the maximum of the function $\omega_i(\nu, \nu_m, \varepsilon_x, \varepsilon_y)$. The characteristic equation (14) has a very complex form even in the simplest cases. For this reason we restrict our investigation by analysing equation (14) only and consider its solutions in the simplest cases.

1.6 Particular cases

1.6.1 The case $\nu_t = \nu_m = 0$ is well-known in the literature [Parker, 1979; Priest, 1982]. The dependence of the increment of the instability on wave numbers are shown in Fig.1. To the absolute maximum of the increment correspond the values of the wave numbers $k_x = \infty$,

$$k_y^2 = \frac{1}{H^2}\left[-\frac{(1+j+3\alpha)}{4\alpha^2(1+\alpha)} + \sqrt{\left(\frac{1+j+3\alpha}{4\alpha^2(1+\alpha)}\right)^2 + (1+\alpha j)[4\alpha(\alpha+j) - j(1+\alpha j)]}\right. \quad (23)$$

The instability of this mode arises for $j < 1 + \alpha$. The equality $k_x = \infty$ corresponds to the so-called problem of the transverse wave number.

1.6.2 The case $\nu_m = 0$, $k_y = 0$ ("kinematic" turbulent viscosity) is defined by the characteristic equation

$$\phi(q) = a_\mu q^\mu + a_2 q^2 + a_0 = \psi^2\left(j+2\alpha+\frac{\mu}{3}\nu n\right) - \frac{n}{\nu}\left(j+2\alpha+\frac{7}{3}\nu n\right)$$
$$+\left[\frac{n^3}{\nu} - \varepsilon_x^2\left(\frac{\delta}{n\nu} + 1\right)\left(\delta + \frac{2}{3}\nu n - j - 2\alpha\right)\right] = 0 \quad (24)$$

$$\psi = q^2 - \left(\varepsilon_x^2 + \frac{1}{4}\right)$$

For $q(n, \varepsilon_x^2)$ we obtain

$$q_1 = -q_2 = \sqrt{\frac{1}{4} + \varepsilon_x^2 + \Psi_-}$$

$$q_3 = -q_4 = -\sqrt{\frac{1}{4} + \varepsilon_x^2 + \Psi_+} \quad (25)$$

where $\Psi_\pm$ are roots of the quadratic equation (24). The dispersion equation which is defined by four boundary conditions ($V_x'(0) = 0$, $V_z'(0) = 0$, $E_x'(0) = 0$ and Re $q_k \leq 0$) may be written clearly but it is complex enough for analytical investigation. Assuming the case of small viscosity $\nu << u.H$ for estimating the transverse scale λ_x it is enough to consider the asymptotic behaviour of the solution for $k_x \rightarrow \infty$, following from the characteristic equation (23) in the case of purely transverse perturbations, for which

$$\omega_{i\perp}^{\nu}(k_x \rightarrow \infty) \rightarrow \frac{(1+\alpha)(1-\alpha-j)}{(j+2\alpha)H^2\nu_t K_x^2} \quad (26)$$

Extrapolating the asymptotic result to the value of the increment at $k_x \rightarrow \infty$ in the case $\nu = 0$ (see Fig. 2) $x_\perp^\nu(k_x \rightarrow \infty) \sim x_\perp^0(k_x \rightarrow \infty)$, we evaluate the transverse scale of the flux tube to be

$$\lambda_x = 2\pi\sqrt{\frac{(j+2\alpha)}{(1+\alpha)(1-\alpha-j)} \cdot \frac{\nu_t u}{g}} \simeq 2\pi\sqrt{\frac{\nu_t u}{g}} \quad (27)$$

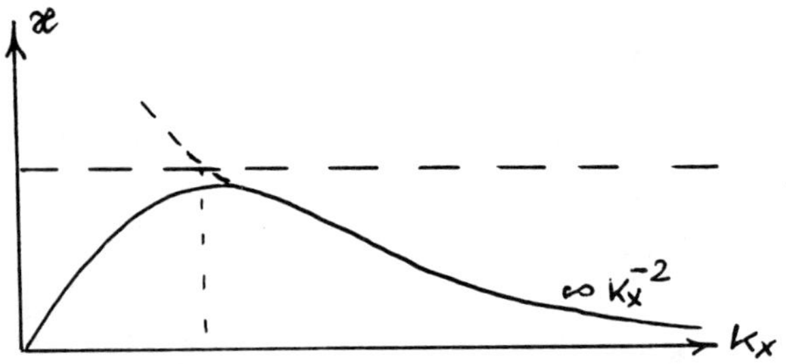

Fig.2.The dependence of the growth rate of instability on k_x for the case 1.6.2.

For purely longitudial perturbations (k_x=0) the increment is maximal for ($\gamma = \delta^2 - j(\delta + \alpha/2)$; $\mu = j+2\alpha$)

$$k_y^2 = \frac{1}{H^2}\frac{\left[-4\alpha j(Y+\frac{1}{8}M^2) + \sqrt{16\alpha^2 j^2 (Y+\frac{1}{8}M^2)^2 + Y(Y+\frac{1}{4}M^2)^2 2\alpha j(\mu-4\alpha)^2}\right]}{2\alpha j(j-2\alpha)^2} \tag{28}$$

For $k_x > 0$ ($k_x \neq 0$, $k_x \neq \infty$) the maximum of the increment will be reached at a wave number between the values (23) and (28). For the solar convective zone where $\alpha << 1$ we take $j \sim 1$ and obtain

$$14.5H = \frac{8\pi H}{\sqrt{3}} = \lambda_y^{(23)} < \lambda_y < \lambda_y^{(27)}$$

$$\frac{4\pi H}{\sqrt{\sqrt{2}-1}} = 19.5H \tag{29}$$

Substituting the values T = 4. 5×10^{4}K, g=2. 7×10^{4}cm/s^2, ν=6.10^{12}cm^2/s (a turbulent value) in the formulae (27) and (29) we evaluate the size of the flux tubes to be

$$\lambda_x \approx H = 1.\ 4\times10^8 \text{cm}$$

$$2\times10^8 \text{cm} < \lambda_y < 2.\ 8\times10^8 \text{cm} \tag{30}$$

The formulae (27) and (29) may be considered as approximate solutions of the equation (22).

1.6.3 The case $\nu_t = 0$, $k_y = 0$ ("magnetic" turbulent viscosity) is defined by the next characteristic equation for the unstable mode

$$\phi_4^{\perp}(n, \nu_m, \varepsilon_x^2, q) = q^4 + \frac{\alpha}{j}q^3 - \left\{\left(1+\frac{2\alpha}{j}\right)\frac{n}{\nu_m} + \frac{h^2}{j} + 2\varepsilon_x^2 + \frac{1}{4} + \frac{\alpha}{2j} + \left(1-\frac{1}{j}\right)\frac{\delta\varepsilon_x^2}{h^2}\right\}q^2 - \frac{\alpha}{j}\varepsilon_x^2 q$$
$$+\left[h\left\{\frac{h^2}{j\upsilon_m} + \left(1+\frac{2\alpha}{j}\right)\left(\varepsilon_x^2+\frac{1}{4}\right)\frac{1}{\nu_m} + \left(1+\frac{\alpha 1}{j}\right)\frac{\delta.\varepsilon_x^2}{n^2\nu_m}\right\} + \varepsilon_x^2\left\{\frac{n^2}{j} + \varepsilon_x^2 + \frac{1}{4} + \frac{\alpha}{2j} + \left(1-\frac{1}{j}\right)\frac{\delta.\varepsilon_x^2}{n^2}\right\}\right] = 0,$$

$$\delta = 1 + \alpha \tag{31}$$

The difference of this equation from the equation (24) is that to the value of n = 0 corresponds not only $\varepsilon_x{}^2 = 0$ but a condition $q^2 = \varepsilon_x{}^2$. Taking into account this condition, the equation (31) for $k_x \to \infty$ gives two asymptotic solutions

$$x_{\perp}^{\nu m}(k_x \to \infty) \to \frac{\mu}{H}\sqrt{\frac{(1+\alpha)(1-\gamma)}{\gamma}};$$

$$x_{\perp}^{\nu m}(k_x \to \infty) \to -\nu'_m k_x^2, \tag{32}$$

the first of which corresponds to the common convective instability without a magnetic field and the second to ohmic dissipation of magnetic perturbations at small scales. With $\nu_m \neq 0$ at small scales ($k_x \to \infty$) the frozen in condition is broken: plasma and magnetic field are uncoupled and the characteristic equation is split into two independent multipliers. The dependence of the growth rate on k_x in that case is shown in Fig. 3. The scales of the flux tubes can be estimated from the formulae (27) and (29) by replacing ν_t by ν_m.

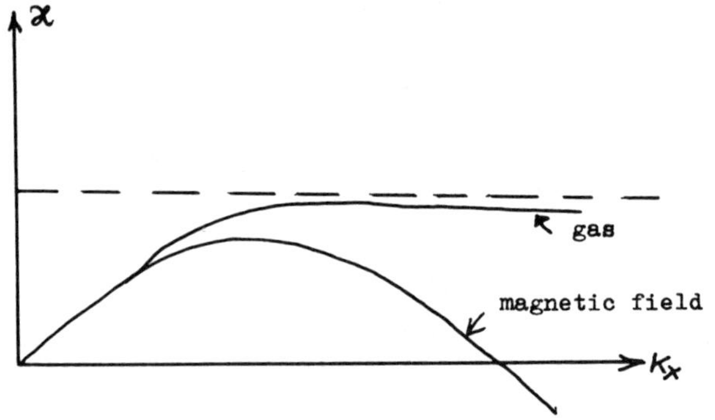

Fig.3.The dependence of the growth rate of instability on k_x for the case 1.6.3.

1.7 Conclusion.

The formulae (27) and (29) define the typical scales of emerging magnetic tubes in the initial stage of fragmentation. These scales (as the numerical estimates show) are in a good agreement with the sizes of the magnetic structures observed as arches and loops.

As the parameters of the medium in the convection zone are strongly variable the fragmentation of the magnetic fields into flux tubes may be most effective in the hydrogen ionization zone where the value of γ is minimum [Pikel'ner, 1966] and the elasticity of the plasma is insufficient to stabilize the longitudinal perturbation instability, connected with bending of the field lines. Taking at the depth $h=3.10^8$cm the parameters $p_0=10^8$dyne/cm^2, $t=2.\ 5\times10^4$K, $\gamma=1.09$ from the model [Baker and Temesvary, 1966] we find that the value of the magnetic field needed for instability equals $B_0 \geq \sqrt{16\pi p_0(\gamma-1)}=2.10^4$G, which is quite acceptable.

2. Global instability of subphotospheric magnetic field and coronal transient origin.

Coronal transients relate to active phenomena on the Sun which have a magnetic origin. Their appearance may be connected with emerging of new magnetic flux and with its penetration into the corona in the form of magnetic tubes [Syrovatskii, 1982]. In the dynamo theory of solar magnetic field generation one supposes the existence of a toroidal magnetic field in the convective zone down to a depth $h=2.10^{10}$cm. The instability of this field gives the origin of emerging magnetic flux. Let us assume that the toroidal magnetic field is distributed in a spherical shell between $r=0.7R_0$ and $r=R_0$ according to same law decreasing with r. The direction of the toroidal magnetic field is practically perpendicular to the axis of rotation, that is to the NS line. This is confirmed by the relative arrangement of leading and following spots, which represent the footpoints of the emerging flux tubes of toroidal field through the photospheric surface. As a rule the leading spot in the north hemisphere is at a higher latitude than the following spot which implies a small angle of inclination of the toroidal field to the equatorial plane. Below for simplicity we restrict our consideration to a two-dimensional pattern in the equatorial plane. The loops of the field, appearing due to the magnetic buoyancy instability, will turn relative to their initial orientation under the action of the coriolis force. The predominant direction of the neutral line of emerging loops is NS. Perhaps, the observed predominant orientation of coronal transients with NS filaments [Trotter et al, 1980] may be connected with this feature.

2.1 Eqilibrium distribution of the toroidal magnetic field.

Hydrostatic equation

$$-\nabla p_0 + \frac{1}{\gamma\pi}\,\mathrm{rot}\,\vec{B}\times\vec{B} - \frac{GM\rho_0}{\tau^2}\cdot\frac{\vec{\tau}}{\tau} = 0 \qquad (33)$$

for $\vec{B}_0 = (0, B_\rho^{(o)}, 0) = B_0(r)\,\vec{e}_\rho$; $\frac{\partial}{\partial\rho} = \frac{\partial}{\partial Z} = 0$ gives

$$-\frac{dp_0}{dZ} - \frac{1}{\gamma\pi}\cdot\frac{B_\rho^{(o)}}{\tau}\frac{\partial}{\partial\tau}(\tau B_\rho^{(o)}) - \frac{GM\rho_0}{\tau^2} = 0 \qquad (34)$$

For simplifications (6) this equation has the following solution ($M \approx M_0$)

$$p_0(\tau) = p_0(R_1)\left(\frac{R_1}{\tau}\right)^{\frac{2}{1+\beta}} \exp\left\{R_*\left(\frac{1}{\tau} - \frac{1}{R_1}\right)\right\} \qquad (35)$$

$$B_\rho^{(o)}(\tau) = \sqrt{\frac{8\pi p_0(\tau)}{\beta}}\ ; \quad \beta = \frac{1}{\alpha} = \frac{2\mu^2}{V_A^2} = \frac{8\pi p_0(\tau)}{B_0^2(\tau)} = \mathrm{const}$$

$$R_* = \frac{\beta GM_0}{(\beta+1)\mu^2} \qquad (36)$$

where R_1 is the internal boundary of the cylinder and $P_0(R_1)$ the plasma pressure at this boundary.

2.2 The equation describing the behaviour of small perturbations.

$$\frac{\partial\vec{B}}{\partial t} = \mathrm{rot}[\vec{v}\times\vec{B}] \qquad (37)$$

$$\rho\frac{d\vec{V}}{dt} = -\nabla p + \frac{1}{4\pi}\,\mathrm{rot}\vec{B}\times\vec{B} - \frac{GM\rho}{\tau^2}\cdot\frac{\vec{\tau}}{\tau} \qquad (38)$$

$$\frac{\partial\rho}{\partial t} + \mathrm{div}\rho\vec{V} = 0 \qquad (39)$$

$$\frac{d}{dt}\left(\frac{p}{\rho\gamma}\right) = 0 \qquad (40)$$

Linearizing the equations (37) - (40) near the equilibrium (35) and substituting a general type of perturbation in cylindrical coordinates in the form f(z) exp ($i\omega t + im\rho$) for small perturbations of the vector potential in the plane z=0 we obtain

$$\eta^4 Q_2(\eta)\frac{\partial^2 A_z'}{\partial\eta^2} + \eta^3 Q_1(\eta)\frac{\partial A_z'}{\partial\eta} + Q_0(\eta)A_z' = 0 \qquad (41)$$

where

$$\eta = \frac{\omega\tau}{m\mu} \quad Q_2(\eta) = \left(\frac{\eta^2}{\gamma} - 1\right)\left[\left(\frac{1}{\gamma} + \frac{\beta}{2}\right)\eta^2 - 1\right];$$

$$Q_1(\eta) = \sum_{K=0}^{K=4} q_{1k}\eta^k \qquad Q_0(\eta) = \sum_{K=1}^{\infty} q_{ok}\eta^k \qquad (42)$$

The coefficients q_{lk} are known functions of the parameters γ, β, $\eta_* = \omega r_*/mu = \omega GM/mu^3$.

The equation (41) is irregular. The coefficient $Q_2(\eta)$ of the highest order derivative equals zero for

$$\eta^2 = \gamma \quad \text{and} \quad \eta^2 = \left(\frac{1}{\gamma} + \frac{\beta}{2}\right)^{-1} \tag{43}$$

The usual method for investigating instability can not be applied in that case[Polovin and Demutskii, 1987]. Because of the inhomogeneity of the system a continuous spectrum of values of ω exists. A special analysis of equation (41) is needed. The curvature of the toroidal magnetic field R_1 in that case is an additional parameter which together with inhomogeneous scale R_* influences the growth rate. The r dependence of the gravity influences equation (41). As the interesting scales of perturbation in the ρ-direction are not large (the size of an active region) then the influence of the toroidal field curvature on the instability must be small and in equation (41) we may to put $m=k_\rho r = 2\pi r/\lambda_\rho >> 1$. In that approximation we adopt the following investigation of the equation (41).

3. Magnetic tubes in the corona: force-free twisted loops.

The equilibrium and instability of twisted loops in the solar atmosphere are considered as a possible description of magnetic field evolution in active regions and as an explanation for the activity exhibited in flares and coronal mass ejections [Colgate, 1978; Yeh, 1982]. The equilibrium of a twisted loop structure in a gravity field is described by the equation

$$-\nabla p + \frac{1}{4\pi} \operatorname{rot} \vec{B} \times \vec{B} + \rho \vec{g} = 0, \tag{44}$$

the solution of which is possible only under some simplifying assumption, relating to the geometry of the loop, values of the forces, boundary conditions and so on [Priest, 1982]. The possible dimensionless parameters, defining the equilibrium of the loop may be as follows: $\beta=8\pi p/B^2$; $q=B_{||}^2/B^2$; $\delta=8\pi\rho g/B^2$; $\varepsilon=r_0/R_0$, where B the value of the magnetic field in the loop, g is the gravitational acceleration, ρ is the density, R_0 and r_0 are the large and small radius of the loop. For the typical values of the magnetic field in the corona $\beta<<1$, $\delta<<1$. For this reason we may put $\beta = 0$, $\delta = 0$ and consider the magnetic field of a twisted loop as being force-free with

$$\operatorname{rot} \vec{B} \times \vec{B} = 0 \tag{45}$$

We will relate the twisted loop to a toroidal configuration and utilize for the description of the magnetic field a quasi-cylindrical orthogonal coordinate system (r, ω, s) with respect to the axis [Solovev et al, 1967]. Equation (45) in these coordinates is of the form

$$\frac{\frac{1}{h_s}\left[\frac{1}{\tau}\frac{\partial}{\partial\omega}(h_s B_s) - \frac{\delta B_\omega}{\delta s}\right]}{B_r} = \frac{\frac{1}{\tau}\left[\frac{\partial}{\partial\tau}(\tau B_\omega) - \frac{\partial B_r}{\partial\omega}\right]}{B_s}$$

$$= \frac{\frac{1}{h_s}\left[\frac{\partial B_r}{\partial s} - \frac{\partial(h_s B_s)}{\partial\tau}\right]}{B_\omega} = \frac{4\pi j}{cB} = \alpha(\rho, \omega, s) \tag{46}$$

where $h_s(r, \omega, s) = 1 - k(s) r\cos\theta = 1 - kr\cos(\omega - \alpha_0(s))$ $= 1 - kr\cos(\varpi - \int_0^s x(s)ds$, k(s) is a curvature of the spatial axis of the loop, $x(s) = \frac{d\alpha_o^o}{ds}$, $\alpha_0(s)$ is the angle of the binormal to the coordinate axis (the twisting of the axis), α (ρ, ω, s) is an arbitary function. The neglect of gravity allows us to suppose spiral symmetry when k(s)=const, x(s)=const and $\frac{\partial}{\partial\omega} = \frac{\partial}{\partial\theta}$; $\frac{\partial}{\partial s} = -x\frac{\partial}{\partial\theta}$. In that case equation (46) gives

$$\frac{1}{Z}\frac{\partial}{\partial\tau}\left[\frac{\tau}{\alpha h_s}\frac{\partial}{\partial\tau}(h_s B_s)\right] + \frac{1}{\tau^2}\frac{\partial}{\partial\theta}\left[\frac{1}{\alpha h_s}\frac{\partial}{\partial\theta}(h_s B_s)\right]$$

$$+ \alpha B_s = 0 \tag{47}$$

However, this configuration of the field is complex enough. First of all we consider the case of axial symmetry, when the twisting of the coordinate axis of the loop equals zero x = 0 ($\partial/\partial s = 0$) i.e. the coordinate axis is a circle in the plane. In that case

$$\vec{B}(\tau, \theta) = \begin{Bmatrix} B_\tau \\ B_\sigma \\ B_\theta \end{Bmatrix} = \begin{Bmatrix} \frac{1}{\alpha h_s \tau}\frac{\partial}{\partial\theta}(h_s B_s) \\ B_s \\ -\frac{1}{\alpha h_s}\cdot\frac{\partial}{\partial\tau}(h_s B_s) \end{Bmatrix} \tag{48}$$

For α=const from (47) we obtain the equation

$$\frac{\partial^2 B_s}{\partial\xi^2} + \frac{1}{\xi^2}\frac{\partial^2 B_s}{\partial\theta^2} + \frac{1}{\xi}\left(\frac{1-2c\cos\theta}{1-c\cos\theta}\right)\frac{\partial B_s}{\partial\xi} +$$

$$\frac{1}{\xi}\frac{c\sin\theta}{(1-c\xi\cos\theta)}\frac{\partial B_s}{\partial\theta} + \left[1 - \frac{B^2}{(1-c\xi\cos\theta)^2}\right]B_s = 0 \tag{49}$$

where $b=k/\alpha=1/\alpha R$; $\xi=\alpha r$. To solve this equation we utilize the approximation of large aspect ratio when $q=kr=b\xi<<1$. Coefficients in equation (49) may be represented as a series

$$\frac{1}{\xi}\cdot\left(\frac{1-2c\xi\cos\theta}{1-c\xi\cos\theta}\right) = \frac{1}{\xi}\left[1 - \sum_{n=1}^{\infty} q^n \cos^n\theta\right]$$

$$\frac{1}{\xi}\cdot\frac{c\sin\theta}{1-c\xi\cos\theta} = \frac{\sin\theta}{\xi^2}\sum_{n=0}^{\infty} q^{n+1}\cos^n\theta \tag{50}$$

$$\frac{c^2}{(1-c\xi\cos\theta)^2} = \frac{1}{\xi^2}\sum_{n=0}^{\infty}(n+1)q^{n+2}\cos^n\theta$$

The solution of the equation (49) may be found in the form

$$B_s(\xi,\theta) = \sum_{n=0}^{\infty} q^n \bar{c}_n(\xi,\theta) = \sum_{n=0}^{\infty} \frac{c_n(\xi,\theta)}{R^n}$$

$$c_n = \tau^n \bar{c}_n = \frac{\xi^n}{\alpha^n}\bar{c}_n \qquad (51)$$

Substitute the formulas (50) and (51) in equation (49) and transform

$$\sum_{k=1}^{\infty} q^k \cos^k\theta \frac{\partial B_s}{\partial \xi} = \sum_{k=1}^{\infty} q^k \cos^k\theta \sum_{n=0}^{\infty} q^n \left\{\frac{\partial B_n}{\partial \xi} + \frac{n}{\xi}\bar{c}_n\right\} =$$

$$\sum_{n=1}^{\infty} q^n \sum_{k=0}^{n-1} \left\{\frac{\partial \bar{c}_k}{\partial \xi} + \frac{k}{\xi}\bar{c}_k\right\} \cos^{n-k}\theta$$

$$\sum_{k=0}^{\infty} q^{k+1}\cos^k\theta \frac{\partial B_s}{\partial \theta} = \sum_{k=0}^{\infty} q^{k+1}\cos^k\theta \sum_{n=0}^{\infty} q^n \frac{\partial \bar{c}_n}{\partial \theta}$$

$$= \sum_{n=1}^{\infty} q^n \sum_{k=0}^{n-1} \frac{\partial \bar{c}_k}{\partial \theta} \cos^{h-k-1}\theta \qquad (52)$$

$$\sum_{k=0}^{\infty} (k+1) q^{k+2} \cos^k\theta B_s = \sum_{k=0}^{\infty} (k+1) q^{k+2} \cos^k\theta \sum_{n=0}^{\infty} \bar{c}_n q^n$$

$$= \sum_{n=2}^{\infty} q^n \sum_{k=0}^{n-2} \bar{c}_k (n-k-1) \cos^{n-k-2}\theta$$

Write the equation (49) in the form

$$\sum_{n=0}^{\infty} q^n F_n = \sum_{n=0}^{\infty} \frac{F_n}{R^n} = 0 \qquad (53)$$

where

$$F_0 = \bar{F}_0 = \frac{\partial^2 c_0}{\partial \xi^2} + \frac{1}{\xi^2}\frac{\partial^2 c_0}{\partial \theta^2} + \frac{1}{\xi}\frac{\partial c_0}{\partial \xi} + c_0;$$

$$c_0 = \bar{c}_0 \qquad (54)$$

$$\bar{F}_{n\geq 1} = \frac{\partial^2 \bar{c}_u}{\partial \xi^2} + \frac{1}{\xi^2}\frac{\partial^2 \bar{c}_n}{\partial \theta^2} + \frac{(2n+1)}{\xi}\frac{\partial \bar{c}_n}{\partial \xi} +$$

$$\left(1 + \frac{n^2}{\xi^2}\right)\bar{c}_n - \bar{g}_n(\xi,\theta)$$

$$\bar{g}_n(\xi,\theta) = \frac{1}{\xi}\sum_{k=0}^{n-1}\left\{\frac{\partial \bar{B}_k}{\partial \xi} + \frac{k}{\xi}\bar{c}_k\right\}\cos^{n-k}\theta +$$

$$\frac{\sin\theta}{\xi^2}\sum_{k=0}^{n-1}\frac{\partial \bar{c}_k}{\partial \theta}\cos^{n-k-1}\theta -$$

$$\frac{1}{\xi^2}\sum_{k=0}^{n-2}\bar{c}_k (n-k-1)\cos^{n-k-2}\theta \qquad (55)$$

The n-approximation $b_n(\xi,\theta)$ is given by the solution of the equation

$$F_n = 0 \qquad (56)$$

In the zeroth-order approximation from equation $F_0=0$ we have a solution for a cylindrical force-free field

$$\vec{B}_0(\xi,\theta) = \begin{Bmatrix} B_r \\ B_s \\ B_\theta \end{Bmatrix} \begin{matrix} \text{Re} \\ \text{Im} \end{matrix} \begin{pmatrix} i\rho T_\rho(\xi) \\ T_\rho(\xi) \\ \dfrac{-\partial T_\rho}{\partial \xi} \end{pmatrix} e^{i\rho\theta};$$

$$\rho = 0, \pm 1, \pm 2, \ldots \qquad (57)$$

The solution in the first-order approximation is defined by equation

$$\frac{\partial^2 B_1}{\partial \xi^2} + \frac{1}{\xi^2}\frac{\partial^2 c_1}{\partial \theta^2} + \frac{1}{\xi}\frac{\partial c_1}{\partial \xi} + c_1 = g_1$$

$$\frac{B_0}{2\alpha}\left[e^{i(p-1)\theta}T_{p-1} - e^{i(p+1)\theta}T_{p+1}\right] \qquad (58)$$

Writing the particular solution in the form

$$c_1^{part}(\xi,\theta) = e^{i(p+1)\theta}W_1(\xi) + e^{i(p-1)\theta}W_2(\xi) \qquad (59)$$

we find

$$W_1(\xi) = \frac{B_0}{4\alpha}\xi T_{p+2}(\xi)$$

$$W_2(\xi) = \frac{B_0}{4\alpha}\xi T_p(\xi) \qquad (60)$$

Thus, the magnetic field B_s in the first approximation is

$$B_s(\xi,\theta) = c_0(\xi,\theta) + \frac{c_1(\xi,\theta)}{R} = C_1 T_p c^{ip\theta} +$$

$$C_2 \frac{\xi}{4\alpha R}\left\{e^{i(p-1)\theta}T_p - e^{i(p+1)\theta}T_{p+2}\right\} \qquad (61)$$

where the constants C_1 and C_2 define the amplitude and boundary conditions. For the p = 0 (fundamental mode) we have

$$B_s = \mathcal{C}_o + \frac{\mathcal{C}_1}{R} = C_1 T_o + C_2 \frac{\cos\theta}{2\alpha R}(\xi T_o - T_1)$$

$$B_r = \frac{1}{\xi}\frac{\partial \mathcal{C}_o}{\partial\theta} + \frac{1}{R}\left\{\frac{\mathcal{C}_o \sin\theta}{\alpha} + \frac{1}{\xi}\frac{\partial \mathcal{C}_1}{\partial\theta}\right\} =$$

$$\frac{\sin\theta}{\alpha R}\left\{\left(C_1 - \frac{C_2}{Z}\right)T_o + \frac{1}{2}C_2\frac{T_1}{\xi}\right\} \qquad (62)$$

$$B_\theta = -\frac{\partial \mathcal{C}_o}{\partial\xi} + \frac{1}{R}\left\{\frac{\mathcal{C}_o \cos\theta}{\alpha} - \frac{\partial \mathcal{C}_1}{\partial\xi}\right\} =$$

$$C_1 T_1 + \frac{\cos\theta}{\alpha R}\left\{C_1 T_o - \frac{C_2}{2}\left(\frac{1}{\xi} - \xi\right)T_2\right\}$$

In paper by Miller and Turner (1981) the constants $C_1$and C_2 are defined by conditions $B_r(0) = 0$, $B_s(0) = B_o$. Approximations of higher order on R may be constructed in a similar manner. Solution (62) may describe loop structures in solar atmosphere with low values of the parameter β, and also configurations of the field of interplanetary magnetic clouds, the formation and ejection of which in the low solar atmosphere is related to flare events, CME's and transients. Taking into account the toroidal effects in coronal loops is important for the analysis of their equilibrium and ejection in interplanetary space [Chen, 1989].

REFERENCES

Baker N H and Temesvary S, *Tables of Convective Stellar Envelope Models*, New York:NASA, 1966.

Chen J, *Astrophys. J.*, 338, 453, 1989.

Colgate S A, *Astrophys. J.*, 221, 1068, 1978.

Kuznetsov V D, Magnetic field fragmentation of flux tubes in subphotospheric regions of the Sun, *Phizika solnechnoi activnosti*, Nauka, Moscow, 1988, 30-53.

Miller G and Turner L, *Phys. Fluids*, 24, 363, 1981.

Parker E N, *Cosmical magnetic fields: Their origin and their activity*, Oxford, 1979.

Pikel'ner S B, *The principles of cosmical electrodynamics*, Nauka, Moscow, 1966.

Polovin R V and Demutskii V P, *The principles of magnetohydrodynamics*, Energoizdat, Moscow, 1987, p.92.

Priest E R, *Solar Magnetohydrodynamics*, D. Reidel Publishing Company, Dirdrecht, Holland, 1982.

Solov'ev L S and Shafranov V D, *Voprosy teorii plasmy*, N5, 1967, Moscow, Atomizdat.

Syrovatskii S I, *Solar Physics*, 76, 3-20, 1982.

Trottet G and MacQueen R M, *Solar Physics*, 68, 177-186, 1980.

Yeh T, *Solar Physics*, 78, 287, 1982.

MAGNETIC RECONNECTION, COALESCENCE, AND TURBULENCE IN CURRENT SHEETS

Manfred Scholer

Instiut für extraterrestrische Physik

Max-Planck-Instiut für Physik und Astrophysik, 8046 Garching, FRG

Abstract. Reconnection in a double periodic current sheet configuration is investigated by means of a two-dimensional compressible MHD code as an initial value problem. The numerical system has a length x of 4π and a height y of 2π, so that the minimum wave number in a Fourier series in x is 1/2 unit. Reconnection is initiated by adding fluctuations in the initial data. The numerical experiments are decay runs; no additional energy is added after time $t = 0$. When the noise is initially only in modes with an integer k_y and k_x with a flat energy spectrum, two islands grow in each current sheet within about 10 Alfvén transit times. When the noise is also distributed into the modes with $|k_x| = 1/2$, a single large sized island grows in each current sheet. In this case kinetic energy and enstrophy (mean square vorticity) production is considerably larger. In a third run, the energy in each of the modes with $|k_x| = 1/2$ has been reduced to 0.25% of the energy in each of the other modes. Two islands grow initially in each current sheet and start to coalesce after about 30 Alfvén transit times into one big island. Maximum kinetic energy and enstrophy production occurs during the final coalescence process. In terms of a Fourier decomposition, the final build-up of one large scale structure can be described as a backtransfer of magnetic excitation of low wavenumbers. It is suggested, that in three-dimensional current sheets the presence of turbulence will drive the growth of small-scale flux ropes, which ultimately merge to a size determined by the largest scale available to the system. The kinetic energy production seems to be independent of the magnetic Reynolds number.

Introduction

Magnetic reconnection is thought to be an important process in determining the magnetic field configuration in astrophysical and space plasmas. The earliest resistive and incompressible models of quasisteady reconnection were developed by Sweet [1958] and Parker [1963] and by Petschek [1964]. In the Sweet-Parker model long current sheets develop and the reconnection rate is rather slow, whereas in the Petschek model the current sheet is limited to a small diffusion region and the outer region contains two pairs of standing slow mode shocks, which deflect and accelerate the incoming plasma, thus allowing for a much larger reconnection rate. Because of the inherent difficulties involved in solving the full resistive problem, including the diffusion layer, one is forced to perform numerical studies of the reconnection process. Such two-dimensional time-dependent simulations of reconnection in an isolated current sheet were performed by a number of authors [e.g., Ugai and Tsuda, 1977; Hayashi and Sato, 1978; Lee and Fu, 1986; Biskamp, 1986]. In addition, reconnection was simulated in configurations appropriate to the magnetotail [e.g., Birn, 1980; Scholer, 1987] or the solar corona [e.g., Forbes and Priest, 1983]. Many of these simulations were devised such that the ensuing reconnection remained smooth, or nonturbulent. In particular, in the driven simulations a highly ordered magnetic field is supplied from the boundaries, while in freely evolving reconnection simulations symmetry conditions are imposed at the midplane.

Matthaeus and Montgomery [1981] were first to point out the importance of turbulence in the reconnection process. In this and a number of subsequent papers Matthaeus and co-workers investigated the effect of turbulence on reconnection by numerically solving the incompressible magnetohydrodynamic (MHD) equations [Matthaeus, 1982; Matthaeus and Lamkin, 1985, 1986]. Turbulence was initiated by including broadband fluctuations in the initial data, and, by using a spectral method for solving the MHD equations, the spectral transfer was followed.

In the present report we will investigate the relation between reconnection, coalescence, and turbulence by numerical simulations using a two-dimensional compressible MHD code. The initial configuration is a periodic sheet pinch consisting of two neutral sheets, and is thus very similar to the one used by Matthaeus and co-workers.

Simulation Model

The 2-dimensional compressible and resistive MHD equations have been solved as an initial value problem by the two step Lax-Wendroff scheme [see, e.g., Scholer,

Geophysical Monograph 58

1987]. We have eliminated all artificial smoothing terms previously added to the scheme. This is possible since contrary to open systems the velocity stays considerably smaller than the Alfvén velocity. The initial configuration is periodic in both dimensions x and y; the sides of the box are 4π in x and 2π in y, i.e., the length is twice the height. Thus, the minimum wave number in a Fourier series in y is one unit, and the minimum wave number in a Fourier series in x is 1/2 unit. Two current sheets are centered at $y = \pi/2$ and $y = 3\pi/2$. The magnetic field is given by

$$B_x = \tanh[(y - 3\pi/2)/\delta] - \tanh[(y - \pi/2)/\delta]$$

where δ is the currrent sheet half thickness. Principally, it would be more appropriate for most space and astrophysical applications to study the behavior of a single current sheet. However, by using a configuration with two current sheets we are able to impose periodic boundary conditions in the x as well as in the y direction. Open boundary conditions are notoriously difficult to handle in numerical simulations and solid wall boundary conditions are rather unphysical.

The density is initially constant and the system is in pressure equilibrium, i.e., $p + B^2 = \text{const}$. The plasma β (ratio of plasma to magnetic field pressure) is assumed to be 1. Velocity is measured in units of the Alfvén speed computed from the unit magnetic field. The time unit is the Alfvén transit times of unit distance, and the dimensionless resistivity, η, is the reciprocal of the magnetic Reynolds number R_m. We do not take into account a finite mechanical Reynolds number. We use a 256 × 128 spatial grid, and the current sheets are 12 cells wide, corresponding to $\delta \approx 0.29$. In the case of tearing mode analysis the magnetic Reynolds number is conventionally scaled by the current sheet half width. Designating this quantity by S, we have $S = \delta R_m$.

In terms of a Fourier representation the double periodic current sheets can be described by the odd modes $k_y = \pm 1, \pm 3, \pm 5, \ldots$, and $k_x = 0$. Reconnection can be initiated by perturbing the initial magnetic field. We will describe in the following five numerical simulation runs. In these five experiments we have selectively enhanced initially various non-sheet pinch Fourier modes.

Simulation Results

In the following five numerical experiments, designated (A), (B), (C), (D), and (E), reconnection was initiated by adding to the initial magnetic and velocity field (initially zero) a low level of random phased perturbations, thus that the perturbation energy is equally divided between magnetic and kinetic energy. For all runs we choose an energy in the magnetic field perturbations of 2% of the initial magnetic field energy. We denote with ϵ the ratio of the energy contained in a mode with $|k_x| = 1/2$ to the energy in a mode with an integer value of k_x. Numerical values for the various run parameters are given in Table 1.

TABLE 1. Parameters of the Simulation Runs

Run	R_m	S	ϵ
A	400	118	0
B	400	118	1
C	1000	294	1
D	400	118	2.5×10^{-3}
E	800	236	2.5×10^{-3}

For case (A) the noise level used to trigger reconnection is divided equally between the Fourier modes with integer wavenumbers ≤ 4, i.e., every Fourier mode with $1 \leq k^2 = k_x^2 + k_y^2 \leq 16$ and with k_x and k_y as integers is excited initially with a flat spectrum in energy and random phase. This leads, for example in the case of the $(k_x, k_y) = (1,0)$ mode, to a magnetic field amplitude of $\sim 2.7 \times 10^{-2}$. The magnetic Reynolds number R_m is 400 ($S \approx 118$). Figure 1 shows in the top part the magnetic field configuration at $t = 20$. The bottom part shows isointensity contours of the current density. As expected, the result is very similar to the results obtained by, e.g, Matthaeus and Lamkin [1986] in the incompressible case.

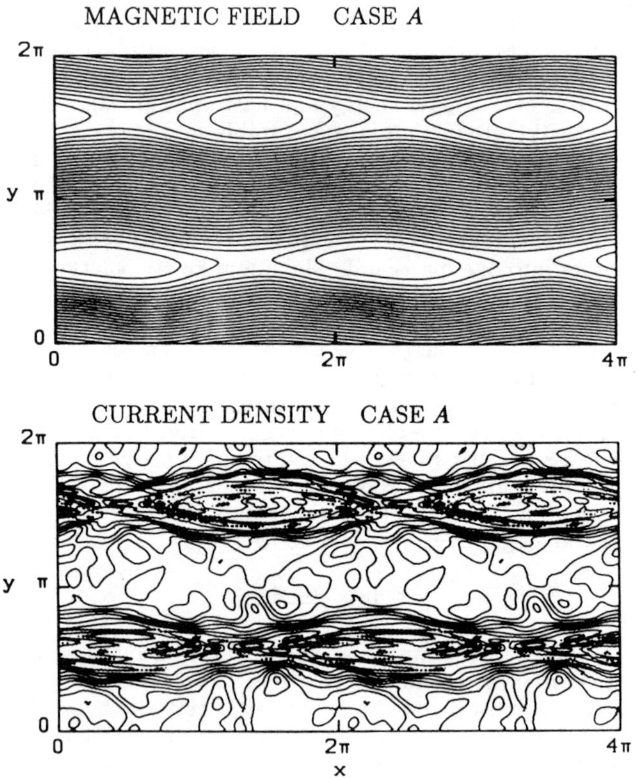

Fig. 1. Magnetic field line configuration (top) and isointensity contours of the current density (bottom) at $t = 20$ for case (A) where the initial noise is only in modes $|k| \leq 4$ with integer values of k_x, k_y. The magnetic Reynolds number is 400.

The only difference here is that two islands form in each current sheet instead of one, simply because the box is twice as long. The randomly chosen phases of the initial perturbations determine the location of the X and O points in the two current sheets. The iso-intensity contours of the current show that the current becomes more and more filamentary, while the magnetic field goes into an ordered state.

For case (B) the initial noise level has been slighly changed. In addition to the integer Fourier modes noise was also added to the $|k_x| = 1/2$ modes such that $1/4 \le k^2 \le 4$. As in case (A), the noise level was equally divided between all Fourier modes, i.e., the energy spectrum is flat. The top panel of Figure 2 shows the initial magnetic field configuration consisting of the two current sheets and the superimposed random phased noise. The magnetic field lines, the contours of constant current density and the contours of constant plasma density at $t = 20$ are shown in the next three panels of Figure 2. Since the lowest possible wave number in x is now also excited ($k_x = \pm 1/2$), a single large island develops in the upper and the lower current sheet. The second panel of Figure 2 shows a high concentration of the current between the islands; the vertical expansion of the large scale islands may already be partly responsible for this. Furthermore, the dent-like structure of the magnetic field in the islands indicates a high speed flow from the reconnection sites into each island. From the bottom panel it can be seen, that although initially constant, the density is also highly structured at later times. The burstiness of the reconnection process leads to knots and clumps within the magnetic islands. This is of course due the compressibility of the fluid and the effect depends on the initial plasma β. The importance of the compressibility will be fully explored in a future publication.

Figure 3 shows in the top panel the time history of the kinetic energy $E_k = \langle \rho \mathbf{V}^2 \rangle / 2$, where ρ is the density, $\mathbf{V}$ the velocity, and $\langle ... \rangle$ denotes a volume average. It can be seen that considerable more kinetic energy is produced during case (B), i.e., in the process of the production of the larger scale island. The bottom panel of Figure 3 shows the time history of the mean square vorticity, or enstrophy $\Omega = \langle \vec{\omega}^2 \rangle / 2$, where $\vec{\omega} = \nabla \times \mathbf{V}$. Since Ω emphasizes the small scale nature of the flow, it can be seen that during case (B) the flow is of considerably smaller scale than during case (A).

The run (B) has been repeated with the same initial conditions, but with a higher magnetic Reynolds number.

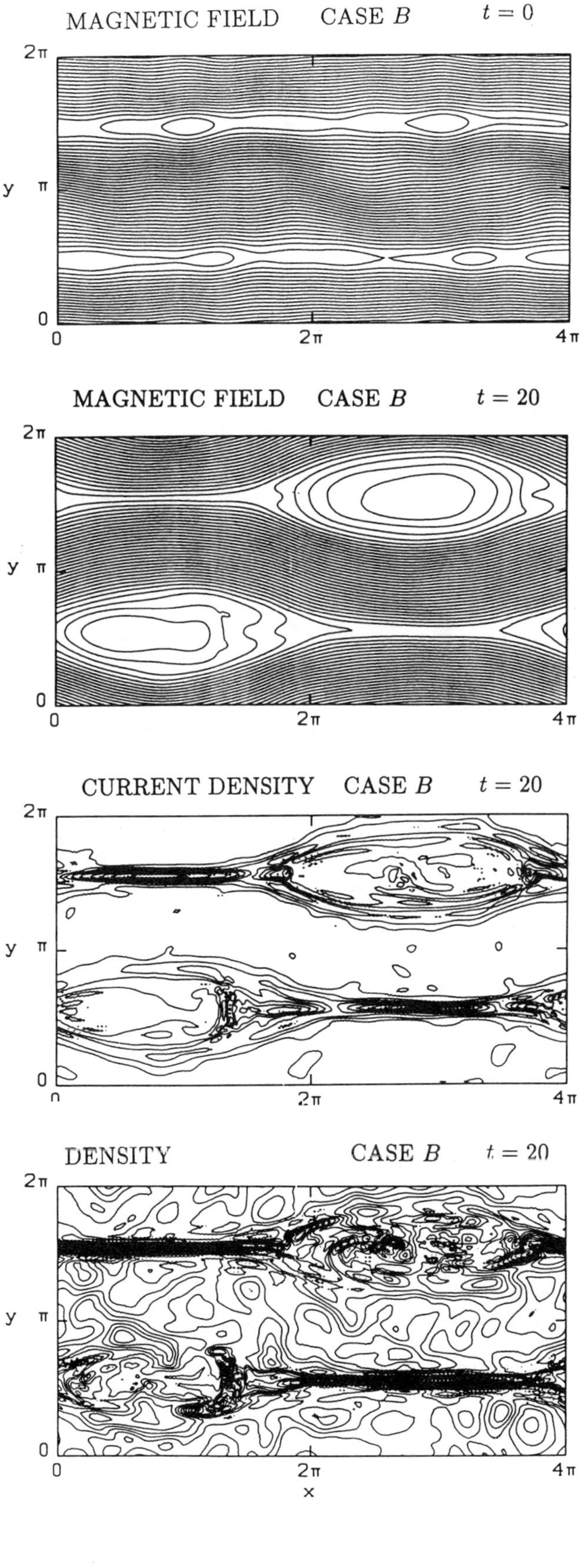

Fig. 2. Top to bottom: Magnetic field line configuration at $t = 0$, magnetic field line configuration, iso-intensity contours of the current density, and plasma density contours, all at $t = 20$, for case (B). In addition to noise in integer modes $|k| \le 4$, noise is initially also in modes $|k_x| = 1/2$. The magnetic Reynolds number is 400.

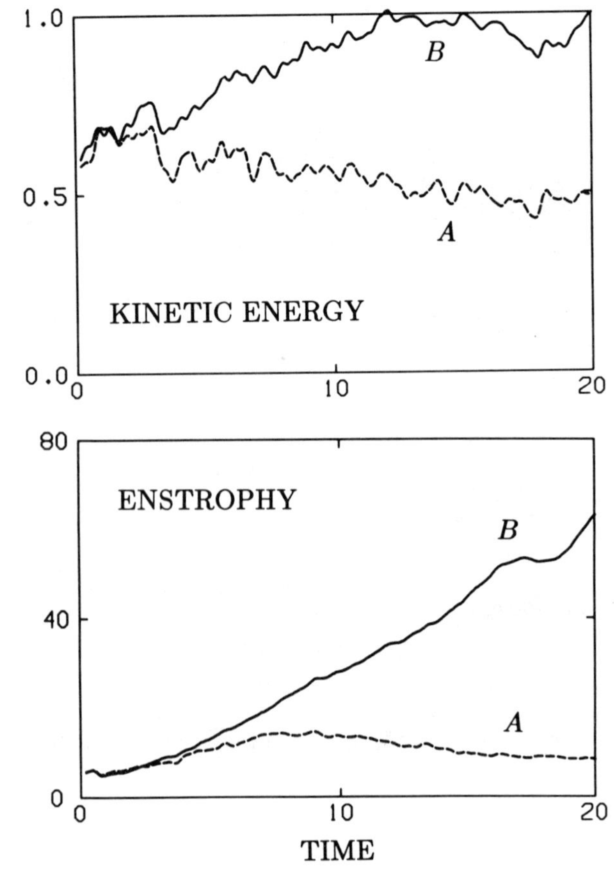

Fig. 3. Time histories of the volume integrated kinetic energy (top) and of the enstrophy (mean square vorticity) (bottom) for cases (A) and (B).

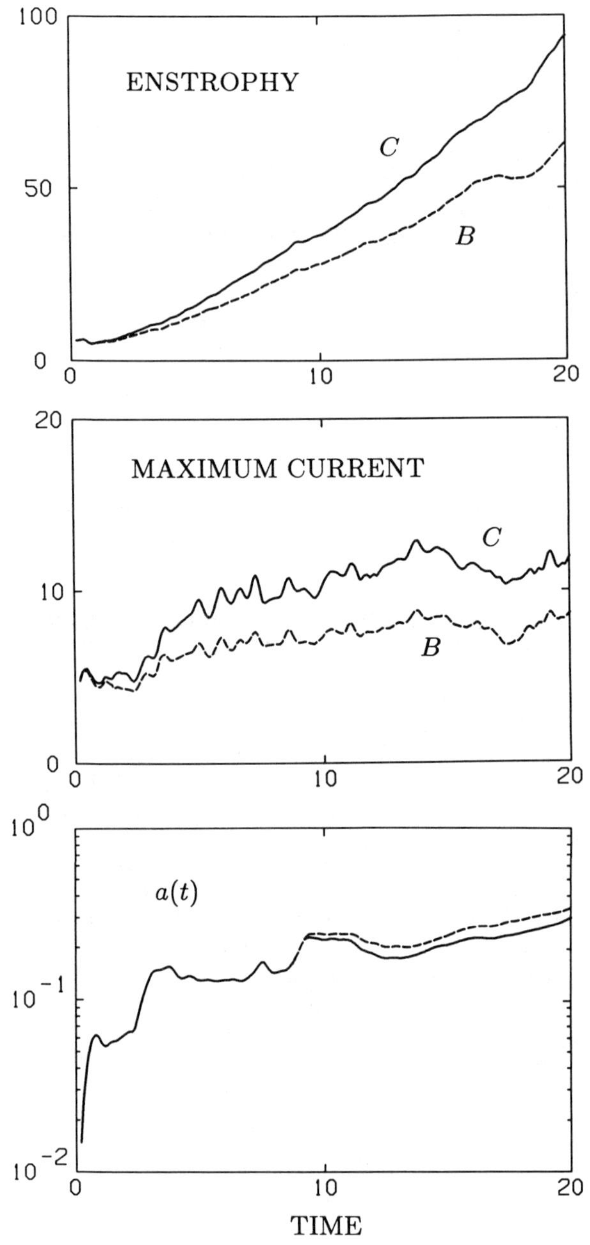

Fig. 4. Time histories of the enstrophy, the maximum current density and global magnetic field reconfiguration represented by $a(t)$ for cases (B) and (C). The initial conditions for case (C) are identical to those of case (B), however the magnetic Reynolds number has been increased to $R_m = 1000$.

We used $R_m = 1000$, corresponding to $S \approx 294$ (case (C)). As a measure for the growth of the instability we have evaluated

$$a(t) = \left[\int (A(x,y,t) - A(x,y,0))^2 dx dy\right]^{1/2}$$

for the two cases (B) and (C). Here, A is the z component of the magnetic vector potential: $\mathbf{B} = \nabla \times A\hat{z}$ with the unit vector $\hat{z}$. Figure 4 shows from top to bottom the time histories of the enstrophy, of the maximum current density in the upper half of the simulation box, and of the quantity a. The maximum current density occurs at the X points and is thus directly proportional to the reconnection electric field. It can be seen that the electric field increases in a burst like fashion. The maximum current density is larger in case (C), indicating stronger current filamentation at higher Reynolds numbers. This is similar to the result reported by Matthaeus and Lamkin [1985]. The global quantity $a(t)$ exhibits initially a sharp rise which is probably due to the initial exchange between the magnetic and kinetic wave energy.

Subsequently, $a(t)$ rises in several steps and on a much larger time scale. It is important to note that the quantity a is almost identical for the two different Reynolds number runs. In the case of the tearing mode for a single current sheet the growth rate is proportional to $R_m^{-5/3}$

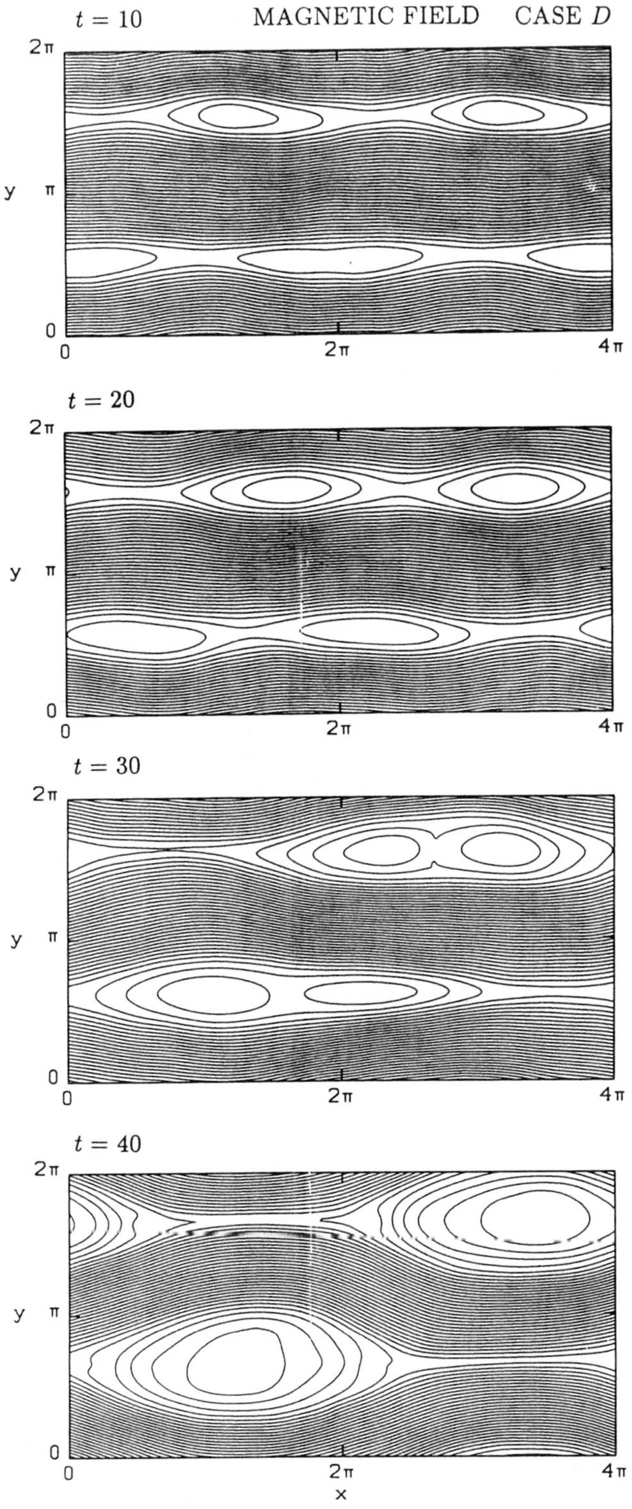

Fig. 5. Evolution of the magnetic field configuration in a case (case (D)), where the energy in each of the $|k_x| = 1/2$ modes has been suppressed to 0.25% of the energy in each of the other modes.

Since the current sheet width 2δ is considerable smaller than the distance $D = \pi$ between the two current sheets, single current sheet tearing mode theory should still be applicable. Figure 4 shows that turbulent reconnection cannot be explained in terms of resistive tearing mode instability.

For case (D) the initial noise level was changed such that only 0.25% of the energy contained in each of the modes with an integer k_x is in each of the modes with $k_x = \pm 1/2$. Figure 5 shows the time development of the magnetic field configuration. Since the initial noise in the $|k_x| = 1/2$ mode is considerably reduced relative to the $|k_x| = 1$ mode, two islands grow initially in each current sheet up to $t = 20$. This run has been continued until $t = 40$. At about $t = 30$ the two islands in each current sheet start to merge and coalesce into one big island, and at $t \approx 40$ the coalescence process is completed. In terms of turbulence this can be considered as a selective decay process, during which energy is transfered into the $(k_x, k_y) = (1/2, 0)$ mode. Other simulation runs have shown that, however small the initial energy in the $(1/2, 0)$ mode, energy is ultimately transferred into that mode. Figure 6 shows a comparison of the kinetic energy and enstrophy time histories of cases (A) and (D). The

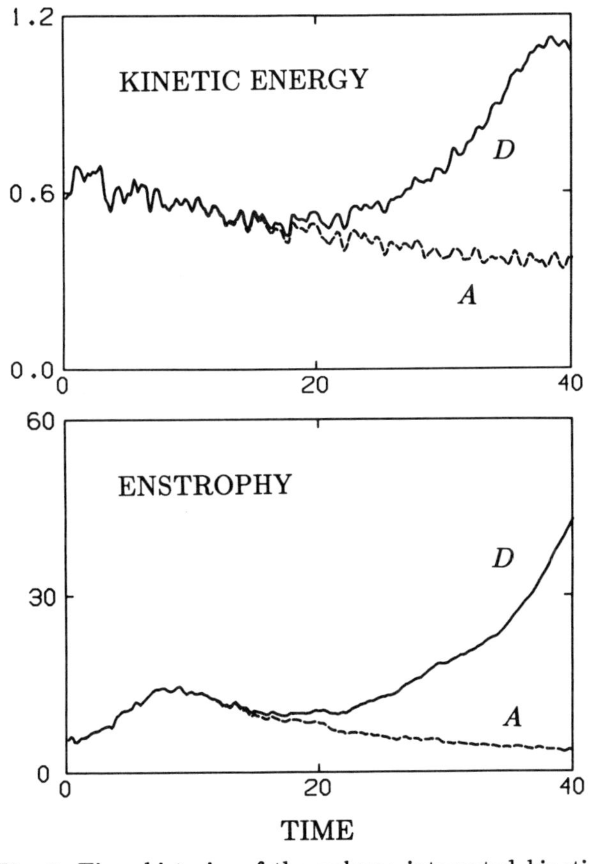

Fig. 6. Time histories of the volume integrated kinetic energy (top) and of the enstrophy (bottom) for cases (A) and (D).

time histories are almost identical up to $t = 20$ and then diverge. Maximum energy and enstrophy production occurs during the final coalescence into one big island. The enstrophy production late in run (D) seems not to be due to the fact that (apart from numerical viscosity) the mechanical Reynolds number is infinite: comparing the bottom panels of Figures 3 and 6 it can be seen, that the time of the enstrophy increase is determined by the energy in the $|k_x| = 1/2$ modes. In the final run presented here (case (E)), the initial noise distribution was the same as in case (D), but the magnetic Reynolds number was increased by a factor two to 800. Figure 7 shows a comparison of the time histories of the kinetic energy and the enstrophy for cases (D) and (E). The kinetic energy seems to be independent of R_m, while the enstrophy increases with increasing R_m. The latter is expected, since filamentation should be larger at larger Reynolds numbers.

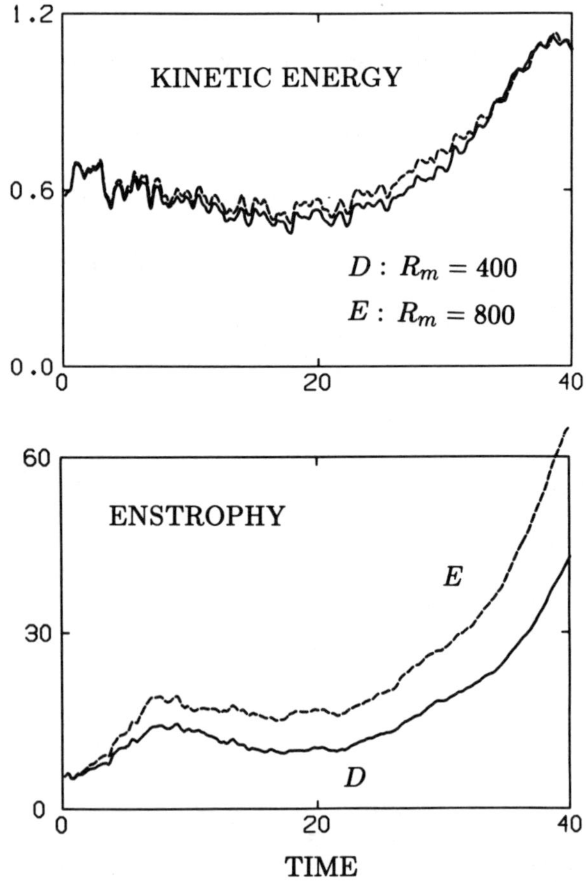

Fig. 7. Time histories of the volume integrated kinetic energy (top) and of the enstrophy (bottom) for cases (D) and (E). The initial conditions for case (E) are identical to those of case (D), however the magnetic Reynolds number has been increased to $R_m = 800$.

Summary

From the limited number of simulations presented in this report we found the following.

1. As has been shown before by Matthaeus and coworkers, islands grow in magnetic current sheets whenever noise or turbulence is present.

2. In the case of a flat turbulence spectrum the island size is determined by the smallest k number present in the initial noise. Maximum kinetic energy and enstrophy production occurs during build-up of the largest scales.

3. When the energy in the mode with the smallest k value is greatly suppressed, the energy is first tranfered into smaller scale structures, which ultimately merge or coalesce into the largest structure allowable in the system. Maximum production of kinetic energy and enstrophy occurs during the final coalescence.

4. The kinetic energy production during the reconnection process seems to be independent of magnetic Reynolds number, while the enstrophy production increases with increasing Reynolds number.

The present simulations are two-dimensional and the magnetic field is purely coplanar. In a three-dimensional configuration with a magnetic field component parallel to the current sheet (i.e., $B_z \neq 0$) the magnetic islands would correspond to magnetic flux ropes. The present two-dimensional results suggest that small scale flux ropes will develop in the presence of turbulence, which ultimalely merge into one large size structure. The size of this large size rope is expected to correspond to the largest scale available to the system. Such a generation of smaller scale flux ropes and their subsequent coalescence may be of particular importance for the magnetopause current layer of the Earth. We note that Song and Lysak [1989] have recently suggested that the formation of the coherent flux transfer event flux tube at the magnetopause may be the result of a self-organization process.

References

Birn, J., Computer studies of the dynamic evolution of the geomagnetic tail, *J. Geophys. Res., 85*, 1214, 1980.

Biskamp, D., Magnetic reconnection via current sheets, *Phys. Fluids, 29*, 1520, 1986.

Forbes, T. G., and E. R. Priest, A numerical experiment relevant to line-tied reconnection in two-ribbon flares, *Solar Phys., 84*, 169, 1983.

Hayashi, T., and T. Sato, Magnetic reconnection: Acceleration, heating, and shock formation, *J. Geophys. Res., 83*, 217, 1978.

Lee, L. C., and Z. F. Fu, A simulation study of magnetic reconnection: Transition from a fast mode to a slow mode expansion, *J. Geophys. Res., 91*, 4551, 1986.

Matthaeus, W. H., Reconnection in two dimensions: Localization of vorticity and current near magnetic X-points, *Geophys. Res. Lett., 6*, 660, 1982.

Matthaeus, W. H., and D. Montgomery, Nonlinear evolution of the sheet pinch, *J. Plasma Phys., 25*, 11, 1981.

Matthaeus, W. H., and S. L. Lamkin, Rapid magnetic reconnection caused by finite amplitude fluctuations, *Phys. Fluids, 28*, 303, 1985.

Matthaeus, W. H., and S. L. Lamkin, Turbulent magnetic reconnection, *Phys. Fluids, 29*, 2513, 1986.

Parker, E. N., The solar flare phenomenon and the theory of reconnection and annihilation of magnetic fields, *Astrophys. J. Suppl. Ser., 8*, 177, 1963.

Petschek, H. E., Magnetic field annihilation, AAS-NASA Symposium on Physics of Solar Flares, *NASA Spec. Publ. 50*, p.425, 1964.

Scholer, M., Earthward plasma flow during near-Earth reconnection: Numerical simulations, *J. Geophys. Res., 92*, 12,425, 1987.

Song, Y., and R. L. Lysak, Evaluation of twist helicity of flux transfer events, *J. Geophys. Res., 94*, 5273, 1989.

Sweet, P. A., The neutral point theory of solar flares, in *Electromagnetic Phenomena in Cosmical Physics*, edited by B. Lehnert, p. 123, Cambridge University Press, London, 1958.

Ugai, M., and T. Tsuda, Magnetic field line reconnexion by localized enhancement of resistivity, 1, Evolution in a compressible MHD fluid, *J. Plasma Phys., 17*, 337, 1977.

WAVE MODES IN THICK PHOTOSPHERIC FLUX TUBES: CLASSIFICATION AND DIAGNOSTIC DIAGRAM

S. S. Hasan[1] and T. Abdelatif[2]

[1]Indin Institute of Astrophysics, Bangalore 560034, India
[2]School for Mathematics, Queen Mary College, London E1 4NS, U.K.

Abstract. We analyse the nature of wave motions in thick photospheric flux tubes. The aim of our investigation is to determine the normal modes of a stratified atmosphere with a vertical magnetic field and to discuss their properties. The results are displayed in the form of a diagnostic diagram. An interesting feature of the solutions is the existence of 'avoided crossings', which occur when adjacent order modes approach each other in the diagnostic diagram. We examine the nature of the modes by decomposing the eigenvectors into longitudinal and transverse components. In general, the character of a mode changes with height in the atmosphere. We apply our results to umbral oscillations and find that the observed oscillations with periods in the range 2-3 min, correspond to low order modes in our calculation. For low horizontal wave number K, the modes, in the photosphere, have almost equal contributions from longitudinal and transverse components. As K increases, the transverse component begins to dominate. In the chromosphere, the modes are essentially transverse and can be identified with slow modes.

Introduction

The magnetic field in the solar photoshere invariably occurs in strong form and confined to discrete elements or flux tubes with different sizes. Sunspots and fibrils are well known examples of magnetic elements corresponding to the extreme limits of thick and thin flux tubes. Oscillations, with periods in a fairly broad range of frequencies, have been extensively observed in sunspots [e.g., Beckers and Schulz, 1972; Abdelatif et al., 1984,1986; Lites and Thomas, 1985; Gurman, 1987; see also review by Moore and Rabin, 1985 for additional references]. However, observations of waves in fibrils are rare, owing to their small horizontal dimensions, apart from those by Giovanelli [1975] and Giovanelli et al. [1978].

Considerable progress has been made in recent years on wave propagation in flux tubes [see reviews by Roberts, 1989; Hollweg, 1989; Thomas, 1989]. Owing to the mathematical complexity of the problem, the analyses have in general been confined to fairly idealised situations, such as neglecting gravity, or invoking the thin flux tube approximation [Defouw, 1976; Roberts and Webb, 1978]. In this paper, we focus our attention on thick flux tubes. We approximate a thick tube as a uniform magnetic field of infinite horizontal extent. Even under this approximation, the mathematical treatment is fairly complicated. Nevertheless, several papers exist on magneto-atmospheric (henceforth abbreviated as MAG) waves in a vertical magnetic field [Ferraro and Plumpton, 1958; Antia and Chitre, 1978; Uchida and Sakurai, 1978; Zhugzhda, 1979; Scheuer and Thomas, 1981; Leroy and Schwarz, 1982 and review by Thomas, 1983 for additional references]. Our investigation differs from previous ones in two important respects: firstly, we present quantitative calculations on the horizontal wave number dependence of frequency i.e., we generate a diagnostic diagram and secondly, we attempt a classification of the modes.

Model and Equations

Let us consider an isothermal stratified atmosphere, with a uniform vertical magnetic field B, which is unbounded in the horizontal direction. In cartesian geometry, the linearised equations for MAG waves can be written in terms of the Lagrangian displacement $\xi \sim e^{i(\omega t - kx)}$ as [Ferraro and Plumpton, 1958]

Geophysical Monograph 58

$$[v_a^2 \frac{d^2}{dz^2} - (c_s^2 + v_a^2)k^2 + \omega^2]\xi_x - ik(c_s^2 \frac{d}{dz} - g)\xi_z = 0 \quad (1)$$

$$[c_s^2 \frac{d^2}{dz^2} - \gamma g \frac{d}{dz} + \omega^2]\xi_z - ik[c_s^2 \frac{d}{dz} - (\gamma - 1)g\ ]\xi_x = 0 \quad (2)$$

where z, the vertical co-ordinate, is measured positive upwards i.e., away from the Sun, k is the horizontal wave number, ω is the frequency, g is gravity, γ is the ratio of specific heats, c_s is the sound speed and v_a is the Alfvén speed. We have implicitly assumed that the propagation and motions of the MAG modes are confined to the $x - z$ plane. This involves no loss of generality.

Equations (1) and (2) were combined into a single fourth order differential equation for ξ_x [Zhugzhda 1979], which we solved in series form using the Frobenius method, similar to Leroy and Schwartz (1982).

Boundary Conditions

In order to determine the normal modes of the system, boundary conditions need to be imposed. Let the atmosphere extend from z=0 to $z = d$ in the vertical direction. We assume rigid boundary conditions and set $\xi_x = \xi_z$=0 at z=0 and $z = d$. The physical reason for this choice is that at the lower boundary, which we assume defines the interface between the photosphere and convection zone, the amplitudes of the motions are fairly small, in view of the large mass density. Let the upper boundary be situated at the base of the transition region. Thus, a sound-like wave propagating upwards will be reflected there. An Alfvén-like wave will probably be reflected much before, since the Alfvén speed increases sharply with height in the atmosphere.

Photospheric Flux Tubes

Diagnostic Diagram

Figure 1 shows the frequency Ω as a function of the horizontal wave number K for different mode orders n, where $\Omega = \omega H/c_s$, $K = kH$ and H is the scale height of the atmosphere. A vertical extension of $20H$ and a ratio of Alfvén to sound speeds v_{a0}/c_{s0}=0.5 at z=0 was used. The curves in Figure 1 correspond to various order solutions, obtained by solving Equations (1) and (2), with rigid boundary conditions. An interesting feature of the solutions is the absence of accidental degeneracy in the frequencies of the various order solutions. We find that when curves of adjacent orders draw close to one another, an 'avoided crossing' occurs. 'Avoided crossings' have been found to occur in the study of global oscillations, but their existence for MAG modes does not ap-

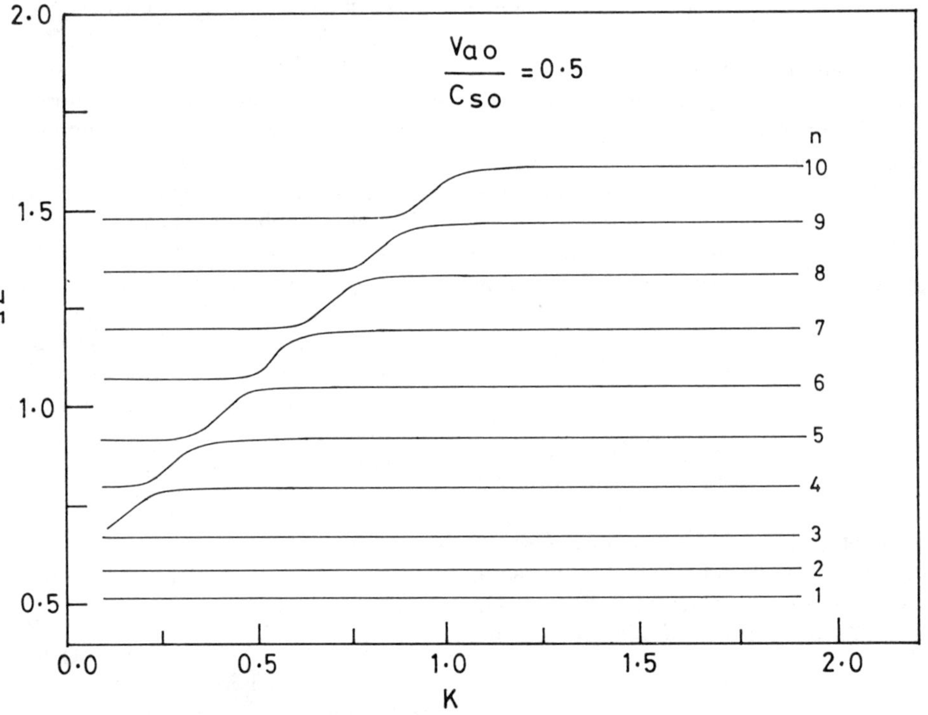

Fig. 1. Variation of Ω with K for v_{a0}/c_{s0}=0.5, where v_{a0} and c_{s0} are the Alfvén and sound speeds respectively at z=0, and n denotes the mode order.

pear to have been noted previously. Physically, this phenomenon is related to the coupling of modes confined to different regions in the atmosphere [Leibacher and Stein, 1981]. Another aspect to note is that the frequency does not depend strongly on the horizontal wave number. For low orders, the frequency spectrum is practically flat for large K.

E_{kin} *vs.* z

Figures 2a and 2b show the kinetic energy density E_{kin} of the lowest three modes as a function of z for $K = 0.1$ and $K = 1.9$ respectively. We find that in both cases, the oscillations have non-negligible energy density over practically the whole vertical extension of the tube. For n=2, the energy in the oscillations is confined to two regions, which we interpret as the photosphere and chromosphere. Increasing n has the effect of increasing the number of energy maxima, with the dominant contribution to E_{kin} moving lower down in the photosphere.

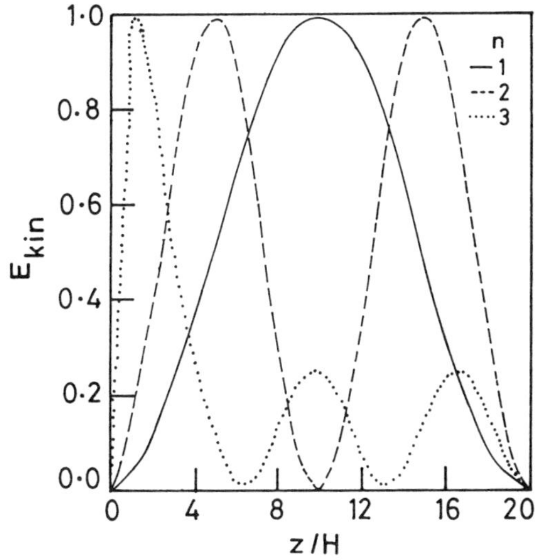

Fig. 2a. E_{kin} vs. z/H for n=1 (solid line), n=2 (dashes) and n=3 (dots), assuming K=0.1.

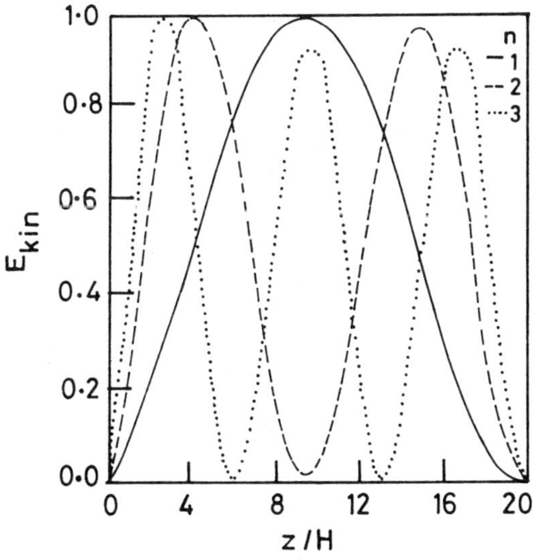

Fig. 2b. E_{kin} vs. z/H for n=1 (solid line), n=2 (dashes) and n=3 (dots), assuming K=1.9.

Mode Classification

We examine the properties of the modes by using Helmholtz' theorem to analyse the eigenvectors in terms of longitudinal (irrotational) and transverse (solenoidal) components. A similar approach was adopted by Aizenman and Smeyers [1977] and Sobouti [1981] in the context of global oscillations and by Hasan and Sobouti [1987] for waves in flux tubes. Thus, we write ξ as follows

$$\xi = \xi_l + \xi_t \tag{3}$$

The longitudinal and transverse components ξ_l and ξ_t can be written in terms of scalars ϕ_l and ϕ_t as

$$\xi_l = -\nabla\phi_l, \qquad \xi_t = \nabla \times \nabla(\mathbf{z}\phi_t)$$

where $\mathbf{z}$ is a unit vector in the z direction. We calculate ϕ_l and ϕ_t from ξ by solving a Poisson type equation, using standard techniques. Figures 3a and 3b depict the z variation of f, where

$$f = \frac{|\xi_l| - |\xi_t|}{|\xi_l| + |\xi_t|}$$

for K=0.1 and K=1.9 respectively. In the former case, we find that the modes in the photosphere have almost equal contributions from both longitudinal and transverse components. However, in the chromosphere, the modes acquire a dominantly longitudinal character. We now establish a rough correspondence with MHD modes in an unstratified medium. Apart from a small region close to z=0, $v_a/c_s \gg 1$ owing to the rapid increase of density with height. Thus, in the upper layers, we find that the mode is of the slow type, since for small K, the slow mode is essentially longitudinal. However, for K=1.9, the modes have a dominantly transverse character, in regions where E_{kin} is large. For $K \gg 1$, the MHD mode, which is transverse, is the slow one.

Umbral Oscillations

Let us at the very outset point out that our aim in this paper is not to present results for detailed comparison with any specific observation, but rather to delineate

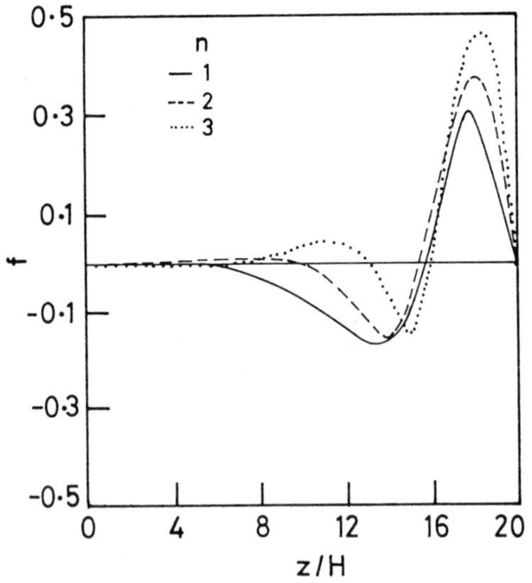

Fig. 3a. f ($f = (|\xi_l| - |\xi_t|)/(|\xi_l| + |\xi_t|)$) vs. z/H for n=1 (solid line), n=2 (dashes) and n=3 (dots), assuming K=0.1.

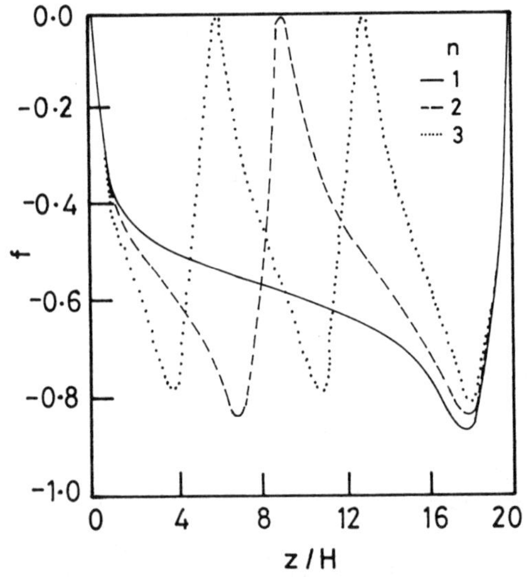

Fig. 3b. f vs. z/H for n=1 (solid line), n=2 (dashes) and n=3 (dots), assuming K=1.9.

the properties of MAG waves in thick flux tubes. Nevertheless, it is tempting to apply out results to sunspots. We have used the same set of parameters as Scheuer and Thomas [1981], who argued that umbral oscillations could be studied in a simplified manner by considering an isothermal layer, with rigid boundaries, and placing the upper boundary high enough from the region where bulk of the energy in the oscillations resides. Approximating, the umbra of a sunspot as a thick vertical flux tube, which is isothermal in the photosphere with a temperature of 4500 K, we find that the three lowest order modes correspond to periods of 208 s, 185s and 159s respectively. Observations suggest that the power spectrum of umbral oscillations is in the 2-3 min range, with most of the kinetic energy trapped in the photosphere [Abdelatif et al., 1984]. These correspond to n=2 and n=3 modes in our calculation. Our results indicate that, in the limit of small horizontal wave number, these modes have a mixed character in the photosphere, but in the chromosphere they resemble slow waves. For large K, the modes have a slow character in both the photosphere and chromosphere. In general, our results are in qualitative agreement with the observations of von Uexküll et al. [1983], who interpreted umbral oscillations in the chromosphere, as slow waves.

References

Abdelatif,T.E., Lites,B.W. and Thomas,J.H.,in *Small-Scale Dynamical Processes in Solar and Stellar Atmospheres*, ed. S.L.Keil, Sacramento Peak Obs., Sunspot,p. 141, 1984.

Abdelatif,T.E., Lites,B.W. and Thomas,J.H., *Astrophys. J. 311*, 1015, 1986.

Aizenman,M.L. and Smeyers,P., *Astrophys. Sp. Sci. 48*, 123,1977.

Antia,H. and Chitre,S.M., *Solar Phys. 63*, 67, 1979.

Beckers,J.M. and Schulz,R.B., *Solar Phys. 27*, 61, 1972

Defouw,R.J., *Astrophys. J. 209*, 355, 1976.

Ferraro,V.C. and Plumpton,C., *Astrophys. J. 129*, 459, 1958.

Giovanelli,R.G., *Solar Phys. 44*, 289, 1975.

Giovanelli,R.G., Harvey,J.W. and Livingston,W.C., *Solar Phys. 59*, 40, 1978.

Gurman,J.B., *Solar Phys. 108*, 61, 1987.

Hasan,S.S. and Sobouti,Y., *Mon. Not. R. astr. Soc. 228*, 427, 1987.

Hollweg,J.V., in *Physics of Magnetic Flux Ropes*, eds. C.T.Russell and E.R. Priest, AGU Monograph, (in press), 1989.

Leibacher,J.W. and Stein,R.F., in *The Sun as a Star*, ed. S.Jordan (Washington:NASA), pp. 263-287.

Leroy,B. and Schwartz,S.J., *Astron. Astrophys. 112*, 84, 1982.

Lites,B.W. and Thomas,J.H., *Astrophys. J. 294*, 682, 1985.

Moore,R.L. and Rabin,D., *Ann. Rev. Astron. Astrophys. 23*, 239, 1985.

Roberts,B. and Webb,A., *Solar Phys. 56*, 5, 1978.

Roberts,B., in *Physics of Magnetic Flux Ropes*, eds. C.T.Russell and E.R. Priest, AGU Monograph, (in press), 1989.

Scheuer,M.A. and Thomas,J.H., *Solar Phys. 71*, 21, 1981.

Sobouti,Y., *Astron. Astrophys. 100*, 319, 1981.
Thomas,J.H., *Ann. Rev. Fluid Mech. 15*, 321, 1983.
Thomas,J.H., *Physics of Magnetic Flux Ropes*, eds. C.T.Russell and E.R. Priest, AGU Monograph, (in press), 1989.
Uchida,Y. and Sakurai,T., *Publ. Astron. Soc. Japan 27*, 259, 1975.
von Uexküll,M., Kneer,F. and Mattig,W., *Astron. Astrophys. 123*, 263, 1983.
Zhugzhda,Y.D., *Sov. Astron. 23*, 42, 1979.

FLUX TUBE WAVES: A BOUNDARY-VALUE PROBLEM

M. P. Ryutova and L. G. Khijakadze

Institute of Nuclear Physics, 630090, Novosibirsk, USSR

Abstract The forced oscillations of an open magnetic flux tube shaken at its footpoints by convective motions are studied as a boundary-value problem of two different types: the one of wave propagation in a semi-infinite tube with a smooth radial profile of plasma parameters and another of standing wave resonance in a tube with step-wise radial profile. It is shown that an excitation of white noise probably leads to the temporal brightening of a flux tube at some height.

1. Introduction

According to observational data all the solar magnetic field has a pronounced filamentary structure. In the photosphere, magnetic field is concentrated in almost vertical flux tubes, usually far removed from each other (about 300 km) with magnetic field strengths of the order of 1-2 kG. In sunspots intense (3-4 kG) flux tubes are assumed to be tightly compressed. As a rule, isolated flux tubes are localized at supergranular cell boundaries extending from the subsurface regions through the photosphere and chromosphere to higher layers, where they form a variety of magnetic structures. These structures, having as a rule longitudinal dimensions much larger than transverse ones, are exposed to the action of the permanently moving convective zone which results in the generation of different kinds of waves and oscillations in such systems. The waves and oscillations of magnetic flux tubes turn out to be one of the most interesting aspects of the structure and dynamics of the solar atmosphere.

In the present paper we analyse the peculiarities of forced oscillations of an open magnetic flux tube which is shaken at its footpoint. We will consider two different boundary problems. One is that of the asymptotic behaviour of the wave amplitude and energy flux at high altitudes in the case when the plasma parameters inside the tube have a smooth radial dependence (Sect. 2). The second problem is related to the situation when the plasma parameters in a flux tube have a sharp longitudinal dependence (Sect. 3). Before proceeding to these problems we recall some properties of flux tube oscillations. We will consider long-wave oscillations whose wavelength $\lambda = 1/k$ is much larger than the radius of the magnetic flux tube R : $kR << 1$. These oscillations are most readily excited by large- scale plasma motions and have a relatively low damping rate. Among these oscillations the most important modes are two: the bending (kink) oscillations which are actually the dipole mode corresponding to the azimuthal wavenumber $m = \pm 1$ and the axisymmetric sausage mode. For an infinitely long homogeneous flux tube (with step-wise radial profiles of density, magnetic field, etc) these oscillations can be described in terms of a simple dispersion relation. For both types of oscillation the frequency scales linearly with wavenumber.

The bending oscillations are an analogue of Alfvén waves. As the external plasma participates in the motion in the vicinity of a tube via the "added mass" effect, its density enters their dispersion relation [Ryutov & Ryutova, 1976]:

$$c_b = \frac{\omega}{k} = \frac{a}{\sqrt{1 + \rho_e/\rho_i}} . \qquad (1)$$

Here subscripts "i" and "e" refer to the tube's interior and its surrounding medium, respectively; ρ is the plasma density, $a = B/\sqrt{\mu\pi\rho_i}$ is the Alfvén velocity.

The sausage mode is a specifically quasi-longitudinal oscillation of a flux tube in which a compression (expansion) of plasma inside the tube is compensated by the decrease (increase) in the longitudinal magnetic field due to a corresponding change in the cross section of the flux tube, so that the sum of gas-kinetic and magnetic pressures is almost not perturbed. Thus the plasma parameters outside the tube have little influence on their dispersion relation (giving the corrections of the order of $(kR)^2$) [Defouw, 1976]:

$$c_T = \frac{\omega}{k} = \frac{a\, s_i}{\sqrt{a^2 + s_i^2}} . \qquad (2)$$

Here $s = \gamma p/\rho$ is a sound speed (γ is a specific heat ratio). Besides the sausage oscillations there are two other axisymmetric modes: the high frequency oscillations in which plasma density and magnetic field are changing in phase. These oscillations are an analogue of fast magnetosonic waves. As their frequency is extremely high (of the order of a/R), they can hardly be excited and experience fast radiative damping. The $m = 0$ torsional oscillations are just Alfvén waves, but their amplitude is very small, namely of the order of $v_{conv}R/h$, where h is the convective cell size and v_{conv} is a characteristic velocity in the convective zone. These oscillations are also usually of less interest.

Geophysical Monograph 58

As to the oscillations with higher azimuthal mode numbers m = ±2,±3,..., they are weakly coupled with the large-scale motions of medium since the matrix elements which determine this coupling contain a small parameter $(kR)^{|m|}$.

In the absence of dissipative processes the only damping mechanism is that of radiation of secondary acoustic waves by oscillating flux tube [Ryutov and Ryutova, 1976; Ryutova, 1981]. The damping rate contains the factor $(kR)^2$ and is typically quite small. The situation becomes quite different in a more realistic case of a smooth radial distribution of plasma parameters (Fig. 1). In this case a new effect appears. Namely, the flux tube oscillations experience the anomalous and strong damping which is caused by a resonance between the phase velocity of tube oscillations and the local Alfvén velocity [Ryutova, 1977]. The physical origin of this effect is somewhat similar to Landau damping. The general features of this damping mechanism in a plasma were studied by many authors [Timofeev, 1970; Tataronis & Grossman, 1973; Chen & Hasegava, 1974].

The quantitative evaluation of the damping rate for the bending oscillations of a flux tube has been studied in a model of a tube with a narrow transition region shown by dashed lines in Fig 1a. The local Alfvén velocity linearly decreases to zero in a narrow region of width εR, where ε is a small parameter. The corresponding damping rate is as follows [Ryutova, 1977]:

$$\frac{\Gamma}{\omega} = \frac{\pi\varepsilon}{4} \frac{\rho_i}{\rho_i + \rho_e} .$$

Qualitatively, on the applicability limit when ε is of the order of unity, the damping rate becomes comparable with the wave frequency. Physically, the nature of the damping is due to the pumping of oscillation energy into the resonance point where the dissipation occurs. The resonance damping, by itself, does not transform the energy of oscillations to plasma heating. It just results in the concentration of the initially smooth eigenfunction near the resonance point, or, in other words, in a conversion of oscillation energy to the Alfvén continuum. And then of course, the usual dissipation mechanisms, like viscosity and magnetic diffusivity turn on and this strongly oscillating distribution damps out (Fig. 2).

This phenomenon appears to be very important in heating processes of the high chromosphere and corona. This subject was explored by many authors. First, as an eigenvalue problem [Ionson, 1978; Wentzel, 1979; Rae & Roberts, 1981] and then as a boundary value problem for a closed coronal loop with both of its footpoints embedded in the convective zone and shaken by the convective motions [Ionson 1982; Heyvaerts & Priest, 1983; Gordon & Hollweg, 1983; Sakurai & Granik, 1984].

In the present paper we will show that there are some mechanisms which lead to the formation of a 1-D resonator in an open flux tube, that is a semi-infinite vertical flux tube can behave as a string with a very small leak of the energy through the upper point.

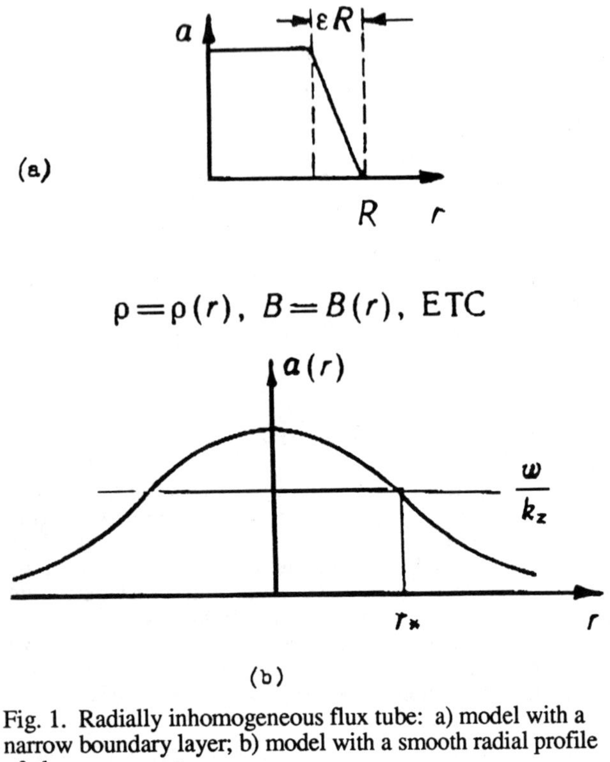

Fig. 1. Radially inhomogeneous flux tube: a) model with a narrow boundary layer; b) model with a smooth radial profile of plasma parameters

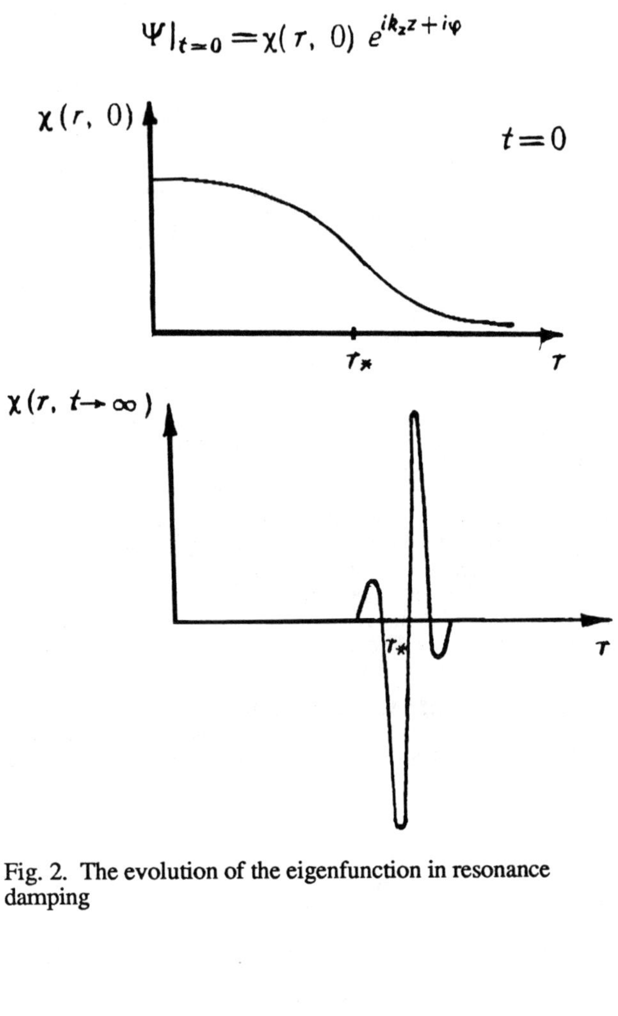

Fig. 2. The evolution of the eigenfunction in resonance damping

2. Flux tube with smooth radial profile of plasma parameters

In this section we consider the most general case of a smooth radial profile of flux tube without any small parameter (Fig. 1b). In this case the weakly damped wave which could be described in terms of a smoothly varying radial eigenfunction does not exist any more. As in a real situation the radial distributions in most cases should be smooth, the question arises as to whether the bending oscillations of a flux tube can still be responsible for energy transfer from the underlying surface to the upper layers of the solar atmosphere. To answer this question one should study a boundary-value problem for the flux tube excited at its foot point. The answer is positive; this is connected with the fact that the resonance damping for a really smooth profile of plasma parameters gives rise to a very peculiar evolution of the radial mode structure of flux tube oscillations in the course of their propagation along the tube from the excitation point (where a rigid $m = 1$ displacement with some given frequency ω is imposed) to higher altitudes. Near the excitation point the radial structure of the perturbation is smooth, while at higher altitudes it becomes more and more spiky. The presence of these small-scale structures results in a fast damping of the wave.

We consider bending and torsional oscillations in the simplest case of an incompressible fluid.

The linearized MHD equations together with the equilibrium condition

$$p_i(r) + \frac{B^2(r)}{8\pi} = p_e$$

lead to the following equations describing the spatial variation of the tube displacement $\vec{\xi}$:

$$-\omega^2 \rho \vec{\xi} = -\nabla \delta P + \frac{B^2}{4\pi}\frac{\partial^2 \vec{\xi}}{\partial z^2} \qquad (3)$$

where $\delta P = \delta p + Bb_z/4\pi$ is a perturbation of total pressure (we use cylindrical coordinates with the z-axis coinciding with tube axis).

To solve these equations one should use the Laplace transform over z, then solve the radial equation, and apply Laplace inversion.

Let us perform these calculations first for a more simple case of a torsional mode ($m = 0$, $\xi_r = \xi_z = 0$). In this case Eqns (3) are reduced to a single equation for ξ_ϕ:

$$-\omega^2 \xi_\phi(r,z) = a^2(r)\, \frac{\partial^2 \xi_\phi(r,z)}{\partial z^2} \qquad (4)$$

Laplace transform

$$\xi_{\phi p} = \int_0^\infty e^{-pz}\, \xi_\phi dz; \quad R_e p > 0 \qquad (5)$$

gives for Eqn (4):

$$\omega^2 \xi_{\phi p}(r) + a^2(r) p^2 \xi_{\phi p}(r)$$

$$= -a^2(r)\left\{\xi_\phi(r,0) + p\frac{\partial \xi_\phi(r,0)}{\partial z}\right\}.$$

So, for $\xi_{\phi p}$ we have:

$$\xi_{\phi p} = -\frac{ia}{2\omega}\left[\xi_\phi(r,0) + p\frac{\partial \xi_\phi(r,0)}{\partial z}\right]$$

$$\left(\frac{1}{p + i\frac{\omega}{a}} - \frac{1}{p - i\frac{\omega}{a}}\right) \qquad (6)$$

Laplace inversion is

$$\xi_\phi(R,Z) = -\int_C e^{pz}\, \xi_{p\phi}(r) dp \qquad (7)$$

where the integration has to be performed along the contour which is a vertical line lying on the right of all the poles of the function $\xi_{\phi p}$. The poles are obviously at the points $p = \pm i\,\omega/a(r)$. Shifting the contour C to the left and taking residues one can easily get

$$\xi_\phi(r,z) = \frac{\xi'_\phi(r,0).a(r)}{\omega}\sin\frac{\omega z}{a(r)} + \xi_\phi(r,0)\cos\frac{\omega z}{a(r)}. \qquad (8)$$

Eqn (8) shows that if at the base of the flux tube $z = 0$, $\xi_\phi(r,0)$ is a smooth function of radius (say, $\xi_\phi(r,0) \sim r$, i.e. a rigid rotation takes place), then at higher altitudes the dependence of ξ_ϕ on r becomes more and more "spiky" (Fig. 3). The characteristic radial scale length of perturbations Δr is of the order of

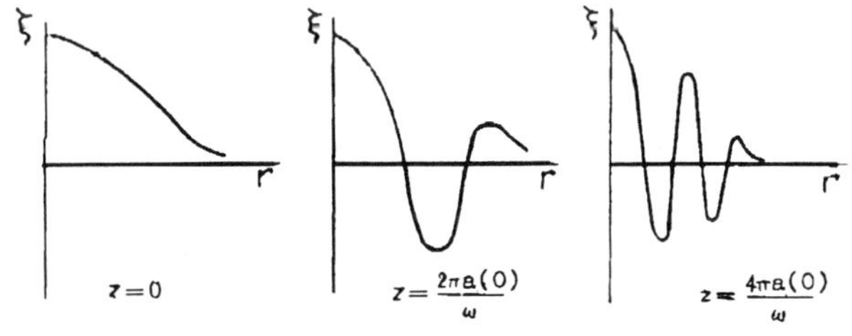

$$\Delta r \sim R\,\frac{a(r)}{\omega z}$$

Fig. 3. The radial structure of perturbations with height in an oscillating flux tube with a smooth radial profile of plasma parameters.

$$\Delta r \sim \frac{a^2(r)}{z\omega a'(r)} \sim \frac{a.R}{\omega z}$$

and diminishes inversely with z. Respectively, the dissipative processes, for instance, viscosity, become more and more important at larger z. At the same time, it is obvious that the energy flux no longer depends on z exponentially.

To obtain a real dissipation one should include the dissipative terms into Eqn (4). Namely, for viscous losses on the right of Eqn (4) we have [Landau & Lifshitz, 1986]:

$$\nu \frac{1}{r^2}\frac{\partial}{\partial r}\left(r^3 \frac{\partial}{\partial r}\frac{\xi_\phi}{r}\right)$$

where ν is the coefficient of kinematic viscosity.

Let us assume that for oscillations with a smooth radial profile (with a scale of the order of the flux tube radius) which initially are generated by convective motions, dissipation is slow, $\omega >> \nu/R^2$. In this case the oscillations will propagate almost without damping up to the heights where the inverse damping rate $\Delta r^2/\nu$ (here $\Delta r \sim aR/\omega z$ is the scale-length of perturbation) becomes equal to the propagation time $z/a(r)$. The corresponding height can be estimated then as

$$z = z_* \sim \frac{a(r)}{\omega}\left(\frac{\omega R^2}{\nu}\right)^{1/3}. \qquad (9)$$

For the volume density of the power Q released by viscous dissipation we have the following estimate

$$\sim \rho\nu \frac{\omega^2\xi^2}{(\Delta r)^2}.$$

As Δr diminshes with height, Q is growing with z. At the altitude given by Eqn (9), Q reaches a maximum and then rapidly decreases (see Fig. 4). Thus, the heating power has a very characteristic shape with a quite pronounced maximum at the altitude z. For comparison, the dotted line in Fig. 4 shows Q for the usual exponentially damping wave.

Thus, we can conclude that even in the case of a smooth radial profile the flux tube oscillations can transfer energy from underlying surface to upper layers of atmosphere.

Let us consider now the bending oscillations with m=1. In a low frequency $\omega << a/R$ the linearised MHD equations can be reduced to the following equation for the r-component of the tube's displacement vector:

$$\frac{\partial^2}{\partial z^2}\left(\frac{B^2}{4\pi}\xi_r + \frac{\partial}{\partial r}\frac{B^2}{4\pi} r \frac{\partial}{\partial r} r\xi_r\right)$$
$$+ \omega^2\left(\rho\xi_r + \frac{\partial}{\partial r}\rho r \frac{\partial}{\partial r} r\xi_r\right) = 0. \qquad (10)$$

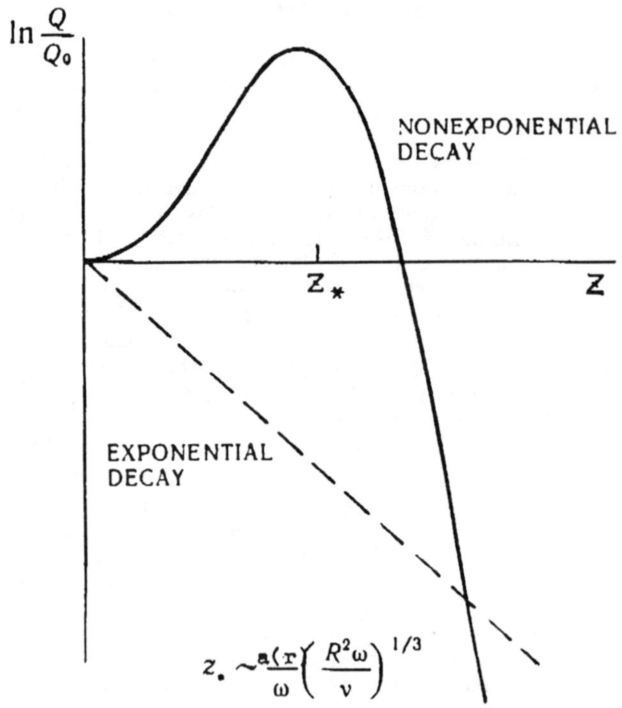

Fig. 4. The volume density of power released by viscous dissipation

The Laplace transform for this equation has the form:

$$\frac{\partial}{\partial r}\rho\left(\omega^2 + p^2a^2\right) r \frac{\partial}{\partial r} r\xi_{rp} + \rho(\omega^2 + p^2a^2)\xi_{rp}$$
$$= - F(0) - pF(0) \qquad (11)$$
$$F(r) = \frac{B^2}{4\pi}\xi_r + \frac{\partial}{\partial r}\frac{B^2}{4\pi} r \frac{\partial}{\partial r} r\xi_r$$

where ξ_{rp} is the Laplace transform for the displacement $\xi_r(z)$. The exact solution of the boundary-value problem in this case is much more difficult than in the previous one. However, some general conclusions on the asymptotic behaviour of the solution as $z \to \infty$ can be obtained in a simpler way. The main contribution to the asymptotic solution comes from those singularities of $\xi_{rp}(p)$ which are nearest the imaginary axis in the complex plane. In other words, we should consider the solution of Eqn (11) in the limit of $Re(p) = \varepsilon \to 0$. In this limit the solution has a strongly oscillating shape near the point where the following condition is fulfilled:

$$a^2(r) = \frac{\omega^2}{(\mathrm{Im}\,p)^2}. \qquad (12)$$

the solution in this region has a form:

$$\xi_{rp} = D(r)\,\ln \frac{1}{r - r_0 + i\dfrac{\varepsilon}{\operatorname{Im} p}\dfrac{a(r)}{a'(r)}} \qquad (13)$$

where D(r) is a slowly changing function p, and r_0 is determined by Eqn (12). As to a(r) and a' = da/dr they are taken at the point $r = r_0$.

From Eqns (9) and (13) for $\xi_{\phi p}$ we have:

$$\xi_{\phi p} \sim \frac{1}{r - r_0 + i\dfrac{\varepsilon}{\operatorname{Im} p}\dfrac{a(r)}{a'(r)}}\,. \qquad (14)$$

It is obvious that the function $\xi_{\phi p}$ has poles near the points p = ± a/ω, i.e. the solution in this case is similar to that for torsional oscillations. The same holds for the asymptotic behaviour of the solution. And the conclusion that even for the smooth radial profiles of flux tube parameters the oscillations (now bending ones) still remain a plausible agent for energy transfer from lower to upper layers of solar atmosphere is valid.

3. Longitudinal resonances in bending oscillations

In this section we consider the problem connected with the situation when a flux tube is "pinned" at some altitude. Then, at this altitude, almost complete reflection of the wave can occur, which means that a resonance structure is formed and standing waves can be excited. In this sense the problem under consideration is similar to that for coronal loops (see References mentioned above) where both ends of the tube are pinned to the underlying surface which provides the formation of a resonance structure. Here we show that such structures can be formed also in the case of open field lines. Below we give two examples of such resonant structures. The first example is the structure which is formed in the case of a loss of radial equilibrium at some height (when the plasma temperature outside the tube is less than that inside). The second example is that of a sharp discontinuity in the vertical temperature distribution. In both cases we have an open resonator with a very small energy leakage.

We consider the flux tube with a stepwise radial profile of density, magnetic field etc. As was mentioned before, in this case weakly damped bending oscillations can propagate along the flux tube (note, that the effects described in the previous section are absent). But now the flux tube's parameters as well as those of its environment depend on z. For the z-dependent case the equation for bending oscillations has the form

$$4\pi R^2(\rho_i + \rho_e)\frac{\partial^2 \xi}{\partial t^2} = \frac{\partial}{\partial z}\left(R^2 B^2 \frac{\partial \xi}{\partial z}\right). \qquad (15)$$

The condition of magnetic flux conservation

$$R^2(z)B(z) = \text{const} \qquad (16)$$

enables us to write down Eqn (14) in the form:

$$4\pi\frac{\rho_i + \rho_e}{B}\frac{\partial^2 \xi}{\partial t^2} = \frac{\partial}{\partial z} B \frac{\partial \xi}{\partial z}. \qquad (17)$$

Let us consider the action of the first mechanism (the loss of radial equilibrium (see, for example, Spruit, 1981)). For simplicity we choose an isothermal atmosphere. Then the z-dependence of plasma parameters obeys the simple barometric law with the inverse pressure scale-height

$$\kappa_{i,e} = \frac{gM}{2T_{i,e}}\,. \qquad (18)$$

Here M is the mass of proton, g is the acceleration of gravity and $T_{i,e}$ is plasma temperature (inside "i" and outside "e" the flux tube).

At $\kappa_e > \kappa_i$, i.e., in the case when the pressure scale- height outside the tube is less than that inside ($T_e < T_i$), at some finite altitude the flux tube rapidly expands and its mass per unit length becomes very large. It is easy to show that at the height

$$z = z_* = \frac{1}{\kappa_e - \kappa_i}\ln\frac{p_{i0} + \dfrac{B_0^2}{8\pi}}{p_{e0}} \qquad (19)$$

(approaching this point from below) the magnetic field tends to zero and according to condition (16), R → ∞. Of course, this assertion is formal, since at the point $z=z_*$ the equilibrium of the flux tube is no more one- dimensional, but the qualitative conclusion that in this region the flux tube rapidly expands, or its inertia becomes infinitely large, remains valid. Respectively, the amplitude of the oscillations becomes zero in this region and Eqn (15) should be solved with the boundary condition $\xi(z_*) = 0$. This is a typical Sturm-Liouville problem and the corresponding solution has a discrete frequency spectrum. This means that resonances are present in the system, and if the spectrum of the exciting force is sufficiently narrow the amplitude of flux tube oscillations can increase at several points along the tube (at the gulf points).

Let us consider now another mechanism for the formation of a one-dimensional resonant structure which is connected with the temperature jump:

$$T = \begin{cases} T_1 & 0 < z < L \\ T_2 & z > L \end{cases} \qquad (20)$$

(now we take the temperatures inside and outside the flux tube to be equal). In the region 0 < z < L Eqn (15) has a form

$$\xi'' - \kappa\xi' + q^2\xi = 0 \qquad (21)$$

where κ is defined by Eqn (18) and

$$q^2 = 4\pi\omega^2 \frac{\rho_i(0) + \rho_e(0)}{B^2(0)} .$$

The solution of Eqn (21) with a given amplitude of oscillation at the base of the flux tube

$$\xi|_{z=0} = \xi_0$$

has a form

$$\xi = e^{\frac{\kappa z}{2}} \left\{ \xi_0 \cos z\sqrt{q^2 - \frac{\kappa^2}{4}} + A \sin z\sqrt{q^2 - \frac{\kappa^2}{4}} \right\}. \tag{22}$$

In the region z > L Eqn (15) is

$$\xi'' - \kappa \frac{T_1}{T_2}\xi' + q^2 \frac{T_1}{T_2}\xi = 0.$$

Its solution is

$$\xi = B\, e^{\frac{\alpha\kappa z}{2}} \exp\left\{ i\sqrt{q^2\alpha - \frac{\kappa^2\alpha^2}{4}} \right\}$$

$$\alpha = T_1/T_2.$$

Arbitrary constants A and B are determined by the boundary conditions at the point z = h:

$$\{\xi\}_{z=L} = 0; \qquad \{\xi'\}_{z=L} = 0.$$

Then simple algebra gives

$$A = -\xi_0 \frac{\frac{\kappa L}{2}\cos x - x\sin x - \left(\alpha\frac{\kappa L}{2} + i\sqrt{\alpha q^2 L^2 - \alpha^2 \frac{\kappa^2 L^2}{4}} \right)}{\frac{\kappa L}{2}\sin x + x\cos x - \left(\alpha\frac{\kappa L}{2} + i\sqrt{\alpha q^2 L^2 - \alpha^2 \frac{\kappa^2 L^2}{4}} \right)} \tag{23}$$

where

$$x = L\sqrt{q^2 - \frac{\kappa^2}{4}} . \tag{24}$$

At $\alpha << 1$ there appear narrow resonances in the system: the imaginary part in the denominator is small and when the real part of denominator tends to zero the amplitude of oscillation becomes infinitely large. Let us consider this case in more detail. To simplify the analysis we can neglect the terms of the order of $(T_1/T_2)^{1/2}$ (and higher) in the numerator and the terms of the order of T_1/T_2 in the denominator. In this approximation we have

$$A = -\xi_0 \frac{\frac{\kappa L}{2}\cos x - x\sin x}{\frac{\kappa L}{2}\sin x + x\cos x - i\sin x\, qL\sqrt{\alpha}} . \tag{25}$$

The resonance is determined by the equation

$$\mathrm{tg}x = -\frac{2x}{\kappa L} . \tag{26}$$

Respectively, for the set of resonance frequencies we have

$$\omega_n^2 = \frac{B^2(0)}{4\pi[\rho_i(0) + \rho_e(0)]} \sqrt{\frac{x_n^2}{L^2} + \frac{\kappa^2}{4}} \tag{27}$$

where x_n are the roots of Eqn (26), n = 1,2,... .

Near the resonance frequency ω_n, that is at $|\omega-\omega_n| << 0$, one can replace the numerator in (25) by its value at the resonance point and expand the denominator over $(\omega-\omega_n)$. This gives

$$A_n = -\xi_0 \frac{\frac{\kappa L}{2} - \frac{2}{\kappa L}x_n^2}{\left(1 + \frac{\kappa L}{2} + \frac{2x_n^2}{\kappa L}\right)\sqrt{1 + \frac{\kappa^2 L^2}{4x_n^2}}} \times$$

$$\frac{\omega_n}{\omega - \omega_n + \frac{i\nu_n}{2}} \tag{28}$$

$$\nu_n = 2\omega_n\sqrt{\alpha}\,\frac{2x_n^2}{\kappa L}$$

$$\frac{1}{\left(1 + \frac{\kappa L}{2} + \frac{2x_n^2}{\kappa L}\right)\sqrt{1 + \frac{\kappa^2 L^2}{4x_n^2}}} .$$

As an example, let us consider the first resonance n=1 which lies in the interval $\pi/2 < x_1 < \pi$. Let us assume that $\kappa L \sim 1$. Then, $x_1 = 1.9$ and

$$\frac{\nu_1}{\omega_1} \approx 1.4\sqrt{\alpha}; \quad A_1 \approx 5\xi_0\left(\omega - \omega_1 + \frac{i\nu_1}{2}\right).$$

The oscillation amplitude can be found from Eqn (22). Near the resonance we can neglect the first term, thus obtaining

$$\xi(L) = A\, e^{\frac{\kappa L}{2}} \sin x_1 .$$

For $\kappa L \approx 1$

$$\xi(L) = 6\xi_0 \frac{\omega_1}{\omega - \omega_1 + 0.7 i \omega_1 \sqrt{T_1/T_2}} .$$

If, for example $T_2 = 10T_1$, the width of the resonance is 20% of the frequency.

We have considered above harmonic waves excited at the base of flux tube. Actually, in the excitation region the flux tube is subject to random displacements produced by the convective motions. So, more adequate is the description of these displacements in terms of the spectral density $(\xi_0^2)_\omega$. The response of flux tube can be presented in the form

$$(\xi_n^2)_\omega = \frac{|F|^2 (\xi_0^2)_\omega}{(\omega-\omega_n)^2 + \frac{\nu_n^2}{4}} \qquad (29)$$

where $(\xi_n^2)_\omega$ is the spectral density of flux tube oscillations averaged over the "length" of the tube, ω_n and ν_n are as before the n-th eigenfrequency and damping rate and F has a meaning of a formfactor which is of the order of unity and depends on the particular longitudinal structure of the eigenfunction. For narrow resonances ($\nu_n << \omega_n$) one can replace $(\xi_0^2)_\omega$ by the constant $(\xi_0^2)_{\omega_n}$ and after integration over frequencies we get the square average of flux tube displacement

$$\overline{\xi_r^2} - \int (\xi_r^2)_\omega d\omega = \frac{2|F|^2 (\xi_0^2)_{\omega_n}}{\nu_n} . \qquad (30)$$

Thus, the displacement of a flux tube at some height can be much larger than at its footpoint. This means that in this region nonlinear dissipation can occur, by formation, for example, of shock waves in the external plasma. So, if somewhere the resonance structure is formed one can expect there quite a large oscillation level and with some probability the amplitude of oscillations can be substantially larger than its time- averaged value. The probability distribution $P(\xi)$ depends on the properties of the random process $\xi_0(t)$. Note, that we do not assume anything as regards the properties of the spectral density of exciting oscillations, and the results obtained are therefore quite general. For example, for a Gaussian process the probability distribution has a familiar exponential form

$$p(\xi) \sim \exp\left(-\xi^2 / 2\overline{\xi^2}\right)$$

which gives a considerable probability for fluctuations during which ξ^2 is 4-6 times larger than its average value $\overline{\xi^2}$ (and $\overline{\xi^2}$ itself is already quite large, see Eqn (30)):

$$\xi^2 \approx (4 - 6)\overline{\xi^2} .$$

In observational data such events could manifest themselves as a temporal brightening of a flux tube region.

Acknowledgement

MPR is grateful to Prof Eric Priest and the American Geophysical Union for invitation, financial support and great efforts to help her attend this meeting.

References

Chen L and Hasegawa A, A theory of long-period magnetic pulsations, *J. Geophys. Res.* 79, 1033, 1974.

Defouw R J, Wave propagation along a magnetic tube, *Ap. J.* 209, 266-269, 1976.

Gordon B E and Hollweg J V, Collisional damping of surface waves in the solar corona, *Ap. J.* 266, 373, 1983.

Heyvaerts J and Priest E R, Coronal heating by phase- mixed waves, *Astr. Ap.* 117, 220-234, 1983.

Ionson J A, Resonant absorption of Alfvénic surface waves and the heating of solar coronal loops, *Ap. J.* 226, 650, 1978.

Ionson J A, Resonant electrodynamic heating of stellar coronal loops, *Ap. J.* 254, 318-334, 1982.

Landau L and Lifshiz E, *Hydrodynamics*, Moscow, Nauka, 1986.

Rae I C and Roberts B, Surface waves and the heating of the corona, *Geophys. Ap. Fluid Dyn.* 18, 197-226, 1981.

Ryutova M P, Anomalous damping of oscillations of a magnetic filament, Proc. of XIII Int. Conf. on Phenomena in Ionized Gases, p 859-860, 1977.

Ryutov D D and Ryutova M P, Sound oscillations in a plasma with magnetic filaments, *Sov. Phys. JETP*, 43, 491-497, 1976.

Sakurai T and Granik A, Generation of coronal electric currents due to convective motions on the photosphere II, *Ap. J.* 177, 404-414, 1984.

Tataronis J A and Grossmann W, Decay of MHD waves by phase mixing, *Z. Phys.* 261, 203-216, 1970.

Timofeev A B, Oscillations in the sheared flows of plasmas and fluids, *Usp. Fiz. Nauk.* 102, 185-210, 1970.

Wentzel D G, Hydromagnetic surface waves, *Ap. J.* 227, 319-322, 1979.

SHOCK WAVES IN THE THIN FLUX TUBE APPROXIMATION[1]

A. Ferriz-Mas

Kiepenheuer-Institut für Sonnenphysik, Schöneckstr. 6,
D-7800 Freiburg, Federal Republic of Germany

Abstract. A discussion of the properties of magnetohydrodynamic (MHD) shock waves in the framework of the thin flux tube approximation (to zeroth order in the expansion) is presented. We consider several simplifying assumptions which make an analytical study possible: Radiative energy exchange is neglected; the equation of state is that of a polytropic ideal gas (i.e., with constant ratio of specific heats), so that ionization effects are ignored; further, the backreaction of the shock on the ambient medium is ignored, so that the theory can be accurate only for not too strong shocks. Despite the limitations introduced by these assumptions, an analytical study provides useful insight into the physics of the problem, thus complementing existing and future numerical investigations under more realistic conditions.

We show that the properties of shock waves confined to flux tubes, as described by the thin flux tube equations, exhibit many analogies with those of slow MHD shocks in extended media. That could be expected on physical grounds: the basic compressive tube wave in the framework of the (zeroth-order) thin flux tube approximation is a slow mode with phase speed given by the *cusp speed* $c_T = c_S\, v_A/(c_S^2 + v_A^2)^{1/2}$ (with c_S and v_A the sound and Alfvén speeds, respectively), and it is precisely the shock resulting from the nonlinear evolution and breaking of this mode that is the subject of our investigation. The analogies and differences with purely hydrodynamic shocks are also pointed out. In particular, it can be shown that the sub- or supercritical character of the flow velocity with respect to the sound, Alfvén and cusp speeds is derivable from thermodynamic considerations only, as for HD shocks, in contrast to general MHD shocks, for which the evolutionary conditions have to be applied.

The theory of shock waves in thin flux tubes is not only of interest in connection with concentrated magnetic structures in the stellar atmospheres. Its understanding is conceptually important from both physical and mathematical point of view since the flux tube provides one of the simplest forms of equations governing the dynamics of a magnetized plasma confined by an external pressure and subject to a permanent constraint (internal gas pressure variations are related to internal magnetic field variations).

1. Introduction

The topic of shock waves in magnetic flux tubes has been a subject of research for some years now. These investigations have been motivated, in part, by the necessity of finding a theoretical explanation for the high temperatures (clearly exceeding the effective temperature) of the upper atmosphere of the Sun and, in general, of all late-type stars. The existence of intense photospheric flux tubes, located in the boundaries of the supergranulation cells, leads to the question of to what extent these flux tubes might provide a channel for carrying mechanical energy from the upper layers of the convection zone to the chromosphere. In particular, magnetohydrodynamic (MHD) shock wave heating resulting from the nonlinear evolution of longitudinal flux tube waves constitutes one of the most promising candidates as a heating mechanism for the chromosphere [e.g. Herbold *et al.*, 1985; Ulmschneider *et al.*, 1987; for a recent review of chromospheric heating theories see Ulmschneider, 1987; Hammer, 1987]. Also, shocks in isolated magnetic flux tubes are believed to play an important role in the so-called syphon flows in connection with the Evershed flow in sunspots [Meyer and Schmidt, 1968; Thomas, 1988; Montesinos and Thomas, 1989; Thomas and Montesinos, 1989]. In addition to numerical simulations, an analytical investigation of the properties of shocks in flux tubes may be of interest in the context of shock wave theory and in the study of nonlinear phenomena in general [Ferriz-Mas and Moreno-Insertis, 1987 (thereafter Ref. 5); Ferriz-Mas, 1988]. The magnetic flux tube provides us with one of the simplest configurations illustrating the physics of a magnetized plasma subject to a permanent constraint: internal pressure variations are related to internal magnetic field variations in such a way that the total pressure within the tube is equal to the confining gas pressure.

In this contribution the transition solution of the physical state across a slow-mode-like magnetohydrodynamic shock of finite strength in a compressible, dissipationless plasma confined to a magnetic flux tube is investigated in the framework

[1]Mitteilungen aus dem Kiepenheuer-Institut Nr. 312

Geophysical Monograph 58

of the "thin flux tube approximation" for magnetic flux concentrations with axial symmetry [Defouw, 1976; Roberts and Webb, 1978, 1979; Ferriz-Mas *et al.*, 1989]. This approximation, consisting essentially of a Taylor series expansion approach in the (scaled) radial coordinate r, is based on the assumption that the radius of the magnetic tube is small compared to all scales of variation along the tube and that the physical quantities have a weak variation over any cross-section of the tube, with the exception of a jump at the boundary. All quantities are expanded in r:

$$\begin{aligned} p(r,z,t) &= p_0(z,t) + O(r^2) \\ \rho(r,z,t) &= \rho_0(z,t) + O(r^2) \\ B_z(r,z,t) &= B_{z0}(z,t) + O(r^2) \\ v_z(r,z,t) &= v_{z0}(z,t) + O(r^2) \\ B_r(r,z,t) &= rB_{r1}(z,t) + O(r^3) \\ v_r(r,z,t) &= rv_{r1}(z,t) + O(r^3), \end{aligned}$$

where p is pressure, ρ is density, and $\mathbf{B} = (B_r, 0, B_z)$ and $\mathbf{v} = (v_r, 0, v_z)$ are magnetic field and flow velocity, respectively; z is the coordinate along the axis of symmetry.
These expansions are introduced in the system of ideal MHD equations written in cylindrical coordinates (r, ϕ, z) with no ϕ−dependence; by retaining only the zeroth- and first-order terms in the set of expanded equations, the following system of nonlinear hyperbolic partial differential equations is obtained [Defouw, 1976; Roberts and Webb, 1978]:

$$\begin{aligned} &\frac{\partial}{\partial t}\left(\frac{\rho}{B}\right) + \frac{\partial}{\partial z}\left(v\frac{\rho}{B}\right) = 0, \\ &\frac{\partial v}{\partial t} + v\frac{\partial v}{\partial z} + \frac{1}{\rho}\frac{\partial p}{\partial z} + g = 0, \qquad (1) \\ &\frac{\partial p}{\partial t} + v\frac{\partial p}{\partial z} = \frac{\gamma p}{\rho}\left(\frac{\partial \rho}{\partial t} + v\frac{\partial \rho}{\partial z}\right); \end{aligned}$$

this system can be closed on application of the condition of instantaneous lateral pressure balance

$$\frac{B^2}{8\pi} + p = p_e, \qquad (2)$$

which constitutes an integral part of the thin flux tube approximation. Here p_e denotes the external pressure at the boundary of the tube. We write $p = p_0$, $\rho = \rho_0$, $B = B_{z0}$ and $v = v_{z0}$ for easiness of notation, but it should be remembered that the quantities appearing in the governing equations (1) and (2) represent values on the axis of symmetry.

2. Shock equations

As usual in the MHD theory we shall assume that on a sufficiently large length-scale a shock wave appears to be a thin transition layer (shock front) across which the fluid variables undergo a jump.

We shall employ the convention that the plasma crosses the shock from the *front side* to the *back side*; the subscript 1 refers to the region in front of the shock (*upstream*) and the subscript 2 to the region behind the shock (*downstream*). In a frame of reference comoving with the shock front the set of ideal MHD equations for a thin vertical flux tube, Eqs. (1) and (2), yields under steady conditions the following conservation equations of mass, momentum and energy across the shock front [Herbold *et al.*, 1985]:

$$\begin{aligned} &\rho_1 v_1 A_1 = \rho_2 v_2 A_2, &(3.a) \\ &A_1(\rho_1 v_1^2 - 2p_e + 2p_1) = A_2(\rho_2 v_2^2 - 2p_e + 2p_2), &(3.b) \\ &\frac{1}{2}v_1^2 + h_1 = \frac{1}{2}v_2^2 + h_2. &(3.c) \end{aligned}$$

Here v is the flow speed of the plasma relative to the shock front and A denotes the cross-sectional area of the tube, related to the zeroth-order longitudinal magnetic field through the condition of magnetic flux conservation:

$$\Phi = A_1 B_1 = A_2 B_2 = \text{const.}$$

The specific enthalpy h may be considered as a given function of p and ρ while A and B are uniquely determined by specifying p and the constants p_e and Φ. The shock relations (3) thus form a system of three nonlinear algebraic equations for the six quantities p_1, ρ_1, v_1, p_2, ρ_2, and v_2. The knowledge of the preshocked state $\{p_1, \rho_1, v_1\}$ together with p_e and Φ completely determines the shock transition [Ref. 5]. The question of the complete determinacy of the shock transition can be extended to the case of an arbitrary frame of reference with respect to which the shock propagates with speed U and the plasma flows with speeds u_1 (front side) and u_2 (back side) such that $u_1 = v_1 + U$ and $u_2 = v_2 + U$. To simplify the analysis we shall consider throughout a reference frame fixed to the shock.

3. The entropy constraint

The symmetry of the jump equations (3) with respect to the variables on both sides of the front is broken when the Second Principle of Thermodynamics is applied: since the entropy production rate in the shock layer must be non negative, we have a further condition: $s_2 \geq s_1$, where s is the specific entropy. This constraint is satisfied if and only if

$$p_2 \geq p_1, \qquad (4)$$

which is equivalent to the condition $\rho_2 \geq \rho_1$ (i.e., the shock wave is compressive). Although the proof of the compresiveness given in [Ref. 5] strictly applies to an ideal gas with constant ratio of specific heats, this result may be proved quite generally for weak shocks [Ferriz-Mas and Moreno-Insertis, in preparation, 1989]. From the inequality (4) the following consequences are obtained:

$$\begin{aligned} &\rho_2 \geq \rho_1, \quad T_2 \geq T_1, \quad c_{S2} \geq c_{S1}, \\ &h_2 \geq h_1, \quad \epsilon_2 \geq \epsilon_1, \quad A_2 \geq A_1, \qquad (5) \\ &B_2 \leq B_1, \quad v_{A2} \leq v_{A1}, \quad |v_2| \leq |v_1|, \end{aligned}$$

where T is the temperature, ϵ is the specific internal energy, c_S is the sound speed and v_A the Alfvén speed. In particular, the result $|v_1| > |v_2|$ corresponds to the general property of all hydrodynamic (HD) and MHD shocks that the effect of the shock is to slow down the flow in the normal direction. In the present case this is accompanied by a weakening of the magnetic field, a feature common to slow MHD shocks. Also, the shock relations (3)-(4) establish that the relative flow speed on the front side is greater than the basic speed c_{T1}, while the flow speed is less than the basic speed c_{T2} behind the shock (the cusp speed $c_T = c_S v_A/(c_S^2 + v_A^2)^{1/2}$ is the basic or characteristic velocity in the thin flux tube context). The physical meaning of this is that information cannot be transmitted ahead of the shock front. This is an extension of the concept of the supersonic character of the flow in front of a hydrodynamic shock and the corresponding subsonic character on the back side. From the theory of MHD shocks we know that the normal upstream (downstream) flow speed relative to the front shock must exceed (must be less than) the characteristic speed on the front side (back side). Nevertheless, notice that in the present case the sub- or supercritical character of the relative upstream and downstream velocities with respect to c_T can be ascertained with only the help of the entropy condition, as for HD shocks, in contrast to general MHD shocks, for which the application of the evolutionary conditions is also required. Moreover, the flow velocity of the unshocked plasma relative to the front shock also has an upper bound: $c_{T1} < |v_1| < v_{A1}$; this double limitation on $|v_1|$, which has no counterpart in HD shocks, is a feature of slow MHD shocks. The above results have been rigorously established in [Ref. 5].

4. Thermodynamics of shocks of arbitrary strength

A useful intermediate variable arising in the theory is

$$< p > \equiv p_e - (p_{m1}\, p_{m2})^{1/2} = p_e[1-(1-\theta_1)^{1/2}(1-\theta_2)^{1/2}], \quad (6)$$

where $p_m \equiv B^2/8\pi$ is the magnetic pressure and $\theta \equiv p/p_e$ is related to the plasma beta through $\beta \equiv 8\pi p/B^2 = \theta/(1-\theta)$. The advantage of introducing $< p >$ is that the thermodynamic relations obtained on eliminating v_1 and v_2 from the jump equations (3) are equally valid for shocks in thin magnetic tubes as well as for HD shocks. Going to the limiting case $\beta \to 0$ (or $\theta \to 0$), i.e., the rigid tube limit, which is formally identical to the one-dimensional HD case, we obtain

$$< p > = p_e[\frac{p_1 + p_2}{2\,p_e} + O(\theta^2)] = < p >_{HD} [1 + O(\theta)], \quad (7)$$

where $< p >_{HD} \equiv (p_1 + p_2)/2$. In terms of $< p >$ the jump in specific internal energy can be written:

$$\epsilon_2 - \epsilon_1 = < p > \left(\frac{1}{\rho_1} - \frac{1}{\rho_2}\right), \quad (8)$$

which expresses in a unified way the results for the magnetic tube and for the HD shocks [cf. Eq. (55.04) in Courant and Friedrichs, 1976]. The interpretation of Eq. (8), for instance, is that the jump in specific internal energy across the shock equals the work done by the mean gas pressure $< p >$ in performing the compression.

For a polytropic ideal gas ($\gamma \equiv$ const.) the specific internal energy is given by $\epsilon = p/[(\gamma-1)\rho]$. On substituting this expression into Eq.(8) we obtain

$$\chi_\rho = G(\chi_p) = \frac{\theta_1\chi_p + r(\theta_1, \theta_1\chi_p)}{\theta_1 + r(\theta_1, \theta_1\chi_p)}, \quad (9)$$

where we employ the notation $\chi_Q \equiv Q_2/Q_1$ and $r(\theta_1, \theta_2) \equiv (\gamma - 1) < p > /p_e$. The function G characterizes a one-parameter family of curves in the (χ_p, χ_ρ)-plane when θ_1 varies in $[0, 1)$. The study of the function $\chi_\rho = G(\chi_p)$ is essential for the understanding of the properties of shocks in thin magnetic tubes. The variable χ_p may vary in $[1, \theta_1^{-1})$, the upper bound resulting from the condition of lateral pressure balance; the curve for $\theta_1 = 0$ corresponds to a HD shock.

The ranges of variation of the different ratios χ's are found to be

$$\begin{aligned} &1 \le \chi_p < 1/\theta_1\,, \\ &1 \le \chi_A < \infty\,, \\ &1 \le \chi_\rho < \chi_\rho^{\max}\,, \\ &1 \le \chi_T < \chi_T^{\max} = \frac{\gamma - 1 + \theta_1}{\gamma\theta_1}\,. \end{aligned} \quad (10)$$

The fact that all ratios χ are ≥ 1 is a consequence of the compressive nature of shocks; the case $\chi = 1$ corresponds to the trivial solution (i.e., no shock). χ_A is a monotonic increasing function of χ_p given by

$$\chi_A^2 = 1 - \theta_1/(1 - \theta_1\chi_p); \quad (11)$$

the value of χ_A may increase arbitrarily (but see the discussion regarding the validity of the thin flux tube approximation and of the other assumptions involved; see also Sect. 5 in Ref. 5). The density jump has an upper bound $\chi_\rho^{\max}$, as in HD shocks — in which case $\chi_\rho^{max} = \gamma - 1/(\gamma+1)$ — and as in all types of MHD shocks.

Nevertheless, there appear some features with no parallel in the HD case: (1) Pressure and temperature cannot increase without bound, which is ultimately due to the constraint set by the confining external pressure. (2) Further, the ratio χ_ρ is a monotonically increasing function of the ratio χ_p for $1 \le \chi_p < \chi_p^*$; χ_ρ attains a maximum value $\chi_\rho^{\max}$ for $\chi_p = \chi_p^* < 1/\theta_1$. But for $\chi_p^* < \chi_p \le 1/\theta_1$ the ratio χ_ρ decreases monotonically from $\chi_\rho^{\max}$ down to $G(1/\theta) = \gamma/(\gamma - 1 + \theta_1)$. The meaning of the decreasing branch of the curve is that it is possible to have two different shocks corresponding to the same compression rate χ_ρ.

The value χ_p^* at which χ_ρ reaches its maximum is determined by $\chi_p^* = \chi_p(\chi_A^+)$, where χ_A^+ is the larger root of the polynomial in χ_A:

$$F(\chi_A) = -(\gamma-1)(1-\theta_1)(\chi_A^2+1) + 2(\gamma - 1 + \theta_1)\chi_A\,. \quad (12)$$

The two roots χ_A^+ and χ_A^- satisfy

$$\chi_A^+\chi_A^- = 1 \Rightarrow \chi_A^+ = \frac{1}{\chi_A^-}, \quad (13)$$

$$\chi_A^+ + \chi_A^- = \frac{2(\gamma - 1 + \theta_1)}{(\gamma-1)(1-\theta_1)} > 2\,. \quad (14)$$

According to (13) both roots have the same sign, and, from (14), they are both positive; further, (13) and (14) imply that the two roots must necesarily obey the inequality $0 < \chi_A^- < 1 < \chi_A^+$, so that only $\chi_A^+ > 1$ is physically acceptable. (Note that a sign error slipped into Eq. (48) of [Ref. 5], which gives the derivative of χ_ρ as a function of χ_A. The correct numerator of Eq. (48) in [Ref. 5], which defines $F(\chi_A)$, should read

$$-(\gamma-1)(1-\theta_1)^2(\chi_A^2+1)+2(\gamma-1+\theta_1)(1-\theta_1)\chi_A\,,$$

and coincides with $F(\chi_A)$ [Eq. (12)] up to a factor $(1-\theta_1)$. Thus, χ_A^- is not negative, as was stated in [Ref. 5], but positive and smaller than unity; nevertheless, it is still physically uninteresting, and therefore the rest of the paper remains unaffected by this error).

We have plotted the graph of $\chi_\rho = G(\chi_p)$ in Fig. 1 for $\theta_1 = 0.2$ and $\theta_1 = 0.3$. In hydrodynamics it is more usual to plot χ_p as a function of the ratio of specific volumes (*Hugoniot curve)*; this is done in Fig. 2, where χ_p is plotted as a function of χ_ρ^{-1} for $\theta_1 = 0.2$ and compared with the hydrodynamic curve, corresponding to $\theta_1 = 0$.

5. Critique and further remarks

The (zeroth-order) thin flux tube approximation is based on the assumption that the radius of the magnetic tube is small compared to all scales of variation along the tube and that the pertinent physical quantities have a weak variation over any cross-section of the tube, with the exception of a jump at the boundary to the medium outside. (In the zeroth-order approximation considered here, the quantities appearing in the governing MHD equations are represented by their values on the axis of symmetry; they can be as well understood as averages over the cross-section of the tube). These assumptions may no longer be valid in the vicinity of the shock region, where the gradients of p, ρ, $B = B_z$ and $v = v_z$ are very large, and hence the radial components B_r and v_r are also large (for example, B_r is related to the gradient of B_z through the condition div $\mathbf{B} = 0$; to first order in r this reads $B_{r1} = \partial B_{z0}/\partial z$). Nevertheless, for not too strong shocks it seems reasonable to assume that the steepening of the wave profile (i.e., the increase of the gradients of pressure, density, temperature and axial components of $\mathbf{B}$ and $\mathbf{v}$) is balanced by the dissipative effects in such a way that the resulting permanent regime profile (i.e., the shock wave) is reached before the radial components of $\mathbf{B}$ and $\mathbf{v}$ increase without bound.

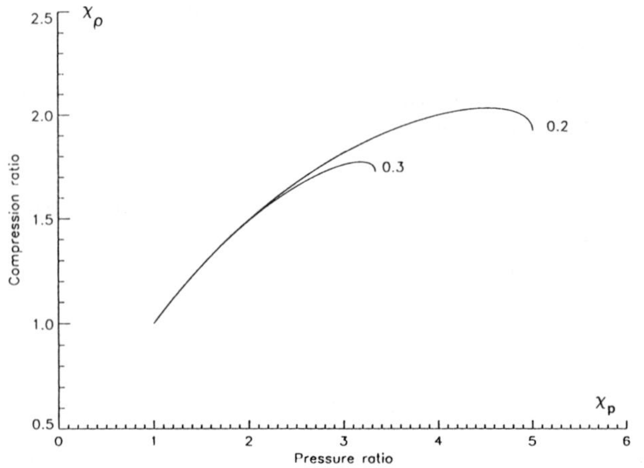

Figure 1. Density jump $\chi_\rho = \rho_2/\rho_1$ versus pressure jump $\chi_p = p_2/p_1$ (with $\gamma = 5/3$) for two different values of the initial pressure ratio $\theta_1 = p_1/p_e$ ($\theta_1 = 0.2$ and 0.3).

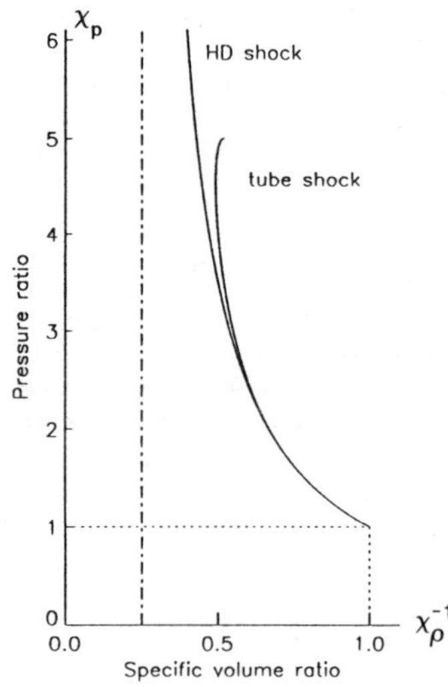

Figure 2. Hugoniot curve for a shock with $\theta_1 = 0.2$ compared to the corresponding curve for a hydrodynamic shock ($\theta_1 = 0$). The latter is a rectangular hyperbola with asymptotes $\rho_1/\rho_2 = (\gamma-1)/(\gamma+1)$ and $p_2/p_1 = -(\gamma-1)/(\gamma+1)$; only the upper part of the curve ($\rho_1/\rho_2 = p_2/p_1 = 1$) has any physical meaning. Here $\gamma = 5/3$.

Apart from the assumptions inherent to the thin flux tube approximation [see also Roberts and Webb, 1978, 1979], the above treatment contains other simplifying assumptions that have to be made in order to obtain analytical results, as, for instance, the neglect of radiative energy transfer and ionization effects. These shortcomings, which would limit the applications in Solar Physics, can be overcome by resorting to numerical techniques [Herbold *et al.,* 1985; Ulmschneider *et al.,* 1987]. A major flaw in the theory is the assumption of a constant-pressure boundary which prevents the transmission of mechanical energy to the surrounding fluid and thus neglects the modulation of motions in the flux tube due to the external wave field (but see [Roberts, 1985] for an example of weakly nonlinear motions in the flux tube assuming linearized perturbations in the surroundings). Indeed, it is reasonable to expect that the shock will cause perturbations in the external medium, which, in turn, could affect the shock strength and structure. For a complete solution to this nonlinear MHD problem it would be necessary to couple the set of equations for the inside of the flux tube with the hydrodynamic equations for the external fluid, but the full problem seems very difficult to treat analytically and, to my knowledge, it has not even been tackled numerically yet. In spite of the limitations of the above model, it is to be hoped that it gives some qualitative insight into what to expect under more realistic assumptions, at least for not too strong shoks.

We have obtained thermodynamic relations in terms of the variable $< p >$ which are valid both for shocks in thin flux tubes and for one-dimensional HD shocks, thus bringing out a formal parallelism between them. But shocks in thin flux tubes have a number of features without a counterpart in the HD case. Although the MHD equations for the thin flux tube approximation are essentially one-dimensional, they implicitly comprise the transverse or radial components of $\mathbf{v}$ and $\mathbf{B}$ (to first order in r) since the cross-sectional area has to vary in response to internal gas pressure variations. Indeed, the longitudinal mode consists of periodic lateral expansions and contractions of the tube. If B_r would vanish, the condition $div\,\mathbf{B} = 0$ would be in contradiction with the possiblity that the zeroth-order axial magnetic field can vary with z. The component v_r can be calculated (to first order) in terms of the density and the longitudinal velocity [Roberts and Webb, 1978, 1979]. Therefore, $\mathbf{B}$ and $\mathbf{v}$ have both radial and axial components. This is equivalent to stating that the effective compressibility of the plasma in the problem under consideration is made up of the gas compressibility plus the *magnetic distensibility* [e.g., Roberts, 1985; Herbold *et al.,* 1985; Ferriz-Mas, 1988], thus reducing the characteristic speed of wave propagation (as compared with the HD case) down to c_T. When the limit $\beta \to 0$ (or $\theta \to 0$) is considered, the magnetic field lines defining the flux tube are rigid and we recover the HD case.

That the properties of shocks in thin magnetic tubes so closely resemble the corresponding ones for slow MHD shocks is not surprising, since the basic compressive tube wave (in the zeroth-order thin flux tube approximation) is a slow mode with phase speed given by c_T. It can be shown that in this zeroth-order limit the fast mode effectively vanishes [Ferriz-Mas *et al.,* 1989].

A final remark seems in order. Recall that actual MHD shocks are not aligned in general, but for MHD shocks in extended media (with the exception of perpendicular shocks) one may always in principle transform to a de Hoffmann-Teller reference frame by an appropriate Galilean transformation, so that the flow velocities are aligned with the magnetic field in the uniform medium on either side of the shock [e.g., Bazer and Ericson, 1959; Priest, 1982] (the inverse transformation can be very involved). This choice is not possible for shocks in magnetically structured media, for which translational invariance over the shock surface is no longer valid due to spatial inhomogeneities. Nevertheless, if we confine ourselves to the thin flux tube approximation to zeroth-order, it turns out that under steady state conditions ($\partial/\partial t \equiv 0$) the plasma velocity is parallel to the magnetic field on both sides of the shock front up to second order in the expansion parameter at least:

$$(\mathbf{v} \wedge \mathbf{B})_r = 0\,,$$
$$(\mathbf{v} \wedge \mathbf{B})_\phi = \frac{1}{2} r \frac{\partial B_{z0}}{\partial t} + O(r^3)\,, \qquad (15)$$
$$(\mathbf{v} \wedge \mathbf{B})_z = 0\,.$$

Thus the comparison of the properties of MHD shocks in thin flux tubes with general slow MHD shocks can be carried out by directly considering the properties of aligned MHD shocks in extended media, as summarized for instance in [Priest, 1982].

Questions and Answers

Thomas: (Comment) The shock jump conditions you have derived within the thin flux tube approximation are quite useful, and in fact, I use them myself in my calculations of siphon flows in thin flux tubes. However, I would like to emphasize the limitations of these relations which you alluded to at the end of your talk. In particular, because of the sharp increase in cross-sectional area across the shock, magnetic curvature forces must be important within the shock. The tube shock is probably highly two-dimensional in structure and the shock front is probably nonplanar. Also, there is no way of calculating the thickness of the tube shock within the thin flux tube approximation. Nevertheless, as long as the shock is relatively thin, the relations you have derived should be reasonably accurate.

Hundhausen: The solutions you have described may be an interesting limiting form for the "upstream-inclined" slow shock geometry suggested for some coronal mass ejections. Could the absence of any fast-shock like solutions be related to your restriction or finding that the post-shock area of the tube is always larger than the pre-shocked area? For a fast shock the field must increase and the area in a "one-dimensional" approximation such as this would have to decrease across the shock.

Ferriz Mas: It has been shown from the shock relations obtained for thin flux tubes and from the law of increase of entropy that the magnetic field B has to decrease across the shock. Now, since the magnetic field is coupled to the cross-sectional area of the tube through the condition of flux conservation ($B \cdot A = \phi_{mag}$), the area A has to increase. A decrease of the cross-sectional area of the tube following the passage of the shock would mean an increase in the magnetic field, and this would correspond to a fast-shock like solution. But these solutions have been excluded from the framework of the thin flux tube approximation used here: the basic compressive tube wave in this approximation is a slow mode. In this limit the fast mode effectively vanishes. The shocks I have just described are essentially slow-mode shocks modified by the presence of a lateral boundary and subject to the condition of lateral pressure balance.

Acknowledgments I would like to thank Dr. J.H. Thomas for comments on the manuscript.

I gratefully acknowledge a travel grant from the *American Geophysical Union*. The investigation reported here has been supported by the *Deutsche Forschungsgemeinschaft* through Grant No. 418 SPA-11.

References

Bazer, J., and W. B. Ericson, Hydrodynamic shocks, Astrophys. J., 129, 758-785, 1959.

Courant, R., and K. O. Friedrichs, "Supersonic Flow and Shock Waves", Springer-Verlag, Berlin, Heidelberg, New York, 1976.

Defouw, R. J., Wave propagation along a magnetic tube, Astrophys. J., 209, 266-269, 1976.

Ferriz-Mas, A., Nonlinear flows along magnetic flux tubes: Mathematical structure and exact simple wave solutions, Phys. Fluids, 31, 2583-2593, 1988.

Ferriz-Mas, A., and F. Moreno-Insertis, An analytical study of shock waves in thin magnetic flux tubes, Astron. Astrophys., 179, 268-276, 1987 [Reference 5].

Ferriz-Mas, A., M. Schüssler, and V. Anton, Dynamics of magnetic flux concentrations: The second order thin flux tube approximation, Astron. Astrophys., 210, 425-432, 1989.

Hammer, R., Waves and thermal instabilities in flux tubes: Their role for the structure of the chromosphere and transition region, in Proceedings of The Role of Fine-Scale Magnetic Fields on the Structure of the Solar Atmosphere, E.-H. Schröter, M. Vázquez and A. A. Wyller (eds.), Cambridge University Press, 255-273, 1987.

Herbold, G., P. Ulmschneider, H. C. Spruit, and R. Rosner, Propagation of nonlinear, radiatively damped longitudinal waves along magnetic flux tubes in the solar atmosphere, Astron. Astrophys., 145, 157-169, 1985.

Meyer, F., and H. U. Schmidt, Magnetisch ausgerichtete Strömungen zwischen Sonnenflecken, Zeitschrift für Angewandte Math. Mech., 48, 218-221, 1968.

Montesinos, B., and J. H. Thomas, Siphon flows in isolated magnetic flux tubes. II. Adiabatic flows, Astrophys. J., 337, 977-988, 1989.

Priest, E. R., "Solar Magnetohydrodynamics", Reidel, Dordrecht, 1982.

Roberts, B., Solitary waves in a magnetic flux tube, Phys. Fluids, 28, 3280-3286, 1985.

Roberts, B., and A. R. Webb, Vertical motions in an intense magnetic flux tube, Solar Phys., 56, 5-35, 1978.

Roberts, B., and A. R. Webb, Vertical motions in an intense magnetic flux tube. III. On the slender flux tube approximation, Solar Phys., 64, 77-92, 1979.

Thomas, J. H.: Siphon flows in isolated magnetic flux tubes, Astrophys. J., 333, 407-419, 1988.

Thomas, J. H., and B. Montesinos, The equilibrium path of an isolated magnetic flux tube containing a siphon flow, (these proceedings), 1989.

Ulmschneider, P.: The present state of heating theories of stellar chromospheres, in Solar and Stellar Activity, Proc. Symp. 11, COSPAR Conference, Toulouse, Adv. Space Res., 1987.

Ulmschneider, P., D. Muchmore, and D. Kalkofen, Acoustic tube waves in the solar atmosphere, Astron. Astrophys., 177, 292-302, 1987.

PROPERTIES AND MODELS OF PHOTOSPHERIC FLUX TUBES

B. Roberts

Department of Mathematical Sciences, University of St Andrews, St Andrews, Fife, Scotland

Abstract. The gross properties of isolated solar magnetic flux tubes, from the intense tube to the sunspot, are summarized, paying particular attention to how these properties determine the basic wave propagation speeds in the structures. An overview of the fundamental modes of oscillations is attempted, considering both thin and wide tubes, and a comparison with observations of oscillations is made. The potential seismological value of such information is stressed.

1. Introduction

The photosphere, that thin layer of the Sun's atmosphere that we see most readily, is the most dynamic of places. Nothing, it would seem, is at rest there! It is through this layer that magnetic field-lines, generated in the lower depths of the Sun (perhaps at the base of the convection zone), emerge like balloons buoyed up in a stratified atmosphere to eventually fill the tenuous coronal atmosphere. It is in the photospheric layers that we can measure magnetic field strengths. Higher in the atmosphere, the field is weaker and the gas is much thinner, so determination of field strength is difficult; lower in the atmosphere it is simply inaccessible to us (at least directly).

The one place that we can measure magnetic field strengths with reasonable accuracy is also the layer in the solar atmosphere that is so dynamic! The amazing thing is that the magnetic field at the photospheric level occurs for the most part in a strongly concentrated form.

The obvious instance of concentrated magnetic field is the sunspot (see Figure 1). The magnetic nature of sunspots has been known for over 80 years (Hale, 1908), yet we still do not understand fully this marvellous phenomenon. A more recent discovery is that, aside from sunspots, much of the remaining magnetic flux occurs not in a diffusive form but in intense flux tubes (see Spruit and Roberts (1983) and Stenflo (1989) for reviews).

These tubes are below the telescopic resolution (some 0.3" or 210km) of current earth-bound telescopes and consequently there is much interest in space missions that aim to resolve these structures. Though small, they are important because they provide a major connection channel linking the energetic but cool photosphere and upper convection zone to the tenuous but hot corona. They are presumably the messengers that provide the presently unknown heating mechanism for the coronal atmosphere. They are capable of transporting stresses and strains induced by slow movements in the photosphere, or the more rapid movements that generate waves.

The main sources of magnetic field in the photosphere, then, are the intense flux tubes (with diameters of order 10^2km) and the sunspots (with diameters of a few $\times$ 10^3km to a few $\times$ 10^4km). Are they connected? Are sunspots nothing more than a grouping of intense flux tubes clumped together in the visible layers but retaining their separate identities just below those layers (Parker, 1979a)?

The photospheric sources of magnetic field find themselves in a moving environment. For the top of the convection zone is the region where the temperature gradient is most strongly superadiabatic, and consequently convection is most pronounced there. It is there that granules and supergranules are seen. The supergranules are the larger of these two convection patterns, with a typical scale of 3×10^4km and flows of 0.5 km s^{-1} or less; they have lifetimes in excess of 2 days (see Wang and Zirin (1989) for a recent discussion). The flux tubes reside in the boundaries between supergranules where downdrafts occur. Granules (see Figure 1) occur near the upper layers of supergranules and have a scale of between 200-2000 km, centred on about 1000 km (see Title, Tarbell, and Topka (1987) and Title et al. (1989) for a recent discussion). There is no typical granule. Granules are moved both randomly and by advection from supergranules (and mesogranules), and have a lifetime of about 5 min. Granules are also strongly influence by magnetic fields: they are generally brighter in magnetic areas than in quiet ones; the advection of granules by steady flows is slower (by a factor of 2 or 3) in magnetic regions than quiet ones (where mean flow speeds are about 0.4 km s^{-1}). Root-mean-square flows within granules are of the order of 1.4 km s^{-1} in quiet regions, but suppressed (about 45% less) in magnetic regions. Granules are also commonly (at least in Title et al's SOUP data) seen to 'explode' (i.e. rapidly expand outwards, at 1-2 km s^{-1}), and such expansions in exploding granules affects other granules and may, indeed, terminate them. Finally, we note a recent study by Brandt et al. (1988) which showed that swirls (vortices) may develop in the granulation pattern; the one studied by Brandt et al. was 5,000 km across and persisted for 1.5 hours. How common such events are is not clear.

The granules and supergranules, then, are the 'sea' in which intense flux tubes are jostled about, moved along granular lanes and generally subject to the flows in this part of the solar atmosphere. Added to this, are sound waves (the

Geophysical Monograph 58

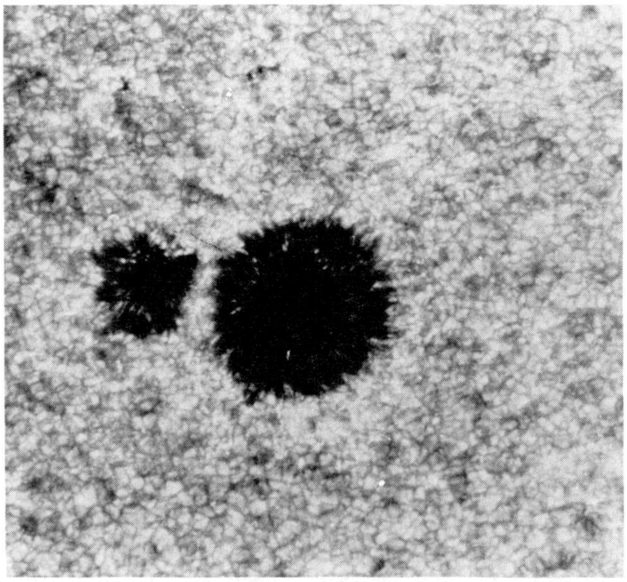

Fig. 1. Sunspots in the photosphere showing the umbral and penumbral structure and the granulation surrounding the spots. (Courtesy R. Muller, Pic du Midi Observatory)

so-called p-modes) which have their strongest amplitudes near the photosphere. The p-modes, too, interact with intense flux tubes. Thus, altogether, the photosphere is a very dynamic place indeed for flux tubes to reside and consequently we may expect the tubes themselves to support waves and be in motion as they respond to their energetic environments. Only if the tubes are sufficiently large, as with sunspots, can we expect them to locally dominate the convection and consequently cool the local photosphere.

Sunspots are large regions of strong magnetic field, the field-lines being broadly vertical as they pierce the solar surface but rapidly expanding outwards to fill the chromosphere and corona above. The inner core of a spot - the umbra - has a field strength of some 3kG and a temperature of 4000°K, two-thirds that of the undisturbed photosphere. Surrounding the umbra of a well-developed sunspot is a region of temperature 5000°K, where the magnetic field-lines are expanding rapidly to become more nearly horizontal. Field strengths in this region, the penumbra, are some 2kG at the inner umbral-penumbral boundary, declining to 1kG at the outer edge of the penumbra. A sunspot may be a monolithic tube or a grouping of smaller tubes, compressed tightly together in the visible layers of a spot but separated from one another below these layers (Parker, 1979a). See Figure 2. Which view provides the more accurate representation is not clear at present.

Just as with intense flux tubes, sunspots reside in a dynamic part of the solar atmosphere. However, they are sufficiently broad as to locally interrupt convective heat transport and thus cool their interiors. The activity associated with sunspots is to be found not in the photospheric layers of the spots but higher in the chromosphere and corona, where the presence of their strong fields (emerging from sunspots residing lower down in the photosphere) gives rise to enhanced heating in the so-called active regions. We thus encounter a paradox. In a sunspot the magnetic field is

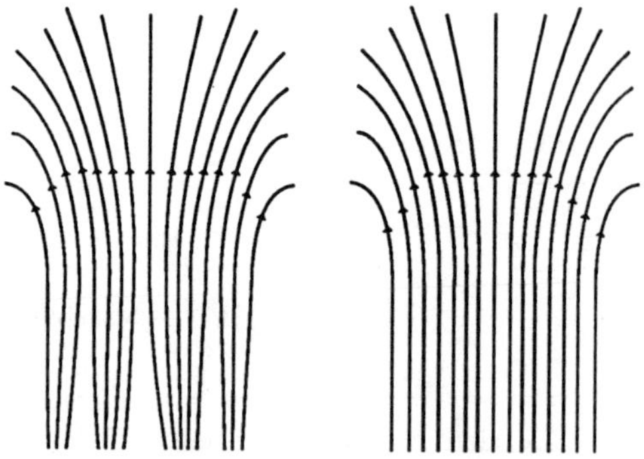

Fig. 2. Two possibilities for the structure of a sunspot, a uniform tube or an ensemble of distinct tubes. (After Parker, 1979a)

responsible for both the coolness of the photospheric layers of a spot and the enhanced heating in the higher atmosphere above the spot. Furthermore, the interior of a spot tends to be dynamically quiet compared with its non- magnetic surroundings. As J M Beckers and H Zirin expressed it, "the quiet Sun was so active, whereas the active Sun was so quiet" (Beckers, 1975). In fact, the "quietness" of a sunspot belies a dynamical nature!

What is so special about a flux tube, a concept going back at least to Faraday? It is commonly said that Nature abhors a vacuum. It would seem, too, that Nature abhors a uniform medium! This is perhaps a pity because so much of what we learned from textbooks written over a decade or so ago may not now be relevant, or at least only partially so. Indeed, so struck were many of us in the solar community by the examples of non-uniform plasmas - the photospheric flux tubes, the coronal loops - that we have worked hard ever since to develop mechanisms for producing a medium that is non-uniform. Indeed, so successful have we been in these efforts that many of us would be in trouble if anyone finds a uniform medium! It seems likely that a non-uniform plasma is the key to understanding so many important solar phenomena, including the enigmatic coronal heating. So, again, we ask what is so special about a flux tube?

A magnetic flux tube is any arbitrarily marked grouping of magnetic field lines intersecting a closed curve; the tube consists of all those field lines that pass through a cross-section S bounded by the closed curve. Since div $\underline{B} = 0$, the magnetic flux

$$\int_S \underline{B}.\underline{dS}$$

through the tube is zero: the flux coming in at one end of the tube leaves by the other end.

A flux tube has a length-scale, associated with it and this immediately marks it out from a uniform medium. In a uniform medium there is no natural length-scale nor natural time-scale. In a tube, we have both: the length-scale defined by the tube's diameter (say), and the time-scale by the time it would take a wave propagating with the Alfvén speed (say) to transverse the tube's cross- section. The occurrence of a length-scale has important implications for the nature of the waves that a tube can support, a topic we take up later in this review.

Flux tubes are building blocks for atmospheric magnetic field. They are the roots of the field that fills the solar corona. Similarly, it seems likely that they are the roots for stellar atmospheric fields. It is evident, then, that they are important to our understanding of coronal (solar or stellar) heating, for it is through flux tubes that the mechanical energy reservoir of the photosphere can be communicated to the atmosphere (chromosphere and corona) above.

Several kinds of flux tubes are likely to occur in astrophysics. In the Sun, the isolated flux tube of the photosphere is a region of density depletion and Alfvén speed enhancement. In the corona, magnetic fields dominate the atmosphere and no isolated tubes can occur. However, regions of density enhancement (with corresponding depressions in Alfvén speed) define non-isolated flux tubes, the coronal loops. Figure 3 illustrates the photospheric and coronal possibilities. Twisted fields may add further complications to this picture.

2. The Structure of Intense Tubes

Because the intense flux tubes of the solar photosphere occur on scales below the currently best available resolution (about 210 km), their properties are particularly difficult to obtain. Consequently, a number of special techniques have been developed in attempts to explore their properties. Progress in this area has been substantial. Nonetheless, we can anticipate even greater progress when high resolution (at about 70 km) space missions, planned for the next decade, and developments currently under way in ground-based facilities finally achieve their aims. An understanding of the structure of a photospheric flux tube calls for developments on all fronts, from observational to numerical and analytical modelling. We give here a summary of the basic properties of photospheric flux tubes, drawing heavily on the excellent review by Solanki (1987).

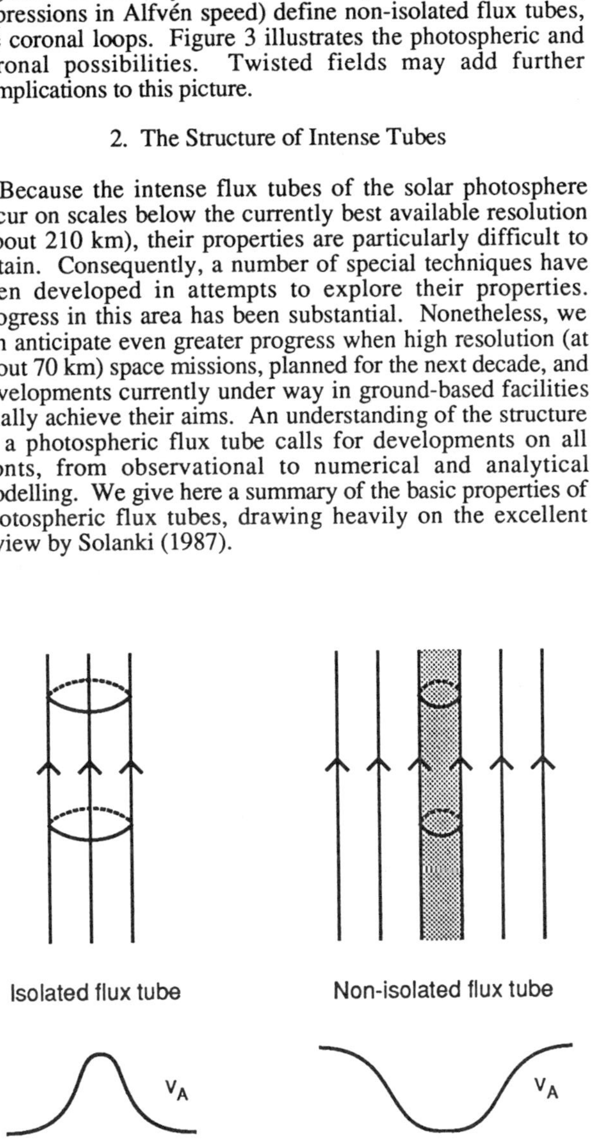

Fig. 3. Two types of magnetic flux tube, the isolated tube of the photosphere (corresponding to a region of enhanced Alfvén speed) and the non-isolated tube of the corona (corresponding to a depletion in Alfvén speed).

a) How strong is B?

The basic question to be asked is how strong are the magnetic fields in isolated intense flux tubes? The first indications that there were strong fields in the photosphere arose in the late 60's; Sheeley (1966,1967) reported field strengths of 0.2-0.7 kG, Beckers and Schröter (1968) gave 0.6-1.4 kG. Shortly after, Howard and Stenflo (1972) and Frazier and Stenflo (1972) suggested that over 90% of the net magnetic flux outside of sunspots resided in strong field form, which Stenflo (1973) gave as 1-2 kG. Thus, the suggestion was that the bulk of the non-spot magnetic field emerging through the photosphere was in a 1-2 kG flux tube form. A direct measurement using the iron-line in the infrared wavelength, reported by Harvey and Hall (1977), gave 1.2-1.7 kG. Similar ranges in field strengths have been reported by Tarbell and Title (1977) from Fourier analysis of the Stokes V profile and by Wiehr (1978) using the line ratio method employed by Stenflo (1973). More recently, a variety of spectral methods have been used obtaining simultaneous information at different wavelengths. Solanki and Stenflo (1984), using a multi-line analysis, determined a field strength of 1.4-1.7 kG. Stenflo, Solanki and Harvey (1987a,b) showed that field strengths declined with height, giving 1.4 kG at optical depth τ = unity outside the tube and 1.1 kG near $\tau = 10^{-2}$. Solanki, Zayer and Stenflo (1988) used Stokes I and V spectra to confirm that field strengths declined with height, giving a strength of 2 kG at $\tau=1$. Zayer, Solanki and Stenflo (1989) have used Stokes V line profiles in the infrared to obtain (for the first time) estimates of the size of the unresolved flux tubes, giving a radius of 150 km at $\tau=1$. They also find that the field strength (2 kG at $\tau=1$) declines with height in a manner consistent with that expected from the thin flux tube approximation (discussed in the next Section).

Altogether, then, there is strong and mounting evidence for 1-2 kG isolated flux tubes with radii of 150 km at $\tau=1$; the field strength declines with height. We should note, however, that not all observers are convinced of the kilogauss field strengths, arguing instead for several hundreds of gauss (see Zirin, 1988). Developments expected in the next few years are likely to resolve this issue one way or another.

b) Density and pressure depletions

Given the existence of kilogauss flux tubes, we may compare the magnetic pressures they sustain with the pressures in the environment of the tubes. A tube of field strength B_0 = 0.2 Tesla = 2 kG has a magnetic pressure $B_0^2/2\mu_0$ of $1.6 \times 10^5 \mathrm{N m^{-2}}$. In c.g.s. units, this is a magnetic pressure (with B_0 in gauss) of $B_0^2/8\pi = 1.6\times10^5$ dynes $\mathrm{cm^{-2}}$. Such a pressure is far in excess of that expected from the dynamic pressure of granular or supergranular flow. A photospheric density of $\rho = 10^{-4} \mathrm{kg\ m^{-3}} = 10^{-7} \mathrm{gm\ cm^{-3}}$ in a flow of v = 0.4 km $\mathrm{s^{-1}}$ produces a pressure of $\frac{1}{2}\rho v^2 = 80$ dynes $\mathrm{cm^{-2}}$, far below the magnetic pressure within an intense tube. In fact, the magnetic pressure is roughly comparable with the gas pressure p_e in the environment of the tube, so that we have

$$\frac{B_0^2}{2\mu_0} >> \frac{1}{2}\rho v^2, \quad \frac{B_0^2}{2\mu_0} \sim p_e. \qquad (1)$$

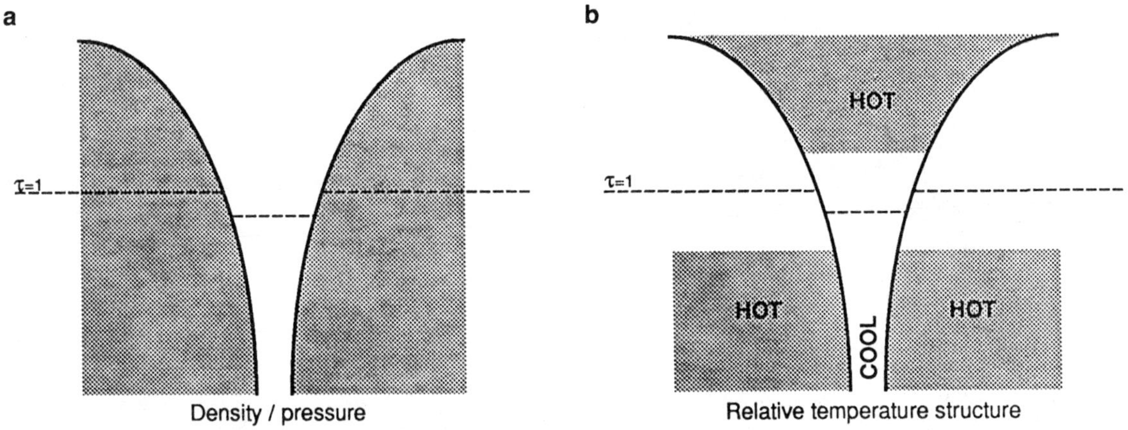

Fig. 4. (a) The photospheric flux tube as a density and pressure depletion (shading denoting excess density and pressure). (b) The thermal structure of a tube is more complicated, showing regions of both coolness and hotness (heavy shading) relative to its immediate surroundings.

We may also expect that thin flux tubes are broadly in transverse pressure balance with their surroundings, so that for a tube of gas pressure p_0 we have

$$p_0 + \frac{B_0^2}{2\mu_0} = p_e. \qquad (2)$$

Combined with Equation (1), this suggests that photospheric flux tubes are regions of gas pressure reduction ($p_0 < p_e$). Unless there are extreme temperature imbalances between the tube and its environment, which is unlikely (see below), this suggests that intense flux tubes are regions of low gas density, being partially evacuated by the presence of the strong magnetic field (see Figure 4). Unless the density evacuation is almost total, which is unlikely, this implies that an intense flux tube is an elastic (not rigid) object with a plasma beta ($\beta \equiv 2\mu_0 p_0/B_0^2$) comparable with unity. (A rigid tube would have $\beta << 1$.) Accordingly, we may expect such elastic tubes to respond to the dynamic pressure fluctuations, produced by granules, supergranules and p-modes, in their environments.

c) Temperature

Since photospheric flux tubes are thin structures we do not expect temperatures in their interiors to depart radically from those in their surroundings, certainly not to the extent exhibited by sunspots (with an umbra at 4000°K compared with an environment of 6000°K). Nonetheless the evacuation of the tube by the magnetic field makes some temperature difference likely: the gas in the tube is less dense and so more readily heated than the denser gas of the surroundings. To determine this temperature difference as a function of height in the tube is a complex modelling exercise, drawing on the observations of line weakening in Stokes I and V profiles (see Solanki (1987) for a detailed discussion). The upshot of this "faculae" modelling is the demonstration that in their higher layers photospheric flux tubes are hotter than their surroundings. Figure 5 shows the run of temperature in various models considered by Solanki (1987), illustrating the fact that at equal optical depth the temperature inside a flux tube is hotter than the non-magnetic Harvard-Smithsonian Reference Atmosphere. However, in interpreting such plots we must remember that the density within a tube is less than that in its surroundings and so we reach an optical depth of unity within a flux tube below the level of optical depth unity in the undisturbed, non-magnetic atmosphere. This is the familiar "Wilson depression" exhibited in observations of sunspots as they approach the solar limb.

Beginning a new generation of flux tube models, Grossmann-Doerth et al. (1988) have presented model calculations of slab magnetic fields incorporating full radiative transfer (grey, LTE). Their results show (among other things) that magnetic sheets are hotter (by up to 400°K) in their upper layers but cooler in their lower layers, the division occurring

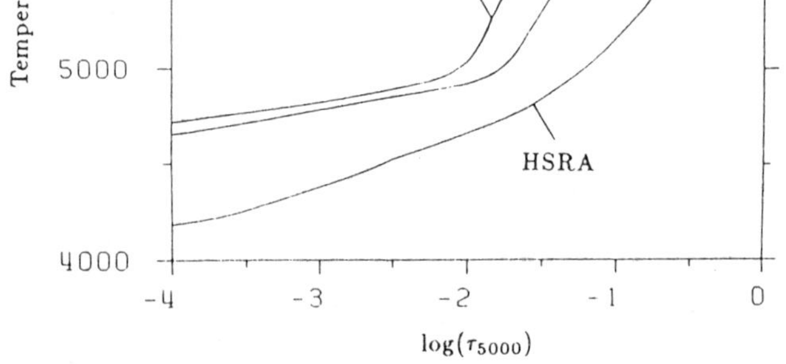

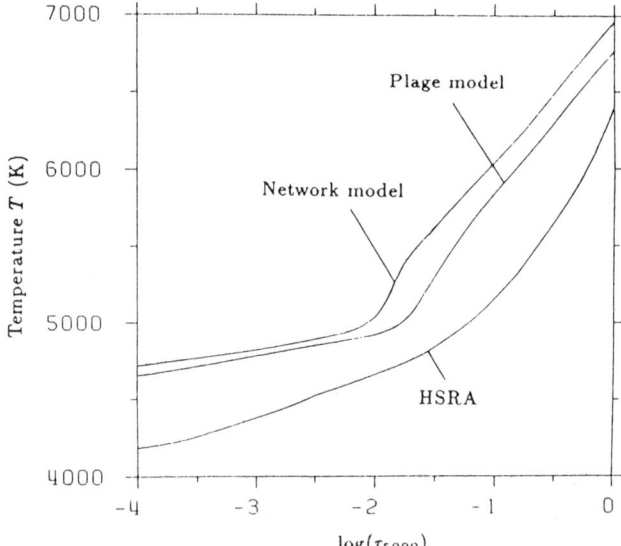

Fig. 5. The run of temperature as a function of optical depth for two flux tube models (network and plage) and the Harvard-Smithsonian Reference Atmosphere. (After Solanki, 1987)

near optical depth $\tau=1$ of the undisturbed atmosphere. The excess heating is due to radiative illumination from the hot lower atmosphere of the magnetic sheet. By contrast, these hot lower layers are in fact cooler than their surroundings at the same geometrical height.

d) Dynamics

We have remarked above that an intense flux tube is an elastic object, subject to pressure fluctuations in its environment. We may thus expect the tube to respond to such fluctuations, producing waves which are guided by the tube and which ultimately dissipate and heat the chromosphere and corona. Additionally, we may expect flux tubes to respond to granular and supergranular motions, transporting the tubes into the downdrafts between neighbouring flows and especially the interstices between several granules. Such bulk movements of tubes are expected to build up stresses in the coronal magnetic field and, through the dissipation of these stresses, lead to heating in the coronal gas (Parker, 1987). The photospheric tube, then, is a dynamical object, forever being moved by bulk flows and at the same time responding to the local variations in the environment's pressure field. A granular velocity of 0.4 km s^{-1} can move a tube a distance equal to its diameter (≈ 300 km) in about 12 minutes (which is about the lifetime of a granule).

Photospheric flux tubes, then, are likely to be surrounded by downdrafts, and possibly also by swirls (Schüssler, 1984,1987) though perhaps smaller in scale than the vortex observed by Brandt et al. (1988). There are two reasons for a downdraft to occur in the immediate neighbourhood of a photospheric flux tube. Firstly, magnetic flux expulsion occurs in convective cells and this transports magnetic field out of the interiors of cells into the boundaries where downdrafts occur. Secondly, the thermal structure of a formed intense tube, with a cool environment, favours the generation of a circulation outside the tube (Deinzer et al., 1984). Thus, taken together, we may expect an enhanced downdraft in the immediate vicinity of an intense tube.

The results obtained by Grossmann-Doerth et al. (1988) from a numerical calculation of a concentrated flux sheet (slab) in which full radiative transfer (grey, LTE) effects are incorporated, show not only the thermal stratification of the sheet (see c) above) but also the development of a circulation (maximum velocity 1 km s^{-1}) in the environment of the sheet.

Given downdrafts in the environment of tubes, are there similar flows within tubes? Early observations (e.g. Giovanelli and Slaughter, 1978) reported the presence of substantial downdrafts within photospheric flux tubes, flows so strong that it became difficult to understand how mass conservation was met. However, a recent detailed analysis by Solanki (1989), concerned with an explanation of the Stokes V asymmetry, convincingly argues that there is little or no flow (< 0.25 km s^{-1}) within the tube, though there is a downdraft of 0.5-1.5 km s^{-1} in its immediate surroundings. The downdraft is of gas which is 250-350°K cooler than in the average quiet Sun, much as produced in the models of Grossmann-Doerth et al. (1988). Furthermore, Solanki argues that a longitudinal wave-like motion (with an amplitude of 1.0-2.5 km s^{-1}) exists within the tube. If substantiated, this provides important evidence for the existence of waves in photospheric flux tubes, developing the earlier observations by Giovanelli, Livingston and Harvey (1978) of waves with 5^m period in intense flux tubes.

In contrast with the observations of waves in intense flux tubes, which are still at an early stage, theoretical developments have proceeded apace and give ever mounting arguments for the occurrence of oscillations. On theoretical grounds, we may expect p-modes to scatter off tubes (Ryutov and Ryutova, 1976; Bogdan and Zweibel, 1987) and to resonantly excite oscillations in the tube (Bogdan, 1989). We may expect resonant absorption effects to locally heat the sides of a tube, just as Hollweg (1989) has argued occurs on the edges of sunspots. And granules will manipulate the boundaries of tubes driving waves within them (Roberts, 1979); 'exploding' granules (Title et al. 1989) are likely to send out shocks which impinge on tubes and also on the magnetic canopy (see Section 3) in the chromosphere. Also, the very act of formation of intense tubes (see Sections 3 and 6) may give rise to oscillations which are overstable because of heat exchange in the tube (Hasan, 1985, 1986; Venkatakrishnan, 1985).

Once generated, what happens to the waves of a flux tube? It seems likely that some will form shocks in the chromosphere (Deinzer et al., 1984; Ferriz-Mas and Moreno-Insertis, 1987) and so contribute to the heating of that layer. Other modes may survive to appear in the H_α-fibrils, in which waves have been observed (Giovanelli, 1975). It is also possible that spicules - thin, pencil-like, jets of gas seen moving steadily outwards (with speeds of about 20 km s^{-1}) in the chromosphere - are driven by wave motions in photospheric flux tubes (Roberts, 1979). A wave generated impulsively in a photospheric flux tube exhibits an oscillatory wake trailing its wave-front (Rae and Roberts, 1982), and this may grow nonlinearly to form a sequence of shocks impinging on the transition layers of the chromosphere (Hollweg, 1982). The numerical investigations by Hollweg (1982) and Sterling and Hollweg (1987) indicate spicule-like formations as a consequence of this wake phenomenon.

Another flux tube candidate for the spicule is the formation of a soliton in a tube (Roberts and Mangeney, 1982; Edwin and Roberts, 1986), though the influence of gravity on the soliton has yet to be worked out. Nor, too, have the consequences of an expanding elastic tube been explored for spicule models: how does a wave (or soliton) propagate from the confines of a thin tube to the expanded portions of the upper atmosphere? Is the motion confined to the edges of the expanded parts - as one might expect for surface waves - or is the expansion in the wave less than the geometrical expansion of the tube, so that the wave motion is concentrated towards the centre of the expanded parts?

3. The Life and Times of a Tube

How are intense flux tubes formed? We know that convective cells (granules and supergranules) expel magnetic field from their interiors, concentrating small-scale fields in their downdrafts. This process organizes the distribution of magnetic field, but it does not define the resulting field strength: as remarked on in Section 2, the dynamic pressure of granulation is far too small to account for kilogauss field strengths. It seems that the explanation for such strong fields is to be found in the fact that they reside in a convectively unstable atmosphere - the solar convection zone - and that atmosphere is most strongly unstable (superadiabatic) in the upper parts of the convection zone, near the photosphere. As a consequence of this instability, a downdraft within a flux tube of moderate field strength, assembled by granulation and supergranulation, leads to the transporting of light gas from the upper layers of the tube to the lower layers. This allows the external field-free gas to compress the tube and thus increase its field strength. The process is stabilized by kilogauss field strengths. This process is referred to as convective collapse (or the superadiabatic effect) and was put forward as an explanation for the observed intense field strengths by Parker (1978, 1979b) and Spruit (1979). A mathematical demonstration of the instability is readily given using the thin flux tube equations (see Section 5), which lead to a second order ordinary differential equation governing vertical motions within the tube (Roberts and Webb, 1978). An analysis of this equation (Webb and Roberts, 1978; Spruit and Zweibel, 1979; Unno and Ando, 1979) allows a detailed description of the instability. We examine this topic briefly in Section 7. For more detailed reviews, see Spruit, 1981a; Spruit and Roberts, 1983; Roberts, 1984a, 1986; Thomas, 1985, 1989; Schüssler, 1987 .

Three dimensional numerical simulations of compressible convection in a magnetic field have been carried out by Nordlund (1983, 1986; and these proceedings) and show the ability of convection to organize concentrations of magnetic flux and for convective collapse to produce intensified magnetic fields. But, of course, the processes of flux expulsion and convective collapse do not operate separately but occur simultaneously. So it may not be possible to observe a vertical flux tube actually collapsing to an intensified form; it may simply be born strong as a consequence of convective forces.

Given the birth of a strong flux tube, what role does it then play in the life of the solar atmosphere? It seems likely that a formed flux tube is in a continual state of excitation! It will be subject to the vagaries of its dynamic environment, shuffled about and buffeted by granules, and interacting with the p-modes of its environment. An intense tube is an elastic object, with a plasma beta of order unity, and so is likely to readily respond to any pressure variations occurring in its environment. These and other considerations make it likely that intense tubes are a continual source of oscillations, waves that propagate up the tube to enter the chromosphere and corona. Some of these waves may form shocks and so heat the chromospheric reaches of the tube (Herbold et al. 1985). Other modes may survive to be manifest as oscillations in H_α fibrils (Giovanelli, 1975), which Edwin and Roberts (1983) classified on theoretical grounds as 'magnetic Love waves' (see also Cally, 1985).

The role, then, of intense flux tubes in the life of the solar atmosphere is likely to be a significant one and it is thus of particular importance to establish observationally the nature of waves in photospheric flux tubes. While the recent results of Solanki (1989) offer much encouragement for our understanding of the dynamical nature of photospheric tubes, it may be that only high resolution studies, perhaps from space, will ultimately reveal the detailed nature of tube oscillations.

If the middle-aged life of a tube is likely to be an exciting one, the death of a tube is enigmatic. How long a tube may survive, before being ultimately removed from the photosphere, is far from clear. One problem is that the vanishing from view of a flux concentration in the photosphere does not necessarily mark the actual demise of the tube: it may simply reappear later! In any case, it seems likely that various combinations of reconnection and tube submergence may act to end the life of a tube, at least in the photospheric layers (see Priest, 1987 for a recent discussion). But precisely how all this operates remains to be discovered.

4. The Geometrical Structure of a Thin Tube

What determines the geometrical structure of a photospheric flux tube? Clearly, the final shape and its rate of expansion depend upon the detailed thermal and dynamical nature of the tube and the environment. But we can give a simple gross answer to this question by considering the statics of a thin tube. The equation of magnetostatics is

$$\text{grad}\left(p + \frac{B^2}{2\mu_0}\right) = -g\rho\hat{\underline{z}} + \frac{1}{\mu_0}(\underline{B}.\text{grad})\underline{B}. \quad (3)$$

for a gas with pressure p, density ρ and magnetic field $\underline{B}$. Gravity, with uniform acceleration g (= 274 m s^{-2}), acts in the negative $\underline{z}$-direction. Equation (3) is to be coupled with the ideal gas law

$$p = \frac{k_B}{m}\rho T, \quad (4)$$

where T is the temperature, k_B is Boltzmann's constant and m the mean particle mass of the gas. Also, the magnetic field is solenoidal:

$$\text{div}\,\underline{B} = 0. \quad (5)$$

Consider now a tube with axis of symmetry aligned with the vertical. If that tube is thin, we can expand variables (pressure, field, etc.) about the axis of symmetry and obtain a reasonable representation of these variables from the first one or two terms in a Maclaurin series. (This was the procedure used by Roberts and Webb (1978) to obtain the dynamical equations for a thin tube with axial symmetry.) Thus, assuming that $\partial/\partial\theta \equiv 0$ in a cylindrical coordinate system r,θ,z, we obtain for a field $\underline{B} = (B_r,0,B_z)$ (of strength $B = (B_r^2+B_z^2)^{1/2}$):

$$\left\{\begin{aligned} p(r,z) &= p_0(z) + rp_1(z) + r^2p_2(z) + \dots, \\ B_r(r,z) &= \qquad\quad rB_{r1}(z) + r^2B_{r2}(z) + \dots, \\ B_z(r,z) &= B_{z0}(z) + rB_{z1}(z) + \dots, \end{aligned}\right\} \quad (6)$$

with similar expansions holding for variables ρ and T.

Examine, first, Equation (5). We have

$$2B_{r1}(z) + \frac{dB_{z0}(z)}{dz} + 0(r) = 0. \quad (7)$$

Hence, we require

$$B_{r1} = -\frac{1}{2}\frac{dB_{z0}}{dz}. \quad (8)$$

Notice that having a predominantly vertical field $\underline{B}$, that is also a function of height z, is not in contradiction to the solenoidal equation; the radial component simply adjusts to the expansion in the vertical component of the field.

Turning now to the radial component of Equation (3), we obtain

$$\frac{\partial}{\partial r}\left(p+\frac{B^2}{2\mu_0}\right)=\frac{1}{\mu_0}\left(B_r\frac{\partial}{\partial r}+B_z\frac{\partial}{\partial z}\right)(rB_{r1}(z)+\ldots). \quad (9)$$

That is

$$\frac{\partial}{\partial r}\left(p+\frac{B^2}{2\mu_0}\right)=0(r),$$

and so

$$p+\frac{B^2}{2\mu_0}=p_0(z)+\frac{B_{z0}^2(z)}{2\mu_0}+0(r^2). \quad (10)$$

In other words, the total pressure (gas p + magnetic $B^2/2\mu_0$) varies only weakly across the tube for small r. The total pressure at the centre (r=0) of the tube is a good approximation to the total pressure at the outer edge of the tube.

But across the outer boundary of the tube we must have total pressure balance. Denoting the gas pressure immediately outside the tube by $p_e(z)$, we thus have

$$p_0(z)+\frac{B_{z0}^2(z)}{2\mu_0}=p_e(z). \quad (11)$$

Consider now the vertical component of Equation (3). To lowest order in r, we have

$$\frac{d}{dz}\left(p_0+\frac{B_{z0}^2}{2\mu_0}\right)=-g\rho_0(z)+\frac{B_{z0}}{\mu_0}\frac{dB_{z0}}{dz},$$

and so

$$\frac{dp_0}{dz}=-g\rho_0. \quad (12)$$

Thus, the gas within the tube is stratified barometrically.

Altogether, then, we have the system of equations

$$\frac{dp_0}{dz}=-g\rho_0,\ \frac{dp_e}{dz}=-g\rho_e,\ p_0+\frac{B_0^2}{2\mu_0}=p_e, \quad (13)$$

where we have written $B_0 \equiv B_{z0}$ (so that the tube's magnetic field is essentially $\underline{B} \approx B_0(x)\underline{z}$) and assumed that the gas in the environment is in hydrostatic equilibrium.

To take Equations (13) much further we must make assumptions about the temperature inside and outside the tube (or else solve an equation of heat transfer). Observations indicate a thermal imbalance between a tube and its surroundings (see Section 2). However, for convenience, we will assume that the tube is in thermal balance with the environment, so that the pressure scale-height $\Lambda_0(z) \equiv p_0/g\rho_0$ inside the tube is the same as that in the environment, $p_e/g\rho_e$. (In other words, T/m is the same inside and outside the tube.) This assumption enables us to display most clearly the geometrical structure of a thin tube, without serious loss in the physics. With thermal balance, Equations (13) admit the integrals

$$\left.\begin{aligned} &p_0(z)=p_0(0)e^{-n},\ p_e(z)=p_e(0)e^{-n},\\ &\rho_0(z)=\rho_0(0)\,\frac{\Lambda_0(0)e^{-n}}{\Lambda_0(z)},\\ &B_0(z)=B_0(0)e^{-n/2},\ A_0(z)=A_0(0)e^{n/2}.\end{aligned}\right\} \quad (14)$$

In Equations (14) we have introduced the cross-sectional area $A_0(z)$ of the tube, requiring that total magnetic flux B_0A_0 be a constant along the tube. Also, n(z) denotes the integrated distance in units of scale-height along the tube:

$$n=\int_0^z \frac{dz}{\Lambda_0(z)}, \quad (15)$$

and z=0 is an arbitrary reference level for variables. (In the case of an isothermal atmosphere (Λ_0 = constant), $n = z/\Lambda_0$.)

It follows immediately from Equations (14) that the area $A_0(z)$ of the tube expands like the reciprocal of the square root of the external pressure field:

$$A_0(z) \propto p_e^{-1/2}(z).$$

The radius of the tube, then, expands like the reciprocal of the fourth root of the external pressure. In an isothermal atmosphere, the external pressure field falls off in height like $\exp(-z/\Lambda_0)$, and so the tube's radius grows like $\exp(z/4\Lambda_0)$. Thus, the tube e-folds in four pressure scale-heights. The shape of the tube, then, is like an exponential horn, opening out to the chromosphere and closing down as we descend into the convection zone. Taking Λ_0 = 125km as representative of the upper photosphere, this indicates that a tube of (say) 100 km radius at z = 0 would have expanded out to a radius of 1000 km at a height of z = 1150 km.

There is another important consequence of Equations (14). Introducing the plasma beta through $\beta = 2\mu_0 p_0(z)/B_0^2(z)$, measuring the relative importance of gas to magnetic pressure, we see that in a thin tube in thermal balance with its surroundings β is <u>independent of depth</u>, and therefore takes on its value at the observable (z=0) level. For a kilogauss tube, this means β is of order unity.

How good is the thin flux tube representation? It is difficult to answer this question in absolute terms since it has not yet proved possible to analyse the behaviour of a stratified flux tube without making approximations. Clearly, the thin tube approximation breaks down as the tube expands into the chromosphere and so some loss of accuracy is to be expected there. Nonetheless, its predictions seem to be qualitatively correct. Calculations of the equilibrium profile of a tube assuming either an expansion procedure (Pneuman, Solanki, and Stenflo, 1986; Hasan, 1988) or generally a finite tube (Pizzo, 1986) give reasonably similar results to the thin tube theory. The behaviour of waves (see Sections 6 and 7) is more difficult to assess, though a detailed study of the unstratified but finite tube (Roberts and Webb, 1979) or of higher order terms in an expansion (Ferriz-Mas, Schüssler, and Anton, 1989) show the thin tube approximation to be a good one within its long-wavelength range of validity. On observational grounds, we note that Solanki et al. (1988) and Zayer et al. (1989) conclude that $|B|$ is almost constant across the tube, with a sudden drop at the tube's edge. Altogether, then, it would seem that the thin tube approximation provides a good description of the structure of photospheric flux tubes but, of course, that description becomes less reliable as the tube fans out with height.

5. Merging of Flux Tubes

We can apply the thin tube representation of photospheric magnetic field to examine approximately the level at which merging of a collection of isolated tubes occurs (Spruit, 1981a). Consider a square of side L, drawn in the solar surface at z=0. The square contains N^2 isolated thin tubes, each with field strength $B_0(z)$ and radius $r_0(z)$, and separated at a height z=0 by a distance d (see Figure 6). The total magnetic flux through the square, made up of N^2 contributions from the individual tubes, is $N^2\pi r_0{}^2(0)B_0(0)$, which we can equate to the flux of an associated mean field B thought of as distributed uniformly throughout the square:

$$\pi r_0^2(0)B_0(0)N^2 = \bar{B}L^2 = \bar{B}N^2d^2, \qquad (16)$$

where we have set L = Nd.

Now, from Equation (14), we see that in an isothermal atmosphere (for which $n = z/\Lambda_0$, with Λ_0 the constant pressure scale-height) an isolated flux tube expands in radius $r_0(z)$ given by

$$r_0(z) = r_0(0)e^{z/4\Lambda_0}. \qquad (17)$$

Such a tube will expand out to a radius of order d/2 before merging with the field from neighbouring tubes. If this occurs at a height $z = z_m$ (the merging height), then Equations (16) and (17) imply (Spruit, 1981a)

$$z_m = 4\Lambda_0 \log_e\left(\frac{d}{2r_0(0)}\right) = 2\Lambda_0 \log_e\left(\frac{\pi B_0(0)}{4\bar{B}}\right). \qquad (18)$$

To illustrate the above we choose $r_0(0) = 100$ km, $\Lambda_0 = 125$ km and $B_0(0) = 1.5$ kG, and then consider two regions accordingly as the mean field $\bar{B}$ is relatively low or high. For a region of low mean field $\bar{B}$, which we may associate with so called quiet regions, taking $\bar{B} = 5$G gives $z_m = 1365$ km and d = 3070 km; this gives $N^2 = 100$ for a square the size of a typical supergranule (L = 30,000 km). Thus, in this scheme, an area the size of a supergranule in a quiet region has within it some 100 isolated kilogauss flux tubes, the magnetic fields of which spread out with height to merge together at a height of about 1365 km. By contrast, an active region, characterized by a higher mean field of (say) $\bar{B} = 100$G, has some 1900 isolated tubes (separated 686 km from one another) in an area the size of a supergranule, and these tubes merge together at a height of 617 km.

Thus, the picture (represented in Figure 6) that emerges from considerations of the expansion of isolated flux tubes is that of a chromosphere permeated by magnetic field arching over an essentially field-free atmosphere below. The height at which the field-free lower atmosphere gives way to a magnetic medium depends upon whether a particular region is a quiet or active one, the two being characterized by the number of isolated tubes that support the atmospheric field. The more active a region, the lower the magnetic atmosphere sits in the medium.

The notion of a magnetic canopy follows quite naturally from the picture of isolated flux tubes expanding in a stratified atmosphere, but its actual presence was first suggested on observational grounds (Giovanelli, 1980; Giovanelli and Jones, 1982). Giovanelli and Jones found nearly horizontal canopies with base heights close to the temperature minimum for active regions or unipolar network regions (see the review by Jones, 1986). This merging height is comparable with the above simple estimate for flux tubes in active regions (and considerably lower than the merging height of flux tubes in quiet regions).

More detailed calculations of merging height have been performed. For example, Anzer and Galloway (1983) have looked at how a potential field expands from localized sources to form a canopy, and Pneuman, Solanki and Stenflo (1986) have considered the role of higher order terms in the thin tube expansion (Section 4).

6. Waves in a Flux Tube

What are the basic modes of oscillation of an isolated magnetic flux tube? What are the basic propagation speeds?

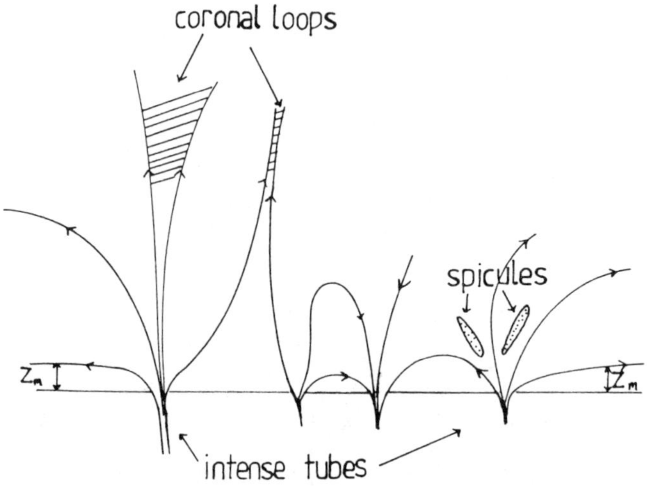

Fig. 6. The merging of magnetic flux tubes in the solar atmosphere (after Spruit, 1981a).

In magnetohydrodynamics there are two fundamental speeds, the sound speed c_s and the Alfvén speed v_A. Consider an isolated flux tube, with magnetic field strength B_0, gas pressure p_0 and density ρ_0, confined by an environmental gas with pressure p_e and density ρ_e. For the moment we will ignore the effect of gravity: the tube and its atmosphere are therefore unstratified. Then p_0, p_e, B_0 are all constants, related by the requirement of pressure equilibrium:

$$p_0 + \frac{B_0^2}{2\mu_0} = p_e. \tag{19}$$

In terms of these equilibrium quantities, we may now define the sound and Alfvén speeds in the tube: $c_s = (\gamma p_0/\rho_0)^{1/2}$ and $v_A = (B_0^2/\mu_0\rho_0)^{1/2}$, where γ is the ratio of specific heats (normally taken to be 5/3). The sound speed in the environment is $c_{se} = (\gamma p_e/\rho_e)^{1/2}$. The Alfvén speed is the characteristic speed with which torsional oscillations - Alfvén waves - of the tube are propagated. Sound waves may also propagate within the tube and they do so at a speed that is lower than c_s. Their characteristic speed is c_T, defined by

$$c_T^{-2} = c_s^{-2} + v_A^{-2}; \tag{20}$$

that is,

$$c_T^2 = \frac{c_s^2 v_A^2}{c_s^2 + v_A^2}. \tag{21}$$

Thus, sound propagates along the tube at a speed which is both sub-sonic and sub-Alfvénic. The reason for the reduction in characteristic speed for sound waves lies in the elasticity of the magnetic flux tube. Compression of the gas within the tube results in changes in the tube's cross-sectional area and consequently must do work against the magnetic field; the result is a reduction in speed, just as occurs for any elastic tube. Indeed, in a blood vessel the characteristic speed of propagation is given by a formula equivalent to Equation (20), with the interpretation that the speed v_A is a measure of the elasticity of the wall of the tube (see, for example, Lighthill, 1978). Of course, in the blood vessel the sound speed is very high and so the tube speed is essentially the speed determined by the elasticity of the wall. A general discussion of the acoustics of tubes has been given by Campos (1986,1987). Waves in magnetic flux tubes are reviewed in Spruit and Roberts (1983), Roberts (1981c, 1984c,1986), Thomas (1985), Hollweg (1986,1989) and Hammer (1987).

The speed c_T is commonly referred to as the tube speed; it is also the cusp speed of the slow magnetoacoustic wave in a uniform medium. The speed c_T arises in the description of slow magnetoacoustic surface waves on a magnetic interface one side of which is field-free (Roberts, 1981a; Miles and Roberts, 1989). Its significance in a tube was brought out in Defouw (1976) and Roberts and Webb (1978).

Typical modes of oscillation of a flux tube are illustrated in Figure 7. Symmetrical modes of oscillation of the tube are referred to as sausage modes. It is also possible for the tube to undergo anti-symmetric oscillations, referred to as kink modes. Kink modes are associated with a second characteristic speed, the kink speed c_k:

$$c_k^2 = \left(\frac{\rho_0}{\rho_0 + \rho_e}\right) v_A^2 . \tag{22}$$

The kink speed c_k is sub-Alfvénic, reflecting the fact that the vibrating tube has to displace about an equal amount of the surrounding gas in its motion, thus adding to the effective inertia of that motion. The speed c_k is also associated with the speed of a surface wave on a magnetic interface (one side of which is field-free) in an incompressible medium (Kruskal and Schwarzschild, 1954). Its significance in a flux tube has been brought out by Ryutov and Ryutova (1976), Ryutova (1988), Parker (1979b, Section 8.10) and Spruit (1981b).

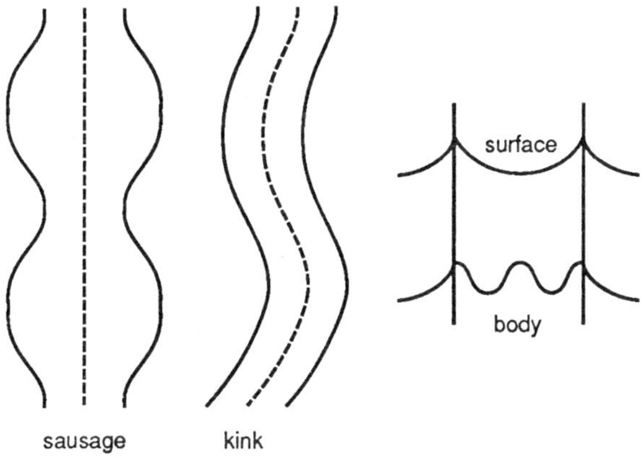

Fig. 7. The modes of oscillation of a magnetic flux tube.

The two speeds c_T and c_k, then, characterize two distinct modes of propagation of waves on a magnetic flux tube. To look more closely at oscillations of a tube, however, requires us to analyse its normal modes by linearizing the usual equations of ideal Mhd (e.g. Parker, 1979b; Priest, 1982) about the equilibrium (19). In a cylindrical coordinate system (r,θ,z), we may write perturbations $f(r,\theta,z,t)$ in the form

$$f(r,\theta,z,t) = \hat{f}(r)e^{i(\omega t + n\theta + kz)}, \tag{23}$$

for amplitude $f(r)$, frequency ω, azimuthal wavenumber n and longitudinal wavenumber k. Then, the linear modes of oscillation of an isolated flux tube of radius a satisfy the dispersion relation (Roberts and Webb, 1978; Wilson, 1980; Spruit, 1982; Edwin and Roberts, 1983; Evans and Roberts, 1989)

$$\rho_0(k^2 v_A^2 - \omega^2) m_e \frac{K_n'(m_e a)}{K_n(m_e a)} + \rho_e \omega^2 m_0 \frac{I_n'(m_0 a)}{I_n(m_0 a)} = 0, \tag{24}$$

where

$$m_e^2 = k^2 - \frac{\omega^2}{c_{se}^2},$$

$$m_0^2 = \frac{(k^2c_s^2 - \omega^2)(k^2v_A^2 - \omega^2)}{(c_s^2 + v_A^2)(k^2c_T^2 - \omega^2)} \qquad (25)$$

and K_n and I_n are modified Bessel functions of order n, with derivatives K_n' and I_n'.

The dispersion relation (24) has been obtained under the assumption that $m_e^2 > 0$, requiring the phase speed ω/k to be less than the external sound speed. This corresponds to motions outside the tube being radially evanescent (non-propagating), declining essentially exponentially to zero as $r \rightarrow \infty$. Otherwise, vibrations of the tube may excite sound waves in the environment which propagate away from the cylindrical tube to infinity. These are commonly referred to as leaky (or radiatively damped) modes (Ryutov and Ryutova, 1976; Roberts and Webb, 1979; Spruit, 1982; Davila, 1985; Cally, 1986).

The dispersion relation (24) is specifically written for the case $m_0^2 > 0$. If $m_0^2 < 0$, Equation (24) is more conveniently rewritten in terms of $n_0^2 = -m_0^2$, and then the expression $m_0 I_n'(m_0 a)/I_n(m_0 a)$ is replaced by $n_0 J_n'(n_0 a)/J_n(n_0 a)$, for Bessel function J_n with derivative J_n'. Solutions of Equation (24) are classified (cf. Roberts, 1981b) as surface modes, since amplitudes decline away from both sides of the tube's edge at r=a. Waves with $m_0^2 < 0$ are classified as body modes, since these solutions of (24) are oscillatory across the tube (though declining outside). In the extreme of a very wide tube, corresponding to ka >> 1, surface modes hug the boundary r=a of the tube, the disturbance hardly penetrating to the tube's interior. Body modes, however, persist throughout the tube's interior, declining only in the environment.

It is clear from Equation (24) that magnetoacoustic waves in a flux tube are dispersive, i.e. their phase-speeds, ω/k, depend upon the wavenumber k. This is a reflection of the fact that a flux tube has an imposed length-scale, the dimension of the tube, by which the wavelength of a disturbance may be measured. This is quite distinct from magnetoacoustic waves in a uniform medium. A uniform medium has no natural lengthscale and consequently its waves are non-dispersive (though they are anisotropic).

The transcendental dispersion relation (24) may be thought of as replacing the usual magnetoacoustic dispersion relation, namely

$$\omega^4 - \kappa^2(c_s^2 + v_A^2)\omega^2 + k^2\kappa^2c_s^2v_A^2 = 0, \qquad (26)$$

where $\kappa^2 = k^2 + k_\perp^2$ is the square of the total wavenumber, made up of the longitudinal component k and the component transverse to the field, $k_\perp$. Equation (26) may be rewritten in the form

$$k_\perp^2 = n_0^2 \equiv -m_0^2, \qquad (27)$$

revealing the complexity of the tube relation (24).

Numerical solutions of Equation (24) are displayed in Figure 8 for two orderings of speeds, namely $c_k < c_T < v_A < c_0 < c_e$ and $c_T < c_0 < c_k < c_e < v_A$, which we take as broadly representative of flux tube conditions. They reveal a complex array of modes, including both fast and slow, surface and body, waves, depending upon the speed orderings.

The dispersion relation (24) and its solutions (depicted in Figure 8) summarize the behaviour of magnetoacoustic waves in a uniform flux tube of arbitrary radius (see Edwin and Roberts, 1983; Cally, 1985, 1986; Abdelatif, 1988; Evans and Roberts, 1989). It is natural to inquire how these modes behave if gravity is included (so the tube is no longer uniform) or if nonlinearities are added. The inclusion of gravity is pursued in the next section. The addition of nonlinearities

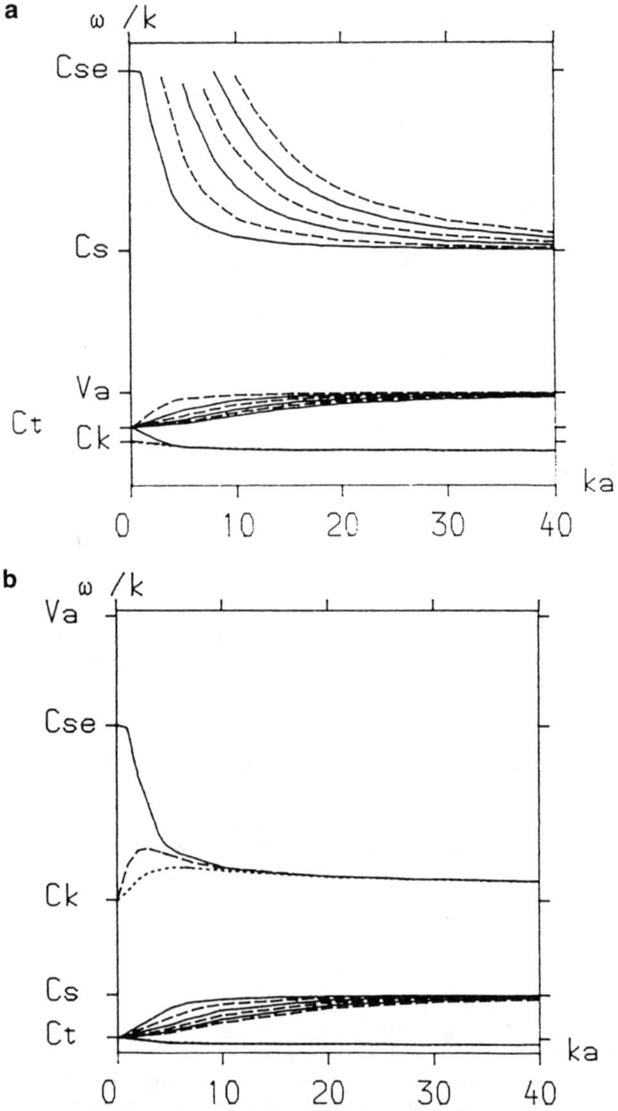

Fig. 8. The phase speeds, ω/k, as a function of dimensionless wavenumber ka of the various modes of oscillation of a flux tube, for two orderings of the basic speeds (after Evans and Roberts, 1989)

raises a number of complexities. We will be content here merely to summarize some of the findings (see also the review by Roberts, 1984b). For example, the slow body modes of a magnetic slab have been investigated by Merzljakov and Ruderman (1986) and shown to form shocks. In contrast, the slow sausage mode of a flux tube may form soliton or solitary wave profiles (Roberts, 1985). For weak (low amplitude) nonlinearity and weak dispersion (corresponding to long wavelength disturbances, with ka << 1 in the linear regime), Roberts (1985) has shown that the sausage mode (with $\omega \approx kc_T$) satisfies the nonlinear integrodifferential equation

$$\frac{\partial v}{\partial t} + c_T \frac{\partial v}{\partial z} + \beta_0 v \frac{\partial v}{\partial z} + \alpha \frac{\partial^3}{\partial z^3} \int_{-\infty}^{\infty} \frac{v(s,t)ds}{[\lambda^2 a^2 + (z-s)^2]} = 0. \quad (28)$$

Equation (28) governs the behaviour of the longitudinal flow v(z,t) in the flux tube when a nonlinear sausage mode propagates. The coefficients β_0 and λ (<1) depend upon the sound and Alfvén speeds of the tube; the coefficient α depends also upon the parameters of the environment and the tube radius, and reflects the dispersive nature of the mode in linear theory.

Of particular interest is the shape of the nonlinearly disturbed tube when a sausage mode propagates along it, balancing the tendency for waves to steepen (due to nonlinearities) with the tendency for waves to disperse (due to the influence of the environment). The shape is readily determined once Equation (28) is solved for a stationary profile. Unfortunately, it has not as yet proved possible to find analytically a solution to equation (28), though a number of its properties have been discovered (see Bogdan and Lerche, 1988). Bogdan and Lerche have termed (28) the Leibovich-Roberts equation.

While an analytical form of the solution of equation (28) is not known, it has proved possible to investigate its structure numerically (Weisshaar, 1989; see also Molotovschikov and Ruderman, 1987). Weisshaar (1989) has shown that its solutions are soliton-like, though whether they are in fact solitons is not clear. However, we know that the equivalent slow sausage mode in an isolated magnetic slab (rather than tube) satisfies the nonlinear integrodifferential equation (Roberts and Mangeney, 1982; Roberts, 1985; Merzljakov and Ruderman, 1985)

$$\frac{\partial v}{\partial t} + c_T \frac{\partial v}{\partial z} + \beta_0 v \frac{\partial v}{\partial z} + \frac{\alpha'}{\pi} \frac{\partial^2}{\partial z^2} \int_{-\infty}^{\infty} \frac{v(s,t)ds}{s - z} = 0, \quad (29)$$

and Equation (29) - called the Benjamin-Ono equation- has the soliton solution

$$v(z,t) = \frac{v_0}{1 + \left(\frac{z - st}{\ell}\right)^2}. \quad (30)$$

Here v_0 is the amplitude of the soliton, ℓ its scale and s its speed; these properties of the soliton are related through

$$s = c_T + \frac{1}{4}\beta_0 v_0, \quad \ell = \frac{4\alpha'}{v_0 \beta_0}. \quad (31)$$

Thus, the larger the amplitude v_0 then the fatter the soliton is, and the faster its propagation speed s. The soliton on a slab of magnetic field resembles a "swollen knee-cap", a bulge in the slab which propagates faster than the tube speed, c_T, along the slab. The solitary wave on a cylindrical flux tube, determined numerically by Weisshaar (1989), has a similar shape and has similar properties.

There have been a number of other nonlinear investigations of closely related interest, pertaining either to a single magnetic interface (Ruderman, 1985; Hollweg, 1987), a slab (Edwin and Roberts, 1986; Sahyouni, Zhelyazkov, and Nenovski, 1988; Zhelyazkov and Nenovski, 1989) or a periodic array of interfaces (Hollweg and Roberts, 1984), but a discussion of these matters would take us too far afield.

Finally, we mention briefly one consequence of dropping the assumption that the tube is a discrete object, with the magnetic field changing abruptly from a high value inside the tube to a low (or zero) value outside. In reality, of course, the magnetic field (and with it the Alfvén speed) will vary continuously (though sharply) from inside to outside and this has consequence for any wave motions that ensue. It is well known that wave motions in continuously structured media reveal that modes tend to undergo resonant absorption, by which strong spatial gradients in the motions build up in regions of sharp inhomogeneity. This suggests that excess heating is likely to take place on the boundary of a flux tube as wave motions are efficiently damped. Ryutova (1981) has suggested that such processes would mean that longitudinal waves would be entirely dissipated before reaching the chromosphere. But it seems more likely to us that this process would simply lead to a boundary layer of excess heating being created on the sides of a tube, though other heat exchange effects may well diminish this form of wave heating somewhat. The effect demands further investigation.

7. Effects of Stratification: The Thin Tube Equations

Gravitational effects are important in the solar photosphere, where the pressure scale-height is perhaps as low as 100km. It is of considerable interest, then, to ask what modifications the inclusion of gravity has on the above description of the modes of oscillation of an isolated flux tube. In particular, we would like to know how dispersion diagrams like Figure 8 are modified in a stratified atmosphere. Unfortunately, it has not proved possible to answer such questions. A chief difficulty is that even the equilibrium configuration of a finite tube is not known, except numerically.

One area in which progress has been possible is that for thin flux tubes. The approach is similar to that employed in Section 4 to describe approximately the shape and physical conditions of a stratified flux tube: all variables are expanded in their Taylor series about the axis (r=0) of the flux tube. Thus, starting from the full nonlinear equations of ideal Mhd, Roberts and Webb (1978) obtained the following system of equations:

$$\frac{\partial}{\partial t}\rho A + \frac{\partial}{\partial z}\rho v A = 0, \quad (32)$$

$$\frac{\partial v}{\partial t} + v\frac{\partial v}{\partial z} = -\frac{1}{\rho}\frac{\partial p}{\partial z} - g, \quad (33)$$

$$\frac{\partial p}{\partial t} + v\frac{\partial p}{\partial z} = \frac{\gamma p}{\rho}\left(\frac{\partial \rho}{\partial t} + v\frac{\partial \rho}{\partial z}\right), \quad (34)$$

$$p + \frac{B^2}{2\mu_0} = \pi_e, \quad (35)$$

$$BA = \Phi_0. \quad (36)$$

These equations had earlier been written down on physical grounds by Defouw (1976); Parker (1974) considered the g=0 case for an incompressible medium. The expansion procedure used by Roberts and Webb (1978) to derive the above has recently been taken to higher order by Ferriz-Mas, Schüssler and Anton (1989).

Equations (32)-(36) are commonly referred to as the thin tube equations. They apply to the sausage (symmetric) mode of oscillation of the tube. The kink mode requires a separate discussion (given below). In the thin tube equations, p(z,t) and ρ(z,t) are the pressure and density within a tube of cross-sectional area A(z,t), and π_e(z,t) is the pressure in the environment of the tube. The longitudinal velocity is v(z,t); the radial component of velocity does not enter into the thin tube equations, though it can be calculated from them (see Roberts and Webb, 1978, 1979). The magnetic flux through the tube is Φ_0, a constant.

We may observe that the thin tube equations for the sausage mode may be grouped into separate parts: Equations (32)-(34), which apply to any elastic tube within which are predominantly longitudinal motions (cf. Lighthill, 1978); and Equations (35) and (36), which effectively relate the magnetic field strength B(z,t) to the area of the tube and the pressure variations π_e in the environment.

Consider the linearized form of Equations (32)-(36) for the equilibrium described in Equation (14). Ignoring pressure variations in the environment of the tube, i.e., setting π_e = p_e(z), the barometrically stratified external gas pressure (see Section 4), allows us to derive a Klein-Gordon equation (Rae and Roberts, 1982):

$$\frac{\partial^2 Q}{\partial t^2} - c_T^2(z)\frac{\partial^2 Q}{\partial z^2} + \Omega_s^2(z)Q = 0. \quad (37)$$

Here c_T(z) is the tube speed in the diverging, stratified tube, defined in terms of the sound and Alfven speeds through equations (21). The quantity Q(z,t) is related to the longitudual velocity v(z,t) through

$$Q(z,t) = \left[\frac{\rho_0(z)A_0(z)c_T^2(z)}{\rho_0(0)A_0(0)c_T^2(0)}\right]^{1/2} v(z,t), \quad (38)$$

and the frequency Ω_s is given by (Roberts and Webb, 1978)

$$\Omega_s^2(z) = \frac{c_T^2}{4\Lambda_0^2}\left[3\Lambda_0' + \frac{9}{4} - \frac{2}{\gamma} + 2\beta\left(\frac{\gamma-1}{\gamma} + \Lambda_0'\right)\right], \quad (39)$$

where $\beta=2\mu_0 p_0/B_0^2$ is the plasma beta of the tube, and Λ_0' is the derivative with respect to height z of the pressure scale-height Λ_0. The Klein-Gordon equation (37) describes the behaviour of slow sausage modes.

If the atmosphere is isothermal (Λ_0 = constant), then the above expression for Ω_s^2 reduces to:

$$\Omega_s^2 = \left[\left(\frac{9}{4} - \frac{2}{\gamma}\right) - \left(\frac{3}{2} - \frac{2}{\gamma}\right)\frac{\beta}{\beta + \frac{2}{\gamma}}\right]\omega_a^2, \quad (40)$$

where $\omega_a \equiv c_s/2\Lambda_0$ is the acoustic cutoff frequency. An expression equivalent to Equation (40) for the tube frequency Ω_s in an isothermal atmosphere was first given by Defouw (1976). The form in which Ω_s^2 is cast in Equation (40) separates the magnetic terms from the non-magnetic ones, and may be interpreted as follows. The first term on the right-hand side of Equation (40) is independent of B_0 and depends simply on the geometry of the tube; it is precisely the frequency squared that would arise in a rigid tube of exponential shape, the cross-sectional area e-folding in two scale-heights of the isothermal atmosphere (see Equation (14)). The second term reflects the elasticity of the flux tube; a rigid tube has β=0. Thus, we may see that the effect of that elasticity is to reduce the frequency Ω_s below its value in a rigid tube.

The frequency Ω_s may not be reduced arbitrarily in size. In fact, we may rewrite our expression for Ω_s^2 to read:

$$\Omega_s^2 = \omega_g^2 + \left(\frac{3}{4} - \frac{1}{\gamma}\right)^2 \frac{c_T^2}{\Lambda_0^2}, \quad (41)$$

where $\omega_g^2 = (g/\Lambda_0)(\gamma-1)/\gamma$ is the buoyancy (Brunt-Väisälä) frequency squared of an isothermal atmosphere. Thus, Ω_s lies above the buoyancy frequency.

So far we have described only the behaviour of sausage modes, but kink modes may be similarly described. In fact, transverse displacements ξ(z,t) of the tube satisfy the equation (Spruit, 1981b)

$$\frac{\partial^2 \xi}{\partial t^2} = c_k^2\frac{\partial^2 \xi}{\partial z^2} + g\left(\frac{\rho_0 - \rho_e}{\rho_0 + \rho_e}\right)\frac{\partial \xi}{\partial z}, \quad (42)$$

where the kink speed $c_k(z)$ is defined in Equation (22). This too may be cast in the form of a Klein-Gordon equation by writing

$$Q(z,t) = e^{-n/4}\,\xi(z,t) \qquad (43)$$

to obtain

$$\frac{\partial^2 Q}{\partial t^2} - c_k^2 \frac{\partial^2 Q}{\partial z^2} + \Omega_k^2 Q = 0, \qquad (44)$$

where the frequency Ω_k of kink waves is given by

$$\Omega_k^2 = \frac{c_k^2}{4\Lambda_0^2}\left(\frac{1}{4} + \Lambda_0'\right). \qquad (45)$$

It may be seen in the above that whereas the speeds c_T and c_k are likely to be broadly comparable, the frequencies Ω_s and Ω_k may differ substantially; in an isothermal atmosphere, Ω_k is much smaller than the acoustic cutoff frequency ω_a whereas Ω_s is comparable with ω_a. The circumstances of a photospheric tube are summarized in Table 1. These results suggest that kink modes are likely to be more readily generated and propagated to higher levels in the atmosphere than sausage modes (Spruit, 1981b).

The significance of the Klein-Gordon equation here is that any impulsively generated wave (sausage or kink) will propagate a wave front at its characteristic speed (c_T or c_k) and behind this front will trail an oscillating wake (with frequency of oscillation Ω_s or Ω_k) (Rae and Roberts, 1982). Such waves could be generated by, for example, shock waves initiated by exploding granules (see Section 1).

What happens when such waves become nonlinear? Hollweg (1982) has argued on the basis of numerical simulations of the nonlinear sound mode in a rigid tube that the effect of the initial wavefront and the subsequent oscillating wake impinging on a transition region is to set the transition region in upward motion and this upward motion is non-ballistic when averaged over several oscillations. The result is a spicule-like motion (Hollweg, 1982, 1989; Sterling and Hollweg, 1988).

Taken together with the earlier discussion of solitons, and the unanswered question of how a nonlinear wave generated in the thin part of an isolated photospheric tube would develop as it propagated into the expanded, non-thin, upper part of the tube, we have raised two suggestions for spicules. Spicules are the result of either solitons or wakes generated in the photospheric or convection zone part of a tube, propagating into the higher parts and there producing the observed 'rosette' pattern.

The behaviour of solitons in a fully stratified atmosphere is unknown as is the behaviour of an elastic tube wake, so neither model can be fully assessed at this stage. But such calculations point to the need for high spatial and temporal resolution studies of flux tubes and spicules to be carried out in the future.

We end this section with a discussion of convective instability. We pointed out earlier (in Section 3) that convective instability in a thin tube may cause it to collapse to higher field strengths (Parker, 1978; Spruit, 1979).

Here we indicate the mathematics of this process as described by the thin tube equations. The Klein-Gordon equation (37) for the sausage mode in a tube may be cast in the form (Webb and Roberts, 1978)

Wave	Speed c	Cut off Ω, $\frac{\Omega}{2\pi}$ (and period)	Amplitude, and e-folding distance
Sausage	c_T (~ 5.3 km s^{-1})	Ω_s~ω_a, 4.8 mHz (208s)	$e^{z/4\Lambda_0}$, 500 km
Sound in exponential rigid tube	c_s (~ 7.5 km s^{-1})	$\Omega \sim \frac{41}{40}\omega_a$, 4.9 mHz (203s)	$e^{z/4\Lambda_0}$, 500 km
Sound in straight rigid tube	c_s(~ 7.5 km s^{-1})	$\Omega = \omega_a \equiv \frac{c_s}{2\Lambda_0}$, 4.8 mHz (208s)	$e^{z/2\Lambda_0}$, 250 km
Kink	c_k (~ 4.5 km s^{-1})	$\Omega_k \sim 0.3\omega_a$, 1.4 mHz (700s)	$e^{z/4\Lambda_0}$, 500 km

TABLE 1. The characteristics of a sausage wave in a magnetic flux tube, a sound wave in a rigid tube with the same cross-section (viz., $A_0(z) = A_0(0)\exp(z/4\Lambda_0)$), a sound wave in a straight, rigid tube, and the kink wave in a magnetic flux tube. We have taken an isothermal atmosphere with $\gamma = 5/3$, $\Lambda_0 = 125$ km, and $v_A = c_s$ (= 7.5 km s^{-1}), all broadly typical of photospheric conditions. After Roberts (1986).

$$\frac{B_0}{\rho_0}\frac{d}{dz}\left(\frac{\rho_0 c_T^2}{B_0}\frac{dv}{dz}\right)+\left\{\omega^2-\omega_g^2\left(\frac{c_T^2}{v_A^2}+\frac{\gamma c_T^2}{2c_s^2}\right)\right\}v=0, \tag{46}$$

where

$$\omega_g^2=\frac{g}{\Lambda_0}\left(\frac{\gamma-1}{\gamma}+\Lambda_0'\right) \tag{47}$$

is the square of the buoyancy frequency in a stratified atmosphere, and ω is the frequency of the mode ($v(z,t) = v(z)e^{i\omega t}$).

Consider Equation (46) under the illustrative boundary conditions

$$v = 0 \text{ at } z = 0 \text{ and } z = -d, \tag{48}$$

giving the vanishing of the flow at two levels of the atmosphere. Equations (46) and (48) constitute a Sturm-Liouville problem, from which we may immediately deduce that a sufficient condition for stability (i.e. for all modes to have $\omega^2>0$) is $\omega_g{}^2>0$. This is the usual Schwarzschild criterion for stability of an atmosphere to convection.

We may obtain a simple guide to the structure of solutions of equation (46) by obtaining a local solution - regarding the coefficients of the differential equation as constants. The result is the solution (Webb and Roberts, 1978)

$$v(z,t)=e^{z/4\Lambda_0}\sin\left(\frac{n\pi z}{d}\right)e^{i\omega t}, \tag{49}$$

vanishing at z=0 and z=-d (for integer n). The local dispersion relation for these modes is

$$\frac{\omega^2}{c_T^2}=\frac{n^2\pi^2}{d^2}+\frac{1}{16\Lambda_0^2}+\left(\frac{1}{v_A^2}+\frac{\gamma}{2c_s^2}\right)\omega_g^2. \tag{50}$$

It is immediately apparent from Equation (50) that modes maybe unstable (i.e. $\omega^2<0$) if $\omega_g{}^2<0$ and the magnetic field is sufficiently weak. Considering the case of large depth d, so that the first term in the right-hand side of Equation (50) may be neglected, we may readily obtain the marginal stability ($\omega^2=0$) limit. Marginal stability occurs for $\beta=\beta_c$, where

$$1+\beta_c=\frac{1}{8\left|\Lambda_0'+\frac{\gamma-1}{\gamma}\right|}=\frac{g}{8\Lambda_0|\omega_g^2|}. \tag{51}$$

It follows, then, that the strength of the superadiabaticity determines the strength of the critical field (for marginal stability). The more superadiabatic is the atmosphere (i.e. the larger is the value of $|\omega_g{}^2|$), then the smaller is the critical plasma beta of the tube. In other words, a high field strength is required to achieve marginal stability in a strongly superadiabatic atmosphere.

Of course, in a real atmosphere a more careful analysis of the critical β_c is required. This has been done for the Sun by Spruit and Zweibel (1979) who find β_c=1.8. This corresponds to kilogauss field strengths in the tube. Further discussion of these aspects may be found in Spruit (1981a), Roberts (1984a), and Thomas (1989). Nonlinear aspects are discussed by Venkatakrishnan (1983) and Hasan (1985).

8. Sunspots

In some ways sunspots would seem to be a perfect place in which to study magnetohydrodynamic effects, for they are clearly regions of the Sun in which magnetic fields play a crucial role and they are sufficiently large as to be observed relatively easily. One would expect to find, then, an extensive body of literature in which theoretical and observational developments go hand in hand, with a deep understanding of the intrinsic nature of sunspots emerging. Unfortunately, this is not the case. Indeed, while we have a straightforward but superficial understanding of a sunspot- it has a cool umbra with a largely vertical magnetic field of some 3kG and a somewhat warmer penumbra with a weaker, largely horizontal, field of some 2kG (declining radially outwards) - we are far from clear as to its sub-surface structure. Is a sunspot a monolithic flux tube or an ensemble of discrete flux tubes pushed close together in the visible layers of the spot but isolated from one another just below those layers (Parker, 1979a)?

Another difficulty to be contended with in any theoretical attempt to understand their structure or dynamics is the fact that sunspots provide us with a good example of a strongly stratified and magnetized plasma. The plasma beta in a sunspot varies from a very low value in the spot's chromosphere to a high value in the sub-photospheric layers. This is in contrast with an isolated thin tube which (at least in the thin tube approximation) has a plasma beta that is constant with depth.

The fact is, then, that well over three and a half centuries after the telescopic observations of sunspots by Galileo, in 1610, and over eighty years after Hale's (1908) establishment of the magnetic nature of spots ,we are still uncertain as to their three-dimensional structure! The geometry of a sunspot is undoubtedly complex.

Nor are sunspots uniform. The penumbra of a spot exhibits a considerable fibrous structure, and umbral dots (small regions of brightness) are to be found throughout the umbra (see Moore and Rabin 1985 for a recent review of sunspots). The role of these non-uniformities and their production in the field (by, presumably, magneto-convective forces) is not clear.

One way forward in understanding sunspots is to analyse their modes of oscillation and use such information as a diagnostic tool - sunspot seismology (Thomas, Cram and Nye, 1982; Zugzda, Locans and Staude, 1983,1987; Zugzda, Staude and Locans, 1984). However, the complexity of a sunspot - its three-dimensionality and high degree of stratification - makes theoretical progress in this area difficult. It is not yet clear how p-modes interact with a sunspot, though simple models involving the neglect of gravity (e.g. Abdelatif and Thomas, 1987) give important clues.

Observations show that sunspots support a variety of modes (see Moore and Rabin, 1985, Lites, 1986 a,b, 1988; Zugzda, Locans and Staude, 1987 for recent results and discussions). There is a 3 minute mode, which is seen across the umbra and at all heights (from photosphere to chromosphere); periods actually lie in the range 100 - 200s. In its strong form, this mode is the umbral flash first reported by Beckers and Tallant (1969). The 3 minute mode seems absent from developing or decaying spots. Its period is apparently not correlated with radius. Amplitudes are low (~0.1 kms^{-1}) in the photosphere and much stronger in the chomosphere. However, because of density stratification, the energy in the motion resides in the low photosphere, suggesting that the 3 minute mode is basically a photospheric phenomenon. Observations by Abdelatif, Lites and Thomas (1984,1986), Lites and Thomas (1985) and Thomas et al. (1987) demonstrate that the kinetic energy density in the oscillations is some 5-10 times larger in the photosphere than in the chromosphere.

In addition to the 3 minute -modes, sunspots exhibit power in their oscillation spectra at 5 minutes (Abdelatif, Lites and Thomas, 1986). This mode is restricted to photospheric levels (and below) but is present across the umbra. It is strongly related to the p-modes of the quiet Sun (which have their power peaked in the vicinity of 5 minutes). Sunspots act as selective filters in admitting some p-mode frequencies but not others, and there is a general shift of power towards longer horizontal wavelengths (Abdelatif, Lites, and Thomas, 1986). A recent discovery is that sunspots seem to act as sinks for p-modes, appearing to absorb as much as half of the incident energy flux in p-modes (Braun, Duvall and Labonte, 1987, 1988).

The penumbra of a sunspot supports its own class of oscillations (see Lites 1988 and references therein). Lites reports frequencies of 3.5 mHz (period 286s) and 2 mHz (period 500s) in the upper photosphere but only the higher frequency (3.5 mHz) modes higher up in the penumbral chromosphere. Motions in the upper photosphere are aligned with the magnetic field. Movies of penumbral oscillations reveal the running penumbra wave, first reported by Giovanelli (1972) and Zirin and Stein (1972): "alternate bands of relative Doppler shift appear to emanate from the umbral-penumbral boundary, then travel radially outward toward the outside edge of the sunspot" (Lites, 1988).

To give a theoretical interpretation of these results is demanding, not least because of the complex geometry of a sunspot with its field diverging with height and becoming roughly horizontal in the penumbra. A variety of approximate models have accordingly been developed from an unbounded vertical field (e.g. Zugzda et al., 1983, 1984, 1987) to one contained in a cylinder (Scheuer and Thomas, 1981; Thomas and Scheuer, 1982). These models include the effects of stratification. A model has recently been investigated by Evans and Roberts (1989), who include the effect of the boundary of the sunspot (flux tube) at the expense, however, of neglecting gravity. In such a model the dispersion curves of Figure 8, presented earlier, are appropriate. In terms of Figure 8, Evans and Roberts argue that the 3 minute oscillation is a slow body mode. The 5 minute oscillation is then identified with the fast body waves of Figure 8. Since fast body modes are unable to propagate above the level $v_A = c_{se}$ without generating modes that leak into the environment of the tube, in this interpretation the 5 minute mode will not reach into the higher levels of sunspots. This is consistent with the observed lack of power at 5 minutes in the upper levels of spots.

Running penumbral waves have been modelled as magnetoacoustic modes in a horizontal magnetic atmosphere, stratified by gravity (Nye and Thomas, 1974, 1976b; Cally and Adam, 1983). Small and Roberts (1984) have regarded

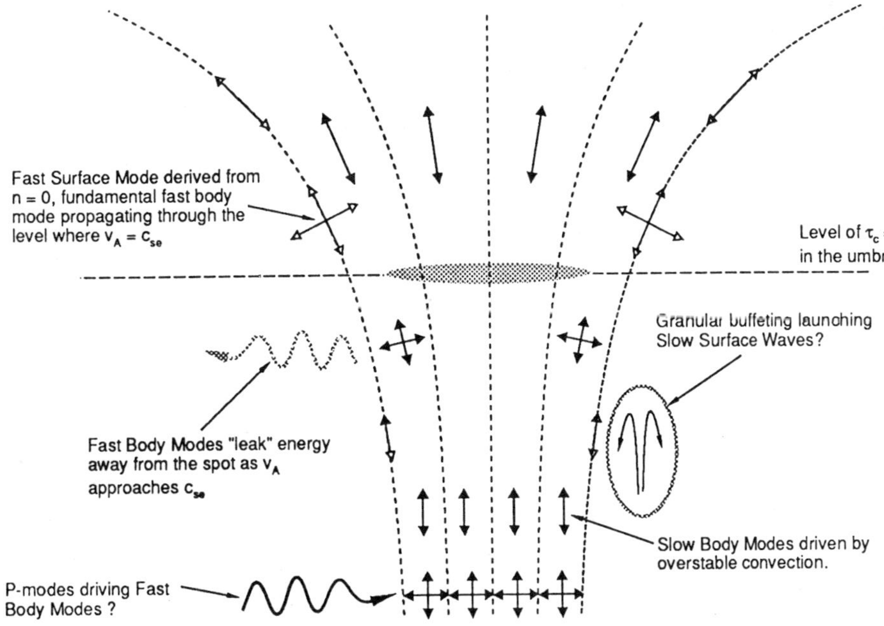

Fig. 9. A sketch of the diversity of waves in a sunspot possible on theoretical grounds (after Evans and Roberts, 1989)

the running penumbral wave as a magnetoacoustic surface mode. Depending upon conditions, two surface modes may exist even in the absence of gravity (Roberts, 1981a; Miles and Roberts, 1989). Gravitational effects on surface waves have only begun to be examined (Small and Roberts, 1984) but are clearly important in a sunspot atmosphere.

Finally, in Figure 9 we sketch the various modes of a diverging (sunspot-like) magnetic flux tube, as suggested on theoretical grounds by Evans and Roberts (1989). It is immediately clear from such a sketch, in which we should stress the effects of stratification are not fully apparent, that the modes of a sunspot are likely to be complex indeed. It is this complexity that makes all theoretical attempts to explain the observed modes of a sunspot such a challenge, with the prospect of obtaining a diagnostic of conditions below the visible surface an added spur.

Questions and Answers

Antiochos: If sunspots are absorbing p-mode energy, how much power would you expect to observe in MHD waves emanating from sunspots?

Roberts: To obtain 10^{25} erg. s^{-1} energy flux in a sunspot of radius 10^4 km (see comment following Hollweg's remark) we may estimate 10^{25} erg $s^{-1} \sim \rho < v^2 > A\, v_A$ for the tube of area A and Alfvén speed v_A. For $v_A = 10$km s^{-1} and a density ρ of 10^{-7}gm cm^{-3} we obtain a mean velocity amplitude of $<v> \sim 0.1$ km s^{-1}. This is quite reasonable for the photospheric layers of a sunspot. Also, Braun et al. (1988) find sunspots absorb 10^7 ergs cm^{-2} s^{-1}, which is consistent with an estimate of 0.1 km s^{-1} velocity amplitude.

Hollweg: Without having numbers in my head, I believe Giovanelli has given a rough estimate of the running penumbral wave energy. This is roughly comparable to the p-mode energy loss at sunspots so the two phenomena may be connected.

Roberts; Yes, Giovanelli et al. (Sp. Phys., 58, 347; 1978) estimated an energy flux in running penumbral waves of very much less than 3×10^6 erg cm^{-2} s^{-1}. Assuming this to be radiated over a spot of radius 10^4 km, Evans and Roberts (1989) estimated a total energy flux of very much less than 10^{25} erg s^{-1}. This is to be compared with Libbrecht's (1989; Ap. J.) estimate of 10^{30} erg s^{-1} as the total energy flux in p-modes, which gives at most 5×10^{25} erg s^{-1} when distributed over a sunspot. So, yes the two phenomena may be connected.

Hollweg: Does the temperature increase with height in a flux tube imply mechanical heating, or is it solely a radiative transfer effect? Is there evidence for mechanical heating? (I believe Solanki has said there is.)

Roberts: In Grossman-Doerth's model calculations, the temperature increase is due to radiative effects. Wave heating may also occur, of course. Computations by Herbold et al. and others show tube waves may form shocks in the upper reaches of the tube, so heating in these layers is to be expected. Solanki has been investigating the role of tube waves in modifying the Stokes line profiles and is inclined towards the view that such waves are necessary to explain the observations; a steady flow outside the tube is not sufficient.

Haerendel: I was brought up to believe that sunspots are cool because of reduced energy supply from below, due to the suppression of convective energy transport by the magnetic field. Obviously, this does not apply to their flux tubes. Are they cooler because of enhanced cooling by the multitude of wave modes that you discussed?

Roberts: Overstable oscillations may cool thin tubes (just as was proposed earlier for sunspots) but further work in this area is needed to assess the effect. Certainly, the inhibition of convection argument as applied to sunspots is not significant in tubes with a scale of 10^2km. In such small tubes, radiative heat exchange with the tube's environment is important.

Schindler: I noticed that you paid little attention to the continuous part of the spectrum. Why do you think that that part does not play an important role in photospheric flux tubes?

Roberts: Certainly one can expect resonant absorption effects near the boundary of tubes, suggesting that those boundaries will be subject to enhanced heating. This may lead to a hot boundary layer on tubes (both thin ones and sunspots). But, of course, other physical effects may add or subtract from such an effect so the net result is not clear in the absence of detailed modeling. However, all these effects are presumable confined to a very thin boundary, so I don't think they have any strong effect on the global structure of tube waves.

Zirin: There is no doubt that flux tubes are discrete. There is only dispute about the strength of the field. Direct measurements give about 100 Gauss, but indirect measurements suggest 1000 Gauss. However, there are problems with the popular "strongfield" model. The brightness of certain K-line points are too great, the size is limited by radiative mean free path, etc. It is to be hoped that the orbiting solar laboratory will have sufficient resolution to settle the question.

Roberts; Some recent work based upon a Fourier transform spectrometer analysis of high quality Stokes V and I spectra (Solanki et al., 1988; Zayer et al., 1989) supports very strong field strengths ($B \approx 2000G$) in the photospheric layer of tubes. Of course, the small size of a tube and its strong stratification make determination of field strengths difficult. So, I agree that OSL will no doubt make an important contribution to this question.

Hundhausen: (1) Are the polarities of flux tubes "clumpy" or random? i.e., do the neighbors of a tube of a given polarity tend to have the same or differing polarity? (2) In your conventional drawing of the canopy structure, what are the field strengths on the field lines that are drawn to be nearly horizontal? If these strengths are high, how is force balance maintained where the field lines are drawn to bend sharply away from the boundary with the fields from the neighboring tube?

Roberts: There are regions of one polarity and also regions of mixed polarity, so both possibilities can be expected. Regarding the canopy structure, I have presented the picture expected on the basis of an expanding thin tube. So, the field field strength declines exponentially fast with height. Pressure balance must exist across the boundary between the almost horizontal field lines and the field-free region below. So, a density drop must occur as we cross from the field-free region into the magnetic canopy field. As I recall, more detailed models, such as carried out by Anzer and Galloway and by Pizzo, give a similar canopy structure.

Zirin: There is direct evidence for canopy spreading in magnetograms near the extreme limit. The magnetic elements appear as a pair of + and - fields, with the opposite polarity on the far side where the lines turn away from us. This is seen for angles $>75°$ from disk center.

Acknowledgements. Discussions with Dr Sami Solanki are greatly appreciated, as are comments from Drs Pat Edwin, Jack Thomas and Rita Ryutova. I am grateful to the American Geophysical Union for financial support towards the cost of travel to this meeting.

References

Abdelatif, T.E., Surface and body waves in magnetic flux tubes, *Astrophys. J.*, 333, 395, 1988.

Abdelatif, T.E., B.W. Lites, and J.H. Thomas, in *Small-scale Dynamical Processes in Quiet Stellar Atmospheres*, S.L. Keil (ed.), National Solar Observatory, Sunspot, NM, p.141, 1984.

Abdelatif, T.E., B.W. Lites, and J.H. Thomas, The interaction of solar p-modes with a sunspot. I. Observations, *Astrophys. J.*, 311, 1015, 1986.

Abdelatif, T.E., and J.H. Thomas, The interaction of solar p-modes with a sunspot. II. Simple theoretical models, *Astrophys. J.*, 320, 884, 1987.

Anzer, U., and D.J. Galloway, A model for the magnetic field above supergranules, *Mon. Not. Roy. Astr. Soc.*, 302, 637, 1983.

Beckers, J.M., New views of sunspots, Sacramento Peak Observatory, Contribution No. 249, 1975.

Beckers, J.M., and E.-H. Schröter, The intensity, velocity, and magnetic structure of the sunspot region. I. Observational technique; Properties of magnetic knots, *Solar Phys.*, 4, 142, 1968.

Beckers, J.M., and P.E. Tallant, Chromospheric inhomogeneities in sunspot umbrae, *Solar Phys.*, 7, 351, 1969.

Bogdan, T.J., Propagation of compressive waves through fibril magnetic fields. III. Waves that propagate along the magnetic field, *Astrophys. J.*, 318, 896, 1987.

Bogdan, T.J., On the resonance scattering of sound by slender magnetic flux tubes, *Astrophys. J.*, in press, 1989.

Bogdan, T.J., and I. Lerche, A note on solitary wave solutions of the Leibovich-Roberts equation, *Quat. Applied Maths.*, XLVI, 365, 1988.

Bogdan, T.J., and E.G. Zweibel, Propagation of compressive waves through fibril magnetic fields, *Astrophys. J.*, 312, 444, 1987.

Brandt, P.N. et al., Vortex flow in the solar photosphere, *Nature*, 335, 238, 1988.

Braun, D.C., T.L. Duvall, and B.J. Labonte, Acoustic absorption by sunspots, *Astrophys. J.*, 319, L27, 1987.

Braun, D.C., T.L. Duvall, and B.J. Labonte, The absorption of high-degree p-mode oscillations in and around sunspots, *Astrophys. J.*, 335, 1015, 1988.

Cally, P.S., Magnetohydrodynamic tube waves: Waves in fibrils, *Aust. J. Phys.*, 38, 825, 1985.

Cally, P.S., Leaky and non-leaky oscillations in magnetic flux tubes, *Solar Phys.*, 103, 27, 1986.

Cally, P.S., and J.A. Adam, On photospheric and chromospheric penumbral waves, *Solar Phys.*, 85, 97, 1983.

Campos, L.M.B.C., On waves in gases. Part I: Acoustics of jets, turbulence, and ducts, *Reviews of Mod. Phys.*, 58, 117, 1986.

Campos, L.M.B.C., On waves in gases. Part II: Interaction of sound with magnetic and internal modes, *Reviews of Mod. Phys.*, 59, 363, 1987.

Davila, J.M., A leaky magnetohydrodynamic waveguide model for the acceleration of high-speed solar wind streams in coronal holes, *Astrophys. J.*, 291, 328, 1985.

Defouw, R.J., Wave propagation along a magnetic tube, *Astrophys. J.*, 209, 266, 1976.

Deinzer, W., G. Hensler, M. Schüssler, and E. Weisshaar, Model calculations of magnetic flux tubes. II. Stationary results for solar magnetic elements, *Astron. Astrophys.*, 139, 435, 1984.

Edwin, P.M., and B. Roberts, Wave propagation in a magnetic cylinder, *Solar Phys.*, 88, 179, 1983.

Edwin, P.M., and B. Roberts, The Benjamin-Ono- Burgers equation: An application in solar physics, *Wave Motion*, 8, 151, 1986.

Evans, D.J., and B. Roberts, The oscillations of a magnetic flux tube and its application to sunspots, *Astrophys. J.*, in press, 1989.

Ferriz-Mas, A., Nonlinear flows along magnetic flux tubes: Mathematical structure and exact simple wave solutions, *Phys. Fluids*, 31, 2583, 1988.

Ferriz-Mas, A., and F. Moreno-Insertis, An analytical study of shock waves in thin magnetic flux tubes, *Astron. Astrophys.*, 179, 268, 1987.

Ferriz-Mas, A., M. Schüssler, and V. Anton, Dynamics of magnetic flux concentrations: the second-order thin flux tube approximation, *Astron. Astrophys.*, 210, 425, 1989.

Frazier, E.N., and J.O. Stenflo, On the small-scale structure of solar magnetic fields, *Solar Phys.*, 27, 330, 1972.

Giovanelli, R.G., Oscillations and waves in sunspots, *Solar Phys.*, 27, 71, 1972.

Giovanelli, R.G., Waves systems in the chromosphere, *Solar Phys.*, 44, 299, 1975.

Giovanelli, R.G., An exploratory two-dimensional study of the coarse structure of network magnetic fields, *Solar Phys.*, 68, 49, 1980.

Giovanelli, R.G., and H.P. Jones, The three-dimensional

structure of atmospheric magnetic fields in two active regions, *Solar Phys.*, 79, 267, 1982.

Giovanelli, R.G., and C. Slaughter, Motions in solar magnetic tubes. I. The downflow, *Solar Phys.*, 57, 255, 1978.

Giovanelli, R.G., W.C. Livingston, and J.W. Harvey, Motions in solar magnetic tubes. II. The oscillations, *Solar Phys.*, 59, 49, 1978.

Grossmann-Doerth, U., M. Knölker, M. Schüssler, and E. Weisshaar, Models of magnetic flux sheets, in *Solar and Stellar Granulation* (NATO Advanced Workshop, Capri), preprint, 1988.

Hale, G.E., On the probable existence of a magnetic field in sun-spots, *Astrophys. J.*, 28, 315, 1908.

Hammer, R., Wave and thermal instabilities in flux tubes: Their role for the structure of the chromosphere and transition region, in *The Role of Fine-Scale Magnetic Fields on the Structure of the Solar Atmosphere*, E-H. Schröter, M. Vazquez and A.A. Wyller (Edits.), Cambridge Univ. Press, p.255, 1987.

Harvey, J.W., and D. Hall, Magnetic Field Observations with FeIλ 15648°A, *Bull. Amer. Astron. Soc.*, 7, 459, 1977.

Hasan, S.S., Convective instability in a solar flux tube. II. Nonlinear calculations with horizontal radiative heat transport and finite viscosity, *Astron. Astrophys.*, 143, 39, 1985.

Hasan, S.S., Oscillatory motions in intense flux tubes, *Mon. Not. Roy. Astr. Soc.*, 219, 357, 1986.

Hasan, S.S., Energy transport in intense flux tubes on the Sun. I. Equilibrium atmosphere, *Astrophys. J.*, 332, 499, 1988.

Herbold, G., P. Ulmschneider, H.C. Spruit, and R. Rosner, Propagation of nonlinear radiatively damped longitudinal waves along magnetic flux tubes in the solar atmosphere, *Astron. Astrophys.*, 145, 157, 1985.

Hollweg, J.V., On the origin of solar spicules, *Astrophys. J.*, 257, 345, 1982.

Hollweg, J.V., Energy and momentum transport by waves, in *Advances in Space Plasma Physics*, B. Buti (edit.), World Scientific, Singapore, p.77, 1986.

Hollweg, J.V., Incompressible MHD surface waves: Nonlinear aspects, *Astrophys. J.*, 317, 918, 1987.

Hollweg, J.V., Resonance absorption of solar p-modes by sunspots, *Astrophys. J.*, 335, 1005, 1988.

Hollweg, J.V., these proceedings, 1989.

Hollweg, J.V., and B. Roberts, Surface solitary waves and solitons, *J. Geophys. Res.*, 89, 9703, 1984.

Howard, R., and J.O. Stenflo, On the filamentary nature of solar magnetic fields, *Solar Phys.*, 22, 402, 1972.

Jones, H.P., The interpretation of spectrum lines formed in small solar structures, in *Small Scale Magnetic Flux Concentrations in the Solar Photosphere*, W. Deinzer, M. Knölker and H.H. Voigt (Edits.), Vandenhoeck & Ruprecht, Göttingen, p.127, 1986.

Kruskal, M., and M. Schwarzschild, Some instabilities of a completely ionized plasma, *Proc. Roy. Soc. London*, A223, 348, 1954.

Lighthill, J., *Waves in Fluids*, Cambridge Univ. Press, 1978.

Lites, B.W., Photoelectric observations of chromospheric sunspot oscillations. III. Spatial distribution of power and frequency in umbrae, *Astrophys. J.*, 301, 992, 1986.

Lites, B.W., IV. The CaIIH line and HeI λ10830, *Astrophys. J.*, 301, 1005, 1986.

Lites, B.W., V. Penumbral oscillations, *Astrophys. J.*, 334, 1054, 1988.

Lites, B.W., and J.H. Thomas, Sunspot umbral oscillations in the photosphere and low chromosphere, *Astrophys. J.*, 294, 682, 1985.

Merzljakov, E.G., and M.S. Ruderman, Long nonlinear waves in a compressible magnetically structured atmosphere. I. Slow sausage waves in a magnetic slab, *Solar Phys.*, 95, 51, 1985.

Metzljakov, E.G., and M.S. Ruderman, II. Slow body waves in a magnetic slab, *Solar Phys.*, 103, 259, 1986.

Miles, A.I., and B. Roberts, On the properties of magnetoacoustic surface waves, *Solar Phys.*, 119, 257, 1989.

Molotovshchikov, A.L., and M.S. Ruderman, IV. Slow sausage waves in a magnetic tube, *Solar Phys.*, 109, 247, 1987.

Moore, R., and D. Rabin, Sunspots, *Ann. Rev. Astron. Astrophys.*, 23, 239, 1985.

Nordlund, A., 3-D Model Calculations, in *Small Scale Magnetic Flux Concentrations in the Solar Photosphere*, W. Deinzer, M. Knölker and H.H. Voigt (Edits.), Vandenhoeck & Ruprecht, Göttingen, p.83, 1986.

Nordlund, A., these proceedings, 1989.

Nye, A.H., and J.H. Thomas, The nature of running penumbral waves, *Solar Phys.*, 38, 399, 1974.

Nye, A.H., and J.H. Thomas, Solar magneto- atmospheric waves. I. An exact solution for a horizontal magnetic field, *Astrophys. J.*, 204, 573, 1976a.

Nye, A.H., and J.H. Thomas, II. A model for running penumbral waves, *Astrophys. J.*, 204, 582, 1976b.

Parker, E.N., Hydraulic concentration of magnetic fields in the solar photosphere. I. Turbulent pumping, *Astrophys. J.*, 189, 563, 1974.

Parker, E.N., Hydraulic concentration of magnetic fields in the solar photosphere. VI. Adiabatic cooling and concentration in downdrafts, *Astrophys. J.*, 221, 368, 1978.

Parker, E.N., Sunspots and the physics of magnetic flux tubes. I. The general nature of the sunspot, *Astrophys. J.*, 230, 905, 1979a.

Parker, E.N., *Cosmical Magnetic Fields*, Clarendon Press, Oxford, 1979b.

Parker, E.N., Why do stars emit X-rays?, *Physics Today*, 40, 36, 1987.

Pizzo, V.J., Numerical solution of the magnetostatic equations for thick flux tubes, with applications to sunspots, pores, and related structures, *Astrophys. J.*, 302, 785, 1986.

Pneuman, G.W., S.K. Solanki, and J.O. Stenflo, Structure and merging of solar magnetic flux tubes, *Astron. Astrophys.*, 154, 231, 1986.

Priest, E.R., *Solar Magnetohydrodynamics*, Reidel, Dordrecht, 1982.

Priest, E.R., Appearance and disappearance of magnetic flux at the solar surface, in *The Role of Fine-Scale Magnetic Fields on the Structure of the Solar Atmosphere*, E.H. Schröter, M. Vazquez, and A.A. Wyller (edits.), Cambridge Univ. Press, p.297, 1987.

Rae, I.C., and B. Roberts, Pulse propagation in a magnetic flux tube, *Astrophys. J.*, 256, 761, 1982.

Roberts, B., Spicules : The resonant response to granular buffeting, *Solar Phys.*, 61, 23, 1979.

Roberts, B., Wave propagation in a magnetically structured atmosphere. I. Surface waves at a magnetic interface, *Solar Phys.*, 69, 27, 1981a.

Roberts, B., II. Waves in a magnetic slab, *Solar Phys.*, 69, 39, 1981b.

Roberts, B., Waves in magnetic structures, in *Physics of Sunspots*, L.E. Cram and J.H. Thomas (edits.), Sacramento Peak, p.360, 1981c.

Roberts, B., The creation of fine structure by magnetic fields, *Adv. Space Res.*, 4, 17, 1984a.

Roberts, B., Solitons in Astrophysics, in *Trends in Physics*, Proc. Gen. Conf. Europ. Phys. Soc., Prague, J. Janta and J. Pantofliceck (Edits.), 177, 1984b.

Roberts, B., Waves in inhomogeneous media, in *The Hydromagnetics of the Sun*, ESA SP-220, 137, 1984c.

Roberts, B., Solitary waves in a magnetic flux tube, *Phys. Fluids*, 28, 3280, 1985.

Roberts, B., Dynamical processes in magnetic flux tubes, in *Small Scale Magnetic Flux Concentrations in the Solar Photosphere*, W. Deinzer, M. Knölker and H.H. Voight (Edits.),Vandenhoeck & Ruprecht,Göttingen, p.169,1986.

Roberts, B., and A. Mangeney, Solitons in magnetic flux tubes, *Mon. Not. Roy. Astron. Soc.*, 198, 7p, 1982.

Roberts, B., and A.R. Webb, Vertical motions in an intense flux tube, *Solar Phys.*, 56, 5, 1978.

Roberts, B., and A.R. Webb, Vertical motions in an intense flux tube. III. On the slender flux tube approximation, *Solar Phys.*, 64, 77, 1979.

Ruderman, M.S., Longitudinal propagation of nonlinear surface Alfvén waves at a magnetic interface in a compressible atmosphere, *Plasma Phys. and Controlled Fusion*, 30, 1117, 1988.

Ryutov, D.D., and M.P. Ryutova, Sound vibrations in a plasma with magnetic filaments, *Sov. Phys. J.E.T.P.*, 43, 491, 1976.

Ryutova, M.P., 'Slow oscillations' of magnetic filaments, *Sov. Phys. J.E.T.P.*, 53, 529, 1981.

Ryutova, M.P., Negative energy waves in magnetic flux tubes, *Sov. Phys. J.E.T.P.*, 94, 138, 1988.

Sahyouni, W., I. Zhelyazkov, and P. Nenovski, Dark envelope solitons of fast magnetosonic surface waves in solar flux tubes, *Solar Phys.*, 115, 17, 1988.

Scheuer, M.A., and J.H. Thomas, Umbral oscillations as resonant modes of magneto-atmospheric waves, *Solar Phys.*, 71, 21, 1981.

Schüssler, M., The interchange instability of small flux tubes, *Astron. Astrophys.*, 140, 453, 1984.

Schüssler, M., Structure and dynamics of small magnetic flux concentrations: Observations versus theory, in *The Role of Fine-Scale Magnetic Fields on the Structure of the Solar Atmosphere*, E-H. Schröter, M. Vazquez and A.A. Wyller (Edits.), Cambridge Univ. Press, p.223, 1987.

Sheeley, N.R., Measurements of Solar Magnetic Fields, *Astrophys. J.*, 144, 723, 1966.

Sheeley, N.R., Observations of small-scale solar magnetic fields, *Solar Phys.*, 1, 171, 1967.

Small, L.M., and B. Roberts, On running penumbral waves, in *Hydromagnetics of the Sun*, ESA SP-220, p.257, 1984.

Solanki, S.K., Structure of magnetic flux tubes as derived from observations with moderate spatial resolutions, in *The Role of Fine-Scale Magnetic Fields on the Structure of the Solar Atmosphere*, E-H. Schröter, M. Vazquez and A.A. Wyller (Edits.), Cambridge Univ. Press, p.67, 1987.

Solanki, S.K., The origin and the diagnostic capabilities of the Stokes V asymmetry observed in solar faculae and the network, *Astron. Astrophys.*, in press, 1989.

Solanki, S.K., I. Zayer and J.O. Stenflo, The internal magnetic field structure of solar magnetic elements in *High Spatial Resolution Solar Observations*, O. von der Lühe (Edit.), 10th Sacramento Peak Workshop, in press, 1989.

Spruit, H.C., Convective collapse of flux tubes, *Solar Phys.*, 61, 363, 1979.

Spruit, H.C., Magnetic flux tubes, in *The Sun as a Star*, S. Jordan (Ed.), Washington, NASA SP-450, p.385, 1981a.

Spruit, H.C., Motion of magnetic flux tubes in the solar convection zone and chromosphere, *Astron. Astrophys.*, 98, 155, 1981b.

Spruit, H.C., Propagation speeds and acoustic damping of waves in magnetic flux tubes, *Solar Phys.*, 75, 3, 1982.

Spruit, H.C., and B. Roberts, Magnetic flux tubes on the Sun, *Nature*, 304, 401, 1983.

Spruit, H.C., and E.G. Zweibel, Convective instability of thin flux tubes, *Solar Phys.*, 62, 15, 1979.

Stenflo, J.O., Magnetic field structure of the photospheric network, *Solar Phys.*, 32, 41, 1973.

Stenflo, J.O., Small-scale magnetic structures on the Sun, *Astron. Astrophys. Rev.*, 1, 3, 1989.

Stenflo, J.O., J.W. Harvey, J.W. Brandt, and S.K. Solanki, 'Diagnostics of solar magnetic flux tubes using a Fourier transform spectrometer, *Astron. Astrophys.*, 131, 333, 1984.

Stenflo, J.O., S.K. Solanki, and J.W. Harvey, Centre-to -limb variation of Stokes profiles and the diagnostics of solar magnetic flux tubes, *Astron. Astrophys.*, 171, 305, 1987a.

Stenflo, J.O., S.K. Solanki and J.W. Harvey, Diagnostics of solar magnetic flux tubes with the infrared line Fe I λ15648.54Å, *Astron. Astrophys.*, 173, 1987b.

Sterling, A.C., and J.V. Hollweg, The rebound shock model for solar spicules: Dynamics at long times, *Astrophys. J.*, 327, 950, 1988.

Tarbell, T.D., and A.M. Title, Measurements of magnetic fluxes and field strengths in the photospheric network, *Solar Phys.*, 52, 13, 1977.

Thomas, J.H., Hydromagnetic waves in the photosphere and chromosphere, in *Theoretical Problems in High Resolution Solar Physics*, H.U. Schmidt (Ed.), Max-Planck-Institut, MPA 212, p.126, 1985.

Thomas, J.H., B.W. Lites, J.B. Gurman, and E.F. Ladd, Simultaneous measurements of sunspot umbral oscillations in the photosphere, chromosphere, and transition region, *Astrophys. J.*, 312, 457, 1987.

Thomas, J.H., these proceedings, 1989.

Thomas, J.H., L.E. Cram, and A.H. Nye, Five minute oscillations as a subsurface probe of sunspot structure, *Nature*, 297, 485, 1982.

Thomas, J.H., and M.A. Scheuer, Umbral oscillations in a detailed model umbra, *Solar Phys.*, 79, 19, 1982.

Title, A.M., T.D. Tarbell, and K.P. Topka, On the relation between magnetic field structures and the granulation, *Astrophys. J.*, 317, 892, 1987.

Title, A.M., et al., Statistical properties of solar granulation derived from the SOUP instrument on Spacelab 2, *Astrophys. J.*, 336, 475, 1989.

Unno, W., and H. Ando, Instability of a thin magnetic tube in the solar atmosphere, *Geophys. Astrophys. Fluid Dynamics*, 12, 107, 1979.

Venkatakrishnan, P., Nonlinear development of convective instability within slender flux tubes. I. Adiabatic flow, *J. Astrophys. Astr.*, 4, 135, 1983.

Venkatakrishnan, P., II. The effect of radiative heat transport, *J. Astrophys. Astr.*, 6, 21, 1985.

Wang, H., and H. Zirin, Studies of supergranules, *Solar Phys.*, 120, 1, 1989.

Webb, A.R., and B. Roberts, Vertical motions in an intense magnetic flux tube: Convective instability, *Solar Phys.*, 59, 249, 1978.

Weisshaar, E., Solitary waves in thin magnetic flux tubes as solutions of the Leibovich-Pritchard- Roberts equation, *Phys. Fluids*, A1, 1406, 1989.

Wiehr, E., A unique magnetic field range for non-spot solar magnetic regions, *Astron. Astrophys.*, 69, 279, 1978.

Wilson, P.R., The general dispersion relation for the vibration modes of magnetic flux tubes, *Astron. Astrophys.*, 76, 20, 1980.

Zayer, I., S.K. Solanki, and J.O. Stenflo, The internal magnetic field distribution and the diameters of solar magnetic elements, *Astron. Astrophys.*, in press, 1979.

Zhelyazkov, I., and P. Nenovski, Nonlinear evolution of large amplitude surface magnetohydrodynamic waves, in SPIG '88, Sarajevo, in press, 1989.

Zirin, H., *Astrophysics of the Sun*, Cambridge Univ. Press, Sect. 6.9, 1988.

Zirin, H., and A. Stein, Observations of running penumbral waves, *Astrophys. J.*, 178, L85, 1972.

Zugzda, Y.D., V. Locans, and J. Staude, The interpretation of oscillations in sunspot umbrae, *Astron. Nachr.*, 308, 257, 1987.

Zugzda, Y., V. Locans, and J. Staude, Seismology of sunspot atmospheres, *Solar Phys.*, 82, 369, 1983.

Zugzda, Y.D., J. Staude, and V. Locans, A model of the oscillations in the chromosphere and transition region above sunspot umbrae, *Solar Phys.*, 91, 219, 1984.

THE STRUCTURE OF PHOTOSPHERIC FLUX TUBES

John H. Thomas

Department of Mechanical Engineering, Department of Physics and Astronomy, and C. E. K. Mees Observatory, University of Rochester, Rochester, New York 14627

Abstract. Basic physical mechanisms for producing the observed intense magnetic flux tubes in the solar photosphere are reviewed. The mechanism of flux expulsion by convective cells can concentrate magnetic flux up to the equipartition field strength, which is only about 200 G at the solar surface for the observed granular convection. Other mechanisms that partially evacuate the flux tube are needed to produce further concentration of magnetic flux to the observed values of 1000-1500 G. Two such mechanisms are discussed: concentration by convective collapse of a vertical flux tube in the superadiabatic layer just below the solar surface, and concentration by a siphon flow in an arched, isolated flux tube.

Introduction

Improved high-resolution observations of the Sun's surface magnetic field over the past twenty years have revealed that almost all of the magnetic flux outside of sunspots is concentrated into localized intense magnetic "elements" with field strengths generally in the range 1000-1500 G, and perhaps as high as 2000 G [see the recent reviews by Solanki, 1987, and Stenflo, 1989]. The standard theoretical picture of a solar magnetic element is an isolated magnetic flux tube emerging from the solar surface. A thin, isolated magnetic flux tube in equilibrium will be in lateral pressure balance with its surroundings, so that $p + B^2/8\pi = p_e$, where p_e is the external gas pressure, p is the internal gas pressure, and $B^2/8\pi$ is the magnetic pressure for a field strength B. The limiting maximum magnetic field strength for a given external pressure is that of a totally evacuated flux tube ($p = 0$), for which $B = (8\pi p_e)^{1/2} \equiv B_p$. At the surface of the Sun, $B_p \approx$ 1700 G (for $p_e = 1.14 \times 10^5$ dynes cm^{-2} at optical depth $\tau_{5000} = 1.0$, according to the model atmospheres of Vernazza, Avrett, and Loeser [1981]). Observed field strengths greater than this have been reported, but this could be due to greater transparency of the gas within the flux tube, meaning that one sees deeper into the flux tube than into the surrounding photosphere, down to where B_p is higher (as in the Wilson depression in sunspots). For example, the value of B_p increases to 2000 G at a depth of only about 60 km below the solar surface, according to the models of Vernazza, Avrett, and Loeser [1981].

In the solar convection zone below the solar photosphere, the magnetic field is expelled from turbulent convective cells and concentrated into flux sheets or tubes through the combined action of turbulent advection and ohmic diffusion. This process, which is discussed in the next section, has been clarified by numerical studies of magnetoconvection of increasing complexity by several researchers. Convective expulsion is capable of creating magnetic flux tubes with field strengths up to an appreciable fraction of the "equipartition" value given by a balance between magnetic pressure and the dynamic pressure of the turbulent convective eddies. The equipartition field strength for solar granular convection is less than the observed values of the magnetic field strength in photospheric flux tubes, however, so apparently there are other mechanisms at work that produce a further increase in the field strength.

In this review I discuss two mechanisms for further intensification of the magnetic field strength in photospheric flux tubes: convective collapse of the flux tube and a siphon flow within an arched flux tube. Each of these mechanisms causes a partial evacuation and decrease in the internal gas pressure in the flux tube, which in turn requires an increase in magnetic pressure in order to maintain lateral pressure balance between the flux tube and the surrounding atmosphere.

Magnetic Flux Expulsion By Convective Eddies

The combined effects of advection and diffusion of magnetic lines of force in a convecting fluid of finite electrical conductivity lead to the expulsion of magnetic flux from the interior of the convective cells or eddies and concentration of the magnetic field into sheets or tubes at the boundaries of the cells or eddies. This basic physical process is now well understood, based on extensive analytical and numerical studies for a Boussinesq fluid [see Parker, 1979, chapter 10, and the review by Proctor and Weiss, 1982] and on numerical studies of fully compressible convection [see Nordlund, 1983, 1986, and the review by Hughes and Proctor, 1988].

In a convecting fluid layer with a high magnetic Reynolds number $R_m = UL/\eta$ (where U is the velocity scale, L is the length scale, and η is the magnetic diffusivity), magnetic flux is swept into regions of converging flow until this advection of magnetic flux is balanced by diffusion of magnetic field lines away from the zone of convergence. This process is illustrated

Geophysical Monograph 58

in several model kinematic calculations that simulate the concentration of magnetic flux at the boundaries between adjacent solar supergranules or granules [e. g., E. N. Parker, 1963; Clark, 1964, 1965, 1966; R. L. Parker, 1966; Weiss, 1966; Clark and Johnson, 1967; Galloway and Proctor, 1983]. Dynamical considerations lead to the expectation that the magnetic field can be concentrated up to about the "equipartition" field strength given by equating the magnetic energy density $B^2/8\pi$ with the kinetic energy density $\rho v^2/2$ of the convective motions. Another way of looking at the equipartition field strength $B_e = (4\pi\rho v^2)^{1/2}$ is that it represents a balance between the magnetic pressure $B^2/8\pi$ and the dynamical pressure $\rho v^2/2$. The equipartition field strength B_e at the solar surface, corresponding to observed convective velocities associated with solar granules, is about 200 G (for $\rho = 2.7 \times 10^{-7}$ gm cm^{-3} and $v \approx 1$ km s^{-1}), well below the observed field strengths of 1000-1500 G in photospheric magnetic flux tubes. Thus, the mechanism of flux expulsion alone does not appear to offer an explanation of the observed intense magnetic flux concentrations in the solar photosphere.

There has been a suggestion by Galloway, Proctor, and Weiss [1977, 1978] that the process of magnetic flux expulsion might lead to field strengths considerably in excess of the equipartition strength B_e, even in a Boussinesq fluid. They argue that in a turbulent convective layer the expulsion process might proceed until there is a balance between the ohmic and viscous dissipation rates. This leads to an estimate of the maximum magnetic field strength given by $B_{max} \sim (\tilde{\nu}/\tilde{\eta})^{1/2} B_e$, where $\tilde{\nu}$ is the turbulent kinematic viscosity and $\tilde{\eta}$ is the turbulent magnetic diffusivity. Thus, if the turbulent viscosity is much larger than the turbulent magnetic diffusivity, field strengths considerably greater than the equipartition value might be achieved by flux expulsion alone. However, because the values of the turbulent diffusivities $\tilde{\nu}$ and $\tilde{\eta}$ appropriate to the solar convection zone are highly uncertain (as, indeed, is their very concept), this suggestion must be viewed as conjectural. Besides, from the point of view of the simple mechanical force balance between the flux tube and its surroundings, it seems unlikely that field strengths much above the equipartition value can be achieved without partial evacuation of the tube in addition to flux expulsion.

Convective Collapse of a Flux Tube

Convective collapse of a photospheric flux tube was first proposed by E. N. Parker [1978] and has been the subject of a number of subsequent studies by several authors. Early work focused on linear stability analysis with adiabatic perturbations. More recent work has extended the analysis to include nonlinear effects and radiative heat exchange. In addition, the three-dimensional numerical simulations of granular convection by Nordlund [1983, 1986] display both magnetic flux expulsion by the granules and convective collapse of the isolated magnetic flux tubes, showing that these two processes are coupled.

Figure 1 illustrates schematically the basic mechanism of convective collapse of a vertical magnetic flux tube in the upper solar convection zone, where the mean temperature gradient is superadiabatic. Assume that initially the temperatures inside and outside the flux tube are equal at every height z; that is, the external and internal temperatures both follow the superadiabatic curve $T(z)$ in the upper convection zone (solid curve). Suppose that a parcel of gas within the tube is displaced vertically by an amount Δz. Because the magnetic field inhibits small-scale convective motions, heat transport within the flux tube is greatly reduced compared to the surroundings, and the displacement takes place nearly adiabatically. To a first approximation the temperature of the displaced parcel follows the adiabatic temperature curve (dashed line) shown in Figure 1. In its displaced position, the parcel has a temperature which differs from that of its surroundings by an amount ΔT, given by

$$\Delta T = -\left[\frac{dT}{dz} - \left(\frac{dT}{dz}\right)_{ad}\right]\Delta z \ .$$

(Note that here z is measured positive upwards and that dT/dz and $(dT/dz)_{ad}$ are each negative in the region of interest.) For a downward displacement ($\Delta z < 0$), the parcel is cooler than its surroundings ($\Delta T < 0$) and, provided the magnetic pressure is not too high, lateral pressure balance implies that the parcel is also more dense than its surroundings. Thus, the displacement is accelerated and a downdraft is created which evacuates the upper part of the tube, causing the tube to collapse to a higher magnetic field strength in order to reestablish the lateral pressure balance. On the other hand, if the parcel is given an initial upward displacement ($\Delta z > 0$), then the parcel will be hotter ($\Delta T > 0$) and less dense than its surroundings and will be accelerated upwards. The tube will expand and weaken and the magnetic field will be dispersed.

If the initial magnetic field strength of the tube is high enough, then we might expect that for an initial downward displacement the lateral pressure balance is maintained largely through an increase in the magnetic pressure and hence the gas pressure in the parcel drops and the density for the parcel may be less than the surroundings, leading to a restoring buoyancy force. In this case the initial equilibrium state of the tube is stable. This suggests that there is a critical magnetic field strength above which a thin flux tube in a superadiabatic layer is stable to adiabatic displacements. This result is confirmed by a linear stability analysis based on the thin flux tube equations, as shown by Webb and Roberts [1978], Spruit and Zweibel [1979], and Unno and Ando [1979]. In all three of these papers it is assumed that in the unperturbed state the plasma beta, $\beta = 8\pi p/B^2$, is constant with depth. This corresponds approximately to equal initial temperatures $T(z)$ inside and outside the tube. (More precisely, as shown by Spruit and Zweibel [1979], it corresponds to equal values of $T(z)/\mu(z)$ inside and outside the tube, where $\mu(z)$ is the mean molecular weight; there is a small effect on $\mu(z)$ due to the decreased gas pressure inside the tube.) The linear stability analysis leads to an eigenvalue problem to determine the critical value of β for a given realistic initial temperature profile $T(z)$. The results depend on the boundary conditions, which must be applied somewhere above and below the superadiabatic layer. For example, Spruit and Zweibel found the critical value $\beta_c = 1.83$ for the convection zone model of Spruit [1977]; flux tubes with $\beta < 1.83$ are stable and flux tubes with $\beta > 1.83$ are unstable to either a downward displacement which leads to collapse or an upward displacement which leads to expansion. The value $\beta_c = 1.83$ corresponds to a field strength of about 1350 G at the solar surface, implying that only flux tubes with field strengths greater than 1350 G are stable. Weaker flux tubes will either collapse to a field strength greater than 1350 G or expand and weaken indefinitely. This picture is in good agreement with observations. The results of Webb and Roberts [1978] also give a value for the critical magnetic field strength consistent

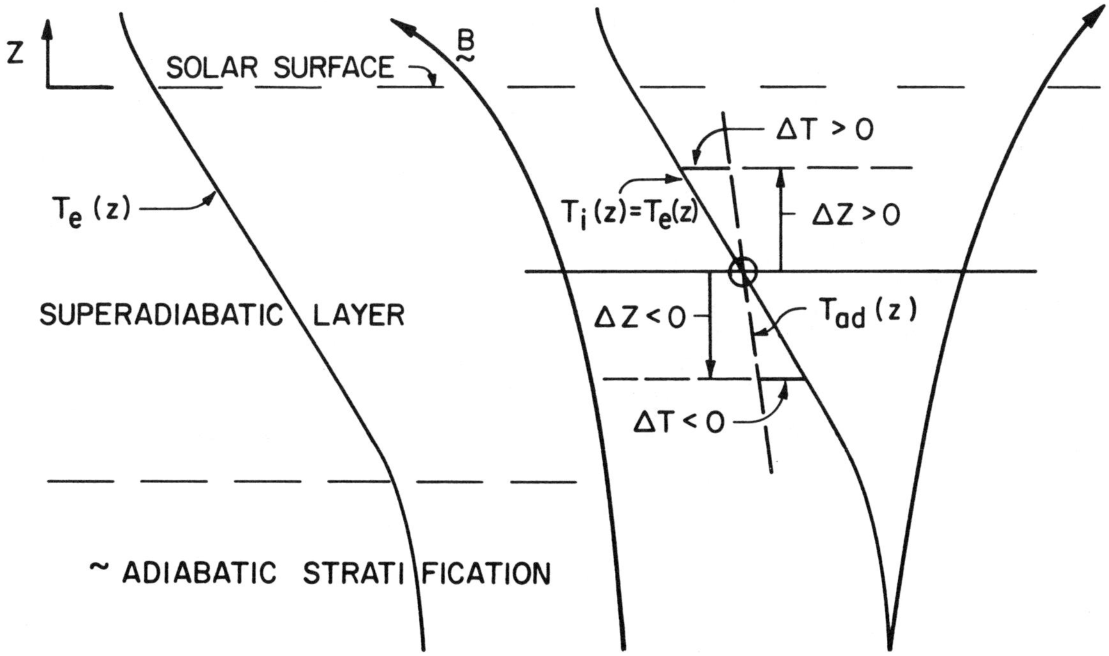

Fig. 1. Illustration of the basic mechanism of convective collapse of a magnetic flux tube in thermal equilibrium in the upper solar convection zone. The mean temperature profile outside the flux tube is superadiabatic just below the solar surface (solid curve). The temperature profile inside the flux tube is initially the same as that outside the tube, but the temperature of a vertically displaced fluid parcel within the flux tube will roughly follow an adiabatic profile (dashed curve) because of the inhibiting effect of the magnetic field on small-scale turbulent heat transport. For a downward displacement, the parcel will be cooler than the surroundings and the tube will collapse and concentrate; for an upward displacement, the parcel will be hotter than the surroundings and the tube will expand and disperse. (See further discussion in the text.)

with observations, but the results of Unno and Ando [1979] do not, most likely due to the unrealistic boundary conditions they use (as pointed out by Spruit and Zweibel [1979]).

What is the consequence of the convective collapse of a flux tube? That is, how does the instability develop nonlinearly and what is the final state? Spruit [1979] showed, by considering nonlinear effects but not the detailed time development of the instability, that for convectively unstable flux tubes there is another equilibrium state, the "collapsed state," which has higher magnetic field strength and is stable. This is illustrated schematically in Figure 2. In the collapsed state the magnetic field strength is higher than the equipartition value given by Boussinesq turbulence theory because the collapse partially evacuates the tube in the upper convection zone as a consequence of the compressibility of the gas. The schematic plot of the total energy $W(\zeta)$ of the system as a function of the displacement ζ shows that the initial state is an unstable equilibrium state whereas the collapsed state is a stable equilibrium state. Spruit speculated that the collapsed state might be overstable if the effects of radiative heat exchange between the flux tube and its surroundings are included. Webb and Roberts [1978] also speculated that overstability might arise if dissipative effects were included.

Venkatakrishnan [1983] and Hasan [1984] studied the nonlinear time development of the convective collapse of a flux tube by solving the nonlinear, time-dependent, thin flux tube equations for adiabatic displacements. Starting with an initial state consisting of a flux tube with a field strength of, say, 700 G in hydrostatic and thermal equilibrium with its surroundings, and then applying a small initial downward displacement inside the tube, Hasan [1984] found that the tube does indeed collapse to a new, collapsed equilibrium state of field strength about 1200 G. However, because of the momentum buildup during the collapse, the displacement overshoots this equilibrium state so that the flux tube oscillates about the collapsed equilibrium state, with velocity oscillations of amplitude 600 m s^{-1} and magnetic field oscillations of amplitude 200 G. The period of this oscillation is about 1250 s. In a subsequent paper, Hasan [1985] added the effects of viscosity and heat exchange between the flux tube and its surroundings, using Newton's law of cooling to model the radiative heat exchange. In this case he found that the oscillations about the collapsed state of the flux tube grow in time; that is, the collapsed equilibrium state is overstable if viscosity and radiative exchange are included (see also Venkatakrishnan [1985]). This confirmed the earlier conjectures of Webb and Roberts [1978] and Spruit [1979]. In very recent work, Massaglia, Bodo, and Rossi [1989] and Hasan (this volume) have extended the calculations of overstable oscillations in a flux tube to include radiative transfer in the Eddington approximation.

Overstable oscillations in a magnetic field in the upper solar convection zone had been studied earlier in other contexts, especially for sunspots, in both the case of Alfvén waves in a

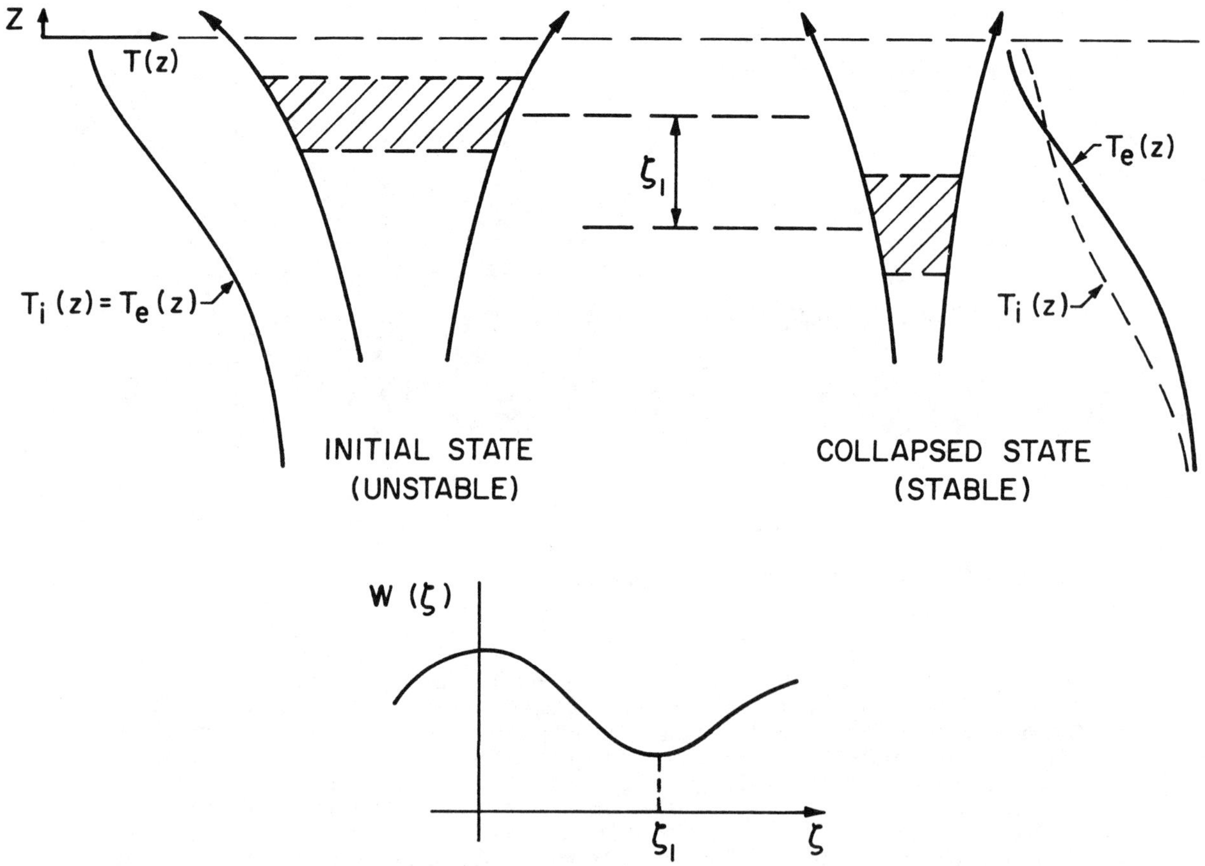

Fig. 2. Schematic diagram of the unstable initial state and the stable collapsed state of a thin flux tube in the upper solar convection zone assuming adiabatic displacements (based on Spruit 1979). In the initial state the flux tube is in thermal equilibrium with its surroundings, whereas in the collapsed state the tube is cooler than the surroundings over most of the superadiabatic layer. The cross-hatched regions correspond to the same masses of gas within the flux tube. The schematic plot of the total energy $W(\zeta)$ of the system as a function of the vertical displacement ζ illustrates the stability properties of the two equilibrium states.

Boussinesq fluid (or, more properly, slow magneto-acoustic waves in the Boussinesq limit, as shown by Cowling [1976]) [Chandrasekhar, 1961; Danielson, 1961; Musman, 1967; Savage, 1969; Moore, 1973; Roberts, 1976] and the fully compressible case [Syrovatskii and Zhugzhda, 1967; Antia and Chitre, 1979]. This earlier work, mostly for a uniform magnetic field, suggested that overstable oscillations might also occur in a thin solar flux tube. Because of the long period of these overstable oscillations (> 1000 s according to Hasan's calculations), it is doubtful that they should actually be observed in a photospheric flux tube because other processes, such as the breakup and reformation of convective granules around the flux tube, may be the limiting factor in the lifetime of the flux tube.

The process of convective collapse will not be as simple on the Sun as it is in the idealized model calculations described above. In the upper solar convection zone the processes of magnetic flux expulsion and convective collapse are occurring simultaneously and continually. This situation is demonstrated clearly in Nordlund's [1983, 1986] three-dimensional numerical simulations of compressible granular convection including radiative transfer and a magnetic field. The magnetic flux is concentrated into the relatively cool, dark intergranular lanes where the flow is downward. This process is eventually retarded by the magnetic pressure force. However, because the magnetic field retards the horizontal flow of gas, there is a reduced transport of heat into the flux tube which causes the flux tube to cool further because of the radiative loss at the surface. This in turn causes a further collapse of the flux tube. Nordlund points out the importance of radiative cooling at the surface for the convective collapse mechanism, an effect that is not present in the simple models based on adiabatic displacements.

Nordlund's simulations also show that the magnetic field pattern continually readjusts in response to changes in the convective granules in field-free regions. Although magnetic flux is not destroyed, it is rearranged by the changing pattern of convection, which means that the effective lifetime of an identifiable magnetic flux tube is comparable to the lifetime of an individual granule, which is of the order of ten minutes. This relates directly to the point made above that such changes may well supersede any growing overstable oscillation in an individual flux tube. More recent results of Nordlund's numerical simulations are reported by him in this volume.

Concentration of Magnetic Flux By Siphon Flows

Siphon flows in isolated, arched magnetic flux tubes are another possible mechanism for producing intense magnetic fields by partially evacuating the flux tube [Thomas, 1984, 1988]. Earlier work on siphon flows, beginning with Meyer and Schmidt [1968], dealt exclusively with the case of a rigid, embedded flux tube in the limit of low plasma beta, appropriate for conditions in the solar corona or chromosphere (see the review by Priest [1981]). For conditions in the solar photosphere or convection zone, with plasma beta of order unity, the flux tube is not rigid and its cross-sectional area and magnetic field strength change in response to changes in internal gas pressure induced by the siphon flow in order to maintain lateral pressure balance with the surrounding atmosphere. The reduced internal gas pressure associated with a siphon flow (the Bernoulli effect) causes an increased magnetic field strength and decreased cross-sectional area of the flux tube relative to the static case.

Both isothermal siphon flows [Thomas, 1988] and adiabatic siphon flows [Montesinos and Thomas, 1989] have been studied for thin, arched flux tubes in an isothermal plane-stratified atmosphere. For a steady isothermal flow along a thin flux tube, the basic equations lead to the following relations between velocity v and height h and between cross-sectional area A and height h :

$$\left(1 - \frac{v^2}{c_t^2}\right)\frac{dv}{v} = \frac{dh}{2H} \quad , \quad \left(1 - \frac{v^2}{c_t^2}\right)\frac{dA}{A} = \left(1 - \frac{v^2}{c_1^2}\right)\frac{dh}{2H} \quad ,$$

where $c_t = [c^2 a^2/(c^2+a^2)]^{1/2}$ is the "tube speed" (i.e, the speed of propagation of axisymmetric distortions of the thin flux tube) and $c_1 = [(\rho_e-\rho)/\rho_e]^{1/2}c = [c^2 a^2/(2c^2+a^2)]^{1/2}$ is another characteristic speed, and where c is the sound speed inside the flux tube, a is the Alfvén speed inside the flux tube, H is the scale height of the external atmosphere, and ρ and ρ_e are the internal and external mass densities. The characteristic speeds are ordered according to $c_1 < c_t < \min(c, a)$. The critical speed for siphon flows in an isolated flux tube is the tube speed c_t, instead of the sound speed c as in the case of a rigid, embedded flux tube in the low-beta limit. A siphon flow can undergo a smooth transition from subcritical speed ($v < c_t$) to supercritical speed ($v > c_t$) only at the top of the arch where the tube is locally horizontal ($dh = 0$).

The qualitative behavior of an isothermal siphon flow can be deduced from the velocity-height and area-height relations above. The maximum speed of the siphon flow depends on the height of the arch and the initial velocity at the upstream footpoint. If the arch height is small enough or the initial velocity is low enough, then the flow remains subcritical throughout. If the maximum velocity is less than the characteristic speed c_1 then the flow behaves as illustrated in Figure 3a; the flow accelerates and the tube expands with increasing height along the upstream leg of the arch, and the flow decelerates and the tube contracts with decreasing height along the downstream leg of the arch. The expansion of the tube with height is less than it would be without the flow, however, due to the decreased internal pressure caused by the flow (the Bernoulli effect). If the maximum velocity exceeds the characteristic speed c_1 but is still less than c_t then the flow

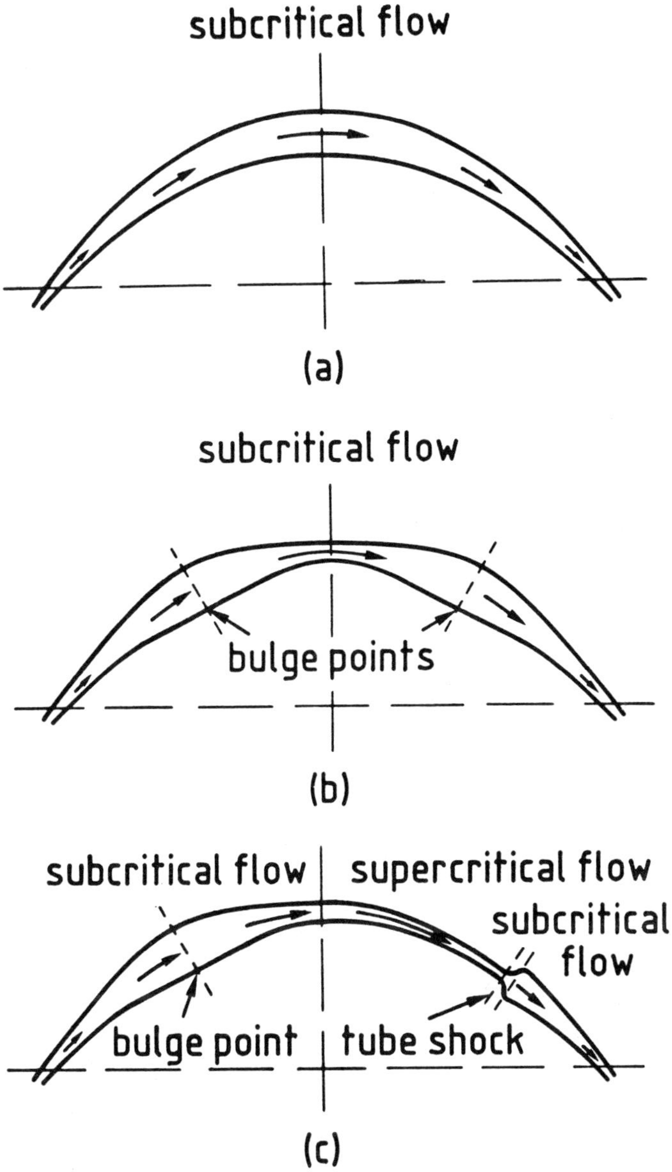

Fig. 3. Schematic diagrams of three types of steady siphon flows in an arched magnetic flux tube: (a) purely subcritical flow without bulge points ($v_{max} < c_1$); (b) purely subcritical flow with bulge points ($c_1 < v_{max} < c_t$); and (c) critical flow with a smooth transition from subcritical flow ($v < c_t$) to supercritical flow ($v > c_t$) at the top of the arch and a standing tube shock in the downstream leg of the arch. (After Thomas 1988; see further discussion in the text.)

behaves as shown in Figure 3b. The flow accelerates and the tube expands up to the point where $v = c_1$ and beyond that point the flow accelerates but the tube contracts with increasing height in the upstream leg of the arch, as required by the sign change of the term $[1 - (v^2/c_1^2)]$ in the area-height relation. This produces a point of local maximum cross-sectional area, or "bulge point," in the tube at the point where $v = c_1$, as shown. In the downstream leg of the arch the flow decelerates symmetrically and there is another bulge point . For a flow like that shown in Figure 3b the Bernoulli effect is so strong that it more than compensates for the decreasing pressure with height outside the tube and causes the tube to contract with height above the bulge point. The siphon flow is thus an effective mechanism for partially evacuating the flux tube and causing a concentration of magnetic flux.

If the arch height is increased or the initial velocity at the upstream footpoint is increased, we reach a "critical flow" in which v $= c_t$ at the top of the arch. In this case the flow can accelerate smoothly to supercritical velocity in the downstream leg of the arch, as shown in Figure 3c. The flow velocity increases and the cross-sectional area decreases with decreasing height in the downstream leg of the arch. However, this supercritical flow requires a very small internal gas pressure at the downstream footpoint which will not be the case in the solar photosphere. For a higher value of the "backpressure" at the downstream footpoint, the flow will decelerate abruptly and the cross-sectional area will increase abruptly across a standing "tube shock" at some point along the downstream leg of the arch. The critical flow is "choked," in the sense that any decrease in the pressure at the downstream footpoint will not increase the mass flow rate along the tube.

A siphon flow within an arched, isolated magnetic flux tube affects the equilibrium path of the flux tube in the surrounding stratified atmosphere. The large-scale mechanical equilibrium of the flux tube involves a balance among the buoyancy force, the net magnetic tension force, and the inertial force due to the siphon flow along curved streamlines (the centrifugal force). In general, the presence of a siphon flow requires that the arch be more highly curved in order that the net magnetic tension force can balance the additional effect of the centrifugal force. Exact equilibrium shapes of thin flux tubes containing siphon flows have been computed by Thomas and Montesinos (preprint).

Figure 4 shows the results of numerical computations of the velocity and cross-sectional area along an arched flux tube for isothermal siphon flows with several different values of the initial velocity. Here the dimensionless height of the arch is $\alpha = h_0/H = 0.550934$, for which the flow is critical for an initial velocity $v_0 = 0.4$. Shown here are four different subcritical flows (for successively greater values of the initial velocity at the upstream footpoint) plus the critical flow with its supercritical downstream leg (dashed line). Note that the faster siphon flows produce smaller cross-sectional areas and thus greater magnetic field strengths in the flux tube. The critical flow and the fastest subcritical flow produce bulge points.

Siphon flows are capable of increasing the magnetic field strength of an isolated flux tube up to as much as 80 or 90% of the limiting vacuum value and thus can, in principle, produce flux tubes with strength 1500 G in the solar photosphere. In general, adiabatic flows produce greater magnetic flux concentration than isothermal flows [Montesinos and Thomas, 1989]. Radiative transfer in a real photospheric flux tube will produce conditions somewhere in between isothermal and adiabatic flow. Siphon flows in arched magnetic flux tubes may the cause some of the intense magnetic fields observed in the solar photosphere outside sunspots. In these cases the strong magnetic field can be highly inclined to the vertical, and observations do indicate that a significant fraction of the photospheric flux tubes are moderately inclined (10° or more) and that a smaller fraction may be highly inclined [Solanki, 1987].

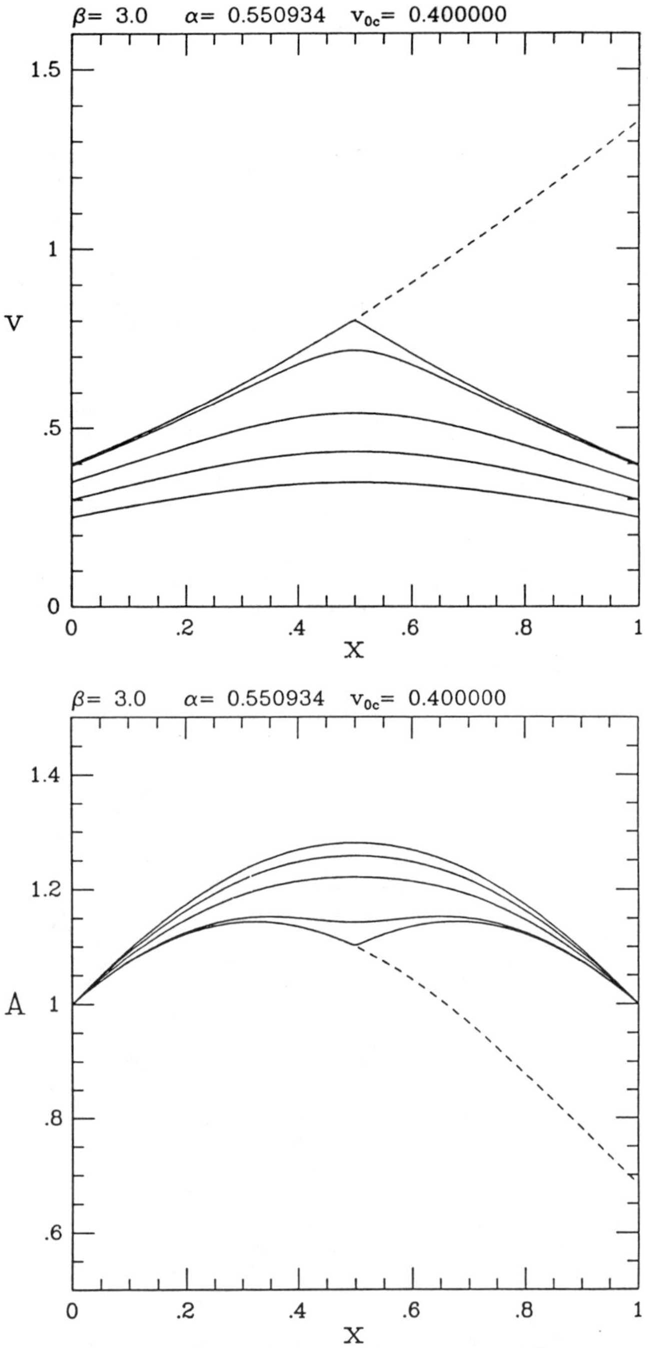

Fig. 4. Plots of the dimensionless flow speed v and the cross-sectional area A as functions of horizontal distance for steady isothermal siphon flows in an arched flux tube with $\beta = 8\pi p_{e0}/B_0^2 = 3.0$ and with dimensionless height $\alpha = h_0/H = 0.550934$. In this case the flow is critical for an initial velocity (at the upstream footpoint) of $v_0 = 0.4$. Four purely subcritical flows are shown, along with the critical flow and its supercritical downstream branch (dashed line). Note the increased concentration of magnetic flux (decreased cross-sectional area) for greater flow speeds, and the presence of bulge points for the critical and nearly critical flows. (From Thomas 1988.)

Questions and Answers

Haerendel: A flux tube does not have to rise much in height before it reaches the temperature minimum. Did you pursue models by which cooling in this region and mass loss by recombination play a role in creating the siphon action?

Thomas: No, I have considered only strict mass conservation in these flows. I assume that you are referring to the possibility of diffusion of neutral mass across field lines and out of the flux tube. I would guess that that process will be slow compared to the time-scale associated with the siphon flows, because the gas within the flux tube is in a highly collisional state.

Ryutova: Did you take into account the interaction between the siphon flow and oscillations of flux tube which inevitably exist in permanently booming atmosphere? My Comments: The point is that this interaction leads to the generation of secondary mass flows along the magnetic field as well as to the generation of the stationary convection across the flux tube. From the other hand this effects results in the two different regions of the tube evolution, namely the splitting of a flux tube in thin tubes and the diffusive broadening of a tube. Certainly, the effects described by you exist and it would be interesting to find the relationship of these effects with those which I mentioned.

Thomas: I have not yet considered the effects of disturbances in the surrounding atmosphere, but I have in mind doing so in the future. The possible effects that you mention are certainly of interest.

Antiochos:Isn't the gas pressure difference between the two ends of a sunspot flux tube opposite to what one would need to drive the Evershed flow?

Thomas: No, I think not. Let me give you an explanation first proposed by Spruit. Imagine that all of the magnetic flux emerging from a sunspot reenters the photosphere at other places in the form of intense flux tubes of strength 1500 G. Then, a flux tube emerging from the umbra of the sunspot, where the field strength is say, 3000G, will have higher magnetic pressure, and hence lower internal gas pressure, at the umbral footpoint of the arched flux tube, compared to the opposite footpoint at the 1500 G magnetic element. For this high-lying arch the siphon flow will be driven inward toward the umbra: this explains the observed inward chromospheric Evershed flow. Now consider a flux tube emerging in the penumbra of the sunspot, where the field strength is only 1000 G. For this low-lying arch, the magnetic pressure is lower and the internal gas pressure is higher at the penumbral footpoint of the arch compared to the other footpoint at the 1500 G magnetic element. Thus, the siphon flow will be driven outward away from the center of the sunspot, in agreement with the observed outward Evershed flow in the penumbral photosphere. We, thus, have an overall picture of siphon flows that is consistent with the observed directions of the chromospheric and photospheric Evershed flows.

Acknowledgments. I am grateful to Bernard Roberts for his comments on the manuscript of this paper and to the American Geophysical Union for paying part of my travel costs for this meeting. My work on siphon flows and the preparation of this review were supported in part by the National Aeronautics and Space Administration under grant NSG-7562.

References

Antia, H. M., and Chitre, S. M. 1979, Waves in the sunspot umbra, *Solar Phys.*, 63, 67-78.

Chandrasekhar, S. 1961, *Hydrodynamic and Hydromagnetic Stability*, (Oxford: Clarendon Press), 652 pp.

Clark, A., Jr. 1964, Production and dissipation of magnetic energy by differential fluid motions, *Phys. Fluids*, 7, 1299-1305.

Clark, A., Jr. 1965, Some exact solutions in magnetohydrodynamics with astrophysical applications, *Phys. Fluids*, 8, 644-649.

Clark, A., Jr. 1966, Some kinematical models for small-scale solar magnetic fields, *Phys. Fluids*, 9, 485-492.

Clark, A., Jr., and Johnson, A. C. 1967, Magnetic field accumulation in supergranules, *Solar Phys.*, 2, 433-440.

Cowling, T. G. 1976, On the thermal structure of sunspots, *Mon. Not. Roy. Astron. Soc.*, 177, 409-414.

Danielson, R. E. 1961, The structure of sunspot penumbras. II. Theoretical, *Astrophys. J.*, 134, 289-311.

Galloway, D. J., and Proctor, M. R. E. 1983, The kinematics of hexagonal magnetoconvection, *Geophys. Astrophys. Fluid Dyn.*, 24, 109-136.

Galloway, D. J., Proctor, M. R. E., and Weiss, N. O. 1977, Formation of intense magnetic fields near the surface of the Sun, *Nature*, 266, 686-689.

Galloway, D. J., Proctor, M. R. E., and Weiss, N. O. 1978, Magnetic flux ropes and convection, *J. Fluid Mech.*, 87, 243-261.

Hasan, S. S. 1984, Convective instability in a solar flux tube. I. Nonlinear calculations for an adiabatic inviscid fluid, *Astrophys. J.*, 285, 851-857.

Hasan, S. S. 1985, Convective instability in a solar flux tube. II. Nonlinear calculations with horizontal radiative heat transport and finite viscosity, *Astron. Astrophys.*, 143, 39-45.

Hughes, D. W., and Proctor, M. R. E. 1988, Magnetic fields in the solar convection zone: magnetoconvection and magnetic buoyancy, *Ann. Rev. Fluid Mech.*, 20, 187-223.

Massaglia, S., Bodo, G., and Rossi, P. 1989, Overstability of magnetic flux tubes in the Eddington approximation, *Astron. Astrophys.*, 209, 399-405.

Meyer, F., and Schmidt, H. U. 1968, Magnetisch ausgerichtete Strömungen zwischen Sonnenflecken, *Zeits. Angew. Math. Mech.*, 48, 218-221.

Montesinos, B., and Thomas, J. H. 1989, Siphon flows in isolated magnetic flux tubes. II. Adiabatic flows, *Astrophys J.*, 337, 977-988.

Moore, R. L. 1973, On the generation of umbral flashes and running penumbral waves, *Solar Phys.*, 30, 403-419.

Musman, S. 1967, Alfvén waves in sunspots, *Astrophys. J.*, 149, 201-209.

Nordlund, A. 1983, Numerical 3-d simulations of the collapse of photospheric flux tubes, in *Solar and Stellar Magnetic Fields: Origins and Coronal Effects, IAU Symposium No. 102*, ed. J. O. Stenflo (Dordrecht: Reidel), pp. 79-83.

Nordlund, A. 1986, 3-d model calculations, in *Small Scale Magnetic Flux Concentrations in the Solar Photosphere*, ed. W. Deinzer, M. Knölker, and H. H. Voigt (Göttingen: Vanderhoeck & Ruprecht), pp. 83-102.

Parker, E. N. 1963, Kinematical hydromagnetic theory and its applications to the low solar photosphere, *Astrophys. J.*, 138, 552-575.

Parker, E. N. 1978, Hydraulic concentration of magnetic fields in the solar photosphere. VI. Adiabatic cooling and concentration in downdrafts, *Astrophys. J.*, 221, 368-377.

Parker, E. N. 1979, *Cosmical Magnetic Fields* (Oxford: Clarendon Press), 841 pp.

Parker, R. L. 1966, Reconnection of lines of force in rotating spheres and cylinders, *Proc. Roy. Soc. London*, A 291, 60-72.

Priest, E. R. 1981, Theory of loop flows and instability, in *Solar Active Regions*, ed. F. Q. Orrall (Boulder, CO: Colorado Associated University Press), pp. 213-275.

Proctor, M. R. E., and Weiss, N. O. 1982, Magnetoconvection, *Rep. Prog. Phys.*, 45, 1317-1379.

Roberts, B. 1976, Overstability and cooling in sunspots, *Astrophys. J.*, 204, 268-280.

Savage, B. D. 1969, Thermal generation of hydromagnetic waves in sunspots, *Astrophys. J.*, 156, 707-729.

Solanki, S. K. 1987, Magnetic fields: observations and theory, in *Proc. Tenth European Regional Astronomy Meeting of the IAU. Vol. 1: The Sun*, ed. L. Hejna and M. Sobotka, Publ. Astron. Inst. Czechoslovakia Acad. Sci., p. 95.

Spruit, H. C. 1977, Ph.D. thesis, University of Utrecht.

Spruit, H. C. 1979, Convective collapse of flux tubes, *Solar Phys.*, 61, 363-378.

Spruit, H. C., and Zweibel, E. G. 1979, Convective instability of thin flux tubes, *Solar Phys.*, 62, 15-22.

Stenflo, J. O. 1989, Small scale magnetic structures on the Sun, *Astron. Astrophys. Rev.*, in press.

Syrovatskii, J., and Zhugzhda, Y. D. 1967, *Astron. Zh.*, 44, 1180-1190 (English translation 1968, Oscillatory convection of a conducting gas in a strong magnetic field, *Sov. Astron.*, 11, 945-952).

Thomas, J. H. 1984, Flow in an isolated magnetic flux tube, in *Small-Scale Dynamical Processes in Quiet Stellar Atmospheres*, ed. S. L. Keil (Sunspot, NM: National Solar Observatory), pp. 276-277.

Thomas, J. H. 1988, Siphon flows in isolated magnetic flux tubes, *Astrophys. J.*, 333, 407-419.

Unno, W., and Ando, H. 1979, Instability of a thin magnetic tube in the solar atmosphere, *Geophys. Astrophys. Fluid Dyn.*, 12, 107-115.

Venkatakrishnan, P. 1983, Nonlinear development of convective instability within slender flux tubes, *J. Astrophys. Astron.*, 4, 135-149.

Venkatakrishnan, P. 1985, Nonlinear development of convective instability within slender flux tubes. II. The effect of radiative heat transport, *J. Astrophys. Astron.*, 6, 21-34.

Vernazza, J. E., Avrett, E. H., and Loeser, R., Structure of the solar chromosphere. III. Models of the EUV brightness components of the quiet Sun, *Astrophys. J. Suppl.*, 45, 635-725.

Webb, A. R., and Roberts, B. 1978, Vertical motions in an intense magnetic flux tube. II. Convective instability, *Solar Phys.*, 59, 249-274.

Weiss, N. O. 1966, The expulsion of magnetic flux by eddies, *Proc. Roy. Soc. London*, A 293, 310-328.

ON THE THIN MAGNETIC FLUX TUBE APPROXIMATION[1]

A. Ferriz-Mas and M. Schüssler

Kiepenheuer-Institut für Sonnenphysik, Schöneckstr. 6,
D-7800 Freiburg, Federal Republic of Germany

Abstract. A large number of theoretical studies on the structure and dynamics of magnetic flux tubes make use of the "thin flux tube approximation", based on an expansion approach about the axis of the tube, which permits the reduction of the full magnetohydrodynamic (MHD) problem to a mathematically more tractable set of equations. In this paper the assumptions underlying this approximation are analyzed.

For a vertical, axisymmetric magnetic flux tube the pertaining physical quantities are expanded in the radial coordinate about the axis of symmetry and the power series are introduced into the MHD equations written in cylindrical coordinates. The assumption of axial symmetry significantly reduces the number of unknowns and equations. The closure of the system is provided by appropriate boundary conditions. By retaining only the zeroth- and first-order terms, the equations of the conventional thin flux tube approximation (Defouw, 1976; Roberts and Webb, 1978) are obtained. To include twisted magnetic fields and azimuthal flows, the expansion has to be extended at least to second order. As an application, the axisymmetric wave modes of a uniform magnetic cylinder are studied. The above formalism is also applied to derive the magnetostatic equations governing the equilibrium structure of an axisymmetric, vertical flux tube embedded in a stratified atmosphere.

Key words: magnetic flux tubes - magnetohydrodynamics - solar and stellar magnetic fields

1. Introduction

The study of the structure and dynamics of magnetic flux tubes in stellar atmospheres, even under the restricting assumption of axial symmetry which makes a reduction to two spatial variables possible, is a complex mathematical problem due to the nonlinear nature of the magnetohydrodynamic (MHD) equations, to the presence of boundary layers and to the inhomogeneities involved. Except in the corona, where the pressure scale-height is very large, the effects of gravity cannot be ignored, thus introducing an additional complexity. Therefore, the entire problem would be one involving three governing forces (pressure gradient force, magnetic Lorentz force and gravity) actuating simultaneously and subject to the presence of lateral boundaries. Thus, unless various simplifying assumptions are made (incompressible limit, neglect of gravity, linearized form of the equations, etc.), the difficulty of solving the whole system of MHD equations is great.

This difficulty can be avoided in some situations by recourse to an approximation based on an expansion of the system of MHD equations in the radial coordinate. The *thin flux tube approximation* was originally introduced by Defouw (1976), who called it the *magnetohydraulic approximation* (the case of an incompressible fluid was first considered by Parker, 1974). In this approximation the scalar variables, density, pressure and temperature, and the axial components of the flow velocity and magnetic field vectors are assumed to be constant over any cross-section of the tube. The resulting *thin flux tube equations* were later derived by Roberts and Webb (1978) considering an expansion in the (scaled) radial coordinate of the complete MHD equations. These equations [see Eqs. (3.2)-(3.5)] only contain zeroth-order expansion coefficients and are essentially a description of the behaviour on the axis of symmetry of the tube; yet, the first-order coefficients of the radial components of magnetic field and velocity are implicitly contained in this approximation (see the discussion in Roberts and Webb, 1979, and also Sect. 3 in this contribution). In this crude approximation, which in many cases allows an analytical treatment, magnetic curvature forces are neglected because the z-component of the Lorentz force is of second order. A generalization of the (zeroth-order) thin flux tube equations for tubes with non-straight axes was presented by Spruit (1981); see also Moreno-Insertis (1986).

We extend this approximation by introducing a general expansion scheme to obtain a fully consistent closed set of magnetohydrodynamic equations in two independent variables, which is particularly useful to describe axially symmetric, time-dependent problems for which all quantities do not vary significantly in the radial direction. This is done by considering the hierarchy of expanded magnetofluid equations in cylindrical coordinates (r, ϕ, z) and equating terms with equal powers in the radial coordinate r. The assumption of axial symmetry

[1]Mitteilungen aus dem Kiepenheuer-Institut Nr. 313

Geophysical Monograph 58

significantly reduces the number of unknowns and equations. The closure of the system is provided by appropriate boundary conditions.

In Sect. 2 we present the set of expanded equations up to second order, which permits the inclusion of twisted magnetic fields and azimuthal flows. By considering only the zeroth-order terms in the expanded equations, the *conventional* or *zeroth-order* thin flux tube approximation is recovered; this is done in Sect. 3. In Sect. 4 we consider the linearized system of expanded MHD equations up to second order; we determine the axisymmetric wave modes of a cylindrical homogeneous flux tube, a configuration for which an exact analytical investigation of linear wave modes is possible (i.e., without resort to an expansion procedure). We compare and discuss the results arising from the expansion procedure with those of the exact treatment. It turns out that body waves in flux tubes are not adequately described by means of expansions truncated at low orders, while long-wavelength surface waves can be well treated. This result extends and complements the analysis of Roberts and Webb (1979), who considered the expansion procedure to zeroth order. The formalism developed in Sect. 2 can also be applied to the magnetostatic equilibrium structure of a flux tube embedded in a stratified atmosphere; this is done in Sect. 5. It turns out that the temperature distribution within the flux tube may be specified *a priori* if no direct reference to an energy equation is made. The magnetostatic equations derived in Sect. 5 encompass special cases studied previously.

A note on the terminology: we will adopt here the term *thin flux tube approximation* to mean the description of the structure and dynamics of magnetic flux tubes by means of a set of expanded equations (truncated at a certain low order) which is mathematically closed on application of appropriate boundary conditions, thus avoiding a two- or three-dimensional MHD problem.

2. Expansion of the magnetohydrodynamic equations

For reasons of simplicity we shall confine ourselves to the framework of ideal MHD (i.e., no magnetic or viscous diffusivities and no heat fluxes) and also assume that the plasma obeys the equation of state of an ideal gas. The system of ideal MHD equations is (in c.g.s units):

$$\begin{aligned}
&\operatorname{div}\mathbf{B} = 0\,,\\
&\frac{\partial \mathbf{B}}{\partial t} = \mathbf{rot}\,(\mathbf{v}\wedge\mathbf{B})\,,\\
&\frac{\partial \rho}{\partial t} + \operatorname{div}(\rho\,\mathbf{v}) = 0\,,\\
&\rho\,[\frac{\partial \mathbf{v}}{\partial t} + (\mathbf{v}\cdot\mathbf{grad}\,)\mathbf{v}] =\\
&\qquad = -\mathbf{grad}\,p + \rho\,\mathbf{g} + \frac{1}{4\pi}\mathbf{rot}\,\mathbf{B}\wedge\mathbf{B}\,,\\
&\frac{\partial p}{\partial t} + \mathbf{v}\cdot\mathbf{grad}\,p = \frac{\gamma\,p}{\rho}\,(\frac{\partial \rho}{\partial t} + \mathbf{v}\cdot\mathbf{grad}\,\rho)\,,
\end{aligned} \tag{2.1}$$

where $\mathbf{B} = (B_r, B_\phi, B_z)$ is the magnetic field, $\mathbf{v} = (v_r, v_\phi, v_z)$ is the flow velocity, p is the plasma pressure, ρ is the mass density and $\mathbf{g} = (0, 0, -g)$ represents the gravitational acceleration.

Assume that the radius of the tube, $R(z)$, is much smaller than *each* characteristic length L of the variation of the physical quantities of the system, and introduce the nondimensional coordinate $\tilde{r} \equiv r/L$, which serves as a small-order parameter. In the following we drop the tilde for ease of notation. The essence of the expansion approach is the assumption that the radial variation of all quantities within the interval $0 \le r \le R(z)$ can be reasonably described by their low-order (zeroth, first, second, ...) radial derivatives taken on the axis ($r = 0$). Consequently, all pertaining quantities are expanded as a Taylor series in r:

$$\begin{aligned}
v_r(r,z,t) &= r v_{r_1}(z,t) + r^3 v_{r_3}(z,t) + \cdots\\
v_\phi(r,z,t) &= r v_{\phi 1}(z,t) + r^3 v_{\phi 3}(z,t) + \cdots\\
v_z(r,z,t) &= v_{z_0}(z,t) + r^2 v_{z_2}(z,t) + \cdots\\
B_r(r,z,t) &= r B_{r_1}(z,t) + r^3 B_{r_3}(z,t) + \cdots\\
B_\phi(r,z,t) &= r B_{\phi 1}(z,t) + r^3 B_{\phi 3}(z,t) + \cdots\\
B_z(r,z,t) &= B_{z_0}(z,t) + r^2 B_{z_2}(z,t) + \cdots\\
p(r,z,t) &= p_0(z,t) + r^2 p_2(z,t) + \cdots\\
\rho(r,z,t) &= \rho_0(z,t) + r^2 \rho_2(z,t) + \cdots\,,
\end{aligned} \tag{2.2}$$

where we have introduced the notation

$$f_k(z,t) \equiv \frac{1}{k!}\frac{\partial^k f}{\partial r^k}(0,z,t)\,, \quad k = 0,1,2,\ldots$$

for the expansion coefficients of any function $f(r,z,t)$ which satisfies the appropriate regularity conditions at $r = 0$.

The form of the expansions (2.2), in which the radial and azimuthal components of $\mathbf{v}$ and $\mathbf{B}$ are odd series in r, while the axial components v_z and B_z as well as the scalar functions p and ρ are even series in r, follows directly from the symmetry of the problem (Ferriz-Mas and Schüssler, 1989).

Inserting the power series (2.2) into the MHD equations (2.1) written in cylindrical polar coordinates and sorting out the contributions corresponding to equal powers in r we obtain a system of expanded magnetofluid equations in the variables z and t (Ferriz-Mas *et al.* 1989; Ferriz-Mas and Schüssler, 1989). Here we present and discuss the set of MHD equations resulting from the expansion up to second order in r (the primes denote partial differentiation with respect to z):

Zeroth-order equations:

- $$\rho_0\left(\frac{\partial v_{z0}}{\partial t} + v_{z0}v'_{z0}\right) + p'_0 + \rho_0 g = 0\,, \tag{2.3}$$
- $$\frac{\partial \rho_0}{\partial t} + (\rho_0 v_{z0})' + 2\rho_0 v_{r1} = 0\,, \tag{2.4}$$
- $$\frac{\partial B_{z0}}{\partial t} + v_{z0}B'_{z0} + 2B_{z0}v_{r1} = 0\,, \tag{2.5}$$
- $$\frac{\partial p_0}{\partial t} + v_{z0}p'_0 - c^2_{S0}\,(\frac{\partial \rho_0}{\partial t} + v_{z0}\rho'_0) = 0\,, \tag{2.6}$$

with $c^2_{S0} = \gamma\,p_0/\rho_0$.

First-order equations:

$$\bullet\ \rho_0\left(\frac{\partial v_{r1}}{\partial t}+v_{z0}v'_{r1}+v_{r1}^2-v_{\phi1}^2\right)+2p_2+$$
$$+\frac{1}{4\pi}\left(-B_{z0}B'_{r1}+2B_{z0}B_{z2}+2B_{\phi1}^2\right)=0\,, \qquad (2.7)$$

$$\bullet\ \rho_0\left(\frac{\partial v_{\phi1}}{\partial t}+v_{z0}v'_{\phi1}+2v_{r1}v_{\phi1}\right)+$$
$$+\frac{1}{4\pi}\left(B_{\phi1}B'_{z0}-B_{z0}B'_{\phi1}\right)=0\,, \qquad (2.8)$$

$$\bullet\ \frac{\partial B_{r1}}{\partial t}+(v_{z0}B_{r1})'-(v_{r1}B_{z0})'=0\,, \qquad (2.9)$$

$$\bullet\ \frac{\partial B_{\phi1}}{\partial t}+(v_{z0}B_{\phi1})'-(v_{\phi1}B_{z0})'+$$
$$+2v_{r1}B_{\phi1}-2v_{\phi1}B_{r1}=0. \qquad (2.10)$$

Second-order equations:

$$\bullet\ \rho_0\left[\frac{\partial v_{z2}}{\partial t}+2v_{r1}v_{z2}+(v_{z0}v_{z2})'\right]-\rho_2\left(\frac{p'_0}{\rho_0}-\frac{p'_2}{\rho_2}\right)+$$
$$+\frac{1}{4\pi}\left(B_{\phi1}B'_{\phi1}+B_{z2}B'_{z0}+B_{r1}B'_{r1}\right)=0\,, \qquad (2.11)$$

$$\bullet\ \frac{\partial \rho_2}{\partial t}+(\rho_0 v_{z2})'+(\rho_2 v_{z0})'+4\rho_0 v_{r3}+4\rho_2 v_{r1}=0\,, \qquad (2.12)$$

$$\bullet\ \frac{\partial B_{z2}}{\partial t}-4v_{z0}B_{r3}-4v_{z2}B_{r1}+$$
$$+4v_{r1}B_{z2}+4v_{r3}B_{z0}=0\,, \qquad (2.13)$$

$$\bullet\ \frac{\partial p_2}{\partial t}+v_{z0}p'_2+v_{z2}p'_0+2v_{r1}p_2-c_{S0}^2\left(\frac{\partial \rho_2}{\partial t}+v_{z0}\rho'_2+\right.$$
$$\left.+v_{z2}\rho'_0+2v_{r1}\rho_2\right)-c_{S2}^2\left(\frac{\partial \rho_0}{\partial t}+v_{z0}\rho'_0\right)=0\,, \qquad (2.14)$$

with

$$c_{S2}^2=\gamma\frac{p_0}{\rho_0}\left(\frac{p_2}{p_0}-\frac{\rho_2}{\rho_0}\right).$$

Equations (2.3) to (2.14) are complemented by the zeroth and second orders of $\operatorname{div}\mathbf{B}=0$:

$$\bullet\ 2B_{r1}+B'_{z0}=0\,, \qquad (2.15)$$
$$\bullet\ 4B_{r3}+B'_{z2}=0\,. \qquad (2.16)$$

Eq. (2.9) reduces to Eq. (2.5) with the help of Eq. (2.15); equation (2.16) in turn permits the elimination of B_{r3} in favor of the unknown B_{z2} in Eq. (2.13). Since equation (2.13) also contains v_{r3}, we have 13 equations for 14 unknowns and the system is not closed yet. In order to close the system we have to specify the kind of problem we wish to consider and the corresponding boundary conditions.

The scheme outlined above is in principle applicable to any axisymmetric MHD configuration (static or dynamic) such that the radial variation of all physical quantities can be well represented by using low-order Taylor expansions. In the present case we shall consider an axisymmetric magnetic flux tube embedded in a non-magnetic medium and confined by external pressure forces.* The boundary between the magnetized plasma inside the flux tube and the external medium represents a *tangential discontinuity;* at this interface we require the continuity of the total (gas plus magnetic) pressure (e.g., Landau and Lifshitz, 1960):

$$p+\left.\frac{|\mathbf{B}|^2}{8\pi}\right|_{r=R}=p_e(R)\,, \qquad (2.17)$$

where $p_e(R)\equiv p_e(R,z,t)$ denotes the external gas pressure at the boundary. This pressure equilibrium is established on a time-scale τ of the order of the travel time of a fast magneto-acoustic wave across the tube; therefore, if the radius of the flux tube is sufficiently small compared to the other length-scales involved ($R/L<<1$), τ can be made arbitrarily small compared to any other dynamical time-scale in the system, so that (2.17) expresses the condition of instantaneous lateral pressure balance (e.g., Spruit, 1981).

To second order in the radial expansion, Eq. (2.17) reads:

$$p_0+\frac{B_{z0}^2}{8\pi}+R^2\left(p_2+\frac{B_{r1}^2+B_{\phi1}^2}{8\pi}+\frac{B_{z0}B_{z2}}{4\pi}\right)=p_e(R)\,. \qquad (2.18)$$

The radius R is determined by the requirement of magnetic flux conservation, which up to second order yields:

$$\pi R^2 B_{z0}=\Phi_{\rm mag}=\text{ const.} \qquad (2.19)$$

Consequently, R is directly related to the instantaneous zeroth-order longitudinal magnetic field.

To determine the instantaneous external plasma pressure we have to solve the hydrodynamic equations for the exterior of the flux tube and match the result to the interior solution so that the conditions of lateral pressure balance, Eq. (2.17), and of continuity of the velocity component normal to the boundary are satisfied. In many analytical as well as numerical studies, the external pressure is simply taken as a given function of z which does not vary in time. This simplification, which is not inherent in the thin flux tube approximation, usually permits a more extensive mathematical investigation, although important phenomena like surface waves (e.g., Roberts and Webb, 1979; Ferriz-Mas et al., 1989) or the formation of solitons by counterbalance between nonlinear effects and the dispersion provided by the inertia of the external medium cannot be then described (Roberts, 1985).

3. The zeroth-order thin flux tube approximation

Consider the zeroth-order equations (2.3)-(2.6) along with (2.15), and the first-order equations (2.7)-(2.10). For an untwisted flux tube ($B_\phi=0, v_\phi=0$), Eqs. (2.8) and (2.10) drop out. With equation (2.15) equation (2.9) reduces to equation (2.5). Note that the second-order coefficient p_2, which appears in the first-order equation (2.7), couples the zeroth- and first-order equations with the second-order contributions. Leaving Eq. (2.7) aside we would end up with 5 equations for 6 unknown functions, viz. ρ_0, p_0, v_{z0}, B_{z0}, v_{r1}, B_{r1}. The unknown v_{r1} may be eliminated between Eqs. (2.4) and (2.5) to yield

$$\frac{\partial}{\partial t}\left(\frac{\rho_0}{B_{z0}}\right)+\frac{\partial}{\partial z}\left(v_{z0}\frac{\rho_0}{B_{z0}}\right)=0\,. \qquad (3.1)$$

* If a magnetic field external to the tube is considered, the magnetic pressure term $B_e^2(z)/8\pi$ has to be added to the external gas pressure.

Dropping all subindices for ease of notation, the system of differential equations describing longitudinal (or axisymmetric) isentropic motions in a vertical, untwisted flux tube are:

$$\frac{\partial}{\partial t}(\frac{\rho}{B}) + \frac{\partial}{\partial z}(v\frac{\rho}{B}) = 0 , \tag{3.2}$$

$$\frac{\partial v}{\partial t} + v\frac{\partial v}{\partial z} + \frac{1}{\rho}\frac{\partial p}{\partial z} + g = 0 , \tag{3.3}$$

$$\frac{\partial p}{\partial t} + v\frac{\partial p}{\partial z} = \frac{\gamma p}{\rho}\left(\frac{\partial \rho}{\partial t} + v\frac{\partial \rho}{\partial z}\right) , \tag{3.4}$$

which is complemented by the zeroth-order contribution of the pressure balance equation at $r = R$,

$$p + \frac{B^2}{8\pi} = p_e . \tag{3.5}$$

This is just the set of equations of the conventional thin flux tube approximation (Defouw, 1976; Roberts and Webb, 1978). Notice that only the values of the variables on the axis of symmetry of the tube appear in Eqs. (3.2)-(3.5). It should be noted, however, that although the azimuthal components of $\mathbf{v}$ and $\mathbf{B}$ have been assumed equal zero, the radial components v_r and B_r *are not ignored* in this model. For if $B_r = 0$ then, as a consequence of $\operatorname{div}\mathbf{B} = 0$, the zeroth-order field $(0, 0, B(z,t))$ would be independent of z. In fact, the condition $\operatorname{div}\mathbf{B} = 0$ permits to determine B_r (to first order) once $B(z,t)$ has been found [Eq. (2.15)]. The radial component v_r can be determined (to first order) from the axial component of the induction equation [Eq. (2.5)], for instance. Thus, the (zeroth-order) thin flux tube approximation does not assume the internal magnetic field to be purely axial and uniform over the cross-section of the tube, as can be read in some papers.

Applying the condition of flux conservation, Eq. (2.19), and setting $A = \pi R^2$ (the cross-sectional area of the tube), Eq. (3.2), which combines the continuity equation and the z-component of the induction equation (to zeroth order) can be cast into

$$\frac{\partial}{\partial t}(\rho A) + \frac{\partial}{\partial z}(\rho v A) = 0 . \tag{3.6}$$

Equations (3.6), (3.3) and (3.4) are identical to the equations of continuity, momentum and energy (isentropic motions) for a (non-ionized) gas in a elastic tube of varying cross-section when the motions $v(z,t)$ are predominantly directed along the tube (Lighthill, 1978). In general, area changes must be related to pressure changes to close the system, i.e., a relationship between $A(z,t)$ and $p(z,t)$ must be specified. When the boundary of the tube is defined by magnetic lines of force (magnetic flux tube), this relationship is taken to be Eq. (3.5). The intrinsic elasticity of the tube is then the *magnetic distensibility*, whose effect can be understood in terms of a new restoring force added to the gas compressibility [see, e.g., Rae and Roberts, 1982; Roberts, 1985; Ferriz-Mas, 1988 (Appendix A)].

4. Linear waves in the second-order thin flux tube approximation

The system of expanded MHD equations to second order derived in Sect. 2 allows for the nonlinear coupling of torsional and longitudinal waves, including the effects of gravitation, yet avoiding a full two- or three-dimensional calculation. A numerical simulation of the nonlinear interaction of a magnetic flux tube with a surrounding whirl flow in the presence of a gravitational field has been carried out by Anton (1989).

By linearizing the system of expanded equations presented in Sect. 2 we can determine longitudinal and torsional Alfvén waves, while transversal waves have been suppressed by assuming $\partial/\partial\phi \equiv 0$. The results can be compared to the exact solutions and may serve as test cases for numerical codes.

As an equilibrium state we consider a homogeneous cylindrical flux tube with uniform axial magnetic field $\bar{B}_{z0} = \text{const.}$ In the absence of gravity a static ($\mathbf{v} \equiv 0$), isothermal equilibrium is described by constant pressure $\bar{p}_0$ and density $\bar{\rho}_0$ within the tube and constant pressure $\bar{p}_e$ and density $\bar{\rho}_e$ outside. The pressure balance condition reads:

$$\bar{p}_0 + \frac{\bar{B}_{z0}^2}{8\pi} = \bar{p}_e . \tag{4.1}$$

We shall consider small departures from the equilibrium state by setting:

$$p_e = \bar{p}_e + \tilde{p}_e \,, p_0 = \bar{p}_0 + \tilde{p}_0 \,, p_2 = \tilde{p}_2 \,, v_{z0} = \tilde{v}_{z0}, \ldots \tag{4.2}$$

where the quantities with a *tilde* denote the perturbations. By substituting (4.2) into Eqs. (2.3)-(2.16) and neglecting squares and products of perturbations the following linearized system results:

Zeroth-order equations:

$$\bar{\rho}_0\frac{\partial \tilde{v}_{z0}}{\partial t} + \tilde{p}_0' = 0 , \tag{4.3}$$

$$\frac{\partial \tilde{\rho}_0}{\partial t} + \bar{\rho}_0\tilde{v}_{z0}' + 2\bar{\rho}_0\tilde{v}_{r1} = 0 , \tag{4.4}$$

$$\frac{\partial \tilde{B}_{z0}}{\partial t} + 2\bar{B}_{z0}\tilde{v}_{r1} = 0 , \tag{4.5}$$

$$\frac{\partial \tilde{p}_0}{\partial t} - \bar{c}_{S0}^2\frac{\partial \tilde{\rho}_0}{\partial t} = 0 , \tag{4.6}$$

$$2\tilde{B}_{r1} + \tilde{B}_{z0}' = 0 , \tag{4.7}$$

with $\bar{c}_{S0}^2 = \gamma\bar{p}_0/\bar{\rho}_0$. Eqs. (4.4) and (4.5) may be combined into:

$$\frac{1}{\bar{B}_{z0}}\frac{\partial \tilde{B}_{z0}}{\partial t} - \frac{1}{\bar{\rho}_0}\frac{\partial \tilde{\rho}_0}{\partial t} - \tilde{v}_{z0}' = 0 . \tag{4.8}$$

First-order equations:

$$\bar{\rho}_0\frac{\partial \tilde{v}_{r1}}{\partial t} + 2\tilde{p}_2 - \frac{1}{4\pi}\left(\bar{B}_{z0}\tilde{B}_{r1}' - 2\bar{B}_{z0}\tilde{B}_{z2}\right) = 0 , \tag{4.9}$$

$$\bar{\rho}_0\frac{\partial \tilde{v}_{\phi 1}}{\partial t} - \frac{1}{4\pi}\bar{B}_{z0}\tilde{B}_{\phi 1}' = 0 , \tag{4.10}$$

$$\frac{\partial \tilde{B}_{\phi 1}}{\partial t} - \bar{B}_{z0}\tilde{v}_{\phi 1}' = 0 . \tag{4.11}$$

Second-order equations:

$$\bar{\rho}_0\frac{\partial \tilde{v}_{z2}}{\partial t} + \tilde{p}_2' = 0 , \tag{4.12}$$

$$\frac{\partial \tilde{\rho}_2}{\partial t} + \bar{\rho}_0\tilde{v}_{z2}' + 4\bar{\rho}_0\tilde{v}_{r3} = 0 , \tag{4.13}$$

$$\frac{\partial \tilde{B}_{z2}}{\partial t} + 4\bar{B}_{z0}\tilde{v}_{r3} = 0\,, \tag{4.14}$$

$$\frac{\partial \tilde{p}_2}{\partial t} - \bar{c}^2_{S0}\frac{\partial \tilde{\rho}_2}{\partial t} = 0\,. \tag{4.15}$$

These equations are complemented by the linearized versions of Eqs. (2.18) and (2.19):

$$\tilde{p}_0 + \frac{1}{4\pi}\bar{B}_{z0}\tilde{B}_{z0} + \bar{R}^2\left(\tilde{p}_2 + \frac{1}{4\pi}\bar{B}_{z0}\tilde{B}_{z2}\right) = \tilde{p}_e(\bar{R})\,, \tag{4.16}$$

$$\bar{R}\tilde{B}_{z0} + 2\tilde{R}\bar{B}_{z0} = 0\,, \tag{4.17}$$

with $\pi\bar{R}^2\bar{B}_{z0} = \Phi_{\text{mag}}$ being the total magnetic flux in the tube.

Eqs. (4.10) and (4.11), which describe the ϕ-components of velocity and magnetic field, are decoupled from the rest of the system and can be combined into a single wave equation which governs the propagation of torsional Alfvén waves. The rest of the system, Eqs. (4.3)-(4.9) and (4.12)-(4.17), is not closed yet.

To consistently include the external pressure field one has to solve the linearized equations for the interior and the exterior of the flux tube and relate the interior ($r < \bar{R}$) and exterior ($r > \bar{R}$) solutions by the boundary conditions of lateral pressure balance, Eq. (4.16), and of continuity of the velocity component normal to the boundary (v_r, in our case). The procedure is similar to that described in Roberts and Webb (1978, 1979), who considered the zeroth-order equations. The governing equations for the field-free exterior medium are the linearized hydrodynamic equations:

$$\begin{aligned}
&\frac{\partial \tilde{\rho}_e}{\partial t} + \bar{\rho}_e \operatorname{div}\tilde{\mathbf{v}}_e = 0\,,\\
&\bar{\rho}_e\frac{\partial \tilde{\mathbf{v}}_e}{\partial t} + \mathbf{grad}\,\tilde{p}_e = 0\,,\\
&\frac{\partial \tilde{p}_e}{\partial t} - c^2_{Se}\frac{\partial \tilde{\rho}_e}{\partial t} = 0\,,
\end{aligned} \tag{4.18}$$

where the perturbations $\tilde{\rho}_e$, $\tilde{p}_e$ and $\mathbf{v}_e = \tilde{\mathbf{v}}_e$ are functions of r, z and t, and $c^2_{Se} = \gamma\bar{p}_e/\bar{\rho}_e$ is the exterior sound speed. Equations (4.18) may be combined into

$$\frac{\partial^2 \tilde{p}_e}{\partial t^2} = c^2_{Se}\left(\frac{\partial^2}{\partial r^2} + \frac{1}{r}\frac{\partial}{\partial r} + \frac{\partial^2}{\partial z^2}\right)\tilde{p}_e\,. \tag{4.19}$$

We consider harmonic plane waves of the form $\tilde{Q}(z,t) = \hat{Q}\exp(i\omega t - ikz)$, where $\tilde{Q}(z,t)$ denotes any disturbed zeroth-, first- or second-order quantity inside the flux tube (e.g., $\tilde{\rho}_0, \tilde{\rho}_2, \tilde{p}_0, \ldots$) and $\hat{Q}$ is the (constant) complex amplitude; ω denotes the angular frequency and k is the wave number in the z direction. For the external perturbations $\tilde{Q}_e$ an Ansatz of the form $\tilde{Q}_e(r,z,t) = \hat{Q}_e(r)\exp(i\omega t - ikz)$ is made, in which the r-dependence of the amplitudes is retained and, in addition to the boundary conditions at $r = \bar{R}$, we shall also require that the disturbances are finite at the origin ($r = 0$) and that they tend to zero as r tends to ∞. The resulting dispersion relation for radially evanescent waves in the exterior is found to be

$$\begin{aligned}
\bar{\rho}_0(\omega^2 - k^2c_T^2)(c_S^2 + v_A^2)m_e\left(1 + \frac{m_0^2\bar{R}^2}{4}\right)K_1(m_e\bar{R}) =\\
= \frac{1}{2}\bar{\rho}_e\bar{R}\omega^2(\omega^2 - k^2c_S^2)K_0(m_e\bar{R})\,,
\end{aligned} \tag{4.20}$$

with

$$\begin{aligned}
m_0^2 &= \frac{(\omega^2 - k^2c_S^2)(\omega^2 - k^2v_A^2)}{(c_S^2 + v_A^2)(k^2c_T^2 - \omega^2)}\,,\\
m_e^2 &= \frac{k^2c_{Se}^2 - \omega^2}{c_{Se}^2}\,.
\end{aligned} \tag{4.21}$$

Here c_S, v_A and c_T are, respectively, the sound, Alfvén and cusp (or tube) speeds referred to the equilibrium state, i.e., $c_S^2 = \gamma\bar{p}_0/\bar{\rho}_0$, $v_A^2 = \bar{B}_{z0}^2/(4\pi\bar{\rho}_0)$ and $c_T^2 = c_S^2v_A^2/(c_S^2 + v_A^2)$. The interior and exterior sound speeds are related through the condition of lateral pressure balance:

$$c_{Se}^2 = \frac{\bar{\rho}_0}{\bar{\rho}_e}(c_S^2 + \frac{\gamma}{2}v_A^2)\,.$$

The signs of the squared radial wavenumbers m_0^2 and m_e^2 determine the nature of the solutions inside and outside the flux tube. The present case of radially evanescent solutions outside the flux tube ($m_e^2 > 0$) corresponds to a vertical phase speed ω/k which is subsonic with respect to the exterior sound speed c_{Se}. The form of the solutions inside the tube depends upon the sign of m_0^2. For $m_0^2 > 0$ the interior solutions are radially evanescent disturbances whose maximum amplitude is reached on the boundary (*surface modes*). For $m_0^2 < 0$ the solutions inside the tube are oscillatory modes (*body modes*). The eigenfunctions inside the tube are proportional to

$$1 + \frac{(m_0r)^2}{4} \quad \text{if} \quad m_0^2 > 0\,, \tag{4.22a}$$

$$1 - \frac{(n_0r)^2}{4} \quad \text{if} \quad m_0^2 \equiv -n_0^2 < 0\,, \tag{4.22b}$$

for $\hat{\rho}(r) = \hat{\rho}_0 + r^2\hat{\rho}_2$, $\hat{p}(r) = \hat{p}_0 + r^2\hat{p}_2$, $\hat{B}_z(r) = \hat{B}_{z0} + r^2\hat{B}_{z2}$, and $\hat{v}_z(r) = \hat{v}_{z0} + r^2\hat{v}_{z2}$, and proportional to

$$m_0r/2 \quad \text{if} \quad m_0^2 > 0\,, \tag{4.23a}$$

$$n_0r/2 \quad \text{if} \quad m_0^2 \equiv -n_0^2 < 0\,, \tag{4.23b}$$

for the radial components $\hat{B}_r(r) = r\hat{B}_{r1}$ and $\hat{v}_r(r) = r\hat{v}_{r1}$.

Turning now to the exact treatment, the dispersion relation for the wave modes of a magnetic cylinder when the disturbances outside the tube are evanescent ($m_e^2 > 0$) is (e.g., Roberts and Webb, 1979):

$$\begin{aligned}
\bar{\rho}_0\,(k^2c_T^2 - \omega^2)(c_S^2 + v_A^2)m_0m_eI_0(m_0\bar{R})K_1(m_e\bar{R}) =\\
\bar{\rho}_e\omega^2\,(k^2c_S^2 - \omega^2)I_1(m_0\bar{R})K_0(m_e\bar{R})\,,
\end{aligned} \tag{4.24}$$

valid for surface modes ($m_0^2 > 0$). For body modes inside the tube ($n_0^2 \equiv -m_0^2 > 0$) the Bessel function I is replaced by J and m_0 is replaced by n_0 :

$$\begin{aligned}
\bar{\rho}_0\,(k^2c_T^2 - \omega^2)(c_S^2 + v_A^2)n_0m_eJ_0(n_0\bar{R})K_1(m_e\bar{R}) =\\
\bar{\rho}_e\omega^2\,(k^2c_S^2 - \omega^2)J_1(n_0\bar{R})K_0(m_e\bar{R})\,.
\end{aligned} \tag{4.25}$$

The eigenfunctions for $0 \le r \le \bar{R}$ are proportional to

$$I_0(m_0r) \quad \text{if} \quad m_0^2 > 0\,, \tag{4.26a}$$

$$J_0(n_0r) \quad \text{if} \quad m_0^2 \equiv -n_0^2 < 0\,, \tag{4.26b}$$

for the perturbations of pressure, density and axial components of the velocity and magnetic field, and proportional to either

$I_1(m_0 r)$ or $J_1(n_0 r)$ for the perturbations of the radial components of $\mathbf{v}$ and $\mathbf{B}$. [Other authors have generalized the problem to describe the wave modes of a homogeneous plasma cylinder embedded in a magnetic environment; a very complex array of wave modes arise. For further references see, e.g., Spruit, 1982; Edwin and Roberts, 1983; Thomas, 1985; Abdelatif, 1988].

The expressions (4.22a) and (4.22b) are the respective expansions of $I_0(m_0 r)$ and $J_0(n_0 r)$ to second order in the argument $[(m_0 r)^2$ or $(n_0 r)^2]$, while (4.23a) and (4.23b) coincide with $I_1(m_0 r)$ and $J_1(n_0 r)$ to first order. As regards the dispersion relations, expanding equation (4.24) in powers of $m_0 \bar{R}$ and retaining only terms up to second order, the dispersion relation (4.20) is recovered; the same holds in the case of body waves replacing m_0 and I by n_0 and J. Now, for surface waves ($m_0^2 > 0$) we have that $k\bar{R} \to 0$ implies $m_0 \bar{R} \to 0$ (Roberts and Webb, 1979), showing that the eigenfunctions (4.22a) and (4.23a) and the dispersion relation (4.20) are a fully consistent approximation to the exact eigenfunctions $I_0(m_0 r)$ and $I_1(m_0 r)$ and to the dispersion relation (4.24). For body waves ($m_0^2 < 0$), on the other hand, $n_0 \bar{R}$ remains finite and non zero for $k\bar{R} \to 0$; more especifically, Roberts and Webb (1979) show that $n_0 \bar{R}$ tends to the roots of the Bessel function J_1 as $k\bar{R} \to 0$ (notice that the dispersion relation (4.25) has indeed an infinite number of roots, which is owed to the oscillatory character of the Bessel functions J_0 and J_1). Therefore, the eigenfunctions (4.22b) and (4.23b) resulting from the expansion procedure can be a good aproximation to the exact eigenfunctions $J_0(n_0 r)$ and $J_1(n_0 r)$ only sufficiently near the axis, where $n_0 r$ can be made arbitrarily small and higher-order terms in the expansions of J_0 and J_1 can be consistently neglected, but not throughout the interval $0 \le r \le \bar{R}$. Thus, it turns out that the expansion procedure applied to a magnetic cylinder of plasma is best suited to approximately describe long-wavelength surface waves, but not body waves.

5. Equations of the magnetostatic structure of a twisted flux tube in a stratified atmosphere

As another application of the general expansion procedure presented in Sect. 2 we now consider the magnetostatic equilibrium structure of a (weakly) twisted vertical flux tube with axial symmetry embedded in a plasma stratified by gravity, which is a problem of current astrophysical interest (see, e.g., Parker, 1979; Sturrock and Uchida, 1981; Browning and Priest, 1982, 1983, 1984; Pneuman et al., 1986). The flux tube is confined by an external plasma pressure which is a function of height: $p_e = p_e(z)$.

Setting $\partial/\partial t \equiv 0$ and $\mathbf{v} \equiv 0$, from the system of expanded equations (2.3) to (2.16) and (2.18) we obtain as a particular case the set of equations which determine the equilibrium structure of a twisted, vertical flux tube (to order r^2):

Zeroth-order:

$$\mathcal{R}\, T_0\, p_0' + g\, p_0 = 0\,, \tag{5.1}$$

$$B_{r1} = -\frac{1}{2} B_{z0}'\,. \tag{5.2}$$

First-order:

$$p_2 + \frac{1}{8\pi}(-B_{z0} B_{r1}' + 2B_{z0} B_{z2} + 2B_{\phi1}^2) = 0\,, \tag{5.3}$$

$$B_{z0} B_{\phi1}' - B_{z0}' B_{\phi1} = 0\,. \tag{5.4}$$

Second-order:

$$p_2' - p_0'\left(\frac{p_2}{p_0} - \frac{T_2}{T_0}\right) + \frac{1}{4\pi}(B_{r1} B_{r1}' - 2B_{r1} B_{z2} + B_{\phi1} B_{\phi1}') = 0\,, \tag{5.5}$$

$$4B_{r3} + B_{z2}' = 0\,, \tag{5.6}$$

$$p_0 + \frac{B_{z0}^2}{8\pi} + R^2\left(p_2 + \frac{B_{r1}^2 + B_{\phi1}^2}{8\pi} + \frac{B_{z0} B_{z2}}{4\pi}\right) = p_e\,, \tag{5.7}$$

where ρ has been eliminated in favor of T through:

$$\rho_0 = \mathcal{R}^{-1} \frac{p_0}{T_0}\,, \qquad \rho_2 = \mathcal{R}^{-1} \frac{p_0}{T_0}\left(\frac{p_2}{p_0} - \frac{T_2}{T_0}\right)\,, \tag{5.8}$$

and $\mathcal{R}$ denotes the universal gas constant divided by the effective molar mass of the plasma.

Since B_{r3} only appears in Eq. (5.6), this equation permits the calculation of B_{r3} once B_{z2} is known. Also, Eq. (5.2) can be used to eliminate B_{r1} from the whole set of equations in favor of B_{z0}. Integration of Eq. (5.4) yields:

$$B_{\phi1}(z) = \frac{B_{\phi1}^*}{B_{z0}^*} B_{z0}(z) \equiv \eta^* B_{z0}(z)\,, \tag{5.9}$$

where the asterisk on a variable denotes its value at a given reference level $z = z_0$. Given the temperature distribution $T_0(z), T_2(z)$, Eq. (5.1) can be integrated as well:

$$p_0(z) = p_0^* \exp\left(-\frac{T_0^*}{H^*}\int_{z_0}^{z} \frac{d\xi}{T_0(\xi)}\right)\,, \tag{5.10}$$

where $H^* = \mathcal{R} T_0^*/g$ denotes the pressure scale-height at the reference level. Also,

$$p_2(z) = \frac{p_0(z) B_{z0}(z)}{B_{z0}^*}\left(\frac{p_2^*}{p_0^*} + \frac{B_{z0}^* T_0^*}{H^*}\int_{z_0}^{z} \frac{T_2(\xi)}{T_0^2(\xi) B_{z0}(\xi)}\, d\xi\right)\,, \tag{5.11}$$

$$B_{z2}(z) = -\frac{4\pi}{B_{z0}} p_2 - \frac{1}{4} B_{z0}'' - \eta^{*2}\, B_{z0}\,. \tag{5.12}$$

Substituting Eqs. (5.2), (5.9), (5.10), (5.11) and (5.12) into the boundary condition Eq. (5.7), a nonlinear differential equation for B_{z0} as function of z is finally obtained:

$$B_{z0}'' - \frac{1}{2}\frac{(B_{z0}')^2}{B_{z0}} + F(B_{z0}) = 0\,, \tag{5.13}$$

with

$$F(B_{z0}) = 2\eta^{*2} B_{z0} + \frac{16\pi}{R^{*2} B_{z0}^*}\left[p_e - (p_0 + \frac{B_{z0}^2}{8\pi})\right]\,, \tag{5.14}$$

where $p_e(z)$ is a given function of height and $p_0(z)$ is defined by Eq. (5.10). This equation has been investigated numerically by Pneuman *et al.*, 1986 [see their Eq.(30)] specifying B_{z0} and B'_{z0} at $z = z_0$ (Pneuman *et al.* also considered a weak external magnetic field, $\mathbf{B}_e$, so that a magnetic pressure term $B_e^2/8\pi$ has to be included in $p_e(z)$.

Although we present here the set of equations up to second order, our treatment is not restricted to any specific truncation order, as in previous studies (see also Ferriz-Mas and Schüssler, 1989).

6. Discussion and final remarks

The expansion procedure presented in Sect. 2 relies on two basic assumptions: axial symmetry and regularity of the pertaining functions at $r = 0$. The assumption of continuous derivatives on the axis guarantees the existence of the expansions (2.2) and the rotational invariance implies that these expansions only involve either even or odd powers of r. The truncated system of expanded equations can be closed on application of appropriate boundary conditions. We have applied this expansion technique to an axisymmetric configuration described by the ideal MHD equations, but it should be noted that the formalism developed in Sect. 2 could be easily extended to include the effects of viscosity, resistivity, radiation damping and heat exchange with the surrounding fluid. In deriving the system of equations (2.3) to (2.16) no assumption has been made as to the nature of the external perturbations, which enter the theory through the boundary conditions that are to be applied at the interface between the flux tube and the surrounding medium, i.e., at $R(z)$, where R is the height-dependent radius of the tube.

In Sect. 3 the zeroth-order approximation has been recovered as a special case of our treatment; even this crude approximation accounts for the radial variation of $\mathbf{B}$ (and, in non-static problems, of $\mathbf{v}$). Higher-order approximations, which take into account the effects of field line curvature, represent corrections to the zeroth-order description and may give more information on the internal variations. We have applied the truncated set of expanded equations (to second order) to a vertical flux tube in magnetostatic equilibrium ($\mathbf{v} \equiv 0$ and $\partial/\partial t \equiv 0$) in a stratified atmosphere. In this case the external pressure at the boundary of the tube is given as a function of height, and the only boundary condition is that of lateral pressure balance. In the time dependent problem, on the other hand, the additional condition of continuity of the velocity component normal to the boundary has to be applied; in this case the external pressure field plays a crucial role in determining the nature of the solutions in the tube.

A precise treatment of the time-dependent problem is possible for the special case of small perturbations (linear theory) excluding the effect of gravity. Depending upon the assumed sign of the squared exterior radial wavenumber m_e^2 different dispersion relations arise. A comparison of the results with those of the exact treatment (i.e., without resort to a radial expansion) shows that the expansion procedure is best applicable to the axisymmetric surface modes ($m_0^2 > 0$) on a flux tube. If we want to investigate the body waves in a flux tube, the expansion scheme is not appropriate (it is like saying that the Bessel function $J_0(x)$ is not well approximated by $1 - x^2/2^2$ within the interval $0 \leq x \leq a$ unless x is small against unity *throughout* the interval). The reason is that we cannot make $n_0\bar{R}$ smaller than a certain value when $k\bar{R} \to 0$.

The approximation of thin flux tubes is valid as long as (A) all physical quantities do not vary *significantly* across the section of the tube and (B) the radius of the flux tube is small compared to the spatial scales of variation of *all* relevant quantities of the tube and its environment. Let us analyze these two conditions in detail.

(A) The first condition, which requires that the radial variation of all physical quantities within $0 \leq r \leq R$ can be well represented by their low-order derivatives taken on the axis ($r = 0$), is the basis of the expansion procedure. In the zeroth-order approximation, for instance, only the constant terms and the terms involving first derivatives taken on the axis are retained in the Taylor expansions of the MHD equations. Therefore, the description of processes involving a significant structure within a cross-section of the tube cannot be accomplished within the *thin flux tube approximation*.

(B) The second condition, which requires that the radius of the tube is small compared to all spatial scales of variation, enters the theory through the boundary condition of *instantaneous pressure balance*. In particular, this condition requires that the radius of the tube is small compared to the different scale-heights of the problem and also to the spatial scales of dynamical processes (e.g., longitudinal and radial wavelengths, scale of variation of flows). Calling L to these scales generically this means that $R << L$; the fulfillment of this condition must be checked for all z and t (for an example of a numerical simulation in which the thinness of the tube is checked throughout the evolution see Moreno-Insertis, 1986).

The conditions (A) and (B) are not mutually independent, as we shall see next.

Consider the zeroth-order thin flux tube approximation, for which the magnetic field is given by

$$\begin{aligned} B_z(r,z) &= B_{z0}(z)\,, \\ B_r(r,z) &= rB_{r1}(z)\,, \end{aligned} \tag{6.1}$$

subject to the divergence-free condition $2B_{r1} + B'_{z0} = 0$. In terms of the scale-height H for the magnetic field ($B'_{z0} \simeq B_{z0}/H$) this last equation may be expressed as

$$B_r \simeq -\frac{r}{2H}B_z\,. \tag{6.2}$$

The Lorentz force per unit volume ($\mathbf{F} = \mathbf{rot\,B} \wedge \mathbf{B}/4\pi$) for a magnetic configuration such as the given by (6.1) takes the form

$$\begin{aligned} F_r &= \frac{1}{4\pi}rB_{z0}B'_{r1}\,, \\ F_z &= -\frac{1}{4\pi}r^2B_{r1}B'_{r1}\,. \end{aligned} \tag{6.3}$$

Now, since we are dealing with the zeroth-order thin flux tube approximation, we require that all terms that are of order r^2 are negligible compared to terms of order r, i.e., $|F_z| << |F_r|$, so that for the approximation to be a consistent one we must have

$$|B_r| << |B_z|\,, \tag{6.4}$$

or, in virtue of (6.2),

$$\frac{r}{H} << 1\,. \tag{6.5}$$

Therefore, the applicability of the thin flux tube approximation requires that $|B_r| << |B_z|$ or $r/H << 1$ even in the magnetostatic case, for which the lateral pressure balance is mantained at all times (regardless how thin the tube may be as compared to the different scale-heights) because the dynamical time-scales are zero. Next we show that, as a consequence of (6.4) or (6.5), the component of the velocity along the tube has to dominate over any transverse motions. The velocity components v_r and v_z are related to the magnetic field components through the z-component of the induction equation (2.5), viz.

$$v_{r1} = v_{z0}\frac{B_{r1}}{B_{z0}} - \frac{1}{2B_{z0}}\frac{\partial B_{z0}}{\partial t}. \qquad (6.6)$$

For a stationary flow ($\partial/\partial t \equiv 0$) we obtain $v_{r1}/v_{z0} = B_{r1}/B_{z0}$, and on application of Eq. (6.5) the desired result follows. In the case of oscillatory motions, taking time averages on both sides of Eq. (6.6) we also obtain $|v_r| << |v_z|$. Thus, for the thin flux tube approximation to be applicable it is necessary that

$$|B_r| << |B_z| \quad \text{and} \quad |v_r| << |v_z|, \qquad (6.7)$$

but this is not a sufficient condition. Here we can link with the discussion in Sect. 4 regarding the applicability of the thin flux tube approximation to the body waves in a flux tube: the condition that $R/L << 1$ for *all* spatial scales of variation L (in particular, the transverse wavelength $\lambda_r = 2\pi/n_0$, which is a dynamical length-scale) is not fulfilled for body waves. The reason is that we cannot make $n_0\bar{R}$ smaller than a certain finite value when $k\bar{R} \to 0$. This example shows that the terms "thin flux tube approximation" and "long-wavelength approximation", although closely related, are not equivalent even in the absence of gravity, when the concept of scale-height becomes meaningless and the only characteristic length-scales present are the longitudinal and the transverse (or radial) wavelengths.

Finally, let us note that in this contribution we have focused on a special case of the *thin flux tube approximation*, namely, that of an axisymmetric configuration. This does not exhaust all possible MHD configurations to which a generalization of the approximation of Roberts and Webb can be applied [for instance, the asymmetric or kink modes of a thin magnetic flux tube in a homogeneous magnetized or unmagnetized fluid (Spruit, 1981, 1982) or the two- or three-dimensional motion of a curved magnetic flux tube embedded in a compressible fluid in the presence of gravity (e.g., Spruit, 1981; Moreno-Insertis, 1986)].

Acknowledgments

Antonio Ferriz Mas thanks the Deutsche Forschungsgemeinschaft for its support of the research reported here under grant 418 SPA-11 and also acknowledges a Travel Grant from the American Geophysical Union.

References

Abdelatif, T. E., Surface and body waves in magnetic flux tubes, Astrophys. J., 333, 395-406, 1988.

Anton, V., Ph.D. Thesis, Universität Göttingen, 1989.

Browning, P. K., and E. R. Priest, The structure of untwisted magnetic flux tubes, Geophys. Astrophys. Fluid Dyn., 21, 237-263, 1982.

Browning, P. K., and E. R. Priest, The structure of twisted magnetic flux tubes, Astrophys. J., 266, 848-865, 1983.

Browning, P. K., and E. R. Priest, The magnetic non-equilibrium of buoyant flux tubes in the solar corona, Solar Phys., 92, 173-188, 1984.

Defouw, R. J., Wave propagation along a magnetic tube, Astrophys. J., 209, 266-269, 1976.

Edwin, P. M., and B. Roberts, Wave propagation in a magnetic cylinder, Solar Phys., 88, 179-191, 1983.

Ferriz-Mas, A., Nonlinear flows along magnetic flux tubes: Mathematical structure and exact simple wave solutions, Phys. Fluids, 31, 2583-2593, 1988.

Ferriz-Mas, A., and M. Schüssler, Radial expansion of the magnetohydrodynamic equations for axially symmetric configurations, Geophys. Astrophys. Fluid Dyn., (in press), 1989.

Ferriz-Mas, A., M. Schüssler, and V. Anton, Dynamics of magnetic flux concentrations: The second order thin flux tube approximation, Astron. Astrophys., 210, 425-432, 1989.

Landau, L. D., and E. M. Lifshitz, Electrodynamics of continuous media, vol. 8 of *Course of Theoretical Physics*, Pergamon Press, chapter VIII, 1960.

Lighthill, M. J., Waves in Fluids, Cambridge University Press, London, 1978.

Moreno-Insertis, F., Nonlinear time-evolution of kink-unstable magnetic flux tubes in the convective zone of the Sun, Astron. Astrophys., 166, 291-305, 1986.

Parker, E. N., Hydraulic concentration of magnetic fields in the solar photosphere. I. Turbulent pumping, Astrophys. J., 189, 563-568, 1974.

Parker, E. N.: Cosmical magnetic fields, Oxford University Press, 1979.

Pneuman, G. W., S. K. Solanki, and J. O. Stenflo, Structure and merging of solar magnetic fluxtubes, Astron. Astrophys., 154, 231-242, 1986.

Rae, I. C., and B. Roberts, Pulse propagation in a magnetic flux tube, Astrophys. J., 256, 761-767, 1982.

Roberts, B., Solitary waves in a magnetic flux tube, Phys. Fluids, 28, 3280-3286, 1985.

Roberts, B., and A. R. Webb, Vertical motions in an intense magnetic flux tube, Solar Phys., 56, 5-35, 1978.

Roberts, B., and A. R. Webb, Vertical motions in an intense magnetic flux tube. III. On the slender flux tube approximation, Solar Phys., 64, 77-92, 1979.

Spruit, H. C., Motion of magnetic flux tubes in the solar convection zone and chromosphere, Astron. Astrophys., 98, 155-160, 1981.

Spruit, H. C., Propagation speeds and acoustic damping of waves in magnetic flux tubes, Solar Phys., 75, 3-17, 1982.

Sturrock, P. A., and Y. Uchida, Coronal heating by stochastic magnetic pumping, Astrophys. J., 246, 331-336, 1981.

Thomas, J. H., Hydromagnetic waves in the photosphere and chromosphere, in Theoretical problems in high resolution solar physics, H.U. Schmidt (ed.), Max-Planck-Institut für Physik und Astrophysik (MPA 212), München, 126-147, 1985.

ON THE EQUILIBRIUM OF A THIN FORCE-FREE MAGNETIC FLUX TUBE IN A STRATIFIED ATMOSPHERE

Prasannalakshmi and M. H. Gokhale
Indian Institute of Astrophysics, Bangalore 560 034, India

Abstract. Earlier we have shown that the equations of a thin, axisymmetric, uniformly twisted, force-free magnetic flux tube in a stratified atmosphere form a closed system of equations up to the second order in the thinness parameter viz, the ratio of the tube radius R to the atmospheric scale height H. In this paper we show that for a given amount of longitudinal magnetic flux and given initial conditions at a point on the axis, the equations lead to a unique geometry of the tube with a unique longitudinal variation of the excess (or deficit) of density (or pressure) at the same horizontal level outside the tube. These do not depend upon either the structure of the surrounding atmosphere or the depth of the initial point in it. The reasons for the difference between the limit of this solution and the solution of the first order system of equations is explained. Limitations on the applicability of the solutions in the two orders are discussed.

Introduction

Theoretical models of the equilibrium of a magnetic flux tube in a stratified atmosphere are essential for studying those parts and properties of solar magnetic flux tubes which are not directly observable. The latter study is important even in the larger context of the stellar magnetic fields and activity.

Parker [1979] derived equations for equilibrium of an isolated, thin, untwisted flux tube in a stratified atmosphere. Spruit [1981] obtained equations for motion of such a flux tube. The equilibrium equations were used by van Ballegooijen [1982] for modeling equilibria of adiabatic flux tubes in the solar convection zone. These investigations were confined to the first order in the thinness parameter, viz. the ratio of the tube radius to the scale height of the surrounding atmosphere. Browning and Priest [1984] extended the theory to study the equilibria of a twisted force free flux tube to the second order in the thinness parameter. We have shown [Prasannalakshmi and Gokhale, 1987] that in the second order the equations for equilibrium of the twisted force-free flux tube form a mathematically closed system of equations.

Here we show that when the twist parameter is prescribed, the flux tube configuration and the longitudinal variation of the differences of plasma density and pressure with respect to the surrounding plasma can be obtained in terms of the initial conditions at a point on the tube plasma. Nor does it depend either upon the external stratification or upon the level of the initial point in it. The external stratification and the level of the initial points are needed only for determining the values of density and pressure on the tube axis.

We explain the cause of the difference in the mathematical nature of the problem between the first and the second orders and discuss the significance of this difference with regard to the applicability of the solutions.

The Second-Order Set of Equations

The balance of the forces due to gravity, plasma stresses and the electromagnetic stresses in a thin isolated axisymmetric magnetic flux tube has earlier yielded the equations of equilibrium in the form [Prasannalakshmi and Gokhale, 1987].

$$\frac{dQ}{ds} = \langle\nabla\rho\rangle\ g\ \pi\ R^2\ \sin\theta \qquad (1)$$

$$Q\frac{d\theta}{ds} = \langle\nabla\rho\rangle\ g\ \pi\ R^2\ \cos\theta \qquad (2)$$

where

$$Q = \pi R^2 \left[\langle\Delta P\rangle + \frac{\langle B_z^2\rangle}{4\pi} - \frac{\langle B^2\rangle}{8\pi} \right]$$

$$\Delta P = P_e - P_i \quad \text{and} \quad \Delta\rho = \rho_i - \rho_e$$

Geophysical Monograph 58

Here, the angular brackets (< >) represent average over the area of cross section (including that of the current sheath), g is the gravitational acceleration, s is the arc length measured along the axis of the tube, B is the magnetic field intensity, B_s is the longitudinal component of the field, R is the radius of the tube at 's', θ is the angle made by the tangent to the axis of the tube with the horizontal ρ and P are the plasma density and pressure, and the suffixes i and e represent values inside and outside the tube respectively.

Following Browning and Priest [1983] the components of a weakly twisted axisymmetric force-free field can be written up to the second order terms in the radial coordinates r, φ, s in a local cylindrical coordinate system as

$$B_s = 2f - (r^2/2)(f''+b^2f),\ B = brf \text{ and } B_r = rf' \qquad (3)$$

where

$$bR << 1,\ f = f(s),\ f' = \frac{df}{ds},\ f'' = \frac{d^2f}{ds^2},\ \text{etc.}$$

Here 2f is the magnetic field on the axis of the tube and b is twice the angle through which the field lines are rotated per unit length of the tube.

Using these expressions we obtain, up to terms in $(R/H)^2$, the following equation to represent the constancy of the longitudinal magnetic flux (F) in the form:

$$\int B_s\, dA = 2\pi R^2 f - \frac{\pi R^4}{4}(f'' + b^2 f) = F \qquad (4)$$

The plasma parameters P_e and ρ_e are functions of z specified by the external stratification and satisfy the equation

$$\frac{dP_e}{dz} = -\rho_e g \qquad (5a)$$

Since the field is force-free inside the tube, P_i and ρ_i are also functions of z, (to be determined), satisfying:

$$\frac{dP_i}{dz} = -\rho_i g \qquad (5b)$$

one must have $P_i = P_{i*}$ at all points along each horizontal line within any cross section of the tube, where P_{i*} represents the value of P_i just inside the boundary at the same horizontal level. The equilibrium of the current-sheath at the boundary requires

$$P_{i*} + B_*^2/8\pi = P_e \qquad (6)$$

where B_* is the value of B just inside the boundary and P_e is the external plasma pressure at the same horizontal level. Dividing the cross section into thin horizontal strips and integrating over the entire cross section, we obtain

$$\langle\Delta P\rangle = \langle P_e\rangle - \langle P_i\rangle = \frac{B_*^2}{8\pi} \qquad (7)$$

where $\langle P_e\rangle$ is the average pressure of the 'displaced' plasma over the cross section. However it can be shown that to the second order in (R/H),

$$\langle\Delta P\rangle = (\Delta P)_a + \frac{R^2\cos^2\theta}{8}\frac{d^2}{dz^2}(\Delta P)_a \qquad (7a)$$

where the suffix 'a' denotes the values on the axis of the tube. Equations (6) and (7) together require:

$$\frac{B_*^2}{8\pi} = (\Delta P)_a + \frac{R^2\cos^2\theta}{8}\frac{d^2}{dz^2}(\Delta P)_a \qquad (8)$$

Equations (1) and (2) can be integrated to give

$$Q = \Lambda \sec\theta \qquad (9)$$

where Λ is the constant of integration. Equation (9) can be simplified using equations (3), (4), and (8) to give

$$\frac{df}{dz} = \pm\left(\frac{16\pi\Lambda}{R^4\cos\theta} + \frac{32f^2}{R^2} - \frac{24Ff}{\pi R^4} - b^2f^2\right)^{1/2}\operatorname{cosec}\theta \qquad (10)$$

We solve equations (1) and (8) to obtain an expression for $(\Delta P)_a$ and simplify the resultant equation to get

$$\frac{dR^2}{dz} = \frac{df}{dz}\frac{A_1}{A_2} \qquad (11)$$

where

$$A_1 = \frac{R^2}{\pi^2}\left(5f^2 + \frac{3\pi\Lambda}{R^2\cos\theta} - \frac{4Ff}{\pi R^2} - 2\pi\,\Delta P_a\right)\left(\frac{F}{\pi R^2} - 3f\right) - \frac{f\Lambda}{\pi\cos\theta}$$

and

$$A_2 = \left(2f^2 - \frac{3Ff}{\pi R^2}\right)\left[\frac{2(\Delta P)_a}{\pi} - \frac{5f^2}{\pi^2} + \frac{4Ff}{\pi^3R^2}\right] + \frac{\Lambda}{2\pi R^2\cos\theta}\left[4\pi(\Delta P)_a - 35f^2 + \frac{38Ff}{\pi R^2} - \frac{12\pi\Lambda}{R^2\cos\theta}\right]$$

Differentiating Equation (10) and using Equation (4) we get

$$\frac{d\theta}{dz} = \frac{R^2\cos\theta}{\pi\Lambda\sin\theta}\left[\frac{df}{dz}\left(-\frac{3f}{R^2}+\frac{F}{\pi R^4}\right)+\frac{dR^2}{dz}\left(\frac{2\pi\Lambda}{R^6\cos\theta}+\frac{2f^2}{R^4}-\frac{3Ff}{\pi R^6}\right)\right] \quad (12)$$

Equations (5a) and (5b) combine to give

$$\frac{d}{dz}(\Delta P)_a = (\Delta\rho)_a g \quad (13)$$

Equation (13) can be used in equation (8) to get an equation for $(\Delta\rho)_a$ in the form

$$d/dz\ (\Delta\rho)_a = \frac{\left[20f^2 - \frac{16Ff}{\pi R^2} + \frac{16\pi\Lambda}{\cos^2\theta} - 8\pi(\Delta P)_a\right]}{\pi g R^2\cos^2\theta} \quad (14)$$

Equations (10) to (14) comprise a system of five first order differential equations in five unknowns, viz f, R, θ, $(\Delta P)_a$ and $(\Delta\rho)_a$.

The Nature of the Solution: Its Independence with Respect to the Atmosphere

It should be noted that the independent variable z is not present explicitly in any of these equations. The values of 'z' will have to be determined by integrating the reciprocal of any of the equations (10) to (14) after solving the whole system (e.g., by Runge Kutta method). Thus, when the initial values of f, R, θ, $(\Delta P)_a$ and $(\Delta\rho)_a$ are specified, the equations lead to a unique shape of the axis and also to a unique longitudinal variation of the differences between the internal and the external density and pressure.

Note that this solution does not depend on the structure of the surrounding atmosphere. The specification of the surrounding atmosphere is needed only for determining the thermodynamical parameters of the plasma along the tube axis using the equations:

$$(P_i)_a = P_e - (\Delta P)_a,$$

$$(\rho_i)_a = \rho_e - (\Delta\rho)_a.$$

The Difference Between the Limit of Such a Solution and the Set of Solutions Obtained from the 'First Order' Set of Equations

The equilibrium of a 'thin' force-free flux tube, to first order of (R/H), should be obtained as a limit of the solution obtained by solving the above system of equations. When the terms in $(R/H)^2$ are neglected, we obtain the system of equations derived by Parker [1979] and van Ballegooijen [1982]. In this limit the mathematical closure of the system of equations is lost because the lateral equilibrium equation, together with the hydrostatic equilibrium equations of the inner and the outer plasma, becomes identical with the equation of equilibrium along the axis. This situation arises from the fact that in neglecting the second-order terms, the terms in "b" and the second order derivatives of f and $(\Delta P)_a$ in equations (4) and (7a) respectively get removed. This amounts to giving up the "force-free" condition. This introduces scope and need for an ad-hoc auxiliary assumption. For a given set of initial conditions, only one of the infinite number of solutions of the 'open' first-order system of equations (corresponding to different auxiliary assumptions) can be expected to be reached as the limit of the unique solution of the 'closed' second order set.

Thus for determining the equilibrium of a force-free tube, however thin, one must first solve the second order set of the equations and then take the limit.

This raises the question whether the first order configuration of a real flux tube (such as a solar magnetic flux tube) will be represented by one of the solutions of the 'open' set of equations in the first order or by the limit of the solution of the closed second order set. The answer will depend upon whether the real flux tube is expected satisfy the "differential" constraint (like "force-free" conditions) more rigorously or less rigorously than the "auxiliary" constraint (like thermodynamical assumption) used in closing the first order set.

References

Browning, P. K. and E. R. Priest, The structure of twisted magnetic flux tubes, Astrophys. J., 266, 848-865, 1983.

Parker, E. N., Cosmical Magnetic Fields, Clarendon Press, Oxford, 1979.

Prasannalakshmi and M. H. Gokhale, Equilibrium of a thin isolated axi-symmetric, force-free magnetic flux tube in a stratified atmosphere, Solar Physics, 114, 75-80, 1987.

Spruit, H. C., Motion of magnetic flux tubes in the solar convection zone and chromosphere, Astron. Astrophys., 98, 155-160, 1981.

van Ballegooijen, A. A., The structure of the solar magnetic field below the photosphere, Astron. Astrophys., 106, 43-52, 1982.

DIAGNOSING THE FINE STRUCTURE OF MAGNETIC FIELDS IN THE PHOTOSPHERIC NETWORK ON THE PERIPHERY OF ACTIVE REGIONS

V. M. Grigoryev and V. L. Selivanov
SibIZMIR, Irkutsk 33, P.O. Box 4206, 664033 USSR

Abstract. In this paper a critical analysis has been made for interpretation of vector-magnetographic data by the line-ratio technique. Magnetic field strengths from lines FeI 5250.2 and 5247.1 Å are determined almost simultaneously, without showing the so-called "saturation effect" for a line with higher magnetic sensitivity. Analysis using a model of the two-component structure of the magnetic field indicates the possible existence of an inhomogeneous structure; however, it is not possible to estimate the filling factor and the inhomogeneous magnetic field strength (of magnetic network elements) because its effect in polarization signals is small as compared with the accuracy of measurements of the single-channel vector-magnetograph signals themselves.

Introduction

According to present understanding, solar photospheric magnetic fields are concentrated in small areas with angular dimensions 0.25"-0.15" and strengths of 1000-2500 G. It should be noted that this conclusion was drawn indirectly by analysing results obtained with the line-ratio technique suggested by Stenflo [1973]. Recent improvements of methods and techniques of magnetographic measurements, nevertheless, have not permitted an unambiguous solution of the question as to the existence of small-scale strong magnetic fields. Thus, no positive result was also obtained by analyzing polarized profiles of spectral lines in papers of Solanki and Stenflo [1984; 1985]. The main difficulty here seems to be in separating the expected "saturation" effect in magneto-sensitive lines from thermodynamical effects leading to changes in profiles of spectral lines which form in different features of the solar atmosphere.

The objective of this paper is to examine some possibilities of studying small-scale magnetic fields using vector-magnetographic data.

Geophysical Monograph 58

Observational Data and Analysis

The analysis used circular and linear polarization measurements in some fragments of an enhanced magnetic network on the periphery of active regions. The measurements were made with the Sayan observatory vector-magnetograph. All data on the magnetograms analyzed are presented in Table 1.

Our magnetograph measures the parameters of relative circular (V) and linear (Q) polarization averaged over the magnetograph exit slits (λ_1 = 38 mÅ, λ_2 = 95 mÅ) in the spectral line wings:

$$V = \int_{\lambda_1}^{\lambda_2} V_0 \, d\lambda / \int_{\lambda_1}^{\lambda_2} I_0 \, d\lambda$$

$$Q = \int_{\lambda_1}^{\lambda_2} \sqrt{Q^2 + U^2} \, d\lambda / \int_{\lambda_1}^{\lambda_2} I_0 \, d\lambda$$

where I_0, Q_0, U_0, and V_0 are the Stokes parameters of the light analyzed.

Points in Figure 1 present results of measured values of the parameter V (the signal of the longitudinal component of magnetic field $H_{||}$) in the line FeI 5250.2 Å, and those in the line FeI 5253.5 Å using magnetograms No. 1 and No. 2 (numbering according to Table 1).

It should be noted that the spread of points from the mean straight line in Figure 1 exceeds the magnetograph r.m.s. noise which usually makes up 0.1%. Under the assumption of a linear calibration dependence of the signal V on the magnetic field strength the following relationship for Doppler calibration values of strength:

$$\frac{H_1^c}{H_2^c} = \frac{g_2\lambda_2}{g_1\lambda_1} \qquad (1)$$

holds, where indices 1 and 2 denote the quantities corresponding to the first and second lines, respectively; and H_i^c, g_i, and λ_i are calibration values of magnetic field strength, Lande factor, and wavelength for the i-th spectral line, respectively.

Tables

TABLE 1. Data on the Active Region Magnetograms Analyzed

No.	Date	AR Center Coordinates	Time, UT	Line λ, Å	Pixel
1	21.05.83	N10W38	05:39 - 06:08	5253.5	4"x4"
2	21.05.83	N10W38	06:16 - 06:45	5250.2	4"x4"
3	26.06.84	S16W32	06:26 - 06:55	5250.2	4"x4"
4	26.06.84	S16W32	07:03 - 07:32	5247.1	4"x4"
5	26.06.84	S16W32	00:34 - 01:00	5250.2	2"x2"
6	26.06.84	S16W32	01:03 - 01:29	5247.1	2"x2"

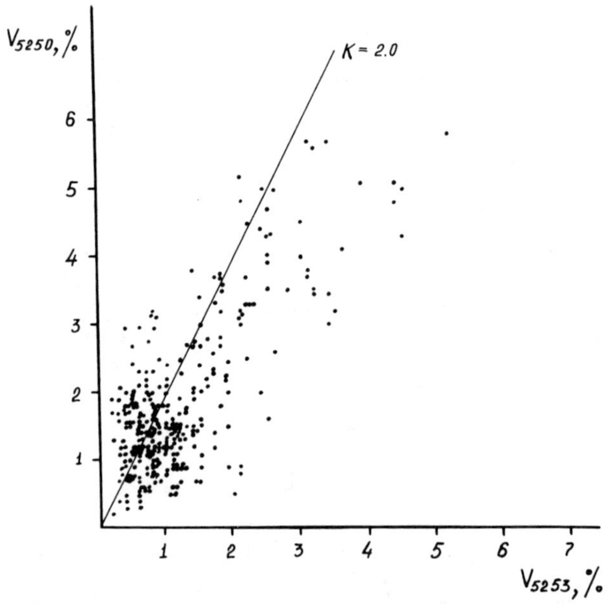

Fig. 1. Circular polarization signals in the magnetic network as measured from lines FeI: λ5250.2 and 5253.5 Å.

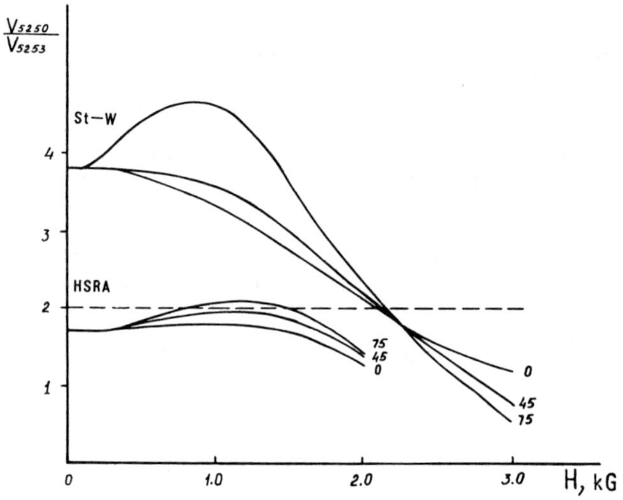

Fig. 2. Calibration dependences V_{5250}/V_{5253} on the magnetic field strength H and on different inclination angles γ for different atmosphere models: photospheric (HSRA) and sunspot model (St-W).

For the quantity H_{50}^c/H_{53}^c, we have the value 2.0. Thus, points lying below the straight line with K = 2.0 in Figure 1 can be referred to magnetic field values at which the signal in line λ5250.2 undergoes saturation and, hence, one can draw the conclusion about the presence of unresolved strong magnetic fields.

Next, we estimate the ratio H_{50}/H_{53} from the calibration functions V(H, γ) (γ being the angle between the strength vector and the line of sight), calculated according to Staude's [1982] program for particular models of the solar atmosphere: photospheric - HSRA [Gingerich et al., 1971] and sunspot model - St-W [Stellmacher and Wiehr, 1975]. Calculation results are presented in Figure 2, showing that it is necessary to take into consideration a more complex dependence for calibration in going from one line to the other than the linear dependence with K = 2.0. A strong dependence of the value of V_{50}/V_{53} on the atmosphere model is obvious, which is due to the high temperature sensitivity of the line λ5250.2 unlike the line λ5253.5. The influence of thermodynamical effects on relative variations of the profiles of a pair of lines λ5250.2 and λ5247.1 must have a much smaller effect because these lines have close-in-value excitation potentials and belong to the same multiplet. This fact is visualized in Figure 3, displaying the dependence V_{50}/V_{47}. Table 2 presents values of the signal ratio V_{50}/V_{47} averaged over all points in definite intervals of the signal V_{50}, using magnetograms No. 5 and No. 6. The third column of the table indicates the localization of the magnetogram points analyzed. The obtained distribution of the ratio V_{50}/V_{47} agrees well with the behavior of the curves presented in Figure 3.

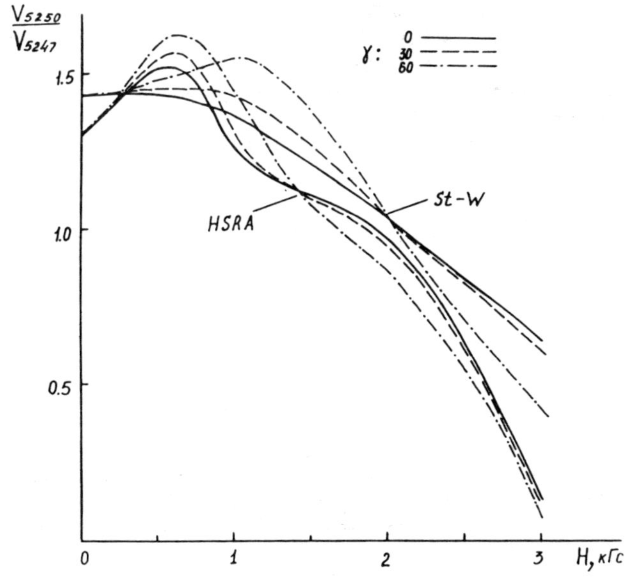

Fig. 3. Calibration dependences V_{5250}/V_{5247} on the magnetic field strength H and on different inclination angles γ for different atmosphere models: photospheric (HSRA) and sunspot model (St-W).

TABLE 2. Values of Circular Polarization Signals Inside and in the Vicinity of a Sunspot as Determined from Lines FeI: λ5250.2 and 5247.1 Å.

V_{50}, %	V_{50}/V_{47}	Localization	H, G
0 - 1	1.321	Magnetic	40
1 - 2	1.397	network	200
2 - 3	1.498		360
3 - 4	1.394		1050
4 - 5	1.388	Penumbra	1100
5 -10	1.303		1340
10-15	1.292		1360
15-20	1.272	Umbra	1420
20-20	1.263		1480

This indicates that it is not necessary to use, for the interpretation of data, inhomogeneous models of the magnetic field with the presence of strong small-scale fields. The fourth column of the table gives magnetic field strength values determined from the calibration curves (Figure 3) for averaged values of γ in corresponding intervals of the table.

Interpretation and Discussion

We have estimated the amount of the effect of a strong small-scale magnetic field in the magnetograph signals within the framework of the two-component field model. In this case the signals of circular and linear polarization recorded by the magnetograph can be expressed as:

$$V_m = V(H_m, \gamma_m) = A_w V(H_w, \gamma_w) + A_s V(H_s, \gamma_s) \quad (2a)$$

$$Q_m = Q(H_m, \gamma_m) = A_w Q(H_w, \gamma_w) + A_s Q(H_s, \gamma_s) \quad (2b)$$

where indices s, w, and m refer to the values of a strong, weak and measured (or mean) magnetic field, respectively. A is the fraction of the area of a corresponding field in the magnetograph entrance aperture. The condition: $A_w + A_s = 1$ is satisfied. Obviously, this system of equations is sufficient only for determining the values of H_m and γ_m, i.e., the parameters of a homogeneous (mean) magnetic field. If the magnetic field distribution represents patches or elements of a strong small-scale field in extended regions without a field (i.e., $H_w = 0$), then for the signal ratio in the two spectral lines (λ5250.2 and λ5247.1), we have

$$\frac{V_{m50}}{V_{m47}} = \frac{V_{50}(H_m, \gamma_m)}{V_{47}(H_m, \gamma_m)} = \frac{V_{50}(H_s, \gamma_s)}{V_{47}(H_s, \gamma_s)} \quad (3)$$

According to Table 2, the magnetic network-averaged ratio V_{m50}/V_{m47} equals 1.4. To this value, according to the calibration curves (Figure 3), there correspond the strengths $H_m \approx 200$-300 G and $H_s \approx 0.8$-1.1 kG. In this case for signals corresponding to a sunspot, values of a homogeneous magnetic field are determined unambiguously from calibration curve data. Thus, it remains to ascertain: Do small-scale magnetic fields with strengths $H_s \approx 0.8$-1.1 kG exist in the photospheric network? For that purpose, it is necessary to estimate the value of the filling factor A_s. Using the approximation [Grigoryev et al., 1985].

$$tg^2\gamma = Q/V^2\ [0.5 + 2(Q + V)^2] \quad (4)$$

for separation of the variables H and γ in calibration functions, and subject to $\gamma_s = \gamma_s^{50} = \gamma_s^{47}$, we have:

$$A_s = \sqrt{\frac{b_{47} - b_{50}}{a_{50} - a_{47}}} \quad (5)$$

where

$$a_{50,47} = \frac{Q_{m50,m47}}{2(V_{m50,m47})^2} \quad (6a)$$

$$b_{50,47} = 4a_{50,47}\ (Q_{m50,m47} + V_{m50,m47})^2 \quad (6b)$$

The values of the quantities calculated in this way for averaged signals for separate fragments of the magnetic network show a spread within 0-0.15. In order to take account of the influence of noise in the magnetograph channels upon the determination of A_s, a calculation was performed with all possible combinations of signals, with a noise amplitude equal to 0.001. In this case, the calculated values of A_s lie in the range 0-0.2. From this it follows that this accuracy (0.001) of magnetographic measurements in our case is insufficient for diagnosing small-scale magnetic fields. Possibly, most of the errors here come from the inaccuracy of overlapping of magnetograms obtained at different times in different spectral lines because ours is a one-channel magnetograph. Also, of importance may be the fraction of errors associated with inaccuracies of the approximation (4) which in the region H = 1-1.5 kG for the inclination angle γ can be about 5-7°.

Conclusions

The conclusion about unresolved small-scale magnetic fields was drawn for the first time by Kiepenheuer [1953], who showed that the magnetograph signal grows as the magnetograph slit is decreased when magnetic fields are being observed in an undisturbed photosphere. This result was later confirmed by Stenflo [1966a,b]. However, the concentration of points on the H_{5250}-H_{5233} and H_{5250}-H_{5247} diagrams along the linear dependence suggested the conclusion [Stenflo, 1973] about a universal distribution of "quantized" strong small-scale magnetic fields in the solar photosphere. Based on the analysis made in Section "Interpretation and Discussion," we believe that this conclusion might be a premature one because it was based on linear calibration relationships. This implied that a simple relationship was used between the value of the observed magnetic field and the unresolved strong magnetic field (a two-component model): $H_m = A_w H_w + A_s H_s$, whereas these values are related by a more complex relationship (2). The existence of a unique magnetic field strength is questioned also by Semel [1985], who argues there is a distribution of strengths, ranging from 600 to 1600 G.

Diagnostics of small-scale magnetic fields by the line-ratio technique, even with the use of theoretical calibration relationships, is presently questionable because of the insufficient, for this purpose, accuracy of magnetographic measurements as well as, possibly, to a greater extent, due to uncertainties in applying particular theoretical models for various photospheric features.

Acknowledgment. The authors express their appreciation to Dr. J. Staude for making the polarization profile calculations available to us.

References

Gingerich, O., R. W. Noyes, W. Kalkofen and Y. Cuny, The Harvard-Smithsonian reference atmosphere, Solar Phys., 18, 347-365, 1971.

Grigoryev, V. M., N. I. Kobanov, B. F. Osak, V. L. Selivanov and V. E. Stepanov, The vector magnetograph of the Sayan solar observatory, NASA Conf. Publ., 2374, 231-256, 1985.

Kiepenheuer, K. O., Photoelectric measurements of solar magnetic fields, Astrophys. J., 117, 447-453, 1953.

Semel, M., Determination of magnetic fields in unresolved features, High Resolution in Solar Physics, (ed. by R. Muller), 178-197, 1985.

Solanki, S. K. and J. O. Stenflo, Properties of solar magnetic fluxtubes as revealed by FeI lines, Astron. Astrophys., 140, 185-198, 1984.

Solanki, S. K. and J. O. Stenflo, Models of solar magnetic fluxtubes: Constraints imposed by FeI and II lines, Astron. Astrophys., 148, 123-132, 1985.

Staude, J., Diagnostics of solar active regions in photosphere and chromosphere by means of the spectral polarimetry of Fraunhofer lines - A compilation of computer programs, HHI-STP Report, 14, 1982.

Stellmacher, G. and E. Wiehr, The deep layers of sunspot umbrae, Astron. Astrophys., 45, 69-76, 1975.

Stenflo, J. O., The Observatory, 86, (Medd. Lunds Astr. Obs. Ser. I, No. 214), 73, 1966a.

Stenflo, J. O., Ark. Astr., 4, (Medd. Lunds. Astr. Obs. Ser. I, No. 215), 173, 1966b.

Stenflo, J. O., Magnetic field structure of the photospheric network, Solar Phys., 32, 41-63, 1973.

DYNAMICAL EFFECTS AND ENERGY TRANSPORT IN INTENSE FLUX TUBES

S. S. Hasan

Indian Institute of Astrophysics, Bangalore 560034, India

Abstract. The aim of the present analysis is to provide a realistic model for conditions within intense flux tubes. In a previous examination, it was demonstrated that convective collapse is a feasible mechanism for generating kilogauss fields in the photosphere. An important finding to emerge was that that the final state of convective collapse is not steady, but oscillatory. In the presence of horizontal heat exchange, overstable oscillations occur. The calculations have now been refined to treat radiative transport in the Eddington approximation and also to allow for convective energy transport within the flux tube. An equilibrium atmosphere in the tube, corresonding to a specified value of $\beta_0 (\beta_0 = 8\pi p_0/{B_0}^2)$ at the top of the tube, is first constructed. This equilibrium is perturbed, by introducing a small downflow, and the subsequent time evolution of the tube is followed. Although oscillatory behaviour is again observed, the nature of the oscillations is different. The flow does not appear to have a simple sinusoidal behaviour, as found earlier, but a fairly complicated one. The uplow and downflow phases do not appear to be symmetric. An important finding is that vertical energy transport through radiation is very important, particularly close to continuum optical depth unity. The observational implications of the calculations are pointed out. A comparison with semi-empirical models shows reasonable agreement.

Introduction

Observations have established that the magnetic field in the solar photosphere is confined to elements with field strengths in the kilogauss range [e.g. Beckers and Schröter 1968, Frazier and Stenflo 1972]. In recent years, much information has become known about these magnetic elements or intense flux tubes (hereafter IFT). Observations of Stokes profiles suggest that flows may exist in IFT's [Giovanelli and Slaughter 1978, Wiehr 1985], but their time-averaged contribution is probably fairly small [Stenflo and Harvey 1985, Solanki 1986].

Geophysical Monograph 58

Numerous quantitative studies have been carried out on IFT's, starting with the static models developed by Spruit [1977]. Subsequently, detailed time-dependent investigations have been made, such as the quasi-1-D simulations by Hasan [1984,1985], the latter hereafter designated Paper I. Model calculations have also been done in 2-D by Deinzer et al. [1984], Hurlburt [1983] and Knölker et al. [1988] and in 3-D by Nordlund (1983). Owing to computational reasons, the spatial resolution of the 2-D and 3-D models was fairly coarse. In addition, the vertical extension that could be treated, was limited to a few scale heights. On the other hand, the 1-D models, though sacrificing information about the horizontal structure, have the important advantage over other models, of allowing a much finer spatial resolution to be achieved in the vertical direction, as well as permitting a larger height range to be treated.

In an earlier paper [Hasan 1988], hereafter called Paper II, a static equilibrium atmosphere in an IFT was constructed. Energy transport due to both radiation and convection was permitted. We now extend the previous analysis to include dynamical effects and examine their influence on the transport of energy in a flux tube. The present investigation differs from the previous one of Paper I in two important respects: radiative transport is treated in a more realistic manner, by solving the transfer equation in the Eddington approximation [following Unno and Spiegel 1966], and convective energy transport is also taken into account.

Model and Equations

Details of the model can be found in Paper I. Briefly, we treat a magnetic flux tube of circular cross-section, extending vertically through the photosphere and convection zone of the Sun. Since the diameter of the tube is fairly small, we employ the thin flux tube approximation.

The relevant equations are essentially identical to those in Paper I, apart from the energy equation, which is

$$\rho C_v(\frac{\partial T}{\partial t} + v\frac{\partial T}{\partial z}) = -\rho C_v(\gamma - 1)T\Delta\frac{\chi_\rho}{\chi_T} + 4\pi\kappa\rho(J - S) - \frac{\partial F_c}{\partial z} \qquad (1)$$

where z, the vertical co-ordinate, is measured positive into the Sun, ρ is the mass density, T is the temperature, v is the vertical component of velocity, $\Delta = (\nabla.\mathbf{v})_{r=0}$, C_v is the specific heat at constant volume,γ is the ratio of specific heats, κ is the Rosseland mean opacity, S is the source function and F_c is the vertical component of the convective flux (we implicity assume that the strong magnetic field suppresses horizontal convective energy transport). Expressions for χ_ρ, χ_T can be found in Paper I. The first term on the right hand side of Equation 1 denotes the contribution due to compressional heating, whereas the second and third terms correspond to energy deposited by radiation and convection respectively. The second term can be related to the radiative flux, defined as $\mathbf{F}_r = -(4\pi/3\kappa\rho)\nabla J$, through the relation

$$\nabla.\mathbf{F}_r = 4\pi\kappa\rho(S - J) \qquad (2)$$

Initial State

We assume that at $t = 0$, the tube is in hydrostatic and energy equilibrium. In the external atmosphere, the model atmosphere of Spruit [1977, pp 26-34] for the convection zone was combined with the Vernazza et al. [1976] model for the photosphere. The atmosphere within the tube was constructed iteratively, by solving the static equations of MHD, keeping β_0 ($\beta_0 = 8\pi p_0/B_0{}^2$) fixed at 1.5, where the subscript o refers to the top boundary. This equilibrium was perturbed by introducing a small downflow velocity (< 50 m s^{-1}) and the subsequent time evolution of the tube was followed by numerically solving Equations 1,2 along with the momentum and continuity equations using an explicit finite difference scheme based upon the Flux Corrected Transport algorithm of Boris and Book [1976]. A finite length of tube, with upper and lower boundaries at $z = -500$ km and $z = 2000$ km was used. The level $z = 0$, corresponds to $\tau_e = 1$, where τ_e is the continuum optical depth in the external atmosphere.

v, F_r and B vs. t

Figure 1 depicts the time variation of the vertical components of velocity v (solid line), radiative flux F_r (dotted line), magnetic field B (dashed line) at $z = 0$. The radiative flux F_r is in units of the observed photospheric flux. Positive values of F_r indicate upward transport of energy. We find that a non-stationary oscillating state develops in the tube, with the amplitude of the oscillations increasing with time. The reasons for this behaviour have already been discussed in Paper I. During the downflow phase, F_r increases since the gas density is reduced. We also discern that F_r and B appear to vary in phase, but some 90° out of phase with v.

F vs. z

Figure 2 shows the vertical energy flux $F = F_r + F_c$ as a function of z, at $t = 0$, 383 s and 657 s. The convective flux was kept constant in time at its initial value. In the

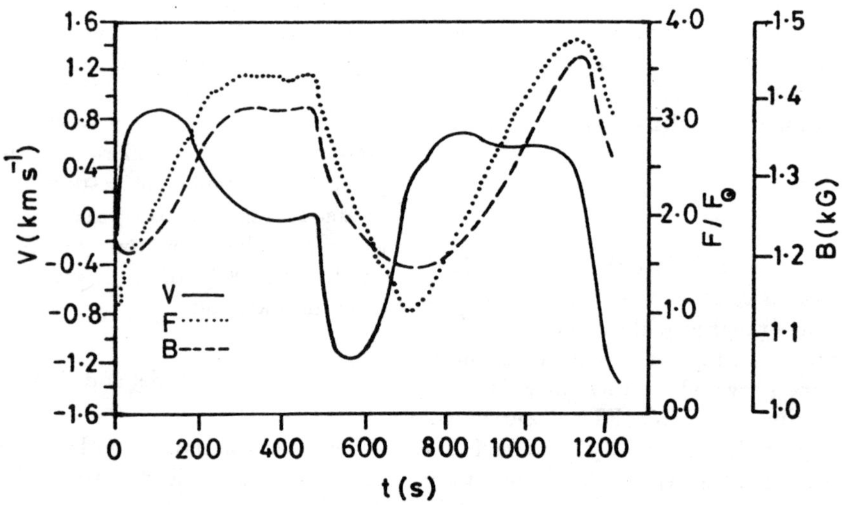

Fig. 1. Variation of the vertical components of velocity v (solid line), radiative flux F_r (dotted line) and magnetic field B (dashed line) with t at $z = 0$ in the tube.

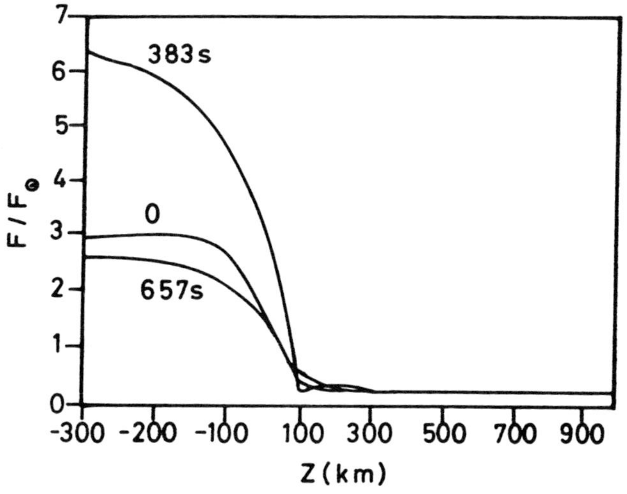

Fig. 2. Depth dependence of the total vertical flux F ($F_r + F_c$) at different times.

upper regions of the tube, where $F = F_r$, the vertical flux can increase to more than double its initial value, owing to evacuation of the surface layers of the tube.

T vs. τ

Figure 3 depicts the variation of T with τ at different epochs: t = 0 (solid line), 123 s (dashed line) and 562 s (dotted line). For purposes of comparison, the temperature variation in the external atmosphere is also shown. During the downflow phase (dashed line) T, at fixed τ, increases (for $\tau < 1$) since the reduced density results in a higher temperature occurring at the same value of τ. A temperature reversal develops for $\tau > 1$, during the downflow. This is probably an effect of our assuming that the convective flux remains constant in time.

Observational Implications

The results indicate that an IFT does not appear to have a stationary equilibrium. All quantities oscillate with an approximate period of some 800 s. We find that v, F and B have typical variations in the range -1.2 km s^{-1} to 1 km s^{-1}, 1 $F_\odot$ to 3.5 $F_\odot$ ($F_\odot$ = -6.284×10^{10} erg cm^{-2} s^{-1}) and 1200 G to 1400 G respectively at the surface (i.e., where τ = 1).These values appear to be compatible with observations [e.g., Stenflo and Harvey 1985]. We also find that at $\tau = 1$, the temperature varies from some 6400 K to 6700 K. A comparison with the semi-empirical model of Solanki [1986], again shows reasonable agreement. The question of whether this model can reproduce the observed V profile asymmetries will be investigated in a forthcoming paper.

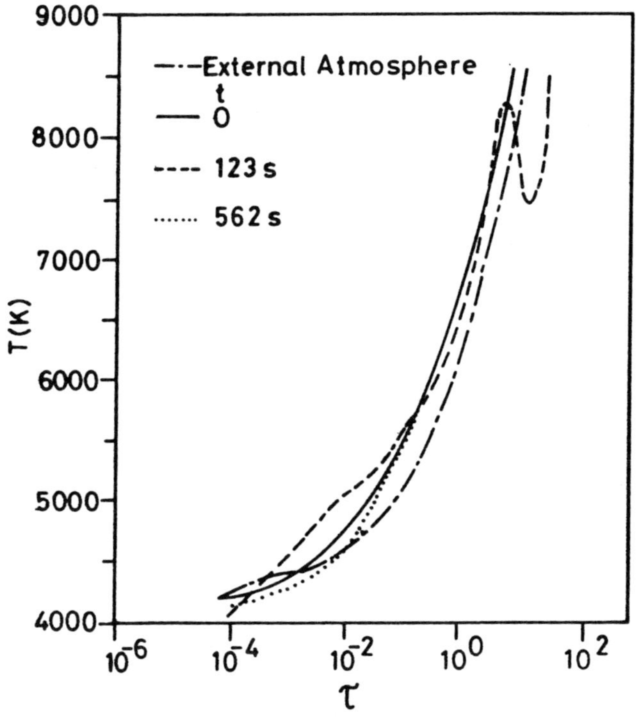

Fig. 3. Temperature in the flux tube T as a function of τ at different times. For comparison, the external temperature is also shown.

References

Beckers,J.M. and Schröter,E.H., *Solar Phys. 4,* 142, 1968.

Boris,J.P. and Book,D.L., *J. Comp. Phys. 20,* 397, 1976.

Deinzer,W., Hensler,G., Schüssler,M. and Weishaar,E. *Astron. Astrophys. 139,* 435, 1984.

Frazier,E.N. and Stenflo,J.O., *Solar Phys. 27,* 330, 1972.

Giovanelli,R.G. and Slaughter,C., *Solar Phys. 57,* 255, 1978.

Hasan,S.S., *Astrophys. J. 285,* 851, 1984.

Hasan,S.S., *Astron. Astrophys. 143,* 39, 1985.

Hasan,S.S., *Astrophys. J. 332,* 499, 1988.

Hurlburt,N., *Ph.D thesis, University of Colorado* 1983.

Knölker,M., Schüssler,M. and Weishaar,E., *Astron. Astrophys. 194,* 257, 1988.

Nordlund,A., *IAU Symp. 102,* 79, 1983.

Roberts,B. and Webb,A., *Solar Phys. 56,* 5, 1978.

Solanki,S,K., *Astron. Astrophys. 168,* 311, 1986.

Spruit,H.C., *Ph.D thesis, Univ. of Utrecht,* 1977.

Stenflo,J.O. and Harvey,J.H., *Solar Phys. 95,* 99, 1985.

Unno,W. and Spiegel,E.A., *Pub. Astr. Soc. Japan 18,* 85, 1966.

Vernazza,J.E., Avrett,E. and Löser,R., *Ap. J. Suppl. 30,* 1, 1976.

Wiehr,E., *Astron. Astrophys. 149,* 217, 1985.

ELECTRIC CURRENTS IN A UNIPOLAR SUNSPOT

A. A. Pevtsov and N. L. Peregud
SibIZMIR, Irkutsk 33, P.O. Box 4026, 664033 USSR

Abstract. A study is made of longitudinal electric currents of a unipolar sunspot (NOAA No. 4744). The pattern of longitudinal currents in the sunspot umbra indicates the presence of a common current system with predominance of the tangential component. The direction of this component, together with the sunspot's polarity, suggests the conclusion that currents flowing in the umbra sustain, at least, the sunspot magnetic field. The Lorentz force that is calculated in cylindrical symmetry is directed largely in radius out of the sunspot and exceeds considerably the horizontal gas pressure gradient. The nature of electric currents is discussed.

Introduction

The longitudinal component $j_{||}$ of the density vector of electric currents can be determined from Maxwell's equation

$$\frac{\partial B_y}{\partial x} - \frac{\partial B_x}{\partial y} = \mu j_{||} \qquad (1)$$

if the B_x and B_y components of the transverse magnetic field are known. Maps of $j_{||}$ distribution were obtained in a number of earlier papers (see, e.g., Kotov, 1970]; however, a complete picture of electric currents in a sunspot is still unavailable to date. Severny [1964] in his pioneering papers demonstrated that regions of longitudinal currents of opposite polarity merge together in a sunspot, while their arrangement resembles a current pattern of an electric motor, in the direction along its axis. Subsequently it became apparent, however, that not all sunspots show such a distribution of currents (for example, Kotov, 1970). Recently Staude and Hofmann [1987] reported the observations of a striking distribution of longitudinal currents in a sunspot in the form of two hills of opposite polarity. Azimuth asymmetry of the transverse magnetic field, and some other features, enabled them to interpret the observed picture in terms of a ring current flowing around one of the flux tubes forming part of the sunspot. In this case it was found that the tube must have a significant inclination with respect to the solar surface such that two hills of longitudinal current were observable.

This paper is a continuation of the study of current structures, seeking to construct a unified system of electric currents of a unipolar sunspot.

Geophysical Monograph 58

Observations and Data Treatment

A unipolar sunspot AR NOAA No. 4744 [Solar Geophysical Data, 1986] was observed on 25 August and from 30 August to 2 September, 1986 using the vector-magnetograph of the Sayan observatory [Grigoryev et al., 1985]. The observations were made in line FeI 525.0 nm with the 2x2" entrance slit and the scanning rate of 2" per second. The observing technique, the calibration procedure and data treatment are described in detail in a paper of Grigoryev et al. [1985]. The seeing was below 1 or 2".

Out of all the observations obtained, we selected six sets of maps featuring the best observing conditions. The table summarizes the appropriate data. The first column indicates the observing date, and the second column gives the times at which the observations were started and terminated. The magnetograms taken during poor atmospheric seeing of 25 August are included in our treatment because they are the only observation of the group to the east of the central meridian.

When calculating the longitudinal current density we encounter two difficulties. One is that in order to determine the B_x and B_y components of the transverse magnetic field $B_\perp$ it is necessary to have an accurate value of its azimuth x. Using the observations it is possible to determine only to an accuracy as high as 180°. In order to eliminate this ambiguity, we have conjectured that the magnetic field of a sunspot (an active region) has, in a rough approximation, a potential structure. The adopted correction procedure is as follows: the longitudinal component $B_{||}$ of the magnetic field is used to calculate, in a potential approximation, the transverse field azimuths which are then compared with actual azimuths. The final value of x was chosen such that its difference from a "potential" azimuth was less than 90°. Such a technique permitted us to have an automatic process of eliminating the ambiguity of x (according to our estimates in a complex sunspot group this method gives 8 to 10% of non-coincidences with "manual" correction using the criterion for comparison with the general character of the field in the active region and 100% of coincidences within a unipolar spot).

The other problem involves the decrease in accuracy when the B_x and B_y components of the transverse field are differentiated. In order to determine the partial derivatives, we chose to use two independent methods, namely the spline-method with a high intrinsic accuracy of differentiation (0.001 $G \cdot km^{-1}$) and the smoothing method of central differences to within 0.033 $G \cdot km^{-1}$. The accuracies indicated have been obtained through modeling of the input data by a radial field. Taking into account the inaccuracies of measurement of $B_\perp$ and x the error of determining $j_{\parallel}$ does not exceed, we believe, 0.14 $G \cdot km^{-1}$. We obtained this estimate by assuming that the error of measurement of the transverse magnetic field $B_\perp$ does not exceed 100 G (the actual noise of $B_\perp$ for the magnetograms used is 50 G), and the error of determining the azimuth of $B_\perp$ is not larger than 2°.

A preliminary analysis of the actual magnetograms using different methods showed that the spline-method gives in details a more complicated pattern of currents as compared with central differences. This is accounted for, in particular, by the higher sensitivity of the splines to random noise in the observational data. For that reason, the initial maps BX and BY were smoothed out by the sliding window 3x3 points. Maps obtained by the different methods agree quite well. However, central differences showed very smoothed $j_{\parallel}$-distributions. The results to be discussed here have been obtained by the spline-method after smoothing.

Analysis of the Results

Figure 1 gives the maps for longitudinal current densities. The isolines are drawn through 0.14 $G \cdot km^{-1}$. Solid lines correspond to a current toward the observer. A dash-dot line indicates the umbra and penumbra of this sunspot. North is to the bottom, and east is to the right. The level of the maximum possible errors (0.14 $G \cdot km^{-1}$) is the first level drawn on the maps. For all observing times $j_{\parallel}$ shows a similar structure, namely two main maxima of opposite sign almost entirely lie in the sunspot umbra, and two secondary maxima are observed in the penumbra. As heliocentric angle θ varies, the main positive maximum decreases, while one of the secondary hills disappears. The structure repeats itself strikingly nicely on maps 2 to 6 (the time interval being slightly more than 30 hours). The observation of 25 August was excluded from this analysis because almost no values of $j_{\parallel}$ above the level of errors remained after smoothing. The map for longitudinal current densities of 25 August that was not smoothed out, differs from the other maps. The difference lies in the fact that the main positive maximum is located in the northern part of the umbra and the negative maximum lies in the southern part (cf. Figure 1). We will now try to explain this difference.

The character of the distributions we have obtained agrees with earlier results [Kotov, 1970; Staude and Hofmann, 1987]. The recurring character and asymmetry of the $j_{\parallel}$ picture with respect to the sunspot center have induced us to describe it in terms of cylindrical symmetry by analogy with the line-of-sight velocity. Two $j_{\parallel}$ maxima with a zero line along the direction towards the disk center can be interpreted in this case as the presence of a large azimuthal component, and asymmetry between the positive and negative hills as the presence of a vertical current. Let $j_{\parallel}$ be represented as

$$j_{\parallel}(r,\varphi)d\varphi \tag{2}$$
$$= \cos\theta j_z(r)d\varphi + \sin\theta\cos\varphi j_R(r)d\varphi - \sin\theta\sin\varphi j_\varphi(r)d\varphi$$

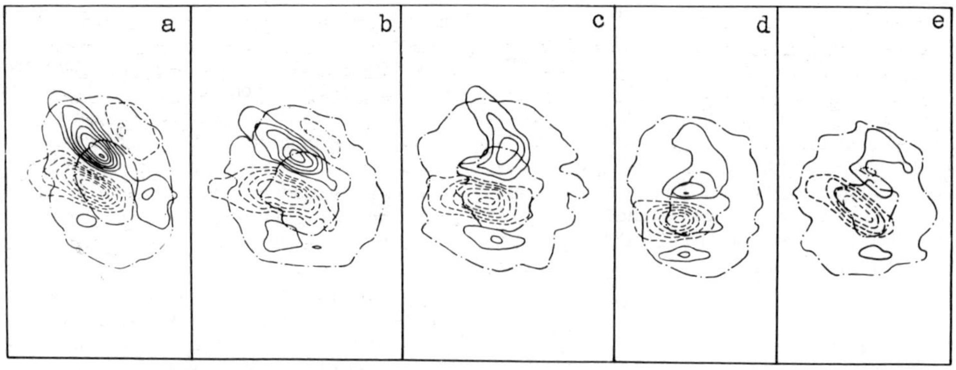

Fig. 1. The maps for longitudinal current density distribution. Letter symbols correspond to the last column of the table.

where j_z, j_R and j_φ are the vertical, radial and tangential components of the current density vector, θ is the heliocentric distance of the sunspot, r is the distance from the sunspot center, and φ is the azimuth of a point inside the sunspot with respect to its center. Integration of (2) gives the expression for j_z, j_R and j_φ (a similar procedure for the line-of-sight velocity was described by Dialetis et al., 1985).

$$j_z(r) = \frac{1}{2\pi\cos\theta} \int_0^{2\pi} j_{||}(r,\varphi)d\varphi$$

$$j_R(r) = \frac{1}{\pi\sin\theta} \int_0^{2\pi} j_{||}(r,\varphi)\cos\varphi d\varphi \qquad (3)$$

$$j_\varphi(r) = \frac{-1}{\pi\sin\theta} \int_0^{2\pi} j_{||}(r,\varphi)\sin\varphi d\varphi$$

We have calculated these three j-components for all maps 2-6. Figure 2 gives an example of the distribution (in 10^{-6} G·m^{-1}) along the sunspot radius for the observation of 1 September, 1986 (map a). In the figure j_z is denoted as W ("+" - upward current), j_R as U ("+" - into the sunspot), and j_φ as V ("+" - clockwise). D/R is the distance from the sunspot center in fractions of its radius. D/R = 0.5 corresponds for this sunspot to the umbra-penumbra boundary.

Thus, the observed distribution of currents $j_{||}$ in a sunspot can be described by a single structure as a reflection of a current system with its components directed downwards, outwards and, to a significant degree, clockwise, i.e., as a downward untwisting spiral. Such a common current system, projected onto the line of sight, will produce two hills $j_{||}$ of different sign. During the passage across the central meridian the positive and negative hills $j_{||}$ must change places merely in virtue of a change in projection direction. It seems likely that this, indeed, becomes apparent when comparing the unsmoothed-out map 1 with other maps.

After calculating the three current density components in cylindrical symmetry we may also calculate the magnitude of the Lorentz force (of course, in the same manner but in cylindrical symmetry).

The only attempt to obtain the distribution of the Lorentz force vector was as yet made by Kotov [1970] because he used magnetograms in two different spectral lines. It is possible that his results were affected by the value of the adopted difference in depth between the levels of line formation. Our method is free from this limitation, but gives only an average picture.

Figure 3 shows the distribution of the Lorentz force in cylindrical symmetry (in N·m^{-3}). Symbols are the same as in Figure 2. The main component of the Lorentz force is a radial one and this is

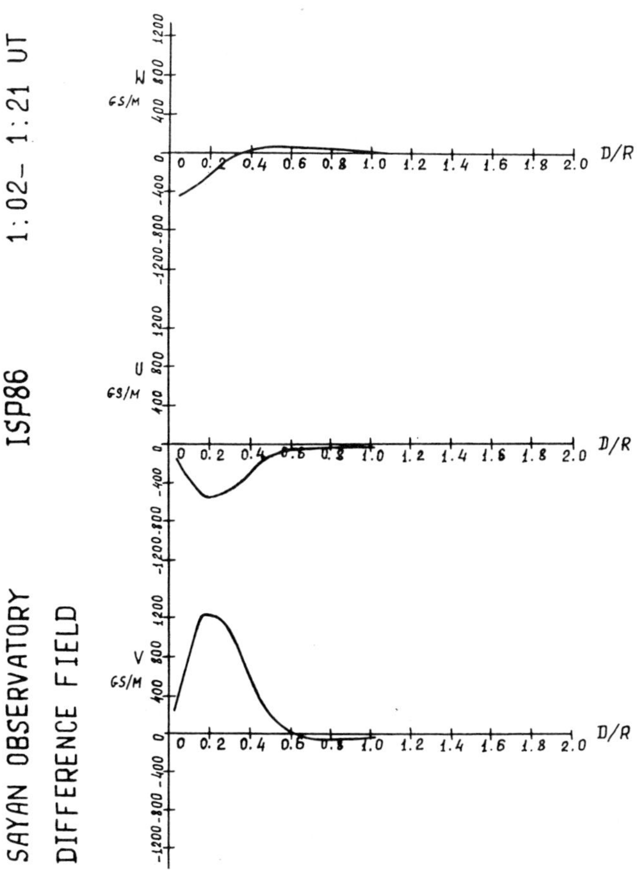

Fig. 2. Three components of the current density vector in cylindrical symmetry in 10^{-6} G·m^{-1}.

natural because the sunspot umbra is dominated by the vertical component of the magnetic field and by the azimuthal component of currents. Maximum values of the radial component of the Lorentz force are given in the table in column 4.

Discussion

The proposed explanation of the obtained picture of longitudinal currents is, of course, not the only possible one. Staude and Hofmann [1987] interpreted a similar current pattern in terms of transverse currents around a very much inclined flux tube. The sunspot considered here differs, however, from that reported in the just cited paper. AR NOAA No. 4744 is a unipolar sunspot, while the sunspot considered by Staude and Hofmann [1987] is the leader in a bipolar group. It is difficult to expect the presence of a large inclination of the flux tube in the unipolar sunspot, whereas for the bipolar group this is quite a logical occurrence. That there is no inclination of the tube in our case is evident from the distribution of the transverse field azimuths as well as the character of behavior of the sunspot longitudinal field neutral line as the sunspot

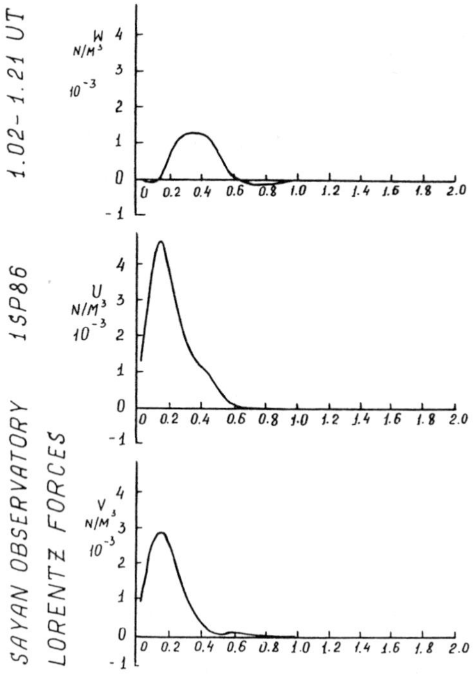

Fig. 3. The Lorentz force components in cylindrical symmetry.

traverses the solar disk. Additional evidence for this is provided by the variation in mutual position of the $j_{\|}$-hills of different sign as the sunspot passes through the central meridian. Therefore, we give preference to our explanation of the observed current picture. The equation of motion in the horizontal plane

$$\rho\ dV/dt = -\nabla P + j \times B + F \qquad (4)$$

assumes that (neglecting viscosity F) in the magnetohydrostatic case ($dV/dt = 0$) the gas pressure gradient ∇P must balance the Lorentz force. It is interesting to compare the obtained value of the Lorentz force jxB with the gas pressure gradient of some generally accepted atmospheric model. As such we chose to use the Stellmacher-Wiehr model [Stellmacher and Wiehr, 1975], and for the photosphere we adopted the convection zone model reported by Spruit [1974]. The gas pressure gradient was determined by linear interpolation for the umbra-penumbra boundary. In Figure 4 the abscissa axis indicates the difference of heights Δh between the $\tau_{5000} = 1$ levels for the photosphere and the sunspot, and the axis of ordinates indicates the pressure gradient ∇P between the $\tau_{5000} = 1$ level and the level in the Spruit model corresponding to it in height (at a given Δh). Dashed lines indicate calculated values of Lorentz forces (on the umbra-penumbra boundary). As is evident, the equality between ∇P and jxB is not reached at any reasonable value of Δh. Possibly, this is a property of the models chosen;

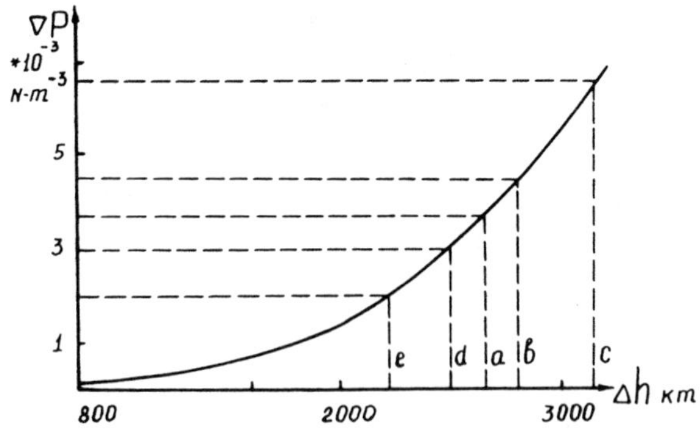

Fig. 4. The dependence of the umbra-photosphere gas pressure gradient on the height difference between levels $\tau_{5000} = 1$ in the umbra and the photosphere. Letter symbols correspond to the last column of the table.

it seems likely, however, that such a discrepancy is an actual one and sustains, for example, the mass outflow from the sunspot (the Evershed effect).

The electric current we have obtained is directed mainly clockwise. Such current must induce a magnetic field of S-polarity. The sunspot involved was precisely of such polarity. From this, one might draw the conclusion about a possible generation of the observed field by electric currents, but the situation is more complicated.

The electric current in the active region is sometimes determined not from equation (1) but from the variation of magnetic flux $\partial\phi/\partial t$ of the vertical field B_z through a certain closed surface. Indeed, according to Ohm's law $j_1 = \sigma E$ (here σ is plasma conductivity, and E is the electric field), and

$$\oint E dS = \frac{-1}{c}\frac{\partial\phi}{\partial t}$$

and the current in such a case is

$$\oint j_2 dS = \sigma\oint E dS = -\frac{\sigma}{c}\frac{\partial\phi}{\partial t} \qquad (5)$$

One should distinguish these two currents. Ampere's equation (1) defines current j_1 which is required for generation of the entire observed magnetic field, and j_2 is induction current, i.e., the current generated by a varying magnetic flux. Between j_1 and j_2 there is a certain relationship. Let us represent the potential magnetic field B_p submerging beneath the photosphere. A decrease in the flux of such a field will generate an electric current which, in turn, produces a magnetic field B_i of the same polarity as the initial field. In

TABLE 1. Maximum Values of Tangential Current (j_{max}) and of the Radial Component of the Lorentz Force (F_{max}) and the Magnetic Field Flux (ϕ) for Observing Intervals

Date	UT	j_{max} $G \cdot m^{-1} \times 10^{-6}$	F_{max} $N \cdot m^{-3}$	ϕ $10^{14}\ G \cdot m^{-2}$	Note
25 AUG 1986	03:59-04:13	-	-	-	-
01 SEP 1986	01:02-02:21	1240	0.0037	6.9	a
01 SEP 1986	03:31-04:04	500	0.0045	13.6	b
01 SEP 1986	06:22-06:55	950	0.0066	13.4	c
02 SEP 1986	05:36-06:10	380	0.0030	11.1	d
02 SEP 1986	07:15-07:49	340	0.0020	4.0	e

virtue of additivity of the field equation (1) is separated

$$\mu j_1 = \nabla \times B_p + \nabla \times B_i = \mu j_2$$

and, as is known, $\nabla \times B_p = 0$.

In this case Ampere's equation (1) yields the distribution of induction current only. Such a relationship of j_1 with j_2, though a more complicated one, exists also for other types of fields. According to our measurements (see Table, column 5), the flux in AR NOAA No. 4744 decreases; this was to be accompanied by generation of a current directed clockwise.

Horizontal currents in the sunspot observed in this case do not contradict the above scheme and can be, mainly, induction currents. The observations available to us suggest, regrettably, only a supposition regarding the nature of electric currents in sunspots. In order to obtain a final answer to this question, it is necessary to have observations of electric currents in an evolving active region in which the increasing magnetic flux must generate a current whose magnetic field is directed opposite to the initial field.

References

Dialetis, D., P. Main and C. E. Alissandrakis, Evershed flow as a steady-state homogeneous phenomenon, Astron. Astrophys., 147, 93-102, 1985.

Grigoryev, V. M., N. I. Kobanov, B. F. Osak, V. L. Selivanov and V. E. Stepanov, Vector-magnetograph of the Sayan solar observatory, NASA Conference publication 2374, 231-256, 1985.

Kotov, V. A., Magnetic field and electric currents of a unipolar sunspot, Izv. Krymsk. astrofiz. obs., 41-42, 67-88, 1970

Severny, A. B., An investigation of the magnetic field and of electric currents of unipolar sunspots, Izv. Krymsk. astrofiz. obs., 33, 34-79, 1965.

Spruit, H. C., A model of the solar convection zone, Solar Phys., 34, 277, 1974.

Staude, J. and A. Hofmann, Electric current density in the sunspot photosphere derived from vector magnetograms, 10th European Regional Astron. Meeting of the IAU, August 24-29, Praha, Czechoslovakia, Ed. by U. Hejna, M. Sobotka, 1, 105-107, 1987.

Stellmacher, G., and E, Wiehr, The deep layers of sunspot umbrae, Astron. Astrophys., 45, 69-76, 1975.

GENERATION OF CURRENTS IN THE SOLAR ATMOSPHERE BY ACOUSTIC WAVES

D. D. Ryutov and M. P. Ryutova

Institute of Nuclear Physics, 630090, Novosibirsk, USSR

Abstract A new mechanism for the generation of currents and magnetic fields in solar plasma by fluxes of acoustic waves is proposed. It is shown that the mechanism is particularly strong in the region where the damping of acoustic waves is caused by nonlinear effects (namely, formation of weak shocks). For example, the upper chromosphere above regions with the strongest convection is the most probable place where our mechanism should manifest itself.

The solar atmosphere is permanently exposed to a flux of acoustic waves generated in the convective zone. These waves have an important impact on the energy balance in the low solar atmosphere [for example Priest, 1982]. A travelling acoustic wave carries a mechanical momentum, and the absorption of the wave by a plasma results in a transfer of momentum to electrons and ions. In other words, the wave acts on each of the plasma species with some force. Generally speaking these forces produce a relative motion of electrons and ions (i.e. a plasma current) and so magnetic fields.

The purpose of our paper is to evaluate the significance of this effect for the solar atmosphere. The fact that absorption of waves can be accompanied by the generation of currents is well known in high-temperature plasma physics and is widely used for current drive in toroidal fusion devices [Parail, 1983] However, the theory of current drive in fusion devices where the particle collision frequencies are quite small is based on collisionless damping mechanisms (similar to Landau damping), so that in this case the absorption of waves is provided by the resonance effects. Under the conditions of the solar atmosphere the wave frequency is much smaller than the electron and ion collision frequencies. Correspondingly, the absorption of the wave is fully governed by the classical collisional effects: thermoconductivity, viscosity and Ohmic losses. This circumstance brings about some peculiarities in the mechanism of current drive.

In the present paper we restrict ourselves to the effects which are caused by the usual acoustic waves (we consider the case of a weak magnetic field in the surrounding medium or of acoustic waves propagating along the field). Prior to quantitative evaluation of the effect let us make some numerical estimates for some typical part of the transition region between chromosphere and corona. Let the density n and plasma temperature T be 10^{+10} cm^{-3} and 10 eV respectively. Then, according to Braginskii [1965] we find that the particle mean free path ℓ is 300 m. For the wavelength Λ of sound oscillations we consider the value $\Lambda = 10^5$m (that is, $\lambda = 2\pi\Lambda$ = 600 km; the wave period for such a wavelength is about 15 seconds). Thus, the particle mean free path satisfies with a great margin not only the condition $\ell << \Lambda$ but also much stronger one

$$\ell << \Lambda \sqrt{\frac{m_e}{m_i}} \tag{1}$$

where m_i and m_e are ion and electron masses. It is easy to verify that the condition (1) automatically guarantees the fact that the time of equalising the electron and ion temperatures is much smaller than the period of the sound wave. In other words, the perturbations of electron and ion temperatures under the condition (1) are equal, i.e. the sound speed has a familiar form valid for a monatomic gas:

$$s = \left(\frac{5}{3}\frac{p}{\rho}\right)^{1/2} \tag{2}$$

where $p = 2nT$ is a pressure and $\rho = m_i n$ is a plasma density.

All these conclusions are of course valid also for the lower chromosphere, where ℓ is much shorter because of the smaller temperature and larger density. They hold also for the longer wave periods.

The absorption of longwave sound is caused by viscosity and thermal conductivity. The damping rate which is defined as

$$\Gamma = -\frac{d \ell n W}{dt} \tag{3}$$

where W is the energy density of oscillations, is given, for example, in Landau and Lifschitz [1986]:

$$\Gamma = \frac{\omega^2}{s^2}\left[\frac{1}{\rho}\left(\frac{4}{3}\eta + \zeta\right) + \frac{2\kappa}{15n}\right] \tag{4}$$

Geophysical Monograph 58

where η and ζ are the hydrodynamic coefficients of viscosity and κ is the thermal conductivity. (Here we consider a harmonic wave of a small amplitude; some of the formulae below have a broader sense.) In our case, the viscosity is determined by ions, while thermal conductivity is provided mostly by electrons; corresponding estimates have a form:

$$\eta \sim \zeta \sim m_i n \ell\, v_{Ti}; \quad \kappa \approx \kappa_e \sim n \ell\, v_{Te}.$$

From Eqn (4) one can see that the main contribution to the damping comes from electron thermal conductivity, i.e.

$$\Gamma \approx \frac{2\omega^2 \kappa_e}{15 s^2 n}. \qquad (5)$$

Here v_{Ti} and v_{Te} are ion and electron thermal velocities. The contribution of ion viscosity and ion thermal conductivity is $\sqrt{(m_i/m_e)}$ times smaller than that of electrons.

Now we proceed to the problem of current drive. In the similar problem for a collisionless plasma the momentum of an absorbed wave is transferred to that of the plasma component which determines the resonance absorption (that is, if the resonance is due to the electron component, this automatically means that just the electrons absorb the wave momentum). In our case of a strongly collisional plasma the answer to the question of which plasma component absorbs the momentum of a sound wave is not so simple. At first sight it seems that the wave momentum has to be gained by electrons since the wave absorption is connected just with the electron thermal conductivity. But the electron mass is small, and another extreme answer is that the whole wave momentum is absorbed by ions (for they are very heavy). As will be shown below the right answer is a bit paradoxical and lies just between these two extremes. Of course, one should bear in mind that the momentum which is gained by electrons finally is transmitted to ions due to friction between the electrons and ions.

Let us write down the equations of motion of a two-component plasma [Braginskii, 1965] choosing the direction of wave propagation to be along the z-axis:

$$m_i n \left(\frac{\partial v_i}{\partial t} + v_i \frac{\partial v_i}{\partial z} \right) = - \frac{\partial p_i}{\partial z} + enE + 0.71 n \frac{\partial T}{\partial z} - \frac{1}{\sigma} enj \qquad (6)$$

$$0 = - \frac{\partial p_e}{\partial z} - enE - 0.71 n \frac{\partial T}{\partial z} + \frac{1}{\sigma} enj. \qquad (7)$$

Here p_i and p_e are the ion and electron pressures, respectively, E is electric field, j is the current density and σ is electrical conductivity. The third term on the right hand side of both equations describes the thermal force, while the last term describes the friction between the electrons and ions. We neglect the ion viscosity in Eqn (6) since its contribution to the wave absorption is small. The current build-up is completely determined by the wave absorption (the perturbation of current in oscillations is negligibly small due to quasineutrality of plasma).

Let us divide Eqns (6)-(7) by n and average them over the space period of the wave without performing the linearization:

$$m_i \frac{\partial}{\partial t} \langle v_i \rangle = - \left\langle \frac{1}{n} \frac{\partial p_i}{\partial z} \right\rangle + e\langle E \rangle - e \left\langle \frac{j}{\sigma} \right\rangle \qquad (8)$$

$$0 = - \left\langle \frac{1}{n} \frac{\partial p_e}{\partial z} \right\rangle - e\langle E \rangle + e \left\langle \frac{j}{\sigma} \right\rangle \qquad (9)$$

(we allow the existence of a mean electric field in a plasma). Adding these equations we get:

$$m_i \frac{\partial}{\partial t} \langle v_i \rangle = - \left\langle \frac{1}{n} \frac{\partial p}{\partial z} \right\rangle \qquad (10)$$

where $p = p_i + p_e$. Eqn (10) expresses momentum conservation in a system and the value on the right can be considered the momentum lost by the sound wave per unit time (and per ion). From (10) one can easily see how the electron part of thermal conductivity leads to the absorption of the wave momentum: this takes place due to the fact that if we take into account the thermal conductivity, p can no longer be expressed as a function of n (as in an isoentropic case) and the value $n^{-1}\, \partial p/\partial z$ is no longer a full derivative over the coordinate. For a travelling sound wave of a small amplitude (not necessarily harmonic) the momentum lost in the unit volume is (see Landau and Lifschiz [1986]):

$$- \frac{1}{s} \frac{\partial W}{\partial t}$$

which gives

$$\langle n \rangle \left\langle \frac{1}{n} \frac{\partial p}{\partial z} \right\rangle = \frac{1}{s} \frac{\partial W}{\partial t}.$$

In our strongly collisional case which is determined by the condition (1) we have $p_e = p_i$ and it is easy to find the mean force acting on the electron gas. This force per electron has a form:

$$F_e \equiv - \left\langle \frac{1}{n} \frac{\partial p_e}{\partial z} \right\rangle = - \frac{1}{2s\langle n \rangle} \frac{\partial W}{\partial t}. \qquad (11)$$

As was already mentioned above, a half of the wave momentum is transferred to electrons and a half to the ions (this directly follows from the condition $p_e = p_i$). Note, that when deriving the Eqn (11), we did not use at all the harmonic dependence of perturbations on coordinates: the equation (11) is valid for an arbitrary periodic travelling wave including a wave which deviates from sinusoidal due to nonlinear effects.

We will obtain the expression for the force in two limiting cases: first, for a purely sinusoidal wave and, second, for a wave in its final stage of nonlinear evolution [Landau and Lifschitz, 1986] when the wave profile reaches a "saw-tooth" form (Fig. 1).

In the first case $\partial W/\partial t = \Gamma W$, where Γ is determined by Eqn (5), and for the force F_e we have

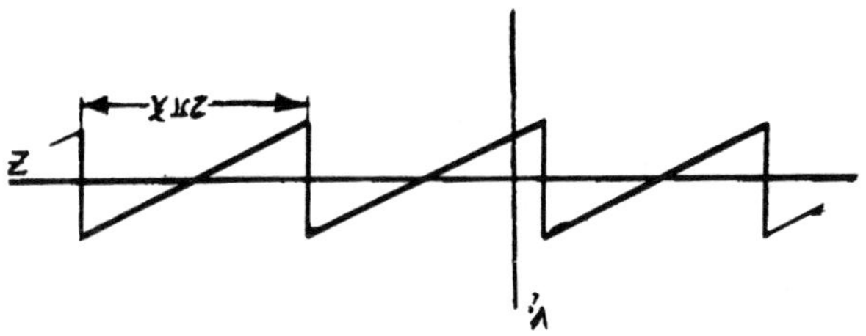

Figure 1. The velocity profile after the formation of weak shocks

$$F_e = \frac{m_i s \kappa_e}{15\Lambda^2 n}\xi \qquad (12)$$

where we introduce the notation

$$\xi = \frac{W}{\mu s^2}.$$

In the second case the energy dissipation occurs at the fronts of weak shocks and doesn't depend on κ_e; according to Landau and Lifschitz [1986] we have:

$$\frac{\partial W}{\partial t} = -\frac{8\sqrt{3}}{\pi}\frac{\rho s^2}{\Lambda}\xi^{3/2} \qquad (13)$$

and

$$F_e = \frac{4\sqrt{3}}{\pi}\frac{m_i s^2}{\Lambda}\xi^{3/2}. \qquad (14)$$

Note that W is the energy density averaged over the wave period.

A rough estimate of the stationary current can be obtained from Eqn (9):

$$j \sim \frac{\sigma F_e}{e}. \qquad (15)$$

Let's apply our results to the transition zone between the chromosphere and corona. In this region, the wave amplitude (due to decreasing plasma density) may achieve values when "overturning" occurs and the wave profile becomes of the "saw-tooth". Respectively, F_e should be evaluated according to Eqn (14). Taking into account the estimate for κ_e [Braginskii, 1965] for the current density we obtain:

$$j \sim \frac{e\, v_{Te}}{\Lambda \sigma_c}\xi^{3/2} \qquad (16)$$

where σ_c is the cross-section for Coulomb scattering of electrons by ions (in the "practical" units $\sigma_c(\mathrm{cm}^2) \sim 3.10^{-13}/T^2\mathrm{eV}$). Assuming now that $\Lambda = 10^7$cm and $T \sim 10$ eV, from the Eqn (16) we find that

$$j \sim 3.10^{-6}\xi^{3/2}\,\frac{A}{\mathrm{cm}^2}.$$

In order to find the spatial and temporal distributions of generated currents and magnetic fields, that is to study nonstationary effects, it is necessary to introduce into the problem the inductive EMF and to take into account the vectorial character of the force F_e. In this case the temporal evolution of the magnetic field is determined by the equation

$$\frac{\partial \vec{B}}{\partial t} = -\,\mathrm{rot}\left(\frac{c^2}{4\pi\sigma}\,\mathrm{rot}\,\vec{B}\right) - \frac{c}{e}\,\mathrm{rot}\,\vec{F}_e \qquad (17)$$

(we used the expression $\vec{j} = \sigma(\vec{E} - \vec{F}_e/e)$ and neglected here the macroscopic motions of the plasma).

As it should be, the generation of current and magnetic field is possible only if the force $\vec{F}_e$ is not curl-free; otherwise an electrostatic potential would be created in the plasma which would exactly compensate the action of the force $\vec{F}_e$ on the electrons. Under the conditions of the solar atmosphere, due to the space variation of intensity (and direction) of the acoustic flux as well as the plasma density and temperature, the vector $\vec{F}_e$ certainly would have a large (of the order of unity) solenoidal component. For illustration, the current pattern is shown for the case when the acoustic flux is generated only in some limiting region of the underyling surface.

The stationary state for which the estimates for the current is determined by Eqn (16) is established during a time of the order of the skin time. At the temperature T ~ 10 eV the skin

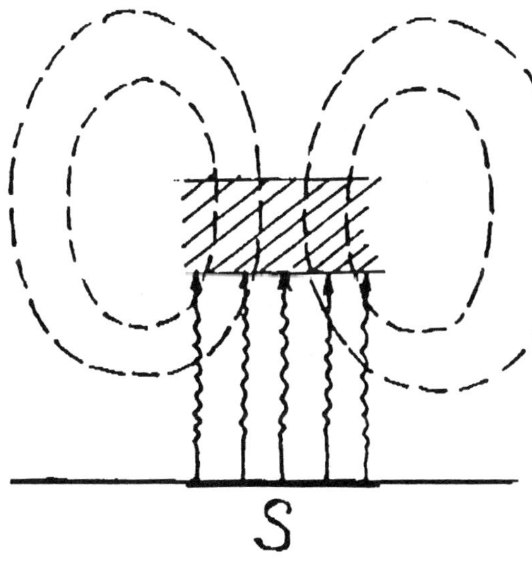

Figure 2. The generation of currents by the acoustic waves radiated from a restricted area S. The region of enhanced absorption (where the weak shocks are formed) is shaded. The current pattern is shown by dashed lines.

time for the scales of 300 km is very large - of the order of 30 years. At smaller times the estimates for the magnetic field (and current) should be made from the equation (17) in which the first term on the right is omitted:

$$\frac{\partial \vec{B}}{\partial t} = -\frac{c}{e} \operatorname{rot} \vec{F}_e. \qquad (18)$$

For a numerical example, let us assume the following orders of magnitude. Let the characteristic scale of variation of the force F_e in space be of the order of 300 km, the characteristic time of its "rearrangement" be ~ 10^7 sec, T ~ 10 eV and Λ ~ 100 km; then for ξ ~ 1 Eqn (18) gives a value for the generated magnetic field of the order of B ~ 100 G.

At a given value of ξ, the force F_e according to Eqn (14) scales as T. This means that at lower altitudes, where the plasma temperature is smaller, the effect is correspondingly weaker.

To some extent, the described effect of magnetic field generation is similar to the effect of the generation connected with the action of a thermal EMF, which takes place in the case when vectors $\underline{\nabla}n$ and $\underline{\nabla}T$ are noncolinear [Arzimovich and Sagdeev, 1965]. Note, however that our effect can work also in the case when they are parallel.

Acknowledgement

MPR is grateful to Prof Eric Priest and to the American Geophysical Union for invitation, financial support and great efforts to help attendance at this meeting.

References

Arzimovich L and Sagdeev R, *Plasma Physics for Physicists*, Moscow, Atomizdat, 309, 1979.

Braginski S I, Transport phenomena in a plasma, *Reviews of Plasma Physics* (Edited by M A Leontovich) Vol. 1, Consultants Bureau, New York, p 183-272, 1965.

Landau L and Lifschiz E, *Hydrodynamics*, Moscow, Nauka, 1986.

Parail V V, Noninductive methods of current-drive in a tokamak, in: *RF Plasma Heating*, Gor'kij, IAP Ed., 253-280, 1983.

Priest E R, *Solar Magnetohydrodynamics*, Dordrecht, D Reidel Publ. Co., 1982.

MAGNETIC FLUX TUBES AND THEIR RELATION TO CONTINUUM AND PHOTOSPHERIC FEATURES

A. Title[1], T. Tarbell[1], K. Topka[1], D. Cauffman[1],
C. Balke[2], G. Scharmer[3]

Abstract. On 29 September, 1988, time sequences of filtergrams of the solar photosphere with high resolution (0.3 to 0.5 arc second) were obtained at the Swedish Solar Observatory (SSO) on La Palma, Canary Islands in an active region near disk center. Every 50 seconds over a 2.5 hour interval, data for a Dopplergram, magnetogram, continuum, and line center image were obtained. We report here on the relationship between photospheric bright points ("filigree"), line center brightness, and magnetic field inferred from sets of individual images and movies. In these images there is a clear difference between two classes of magnetic regions. In the first, the granulation pattern is normal, the vertical velocity field is average, and the magnetic field is largely confined in narrow lanes. In the second, the granulation pattern scale is much smaller and "abnormal" in contrast, the vertical velocity is lower, and the magnetic field is relatively less compact. In the locales where the granulation is normal, there is excellent spatial correlation between bright continuum, line center, and the magnetic field fine structure. In the regions of abnormal granulation, the correspondence between bright structures and magnetic field is much more complex.

Introduction

The Swedish Solar Observatory at the Observatorio del Roque de los Muchachos on La Palma in the Canary Islands is located at an altitude of 7,900 ft. at a site that often has exceptional seeing (0.3 arc second or better). The optical quality of the 50 cm refracting telescope is such that interferometric measurements using stars show a wavefront error of $\lambda/15$ (Scharmer, et al., 1985, Scharmer, 1989). Using an engineering evaluation model of the Coordinated Instrument Package tunable filter system under development for the NASA Orbiting Solar Laboratory, observations were obtained from 7 September to 1 October 1988. On 29 September, nearly continuous data were collected for approximately 2.5 hours in bright plage of NOAA active region 5168 located at North 18, West 0. This beta type active region was born on the back side of the sun and covered 240 millionths of the surface on the day of observation. It had been stable in area for 2 days. Data of comparable quality were obtained on 7 other days. A data analysis system has been developed at Lockheed Palo Alto Research Laboratory to reduce the time sequences of images. We report on the first stages of the analysis effort. A previous study of this type was based on film filtergrams of somewhat lower spatial resolution at a single time (Title, et al., 1987).

Data Collection and Processing

Figure 1 shows the optical layout employed to make the observations. The instrument consists of a high-speed steering mirror for image stabilization, reimaging optics, a polarization analyser, a blocking filter wheel, a narrowband tunable filter from the NASA Solar Optical Universal Polarimeter (SOUP) instrument, and a 1024 × 1024 pixel CCD camera that uses a Texas Instruments virtual phase detector. During periods of the best seeing, only data from a 512 × 512 area of the array were recorded to increase the speed of the data cycle. Data recording is limited by the tape drive to 246 kilobytes/s, and a 512 × 512 image can be acquired and stored every 4 seconds. Each data tape (8 mm cassette) can store about 2.3 gigabytes of data or about 2.5 hours of data. Brief pauses in the observations were required to change data tapes.

The tunable filter bandpass was about 80 milliangstroms, slightly narrower than the solar absorption

[1] Solar and Astrophysics Laboratory, Lockheed Palo Alto Research Laboratory, Palo Alto, CA 94304, USA
[2] Utrecht Astronomical Institute, Utrecht, The Netherlands
[3] Royal Swedish Academy of Sciences, Stockholm Observatorium, S-13300, Saltsjobaden, Sweden

Geophysical Monograph 58

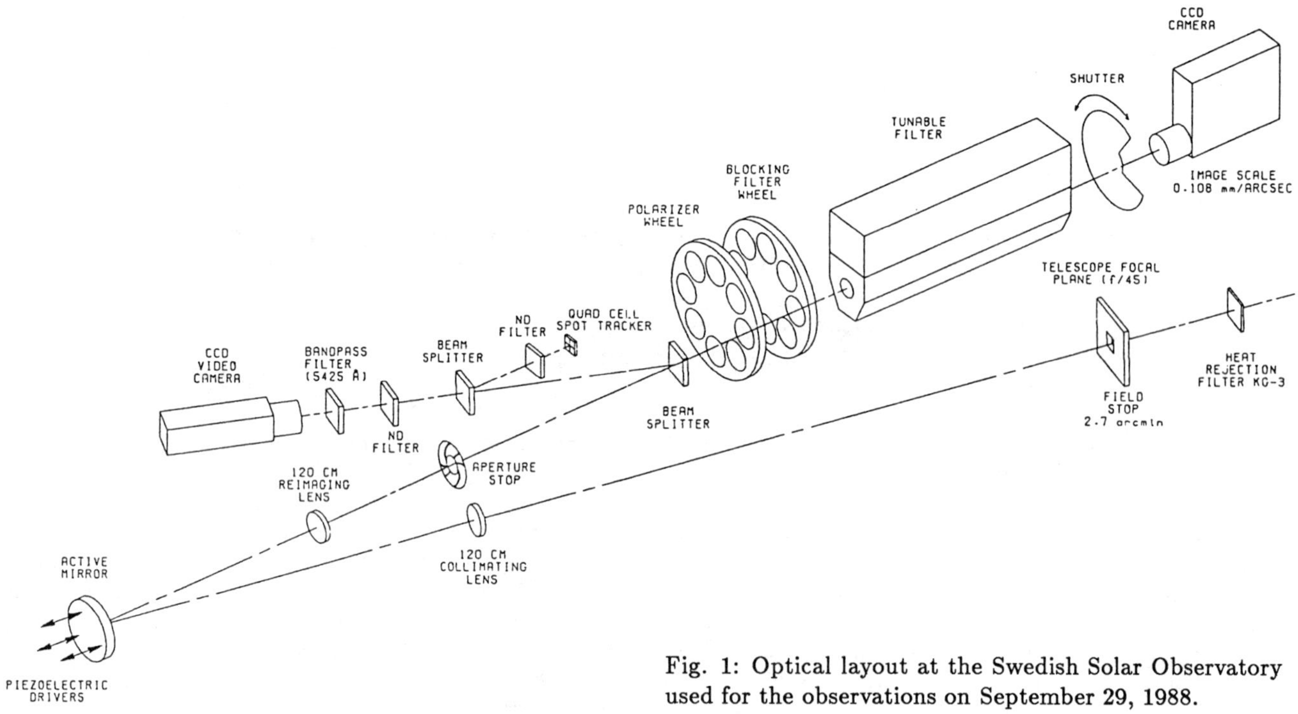

Fig. 1: Optical layout at the Swedish Solar Observatory used for the observations on September 29, 1988.

lines. The image scale was adjusted so that one pixel corresponded to 0.166 arc second (120 km) on the solar surface, for a total field-of-view of 85 arc seconds (62,000 km) on a side. The observing program for the data described here consisted of continuous repetitions every 50 seconds of a cycle of filtergrams. Each cycle consisted of: four images, two each in the right and left circular polarization (for magnetograms), in Fe I 6302.5 angstroms (g=2.5) at -60 milliangstroms from line center; two images in the continuum at 6768.5 angstroms; and four, without polarization optics, in Ni I 6767.8 angstroms at -90, -30, +30, and +90 milliangstroms from line center (for Dopplergrams). The exposure times were approximately 300 milliseconds and where taken on a fixed time base. In addition to the narrowband filtergrams, the realtime video image selection system at the observatory was used to collect the "best" white light image (a 50 angstrom band centered at 4650 angstroms) in every 8 second interval.

The Fe I 6302 images are used to make a longitudinal magnetogram, an image of the fractional circular polarization in the blue wing of the Zeeman-sensitive line. The magnetogram signal is approximately proportional to the line-of-sight component of the magnetic field vector, averaged over the spatial resolution element of the image. In general, the calibration factor which converts percent polarization to Gauss depends in a complex way on the size, field strength, and thermal structure of each solar feature. We use an average value based on simple models of the line profile, and so the Gauss scale could be off by as much as 50% in some features.

The four measurements in Ni I 6768 are used to obtain the line-of-sight velocity using a Fourier fitting method. This method is more accurate and suffers from fewer saturation effects than does the traditional 2-wavelength Dopplergram method. Once the velocity is known, the four intensities are interpolated to obtain the intensity at each pixel at the central wavelength of the line, corrected for the Doppler shift. We refer to this image below as the "Ni 6768 line-center".

To create movies from these data, we first correct the images for measured variations of gain and dark current of the array detector and reorder the data so that all images at a specific wavelength are sequential. Once movies at the various wavelengths are created, each image sequence is derotated. (The SSO telescope has an altitude-azimuth mount so that the image rotation rate varies as the sun moves through the sky.) The derotated images are precisely reregistered to correct for residual guiding errors and then destretched to remove atmospheric image distortions, differential image motion across the field of view (Topka, et al., 1986; November, 1986). Using the aligned and destretched images, Dopplergrams, magnetograms, and 6768 line center images are constructed at every time step. It is possible to register and destretch all of the different wavelengths onto each other because they have sufficient solar structures in common (such as the residual granulation pattern and small pores). Thus, we are able to create a series of movies of the magnetic field, vertical velocity field, continuum intensity, and line center that are essentially simultaneous and aligned to better than 1/5 pixel accuracy, or 0.05 arc second.

The data analysis described above is time consuming. Each destretched movie requires about 30 hours of computer time on a Vaxstation 3200.

Comparison of Regions of Differing Magnetic Flux

The movies and still frames show a complex superposition of many different phenomena in the solar atmosphere. Figures 2, 3, 4, 5, and 6 show, respectively, a set of aligned hydrogen alpha, continuum, magnetogram, Dopplergram, and line center images. The hydrogen alpha picture (Figure 2) was taken just before the observing run began. The others are all from a single 50 second cycle about 1.5 hours later. The continuum image (Figure 3) shows a variety of granules, pores, and small sunspots. In the magnetogram (Figure 4), white represents positive polarity regions; black negative; and gray, weak or no magnetic field. The noise level in the magnetic signal is about 30 gauss on each pixel. Two distinct overall regions of opposite polarity are observed on the left and center of the image. But no mixed polarity is observed outside the boundary regions. In the Dopplergram (Figure 5), white and black areas are moving toward and away from the earth, respectively, along the line of sight. The noise level in the Dopplergram is about 60 m/s on each pixel.

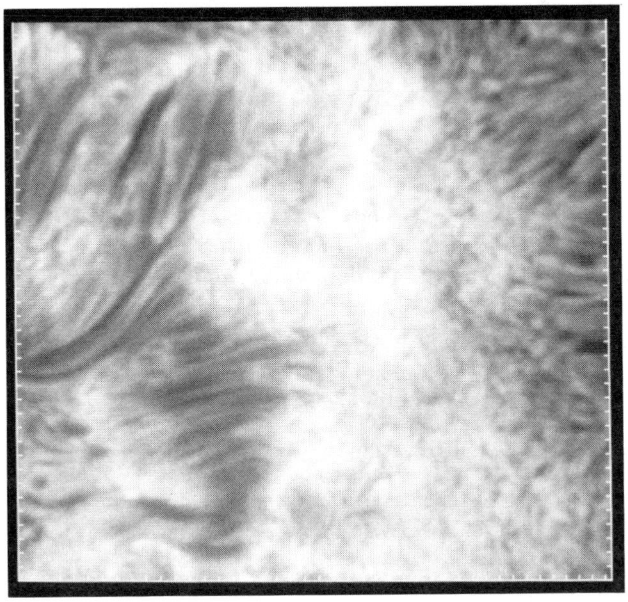

Fig. 2: Filtergram in the hydrogen alpha line taken 1-1/2 hours before the data shown in the remainder of the figures was collected. The region shown is 85 by 85 arcseconds, with tics each 2 arcseconds. It was located near sun center at North 18, West 0. (The same region is shown in Figures 3 through 6.)

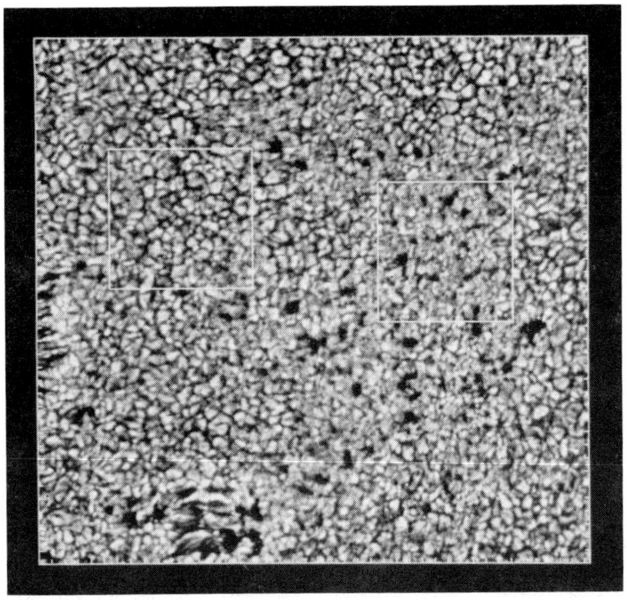

Fig. 3: Filtergram in the continuum at 6768.5 Angstroms, of the same region shown in Figure 2. Figures 3 – 6 were all taken within 50 seconds and depict the same region. The boxes show subregions corresponding to Figures 10 – 13.

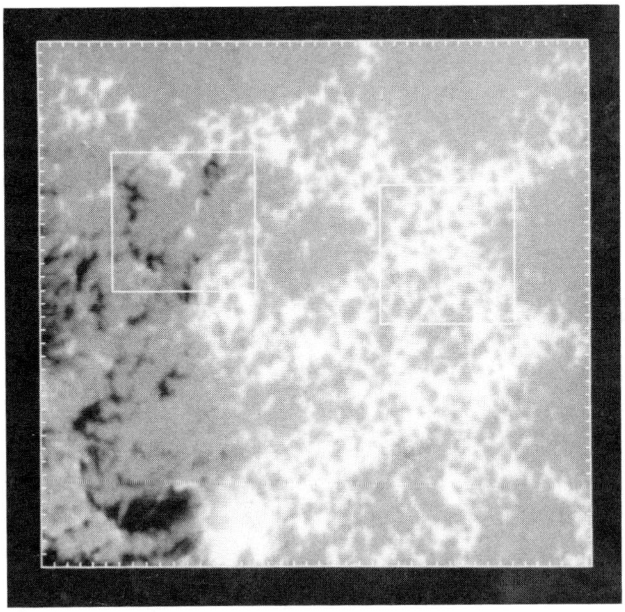

Fig. 4: Magnetogram constructed from measurements of the Fe I 6302.5 Angstrom line. White represents positive polarity and black, negative.

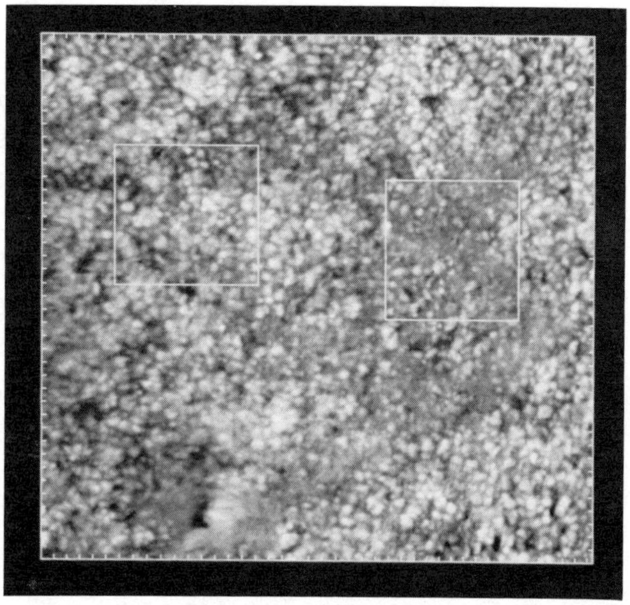

Fig. 5: Dopplergram constructed from four measurements of the Ni I 6767.8 Angstrom line. White represents motion towards the earth along the line of sight, and black, away.

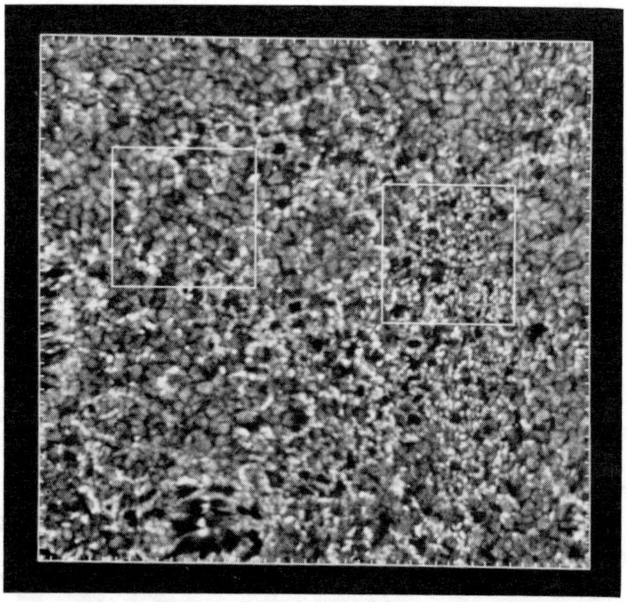

Fig. 6: Filtergram showing the image constructed of the line center of the Ni I 6767.8 Angstrom line, using four measurements across the line.

The magnetogram has been used to make masks (Figure 7) for distinguishing high and low field regions. The high-field mask is unity when the magnetogram signal is greater than 150 gauss and zero elsewhere, and the low-field mask is its complement. Note that the strong field criterion actually depends on the product of the field component and filling factor. Overlaid on Figure 7 are two boxes outlining regions that are predominantly low and high field, respectively. Figures 8 and 9 are continuum images multiplied by the low- and high-field masks, respectively, in the two indicated boxes. Figure 8 reveals the granulation pattern in weakly magnetic regions, while Figure 9 shows it in strongly magnetic areas.

The differences between high and low magnetic areas are significant, even in these still frames. The granules observed in Figure 8 have about 2 arc second diameters with well-defined high contrast lanes, while in Figure 9 the granules are "abnormal" (Dunn and Zirker, 1973, hereafter referred to as DZ) and have diameters of 1 arc second or less with indistinct low contrast lanes. Note that in the high magnetic field regions the lane widths are near the limit of seeing for these observations. Although they are larger than the image pixel size, it is possible that the lanes are well defined but not fully resolved. Nevertheless, the differences in scale are readily apparent and are remarkably consistent in these observations.

The movies emphasize the difference between the two types of regions. In the continuum, Dopplergram, and line center movies, the 5-minute (f- and p-mode) oscillations are clearly stronger in the low magnetic regions, as is the vigor of the dynamics in the granulation (Title, et al., 1989; Tarbell, et al., 1988).

Observations of Bright Points

Filigree are bright structures seen in the photosphere which are associated with fine scale magnetic fields (DZ). They range from points to linear structures, typically one or two arc seconds in length, one quarter arc second or less in width, and often "crinkled" on an arc second scale. Modern analysis of various spectroscopic observations (Stenflo, 1973; see reviews by Muller, 1985, and Stenflo, 1989; see Zirin, 1988, for a dissenting view) suggests that nearly all magnetic fields in the photosphere are in flux tubes too small to be observed directly, with kilogauss field strengths and diameters of order 100 km (0.14 arc seconds). It has been suggested (DZ; Mehltretter, 1974) that the filigree bright points are the elementary flux tubes, but some observations appear to show the magnetic fields covering a much larger area (Simon and Zirker, 1974).

The 29 September data show many examples of filigree in the continuum and Ni I 6768 line center images. In Figure 10 a portion of the line center image is enlarged. The region is in the boundary between the opposite polarities seen in the magnetogram and is mostly free of strongly magnetic structures. The location is marked by boxes overlaid on figures 2 through 8. In this enlargement, the filigree appears as bright chains.

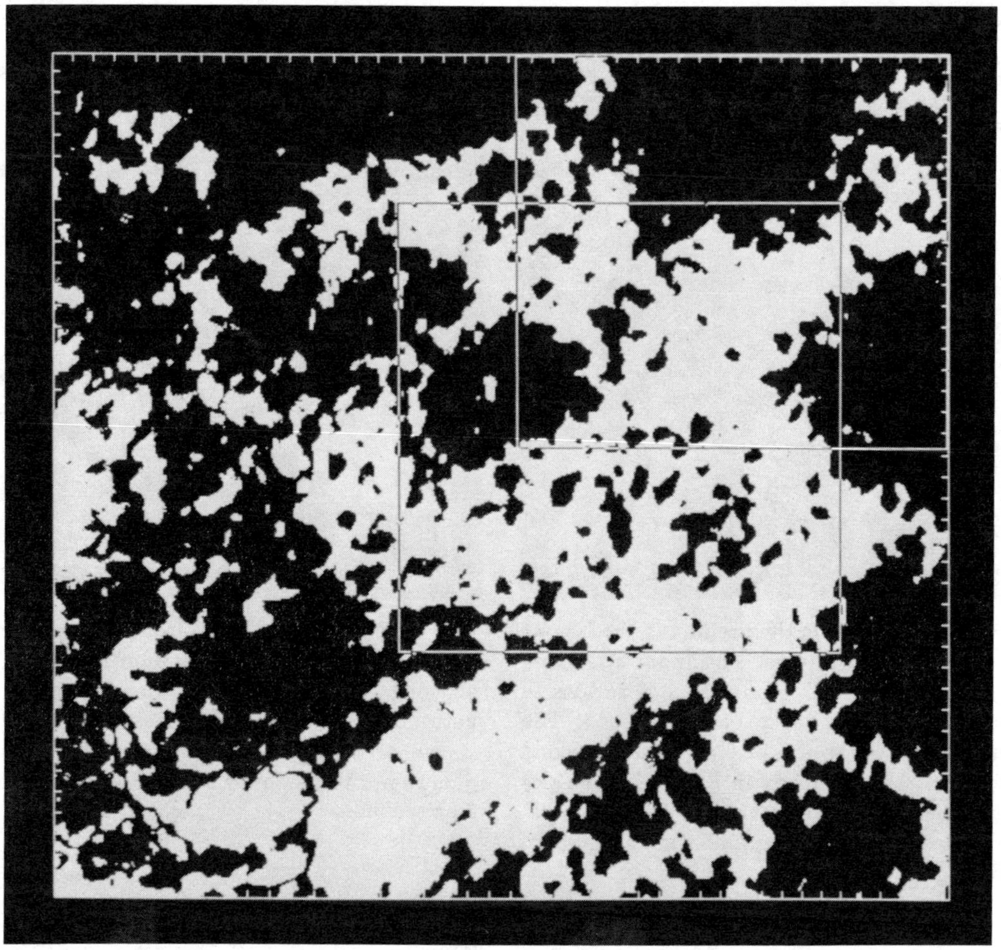

Fig. 7: Mask made from Figure 4 by coding magnetogram signals of greater than 150 Gauss of either polarity as 1 (white) and smaller fields as 0 (black). The boxes on this figure correspond to the subregions depicted in Figures 8 and 9.

In magnetograms taken within 20 seconds of the line center images, we can identify the magnetic counterparts of some of these structures. To show the relationship of these bright points to magnetic field structures, contours were drawn around the brightest 25% of the pixels in the line center image. In Figure 11 these contours are shown superposed upon the magnetogram image of the same region as is shown in Figure 10. The contours in nearly every case overlay the magnetic field structures regardless of polarity. Furthermore, the magnetic and brightness contours are both about the same spatial scale (1 arc second or less in diameter). There are, however, many magnetic locations that are not surrounded by bright point contours.

Continuum images have tiny bright structures at the same locations as the filigree observed in the line center images. These are readily seen by eye using blink comparison of the frames. However, the observed intensities of these intergranular bright points are not large compared with the values in non-magnetic granulation, and so a simple contour or threshold procedure does not show them clearly.

In regions of higher magnetic flux, a quite different pattern is observed. Figure 12 shows the line center intensity image in a region which is largely filled with high magnetogram signals. (It's location is also indicated by overlays in figures 2 through 8.) In this image the bright points do not tend to form lines, but rather are grouped in clusters. Figure 13 shows the magnetogram for this region with the bright point contours superposed. The contrast between figures 11 and 13 is striking. In Figure 13, the bright points do not center on, nor cover the magnetic regions.

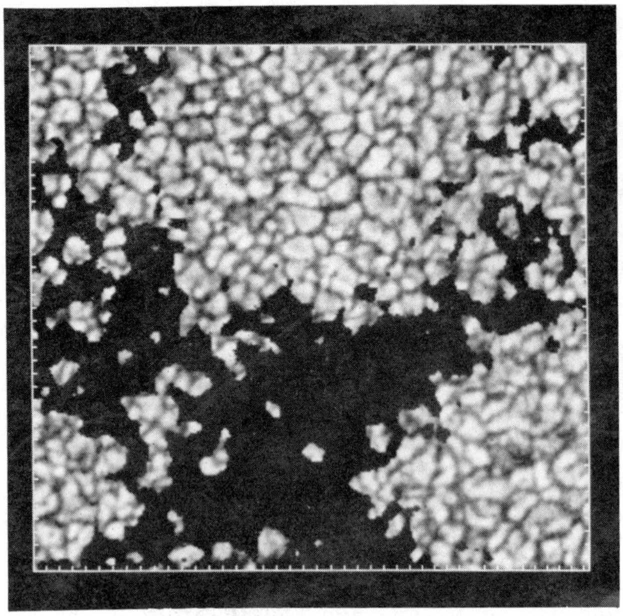

Fig. 8: Filtergram formed from the product of the inverse of the mask of Figure 7 with the continuum filtergram of Figure 3, showing "normal" granulation in regions of low magnetic flux. High flux regions are blackened. The region shown corresponds to the 42.5 by 42.5 arcsecond subregion shown in the upper box in Figure 7. The tics are each 2 arcseconds. The comparison of Figures 8 and 9 shows the contrast between the granulation patterns in weakly and strongly magnetic areas.

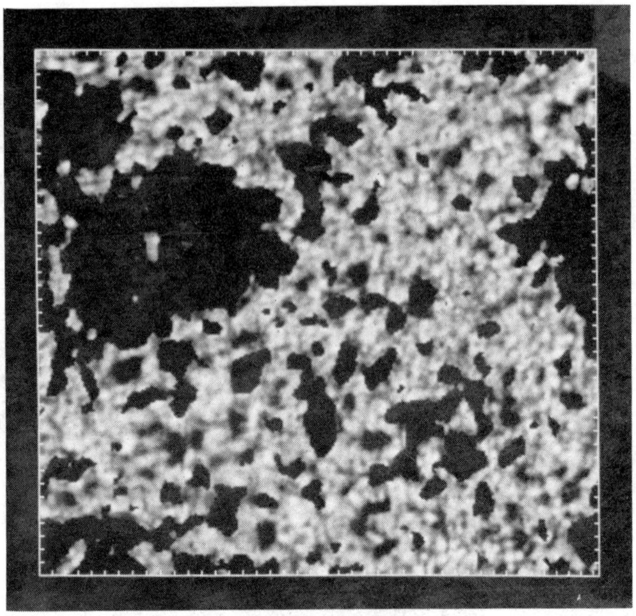

Fig. 9: Filtergram formed from the product of the mask of Figure 7 with the continuum filtergram of Figure 3, showing "abnormal" granulation in regions of high magnetic flux. Low flux regions are blackened. The region shown corresponds to the 42.5 by 42.5 arcsecond subregion shown in the lower box of Figure 7. The tics are each 2 arcseconds.

The magnetic structures are typically larger than an arc second, while many of the bright structures are smaller than an arc second.

The relationship of bright point occurrence to magnetic flux for these data is presented more quantitatively in Figure 14. Bright points are defined as pixels with line center brightness greater than the 75th percentile level. In this figure, the ordinate gives the probability that a pixel with the magnetogram signal identified on the abscissa is also a bright point. If these two properties were unrelated, the curve should follow the horizontal straight line at 0.25. Instead, the bright points are less probable than randomly expected below about 150 Gauss, and are more probable from 150 to about 600 Gauss. Above 600 Gauss, the photosphere begins to darken, eventually forming a pore, and so the association between bright points and magnetic field disappears.

The data allows us to establish the four-dimensional relationships between magnetogram signal, Doppler velocity, continuum, and line center intensity. Work on establishing these is now in progress and will be published elsewhere.

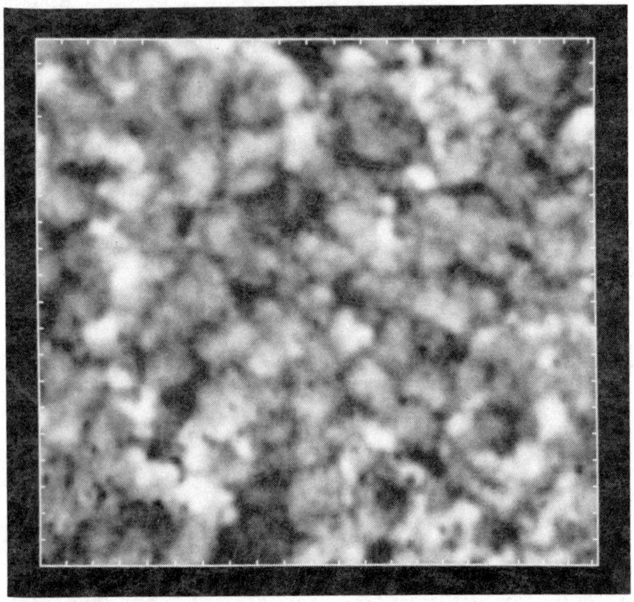

Fig. 10: An enlarged 21.25 x 21.25 arcsecond portion of the Ni I line center image shown in the leftmost box in Figure 6. This area is relatively free of magnetic flux. Tics are each 1 arcsecond.

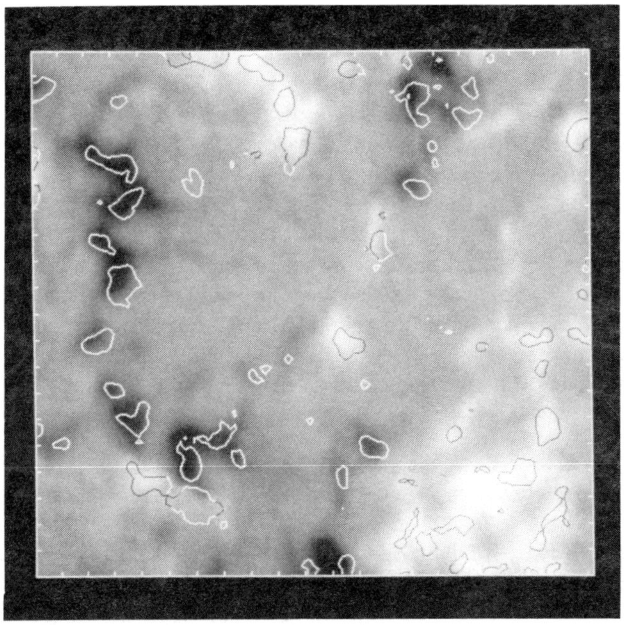

Fig. 11: Contours outlining the brightest 25% of the Ni I line center filtergram shown in Figure 10, superposed upon the same region of the magnetogram (leftmost box in Figure 4). This Figure illustrates the spatial correlation observed between the bright points and magnetic structures in an area of low magnetic flux.

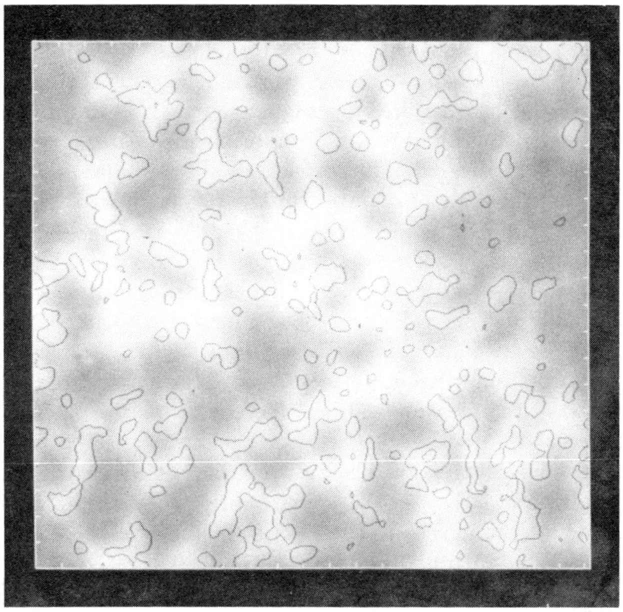

Fig. 13: Contours outlining the brightest 25% of the Ni I line center filtergram, corresponding to the rightmost box in Figure 6, superposed upon the same region of the magnetogram (rightmost box in Figure 4). This Figure illustrates the more complex relationship of bright points and magnetic structures in a higher magnetic flux region, compared to the low flux case illustrated in Figure 11.

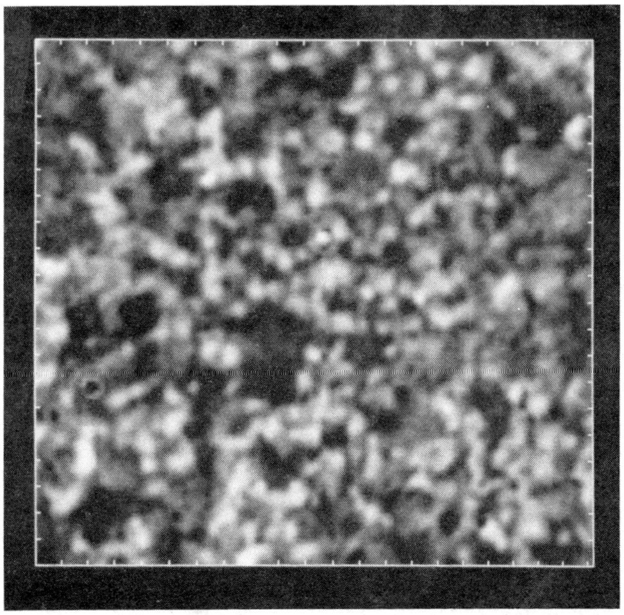

Fig. 12: An enlarged 21.25 x 21.25 arcsecond relatively high magnetic flux portion of the Ni I line center image shown in the rightmost box in Figure 6. Tics are each 1 arcsecond.

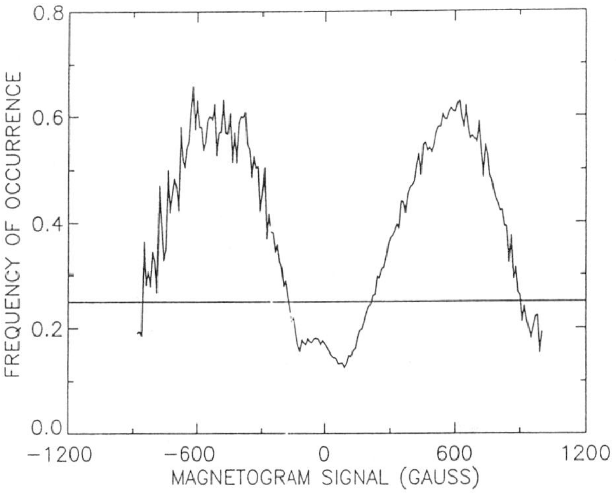

Fig. 14: Probability that a pixel associated with a given magnetogram signal (Figure 4) is in the top 25% of brightness in the Ni I line center (Figure 6). The horizontal line represents the distribution that would be expected if the parameters were unrelated.

It has long been known that bright filigree evolve significantly on time scales of 5 to 15 minutes (DZ; Muller, 1983). However, lower resolution magnetograms have not shown the vast majority of magnetic structures to evolve so rapidly. Indeed, it would be quite surprising if the magnetic field appeared and disappeared in a unipolar region on the time scale of minutes. In the very high resolution magnetic movies made with this data, however, magnetic structures are observed both to move and to change shape on such time scales. Although the total magnetic flux is probably not changing, the detailed magnetic configuration evolves on a time scale of minutes. If the magnetograms are averaged over 45 minutes, the entire interior region of this dense plage becomes covered with field because of the motion of the individual structures. Some magnetic features are observed to move with velocities approaching 3 km/s. We are currently studying the detailed interactions using the movies and the corresponding sets of coaligned digital data.

Conclusions

Evidence has been shown that in the 29 September La Palma data set:
(1) Magnetic regions have an "abnormal" granulation as seen in the continuum that is distinctly different from the granulation pattern seen in low magnetic field regions. Granules and the lanes between them are smaller in the abnormal regions. In the photosphere above the abnormal granules the vertical component of the velocity field is lower on average. This reduction occurs both in the 5 minute oscillation and in the granulation velocities.
(2) The bright points observed in the Ni I 6768 line center are related to magnetic field structures but the converse is not always true. In magnetic regions with a low average flux density. the bright points are co-spatial with equally fine scale magnetic structures. In areas of higher average magnetic flux, bright points lie within magnetic structures and are significantly smaller than the magnetic elements.
(3) Bright points in the Ni I 6768 line center are also usually seen as weak local maxima in the continuum intensity. In stronger magnetic areas, the continuum intensity is usually average or below average. The relation between line center and continuum varies considerably, depending on the magnetic flux, indicating varying intrinsic absorption depth of the Ni I line.
(4) In movies, both the bright points and the magnetic field structures evolve on comparable time scales. However, this does not mean that magnetic field disappears on a time scale of tens of minutes. Rather, magnetic field elements move and change their local configuration.

Both the subjective impression obtained from observing the movies and the detailed examples from sets of images presented above indicate fundamental differences between the photosphere in weakly and strongly magnetic regions. At present we do not have a precise characterization of a strongly magnetic region because the true field strengths and filling factors are not yet known. However, inside such a region the convective processes in the surface are highly modified, the vertical velocity in the photosphere is much lower, and the global oscillations are suppressed.

Nothing in these observations is inconsistent with the identifying the bright points as candidate "elementary magnetic flux tubes". However, much of the magnetic flux in the regions of dense magnetic field is not associated with bright points. Further, our magnetogram signals often can 800 gauss or more in these areas, which strongly suggests that these flux tubes are not much smaller than 200 to 300 km diameter.

Acknowledgments. This work has been supported by NASA contracts NAS8-32805 (SOUP) and NAS5-26813 (OSL), the Lockheed Independent Research Fund, and the Royal Swedish Academy of Sciences. The observations were obtained at the Observatorio del Roque de los Muchachos of the Instituto de Astrofisica de Canarias.

References

Dunn, R.B. and Zirker, J.B., The Solar Filigree, *Solar Phys.*, **33**, 281–304, 1973.

Mehltretter, P., Observations of Photospheric Faculae at the Center of the Solar Disk, *Solar Phys.*, **38**, 43, 1974.

Muller, R., The Dynamical Behavior of Facular Points in the Quiet Photosphere, *Solar Phys.*, **85**, 113, 1983.

Muller, R., The Fine Structure of the Quiet Sun, *Solar Phys.*, **100**, 237, 1985.

November, L.J., Measurement of Geometric Distortion of a Turbulent Atmosphere, *Appl. Optics*, **25**, 392, 1986.

Scharmer, G.B., Brown, D.S., Pettersson, L. and Rehn, J., Concepts for the Swedish 50 cm Vacuum Solar Telescope, *Appl. Optics*, **24**, 2558, 1985.

Scharmer, G.B., High Resolution Granulation Observations from La Palma: Techniques and First Results, in Solar and Stellar Granulation, eds. Robert J. Rutten and Giuseppe Severino, Kluwer Academic Publishers, Dordrecht, The Netherlands, p. 161, 1989.

Simon, G.W. and Zirker, J.B., A Search for the Footpoints of the Solar Magnetic Field, *Solar Phys.*, **35**, 331, 1974.

Stenflo, J.O., Magnetic Field Structure of the Photospheric Network, *Solar Phys.*, **32**, 41, 1973.

Stenflo, J.O., Small Scale Magnetic Structures on the Sun, *Astron. Astrophys. Rev.*, in press, 1989.

Tarbell, T.D., Peri, M., Frank, Z., Shine, R. and Title, A., Observations of F- and P-Mode Oscillations of High Degree ($500 < \ell < 2500$) in Quiet and Active Sun, in Seismology of the Sun and Sun-Like Stars, eds. Domingo, V. and Rolfe, E.J., European Space

Agency SP-286, (ESTEC, Noordwijk, The Netherlands), 1988.

Title, A.M., Tarbell, T.D., and Topka, K.P., On the Relation Between Magnetic Field Structure and Granulation, *Ap. J.*, **317**, 892, 1987.

Title, A.M., Tarbell, T.D., Topka, K.P., Ferguson, S.H., Shine, R.A. and the SOUP Team, Statistical Properties of Solar Granulation Derived from the SOUP Instrument on Spacelab 2, *Ap. J.*, **336**, 475, 1989.

Topka, K.P., Tarbell, T.D. and Title, A.M., High-Resolution Observations of Changing Magnetic Features on the Sun, *Ap. J.*, **306**, 304, 1986.

Zirin, H., Astrophysics of the Sun, (Cambridge Univ. Press, Cambridge), pp. 131-5, 1988.

WAVES IN SOLAR PHOTOSPHERIC FLUX TUBES AND THEIR INFLUENCE ON THE OBSERVABLE SPECTRUM

S. K. Solanki and B. Roberts
The Mathematical Institute, University of St. Andrews, St. Andrew, KY16 9SS, Scotland

Abstract: Linear calculations of undamped magnetoacoustic waves in thin solar magnetic flux tubes are presented and their influence on the Stokes V profiles of various iron lines is studied. This is a necessary first step for the diagnostics of the properties of flux tube waves, in particular the amount of energy transported by them into the upper atmosphere. It is shown that, with sufficiently high spatial resolution, observations can distinguish between standing and propagating waves on the basis of line parameters of photospheric spectral lines alone. Particular attention is given to exploring quantitative diagnostics for the wave amplitude, since it is currently the most important unknown parameter determining the energy flux carried by the waves. It is found that although this parameter can be derived relatively simply if the thermal fluctuations produced by the wave are ignored (i.e. for an isothermal wave), the task becomes much more complex for the more realistic case of a coupled variation of temperature and velocity.

1. Introduction

Physical considerations lead us to expect a rich variety of wave modes in solar magnetic flux tubes, including longitudinal tube waves [cf. Spruit and Roberts, 1983; Roberts, 1984, 1986; Thomas, 1985]. Indirect observational evidence for the presence of large amplitude longitudinal waves in solar flux tubes is also mounting [Solanki, 1986, 1989], although direct evidence is still lacking. Such waves are expected to play an important role in heating the outer solar atmosphere. So far investigations of waves in flux tubes have either concentrated on the purely MHD aspects, which have been dealt with in great detail by various authors [e.g., Defouw, 1976; Roberts and Webb, 1978, 1979; Webb and Roberts, 1980; Spruit, 1981; Rae and Roberts, 1982; Herbold et al., 1985; Musielak et al., 1989], or on the purely observational. No attempts have been made to combine the two approaches quantitatively. Indeed, we have only a rough qualitative idea of what the observational signatures of flux tube waves are. Here, we describe the first investigation of the influence of longitudinal flux tube waves on spectral line profiles, in particular on Stokes V profiles which are the main carriers of information on magnetic elements or small flux tubes. We hope to identify the observations best suited to deriving information on flux tube waves, and to develop methods for their analysis. In a future step presently available observations may also be used to diagnose some interesting parameters, e.g. to set limits on the energy flux being carried by such waves into the chromosphere. We do not attempt to reproduce directly the observations at this stage.

Geophysical Monograph 58

2. Longitudinal Flux Tube Waves

2.1. Summary of Hydrodynamics

The basic assumptions of the model are: 1) The flux tube structure is described by the thin tube approximation [Defouw, 1976; Roberts and Webb, 1978, 1979; Parker, 1979], which agrees very well with observations [Zayer et al., 1989]. 2) The waves considered are linear. This assumption allows us to vary wave parameters easily, thus permitting us to assess the influence of waves over as wide a range of properties as possible. 3) Radiative damping is neglected. Its influence will be studied in a future publication. 4) The gas inside the flux tube is not coupled to the field free surroundings; i.e. the calculated flux tube waves do not excite disturbances in the field-free atmosphere surrounding the flux tube.

For the calculation of the flux tube wave the following differential equation for the normalized velocity, Q, must be solved (see Roberts and Webb [1978] for a derivation):

$$\frac{d^2Q}{dz^2} + \left(\frac{\omega^2 - \Omega^2(z)}{c_T^2}\right) Q = 0; \tag{1}$$

$Q(z)$ is related to the local longitudinal velocity, v, of the oscillations due to the wave via

$$Q(z) = \sqrt{\frac{\rho_0(z) A_0(z) c_T^2(z)}{\rho_0(0) A_0(0) c_T^2(0)}}\, v(z). \tag{2}$$

Subscript zero denotes quantities related to the undisturbed, stationary atmosphere. ρ is the gas density, A is the cross-sectional area of the flux tube, z is the height in the atmosphere (the height scale has been chosen such that $\tau_{5000} = 1$ in the quiet sun corresponds to $z = 0$) and c_T is the tube speed defined as $c_T^2 = c_s^2 v_A^2/(c_s^2 + v_A^2)$, where c_s is the sound speed and v_A is the Alfvén speed within the flux tube. Eq. (2) determines v once Q is known from a solution of Eq. (1).

In Eq. (1) ω is the frequency of the wave and

$$\begin{aligned}\Omega^2 = & c_s^2 \left(\frac{1}{2}(\ln \gamma\zeta)'' + \frac{1}{4}\left((\ln \gamma\zeta)'\right)^2 + \frac{g}{c_s^2}\left(\left(\ln \frac{\rho_0}{p_0}\zeta\right)' + \frac{g}{c_s^2}\right)\right) \\ & - g\left((\ln \rho_0)' + \frac{g}{c_s^2}\right).\end{aligned} \tag{3}$$

Here g is accelaration due to gravity, γ is the ratio of heat capacities, $\zeta = p_0 B_0^3/(4\pi\gamma p_0 + B_0^2)$, p is the gas pressure and B is the magnetic field strength. Once ω and v are prescribed (the latter at two different heights in the atmosphere), Eq. (1) can be solved numerically. With the now known velocity the fluctuations to first order of the rest of the atmospheric parameters required for the calculation of line profiles can be derived from the thin tube equations in a straightforward manner.

2.2. Summary of Radiative Transfer

Since we are primarily interested in basic effects and (at the moment) not in a direct comparison with the data, we have chosen to use hypothetical lines of Fe I and II which can be selected to give an optimum coverage of line strength and excitation potential ranges with a minimum number of lines. The equivalent width W_λ and excitation potential χ_e of the chosen lines are given in Table 1. The Stokes profiles are calculated numerically. The influence of the waves is quantified by considering specific line parameters, like wavelengths, line widths, asymmetries.

numbers listed in these columns are the wave amplitudes as derived from the simulated observations. Although they are reasonably accurate for the wave with the longest wavelength, they can be up to 10 times too small for the λ_w = 150 km wave. This implies that the energy flux transported by flux tube waves can be underestimated by up to a factor of 100, and possibly even more for waves of even smaller λ_w, from direct observations of wavelength shifts, if no correction is made for radiative transfer effects.

There are two main reasons for the behaviour of $v_V(z_F)$ seen in Table 1. If $\lambda_w \lesssim$ the order of the width of a typical Stokes V contribution function (see Van Ballegooijen [1985] and Grossmann-Doerth et al. [1989] for definitions of the Stokes V contribution function) considerable velocity gradients occur over the width of the contribution function at certain phases, leading to the reduction of the signal in the zero-crossing wavelength. As

TABLE 1. Hypothetical Spectral Lines, Heights of Formation and Wave Amplitudes

Line No.	Ion	W_λ (mÅ)	χ_e (eV)	z_F (km)	$\log \tau_F$	$v_a(z_F)$ (km s^{-1})	$v_V(z_F)$ for λ_w = 900 km (km s^{-1})	$v_V(z_F)$ for λ_w = 300 km (km s^{-1})	$v_V(z_F)$ for λ_w = 150 km (km s^{-1})
1	Fe I	55	0	150	−2.2	1.3	1.15	0.8	0.25
2	Fe I	15	0	145	−2.2	1.27	1.15	0.75	0.25
3	Fe I	110	0	235	−2.75	1.5	1.55	1.3	0.3
4	Fe I	55	4	60	−1.45	1.1	1.0	0.65	0.1
5	Fe II	55	3	75	−1.6	1.13	1.05	0.8	0.35

The perturbations produced in the horizontal direction (due to the elasticity of the flux tube) by a longitudinal flux tube wave is observed to be small compared to the perturbations in the vertical direction. Therefore, such waves are not expected to provide a sizeable signal near the limb and cannot explain the large line widths observed there [Pantellini et al., 1988]. Accordingly, the calculations presented here are restricted to solar disk centre, so that Stokes Q and U can be neglected. Therefore, in the present contribution we consider Stokes I (the unpolarised spectrum) and Stokes V (the difference between right hand circularly polarized light and left hand ciruclarly polarized light) profiles only.

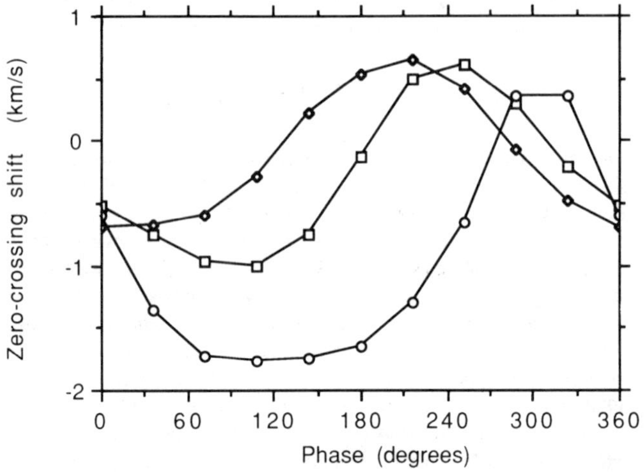

Fig. 1. Variation of zero-crossing wavelength shift (in km s^{-1}) of synthetic Stokes V profiles over a full period of a propagating wave with λ_w = 300 km and $v_a(0)$ = 1 km s^{-1}. Squares: line 1, circles: line 3, diamonds: line 4.

3. Results

3.1. Zero-Crossing Wavelength

Figure 1 shows v_V, the zero-crossing wavelength shift of Stokes V in velocity units vs. phase of the wave for lines No. 1, 3 and 5 of Table 1. Portrayed is the influence of a propagating wave with $v(z = 0) = 1$ km s^{-1}, a wavelength of 300 km and a period of approximately 80 s. Note that the different heights of formation of the cores of the three lines are reflected by the different times or phases at which the largest redshifts for the various lines occur (arrows in Fig. 1). It should, therefore, in principle be possible to distinguish between propagating waves and standing waves or oscillations, and possibly even to deduce propagation velocities from high spatial resolution observations of purely photospheric spectral lines. Conversely, we can derive rough heights of formation from this diagram, since we know the height of the downflowing peak of the wave at a given phase.

Table 1 lists the heights of formation, z_F, and the corresponding optical depths, $\log \tau_F$, of the cores of the 5 lines, the true wave amplitudes $v_a(z_F)$ at these heights and the "wave amplitudes" derived directly from the oscillation amplitudes of the zero-crossing wavelength $v_V(z_F)$, i.e. the signal shown by simulated observations. λ_w refers to the wavelengths of the waves. All three waves have amplitudes of 1 km s^{-1} at $z = 0$. The

expected this effect is seen to increase dramatically with decreasing wavelength of the wave. For Stokes I a similar effect has previously also been noticed for simple sine waves in the quiet photosphere [e.g. Keil and Marmolino, 1986].

The second effect which reduces the oscillation amplitude of the zero-crossing wavelength for a propagating tube wave (or any other propagating acoustic or magnetoacoustic wave) is the variation in temperature with phase. Whereas (in the absence of radiative damping) for standing waves the phase difference between velocity and pressure, temperature etc. is 90°, for propagating waves the phase difference is close to 180°. Therefore the temperature in the upflowing phase is lower than in the downflowing phase. As a result the temperature sensitive low excitation Fe I lines weaken considerably during the downflowing phase and strengthen during the upflowing phase, leading to a reduction of the zero-crossing oscillation amplitude and a blueshift of the line (averaged over a full oscillation period).

3.2. *Line Widths*

Another interesting diagnostic parameter is the line width. In time averaged observations of flux tubes strongly broadened Stokes V profiles are seen. In the past such line widths have been modelled using either macroturbulence or a mixture of macro- and microturbulence.

When the synthetic V profiles are considered as a function of time, or phase of the wave, then the widths of lines 2, 4 and 5 do not change appreciably with phase. Lines 2 and 3, on the other other hand, have larger widths during the cool phase than during the warm phase. Measurements of line widths of well chosen lines as a function of time may, therefore, also be used to diagnose flux tube waves. Such measurements can help to overcome the ambiguities faced when attempting to derive wave amplitudes from zero-crossing shift measurements, in particular when combined with measurements of Stokes I line depth or Stokes V strength (e.g. areas of the Stokes V wings) which can also vary strongly with phase.

One interesting numerical experiment is to test the reliability of the "Gaussian micro- and macroturbulence" approximation used to describe the time averaged line broadening in flux tubes in the past [e.g. Solanki, 1986; Pantellini et al., 1988]. To this end we have used isothermal propagating waves which vary purely sinusoidally with height, thus coming closest to the generally used height independent macro- or microturbulence velocity. We find that if the wavelengths of the waves are sufficiently large (e.g. much larger than the width of the Stokes V contribution function) the line widths and shapes behave qualitatively as expected from the macroturbulence approximation, i.e. the lines become broader and more "V" shaped, while waves with small wavelengths produce strongly "U" shaped lines, a clear signature of microturbulence. Also, such waves affect the width of line 3 (lying on the horizontal part of the curve of growth) more strongly than the weaker lines, also in good agreement with classical microturbulence theory. However, there are considerable quantitative differences between the effects of the waves and of the turbulence velocities, due to the generally assumed Gaussian shape for the turbulence velocity distribution which differs considerably from the distribution produced by a sinusoidal wave. This can give rise to a wrong upper limit on the mechanical flux carried by flux tube waves if the line widths are analysed using turbulence velocities.

As long as there are no temperature fluctuations, standing and propagating waves give rise to reasonably similar line broadenings. In particular, the relative broadenings of the various lines are not affected. As soon as we let the temperature vary with phase then large differences between standing and propagating waves become evident. Due to the particular phase relationship between temperature and velocity, the widths of lines formed in the presence of standing waves are not affected at all by temperature fluctuations. Widths of lines formed in the presence of propagating waves may, on the other hand, change strongly. Fig. 2 shows the line width vs. $v(0)$ for isothermal (Fig. 2a) and for moderately non-isothermal (Fig. 2b) waves (λ_w = 300 km, period of approximately 80 s). Note that lines 1 and 2 differ extremely in the two cases, being much narrower for the wave with temperature fluctuations included. The widths of these lines may even decrease somewhat again as the wave amplitude is increased. Line 3 differs only slightly and the Fe II line (No. 5) appears completely unaffected.

The observed behaviour is due to the alternate weakening and strengthening of the lines in the up- and downflowing phases, respectively. For the weak lines this implies mainly a change in line depth, while for the strong line it means mainly a change in line width. Therefore, the weak line almost disappears in the hot phase, and gives very little contribution to the average profile which thus remains quite narrow corresponding to the width of the line in the upflowing phase. For the strong line, due to its increasingly prominent wings, the increase in width during the cool phase manages to offset its decrease during the hot phase.

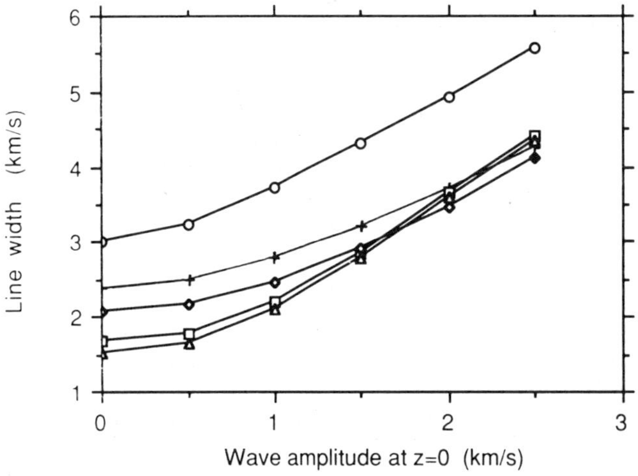

Fig. 2a. Line widths of all 5 lines as a function of wave amplitude at z = 0 for isothermal propagating waves. Squares: line 1, triangles: line 2, circles: line 3, diamonds: line 4, plusses: line 5.

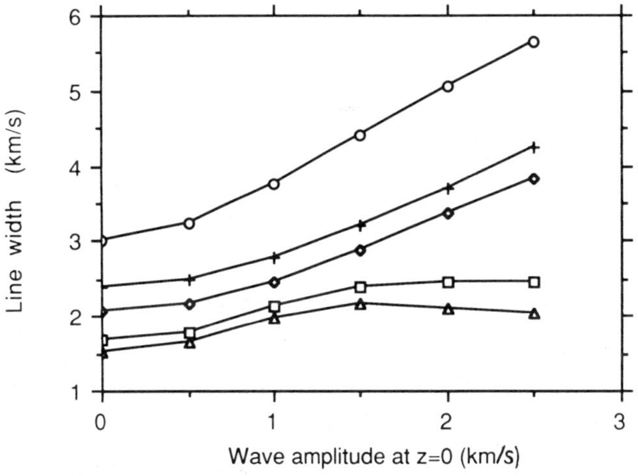

Fig. 2b. The same as Fig. 2a for non-isothermal waves.

It is important to note that in the presence of non-isothermal waves even the line width of temperature sensitive lines can become an unreliable indicator of the wave amplitude. As seen in such lines non-isothermal short period tube waves "disappear", except for their contribution to the V asymmetry.

3.3. *Stokes V Asymmetry*

Finally, let us consider how the tube waves affect the Stokes V asymmetry. Firstly, standing waves, or isothermal propagating waves do not produce any net asymmetry in the Stokes V profile averaged over a full period. Non-isothermal propagating waves also produce only negligible area asymmetry δA (defined as $(A_b - A_r)/(A_b + A_r)$) in agreement with the much simpler calculations of Solanki [this volume]. δa, the amplitude asymmetry defined similarly, on the other hand, can be quite substantial (of the order of 10%) in the 2-D case. Although exciting, this result is based on only a few 2-D calculations. More calculations, in particular such which consider the combination of a wave and an external downflow whose presence is suggested by recent empirical investigations [cf. Solanki, this volume], are planned.

4. Conclusions

We have presented linear calculations of undamped magnetoacoustic waves using the thin tube approximation and have tested their influence on the Stokes I and V line parameters of a set of hypothetical spectral lines of iron. The dependence of the line parameters on waves parameters has been studied, with a view of improving the meagre observational diagnostics of flux tube waves available so far. In particular, the importance of at least linear MHD wave calculations has been demonstrated. For example, propagating magnetoacoustic waves produce an amplitude asymmetry of Stokes V, of the correct sign and approximate magnitude for a correct reproduction of the observations of some lines [cf. Solanki, this volume]. MHD calculations are also needed to interpret line widths properly. Further calculations are underway and improvements to the description of the waves are planned.

References

Defouw, R.J., Wave Propagation Along a Magnetic Tube, *Astrophys. J.* **209**, 266–269, 1976.

Grossmann-Doerth, U., Larsson, B., and Solanki, S.K., Contribution and Response Functions for Stokes Line Profiles Formed in a Magnetic Field, *Astron. Astrophys.* **204**, 266–274, 1988.

Herbold, G., Ulmschneider, P., Spruit, H.C., and Rosner, R., Propagation of Nonlinear Radiatively Damped Longitudinal Waves Along Magnetic Fluxtubes in the Solar Atmosphere, *Astron. Astrophys.* **145**, 157, 1985.

Illing, R.M.E., Landman, D.A., and Mickey, D.L., Broad-Band Circular and Linear Polarization in Sunspots: Spectral Dependence and Theory, *Astron. Astrophys.* **41**, 183, 1975.

Keil, S.L., and Marmolino, C., Diagnostic for Propagating Waves in the Solar Photosphere, *Astrophys. J.* **310**, 912–926, 1986.

Musielak, Z.A., Rosner, R., and Ulmschneider, P., On the Generation of Flux Tube Waves in Stellar Convection Zones. I. Longitudinal Tube Waves Driven by External Turbulence, *Astrophys. J.* **337**, 470–484, 1989.

Pantellini, F.G.E., Solanki, S.K., and Stenflo, J.O., Velocity and Temperature in Solar Magnetic Fluxtubes from a Statistical Centre-to-Limb Analysis, *Astron. Astrophys.* **189**, 263–276, 1988.

Parker, E.N., *Cosmical Magnetic Fields*, Clarendon Press, Oxford, 1979.

Rae, I.C., and Roberts, B., Pulse Propagation in a Magnetic Flux Tube, *Astrophys. J.* **256**, 761–767, 1982.

Roberts, B., Waves in Inhomogeneous Media, in *The Hydromagnetics of the Sun*, T.D. Guyenne and J.J. Hunt (Eds.), *Proc. Fourth European Meeting on Solar Physics*, ESA SP–220, 137–145, 1984.

Roberts, B., Dynamical Processes in Magnetic Flux Tubes, in *Small Scale Magnetic Flux Concentrations in the Solar Photosphere*, W. Deinzer, M. Knölker, H.H. Voigt (Eds.), Vandenhoeck & Ruprecht, Göttingen, p. 169–190, 1986.

Roberts, B., and Webb, A.R., Vertical Motions in an Intense Magnetic Flux Tube. I, *Solar Phys.* **56**, 5–35, 1978.

Roberts, B., and Webb, A.R., Vertical Motions in an Intense Magnetic Flux Tube. III. On the Slender Flux Tube-Approximation, *Solar Phys.* **64**, 77–92, 1979.

Solanki, S.K., Velocities in Solar Magnetic Fluxtubes, *Astron. Astrophys.* **168**, 311–329, 1986.

Solanki, S.K., The Origin and the Diagnostic Capabilities of the Stokes V Asymmetry Observed in Solar Faculae and the Network, *Astron. Astrophys.* in press, 1989.

Spruit, H.C., Propagation Speeds and Acoustic Damping in Waves in Magnetic Flux Tubes, *Solar Phys.* **75**, 3–17, 1982.

Spruit, H.C., and Roberts, B., Magnetic Flux Tubes on the Sun, *Nature* **304**, 401–406, 1983.

Thomas, J.H., Hydromagnetic Waves in the Photosphere and Chromosphere, in *Theoretical Problems in High Resolution Solar Physics*, H.U. Schmidt (Ed.), Max Planck Inst. f. Astrophys., Munich, p. 126, 1985.

Van Ballegooijen, A.A., in *Measurements of Solar Vector Magnetic Fields*, M.J. Hagyard (Ed.), NASA Conf. Publ. 2374, p. 322, 1985.

Webb, A.R., and Roberts, B., Vertical Motions in an Intense Magnetic Flux Tube. V. Radiative Relaxation in a Stratified Medium, *Solar Phys.* **68**, 87–102, 1980.

Zayer, I., Solanki, S.K., and Stenflo, J.O., The Internal Magnetic Field Distribution and the Diameters of Solar Magnetic Elements, *Astron. Astrophys.* in press, 1989.

QUANTITATIVE EXPLANATION OF STOKES V ASYMMETRY IN SOLAR MAGNETIC FLUX TUBES

S. K. Solanki
Department of Mathematical Sciences, University of St. Andrews, St. Andrews, KY16 9SS, Scotland

Abstract: Stokes profiles of four spectral lines with very different properties are calculated in a two dimensional flux tube model of a solar magnetic element. The model has empirically derived temperature and magnetic field strength values within the magnetic element and satisfies pressure balance. The considered model can reproduce the asymmetry between the blue and red Stokes V wings, as well as other line parameters observed near disk centre in solar active region plages and the network if it incorporates the following three features: 1) A downflow of 0.5–1.5 km s^{-1} in the immediate surroundings of the flux tube (but not inside it). 2) A 250–350 K lower temperature in the downflowing non-magnetic atmosphere than in the average quiet sun. 3) A longitudinal wave-like or oscillatory motion with an amplitude of between 1 and 3 km s^{-1} within the magnetic element. The Stokes V asymmetry is thus seen to be a natural outcome of the current picture of magnetic elements embedded in cool downflowing intergranular lanes and of the presence of relatively large amplitude non-stationary mass motions within magnetic elements. The observations also suggest that the upflowing and the downflowing phases of the waves differ in some important respects.

1. Introduction

One of the outstanding features of the Stokes V profiles observed in active region plages and in the quiet solar network is their pronounced asymmetry [e.g. Stenflo et al., 1984, Wiehr, 1985]. Near solar disk centre almost all spectral lines have Stokes V profiles whose blue wings are stronger than their red wings [e.g. Solanki and Stenflo, 1984, 1985], i.e. $\delta A > 0$ and $\delta a > 0$. δA is the relative area asymmetry defined as $\delta A = (A_b - A_r)/(A_b + A_r)$ with A_b and A_r being the absolute values of the areas of the blue and red wings of Stokes V. δa is the relative amplitude asymmetry defined as $\delta a = (a_b - a_r)/(a_b + a_r)$, where a_b and a_r are the absolute blue and red Stokes V amplitudes.

Until recently the source of this asymmetry was unknown. Although various mechanisms have been proposed to explain these observations, none has so far reproduced the data consistently within the framework of a physically reasonable model.

Van Ballegooijen [1985] first pointed out that downflows outside magnetic elements can produce a Stokes V asymmetry having the correct sign of δA if the expansion of the magnetic elements with height is taken into account. Grossmann-Doerth et al. [1988, 1989] have shown that in this model, when B and v are nowhere cospatial, Stokes V profiles can be asymmetric without exhibiting *any* zero-crossing wavelength shift, thus overcoming one of the key observational hurdles without compromising physical consistency. However, no attempt was made by the above authors to quantitatively reproduce, e.g., the asymmetry of more than one Stokes V profile, or the correct ratio of δa to δA. In the present paper I, therefore, attempt a first simple quantitative fit to the data with this model.

Non-stationary mass-motions within the magnetic elements have also been proposed as a source of the Stokes V asymmetry [e.g. Solanki and Stenflo, 1984], but their influence has never been studied quantitatively. I explore how the Stokes V asymmetry compares with the data, with the help of a very simple two time-components model of wave-like or oscillatory motions. Particular emphasis is placed on the combined effect of oscillations within the magnetic elements and downflows external to them. A physically much more consistent model of flux tube waves is investigated in greater detail in a separate paper [Solanki and Roberts, this volume].

2. Description of Model Flux Tube

The model is composed of a magnetic flux tube which expands with height. It is surrounded by and partly overlies a downflowing field free atmosphere.

The stratification of the magnetic field strength, B, within the flux tube is due to exact horizontal pressure balance, i.e. it is calculated using the thin tube approximation [e.g. Roberts and Webb, 1978, 1979]. $B = 2000$ G has been chosen at $\tau = 1$ within the flux tube, in accordance with observations [e.g., Zayer et al., 1989]. The atmosphere outside the flux tube is initially represented by a model of the quiet sun, the HSRASP [Chapman, 1979]. Later, other temperature profiles are also used. These are constructed by increasing or decreasing $T(\tau)$ of the modified HSRASP, without changing the temperature gradient. τ signifies the continuum optical depth at 5000 Å. The empirically derived network flux tube model of Solanki [1986] represents the atmosphere inside the flux tube.

The flow outside the flux tube is assumed to be purely vertical, independent of height and directed downwards. The influence of internal waves and oscillations is modelled in a very simple manner by assuming a two time-component model, one component each for the upflow and the downflow phases. The velocity is kept constant for the duration of each phase and is also independent of height in the atmosphere. Both phases have equal but opposite velocities.

The model is intersected at various distances from the symmetry axis by a number of vertical rays. Along each of these the continuum optical depth is determined and the Stokes profiles are calculated numerically in LTE, using a code based on the numerical method of solution of the Unno-Rachkovsky equations first described by Beckers [1969]. Next, each constituent profile is weighted according to the solar surface area represented by the ray along which it is formed and the average is calculated of all the profiles formed

Geophysical Monograph 58

within a given radius from the centre of the flux tube. For a comparison with the data I have selected four of the spectral lines used in an earlier study of Stokes V asymmetry by Solanki and Pahlke [1988]. These include three Fe I lines (5083.3 Å, 5127.7 Å and 5250.2 Å) having similar (low) excitation potentials, but widely different equivalent widths in the quiet sun, so that they are strongly weakened by higher temperatures and exhibit very different amounts of saturation. In addition, Fe II line (5197.6 Å) has been selected which reacts quite differently to temperature due to its much higher combined ionisation and excitation energy. Solanki and Pahlke [1988] failed to even remotely reproduce the Stokes V asymmetry of these four lines with 1-D models which were otherwise consistent with observations, so that this set of lines is sufficiently sensitive to the atmospheric structure to allow at least some models to be ruled out.

The calculated line parameters are compared with observations obtained near solar disk centre using the Fourier transform spectrometer (FTS) and the McMath telescope in 1979. Stenflo et al. [1984] give a detailed description of the data. δA and $\delta a/\delta A$ values observed in an active region are listed for the four lines in Table 1.

3. Results of Models with Flows Outside the Magnetic Elements Only

In this section only those models are considered which have no velocities within the magnetic elements (i.e. $v_{int} = 0$). First I consider a basic model which has a $T_{ext}(z) = T_{HSRASP}(z)$.

Fig. 1 shows various Stokes V line parameters of Fe I 5250.2 Å calculated with this model. In Fig. 1a the relative area asymmetry δA, and in

TABLE 1. Observed Stokes V Asymmetry

Ion	$\lambda_\odot$ (Å)	δA (%)	$\delta a/\delta A$
Fe I	5083.345	5.35 ± 0.84	3.64 ± 0.62
Fe I	5127.684	-2.06 ± 4.14	-1.79 ± 4.98
Fe II	5197.574	4.43 ± 0.97	4.17 ± 0.98
Fe I	5250.217	5.21 ± 0.42	1.84 ± 0.22

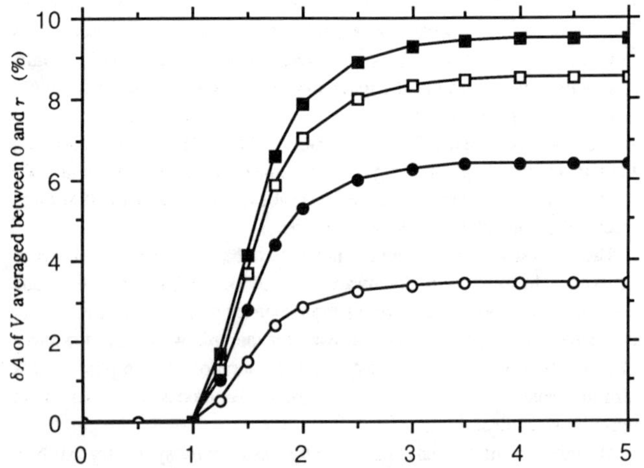

Fig. 1a. Relative area asymmetry, δA, of Fe I 5250.2 Å in % vs. r, the distance from the axis of the flux tube. The δA values plotted at any given r in the figure belong to the average of the Stokes V profiles formed between the axis of the tube and r. External downflow velocity, $v_{ext} = 0.5$ km s^{-1} (open circles), 1 km s^{-1} (filled circles), 1.5 km s^{-1} and 3 km s^{-1} (open squares), and 2 km s^{-1} and 2.5 km s^{-1} (filled squares).

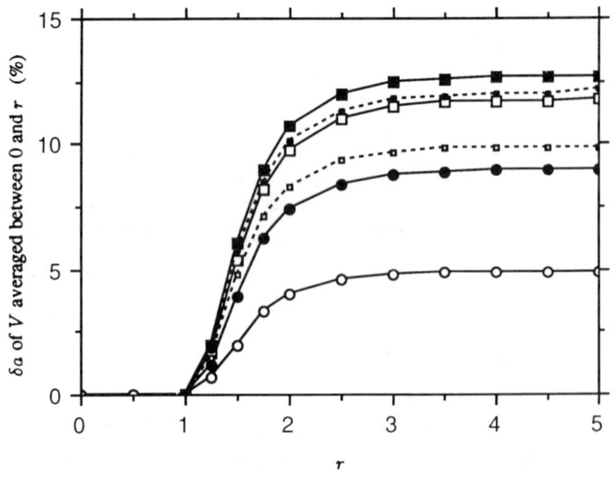

Fig. 1b. Relative amplitude asymmetry, δa, of the averaged Stokes V profiles of Fe I 5250.2 Å in % vs. r. The symbols refer to the same v_{ext} as in Fig. 1a. However, now the curves for $v_{ext} = 2.5$ km s^{-1} and 3 km s^{-1} are plotted separately (dashed) with smaller symbols.

Fig. 1b the relative amplitude asymmetry δa are plotted vs. radius r. The δA and δa values plotted at a given r belong to the area weighted mean Stokes V of all the profiles formed along rays between the flux tube axis and r. δA is shown for different values of v_{ext} (see Figure caption). δA increases steadily with v_{ext} up to $v_{ext} \approx 2.25$, but decreases for larger v_{ext} (cf. Grossmann-Doerth et al. [1989] for a more detailed analysis of this effect).

δA also increases with increasing r. For Fe I 5250.2 Å the main contribution to δA comes in the range of radii between 1 and 2, reaching approximately 80% of the δA value at large r by $r = 2$. At a greater distance from the flux tube axis the contribution of any individual ray to Stokes V is extremely small, despite the large area represented by that ray, and δA becomes essentially constant. Even with reasonable values of v_{ext} ($0.5 \lesssim v_{ext} \lesssim 1.5$ km s^{-1}) δA values comparable to the observations are produced, although only the "canopy" profiles are asymmetric. (The canopy refers to the part of the flux tube overlying field free regions.) This implies that the Stokes V profiles produced in parts of the canopy must be extremely asymmetric. Indeed, closer examination shows that individual Stokes V profiles can have δA and δa values of over 80% (for Fe I 5083.3 Å δA values of well over 90% are seen). Such profiles are almost completely composed of only one Stokes V wing.

As can be seen from Fig. 1b, δa shows a qualitatively similar behaviour to δA. For all the calculated cases $\delta a > \delta A$, but the ratio $\delta a/\delta A$ decreases markedly as v_{ext} is increased, being 1.42 for $v_{ext} = 0.5$ km s^{-1}, but only 1.17 for $v_{ext} = 3.0$ km s^{-1}. The major part of this decrease in $\delta a/\delta A$ occurs above $v_{ext} = 1.5$ km s^{-1}. Although they are asymmetric, all the Stokes V profiles calculated in this section, where velocity within the tube is zero, have zero-crossing wavelengths corresponding to their rest wavelengths. This result has been rigorously proved by Grossmann-Doerth et al. [1988, 1989].

Although the observed δA and zero-crossing wavelength of a single Stokes V profile may be reproduced very simply with the "basic" model used so far, the δA of the four chosen lines cannot be reproduced by a single reasonable v_{ext} value. In addition, the calculated $\delta a/\delta A$ ratios for Fe I 5083.3, 5250.2 and Fe II 5197.6 Å are too small, as are the widths of the average Stokes V profiles of all four lines. Obviously the model must be modified if it is to explain the observations quantitatively.

One successful modification has been to change the temperature of the surroundings of the model flux tube in the manner described in Sect. 2. A series of models with ΔT_{ext} values ranging from −600 K to +300 K have been calculated. $\Delta T_{ext} = 0$ corresponds to the temperature structure of the HSRASP. The resulting δA of all four lines is plotted as a function of ΔT_{ext} in Fig. 2 for $v_{ext} = 1$ km s^{-1}. δA of each of the lines has a distinctive dependence on ΔT_{ext}. δa behaves quite similarly to δA, so that $\delta a/\delta A$ is not significantly affected by changing ΔT_{ext}.

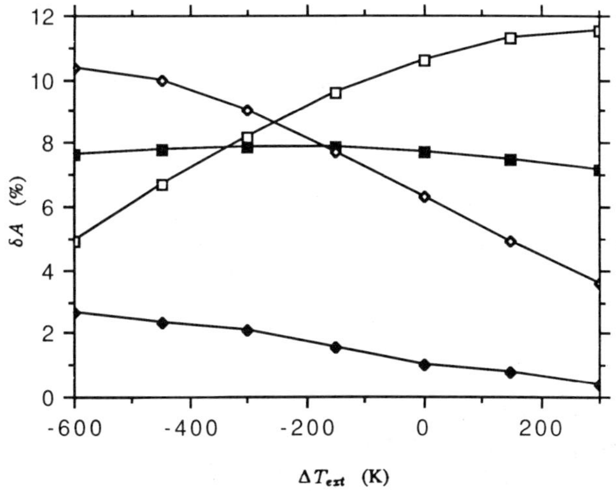

Fig. 2. δA in % of V profiles averaged between $r = 0$ and 4 (in the units used in Fig. 1) of Fe I 5083.3 Å (open squares), 5127.7 Å (filled diamonds), Fe II 5197.6 Å (filled squares) and Fe I 5250.2 Å (open diamonds) vs. ΔT_{ext}, the temperature difference between the non-magnetic atmosphere surrounding the flux tube and the HSRASP. $v_{ext} = 1$ km s^{-1}.

The best fit to δA of all four lines (within the observational uncertainty) is obtained for ΔT_{ext} between approximately −250 K and −350 K, i.e. when the surroundings of the magnetic elements are rather cool compared to the average quiet sun. This agrees well with the picture of magnetic elements being concentrated in dark intergranular lanes [e.g. Dunn and Zirker, 1974; Mehltretter, 1974; Title et al., 1987].

4. Results of Models with Velocities Inside and Outside the Magnetic Elements

In this section I investigate the influence of non-stationary mass-motions within the magnetic elements on the Stokes V asymmetry in combination with external flows. Only models with an external atmosphere having $\Delta T_{ext} = -300$ K are considered, in agreement with the results of Sect. 3.

A very simple two time-component model of a flux tube wave or oscillation, whose basics are described in Sect. 2. has been constructed. For the calculation of the line profiles the same atmosphere is used in both phases. However, when adding the two phases together to produce the time-averaged Stokes V (which is the proper Stokes V profile to compare with time averaged or low spatial resolution data), the profiles resulting from the two phases can be weighted differently to reflect, in a crude manner, e.g., differences in thermal structure between the upflowing and downflowing phases. It can be easily shown that adding two unequally weighted and shifted Stokes V profiles can easily produce a resulting Stokes V profile with a significant *amplitude* asymmetry δa, even if both the original profiles are antisymmetric. The calculations presented in this section should therefore influence δa considerably. The most important result of this section is illustrated in Fig. 3, where the ratio $\delta a/\delta A$ of Fe I 5250.2 Å and 5083.3 Å is plotted vs. the weight, w_u, given to the Stokes V profile resulting from the upflow phase. w_u is normalised such that the sum of the weights of both phases is always unity. The "wave amplitude" is 1 km s^{-1}, i.e. $v_{int} = \pm 1$ km s^{-1}, and v_{ext} is 0.5 km s^{-1}. The curves have only been plotted for those values of w_u for which $\delta a/\delta A \geq 0$ and $\delta A > 0$. Both lines exhibit the same qualitative behaviour, although $\delta a/\delta A$ of Fe I 5083.3 Å can be enhanced much more than of Fe I 5250.2 Å, in agreement with the data. The larger $\delta a/\delta A$ values are always produced when the upflow component is more strongly weighted.

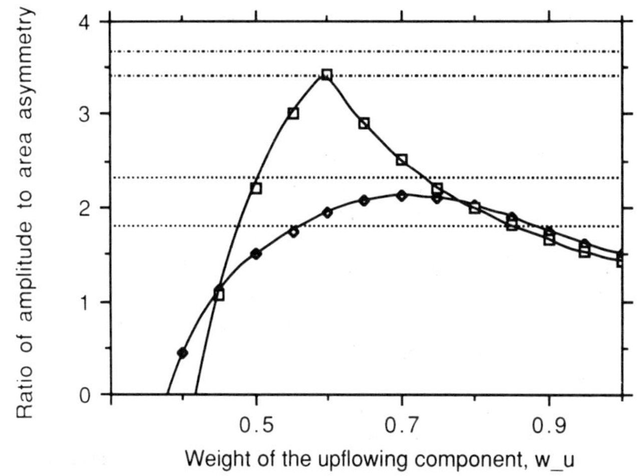

Fig. 3. The ratio $\delta a/\delta A$ of the Stokes V profile of Fe I 5250.2 Å (diamonds) and 5083.3. Å (squares) averaged between $r = 0$ and $r = 4$ vs. w_u, the weight of the internal upflow velocity component. $w_u = 1$ signifies that only an upflow is present, while $w_u = 0.5$ means that up- and downflows have equal weight. $v_{int} = \pm 1$ km s^{-1} and $v_{ext} = 0.5$ km s^{-1}. Only those points have been plotted for which both δa and δA are positive. The horizontal dashed lines represent the observations, in a plage and network region, of 5250.2 Å, while the dot-dashed lines mark the observations of 5083.3 Å.

The observations are represented by the horizontal dashed lines for 5250.2 Å and the dot-dashed lines for 5083.3 Å. The $\delta a/\delta A$ observations of both lines can be approximately reproduced by the chosen model parameters between $w_u \approx 0.55$ and $w_u \approx 0.65$. Unfortunately, the zero-crossing wavelength is no longer conserved when the up- and downflow phases are not equally weighted. However, zero-crossing wavelength shifts are minute when the weighting of the two phases is nearly the same. The observational upper limit of ±250 m s^{-1} on the zero-crossing shift [Solanki, 1986] implies that only calculations with w_u values between approximately 0.4 and 0.6 need be considered further, as far as a comparison with the observational data is concerned. For Fe I 5250.2 Å the profile with the $\delta a/\delta A$ value closest to the observations is blueshifted by approximately 150–250 m s^{-1} and lies just below the upper limit set by the observations. For Fe I 5083.3 Å, the blueshift of the corresponding profile is considerably smaller (50–100 m s^{-1}). As expected, the Stokes *V* line widths also increase with increasing "wave" amplitude. An external downflow of 0.5–1.5 km s^{-1} and an internal wave of amplitude between 1 km s^{-1} and 1.5 km s^{-1} with a weight of the upflow phase, w_u, of 0.55–0.6 are required to reproduce of the plage data best.

5. Conclusions

In the present paper the origin of the Stokes *V* asymmetry observed near disk centre in solar active regions and in the quiet network is studied with the help of a 2-D flux tube model of solar magnetic elements. It is shown that the observed relative area and amplitude asymmetry, the zero-crossing wavelengths and the widths of four Fe I and II Stokes *V* profiles belonging to lines with widely different properties can be reproduced relatively well within the framework of this model if it incorporates the following three features. Firstly, a downflow is present in the immediate surroundings of the magnetic elements. Secondly, this downflowing region is cooler than the quiet sun. Thirdly, an oscillatory or wave-like motion is present within the magnetic elements. Since the thermal and magnetic structures of the model are empirically derived, this implies that the same model effectively also reproduces a host of other line parameters (e.g. magnetic and thermal line ratios, Zeeman splitting of infrared lines etc.).

The following quantitative results are derived from the present calculations. The immediate surroundings of magnetic elements are approximately 250–350 K cooler than the average quiet sun and contain a downflow of 0.5–1.5 km s^{-1} near the walls of the magnetic elements. The waves or oscillations inside magnetic elements have a velocity amplitude of 1–1.5 km s^{-1}, derived on the basis of a simple two time-component model. It is expected that the true wave amplitude is larger, since it must compensate for the larger weighting of small velocities in a sinusoidal wave, so that more realistically the wave amplitude lies between 1 and 3 km s^{-1}. The observations also suggest that the upflowing and downflowing phases probably differ in some important respects.

The picture of solar magnetic elements emerging from the present use of the Stokes *V* asymmetry as a diagnostic is highly appealing, since it fits in very well with our present theoretical understanding of such structures [cf. Spruit and Roberts, 1983; Schüssler, 1986; Solanki, 1987, Roberts, this volume, for reviews].

References

Beckers, J.M., The Profiles of Fraunhofer Lines in the Presence of Zeeman Splitting. I. The Zeeman Triplet, *Solar Phys.* **9**, 372–386, 1969.

Chapman, G.A.: New Models of Solar Faculae, *Astrophys. J.* **232**, 923–928, 1979.

Dunn, R.B., and Zirker, J.B., The Solar Filigree, *Solar Phys.* **33**, 281–304, 1973.

Giovanelli, R.G., Livingston, W.C., and Harvey, J.W., Motions in Solar Magnetic Tubes. II. The Oscillations, *Solar Phys.* **59**, 49–64, 1978.

Grossmann-Doerth, U., Schüssler, M., and Solanki, S.K., Unshifted, Asymmetric Stokes *V*-Profiles: Possible Solution of a Riddle, *Astron. Astrophys.* **206**, L37–L39, 1988.

Grossmann-Doerth, U., Schüssler, M., and Solanki, S.K., Stokes *V* Asymmetry and Shift of Spectral Lines, *Astron. Astrophys.* in press, 1989.

Mehltretter, J.P., Observations of Photospheric Faculae at the Center of the Solar Disk, *Solar Phys.* **38**, 43–57, 1974.

Roberts, B., and Webb, A.R., Vertical Motions in an Intense Magnetic Flux Tube. I, *Solar Phys.* **56**, 5–35, 1978.

Roberts, B., and Webb, A.R., Vertical Motions in an Intense Magnetic Flux Tube. III. On the Slender Flux Tube-Approximation, *Solar Phys.* **64**, 77–92, 1979.

Schüssler, M., MHD Models of Solar Photospheric Magnetic Flux Concentrations, in *Small Scale Magnetic Flux Concentrations in the Solar Photosphere*, W. Deinzer, M. Knölker, H.H. Voigt (Eds.), Vandenhoeck & Ruprecht, Göttingen, p. 103–120, 1986.

Solanki, S.K., Velocities in Solar Magnetic Fluxtubes, *Astron. Astrophys.* **168**, 311–329, 1986.

Solanki, S.K., Magnetic Fields: Observations and Theory, in *Proc. Tenth European Regional Astronomy Meeting of the IAU. Vol. 1: The Sun*, L. Hejna, M. Sobotka (Eds.), Publ. Astron. Inst. Czechoslovak Acad. Sci., p. 95–102, 1987.

Solanki, S.K., and Pahlke, K.D., Can Stationary Velocity Fields Explain the Stokes *V* Asymmetry Observed in Solar Magnetic Elements?, *Astron. Astrophys.* **201**, 143–152, 1988.

Solanki, S.K., and Stenflo, J.O.: Properties of Solar Magnetic Fluxtubes as Revealed by Fe I Lines, *Astron. Astrophys.* **140**, 185–198, 1984.

Solanki, S.K., and Stenflo, J.O., Models of Solar Magnetic Fluxtubes: Constraints Imposed by Fe I and II Lines, *Astron. Astrophys.* **148**, 123–132, 1985.

Spruit, H.C., and Roberts, B., Magnetic Flux Tubes on the Sun, *Nature* **304**, 401–406, 1983.

Stenflo, J.O., Harvey, J.W., Brault, J.W., and Solanki, S.K., Diagnostics of Solar Magnetic Fluxtubes Using a Fourier Transform Spectrometer, *Astron. Astrophys.* **131**, 333–346, 1984.

Title, A.M., Tarbell, T.D., and Topka, K.P., On the Relation Between Magnetic Field Structures and the Granulation, *Astrophys. J.* **317**, 892–899, 1987.

Van Ballegooijen, A.A., Discussion Remarks in *Theoretical Problems in High Resolution Solar Physics*, H.U. Schmidt (Ed.), Max Planck Inst. f. Astrophys., Munich, p. 177, 1985.

Wiehr, E., Spatial and Temporal Variation of Circular Zeeman Profiles in Isolated Solar Ca^+ K Structures, *Astron. Astrophys.* **149**, 217–220, 1985.

Zayer, I., Solanki, S.K., and Stenflo, J.O., The Internal Magnetic Field Distribution and the Diameters of Solar Magnetic Elements, *Astron. Astrophys.* in press, 1989.

A BRIEF INTRODUCTION TO CORONAL "LOOPS"

Robert Rosner
Enrico Fermi Institute and Dept. of Astronomy and Astrophysics
The University of Chicago

Abstract. I discuss the physics of coronal loop structures from first principles, focussing on their geometry, their energetics, and interpretational problems.

Introduction

In this review, I will attempt to focus on the basic principles governing the stucturing of the Sun's outer atmosphere, without any attempt at covering the latest "wrinkles" in the field. In particular, I will focus on the quasi-static atmosphere, or, more properly put, on those aspects of the outer atmosphere for which a quasi-static description gives an adequate account of the observations.

First, a bit of history. Prior to the flights of imaging instruments functioning at UV and shorter wavelengths in the late 1960's and early 1970's, the commonly accepted view of the Sun's outer atmosphere held that to first approximation, this atmosphere was homogeneous; but that significant — but occasional — structures were present. Coronal studies of the period can thus be subdivided into work focussing on the homogeneous aspects; and work focussing on the occasional interlopers. The first category of studies typically led to so-called "reference atmospheres", which were regarded as the experimental touchstone for theoretical work. All this changed completely once it became possible to image the outer atmosphere on the disk; and the collection of telescopes on the Skylab mission of 1973-4 arguably contributed more than any other experiments to changing the perception of what the Sun's outer atmosphere is really like.

This change of view is commonly summarized by showing one of the classic full-disk images obtained from the x-ray telescopes onboard Skylab; but I won't do this here. Instead, I ask you to inspect Figure 1, which shows a small portion of the solar disk, where a particular region of heightened activity has emerged from beneath the Sun's surface. The key issue is a point often lost in the full-disk images, namely that essentially all of the emission seen in this figure comes from what has been called "loop structures", or more simply, "loops". There is no diffuse emission, in the sense that even if there were, it hardly contributes to what is actually seen, and therefore — without any further quibbling — we are justified in asserting that from an energetic standpoint, diffuse emission, if present, is simply irrelevant; and what we have to explain when we are talking about the "coronal heating problem" is exactly why one sees these filamented structures.

"Loops" And The Coronal Magnetic Field

To begin with, let us focus on the problem of how such a loop might "work". It is a straightforward exercise to obtain magnetograph images of the photospheric regions which underlie the loops seen in Figure 1, and using the valued of the line-of-sight magnetic field at the photospheric level, one can compute estimates of the magnetic field strength and configuration in the overlying atmosphere. Of course, there

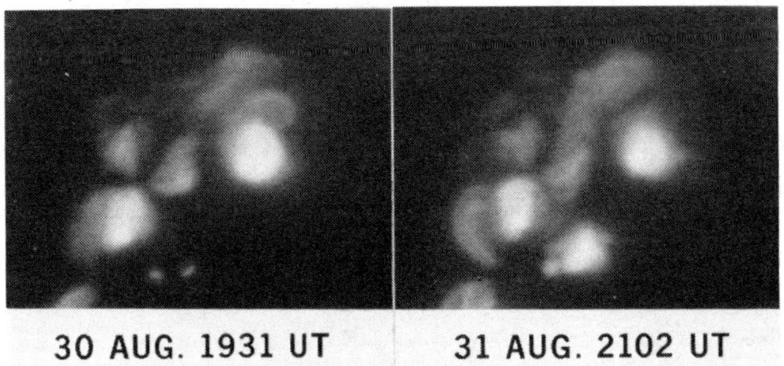

Fig. 1. Two images of a small active region complex, taken by the Skylab S-O54 x-ray telescope. Note the dominance of loop structures, and the virtual absence of any evidence for diffuse emission from surrounding gas.

Geophysical Monograph 58

are a variety of assumptions that go into such a calculations, some of which are surely suspect (for example, that the overlying atmosphere is current-free); but the basic result which emerges is not at all sensitive to whether these assumptions are valid or not: that is, one finds that the magnetic pressure completely dominates the gas pressure in the regions seen in emission in Figure 1. This result has given rise to the commonly-held notion that the loops are nothing but "magnetic bundles", or ropes or tubes, which constrain the high-temperature gas from leaving the Sun's vicinity by virtue of the fact that the ions and electrons in this gas are line-tied to the field. This picture is only partially correct.

To understand the basic problem, note first that the time scales for pressure equilibration are far shorter than any other times of interest (such as, for example, typical cooling times in these loop structures). This means that, for our purposes, the loop structure are in pressure balance with their surroundings. But the surroundings are dark — and it is readily shown that this implies a decreased density, and decreased gas pressure, outside the bright loops. Since the total pressure inside and outside a given loop must balance, and since the magnetic pressure contribution is positive definite, it must be that the loop interior has a weaker magnetic field than the gas just outside. Thus, the loops one sees in x-rays are in some sense anti-flux ropes or anti-tubes.

Loop Energetics

As long as we are discussing quasi-static loop structures, or, more precisely, loops which are not flaring on typical sound crossing times, one can discuss the energetics in a relatively straightforward manner. To begin with, the loop has basically two ways in which it can transfer energy to its surroundings — by radiation and by thermal conduction. It must be heated. And it may also have internal processes (such as flows along and across the confining magnetic fields) which carry enthalpy, and hence may also contribute to the energetic balance. However, let us keep things simple (while keeping in mind that "real" loops may not care one wit about a theorist's desire to maintain simplicity), and assume that there are no flows at all. In that case, energy balance can be looked at schematically as in Figure 2.

To understand what happens, it is useful to recall that at coronal temperatures, thermal conduction is extremely efficient. For that reason, it turns out not to make too much of a difference exactly how uniform the heating is along the loop's length [cf. Craig, McClymont, and Underwood 1978], aside from some interesting dynamics which arises if the heating is not symmetric about the loop apex; I will not discuss this latter issue here at all. For the same reason, much of the coronal portion of the loop is at roughly the same temperature; and if we further suppose that the loop size is smaller than the density scale height at coronal temperatures, then the density variation will be small as well. All of

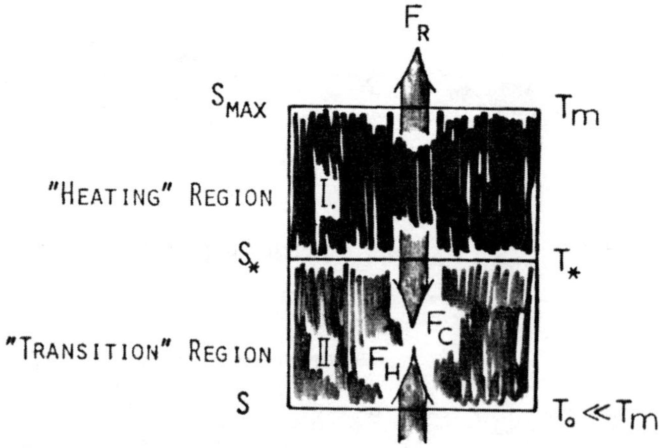

Fig. 2. Sketch of the energetic balance of a schematized coronal loop. Note that we are assuming that thermal conduction acts essentially as an energy redistribution mechanism, and not as a loss process from the loop base.

this is to say that, first, we might as well assume that the heating is spatially uniform; and, second, that the radiative losses (which scale as the square of the density times a weak function of temperature) are also roughly constant along

$$div\ F_c \ = \ n^2 P(T) + E_H \tag{1}$$

by assumption (F_c is the heat flux, n is the number density. $P(T)$ is a known function of temperature, and E_H is the heating function), our argument leads us to the conclusion that, roughly speaking, the divergence of the heat flux in the corona is comparable to the radiative losses,

$$|div\ F_c\ | \ \approx \ n^2 P(T)\ . \tag{2}$$

This immediately allows us to scale the coronal temperature at the loop apex T_m to the loop length L_c, since equation (2) yields the relation

$$10^{-6}\ T_m^{7/2}/L_c \ \approx \ n^2 P(T_m), \tag{3}$$

where we used the fact that for $T > 10^{5.5}\,K$, $P(T) \approx T^{-1/2}$. Solving for T_m, and using the fact that the gas pressure p $(= 2nkT)$ is roughly constant, we arrive at the scaling law

$$T_m \ = \ T_m(p, L_c) \ \approx \ (pL_c)^{1/3}; \tag{4}$$

using the fact that in the corona, E_H must balance both conduction and radiation, we can also derive a scaling law connecting the loop pressure to the heating rate and loop length,

$$p \ = \ p(E_H, L_c) \ \approx \ E_H^{6/7} \cdot L_c^{5/7} \tag{5}$$

A more detailed version of this argument was first presented by Rosner, Tucker, and Vaiana (1978), and has been rederived in various, elegant ways by many other authors since. Indeed, very soon after, Hood and Priest (1979) pointed out that there is another solution to equation (1), the so-called "cool" loop solution; these solutions were worked out in detail recently by Antiochos and Noci (1986), who pointed out that one can obtain their second type of scaling relation by simply assuming the alternative assumption to equation (2) above, namely that

$$E_H \approx n^2 P(T) \; . \tag{6}$$

The key observation here is that thermal conduction becomes irrelevant for sufficiently cool loops, which is precisely the case here. We will return to the issue of cool loops in a moment.

An immediate question is whether these scaling relations actually hold up. This question was addressed early on by Peres et al. (1982) in the course of developing hydrodynamic codes able to handle solar loop flares. The obvious issue was: equation (5) states essentially that for a given loop (and hence a given loop size), the loop pressure is essentially determined by the heating rate; and, via equation (4), that the loop temperature then follows as well. Can these effects be verified in a simulation? The answer is that they can, indeed, demonstrating at least the cognency of the original scaling arguments, if not the fact that the Sun really does behave like that.

The Problem Of The Differential Emission Measure

An important aspect of modeling loop structures is to determine how much mass there is at any given temperature in the outer solar atmosphere. This information is gotten in a straightforward manner by starting with the relation connecting a line (or filter) intensity $I(\lambda,\lambda+\Delta\lambda)$ with the density and temperature distributions,

$$I(\lambda,\lambda+\Delta\lambda) = \frac{1}{4\pi d^2}\int dV \; n^2 \cdot P(T;\lambda,\lambda+\Delta\lambda) \; ,$$

or

$$I(\lambda,\lambda+\Delta\lambda) = \frac{1}{4\pi d^2}\int dA \int d(\ln\, T) \cdot DEM(T) \cdot P(T;\lambda,\lambda+\Delta\lambda) \; , \tag{7}$$

where the differential emission measure is defined as

$$DEM(T) \equiv \frac{n^2\; T}{\frac{dT}{dr}} \; . \tag{8}$$

Since the earliest analyses of the Skylab data, it was clear that simple models of the corona and transition region simply could not match the differential emission measure at temperatures below $\approx 10^{5.3}$ K (and Antiochos and Noci 1986, among others, have shown that superpositions of the hot loop solutions discussed above will not fix the discrepancy either). Typically, the observed *DEM* exceeds the computed *DEM* by a factor of up to 10 at temperatures of 10^5 K; that is, one observes more material at low temperatures than one would expect from model calculations (such as in observations of lines from lower ionization states of Calcium).

A number of solutions have been proposed. These range from the straightforward (e.g., allow spicules to contribute to the flux from lower-temperature material, cf. Athay 1984; or hypothesize multi-temperature structures perpendicular to the field due to localized current heating, Rabin and Moore 1984) to the very complex (e.g., if local heat transport is anomalous — involving hot electrons with very long ranges — the effect is to decrease the effective temperature gradient, and hence to artificially increase the *DEM*; cf. Shoub 1983) However, the most straightforward solution is to acknowledge that the theory for scaling laws of hot loops also allows for cool loops, which by necessity must be physically small. These cool and small loops then offer the possibility of resolving the *DEM* problem by simply enlarging the definition of what a "loop" is [Antiochos and Noci 1986; Dowdy, Emslie, and Moore 1987]. At present, it is not feasible to directly test this idea, but future missions with UV imaging capability at high spectral and spatial resolution, such as the OSL, should be able to resolve this problem.

So, Why Are There Loops Anyway?

A key point we have managed to evade for a number of pages so far is the question that any untutored eye gazing at Figure 1 would ask: why are there loops on the Sun in the first place? To see the problem, recall that for the simplest magnetic model of an ensemble of loops, one finds immediately that field extrapolation from some given magnetic surface field distribution into the overlying corona leads to coronal field structures which are rather smooth, and in particular do not show the kind of transverse structuring that makes the x-ray and EUV images of the Sun so striking. The previous discussion makes plain what must be going on: since thermal conduction is so efficient in carrying heat along the coronal magnetic fields, it must be that the heating itself is highly inhomogeneous.

Now, before we embark on a more considered discussion, let me point out that on a certain level, loops can be regarded as mere surface blemishes. That is to say, since the plasma beta ($\equiv p_{gas}/p_{magnetic}$) for typical active region loops is rather small (certainly less that 0.1), the local perturbations in the dominant pressure term (due to the fields) are very small. Thus, rather minor changes in coronal magnetic field configurations can lead to extremely dramatic changes in the appearance of the luminous corona at x-ray wavelengths; and, conversely, dramatic changes in the x-ray appearance of the corona may be associated with rather minor rearrangements of coronal magnetic fields. In this sense, coronal x-ray emission can be regarded as a sensitive amplifier, or probe, of magnetic field configuration changes.

There are basically only 3 generic ways of producing filamentary structues in a magnetized gas: (i) localized heating; (ii) instability of the existing gas, at a time when it was homogeneous; and (iii) instability of a "pre-coronal" gas (in this view, there never was, and never will be, a homogeneous coronal atmosphere). I will simply discount the second possibility, mostly on the basis of Heyvaert's work showing that for typical coronal conditions, the hot gas is in fact stable to filamenting instabilities.

The first possibility is basically what one would expect from heating models founded on the idea of highly transient, localized heating [cf. discussion in Parker 1979]. The difficulty with this idea is the localization itself. That is, the typical physical process involved in the energy release is reconnection, which is intrinsically very small scale (e.g., the interesting dynamics can take place on scales as small as centimeters). The problem is then how the energy released on scales of 0.1-1 km or less can spread transverse to the field over dimensions of order 10^8 cm or larger. I regard this issue as still unresolved.

The third possibility is based on the appealing notion that if one fixes attention on a given loop structure, there is a point during which the flux bundle must pass through the photosphere and above; at this point, the material which is coming along for a ride is in rough thermal equilibrium with the immediate surroundings (and so is relatively cold), and the question is then whether this material is thermally unstable. This has been a recently active research area, in which I have been collaborating with A. Ferrari, G. Bodo, and S. Massaglia at the Univ. of Torino; the principal result to date is that in a gravitationally stratrified atmosphere, very large spatial scale (along $\vec{B}$) modes are easily destabilized. It is amusing that the transverse spatial scales involved in this case are rather similar to those involved in the localized heating model; and for similar reasons, a definitive test will not be feasible unless we are able to directly image loop structures at sub-arcsecond spatial resolution (sub-100 km structures are of course not not likely to be resolved in the EUV or at x-ray wavelengths in the foreseeable future).

Finally, I note that wave heating mechanisms based on the idea of resonant absorption of waves [cf. Hollweg 1984] can in principle resolve the problem of localized dissipation and ultimate dispersal of the released energy. However, it remains unclear to me how such heating naturally leads to the appearance of tubular structures, rather than other types of geometries for the heated regions. In particular, it is not obvious by what mechanism the transverse spatial scale for loop structures in images such as shown in Figure 1 is fixed.

Summary

I have given a brief tour through the elementary aspects of the physics of coronal loops, with the aim of demonstrating that — once one has a clear idea of what exactly one is modeling — the mysteries of coronal loops — the why and wherefores of their temperatures and pressures — drop away. Interestingly enough, the ultimate question, namely why are there loops in the first place, remains unanswered. I firmly believe that in this case too the solution will only come about once we have in hand the appropriate data, which in this case is high (arcsecond or better) angular resolution, high contrast, images of loop structures. I have also not addressed the heating problem in any detail; this is simply because I believe that while a number of heating processes for coronal structures can be eliminated (such as purely acoustic heating), the number of viable alternatives is still large, and essentially unconstrained (or rather, undifferentiated) by currently available data. This will change shortly, as new instruments are brought to bear on this problem, principally the SOHO mission, with its capability of sensing motions in the corona. Finally, it is easy to see that in order to model solar data in any detail, the simple kind of model discussed here will not do; this was in fact the point of the above discussion of the loop differential emission measure. These complications I leave for the subsequent speakers to discuss.

Questions and Answers

Drake: You argue that a cold corona would be unstable to the super-heating instability. This implies that Ohmic heating can balance radiation in this cold state. Is this true?

Rosner: Yes, if the initial state is cold ($T \leqq 10^8$ cm) and stratified, so that the density at coronal heights ($\geqq 10^8$ cm) is extremely low. I assume what you're getting at is that as the density at coronal heights increases (and the temperature increases as well), one does reach the point where classical current heating will no longer be able to balance the losses - that is correct, but is a regime our calculations explicitly do not apply to.

Haerendel: You showed the relationship of x-ray loops and EUV emissions from the transition region. What does one see in the lower chromosphere in H_α at the end points of the x-ray loops?

Rosner: I regret not having shown the figure that can show the answer directly: Indeed, G. S. Vaiana and collaborators demonstrated many years ago that there are bright H_α points at the footpoints of x-ray loops.

Forbes: What role does enthalpy flux have in the energetics of coronal loops? I know it's important during the formation of the loop, but is it of any importance once the loop reaches an equilibrium state?

Rosner: It depends on the type of loop. For typical hot coronal loops, the enthalpy flux is very likely irrelevant, in the sense that most features of such loops are captured by static models. This is however, not the case for cool loops. The classic example of the latter is the sunspot plume, which is an EUV feature studied first extensively with data from the Skylab S-0SS instrument; sunspot plumes characteristically show emission in EUV lines at heights well above the sale height corresponding to the temperatures at which these lines are formed. For this reason, the Harvard group suggested that these plumes reflect matter which is falling and simultaneously cooling: there the enthalpy flux plays an essential role in the energetics.

Roberts: You have not mentioned transport processes across the magnetic field, such as perpendicular thermal conduction and electrical conductivity. Such effects produce very small scales across the structure, assigning the heating physics to these small scales but leaving a more amorphous, smoothed out structure across a loop. Can you comment on this?

Rosner: Actually, the calculations of filamenting instabilities I just discussed depend essentially on cross-field transport and, as you say, the effect is to broaden very thin layers. However, classical cross-field transport cannot hope to explain coronal (i.e., $T \geqq 10^6$ K)transverse scales of the sort actually observed, that is, transverse loop dimensions well in excess of 10^3 km.

Kundu: I have two questions: (1) When we apply the scaling laws that you derive from x-ray observations to the radio data (for example, microwave loops observed with the VLA), we find that we've to adjust the scaling parameters to fit the radio data. This may be primarily because in radio (microwaves), we also deal with gyro-resonance radiations (at low harmonies of the gyro-frequency) specially near the footpoints where magnetic field is higher in addition to thermal bremsstrahlung. Did you ever look into the situation? (2) With regard to "hot loop" model versus "cold loop" model, there are two theoretical pieces of work - one due to Zheleznyakov who showed that for hot loops, the radio signature will be in the form of electron radiation at low harmonis of ω_H , $2\omega_H$, $3\omega_H$, etc. The other (cold loop) due to Syrovatskii predicts that the radio manifestation will be primarily in the millimeter domain.

Rosner: (1) No, I have not thought through how these static loops would "look" to the VLA; that is a useful thing to do, however, because radio observations are nowadays much more readily carried out than space-based observations at short wavelengths. The trade-off is of course that the radio data is more difficult to interpret than the optically-thin emission in the EUV and at higher energies.

Kahler: We have seen examples of hot x-ray loops and cool XUV loops. Are there two temperature classes of loops rather than a broad continuum?

Rosner: Yes, in the sense that these exists a clear excess in differential emission measure for $T \leqq 10^5$K over what one would expect from standard loop models, this point has been extensively discussed by Hood and Priest (1980), Antiochos and Noci (1986) and, more recently, by Dowdy, Emslie and Moore (1987).

Nordlund: You see a problem in having a very localized heat source effect the heating of a "fat" loop. This is a problem only if one assumes the magnetic field in the loop is nicely ordered. If the field in the loop is "chaotic" (braided; cf., Parker's braiding theory) this is not a problem. The size scale is then given by the braiding "length" over the time scale of the existence of the loop, or alternatively the time scale for dissipation (destruction) of the braiding (cf., the time-scales argued by Parker and typical velocities of footpoint random motion).

Rosner: I agree that "chaotic" fields are a way of getting around the cross-field transport problem; in fact, that was exactly the point made by Tsinganos, Rosner and Distler (1986). However, that still doesn't answer the question I posed, namely, what process determines the transverse loop dimension, R.

Martens: In all the heating models it is sort of assumed that the heating is stationary, while in fact, line-observations by, for example, Sheeley and Golub show strong variability? What is your comment on that?

Rosner: You are right - as I showed, one sees variability in soft x-rays on all time scales, down to the radiative cooling time (and shorter yet for some flare loops). The static models obviously can't hope to shed light on this behavior. The key here must be the process that heats the loops - the process that leads to both temporal and spatial intermittency.

Acknowledgments. I would like to thank L. Golub and G. S. Vaiana for many useful discussions on this topic over a number of years. Much of the work discussed here was supported by NASA, most recently via its Solar-Terrestrial Theory Program at the University of Chicago.

References

Antiochos, S. K., and G. Noci, The Structure of the Static Corona and Transition Region, *Astrophys. J.,* **301**, 440-447, 1986.

Athay, R. G., The Origin of Spicules and Heating of the Lower Transition Region, *Astrophys. J.,* **287**, 412-417, 1984.

Bodo, G., S. Massaglia, R. Rosner, and A. Ferrari, *M.N.R.A.S.,* submitted, 1989.

Craig, I. J. D., A. N. McClymont, and J. H. Underwood, The Temperature and Density Structure of Active Region Coronal Loops, *Astron. Ap.,* **70**, 1-11, 1978.

Dowdy, Jr., J. F., A. G. Emslie, and R. Moore, On the Inability of Magnetically Constricted Transition Regions to Account for the 10^5 to 10^6 K Plasma in the Quiet Solar Atmosphere, *Solar Phys.,* **112**, 255-279, 1987.

Hollweg, J. V., Resonances of Coronal Loops, *Astrophys. J.*, **277**, 392-403, 1984.

Hood, A. W., and E. R. Priest, The Equilibrium of Solar Coronal Magnetic Loops, *Astron. Ap.*, **77**, 233-251, 1979.

Parker, E. N., *Cosmical Magnetic Fields*, Clarendon, Oxford, 1979.

Peres, G., R. Rosner, S. Serio, and G. S. Vaiana, Closed Coronal Structures. IV. Hydrodynamical Stability and Response to Heating Perturbations, *Astrophys. J.*, **252**, 791-799, 1982.

Rabin, D., and R. Moore, Heating the Sun's Lower Transition Region with Fine-scale Electric Currents, *Astrophys. J.*, **285**, 359-367, 1984.

Rosner, R., W. H. Tucker, and G. S. Vaiana, Dynamics of the Quiescent Solar Corona, *Astrophys. J.*, **220**, 643-665, 1978.

Shoub, E., Invalidity of Local Thermodynamic Equilibrium for Electrons in the Solar Transition Region. I. Fokker-Planck Results, *Astrophys. J.*, **266**, 339-369, 1983.

FORMAL MATHEMATICAL SOLUTIONS OF THE FORCE-FREE EQUATIONS, SPONTANEOUS DISCONTINUITIES, AND DISSIPATION IN LARGE-SCALE MAGNETIC FIELDS

E. N. Parker
Enrico Fermi Institute, Departments of Physics and Astronomy
University of Chicago, 933 E. 56th St., Chicago, Illinois 60637

Abstract. Direct integration of the force-free field equation $\nabla \times \mathbf{B} = \alpha\mathbf{B}$, in the simple case of the local deformation of a laminar field, produces field configurations containing tangential discontinuities (current sheets). Whereas continuous solutions allow only restricted field topologies, the discontinuities provide the necessary release from those restrictions in more general topologies.

Magnetic fields in nature are strongly deformed by convection, so as to contain significant internal discontinuities. The bipolar magnetic fields containing the active x-ray corona of the sun are a case in point. It appears that the dissipation caused by the discontinuities may be the primary heat source producing the x-ray corona.

Introduction

A magnetic field in a nonresistive fluid (e.g. an extremely hot, and perhaps very tenuous, plasma) represents an unusual elastic medium. As is well known, the magnetic field $\mathbf{B}(\mathbf{r})$ provides an isotropic pressure $B^2/8\pi$ and a tension $B^2/4\pi$ in the direction of $\mathbf{B}$. The balance of these forces in static equilibrium with a gas pressure p is described by the familiar equation $\nabla(p + B^2/8\pi) = (\mathbf{B}\cdot\nabla)\mathbf{B}/4\pi$. In the absence of any significant variation in the fluid pressure p, the equilibrium equations reduce to the well known force-free form $\nabla \times \mathbf{B} = \alpha\mathbf{B}$. An essential feature of the force-free field equation is that the torsion coefficient, or local magnetic helicity, $\alpha = \mathbf{B}\cdot\nabla\times\mathbf{B}/B^2$ is rigorously constant along each line of force ($\mathbf{B}\cdot\nabla\alpha = 0$ as a consequence of $\nabla\cdot\mathbf{B} = 0$). The curious feature of this static stress balance is its creation of internal tangential discontinuities (shear planes, current sheets) in almost all field topologies. The discontinuities play an essential role in many astrophysical settings, providing intense dissipation of magnetic energy into thermal (or suprathermal) energy of the gas, whereas there would be little or no such dissipation if the fields were continuous (Syrovatskii, 1971, 1978, 1988; Parker, 1972, 1979, 1981a,b, 1982, 1983a,b,c; 1986a,b, 1987a, 1989a,b,c,d,e; Tsinganos, 1982; Tsinganos, Distler and Rosner, 1984).

Nowhere is this more evident than in the X-ray corona of the sun, where Rosner, Tucker, and Vaiana (1978) have shown that there is a direct equation between heat input (X-ray output) and magnetic field strength, more or less independent of the scale of the bipolar fields enclosing the superheated gas (over a range 4-200x10^3 km). This fact poses the fundamental question of the means by which energy is transferred from the magnetic field into thermal energy of the local gas. No MHD waves of sufficiently short period ($\sim$ 1 sec) to heat the smaller bipoles are observed, or expected on theoretical grounds. In the absence of internal discontinuities, the estimated resistive dissipation for the fields is much too slow ($> 10^4$ years) to be of interest. It appears that the internal discontinuities must provide the dissipation that is responsible for heating the X-ray emitting coronal gas (Parker, 1979; 1981a,b, 1983a,b).

The discontinuities appear in an otherwise continuous field in response to slow (quasi-static) continuous deformation of the field by the motion of the gas. For the solar X-ray corona it is the photospheric convection (the granules) out of which there extend the bipolar magnetic fields (10^2 gauss) containing the X-ray emitting gases (10^{10} atoms/cm^3, 2-3x10^6 $^\circ K$, 10^7 ergs/cm^2 sec output). The coronal gas pressure (6-10 dynes/cm^2) is approximately 2x10^{-2} of the magnetic pressure (4x10^2 dynes/cm^2); the speed of sound (3x10^7 cm/sec) is 0.15 of the Alfven speed of 2x10^8 km/sec. Hence the coronal magnetic fields are essentially force-free, and it is the force-free field that we study to determine the general properties of those fields.

The essential point (Parker, 1972, 1979, 1981a,b 1983a,b) is that the random photospheric convection of the foot-

Geophysical Monograph 58

points of bipolar magnetic fields in the corona leads to a random wrapping and interweaving of the lines of force of the fields in the manner sketched in Fig. 1, which can be considered to be a picture of the topology of the lines of force in a typical segment of a bipolar coronal magnetic field. To treat a particularly striking example, denote by B_0 the magnitude of the mean field (in the z-direction) and consider the magnitude of the transverse gradients of the transverse components of the field, i.e. $\partial B_x/\partial x, \partial B_x/\partial y, \partial B_y/\partial x, \partial B_y/\partial y$. The mean value of any one of these, say $\partial B_y/\partial x$, along a line of force through one local swirl of transverse scale ℓ is of the order of $B_\perp/\ell$ or less, where $B_\perp$ is the rms transverse field $(B_x^2+B_y^2)^{\frac{1}{2}}$. The mean may be either positive or negative, of course, with equal apriori probability.

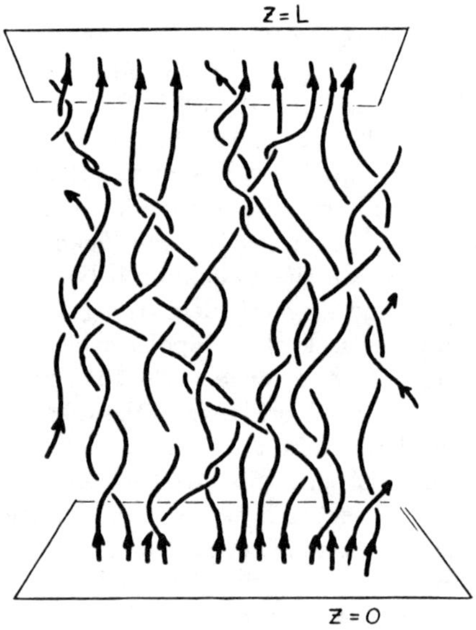

Fig. 1: A sketch of the random winding and interweaving of the magnetic lines of force at any given location within the region of magnetic field.

Now the individual field line passes through many such random, unrelated swirls. The mean value of $\partial B_y/\partial x$, computed along any given line of force passing through n uncorrelated swirls in succession, is the sum of the n mean values of the individual swirls, divided by n. Recalling that the individual mean values are $O(B_\perp/\ell)$ or less, with equal probability of being positive or negative, it follows that the mean of n correlated successive swirls is $O(B_\perp/\ell n^{\frac{1}{2}})$, or less, on almost all lines of force. Hence, in the limit of large n, the mean value along the line of force declines asymptotically to zero. Similarly the mean values of $\partial B_x/\partial x, \partial B_x/\partial y$, etc. are of comparable magnitude and decline asymptotically to zero in the limit of large n. It follows that the mean value of $(\nabla \times \mathbf{B})_z = \partial B_y/\partial x - \partial B_x/\partial y$ declines asymptotically to zero in the limit of large n.

Consider, then, the mean value of the z-component of the field equation $\nabla \times \mathbf{B} = \alpha \mathbf{B}$ along the given line of force. The left hand side tends to zero in the limit of large n. The torsion coefficient α is rigorously constant along the line, while B_z has some mean value of the same order as the mean value B_0 of the field at large. Thus the mean of the field equation yields $\alpha B_0 \sim 0$ in the limit of large n. It follows that α vanishes in the limit of large n on almost all field lines. But if $\alpha \sim 0$, the field equation itself reduces to $\nabla \times \mathbf{B} \sim 0$. The magnetic field becomes a potential field $\mathbf{B} \sim -\nabla\phi, \nabla^2\phi = 0$ in the limiting case.

The continuous solutions of Laplace's equations are well known, of course, and contain no forms with the complicated topology sketched in Fig. 1. On the other hand, such random sequences of winding and interweaving of the lines of force can be produced by the physical mixing and swirling of the footpoints of the field at $z = 0, L$. (The mathematics of the imposed fluid motions is described in Parker, 1986a). Hence the appropriate solutions of Laplace's equations are discontinuous, with the field satisfying Laplace's equation throughout the finite spaces between the surfaces of tangential discontinuity (Parker, 1989a). The hydrodynamic analogue is the irrotational flow of an inviscid incompressible fluid (described by the stationary Euler equation) in which there are stationary vortex sheets. It is interesting to note that Arnold (1965, 1966, 1974) pointed out many years ago on purely formal mathematical grounds that almost all solutions to the stationary Euler equation $\nabla(\frac{1}{2}v^2 + p/\rho) = (\nabla \times \mathbf{v}) \times \mathbf{v}$ contain tangential discontinuities. Moffatt (1985, 1986) has emphasized the exact mathematical analogy to the magnetostatic equation, as well as treating the general stability of the magnetic field and instability of the hydrodynamic field.

Now the occurrence of tangential discontinuities does not require a long sequence of winding patterns along the field. The case that $n \to \infty$ was treated only because the results are spectacular. In fact the physical necessity for tangential discontinuities arises from the simple fact that, in all but the most carefully constructed magnetic field configurations, magnetic flux bundles are called upon to spiral in a right handed sense about the neighboring flux at one location along the bundle, and in a left hand sense somewhere else along the bundle. But the torsion coefficient α within the flux bundle is rigorously constant along the bundle, with the result that the internal spiralling of the field cannot be made to conform to the neighboring field around which the bundle is wrapped first in one sense and then in the other. Nor can the neighboring field be expected to have the correct values of α to conform to the given flux bundle without being in jeopardy somewhere else along the field. The field accommodates by developing tangential discontinuities where the torsional differences are made up in layers of vanishing thickness, containing no magnetic flux (Parker, 1972, 1979, 1986a,b 1989a). In this way the restriction imposed by $\mathbf{B} \cdot \nabla \alpha = 0$ on the continuous magnetic field is evaded at the discontinuities, so as to accommodate alternating shear in the arbitrary winding pattern of the field. The field consists of finite size bundles of flux bounded along their sides by surfaces of tangential discontinuity, with α constant along each line of force throughout each bundle. As already noted, α declines to zero in the field extending through a large number n of random winding patterns or swirls, and the field within each bundle of continuous field approaches a potential form.

In summary, any continuous manipulation of the footpoints of the field introducing both right and left handed torsion into the winding of a flux bundle about its neighbors is expected to produce tangential discontinuities within the field. Simple deformations of most continuous force-free fields produce tangential discontinuities because for reasons of symmetry the deformations introduce opposite torsion or opposite side of the region of deformation. An impressive repertoire of special examples of the production of discontinuities in simple field configuration has accumulated in the literature (cf. Parker, 1972, 1979, 1981a,b, 1983c, 1987a; Rosenbluth, Dagazian and Rutherford, 1973; Park, Monticello, and White, 1983; Low, 1987; Moffatt, 1987; Low and Wolfson, 1987; Strauss and Otani, 1988; Otani and Strauss, 1988). In the next section we show by formal integration of the force-free field equations that a single localized compression of a laminar force-free field with $\alpha = constant$ produces tangential discontinuities.

Formal Integration of the Field Equations

The preceding discussion points out the severe limitations imposed by the continuity of a magnetic field in force-free equilibrium, from which we infer the necessity for internal tangential discontinuities. In this section we integrate the field equation to demonstrate the appearance of the discontinuities as an essential part of the solution (Parker, 1989e). Consider, then, the primitive force-free field with constant $\alpha = q$,

$$B_x = +B_0 \cos qz, B_y = -B \sin qz, B_z = 0 \qquad (1)$$

representing a simple laminar field. The lines of force lie in the planes $z = constant$, with the field uniform at an angle qz to the x-direction in each such flux surface. In polar coordinates (ϖ, φ, z) the field is

$$B_\varpi = +B_0 \cos(qz+\varphi), B_\varphi = -B_0 \sin(qz+\varphi), B_z = 0 \qquad (2)$$

We choose this configuration as the prototype force-free field because it represents the basic pattern of almost all local portions of a general continuous force-free field.

Consider a layer of field of thickness $2h(-h < z < +h)$, confined initially by a uniform pressure $B_0^2/8\pi$. Fix the field in the distant infinitely conducting rigid cylindrical boundary $\varpi = R$ and consider what happens throughout the interior of $\varpi = R$ when a local region of scale ℓ at the origin is squeezed. We work in the limit of large R/ℓ and assume that both ℓq and hq are O(1). In the neighborhood of the origin the pressure changes to $B_0^2 H^2(x,y)/8\pi$ with $H(x,y)$ significantly greater than one over a region of scale ℓ, dropping smoothly to one outside that region. The initial field (B_ϖ, B_φ, O) described by equation (2) becomes $(B_\varpi + b_\varpi, B_\varphi + b_\varphi, b_z)$ as a consequence of $H(\varpi,\varphi) \neq 1$. The torsion coefficient becomes $\alpha = q[1+A(\varpi,\varphi)]$. In the neighborhood of the origin b_ϖ, b_φ and b_z may be comparable in magnitude to B_0, since the field is strongly deformed by H. The boundary conditions yield b_ϖ identically equal to zero at $\varpi = R$. Now the change qA in the torsion coefficient is spread uniformly along the entire length of each line of force. Hence the finite total twist or shear produced by a strong deformation over a length ℓ spreads out uniformly over the total length $2R$ so that $A \lesssim O(\ell/R)$. As a matter of fact, it turns out later that A is small to second order in ℓ/R.

Noting that $(B_\varpi, B_\varphi, 0)$ satisfies the force-free field equation with $\alpha = q$, the field equation for the deformed field $(B_\varpi + b_\varpi, B_\varphi + b_\varphi, b_z)$ reduces to

$$\frac{1}{\varpi}\frac{\partial b_z}{\partial \varphi} - \frac{\partial b_\varphi}{\partial z} = qA\, B_\varpi \qquad (3)$$

$$\frac{\partial b_z}{\partial \varpi} = q(1+A)b_\varphi + qAB_\varphi \qquad (4)$$

$$\frac{1}{\varpi}\frac{\partial}{\partial \varpi}\varpi b_\varphi = q(1+A)b_z \qquad (5)$$

Consider equation (5) in the neighborhood of $\varpi = R$. If b_φ falls off asymptotically as ϖ^{-s}, and $A << 1$, the result is

$$b_z \sim (1-s)b_\varphi/qR \qquad (6)$$

at $\varpi = R$, indicating that b_z is smaller than b_φ by one power of $(qR)^{-1}$. Noting that $\partial/\partial z = O(q), \partial/\partial\varphi = O(1)$ and $q\ell = O(1)$, equations (3) and (4) can be written

$$\frac{1}{q}\frac{\partial(b_\varphi/B_0)}{\partial z} + A \cos u = 0\left(\frac{b_\varphi}{B_0}\frac{1}{q^2R^2}\right), \qquad (7)$$

$$\frac{b_\varphi}{B_0}(1+A) - A \sin u = 0\left(\frac{b_\varphi}{B_0}\frac{1}{q^2R^2}\right) \qquad (8)$$

at $\varpi = R$, where $u = qz + \varphi$. The right hand sides of both equations are small to second order and so may be neglected. Equation (8) yields the relation

$$b_\varphi/B_0 = [A/(1+A)]\ sin\ u. \tag{9}$$

Substituting this into equation (7) and integrating, we obtain

$$[A(2+A)/(1+A)^2]\ sin^2u = C \tag{10}$$

where C is at most a function of φ. Solving for A yields

$$A = -1 \pm\ sin\ u/(sin^2u - C)^{\frac{1}{2}}. \tag{11}$$

It follows from (9) that

$$b_\varphi/B_0\ =\ sin\ u \mp (sin^2u - C)^{\frac{1}{2}}. \tag{12}$$

To demonstrate the character of equations (11) and (12), hold φ fixed and vary u from $-\pi/2$ to $+\pi/2$. Then C is constant, since it is at most a function of φ. Suppose in the first instance that $C < 0$, writing $-C = D > 0$. Then $sin^2u - C = sin^2u + D$ has no zeros, and hence no branch point at which $\pm$ can change sign. We may choose either + or -, but there is no possibility to change from one to the other anywhere in the region. So consider the upper sign. Then at $u = \frac{1}{2}\pi$, it follows that

$$A = -1 + 1/(1+D)^{\frac{1}{2}}, b_\varphi/B_0 = 1 - (1+D)^{\frac{1}{2}}$$

But at $\varpi = R$ A and b_φ/B_0 are both small $O(1/q^2R^2)$ in the limit of large qR, which requires that D be comparably small, with $A \cong b_\varphi/B_0 \cong -\frac{1}{2}D$ to lowest order. With this small value of D, the result is

$$A \cong b_\varphi/B_0 \cong -2$$

at $u = -\pi/2$, so that A and b_φ/B_0 are O(1) rather than $O(1/q^2R^2)$. Using the lower sign simply reverses this lop-sided solution, which is not at all like the real physical situation.

Evidently, then, C is not negative. If $C = 0$, it follows from equation (10) that there are four roots, $A = 0, -2$, and $\pm\infty$. The first root $A = 0$ represents the initial field, with $b_\varphi/B_0 = 0$, without the applied deformation. The other roots are inapplicable to the problem wherein A is small $O(1/q^2R^2)$ in the limit of large qR. This leaves $C > 0$, in which case $sin^2u - C$ has two zeroes as u varies from $-\frac{1}{2}\pi$ to $+\frac{1}{2}\pi$. In particular, $(sin^2u - C)^{\frac{1}{2}}$ is imaginery in the interval $-C^{\frac{1}{2}} <\ sin\ u < +C^{\frac{1}{2}}$, centered on $u = 0$. The upper sign may be used in $sin\ u > +C^{\frac{1}{2}}$ and the lower sign in $sin\ u < -C^{\frac{1}{2}}$, so that the solution has the expected symmetry about $u = 0$, and a gap $(-C^{\frac{1}{2}}, C^{\frac{1}{2}})$ where A is complex.

Consider the solution in $C^{\frac{1}{2}} < u < \frac{1}{2}\pi$, using the upper sign. Evidently $C << 1$ if A and b_φ/B_0 are small. Note, then, that for $1 \gtrsim sin^2u >> C$ equations (11) and (12) reduce to

$$A \cong \frac{1}{2}C/sin^2u + \frac{3}{8}(C/sin^2u)^2 + ... \tag{13}$$

$$\frac{b_\varphi}{B_0} = sin\ u[\frac{1}{2}C/sin^2u - \frac{1}{8}(C/sin^2u)^2 + ... \tag{14}$$

so that both are small, with $C = O(1/q^2R^2)$ in order that b_φ is $O(1/q^2R^2)$. But as sin^2u declines toward C, it is evident that A increases without bound producing a spike at $sin^2u = C$ with an area $C^{\frac{1}{2}} = O(1/qR)$ under the curve. The characteristic width of the spike is $C^{\frac{1}{2}}$. Thus, in the limit of large qR, the spike becomes a delta function and moves to the origin $u = 0$. A similar result obtains for $sin\ u < -C^{\frac{1}{2}}$ in the limit of large qR using the lower sign. The net result is a delta function $C^{\frac{1}{2}}[\delta(u - C^{\frac{1}{2}}) + \delta(u + C^{\frac{1}{2}})]$, or

$$A = 2C^{\frac{1}{2}} \sum_m \delta(u - m\pi) + O(1/q^2R^2) \tag{15}$$

in the limit of large qR. Note that b_φ/B_0 remains small across $u = 0$, in the limit of large qR.

It follows that

$$\alpha = q(1+A) = q[1 + 2C^{\frac{1}{2}} \sum_m \delta(u - m\pi) + O(1/q^2R^2)] \tag{16}$$

at $\varpi = R$. We recall that α is constant along the lines of force, which at $u = 0$ or π extend all the way across the region. Hence, this value of α applies all the way across the region. The delta function represents a tangential discontinuity (Parker, 1989e). Except at the surfaces of discontinuity ($u = m\pi$) α deviates from q only by terms $O(1/q^2R^2)$.

This is just the effect calculated from the optical analogy (Parker, 1989b,c), showing that the lines of force of each flux surface are "refracted" by the index of refraction $H(\varpi, \varphi)$ so as to pass around, rather than through, the region of enhanced pressure ($H > 1$) localized about the origin. The result is a gap in each flux surface extending outward along the lines of force from the neighborhood of the origin to where the field is fixed at $\varpi = R$, sketched in Fig. 2. The angle subtended at $\varpi = R$ by the pie shaped gap is $O(\ell/R) = O(1/qR)$. The gap in any given flux surface permits the flux surfaces on either side (with orientations differing by their difference in qz) to press against each other, thereby creating a directional discontinuity in the amount $O(\ell/R) = O(1/qR)$ for $\varpi = O(R)$. The precise magnitude and form of the discontinuity depends upon the precise magnitude and form of $H(\varpi, \varphi)$ (see detailed examples in Parker, 1989c). The present integration of the equations at $\varpi = R$ does not consider anything more than the order of magnitude ℓ and $1/q$ of the displacement of field in the neighborhood of the origin. Hence the result is only that the area under the delta

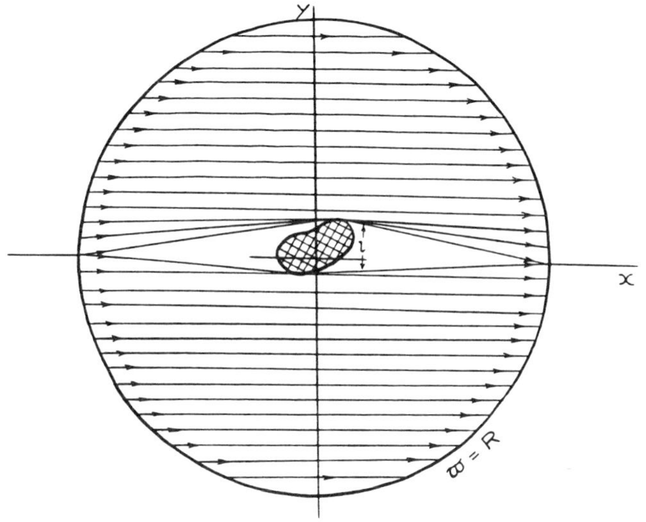

Fig. 2: A sketch of the flux surface $qz = 0$, showing the gap produced by the pressure enhancement of scale ℓ (cross hatched) at the origin.

function is $O(1/qR)$. The essential point of the calculation is to demonstrate a tangential discontinuity by formal conventional integration of the field equations. The optical analogy provides a more powerful mathematical approach, of course, but it does not constitute a "conventional" method. A more detailed presentation of the mathematics is given elsewhere (Parker, 1989e).

Discussion

It is instructive to go back to the field equations with $\alpha = q$ (as in the foregoing example) to explore the continuous solutions. In this case the force-free field equations can be written

$$\nabla^2 B_z + q^2 B_z = 0 \tag{17}$$

with

$$\left(\frac{\partial^2}{\partial z^2} + q^2\right) B_\varpi = \left(\frac{\partial^2}{\partial \varpi \partial z} + \frac{q}{\varpi}\frac{\partial}{\partial \varphi}\right) B_z \tag{18}$$

$$\left(\frac{\partial^2}{\partial z^2} + q^2\right) B_\varphi = \left(\frac{1}{\varpi}\frac{\partial^2}{\partial \varphi \partial z} - q\frac{\partial}{\partial \varpi}\right) B_z \tag{19}$$

Note, then, that equation (17) is fully elliptic. The hyperbolic characteristics of the general force-free field equation vanish in the present case that $\nabla\alpha = 0$. Separating the variables gives the general solution

$$B_z = \sum_{k,\nu} C_{k\nu} Z_\nu(\lambda\varpi)\; exp(\pm ikz \pm i\nu\varphi)$$

where $\lambda^2 = q^2 - k^2$ and Z_ν is any solution to Bessel's equation. In the present case, with the field at the boundaries depending on $u = qz + \varphi$, it is necessary to put $k = q$ and $\nu = 1$, so that

$$B_z = B_0 Q \varpi^{-1}\; exp\; iu$$

where Q is an arbitrary constant. We can now go back to equations (18) and (19) to work out B_ϖ and B_φ, to which we apply the boundary condition that B_ϖ is fixed at $\varpi = R$ at the value given by equation (2). The calculation is straightforward but tedious and need not be repeated here (see Parker, 1989e for details). The net result is that a unique solution follows from application of the boundary conditions, and the unique solution is qualitatively unlike the field described by equations (1) and (2). In particular, the field of (1) and (2) has been deformed only in the locality of the origin. Clearly it is bounded and preserves its simple topological form, connecting straight across the entire region $\varpi \lesssim R$. The continuous solution has none of these properties. Once again, it is evident that the singularities that appear in the general integration of the equations in the preceding section are an essential part of the release of that solution from the restrictions of universal continuity.

In the present calculation the spiral surfaces $(u = 0, \pi)$ of discontinuity free the mathematical solution from the requirement that $\nu = 1$ (in order that the field be a continuous single valued function of φ) since the continuous solution is terminated in both directions in φ by the discontinuities. Similarly the solutions are freed from the requirement that $k = q$, because the continuous solution is terminated in both directions in z.

A recent theoretical development further illuminates the restrictions on the continuous field, providing a second magnetic-hydrodynamic analogy of some interest. Consider an initially uniform field B_0 mixed and interwoven by arbitrary motions of the footpoints at $z = 0, L$, so that it has the formed sketched in Fig. 1, but stretched out in the z-direction so that the field perturbations $\Delta\mathbf{B}$ are small $O(\epsilon B_0)$. We showed (Parker, 1972, 1979) that there are continuous solutions of the field equations only if the field is invariant in the z-direction $(\partial/\partial z = 0)$ of the mean field. The result follows to all orders in ϵ. The torsion coefficient α, constant along each line of force, is also independent of z in this case, to all orders in ϵ. Recently van Ballegooijen (1985) has extended this result, thereby providing a broader base for the continuous solutions, by explicitly recognizing the transverse scale $\ell_\perp$ (that is small $O(\epsilon)$ compared to the characteristic scale $\ell_\parallel$ along the field) in the perturbation scheme. This reduces the z-component of the first order perturbation equations to second order, thereby permitting continuous solutions if the torsion coefficient α varies in the z-direction in the same manner that with vorticity $\omega = (\nabla \times \mathbf{v})_z$ of a two dimensional ideal fluid flow $\mathbf{v}$ varies with time. Since α is a direct measure of $\nabla \times \mathbf{B}$ in the perturbation scheme, the result is equivalent to the statement that the first order perturbation field $\Delta\mathbf{B}$ varies with z in the same manner that the analogous two dimensional fluid velocity $\mathbf{v}$ varies with time. In particular, α is constant along each field line and ω is constant on the world line of each moving element of inviscid incompressible fluid. Note that this direct analogy is in addition to the analogy between the three-dimensional magnetostatic field $\mathbf{B}$ and a three-dimensional stationary flow $\mathbf{v}$ of an ideal fluid. Combining the two analogies leads to the statement that the first order deviation $\Delta\mathbf{u}$ of a stationary Beltrami flow from a uniform flow in the z-direction varies with z in the same manner that a two dimensional flow varies with time.

The more restrictive case, that $\partial\Delta\mathbf{B}/\partial z = 0$, is a special case of the general condition that $\Delta\mathbf{B}$ varies with z in the same manner that a two dimensional flow varies with time. The final conclusion is, of course, that a field without internal discontinuities cannot be achieved if the field topology contains no reversals in torsion along each and every line of force, just as a continuous two dimensional flow of ideal fluid can have only those topologies that involve no reversals in vorticity along the world line of each and every element of fluid. In fact it requires nothing so extreme as a reversal. Any topology that requires a variation in torsion or vorticity along any field line or world line precludes a continuous field. Hence almost all winding patterns (e.g. Fig. 1) preclude a continuous field.

In summary, it appears that almost all quasi- equilibrium fields in nature possess tangential discontinuities (Parker, 1972, 1979). New discontinuities, or current sheets, are created in almost all cases when the field is further deformed (1987a, 1989e). For instance, we expect this formation of discontinuities to occur in the terrestrial magnetosphere when there is a significant change in the strength, direction, and perhaps temperature and density, of the solar wind. We expect tangential discontinuities to be the normal state of affairs in the bipolar magnetic fields of active regions on the sun and other stars, as a consequence of the continual mixing and swirling of the footpoints of the field in the photospheric convection. We conjecture (Parker, 1979, 1981b, 1983b) that it is the dissipation at the tangential discontinuities that is the principal heat source responsible for the X-ray corona. If this is correct, it is the principal cause of the X-ray emission from almost all stars. It follows that the X-ray corona (Parker, 1988, 1989d) and much of the region of a stellar flare (Parker, 1987b) represent a cloud of small reconnection events (nanoflares) in the ambient tangential discontinuities.

Similar creation of discontinuities is expected in the magnetic fields of our galaxy and of other galaxies, and in intergalactic space within clusters of galaxies. However, the vigorous suprathermal activity of most galaxies, as evidenced by the copious production of cosmic rays in supernovae, active galactic nuclei, and other compact objects, appears to overwhelm the effect on galactic scales, so that it may be of little practical interest. So far the most conspicuous consequence of the spontaneous tangential discontinuities seems to be the stellar flare and the universal emission of X-rays by ordinary stars like the sun.

Questions and Answers

Antiochos: Is the conclusion that the heating increases as the resistivity decreases valid for the solar corona which has an unbounded volume and consequently, $B_\perp$ may have an upper bound?

Parker: I think the conclusion is valid. The $B_\perp$ *involved in heating the corona is well below any level that might cause an eruption into space. Or, to put it differently, I am concerned with the effects of* $B_\perp$ *short of eruption. Coronal eruptions (coronal mass ejections) are another subject, beyond the scope of the present theory.*

Chen: A field line does not stop in the photosphere; it presumably continues into the subphotospheric regions where the resistivity is considerably higher (weak-ionization). Thus, as the photospheric footpoints wander around, field lines in the subphotospheric regions are also twisted up. In addition, two field lines which may reconnect in the corona are separated by shorted distances ($\sim B^{-\frac{1}{2}}$) in the subphotospheric regions. It seems that reconnection, if it occurs, would take place faster in the much more resistive medium, rather than in the corona where the resistivity is very small.

Parker: The answer to your question is that beneath the photosphere the fields are dominated by the convection, so that current sheets do not form as they do in the corona where the field completely dominates the fluid. There is, of course, a photospheric reconnection rate which can cut across a typical granule scale in 10^5 *seconds or so. So, there is not time to twist up coronal flux tubes much longer than about* 10^5 *km, because it would require longer than the available* 10^5 *seconds. It is interesting to note that coronal x-ray loops do not exist on scales longer than about* 10^5 *km.*

Roberts: Your arguments for heating stellar atmospheres are applicable to active regions, possessing re-entrant fields and so a length-scale L. That leaves the problem of coronal holes (with open field lines). Presumably, whatever heats coronal holes (perhaps Alfvén waves) is also operative in active regions, perhaps even more so. So, whatever heats coronal holes to 1.5 x 10^6K would also heat active regions to much the same temperature, suggesting that your mechanism is responsible for the higher temperature (2 - 3 x 10^6K) of active regions. In other words, even if your mechanism was not operating, we would still have a 1.5 x 10^6K atmosphere, heated by whatever is responsible for heating coronal holes.

Parker: I think that one cannot compare coronal holes with the coronal x-ray loops of finite ($\leq 10^6$ km) length. Not only does the dense gas of the x-ray loop require 20 times more energy to make it hot (so that an energy source adequate to a coronal hole would have negligible effect in a coronal x-ray loop), but the energy deposition in a coronal hole may be spread out over 10^6km, whereas, in a small coronal loop the deposition must be accomplished in a distance as small as 4×10^3 km. The requirements are simply qualitatively different for the two cases.

Acknowledgements. This work was supported in part by the National Aeronautics and Space Administration under NASA grant NGL 14-001-001.

References

Arnold, V. "Sur la topologie des ecoulements stationnaires des fluids parfaits," *C.R. Acad. Sci. Paris* **261**, 17-20, 1965.

Arnold, V. "Sur la geometric differentielle des groups de Lie de dimension infinie et ses applications a l'hydrodynamique des fluides parfaits," *Ann. Inst. Fourier Grenoble* **16**, 361-391, 1966.

Arnold, V. "The asymptotic Hopf invariant and its application (in Russian) in *Proc. Summer School in Differential Equations*, Erevan, Armenian SSR Acad. Sci., 1974.

Low, B.C., "Electric current sheet formation in a magnetic field induced by continuous footpoint displacements," *Astrophys. J.* **323**, 574-581, 1987.

Low, B.C. and Wolfson, R., "Spontaneous formation of electric current sheets and the origin of solar flares," *Astrophys. J.* **324**, 574-581, 1987.

Moffatt, H.K. "Magnetostatic equilibria and analogous Euler flows of arbitrarily complex topology. I. Fundamentals," *J. Fluid Mech.* **159**, 359-378, 1985.

Moffatt, H.K. "Magnetostatic equilibria and analogous Euler flows of arbitrarily complex topology. II. Stability considerations," *J. Fluid Mech.* **166**, 359-378, 1986.

Moffatt, H.K. Advances in Turbulcncc, ed. G. Comte-Bellot and J. Mathieu, Springer-Verlag, Berlin, 240-241, 1987.

Otani, N.F. and Strauss, H.R., "Current-driven resistive ballooning modes in axially bounded solar flare plasmas," *Astrophys. J.* **325**, 468-475, 1988.

Park, W., Monticello, D.A. and White, R.B., "Reconnection rates of magnetic fields including the effects of viscosity," *Phys. Fluids* **27**, 137-149, 1983.

Parker, E.N. "Topological dissipation and small-scale fields in turbulent gases," *Astrophys. J.* **174**, 499-510, 1972.

Parker, E.N. "Cosmical magnetic fields," Clarendon Press, Oxford, pp. 359-391, 1979.

Parker, E.N. "Dislocation and flattening of inhomogeneous magnetic fields and the problem of coronae. I. Dislocation and flattening of flux tubes, *Astrophys. J.* **244**, 631-643, 1981a.

Parker, E.N. "Dislocation and flattening of inhomogeneous magnetic fields and the problem of coronae. II. The dynamics of dislocated flux tubes," *Astrophys. J.* **244**, 644-652, 1981b.

Parker, E.N. "The rapid dissipation of magnetic fields in highly conducting fluids," *Geophys. Astrophys. Fluid Dyn.* **22**, 195-218, 1982.

Parker, E.N. "Magnetic neutral sheets in evolving fields. I. General theory," *Astrophys. J.* **264**, 635-641, 1983a.

Parker, E.N. "Magnetic neutral sheets in evolving fields. II. Formation of the solar corona," *Astrophys. J.* **264**, 642-647, 1983b.

Parker, E.N. "Absence of equilibrium among close packed twisted flux tubes," *Geophys. Astrophys. Fluid Dyn.* **23**, 85-102, 1983c.

Parker, E.N. "Equilibrium of magnetic field with arbitrary interweaving of the lines of force. I. Discontinuities in the torsion," *Geophys. Astrophys. Fluid Dyn.* **34**, 243-264, 1986a.

Parker, E.N. "Equilibrium of magnetic field with arbitrary interweaving of the lines of force. II. Discontinuities in the field," *Geophys. Astrophys. Fluid Dyn.* **35**, 277-301, 1986b.

Parker, E.N. "Magnetic reorientation and the spontaneous formation of tangential discontinuities in deformed magnetic fields," *Astrophys. J.* **318**, 876-887, 1987a.

Parker, E.N. "Stimulated dissipation of magnetic discontinuities and the origin of solar flares," *Solar Phys.* **111**, 297-308, 1987b.

Parker, E.N. "Nanoflares and the solar X-ray corona," *Astrophys. J.* **330**, 474-479, 1988.

Parker, E.N. "Tangential discontinuities and the optical analogy for stationary fields. I. Force-free fields, potential fields, and discontinuities," *Geophys. Astrophys. Fluid Dyn.* (in press), 1989a.

Parker, E.N. "Tangential discontinuities and the optical analogy for stationary fields. II. The optical analogy," *Geophys. Astrophys. Fluid Dyn.* (in press), 1989b.

Parker, E.N. "Tangential discontinuities and the optical analogy for stationary fields. III. Zones of exclusion," *Geophys. Astrophys. Fluid Dyn.* (in press), 1989c.

Parker, E.N. "Tangential discontinuities and the optical analogy for stationary fields. IV. High speed sheets of fluid," *Geophys. Astrophys. Fluid Dyn.* (submitted for publication), 1989d.

Parker, E.N. "Tangential discontinuities and the optical analogy for stationary fields. V. Formal solution of the force-free field equations," *Geophys. Astrophys. Fluid Dyn.* (submitted for publication), 1989e.

Rosenbluth, M.N., Dagazian, R.Y. and Rutherford, P.H., "Nonlinear properties of the internal m=1 kink instability in the cylindrical tokamak," *Phys. Fluids* **16**, 1894-1902, 1973.

Rosner, R., Tucker, W.H., and Vaiana, G.S., "Dynamics of the quiescent solar corona," *Astrophys. J.* **220**, 643-665, 1978.

Strauss, H.R. and Otani, N.F., "Current sheets in the solar corona," *Astrophys. J.* **326**, 418-424, 1988.

Syrovatskii, S.I., "Formation of current sheets in a plasma with a frozen-in strong magnetic field," *Soviet Phys. JETP* **33**, 933-940, 1971.

Syrovatskii, S.I., "On the time evolution of force-free fields," *Solar Phys.* **58**, 89-94, 1978.

Syrovatskii, S.I., "Pinch sheets and reconnection in astrophysics," *Ann. Rev. Astron. Astrophys.* **19**, 163-229, 1981.

Tsinganos, K.C., "Magnetohydrodynamic equilibrium. IV. Nonequilibrium of nonsymmetric hydrodynamic topologies," *Astrophys. J.* **259**, 832-843, 1982.

Tsinganos, K.C., Distler, J., and Rosner, R., "On the topological stability of magnetostatic equilibria," *Astrophys. J.* **278**, 409-419, 1984.

STRUCTURE AND FLOWS IN CORONAL LOOPS

Spiro K. Antiochos
E. O. Hulburt Center for Space Science
Naval Research Lab, Washington, DC 20375

Abstract. The canonical model for the structure of the solar coronal plasma is a collection of loops. Each loop is believed to correspond to a magnetic flux tube in which the field dominates the plasma. In this review we discuss plasma flows in solar loops. The field is approximated as completely rigid since the plasma beta in the corona is very low, and the observed time scales of the motions are typically much longer than the Alfven time scales.

First, a brief overview of observations is presented; in particular the redshifts observed in UV emission lines. Next we review the work on steady-state flows in coronal loops. An important point is that static models, although widely used, are not totally valid since any asymmetry in the loop geometry, in the coronal heating, or in the chromospheric boundary conditions will result in a "siphon" flow along the loop. We discuss whether such flows can account for the observed redshifts, and conclude that the siphon flow models require extreme and contrived assumptions on the heating process in order to fit the data. We also discuss the possible importance of non-equilibrium ionization in these models, and conclude that the velocities deduced from the observed line shifts imply ionization non-equilibrium in the lower transition region, $T < 10^5$ K. Finally, we review work on flows due to impulsive heating in coronal loops. We argue that implusive heating may be able to account for the observed redshifts, but conclude that much more work is needed in order to verify or disprove this conjecture.

Introduction

Although static models have been widely applied to the solar corona, and have been found to explain many of the observed features of the coronal plasma, it is evident from the observations that the corona is never truly static. Motions are always evident in the data, especially in the transition region, $10^4 < T < 10^6$ K. The observed dynamics of the coronal plasma are very important because, as we will demonstrate below, they may provide strong constraints on the possible heating mechanisms.

Observations with high spatial and and high temporal resolution show a hodge-podge of dynamic phenomena, such as spicules, coronal bullets, etc. [e.g. Brueckner 1981]. The mechanism for these transient events is not known, but since they generally involve supersonic velocities, it is likely that they are magnetically driven. For investigating the heating it seems most useful to study only the motions that are quasi-steady or, at least, long-lived compared to the lifetimes of coronal loops (of order hours). When the observations are averaged over a large area or over a long time, then usually only two types of velocity signature remain, and both are subsonic. One is a redshift of lines formed at lower-transition region temperatures $T \sim 10^5$ K. Interpreting this redshift as due to a downflow, the velocities required to produce the line shifts are of order 10 km/sec. Note that the sound speed at these temperatures is 50 km/sec. Athay and Dere (1989) find from a compilation of available data that the redshift velocity is given by,

$$V \approx 10(T/10^5)^{1.5} km/s, \qquad 1.6 \times 10^4 \leq T \leq 10^5 K \quad (1)$$

The other velocity signature is a broadening of the lines above their thermal widths. This is commonly interpreted as due to a distribution of randomly oriented velocities. For the "turbulent" width, ξ, Athay and Dere (1989) find

$$\xi \approx 20(T/10^5)^{0.5} km/s, \qquad 10^4 \leq T \leq 3 \times 10^5 K \quad (2)$$

Both types of motions can be seen in Figure 1, which presents results from a recent (November 1988) rocket flight of the High Resolution Spectrograph and Telescope (HRTS) [Brueckner and Bartoe 1983]. The Figure shows a horizontal strip of the solar disk aproximately 30,000 km wide by 700,000 km long. (The strip is broken up into two parts in order to fit on the Figure.) This section of the Sun was selected so that it would contain no active regions or coronal holes, hence, it should be representative of the average Sun. The Figure contains three panels, the bottom panel shows the absolute intensity of the C IV 1550 Å line, the middle panel shows the wavelength shift of the center of this line with red indicating a shift to larger wavelength and blue to shorter, and the top panel shows the width of the line.

Several important features of the structure and dynamics of the solar plasma appear in Figure 1. First, it is evident that the emission is strongly concentrated in small structures, (below the resolution of the photograph), that

Geophysical Monograph 58

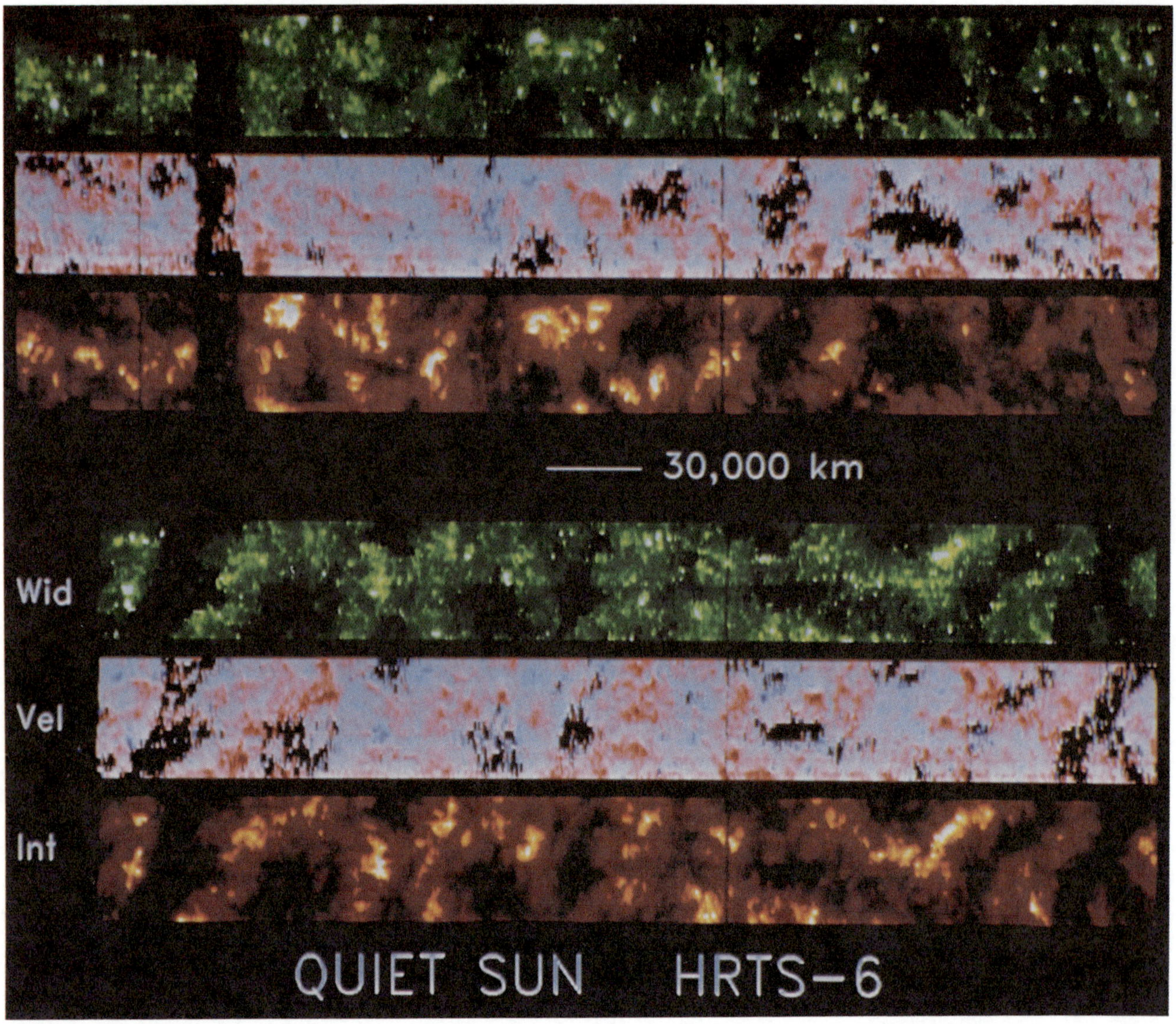

Fig. 1. A 30,000 by 700,000 km strip of the quiet sun as observed by HRTS (provided by K. P. Dere). The bottom panel shows intensity in C IV 1550 Å. The middle shows line shift and the top shows line width.

occur in the supergranule network. We believe that these structures are small loops, perhaps of the type described by Antiochos and Noci (1986). The network is the place where one would expect small loops to occur since the magnetic field has a fine-scale mixed polarity structure there [Dowdy *et al.,* 1986]; but the HRTS observations are insufficient to verify this conjecture. Second, it can be seen that a large fraction of the emission is redshifted. Blue shifts do occur as well, but they are much less extensive. If one integrates over the whole field-of-view, the line has a net redshift. Third, there does not appear to be a strong correlation between the intensity of the emission and the magnitude of the line shift.

It should be noted that since there was no absolute wavelength source on this or on other solar instruments, the line shifts are obtained by measuring the position of the C IV line relative to lines formed in the deep chrompsphere and photosphere. The redshifts actually indicate only the relative velocity between the transition region and the photosphere. The solar observations could, in principle, be interpreted as indicating that the photosphere is moving up. Confirmation of a true downflow comes, interestingly enough, from stellar observations. Data obtained with the IUE spacecraft, which does have an absolute wavelength source, indicate that all nearby solar-like stars have redshifted UV emission lines, and that the magnitude of the shifts are very similar to those of the Sun's [e.g. Ayres 1984]. It appears that downflows are a feature common to all stellar coronae.

A controversial question concerns the center-to-limb variation in the redshifts. If the flows are radially downward, then one would expect to see a marked decrease in the redshifts toward the limb where the velocities become perpendicular to the line-of-sight. Even if the flows had a large horizontal component the redshifts should still disappear at the limb. Since the UV emission lines are, for the most part, optically thin, flows toward and away from the observer should produce cancelling line shifts, so that only a broadening remains. A very interesting study by Feldman *et al.*, (1982), however, seems to indicate that although they may decrease by a factor of approximately two near the limb, the redshifts do not vanish there. On the other hand, other studies do seem to show the expected center-to-limb variation of the redshifts [Athay and Dere 1989]. Due to this uncertainty in the observations, we will not use it as a test of the models. Of course, the lack of a center-to-limb variation would be a difficult result for any model to explain. We have argued [Antiochos 1984] that it requires the presence of very cool material $T < 5 \times 10^3$ K lying above the transition region and absorbing the UV emission.

Another feature of Fig. 1 that all the observable emission has nonthermal broadening. A key property of this broadening is that its magnitude appears to be largely independent of the spatial resolution of the observations; hence, it must be due to motions with a very small scale, < 2000 km, Dere (1989). The origin of this broadening is not known. Flare observations suggest that the line broadening is associated with the energy release process. Flare X-ray line profiles exhibit a large broadening but only at the impulsive phase of a flare when, presumably, the heating is greatest [Antonucci *et al.*, 1984]. Most theories for the heating in flares and in quiet regions require that the energy release occurs in very small regions, which should exhibit large mass motions. Our belief is that an understanding of the line broadening must await an understanding of the coronal heating process.

Steady-State Siphon Flow Models

Several models have been proposed to explain the transition region redshifts. Since the flows are long-lived and subsonic, they all assume that gas pressure is the driving mechanism.

An important contraint on the models is that the amount of downflowing material is large; it represents a mass flux sufficient to empty the corona in a few days. Consequently there must be a mass flow upward into the corona equal to the observed downflow. This upflow will produce blueshifted emission. The observations, Figure 1, indicate that there is unshifted emission as well. We expect, therefore, that a line such as C IV 1550 Å consists of three components, two oppositely shifted ones and a rest compnent. The observation of a net redshift in a line requires that the emission from downflowing plasma at a given temperature dominate that of upflowing material at the same temperature. An estimate of the contribution of each shifted component to the observed line shift is given by

$$\Delta\lambda \sim JNH_T \tag{3}$$

where N is the electron number density, J is the particle current $J = NVA$, and H_T is the temperature scale height, $H_T = T\,ds/dT$. Note that the lines are optically thin so that the emissivity varies as N^2. In order to account for the observations, then for temperatures $\sim 10^5$ K either the current (the integrated mass flux), the density, or the temperature scale height must be significantly different in the upflowing and downflowing plasma.

Perhaps the simplest explanation for the redshifts is that they are due to a steady siphon flow along a coronal loop. It should be emphasized that such a flow is the natural state of a coronal loop. The static models are, in fact, somewhat unphysical since they assume that the loop has a symmetry about its apex. There are at least three quantities that almost certainly will break such as a symmetry: the magnetic geometry, the energy input rate, and the chromospheric boundary conditions at the two footpoints of the loop. In particular, if there is a pressure difference between the two footpoints a flow must result.

Pressure difference flows have been discussed by a number of authors [e.g. Cargill and Priest 1980; Glencross 1980; Antiochos 1984] The primary conclusion is that large velocities can readily be obtained by this mechanism; indeed, there is observational evidence in C IV [K. Dere, private communication, 1989] that highly supersonic flows are driven into sunspots by the difference in plasma pressure between the footpoints of flux tubes that have one of their ends in the spot umbra. However, as discussed below, it appears unlikely that such flows can produce redshifted emission lines. Flows driven by asymmetries in the energy input and in the loop geometry have also been investigated [e.g. Boris and Mariska 1982; McClymont and Craig 1986, 1987; Mariska 1988]. In this case the main difficulty appears to be obtaining sufficiently large velocities at transition region temperatures to match the observations. These results are also discussed below.

The equations commonly used to describe the siphon models are the 1-d, steady-state, single-fluid equations for a fully-ionized hydrogen plasma:

$$NVA(s) = J = const, \tag{4}$$

$$m_pNV^2 + dP/ds = m_pNG(s), \tag{5}$$

$$\frac{d}{ds}\left(A\left(\frac{m_pNV^3}{2} + \frac{5PV}{2} - 10^{-6}T^{5/2}\frac{dT}{ds}\right)\right) = A\left(E - N^2\Lambda(T)\right) + m_pJG. \tag{6}$$

In Eqtns (4) – (6) m_p is the proton mass, (the mass due to electrons and heavy elements is usually neglected); $A(s)$is the cross-sectional area of the loop, $(AB = magnetic flux = const.)$; $G(s)$ is the component of gravity along the field; $\Lambda(T)$ is the radiative loss coefficient for optically-thin emission [e.g. Gaetz and Salpeter 1984]; and E is an energy source term representing coronal heating. It is simply parametrized as a function of position, density, and temperature, $E = E(s, N, T)$. A number of different sets of boundary conditions are possible for this system. One of the most straightforward is to specify at the upflowing footpoint the temperature and pressure. In addition the heat flux is usually taken to vanish at both footpoints, since the heat flux into the chromosphere must be small [Antiochos and Sturrock 1978].

Flows Due to a Footpoint Pressure Difference

Let us consider the lineshifts that would be produced by a steady flow driven by a pressure difference at the chromospheric footpoints. If the flows are steady then the mass flux is the same in the upflowing and downflowing components to the emission Eqtn (2), so that a net shift in a line must be due to a difference in either the density or the temperature scale height. However, we do not expect a large difference in the density because the pressure variations along the loop cannot be more than of order unity. The reason is that the observed velocities are subsonic and the effects of gravity are small because the gravitational scale height in the corona $H_P = (2kT)/(m_p G) \sim 10^{10}$ cm for $T \sim 10^6$ K, is much larger than typical loop heights $\sim 10^9$ cm. Hence, the temperature scale height, i.e. the width of the transition region, is the only parameter that is likely to vary significantly between the up and down leg and, thereby, to produce large differences between the red and blue emission.

In order to determine the transition region widths in the siphon flow model, let us consider a simple order-of-magnitude analysis of the relevant terms in Eqtns. (4) – (6). Note that the temperature gradients are determined by the energy equation (6). In the transition region of the static loop models it is well known that the energy balance is primarily between thermal conduction from the corona and radiation losses. The coronal heating rate is small compared to those terms there [e.g. Craig *et al.*, 1978; Vesecky *et al.*, 1979]. The question is: What is the effect of the flows on the energy balance? We rewrite (6) in the form:

$$\frac{10^{-6}T^{7/2}}{{h_T}^2} = A(N^2\Lambda(T) - E) \pm \frac{(m_p N V^3 + 5PV)}{2h_T} \pm m_p J G. \qquad (7)$$

where + (-) refers to the up (down) flowing legs of the loop respectively.

The key feature of this equation is the change in sign in the enthalpy term between the up and down flowing legs. In the upflowing leg the velocities act as a heat sink because plasma must be heated and accelerated as it flows from low to high temperatures; but, in the downflowing leg the flows are a heat source since the plasma cools and decelerates. The gravitational term acts in the same way; however, it is usually negligible compared to the enthalpy term. Now let us assume that the area variation is small and that the energy input has no explicit spatial dependence. In this case the energy input term and the radiative loss term depend only on pressure and temperature. We argued above that the pressure variation is small, consequently these two terms must be nearly equal at any given temperature in the up and down legs. Therefore the heat flux must be smaller in the upflowing leg than in the downflowing one due to the change in sign in the enthalpy. Since the temperature scale height is inversely proportional to the flux, we conclude that the effect of the flows is to decrease the temperature scale height in the downflowing leg and increase it in the upflowing one. This is the exact opposite of what is needed to explain the redshifts; siphon flows due to footpoint pressure differences tend to produce blueshifts.

Let us estimate the magnitude of the enthalpy term compared to the radiative term. This will determine the importance of the enthalpy term for the heat flux and the transition region widths. Taking a velocity of 10 km/s and a typical quiet sun plasma pressure of 0.1 $ergs/cm^3$, yields an enthalpy flux $5PV/2$ of 10^5 $ergs/cm^3/s$. We can neglect the kinetic energy flux since the velocities are clearly subsonic. Balancing this flux by a conductive flux, $10^{-6}T^{7/2}/h_T \approx 10^5$, implies a size scale h_T of the 10^5 K plasma of 30 km. The radiative flux $N^2\Lambda(10^5)h_T$ now turns out to be also $\approx 10^5$. Hence, the mass motions can be the major effect determining the magnitude of the heat flux at 10^5 K. We found [Antiochos 1984] that there can be over an order of magnitude difference between the fluxes in the up and down legs. In this case the C IV emission would be strongly blue shifted. This conclusion is confirmed by detailed calculations, presented below.

Effects of Non-Equilibrium Ionization

In this section we calculate the detailed emission line profiles from Carbon ions for a loop undergoing siphon flow. One assumption that is commonly made in the loop models is that the plasma is in ionization equilibrium. For the static models this is obviously valid; however, it is likely to be invalid in the presence of flows due to the large temperature gradients in the transition region. We estimated above that the scale heights at 10^5 K could be as small as tens of kms. If so then a velocity of 10 km/s implies that the time scale for the plasma to change temperatures is order 10 sec. For coronal parameters, this time scale can be significantly smaller than the time scales for ionization and recombination. Hence, as the plasma flows through the transition region it may end up out of ionization equilibrium. Note that this effect would be important for non-steady flows as well. It has been investigated by several authors [e.g. Dupree *et al.*, 1979; Joselyn *et al.*, 1979; Borrini and Noci 1982; Mariska and Boris 1983], especially for solar wind outflow in open flux tubes and for supersonic flows.

In Figure 2 we show the results of a calculation by Noci et al (1989) of ionization non-equilibrium in a siphon flow loop. The model assumes a loop of uniform cross-section and energy input of 10^{-3} $ergs/cm^3/s$, and neglects the effect of gravity. The temperature is taken to be 3×10^4 K at both ends of the loop and the heat flux is assumed to vanish at the upflowing end. The loop has a constant mass flux of 5.0×10^{-9} $g/cm^2/s$ and constant momentum flux of .40 $ergs/cm^3$. The temperature, density and velocity profiles are shown in Figures 2a and 2b [from Noci *et al.*, 1989].

Given these profiles, the evolution of the carbon ionization as this element flows along the loop can be calculated [Noci *et al.*, 1989]. We assumed collisional ionization, and radiative and dielectronic recombination with the boundary condition that the carbon is in ionization equilibrium when it enters the loop at the upflowing footpoint. The densities obtained for the C I through C VII ions in the up and down flowing legs are shown in Figures

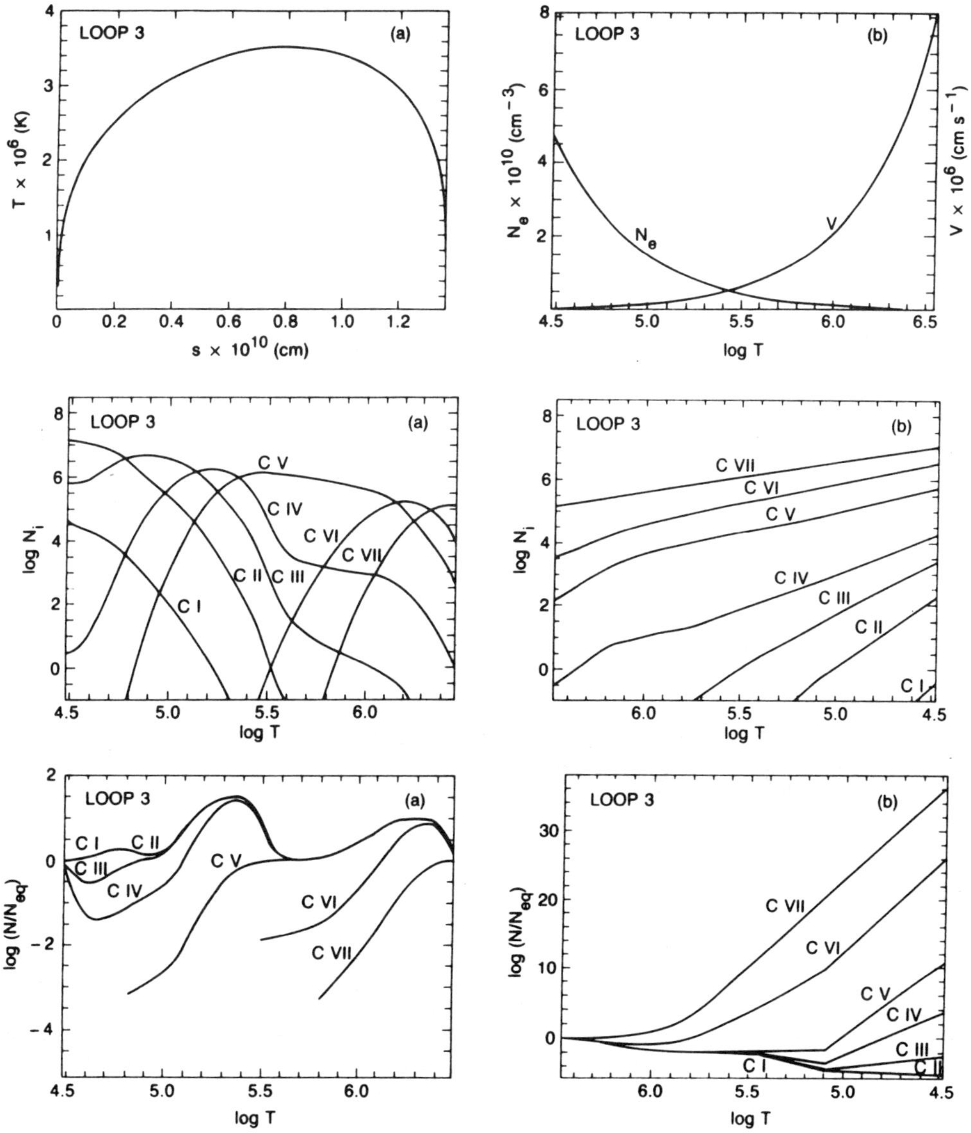

Fig. 2. The C I – C VII ionization balance in a coronal loop undergoing steady siphon flow, (from Noci *et al.*, 1989). (a)Plasma temperature as a function of position. (b)Velocity and electron density as a function of temperature along the loop. (c)Density of Carbon ions as a function of temperature in the upflowing leg. (d)Density of Carbon in downflowing leg. (e)Ratio of ion density to the equilibrium value at that temperature in the upflowing leg. (f)Ratio in downflowing leg.

2c and 2d. The departures from equilibrium are shown in 2e and 2f. It is evident that the flows produce large departures from ionization equilibrium.

The profile of the dominant emission line from each ion has been calculated [Spadaro *et al.*, 1988] and our results shown in Figure 3. For this calculation the loop is assumed to lie at the center of the disk so that it is observed from directly above. Note that all the lines have their center shifted to the blue rather than the red. This is a consequences of two effects. One is that, as discussed above, the thickness of the transition region is greater in the upflowing leg. For ions such as C IV there is another effect due to the non-equilibrium ionization that favors the blueshifted emission. It can be seen from Figures 2c and 2d that over most of the loop temperature range, 10^5 K – 10^6 K, the density of C IV ions is substantially lower in the downflowing leg. Only near the downflowing footpoint does the C IV density become significant. However, the velocities and the size scales there are small so that very weak redshifts are produced.

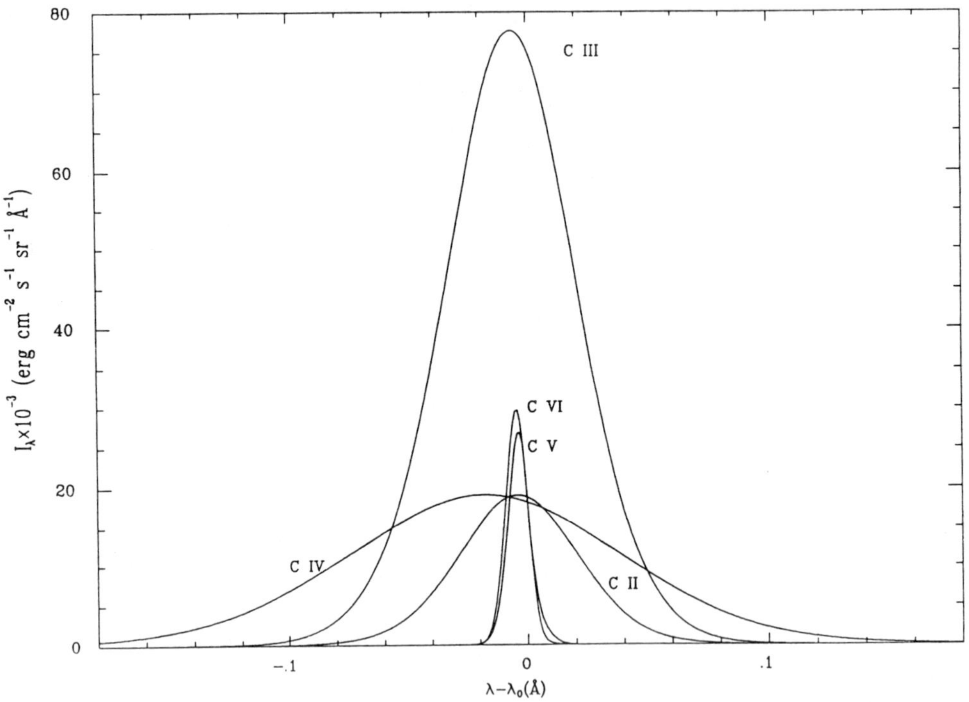

Fig. 3. The profiles of the main Carbon lines as would be observed from the loop of Figure 2 if it were at disk center (from Spadaro *et al.*, 1989). The lines are: C II 1335.3, CIII 977.03, C IV 1548.20, C V 40.27, and C VI 33.77.

We conclude that siphon flows, at least those driven by a footpoint pressure difference, cannot account for the redshifts.

Effects of a Spatially–Varying Coronal Heating

Although the results above are clearly not encouraging for explaining the redshifts with siphon flows, the possibility remains that a spatially dependent energy input rate or cross-sectional area may lead to different results. This question has been examined by Boris and Mariska (1982), McClymont and Craig (1986, 1987), and Mariska (1988).

One important result of their work is that for the magnitudes of area variations expected for coronal flux tubes, the flow speeds produced are much smaller than the observed velocities. Hence, the area variation does not appear to play a significant role; the main effect must come from the spatial variation in $E(s)$. In order to obtain redshifts, the width of the downflowing transition region must be larger than that of the upflowing. But, one has to overcome the large effect of the enthalpy term in Eqtn (7) which acts in the opposite sense. This requires a very strong spatial dependence in $E(s)$. However, since the coronal heating process is not known it is possible that such a dependence does exist.

McClymont and Craig (1987) and Mariska (1988) argue that forms of $E(s)$ can be found that produce the correct line shifts. The requirements are that $E(s)$ be highly localized so that essentially all the heating in the loop occurs in a very narrow width, < 100 km, and that this heating is located very near one of the loop footpoints, within ~ 100 km of the chromosphere. In addition, the energy input and the loop size must be such the the maximum temperature in the loop be quite low, $\sim 2 \times 10^5$ K. Under these conditions redshifts will be produced at the temperature of 10^5 K.

The explanation for this result is that the upflowing transition region must lie between the heating region where the temperature reaches its maximum value and the "near" footpoint where the temperature is set to the chromospheric value. By placing the heating in a small region very near one footpoint, the upflowing transition region is forced to be very narrow. In principle, one could make the width of the 10^5 K upflowing region arbitrarily small by placing a delta-function heating arbitrarily close to the footpoint boundary. The width of the downflowing region, however, is set by the length of the loop and by the maximum temperature, so that it stays approximately constant as the heating is placed nearer the base. Hence, it is possible to choose the heating so that the downflowing emission dominates the upflowing.

One point of uncertainty in this result is that the calculations so far have all assumed ionization equilibrium. We found that non-equilibrium can strongly affect the UV emission in the previous siphon flow models with uniform heating. We expect that the effects would be even stronger in non-uniform heating models since the temperature gradients are more severe, and we are presently investigating this issue.

Although non-uniform heating appears to be able to produce redshifts, this model requires extremely restrictive conditions on the form of coronal heating. Both the location and the width of the heating are constrained to a small region, much smaller than the scale of the loop, i.e. the scale of the magnetic field. On the one hand, it can be argued that this is a positive feature of the model because it yields strong constraints on the form of the heating. We believe, however, that the constraints are simply too severe. Note that many coronal heating theories involve thin energy-release layers such as current sheets, but the small scale in this case is the cross-field scale. Parallel to the field the relevant scale is usually the scale of the field, which is the scale of the loop. The redshift model requires that the scale of the heating along the field be much smaller than the magnetic scale. Another difficulty is that in order to obtain the observed magnitude of the velocities and the line intensity, both the magnitude of the heating and the size of the loops are highly constrained [McClymont and Craig 1987; Mariska 1988].

We conclude that in view of all the extreme assumptions required in order for steady-state flow models to match the observations, it is not the likely explanantion for the redshifts.

Transient Flows

If steady flows are not the answer, the remaining possibility is that the flows are transient in nature. Again there must be is a difference between the upward and downward fluxes, but now it is possible for the mass flux at a given temperature to have this difference. This is the gist of the model advocated by some authors [Athay 1981; Athay and Holzer 1982; Pneuman and Kopp 1977]. Athay proposes that the transition region plasma responsible for the redshifts is first raised into the corona in the form of spicules, which are well-known to consist primarily of cool chromospheric plasma. This plasma is then assumed to heat to transition region or even coronal temperatures, and then to fall back onto the chromosphere. In this scenario no blueshifts would be observed in the UV lines since the plasma is rising only when it is very cool, $T < 10^4$K. There may be a stationary component, but this could be weak compared to the redshifted one.

One obvious difficulty with this picture is that it also requires seemingly contrived conditions on the heating process. The heating to transition region temperatures would have to occur only after the plasma has been elevated to its maximum height. Neither the process for lifting the material nor for the subsequent heating is identified. Hence, the model must be considered as highly speculative.

Another possibility is that the shifts are due to transient coronal heating. In this scenario the corona is heated by small flare-like events, so-called microflares, rather than by a steady process. Impulsive heating has had considerable observational support recently [e.g. Lin *et al.*, 1985, Porter *et al.*, 1987] although it is far from verified as the dominant heating mechanism. The dynamic processes with impulsive heating are simply those discussed in flares, chromospheric evaporation during the rise in heating and coronal condensation during the decay. In the impulsive heating model the redshifts would be produced by loops undergoing cooling.

Order-of-magnitude analysis suggests some support for this model as an explanation of the redshifts. Consider a coronal loop that is initially in static equilibrium, and then suddenly the heating is turned off. The loop begins to cool due to the radiative losses that are now unbalanced. At the time that the heating is just turned off the radiative cooling time for plasma at different temperatures along the loop is given by

$$\tau \sim \frac{P}{N^2\Lambda(T)} \sim \frac{T^2}{\Lambda(T)}$$

since the pressure is approximately constant along the loop. This result implies that the plasma at the temperature where Λ/T^2 is maximum, $\sim 10^5$ K, cools most rapidly, creating pressure gradients which drive downflows from the corona into this region. Hence, the maximum velocities should occur near 10^5 K as is observed. An estimate for the magnitude of these velocities is given by $V \sim h_T/\tau$. In the usual loop models [e.g. Craig *et al.*, 1978, Vesecky *et al.*, 1979] this is given by the balance between conduction and radiation,

$$10^{-6}T^{7/2}h_T^{-2} \approx N^2\Lambda(T) \quad (8)$$

so that we find

$$V \approx 10^6(T/10^5)^{.75}\sqrt{\Lambda(T)/\Lambda(10^5)} \quad (9)$$

This result appears promising. First we note that the magnitude of the velocities depends only on the temperature; it is independent of density. This would explain the observed lack of correlation between the intensity of the emission and the magnitude of the redshifts. Second, there are no adjustable parameters in this model as there were in the siphon flow models. The magnitude of the downflows is determined by the magnitude of Λ. Therefore, the fact that the velocity we obtained for the 10^5 K plasma agrees well with the observations is convincing support for this model. Note that in the siphon flow model the velocities at 10^5 K were essentially arbitrary depending on the parameters chosen for the footpoint pressure-difference or the heating asymmetry.

There are some discrepancies, however, between our result, Eqtn (9), and the observations. In particular, the dependence of V on T at low temperatures is stronger, ($V \sim T^{2.25}$ for $\Lambda \sim T^3$) than is observed, ($T^{1.5}$). There is considerable uncertainty, however, in both the observations and the form of Λ [Cook *et al.*, 1989]. Note also, that (9) applies only to the early cooling phase when the scale height is given by (8). It is not apparent how the scale height evolves after significant cooling occurs; detailed numerical simulations of the complete cooling evolution are needed. The only investigation so far has been by Mariska (1987). His results were somewhat mixed. He found that impulsive cooling produced velocities in the transition region of the correct magnitude, but that they ocurred only after the intensity of the emission became low compared to the initial emission of the static loop. We believe that this may be a result of the particular loop parameters that he chose, and in particular, we feel that cooling in loops with high density and pressure may be able to fit the observations better. More simulations are needed to test this conjecture.

In summary, we feel it is fair to conclude that no convincing model for the redshifts has been developed so far.

The observed redshifts remain as one of the major unresolved issues of the solar transition region. We believe that they may be providing us with key information on the structure and heating of the solar atmosphere, and that more observational and theoretical work is required on this interesting problem.

References

Antiochos, S K., A Dynamic Model for the Solar Transition Region, Astrophys. J., 280, 416, 1984.

Antiochos, S.K., and Noci, G., The Structure of the Static Corona and Transition Region, Astrophys. J., 301, 440, 1986.

Antiochos, S.K., and Sturrock, P.A., Evaporative Cooling of Flare Plasma, Astrophys. J., 220, 1137, 1978.

Antonucci, E., Gabriel, A.H., and Dennis, B.R., The Energetics of Chromospheric Evaporation in Solar Flares, Astrophys. J., 287, 917, 1984.

Athay, R.G., Chromosphere-Corona Transition Region Models with Magnetic Field and Fluid Flow, Astrophys. J., 249, 340, 1981.

Athay, R.G. and Holzer, T.E., The Role of Spicules in Heating the Solar Atmosphere, Astrophys. J., 255, 743, 1982.

Athay, R.G. and Dere, K.P., Temperature and Center-Limb Variations of Transition Region Velocities, Astrophys. J., December, 1989.

Ayres, T.R., The Many Faces of Capella, Astrophys. J., 284, 784, 1984.

Boris, J.P. and Mariska, J.T., An Explanation for the Sytematic Flow of Plasma in the Solar Transition Region, Astrophys. J. (Letters), 258, L49, 1982.

Borrini, G. and Noci, G., Non-Equilibrium Ionization in Coronal Loops, Solar Phys., 77, 153, 1982.

Brueckner, G.E., in Solar Active Regions, (ed. F.Q. Orrall), Colo. Assoc. Univ. Press., Ch. 5., 1981.

Brueckner, G.E. and Bartoe, J.-D.F., Observations of High-Energy Jets in the Corona Above the Quiet Sun, Astrophys. J., 272, 329, 1983.

Cargill, J.P. and Priest, E.R., Siphon Flows in Coronal Loops: I. Adiabatic Flow, Solar Phys., 65, 251, 1980.

Cook, J.W., Cheng, C.-C., Jacobs, V.L., and Antiochos, S.K., Effect of Coronal Elemental Abundances on the Radiative Loss Function Astrophys. J., 338, 1176, 1989.

Craig, I.J.D., McClymont, A.N., and Underwood, J.H., The Temperature and Density Structure of Active Region Coronal Loops, Astron. Astrophys., 70, 1, 1978.

Dere, K.P., Turbulent Power and Dissipation in the Solar Transition Region, Astrophys. J., 340, 599, 1989.

Dupree, A.K., Moore, R.T., and Shapiro, P.R., Non-Equilibrium Ionization in Solar and Stellar Winds, Astrophys. J. (Letters), 299, L101, 1979.

Dowdy, J.F., Jr., Rabin, D. and Moore, R.L., On the Magnetic Structure of the Quiet Transition Region, Solar Phys., 105, 35, 1986.

Feldman, U., Cohen, L., and Doschek, G.A., Doppler Wavelength Shifts of Ultraviolet Spectral Lines in Solar Active Regions, Astrophys. J., 255, 325, 1982.

Gaetz, T.J. and Salpeter, E.E., Line Radiation from a Hot, Optically-Thin Plasma: Collision Strengths and Emissivities, Astrophys. J. (Suppl.), 52, 155, 1983.

Glencross, W.M., Plasma Flow along Sheared Magnetic Arches within the Solar Corona, Astron. Astrophys., 83, 65, 1980.

Joselyn, J., Munro, R.H., and Holzer, T.E., Mass Flow and the Validity of Ionization Equilibrium in the Sun, Solar Phys., 64, 57, 1979.

Lin, R.P., Schwartz, R.A., Kane, S.R., Pelling, R.M., and Hurley, K.C., Solar Hard X-Ray Microflares, Astrophys. J., 283, 421, 1984.

Mariska, J.T., Solar Transition Region and Coronal Response to Heating Rate Perturbations, Astrophys. J., 319, 465, 1987.

Mariska, J.T., Observational Signatures of Loop Flows driven by Asymmetric Heating, Astrophys. J., 334, 489, 1988.

Mariska, J.T. and Boris, J.P., Dynamics and Spectroscopy of Asymmetrically Heated Coronal Loops, Astrophys. J., 267, 409, 1983.

McClymont, A.N. and Craig, I.J.D., An Explanation for Fast Downflows on the Sun, Nature, 324, 128, 1986.

McClymont, A.N. and Craig, I.J.D., Fast Downflows in the Solar Transition Region Explained, Astrophys. J., 312, 402, 1987.

Noci, G., Spadaro, D., Zappla, R.A., and Antiochos, S.K., Mass Flows and the Ionization States of Coronal Loops, Astrophys. J., 338, 1131, 1988.

Porter, J.G., Moore, R.L., Reichmann, E.J., Engvold, O., and Harvey, K.L., Microflares in the Solar Magnetic Network, Astrophys. J., 323, 380, 1987.

Pneuman, G.W. and Kopp, R.A., Downflows of Spicular Material and Transition Region Models, Astron. Astrophys., 55, 305, 1977.

Spadaro, D., Noci, G., Zappala, R.A., and Antiochos, S.K., in The Solar Interior and Atmosphere, Tucson, AZ, Nov. 15 - 18, 1988.

Vesecky, J.F., Antiochos, S.K., and Underwood, J.H., Numerical Modeling of Quasi-Static Coronal Loops, Astrophys. J., 233, 987, 1979.

DYNAMICS OF AXISYMMETRIC LOOPS

R. S. Steinolfson

Department of Space Sciences
Southwest Research Institute, San Antonio, TX 78228

Abstract. The evolution of a magnetic loop in response to rotation within circular sections at the loop ends is investigated using time-dependent, two-dimensional MHD simulations in cylindrical geometry. The magnetic field components at the loop ends are modified by the applied rotating flow in a manner consistent with the generated electric fields. The axisymmetric evolution of the magnetic field and plasma flow within the loop can be characterized as passing through several distinct and identifiable stages. Early on, the magnetic energy increases with the square of time, the current and magnetic field are parallel (force-free), the average kinetic energy remains constant and is negligible compared to the magnetic energy increase, and the solution fluctuates at a characteristic loop frequency. The average kinetic energy then increases rapidly and levels off at a higher value, which is still much less than the magnetic energy increase. The magnetic energy now increases exponentially with time and deviations from a force-free field begin to appear, particularly near the loop axis and ends. In the final phase, the current and field are no longer force-free, and the solution becomes highly nonlinear. The fluctuations that appeared early on continue with ever increasing amplitude into this last stage. Finally, a critical shear level is exceeded and the oscillatory behavior stops and is replaced by an expansion propagating radially outward at large radii, a large radial inflow toward the loop axis within the outward travelling expansion, and a strong cylindrical shock around the axis that brings the inflowing plasma to rest. It is suggested that this shock may produce the heating and particle acceleration in compact loop flares.

I. Introduction

The evolution of the magnetic field and plasma velocity in magnetic loops (flux tubes) as a result of a localized rotation of cylindrical sections of the axial loop ends is investigated using numerical solutions of the MHD equations. The magnetic field is assumed to be firmly anchored in the end plates so the field within the loop must adjust to azimuthal motion of the footpoints attached at the ends. The resulting twist of the loop field generates currents as the energy level within the loop increases above that in the initial, potential state. A possible application of this study is to the buildup of nonpotential magnetic energy in coronal magnetic loops driven by photospheric motion. The high beta (ratio of plasma thermal pressure to magnetic pressure) and high inertia of the photosphere causes magnetic fields in this lower part of the solar atmosphere to be convected with the photospheric motion. By contrast, due to the relatively low beta and low inertia of the corona, the response of coronal field lines connected to the photosphere dominates the coronal plasma dynamics and produces velocities in the corona. The increase of stored, nonpotential magnetic energy in the twisted loop may provide the energy source for compact flares, which is one of the two classes of flares identified during the Skylab Workshop [Pallavicini et al., 1977]. It should be noted that post-Skylab observations have led Bai and Sturrock [1989] to suggest an alternate flare classification scheme. The present computations not only simulate the slow, preflare energy buildup phase but show a much more rapid evolution once the field twist exceeds a critical level. The behavior during the rapid phase suggests possible physical mechanisms for the heating (and subsequent continuum and line emission) and particle acceleration during the flare impulsive phase.

Zweibel and Boozer [1985] analytically studied a problem similar to that posed here using the assumptions that the coronal motion could be neglected, the field remained force-free, and the twist per unit length was small. Work on the present problem has been initiated by Steinolfson and Tajima [1987, hereafter referred to as Paper I], who published results for only one special case. Their particular example has only limited application to coronal loops due to the very small loop aspect ratio (length/radius = 2) that was used. Steinolfson and Tajima show that their results agree with those of Zweibel and Boozer during the early part of the field evolution.

The buildup of energy in magnetic arcades by photospheric motion has also been studied. Energy buildup in an arcade is in one important aspect fundamentally different from that in a loop, since the arcade is an open system whereas the loop is essentially a closed system. As for loops, in some investigations the coupling between the coronal magnetic field and the fluid motion is neglected, and the corona is assumed to evolve through a sequence of static equilibrium status [e.g., Birn et al.,1978; Low, 1982; Zwingmann, 1987], while others include coupling to the fluid motion [e.g., Wu et al., 1983; Mikic et al.,1986]. The above studies indicate that there is a critical shear beyond which a nearby equilibrium does not exist and the field begins to evolve more rapidly (analogous to the preflare buildup and the flare impulsive phases). More recent work, however, does not support the above loss of equilibrium picture and shows that the coronal field continues to evolve slowly to increased photospheric shear [Biskamp and Welter, 1989; Klimchuk et al., 1988]. Additional simulations using more physically realistic models are needed to resolve the loss of equilibrium issue for arcades.

Any conclusions regarding a critical shear in arcades cannot be directly applied to loops due to the basic difference between the two magnetic geometries, as mentioned above. The present results suggest that a critical shear, beyond which a much more rapid evolution occurs, does exist in magnetic loops. However, it must be kept in mind that the geometrical and physical assumptions made in this model

Geophysical Monograph 58

(see Sec. II) severely limit the possible application of the results to actual magnetic loops in the corona. The model has been simplified to identify basic physical processes, as discussed in Sec. III. It remains to determine whether the evolution picture developed here obtains when the geometric simplifications are relaxed and more physics is included in the model. Our results are placed in their proper context in the final section where some suggested model improvements are also discussed.

II. Model and Numerics

The coronal loop is approximated by a cylinder with an initially uniform magnetic field in static equilibrium parallel to the cylinder axis. It is further assumed that gravitational effects are negligible, the evolution is axisymmetric (all three components of vector quantities are included, however), and that thermal pressure gradients are insignificant ($\beta \ll 1$). The latter assumption removes sound waves. Finally, the density evolution is not computed self consistently from the continuity equation, or equivalently, density changes are assumed to be small relative to the initial density. With these assumptions, the dynamics uncouple from the energetics, and the equations that govern the evolution become (in dimensionless form)

$$\frac{d\vec{v}}{dt} = \left(\nabla \times \vec{B}\right) \times \vec{B} + \frac{1}{R}\nabla^2 \vec{v}, \tag{1}$$

$$\frac{\partial \vec{B}}{\partial t} = \nabla \times \left(\vec{v} \times \vec{B}\right) - \frac{1}{S}\nabla \times \left(\nabla \times \vec{B}\right). \tag{2}$$

The velocity $\vec{v}$ is normalized to the initial Alfvén velocity ($v_A = a/\tau_A$), magnetic field $\vec{B}$ to the initial uniform field, time t to the initial Alfvén time τ_A, and distance to the maximum radius (a) to which the rotation of the loop base extends (radial extent from the loop axis of the photospheric shear motion). The relevant times in the problem are the Alfvén τ_A, resistive τ_r, and viscous τ_ν times defined by

$$\tau_A = \frac{a\sqrt{4\pi\rho}}{B}, \qquad \tau_r = \frac{4\pi a^2}{c^2\eta}, \qquad \tau_\nu = \frac{a^2\rho}{\nu},$$

where c is the light speed, η the resistivity, ρ the density, and ν is the viscosity. The only physical parameters in Eqs. (1) and (2) are the Lundquist number $S = \tau_r/\tau_A$ and a dimensionless quantity involving the viscosity, $R = \tau_\nu/\tau_A$.

The only place the density would enter into the defining equations is to multiply the inertia term on the left-hand side of Eq. (1). As long as dynamic processes are relatively slow compared to the Alfvén time, the precise value of the density should not significantly affect the evolution. During the later stages of the present simulation, however, the solution evolves rapidly relative to the local Alfvén time and the constant density assumption may no longer be even approximately valid. Further work is necessary to determine whether relaxation of this assumption would more than quantitatively alter the general behavior computed from Eqs. (1) and (2). Note that the velocity components calculated using Eq. (1) will not necessarily satisfy the constant density continuity equation $\nabla \cdot \vec{v} = 0$. This approximation has also been used in numerical studies of energy buildup in arcades [Mikic et al., 1988; Biskamp and Welter, 1988].

Equations (1) and (2) are solved numerically using the semi-implicit method of Harned and Schnack [1986]. This differencing scheme was developed to remove the shear Alfvén CFL restriction on the allowable time step for which the solution remains numerically stable. It has been successfully applied in simulations of sheared arcades where time steps substantially in excess of the CFL value are used, particularly during the early quasistatic evolution [e.g., Mikic et al., 1988; Biskamp and Welter, 1988]. The algorithm was not as useful for the present application to loops where it proved necessary to resolve features on the shear Alfvén time scale. Hence, the maximum time step was restricted to about twice the CFL time step. Given the increased numerical complexity of the semi-implicit method, it probably did not provide any significant improvement in the required CPU time over a simple explicit method.

We assume that the loop ends are rotated by the same amount in opposite directions. All physical variables then become either symmetric or antisymmetric midway along the loop axis, and the simulation only has to be performed in one-half the total loop length. The only velocity allowed at the loop base is the applied rotation (azimuthal) velocity. Solid-body rotation with a linear increase in shear velocity is assumed to exist out to $r = 0.75$ (in dimensionless radius) followed by a linear decrease in the velocity to zero at $r = 1$. The electric field at the previous time step is used to update the azimuthal magnetic field, and zero-order extrapolation provides the radial component. The axial field is held fixed at the loop ends so the total magnetic flux into the loop remains constant. These boundary conditions are consistent with the nondissipative (hyperbolic) form of Eqs. (1) and (2), since only two characteristics transport information to the loop base.

The numerical procedure used previously for this problem (Paper I) has been improved by transforming the radial coordinate using $r = c\sinh^{-1} y$ where c is a constant. This transformation provides good numerical resolution within $r = 1$ while placing the outer radial boundary so far away that it has no effect on the solution. The outer boundary is typically located at $r = 27$, and zero-order extrapolation is certainly adequate at this distance. The simulations discussed in the following section have at least 100 radial and 60 axial grid points.

Since the pressure (temperature) has been removed in our simplified analysis, the only relevant physical quantity in the initial atmosphere (other than the dissipative parameters) is the Alfvén velocity or the ratio $B/\rho^{1/2}$, which determines the speed at which information is transmitted. The important geometric factor is the loop aspect ratio AR. For a fixed radial shear profile, the shear is completely specified by the peak shear velocity v_{sh} or, more importantly, by the ratio of this velocity to the Alfvén velocity. An additional parameter regarding the applied shear is the manner in which the shear increases to its maximum value and the time interval $\Delta\tau_s$ of the increase. We generally assume that the shear velocity is applied instantaneously ($\Delta\tau_s = 0$), but have also computed results for a linear increase over a finite time ($\Delta\tau_s > 0$). Therefore, a unique solution is given by the values of v_A, AR, S, R, v_{sh}, and $\Delta\tau_s$.

III. Numerical Results

The temporal behavior of the kinetic energy and the increase in magnetic energy, obtained by integrating over the entire computational box, are shown in Fig. 1 for the following parametric values: $v_A = 727\,\mathrm{km\,s^{-1}}$, $AR = 10$, $S = 10^6$, $R \to \infty(\nu = 0)$, $v_{sh} = 10\,\mathrm{km\,s^{-1}}$, $\Delta\tau_s = 0$. The Alfvén velocity in the twisted loop will, of course, be much larger than the above initial reference value. The solution is computed for $0 \le z \le 3$, $0 \le r \le 27$. The total initial magnetic energy is 5.1×10^{31}ergs. Note that the energies in the figure are scaled by a factor of 10^{25}. The initial Alfvén time (for $a = 10^4$km) is 13.8 sec so this run extends over about 925 Alfvén times, which required 28,000 computational time cycles.

As for the example discussed in Paper I, the evolution occurs in several distinct and identifiable stages marked on Fig. 1. In the first phase, the current and magnetic field are parallel throughout the loop, the kinetic energy is negligible, and the induction equation can be approximated by

$$\frac{\partial B}{\partial t} = \nabla \times \left(\vec{v}_{sh} \times \vec{B}_0\right), \tag{3}$$

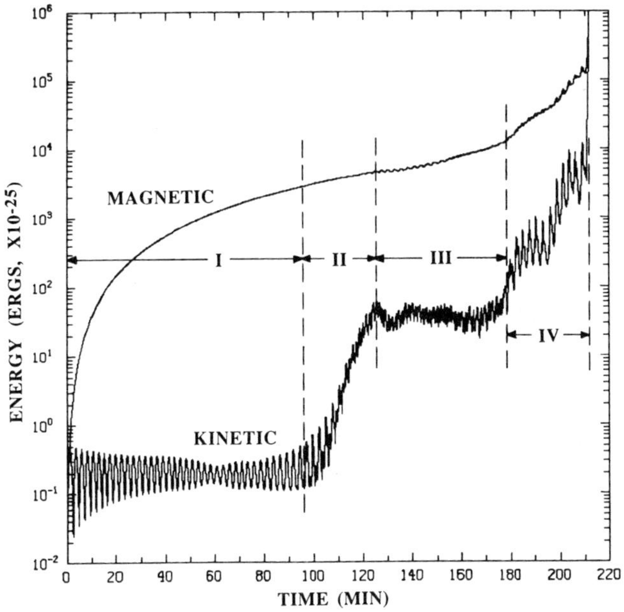

Fig. 1. The temporal behavior of the kinetic energy and the increase in magnetic energy integrated over the entire computational cylinder ($0 \leq r \leq 27$, $0 \leq z \leq 5$) for the following parameter values: $v_A = 727\,\mathrm{km\,s^{-1}}$, $AR = 10$, $S = 10^6$, $\nu = 0$, $v_{sh} = 10\,\mathrm{km\,s^{-1}}$, $\Delta\tau_s = 0$. The vertical dashed lines separate different stages in the solution that are discussed in the text.

which predicts that the magnetic energy increases as t^2 in excellent agreement with results in the figure. The solution for the magnetic field configuration during this period agrees with that given by Zweibel and Boozer [1985]. The period of the kinetic energy oscillation equals the time required for a disturbance to travel the length of the loop. There are also similar oscillations in the magnetic energy with the same period, but the amplitude is so small compared to the magnetic energy put into the loop that they do not appear on the figure. The oscillations result from a standing wave within the cylinder. The kinetic energy during this early stage is primarily in the azimuthal component. The average value of kinetic energy varies linearly with the applied shear velocity, and the oscillation magnitude varies with $\Delta\tau_s$.

The average kinetic energy increases rapidly during the second phase. The kinetic energy rise coincides with an increase in the axial component of magnetic energy, which had been negligible throughout the first stage. This suggests that the magnetic field is no longer simply being twisted up, as it would be if all the energy went into the azimuthal component, but that the configuration is changing.

The average kinetic energy again remains constant during phase III on Fig. 1, but now the perturbed magnetic field becomes comparable to the initial field. The current and magnetic field are still approximately parallel throughout most of the loop except for regions near the loop base and axis. The average magnetic field evolution is now more accurately represented by

$$\frac{\partial B}{\partial t} = \nabla \times (v_{sh} \times B)$$

rather than Eq. (3), which gives an exponential increase in magnetic energy. As seen in the figure, the magnetic energy does increase exponentially over about the last two thirds of stage III.

In the last phase on the figure, the solution becomes highly nonlinear and the magnitudes of the oscillations grow until both energies increase simultaneously and the solution terminates. The magnetic configuration changes dramatically just prior to the final simultaneous increase in energies and will be discussed in more detail below. Although the final rise appears as a discontinuity in the figure, the time step is small enough that it is well-resolved temporally (it occurs over hundreds of time steps). Other features of the evolution prior to the above change in configuration have been discussed in Paper I. We now go on to consider results for a different set of parameters and concentrate on the final rapid change in the magnetic configuration.

The energies for the parametric values of $v_A = 230\,\mathrm{km\,s^{-1}}$, $AR = 6$, $S = 10^6$, $R \to \infty(\nu = 0)$, $v_{sh} = 10$ km s^{-1}, and $\Delta\tau_s = 0$ are shown in Fig. 2. The various phases identified for the solution in Fig. 1 can also be seen in this solution. The oscillatory period is now somewhat longer due to the smaller Alfvén velocity, although the shorter loop acts to decrease the period. The smaller figure at the top shows a blow-up of the energies near the end of the solution. The vertical wavy line represents the approximate time t_c at which the solution changes from having an oscillatory behavior to one that evolves monotonically. To illustrate the nature of this change, we now consider details of the solution at several times just prior to and following the time t_c.

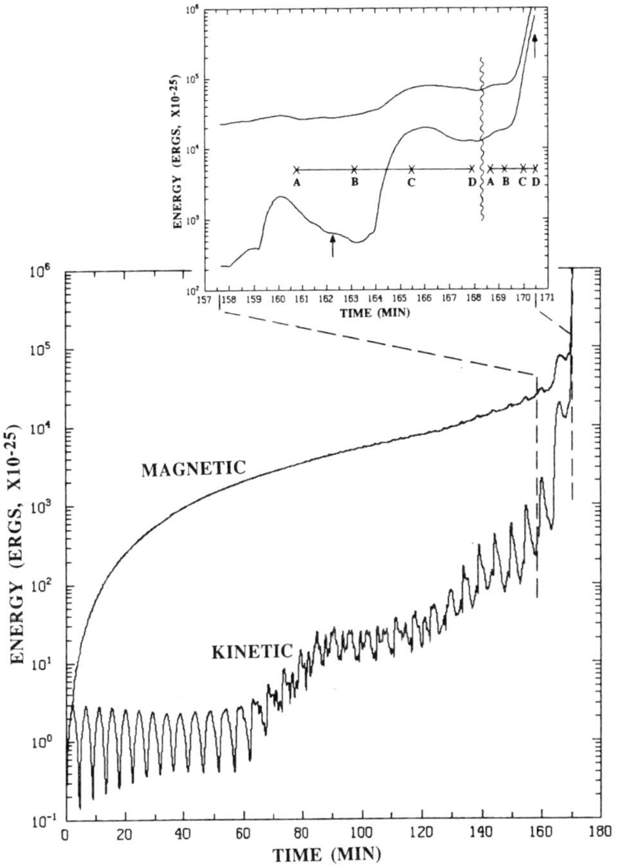

Fig. 2. The temporal behavior of the kinetic energy and the increase in magnetic energy integrated over the entire computational cylinder for the same parameters as in Fig. 1, except that $v_A = 230\,\mathrm{km\,s^{-1}}$ and $AR = 6$. The upper frame shows a blow-up of the energies near the end of the calculation.

A few of the magnetic field lines at the two times marked by the vertical arrows in Fig. 2 are shown in three different views of the same field lines in Fig. 3. The field lines were computed by following the field lines, starting from the same radial locations at the loop base at both times, until they reached the loop midpoint. The top panels show the lines as viewed from the side of the loop perpendicular to the r–z plane on which the fields emerged through the base [$z = 0$ in Figs. 3(a)]. The view from the loop end is given in the center panels, and the field lines projected into the r–z plane (the azimuthal component is set to zero) are shown in the bottom panels. The two field lines that appear to terminate at the latter time in Fig. 3(c) are being wrapped around the axis so often ($B_z \approx 0$ for these two field lines) that the allowable field line length in the numerical code was exceeded. As seen at the earlier time in the figure some of the

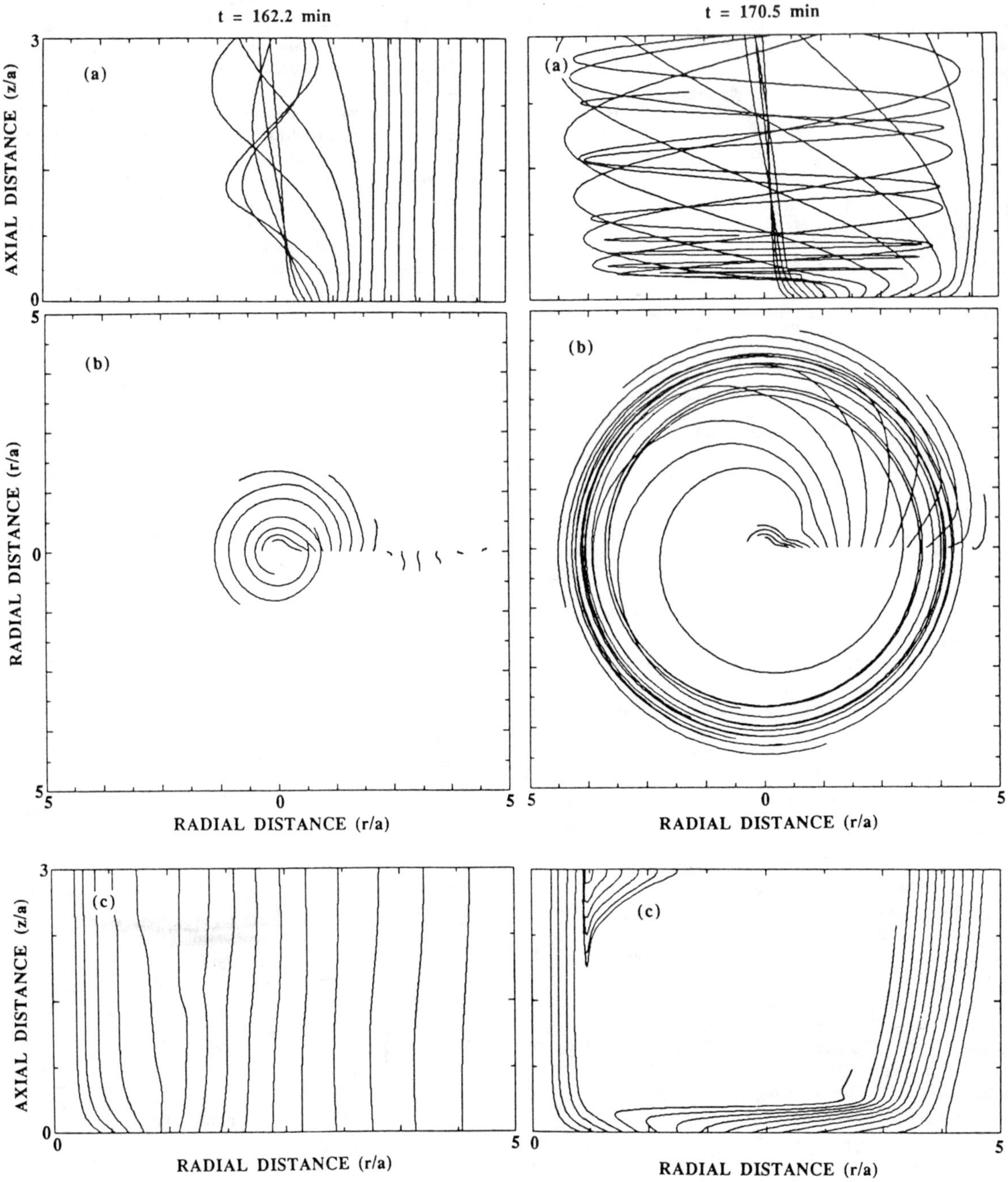

Fig. 3. Three different views of the same field lines in one half of the loop at the times marked with the vertical arrows ($t = 162.2$ min and $t = 170.5$ min) in the upper frame in Fig. 2. The view from the side of the cylinder is shown in (a), that from the end in (b), and the field lines projected into the r–z plane (B_θ set to zero) are shown in (c).

field lines are just being wrapped around the axis once between the base and the loop midpoint at $z = 3$ (the total loop length is $z = 6$). In addition, before the configuration change, most of the adjustment of the loop field takes place in the field lines that intersect the base within or just beyond the radius of the applied shear (i.e., for approximately $r < 2$). All field lines emerging from the sheared region converge toward the axis while the outer field lines diverge away from the loop axis (see Paper I for an explanation). At the later time in Fig. 3 (beyond the indicated time of the configuration change in Fig. 2) the inner converging field lines intersecting the sheared region do not change significantly, although they are forced closer to the axis. The field lines emerging from the base beyond the applied shear do change substantially, however, as they are bent radially away from the axis with some field lines being wrapped several times around the axis. As shown in Fig. 3(c) at $t = 170.5$ min, a new magnetic island also forms.

The axial magnetic field after the configuration change is approximately zero over a large cylindrical annulus extending over almost the entire loop length with the exception of small regions near the ends. This can be seen in Fig. 3(c), but is shown more clearly in Fig. 4(a). The axial current density at this time in Fig. 4(b) is confined to a narrow channel around the loop axis in one direction with two return annular channels just beyond the current on axis (centered at $r \approx 1$) and outside of the region of vanishing axial field (centered at $r \approx 4.5$). Note that the solution is only being shown out to $r = 5$ although the computation actually extends to $r = 27$ so the outer radial boundary is not restricting the radial motion.

Another informative view of the behavior of the solution for the time interval in the top of Fig. 2 is shown in Fig. 5. These plots give the radial variation of several physical quantities at a fixed axial location near the axial center of the loop (at $z = 2.5$ in Fig. 4). The plots in the left column are for the times identified by A, B, C, and D in the top plot in Fig. 2 prior to t_c, and the right column contains values at times A, B, C, and D during the monotonically growing phase after t_c. A shock forms around the axis at $r \approx 1$ near the maximum of the oscillatory energy before t_c. However, it again becomes weaker before strengthening once more after t_c and then continues to converge toward the axis and increase in strength until the solution terminates. A large super-Alfvénic and supersonic flow towards the loop axis develops and is brought to rest by the shock wave. The region with the large inflow velocity coincides with the development of large azimuthal field and negligible axial field. Hence the axial field provides the source for this rapid change in configuration. The axial field, however, remains strong and dominant within the loop core inside the shock. Since the thermodynamics are not included here, this shock is not a true Rankine-Hugoniot shock wave. Nonetheless, the equations we are using still permit the existence of a shock-type discontinuity.

During the oscillatory part of the solution before t_c, the imploding behavior shown at the earlier time in Fig. 5 only extends over approximately the central half of the total loop length. When the monotonic behavior after t_c begins, the length of the implosion region expands axially until it reaches the loop ends at about $t = 170.0$ min and the solution looks similar to that in Fig. 4. Prior to 170 min the energy increase within the loop is approximately equal to the energy flux through the loop ends (to within numerical accuracy of a few percent). After 170 min, however, the loop energy exceeds (by ever increasing margins) the energy flux into the loop. A narrow boundary layer develops near the ends, and the interior region evolves independent of the applied boundary conditions. In essence, the poynting flux into the loop is no longer given by the applied value at the boundary ($v_\theta B_\theta B_z$) but by the much larger value at grid points adjacent to the

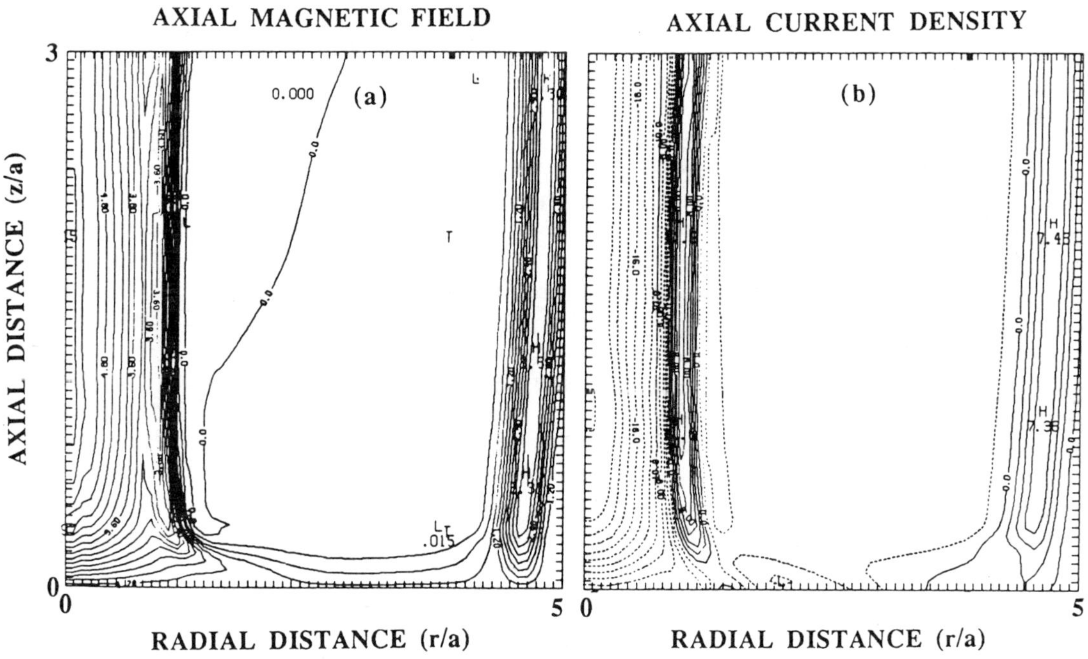

Fig. 4. Contour plots in half the loop length of the axial components of the magnetic field (a) and current density (b) at $t = 170.5$ min for the solution in Fig. 2. Dashed contours of the current indicate flow toward the base ($z = 0$) and solid contours identify flows away from the base. The horizontal dashed line in (a) marks the axial location of the radial scans in Fig. 5.

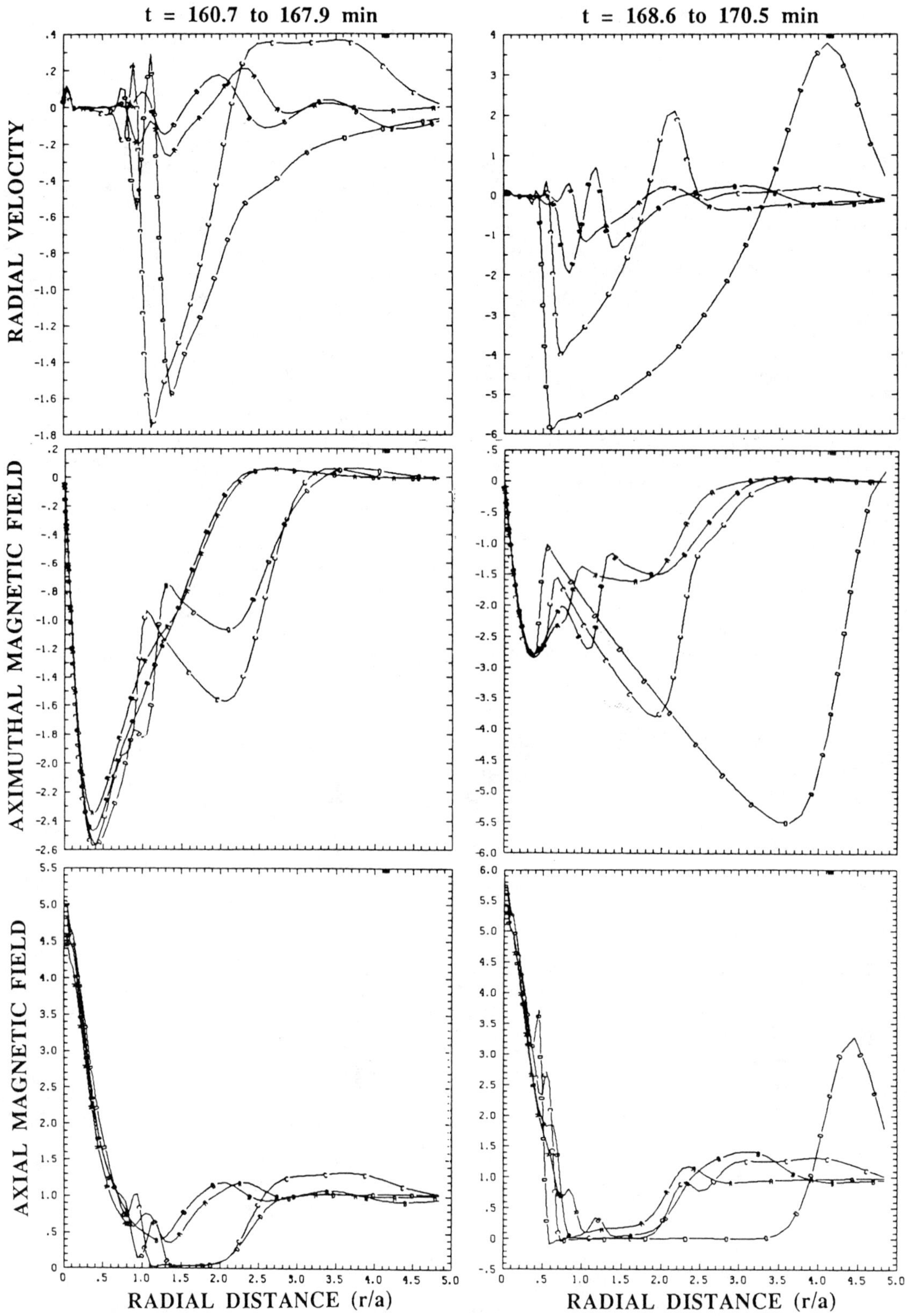

Fig. 5. Radial variation at the axial location marked by the dashed line in Fig. 4(a) of the radial velocity (b), azimuthal magnetic field (b), and axial magnetic field (c). The values in the left column are for the four times from t = 160.7 min to 167.9 min marked by A, B, C, and D in the top part of Fig. 2. The right column contains values from t = 168.6 min to 170.5 min identified also by A, B, C, and D in Fig. 2.

boundary ($v_r B_r B_z$). Hence, once the imploding region expands to the loop ends, the solution is no longer physically realistic. However, until that time the solution remains valid. Thus, the oscillatory behavior before t_c involving just the central part of the total loop length, the change in configuration at t_c, and the growth of the imploding region to fill the entire loop must be considered physically realistic predictions of this model.

Since the oscillations appear to play a crucial role in the loop evolution, two additional computations with parameters selected to attempt to suppress the oscillations have been carried out, although the results are not shown here due to space limitations. The selected parameters were the same as those used for the results in Fig. 2, except that finite viscosity ($R = 10^4$) was used in one case and in the other the shear velocity was gradually increased ($\Delta\tau_s = 30\,\text{min}$) to the maximum value. In both examples the oscillations are essentially suppressed during the first stage. They eventually reappear, however, and the solutions have the same qualitative behavior as for the two shown here.

IV. Discussion

Numerical simulations of the buildup of nonpotential magnetic energy in magnetic loops due to rotation of sections of the loop end plates, initiated by Steinolfson and Tajima [1987], Paper I, have been extended to more realistic physical conditions (for application to coronal loops), the numerics have been improved, and a detailed study of the rapid evolution when the twist exceeds a critical value (which may explain heating and particle acceleration in compact flares) has been carried out. The results of the studies presented here have the same average, qualitative behavior as the simulation in Paper I. When the results are considered in terms of the temporal evolution of the total kinetic energy and the total increase in magnetic energy within the loop, the solutions can be characterized as evolving through four identifiable stages. This same general behavior obtains for all simulations that have been carried out to date, including approximately thirty whose results are not discussed here. To summarize, in the first stage the magnetic energy increases at t^2, the current and magnetic field are parallel throughout the loop, the energies oscillate with a period equal to the time it takes an Alfvén wave to traverse the length of the loop, and the kinetic energy is negligible. The kinetic energy rises rapidly during the second phase (maintaining the oscillations) but is still negligible relative to the increase in magnetic energy. During the third stage, the kinetic energy levels off at a new average value but continues to oscillate, the magnetic energy now increases exponentially, and deviations from parallel current and field begin to appear near the loop axis and ends. The energies continue to oscillate in the final stage, but the solution becomes highly nonlinear and the current and field are no longer parallel. The oscillations grow in magnitude during this final phase until the magnetic configuration changes substantially, the solution evolves monotonically, and finally terminates due to numerical problems. The nature of the evolution just before and after the configuration change is studied in some detail since it is suggested that this stage may explain the physical processes involved in the heating (and associated radiation) and particle acceleration during the impulsive phase of compact (loop) flares. In one-half of the oscillation period a large radial flow toward the axis develops in an annulus around the axis in about the central half of the total loop length. The flow rebounds outward during the other half of the cycle. The energy during the oscillations increases in magnitude until finally only the inflow survives and the flow collapses monotonically toward the loop axis. The region involved in the inflow expands axially until it extends over the entire loop except for a narrow boundary layer at the ends. Once the inflow in the annular region expands to the loop ends, the poor numerical resolution produces numerical problems and the solution no longer conserves energy. The general behavior up until this time remains valid and must be considered physically realistic predictions of the model.

The converging flow is brought to rest by a cylindrical shock surrounding the loop axis. The heating produced by this shock could produce the thermal electromagnetic radiation observed in compact solar flares [e.g., Kane et al., 1980]. The short time scales of quasi-periodic emission features prior to the main impulsive phase and those during the impulsive phase can also be easily explained by this model. Although nonthermal effects are not included in this model, the shock is a quasi-perpendicular shock and should be capable of producing the particle acceleration associated with such flares. More extensive analysis is required to quantify the anticipated radiation and particle acceleration signatures for comparison to observations.

An interpretation of the characteristic loop response is that the loop has been sheared beyond a critical value at which the oscillating solutions prior to the configuration change can occur. Hence, the plasma and magnetic field are simply seeking out a new solution, with a substantially different magnetic configuration. Since the azimuthal magnetic field component becomes comparable to or larger than the axial component before the solution becomes monotonic, it is conceivable that a kink instability could occur at some critical shear level providing the axisymmetric approximation was removed, which may result in a much different adjustment than that computed herein. It is definitely necessary to do a nonaxisymmetric calculation, but there are a couple of reasons why a kink instability may not occur. First of all, the solution is extremely dynamic in the oscillatory phase, and it is not clear whether a coherent kink could develop fast enough so as not to be dominated by the velocity fluctuations. In addition, the axial magnetic field exceeds the azimuthal component in a region near the loop axis and may act as a stabilizing effect against the kink.

Before this model can be applied with confidence to the quantitative observed emission characteristic in compact flares, a number of the simplifying assumptions must be relaxed. Among the more important modifications that could easily be implemented are the following: permit nonaxisymmetric evolution, include compressibility, and include an energy equation. Additional changes that may be important, but which would be more difficult, are to allow loop curvature and include gravity.

Acknowledgments. The author appreciates beneficial conversations with T. Tajima. This research was supported by NASA Grants NAGW-846 and NAGW-1324, and the majority of it was performed while the author was at the University of Texas at Austin. The use of the computing facilities at the San Diego Supercomputer Center and at the National Center for Atmospheric Research, which is sponsored by the National Science Foundation, is gratefully acknowledged.

References

Bai, T., and P. A. Sturrock, Classification of solar flares, *Ann. Rev. of Astron. and Astrophys.*, in press, 1989.

Birn, J., H. Goldstein, and K. Schindler, A theory of the onset of solar eruptive processes, *Solar Phys.*, *57*, 81–101, 1978.

Biskamp, D., and H. Welter, Magnetic arcade evolution and instability, *Solar Phys.*, *120*, 49–77, 1989.

Harned, D. S., and D. D. Schnack, Semi-implicit method for long time scale magnetohydrodynamic computations in three dimensions, *J. Comp. Phys.*, *65*, 57–70, 1986.

Kane, S. R., and ten co-authors, Impulsive phase of solar flares, in *Solar Flares*, edited by P. A. Sturrock, 187–226, Colorado Assoc. Univ. Press, Boulder, Colorado, 1980.

Klimchuk, J. A., P. A. Sturrock, and W.-H. Yang, Coronal magnetic fields produced by photospheric shear, *Astrophys. J.*, *335*, 456–467, 1988.

Low, B. C., Nonlinear force-free fields, *Rev. Geophys. and Space Phys.*, *20*, 145–159, 1982.

Mikic, Z., D. C. Barnes, and D. D. Schnack, Dynamical evolution of a solar magnetic field arcade, *Astrophys. J.*, *328*, 830–847, 1988.

Pallavicini, R., S. Serio, and G. S. Vaiana, A survey of soft x-ray limb flare images: the relation between their structure in the corona and other physical parameters, *Astrophys. J.*, *216*, 108–114, 1977.

Steinolfson, R. S., and T. Tajima, Energy buildup in coronal magnetic flux tubes, *Astrophys. J.*, *322*, 503–511, 1987 (Paper I).

Wu, S. T., Y. Q. Hu, Y. Nakagawa, and E. Tandberg-Hanssen, Induced mass and wave motions in the lower solar atmosphere, I. Effects of shear motion of flux tubes, *Astrophys. J.*, *266*, 866–881, 1983.

Zweibel, E. G., and A. H. Boozer, Evolution of twisted magnetic fields, *Astrophys. J.*, *295*, 642–647, 1985.

Zwingmann, W., Theoretical study of onset conditions for solar eruptive processes, *Solar Phys.*, *111*, 309–324, 1987.

TWISTED FLUX ROPES IN THE SOLAR CORONA

P. K. Browning

Department of Pure and Applied Physics, UMIST, P. O. Box 88, Manchester M60 1QD, England

Abstract Loop structures, which are essentially magnetic flux tubes, often with twist, are common in the solar corona. This paper considers the magnetic equilibrium of twisted coronal loops. Ignoring curvature, as a loop is twisted at the photospheric footpoints, longitudinal structure develops with the centre of the loop tending to expand radially. Results of a 2D numerical code show that nearly all of the expansion occurs in narrow boundary layers near the photosphere, and most of the loop is approximately a straight cylinder. If a small amount of dissipation is allowed, the field relaxes to a minimum energy state $\nabla \times \underline{B} = \mu \underline{B}$. In the limit of immediate relaxation, the loop evolves through a sequence of such states; the loop surface is free, the radius being determined by pressure balance and flux conservation. A critical point is reached if the loop is strongly twisted, whereby the local current $\mu = \mu_0 j/B$ cannot be further increased. The heating vanishes in this immediate relaxation limit, but for finite relaxation times sufficient energy can be dissipated to heat the corona.

Introduction

The solar corona contains a large number of loops structures which may be seen in X-rays or EUV; it is widely accepted that these loops outline magnetic field structures. In many cases, loops show evidence of twisted or helical fields. Since the coronal magnetic field is very strong and the pressure is low ($\beta \ll 1$), the equilibrium fields must be approximately force-free ($\underset{\sim}{j} \times \underset{\sim}{B} = \underset{\sim}{0}$). In this case, twisted loops correspond to non-potential fields ($\underset{\sim}{j} \neq 0$). Extrapolations of the coronal field from line-of-sight photospheric measurements show that observed structures, particularly in active regions, are best modelled by non-potential force-free fields. Thus coronal loops can be generally thought of as twisted flux ropes.

Most of the physical processes in the corona are dominated by the magnetic field. It is thus very important to understand the basic properties of magnetic loops. Except in relatively rare events such as flares, coronal loops change rather slowly, on a time-scale much slower than an Alfven wave travel time: hence the magnetic fields are force-free equilibria, evolving only quasi-statically. Therefore we need an understanding of the possible equilibria of twisted coronal loops. Other important phenomena, such as wave propagation, plasma flows, non-equilibrium and eruption, instabilities and thermal properties may be studied, given an equilibrium as a basis. Coronal fields are complex, loops consisting of curved twisted structures in 3-D, with ends embedded in the dense photosphere, surrounded by the ambient coronal field and plasma. A wide range of models have been developed, concentrating on different features. Parker [1979], using a 1D description, shows that a loop tends to expand if it is strongly twisted; our 2D results confirm this. We present in Section 2 a model whose main aim is to take account both of the line-tying at the photosphere and of the free boundary between the loop and the background coronal field. This extends previous linearised models of Zweibel and Boozer [1985] and Lothian and Hood [1989] to the strong twist regime. We find that the cross-section and structure of the loop is invariant along its length, except in narrow layers near the photosphere: a 1D model of the loop is then developed.

An important feature of twisted flux ropes is that they store energy, as they contain excess magnetic energy above a potential state. This energy may be released in principle by dissipating the currents. In the corona, the stored energy may be released suddenly, as in a flare, or continually (or perhaps in small bursts) in order to heat the corona. We discuss here the possibility of heating the corona by dissipation of equilibrium currents, generated by slow footpoint motions, using small-scale magnetic reconnection. Section 3 describes this process in a twisted flux tube using relaxation theory, which

Geophysical Monograph 58

predicts that the field tends to relax towards a minimum energy state. The slow evolution through relaxed equilibria is described using the ideas developed in Section 2, with the 1-D model appropriate for the central part of the loop. The heating rate is calculated for specified twisting motions.

The Equilibrium of a Twisted Loop in an External Field

In this section we study a twisted coronal loop whose footpoints are line-tied in the dense photosphere and which is embedded in the ambient potential coronal field. A loop is thus viewed as a twisted flux rope which is simply a bundle of fieldlines within a global field system. The flux rope is assumed straight, ignoring curvature. Pressure gradients are also neglected, since the equilibrium is likely to be only slightly modified by the effects of small finite β (nevertheless, a loop may well have enhanced pressure, and the effects of this are mathematically similar to those of twist and will be studied in future).

Consider an initially uniform field between the planes z=0,L (representing the photospheric surface), subjected to slow axisymmetric twisting motions of the footpoints. Line-typing ensures that the normal field at the photosphere remains constant, and the field evolves through a sequence of 2-D force-free states with the azimuthal field determined by the angular footpoint displacement if resistive effects are neglected. The twisting is confined within a radius r_b from the symmetry axis; outside the loop is potential field which adjusts to the changing loop shape but remains straight and uniform at large distances (Figure 1). No assumption is made concerning the aspect ratio or the magnitude of the twist; this model is thus an extension to the non-linear regime of the linearised perturbation models of Zweibel and Boozer [1984] and Lothian and Hood [1989], who both study large aspect-ratio, weakly twisted loops.

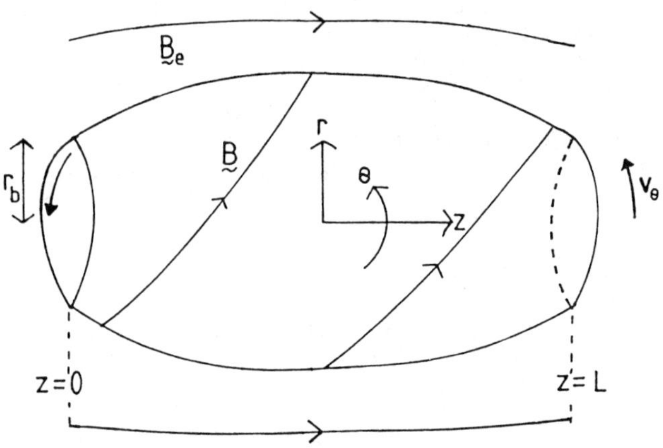

Fig.1 A sketch of the model for a twisted coronal loop, line-tied in the photosphere at z=0,L

A numerical code is used to solve the Grad-Shafranov equation

$$\frac{\delta^2\psi}{\delta r^2} - \frac{1}{r}\frac{\delta\psi}{\delta r} + \frac{\delta^2\psi}{\delta z^2} = -G\frac{dG}{d\psi} \qquad (1)$$

where the azimuthal field is

$$rB_\theta = G(\psi)$$

with a variety of current profiles $G(\psi)$. On the planes z=0,L the flux function is taken to be $\psi = cr^2$, in order to model line-tying with an initially uniform field. The footpoint displacement is calculated a posteriori from

$$\Phi = \int_0^L (B_\theta/rB_z)dz \qquad (2)$$
$$[\psi = \text{const}]$$

where the integral is taken following a fieldline. Results are shown in Figure 2 for the fieldlines and the footpoint displacments; the profile used is

$$G = \begin{cases} \mu_a(1-\psi/\psi_b), & \psi < \psi_b; \\ 0, & \psi \geq \psi_b; \end{cases} \qquad (3)$$

where ψ_b is the flux within the loop (see Browning and Hood [1989] for further details).

It can be seen from Figure 2 that the coronal field develops a very inhomogeneous structure as the footpoints are twisted. An inner core of very strong longitudinal field with weak azimuthal field is formed, with the flux surfaces compressing inwards from their untwisted positions. Then there is a rather wide sheath with large B_z but weak B_θ. The surface of the loop (at which $B_\theta = 0$) expands outwards as the loop is twisted, in accordance with the predictions of Parker [1979]. This expansion only appears in the non-linear (strong twist) regime; Lothian and Hood [1989], for example, find that the loop radius is unchanged by twisting. Also note that if B_θ does not fall to zero at a finite radius, as in some of Zweibel and Boozer's models, then all fieldlines contract <u>inwards</u>.

A significant feature apparent from these 2-D numerical results is that the flux surfaces are essentially cylindrical and z-independent for most of the loop's length, with 2-D effects apparent only in rather narrow boundary layers near the photosphere [Parker, 1972]. This result was pointed out for the linearised weak twist case by Lothian and Hood, and can be proved for large aspect ratio loops ($r_b/L \ll 1$). Coronal flux ropes are indeed of large-aspect ratio (typically $r/L \sim 1/10$), and in fact our results show that the central part of the loop is well described by a 1-D model even for rather fat loops (e.g. $r_b/L \sim 1/3$). It is thus appropriate to develop a simple model taking account of the essentially 1-D nature of most of the loop [Browning and Hood, 1989].

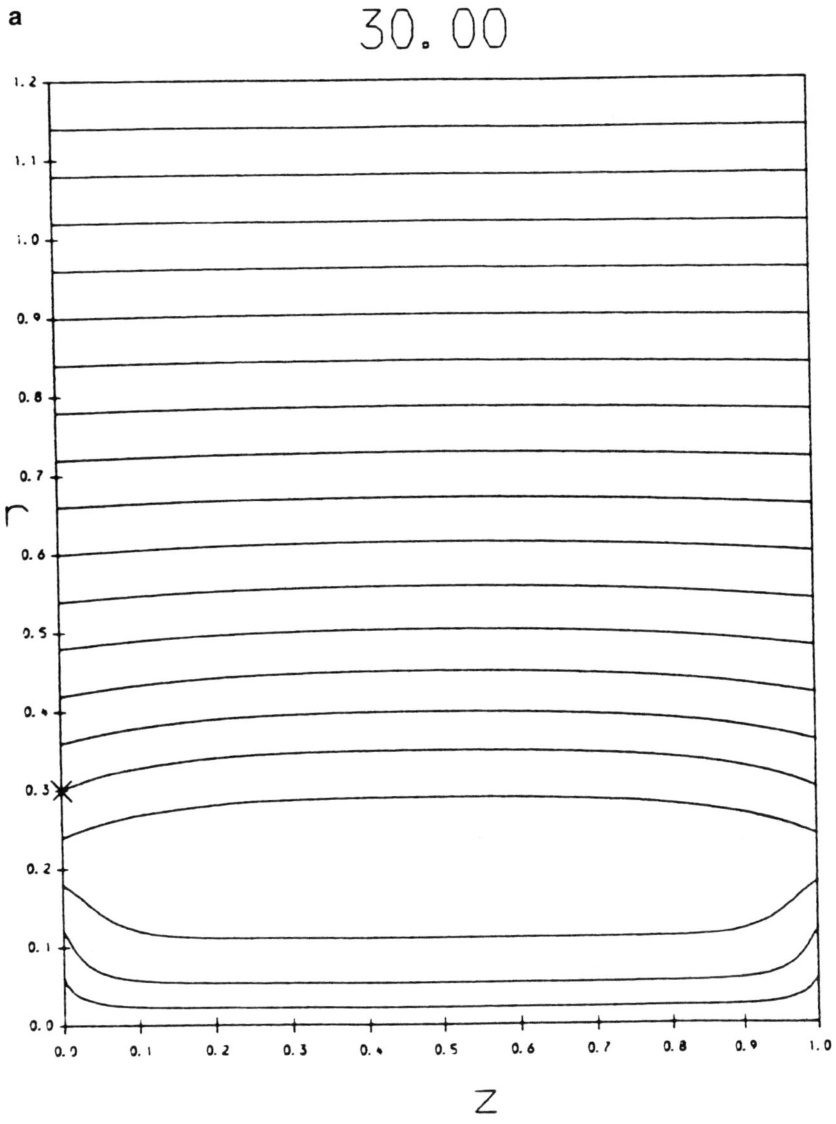

Fig.2
(a) The fieldlines projected in a plane θ = constant for the current profile (3) with μ on axis $\mu_a = 30.0$: the inverse aspect ration r_b/L is 0.3. Lengths are non-dimensionalised with respect to the loop length L.
(b) The angular footpoint displacement as a function of the photospheric radius r for this field profile.

Ignoring differences in the fieldline rotation in the boundary layers, it is possible to specify directly the footpoint angular displacement using this 1-D prescription. The flux function is matched to its known form at the loops ends (accounting for line-tying), giving an ordinary differential equation for ψ in the central part of the loop

$$(1 + r^2\Phi^2/L^2)\frac{d^2\psi}{dr^2} + \left(-1/r + \frac{d}{dr}(r^2\Phi^2/2L^2)\right)\frac{d\psi}{dr} = 0 \qquad (4)$$

in terms of a given twist profile $\Phi(\psi)$. The solution of (4) may then be used to determine the right-hand side $-GdG/d\psi$ of the Grad-Shafranov equation, which may be solved using the numerical code to give a full 2-D field whose footpoint displacement matches very closely the prescribed profile $\Phi(r)$. Typical results of this procedure are shown in Figure 3, for the profile

$$\Phi(r) = \begin{cases} \mu_a L(1-r^2/r_b^2), & r \le r_b; \\ 0, & r > r_b. \end{cases}$$

Notice that the twist profile determined from final 2-D calculation, using $G(\psi)$ from the 1-D

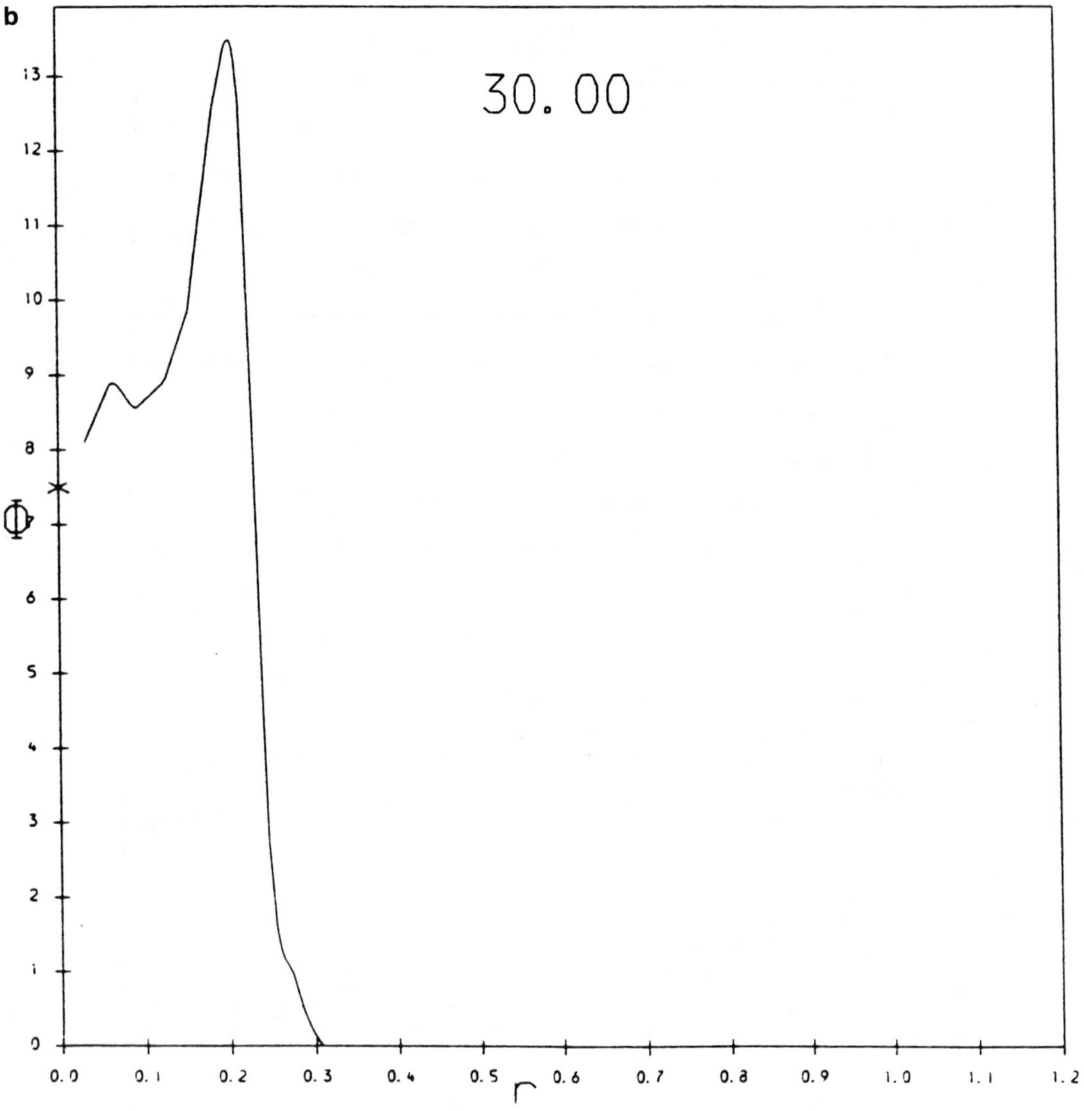

model, agrees very well with the predicted form, except close to the axis. Inaccuracies in Φ(r) arise for small r because these fieldlines compress inwards strongly, so that the value of r in the main straight portion of the loop is very small indeed.

This simple 1-D model may be used to study the effects of various twist profiles on the flux rope's shape and structure. Work is in progress investigating the possible onset of non-equilibrium when the twist or pressure is increased beyond some threshold.

Coronal Heating in Twisted Loops

A major unanswered question in solar physics is how the corona is heated to temperatures of millions of Kelvin. It is widely accepted that a magnetic heating mechanism must balance the radiative and conductive energy losses, energy being supplied by the turbulent motions of the photospheric footpoints of the coronal field [Priest, 1982; Pneuman and Orall, 1986; Parker, 1987]. One class of mechanisms considers slow motions, so that the coronal field evolves approximately through a sequence of equilibria: excess energy is stored by field-aligned currents, generated if the footpoints are twisted or sheared, which may dissipate and thus heat the corona. Dissipation on reasonable time-scales is difficult due to the very low resistivity and viscosity, but a plausible candidate for rapid dissipation is magnetic reconnection in thin current sheets.

We discuss here a theory which quantifies the release of stored energy due to small-scale dissipation in terms of the driving footpoint motions. This is an extension of Taylor's hypothesis [Taylor, 1974], which states that in a

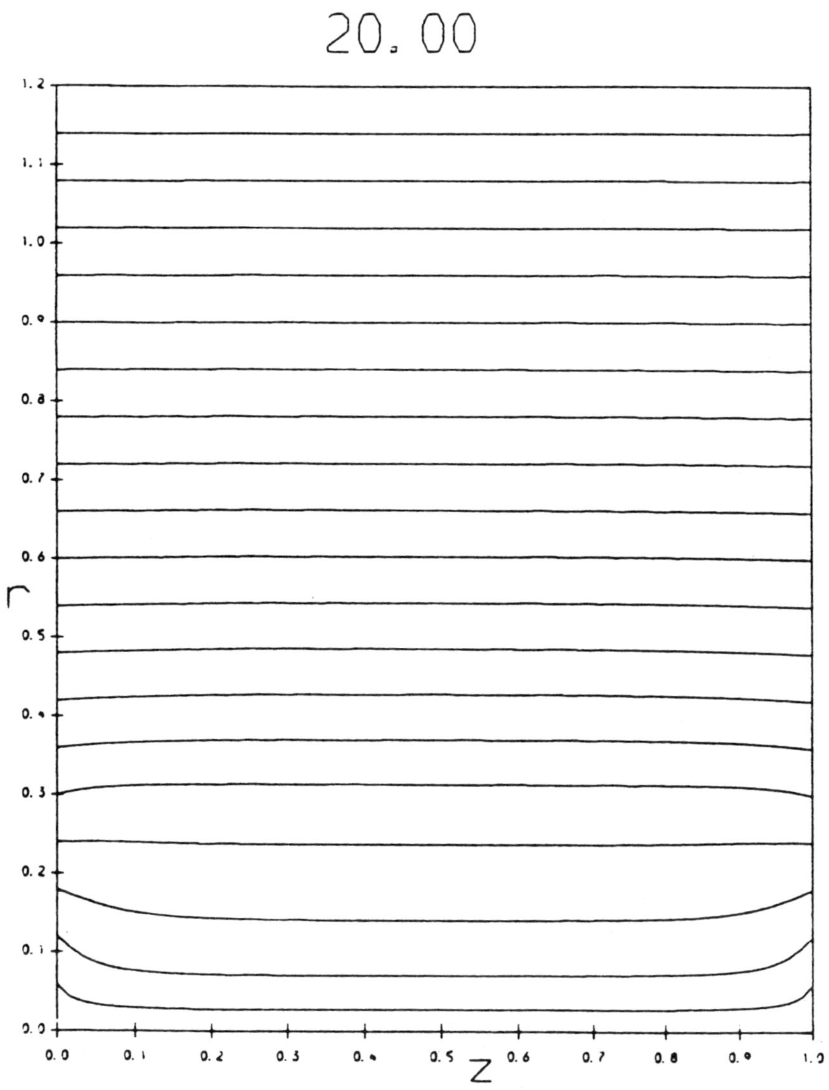

Fig.3 (a) The projected fieldlines for the 2D field, with the azimuthal field profile calculated from the 1D solution of (4) with $\Phi = \alpha(1 - r/r_b)$, $r \leq r_b$
(b) The angular footpoint displacement as a function of photospheric radius r_b for the 1D field as imposed (shown dotted) and for the 2D field calculated from this (shown solid).

slightly dissipative magnetised plasma, the field will tend to relax to a state of minimum magnetic energy, subject to the constraint that the global magnetic helicity

$$K = \int_V \underline{A}.\underline{B} dV, \qquad \underline{\nabla} \times \underline{A} = \underline{B},$$

is conserved. The minimum energy state turns out to be a constant-μ or linear force-free field

$$\underline{\nabla} \times \underline{B} = \mu \underline{B} \quad (\mu \text{ constant}).$$

The relaxation process involves small-scale turbulence and magnetic reconnection which anomalously dissipate energy but not helicity. As a loop is twisted, the field profile is initially determined by the footpoint connections as described in section 2, with μ (where $\mu = dG/d/)$ in general varying across the field. Dissipation processes within the loop then make the field evolve towards a minimum energy state resulting in a uniform μ profile across the loop. We assume here that the relaxation process takes place only within the loop, so that a sheet reversed current remains between the loop field and the ambient untwisted field (in practice this

b

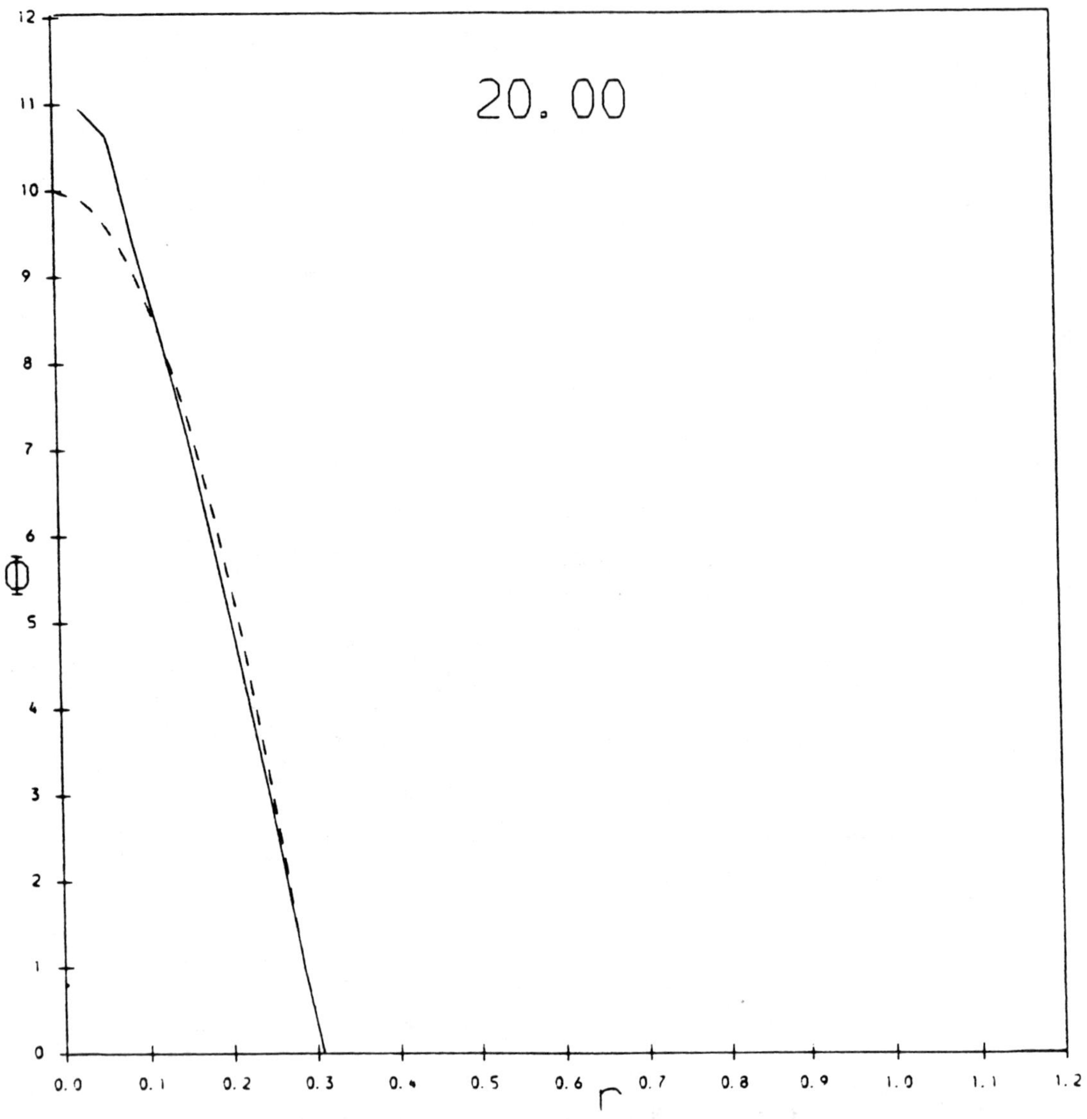

discontinuity would clearly be spread over a finite sheath).

In order to apply Taylor's theory (first developed for laboratory Reversed Field Pinch experiments) to the corona, allowance must be made for the fieldlines which are rooted in the photosphere, and the supply of energy by footpoint motions [Heyvaerts and Priest, 1984]. These motions inject both energy and helicity into the corona, with only the free energy above a constant-μ state available for heating. In the limit of very fast relaxation ($t_r \ll t_v$, where t_r and t_v are the time-scales of relaxation/reconnection and of the velocity field, respectively), the field evolves quasi-statically through a sequence of relaxed states

$$\underline{\nabla} \times \underline{B} = \mu(t)\underline{B}$$

The time dependence is given by the helicity injection equation

$$\frac{dK}{dt} = \int_s (\underline{A}.\underline{v} + g)\underline{B}.\underline{ds}; \quad \frac{\partial \underline{A}}{\partial t} = \underline{v} \times \underline{B} + \underline{\nabla} g \quad \text{on } s. \tag{5}$$

In this case there is no heating as no free energy is built up [Browning and Priest, 1986]: however, if the relaxation takes place more slowly, the field is driven into a non-linear force-free state, and then relaxes, releasing some magnetic energy as heat, to a minimum energy state. This process of energy input by motions, with a sequence of small relaxation events releasing energy, may then heat the corona.

This model may be considered in detail for the case of a coronal loop subject to twisting. Consider an axisymmetric flux rope with twisting motions inside a radius r_b, surrounded by a constant ambient external field: the boundary between the loop and the external medium is free. This is the situation modelled in Section 2, except that there is a current sheet at the loop surface. Following Section 2, it may be noted that if the loop length is much longer than the radius, then all z-dependence is confined in narrow layers near the photosphere, where the fieldlines join their line-tied positions. It is thus valid to treat the flux tube as a straight cylinder, with field components B_z and B_θ and radius R, and we therefore calculate the quasi-static evolution of a twisted cylindrical flux rope with a free surface, taking account of relaxation (see Browning [1988] for further details).

In the limit of immediate relaxation, the loop evolves through a sequence of constant-μ states

$$B_z = B_0 J_0(\mu r), \quad B_\theta = B_0 J_1(\mu r) \tag{6}$$

This evolution is subject to the constraints of conservation of flux

$$\Phi \equiv B_0 R J_1(\mu R)/\mu = \text{constant} \tag{7}$$

and constant pressure at the curved free surface

$$2\mu_0 p \equiv B_0^2(J_0^2(\mu R) + J_1^2(\mu R)) = \text{constant} \tag{8}$$

The time-dependence is determined by the footpoint motions through the helicity injection equation (5). Substituting the field (6) into (5), making use of the constraints, leads to

$$\frac{dK}{dt} = \frac{8\pi B_0^2}{\mu}\int_0^R rJ_0(\mu r)J_1(\mu r)v_\theta(r)dr \tag{9}$$

where

$$K = \frac{4\pi B_0^2 L}{\mu}\left[R^2(J_0^2(\mu R) + J_1^2(\mu R)) - \frac{2R}{\mu}J_0(\mu R)J_1(\mu R)\right]$$

[Browning, 1988]. This may also be transformed into an ODE for $\mu(t)$

$$\frac{d\mu}{dt} = \frac{1}{f}\int_0^R rJ_0(\mu r)J_1(\mu r)v_\theta dr,$$

where $f(\mu,R,B_0)$ is a rather complicated function, given by Browning [1988]. This may be integrated in time to determine the flux tube evolution.

Some interesting features arise from considering the constraints (7) and (8) which predict the loop radius. The graph of $R(\mu)$ is shown in Figure 4a. As μ is increased from zero, twisting up an initially potential tube, at first R increases, but only by a very small amount as expected from Section 2. As μ increases further, the tube expands more substantially, until a critical point $\mu*$ is reached at which the curve $R(\mu)$ bends back on itself. There are no neighbouring solutions for $\mu>\mu*$, although there are separate higher branches of the solution curve: this is in some sense a non-equilibrium point. However,, the curve of $K(\mu)$ has a similar shape (Figure 4b), and in fact for each K there exists a unique value of μ on the first solution branch (there are higher values of μ for given K, on the higher solution branches, but these are not minimum energy states). Thus if K is increased by twisting the footpoints, as described by equation (9), the field evolves continuously through a sequence of constant-μ states: when the value $\mu*$ is reached, the tube starts to expand strongly, and the local current ($\mu = \mu_0 j/B$) actually falls as further twisting takes place. It is interesting to note that the form of the curves $R(\mu)$ is independent of the fact that here μ is constant (since the tube is always in a minimum energy state). Thus we expect qualitatively similar behaviour if a loop is twisted up by general footpoint motions in the ideal MHD limit: if the azimuthal field $G(\psi)$ is increased beyond a critical magnitude, no equilibrium is possible. Preliminary numerical results, both 2D and 1D, confirm this prediction. However, note that while non-equilibrium may be reached by increasing $G(\psi)$, it is likely that solutions always exist for the more physically realistic problem where the footpoint displacement Φ is imposed.

In this immediate relaxation limit, the heating rate vanishes. In order to calculate a non-zero dissipation rate, finite relaxation time t_r must be assumed [Heyvaerts and Priest, 1984]. First a relaxed state is subject to an ideal displacement, producing a non-linear force-free field. This subsequently relaxes, conserving helicity but releasing some free energy δW_h as heat, to a new minimum energy state. This process continually repeats as the field evolves (Figure 5). By calculating this in detail using a second order perturbation expansion in the small quantity t_r/t_v about a relaxed state, it is possible to quantify the dissipation. Browning [1988] performs this calculation for fields which are close to potential, and it is shown that for a quadratic velocity profile $v_\theta = 4v_0 r(R-r)/R$ the heating rate is

$$F = \frac{8}{75}\frac{B_0^2}{\mu_0}v_0\frac{t_r}{t_v}\frac{R}{L}.$$

It may be seen that the heating rate increases with t_r/t_v: although our calculation is strictly valid only when this parameter is small, we expect the heating to be optimal when $t_r = t_v$ - in other words, when the dissipation time-scale matches the driving time-scale. Indeed, the reconnection rate may well adjust itself to the driving in order to approximately ensure this equality. Taking typical values for the other parameters (B_0 = 100G, v_0 = 1km/s, R/L = 1/10) we find the heating rate to be 1700 W/m². This is more than adequate for the quiet corona but too

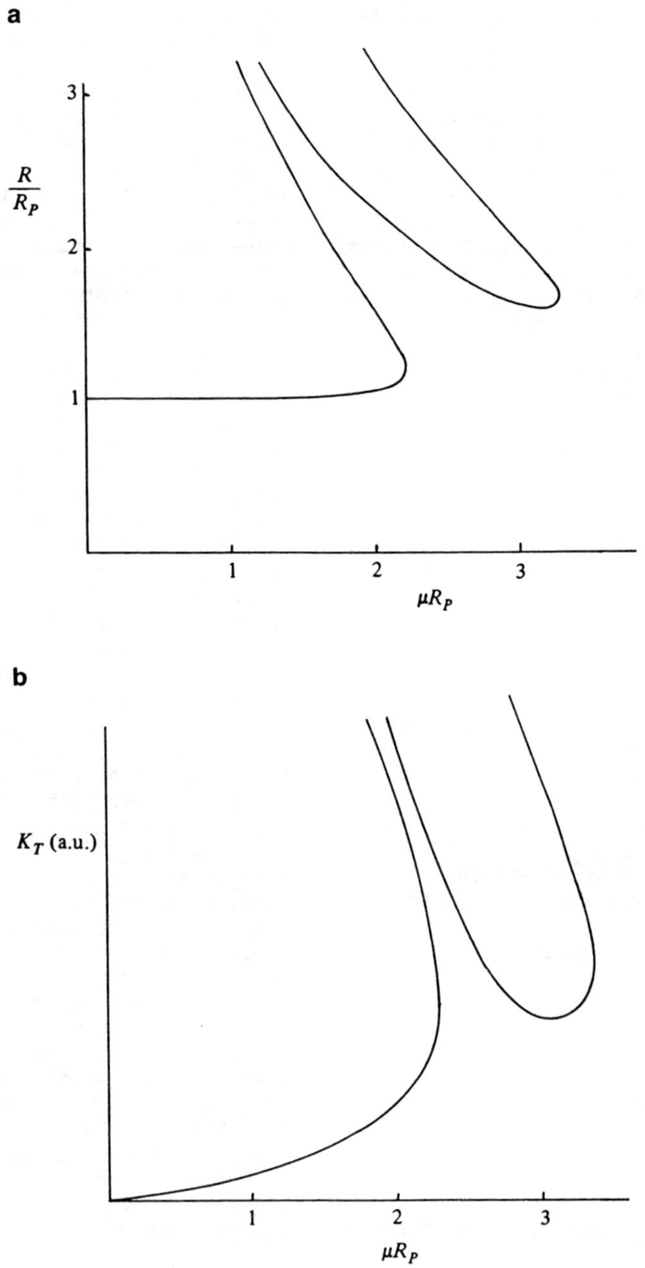

Fig.4 (a) The radius (R) as a function of the ratio of field to current (μ) for a twisted coronal loop with a free surface in a constant ambient medium, evolving through a sequence of relaxed states. Both quantities are non-dimensionalised with respect to the untwisted tube radius (R_p).
(b) The helicity (K), in arbitrary units, as a function of μ.

low to heat an active region. However, Browning and Priest [1986] have shown that the heating rates in fields which are strongly sheared are far greater than in fields that are close to potential. So it is likely that if we considered a loop with significant twist rather than a near potential loop, we would find much more effective heating. Active regions, which require stronger heating, do indeed consist of more complex fields.

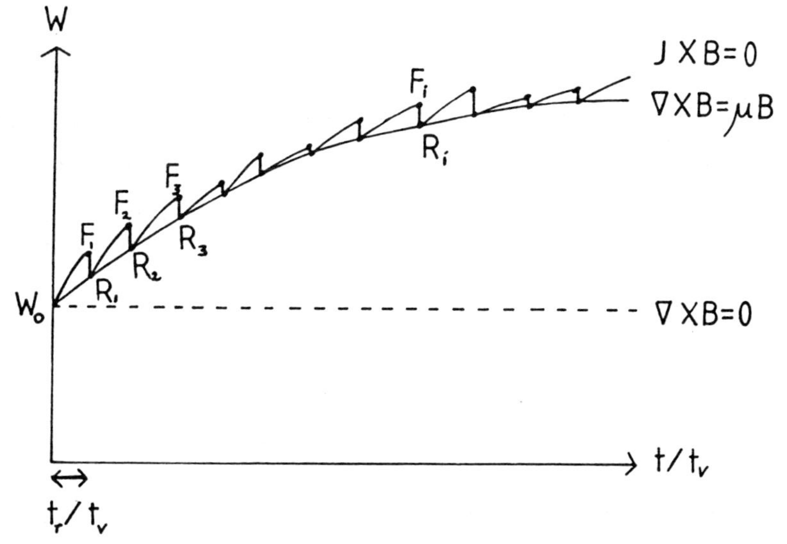

Fig.5 The evolution of a field subject to footpoint motions on a time-scale t_v. The field, in a relaxed state R_i, is moved by an ideal displacement into a non-linear force-free configuration F_{i+1}; this then relaxes, releasing some of the energy input but conserving helicity, into a new relaxed state R_{i+1}. This process continues, each relaxation event lasting a time of the order t_r.

Summary

Some properties of force-free twisted coronal loops have been discussed. Section 2 investigated the shape of a magnetic loop in force-free equilibrium. A coronal loop was modelled as a bundle of twisted field lines within a global potential field, whose footpoints are fixed at the photosphere. The effects of slowly twisting up an initially straight uniform field were studied, using a 2D numerical code to solve the Grad-Shafranov equation (with the azimuthal field profile specified).

The line-typing causes the coronal radius and field structure to differ from those at the footprints. The photospheric normal field is constant, but in the central part of the loop considerable radial inhomogeneity develops as the twisting proceeds. A core of strong longitudinal field and current is formed: for example, with the peak angular footpoint displacement of just over 2π radians, the value of B_z at the loop axis is about three times its untwisted value. This enhancement of field inhomogeneities and currents may increase the likelihood of reconnection and hence improve the efficiency of current dissipation and heating. Outside this core is a sheath of strong azimuthal but weak longitudinal field. The loop surface may either expand or contract. If the field is twisted outside the loop, then all fieldlines contact inwards, but if, more realistically, fieldlines outside the loop are untwisted (so that the net loop current is zero), then the surface expands outwards with twisting. This expansion is very slight unless the field is quite strongly twisted.

These results show that if the loop is of fairly large aspect ratio the structure is longitudinally invariant except in narrow photospheric boundary layers. This permitted a simplified 1D description to be developed, with matching to the line-tied footpoints. Using this idea, it was possible to calculate easily the field structure in the central part of the loop for given footpoint angular displacements. The results of this 1D calculation could then be used to determine the appropriate form of the azimuthal field profile in order to solve the 2D Grad-Shafranov equation. The effects of a variety of different twist profiles and also pressure gradients may then be investigated. Work is in progress using this method to determine the possible onset of non-equilibrium.

Twisted flux tubes store magnetic energy, which may be released as thermal energy in order to heat the corona. In Section 3, a model was described which assumed that this energy was dissipated by small-scale turbulence and reconnection, and that the amount of free energy available for heating is determined by Taylor's hypothesis, which predicts that the accessible minimum energy state satisfies $\nabla \times \mathbf{B} = \mu\mathbf{B}$. The evolution of a coronal loop, subject to slow twisting motions, through a sequence of relaxed states was calculated: in accordance with the results of Section2, a 1D description was used, which is valid outside narrow boundary layers. In the limit of immediate relaxation, the loop evolution is determined through a helicity injection equation, and a formulation was given for the time dependence of the field in terms of arbitrary footpoint motions. For small twists, the loop gradually expands and μ increases, but if the field is twisted beyond a critical point the tube begins to expand significantly while μ actually falls.

The heating rate can be quantified by considering a small departure from a relaxed state and calculating, correct to second order, the energy changes. It was shown that heating rates adequate for the coronal energy balance can be obtained.

Twisted loops of magnetic flux are the major building block of the coronal magnetic field, and many properties of the corona can be explained in terms of loops. We have touched here on some aspects related to their magnetic equilibrium, but the behaviour of twisted flux ropes in the solar corona is a vast subject. Some other topics are dealt with elsewhere in this volume - many more phenomena remain to be understood!

References

Browning, P.K., Helicity injection and relaxation in a solar-coronal magnetic loop with a free surface. J.Plas.Phys., 40, 263, 1988.

Browning, P.K. and Hood, A.W. The shape of twisted line-tied coronal loops. Submitted to Sol. Phys. 1989.

Browning, P.K. and Priest, E.R. Heating of coronal arcades by magnetic tearing turbulence, using the Taylor-Heyvaerts hypothesis. Aston. Astrophys., 159, 129, 1986.

Heyvaerts, J. in Unstable Current Systems and Instabilities in Plasmas (ed. M.R.Kundu and G.D. Holman), Reidel, p95, 1985.

Heyvaerts, J. and Priest, E.R. Coronal Heating by Reconnection in DC current systems: a theory based on Taylor's hypothesis. Astron. Astrophys., 137, 63, 1984.

Lothian, R. and Hood, A.W. Twisted magnetic flux tubes: the effects of small twist. To appear in Sol. Phys. 1989

Pneuman, G.W. and Orrall, F.Q. in Structure, dynamics and heating of the solar atmosphere in Physics of the Sun, Vol.2, (ed. P.A.Sturrock, T.E. Holzer, D.M.Mihalas and R.K. Ulrich), Reidel, 1986.

Parker, E.N. Astrophys.J., 74,499, 1972.

Parker, E.N. Cosmical Magnetic Fields, Oxford University Press, 1979.

Parker, E.N. Magnetic neutral sheets in evolving fields. Astrophys.J., 264, 1983.

Parker, E.N. Why do stars emit x-rays? Physics Today, 40, 3, 1987.

Priest, E.R. Chapter 6 in Solar Magnetohydrodynamics, Reidel, 1982.

Taylor, J.B. Relaxation of toroidal plasma and the generation of reverse field. Phys.Rev.Lett., 33, 1139, 1974.

Zweibel, E.G. and Boozer, A.H. Evolution of twisted fields, Astrophys.J., 295,642, 1985

THE OBSERVATION OF POSSIBLE RECONNECTION EVENTS IN THE BOUNDARY CHANGES OF SOLAR CORONAL HOLES

S. W. Kahler

Physics Research Division, Emmanuel College, 400 The Fenway, Boston, MA 02115

J. D. Moses

American Science and Engineering, Inc., Ft. Washington, Cambridge, MA 02139

Abstract. Coronal holes are large scale regions of magnetically open fields which are easily observed in solar soft X-ray images. The boundaries of coronal holes are separatrices between large-scale regions of open and closed magnetic fields where one might expect to observe evidence of solar magnetic reconnection. Previous studies by Nolte and colleagues using Skylab X-ray images established that large scale ($\geq 9 \times 10^4$ km) changes in coronal hole boundaries were due to coronal processes, i.e., magnetic reconnection, rather than to photospheric motions. Those studies were limited to time scales of about one day, and no conclusion could be drawn about the size and time scales of the reconnection process at hole boundaries.

We have used sequences of appropriate Skylab X-ray images with a time resolution of about 90 min during times of the central meridian passages of the coronal hole labelled "Coronal Hole 1" to search for hole boundary changes which can yield the spatial and temporal scales of coronal magnetic reconnection. We find that 29 of 32 observed boundary changes could be associated with bright points. The appearance of the bright point may be the signature of reconnection between small-scale and large-scale magnetic fields. The observed boundary changes contributed to the quasi-rigid rotation of Coronal Hole 1.

Introduction

Coronal holes are regions of unusually low density and temperature in the solar corona. They are present at all phases of the solar cycle, but reach their maximum extent in the two or three years before solar minimum. Over a decade ago Krieger (1977) in his review of the temporal behavior of coronal holes posed several fundamental questions about the evolution of holes that have yet to be completely answered. In particular, he asked: 1) What is the relationship between the stochastic diffusion of photospheric magnetic flux and the large-scale boundary changes? 2) What is the characteristic time scale for coronal hole boundary changes? 3) What is the role of emerging flux? 4) Are the large-scale boundary shifts cases of field line reconnection or of the evacuation of previously opened field lines?

Geophysical Monograph 58

An examination of the boundary changes of coronal holes was carried out by Nolte and colleagues (Nolte et al., 1978 a,b,c) using Skylab X-ray images from the period of May to November 1973. For each central meridian passage (CMP) of the Skylab coronal holes they compared the boundaries observed in three X-ray images: an image at CMP, an image 1 day earlier, and an image 1 day later. This procedure allowed them to study boundary changes with a time resolution of 1 day. Because of a concern with the possibility that the boundaries could move as a result of the diffusive motion of the field lines, they considered two classes of changes. Small-scale changes ranged from ~ 1.2×10^4 km, the smallest changes they could measure, to 9×10^4 km. Large-scale changes were those exceeding 9×10^4 km, ~ 3 times the average supergranulation cell length. This criterion was used to preclude the possibility that large-scale changes could arise from the chance association of random motions. Nolte et al. (1978a) found statistically that about 38% of the boundary lengths showed a significant change over 1 day. The small-scale changes accounted for 70% of this total, and the large-scale changes for the remaining 30%.

In their second paper Nolte et al. (1978b) inferred that the large-scale changes (which they referred to as "sudden") must involve a process different from that of at least some of the small-scale changes because the large-scale changes were found to account for most of the long-term (rotation-to-rotation) changes in coronal hole areas whereas the small-scale changes seemed poorly correlated with the long-term changes.

In the third paper Nolte et al. (1978c) studied the specific coronal structures which seemed to play roles in the growth and decay of coronal holes. They found a general agreement with the hypothesis that holes are born and grow in conjunction with active regions. They also found evidence that holes decayed when the number of X-ray bright points in the longitude bands containing the holes was relatively high. X-ray bright points are pointlike X-ray emitting features associated with small bipolar magnetic features (Golub et al., 1974).

We might expect that the detailed studies of Nolte et al. (1978 a,b,c) would have explored Krieger's (1977) questions to the limit of the X-ray observations. However, those studies were based only on comparisons of X-ray images obtained at 1-day intervals. Appropriate X-ray images were regularly obtained at roughly 6-hr intervals through most of the Skylab mission and in some cases, which we discuss here, the observations were made at least once per orbit (~ 90 min) for sequences of 3 to 7 consecutive orbits. We use these images to study coronal hole boundary changes on this substantially shorter time scale.

Analysis

The X-ray spectrographic telescope built by American Science and Engineering, Inc. flew on the Skylab spacecraft in 1973 and 1974. During the 8-month operational lifetime of the mission soft X-ray images of the sun were recorded on film with a spatial resolution of ~ 2 arc sec. Six different broad-band filters and a large dynamic range of exposure times were used to image various solar features and provide effective temperature diagnostics. The instrument has been described in detail by Vaiana et al. (1977), and an atlas of daily full-sun images of the X-ray corona was published by Zombeck et al. (1978).

The optimum images for studying the faint features of coronal holes are those obtained with the largest X-ray fluence. These are the 256 s exposures taken through the thinnest filter (filter 3) with passbands of 2-32 and 44-54 A. Usable images in this mode were obtained from 1973 May 28 to November 21 (Nolte et al., 1976). We examined a catalog of all such sun-centered images obtained during 5-day periods centered on the CMPs of low-latitude coronal holes determined by Nolte et al. (1976) to look for images in three or more consecutive orbits. We restricted the images to those with coronal holes near CMP and limited the regions of interest to latitudes of $\pm$ 40° to minimize the projection effects of optically thin structures at hole boundaries (Nolte et al., 1976). Since we wanted to study hole boundary changes, we sought large area holes with extensive boundaries. For that reason we eliminated the sequences of images of coronal holes 2 (on May 29 and August 18) and 3 (on August 12 and 13) because of their small areas (Nolte et al., 1976) and concentrated on Coronal Hole 1 (hereafter CH 1, following the designation used by Timothy et al. (1975) for the first coronal hole observed during the Skylab mission). The only images satisfying our requirements were obtained on 3 consecutive orbits on June 2, 7 orbits on August 19, 4 orbits on August 20, and 4 orbits on August 21.

Full-disk X-ray images of CH 1 at each CMP have been published by several authors (i.e., Figure 9 of Timothy et al. (1975); Figure 1 of Nolte et al. (1978c); and Figure 1 of Maxson and Vaiana (1977)) and will not be repeated here. It was the largest of the Skylab coronal holes, extending from the north pole to about S 20° with a width of order 15° at the equator. The extensive boundaries of the hole allow us a good opportunity to study the details of the boundary changes.

The changes of the hole boundaries were examined by a visual comparison of second-generation transparencies with a disk diameter of 10.8 cm. Since the positions of the boundaries and the changes in those positions over several consecutive orbits involves a subjective determination, we first listed all suspected boundary changes in all sets of images and then repeated the effort to get only the clearest examples. We eliminated cases where the area changes were so small as to be questionable or where the brightness change of a boundary feature was not sufficient to cause one to redraw the boundary. Although they claimed that coronal hole boundaries are sharp, Maxson and Vaiana (1977) presented cross sections of photographic density through the filter 3, 256 s images that clearly display the low spatial gradients of brightness at the boundary that render the boundary determination uncertain by perhaps 10-30 arc sec. Our boundary changes, characterized as one-dimensional features, ranged from ~ 10 arc sec (~ 7 x 10^3 km) to ~ 1 arc min (4.3 x 10^4 km). Our lower limit is slightly less than that (~ 1.2 x 10^4 km) of Nolte et al. (1978a) who examined hole boundary shifts on a time scale of 1 day.

In the examination of the boundary changes it was immediately apparent that bright points played an important role. This can be seen in Figure 1, which shows the sequences of filter 3, 256 s images of CH 1 during the times of 7 consecutive orbits on August 19. In the figure black arrows point to the bright points associated with coronal hole expansions and white arrows point to the bright points associated with coronal hole shrinkages. One case of a hole shrinkage with no bright point association is shown with the lower white arrow at 0651 UT in Figure 1. In the images of August 20 and 21 there were two cases of coronal hole expansions without any observed associated bright point. In the images of all four dates we found 32 boundary changes of which 29 could clearly be associated with bright points.

The most common kind of boundary change is simply the appearance of a new bright point or the disappearance of a pre-existing bright point at the coronal hole boundary in such a way as to cause an apparent shift in the boundary by about the dimension of the bright point itself. Most of the boundary changes of Figure 1 are of this type. In some cases an X-ray region somewhat more extensive than just the bright point itself will brighten or dim. Two examples in Figure 1 are shown by the lower white arrow at 0226 UT, in which a relatively large X-ray structure in the hole attaches itself to the boundary, and by the black arrow at the eastern boundary at 0415 UT, where a large bright region surrounding the bright point slowly fades after the transient appearance of the bright point.

A summary of the time and size scales of the three kinds of boundary changes is given in Table 1, where we have averaged the measured sizes and the number of orbits over which the brightening or dimming of the X-ray structure was observed. Since the time resolution is ~ 90 min, the actual time scales could be significantly less than the observed values, and we have given them as upper limits. To compare these boundary changes with those expected from supergranulation motions, we can use the time and size scales to calculate a characteristic speed for the boundary changes of $\geq$ 6 x 10^3 $km{\cdot}hr^{-1}$ for all categories of changes. Assuming a supergranulation cell size of 3.2 x 10^4 km and cell lifetime of 20 hr, we see that the speeds of the boundary changes exceed the supergranulation speed of 1.6 x 10^3 $km{\cdot}hr^{-1}$ by at least a factor of 4. These boundary changes are therefore not due to supergranulation motion.

The dimensions of the 32 observed boundary changes ranged from 7 x 10^3 km to 4.5 x 10^4 km with an average of 1.7 x 10^4 km; only one event exceeded 3.2 x 10^4 km. We therefore find no evidence for large scale boundary shifts of a size exceeding three times the supergranulation cell size (~ 9 x 10^4 km) discussed by Nolte et al. (1978a).

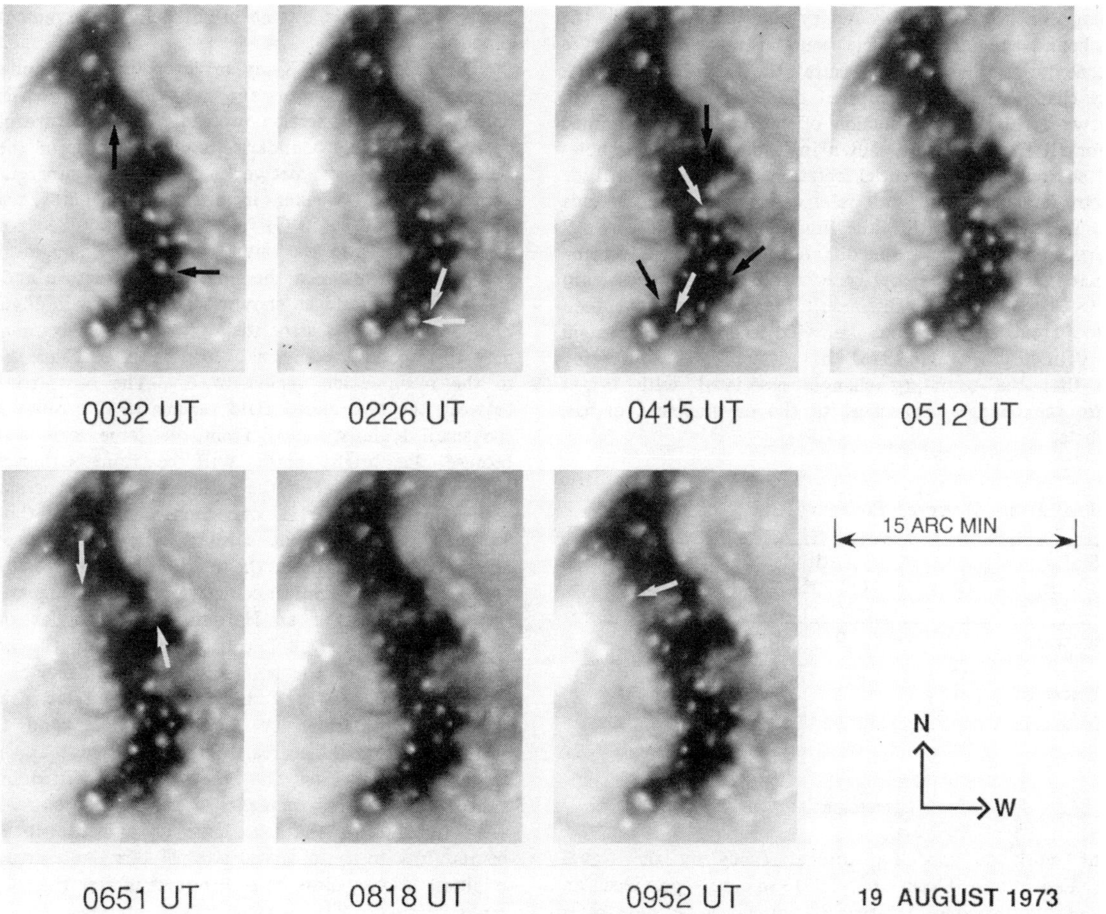

Fig. 1. Skylab X-ray images of CH 1 during seven consecutive orbits on 1973 August 19. The five bright points which were associated with expansions of the hole area are shown by black arrows; the six bright points associated with hole shrinkage by white arrows. One case of a hole shrinkage with no obvious bright point association is shown by the lower white arrow at 0651 UT.

Most of the bright points associated with the boundary changes are much fainter than those used by Golub et al. (1974) for bright point statistics studies. Those authors used bright points visible on 4 s exposures, while we have used 256 s exposures. Comparing bright point counts in coronal holes on 4 s and 256 s images, Golub et al. found about 100 times more bright points visible on the longer exposures. They also found a correlation between the maximum areas and the lifetimes of bright points. The bright points we have observed are generally small in area ($< 20 \times 10^7$ km^2) and short lived (1-5 hrs), consistent with this correlation.

The tendency of CH 1 to rotate quasi-rigidly rather than to participate in the solar differential rotation was discussed by Timothy et al. (1975). To see whether the 32 boundary changes we found in the sequences of images contributed to that quasi-rigid rotation, we establish two categories of boundary changes. X-ray brightenings on the western boundary and dimmings on the eastern boundary of the hole result in an eastward shift of the hole boundary. Conversely, X-ray

TABLE 1. Time and Size Scales of Boundary Changes

Boundary Change	Dimming	Brightening
Bright Point Only	11 cases 1.3×10^4 km ≤ 2.3 hr	9 cases 1.4×10^4 km ≤ 1.8 hr
Bright Point and Extended Structure	6 cases 2.2×10^4 km ≤ 3.0 hr	3 cases 2.3×10^4 km ≤ 4.0 hr
Extended Structure Without Bright Point	2 cases 2.3×10^4 km ≤ 3.0 hr	1 case 2.0×10^4 km ≤ 3.0 hr

brightenings on the eastern boundary and dimmings on the western boundary shift the hole boundaries westward. We used Stoneyhurst disks to measure the latitude of each boundary change and then compared the eastward shifts with the westward shifts as a function of latitude. The summed results for all four dates are shown in Table 2. Coronal holes will be sheared by differential rotation as the low-latitude regions are shifted westward relative to the high-latitude regions. We see that within the limited statistics of Table 2 the observed shifts oppose the differential rotation by being predominately eastward at low (≤ 20°) latitudes and westward at high (> 20°) latitudes. This result may perhaps have been anticipated from our previous knowledge of the quasi-rigid rotation (Timothy et al., 1975), but it provides supporting evidence that the boundary changes associated with bright points are the changes important to the development of the coronal hole.

TABLE 2. Observed Boundary Shifts of CH 1

Shifts	Latitude			
	01°-10°	11°-20°	21°-30°	31°-40°
Eastward	4	9	2	1
Westward	3	4	6	3

Discussion

Recent work on modeling coronal fields by the Naval Research Lab group (Nash et al., 1988) has provided an explanation for the rigid rotation of coronal holes near solar minimum. Using a potential field model with differential rotation, diffusion and meridional flow, they found that the outer coronal field rotates more rigidly than the underlying photospheric field because it depends on only the lowest-order harmonic components. The motion of the hole boundary is uncoupled from that of the underlying photospheric flux elements by continual reconnection of magnetic field lines. The details of the reconnection process are not specified. One possibility is that this reconnection occurs in the high corona. The time scale of the boundary changes (~ 1-5 hrs) is consistent with this, but no bright point involvement would be expected.

The appearance of X-ray bright points in boundary changes suggests that we examine the weak photospheric fields for the source of the reconnection process. The structure of the magnetic fields at hole boundaries is characteristic of the quiet sun fields consisting of network clusters at supergranular cell vertices and of weaker intranetwork fields (Zwaan, 1987). The latter weak (< 50 G) fields consist of mixed polarities and do not extend into the outer corona. We suggest that reconnection occurs between the small-scale structure and the larger scale magnetic field as shown schematically in Figure 2. The X-ray bright points associated with the hole boundary changes may correspond to the small loop in A or C of the figure or to the reconnection region in B. The separatrix is drawn between the two closed field regions in C because it separated the small scale structure from the large scale structure and because the bright point will be faint either before the sequence C,B,A or after the sequence A,B,C. The size and time scales of the proposed reconnection scenario are those given in Table 1. A somewhat similar schematic was proposed by Marsh (1978) to explain the relationship between bright point flares and supergranulation network flux elements. He observed several cases of an H-alpha brightening at the network element followed by a fibril system linking the network element with one of the poles of the bipolar region.

Nolte et al. (1978c) found a statistical relationship between the bright point density in coronal holes and the rate of shrinkage of the hole area. They suggested that this was due to two reasons. First, the hole was being filled in by X-ray-emitting closed-field remnants of the bright points. A problem with this idea is that we have no evidence that the bright points grow to the observed sizes of large-scale structures. The brightest bright points have lifetimes of less than a day (Golub et al., 1974). The second reason proposed by Nolte et al. (1978c) was that the bright points enhanced the rate of reconnection of open field lines at the hole boundaries. However, if a bright point reconnects with an open field line, one end of the bright point bipole must also be open after the reconnection process. Thus the proposed reconnection scenario will not result in a net closing of large-scale open field lines. In contrast, in our Figure 2 we see that the bright point in C interacts with adjacent closed field line flux to produce a shrinking of the hole area in the C,B,A sequence by motions of previously closed field lines. A further observational

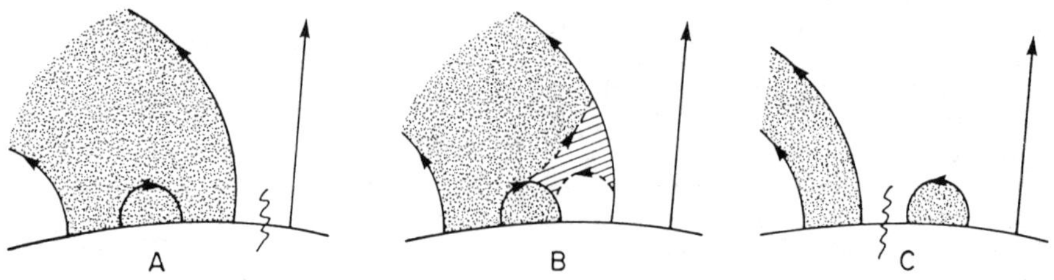

Fig. 2. Schematic for reconnection of magnetic fields at coronal hole boundaries. Dotted regions are closed fields; the wavy line is the separatrix between open and closed fields. Reconnection occurs in B in the shaded region. The sequence A,B,C corresponds to an expansion of the hole area; C,B,A corresponds to a shrinking of the hole area.

problem with the Nolte et al. idea is that a more detailed examination of the bright point densities in coronal holes by Davis (1985) showed no association between bright point density and the rates of hole growth or decay.

At the time of their discovery it was obvious that bright points were bipolar magnetic structures (Golub et al., 1974). They were interpreted as regions of emerging flux by Golub et al. (1974) and others. This view was challenged by Harvey (1985), who used He I 10830 A dark points as a proxy for X-ray bright points and found that about two-thirds of the dark points were associated with chance encounters of features of opposite magnetic polarity. In a recent study Webb and Moses (1988) compared bright points observed in rocket solar X-ray images with bipoles observed in simultaneous videomagnetograms. The great majority of bipoles were not associated with X-ray bright points, but 11 of 16 observed X-ray bright points were associated with cancelling bipoles and only one with an emerging bipole. Webb and Moses concluded that their results were consistent with the Harvey (1985) interpretation that most bright points are associated with encounters of opposite polarity features. Our observations suggest that the bright points form due to coronal heating at some time during the reconnection process. X-ray bright points are known to flare on a time scale of minutes (Golub et al., 1974), but it is not clear how the flare event or the formation and disappearance of the bright point are related to the reconnection scenario of Figure 2.

Questions and Answers

Pizzo: Are there a sufficient number of bright points (as you define them) scattered about the solar surface that any significant change in the boundary will by default occur near a bright point? (i.e., is there necessarily a physical connection or just random correlation?)

Kahler: There are many faint bright points in and around the coronal hole. However, relatively few are close to the boundary itself. This is clear from looking at the photographic prints.

Kane: I am trying to understand the scale of this phenomenon in terms of its effect on the overall coronal hole. The title of the paper seems to imply overall changes in the boundary of the coronal hole. Is this true? Or is it only a small change in a small part of the coronal hole, interesting by itself but not of any significant importance to the structure of a coronal hole?

Kahler: I think these changes are the means by which the entire coronal hole boundaries evolve. I stated at the beginning that we have looked only at one hole and in one epoch, but we have no other information at this point regarding the general details of how the boundaries change. The changes occurred at all latitudes and on the leading and trailing hole boundaries.

Nordlund: There is something trivial I don't understand here. Are you saying (1) that the corona hole rotates rigidly until bright points pass through the boundary and the boundary then "pops," or (2) that the CH rotates differentially until bright points come along and "restore" the boundary to the rigid rotation location?

Kahler: The hole would rotate differentially if no reconnection process modified the boundaries. I see the bright points as the signatures of the reconnection process that restores the quasirigid rotation. It is not clear whether the bright points involved in boundary changes have an "independent" existence unrelated to the hole boundary changes or are produced only because of the reconnection process driven by some other process.

Antiochos: Do you really need reconnection in order to understand how bright points cause coronal holes to expand? Could it not simply be that when bright point erupt near a hole boundary, the heating input into the corona increases which results in opening closed field lines.

Kahler: The size of loops that open are very large, so a substantial heating would probably be required to open the loop. The bright points involved here are very small, and it is not clear they do any heating outside the BPs themselves. A real problem for your idea is that the BPs are also associated with a shrinking of the coronal hole as well as the expansion.

Kundu: Approximately 10% of flaring bright points are associated with type III radio bursts, implying that BPs are associated with open field lines. This would be consistent with the open field lines that you produce after reconnection. But, then we'd expect type III's and by extrapolation the flaring BPs only in the third stage of your schematic diagram. Is that the case?

Kahler: The BPs discussed here are all fainter and shorter-lived than the flaring BPs discussed by Golub et al. It is not clear that these features ever undergo flaring, but even if they did, I would expect any particle production to be too faint to be observed as type III bursts.

Webb: To comment on Mukul Kundu's comment: I believe the bright points Steve examined were there before and after the boundary change. Therefore, they could not be flaring bright points because during Skylab flaring bright points always disappeared after flaring.

Acknowledgements. We thank D. Webb, E. Hildner, R. Moore, A. Nash, and N. Sheeley, Jr. for helpful comments. This research was supported at Emmanuel College by AFGL contract F19628-87-K-0033 and at AS&E by NASA contract NAS5-25496.

References

Davis, J.M., Small-scale flux emergence and the evolution of equatorial coronal holes, Solar Phys., 95, 73-82, 1985.

Golub, L., A.S. Krieger, J.K. Silk, A.F. Timothy, and G.S. Vaiana, Solar X-ray bright points, Ap.J., 189, L93-L97, 1974.

Harvey, K.L., The relationship between coronal bright points as seen in He I 10830 and the evolution of the photospheric network magnetic fields, Aust. J. Phys., 38, 875-883, 1985.

Krieger, A.S., Temporal behavior of coronal holes, in Coronal Holes and High Speed Wind Streams, edited by J.B. Zirker, Colorado Associated University Press, Boulder, 71-102, 1977.

Marsh, K.A., Ephemeral region flares and the diffusion of the network, Solar Phys., 59, 105-113, 1978.

Maxson, C.W., and G.S. Vaiana, Determination of plasma parameters from soft X-ray images for coronal holes (open magnetic field configurations) and coronal large-scale structures (extended closed-field configurations), Ap.J., 215, 919-941, 1977.

Nash, A.G., N.R. Sheeley, Jr., and Y.-M. Wang, Mechanisms for the rigid rotation of coronal holes, Solar Phys., 117, 359-389, 1988.

Nolte, J.T., A.S. Krieger, A.F. Timothy, G.S. Vaiana, and M.V. Zombeck, An atlas of coronal hole boundary positions May 28 to November 21, 1973, Solar Phys., 46, 291-301, 1976.

Nolte, J.T., A.S. Krieger, and C.V. Solodyna, Short term evolution of coronal hole boundaries, Solar Phys., 57, 129-139, 1978a.

Nolte, J.T., M. Gerassimenko, A.S. Krieger, and C.V. Solodyna, Coronal hole evolution by sudden large scale changes, Solar Phys., 56, 153-159, 1978b.

Nolte, J.T., J.M. Davis, M. Gerassimenko, A.S. Krieger, C.V. Solodyna, and L. Golub, The relationship between solar activity and coronal hole evolution, Solar Phys., 60, 143-153, 1978c.

Timothy, A.F., A.S. Krieger, and G.S. Vaiana, The structure and evolution of coronal holes, Solar Phys., 42, 135-156, 1975.

Vaiana, G.S., L. van Speybroeck, M.V. Zombeck, A.S. Krieger, J.K. Silk, and A. Timothy, The S-054 X-ray telescope experiment on Skylab, Space Sci. Instr., 3, 19-76, 1977.

Webb, D.F., and J.D. Moses, The correspondence between small-scale coronal structures and the evolving solar magnetic field, Advances Space Res., in press, 1988.

Zombeck, M.V., G.S. Vaiana, R. Haggerty, A.S. Krieger, J.K. Silk, and A. Timothy, An atlas of soft X-ray images of the solar corona from Skylab, Ap.J. Supple., 38, 69-85, 1978.

Zwaan, C., Elements and patterns in the solar magnetic field, Ann. Rev. Astron. Astrophys., 25, 83-111, 1987.

QUASI-STATIC EVOLUTION OF A THREE-DIMENSIONAL FORCE-FREE MAGNETIC FLUX TUBE OR ARCADE

J. J. Aly

Service d'Astrophysique—CEN Saclay—91191 Gif-sur-Yvette–France

Abstract. We consider in the perfectly conducting half-space $\{ z > o \}$ a simple topology line-tied force-free magnetic field $\tilde{B}$ and its quasi-static evolution driven by motions imposed to the feet of its lines on the boundary $\{z = o\}$. We discuss in particular : i) the possibility of a global Euler representation of $\tilde{B}$; ii) the equations satisfied by the field; iii) the existence and stability of solutions of these equations; iv) their asymptotic behaviour at large time. We also discuss for the particular case of an axisymmetric arcade the possibility of a transition by reconnection to a lower energy state when the topology of the lines is allowed to change (the assumption of an ideal embedded plasma being partly released).

Introduction

The problem of the quasi-static evolution of a force-free magnetic field embedded in a perfectly conducting plasma and having the feet of its lines on a boundary submitted to given motions, has long been recognized as a central problem of the physics of the solar corona. In the last few years, we have pursued an analytical study of this problem, concentrating most of our attention on the case of a 2D x-invariant arcade field in the half-space $\{ z > o \}$(see e.g. [Aly, 1987]). In this Communication, we would like to report some preliminary results we have obtained in an attempt to extend our 2D results to a fully 3D situation in which one has in $\{ z > o \}$ an evolving force-free field whose lines have a simple topological pattern - i.e. for which it is possible to define a "covering" set of nested magnetic surfaces, which are either "tube-like" or "arcade-like".

Representation of the Field

Let us consider in $\Omega = \{ z > o \}$ a magnetic field $\tilde{B}$ which is such that:

i) $|\tilde{B}| \neq 0$ in Ω ; ii) $\lim_{r\to\infty} |\tilde{B}|\ r^2 = 0$, say ; iii) on $\partial\Omega = \{z = o\}$, the parts $\partial\Omega^+$ and $\partial\Omega^-$ on which $B_z > 0$ and $B_z < 0$, respectively, are separated by a simple curve L (neutral line) which is either closed or open; in the former case, we assume, without loss of generality, that L encloses $\partial\Omega^+$ (Figure 1); iv) all the field lines of B cut twice $\partial\Omega$; there are no lines meeting $\partial\Omega$ tangentially on L.

With these assumptions, the lines of $\tilde{B}$ establish a one-to-one correspondance (magnetic mapping) between $\partial\Omega^+$ and $\partial\Omega^-$ and $\tilde{B}$ may be shown to admit the global Euler representation

$$\tilde{B} = \nabla u \times \nabla v, \tag{1}$$

where: i) the potential u determines a space covering set of nested magnetic surfaces Σ_u (on Σ_u, $u(r) =$ const.) which are tube-like (resp. arcade-like) when L is open (resp. closed); each Σ_u intersects $\partial\Omega^+$ along a closed curve C_u^+ enclosing a magnetic flux just equal to u; note that we may construct our representation of $\tilde{B}$ by choosing arbitrarily the set $\{ C_u^+ \}$; ii) v is a multivalued function which changes by one unit when crossing a particular surface S.

It is worth noticing that the relative helicity of $\tilde{B}$ [Berger and Field, 1984], which is an important topological quantity, may be expressed as (X_t denoting, quite generally, the component of $\tilde{X}$ parallel to $\partial\Omega$)

$$H = \int_{\partial\Omega^-} u\ u_o\ (\nabla_t\ v \times \nabla_t\ v_o).\hat{z}\ d\sigma \tag{2}$$

where we have set $\tilde{B}_o = \nabla u_o \times \nabla v_o$ ($\tilde{B}_o$ is defined by $\nabla \times \tilde{B}_o = 0$ in Ω and $B_{oz} = B_z$ on $\partial\Omega$) and choosen for that potential field, also assumed to satisfy our topological assumptions, the same set of curves $\{C_u^+\}$ as for $\tilde{B}$.

Equations Describing the Quasi-Static Evolution of a Force-Free Field

Let us now assume that Ω is filled up with a perfectly conducting plasma and that a potential

Geophysical Monograph 58

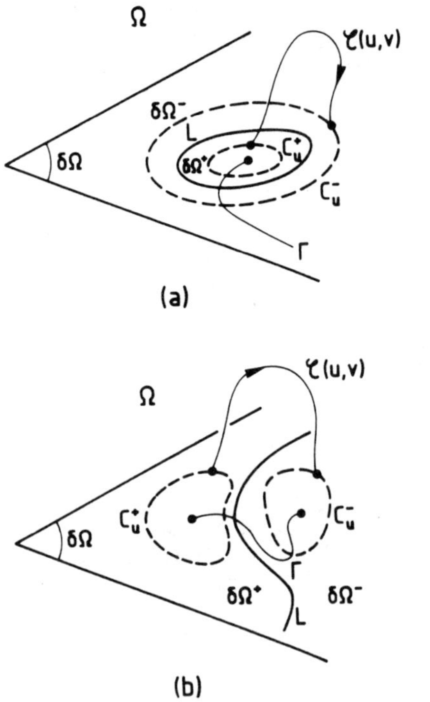

Fig. 1. Topology of the field lines.
(a) Arcade topology (neutral line L closed); (b) tube topology (L open).

field $\underset{\sim}{B}_o$ satisfying the conditions of § 2 is made for $t \geqslant 0$ evolving quasi-statically through a sequence of force-free configurations as a consequence of a smooth velocity field $\underset{\sim}{c}$ (with $c_z = 0$ and $|\underset{\sim}{c}|$ decreasing fast to zero at infinity) imposed on the boundary $\partial\Omega$. Then:

i) at each time t, the potentials u and v satisfy [Barnes and Sturrock, 1972]

$$\nabla.\{\nabla u x(\nabla u x \nabla v)\} = \nabla.\{\nabla v x(\nabla u x \nabla v)\} = 0 \qquad (3)$$

ii) the boundary conditions (u,v) (x,y,o,t) = (U,V)(x,y,t) for (3) are obtained by solving

$$\frac{dU}{dt} = \frac{\partial U}{\partial t} + \underset{\sim}{c}.\nabla_t U = \frac{dV}{dt} = \frac{\partial V}{\partial t} + \underset{\sim}{c}.\nabla_t V = 0 \qquad (4)$$

$$(U,V)(x,y,o) = (U_o, V_o)(x,y) \qquad (5)$$

where (U_o, V_o) are the boundary values of the potentials of the initial field;

iii) one has the "asymptotic" condition

$$\int_\Omega B^2 \, d\underset{\sim}{r} = \int_\Omega |\nabla u \times \nabla v|^2 \, d\underset{\sim}{r} = \int_o^{u_m} du \int_{C_u} v \, \underset{\sim}{B}_t \cdot d\underset{\sim}{s} < \infty \qquad (6)$$

where we have introduced for the energy of a force-free field a new formula in which C_u is a closed oriented curve constituted of C_u^+, C_u^- and the two sides of that part Γ_u of Γ (on which v = 0, 1, ...) joining them (u_m represents the total flux through $\partial\Omega^+$).

iv) the topology of the lines of $\underset{\sim}{B}$ is conserved during the evolution (frozen-in law).

Methods of Solution

Two methods have been used to try to prove that the shearing problem stated in § 3 has a solution.

Variational method

One tries to minimize the functional

$$C[u,v] = \int_\Omega |\nabla u \times \nabla v|^2 \, d\underset{\sim}{r} \qquad (7)$$

- which admits (3) as its Euler - Lagrange equations - over the set of functions (u,v) satisfying the constraints (ii) - (iv) above. A possible way to effect the minimization is as follows. One first fixes a set of magnetic surfaces Σ_u meeting the given curves C_u^- and one minimizes with respect to v. This first step amounts to solve the linear equation

$$\nabla_t \cdot \{ |\nabla u| \nabla_t v \} = 0 \qquad (8)$$

on each Σ_u, with v being given on $\partial\Sigma_u$ and a "twist number" $\tau(u)$ being imposed. One then gets a unique solution for v[u], which determines the optimum shape for the field lines on Σ_u. Reporting this v[u] into (7), one obtains a functional of u alone which is still under study.

Perturbation method

One starts from a known equilibrium (u,v) (e.g. a potential field). Thus one changes slightly the boundary conditions $[(U,V) \rightarrow (U + \varepsilon U_1, V+\varepsilon V_1)]$ and looks for a new solution of the form $(u+\varepsilon u_1, v+\varepsilon v_1)$. Then (u_1, v_1) must be solution of

$$N(\underset{\sim}{r})\,(u_1, v_1) = \varepsilon M(\underset{\sim}{r}, u_1, v_1, \varepsilon) \qquad (9)$$

$$(u_1, v_1)(x,y,o) = (U_1, V_1)(x,y) \qquad (10)$$

where N (resp. M) in a second order linear (resp. non-linear) operator. Unfortunately, N turns out to have some bad properties which preclude (9) - (10) to be solved by standard iterative schemes and it is necessary to appeal to uneasy techniques. The calculations have not yet been done successfully for the half-space problem, but they have been developed to prove the existence of solutions (in the neighbourhood of a uniform field $B_o = B_o\hat{z}$) for the well known Parker's problem, in which the configuration is confined between the two planes { z = o } and { z = h }.

Thus the existence of solutions for the 3D shearing problem in Ω is still an open problem, even at small shear. It seems likely that a solution B_t exists, in some sense, for $0 \leq t \leq \infty$. However, B_t may be not smooth, developing e.g. current sheets when sufficiently stressed (Parker, these Proceedings). Numerical computations, using empirically the two methods above, have seldom been attempted (see, however, [Sakurai, 1979]) and should be worth developing.

Stability of the Solutions

To try to understand better the variational principle stated in § 4, we have considered in some details the behaviour of the second variation of the energy $\delta^2 W$ near a given line-tied equilibrium configuration. We have yet been able to determine conditions implying $\delta^2 W > 0$, and thus sufficient criteria for linear ideal MHD stability. Let us set: $\mathcal{C}(u,v)$ = field line labelled by the values u and v of the potentials; $\ell(u,v)$ = length of $\mathcal{C}(u,v)$; $\alpha(u,v)$ = constant value of the parameter $\alpha(\nabla \times B = \alpha B)$ on $\mathcal{C}(u,v)$. Then a field is linearly stable if

$$2\alpha(u,v)\ell(u,v) < \left[\int_0^1 {}_{\mathcal{C}(u,v)} d\zeta \frac{|\nabla u|^2}{B} \int_0^1 {}_{\mathcal{C}(u,v)} d\zeta \frac{|\nabla v|^2}{B} \right]^{-1} \quad (\leq 1) \tag{11}$$

on any line $\mathcal{C}(u,v)$ ($d\zeta = ds/\ell(u,v)$), or if (global criterion) [Aly, 1989]

$$\left[\int_\Omega |\alpha|^3 d\underset{\sim}{r} \right]^{1/3} + \left[\int_\Omega |f|^{3/2} d\underset{\sim}{r} \right]^{1/3} < \frac{\sqrt{3}}{2} \pi^{2/3} \tag{12}$$

where

$$f = f(u,v) = \int_{\mathcal{C}(u,v)} |\nabla\alpha|^2 B^{-1} ds \times \int_{\mathcal{C}(u,v)} B\, ds \tag{13}$$

It is worth emphasizing the fact that our conditions (11) - (12) are sufficient for stability to hold, but not necessary: thus their violation does not imply instability. In the interesting case when B and α have similar length-scales d, they allow to conclude at the stability of the field when $|\alpha|_m d < C_{stab} = O(1)$ ($|\alpha|_m = \sup |\alpha|$). Comparing this condition with that one: $|\alpha|_m d \leq C_{exist} = O(1)$ which just results from B being force-free in Ω [Aly, 1984], one may reasonably conjecture that such fields B are always stable.

Asymptotic Behaviour

For definiteness, let us assume that the shearing velocity field $\underset{\sim}{c}$ vanishes on $\partial\Omega^-$, is stationnary, and admits current lines forming a nested set of curves on $\partial\Omega^+$. Then, if we use all these curves to constitute our set $\{C_u^+\}$ (choosing arbitrary curves in the annular parts of $\partial\Omega^+$ where $\underset{\sim}{c} = 0$, if any), we see at once that the field lines on some of the tube or arcade-like Σ_u will be indefinitely sheared, and the question naturally arises of the asymptotic states which may be reached by the field when $t \to \infty$. Actually, by following techniques quite similar to those used in the 2D case (e.g. [Aly,1987]), one may show that the 3D field (assuming a solution of the shearing problem to exist for all t) opens too, the electric currents concentrating into a sheet (generally moving). Two key ingredients necessary to reach that result are the following ones:

i) for B_z fixed on ∂Ω ($B_z|_{\partial\Omega} = g$),the energy of any finite-energy field which is force-free in Ω, is bounded from above by a number $C_o[g]$ [Aly, 1984]; this implies for our problem (in which g changes in time) the existence of a upper bound $C^m[g(t=0), \underset{\sim}{c}]$ for the energy ;

ii) one has the inequality ($v(u)$ = volume bounded by Σ_u and ∂Ω)

$$\int_0^u du' \int_0^1 dv' \; \ell(u',v') \leq [v(u)C^m]^{\frac{1}{2}} \tag{14}$$

which shows at once that $v(u) \to \infty$ (and the field opens) if $\ell(u',v') \to \infty$ on a non-negligible set of lines; but, when $t \to \infty$, one has indeed $\ell(u',v') \to \infty$ on any sheared surface $\Sigma_{u'}$ (this is not completely obvious, as one could imagine a priori that ℓ stays finite, $\Sigma_{u'}$ pinching towards the magnetic axis - i.e. the singular surface Σ_o; but this possibility is excluded, as it may be shown to lead - as a consequence of flux conservation inside Σ_u - to an indefinite increase of the energy).

A few years ago [Aly, 1984], we have conjectured that the best possible value for the number $C_o^m[g]$ introduced above is just equal to the energy $C_{op}[g]$ of the totally open field associated with g (for particular classes of g, we have actually shown explicitely $C_{ff}\{g\} \leq C_{op}[g]$; compare with [Barnes and Sturrock, 1972]): clearly, the result above on the opening of the field strongly supports our conjecture.

Reconnection

The previous result strongly suggests that, if we allow the plasma to have a small resistivity, the field must become unstable with respect to reconnection when the shear becomes large enough. Up to now, we have been able to gain some insight into this conjecture only in the particular situation where $\underset{\sim}{B}$ is axisymmetric around $\hat{z}$. In that case, we may reasonably admit that 2D reconnection, acting on a time scale much smaller than the diffusion time scale of the field, can change the topology of the lines, but not the "distribution of the magnetic fluxes": i.e. the range of values of the "flux function" u, as well as the amount of toroidal flux between any two surfaces Σ_{u1} and Σ_{u2}, are conserved during a reconnecting transition. Then reconnection is possible in an arcade configuration only if there exists a field related to it by these flux constraints, but having a more complex topology and a lower energy. We have shown that such fields start existing at some critical time t_c where the shear exceeds a critical value. For $t > t_c$, the arcade is metastable with respect to reconnection and is able to effect an explosive transition (flare) towards a lower energy state under the action of a finite perturbation (figure 2).

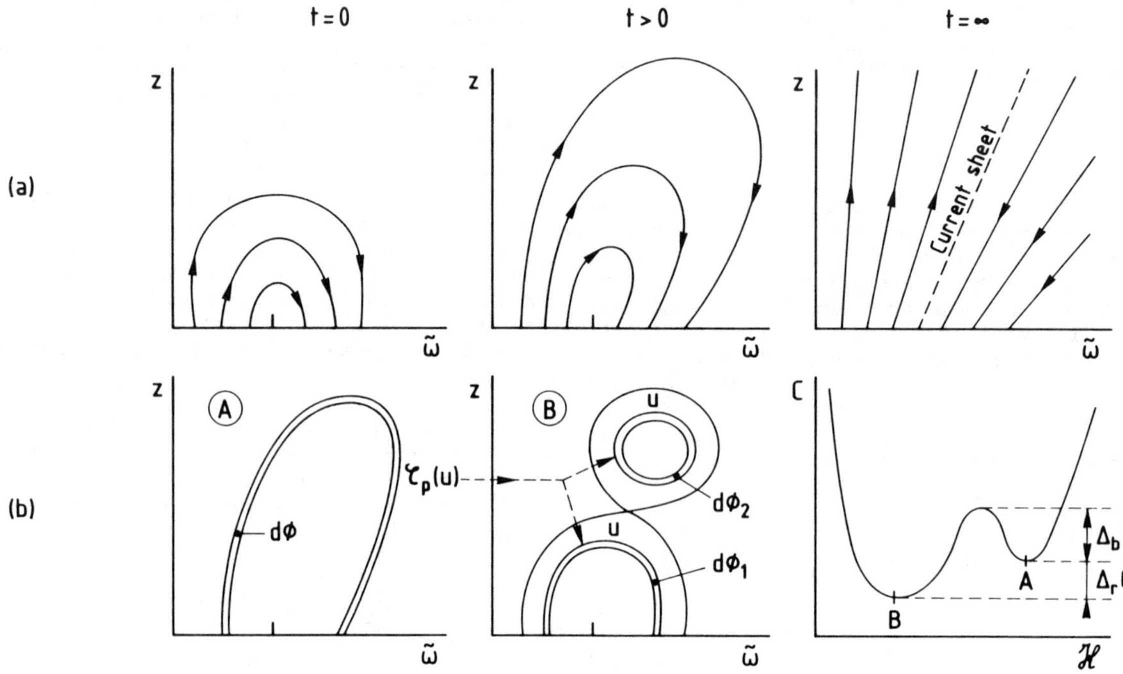

Fig. 2. Evolution of an axisymmetric force-free field (poloidal structure: the lines represent the intersections of magnetic surfaces by the half-plane { Φ = o }). (a) the plasma is perfectly conducting, the field evolves gently towards an open field; (b) fast reconnection is allowed: at $t_r > t_c$, the arcade (A) is submitted to a finite perturbation, the "energetic barrier" $\Delta_b C$ in the space $\mathcal{H}$ of admissible fields is overpassed, and there is an explosive transition towards (B) (with flux being conserved: here, e.g., $d\phi = d\phi_1 + d\phi_2$); an energy $\Delta_r C$ is released.

Conclusion

We have been able to extend to a 3D situation some of the analytical results previously obtained in a study of the quasi-static evolution of a 2D force-free field. In particular, we have shown that a 3D force-free field may approach open configurations when indefinitely submitted to stationnary boundary motions. This result rests, however, on the as yet unproved assumption that the field may continuously adjust to new equilibria up to arbitrarily large time. Work is currently being done to prove that this must be indeed the case.

Questions and Answers

Martens: You showed a transparency demonstrating the formation of a plasmoid from an arcade, the plasmoid having a lower net energy than the original arcade. How do you decide which field lines reconnect (to the plasmoid) and which remain in the original arcade?

Aly: What I have shown is that, when the constraint of topology conservation is given up, then, for any large value of the shear, the configuration of minimum energy has a greater topological complexity and a lower energy than the corresponding arcade. For this configuration, it is just the mathematics which decide which field lines reconnect. This result just proves that reconnection becomes energetically favorable at large shear. Now, if reconnection actually occurs in an arcade, which lines reconnect is certainly determined by the details of the physical processes (large amplitude perturbation, diffusion ...) triggering the onset of this reconnection. Of course, the final state will not be generally the true minimum energy state, but will stay between it and the original arcade.

Mikic: Does the result that the axisymmetric force-free field has a minimum energy (and hence stable) equilibrium at any specified shear depend on the fact that axisymmetry is enforced? Do you have any conjecture or speculation on what might happen if the assumption of axisymmetry were relaxed?

Aly: The problem of producing an axisymmetric equilibrium at any specified shear by minimizing the energy has been considered indeed by varying the energy over the set containing only the axisymmetric functions satisfying the boundary conditions. If we extend the set of test functions to include also 3-D fields, we do not expect any change at moderate values of the shear (because of our stability criteria, which allow for 3-D perturbations). On the other end, our asymtatic result shows that the asymptotic state (approached for large shears) has to be necessarily the axisymmetric open field. These statements may be taken as an indication that bifurcation to a 3-D configuration should not occur at all, but of course the problem has to be considered as open.

Antiochos: Comment: For the case of a twisted flux tube between two plates, although the field along the axis contracts due to the twist, the flux tube as a whole does expand outward from its initial radius.

Aly: The difference with a twisted flux tube in the half-space is that in this last case, the magnetic surfaces located just near the axis also expand; pinching never dominates and a filament singularity cannot be produced.

REFERENCES

Aly, J.J , On some properties of force-free magnetic fields in infinite regions of space, Astrophys. J., 283, 349, 1984.

Aly, J.J., Evolving magnetostatic equilibria, in Interstellar Magnetic Fields, R. Beck and R. Gräve, Ed., Springer Verlag, Berlin, p. 240, 1987.

Aly, J.J., The stability of a line-tied force-free magnetic field in an unbounded region of space, Phys. of Fluids (in press), 1989.

Barnes, C.W., and Sturrock, P.A., Force-free magnetic structures and their role in solar activity, Astrophys. J., 174, 659, 1972.

Berger, M.A., and Field, G.B., The topological properties of magnetic helicity, J. Fluid. Mech., 147, 133, 1984.

Sakurai, T., A new approach to the force-free field and its application to the magnetic field of solar active regions, Publ. Astron. Soc. Japan, 31, 209, 1979.

THE QUASI-STATIC EVOLUTION OF MAGNETIC CONFIGURATIONS ON THE SUN AND SOLAR FLARES

Yu G. Matyukhin and V. M. Tomozov

SibIZMIR, Irkutsk 33, P.O. Box 4026, 664033 USSR

Abstract. We consider the problem of quasi-static evolution of a magnetic configuration on the Sun under the action of shear motions at footpoints of field lines on the photosphere. It is shown that an equilibrium system (a magnetic configuration), with a certain value of shear, changes its structure qualitatively, i.e., a topological reconstruction occurs. The character of the transition from one equilibrium magnetic field configuration to another one depends strongly on the flow structure at field line footpoints and on the boundary conditions imposed on the original magnetic configuration. It is shown that for a loop arcade a shear motion such as a viscous boundary layer leads to its topological reconstruction, with the energy released comparable with that released in large solar flares. The topology of the original arcade configuration, when a critical value of shear is reached, becomes an open one.

We consider the problem of quasi-static evolution of a magnetic configuration on the Sun as a loop arcade under the action of shear motions of a certain kind at footpoints of field lines on the photosphere. It is shown that such a formulation of the problem is possible if the velocity of motions of field line footpoints v is much smaller than the Alfven velocity v_A. Thus, equations describing the quasi-static evolution of a magnetic structure are derivable from a system of magnetohydrodynamic equations by the method of expansion in the small parameter $\alpha = v/v_A \ll 1$. Such a system of equations, for a force-free field, has the form:

$$\mathrm{rot}\,[\vec{B}] \times \vec{B} = 0, \quad \partial\vec{B}/\partial t = \mathrm{rot}\,[\vec{v} \times \vec{B}] \qquad (1)$$

and field $\vec{B}$ is given as (a two-dimensional configuration):

$$\vec{B} = \mathrm{grad}\, A \times \vec{e}_z + B_z\vec{e}_z \qquad (2)$$

Geophysical Monograph 58

(A(x,y) is a scalar potential). On replacing the equation for "frozen-in" condition with the equation for field lines [Parker, 1979], we obtain the following system of equations of quasi-static evolution of a magnetic structure:

$$2\Delta A = (\partial/\partial A)\, B_z^2(A), \quad dx/B_x = dy/B_y = dz/B_z \qquad (3)$$

In the case of a slow ($v \ll v_A$) motion of field line footpoints the second equation of the system can be related to $B_z(A)$ [Birn and Schindler, 1981; Aly, 1984] by

$$\xi(A) = v(A)\cdot t = B_z(A) \int_{A=\mathrm{const}} \frac{dS}{B_p} \qquad (4)$$

where dS is an element of the arc of projection of the field line with A = constant onto the plane (x,y), and

$$B_p = \left[\left(\frac{\partial A}{\partial x}\right)^2 + \left(\frac{\partial A}{\partial y}\right)^2\right]^{1/2}$$

$\xi(A)$ is the displacement of field lines; t is time. The first equation of the system (3) is parametrized

$$\frac{1}{2}\frac{\partial}{\partial A} B_z^2(A) = \lambda f(A) \qquad (5)$$

then

$$\xi(A) = \left[2\lambda \int_0^A f(A)dA\right]^{1/2} \int_{A=\mathrm{const}} \frac{dS}{B_p} \qquad (6)$$

and the parameter λ is always non-negative, and f(A) is an arbitrary positively-defined function. With a given function f(A) and fulfillment of the condition

$$\int_{A=\mathrm{const}} \frac{dS}{B_p} = \int_{A=\mathrm{const}} \frac{dS^\circ}{B_p^\circ}$$

where

$$\int_{A=\mathrm{const}} \frac{dS^\circ}{B_p^\circ} = V_0(A)$$

is a specific volume of the flux tube for $\lambda = 0$, the value of displacement $\xi(A)$ and the parameter λ will be uniquely related. Currently two classes of boundary value problems (BVP 1 and 2) of the system of equations (5), (6) are usually considered for astrophysical applications [Priest, 1982]. Priest and Milne [1980] considered BVP 1, in terms of which it was found that the system of equations (5), (6) can have two solutions for the same displacement $\xi(A)$. In this paper, unlike those by Priest and Milne [1980], we investigate BVP 1 of the system (5), (6) which may admit the existence of three solutions for the same value of the parameter λ. In order for these solutions to satisfy the Priest [1982] requirement, i.e., to correspond to the same value of displacement $\xi(A)$, an additional boundary condition

$$\int_{A = \text{const}} \frac{dS}{B_p} = \int_{A = \text{const}} \frac{dS^\circ}{B_p^\circ}$$

is introduced. Thus, the relationship between the parameter λ and the value of displacement $\xi(A)$ becomes unambiguous.

In order to solve the problem, we have constructed an iteration scheme based on the Courant-Gilbert method (the method of upper and lower solutions) which makes it possible to obtain a full set of solutions of the nonlinear equation with the Neuman boundary conditions for an arbitrary value of the parameter λ. The choice of the function f(A) was due to the requirement that at field line footpoints of the magnetic structure the velocity profile should be identical to that of a viscous boundary layer (for more details, see a paper of Golovko et al. [1985, 1988]). Physically, this means that with such a choice of the velocity profile the function f(A) in the vicinity of $A \to 0$ has the form $f(A) \sim A^s$, where $s \geq 3$.

For values of λ lying within $0 \leq \lambda < 1.97$ the equation $\Delta A = -\lambda f(A)$ with the Neuman boundary conditions has a unique solution; it has three solutions in the range $1.97 < \lambda < 4$ and two solutions for $\lambda = 1.97$ and $\lambda = 4$. We emphasize that multiple solutions (solutions with the same λ) describe different magnetic topologies. Note that the linear stability of solutions to disturbances which do not alter the field line topology, proves to be different, i.e., for double and triple solutions one of them at any λ becomes linearly unstable, while unique solutions always are linearly stable.

In order to describe the evolution of a magnetic configuration under the action of shear motions of its footpoints (such as a viscous boundary layer), it is necessary to investigate the dependence of the full magnetic energy of a given structure on the value of shear, i.e., on the parameter λ. Figure 1 gives the dependence of the full magnetic energy of the configuration W_{tot} on the value of λ for stable solutions of the equation $\Delta A = -\lambda f(A)$. Here W_0 is the magnetic configuration energy for $\lambda = 0$. Curve 1 in Figure 1 represents the magnetic energy behavior of a topologically closed magnetic structure versus λ, and curve 2 describes that of a topologically open one. It should be stressed that solutions of the equation for a closed topology of field lines become linearly unstable when $\lambda = \lambda_{cr}^2$, while solutions for an open topology of field are linearly unstable when $\lambda = \lambda_{cr}^1$.

Thus, in order to treat the quasi-static evolution of a magnetic configuration as an evolutionary process, it is necessary to suppose that depending on the parameter λ which increases monotonically with time, the magnetic configuration is able to be in stable states only. Such a behavior of the full magnetic energy of the configuration versus shear and the assumptions made above (the parameter λ) permit us to suggest the following scenario of quasi-static evolution of a magnetic arcade. Under the action of a shear motion the originally closed equilibrium magnetic structure necessarily evolves to such an equilibrium state that it becomes linearly unstable so that if the value of magnetic field shear reaches a critical value $\lambda = \lambda_{cr}^2 = 4$, then the magnetic structure must of necessity reach a new stable equilibrium state with a change of the field topology (since for $\lambda > \lambda_{cr}^2$ there exist no solutions with a closed topology). During the transition there occurs a change of the magnetic field topology from closed to open, and this process is accompanied by the release of the accumulated free energy of the magnetic field, ΔW_f:

$$\Delta W_f = \int \frac{B_z^2}{8\pi} dV \approx \int \frac{\xi^2 B_p^2 \lambda}{8\pi L^2} dV \qquad (7)$$

For a length of the arcade along its symmetry axis $L = 2 \cdot 10^{10}$ cm, a particular calculation gives a value of $\Delta W_f \approx 10^{32}$ ergs. The fact that the transition is accompanied by a change of the magnetic structure topology and by the release of the field free energy ΔW_f, indicates that the transition has a dissipative character, during which the equation

$$\partial \vec{B}/\partial t = \text{rot}\,[\vec{V} \times \vec{B}]$$

fails to be satisfied. Note that such a scenario of magnetic arcade evolution on the Sun explains quite well the main energy characteristics of large solar flares and predicts the origins of open magnetic structures in active regions of the Sun after flares. Notice also that the above-mentioned topological reconstruction of a magnetic configuration under the action of shear motions turns out to be equivalent to a phase transition of the first kind in liquid crystals during their deformation.

Let us consider the last question in greater detail. First of all, attention is attracted by the fact that in the region of values of the parameter λ, where multiple solutions of the equation $\Delta A = -\lambda f(A)$ exist, the magnetic field topology described by these solutions is different. This very much resembles the change in the character of symmetry in liquid crystals under the external influence upon them. The above similarity can be characterized as follows. The set of solutions of the equilibrium equation depending on the parameter λ involves the following:

1. $\lambda < \lambda_{cr}'$ - there exists a unique solution with a closed topology of field lines and it is topologically stable, with $\lambda = 0$ corresponding to a potential magnetic structure.
2. $\lambda_{cr}' < \lambda < \lambda_{cr}^2$ - there exists a triple solution with a different topology of field lines, and there are always two solutions describing a different topology (open and closed) which are topologically stable.
3. $\lambda = \lambda_{cr}'$ and $\lambda = \lambda_{cr}^2$ - there are double solutions with a different topology, one of which always is topologically stable, and the other is neutrally stable.
4. $\lambda > \lambda_{cr}^2$ - there exists one topologically stable solution.

Now, using the dependences $W_{tot}(\lambda)$ given in Figure 1 and a classification of solutions of the equation $\Delta A = -\lambda f(A)$, one can construct the function of state of the system describing the quasi-static evolution of a magnetic configuration following Gilmore [1981]. It will be assumed that X is a variable of the state of the system; then from the character of the evolution it becomes obvious that it must be characterized by two main attributes, namely a total energy of the magnetic configuration and the character of its topology. These two attributes fully describe a magnetic configuration with a given λ, i.e., if the topology

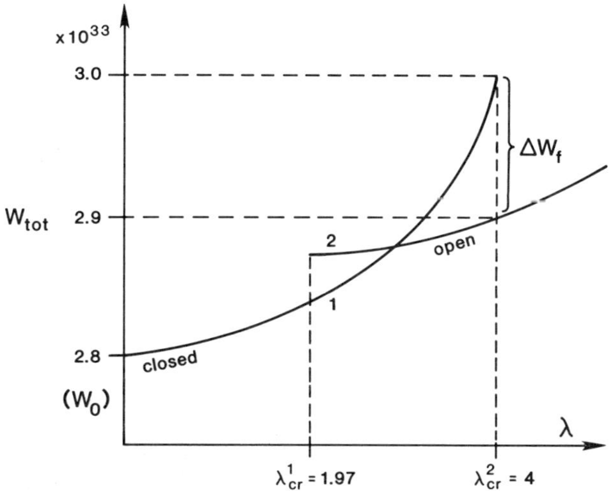

Fig. 1. The dependence of the total magnetic energy of the configuration W_{tot} on the value of the parameter λ.

can be characterized by the function $\theta(\lambda)$, and the magnetic energy, by $W_{tot}(\lambda)$, then the configuration will be described, provided that for both attributes their realization is known, i.e., $X(\lambda) = \theta(\lambda)W_{tot}(\lambda)$ (operation of logical multiplication). It should be stressed that the topology variability is determined, for a given magnetic configuration, by the value $\Delta\lambda_{cr} = \lambda_{cr}^2 - \lambda_{cr}'$. Suppose that there exists a function $\phi(X, \lambda, \Delta\lambda_{cr})$ which permits the state of a system to be described from the given parameters $\Delta\lambda_{cr}$, λ and X, and minima of this function correspond to stable equilibrium states of the magnetic configuration, i.e., the surface $\partial\phi/\partial X = F(X, \lambda, \Delta\lambda_{cr}) = 0$ is a surface of equilibrium stable states of the system in the space of driving parameters, $\Delta\lambda_{cr}$ and λ. Note that the quantity $\Delta\lambda_{cr}$ for a particular magnetic configuration is a fixed parameter. From the above classification of solutions, it is easy to see that the function $\phi(X, \Delta\lambda_{cr}, \lambda)$ can contain no more than two minima. A classification of the functions of state reported by Gilmore [1981] shows that all of the above-mentioned requirements to the function of state are satisfied by the following normal form:

$$\phi(X, \lambda, \Delta\lambda_{cr}) = \tfrac{1}{4}x^4 + \tfrac{1}{2}ax^2 + bx \qquad (8)$$

where $a = a(\lambda, \Delta\lambda_{cr})$ and $b = b(\lambda, \Delta\lambda_{cr})$ are the driving parameters, and the function $\theta(\lambda)$ that determines the character of field topology, is chosen in the form:

$$\theta(\lambda) = \begin{cases} +1 & \lambda < \Delta\lambda_{cr} - (\Delta\lambda_{cr})^{3/2} \quad \text{closed field topology} \\ -1 & \lambda > \Delta\lambda_{cr} - (\Delta\lambda_{cr})^{3/2} \quad \text{open field topology} \end{cases} \qquad (9)$$

Since for the particular magnetic configuration, a normal field component $\vec{B}_n$ is specified, $a = a(\Delta\lambda_{cr})$ is then independent of λ, and the value of a can be chosen from the condition of equality of a total field energy W_{tot} for the upper and lower solutions. From the condition $\partial\phi/\partial X = 0$ we find the value of the parameter $a = \Delta\lambda_{cr}$. The existence of such unique stable solutions only for one field topology indicates that the parameter b can assume both positive and negative values. Since, when $b = 0$, the function $\phi(X, \lambda, \Delta\lambda_{cr})$ is symmetric and the separatrix on the region of driving parameters has the form $a^3/3 = -b^2/2$, the value of $b = \lambda - \Delta\lambda_{cr}$. Note that by the separatrix we understand a curve on the surface of driving parameters, in the vicinity of which the function of state undergoes qualitative changes (for example, to the right and to the left of the separatrix the function of state can have two or one stable states of equilibrium). Now the value of X must be normalized so that, with a given topology, the value of $|X|$ corresponds to the total magnetic energy of the field of a given state of the configuration. Such a procedure is performed with the help of recalculation transformations reported by Gilmore [1981]. Next, on determining the form of the function $\phi(X, \lambda, \Delta\lambda_{cr})$ and of the driving parameters a and b, one can consider the way in which the evolution of the magnetic configuration will proceed under the action of shear motions of field line footpoints on the photosphere. Assuming that the evolution of the system starts from a potential configuration of the magnetic field with $\lambda = 0$ and, under the action of shear motions, the system accumulates the energy and its topology becomes similar to a potential one (namely, a closed one with $\theta = 1$) up to

$$b = -\left(\frac{2|a|^3}{3}\right)^{1/2} = -(\Delta\lambda_{cr})^{3/2}$$

i.e., the system of equations (3) has a unique solution, and the function of state has a unique minimum. When $b > -(\Delta\lambda_{cr})^{3/2}$, the system of equilibrium equations has two stable solutions, and the function $\phi(X, \lambda, \Delta\lambda_{cr})$ has two minima. It must be pointed out that according to the causality principle, the system will remain in that minimum in which it has been before the appearance of a new stable equilibrium state. When $b > (\Delta\lambda_{cr})^{3/2}$, the system (3) again has a unique solution, but when $b = (\Delta\lambda_{cr})^{3/2}$ and $\theta = -1$ the minimum of the function of state where the configuration had been before the value of shear reached this value, degenerates and, therefore, near this minimum there is no equilibrium solution such that the magnetic structure must necessarily go to a new equilibrium state in the case of increasing shear. Note that the transition of the structure from one equilibrium state to another occurs in accordance with the principle of maximum duration [Gilmore, 1981] which implies that the state of the system is determined by a stable or metastable minimum until such a minimum exists.

Thus, the evolution of a magnetic configuration under the action of given shear motions at field line footpoints on the photosphere can be regarded as a phase transition of the first kind in which the most important physical attributes of such a transition are manifested, namely modality - the function of state at some values of driven parameters has a non-unique local minimum; nonreachability - the presence of topologically unstable solutions which do not correspond to the causality principle; catastrophic jump - dissipative transition of the magnetic configuration from one topological state to another; and finally, divergence - the system of equations (3) does not have stable multiple solutions for $\lambda < \lambda_{cr}^1$ and $\lambda > \lambda_{cr}^2$.

Thus, the above quasi-static evolution of a magnetic structure caused by shear motions, with a jump-like change of the magnetic field topology (from closed to open) and with the release of free energy, in terms of catastrophic theory, turns out to be equivalent to a normal form of the "cusp" type [Gilmore, 1981].

Acknowledgments. We are profoundly grateful to Prof. E. R. Priest for his attention to this work and to Dr. T. Amari for helpful comments. Thanks are also due to Mr. V. G. Mikhalkovsky for his assistance in preparing the English version of the manuscript and for typing the text.

References

Aly, J. J., On some properties of force-free magnetic fields in infinite regions of space, Astrophys. J., 283, 349-362, 1984.

Birn, J., and K. Schindler, Two-ribbon flares: Magnetostatic equilibria, in Solar Flare Magnetohydrodynamics, edited by E. R. Priest, pp. 337, Gordon and Breach, New York, 1981.

Gilmore, R., Catastrophic Theory for Scientists and Engineers, A. Wiley-Interscience Publication, New York, 1981.

Golovko, A. A., G. V. Kuklin, A. V. Mordvinov, and V. M. Tomozov, The role of large-scale velocity fields in producing a pre-flare situation, in Solar Maximum Analysis Workshop, Abstracts of papers, June 17-24, Irkutsk, USSR, pp. 62, 1985.

Golovko, A. A., G. V. Kuklin, A. V. Mordvinov, and V. M. Tomozov, The role of large-scale velocity fields in producing a pre-flare situation, in Solar Maximum Analysis. Additional Issue, edited by V. E. Stepanov, V. N. Obridko and G. Ya. Smolkov, Proceedings of the International Workshop held in Irkutsk, USSR, June 17-24, 1985, pp. 278-290, Nauka, Novosibirsk, 1988.

Parker, E. N., Cosmical Magnetic Fields: Their Origin and Their Activity, Clarendon Press, Oxford, 1979.

Priest, E. R., and A. M. Milne, Force-free magnetic arcades relevant to two-ribbon solar flares, Solar Phys., 65, 315-346, 1980.

Priest, E. R., Solar Magnetohydrodynamics, D. Reidel Publishing Company, Dordrecht, Holland, 1982.

QUASI-POTENTIAL-SINGULAR-EQUILIBRIA AND EVOLUTION OF THE CORONAL MAGNETIC FIELD DUE TO PHOTOSPHERIC BOUNDARY MOTIONS

T. Amari* and J. J. Aly**

*Applied Mathematics Division, The University, St Andrew, Fife, Scotland
**Service d'Astrophysique, Centre d'etudes Nucléaires de Saclay, 91191 Gif-sur-Yvette Cedex, France

Abstract. We consider the structure of a particular class of 2D-x-invariant magnetostatic equilibria (in which the magnetic field is potential everywhere in some domain of space but on some singular surfaces - current sheets - carrying a non-zero x-current) for which only particular examples were selected in all the previous models. We present some new general properties of configurations of this type. We prove in particular that the general condition of equilibrium at the extremities of the Current Sheet which has never been analysed before implies a new constraint on the magnetic field. We also present a variational principle and results of existence and stability of such singular equilibria when they arise in a physical context. We consider in detail the situation in which such singular states are obtained asymptotically by an arcade like x-invariant force free field in {z > 0} when indefinitely sheared. We present a method which allows us to compute analytically these asymptotic states as the solutions of well set boundary value problems even in the slightly non-symmetric case for which no analytical model has ever been proposed before.

1. Introduction

The process of spontaneous formation of current-sheets (CS) during the quasi-static evolution of magnetohydrostatic equilibria embedded in perfectly conducting plasma is of rather general occurrence. There have been many elaborated models since the simplest one of Dungey (1953). The appearance of the CS is related to the necessity of preserving the field line topology (because of the frozen-in law) when the equilibrium evolves slowly as a consequence of the boundary motions imposed to the feet of the field lines. However, in all the previous models only particular examples were selected for reasons of mathematical convenience.

In this communication we present some new results concerning the structure of a particular class of 2D x-invariant magnetostatic equilibria - Quasi-Potential-Singular-Equilibria (QPSE) - in which the magnetic field is potential everywhere in some domain of space except on some singular surfaces - CS - carrying a non-zero x-current. We derive the conditions which have to be satisfied for a configuration to be a QPSE. We then present methods for computing explicitly in a direct way QPSE which may also appear in a different context as asymptotic limits of a sequence of arcade like x-invariant force free fields in {z > 0} which are indefinitely sheared parallely to x. The details of the calculations and references to the current literature on the subject may be found in Amari (1988), Aly and Amari (1989), Amari and Aly (1989a).

2. Quasi-Potential-Singular Equilibria and General Properties

2.1 Assumptions

In this section we introduce the complex functions formalism which will be of great use in the following sections.

Geophysical Monograph 58

(i) Let Ω be a simply-connected open (bounded or unbounded) domain of the complex plane with regular boundary $\partial\Omega$ and $\{\Gamma_k\}$ is a finite set of differentiable arcs of $\bar{\Omega} = \Omega\cup\partial\Omega$. The endpoints $\eta_k^{\pm}$ of Γ_k are the only points of this arc which may belong to some Γ_j ($j \neq k$) or to $\partial\Omega$ if only one of the $\eta_k^{\pm}$ is on $\partial\Omega$ (we choose it to be η_k^-). Each Γ_k is oriented from η_k^- towards η_k^+ and admits the parametric representation $\eta_k = \eta_k(s_k)$ where s_k is the arc length on Γ_k. $\tau_k = d\eta_k/ds_k = \tau_{ky}+i\tau_{kz}$ and $\eta_k = i\tau_k$ are the tangent and normal unit vectors defined at each point of $\Gamma_{kc} = \Gamma_k/\{\eta_k^{\pm}\}$, respectively. We set $(\Gamma = \cup_k \Gamma_k, \Gamma_c = \cup_k \Gamma_{kc}, \Omega_c = \Omega/\Gamma$.

(ii) Let us consider in R_x ($= \{-\infty < x < +\infty\}$) $\times\, \Omega$ the x-invariant magnetic field

$$B(\eta,\bar{\eta}) = (B_y - iB_z)(\eta,\bar{\eta}) \tag{1}$$

which satisfies

$$\frac{\partial B}{\partial \bar{\eta}} = -\frac{2i\pi}{c}\, j\, \delta_\Gamma \text{ in } \Omega \tag{2}$$

where j is the density of surface current flowing parallely to x on $\Sigma = R_x \times \Gamma$ (which appears as a CS).

(iii) B is holomorphic in Ω_c ($\partial B/\partial\bar{\eta} = 0$ in Ω_c).

(iv) On Γ_c, $[B\tau]_{\Gamma_c} = -4\pi/c\ j$ (where $[X] = X^+ - X^-$) express the fact that B_n is continuous across Σ while B_t suffers a discontinuity owing to the singular current.

(v) Σ is submitted to the Lorentz force whose surface density is given by

$$f = f_y - if_z = [B^2]_{\Gamma_c} \frac{n}{8\pi} = -\frac{i}{c}\, jB|_{\Gamma_c} \tag{3}$$

(vi) $F(\eta) = (V+iA)(\eta)$ is the potential associated to B in Ω_c which satisfies $B(\eta) = dF(\eta)/d\eta$.

2.2 <u>Conditions of equilibrium</u>

(i) The Lorentz force has to vanish. Thus

$$[B^2]_{\Gamma_c} = 0 \ (\text{or } 2B|_{\Gamma_c} = (B^+ + B^-)|_{\Gamma_c} = 0)$$

at any point on Γ_c, which implies: (4)

$$A|_{\Gamma_{kc}} = A^{\pm}|_{\Gamma_{kc}} = \gamma_k = \text{cte}. \tag{5}$$

(ii) Let us now consider an endpoint η_I (at which p arcs meet). Then if $p \geq 2$ from the analysis of the behaviour of B near η_I (see Aly and Amari 1989 for the details) and (4) it results that the condition

$$B(\eta_I) = 0 \tag{6}$$

has to hold.

If p=1, one may have near η_I

$$B \sim c_I(\eta-\eta_I)^{-1/2} + (\eta-\eta_I)^{1/2}\psi_I(\eta) \tag{7}$$

with ψ_I holomorphic and c_I a constant. Then, by computing the force $\mathcal{F}$ acting on a small cylinder whose base is centred on η_I ($\mathcal{F} = 1/4c_I^2$) it comes the new equilibrium constraint $c_I = 0$ on the field and (6) still holds.

2.3 <u>General properties of QPSE</u>

We now report some properties whose details of proof may be found in Aly and Amari 1989.

(i) B^2 is holomorphic in Ω.

(ii) A functional form of $B(\eta)$ may be obtained in general from the expression

$$B(\eta) = \prod_k [(\eta_k^+ - \eta)(\eta - \eta_k^-)]^{1/2} H(\eta) \stackrel{\text{def}}{=} Q(\eta)H(\eta) \tag{8}$$

where $H(\eta)$ is holomorphic in Ω and when $\eta_k^{\pm}$ belongs to Ω and to only one Γ_k. (8) has to be slightly adapted for different situations ($\eta_k^{\pm}$ are common to several arcs, $\eta_j^+ = +\infty$ for some j, or $\eta_\ell^{\pm} \in \partial\Omega$ for some ℓ). Moreover H has to satisfy: $I_m\{Q^+H\tau\} = 0$ on Γ.

(iii) in a QPSE the CS are analytical curves and reciprocally an analytical curve may be always considered as a CS embedded in a QPSE.

(iv) a QPSE (Ω',Γ',B') is transformed into a new QPSE (Ω,Γ,B) by a conformal mapping G, with $\Omega = G(\Omega')$, $\Gamma = G(\Gamma')$, $B(\eta) = B'[G^{-1}(\eta)]dG^{-1}(\eta)/d\eta$ (or equivalently $F[\eta] = F'[G^{-1}(\eta)]$). This property has been used to construct non y-symmetric models of magnetic support of prominences in potential magnetic fields. (Aly and Amari 1988.)

(v) The x-current

$$I_k = \int_{\Gamma_k} j \, ds$$

flowing in Γ_k and the magnetic energy

$$w_m = \frac{1}{8\pi} \int_{\Omega} B\bar{B} \, dy \, dz$$

in Ω are left invariant when the QPSE (Ω', Γ', B') is submitted to an arbitrary conformal mapping G.

(vi) Applying the Gauss theorem

$$\int_D \frac{\partial h(\eta,\bar{\eta})}{\partial \bar{\eta}} dydz = \frac{1}{2i} \oint_{\partial D} h(\eta,\bar{\eta}) d\eta \tag{9}$$

successively for $D = \Omega$ and $h = B^2$, $h = \eta B^2$, $h = \bar{\eta} B^2$ and for $D = \Omega_c$, $h(\eta,\eta) = B(\bar{F}-F)$ one may derive several integral relations expressing the global mechanical equilibrium, virial relations and the total magnetic energy w_m which may be shown to reduce to

$$w_m = \int_{\partial\Omega} (\gamma_1 - A) B_\tau ds$$

in the particular case of only one CS Γ_1.

2.4 Quasi-static evolution and a variational Principle

Let us now discuss how QPSE arise in the following physical context. In Ω (assumed filled of a perfectly conducting plasma), let us start at time t=0 with a field in the potential state $\underline{B}_0 = \underline{\nabla A}_0 \times \hat{x}$ corresponding to $A_0|\partial\Omega = g_0(s)$. Then for $t \geq 0$ the feet of the field lines are moved (by a regular velocity field $\underline{v} = v(s,t)\hat{t}$) parallely to $\partial\Omega$ but normally to $\hat{x}$. The field is then brought into an evolution (that is assumed quasi-static) through a sequence of force-free states for $t \geq 0$. The field $\underline{B}_t = \underline{\nabla A}_t \times \hat{x}$ is then force free and $A_t|\partial\Omega = g_t$ where g_t is the solution of $\partial_t g_t + v\partial_s g_t = 0$ with $g_{(t=0)} = g_0$. Moreover, the frozen-in law imposes: $A_t[\underline{r} + \underline{X}(\underline{r},t)] = A_0(\underline{r})$ for some field of displacement $\underline{X}(\underline{r},t)$ which coincides on $\partial\Omega$ with the displacement resulting from $\underline{v}$, and satisfies a "Jacobian Condition".

Then, if we assume that this evolving boundary value problem has a solution $\underline{B}_t$ of finite energy (i.e. $C(A_t) < \infty$) where

$$C(A) = \int_{\Omega} |\nabla A|^2 d\underline{r} \tag{10}$$

it is then reasonable to assume that for all t, A_t makes C(A) an absolute minimum over some functional space $\mathcal{H}_t$ that satisfies the right constraints, and we may then determine A_t by solving the minimisation problem:

$$C[A_t] = \operatorname*{Inf}_{\mathcal{H}_t} C(A), \quad A_t \in \mathcal{H}_t.$$

(i) We may then prove that this variational principle admits a solution in $\mathcal{H}_t$.

(ii) Moreover we may develop heuristic arguments tending to prove that this solution should be unique and should coincide with a QPSE. In particular we can prove the linear stability of any QPSE with respect to 2D perturbations.

(Details of the definition of $\mathcal{H}_t$ and the discussion may be found in Aly and Amari 1989.)

3. QPSE as Asymptotic Limits of Indefinitely Sheared Force Free Fields

3.1 Obtention of the QPSE : the shearing problem

Let us start in {z>0} with an x-invariant potential magnetic field $\underline{B}_0 = \underline{\nabla A}_0 \times x$ where A_0 is the solution of

$$\Delta A_0 = 0 \tag{11}$$

$$A_0(y,0) = g(y) \tag{12}$$

$$\lim_{r\to\infty} A_0 = 0 \tag{13}$$

and g satisfies (i) $\inf g = 0 \leq g \leq A^m = \sup g$, (ii) $y\dot{g}(y) < 0$ for $y \neq 0$, these two conditions ensuring that B_0 has an arcade like topology. Then, if we impose at t=0 an increasing shear with time $X_t(A) = t\,\xi(A)$ (stationary shearing motions whose velocity is parallel to x, where X_t is the difference between the x-position of the left and right feet, and $A^1 < A^m = \sup g$ is the smallest value of A such that $\xi(A)=0$ in $[A^1,A^m]$) to the feet of the field lines of $\underline{B}_0$, the field will evolve through a sequence (that we assume to be quasi-static) of force free fields $\underline{B}_t = \nabla A_t \times \hat{x} + B_{tx}[A_t]\hat{x}$ with the same topology as $\underline{B}_0$ (because of the frozen-in law) where A_t is solution of a nonlinear and non-local problem. As it is proved in Aly (1985,1987) and Aly and Amari (1985) this problem admits a solution for arbitrary large values of t and when $t \to \infty$ this solution converges asymptotically towards a QPSE which consists of a semi-open configuration A_∞ in which the CS extends above an arcade like region constituted of closed field lines $A^1 < A \leq A^m$. A_∞ is solution of the following free boundary value problem which depends only on g(y) and A^1:

$$-\Delta A_\infty = 0 \text{ in } \Omega/\Gamma \tag{14a}$$

$$A_\infty(y,0) = g(y) \tag{14b}$$

$$A_\infty|_\Gamma = A_1 \tag{14c}$$

$$[|\nabla A_\infty|^2]_\Gamma = 0 \tag{14d}$$

$$\nabla A_\infty(P^-) = 0 \tag{14e}$$

where Γ is a curve of extremities P^- and the point at infinity, whose position has to be determined as a part of the problem. One may notice that (14c-e) are nothing more than the conditions of equilibrium derived in 2.2 in order to have a QPSE.

Let us now consider the possibility of computing analytically A_∞ (noted A hereafter for simplification), in the following situations.

3.2 The Symmetric Case : g(y) = g(-y)

In this case it is possible to prove (see Amari & Aly 1989a) that the solution of (14a-e) is symmetric (which is not an immediate consequence of g(y) = g(-y)). Therefore, the CS Γ has to be located on the z-axis and only A and P^- remain the only unknowns for the QPSE solution of the problem (14a-e). This solution may be then obtained by reformulating (14a-e) as the problem of determining in the quarter-plane {y>0, z>0} an holomorphic function $B(\eta) = B_y - iB_z$ which satisfies the boundary conditions: (i) $\operatorname{Im} B(y) = g(y) = dg/dy$ for $y \in \mathbf{R}_+$, (ii) $\operatorname{Im} B(iz) = 0$ for $0 < z < z_1$, (iii) $\operatorname{Re} B(iz) = 0$ for $z_1 < z < \infty$, (iv) $B(iz_1) = 0$, (v) $\lim_{\eta\to\infty} B(\eta) = 0$, where z_1 is a constant determined by the "flux condition"

$$\int_0^{z_1} By(o,z)\,dz = -(A^m - A^1).$$

Effecting then the conformal mapping $\xi = \eta^2$, which maps the quater plane {y>0, z>0} onto the half plane {z>0}, the problem is then reduced to determining in {z>0} a bounded holomorphic function $H(\xi) = iB(\sqrt{\xi})$ whose either the real or imaginary parts are given on segments of the real axis. The solution of this problem is then given by the well known "Keldish-Sedov Formulae". Going back to the quater plane one gets the solution of (14a-e) in the closed form

$$B(\eta) = \frac{2}{\pi}(\eta^2+z_1^2)^{1/2} \int_0^\infty \frac{\dot{g}(y')}{(y'^2+z_1^2)} \frac{y'}{y'^2-\eta^2}\,dy' \tag{15}$$

where z_1 is the unique solution of the equation given by the flux condition which may be written:

$$A^m - A^1 = -z_1 \int_0^{\pi/2} \dot{g}(z_1 t_g t)\,\frac{dt}{1+\sin t}\,. \tag{16}$$

One may notice that (15) represent effectively an adapted form of the general expression (8) when there is only one CS and $\eta^+ = \infty$ (by suppressing the factor $(\eta - \eta^+)$ from the corresponding square root in (8)).

3.3 The Asymmetric Case

(i) In this case we cannot guess a priori the location of the CS Γ. Nevertheless we may content ourself in a first step of constructing asymmetrical equilibrium configurations when the constraint (14b) is relaxed, (whereas for a true solution of the whole problem the value of A and of B_z on {z=0} is fixed). This may be achieved by using the general theorem 2.3 iv). Assuming then that we have a known symmetric configuration

(Ω', Γ', B') in Ω, we may consider an arbitrary conformal transform $\xi = G(\eta)$ which maps Ω onto $\Omega_1 \subset \Omega$ with $\Omega_1 \supset \Gamma$, leaves the point at infinity invariant and transforms the y axis into a curve C of Ω ($C = \partial\Omega_1$). The QPSE (Ω', Γ', B') is then transformed into a new asymmetric QPSE (Ω, Γ, B) given by 2.3 iv). It may be worth noticing that the restriction $\Omega_1 = G(\Omega) \subset \Omega$ (or $C = G(\{z=0\}) \subset \Omega = \{z \geq 0\}$ can be partly relaxed because B may be analytically extended into a part Ω^- of $\{z<0\}$ (see Amari and Aly (1989)).

(ii) Let us now reconsider the full problem (14a-e) when g(y) is a given asymmetric function (i.e. $g(y) \neq g(-y)$). The method we propose is then the following one. We start from a known QPSE (Ω, Γ, B) and then try to determine a conformal transform $\xi = G(\eta)$ of Ω onto some Ω_1, such that a solution (Ω', Γ', B') of our problem may be obtained from the QPSE (Ω, Γ, B) by G, which is determined if it exists by the "boundary relation" on $\{z=0\}$ (R):$A(G(y)) = g(y+\lambda)$ where λ is introduced to account for the arbitrary in the choice of the origin on the y-axis. The problem is reduced to solving a so called "nonlinear Riemann-Hilbert problem" whose existence of solutions is far from being guaranteed a priori.

However, in the case of weak asymmetry g may be written as $g(y) = g_S(y) + \delta g(y)$ where $g_S = 1/2\ [g(y)+g(-y)] \geq 0$ is a y-symmetric function and $\delta g(y) = 1/2\ [g(y)-g(-y)]$ is a y-antisymmetric one. Let us then take for initial QPSE, the symmetric solution corresponding to (g_S, A^1) and set $G(\eta) = \eta + 2i\phi(\eta)$ with $\phi(\eta) = u+iv$ where $\phi(\eta)$ is a small correction. Linearising (R) and introducing in $\mathbf{R}^2$ the sectionally holomorphic function $\psi(\eta)$ defined by i) $\psi(\eta) = \phi(\eta)$ in $z>0$ and ii) $\psi(\eta) = -\phi(\eta)$ in $z<0$, ψ satisfies on the y axis

$$\psi^+(y) = \Lambda(y)\psi^-(y) + C(y) \qquad (17)$$

where $\Lambda(y) = (\bar{B}/B)(y)$,

$C(y) = [\delta g(y) + \lambda g_S(y)]/B(y)$ and

$\psi^{\pm}(y) = \lim_{\varepsilon \to 0} \psi(y \pm i\varepsilon)$.

(17) is then nothing more than a so-called "Hilbert Pivalov Problem", whose solution may be completely determined analytically in terms of the known quantities B, δg and g_S since λ is determined by the asymptotic condition $\psi(\infty) = 0$ (see Amari and Aly 1989a for more details).

4. Conclusion

We have reported here some of the results we have recently obtained in our study of the general properties of current sheets in equilibrium in potential magnetic fields (QPSE) for which in the previous models only particular examples were selected. It is proved in particular that the condition of equilibrium at the extremities of the current sheet (which has never been explicitly analysed before) implies a new constraint on the magnetic field. We have proved a property of covariance of QPSE under conformal mappings which allows us to construct new non-symmetric QPSE and which is also of great importance for building non-symmetric models of magnetic support of prominences (Aly and Amari, 1988). We have also presented a variational principle which admits always a solution which is a QPSE and which is stable with respect to 2D perturbation. We have also presented methods which allow us to compute analytically such configurations (as the solutions of well set boundary value problems) when they appear asymptotically as limits of indefinitely sheared x-invariant arcade like force free fields with either symmetric or slightly non-symmetric (a situation for which no analytical model has ever been proposed before) boundary conditions on $\{z=0\}$. Moreover, the methods developed here are also relevant to compute analytically a sequence of QPSE solution of a quasi-static evolution problem when a potential magnetic configuration in $\{z>0\}$ with complex topology (having separatrices, as in Low 1987) is submitted to boundary motions whose velocity is parallel to y. More recently we have been able to compute such sequences, as well as a relaxation mechanism relevant for the heating of the corona by magnetic dissipation, in which the previous sequence dissipate successively its free energy by reconnection by "falling back" to the potential state of lower energy at each step (Amari and Aly, 1989b).

Questions and Answers

Antiochos: What produces the current concentration needed for your prominence model?

Amari and Aly: To reinterpret our second model in the case $\mathbf{E} = +1$*, we need to produce a 2-D configuration containing a flux tube (a bundle of field lines not connected to the boundary) in which one has a concentration of currents flowing in the positive direction. Such a configuration may be certainly produced by a reconnection process developing in an arcade field (see my talk on "Quasistatic evolution of force-free fields). It may also result if a flux tube emerges from below the photosphere (some reconnection being necessary in that case, too).*

van Ballegooijen: Are the current sheets in your model located at fixed positions in space, or are they allowed to vary in position?

Amari: In the first part of my talk, we deal with the general properties of current sheets in equilibrium in a potential magnetic field (what we call a QuasiPotential Singular Equilibrium). We do not compute the positions of the current sheets which can be arbitrary.

In the second part, we first consider the QPSE corresponding to symmetric boundary conditions: we then know that the current sheet has to be located on the vertical z-axis, and we can then compute the magnetic field. In the case of arbitrary asymmetric boundary conditions, we do not present an analytical way to solve the free boundary value problem, but we propose an iterative scheme to solve it, the current sheet varying in position until it reaches its equilibrium position corresponding to the state of minimum energy.

In the case of weak asymmetry, we can compute analytically the QPSE even if the position of the current sheet cannot be guessed a priori, the position being solution of the boundary-value problem.

References

Aly, J. J., Quasi-static evolution of sheared force-free fields and the solar flare problem, Astron. Astrophys. 143, 19, 1985.

Aly, J. J., Evolving magnetostatic equilibria, in "Interstellar Magnetic Fields", ed R. Beck and R. Grave, Springer-Verlag, Berlin, p.240, 1987.

Aly, J. J. and Amari, T. : 1985, Some new results in the theory of two-dimensional magnetostatic equilibria, in Theoretical Problems in High Resolution Solar Physics, ed H. U. Schmidt, MPA 212, p.319.

Aly, J.J. and Amari, T., Two-dimensional non-symmetric models of quiescent prominences in potential magnetic fields, Astron. Astrophys. 207, 154, 1988.

Aly, J. J. and Amari, T., Current sheets in two dimensional potential magnetic fields : I - general properties, Astron. Astrophys. (in Press), 1989.

Amari, T., Theorie des equilibres magnetostatiques et applications a l'astrophysique, Ph.D. Thesis, University of Paris VI, 1988.

Amari, T. and Aly, J. J., Current sheets in two dimensional potential magnetic fields : II - Asymptotic limits of indefinitely sheared force-free fields, Astron. Astrophys., in press, 1989a.

Amari, T. and Aly, J. J., Current sheets in two dimensional potential magnetic fields : III - Formation in complex topology configurations and application to coronal heating, Preprint, 1989b.

Dungey, J. W., Condition for the occurrence of electrical discharges in astrophysical systems, Philos. Mag. (17) 44, 725, 1953.

Low, B. C., Electric current sheet formation in a magnetic field induced by continuous magnetic footpoint displacements, Ap. J. 323, 358, 1987.

BRAIDED FLUX ROPES AND CORONAL HEATING

Mitchell A. Berger[1]

Department of Mathematical Sciences, University of St Andrews, Fife, Scotland

Abstract. We model a coronal loop as a set of N flux tubes braided about each other. The braiding is generated by random motions of the photospheric foot points. For one or two tubes the magnetic energy grows linearly with time as the motions proceed. For three or more tubes, however, energy grows quadratically with time. A method is given which provides a qualitative description of the flux rope structure for $N = 3$.

In Parker's (1983) coronal heating model, randomly braided fieldlines undergo reconnection; the magnetic energy associated with the braiding structure is then released as a source of heat. Calculating the energy storage and release is a difficult problem. Using a simple model for the structure of three braided flux tubes, a Monte Carlo simulation yields an average energy growth rate significantly smaller than that estimated in the Parker model. However, higher values of N will be needed before reliable comparisons can be made.

1. Introduction

Solar coronal loops are rooted in a dense and turbulent plasma below the photosphere. Random motions of field line endpoints at the photosphere can twist and braid the field lines above. Because of this, coronal loops can acquire a considerable amount of internal structure. Dissipation of the magnetic energy associated with the random component of the field has been suggested as a source for coronal heating [Sturrock and Uchida 1981; Parker 1983]. In order to evaluate this possibility, we need to have some idea of how fast magnetic energy builds up due to random boundary motions.

A quiescent loop satisfies a nonlinear equilibrium equation ($\vec{J} \times \vec{B} = 0$ or $\vec{J} \times \vec{B} = \nabla p$). For a given field topology there will be some minimum energy field which solves one of these equations. Unfortunately, three-dimensional solutions generally possess either current sheets or very thin current layers [Parker 1972; van Ballegooijen 1987; Mikić et al 1988]. For this reason highly structured loops are very difficult to model numerically or analytically. However, for the purposes of estimating energy storage it is not necessary to have an exact equilibrium solution. A magnetic field satisfying the same boundary conditions and topological constraints as an equilibrium field will have a higher energy than that equilibrium field. If such a field is available, upper limits can be placed on the equilibrium magnetic energy. This report describes how such fields might be constructed, given an arbitrary braiding pattern.

2. Braided Loop Model

We model a coronal loop as a straight cylinder of length L and radius R. The volume inside the cylinder is filled by N distinct flux tubes braided about each other. Near the ends of the cylinder (the photosphere) there is a very thin transition region where the flux tubes narrow down to points. Thus at the ends the magnetic flux is contained in N footpoints, each footpoint having a negligibly small radius. This model approximates the actual solar conditions, where the flux at the photosphere is highly localised [Frazer and Stenflo 1972; Zayer et al 1989]. Initially, the N tubes are straight and parallel. The footpoints then random walk about each other. In the absence of reconnection the flux tubes above become braided due to this boundary motion. An important parameter is the number of footpoints per unit area, $1/d^2 = N/\pi R^2$. The length d gives a typical distance between footpoints. The footpoints always stay within radius R; if a footpoint approaches the boundary at R, it simply bounces off elastically. This keeps d constant.

If we make N and R large enough (for a given d), the effects of the boundary conditions at R become unimportant (relatively few footpoints actually bounce off the wall at R). Furthermore, the shape of a flux tube depends upon its near neighbours (it must snake about them), but does not depend upon far away flux tubes. Intuitively, then, the magnetic energy per unit area should be independent of N for large N (given a fixed d). In

[1] Present Address: Department of Mathematics, University College London.

Geophysical Monograph 58

practice, if we have a good algorithm for approximating the minimum energy field, we may not have to go to a very large N in order to obtain a reasonable estimate of the energy stored in a braided field. This research report describes the situation for $N \leq 3$. Even for N as low as 3, there are interesting effects due to non-trivial braiding. Eventually, however, higher N must be examined if we are to accurately simulate tangled coronal loops.

3. One or Two Tubes

First consider $N = 1$. A single flux tube fills the entire cylinder (we ignore the thin transition regions where the tube narrows to a point at the photosphere). The free magnetic energy comes from field lines twisting about the central axis. Let T be the number of turns through 2π taken by a field line. The mean twist $\overline{T}$ is zero, but the mean-square twist $\overline{T^2}(t)$ grows linearly with time [Sturrock and Uchida 1981]: starting with $\overline{T^2}(t) = 0$ at $t = 0$ and assuming infinite conductivity one finds [Berger 1989]

$$\overline{T^2}(t) = 3\tau_c t \overline{V^2}/2\pi^2\lambda^2. \quad (1)$$

Here $\overline{V^2}$ is the mean square photospheric velocity, τ_c is the (Lagrangian) correlation time for the photospheric flow, and λ is the velocity correlation length for the flow. The mean energy also grows linearly with time. For granulation, if we use the parameters $\lambda = 800$ km, $V \equiv (\overline{V^2})^{1/2} = 1$ km s^{-1}, and $\tau_c = \lambda/V = 800$ s, then $\overline{T^2}(t) = t/1.5$ hours, implying an rms twist of about four turns after 24 hours.

The situation for $N = 2$ is similar: the two tubes have internal twists T_1 and T_2, but also wind about each other through an angle θ_{12}. This angle is called the *winding number*, and has attracted considerable interest in random walk theory [Spitzer 1958; Berger and Roberts 1988]. The mean winding number is zero but the mean-square winding number increases linearly with time (this will be explained in more detail in the next section). Coronal loops consisting of two intertwined fluxtubes have been discussed by Glencross [1975].

Suppose magnetic energy dissipates with a timescale τ_d. For both $N = 1$ and $N = 2$ the power input (energy input per unit time) is independent of t, and hence independent of τ_d. In other words, the power input does not depend on the amount of structure already built up within the cylinder, or for that matter on how much structure has been lost to dissipation. Consequently, the extremely difficult details of reconnection in the line-tied corona (e.g. Velli and Hood 1989) can be ignored when calculating the heating rate. Note that τ_d may be much shorter than the resistive time because of rapid reconnection events. The heating can be considered stochastic (and independent of τ_d) as long as $\tau_d > \tau_c$.

4. Three or More Tubes

The situation becomes more complex and more interesting for $N > 2$. Here the tubes can be *tangled* about each other. By tangled, we mean that successive braiding patterns do not commute with each other; in other words, the order in which footpoints wind about each other is important (figure 1). To see this, consider chopping a braid into a number of pieces X, Y,

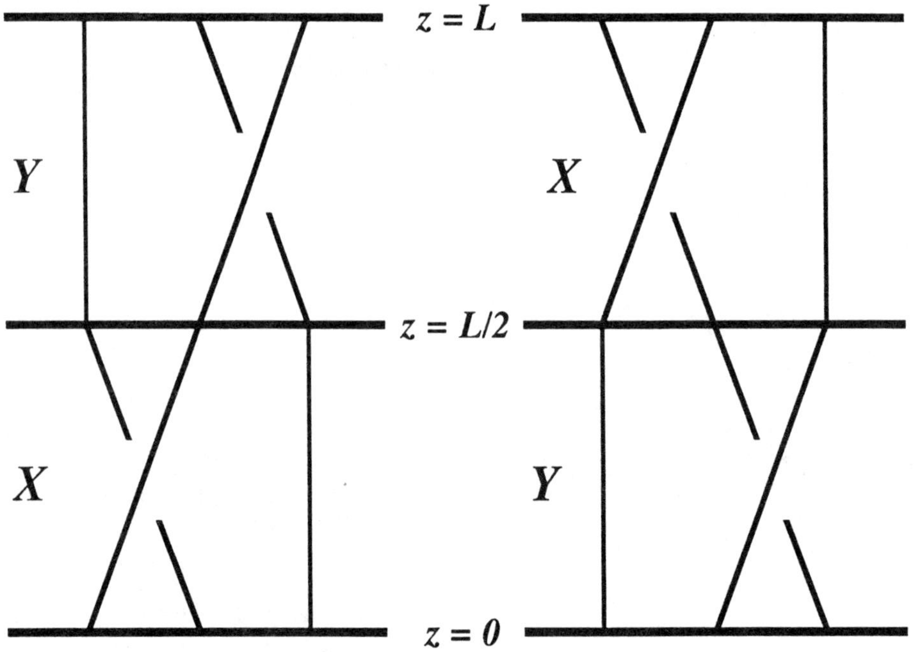

Fig. 1. The braid XY on the left is not equivalent to YX on the right. Thus X and Y do not commute.

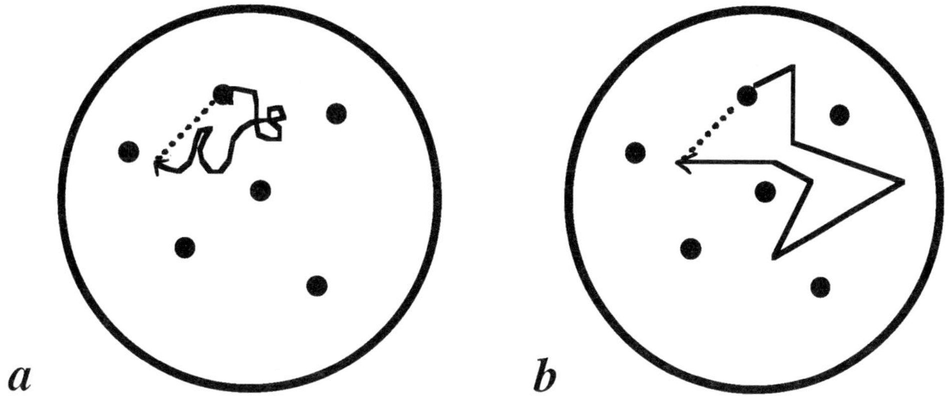

Fig. 2. The paths of footpoints at the photosphere. a: If a path jiggles, there is little net change in the coronal field structure. b: Because of backtracking, the solid line and the dashed line yield equivalent paths.

Z, ..., and look at how these pieces interact. In general a braid XY consisting of pattern X between $z = 0$ and $z = L/2$, and pattern Y between $z = L/2$ and $z = L$, is not topologically equivalent to the braid YX with pattern X above pattern Y. Furthermore, every braid X has an inverse X^{-1} where, if X^{-1} is placed above X, the whole thing unravels and we are left with a set of parallel tubes ($XX^{-1} = I$ = the *trivial braid.*)

Let us go back to two tubes for a moment. A braid between two tubes (call the tubes A and B) is completely described by the internal twists T_A, T_B, and the mutual winding θ_{AB}. To simplify the present discussion, let us assume that these twists and windings come in discrete pieces, say in units of one turn (in reality T_A, T_B, and θ_{AB} evolve continuously as footpoint motions proceed). Let us call a right handed twist of one turn on tube A γ_A with inverse (left handed twist) γ_A^{-1}. The twists on tube B are γ_B or γ_B^{-1}, and mutual windings between the tubes are γ_{AB} or γ_{AB}^{-1}. *All these twists and windings commute with one another.*

A sequence of random motions of the footpoints of the tubes will produce a braid which can be described by a sequence of these symbols. For example, the motions might produce something like

$$\gamma_A \gamma_B^{-1} \gamma_A^{-1} \gamma_{AB} \gamma_B^{-1} = \gamma_{AB} \gamma_B^{-2}. \qquad (2)$$

The number T_A equals the number of γ_A elements minus the number of γ_A^{-1} elements, etc. Thus this braid has $T_A = 0, T_B = -2$, and $\theta_{AB} = 1 \cdot 2\pi$. The numbers T_A, T_B, and θ_{AB} grow like a one-dimensional random walk on the integers with mean zero and mean-square proportional to the number of steps.

Now consider a braid with three or more tubes. Some of the braid patterns do not commute with each other. Thus the braid pattern XYX^{-1} cannot be reduced; X and X^{-1} cannot cancel each other out because Y is in the way. A sequence of patterns X, X^{-1}, Y, and Y^{-1} may have some cancellations, but in general the length of the sequence will grow linearly with the number of patterns.

5. The Topology of 3-Braids

Tangling for three braided curves can be quantified by a *tangle number* $\mathcal{T}$ (see figure 3). At any plane $z = constant$ between the boundaries $z = 0$ and $z = L$, the three curves intersect the plane in three points. The tangle number equals the minimum number of planes where the three points are co-linear. These planes can be labelled A, B, or C depending on which point is in the middle. Thus there is a sequence (e.g. $ACBCBABA$...) of planes, reading from bottom to top.

Let us phrase this in terms of the discussion in §4. Slice the braid into a number of sections, each slice being parallel to the $x - y$ plane. Each section should contain a single plane where the three points are co-linear. A section then consists of a simple braiding pattern which we may also label A, B, or C. Note that $A = A^{-1}$, etc. Thus the sequence of braid patterns (ACB in figure 3) does not contain repeating letters (like AA). Furthermore, the patterns A, B, and C do not commute. For this reason the sequence for any braid is unique. A pigtail braid follows a sequence like $ACBACBACB$.... The number of letters in a sequence equals the tangle number $\mathcal{T}$.

We can find a simple 'smooth braid' configuration for three curves as follows: First subtract the global twist. Second slice the cylinder into $\mathcal{T}$ sections, each of height $L/\mathcal{T}$. At the bottom and top planes of each section the intersection points of the curves form an equilateral triangle. The size of the triangle is chosen so that the vertices have radius $r = 2R/3$, as explained below. At the center plane of each section the intersection points are co-linear, one curve having moved in between the other two.

We wish to consider the three curves as representing the center field lines of three magnetic flux tubes (of equal flux). These tubes should completely fill the cylinder of radius R. If the field were parallel to $\hat{z}$ then each tube would (in the $x - y$ plane) take the shape of a pie slice, with geometric center at $r = 2R/3$.

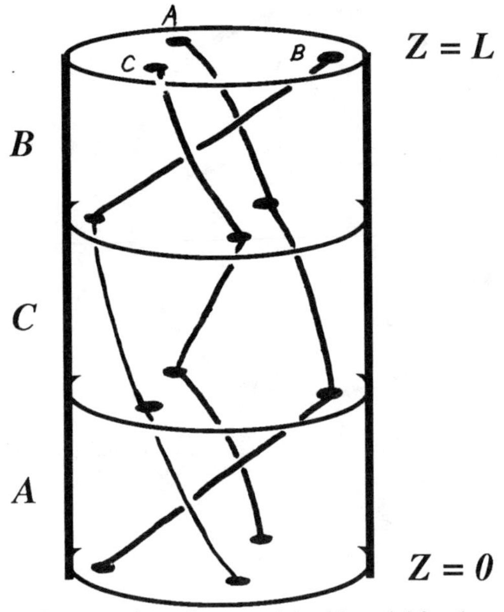

Fig. 3. Construction of a smooth braid model for three intertwined flux tubes. In each piece one tube moves in between the other two.

Figure 4a shows a projection of three curves which have random walked about each other. Figure 4b displays the smooth braid as described in the previous paragraphs. In figure 4c the twist has been put back. The curves in 4c are topologically equivalent to 4a.

6. The Energy of a Tangled Field

Parker [1983] estimated the heating rate of a tangled field in a Cartesian geometry, with a field extending between the planes $z = 0$ and $z = L$. The endpoints of initially vertical field lines random walk for a time t with a typical velocity V. He suggested that the transverse length ℓ of a field line (i.e. the length of the line as projected onto the $x - y$ plane) should be approximately $\ell \approx Vt$. The reason is that each flux tube gets so tangled about the other tubes that it cannot straighten itself out. Its total transverse length ℓ is then the total distance travelled, Vt. The transverse magnetic field $B_\perp$ then grows as $B_\perp = B_0 Vt/L$ where B_0 is the axial field strength. The energy now grows *quadratically* in time. As a consequence, the heating rate grows *linearly* with τ_d, i.e. the more the magnetic energy stored, the greater the heating rate.

Certain effects, however, can reduce the transverse length, and hence the transverse field strength and the free magnetic energy (see figure 2). First, even when the footpoints are closely packed, the path of an individual point will still have some jitter which can be smoothed out. Secondly, there is some probability for back-tracking, where a point reverses its direction so as to decrease the tangling. If the probability for back-tracking is p, then the transverse length of a tube reduces to $\ell \approx (1-2p)Vt$. A first guess might be that $p \sim 1/3$ or $1/4$, since there will typically be 3 or 4 different paths a point might take about its neighbours. Third, and more subtly, some of the braiding will go into coherent large-scale twists of three or more tubes about each other. Such twists do commute with other forms of braid-

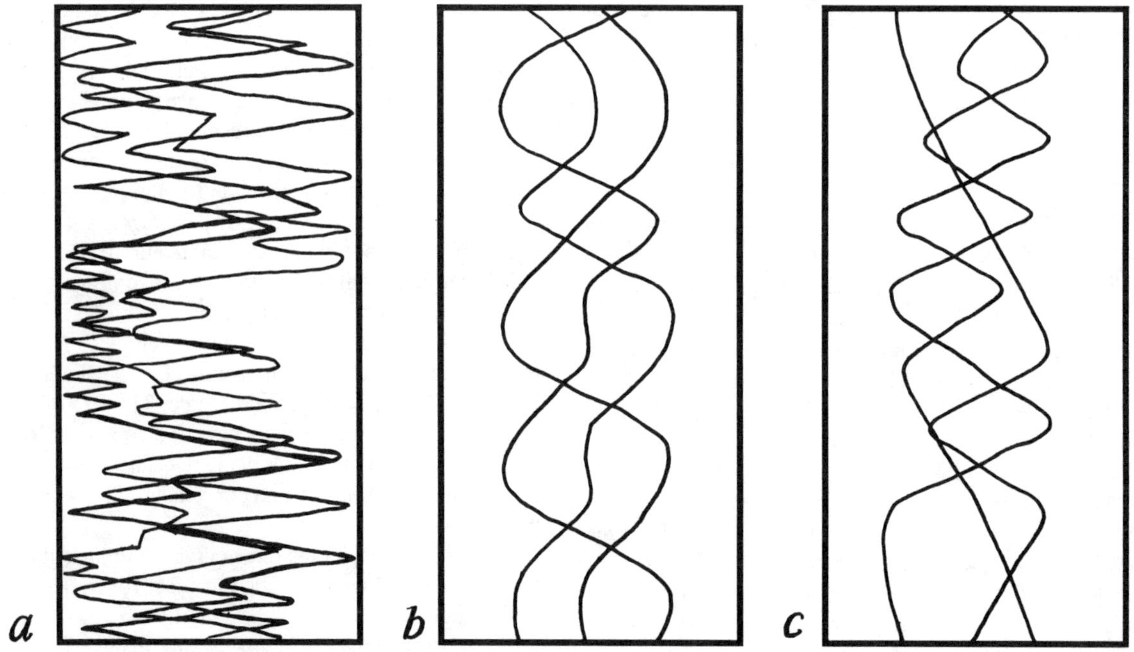

Fig. 4. Reductions of a random braid to a smooth one. a: three points have random walked 36 steps each. The x coordinate is plotted vs. time. b: Global twist has been subtracted and the smooth braid constructed, with $\mathcal{T} = 9$ ($x - z$ projection). c: Global twist has been put back.

ing; twists in opposite directions at widely separated times can cancel each other out. As with $N = 1$ or $N = 2$, the commuting portion of braiding leads to a linear energy growth rate; but the new element of non-commuting tangling leads to a quadratic growth rate.

Given these considerations, one might expect the mean square transverse length $\overline{\ell^2}$ to scale as

$$\overline{\ell^2} = at + b\frac{\lambda^2}{d^2}(Vt)^2 \tag{3}$$

where a and b are numerical factors and d is the typical distance between footpoints. The first term takes into account the Sturrock-Uchida twisting of field lines within tubes, as well as large scale commuting twist between tubes. We will not try to estimate a here. The second term represents non-commuting braiding (tangling). The constant b allows for the effects of backtracking and the fact that some of the braiding motion contributes to the first term (twisting rather than tangling). The factor λ^2/d^2 allows for the time it takes for a randomly moving footpoint to go around another footpoint. We will ignore the first term from now on.

If $B_\perp/B_0$ were constant over the length of a field line then $B_\perp = \ell B_0/L$. A tangled field contains a mean free energy density $(\overline{B_\perp^2}/8\pi)$ which scales as

$$W_f = w\left(\frac{\overline{\ell^2}}{L^2}\right)\left(\frac{B_0^2}{8\pi}\right) \tag{4}$$

$$= \left[wb\frac{\lambda^2}{d^2}\right](Vt/L)^2\,(B_0^2/8\pi)\,. \tag{5}$$

Here w is a numerical factor of order unity: estimating the energy from a knowledge of $\overline{\ell^2}$ is only approximate. To calculate the power per unit area P flowing into a closed coronal loop, remember that there are two photospheric footpoints: ℓ increases owing to motions at both ends so multiply $\overline{\ell^2}$ by 4; integrate from $z = 0$ to $z = L$; and divide by two because two footpoint areas are included in "per unit area". For $t = \tau_d$ the power in $\text{erg cm}^{-2}\,\text{s}^{-1}$ is

$$P = 2LdW_f/dt = 1.4\times 10^8\left[wb\frac{\lambda^2}{d^2}\right]B_2^2\,V_5^2\,L_{10}^{-1}\,\tau_{d1} \tag{6}$$

($B_2 = B_0/10^2 G$, $V_5 = V/10^5\,\text{cm s}^{-1}$, $\tau_{d1} = \tau_d/1\,\text{day}$, and $L_{10} = L/10^{10}\,\text{cm}$). A power of $P = 10^7\,\text{erg cm}^{-2}\,\text{s}^{-1}$ is sufficient to heat an active region [Withbroe and Noyes 1977]. The extra terms within the brackets modify Parker's original model [Parker 1983]. If we ignore these terms and set $V = 0.4$ km s^{-1}, then a decay time of $\tau_d = 11$ hours will provide enough heating power.

7. Monte Carlo Simulation for $N = 3$

We have examined how the tangle number evolves due to random boundary motions in a Monte Carlo simulation. Three points random walk about each other at $z = 0$ for 200 steps, staying inside a disk of radius R (as described in §2). The step size (s) is constant. This generates a set of three curves (the field lines connecting to the three points) inside the cylinder. A numerical algorithm reduces the curves to the smooth braid pattern as pictured in figure 4. For the curves in 4b

$$\overline{\ell^2} = 0.9R^2\mathcal{T}^2. \tag{7}$$

From equation (4) the energy is

$$W_f = 0.9w\left(\frac{R^2\mathcal{T}^2}{L^2}\right)\left(\frac{B_0^2}{8\pi}\right). \tag{8}$$

Does the step size s for a random walk correspond to the velocity correlation length λ for a smooth flow (assuming $V\tau_c = \lambda$)? In the random walk simulation, each footpoint moves independently of the other two. For a random velocity field, on the other hand, neighboring footpoints will have correlated motions if their mutual distance is less than the coherence length λ. Strictly speaking, the step size s for the random walk corresponds to λ as long as the typical distance between footpoints satisfies $d >> \lambda$. In the simulation, a dependence $\mathcal{T} \sim (s/d)^2$ was found, consistent with the $(\lambda/d)^2$ dependence in equations (3) and (5).

For $R = s$ (analogous to one photospheric flux element per granule) the tangle number was, on average, about 1/4 the total number of steps taken by each footpoint ($\overline{\mathcal{T}} \approx 46$ for 200 steps). The coefficient b in equations (3), (5), and (6) was found to be $b \approx 0.05$. This suggests, for $N = 3$, an energy growth rate an order of magnitude smaller than in the Parker model. Equivalently, a decay time τ_d an order of magnitude greater would be needed to provide the same coronal heating power. It is possible that the coefficient b, and hence the energy growth rate, will increase with N.

8. Conclusions

The general situation of N flux tubes braided about each other within a flux rope has been considered. If the braiding is generated by random boundary motions, then the magnetic energy evolution will have two terms (equations (3) and (4)). The term linear in time represents motions such as rotations of individual tubes and large scale uniform rotations; these motions create braiding structures which commute with each other, as described in §4. The quadratic term represents non-commutative braiding, or tangling.

A method has been given in §6 for constructing a regular 'smooth braid' configuration for any set of three braided flux tubes (as pictured in figure 4). This configuration does not precisely satisfy the magnetostatic equilibrium equations; because of its simplicity it may nevertheless be useful in modelling braided flux ropes (reducing a set of three random walks of 200 steps to a smooth braid pattern takes less than a second of VAX time). The energy of the smooth braid is significantly less than in the Parker approximation. However, generalization to higher numbers of flux tubes remains uncertain.

Questions and Answers

Birn: It seems that your definition of the "tangle numbers" relies on the orientation of arbitrary boundaries and is, thus, not an invariant property of the magnetic field itself. Is that correct?

Berger: Imagine projecting the curves onto the cylindrical boundary. The tangle number is essentially equal to the number of crossovers of these projected curves. This number does not change if the boundary orientation is chosen differently. The tangle number is an invariant of a set of braided curves.

Acknowledgments I am grateful to Eric Priest and Joe Hollweg for important suggestions.

References

Berger, MA 1989, Generation of magnetic field structure by random boundary motions: mean square twist and current density , (preprint).

Berger, MA and Roberts, PH 1988, On the winding number problem with finite steps *Adv. Applied Probability* **20**, 261.

Frazer, EN and Stenflo, JO 1972, On the small scale structure of solar magnetic fields *Solar Physics* **27**, 330.

Glencross, WM 1975, Heating of coronal material at X-ray bright points *Astrophysical J.* **199**, L53.

Mikić, Z, Schnack, DD, and Van Hoven, G 1988, Creation of current filaments in the solar corona , (preprint).

Parker, EN 1972, Topological dissipation and the small-scale fields in turbulent gases *Astrophysical J.* **174**, 499.

Parker, EN 1983, Magnetic neutral sheets in evolving fields. II. Formation of the solar corona *Astrophysical J.* **264**, 642.

Spitzer, F 1958, Some theorems concerning 2-dimensional Brownian motion *Amer. Math. Soc. Trans.* **87**, 187.

Sturrock, PA and Uchida, Y 1981, Coronal heating by stochastic magnetic pumping *Astrophysical J.* **246**, 331.

Withbroe, GL, and Noyes, RW 1977, Mass and energy flow in the solar chromosphere and corona *Annual Review of Astronomy and Astrophysics* **15**, 363.

van Ballegooijen, AA 1987, *Proc. Ninth Sacramento Peak Workshop on Solar and Coronal structure and Dynamics* , Sunspot, New Mexico.

Velli, M and Hood, AW 1989, Resistive tearing in line-tied magnetic fields: slab geometry *Solar Physics* **119**, 107.

Zayer, I, Solanki, SK, and Stenflo, JO 1989, The internal magnetic field distribution and the diameters of solar magnetic elements *Astronomy and Astrophysics* , to appear.

CORONAL LOOP HEATING BY RESONANT ABSORPTION

Stefaan Poedts[1], Marcel Gooseens[1], and Wolfgang Kerner[2]

[1]Astronomisch Instituut, Katholieke Universiteit Leuven
Celestijnenlaan 200 B, B—3030 Heverlee, Belgium
[2]Max—Planck—Institut für Plasmaphysik, Euratom Association
Boltzmannstraβe 2,D—8046 Garching bei München, FRG

Abstract. The heating of coronal loops by resonant absorption of Alfvén waves is studied in compressible, resistive magnetohydrodynamics by means of numerical simulations in which the loops are approximated by straight cylindrical, axisymmetric plasma columns. The incident waves, which excite the coronal loops, are modelled by a periodic external driver. The efficiency of the heating mechanism and the localization of the heating strongly depend on the characteristics of both the external source and the equilibrium. The numerical results show that resonant absorption is very efficient for typical coronal loop parameter values. A considerable part of the energy supplied by the external driver, is actually dissipated ohmically and converted into heat.

1. Introduction

The heating of the solar corona is a major problem in solar and stellar physics since many classes of stars have a corona like the sun [see e.g. Rosner et al., 1985 ; Jordan and Linsky, 1987]. Satellite observations (e.g. Skylab, 1973) of the solar corona in soft X – rays have revealed the dominant influence of the magnetic field on the structure of the corona and there is now general consensus that the heating of the corona is magnetic in nature. The corona consists of open and closed magnetic structures. The open regions are known as coronal holes and give the most important contributions to the solar wind. The regions in which the magnetic field is mainly closed consist of myriads of loops. These hot and dense loops are outlined by the magnetic field and show up as bright structures in X – ray observations. By consequence, the loop heating mechanism is of primordial importance.

A general photospheric magnetic field disturbance produces waves of several types such as fast and slow magnetosonic waves and Alfvén waves. The magnetosonic waves steepen into shocks and dissipate in a similar manner to pure acoustic waves. The strong inhomogeneity of the solar corona magnetic field and its strong concentration in intense kilogauss fields, leads to the suggestion that Alfvén waves play a dominant role in the heating of the solar corona [Hollweg, 1979, 1981]. Due to the strong magnetic fields, Alfvén waves can propagate very easily in the solar atmosphere with hardly any attenuation at all. In fact, the problem with Alfvén waves is not to explain how they can reach the corona, but rather how they are dissipated.

In an inhomogeneous plasma magnetic waves and Alfvén waves in particular can be dissipated very efficiently by means of resonant absorption. The spectrum of oscillation frequencies of an inhomogeneous plasma contains two continuous parts in linear ideal MHD : an Alfvén continuum and a slow magnetoacoustic continuum. Consider a diffuse cylindrical axisymmetric plasma with equilibrium quantities only varying in the radial direction. The local Alfvén frequency – given by the dispersion relation $\omega_a(r) = k(r)\, v_a(r)$, with k the wave number along the magnetic field and v_a the Alfvén speed – is then a function of the radial coordinate r. When the plasma is excited periodically by an external source with a frequency, say ω_0, within the range of the Alfvén continuum, a resonance occurs on the magnetic surface r_s where $\omega_0 = \omega_a(r_s)$. The radial velocity component has a logarithmic singularity but the singularity manifests itself in a much stronger way in the tangential components of the velocity field which possess $(r - r_s)^{-1}$ – singulatities which are non – square integrable [see e.g. Goedbloed, 1983]. As time progresses, the plasma energy will, therefore, accumulate unbounded (in ideal MHD) in an ever diminishing plasma layer around the resonance point. However, due to dissipative effects, the system attains a stationary state after a finite time. In this stationary state all physical quantities oscillate in time with the frequency of the driver, ω_0, and the energy dissipation rate in the resonant layer exactly balances the power of the external driver. The absorbed energy is thermalized by the dissipation.

Geophysical Monograph 58

Resonant absorption of Alfvén waves was first studied in the context of controlled thermonuclear fusion research as a supplementary heating mechanism for bringing laboratory plasmas in the ignition regime. It was first proposed as a heating mechanism for coronal loops by Ionson [1978]. Since then this heating mechanism has been investigated by many authors in this context [see e.g. Ionson, 1982, 1984 ; Rae and Roberts, 1981, 1982 ; Heyvaerts and Priest, 1983 ; Nocera et al., 1983 ; Sakurai, 1985 ; Sakurai and Granik, 1984 ; Mok and Einaudi, 1985 ; Lee and Roberts, 1986 ; Grossmann and Smith, 1988 ; Hollweg, 1987a, 1987b ; Hollweg and Yang, 1988]. Poedts et al. [1989a] have shown that resonant absorption can be very efficient for typical coronal loop parameter values. The aim of the present paper is to investigate the dependency of the localization of the heating and the efficiency of the heating mechanism on the characteristics of the external source and the equilibrium. The physical model we consider for this purpose is discussed in the next Section. The results are presented in Section 3 and the conclusions are drawn in Section 4.

2. Physical Model

2.1 Configuration

The calculations presented in this paper are obtained with a computer code in which a cylindrical plasma with radius r_p is considered, surrounded by a vacuum and by a perfectly conducting wall with radius r_w [see Poedts et al., 1989b]. The linear displacements of the plasma column are assumed to be excited by an idealized external antenna situated in the vacuum region, with radius r_a ($r_p \leq r_a \leq r_w$). Of course, coronal loops are not surrounded by a vacuum region and a perfectly conducting wall neither are they excited by a current in external coils. In fact, the power spectrum of the waves that are incident on the loops is not known. Therefore, we restrict ourselves to the determination of the intrinsic energy dissipation rate in coronal loops [cfr. Grossmann and Smith, 1988]. For the determination of the intrinsic energy dissipation rate it is assumed that the power spectrum of the external driver is uniform, i.e. all the waves that are incident on the coronal loops are assumed to have the same amplitude. The actual heating of the loops can only be determined when the power spectrum of these waves is known. The intrinsic energy dissipation can be computed by placing (in the numerical code) the driver at the plasma – vacuum interface ($r = r_p = r_a$) and by taking the wall away from the plasma column ($r_w \rightarrow \infty$). In this set – up the periodic driver simulates a wave with particular wave numbers and frequency that is incident on the loop. The solution is normalized by taking a unit current in the antenna. This procedure yields the source impedance which is a very useful quantity in a system which is excited by an external source. The impedance is proportional to P, the power emmited by the external source. The real part of P indicates the dissipative damping while the imaginary part of P represents the fraction of the total energy circulating in the system (see Section 2.3).

2.2 Resistive MHD equations

We consider a onedimensional cylindrically – symmetric plasma column with equilibrium quantities varying only in the radial direction. The resistive MHD equations that govern linear displacements about this ideal static equilibrium are :

$$\rho \frac{\partial \mathbf{v}}{\partial t} = -\nabla p + (\nabla \times \mathbf{B}) \times \mathbf{b} + (\nabla \times \mathbf{b}) \times \mathbf{B} , \tag{1}$$

$$\frac{\partial p}{\partial t} = -\mathbf{v}.\nabla P - \gamma P \nabla.\mathbf{v} , \tag{2}$$

$$\frac{\partial \mathbf{b}}{\partial t} = \nabla \times (\mathbf{v} \times \mathbf{B}) - \nabla \times (\eta \nabla \times \mathbf{b}) . \tag{3}$$

with P, ρ, and $\mathbf{B}$ respectively the equilibrium plasma pressure, density and magnetic field (magnetic induction) and $\mathbf{v}$, p, and $\mathbf{b}$ the Eulerian variation of respectively the velocity, the plasma pressure and the magnetic field. η is the electric resistivity and γ the adiabatic index. Equations (1) – (3) are expressed in dimensionless units. The distance is normalized to the plasma radius r_p and the time to the Alfvén – transit – time $t_a = r_p / V_a$, where V_a is the Alfvénspeed given by the longitudinal magnetic field and the density on the axis ($r = 0$). Equation (1) is the momentum equation for a non – viscous plasma. Equation (3) is the induction equation which includes the ohmic term (due to the finite electric conductivity of the plasma). Equation (2) is the ideal MHD version of the equation for the variation of the internal energy. Notice that there is no restriction to incompressible plasmas. This enables us to describe the coupling of the driver to the plasma. In the process of resonant absorption the energy of the driver is transported by means of fast magnetosonic waves which only appear in a compressible plasma. The coupling of the driver to the plasma is an important aspect of the heating mechanism since this coupling determines the fraction of the supplied energy that is actually converted into heat, in other words the efficiency of the process of resonant absorption.

The solutions of the system (1) – (3) have to satisfy appropriate boundary conditions : the solutions have to be regular in $r = 0$ and the total pressure and all three components of the magnetic field have to be continuous at the plasma – vacuum interface.

Since the equilibrium quantities do not depend on θ and z, the perturbed quantities can be Fourier analyzed in these coordinates and each Fourier term can be studied separately. The following separation ansatz is made for the perturbed quantities f :

$$f(r, \theta, z ; t) = f(r ; t) \ e^{i(m\theta + nkz)} . \tag{4}$$

The number $k = \pi / L$ defines a quantization factor to allow an integral number of half – wavelengths on the column (the ends of the loop are tied in the dense photosphere) with L the lenght of the loop in the z – direction ($L = \pi\epsilon$, with ϵ the the aspect ratio of the loop). The cylindrical geometry closely approximates that of a large aspect ratio loop. For coronal loops ϵ is typically 20. n is the longitudinal mode number and m is the mode number in the θ – direction. When the dissipative system is excited by an external periodic source, it reaches a stationary state in which all physical quantities vary periodically in time with the frequency of the external driver. The energy dissipation rate in this stationary state can be computed by substituting the temporal dependence

$$f(r\,;\,t) = f(r)\ e^{i\omega_p t} \tag{5}$$

in equations (1) – (3), where ω_p is the frequency of the external source. With the substitutions (4) and (5) the system (1) – (3) reduces to a set of ordinary differential equations of order 6 in the radial coordinate r. This system is solved numerically with the finite element technique [see Poedts et al., 1989a, 1989b for details].

2.3 Energetics

When the system (1) – (3) is solved the energy dissipation rate is calculated in order to determine the efficiency of the heating proces. Moreover, the complete picture of energy conservation in the system provides us with an excellent test for the numerical accuracy of the code and for the validity of the physical assumptions made. The linearized resistive MHD equations yield the following energy equation

$$-\int_{V_p} \nabla.(\mathbf{e}^*\times\mathbf{b})\,dV = \underbrace{\int_{V_p} \rho\mathbf{v}^*.\frac{\partial\mathbf{v}}{\partial t}\,dV}_{\equiv K} + \underbrace{\int_{V_p} \eta j^2\,dV}_{\equiv OD} + \underbrace{\int_{V_p}\left\{\mathbf{v}^*.\nabla p - \mathbf{v}^*.\mathbf{J}\times\mathbf{b} + \mathbf{b}.\frac{\partial\mathbf{b}^*}{\partial t}\right\}dV.}_{\equiv W_p} \tag{6}$$

e and j are the Eulerian perturbations of respectively the electric field and the electric current density. The * denotes the complex conjugate and V_p is the volume of the plasma. According to equation (6) an inflow of electromagnetic energy produces a rise of the kinetic energy of the plasma (K), a change of the potential energy of the plasma (W_p) and heat by ohmic dissipation (OD). The left hand side term of equation (6) is related to P, the power emitted by the antenna [see Poedts et al., 1989b]

$$P = -\int_{V_v} \mathbf{j}_{ant}.\mathbf{e}_v^*\,dV.$$

V_v is the volume of the vacuum region and $\mathbf{j}_{ant}$ is the electric current density in the antenna. In the stationary state all physical quantities oscillate with the frequency of the driver and the resistive energy balance reads

$$\mathcal{Re}(P) = OD\,. \tag{7}$$

This means that the supply of energy from the antenna exactly balances the rate at which energy is converted into heat by ohmic dissipation.

The resistive energy equation (6) and the resistive energy balance (7) provide excellent tests for the numerical accuracy of the code and for the validity of the physical approximations made (ideal static equilibrium, adiabatic law). These approximations are good as long as the equations (6) and (7) are satisfied within reasonable bounds. The numerical accuracy is determined by the distribution of the meshpoints and can be raised to any level by enlarging the number of gridpoints and/or mesh accumulation in the nearly – singular layer and at the plasma boundary. The relative error in the energy balance, made by the physical assumptions, is typically 0.001 %. These assumptions yield, therefore, a very good approximation.

3. Results

In this Section we investigate how the efficiency of resonant absorption and the localization of the intrinsic heating depend on the characteristics of the external driving source and the equilibrium. We consider a class of equilibria given in dimensionless units by the profiles

$$B_z = 1,$$

$$J_z = j_0\,(1 - r^2)^{\,\nu},$$

$$\rho = 1 - (1 - d)\,r^2.$$

The distance is normalized with respect to the radius of the plasma colum (r_p), and the magnetic field and the density are normalized, respectively, by $B_0 = B_z(0)$ and $\rho_0 = \rho(0)$. The constants d, j_0, and ν are free parameters with $j_0 = 2k/q_0$ (where k is the inverse aspect ratio of the loop and q_0 the safety factor on the axis, $q(r) = rkB_z/B_\theta$). d denotes the density at the plasma boundary, $d = \rho(1)$.

The efficiency of the resonant absorption process is expressed in terms of the fractional absorption, f_a, which is defined as

$$f_a = \frac{\text{ohmic dissipation rate}}{\text{total power input}}\,.$$

Clearly, $0 \leq f_a \leq 1$, and the closer f_a approximates 1 the better the coupling between the driver and the plasma is, *i.e.* the larger the fraction of the supplied energy that is actually converted into heat by ohmic dissipation. The localization of the intrinsic dissipation is studied by adding up the contributions $\eta j^2(r)$ for 50 driving frequencies in the range of the ideal continuum (equidistant) and normalizing this sum such that its integral over the plasma volume equals 1. Here η is the electrical conductivity (assumed constant) and $j(r)$ is the current density at the radial position r.

3.1 External source

Poedts et al. [1989a] have shown that the fractional absorption (f_a) is very sensitive to the frequency of the external driver. Extremely high efficiency is obtained when the driving frequency is close to the frequency of a so – called 'collective mode' [see e.g. Balet et al., 1982 ; Sedlacek, 1971]. For a bandwidth around the optimal driving frequency $f_a \approx 1$. For very smooth equilibrium profiles this 'efficient' bandwidth is narrow but for realistic variations of the equilibrium quantities it comprises almost the complete range of the ideal continuum.

The efficiency of resonant absorption depends strongly on the wave numbers of the waves that are incident on the coronal loops. Figure 1 shows the fractional absorption versus the driving frequency for $n = 1$ and $m = 1$ to 5, where the other parameters are fixed. The range of the ideal Alfvén continuum is different for different values of m which means that the bandwidth of frequencies that can be resonantly excited depends on m. By consequence, many different driving frequencies contribute to the heating of the same plasma layers as the corresponding resonant layers range from $r = 0$ to $r = 1$ for each value of m. The $m = 1$ mode yields the strongest coupling and so the most efficient heating. For this mode $f_a \geq 90$ % for a large bandwidth. The fractional absorption gradually decreases as m increases and for $m = 5$, f_a amounts to 0.26 for the optimal driving frequency. When the wave number m is fixed and n is varied, it is found that the maximal value of f_a (for the optimal driving frequency) does not depend on the value of n when m is positive. For negative values of m, however, the optimal efficiency is higher for the higher values of n.

3.2 Equilibrium

η dependence. The ohmic dissipation rate in the stationary state is almost independent of η for the relevant (very small) values of this parameter [Poedts et al., 1989a]. For a given driving frequency the heating is localized in a resonant layer around the 'singular' surface. The thickness of this resonant layer, Δ_{res}, depends on the resistivity of the plasma and the variation of the equilibrium profiles. A parametric study yielded the following result :

$$\Delta_{res} \sim [\eta \,/\, \omega_a'(r_s)]^{1/3},$$

where $\omega_a(r_s)$ is the local Alfvén frequency in $r = r_s$ and the prime denotes the derivation with respect to r.

The ρ – profile. Varying the parameter d $(= \rho(r = 1))$ also affects the efficiency of the heating mechanism. This is illustrated in figure 2 where the fractional absorption is displayed versus the driving frequency for $d = .10$ to $.30$. Resonant absorption is more efficient for stronger density variations in the coronal loop plasma, i.e. for lower values of the parameter d. Notice also in figure 2 that the range of the ideal Alfvén continuum is larger for a stronger variation of the density, which means that the bandwidth of frequencies that are eligible for resonant absorption, is larger. The intrinsic dissipation is localized in the outer layers of the plasma column and so the heating mechanism is most efficient in these layers. It is found that the parameter d has little influence on the localization of the

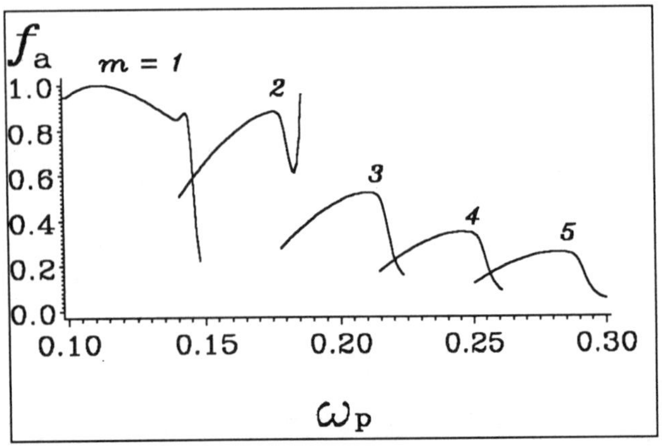

Fig. 1. The fractional absorption versus the frequency of the driver for $\nu = 2$, $d = 0.2$, $j_0 = 0.1$, $\eta = 10^{-8}$, $n = 1$, and $m = 1, 2, 3, 4$, and 5.

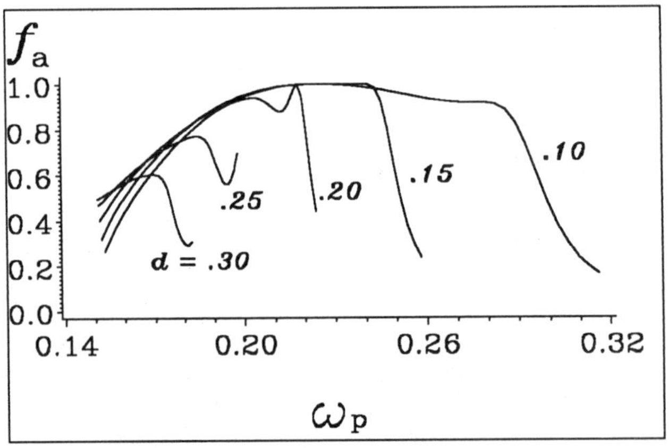

Fig. 2. The fractional absorption versus the frequency of the driver for $\nu = 1$, $j_0 = 0.1$, $\eta = 10^{-8}$, $m = 2$, $n = 1$, and $d = 0.10, 0.15, 0.20, 0.25$, and 0.30.

intrinsic heating. The ohmic dissipation is slightly more localized in the outer regions of the coronal loop for the higher values of the parameter d, i.e. for the smoother density variations.

The J_z – profile. The variation of the equilibrium current density is determined by the parameters ν and j_0 ($= 2k/q_0$). A peaking of the current profile increases the efficiency of the heating mechanism and affects the localization of the intrinsic dissipation substantially as illustrated in figure 3. Figure 3 shows the variation of the (intrinsic) ohmic dissipation in the radial direction of the plasma column for $\nu = 0.5$, 1, 2, and 4. For smooth current density profiles (e.g. $\nu = 0.5$) the ohmic dissipation is almost uniformly spread over the coronal loop. A peaking of the current results in a localization of the intrinsic heating in the outer plasma layers.

For a given current profile ($\nu = $ cst) the efficiency is higher when the current on the axis ($r = 0$) increases. Figure 4 illustrates this by showing the fractional

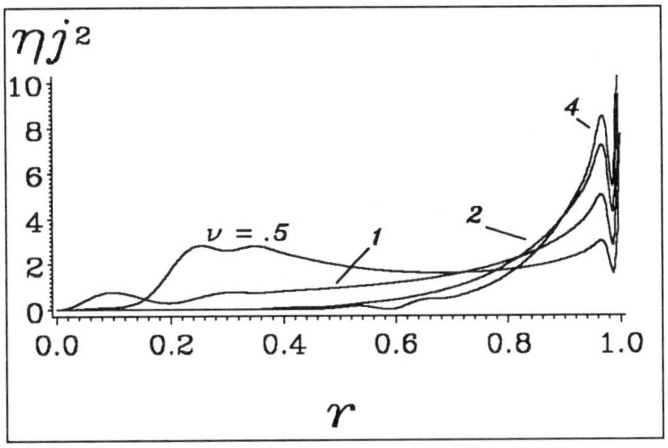

Fig. 3. The ohmic dissipation rate versus the radial coordinate for $d = 0.2$, $j_0 = 0.1$, $\eta = 10^{-7}$, $n = 1$, and $\nu = 0.5$, 1, 2, and 4.

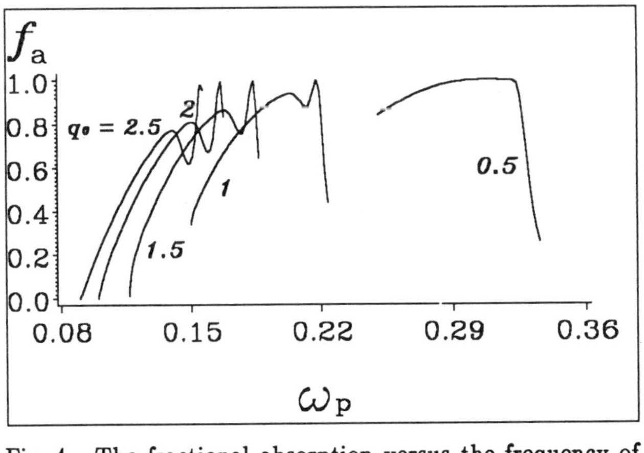

Fig. 4. The fractional absorption versus the frequency of the driver for $\nu = 1$, $d = 0.2$, $\eta = 10^{-8}$, $n = 1$, $m = 2$,and $j_0 = 0.04$ to 0.20 ($q_0 = 0.5$, 1, 1.5, 2, and 2.5).

absorption versus the driving frequency for $j_0 = 0.2$ to 0.04 (i.e. $q_0 = 0.5$ to 2.5). For all these parameter values the intrinsic dissipation is localized in the outer regions of the plasma. This localization of the ohmic dissipation in the stationary state is not very sensitive to the value of j_0 and is only slightly more pronounced for the lower values of this parameter.

4. Conclusions

We have investigated the heating of solar coronal loops by resonant absorption of Alfvén waves in the framework of linearized compressible resistive MHD. The resonant absorption of the waves that are incident on the coronal loops is numerically simulated in straight cylindrical, axisymmetric loop models, externally excited by a periodic source. The power spectrum of the external source is not known and so not specified in the numerical simulations (where it is assumed to be uniform). This reduces the calculations to the determination of the 'intrinsic' dissipation. The actual heating of a coronal loop can only be determined when the driving spectrum of the loop is known, namely by taking the convolution of this input spectrum with the intrinsic dissipation.

Resonant absorption of Alfvén waves is a very efficient heating mechanism for coronal loops. The fractional absorption in the stationary state is very high for typical coronal values of the equilibrium parameters. The intrinsic dissipation is localized in the outer plasma layers indicating that the absorption process is most efficient in the outer regions of the coronal loops. We conclude that resonant absorption is a viable heating mechanism for coronal loops provided the power spectrum of the external source includes continuum frequencies.

Questions and Answers

Pizzo: Physically, why do you get the result that the ohmic heating is independent of the Reynolds number, when the Reynolds number is sufficiently large?

Poedts: Because as the width of the resonant layer decreases for larger magnetic Reynolds numbers, the amplitudes in this layer increase, therefore, the integral of $\eta j^2(r)$ over the resonant layer is almost independent of the magnetic Reynolds number.

Nordlund: These are nondimensional frequencies. What are the actual frequencies for typical loops?

Poedts: For typical coronal loop values the continuum frequencies are typically 0.01Hz. The range of frequencies that are eligible for resonant absorption agrees, therefore, very well with the observed motions in the solar atmosphere which have periods of typically 3 sec.

Ryutova: I think that if you consider the boundary-value problem for an inhomogeneous loop you will find that the resonant absorption becomes less efficient than "phase mixing" since in this case dissipation becomes important providing the heating of the loop region at a definite height. As to 90%

efficiency, do you mean that you know the input of energy and the real structure of a loop?

Poedts: No, I don't know the power spectrum of the incident waves. I wish I did? I just say that 90% of the energy supplied by the external source is actually dissipated and converted into heat, for typical coronal loop parameter values. In my 1-D model the dissipation is uniform along the loop. To explain the temperature difference between the footpoints and the top of the loop, one has to solve a 2-D problem (the "boundary-value problem" as you called it). Phase mixing and resonant absorption involve the same physical mechanism but differ in the way they couple the energy of the source to the continuum Alfvén waves. In phase mixing, the coupling is direct, while in resonant absorption, a "collective mode" is needed to transport the energy across the magnetic surfaces to the resonant layer.

Roberts: Parker has pointed out that whatever heats active regions must work efficiently at scales (loop lengths) varying from a few times 10^3 km up to 10^5 km. How does the efficiency of resonant absorption, as calculated in your model, scale with the length of a loop?

Poedts: The only place where the length of the loop enters in the calculations is in the quantization factor of the longitudinal mode number. This quantization factor is equal to the inverse aspect ratio of the loop. Given the fact that the longer loops have larger radii, the aspect ratio is roughly the same for all loops (cf., the tutorial paper of R. Rosner). Therefore, the loop length does not affect my results and the heating mechanism is equally efficient for all loops.

Martens: One of the often heard objections against wave heating theories is that at coronal values for the resistivity, the required amplitude of the waves (in order to reproduce the observed heating rates) is much higher than can be inferred from observations. Have you resolved this problem?

Poedts: No, I did not resolve this problem. My results are all in dimensionless units and I did not estimate the required amplitudes of the waves. According to Hollweg and Yang (J. Geophys. Res., 93, 5423, 1988), however, the required amplitudes are about 40 km s- and this estimate is very close to the measured "turbulent" velocities for a quiescent coronal active region.

Acknowledgements. S. Poedts is Research Assistant of the Research Council of the K.U.Leuven whose support is greatly acknowledged. The K.U.Leuven – MPIPP collaboration is supported by the E.E.C. Scientific Cooperation Contract N° ST2J – 0275 – C.

References

Balet, B., Appert, K., and Vaclavik, J. : 1982, *Plasma Phys.*, **24**, 1005.

Goedbloed, J.P. : 1983, Rijnhuizen Report 83 – 145.

Grossmann, W. and Smith, R.A. : 1988, *Ap. J.*, **332**, 476.

Heyvaerts, J. and Priest, E. : 1983, *Astron. & Astrophys.*, **117**, 220.

Hollweg, J.V. : 1979, *Solar Phys.*, **62**, 227.

Hollweg, J.V. : 1981, *Solar Phys.*, **70**, 25.

Hollweg, J.V. : 1987*a*, *AP. J.*, **312**, 880.

Hollweg, J.V. : 1987*b*, *AP. J.*, **320**, 875.

Hollweg, J.V. and Yang, G. : 1988, *J. Geophys. Res.*, **93**, 5423.

Ionson, J.A. : 1978, *Ap. J.*, **226**, 650.

Ionson, J.A. : 1982, *Ap. J.*, **254**, 318.

Ionson, J.A. : 1984, *Ap. J.*, **276**, 357.

Jordan, C. and Linsky, J.L. : 1987, in *'Exploring the Universe with the IUE Satellite'* (*ed.* Kondo, Y.), 259.

Kerner, W., Lerbinger, K., Gruber, R. and Tsunematsu, T. : 1985, *Computer Physics Communications*, **36**, 225.

Lee, M.A. and Roberts, B. : 1986, *Ap. J.*, **301**, 430.

Mok, Y. and Einaudi, G. : 1985, *J. Pl. Phys.*, **33**, 199.

Nocera, L., Leroy, B. and Priest, E.R. : 1983, *Astron. & Astrophys.*, **133**, 387.

Poedts, S., Goossens, M. and Kerner, W. : 1989*a*, *Solar Phys.*, in press.

Poedts, S., Kerner, W. and Goossens, M. : 1989*b*, *J. Plasma Phys.*, in press.

Rae, I.C. and Roberts, B. : 1981, *Geophys. Astrophys. Fluid Dyn.*, **18**, 197.

Rae, I.C. and Roberts, B. : 1982, *Mon. Not. R. Astron. Soc.*, **201**, 1171.

Rappaz, J. : 1975, *Num. Math.*, **28**, 15.

Rosner, R., Gollub, L. and Vaiana, G.S. : 1985, *An. Rev. Astron. & Astrophys.*, **23**, 413.

Sakurai, T. : 1985, in H.U. Schmidt (*ed.*), Proc. of the Workshop MPA/LPARL *'Theoretical Problems in High Resolution Solar Physics'*, Max – Planck – Institut für Astrophysik, Munich, 263.

Sakurai, T. and Granik, A. : 1984, *Ap. J.*, **277**, 404.

Sedlacek, Z. : 1971, *J. Plasma Physics*, **5**, 239.

LINEAR EVOLUTION OF CURRENT SHEETS IN SHEARED FORCE-FREE MAGNETIC FIELDS WITH DISCONTINUOUS CONNECTIVITY

Richard Wolfson

Department of Physics, Middlebury College, Middlebury, VT 05753

Abstract. The formation of thin current sheets in tenuous, magnetized plasma may be an important mechanism for the buildup and subsequent release of energy in astrophysical situations, especially in the solar corona. Such sheets might arise from the random but continuous motion of magnetic footpoints associated, for example, with photospheric velocity fields. Whether such sheets can form when all features of the magnetic and velocity fields are strictly continuous remains controversial. However, current sheet formation in situations incorporating some sort of discontinuity in field line connectivity is expected, although few quantitative examples exist. This paper develops a model for studying the quasistatic evolution of current sheets due to shearing of footpoints in a highly idealized geometry incorporating an abrupt jump in field line connectivity. Although the model is limited to the linear regime in which the field deviates only slightly from its initial potential state, it nevertheless shows clearly the formation of thin current layers, and permits surprisingly large shearing motions before the linear approximation is violated. Finally, calculations of energy buildup, scaled to the size of a typical solar flare region, show that excess energy comparable to that released in a flare can be stored in the sheared field.

Introduction

The formation and subsequent dissipation of electric current sheets in tenuous, highly conducting astrophysical plasmas may provide a mechanism for the buildup and sudden release of energy that characterizes such events as solar flares and coronal mass ejections. More frequent dissipation of weaker current sheets may also contribute to coronal heating (Parker 1986a, 1988). Scenarios for current sheet formation generally involve the motion of magnetic footpoints anchored in a high-beta gas such as the solar photosphere; the overlying field is generally taken force free, a reasonable approximation in the tenuous, magnetically dominated solar corona.

Whether and how such current sheets actually form is presently a matter of controversy. Parker (1972, 1983, 1986b, 1987, 1989) has argued that the evolution of an initially continuous force-free magnetic field almost invariably leads to the formation of current sheets. Parker's suggestion has been verified for some special cases by Tsinganos *et al.* (1984), Moffat (1985), Aly (1986), Low and Wolfson (1988), Low (1989), Wolfson (1989) and others. On the other hand, Van Ballegooijen (1985, 1986, 1988), Antiochos and Karpen (Antiochos 1987; Antiochos and Karpen 1989), and Field (1989) argue that current-sheet formation is not a necessary consequence of force-free evolution, and indeed may not be possible under the most general conditions in which all aspects of the initial magnetic field and of the velocity field associated with the motion of magnetic footpoints are continuous.

Recently Low and Wolfson (1988; hereafter LW) have suggested that current sheets must arise when quadrupolar magnetic field configurations are sheared, and numerical calculations (Wolfson 1989) have helped verify that suggestion. The essential feature of the quadrupolar configurations considered by LW is that the field line connectivity be discontinuous, in that there exists a pair of field lines whose footpoints at one end are infinitesimally close, while those at the other end are separated by a finite distance. Configurations with this property might arise if two dipolar regions approached to form a quadrupolar configuration. Or they might occur because solar magnetic flux emerges predominantly in the network, so that two field lines adjacent in the corona could be rooted at distant points in the photosphere.

Although the arguments of LW regarding sheared quadrupolar fields appear sound, and seem supported by numerical calculations, it is hard to make those numerical studies fully convincing because of the difficulty in dealing with the discontinuity represented by a current sheet. With the quadrupolar fields considered by LW, the discontinuity is abrupt and cannot be smoothed over; that is, the geometry precludes any possibility of approximating the discontinuity in field line connectivity with a steep but continuous function.

Theory

To circumvent this difficulty, and to isolate the essential features of current sheet formation due to shearing, we consider a highly artificial situation that appears to be one of the simplest in which current sheet formation is definitely required. In this simplified situation, an abrupt change in field line connectivity is achieved by having field lines anchored to an irregular boundary. Figure 1 shows schematically such a boundary, along with field lines of a potential field in the irregular domain. Here the potential field is parallel to the yz plane, and x is an ignorable coordinate.

Now consider shearing the field of Fig. 1, by moving the left-hand footpoints in the x direction, perpendicular to the plane of the figure. The shear could be any continuous function of z, but suppose for simplicity that it has the constant value Δx. The shearing introduces an x component of the magnetic field, along with associated currents; the field is no

Geophysical Monograph 58

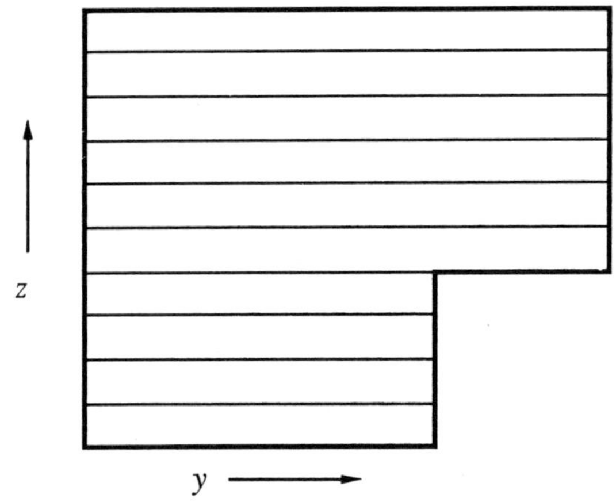

Fig. 1. Irregular domain with potential field lines. Dimensions are those used in the actual calculations, except that here the transition is a true discontinuity.

longer potential. It is well known (see Priest 1982) that a force-free field with x ignorable is described by the equation

$$\nabla^2 A + B_x \frac{dB_x}{dA} = 0 \quad , \tag{1}$$

where the scalar functions B_x and A are each constant on a field line. A is simply the x component of the vector potential, so the y and z components of the field are given by $B_y = \partial A/\partial z$, $B_z = -\partial A/\partial y$. If the shear Δx is specified, integration over a field line then shows that B_x must satisfy the equation

$$\Delta x = B_x \int \frac{ds}{|\nabla A|} \quad , \tag{2}$$

where ds is the element of arc length along the field line, and where the integration goes between footpoints. Equations (1) and (2) form a nonlinear integro-differential system for determining A.

Here we approach the nonlinear problem by considering the linear case in which the function A is only slightly different than in the potential field case. As we show, however, this approximation does not preclude the formation of thin current layers.

In the linear case, the function $A(y,z)$ for the potential field is used to calculate B_x in equation (2). The resulting B_x is therefore a known function of y and z, and equation (1) becomes linear. Solving this linear equation then gives an approximate solution for the sheared field. Using the result again in equation (2) would give an iterative procedure useful in the nonlinear regime. Here, however, we consider only the linear approximation.

To handle the problem numerically, we replace the boundary discontinuity shown in Fig. 1 with a steep but continuous function, describing the right-hand edge of the domain by $y = y_b(z)$. If we let $A = B_0 z$ describe the potential solution, where B_0 is the magnitude of the initial uniform potential field, then it is easily shown that in the linear approximation equation (2) yields an x component of the sheared field given by

$$B_x = \frac{B_0 \Delta x}{y_b(z)} \quad , \tag{3}$$

where Δx is the shear, here assumed independent of z. Using this result in equation (1) gives

$$\nabla^2 A - \frac{B_0 (\Delta x)^2}{y_b^3(z)} \frac{dy_b(z)}{dz} = 0 \tag{4}$$

Equation (4) is to be solved on an irregular boundary specified by $y_b(z)$. The boundary can be made regular by introducing a new variable $w(y,z)$, given by $w = y/y_b(z)$. Furthermore, the inhomogeneous term in equation (4) becomes large where the boundary shape changes rapidly. This situation can be handled by introducing a new variable $v = v(z)$, scaled to put more grid points in the transition region, whose approximate width is specified by Δz. An appropriate function is given by

$$\frac{dz}{dv} = 1 - \sigma f(z) \quad ,$$

where the function f is

$$f(z) = -\frac{B_0 (\Delta x)^2}{y_b(z)} \frac{dy_b}{dz} \quad ,$$

and where σ is a constant chosen so that one-fourth of all grid points lie in the transition region, no matter how narrow. In these new variables w,v, the differential equation becomes

$$\frac{\partial^2 A}{\partial w^2} + y_b^2 \left(\frac{dv}{dz}\right)^2 \frac{\partial^2 A}{\partial v^2} - \sigma\, y_b^2 \frac{df}{dz} \frac{\partial A}{\partial v} + f(z) = 0 \quad . \tag{5}$$

This equation can be solved using standard codes for linear PDE's. The field geometry can be determined by contouring the solution, and current density and other physical parameters are readily calculated. Figure 2 shows a typical grid–that is, lines of constant w and v–imposed on the irregular domain in yz space.

Results

For all calculations, we use a domain with $0<z<1$, and with y ranging from 0 to a maximum of 0.8 at $z=0$ to 1.2 at $z=1$. The boundary transition is centered at z coordinate of 0.4, and its width is designated by Δz. These boundary parameters are established using a boundary-shape function $y_b(z)$ given by $y_b = \alpha + \beta \tan^{-1}[\gamma(z - z_1)]$, where the constants α, β, and γ are chosen to fix the width Δz of the transition region and the magnitude of the change in the boundary location (i.e., from $y=0.8$ at $z=0$ to $y=1.2$ at $z=1$), and where $z_1=0.4$ is the z coordinate at the center of the transition region. Strictly speaking, a true current sheet exists only in the limit $\Delta z \rightarrow 0$; our calculations show how that limit is approached in a sequence of finite current layers. We also show the effect of varying the shear Δx.

Figure 3 shows the steepening and narrowing of the current layer as the transition region thickness Δz is decreased

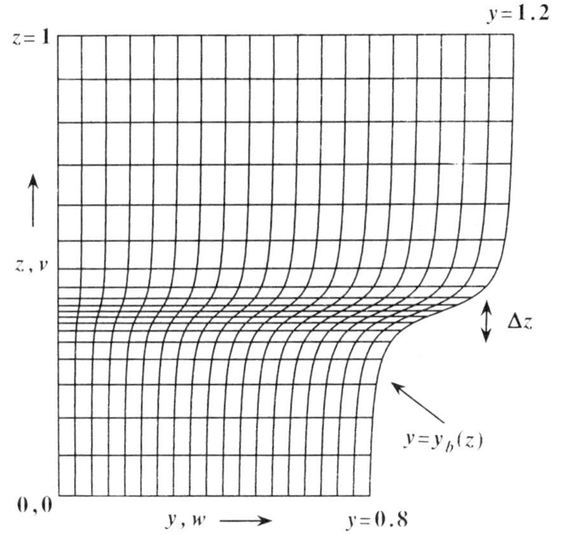

Fig. 2. *wv* grid lines shown in physical (*yz*) domain. Domain width in *y* direction (horizontal) varies from 0.8 at bottom to 1.2 at top; domain is bounded on the right by the curve designated $y=y_b(z)$. For the case shown, half-width Δz of the transition region is 0.05. Note the bunching of grid points near the transition region. Actual solutions are done on a much denser 128x128 grid.

from 0.12 to 0.08 to 0.04, with shear Δx held constant at 0.15. Although the detailed field configuration in the immediate vicinity of the transition changes significantly as Δz is varied, note that the general shape of the field, and in particular of the field lines within the sheet, remains nearly the same despite the factor-of-three variation in the transition region width Δz. This similarity is also evident in the current sheet profiles, where the general curvature of the current sheet does not change significantly with decreasing Δz. The height of the sheet does, of course, increase, since the total current needed to effect the nearly constant change in field magnitude across the sheet remains the same in all cases, even as the width of the region in which significant current flows is decreasing. (The change in the field magnitude is due almost entirely to the shear-associated component B_x, which is not evident in Fig. 3 since it is perpendicular to the plane of the figure.) The fact that the general configuration of the field and associated current system does not change significantly with decreasing Δz suggests that the use of a finite-width current layer probably represents a good approximation to the limit of a true current sheet, in which Δz=0.

Figure 4 shows the effect of increasing the shear Δx while the transition thickness Δz is held constant. Here the current sheet does change in the curvature of its profile as well as in height, indicating that the changing shear has a real physical effect. In particular, increasing magnetic pressure associated with the growing field component B_x perpendicular to the plane of the figure causes the current sheet to bulge. The direction of this bulging is upward because, as equation (3) shows, the shear-associated component of the field is greatest where the domain width y_b is smallest–that is, in the region below the current layer.

The results summarized in Figs. 3 and 4 represent numerical solutions to equation (5), which arises from the integro-differential system of equations (1) and (2) for the case when the unsheared potential field is used to calculate the shear-associated field component B_x that rightfully should be determined self-consistently. Thus the results are valid only in linear regime in which deviations from the potential field are small. For the case Δx=1.5 shown in all frames of Fig. 3, the maximum deviation in *A* from its potential-field value is about 30 per cent, so the linear approximation is at best marginal. (Significant deviation from the potential-field values is, however, confined to the immediate vicinity of the transition–that is, only near the right-hand side of the current layer). That this deviation remains essentially the same for all three frames of Fig. 3–that is, for the three different values of transition width dz–is further indication that the finite-layer approximation used here is close to representing the limit of a true current sheet. For the first two cases shown in Fig. 4, the deviation in *A* from its potential values is much smaller, so the linear approximation should be satisfied. Note that even in these two cases the shear Δx is not small; it corresponds to displacement of field line footpoints in the *x* direction by 50 and 100 per cent of the domain dimensions in the *yz* directions. Thus the linear approximation can be used for significant shears.

Although the model used here is highly idealized, it is instructive to calculate the energy buildup associated with the sheared field. In the linear approximation, we may neglect the *z* component of the field (that component associated with the curvature of the field lines), so the energy per unit length in the *x* direction is given approximately by

$$\varepsilon = \frac{1}{8\pi}\int \left(B_x^2 + B_y^2 \right) dA = \frac{B_0^2}{8\pi}\int_0^1 \left[1 + \left(\frac{\Delta x}{y_b(z)} \right)^2 \right] y_b(z)\, dz,$$

where B_0 is the strength of the initial uniform potential field, through which the potential-field *A* is specified as a function of *z* by $A=B_0 z$. The second term in the square brackets is associated with the nonpotential effects–that is, with excess energy arising from the shear. Designating this excess energy per unit length in the *x* direction by $\Delta\varepsilon$, we have

$$\Delta\varepsilon = \frac{B_0^2 (\Delta x)^2}{8\pi} \int_0^1 \frac{dx}{y_b(z)} = \frac{B_0^2 (\Delta x)^2}{8\pi} \frac{z_0}{\langle y_b \rangle},$$

where $\langle y_b \rangle$ is average value of $y_b(z)$ over the right-hand boundary; in our case, $\langle y_b \rangle$ is of order unity. Thus the energy per unit length is given approximately by

$$\Delta\varepsilon = \frac{B_0^2 (\Delta x)^2}{8\pi} .$$

Scaling the domain to the size of a typical flare region, taken as a cube 5x10^4 km on a side, and using B_0=500 G for a typical field strength, this equation gives excess energy buildup of nearly 3x10^{33} erg–well over that needed for a large flare–for the maximum shear Δx=1.5. The energy buildup as a function of shear is shown in Fig. 5; clearly the buildup of flare energies is possible even with the more modest shears Δx<1 appropriate to these linear calculations.

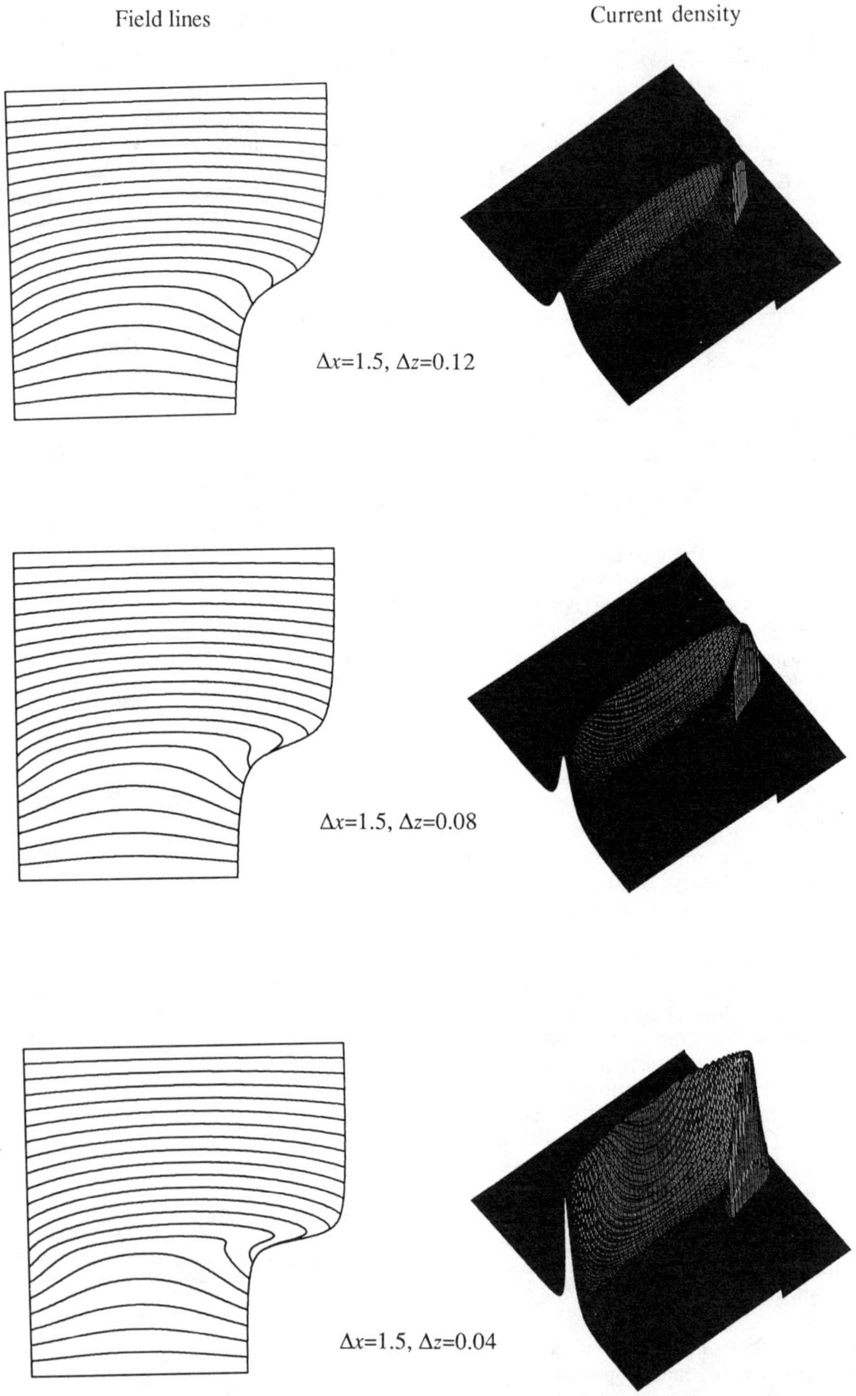

Fig. 3. Effect of decreasing width Δz of the transition region, for fixed shear Δx=1.5 (i.e., moving footpoints in the x direction by one and one-half times the height of the domain). From top to bottom, Δz=0.12, 0.08, 0.04, compared with a total extent from 0 to 1 in the z direction. Left-hand plots are projections of field lines on the yz plane, plotted as contours of constant A. Right-hand plots show current density, with vertical scale the same for all three plots. Depressed areas at lower right mark regions outside the irregular domain in which the boundary-value problem is posed. Note that the general shape of the current layer is the same in all cases, indicating that these narrow layers may approach closely the actual configuration of an infinitesimally thin current sheet.

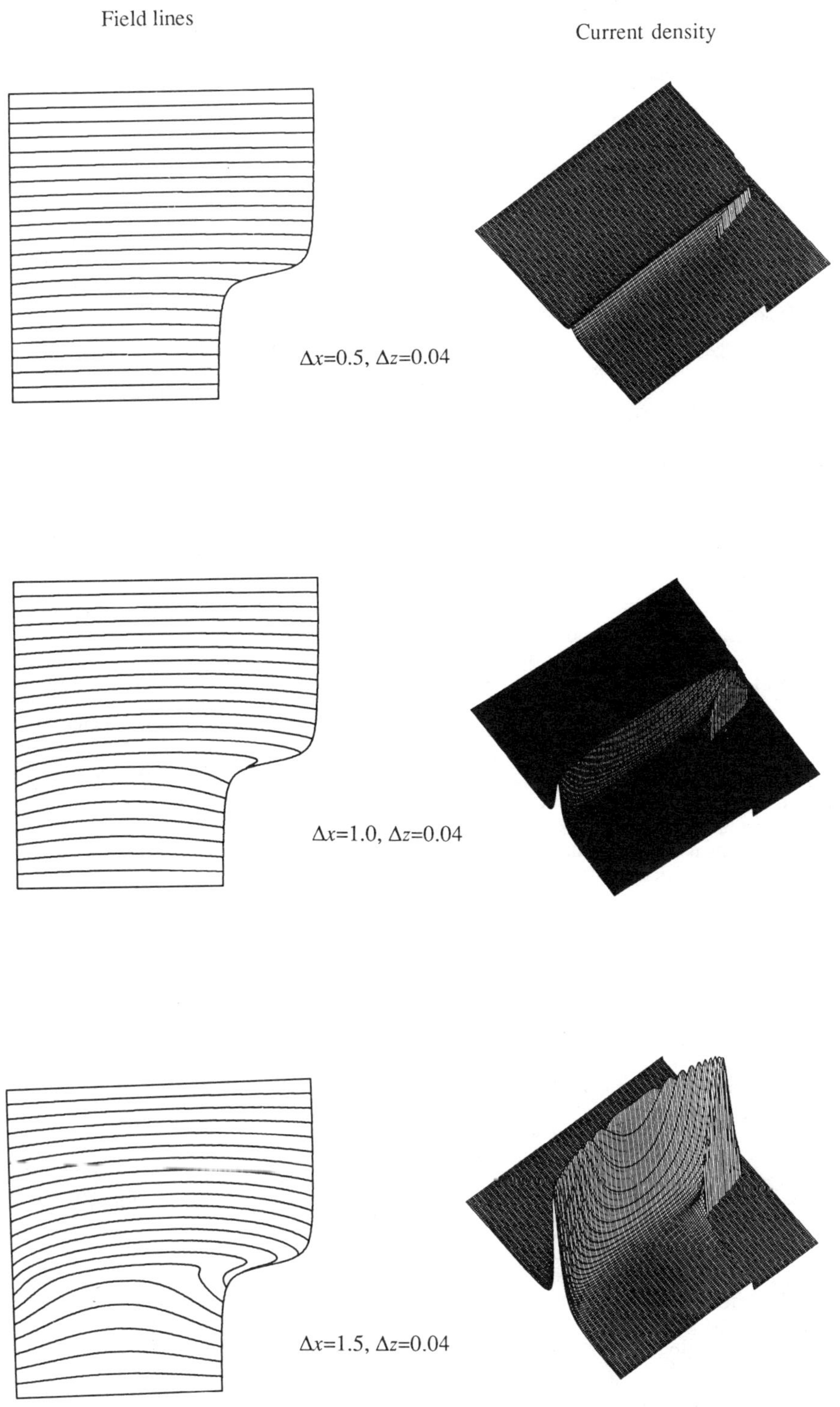

Fig. 4. Effect of increasing shear Δx at fixed transition region thickness Δz=0.04. From top to bottom, Δx=0.5, 1, 1.5. Note the increasing curvature of the current layer, due to magnetic pressure associated with the increasingly sheared field.

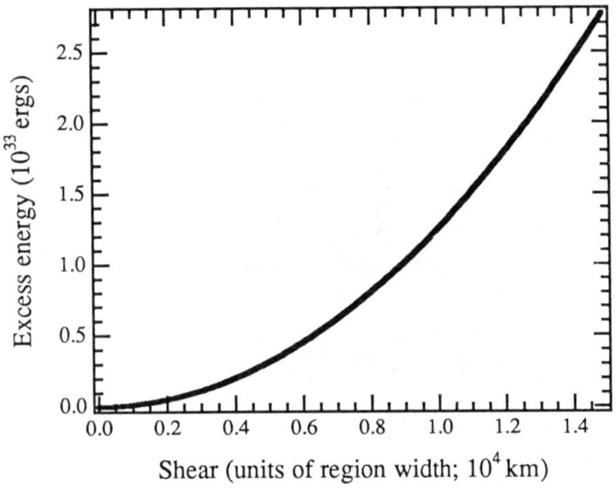

Fig. 5. Buildup of excess energy, above that in the initial potential field, in a region roughly $5x10^4$ km on a side, with initial magnetic field B=500 G.

One question left unanswered by the linear approached used in this paper is whether equilibrium solutions remain possible for all values of shear. In the linear model, iteration to self-consistency is not done, and there is no question of convergence; a solution of equation (5) is assured. When a fully nonlinear model is developed, it will be important to study the question of the existence of solutions as a function of shear. The absence of equilibrium solutions may shed some light on conditions necessary for the sudden onset of dynamic behavior in previously quasistatic force-free magnetic structures. Such behavior might model aspects of the eruption of solar flares, eruptive prominences, and coronal mass ejections.

Conclusion

We have demonstrated the formation of narrow current layers in a sheared, force-free magnetic field whose field line connectivity exhibits an sharp change due, in this case, to an irregular boundary. Since the adaptive gridding scheme used in the model permits arbitrarily sharp transitions, the model shows how a true current sheet may be considered the limit of an arbitrarily thin layer that is in force-free equilibrium throughout. Although our model is highly idealized, it may correctly describe the physics involved in the formation of current layers through shearing of more realistic configurations like multipolar regions or network field on the sun. Simple scaling arguments show that even in the linear regime, shearing may build up the energy necessary for a solar flare.

Acknowledgments. This work was supported primarily by NASA Solar Maximum Mission Guest Investigator Grant NAG 5-762 to Middlebury College. Computer time was provided by the National Center for Atmospheric Research, which is sponsored by the National Science Foundation, under NSF grant AST 85-20557 to Middlebury College.

Questions and Answers

Antiochos: For the case of the quadrupole field, I would argue that current sheets are not formed because the line-tying boundary condition must be dropped at the central neutral line.

Wolfson: I agree, but I don't think that argument applies to the configuration shown here, and while your objection may apply to the real corona/photosphere boundary, it is still important to answer the theoretical question of whether continuous footpoint motions in a perfectly conducting boundary can lead to current sheet formation.

References

Aly, J.J. 1986, in Proceedings of the Workshop on Interstellar Magnetic Fields, Schloss Ringberg (RFA), September 8-12, 1986.

Antiochos, S.K. 1987, *Ap. J.*, **312,** 886.

Antiochos, S.K. and Karpen, J.T. 1989, *Bul. Am. Astron. Soc.*, **20,** 1029.

Field, G.B. 1989, *Bul. Am. Astron. Soc.*, **20,** 979.

Low, B.C. 1989, "Spontaneous formation of electric current sheets by the expulsion of magnetic flux," *Ap. J.*, in press.

Low, B.C. and Wolfson, R. 1988, *Ap. J.*, **324,** 574.

Moffat, H.K. 1985, *J. Fluid Mech.*, **159,** 359.

Parker, E.N. 1972, *Ap. J.*, **174,** 499.

Parker, E.N. 1983, *Geophys. Astrophys. Fluid Dynamics*, **23,** 85.

Parker, E.N. 1986a, in *Coronal and Prominence Plasmas* (NASA Conf. Publ. 2442, A. Poland, ed).

Parker, E.N. 1986b, *Geophys. Astrophys. Fluid Dynamics*, **35,** 277.

Parker, E.N. 1987, *Ap. J.*, **318,** 876.

Parker, E.N. 1988, *Ap. J.*, **330,** 474.

Parker, E.N. 1989, "Formal mathematical solutions of the force-free equations, spontaneous discontinuities, and dissipation in large-scale magnetic fields," preprint.

Priest, E.R. 1982, *Solar Magnetohydrodynamics* (Dordrecht: D. Reidel Publishing Co.), p. 143.

Tsinganos, K.C., Distler, J., and Rosner, R. 1984, *Ap. J.*, **278,** 409.

Vainshtein, S.T. and Parker, E.N. 1986, *Ap. J.*, **304,** 821.

Van Ballegooijen, A.A. 1985, *Ap. J.*, **298,** 421.

Van Ballegooijen, A.A. 1986, in *Coronal and Prominence Plasmas* (NASA Conf. Publ. 2442, A. Poland, ed).

Van Ballegooijen, A.A. 1988, *Geophys. Astrophys. Fluid Dynamics*, **41** 181.

Wolfson, R. 1989, "Current Sheet Formation in Sheared Force-Free Magnetic Field," *Ap.J.*, in press.

DYNAMICS, CATASTROPHE AND MAGNETIC ENERGY RELEASE OR TOROIDAL SOLAR CURRENT LOOPS

James Chen
Space Plasma Branch
Naval Research Laboratory, Washington, DC 20375–5000

Abstract. An earlier paper has shown that toroidal current loops can exhibit a wide range of equilibrium and dynamical behavior under the action of toroidal forces; slow or rapid (supersonic) mass motion accompanied by a corresponding range of magnetic energy dissipation. In this paper, further dynamical properties are explored, including nonzero ambient field B_s previously excluded from the dynamics. The possible influence of subphotospheric current/flux structures on the loop dynamics is discussed. Distinct mechanisms for in situ (coronal) and subphotospheric magnetic energy storage are considered. A scenario of subphotospheric trigger for eruptive processes and magnetic energy release is discussed. Quasi-equilibrium evolution of an isolated toroidal current loop is studied with respect to specifying the poloidal flux of the loop. It is found that no bifurcation results if the infinitely conducting photosphere assumption is imposed but that a cusp catastrophe is obtained if a subphotospheric flux structure is included.

Introduction

Localized current and magnetic structures have been extensively studied, both observationally and theoretically, in solar, planetary and astrophysical plasmas. They are variously referred to as flux ropes, flux tubes and loops, and are presumed to be organized by magnetic fields and currents. Flux transfer events (FTEs) at the earth's dayside magnetopause and solar coronal loops are but a few examples for which loop structures have been studied. In this paper, we describe a theoretical study of the dynamical behavior of model current loops embedded in a background plasma such as the solar corona which may be magnetized or unmagnetized. The study is confined within the MHD framework. Thermal and radiative properties are not addressed.

Geophysical Monograph 58

Figure 1, reproduced from Chen (1989), describes schematically a solar current loop which is initially in equilibrium, showing the components of the current density $\underline{J}$ and magnetic field $\underline{B}$. The subscripts "t" and "p" refer to the toroidal and poloidal directions, respectively. Such a current loop experiences "toroidal forces" in the major radial (R) direction, an action of $\underline{J} \times \underline{B}$ and ∇p in a toroidal segment. By toroidal, we do not refer to an actual torus but to a plasma structure which can be approximated as a section of a torus. These forces and the equilibrium and dynamical behavior determined by them are described in detail in Chen (1989), hereafter referred to as Paper 1, which contains a more extensive bibliography.

Paper 1 shows that current loops can exhibit a wide range of behavior under the action of toroidal forces: A loop can be in stable equilibrium along both the major (R) and minor (a) radii, expand subsonically or supersonically, expand and reach a second equilibrium, oscillate, or various combinations thereof. It can also undergo quasi-equilibrium evolution. The motion of the loop constitutes a mechanism to convert magnetic energy into kinetic and thermal energy via drag heating. The time scale of motion and magnetic energy release is found to be minutes to tens of minutes. Examples were given, showing magnetic energy release of ~10^{29} erg to ~10^{32} erg in ~30 min. More slowly expanding loops release magnetic energy at lower rates. It was suggested that slowly moving loops, if numerous enough, can contribute to more quiescent coronal heating, an alternative to coronal heating mechanisms based on Alfven wave damping or reconnection. The major radial expansion velocity scales as $I_t^2/\bar{n}$. A novel aspect of Paper 1 is the inclusion of subphotospheric current in considering the dynamics and the suggestion that rapid motion and energy release can be triggered by subphotospheric changes in the topology of the underlying current and magnetic structures.

In the next section, we briefly review the model described in Paper 1. In the subsequent sections, we provide a summary of some recent results regarding the dynamical behavior of the

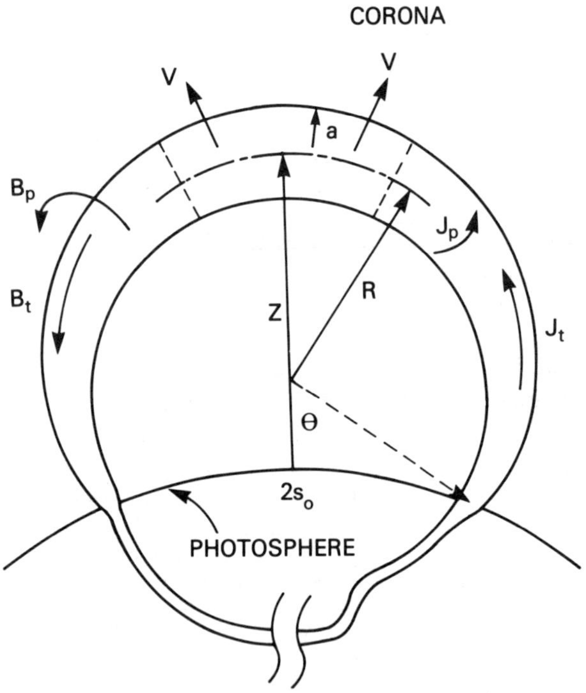

Fig. 1. Schematic drawing of a model solar current loop. Reproduced from Chen (1989). The major radius is R and the minor radius is a. Quantities inside the loop are denoted by bars ($\bar{p}$, $\bar{n}$, etc.). A subscript a denotes an ambient quantity.

model current loop. We then consider quasi-static evolution of a toroidal current loop with respect to increasing the total poloidal flux. We find that the apex height as a function of specified total flux has no bifurcation if the current is assumed to be closed entirely in the photosphere but that a cusp catastrophe is obtained if a subphotospheric flux reservoir is included. Finally, we discuss a number of issues for future research.

Toroidal Current Loops

In order to satisfy current conservation, the model allows the current to close in or below the photosphere (Fig. 1). The only requirement is that the total current be conserved; no particular photospheric or subphotospheric current structure is specified. The footpoint separation $2s_o$ is taken to be fixed because of the dense subphotospheric plasmas (but ideal MHD line-tying is not invoked). As a simplified geometry, the major radius R is taken to be related to the height of the apex Z by $R(Z) = (Z^2 + s_o^2)/2Z$. The force acting on plasma elements in the loop is given by $\underline{f} = (1/c)\underline{J} \times \underline{B} - \underline{\nabla}p$ with $\underline{J} = (c/4\pi)\underline{\nabla} \times \underline{B}$. The motion of the center of mass of a section of the loop, located on the dash-dot line in Fig. 1, is determined by the total force integrated over the given section:

$$F_R = \frac{I_t^2}{c^2R}\left[\ln\left(\frac{8R}{a}\right) + \frac{1}{2}\beta_p - \frac{1}{2}\frac{B_t^2}{B_p^2} + 2\left(\frac{R}{a}\right)\frac{B_s}{B_p} - 1 + \frac{\xi_i}{2}\right] - F_d , \qquad (1)$$

where $F_R = Md^2Z/dt^2$ with $M = \pi a^2\bar{n}m_i$, $\beta_p \equiv 8\pi(\bar{p} - p_a)/B_p^2$ and ξ_i is the internal inductance. Here, F_R is the force per unit length of the loop in the major radial direction, $\bar{p}$ is the average pressure inside the loop, p_a is the ambient pressure and $B_p = B_p(a)$. The ambient pressure p_a is assumed to decrease exponentially as $\exp(-Z/H)$ where $H = 2kT_a/m_ig$ is the gravitational scale height of the corona, with g calculated at each Z. However, for most of the examples used, the gravitational force acting on loops is relatively small in comparison with the Lorentz force and is neglected in the present paper. The term F_d is the drag force per unit length. The model uses a simple form

$$F_d = C_d(n_am_iaV^2).$$

[Equation (32) of Paper 1 contains an erroneous factor of $2/\pi$, which did not propagate.] The poloidal component $B_p(r)$ decreases for $r > a$ with $B_p \sim r^{-1}$ near the loop. We have included an ambient field component B_s, which is normal to the plane of the loop (i.e., the paper in Fig. 1). In general, there is another component B_{et} in the toroidal direction (Paper 1), which is not included in the present discussion. The scale lengths of the ambient field are taken to be comparable to or larger than the loop. The components B_p and B_t are the self-field components of the current loop. The actual field is the superposition of these fields and the total field is sheared. (Applied to prominences, $B_s < 0$ corresponds to the normal polarity.) The minor radial evolution a(t) is described by

$$\frac{d^2a}{dt^2} = \frac{I_t^2}{Mc^2a}\left(\frac{B_t^2}{B_p^2} - 1 + \beta_p\right) . \qquad (2)$$

Note that eqs. (1) and (2) contain the full contribution of $\underline{J} \times \underline{B}$, including the magnetic pressure and tension forces.

Given a current loop, we can define the toroidal flux $\Phi_t \equiv B_ta^2$ and the total poloidal flux $\Phi_T \equiv L_T I_t$ where L_T is the total self-inductance of the entire current distribution, including the subphotospheric structure. If we further define Φ_p and Φ_s to be the poloidal fluxes above and below the photosphere, respectively, then we have $L_p \equiv \Phi_p/I_t$ and $L_s \equiv \Phi_s/I_t$ with $L_T \equiv L_p + L_s$ and $\Phi_T \equiv \Phi_p + \Phi_s$. Note that Φ_s is a physically definable, albeit not necessarily measurable, quantity. In equilibrium, we define

$$\varepsilon \equiv \frac{\Phi_p}{\Phi_T} = \frac{L_p}{L_T} . \tag{3}$$

This quantity is a measure of the size of the loop above the photosphere relative to the entire current structure in equilibrium. During rapid expansion, ε must be re-interpreted (see the last section). For a given loop in the corona, ε parametrizes the subphotospheric structure. Paper 1 studied the properties of toroidal loops as determined by equations (1) and (2), subject to the global constraints that the total poloidal flux Φ_T is conserved and that the adiabatic gas law applies inside the loop. (These constraints may be replaced by others for different systems.)

Because Φ_p and Φ_s are not assumed to be separately conserved, flux exchange between the corona and subphotospheric region is allowed. If the current is entirely closed in the photosphere (e.g., infinitely conducting), then $\varepsilon = 1$. In this case, there is no flux (i.e., magnetic energy) exchange across the photosphere. If $\varepsilon < 1$, then flux and magnetic energy can be exchanged. It was suggested that concentrated magnetic flux tubes may extend below the photosphere and may be a conduit for transport of flux and energy via electromagnetic and other processes. This point will be discussed further in the last section. No specific transport properties have been analyzed.

The idealized model loop of Fig. 1 is not the most general three-dimensional configuration. It has a uniform aspect ratio R/a and the expansion is subject to a simplifying constraint R(Z) described above. The belief is that this model geometry can capture certain essential effects of the toroidal forces. Given these approximations, the coupled differential equations (1) and (2) describe the major and minor radial evolution of the model toroidal current loop. The degree to which the underlying simplifications and assumptions may be valid will be determined by future calculations of specific predictions implied by the model and comparing them with observations.

Current Loop Dynamics

In equilibrium, we demand that $F_R = 0$ and $d^2a/dt^2 = 0$. These conditions are discussed in Paper 1. The simplest case has $B_s = 0$. Then, the equilibrium condition requires $\beta_p < 0$ (Xue and Chen, 1983). Paper 1 studied loop dynamics in this limiting case in order to illustrate the essential physics of toroidal forces. In this section, we will consider some examples with $B_s \neq 0$ (hence $\bar{p} > p_a$) and explore the dynamical behavior implied by equations (1) and (2). The essential physics is not altered but $B_s \neq 0$ allows a broader range of behavior. The examples here correspond to heavier loops than those of Paper 1.

Figure 2(a) shows the height-time curve and expansion velocity profile for a model loop. This loop is in equilibrium with $B_{so} = -6.0$ G. The Z-dependence of B_s is not modelled self-consistently and an ad hoc profile, $B_s = B_{so}(Z/Z_o)^{-2}$, is used. The toroidal current is $I_t \simeq 3.5\times10^{11}$ A with the self-field components $B_p \simeq 27$ G and $B_t \simeq 27$ G. The initial height is $Z_o = 10^5$ km, $Z_o/s_o = 0.8$ and the density is $\bar{n} \simeq 10^{10}$ cm^{-3} with the total mass

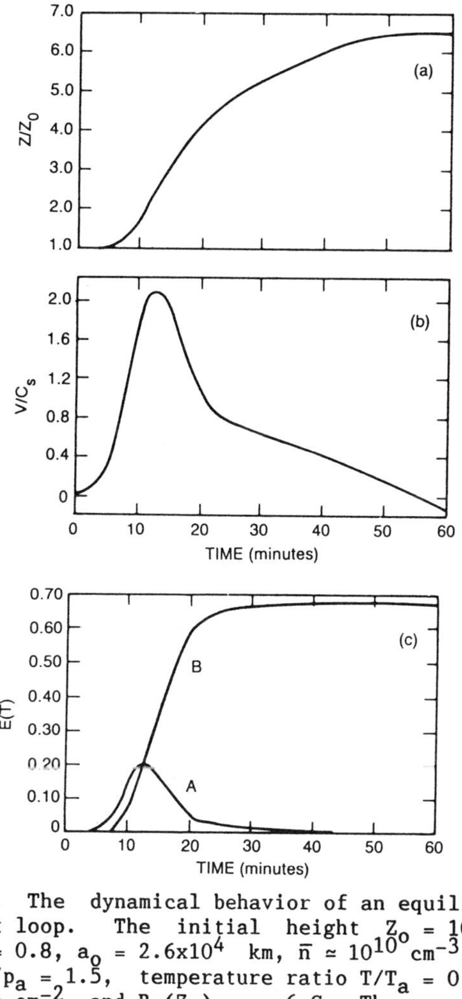

Fig. 2. The dynamical behavior of an equilibrium current loop. The initial height $Z_o = 10^5$ km, $Z_o/s_o = 0.8$, $a_o = 2.6\times10^4$ km, $\bar{n} \simeq 10^{10}$ cm^{-3}, $\bar{n}/n_a = 3$, $\bar{p}/p_a = 1.5$, temperature ratio $T/T_a = 0.5$, $p_a = 2$ dyn cm^{-2}, and $B_s(Z_o) = -6$ G. The current is closed entirely in the photosphere: $\varepsilon = 1$. (a) Height Z versus time T; (b) Apex velocity versus time. $C_s = 235$ km s^{-1}; (c) Curve A: Loop kinetic energy. Curve B: Time-integrated drag heating. Energy in units of 2.6×10^{31} erg.

of $M_T \simeq 1.3\times10^{16}$ g. For this example, we have set $\varepsilon = 1$ so that there is no transport of flux or magnetic energy across the photosphere. For this example, F_R is ~10 times greater than gravity. We have chosen $d|B_s|/dz < 0$ so that the loop is unstable. If $d|B_z|/dZ > 0$, then the loop is stable. We see that the loop reaches a second equilibrium near $Z \simeq 6.5\times10^5$ km. The apex attains a velocity of ~500 km s^{-1} in about 12 minutes [Fig. 2(b)]. Subsequently, the loop slows down and executes damped oscillation. In Fig. 2(c), Curve A shows the kinetic energy of the loop. At $t \simeq 12$ min, the loop has $\sim5\times10^{30}$ erg in kinetic energy, which is dissipated by drag heating. Curve B shows the time-integrated energy dissipation by drag. The loop releases about 2×10^{31} erg in ~30 min. The initial poloidal magnetic energy of the loop is $\sim7\times10^{31}$ erg. In general, an unstable loop ($d|B_s|/dZ < 0$) with $\varepsilon = 1$ cannot expand indefinitely. Also see the discussion of Fig. 4 below.

In the preceding example, we have used an ad hoc profile for B_s. For more complete modelling, the dynamics of the ambient magnetized plasma must be calculated. However, because of the three-dimensional geometry of the apex, the ambient field B_s away from the apex would have a tendency to be pushed aside by an expanding loop. It is also unlikely that a steady-state stagnation point would form in front of the apex. Nevertheless, there is additional work the loop must do to displace magnetized plasma. This would enter the analysis in the form of "mass loading" or "wave loading" and modify the drag coefficient C_d. The ambient toroidal field B_{et}, but not the self-field B_t, has been neglected in order to simplify the discussion. Equation 2(a) of Paper 1 provides a more complete equation including B_{et}. For a discussion on the energy budget for magnetic field components, see eqs. (25) - (29) of Paper 1.

Figure 3 describes an example with $\varepsilon = 0.1$ so that flux/magnetic energy is transported from below the photosphere. The size of the initial loop, mass and pressure profile are the same as in Fig. 2 except that $B_{so} = -2$ G. This loop is out of equilibrium, perhaps following subphotospheric changes. (This loop is such that it would be in equilibrium with $B_{so} \simeq -6$ G.) The current is $I_t \simeq 3\times10^{11}$ A with $B_{so} = -2$ G, $B_p \simeq 23$ G and $B_t \simeq 22$ G. The figure shows that the apex velocity attains a maximum value of $V \simeq 1000$ km s^{-1} in ~10 minutes [Fig. 3(b)]. At 60 minutes, the expansion speed is ≃260 km s^{-1} and is decreasing. In Fig. 3(c), the magnetic energy tranport profile implied by the choice of $\varepsilon = 0.1$ is shown. The energy transport rate has a maximum of $\sim6\times10^{29}$ erg s^{-1} at $t \simeq 7$ minutes. Note that transport rates lower than 6×10^{29} erg s^{-1} but lasting longer could also produce similar peak velocities but that the acceleration would be smaller. Thus, a wide range of dynamical behavior is possible for different energy transport profiles. Figure 3(d) shows that $\sim2\times10^{32}$ erg of magnetic energy is dissipated by drag heating of the ambient coronal gas in ~20 minutes (Curve B). The peak kinetic energy of the loop is $\sim2\times10^{31}$ erg (Curve A). This particular example has been found to continue expanding at 10 solar radii where the solar wind can become an added (perhaps dominant) driver. We suggest that such a loop, driven into the interplanetary space by toroidal forces, may be observed as a magnetic cloud at 1AU (Burlaga et al., 1981; Burlaga, 1988).

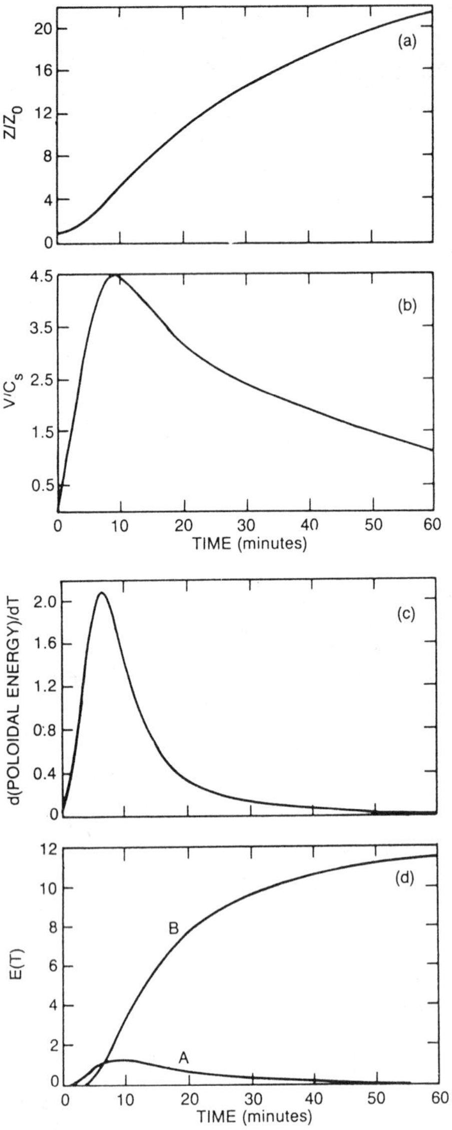

Fig. 3. The dynamical behavior of a loop following loss of equilibrium. The size, mass and pressure are the same as in Fig. 2 but $B_{so} = -2$ G. (a) Height; (b) Apex velocity. (c) Rate of poloidal magnetic energy transported from below the photosphere, implied by the choice of $\varepsilon = 0.1$. In units of 3×10^{29} erg s^{-1}; (d) Curve A: Kinetic energy of the loop. Curve B: Time-integrated drag heating. In units of 1.9×10^{31} erg.

It has been analytically shown in Paper 1 that, if $\varepsilon = 1$, a current loop is stable even in the absence of the restoring field B_s. The reason is that, for $\varepsilon = 1$, conservation of the flux $L_p I_t$ requires that I_t, hence the upward driving force, decrease as the loop expands, causing L_p to increase. For $\varepsilon = 1$, this decrease in I_t occurs faster than the decrease in the restoring force. In Fig. 4, we show the apex position of a current loop, initially in equilibrium with $Z_0 = 10^4$ km and $Z_0/s_0 = 0.8$. Here, we have set $\varepsilon = 1$ and $B_s = 0$. The loop density is $\sim 2\times10^{10}$ cm^{-3} and the loop field components are $B_p = 1.5$ G and $B_t = 2.7$ G. At $t = 0$, we introduce an upward flow in the ambient gas at ~350 km s^{-1}, lasting for 10 minutes and decreasing to zero in 2 minutes. In effect, the loop is "blasted" from below. The figure shows that the loop is carried by the upflow for the duration of the peak flow and returns to its original equilibrium position whereupon it executes damped oscillation. The velocity amplitude ranges from ~37 km s^{-1} at $t = 14$ min, decreasing to ~11 km s^{-1} at $t = 90$ min. The period is ~17 min initially and decreases to ~15 min later (not shown) when the amplitude becomes small. The linear oscillation period can be calculated from $\Gamma(t = 0)$, eq. (21) of Paper 1. For this particular example, we find $2\pi(\Gamma)^{-1/2} \sim 15$ min. If a restoring magnetic field is included, $B_s < 0$ with $d|B_s|/dZ > 0$ (stable), then similar oscillatory behavior occurs. If $d|B_s|/dZ < 0$ (unstable) or if the upflow drives the loop from a stable to unstable region, then the loop expands to and oscillates about a new equilibrium height which is greater than the initial height. Note that the ambient flow profile is ad hoc, intended only to illustrate a generic effect. Different upflow profiles affect the amplitude but not the frequency of oscillation, which is determined by the current, mass, the geometry and the pressure profile of the loop (see Paper 1).

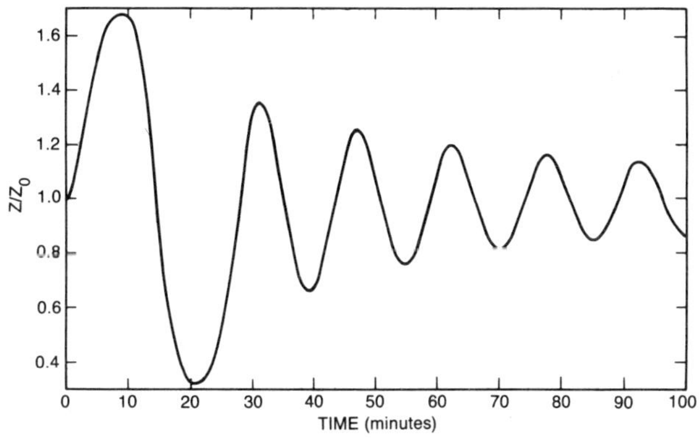

Fig. 4. Behavior of a loop "blasted" from below. $\varepsilon = 1$. The initial loop has $Z_0 = 10^4$ km, $Z_0/s_0 = 0.8$, $a_0 = 1.3\times10^3$ km, $R_0/a_0 = 10$, $\bar{n} = 2\times10^{10}$ cm^{-3}, $\bar{n}/n_a = 6$, $B_s = 0$ (no restoring field), and $\bar{p}/p_a = 0.6$. The upflow velocity is ~350 km s^{-1}, lasting from $t = 0$ to $t = 10$ min and decreasing to zero linearly at $t = 12$ min.

This behavior of the model loop may be the underlying reason for the robustness of certain filaments which oscillate near their initial positions after being blasted from below (e.g., Martin 1989). Ramsey and Smith (1966) reported on several flare-initiated oscillating ("winking") filaments. The oscillations were damped and had periods of 6 ~ 40 minutes. The oscillation profiles were inferred from Doppler shift measurements with $H_\alpha \pm 0.5$Å filters. They also pointed out that filaments tended to exhibit characteristic oscillation frequencies when excited by different flare waves. That the model loops possess characteristic frequencies, as in this example, is consistent with this observation. We believe that the behavior described above also applies to more general three-dimensional loops.

Accompanying the motion of loops is heating of the ambient plasma due to turbulent drag. For supersonic expansion, shock heating is possible (Paper 1). Slowly expanding loops, if numerous enough, can contribute to coronal heating. This mechanism is an alternative to the coronal heating mechanisms based on wave energy dissipation or reconnection. Book (1981) proposed a coronal heating mechanism in which a magnetically expanding straight cylinder does thermodynamic work against the ambient plasma. It was shown in Paper 1 that the minor radial expansion is typically 1/10 of the major radial expansion. This is true for slow or fast expansion. The model of Paper 1 is based on the major radial expansion which releases a much greater amount of magnetic energy than the mechanism based on minor radial expansion proposed by Book (1981).

Quasi-Equilibrium Evolution

In Paper 1, it has been pointed out that a loop can expand slowly if the toroidal current I_t or internal pressure $\bar{p}$ is increased slowly. During such evolution, the loop typically expands and exhibits small-amplitude oscillations in major and minor radii. (The oscillatory behavior will be discussed in detail elsewhere due to space limitations.) It is of interest to note that, under the perfectly conducting photosphere assumption, 2D arcade equilibria have been found to develop no bifurcations with respect to increasing the footpoint displacement (Zwingmann, 1987; Finn and Chen, 1989), consistent with the suggestion of Jockers (1978). Furthermore, (Finn and Chen, 1989) found that no bifurcations occur if the entropy, rather than pressure, is specified. They argued that pressure is not a specifiable quantity and that the pressure-based bifurcation obtained by Zwingmann (1987) is not physically obtainable from a given initial equilibrium. (Note that in Paper 1, I_t and $\bar{p}$ are varied but not specified.) Mikic et al. (1988) have carried out MHD simulations showing that increasing footpoint shear of a double arcade can lead to instability and plasmoid ejection.

Biskamp and Welter (1988) have argued that this is an artifact due to the fact that the arcades are restricted horizontally and symmetric. With three nonsymmetric neighboring arcades, Biskamp and Welter (1988) have found that increasing the photospheric shear can lead to plasmoid ejection from the central arcade. We note that the central arcade is now restricted by the neighboring arcades. Thus, both models are based on restricting the horizontal extent of arcades.

In the above papers, two-dimensional structures were considered. Browning and Priest (1984) considered a thin twisted flux tube, a three-dimensional structure, with a constant aspect ratio and found bifurcations as the field-line twist was increased. In Paper 1, it was argued that field-line twist in the corona is not equivalent to the footpoint twist. In this paper, we show an example of the quasi-equilibrium behavior of an isolated toroidal loop with respect to a slow increase in the poloidal flux Φ_T. The footpoint separation $2s_0$ is held fixed but the loop itself is not restricted in horizontal or vertical direction. The twist of field lines in the photosphere and corona is not specified. In Fig. 5, we have shown two curves, one for $\varepsilon = 1$ and the other for $\varepsilon = 0.01$. For both cases, the starting equilibrium configuration has $Z_0 = 10^4$ km, $R_0/a_0 = 5$, $Z_0/s_0 = 0.6$, $\bar{n} \simeq 7\times10^9$ cm^{-3}, $\bar{p}/p_a = 0.9$ with $B_s = 0$, and $I_t = 3\times10^9$ A with $B_p = 1.6$ G and $B_t = 2.7$ G. In the quasi-equilibrium sense, ε has the same meaning as in equilibrium, defined by eq. (3). If the current is closed entirely in the photosphere, then $\varepsilon = 1$ and the height Z simply increases monotonically with increasing Φ_T. If we allow the coronal current loop to be a part of a larger current structure, then $\varepsilon < 1$. For $\varepsilon = 0.01$, the Z versus Φ_T curve exhibits a cusp catastrophe. However, we argue that this catastrophe in and of itself with no subphotospheric loss of equilibrium and flux injection need not lead to an eruptive process because, depending on the subphotospheric current distribution, rapid motion can cause the loop to act as if $\varepsilon \simeq 1$ on the dynamical time scale whether or not the photosphere is infinitely conducting. The motion of the loop is critically dependent on how much flux is accessible on the Alfven transit time along and inside the flux tubes. If the underlying current is large with a significant amount of flux near the photosphere, then $\varepsilon \ll 1$ is possible. If, on the other hand, the underlying current structure is deep, then $\varepsilon \simeq 1$. In the latter case, the quantity ε_{cr} [eq. (23a) of Paper 1] and the condition of stability or instability given by ε_{cr} [eq. (23b) of Paper 1] are not applicable. (See below for further discussion on this point.) During the quasi-static evolution, $\varepsilon < 1$ is possible because the evolution is assumed to be slower than any other time scales. Quasi-equilibrium evolution of toroidal loops will be reported in detail elsewhere.

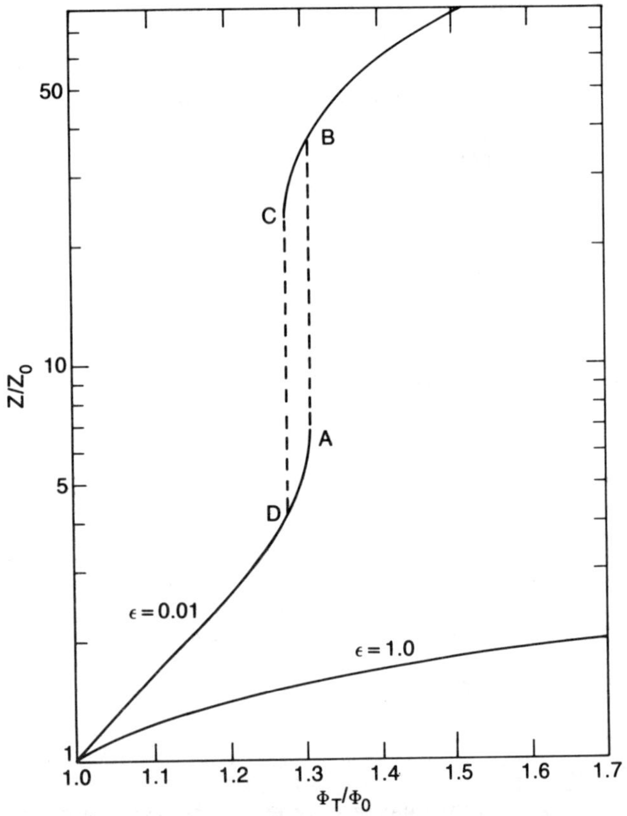

Fig. 5. Quasi-equilibrium evolution of an isolated toroidal current loop. $B_{so} = 0$. Equilibrium height versus the total poloidal flux. $\varepsilon = 1$: infinitely conducting photsphere. $\varepsilon = 0.01$: "large" subphotospheric flux reservoir.

Discussion

A broad range and a wide variety of properties of model toroidal current loops compatible with solar conditions have been described in Paper 1 and the preceding sections. Magnetic energy storage both in the corona and below the photosphere is allowed. Figure 2 gives an example in which the dissipated energy is stored entirely in the corona. This example may be applicable to a current loop in a restoring field $B_s < 0$ which increases in magnitude ($d|B_s|/dZ > 0$, stable) up to $Z = Z_\star$, decreasing for $Z > Z_\star$ ($d|B_s|/dZ < 0$, unstable). If the loop, initially at $Z < Z_\star$, rises past $Z_\star$, possibly by quasi-equilibrium evolution (Fig. 5) or by an impulsive event, then Fig. 2 would describe the subsequent motion ($Z > Z_\star$). This scenario allows slow evolution leading to eruption or expansion by impulsive coronal "triggers". (Note that the catastrophe described in Fig. 5 does not lead to sudden expansion.) The onset condition is determined by the spatial and/or temporal profile of B_s. Figure 3 corresponds to an example with $\varepsilon < 1$ in which a fraction of the dissipated energy is transported ("injected") from below the photosphere in the form of current and magnetic field. Both examples show that loops can expand rapidly. However, with no energy injection (ε = 1), a loop cannot expand indefinitely with or without $B_s \neq 0$ while, with sufficient energy injection ($\varepsilon << 1$), a loop can expand into the interplanetary space.

Figure 4 shows that simply "blasting" a loop upward from below by an impulsive event without changing the flux Φ_p of the loop or without driving it into a region with $d|B_s|/dZ < 0$ does not trigger sustained expansion of the loop.

Brueckner (1976) reported an association of high speed (~400 km s^{-1}) mass motion ("bullet") with a hot Fe XXIV cloud above the bullet in the explosive phase of the 15 June 1973 flare. It was suggested that electrons and ions ejected as non-thermal beams and fast moving clouds can rapidly heat the corona above. The apex of an expanding loop may be an underlying structure for a fast moving cloud. If this is the case, then the drag heating suggested in Paper 1 may be relevant. A number of researchers have suggested that global MHD instabilities may be important for eruptive processes and flares. For example, Kahler et al. (1988) have argued that filament eruption and onset of impulsive phase of solar flares may be due to a common process, some global MHD instability. The initial exponential increase in height shown in Figs. 2 and 3 and in the examples of Paper 1 is consistent with the power-law fit of acceleration versus time given by Kahler et al., with the difference that the basic process may be loss of equilibrium rather than an instability. The present loop model is compatible with this observation. In Paper 1, an additional suggestion has been made that dynamical and energetic processes may be triggered by subphotospheric events, one possibility being loss of equilibrium due to topological changes in the underlying current structure. In the ensuing relaxation to a new equilibrium, electromagnetic processes and electromotive forces can be generated, leading to flux and energy transport along flux tubes. In this scenario, the loop expansion and magnetic energy release in the corona are an integral part of this relaxation process. Motion and magnetic energy release are important components of eruptive and energetic processes such as flares.

During expansion, eq. (3) loses meaning if the initial loop has $\varepsilon < 1$ and ε must be re-interpreted since the subphotospheric flux cannot be "dragged" up freely. Due to the large inertia, the characteristic speed of energy transport below the photosphere is much slower than that (Alfven speed) in the coronal loop. Thus, Φ_p and Φ_s are effectively decoupled on the coronal dynamic time scales and the current loop should act as if $\varepsilon \simeq 1$. However, if the subphotospheric structure itself loses equilibrium, then flux can be forced into the corona (i.e., increased current). So far, we have included ε only as a "free" parameter, determining the implied energy transport profile. However, this profile should be determined by subphotospheric processes and imposed. The problem can be reformulated with an imposed $\varepsilon_i(t)$. This work will reported elsewhere.

We have argued that the long-time loop dynamics can be critically affected by flux transport properties. This also implies that the overall expansion can be limited by flux and energy transport from below the photosphere. Although we have not calculated transport properties for postulated subphotospheric current structures, we briefly consider the implications in a heuristic, order-of-magnitude sense. Consider the example described by Fig. 3. The loop mass is $M_T \simeq 1.3 \times 10^{16}$ g and the magnetic field is $B_t \simeq 22$ G and $B_p \simeq 23$ G. If the minor radius in the photosphere decreases by a factor of 10 to $a \simeq 2 - 3 \times 10^3$ km, then the magnetic field is $B \simeq 2$ kG. It is also possible that the footpoints consist of thin flux tubes with magnetic fields of ~2kG. Estimating the characteristic speed of transport by the Alfven speed inside flux tubes, $v_A \simeq 14$ km s^{-1} where $B \simeq$ 2kG, $n \simeq 10^{17}$ cm^{-1} and $m = 1.7 \times 10^{-24}$ g. (Note that v_A outside flux tubes is expected to be slower.) We find that the magnetic energy accessible based on the above Alfven speed estimate gives a rate of a few times 10^{29} erg s^{-1} for sub-Alfvenic flux transport along flux tubes. (Mass motion at ~v_A is not implied.) The peak energy transport rate (~6×10^{29} erg s^{-1}) in Fig. 3(c) is somewhat larger than this value. The initial poloidal magnetic energy of the loop above the photosphere is ~6×10^{32} erg. This suggests that 1000 km s^{-1} may be roughly the maximum velocity to which large and heavy loops ($Z_o = 10^5$ km and $M_T = 10^{16}$ g) can be accelerated in short periods of time (in ~10 min in this example) by toroidal forces. This order-of-magnitude estimate is compatible with the fact that sustained mass (~10^{16} g) motion significantly faster than 1000 - 1200 km s^{-1} is rarely reported. More careful calculations of possible transport mechanisms are needed to quantify these points. The above energy estimates are based on the existence of flux tubes in which B ~2 kG and $n \simeq 10^{17}$ cm^{-3}. The plasma properties inside the flux tubes may be different from the surrounding regions. If the current is completely diffuse below the photosphere, then magnetic energy is not concentrated (one traditional view) and we expect no significant flux or energy transport on the dynamical time scales (e.g., Fig. 2). This corresponds to $\varepsilon \simeq 1$, a possible limiting case included in the model.

Questions regarding the energy storage site (in situ versus remote) and the mechanisms for triggering energy release have been a long-standing problem (e.g., Spicer et at., 1986). It was shown (Xue and Chen, 1983) that the magnetic energy which an isolated toroidal current loop can store in the corona in the absence of ambient fields ($B_s = B_{et} = 0$) is limited by the low coronal pressure (e.g., ~2 dyn cm^{-2} in active regions) to ~10^{28} erg. In situ storage of significantly greater amounts can be achieved if $B_s \neq 0$ (Fig. 2). In Paper 1 and the present paper, we have suggested specific and distinct scenarios for both in situ and subphotospheric storage. Possible "trigger" mechanisms or onset conditions have been discussed for the respective scenarios. For coronal storage and trigger, the profile of B_s plays a critical role while for subphotospheric storage and trigger, the topological changes in the subphotospheric current/flux structure play the key role. These scenarios will be disccused further in the future.

Toroidal current loops need not be limited to the sun. For example, in the model of FTEs by Russell and Elphic (1978), elbow-shaped flux tubes are presumed to interconnect the earth and the solar wind. Sonnerup (1987) investigated the Lorentz force acting on such isolated flux tubes.

In the preceding sections, we have pointed out certain phenomena where the model of current loops under the action of toroidal forces may be potentially applicable. However, a number of approximations have been made in the calculations. For example, the uniform aspect ratio R/a and toroidal expansion R(Z) are the major geometrical approximations. Inclusion of an ambient toroidal field B_{et} would give a more complete description of the coronal environment. The model equations used in Paper 1 are amenable to such refinements. Other realistic effects should also be added. For example, a more detailed treatment of the interaction between the corona and an expanding loop (waves and other disturbances) would modify the drag coefficient. In the examples presented here, gravity has been neglected, which is valid if the current is large such that $\tau_A \ll \tau_g$ where $\tau_A = R/v_A$ is the Alfven transit time inside the loop and $\tau_g = (R/g)^{1/2}$ is the free-fall time scale. [See eq. (33) of Paper 1.] For the examples of Figs. 2 and 3, F_R is ~10 times greater than gravity while for Figs. 4 and 5, F_R is comparable to gravity. The behavior of loops is certain to be affected by these approximations but the basic physics of toroidal forces should not be altered. See Paper 1 for more detailed discussion on the major approximations. The objective of the paper has been to elucidate the essential physics of toroidal forces. The precise applicability and relative importance of toroidal forces in specific observed phenomena have not been quantitatively assessed. Nevertheless, the model current loops under the action of toroidal forces can mimic a range of dynamical effects (motion, energy dissipation, oscillation, quasi-static evolution and so forth) in the solar environment.

The model does not predict what kind of topological changes take place below the photosphere nor does it require any particular structure. (Recall that the model allows both in situ and subphotospheric energy storage and "triggers".) Although presently unobservable, different subphotospheric structures or events can lead to different behaviors of loops in the corona. If so, it is desirable to understand the influences and coronal manifestations, if any, of subphotospheric events. Work is under way to quantify flux (and magnetic energy) transport issues. The present model, with further development, can potentially allow one to calculate such manifestations quantitatively and compare with observations.

Questions and Answers

Mikic: If we assume that the photosphere provides line-tied boundary conditions on the field (which is a good approximation) then how can the behavior of a loop depend on what happens below the photosphere?

Chen: In equilibrium or quasistatic evolution, there is presumed to be an "infinite" amount of time. Thus, the loop "knows" about the subphotospheric structure so that $E<1$ is possible, unless the photosphere is rigorously perfectly conducting, in which case $E=1$. In quasiequilibrium evolution, it is useful to think of a three-step (imagined) process. Start with an equilibrium with $E<1$. Increase the poloidal flux (i.e., expand) (step 2). The loop size, and hence the inductance, increases, reducing the current while converting the flux. This takes place on the coronal Alfvén time scale. Now the flux distribution across the photosphere is out of equilibrium since the characteristic speed is much slower in and below the photosphere. Therefore, further relaxation to equilibrium occurs across the photosphere on a slower time scale, which entails (slow) transport of flux to the corona (step 3). The system repeats these processes until a new equilibrium is reached. In this regard, it should be pointed out that the quasistatic evolution is assumed to be comparable to or slower than the characteristic speeds below the photosphere. In the present calculation, the subphotospheric structure has been assumed to undergo much smaller fractional changes in geometry than the changes in the corona. Thus, the footpoints are taken to be essentially stationary. The communication between the coronal and subphotospheric structures is accomplished by $\partial \mathbf{B}/\partial t$, etc., not necessarily by footpoint motion.

Antiochos: What is the value for E for the solar photosphere?

*Chen: The parameter **E** does not have a unique value for the solar photosphere. The model calculation, at the present level of sophistication, treats **E** as an input parameter. In a more complete model, I would expect the following: In equilibrium, **E** is the ratio Φ_p/Φ_T as before. If the sun is rigorously infinitely conducting in and below the photosphere, or if the current is closed entirely in the photosphere, then $\mathbf{E}=1$. Otherwise, $\mathbf{E}<1$. In a dynamic situation, the meaning of **E** is modified; it is now the effective "circuit" which the loop in the corona can "see" on the time scale of motion. The faster the loop motion, the larger $\mathbf{E}<1$ should be. Given the slow speed (slower than the coronal Alfvén speed inside the loop) and the large mass density below the photosphere, I expect **E** to be close to unity for moderate to rapid motion in the corona. In a dynamic situation, energy/flux transport into the corona from subphotospheric regions plays the role of **E**. I have not done specific calculations regarding these transport properties in and below the photosphere. Issues pertaining to possible transport* *mechanisms are important. A simple order-of-magnitude discussion has been given in the presentation.*

Acknowledgments. I would like to thank Dr. S. K. Antiochos and Dr. S. F. Martin for useful discussisons. This work supported by ONR.

References

Biskamp, D. and H. Welter, Magnetic arcade evolution and instability, Solar Phys., 120, 49, 1988.

Book, D. L., A mechanism for heating the solar corona, Comments Plasma Phys. Controlled Fusion, 6, 193, 1981.

Browning P. K. and E. R. Priest, The magnetic non-equilibrium of buoyant flux tubes in the solar corona, Solar Phys., 92, 173, 1984.

Brueckner, G. E., A.t.m observations on the X u.v. emission from solar flares, Phil. Trans. R. Soc. Lon., A281, 443, 1976.

Burlaga, L. F., E. Sittler, F. Mariani, and R. Schwenn, Magnetic loop behind an interplanetary shock: Voyager, Helios, and IMP 8 observations, J. Geophys. Res., 86, 6673, 1981.

Burlaga, L. F., Magnetic clouds and force-free fields with constant alpha, J. Geophys. Res., 93, 1988.

Chen, J., Effects of toroidal forces in current loops embedded in a background plasma, Astrophy. J., 338, 453, 1989 (Paper 1).

Finn, J. M. and Chen, J., Equilibrium of solar coronal arcades, in press, Astrophys. J., 1989.

Jockers, K., Birfurcation of force-free solar magnetic fields: A numerical approach, Solar Phys., 56, 37, 1978.

Kahler, S. W., R. L. Moore, S. R. Kane and H. Zirin, Filament eruptions and the impulsive phase of solar flares, Astrophys. J., 328, 824, 1988.

Martin, S. F., Solar Phys., in press, 1989.

Mikic, Z., D. C. Barnes, and D. D. Schnack, Dynamical evolution of solar coronal magnetic field arcade, Astrophys. J., 328, 830, 1988.

Ramsey, H. E. and S. F. Smith, Flare-initiated filament oscillations, Astrophys. J., 71, 197, 1966.

Russell, C. T. and R. C. Elphic, "Initial ISEE magnetometer results: Magnetopause observations", Space Sci. Rev., 22, 681, 1978.

Sonnerup, B. U. O., On the stress balance in flux transer events, J. Geophys. Res., 92, 8613, 1987.

Spicer, D. S., Mariska, J. T., and Boris, J. P., Magnetic energy storage and conversion in the solar atmosphere, in Physics of the Sun, Vol. II, edited by P. A. Sturrock, p. 181, D. Reidel Pub. Co., Dordrecht-Holland, 1986.

Xue, M. L. and Chen, J., MHD equilibrium and stability properties of a bipolar current loop, Solar Phys., 84, 119, 1983.

Zwingmann, W., Theoretical study of onset conditions for solar eruptive processes, Solar Phys., 111, 309, 1987.

AN ELECTRODYNAMICAL MODEL OF SOLAR FLARES

A. I. Podgorny

P. N. Lebedev Physical Institute, USSR Academy of Sciences
117924 Leninsky Prospect 53, Moscow, USSR

I. M. Podgorny

Space Research Institute, USSR Academy of Sciences,
Profsoyuznaya Str., 84/32, Moscow, USSR

Abstract. An electrodynamical model of solar flares is proposed which takes into account the energy transport into a current sheet, its release and the generation of field-aligned currents mapping to the photosphere. The analogy between solar flares and magnetospheric substorms is discussed. The energy of the X-ray radiation in flare kernels, and the maximum energy of particles accelerated in the current sheet are estimated.

1. Energy Accumulation in the Current Sheet

Observations (see, for example, de Jager, 1985; Kahler et al., 1986) indicate that the initial energy release in solar flares takes place at a high altitude in the solar corona. The problem considered here is how to accumulate a large quantity of energy ($\sim 10^{32}$ erg) in such a way as to be able to release it quickly (within $\sim 10^3$ s). Magnetic energy can be accumulated in the vicinity of a magnetic field singular line (Syrovatskii, 1978; Baum and Bratenahl, 1980; Priest, 1985, 1989) by converging MHD-disturbances, and a singular line can appear in the solar corona for a corresponding configuration of magnetic field sources in an active region. The simplest examples of a singular line which have the main properties of the general type of singular line are the zero x-type lines (in its vicinity the magnetic field is $B = \{-h_0y, -h_0x, 0\}$) and the x-type lines with longitudinal magnetic fields $B = \{-h_0Y, -h_0X, B_{0z}\}$. Disturbances are focused by the deformation of the frozen magnetic field in a plasma which flows into the vicinity of a singular line along one axis (for example, the y-axis) and flows out of the vicinity along another axis (for example, the x-axis). As a result of the field deformation, the current density $\bar{j}$ is increased, and therefore, the magnetic force, $1/c\ \bar{j}\times\bar{B}$, is increased, too. This force further propels plasma in the direction of the pre-existing plasma flow and, therefore, it increases the magnetic field deformation.

Geophysical Monograph 58

The plasma flow elongates the magnetic field along one coordinate axis (x) but compresses it across the other (y), and so the magnetic field is deformed into a configuration corresponding to a current sheet (along the x-axis). The magnetic energy accumulation ceases when the current sheet thickness becomes so small that the frozen-in condition is violated, and all the magnetic field carried by the plasma flow into the sheet is dissipated. Hence,

$$a = \frac{\nu_m}{V_{in}} \tag{1}$$

where V_{in} is the plasma inflow velocity to the sheet, a is the sheet thickness, and $\nu_m = c^2/4\pi\sigma$ is the magnetic diffusivity for the conductivity σ. Physically, (1) means that the electric field near the sheet ($E_{ns} = 1/c\ V_{in} B_s$, where B_s is the value of the magnetic field near the sheet) is equal to that in the sheet ($E_s = j/\sigma = B_s/4\pi a\sigma$, Parker, 1957).

It is important to explain why an instability does not occur during the accumulation of magnetic energy as the current sheet slowly forms and develops, and why the latter instability does occur which leads to a rapid energy release. The explanation proposed here is that the created quasi-stationary current sheet slowly develops in time. During the slow evolution the current sheet is transformed from a stable state to an unstable one.

A numerical solution of the MHD-equations by Podgorny and Syrovatskii [1981] showed that a current sheet can be created by driving a disturbance wave from the boundary of the numerical domain towards the singular line. A schematic picture of the current sheet vicinity is shown in Figure 1. The plasma pressure increases in the

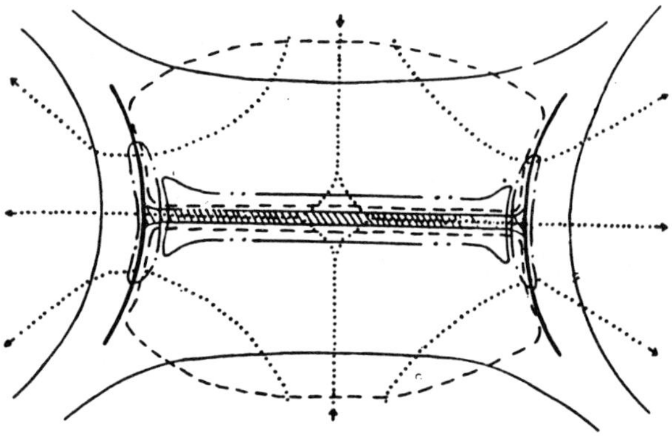

Fig. 1. A schematic picture of plasma and magnetic field behavior in the vicinity of a current sheet. Magnetic field lines are depicted by thin lines, while arrows show the directions of the plasma motion along the dotted lines. Standing fronts of fast shock waves are depicted by thick lines and downstream of these shocks are regions of strongly increased plasma density and magnetic field. The region of the increased plasma pressure in the current sheet is shaded. Regions bounded by other lines are as follows: -..-..-..- the region of the increased current density; - . - . - . regions of oppositely directed current created by compression of the magnetic field by plasma flow coming out of the sheet; - - - - - - regions of low plasma density near the sheet (after Podgorny and Syrovatskii, 1981).

sheet mainly because it must balance both the magnetic pressure and the inflowing plasma. Since it is propelled chiefly by the magnetic tension force the plasma moves along the sheet, and it can compress the magnetic field component perpendicular to the sheet (i.e. B_y) if the outflowing plasma slows near the endpoints of the current sheet. The plasma velocity at the endpoints reaches a value 0.4 $V_A (V_A = B_r/\sqrt{4\pi\rho r}$ is the Alfven velocity and B_r and ρ_r are the magnetic field and density in a coronal active region). Since this velocity exceeds the local Alfven velocity, fast shockwaves appear at the sheet's endpoints (see also Forbes [1986]). The increased normal magnetic field at the endpoints corresponds to reverse currents which are directed opposite to the current in the sheet (see also Biskamp [1986]).

In the quasi-stationary current sheet all the values slowly change with the course of time (i.e. the sheet becomes thinner, the current density increases), but among all of these changes the total plasma mass in the sheet undergoes the most rapid change - it decreases due to a fast plasma outflow. The total plasma mass in the sheet decreases with time in such a way that the plasma density in the middle of the sheet (where the plasma pressure cannot be less than the magnetic) is practically unchanged with time, while the plasma density near a sheet boundary decreases.

2. Magnetic Field Reconstruction

The term "reconstruction" means a change in the magnetic field configuration due to a fast field diffusion in the sheet. The total magnetic field energy ($\int(B^2/8)dV$) at the final state becomes less than that in the initial state.

Numerical calculations show that for a sufficiently high plasma conductivity ($R_m = IV_A/\nu_m \sim 300$, where I is the region size) after the long-time evolution of the quasi-stationary current sheet a fast reconstruction of the magnetic field and plasma flow takes place which releases a substantial part of magnetic energy. During the reconstruction the plasma velocity out of the sheet sharply increases, so that the velocity exceeds the Alfven velocity.

To explain the above process A. I. Podgorny [1988] has developed a model of the current sheet which is based, like the models of Parker and Sweet [Parker, 1956] and of Syrovatskii [1976], on relationships obtained by integrating the MHD-equations around the boundary of the current sheet. However unlike the models of Parker and Sweet and Syrovatskii, Podgorny's model does not assume equality between the plasma flowing into and out of the sheet (instead, the total plasma mass in the sheet can slowly change with time as has occurred in the numerical simulations [Podgorny and Syrovatskii, 1981]. Other new elements of the model are the propelling of the plasma along the sheet mainly by the magnetic tension force (rather than by the plasma pressure) and the counteracting of the magnetic field pressure at the sheets' endpoints by the plasma flow out of the sheet.

A steady state magnetic field (rotE = 0) in the two-dimensional case means that E = constant. Equalizing the electric fields in the sheet and near the sheet, we obtain equation (1), while equalizing the electric fields in the current sheet and at its endpoints ($1/c\ V_{out}B_n$, where V_{out} is the plasma outflow velocity from the sheet and B_n is the magnetic field component normal to the sheet) we obtain:

$$\frac{\nu_m B_s}{a} = V_{out}B_n \qquad (2)$$

This equation has also a magnetohydrodynamic interpretation, namely the diffusion of the normal magnetic component into the sheet is equal to the magnetic field flowing out of the sheet with the plasma.

The plasma flow along the sheet is propelled mainly by the magnetic tension force $1/c\ jB_n$. This means that:

$$\rho_s V_x \frac{dV_x}{dx} = \frac{B_y(x)\ B_x}{4a} \qquad (3)$$

where ρ is the plasma density in the sheet. Assuming a linear dependence of B_y on x of the form

$B_y(x) = B_n x/b$ (where b is the size of the sheet), we can integrate (3) and obtain:

$$V_{out} = \frac{b}{a} \frac{B_n B_s}{4\pi\rho_s} \tag{4}$$

Our assumption that B_y varies as B_nX/b is consistent with the previous numerical experiment.

There is a substantial deceleration of plasma flow out of the sheet by the magnetic field pressure. Therefore, if one approximates the value of the magnetic field at the sheet's endpoints by the initial hyperbolic field at this point (h_0b), then:

$$\frac{\rho_s V_{out}^2}{2} = \frac{h_0^2 b^2}{8\pi} \tag{5}$$

If ρ_s is known, then using (1), (2), (4) and (5), we can express a, b, V_{out} and B_n as functions of V_{in} and B_s. That is:

$$a = \frac{\nu_m}{V_{in}} \tag{6}$$

$$b = \frac{B_s^{3/2} V_{in}}{h_0^{3/2} V_{AS}^{1/2} \nu_m^{1/2}} \tag{7}$$

$$B_n = \frac{h_0^{1/2} \nu_m^{1/2} B_s^{1/2}}{V_{AS}^{1/2}} \tag{8}$$

$$V_{out} = \frac{B_s^{1/2} V_{in} V_{AS}^{1/2}}{h_0^{1/2} \nu_m^{1/2}} \tag{9}$$

where $V_{AS} = B_s/4\pi\rho_s$.

It is difficult to obtain a theoretical estimate for ρ_s because we do not quite know all the mechanisms of the plasma's cooling. Using the balance between gas and magnetic pressures in the sheet, gives

$$\rho_s T_s = B_s^2/8\pi \tag{10}$$

where T is the temperature in the sheet. Using the value of plasma density observed in coronal arch-type structures and relationship (10), we estimate that $\rho_s/\rho_r \sim 10^3 - 10^6$, (where ρ_r, the density in corona is $\sim 10^8$ cm^{-3}).

Combining (6)-(9) we obtain the criterion for the decrease of the plasma mass in the current sheet with time:

$$\frac{V_{in} \cdot b \cdot \rho_r}{V_{out} \cdot a \cdot \rho_s} = \frac{V_{in}}{V_A} R_m \sqrt{\frac{r}{s}} < 1 \tag{11}$$

Using the linearized MHD-equations we have analyzed the stability of the current sheet with respect to small perturbations. The physical meaning of instability is as follows. A local increase of the plasma flowing into the sheet causes an increase of the magnetic field normal to the sheet. In turn, this increase causes an increase of the magnetic field tension which propels plasma out of the sheet. The additional increase of the magnetic tension is caused by the increase of the current density. The decrease of the plasma density, and consequently of the plasma pressure in the sheet leads to an increase of the inflow velocity. Thus, plasma flowing into the sheet hinders the decrease of the plasma density in the sheet, and thus the sheet is stabilized. The plasma flowing into the sheet, $\rho_{ns}V_{in}$, depends on the plasma density near the sheet, ρ_{ns}. So, our conclusion that the instability of the sheet appears after a decrease of the plasma density near the sheet can be understood from a physical point of view.

For a disturbance wave length comparable with the size of the sheet, it is possible to release a substantial part of the sheet's energy during the nonlinear phase of the instability. For this case the mode with the maximum growth rate, γ_{max} is periodic (Im $\gamma = 0$) and given by:

$$\gamma_{max} = \frac{1}{2} R_m^{-1} \left(\frac{V_{in}}{V_A}\right)^{-2} + \sqrt{\left(\frac{1}{2} R_m^{-1} \left(\frac{V_{in}}{V_A}\right) \frac{-2\rho_r}{\rho_{ns}}\right)^2 + R_m^{-1} \left(\frac{V_{in}}{V_A}\right)^{-2} \left(\frac{\rho_r}{\rho_{ns}}\right)^{1/2} - 2\frac{\rho_r}{\rho_s}} - \sqrt{\frac{\rho_r}{\rho_s}} \tag{12}$$

The condition for instability, $\gamma_{max} > 0$, is

$$R_m \left(\frac{V_{in}}{V_A}\right)^2 \sqrt{\frac{\rho_r}{\rho_s}} \frac{\rho_{ns}}{\rho_r} < \frac{1}{2} \tag{13}$$

Initially, when the current sheet is first formed, the situation in the solar corona can be such that the instability condition (13) is not fulfilled while the condition (11) for the plasma mass to decrease is fulfilled. But after a while, the plasma density may slowly decrease at the sheet boundary so that condition (13) becomes fulfilled and the above instability becomes triggered. Once triggered, the instability can provide the magnetic energy to drive the flare.

A similar MHD-instability in the Earth's magnetospheric trail can cause a substorm. However, the growth rate and criterion for the instability in the magnetospheric tail are likely to differ from Equations (12) and (13), since these have been derived using the relationships for the sheet model (6)-(9). It is not clear to what extent our current sheet formation model is applicable to the magnetospheric tail, since the current sheet in the tail is formed by the pull of the powerful directional flow of the solar wind. For the rapid reconstruction of the current sheet, however, the energy release process should still follow the same scenario.

3. Field-aligned currents and particle acceleration

To build an electrodynamical model of a solar flare which explains the main observational facts, let us use the analogy between magnetospheric substorms and solar flares. Two of the principal phenomena during substorms, are the bright auroras in midnight sectors of the Northern and Southern auroral ovals and the accelerated plasma flows which appear in the current sheet in the geomagnetic tail. In the geomagnetic tail the current sheet instability manifests itself initially as a sharp decrease in the sheet thickness [Fairfield et al., 1981] and an increase in the 1/c jxB force directed towards the Earth. The accelerated plasma is injected to the inner magnetosphere, and concurrently, the western electrojet (i.e. the Hall current along the polar oval) grows rapidly in the polar oval at the night side. The electrojet flows in the ionosphere between two layers of field-aligned currents (upward and downward).

The projection of field-aligned currents along magnetic field lines shows that the current source is located in the current sheet of the magnetospheric tail [Podgorny, I. M., 1988]. This source generates an electrical circuit which is closed through field-aligned currents and the Pedersen currents in the ionosphere. The flowing of both Hall and Pedersen currents becomes possible only at low altitudes (about 100 km) where particle collisions become significant. The upward field-aligned current is mainly caused by electrons with energies of 1-10 keV. The downward electron acceleration occurs either in electrical double layers or in the regions with anomalous resistivity. The current velocity of electrons is comparable to their thermal velocity at an altitude of about 10,000 km. The accelerated electrons precipitate and initiate auroral arcs in the Northern and Southern polar ovals. It has been shown by Podgorny, I. M., et al. [1988] that field-aligned currents can be generated because the current in the tail's current sheet is transferred by electrons and the 1/c jxB force is applied to the electron gas. Then plasma ions in the plasma (i.e. current) sheet of the geomagnetic tail are accelerated by the Hall electric field:

$$E = \frac{1}{enc} \, j \times B \qquad (14)$$

directed along the Earth-Sun axis. Figure 2 shows

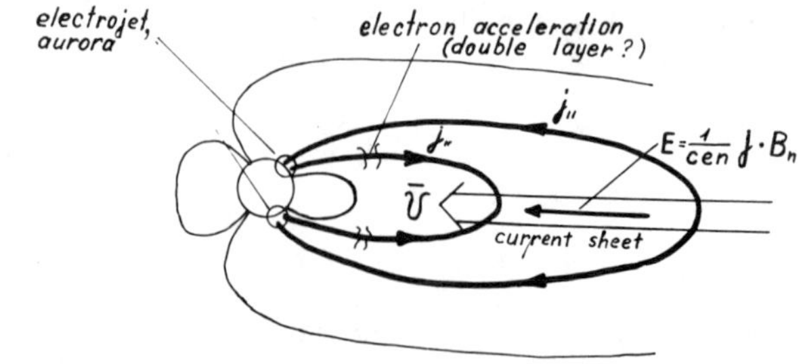

Fig. 2. A schematic diagram of a field-aligned current generator located in the current sheet of the Earth's geomagnetic tail. Arrows show directions of the currents and the electric field.

the distribution of fields and currents during the development of the instability in the geomagnetic tail.

The current sheet of the Earth's magnetosphere is also a source of particles with a much higher energy of about 1 MeV [Krimigis and Sarris, 1979], but the origin of these particles probably is not associated with field-aligned currents. More likely, these particles are accelerated during the development of the instability in the tail's current sheet around neutral points of X- or O-type.

The basic elements of the electrodynamical model, discussed above for a substorm, correspond well to the main events observed during solar flares. Figure 3 presents the circuit of field-aligned currents and connecting Pedersen currents which should be generated during the rapid release of magnetic energy in a solar flare. Here we assume that the energy is released by the development of the current sheet instability. A current sheet is formed under the magnetic arch

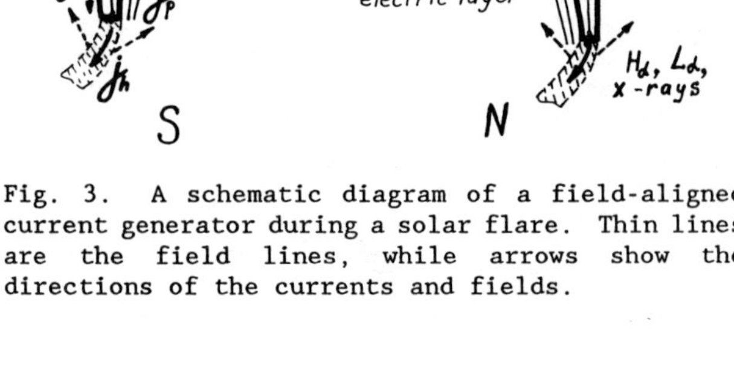

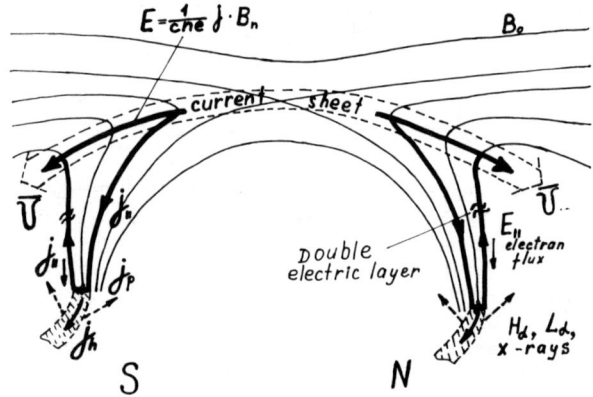

Fig. 3. A schematic diagram of a field-aligned current generator during a solar flare. Thin lines are the field lines, while arrows show the directions of the currents and fields.

whose field is opposite to the field of the other sources. Since the energy released by reconnection is transferred to kinetic energy of plasma flow, field-aligned currents (cf. Figure 3) are generated in a manner which is similar to that in the Earth's magnetospheric tail. If the current in the sheet is transported by electrons, ions are accelerated by the electric field Ene = 1/nec jxB.

Let us use the relationships (6)-(9) to estimate the energy of particles accelerated by the electric field (14) which is generated in the current sheet and which maps along field lines to the photosphere. Using B_s = 300 Gauss, $V_{in} = 0.5 \times 10^3$ cm/s, $\rho_r = 10^8$ cm^{-3}, $\rho_s = 10^{14}$ cm^{-3}, $\nu_m = c^2/4\pi\sigma = 0.5 \times 10^4$ cm^2/s (T = 10^{40} K) for the parameters in the corona, the energy of the accelerated particles will be about 1 to 3 keV. Numerical calculations show that at the nonlinear stage of the instability the current density in the sheet, j, and the value of the normal magnetic component, B_n, are increased by several times. So during the nonlinear phase of the flare this energy of the accelerated particles will be about 30 keV. Such energies are typical for X-ray quanta whose emission was observed by the SMM in flare kernels located along each side of the magnetic field polarity interface [de Jager, 1985].

If during the development of a solar flare the field-aligned current density reaches its critical value, the major part of the potential difference Eb will turn out to be applied along the magnetic field. Unfortunately, the observational data on solar flares makes it impossible to estimate the density of field-aligned currents. High-energy particles could be generated during a solar flare by acceleration along the X-line. The current sheet of a solar flare can be an effective mechanism for charged particle acceleration (see e.g. Chupp, 1988) since a strong inductive electric field appears during a reconstruction of the magnetic field. The maximum velocity of the plasma inflow into the sheet during reconstruction can approach the Alfven speed. Estimating the maximum electric field as $E_{max} = 1/c\, V_A B$ and setting B ~ 300 Gauss, $V_A = B/\sqrt{4\pi\rho}$, $\rho = m_i n$, $m_i = 1.6 \times 10^{-24}$ g is the ion mass, n = 10^8 cm^{-3} is the density of the corona, and assuming the sheet width equal to 10^{10} cm we can obtain the upper limit for the energy of accelerated charged particles. That is:

$$W_{max} = elE_{max} \sim 10^{15} \text{ eV}$$

References

Baum, P. J. and A. Bratenahl, Flux linkages of bipolar sunspot groups: A computer study, Solar Phys., 67, 245-258, 1980.

Biskamp, D., Magnetic reconnection via current sheets, Phys. Fluids, 29, 1520-1531, 1986.

Chupp, E.L., Solar neutron observations and their relation to solar flare acceleration problems, Solar Phys., 118, 137-154, 1988.

Fairfield, D. H., R. P. Lepping, E. H. Hones, Jr., S. J. Bame and J. R. Asbridge, Simultaneous measurements of magnetic-tail dynamics by IMP spacecraft, J. Geophys. Res., 86, 1396-1414, 1981.

Forbes, T. G., Fast-shock formation in line-tied reconnection models of solar flares, Astrophys. J., 305, 553-563, 1986.

Krimigis, S. M. and T. E. Sarris, Energetic particle bursts in the Earth's magnetotail, Dynamics of the Magnetosphere, ed. by Akasofu, p. 599, Reidel, Higham, Mass., 1979.

de Jager, C., Coronal explosions, Solar Phys., 96, 143-156, 1985.

Parker, P. A., Sweet's mechanism for merging magnetic fields in conduction fluids, J. Geophys. Res., 62, 509-520, 1957.

Podgorny, A. I., Model of quasistationary current sheet in the solar corona, Kosmicheskiye Issledovaniya, 26, 910-916, 1988.

Podgorny, A. I. and S. I. Syrovatskii, The creation and development of the current sheet for different magnetic viscosities and gas pressures, Phys. Plazmy USSR, 7, 1055-1063, 1981.

Podgorny, I. M., E. M. Dubinin, P. L. Izrailevich and N. S. Nikolaeva, Large-scale structure of the electric field and field-aligned current in the auroral oval from the Intercosmos-Bulgaria-1300 satellite data, Geophys. Res. Lett., 15, 1538-1540, 1988.

Priest, E. R., The magnetohydrodyanmics of current sheets, Rep. Prog. Phys., 48, 955-1090, 1985.

Priest, E. R. and T. G. Forbes, Steady magnetic reconnection in three dimensions, Solar Phys., 119, 211-214, 1989.

Syrovatskii, S. I., Characteristics of the current sheet and the heating trigger of solar flares, Astron. J. USSR Lett., 2, 35-38, 1976.

Syrovatskii, S. I., Freezing-in condition for a magnetic field and current sheet in plasma, Astrophys. Space Sci., 56, 3-12, 1978.

THE FLARE AS A RESULT OF CROSS-INTERACTION OF LOOPS

A. M. Uralov

SibIZMIR, Irkutsk 33, P.O. Box 4026, 664033 USSR

Abstract. We here consider a normal mode analysis for the instability of a straight flux tube and suggest that emerging curved loops may become unstable. A scenario for reconnection of emerging loops with an overlying arcade is put forward to explain two-ribbon solar flares, similar to the Emerging Flux Model of Heyvaerts, Priest and Rust [1977].

Is the Flare Energy Release Possible in a Single Loop?

The model of flare energy release in a single solar loop is based on the apparent analogy with processes occurring in Tokamak. In order to realize a kink instability, a trial disturbance (the energy approach) inside a cylinder with a current is usually specified in a combined form, for example,

$$\xi = \xi(r)\ \sin(\pi z/\ell)\ \exp(iwt + im\varphi + ikz)$$

[Priest, 1982]. With such a combination, it becomes possible to achieve a simultaneous fulfillment of the resonance condition $KB_z + mB_\varphi/r = (\vec{k}\cdot\vec{B}) = 0$ and of the condition for rigid fastening of the boundaries. However, a disturbance satisfying these conditions simultaneously, is, actually, not an eigen-function of the problem (a standing wave). Of course, for a variational analysis it does not need to be. This is already evident, by considering an example of a homogeneous magnetic cylinder with a homogeneous longitudinal current: ξ = constant, B_z = constant, $B_\varphi = \alpha r$, and α = constant. The equation for the variation p of a total (magnetic plus gas) pressure is in this case thus:

$$(D^2L)p - \partial^2/\partial z^2\ (F^2-D^2)p = 0;\ \text{div}\ \xi = 0 \qquad (1)$$

where $D = -\rho\omega^2+(\hat{k}\vec{B})^2/4\pi$; $F^2 = B_\varphi^2(\hat{k}\vec{B})^2/4\pi^2r^2$; $L = (1/r)(\partial/\partial r)(r\cdot\partial/\partial r) - m^2/r^2$; $\vec{B} = B_z\vec{e}_z + B_\varphi\vec{e}_\varphi$; $\hat{k} = -\vec{e}_z i(\partial/\partial z) + \vec{e}_\varphi(m/r)$.

Geophysical Monograph 58

Because the operators D and F are functions of $\partial/\partial z$ only, equation (1) is solved by the method of separation of variables: p = constant, $R(r)Z(z)\exp(i\omega t + im\varphi)$, $LR + \lambda^2R = 0$, λ^2 = constant:

$$\partial^2/\partial z^2\ (F^2 - D^2)\ Z + \lambda^2D^2Z = 0 \qquad (2)$$

A known solution describing waves in a free incompressible cylinder; $R(r) = J_m(r\sqrt{\lambda^2})$; $Z(z) \sim \exp(i\kappa z)$, corresponds to these equations. For fixed values of ω^2 and m there are only two such waves (despite the sixth order of the initial equation. Pairs of waves corresponding to a fast and slow magnetosonic wave in this special case div $\xi = 0$ are absent), forward and inverse: $\kappa = K_1$, $\kappa = K_2$, $K_{1,2} = R_e$, $K_{1,2} > 0$. In this case $\partial/\partial z \rightarrow i\kappa$ in the expressions for F and D: $(\hat{k}\vec{B}) \rightarrow \kappa B_z + \alpha m$; $F_{1,2}^2 = F^2\{(\vec{k}\vec{B})_{1,2}^2\}$; $D_{1,2} = D\{w^2, (\vec{k}\vec{B})_{1,2}^2\}$, $(\vec{k}\vec{B})_1 = k_1B_z + \alpha m$ = constant; $(\vec{k}\vec{B})_2 = -k_2B_z + \alpha m$. The condition λ^2 = constant for these solutions implies fulfillment of the equality:

$$\lambda^2 = k_1^2\ (F_1^2/D_1^2 - 1) = k_2^2\ (F_2^2/D_2^2 - 1) \qquad (3)$$

Therefore $Z(z) = C_1\exp(ik_1z) + C_2\exp(-ik_2z)$. The combined use of (3) and of the boundary conditions for the function p in coordinates z and r must determine the dispersion properties of the problem. We are, however, interested in the form of the eigen-disturbance only.

If $p(z=0,\ell) = 0$ is the boundary condition, then $c_1 = -c_2$; $k_1 + k_2 = 2\pi n/\ell \equiv 2k_n$. Using trigonometric identities one can find:

$$R_e\ \{\exp(im\varphi)\ Z(z)\} = -2\ \sin(m\varphi - k'z)\ \sin k_nz$$

where $k' = K_2 - K_n$. In the limiting cases $B_z = 0$, $B_\varphi \neq 0$, and $B_\varphi = 0$, $B_z \neq 0$; $k_1 = k_2 = k_n$, $k' = 0$ immediately follows from (3). In intermediate cases it outwardly resembles a combined function which, however, satisfies the condition $(\vec{k}'\bullet\vec{B}) = 0$. It is easy to show, using (3), that fulfillment of an analogous equality $(-k'B_z + \alpha m) = 0$ (in this case $F_1^2 = F_2^2$; $D_1^2 = D_2^2$) is possible only when $B_\varphi = 0$. The eigen-disturbance when B_φ, $B_z \neq 0$ is a weakly helical one such that this resonance condition is never reached.

It is more customary to investigate the problem in terms of equations for plasma displacements ξ. These are expressed, in the case div $\xi = 0$, in terms of the relationships involving $\partial p/\partial z$, $\partial p/\partial \varphi$, $\partial p/\partial r$, and p. Since equation (1) is solved by the method of separation of variables, the dependence of these quantities on z is determined by the dependence p(z).

If, however, a combined disturbance $((\vec{k}\cdot\vec{B}) = 0)$ is, nevertheless, specified from outside (the Cauchy problem), it, without having time to grow, decays into two disturbances running away from each other (an analog of D'Alambere's solution). There arises an oscillatory process. Therefore, the conclusion about the possibility of a linear growth (near the condition $(\vec{k}'\bullet\vec{B}) = 0$) of helical modes in a solar loop with anchored footprints is premature.

The development of a helical bend is, however, possible at times of emergence of magnetic loops, which is equivalent to imposing an edge mode [Uralov, 1987]. The solution of the problem of propagation of the edge mode always involves a term which represents a wave excited at one end of the cylinder, independently of the influence of the other end as if the cylinder were an infinite one. This last circumstance implies a possibility for development of unstable helical modes. These modes are eigen-functions of the cylinder with identified ends. Their growth rate is determined by the rate of emergence of new parts of the loop from beneath the photosphere.

In the discussion to follow we will be concerned with a helical bend of the loop as a whole (external mode m = 1). As is known, it is extremely difficult to subject to stabilization this large-scale mode, whereas all the other ones can be suppressed.

Bending of the Emerging Loop

A direct current i placed into an external magnetic field $\vec{B}$ is twisted (Figure 1a) by the right-hand screw rule. In the case of a totally screened (Figure 1b) current $J = i + i_s = 0$, the sign of helical turn-over is determined by the reverse skin current i_s, whose magnetic moment is slightly larger than the direct current moment of the finite radius of the tube.

A helical bend of the magnetic loop emergent from beneath the photosphere is shown in Figure 2. Electric current parallel to the magnetic field flows along the loop. Again, the right-hand screw rule is applied. The amplitude of bend increases with the length of the loop. (a) the upper turn-over situation (UT), the most simple and stable one. (b) lower turn-over. Situation (b") can go unstable with respect to a change to state (a) (a prominence ejection?).

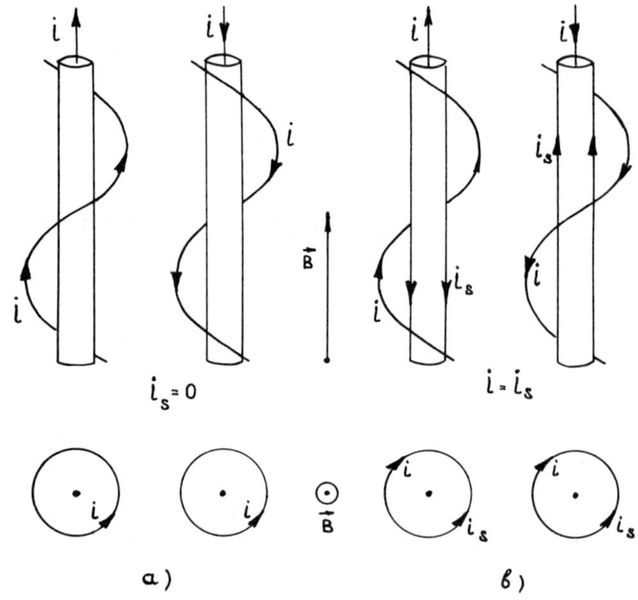

Fig. 1. Bending of direct current.

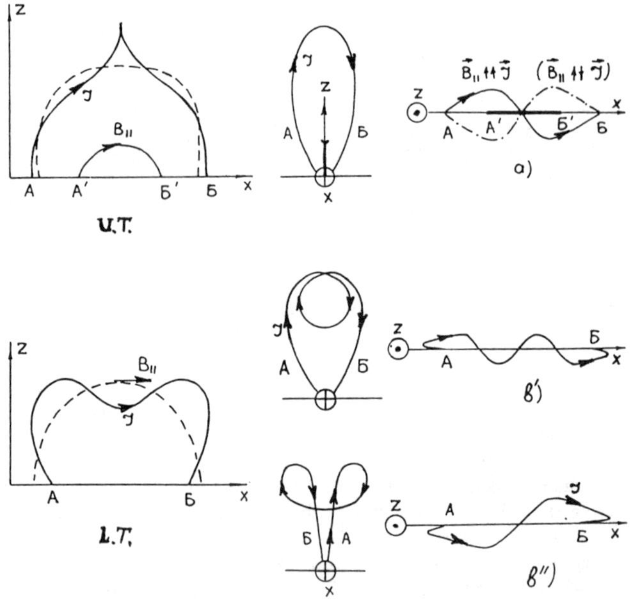

Fig. 2. Bending of the emerging loop.

The Evolution of Equilibrium of an Emerging Loop Arcade

While turning over individually, for example, as in Figure 2a, the loops of a growing arcade will begin interfering with each other. This situation is shown in Figure 3. The amplitude of bend here is exaggerated because the loops themselves are drawn too thin and their initial turn-over to the zero line is chosen to be large. (1) initial position of the loops with longitudinal force-free current. (2) rise and bending of the loops. (3) rise and the turning-around loops and their sticking-together in knots (cross). The loops take on a helmet-like shape. (3') rupture of the reconnected loops. There arises a long magnetic

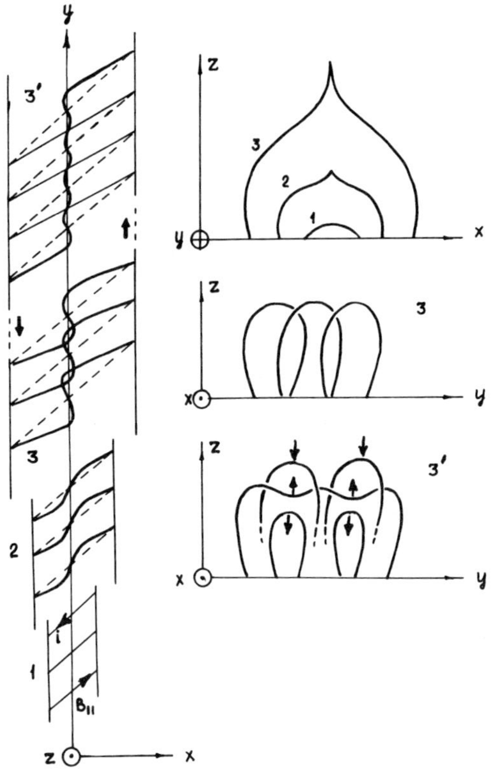

Fig. 3. Interaction of loops in the arcade. Formation of a magnetic filament.

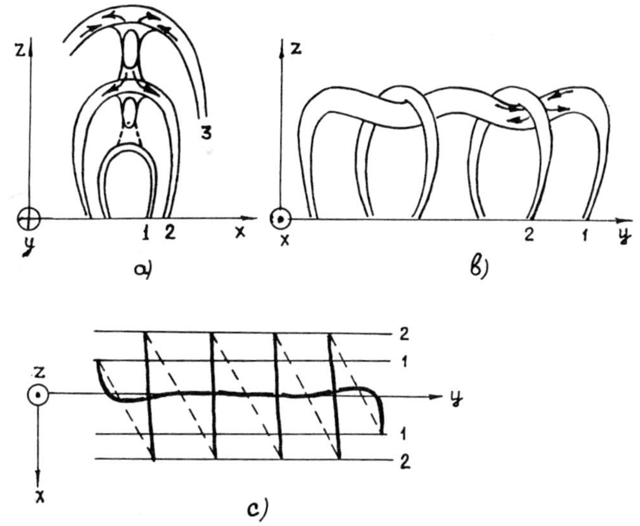

Fig. 4. Magnetic "knots" (of a prominence?). (Flare) decay of the filament - dashes.

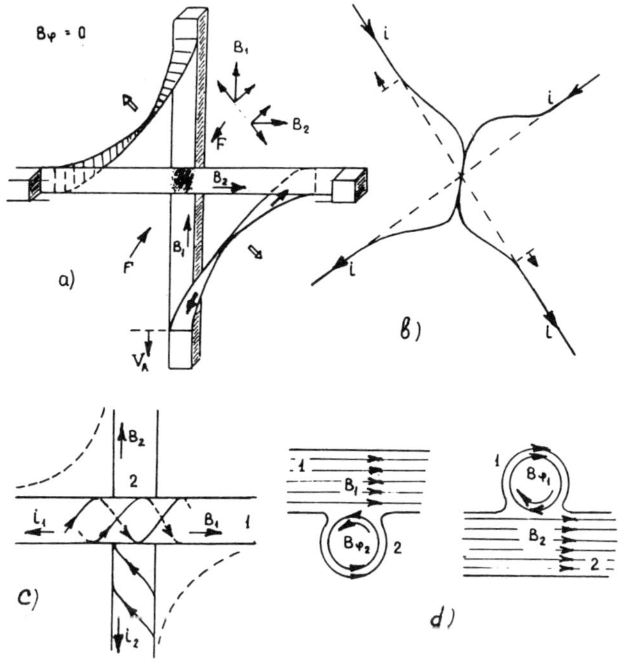

Fig. 5. Merging and rupture of magnetic tubes in the "knots".

filament and short, nearly flat loops. The magnetic filament always is in a state of pulling together with short loops of the preceding arcade. Heavy arrows indicate plasma flow toward the bend.

A magnetic filament, while being attracted to short loops of the preceding arcade, exchanges plasma with them (Figure 4 a,b). (c) flare decay of the filament and return to a situation (dashed) close to the original one. Magnetic surfaces 1 and 2 in this case become connected. The plasma accumulated in the filament is "shot" into short loops. Energetic particles also escape along the horizontally lying filament. A disturbance generated by the reconnection act in one knot, propagates along the filament as a waveguide and initiates a flare process in the neighboring knots (a two-ribbon flare). As new magnetic flux emerges from beneath the photosphere, the picture described here can result.

Cross-Interaction of Identical Magnetic Tubes

Cross-interaction of loops in magnetic "knots" is illustrated by the scheme in Figure 5. (a) forced reconnection of two thin magnetic stripes which compose each magnetic bar. There is no longitudinal electric current present. Total reconnection results in the formation of two bent (with twisting at the place of bending), diverging tubes, (b) bending of the tubes with a strong longitudinal electric current and spontaneous sticking-together of the portions carrying parallel currents. (d) current-carrying tubes ($B_\varphi(a) \approx B_{\parallel}(a)$) arranged as shown in Figure 5c, stick together to form a knot with the subsequent possible rupture. (If tube 2 is placed above tube 1, then instead of sticking together, they push apart). A total time of reconnection

$$\tau \approx (L/V_e)^{1/2} (a^2/\nu_i)^{1/2} \rho_e/\rho_i,$$

where ν_i is magnetic viscosity in a diffusion layer of size L and thickness σ. ρ_i/ρ_e is the compression of the plasma flowing into the layer. The values L = a (a is tube diameter) and L = σ are limiting ones. In the latter case the heat released is a minimum and the rate of energy released is a maximum: $\tau = a/V_e$.

A more detailed treatment of the issues considered here is contained in a paper of Uralov [1987].

Acknowledgment. I am grateful to Mr. V. G. Mikhalkovsky for his assistance in preparing the English version of the manuscript and for typing the text.

References

Heyvaerts, J., E. R. Priest, and D. M. Rust, An emerging flux model for the solar flare phenomenon, Astrophys. J., 216, 123-137, 1977.

Priest, E. R., Solar Magnetohydrodynamics, D. Reidel Publishing Company, Holland, 1982.

Uralov, A. M., The flare as a result of cross-interaction of loops: Causal relationship with a prominence, Preprint SibIZMIR, No. 28, 1987.

EFFECTS OF PLASMA MASS FLOW ON ALFVEN WAVE PHASE MIXING IN CORONAL LOOPS

M. Peredo and J. A. Tataronis

University of Wisconsin-Madison, Madison, WI 53706

Abstract. Ground-based and space observations suggest that flow phenomena are intrinsic to the solar corona loops. The present analysis addresses flow effects on Alfvén wave behavior in coronal loops. Specifically, we investigate the effect of flow on the shear Alfvén wave continuum and the associated damping process resulting from phase-mixing. Flow Doppler-shifts the Alfvén wave continuum, thus displacing the range of frequencies for optimum heating. The present study explores Alfvén resonance effects on surface waves propagating along coronal loops that are under the influence of flow. Coupling to the Alfvén resonance leads to damping of the surface waves. We compute the damping rate in the presence of sub-Alfvénic flows. Flow increases the damping rate of certain modes, thus enhancing the heating effectiveness of those modes. Therefore, our calculations suggest that flow is potentially an important ingredient for an accurate evaluation of heating the solar corona by dissipation of Alfvén waves via phase-mixing.

Introduction

It has been speculated in the past that the high temperature levels of the corona are sustained by shear Alfvén waves, which transport energy from the chromosphere along plasma loops formed by the coronal magnetic field. In the course of time, the shear Alfvén wave disturbances undergo phase-mixing and eventually damp, leading to a conversion of coherent wave energy into plasma thermal energy through viscous effects that are intrinsic to the ionized medium of the corona. Previous studies of coronal heating via phase-mixing of Alfvén waves have neglected equilibrium plasma flow [Heyvaerts and Priest, 1983; Nocera et al., 1984]. However, studies of the solar corona suggest that a variety of flow processes is inherently present in the loops [Foukal, 1978; Priest, 1982, p.45 and p. 242]. Specifically, observations have revealed both upflows and downflows along the legs of the loops with characteristic speeds in the range of 1-45 km/s [Foukal, 1978; Nicolas et al., 1982; Kingston et al, 1982; Gurman and Athay, 1983; Habbal et al., 1985; Kopp et al., 1985]. Plasma flow can affect the characteristics and behavior of the loops by modifying the spectral properties and stability of wave phenomena. A fundamental consequence of flow is a Doppler-shift of the Alfvén wave continuum. Therefore, an accurate estimate of the heating effectiveness requires the inclusion of flow effects. Past investigations of flow effects on the continuum, and on the stability of waves, have emphasized laboratory plasmas [Tataronis and Mond, 1987; Bondeson et al., 1987]. In this paper, we explore the effects of plasma flow on Alfvén wave dynamics in coronal loop structures. Specifically, we address flow effects on the shear Alfvén wave continuum, and the associated damping process resulting from phase mixing.

Geophysical Monograph 58

Coronal Loop Model

In order to isolate the effects of flow on Alfvén wave phenomena in coronal loops, we consider an idealized configuration. Namely, we model a coronal loop as an infinite, free boundary, cylindrical plasma column, embedded in an infinite ambient plasma medium. Figure 1 illustrates schematically the loop-ambient plasma configuration. Clearly, the geometry of the loops limits the validity of our model. Specifically, real loops are curved structures of finite length. Thus, our analysis is relevant for loops with characteristic length much larger than both the Alfvén wavelength and the minor radius. A key assumption in our analysis is that the equilibrium flow velocity is aligned parallel to the magnetic field, $\mathbf{V} = \alpha\mathbf{B}$, where α is a scalar. Our analysis is based on the time-dependent equations of magnetohydrodynamics (MHD). An additional assumption made here is that the equilibrium mass density, ρ, is constant throughout the plasma. The assumption of constant mass density is consistent with EUV and Thomson scattering observations [Foukal, 1978], which suggest a reasonably uniform density within the loops. These observations also show that the pressure within the loops increases dramatically from its value at the center of the loop through a sheath region connecting the interior of the loop to the ambient plasma. Therefore, we assume a magnetic field profile consistent with the observed structure. We consider constant axial fields in the interior region, and the ambient plasma, joined through a sheath region where the field is linear in r, as illustrated in Figure 2. In the presence of such a sheath, surface waves travelling along the loop become increasingly out of phase, and eventually are damped via phase-mixing. Tataronis and Grossmann [1973a, b] studied the decay of MHD surface waves by phase-mixing for static configurations. We use their technique, modified to allow flow in the equilibrium state. The damping process can be understood as follows. As the surface waves propagate along the loops, there are magnetic surfaces in the plasma at which the frequency of the surface waves matches the local frequency of Doppler-shifted shear Alfvén waves. When this matching holds, energy accumulates about the Alfvén resonance leading to damping of the surface waves. It must be emphasized that as the region of non-uniformity vanishes, the surface waves propagate undamped according to the dispersion relation for the sharp boundary configuration characterized by a jump in the axial field at the surface of the loop. Mathematically, the absence of damping results from condensation of the continuous spectrum into the discrete eigenvalues of the sharp boundary configuration.

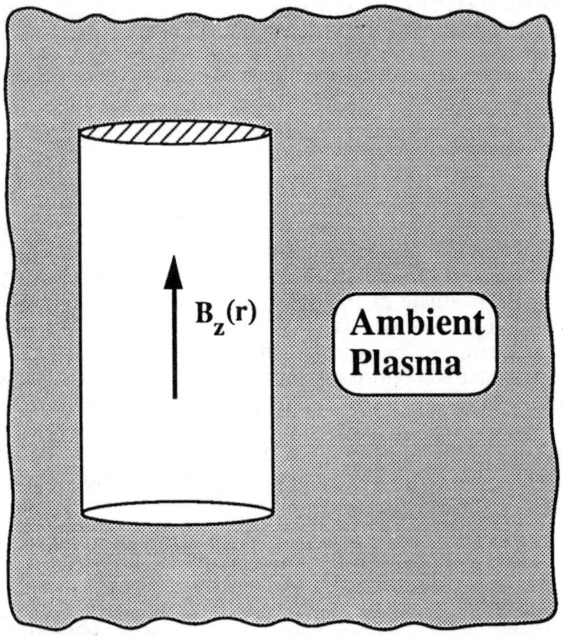

Fig. 1. Schematic description of Coronal Loop Model. The loop is represented by a cylindrical plasma column embedded in an infinite ambient plasma medium.

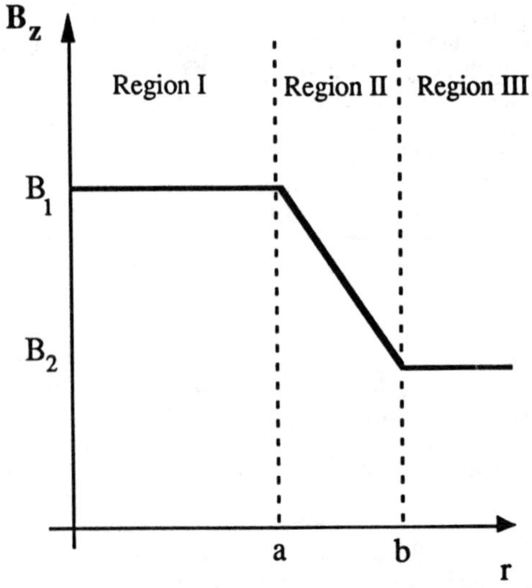

Fig. 2. Radial Variation of the Axial Magnetic Field. The field is uniform in the loop interior ($0 < r < a$), varies proportional to r in the sheath region ($a < r < b$), and is uniform in the ambient plasma ($b < r$).

Governing MHD Equations

Basic Equations

Our description of the dynamics of the coronal plasma is based on the time-dependent MHD equations with scalar pressure, infinite conductivity and zero viscosity. Under these conditions, the MHD equations read as follows,

$$\frac{\partial \rho}{\partial t} + \nabla\cdot(\rho \mathbf{V}) = 0 \qquad (1)$$

$$\rho \frac{\partial \mathbf{V}}{\partial t} + \rho(\mathbf{V}\cdot\nabla)\mathbf{V} = \frac{1}{\mu_0}(\nabla \times \mathbf{B}) \times \mathbf{B} - \nabla P \qquad (2)$$

$$\frac{\partial \mathbf{B}}{\partial t} - \nabla \times (\mathbf{V} \times \mathbf{B}) = 0 \qquad (3)$$

$$\nabla\cdot\mathbf{B} = 0 \qquad (4)$$

where ρ, P, $\mathbf{V}$ are, respectively, the fluid mass density, pressure and velocity; $\mathbf{B}$ is the magnetic field; and μ_0 is the vacuum permeability. Closure of the system requires an equation of state. We restrict our analysis to the case of incompressible plasma motions characterized by the condition,

$$\nabla\cdot\mathbf{V} = 0 \qquad (5)$$

Aligned Flow Equilibria

Observations reveal that coronal loop flows largely follow the magnetic field geometry. Therefore, we consider equilibria with mass flow aligned parallel to the magnetic field,

$$\mathbf{V_0} = \alpha \mathbf{B_0} \qquad (6)$$

where the subscript zero designates equilibrium variables. Note that Alfvénic flow corresponds to $\alpha = \alpha_c \equiv (\rho\mu_0)^{-1/2}$. Coronal observations reveal flows with speeds in the range of 1-45 km/s. On the other hand, characteristic values for the Alfvén speed in coronal loops fall in the range 10^2-10^3 km/s; thus, observations suggest that coronal loop flows generally satisfy the sub-Alfvénic criterion $\alpha \leq \alpha_c$, and we restrict our analysis to this regime. Furthermore, present knowledge of loop flows is not detailed enough to reveal if α depends on spatial location along the loop. Therefore, our study is limited to the case of uniform α. The specific configuration we explore consists of an axial magnetic field of the form,

$$\mathbf{B_0} = B_z(r)\,\hat{\mathbf{z}} \tag{7}$$

where $\hat{\mathbf{z}}$ is a unit vector in the axial direction. Equilibrium pressure balance follows from the radial component of Eq. (2) with the time derivatives set equal to zero,

$$\frac{d}{dr}\left(P + \frac{B^2}{2\mu_0}\right) = 0 \tag{8}$$

Linearized Equations

We consider perturbations about the equilibrium state described in the previous section. Substituting in Eqs. (1)-(5) the expansions

$$\mathbf{V} = \mathbf{V_0} + \mathbf{v}; \quad \mathbf{B} = \mathbf{B_0} + \mathbf{b}; \quad P = P_0 + p \tag{9}$$

and retaining only first-order terms in the small fluctuating quantities **v**, **b**, and p, the following system of linear equations results:

$$\frac{\partial \mathbf{v}}{\partial t} + \alpha(\mathbf{B_0}\cdot\nabla)\mathbf{v} + \alpha(\mathbf{v}\cdot\nabla)\mathbf{B_0} - \frac{1}{\rho\mu_0}(\mathbf{B_0}\cdot\nabla)\mathbf{b} - \frac{1}{\rho\mu_0}(\mathbf{b}\cdot\nabla)\mathbf{B_0} = -\frac{\nabla(\delta\pi)}{\rho} \tag{10}$$

$$\frac{\partial \mathbf{b}}{\partial t} - (\mathbf{B_0}\cdot\nabla)\mathbf{v} + (\mathbf{v}\cdot\nabla)\mathbf{B_0} - \alpha(\mathbf{b}\cdot\nabla)\mathbf{B_0} + \alpha(\mathbf{B_0}\cdot\nabla)\mathbf{b} = 0 \tag{11}$$

$$\nabla\cdot\mathbf{v} = 0 \tag{12}$$

where $(\delta\pi)$ corresponds to the linearized part of the total pressure $\pi = P + B^2/(2\mu_0)$. We follow Tataronis and Mond [1987], who use cylindrical coordinates (r, θ, z), and assume a harmonic dependence of the form exp $i(\omega t - m\theta - kz)$, where ω is the wave frequency, m is the poloidal mode number, and k is the wavelength. We must point out that a harmonic dependence of the form assumed is valid only in the infinite configuration considered. The more realistic, bounded configuration requires a different approach, which will be the subject of a future investigation. With the assumed harmonic dependence, Tataronis and Mond [1987] have reduced Eqs. (10)-(12) to a second-order ordinary differential equation for the quantity, $\Phi = (r\,v_r)$:

$$\frac{d}{dr}\left[\frac{r\,\Delta(r)}{m^2 + k^2r^2}\frac{d\Phi}{dr}\right] - \frac{\Delta(r)}{r}\Phi = 0 \tag{13}$$

where the coefficient Δ(r) is defined as follows,

$$\Delta(r) = \frac{\left[\omega + f(r)(\alpha_c-\alpha)\right]\left[\omega - f(r)(\alpha_c+\alpha)\right]}{\alpha^2 - \alpha_c^2} \tag{14}$$

In Eq. (14), f(r) designates the quantity $kB_z(r)$. Equation (13), along with appropriate homogeneous boundary conditions, constitutes a generalized eigenvalue problem for the frequency ω. Solutions of this eigenvalue problem correspond to natural modes of the system. If the coefficient Δ is free of zeros in the domain of integration, surface waves propagate undamped along the loop. This is a direct consequence of the absence of dissipative terms in the original equations. However, if Δ vanishes somewhere in the domain of integration, Eq. (13) is singular near zeros of Δ. These singularities are associated with the continuous spectrum of the differential operator in Eq. (13). In the presence of inhomogeneities, the coupling between the Alfvén resonance and the surface waves leads to complex frequencies corresponding to damped surface waves as shown by Tataronis and Grossmann [1973a, b].

Effects of Flow on the Shear Alfvén Wave Continuum

Zeros of the coefficient of the highest derivative in Eq. (13) lead to non-square integrable singularities of the differential equation. These singularities correspond to the shear Alfvén wave continuum. In the presence of inhomogeneities, these singularities are associated with accumulation of energy about the Alfvén resonance layer. This accumulation of energy about the Alfvén resonance layer is the source of coronal heating in our model. Singularities of Eq. (13) occur at zeros of Δ(r), i.e. at points satisfying one of two possible conditions,

$$\text{(i)}\ \omega = -f(\alpha_c - \alpha); \qquad \text{(ii)}\ \omega = f(\alpha_c + \alpha) \tag{15}$$

The frequency $f\alpha_c$ corresponds to an Alfvén wave propagating in a stationary plasma. Consequently, Eq. (15) represents Alfvén wave frequencies Doppler shifted by the flow term fα. This shift of the continuum leads to a criterion for effective heating by dissipation of Alfvén waves. A necessary condition for effective heating is that the frequency of the convective motions at the photosphere, which generate the surface waves, lie in the flow-shifted Alfvén wave continuum.

Damping of Surface Waves via Phase Mixing

We consider Eq. (13) with the magnetic field profile illustrated in Fig. 2. Furthermore, in accordance with the technique of Tataronis and Grossmann [1973a, b], we assume the sheath region to be thin compared with the radial dimension of the loop. Under these assumptions, we solve Eq. (13) separately in the three regions identified in Fig. 2.

Region I: $0 \le r \le a$

In this region B_z is constant, implying that Δ is independent of r. It is readily demonstrated that the following function satisfies Eq. (13),

$$\Phi = C_1\, kr\, I_m'(kr) + C_2\, kr\, K_m'(kr) \tag{16}$$

where I_m and K_m are modified Bessel functions of order m, the prime represents derivative with respect to the argument, and C_1 and C_2 are constants of integration determined by boundary conditions.

Region II: $a \leq r \leq b$

In the sheath region, B_z depends linearly on r. Zeros of $\Delta(r)$ occur at the following radial points,

$$r_0 = \frac{(f_2 a - f_1 b)}{(f_2 - f_1)} \pm \frac{\omega\,(b - a)}{(f_2 - f_1)\,(\alpha_c \pm \alpha)} \tag{17}$$

where $f_1 = kB_1$ and $f_2 = kB_2$ correspond, respectively, to the values of f(r) in the interior of the loop and the ambient plasma. It must be emphasized that r_0 is complex if the eigenvalue ω is complex. Caution must be exercised in order to correctly locate the position of the eigenvalues in the complex-ω plane. In general, these roots do not lie on the principal sheet of the complex-ω plane, and analytic continuation under the proper branch cut is necessary to expose them. In the process of locating the eigenvalue, as ω is analytically continued into the principal sheet, a zero of Δ crosses the real-r axis within the domain of integration. Details of this analytic continuation process appear in Tataronis and Grossmann [1973a, b]. Therefore, in the sheath region, we expand the coefficients of the differential equation about this zero of Δ and solve. We find the general solution in Region II to be,

$$\Phi = C_3\, I_0\big[\delta(r-r_0)\big] + C_4\, K_0\big[\delta(r-r_0)\big] \tag{18}$$

where I_0 and K_0 are zeroth order modified Bessel functions, C_3 and C_4 are integration constants determined by boundary conditions, and δ is given by,

$$\delta^2 = \frac{\left(m^2 + k^2 r_0^2\right)}{r_0^2} \tag{19}$$

Region III: $b \leq r$

Here B_z is again constant and we have:

$$\Phi = C_5\, kr\, I_m'(kr) + C_6\, kr\, K_m'(kr) \tag{20}$$

with two additional integration constants. In order to determine the six integration constants C_1 to C_6, we impose boundedness, continuity and jump conditions on Φ. Specifically, we require Φ to be bounded as $r \to 0$ and $r \to \infty$; we also impose continuity of Φ, and $[r\,\Delta/(m^2 + k^2r^2)]\, d\Phi/dr$ at r=a and r=b. The boundedness conditions imply $C_2 = C_5 = 0$, while the conditions at r=a give the dispersion relation:

$$\big[\omega + f_1(\alpha_c - \alpha)\big]\big[\omega - f_1(\alpha_c + \alpha)\big]\,\varepsilon_1 = 2\,\omega\,\alpha_c \frac{(f_1 - f_2)}{(b - a)}$$

$$\times\ \delta(a-r_0)\,\frac{\sigma\, I_1\big[\delta(a-r_0)\big] + K_1\big[\delta(a-r_0)\big]}{\sigma\, I_0\big[\delta(a-r_0)\big] + K_0\big[\delta(a-r_0)\big]} \tag{21}$$

where $\sigma = C_3/C_4$, is determined from the two continuity conditions at r=b,

$$2\,\omega\,\alpha_c \frac{(f_1 - f_2)}{(b - a)}\ \delta(b-r_0)\,\frac{\sigma\, I_1\big[\delta(b-r_0)\big] + K_1\big[\delta(b-r_0)\big]}{\sigma\, I_0\big[\delta(b-r_0)\big] + K_0\big[\delta(b-r_0)\big]} =$$

$$\big[\omega + f_2(\alpha_c - \alpha)\big]\,\big[\omega - f_2(\alpha_c + \alpha)\big]\,\varepsilon_2 \tag{22}$$

In Eqs. (21)-(22), we have introduced the quantities ε_1 and ε_2 defined by,

$$\varepsilon_1 = \frac{(m^2+k^2a^2)\, I_m(ka)}{ka^2\, I_m'(ka)};\quad \varepsilon_2 = \frac{(m^2+k^2b^2)\, K_m(kb)}{kb^2\, K_m'(kb)} \tag{23}$$

We now make the thin sheath, or equivalently, steep gradient approximation, i.e. $|b-a| \ll a$; thus, $(a-r_0)$ and $(b-r_0)$ are small quantities, and we may expand the Bessel functions as,

$K_0(z) \sim -\ln z$; $\quad K_1(z) \sim -1/z$; $\quad I_0(z) \sim 1$; $\quad I_1(z) \sim 0$

In this limit, Eqs. (21) and (22) reduce to:

$$\big[\omega + f_1(\alpha_c - \alpha)\big]\,\big[\omega - f_1(\alpha_c + \alpha)\big]\,\varepsilon_1 =$$

$$+\,2\,\omega\,\alpha_c \frac{(f_1 - f_2)}{(b - a)}\ \delta\,(a - r_0)\ \frac{-\,1/\,\delta(a - r_0)}{\sigma - \ln\big[\delta(a - r_0)\big]} \tag{24}$$

and

$$2\,\omega\,\alpha_c \frac{(f_1 - f_2)}{(b - a)}\ \delta(b - r_0)\ \frac{-\,1/\,\delta(b - r_0)}{\sigma - \ln\big[\delta(b - r_0)\big]} =$$

$$\big[\omega + f_2(\alpha_c - \alpha)\big]\big[\omega - f_2(\alpha_c + \alpha)\big]\,\varepsilon_2 \tag{25}$$

Replacing σ and r_0 from Eqs. (25) and (17) respectively, the dispersion relation reads,

$$\big[\omega + f_1(\alpha_c - \alpha)\big]\big[\omega - f_1(\alpha_c + \alpha)\big]\,\varepsilon_1 =$$

$$\big[\omega + f_2(\alpha_c - \alpha)\big]\big[\omega - f_2(\alpha_c + \alpha)\big]\,\varepsilon_2\, q \tag{26}$$

where we have set

$$\frac{1}{q} \equiv 1 + (b-a) \ln\left[\frac{f_2(\alpha_c+\alpha) - \omega}{f_1(\alpha_c+\alpha) - \omega}\right]$$

$$\times \frac{\left[\omega + f_2(\alpha_c-\alpha)\right]\left[\omega - f_2(\alpha_c+\alpha)\right] \varepsilon_2}{2\,\omega\,\alpha_c\,(f_1 - f_2)} \tag{27}$$

Notice that for $b = a$, q reduces to unity, and we obtain the dispersion relation for the sharp boundary configuration,

$$\left[\omega_0 + f_1(\alpha_c-\alpha)\right]\left[\omega_0 - f_1(\alpha_c+\alpha)\right] \varepsilon_1 =$$

$$\left[\omega_0 + f_2(\alpha_c-\alpha)\right]\left[\omega_0 - f_2(\alpha_c+\alpha)\right] \varepsilon_2 \tag{28}$$

In Eq. (28), a zero subscript identifies the frequency corresponding to the sharp boundary case. It must be emphasized that ω_0 is real, an expected result due to the absence of dissipation. It is the inhomogeneity in the sheath region that causes an imaginary component of frequency to arise from the full dispersion relation, Eq. (26), when q differs from unity.

Next, we examine the logarithm term in detail in order to avoid singular behavior. We write

$$J = \ln\left[\frac{f_2(\alpha_c+\alpha) - \omega}{f_1(\alpha_c+\alpha) - \omega}\right] \tag{29}$$

Notice that J has branch point singularities in the complex-ω plane at the points

$$\omega = \omega_1 \equiv f_1(\alpha_c+\alpha); \qquad \omega = \omega_2 \equiv f_2(\alpha_c+\alpha) \tag{30}$$

Analytic continuation past the singular points is achieved by defining,

$$J = \begin{cases} P \ln\left[\dfrac{\omega - \omega_2}{\omega - \omega_1}\right] & \mathrm{Im}\,\omega < 0 \\ \ln\left[\dfrac{\omega - \omega_2}{\omega - \omega_1}\right] + i\pi & \mathrm{Im}\,\omega = 0 \\ P \ln\left[\dfrac{\omega - \omega_2}{\omega - \omega_1}\right] + 2\pi i & \mathrm{Im}\,\omega > 0 \end{cases} \tag{31}$$

where P refers to the value of the logarithm on the principal sheet. With this definition in mind, we expand ω in Eqs. (26) and (27), as $\omega = \omega_r + i\,\omega_r$ and find:

$$\omega = \omega_0 + \tilde{\omega}_r + i\,\tilde{\omega}_i \tag{32}$$

where ω_0 is the frequency corresponding to the sharp boundary case $b = a$; $\tilde{\omega}_r$ and $\tilde{\omega}_i$ are corrections of order (b-a) due to the inhomogeneity in the sheath region. Specifically, for $\mathrm{Im}\,\omega > 0$

$$\omega_0 = \frac{\alpha(f_1\varepsilon_1 - f_2\varepsilon_2)}{(\varepsilon_1 - \varepsilon_2)}$$

$$\pm \frac{\sqrt{\alpha^2(f_1\varepsilon_1-f_2\varepsilon_2)^2 + (\varepsilon_1-\varepsilon_2)(f_1^2\varepsilon_1-f_2^2\varepsilon_2)(\alpha_c^2-\alpha^2)}}{(\varepsilon_1 - \varepsilon_2)} \tag{33}$$

$$\tilde{\omega}_r = -(b-a)\,\frac{\left[\omega_0^2\,\varepsilon_2 - 2\,\omega_0\alpha f_2\varepsilon_2 - f_2^2\varepsilon_2\,(\alpha_c^2-\alpha^2)\right]^2}{4\,\omega_0\,\alpha_c\,(f_1 - f_2)\left[(\omega_0-\alpha f_1)\varepsilon_1 - (\omega_0-\alpha f_2)\varepsilon_2\right]}$$

$$\times P \ln\left[\frac{\omega_0 - \omega_2}{\omega_0 - \omega_1}\right] \tag{34}$$

$$\tilde{\omega}_i = -\pi\,(b-a)$$

$$\times \frac{\left[\omega_0^2\,\varepsilon_2 - 2\omega_0\alpha f_2\varepsilon_2 - f_2^2\varepsilon_2\,(\alpha_c^2 - \alpha^2)\right]^2}{2\,\omega_0\,\alpha_c\,(f_1 - f_2)\left[\omega_0(\varepsilon_1-\varepsilon_2) - \alpha(f_1\varepsilon_1-f_2\varepsilon_2)\right]} \tag{35}$$

Flow modifies the damping rate, $\tilde{\omega}_i$, explicitly through the terms involving α, and implicitly, through the influence of flow on the sharp boundary frequency, ω_0. Detailed study of flow effects on the damping rate require a numerical study of Eq. (35). However, observed coronal flows are weak compared to the Alfvén speed, i.e. $\alpha \ll \alpha_c$. We discuss the weak flow limit here, and reserve a detailed numerical study of Eq. (35) to a future communication. In the weak flow limit, the effect of flow on the damping rate can readily be seen since Eqs. (33) and (35) reduce to:

$$\omega_0 = \frac{\alpha(f_1\varepsilon_1 - f_2\varepsilon_2)}{(\varepsilon_1-\varepsilon_2)} \pm \sqrt{\frac{(f_1^2\varepsilon_1 - f_2^2\varepsilon_2)}{(\varepsilon_1 - \varepsilon_2)}}\;\alpha_c \tag{36}$$

and

$$\tilde{\omega}_i = \pi\,(b-a)\,\frac{\varepsilon_1^2\,\varepsilon_2^2\,(f_2^2 - f_1^2)^2\,\alpha_c}{2\,(f_1-f_2)\,(\varepsilon_1-\varepsilon_2)^2\,(f_1^2\varepsilon_1-f_2^2\varepsilon_2)}\left[1 \pm Z\,\frac{\alpha}{\alpha_c}\right]$$

$$+\,O\!\left(\frac{\alpha^2}{\alpha_c^2}\right) \tag{37}$$

where the coefficient Z is given by,

$$Z = \sqrt{\frac{16(f_1^2\varepsilon_1 - f_2^2\varepsilon_2)}{(f_2 + f_1)^2\,(\varepsilon_1 - \varepsilon_2)}}$$

$$- \sqrt{\frac{(f_1\varepsilon_1 - f_2\varepsilon_2)^2}{(f_1^2\varepsilon_1 - f_2^2\varepsilon_2)\,(\varepsilon_1 - \varepsilon_2)}} \qquad (38)$$

The two signs in Eq. (37) correspond to the two natural modes of the sharp boundary case described by Eq. (33). At this point, it is apparent from Eq. (36), that flow Doppler-shifts these modes. Furthermore, Eq. (37) reveals that the damping rate of one mode increases with increasing flow, while that for the other decreases. The fundamental effect of flow on the two modes may be understood as follows. The resonant surfaces for the two natural modes coincide in the static case, since the modes have frequencies of equal magnitude. Furthermore, the damping rate is identical for both modes. However, in the presence of flow, the frequencies of shear Alfvén waves and the two modes of surface waves experience a Doppler-shift. These frequency shifts modify the location of the resonant surfaces. The displacement of the resonant surfaces is such that one of the natural modes is closer to resonance while the other is farther from resonance relative to the static case. Under these conditions, one mode damps more efficiently than in the static case, while the opposite holds for the other mode.

We evaluate the specific effect of flow on the damping rate assuming the following parameters appropriate to the corona,

$$B_1 \sim 50\text{ G}; \qquad B_2 \sim 10\text{ G}; \qquad \rho \sim 10^{-11}\text{ kg/m}^3$$

$$a \sim b \sim 5 \times 10^3\text{ Km} \qquad ka \sim kb \sim 0.05$$

In addition, we assume perturbations with poloidal mode number m = 1. We find the effect of flow on the sharp boundary frequency is given by,

$$\omega_0 = \pm\, 10.\,1693 + 8.4609\left(\frac{\alpha}{\alpha_c}\right) \qquad (39)$$

whereas, the effect on the damping rate is,

$$\tilde{\omega}_i = 1.2301 \times 10^{-3}\,(b\text{-}a)\left[1 \pm 3.2353\left(\frac{\alpha}{\alpha_c}\right)\right] \qquad (40)$$

Thus, if the equilibrium flow is 10% of Alfvénic flow, the heating effectiveness of one mode experiences a 32% enhancement relative to the static case.

Summary

We have explored the effects of equilibrium mass flow on the shear Alfvén wave continuum, and on the heating of coronal loops via phase-mixing of Alfvén waves. Specifically, we have shown that flow Doppler-shifts the shear Alfvén wave continuum, displacing the location of the range of frequencies for optimal heating. We have also shown that flow affects the damping rate associated with phase-mixing of the Alfvén waves. In particular, flow enhances the damping effectiveness of certain modes, and thus must be included in any realistic attempt to determine the effectiveness of coronal heating by means of Alfvén wave phase-mixing. An important shortcoming of the present treatment is the assumption of incompressibility. When compressible motions are allowed, supersonic flows lead to the formation of shocks [Cargill and Priest, 1980; Noci, 1981]. The presence of shocks may lead to localization of the Alfvén waves due to reflections at the shock fronts; this in turn may alter significantly the heating effectiveness. A study of flow effects in the presence of shocks is the subject of a future investigation. Lastly, we must point out that the problem considered in this paper is an initial value problem where the wavenumber k is assumed real, and we solve for the (complex) frequency ω. A more appropriate description of coronal loops would be the boundary value problem where the (real) frequency is given, and one must solve for the (complex) wavenumber. The boundary value problem is mathematically more complicated and will be the subject of a future communication.

Acknowledgements. This work was supported in part by the National Science Foundation under grant PHY-8521231, and in part by the National Aeronautics and Space Administration under grant NAGW-1778.

References

Bondeson, A., Iacono, R., and Bhattacharjee, A., Phys. Fluids, **30**, 2167-2180, 1987.

Cargill, P. J., and Priest, E. R., Solar Phys., **65**, 251-269, 1980.

Foukal, P., Astrophys. J., **223**, 1046-1057, 1978.

Grossmann, W., and Tataronis, J. A., Z. Physik, **261**, 217-236, 1973b.

Gurman, J. B., and Athay, R. G., Astrophys. J., **273**, 374-380, 1983.

Habbal, S. R., Ronan, R., and Withbroe, G. L., Solar Phys., **98**, 323-340, 1985.

Heyvaerts, J., and Priest, E. R., Astronomy and Astrophysics, **117**, 220-234, 1983.

Kingston, A. E., Doyle, J. G., Dufton, P. L., and Gurman, J. B., Solar Phys., **81**, 47-58, 1982.

Kopp, R. A., Poletto, G., Noci, G., and Bruner, M., Solar Phys., **98**, 91-118, 1985.

Nicolas, K. R., Kjeldseth-Moe, O., Bartoe, J.-D. F., and Brueckner, G. E., Solar Phys., **81**, 253-280, 1982.

Nocera, L., Leroy, B., and Priest, E. R., Astronomy and Astrophysics, **133**, 387-394, 1984.

Noci, G., Solar Phys., **69**, 63-76, 1981.

Priest, E. R., Solar Magnetohydrodynamics, D. Reidel Publishing Company, Dordrecht, Holland, 1982.

Tataronis, J. A., and Grossmann, W., Z. Physik, **261**, 203-216, 1973a.

Tataronis, J. A., and Mond, M., Phys. Fluids, **30**, 84-89, 1987.

BASIC PROPERTIES AND MODELS OF SOLAR PROMINENCES

T. G. Forbes

Institute for the Study of Earth, Oceans, and Space
University of New Hampshire, Durham 03824

Abstract. Prominences are relatively cool (10^4 K) and dense (10^{11} cm^{-3}) plasma clouds which may persist for 100 days or more in the midst of the much hotter (10^6 K) and more tenuous (10^9 cm^{-3}) corona. Many observations imply that the magnetic field in and around prominences is responsible both for isolating prominences from the corona and for supporting them against gravity. It is not at all obvious how the magnetic field can do both these tasks, but the limited theoretical models that are available suggest that a magnetic-flux rope is involved. Here we argue using a new analytical model that the flux rope could also play a key role in the eruption of a prominence by supplying the magnetic energy necessary to drive the prominence outwards.

1.0 Introduction

As magnetic-flux ropes, prominences have several unique features. First, they have been scientifically observed for over 250 years, whereas almost all other known flux-ropes in the solar system have been observed only since the advent of the space age (the exception being comet tails). Second, prominences also have the advantage that they are observable from both global and local perspectives. In the lower solar corona they exist as telescopic phenomena observable only by remote sensing, but in the interplanetary medium remnants of prominences can be directly sampled by spacecraft. Finally, prominences are unique as magnetic-flux ropes because their existence depends on the radiative and gravitational dynamics of the solar atmosphere. By contrast, radiation and gravitation are completely unimportant for flux ropes in planetary magnetospheres.

Since this paper is a tutorial review, recent theoretical and observational developments are not considered in any detail. For reviews along these lines, the reader should see Schmieder (1989), Malherbe (1989), Hood (1989), and Zirker (1989).

1.1 *Historical Perspective*

Because prominences can be seen with the naked eye during solar eclipses, prominences have been casually observed for millennia, but the earliest recorded scientific observations are those of the Swedish astronomer Vassenius. During the eclipse of May 2, 1733 in Gothenburg, Sweden, Vassenius saw several prominences which he interpreted as clouds of fire in a hypothetical lunar atmosphere (cf. Table 1). Unfortunately, Vassenius's observations were ignored, so that the appearance of prominences during the solar eclipse of July 8, 1842 in France and Italy came as a complete surprise to astronomers. These included Airy, the Astronomer Royal of Great Britain, who thought that prominences might be solar mountains (Tandberg-Hanssen, 1974).

TABLE 1. Some Observational High Points

1733	- Start of eclipse observations. Vassenius sees 'lunar fires'.
1842	- Rediscovery of prominences by Bailey and Airy – 'solar mountains'.
1868	- Spectroscopy starts. Secchi and de la Rue discover helium.
1889	- Continuous observations start. Hale invents spectroheliograph.
1908	- Hale applies Zeeman splitting to the photosphere.
1930	- Lyot makes the first workable coronagraph.
1952	- Babcocks develop the magnetograph

The first real progress in deciphering the nature of prominences came with the beginning of spectroscopic observations by Secchi and de la Rue in 1868. During the eclipse of August 18th in India and Malacca, they discovered an emission line which could not be identified with any known terrestrial element. After considering several possibilities, Secchi and de la Rue correctly deduced that they had discovered a new element which they christened 'helium'.

The invention of the spectroheliograph by Hale in 1889 was especially important for prominence research, because it made continuous observations of prominences possible for the first time. In the Hα Balmer line of the spectroheliograph, prominences appear as dark 'filaments' on the surface of the sun, and it was not immediately recognized that these dark filaments were indeed the same objects as the bright prominences seen against the dark sky of an eclipse. To this day observers still tend to use the word 'filament' to refer to prominences seen on the disk, although it is now quite acceptable to use the two words 'filament' and 'prominence' interchangeably. That filaments were indeed the same objects as

Geophysical Monograph 58

the prominences observed during eclipses, became obvious after Lyot's invention of the coronograph in 1930 which freed astronomers from the necessity of waiting for eclipses in order to see prominences on the limb of the sun (Tandberg-Hanssen, 1974).

The importance of magnetic fields in prominences was not recognized until Hale's application of Zeeman splitting to the solar photosphere in 1908, and the realization that prominences were probably flux ropes followed from the development of the solar magnetograph in 1952 by the Babcocks (see Zirin, 1988). Direct measurement of the vector magnetic fields within prominences was recently accomplished by Leroy et al. (1983), but the coronal magnetic fields surrounding prominences are still unobservable.

1.2 *Standard Definition*

Prominences are traditionally defined as any material visible in Hα above the solar surface (specifically the chromosphere). Since Hα is excited only if the hydrogen plasma is in the range of 6,000 to 11,000 degrees Kelvin, this observational definition categorizes prominences as plasma clouds which are roughly 100 times cooler than the coronal plasma (10^6 K) surrounding them.

1.3 *Classification*

Table 2 shows the most commonly used classification scheme for prominences based on the above definition (Zirin, 1988). There are two distinct classes of prominences known as Active and Quiescent whose members are clearly separated by their time behavior. Active prominences are transient phenomenon which exist on relatively short time-scales ranging from minutes (e.g. surges) to hours (e.g. flare loops), and they typically occur during or after flares. By contrast, quiescent prominences exist on long time-scales ranging from a half a day to hundreds of days. As a rule, active prominences do not show any structure indicating an obvious association with magnetic flux ropes, but quiescent prominences do show such structure. Consequently, in this paper we will concentrate on the quiescent prominences which are the objects one usually thinks of when the word prominence is mentioned.

Quiescent prominences can be further divided into prominences which occur in quiescent regions (Quiescent Region Prominences or QRP's) and active regions (Active Region Prominences or ARP's). The ARP sub-class should not be confused with the Active class, since ARP's are themselves quiescent, even though they occur in active regions. There is no sharp distinction between QRP's and ARP's, but rather a gradual blending from one type to the other. In fact a prominence which extends from an active region into a quiescent region may have parts belonging to both classes. From this property it is inferred that the physical differences between QRP's and ARP's are due primarily to the difference in magnetic field strengths between quiescent and active regions.

TABLE 2. Prominence Classification

I. Active	II. Quiescent
Flare Loops	Quiescent Region Prominence (QRP)
Surges	Active Region Prominence (ARP)
Coronal Rain	
Sprays	

1.4 *Morphology*

Figure 1 diagrams the principal features of quiescent prominences. The prominence occurs as a thin, vertically oriented slab of plasma which is suspended in the corona and which has one or more arch-like structures that rise out of the chromosphere and then descend back down. Each arch-like structure constitutes a section, whose chromospheric terminations are usually referred to as 'legs'. ARP's usually have only a single arch, while QRP's may have several.

The coronal cavity is a low density, cylindrically shaped region which surrounds the prominence as shown in Figure 1. Often the cavity is in turn contained within a helmet streamer, and although the helmet streamer is not always present, the polarity of the magnetic field in the chromosphere is always of the type associated with helmet streamers. That is, the field in the regions lying adjacent to the prominence are always of opposite polarity, and together they form the 'magnetic channel' of the prominence.

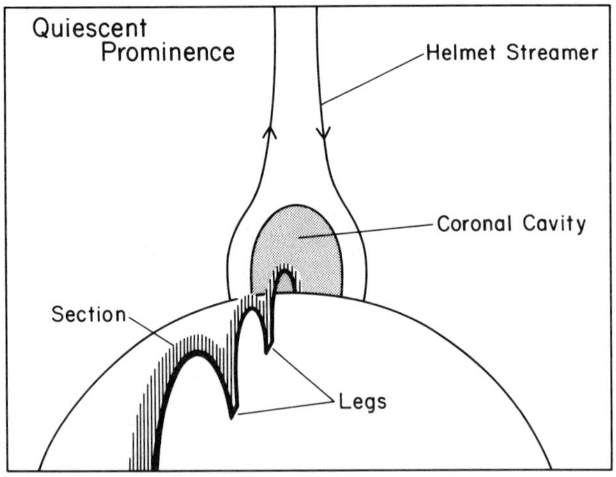

Fig. 1. Principle features of quiescent prominences. The prominence legs rise out of the chromosphere to form the filament sections, which are imbedded within the coronal cavity and helmet streamer.

2.0 Static Properties

Table 3 list the basic properties of quiescent prominences. Both QRP's and ARP's have similar densities and temperatures, and so do the cavities which surround them. Because the cavity is less dense than the corona, it is tempting to assume that the denser plasma of the prominence comes directly from the coronal cavity. However, many observers believe the missing mass of the cavity is insufficient to supply the prominence by about an order of magnitude, and it is assumed that the prominence mass comes from the chromosphere (Tandberg-Hanssen, 1974). However, Engvold (1989) has recently argued that the inferred prominence mass is nearly the same as the mass that would have originally been contained in the cavity prior to its formation, and so there is no general consensus on this point at the present time.

2.1 *Magnetic Field*

QRP's occur at higher altitudes and are much longer lived than their active region counterparts. These differences are

TABLE 3. Static Properties

a. Filament versus Cavity

	Filament	Cavity
Density	10^{11} cm^{-3}	10^{8} cm^{-3}
	(100 × coronal)	(1/10 × coronal)
Temperature	≥ 6000 K	10^{6} K
	(1/100 × coronal)	(equals coronal)

b. ARP versus QRP

	ARP	QRP
Height	$\leq 4 \times 10^4$ km	$\leq 2 \times 10^4$ km
Length	2×10^5 km	5×10^4 km
B Field Strength	5 - 10 Gauss	100 - 200 Gauss
B Polarity (Ave.)	Inverse	Normal

thought to be due solely to the difference in magnetic field strength between the quiescent (5 to 10 Gauss) and active (100 to 200 Gauss) regions of the photosphere. However, QRP's typically have a magnetic topology which is different from the magnetic topology of ARP's . This difference in topology is inferred from comparisons of the vector magnetic field in the prominence with the line-of-sight magnetic field in the photosphere, and it is expressed in terms of the prominence's magnetic polarity.

Figure 2 shows an example of a prominence with normal magnetic polarity. Vector field measurement based on the Hanlé effect indicate that the magnetic field component parallel to the axis of the prominence is from 5 to 10 times stronger than the horizontal field component transverse to the axis (Leroy et al., 1983, see also Tandberg-Hanssen and Anzer, 1970). If the weaker transverse component lies in the same direction as the field component obtained by current-free extrapolation of the photospheric field, then the prominence has normal polarity. However, if the transverse field lies in the reverse direction of the extrapolated photospheric field, then the prominence has inverse polarity.

In a survey of both QRP's and ARP's Leroy et al. (1983) found that about 75% of QRP's had inverse polarity, while about 75 % to 50% of ARP's had normal polarity. (The lower altitudes of ARP's makes it more difficult to measure their transverse field, and the polarity of ARP's is therefore more uncertain.)

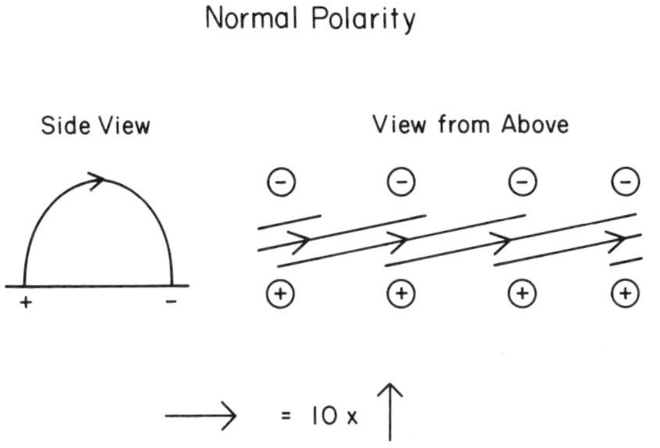

Fig. 2. A prominence with normal magnetic polarity has its transverse component in the same direction as the transverse component of the potential field extrapolated from the photosphere. In general the transverse field is 5 to 10 times smaller than the field parallel to the prominence axis.

Because most modelers had assumed prominences always to have a normal polarity, these results came as somewhat of a surprise.

2.2 *Support Models*

Following earlier work by Menzel (Tandberg-Hanssen, 1974) and Dungey (1953), Kippenhahn and Schlüter (1957) constructed what is now the best known mathematical model for prominences with normal polarity. This model assumes the configuration shown in Figure 3a and gives a description of the field and plasma only within the vicinity of the prominence itself. Kippenhahn and Schlüter's model is not really a global model since it says nothing about how the field lines are anchored to the photosphere. Their model also ignores the very strong field component that exists perpendicular to the plane of the figure. Assuming the horizontal magnetic field component, B_x, to be constant in the vicinity of the prominence, Kippenhahn and Schlüter found the vertical component, B_y, to be

$$B_y = B_y(\infty) \tanh\left[\frac{B_y(\infty)}{B_x}\frac{x}{2H}\right], \tag{1}$$

with the corresponding density distribution

$$\rho = \frac{mB_y(\infty)^2}{8\pi k T}\left[1 - \tanh^2\left(\frac{B_y(\infty)}{B_x}\frac{x}{2H}\right)\right]. \tag{2}$$

Here $B_y(\infty)$ is the vertical magnetic field at $x = \infty$, and H is,the atmospheric scale height kT/mg within the prominence.

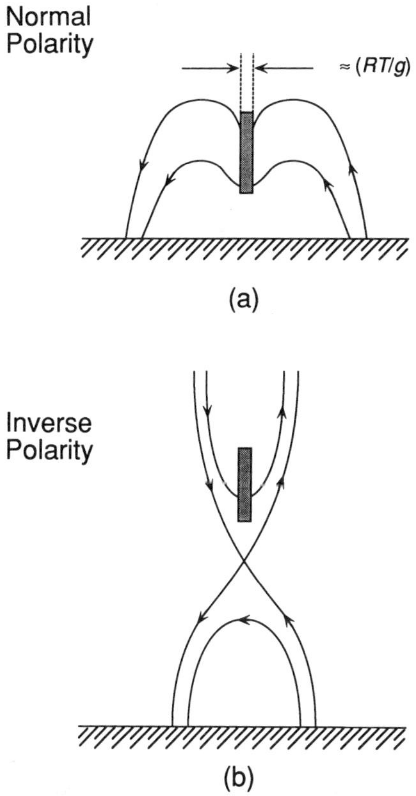

Fig. 3. Early gravitational support models for quiescent prominences (a) Kippenhahn-Schlüter with normal polarity and (b) Kuperus-Raadu with inverse polarity.

More generalized Kippenhahn-Schlüter-type configurations can be found in Low (1981).

Following the discovery that most quiescent region prominences have inverse polarity, Kuperus and Raadu (1974) proposed the inverse polarity configuration shown in Figure 3b, and their work has subsequently been developed and generalized by other modelers (e.g. Malherbe and Priest, 1983; Pneuman, 1983). Because the field lines threading the prominence are disconnected from the chromosphere, the prominence must be supported by field lines anchored only in the corona (Anzer, 1985).

Neither the Kippenhahn-Schlüter nor the Kuperus-Raadu models address the thermodynamics of prominences, nor do they address the global aspects of gravitational support and mechanical stability of the overall configuration. Recently, some efforts have been made to construct more realistic static models of prominences, and we refer the reader to the review of static support models by Anzer (1989). (Also see the paper by Anzer in this volume.)

3.0 Dynamical Properties

It is convenient to consider separately the dynamics of prominences during their development and evolution from those during their disappearance. The development and evolution of prominences is controlled by the slow evolution of the magnetic fields in the photosphere and the convection zone. By contrast the disappearance of prominences is due to the rapid evolution of the magnetic fields in the corona as a result of a loss of mechanical or thermal equilibrium.

3.1 *Global Movement*

Quiescent prominences have life-times ranging from a half a day to 100 days or more, and during their life-time they undergo slow, migratory movements in the photosphere. QRP's in particular move slowly polewards to form 'polar crowns' around the coronal holes at the sun's rotational poles. In addition both QRP's and ARP's are thought to become increasingly aligned parallel to the sun's equator as a results of differential rotation (Glackin, 1974). Because the photosphere is rotating faster at the equator than at the poles, the end of the prominence closest to the pole lags farther and farther behind the other end.

Recent studies have cast some doubt as to whether most prominences respond to the differential rotation in this simple way. Examination of prominence migratory movements suggests that many prominences do not respond to the differential motion seen in the photosphere, and instead the photosphere flows around them (Soru-Escaut et al., 1985; Mouradian et al., 1987; see also Zirin, 1988). This anomalous motion suggests that prominences are somehow anchored differently to the convection zone than are sunspots.

3.2 *Disparition Brusque - Dynamic*

The French term 'disparition brusque' refers to the sudden disappearances of prominences. These disappearances are of two types. The first, and better known, is the dynamic disappearance caused by the loss of mechanical equilibrium, while the second, and lesser known, is the thermal disappearance caused by heating of the prominence material to coronal temperatures. These two types of disappearances are illustrated in Figure 4.

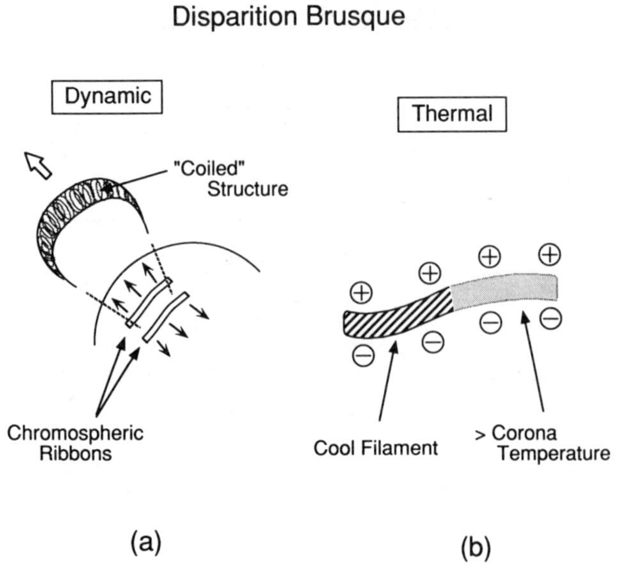

Fig. 4. Disparition Brusque. In dynamic eruptions (a) the magnetic field changes dramatically and a coronal mass ejection is propelled outwards. At the same time chromospheric emission ribbons visible in Helium and Hα lines, propagate away from the original location of the prominence. In thermal disruptions (b) no obvious magnetic field changes are observed and all or part of the filament is simply replaced by a region (grey shading) at a somewhat higher temperature than the corona. Circles with + and – signs indicate the magnetic channel.

In a dynamic disappearance the entire magnetic field structure surrounding the prominence is disrupted and the prominence is ejected outwards into interplanetary space as a coronal mass ejection. Just before an eruption, prominences typically become 'activated' meaning that they darken and have rapid oscillatory motions (Martin, 1980). Once the eruption is in progress, thin threads exhibiting a 'coiled' structure become visible within the prominence. This 'coiled' structure often looks like a very tightly wound spiral with dozens of turns, and it has sometimes been assumed that the magnetic field must be wound up in the same way. However, such coiling would imply that the transverse field in the prominence is much greater than the field parallel to the prominence axis – exactly opposite to what is observed prior to the eruption. Thus, the 'coiled' structure seems more likely to be due to fragmentation of the prominence material during the course of the eruption.

When a prominence erupts, two parallel ribbons of enhanced emission in Helium (and sometimes in Hα also) appear in the chromosphere as illustrated in Figure 4a (K. Harvey, private communication). The two ribbons first appear as a single wide ribbon just beneath the original location of the prominence, but they become visible as separate ribbons within a few minutes. Over the course of many hours, the ribbons continually move farther and – apart. It is clear from Doppler-shift analysis and examination of small-scale structures within the ribbons, that the motion of the ribbons is not due to a bulk flow of the chromosphere, but rather to the progressive brightening of the chromosphere at the outer edges of the ribbons, and the continual dimming of the chromosphere at the inner edges of the ribbons. As in the case of two-ribbon flares, most researchers interpret this wave-like propagation of the ribbons as evidence that reconnection is taking place somewhere in the corona on the field lines which map from the corona to the ribbons (Pneuman and Orrall, 1986).

3.3 *Disparition Brusque - Thermal*

Sometimes prominences disappear simply by fading in place without any obvious eruption or change in their shape, and after the prominence has faded, high temperature plasma is seen in the same region previously occupied by the prominence (Schmahl et al., 1982; and Mouradian et al., 1986). Because the temperature of this plasma is slightly higher than that of the ambient corona, it is most easily observed in extreme ultra-violet emission lines as a kind of ghost image of the original prominence (cf. Figure 4b). This kind of disappearance is thought to result from a heating of the prominence by an external agent such as a nearby flare. The cool prominence can also reappear and disappear several times in the same magnetic channel, and only part of the prominence may disappear while the rest remains at cooler temperatures.

4.0 Thermal Equilibrium

In prominence research two key questions are: How are cold, heavy prominences kept from heating up in the hot corona, and how are they supported against gravity? The former question is the subject of this section, and to answer it requires an understanding of how the corona loses thermal energy to radiation.

The thermodynamics of the corona can be described by an energy equation of the form

$$\rho c_p \, DT/Dt - Dp/Dt = -L, \qquad (3)$$

where D/Dt is the convective derivative, T is the temperature, p is the pressure, ρ is the density, and c_p is the specific heat at constant pressure (Priest, 1982). The quantity L is the energy loss function

$$L = -\nabla \cdot \mathbf{q} - j^2/\sigma + L_r - h_c \rho, \qquad (4)$$

where $\mathbf{q}$ is the heat flux due to particle conduction, j^2/σ is the ohmic dissipation, L_r is the radiative loss function, and $h_c \rho$ represents the sum of all other heating sources. If thermal conduction and ohmic dissipation are ignored, than thermal equilibrium occurs when

$$L_r = h_c \rho. \qquad (5)$$

The two most important heating sources for quiescent prominences are photospheric radiation and magnetic dissipation or waves (Tandberg-Hanssen, 1974; Zirin, 1988). Photospheric photons contribute to the temperature of a prominence by photo-excitation. In radiative equilibrium the prominence absorbs 2π steradians of incoming photons from the photosphere, and then re-emits these as outgoing photons over 4π steradians. Thus the prominence temperature T_{prom} is very approximately

$$2\pi\sigma_s \, T_{phot}^4 = 4\pi\sigma_s \, T_{prom}^4 \, . \qquad (6)$$

where σ_s is the Stefan-Boltzman constant. Since the effective temperature of the photosphere, T_{phot}, is 5770 K, the prominence temperature is approximately 4850 K. Somewhat lower temperatures are predicted when corrections are made for the transparency of the prominence in the continuum and for the radiation of the photons in hydrogen and helium lines. Since actual prominence temperatures are considerably hotter, lying in the range of 5000 to 6000 K, a supplementary heating source is necessary (Zirin, 1988).

Heating by photospheric irradiation results in a heating term which is approximately constant per unit mass once thermal equilibrium is established, and the heating coefficient h_c in the energy equations is approximately constant. However, if thermal equilibrium is lost or not yet established, a full radiative transfer analysis is required. For this tutorial we will assume, as is often done, that photospheric irradiation and the supplementary heating are both constant per unit mass.

4.1 *Radiative Loss Function*

Parker (1953) was one of the first to point out that the form of the radiative loss function makes the corona thermally unstable in the absence of heat conduction. Because the corona is optically thin, the radiative loss function L_r can be expressed as

$$L_r = (\rho/m)^2 \, Q(T) \, , \qquad (7)$$

where m is the proton mass and $Q(T)$ is the function shown in Figure 5 (Priest, 1989). This function gives the radiative loss of the corona in ergs s^{-1} cm^3 as a function of temperature, and it has a maximum at about 1.4×10^5 K. To the right of the maximum is a temperature range where the radiation loss decreases as the temperature increases – just the opposite of what happens with a black body, where

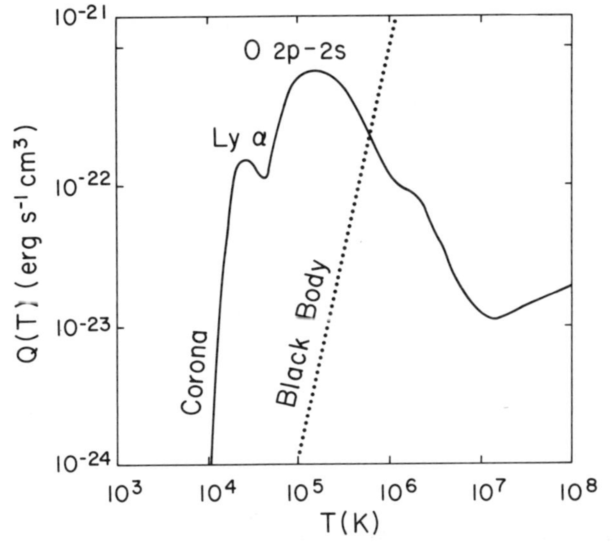

Fig. 5. The radiative loss function for the corona (after Priest, 1982). For temperatures below 4×10^4 K the corona behaves approximately like a black body (dotted curve), but above this temperature the corona actually radiates less energy as the temperature increases.

$$L_r = \sigma_s T^4 / L, \qquad (8)$$

and L is a scale length.

Compared to this black-body radiation loss function, the coronal radiation loss function is obviously quite different above 10^5 K (cf. Figure 5 which plots $L_r m^2/\rho$ versus T). This difference is due to the fact that the coronal radiation comes from atomic line transitions which disappear with the increasing ionization at higher temperatures. For example, the decreases at 2×10^4 K and 3×10^5 K are due to the disappearance of the Lyman α transition in hydrogen and the 2p-2s transition in oxygen. The radiation loss does not start to increase again until x-ray bremsstrahlung radiation becomes significant above 10^7 K.

4.2 *Thermal Instability*

For illustrative purposes let's assume that the pressure in the energy equation is constant and that

$$Q(T) = m^2 \chi T^\alpha . \qquad (9)$$

With these assumptions and the specific heating function of (5), the energy equation is

$$c_p \, \partial T/\partial t = h_c - \chi \rho T^\alpha , \qquad (10)$$

and the condition for thermal equilibrium is

$$0 = h_c - \chi \rho_o T_o{}^\alpha , \qquad (11)$$

where ρ_o is the equilibrium density at the equilibrium temperature, T_o. The energy equation can now be rewritten as

$$\partial T/\partial t = (\chi \rho_o T_o{}^\alpha / c_p) \, [1 - (T/T_o)^{\alpha-1}] , \qquad (12)$$

which, for a small departure ΔT from the equilibrium temperature T_o, becomes

$$\partial T/\partial t = \tau_r^{-1} (1 - \alpha) \, \Delta T , \qquad (13)$$

where

$$\tau_r = c_p / (\chi \rho_o T_o{}^{\alpha-1}) , \qquad (14)$$

is the linear radiative cooling time. For typical coronal conditions τ_r is in the range 10^4 to 10^5 s (Priest, 1982).

For heating proportional to density, an isobaric temperature perturbation decreases with time if $\alpha > 1$, and the equilibrium is stable; but if $\alpha < 1$, then an isobaric perturbation increases with time, and the equilibrium is unstable. (If the heating is constant per unit volume, then $\alpha < 2$ gives instability.) When the equilibrium is unstable, the temperature decreases if the initial perturbation is negative but increases if the initial perturbation is positive. Thus a positive perturbation causes a thermal run-away to a higher temperature, while a negative perturbation causes a thermal collapse to a lower temperature. The plot of $Q(T)/T$ versus T in Figure 6 indicates the temperature range where an optically thin plasma with heating proportional to density is unstable in the absence of thermal conduction. The corona, which has a temperature of about 10^6, is thermally unstable if thermal conduction is inhibited.

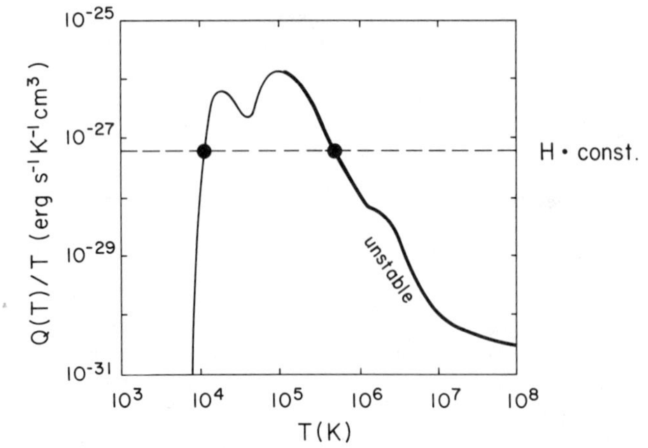

Fig. 6. Thermodynamic equilibrium occurs at constant pressure when $Q(T)/T$ is constant, but the equilibria at the higher temperatures (thick line) are unstable.

Thermal conduction can stabilize the corona by removing or adding heat at a rate that is faster than the growth rate of the thermal instability (Field, 1965). When thermal conduction is included, the energy equation at constant pressure is

$$c_p \, \partial T/\partial t = \rho^{-1} \nabla\cdot(\kappa \nabla T) - \chi\rho T^\alpha + h_c , \qquad (15)$$

where κ is the anisotropic thermal conduction tensor. Thermal conduction is much stronger along the magnetic field then across it, so if we replace the ∇ operator by $1/L$ and use $\lambda T^{5/2}$ for the thermal conduction coefficient along the field, we get

$$c_p \, \partial T/\partial t = [(1 - \alpha) \, \chi\rho_o T_o{}^{\alpha-1} - 9\lambda \, T_o{}^{5/2}/(2\rho_o L^2)] \, \Delta T. \qquad (16)$$

Whereas the first term is positive when $\alpha < 1$, the second term is always negative. Therefore, thermal conduction opposes any change in temperature, and can stabilize the plasma if L is sufficiently small.

Whether or not thermal conduction really prevents thermal instability, depends very much on the magnetic field geometry. If the field lines are straight as shown in Figure 7a, then any small perturbation of scale-size L will damp out. However, if the field lines are circular as shown in Figure 7b, then plasma in the central region is thermally isolated and small perturbations can grow. A full stability analysis requires knowledge of the magnetic field geometry of the prominence and understanding of how the field interacts with the fluid dynamics of the plasma (see Sparks and Van Hoven, 1985, 1988; and Van Hoven et al., 1986).

Several numerical simulations have been done by various researches in order to explore the complex, dynamical evolution of the thermal instability in various magnetic geometries (e.g. Hildner, 1974; Oran et al., 1982; Van Hoven et al., 1987). In general, these simulations show that sheared magnetic field structures can indeed strongly inhibit thermal conduction along the magnetic field and thereby create a thermally unstable environment. However, thermal conduction perpendicular to the magnetic field, and the boundary conditions at the chromospheric footpoints of the field lines, are also impor-

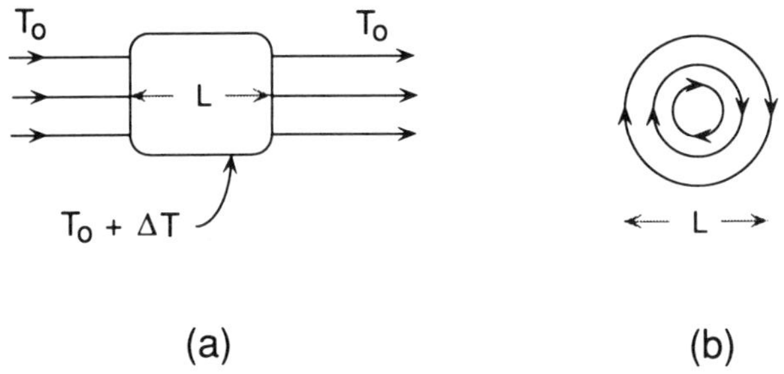

Fig. 7. High temperature equilibria at constant pressure can be stabilized against small temperature perturbations, ΔT, if their scale length, L, is small, and if they have a magnetic topology like that shown in (a). If their magnetic topology isolates them from the rest of the corona, such as shown in (b), then thermal conduction does not necessarily stabilize the equilibria.

tant factors in determining whether a particular configuration is stable or unstable (see the paper by G. Van Hoven in this volume).

5.0 Eruption Dynamics

5.1 *Energetics*

From the viewpoint of solar-terrestrial interactions, the eruption of a quiescent prominence is one of the most important solar phenomenon influencing the Earth's magnetosphere. The kinetic energy of the mass ejection in a large eruption is on the order of 10^{32} ergs (Schmahl and Hildner, 1977; Webb et al, 1980), and the kinetic energy often exceeds the radiated energy by a factor of two or more (Canfield et al., 1980). One of the principal goals of present research is to understand how this enormous kinetic energy is generated.

The only feasible source is the energy stored in the coronal magnetic field in and around the prominence. Prior to eruption, the incipient mass ejection occupies a volume on the order of 10^{15} km^3, which is many times larger than the volume of the prominence alone, since the prominence itself occupies a small fraction of the mass ejection volume. If the magnetic field in this volume is 100 Gauss, then only a 14% decrease in field strength is needed to generate the 10^{32} ergs required for a mass ejection. By comparison, a total annihilation of the magnetic field would release 5.3×10^{33} ergs – enough energy for 50 coronal mass ejections.

During the last few years the relationship between coronal mass ejections and flares has been clarified by several studies using data from the Solar Maximum Mission spacecraft (e.g. Harrison, 1986; Hundhausen, 1987). And it now seems likely that a flare associated with coronal mass ejections is not so much the cause of the ejection as a consequence of it. The flares associated with coronal mass ejections are usually proceeded by the eruption of an ARP whose mass ends up as part of the ejected material (Martin, 1980).

Many flares, especially smaller ones, do not produce coronal mass ejections at all, and likewise, many mass ejections do not produce flares, so the correlation between the two is not strong. However, a much better correlation exists between coronal mass ejections and the eruptions of QRP's. Like flares, QRP eruptions produce chromospheric ribbons albeit they are much fainter than flare ribbons (Munro et al., 1979). Thus, it is tempting to think of coronal mass ejections, prominence eruptions, and at least some flares as different manifestations of a single phenomenon.

Solar flares and magnetospheric substorms are often thought of as very similar phenomena because both involve the rapid conversion of magnetic energy into kinetic and thermal energy. Yet there could very well be a fundamental difference between the energy conversion process in a flare driven by a coronal mass ejection and the energy conversion process in a substorm. A substorm derives its energy from the magnetic energy stored in the geomagnetic tail by the solar wind as illustrated in Figure 8a. The magnetic energy resides in the compressed and stretched field which generates a tail current sheet, and during the course of the substorm the total current in the sheet weakens as the magnetic energy is released and channeled into the ionosphere. On the other hand, a solar flare driven by a coronal mass ejection may actually create a current sheet during the course of its evolution as shown in Figure 8b. Many of the magnetic field lines in the coronal mass ejection are anchored into the photosphere, so that when the ejection moves outwards the field lines are stretched and become extended to form a field structure analogous to the geomagnetic tail (Carmichael, 1964; Sturrock, 1968; Kopp and Pneuman, 1976). In this scenario a coronal mass ejection would actually store magnetic energy in a current sheet, while at the same time it decreases the overall magnetic energy of the system (see Sturrock, 1987).

Barnes and Sturrock (1972) tried to demonstrate the principle that an extended open field could be produced while reducing the magnetic energy of the overall configuration (see also Mikic et al., 1988; Biskamp and Welter, 1989; van Ballegooijen and Martens, 1989), but a few years ago Aly (1984) found a misconception in their analysis which puts this principle in doubt. Because the corona has a strong magnetic field (plasma β of 10^{-3}), the currents in the corona must be either force-free or confined to very thin current sheets or threads. Thus, prominences are thought to be force-free magnetic flux ropes. Yet Aly found, using a general theoretical analysis, that an open magnetic field with the same boundary conditions as a corresponding force-free field, always has the greater magnetic energy of the two.

Although Aly's theoretical analysis greatly restricts the mechanisms one might consider for driving coronal mass ejections magnetically, there are still several possibilities remaining. First, the initial field configuration in the corona might have current sheets or threads and therefore, not be force-free. Second, resistive processes, such as magnetic reconnection, might dissipate the current sheet rapidly enough to eliminate the energy build-up it produces. And third, the mass ejection might move from a low altitude equilibrium to a high altitude equilibrium without ever opening the field lines to infinity. If the second equilibrium is higher than about two solar radii, then the solar wind will carry the mass ejection into interplanetary space and thereby open the field.

In the following subsection we present a two-dimensional analytical model which demonstrates that it is indeed possible for a coronal mass ejection to form an extended, open field with a current sheet, while lowering the magnetic energy of the overall configuration. The model does not contradict Aly's results for two reasons. First, its initial field is not force-free, and second, the field lines, though greatly extended, never become truly open in a strict mathematical sense.

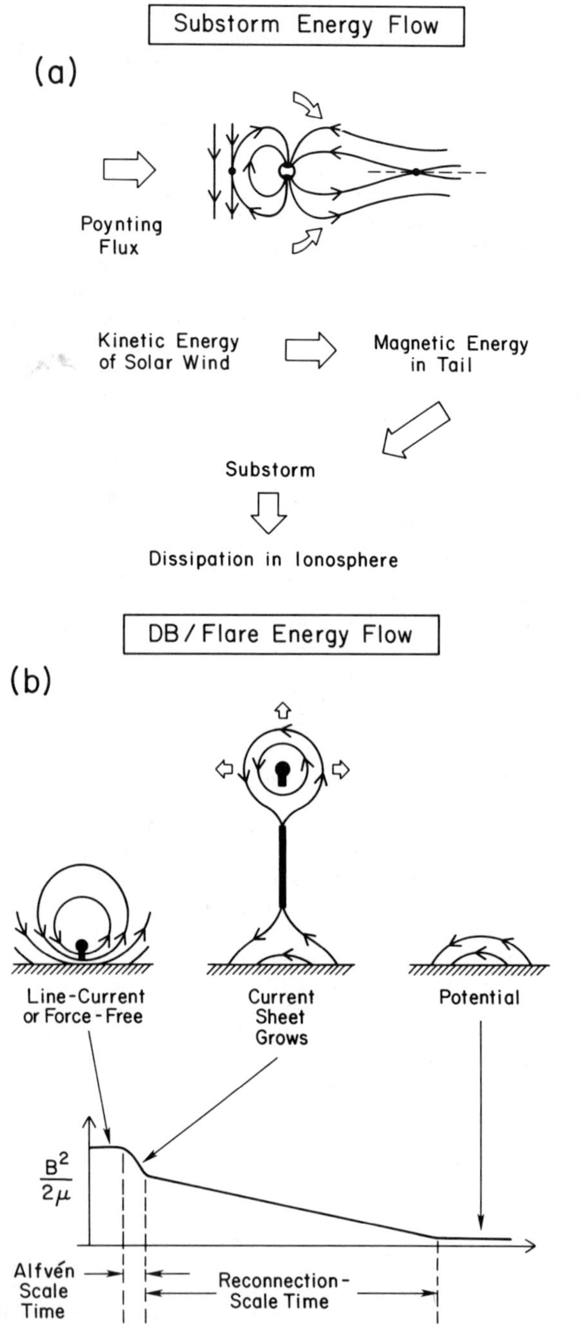

Fig. 8. The energy transfer in (a) substorms and (b) prominence eruptions. Unlike a substorm in the geomagnetic tail, a prominence eruption may act to extend the magnetic field, rather than to relax it.

5.2 *An Analytical Model*

In the limit of zero plasma β, currents in two-dimensional, ideal MHD become concentrated in lines and sheets which can be represented as poles and branch cuts in the complex plane. To model an erupting current filament with a current sheet below it we choose a field of the form:

$$B_y + iB_x = iA\,\frac{\sqrt{(z^2+p^2)(z^2+q^2)}}{z^2\,(z^2+h^2)}, \tag{17}$$

which corresponds to a line current at $z = ih$, a two-dimensional dipole field at $z = 0$, and a current sheet stretching from $z = ip$ to $z = iq$ in the $z \geq 0$ plane as shown in Figure 9 (from Priest and Forbes, manuscript submitted to Solar Phys., 1989). Below $z = 0$ there are corresponding image currents for the filament and the current sheet. The three parameters A, p, and q are fixed by the following ideal-MHD constraints:

(*i*) Near $z = ih$, the field behaves like $I(h)/(z-ih)$, where $I(h)$ is the current in the filament times $2\pi/\mu$.

(*ii*) Near $z = 0$, the field behaves like im/z^2, where m is constant.

(*iii*) The magnetic flux between the dipole at $z = 0$ and the bottom edge of the current sheet at p is constant.

(*iv*) The magnetic flux between the filament at $z = ih$ and the top edge of the current sheet at q is also constant.

Applying conditions (*i*) and (*ii*) gives

$$A = mh^2/pq\,, \tag{18}$$

and

$$2hpq\sqrt{(h^2-p^2)\,(h^2-q^2)} = m/I\,(h). \tag{19}$$

where $I(h)$ is a yet to be determined function of h.

When the background dipole is placed at a depth, h_b, below the photosphere, the equilibria occur at

$$h_o = m/I_o - h_b \pm [(m/I_o)^2 - 2h_b(m/I_o)]^{1/2}\,. \tag{20}$$

as shown in Figure 10 for $h_b = 1$. This set of equilibria has both

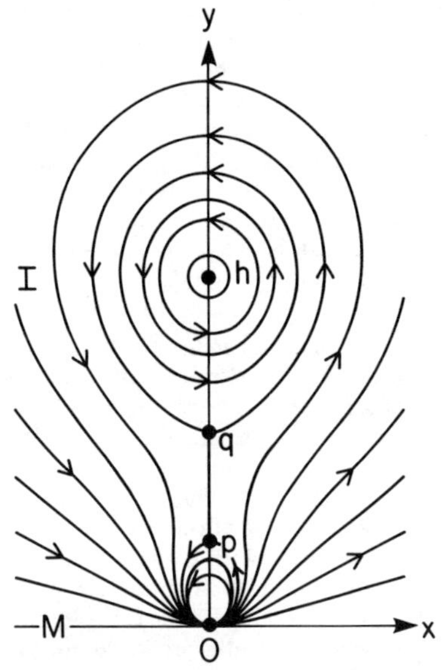

Fig. 9. Magnetic field configuration in an ideal, low β plasma for an infinitely thin line filament with a current sheet below it. The current sheet's parameters are self-consistently determined by frozen field conditions.

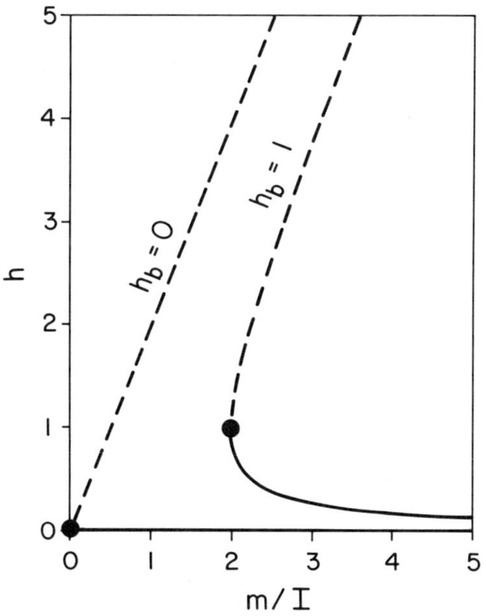

Fig. 10. The equilibria curve for the field configuration shown in Fig. 9 ($h_b = 0$) compared to the equilibria curve for a similar field configuration but with the background dipole at a distance h_b below the line-tying surface. Solid and dashed sections of the curves correspond to stable and unstable equilibria respectively.

a stable and an unstable branch which meet at $m = 2I_o h_b$, $h_o = h_b$, and there are no equilibria at all for $m < 2I_o h_b$. As h_b goes to zero the unstable branch goes to $h_o = 2m/I_o$, while the stable branch goes to $h_o = 0$.

The equilibrium curve in Figure 10 suggests the following scenario for a prominence eruption: If the magnetic field threading the photosphere slowly decreases with time, then a filament located on the stable branch will lose its equilibrium when the field falls below a critical value. In the model of (20) this corresponds to a critical dipole strength of $m = 2I_o h_b$. Once the dipole strength falls below this value, the filament is pushed upwards by the imbalance in magnetic forces (cf. Figure 11).

Unfortunately, a solution has not yet been found for the general case of arbitrary h_b, but only for the case of $h_b = 0$. In order for the filament to be initially in equilibrium for this specific case, $h_o = 2m/I_o$, and since there is no current sheet initially, $p_o = q_o = h_o/\sqrt{5}$

To determine the function $I(h)$, we use condition (*iv*) which requires the flux between the top of the current sheet at q and the filament at h to be constant. That is

$$\int_q^h B_x(0,y,h)\,dy = \int_{q_0}^{h_o} B_x(0,y,h_o)\,dy \quad . \tag{21}$$

Upon application of condition (iii) and substitution of (17) through (19), this gives (Priest and Forbes, manuscript submitted to Solar Phys., 1989):

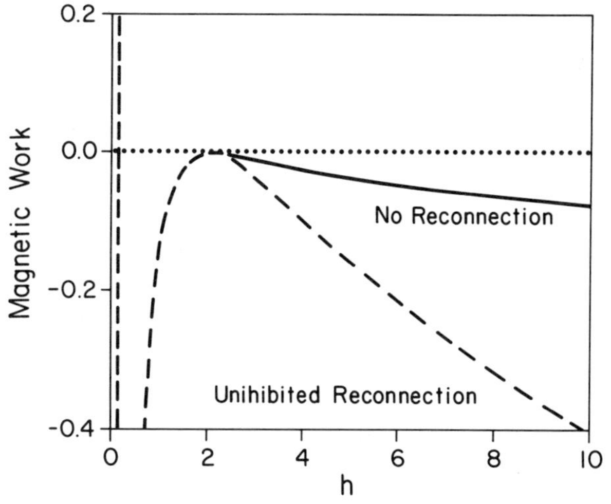

Fig. 11. Magnetic work versus height, h, for no reconnection (solid curve) and uninhibited reconnection (dashed curve) below the filament. Even when no reconnection is allowed, an upwards perturbation still causes the filament to move outwards at an ever increasing speed.

$$\lim_{R_p \to 0} \int_q^{h-R_p} \frac{1}{y-h}\left[1 - \frac{h_o h^2 \sqrt{(y^2-p^2)(y^2-q^2)}}{2pqy^2\,(y+h)\,(I_o/I)}\right] dy$$

$$= \ln\left[\frac{(\sqrt{5}-1)}{(\sqrt{5}+1)}\,\frac{2h_o}{(h-q)}\right] + \frac{1-\sqrt{5}}{2} \;, \tag{22}$$

where R_p is the radius of the current filament. The integral here is expressed as a limit for infinitesimal R_p because of the indeterminate nature of the integrand as $y \to h$. Solving (22) for I/I_0 gives

$$I/I_0 = 1 - \lim_{R_p \to 0} \frac{G(h) + (\sqrt{5}-1)/2}{\ln[(h-q)/R_p]} \;, \tag{23}$$

where

$$G(h) = \int_q^h \frac{1}{y-h}\left[1 - \frac{h_o h^2 \sqrt{(y^2-p^2)(y^2-q^2)}}{2pqy^2\,(y+h)}\right] dy$$

$$- \ln\left[\frac{(\sqrt{5}-1)}{(\sqrt{5}+1)}\,\frac{2h_o}{(h-q)}\right] \;. \tag{24}$$

Numerical evaluation of $G(h)$ with q and p determined by the previous constraints shows that $G(h)$ is a finite function which increases from -0.618 at $h = h_o = 2$, to 0.139 at $h = 4.6$.

In the limit that R_p tends to zero, the solution (23) gives $I = I_o$, but this is only true in the limit. For a non-zero R_p the filament current must decrease in order to conserve flux and energy. However, in the limit of $R_p = 0$, the required decrease is infinitesimally small. This infinitesimally small change makes a finite contribution to the overall flux balance, and one cannot simply set $I = I_o$ when evaluating (22). Instead, the explicit expression following the limit in (23) must be used in the integrand before the limit is taken.

Figure 9 shows the field configuration described by equation (17) and the constraints (*i*) through (*iv*) when $h/h_o = 1.5$. Here the lower and upper edges of the current sheet are at $p/h_o = 0.388$ and $q/h_o = 0.792$, respectively. There is an upward force F on the filament of

$$F = I_o \lim_{y \to h} \left[B_x(0,y) + \frac{I_o}{y-h} \right] \qquad (25)$$

which reduces to

$$F = 0.0614\, I_o^2/h \qquad (26)$$

in the limit of large h. By numerically integrating the force one finds that the magnetic energy of the system decreases with filament height as shown in Figure 11. As also shown in the figure, the energy loss when there is no reconnection (i.e. the ideal MHD case) is much less than the energy loss when the reconnection is uninhibited (i.e. the vacuum case). When there is no reconnection, 87.7 % of the available magnetic energy becomes locked up in the current sheet as $h \to \infty$, so that only 12.3 % is left to accelerate the filament.

Equation (26) seems to imply that the force on the filament will move the filament upwards and open all of the field lines except those below p. However, this equation is not valid at $h = \infty$, if R_p is non-zero, since,flux conservation requires the filament current to go to zero as h goes to ∞. Although the above analytical model cannot prove it, it seems likely that instead of moving outwards indefinitely, a filament of non-zero radius moves outwards only to a second closed-field equilibrium at a higher altitude than the first one. As R_p becomes progressively smaller, the altitude of the second equilibrium becomes progressively higher, and the field lines become more and more extended without really ever being opened in a strict mathematical sense.

Large flares and quiescent prominence eruptions continue to release energy for many hours after the initial energy outburst with which they are identified. The above analytical model implies that if reconnection occurred very slowly, then most of the energy release would occur long after the initial energy outburst. On the other hand, if the reconnection occurs rapidly, on an Alfvén scale-time, then all of the available magnetic energy would be released rapidly.

6.0 Conclusion

Although prominences have been recognized as magnetic phenomena since the early 1900's, there is still no generally accepted model of prominences which quantitatively explains how the magnetic field supports prominences against gravity, and how the magnetic field thermally isolates them from the hot corona. Similarly, there is no generally accepted model which explains why prominences erupt outwards into the outer corona instead of simply collapsing into the chromosphere.

According to the model proposed by Sturrock (1968, 1987), a prominence should create an extended field configuration if it erupts to form a coronal mass ejection. However, Aly (1984) has argued that it is energetically impossible to create a truly open field from an initially closed force-free field with the same photospheric boundary conditions. Using a new analytical model based on ideal MHD we find that the rapid formation of an extended field and a current sheet is possible in principle as long as field lines are not required to be truly opened in a strict mathematical sense. The model suggests that a prominence eruption occurs when a magnetic-flux-rope's anchorage in the photosphere is eroded below a critical level. When the eruption occurs, magnetic energy stored in the flux-rope is released, and the flux-rope moves outwards as a coronal mass ejection.

Questions and Answers

Antiochos: Is it not true that the bulk of the flare energy is released during the impulsive phase in the form of the mass ejection, which seems to be opposite to what your model predicts?

Forbes: Yes, this simple model predicts more energy release during the gradual phase than the impulsive phase if it is assumed that no reconnection occurs during the impulsive phase. However, if reconnection were to occur rapidly, say in a few Alfvén scale-times, then virtually all of the energy would be released in an impulsive phase. In any case, this model is too simple to compare in a quantitative way to observations. What is more important is that the model demonstrates the principle, that field lines can be opened, while the magnetic field decreases.

Zirin: Why do flares only erupt upwards? What is the buoyancy effect?

Forbes: In the scenario I describe, the upward force is produced by the compression of the magnetic field between the current in the filament and the surface of the photosphere. The filament is only prevented from flying upwards by those magnetic field lines tied to the photosphere. As these field lines are disconnected by reconnection in the photosphere, the filament reaches a point where it suddenly becomes unbalanced and the compressed field below it expands and pushes it upwards.

Chen: You mentioned that prominences may in fact have deep "anchorage" below the photosphere. If it is true, which I think is very plausible, then it is possible that subphotospheric events, changes etc., can influence the behavior of prominences and loops. Then, disruption, flares, and other energetic processes may occur as surface manifestations of subphotospheric events. An attempt is under way to develop a model along this line (Chen, Ap. J., 338, 453, 1989).

Forbes: Yes, it should be possible, in principle, to deduce information about subphotospheric flows from the long-term time behavior of prominences, and it is good to see that people are interested in this problem.

Kane: I do not understand your division of the magnetic field energy release between short time- and long time-scales in a flare. As I understand, most of the flare energy is released during the first ~10 minutes of a flare. Emission lasting for one or more days after a flare can hardly be considered part of a flare.

Forbes: For very large flares there is a low level power output which lasts for 10 to 20 hours and although this output is only about 10^{-2} to 10^{-3} of the output during the impulsive and flash phases, the total energy output is of the same order of magnitude as the total energy output of impulsive and flash phases. This low level output during the gradual phase comes from the x-ray flare loops which are a direct consequence of the flare and, therefore, in my view, the energy output by these loops should be considered as part of the energy released by the flare.

Antiochos: Why is the width of quiescent prominences much less than their other scales?

Forbes: Well, the answer given by Kippenhahn and Schlüter is that the width is fixed solely by the gravitational scale-height of a few hundred kilometers. However, this really doesn't explain why the prominences tend to have a slab-like shape. To explain the prominence width also requires a realistic magnetic field model.

Webb:

(1) Many believe that mass ejecta continue into the "gradual" phase of flares. This is especially true if, as in some models, reconnection continues for long times, thus coupling the rising mass ejecta (CME) to the surface heated "postflare" loops.

(2) Terry, if, as you show in your model on the screen, the prominence is on the unstable branch and continues rising, would you expect all the supporting field lines to reconnect and form a closed upper system and/or would the current sheet continue indefinitely to connect the prominence with the surface field?

Forbes: Whether the field lines reconnect quickly, or whether they do not, depends on the rate of reconnection, and the model does not address that question. However, I believe the existence and movement of the flare ribbons shows that reconnection occurs quickly during the first 5 to 15 minutes of a flare, but then slows considerably so that complete reconnection takes many hours.

Martens:

(1) You made a clear distinction between quiescent region prominences and active region filaments with respect to the normal polarity. The quiescent prominences having inverse polarity, and the active region filaments normal polarity. Aad van Ballegooijen and myself have studied the papers by Leroy and Coworkers, and we have come to the conclusion that for active region filaments, really both polarities tend to be observed. What is your comment on that?

(2) In your transparency for an erupting filament, you propose an ideal MHD impulsive phase in which about 12% of the energy is released. However, in ideal MHD all of the released energy would have to go into kinetic energy of the filament, since you assume ideal MHD, while observations do show a lot of x-ray emission in the impulse phase. How do you reconcile that?

Forbes: Response to Q. (1): To a certain extent that is true, since the Leroy and Bommier results for prominence polarities are statistical in nature. As I recall off hand, they find that for QRP's about 75% have inverse polarity and 25% have normal polarity, while for ARP's they find the reverse. So, it is true that for ARP's both polarities can be found.

Response to Q. (2): Yes, in the absence of discontinuities, all of the released magnetic energy in ideal MHD would be converted into the kinetic energy of the bulk flow, but in the ideal impulsive phase I am discussing, a shock is formed almost at once and so the system is no longer strictly ideal once the shock has formed. The important thing here is that the energy release begins as an ideal process occurring on the Alfvén-time scale of the system, rather than as a resistive process occurring on the reconnection-time scale of the system.

Acknowledgements. This work was supported by NASA Grant NAGW-76 and NSF Grant ATM-8711089 to the University of New Hampshire. The author also benefitted from his participation in the 1988 NOAA/SMM Workshop on Solar Events and Their Influence on the Interplanetary Medium.

References

Aly, J.J., On some properties of force-free magnetic fields in infinite regions of space, *Astrophys. J, 283*, 349,1984.

Anzer, U., The global structure of magnetic fields which support quiescent prominences, in *Measurements of Solar Vector Magnetic Fields*, p. 101, NASA CP2374, 1985.

Anzer, U., Structure and equilibrium of prominences, in *Dynamics and Structure of Quiescent Solar Prominences*, edited by E.R. Priest, p. 143, Kluwer Academic, Boston, 1989.

Barnes, C.W. and P.A. Sturrock, Force-free magnetic-field structures and their role in solar activity, *Astrophys. J, 174*, 659, 1972.

Biskamp, D. and H. Welter, Magnetic arcade evolution and instability, *Solar Phys. 120*, 49, 1989.

Canfield, R.C., C.-C. Cheng, K.P. Dere, G.A. Dulk, D.J. McLean, R.D. Robinson, Jr., E.J. Schmahl, and S.A. Schoolman, Radiative energy output of the 5 September 1973 flare, in *Solar Flares*, edited by P.A. Sturrock, p. 451, Colorado Assoc. Univ. Press, Boulder, 1980.

Carmichael, H., A process for flares, in *AAS–NASA Symposium on the Physics of Solar Flares*, edited by W.N. Hess, p. 451, NASA SP-50, 1964.

Dungey, J.W., A family of solutions of the magneto-hydrostatic problem in a conducting atmosphere in a gravitational field, *Monthly Notices Roy. Astron. Soc. 113*, 180, 1953.

Engvold, O., Prominence environment, in *Dynamics and Structures of Quiescent Solar Prominences*, edited by E.R. Priest, p. 47, Kluwer Academic Publishers, Dordrecht, 1988.

Field, G.B., Thermal instability, *Astrophys. J. 142*, 531, 1965.

Glackin, D.L., Differential rotation of solar filaments, *Solar Phys. 36,* 51, 1974.

Harrison, R.A., Solar coronal mass ejections and flares, *Astron. Astrophys. 162,* 283, 1986

Hildner, E., The formation of solar quiescent prominences by condensations, *Solar Phys. 35,* 1974.

Hood, A., Stability and eruption of prominences, in *Dynamics and Structure of Quiescent Solar Prominences,* edited by E.R. Priest, p. 167, Kluwer Academic, Boston, 1989.

Hundhausen, A.J., The origin and propagation of coronal mass ejections, in *Proceedings of the Sixth International Solar Wind Conference,* edited by V.J. Pizzo, T.E. Holzer, and D.G. Sime, p. 181, NCAR/TN-306+Proc, Boulder,1987.

Kippenhahn, R. and A. Schlüter, A theory of solar prominences, *Zs. Astrophys. 43,* 36, 1957.

Kopp, R.A. and G.W. Pneuman, Magnetic reconnection and the loop prominence phenomenon, *Solar Phys. 50,* 85, 1976.

Kuperus, M. and M.A. Raadu, The support of prominences formed in neutral sheets, *Astron. Astrophys. 21,* 189, 1974.

Leroy, J.L., V. Bommier, S. Sahal-Brechot, The magnetic field in the prominences of the polar crown, *Solar Phys. 83,* 135, 1983.

Low, B.C., The field and plasma configuration of a filament overlying a solar bipolar region, *Astrophys. J. 246,* 538, 1981.

Malherbe, J.M., The formation of solar prominences, in *Dynamics and Structure of Quiescent Solar Prominences,* edited by E.R. Priest, p. 115, Kluwer Academic, Boston, 1989.

Malherbe, J.M. and E.R. Priest, Current sheet models for solar prominences, I. Magnetohydrostatics of support and evolution through quasi-static models, *Astron. Astrophys. 123,* 80, 1983.

Martin, S.F., Preflare conditions, changes, and events, *Solar Phys. 68,* 217, 1980.

Mikic, Z., D.C. Barnes, and D.D. Schnack, Dynamical evolution of a solar coronal magnetic field arcade, *Astrophys, J. 328,* 830, 1988.

Mouradian, Z., M.J. Martres, and I. Soru-Escaut, The heating of filaments as a disappearance process, in *Coronal and Prominence Plasmas,* edited by A.I. Poland, p. 221, NASA CP-2442, 1986.

Mouradian, Z., M.J. Martres, I. Soru-Escaut, and L. Gesztelyi, Local rigid rotation and the emergence of active centers, *Astron. Astrophys. 183,* 129, 1987.

Munro, R.H., J.T. Gosling, E. Hildner, R.M. MacQueen, A.I. Poland and C.L. Ross, The association of coronal mass ejection transients with other forms of solar activity, *Solar Phys. 61,* 201, 1979.

Oran, E.S. J.T. Mariska, and J.P. Boris, The condensational instability in the solar transition region and corona, *Astrophys. J. 254,* 349, 1982.

Parker, E.N., Instability of thermal fields, *Astrophys, J. 117,* 431, 1953.

Pneuman, G.W., The formation of solar prominences by magnetic reconnection and condensation, *Solar Phys. 88,* 219, 1983.

Pneuman, G.W. and F.Q. Orrall, Structure, dynamics, and heating of the solar atmosphere, in *Physics of the Sun, Vol. II,* edited by P.A. Sturrock, T.E. Holzer, D.M. Mihalas, and R.K. Ulrich, p. 71, Reidel, Dordrecht, 1986.

Priest, E.R., *Solar Magnetohydrodynamics,* Reidel, Dordrecht, 1982.

Priest, E.R., Introduction to quiescent solar prominences, in *Dynamics and Structure of Quiescent Solar Prominences,* edited by E.R. Priest, p. 1, Kluwer Academic, Boston, 1989.

Schmahl, E.J., and E. Hildner, Coronal mass ejections-kinematics of 19 December 1973 event, *Solar Phys. 55,* 473, 1977.

Schmahl, E.J., Z. Mouradian, J.J. Martres, and I. Soru-Escaut, EUV arcades: signatures of filament instability, *Solar Phys. 81,* 91, 1982.

Schmieder, B., Overall properties and steady flows, in *Dynamics and Structure of Quiescent Solar Prominences,* edited by E.R. Priest, p. 15, Kluwer Academic, Boston, 1989.

Soru-Escaut, I. M.J. Martres, Z. Mouradian, Singularity of solar rotation and flare productivity, *Astron. Astrophys. 145,* 19, 1985.

Sparks, L. and G. Van Hoven, The physics of thermal instability in two dimensions, *Solar Phys. 97,* 283, 1985.

Sparks, L. and G. Van Hoven, Thermal instability of a radiative and resistive coronal plasma, *Astrophys. J. 333,* 953, 1988.

Sturrock, P.A., A model of solar flares, in *Structure and development of solar active regions,* edited by K. Kiepenheuer, p. 471, IAU, Paris, 1968.

Sturrock, P.A., Solar flares and magnetic topology, *Solar Phys. 113,* 13, 1987.

Tandberg-Hanssen, E. *Solar Prominences,* Reidel, Dordrecht, 1974.

Tandberg-Hanssen and U. Anzer, Orientation of magnetic fields in quiescent prominences, *Solar Phys. 15,* 158, 1970.

van Ballegooijen and P.C.H. Martens, Formation and eruption of solar prominences, *Astrophys. J.* in press, 1989.

Van Hoven, G. L. Sparks, and T. Tachi, Ideal condensations due to perpendicular thermal conduction in a sheared magnetic field, *Astrophys. J. 300,* 249, 1986.

Van Hoven, G., L. Sparks, and D.D. Schnack, Nonlinear radiative condensation in a sheared magnetic field, *Astrophys. J. Letts. 317,* L91, 1987.

Webb, D.F., C.-C. Cheng, G.A. Dulk, S.J. Edberg, S.F. Martin, S. McKenna Lawlor, and D.J. McLean, Mechanical energy output of the 5 September 1973 flare, in *Solar Flares,* edited by P.A. Sturrock, p. 471,Colorado Assoc. Univ. Press, Boulder, 1980.

Zirin, H., *Astrophysics of the Sun,* Cambridge Univ. Press, New York, 1988.

Zirker, J.B., Quiescent Prominences, *Solar Phys. 119,* 341, 1989.

STRUCTURE AND STABILITY OF PROMINENCES

U. Anzer

Max-Planck-Institut für Astrophysik, Karl-Schwarzschild-Str. 1, 8046 Garching, FRG

Abstract. This review discusses the basic questions about magnetohydrostatic equilibria of quiescent prominences. First the notion of field configurations with normal and inverse magnetic polarities is outlined. Then some of the principal models for both types of configuration are described. After this a new model based on a twisted flux tube is presented. Finally the very few results concerning the stability of these prominence models are discussed.

1. Introduction

Quiescent prominences are thin, sheet like structures which extend vertically into the solar corona. They have typically the following dimensions: length 200 000 km, height 60 000 km and a width of only 5 000 km (details about observations and models can be found in the books by Tandberg-Hanssen [1974], Poland [1986] and Priest [1989]). One may therefore ask the question what have such sheets to do with magnetic flux ropes, which is the topic of this conference. Before I come to this question in section 3 I shall give, in the next section, a brief summary of the existing two-dimensional models. Then in section 4 I shall present the scarce results which have been obtained on the stability of those model prominences. Section 5 is devoted to some conclusions and outlooks.

2. Two-dimensional models

In this section models for 2-d magnetohydrostatic equilibria for quiescent prominences will be presented. But before going into any details I would like to describe the concept of normal ($N-$) and inverse ($I-$) magnetic polarity. One can study the orientation of the magnetic field in prominences in its relation to the underlying photospheric fields. There are the following two possibilities:

1) the field emerges from the photosphere on one side of the prominence, traverses the prominence and returns to the photosphere on the other side. In this case the topology of the field is that of a potential field and such configurations have been termed $N-$ polarity. They are only potential-like because currents flow in the prominence and possibly also in the surrounding corona.
2) the field in the prominence has opposite direction to that in case 1). These $I-$ polarity configurations require more complex topologies with magnetic islands (or helical structures in 3-d configurations).

The observations show that both configurations with $N-$ and $I-$ polarity actually exist on the sun. A detailed review of the existing magnetic field observations is given by Leroy [1989, and references therein].

For the description of the different models we use a cartesian coordinate system with $x = 0$ being the prominence plane, $z = 0$ the photosphere and the y-direction is along the prominence. The 2-d models are all independent of y. The magnetic field outside the prominence is taken either as a potential field or a force free field. Inside the prominence the equilibrium requires a balance between gravity and Lorentz force, i.e.

$$\underline{j} \times \underline{B} = -\rho \underline{g} \quad . \qquad (1)$$

Most models treat the prominences as infinitely thin sheets of matter and current with densities ρ^s and j^s respectively. Using Maxwell's equation for a sheet current

$$j^s = \frac{1}{\mu}[B_z] \qquad (2)$$

one then obtains

$$\frac{1}{\mu}[B_z]\, B_x = \rho^s g \quad , \qquad (3)$$

where $[B_z]$ is the jump of B_z accross the sheet. Some particularly simple models use a line current, I, and a line distribution of mass, m. Then the equilibrium requires

$$I B_x = mg \quad . \qquad (4)$$

There are also some 2-d models which allow for volume currents inside the prominence. Such models can be found in an early paper by Brown [1958]. Although useful for the description of the internal prominence structure they are of limited value for the representation of the global field which is associated with prominences.

a) Models with normal polarity.

Geophysical Monograph 58

The first sheet models for this case were developed by Kippenhahn and Schlüter [1957)] One of these is represented by the field

$$\underline{B} = \left(\frac{x+1}{(x+1)^2+z^2} - \frac{x-\alpha}{(x-\alpha)^2+z^2}, \quad 0, \right.$$
$$\left. z\left(\frac{1}{(x+1)^2+z^2} - \frac{1}{(x-\alpha)^2+z^2} \right) \right) \tag{5}$$

for $x < 1$ and

$$\underline{B} = \left(-\frac{x-1}{(x-1)^2+z^2} + \frac{x+\alpha}{(x+\alpha)^2+z^2}, \quad 0, \right.$$
$$\left. -z\left(\frac{1}{(x-1)^2+z^2} - \frac{1}{(x+\alpha)^2+z^2} \right) \right) \tag{6}$$

for $x > 1$.

Here $\alpha < 1$ is a free parameter. The field configuration is shown in Fig. 1. This field can support a prominence with the following mass distribution

$$\rho^s = \frac{2z}{\mu g} \left(\frac{1}{1+z^2} + \frac{\alpha}{\alpha^2+z^2} \right)$$
$$\left(\frac{1}{\alpha^2+z^2} - \frac{1}{1+z^2} \right) \tag{7}$$

Such a prominence would extend from the photosphere to infinite heights and one obtains for the total mass per unit length

$$M = \int_0^\infty \rho^s dz < \infty$$

$2-d$ potential fields can also be obtained from complex-variable functions $\phi(\xi)$ if one takes $\xi = x + iz$, $B_x = Im(\phi)$

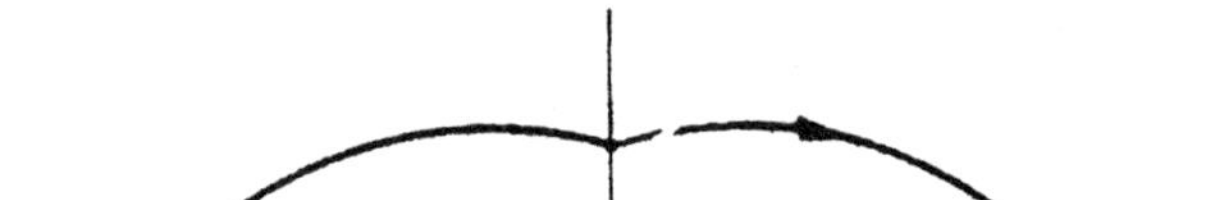

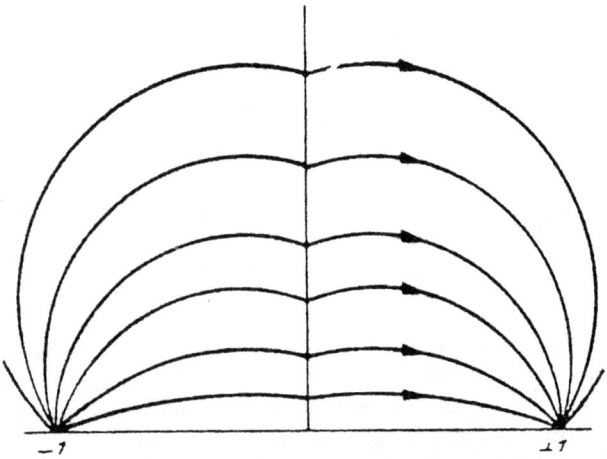

Fig.1: Current sheet model of a prominence extending from $z = 0$ to $z \to \infty$ [Kippenhahn and Schlüter, 1957].

and $B_z = Re(\phi)$. If one introduces cuts along the z-axis one can then produce fields with current sheets. This method was used by Malherbe and Priest [1983]. One typical model of their computations is given by

$$\phi(\xi) = -\frac{B_1}{\xi} - B_0 \frac{\sqrt{(p^2+\xi^2)(q^2+\xi^2)}}{\xi(\xi+i)^2} \quad . \tag{8}$$

The field is shown in Fig. 2. In this case the prominence extends from $z = p$ to $z = q$. The advantage of this procedure is that one can use many different kinds of complex functions to produce a variety of models.

b) Models with inverse polarity

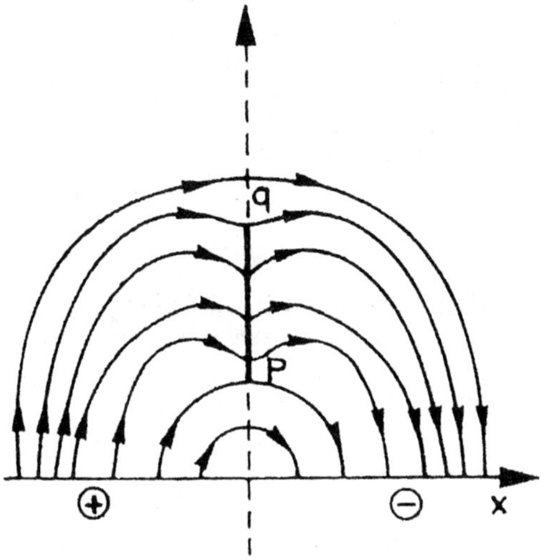

Fig.2: Current sheet obtained from a particular complex function; the sheet reaches from $z = p$ to $z = q$ [Malherbe and Priest, 1983].

Since such configurations are more complicated than those with $N-$ polarity most of the modelling in this case is done by describing the prominence by a line current. For such configurations $I-$ polarity refers to the situation where the field just underneath the singular current line is directed oppositely to the potential field. Such models were initiated by Kuperus and Raadu [1974] and a sketch of the configuration which they had in mind is shown in Fig. 3. These models require that the current in the prominence flows in opposite direction to the one in $N-$ polarity models. This then leads to a downward Lorentz force which amplifies the gravitational force. Therefore these configurations are initially completely out of equilibrium. Kuperus and Raadu suggested that the resulting downward motion of the prominence then induces strong currents in the photosphere. These induced currents are repulsive with respect to the prominence current. One finds that equilibria can be obtained once the prominence has moved downward

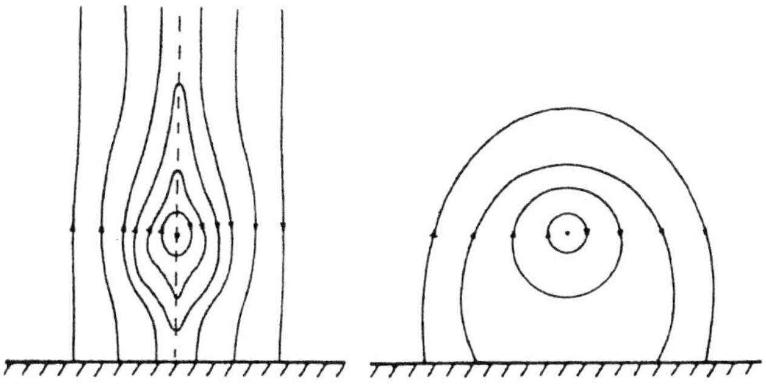

Fig.3: Sketch of the $I-$ polarity configurations proposed by Kuperus and Raadu [1974].

a sufficient distance. Anzer [1984] and Anzer and Priest [1985] have studied these effects in more detail. They also tried to extend this approach to models with curent sheets. They were able to recover the global equilibria described above but could not find any sheet configuration which is in local equilibrium everywhere. All models showed downward Lorentz forces in the upper parts of the sheet. This is a natural consequence of the pinching effect of extended current distributions. Malherbe and Priest [1983] applied the complex variable technique also to configurations with $I-$ polarity. They encountered similar difficulties. In their model of a finite current sheet a singularity occurs at the upper edge which has a downward force of infinite density. But even more seriously is the fact that in their model the forces integrated over the entire sheet give a net downward pull.

c) Models with sheared magnetic fields

The observations indicate that the magnetic field in the neighbourhood of prominences is highly sheared. One can try to take this into account in the models described above by superimposing a constant field in y-direction, $\underline{B}_y^0$. Since all currents flow in y-direction adding a constant $\underline{B}_y^0$ does not change the equilibrium. But since the constraint of $\underline{B}_y^0$ being constant is very severe such a procedure is not very satisfactory.

A better way is to replace the potential fields outside the prominence by force free fields, defined by

$$\nabla \times \underline{B} = \alpha \underline{B} \tag{9}$$

Amari and Aly [1989a] studied such configurations. They assumed that α is constant and represented the prominence by a line current I at $x = 0$, $z = h$. They constructed a class of special perodic fields given by

$$\underline{B} = \nabla A \times \underline{e}_y + \alpha A \underline{e}_y \tag{10}$$

with

$$A = B_0 L \cos\left(\frac{\pi x}{L}\right) e^{-\gamma_1 z} + \frac{\mu I}{L} \sum_{p=0}^{\infty} \frac{1}{\gamma_{2p+1}} \cos\left((2p+1)\frac{\pi x}{L}\right) \left(e^{-\gamma_{2p+1}|z-h|} - e^{-\gamma_{2p+1}(z+h)}\right) \tag{11}$$

and

$$\gamma_{2p+1} = \sqrt{(2p+1)^2 \frac{\pi^2}{L^2} - \alpha^2} \quad . \tag{12}$$

Equilibrium then requires

$$I\left(\frac{\mu I}{L} \sum_{p=0}^{\infty} e^{-2\gamma_{2p+1} h} - B_0 L \gamma_1 \quad e^{-\gamma_1 h}\right) = mg \quad . \tag{13}$$

This equation describes equilibria of both $N-$ and $I-$ polarity configurations. In this conference Amari and Aly [1989b] announced that they have now extended this model to incorporate current sheets. They used the same photospheric flux distribution and again took a force free field with constant α. The current sheet which they use extends from $z = a$ to $z = b$ and has the following form:

$$j^s = I \frac{6(b-z)(z-a)}{(b-a)^3}$$

which is the same as previously used by Anzer [1985]. They found for these configurations that in all cases with inverse polarity the upper parts of the sheet show downward Lorentz force. Therefore even the inclusion of shear cannot provide equilibrium models of $I-$ polarity.

Attempts to do 3-d modelling were made by Wu and Low [1987]. They took 2-d current sheets and superimposed 3-d potential fields. Obviously the resulting solutions are very special and it is not clear how useful they are for describing real prominence configurations.

3. Flux tube model

Priest et al. (1989) developed a different model. They noticed that only very special fields allow the formation of a prominence. There must be initially at least some field lines which have a dip. At the location of these dips then material which condenses out of the corona can accumulate. This will then further increase the dip and the prominence can grow. The question arises how to form such field configurations in a natural way. Suppose that there is one giant flux tube without twist, as sketched in Fig. 4a. This initial field has downward curvature everywhere. But if one rotates the foot points and twists the flux rope dips will form in the fieldlines as indicated in Figs. 4b and 4c. At those places then a prominence can form. The condition for a dip is given by

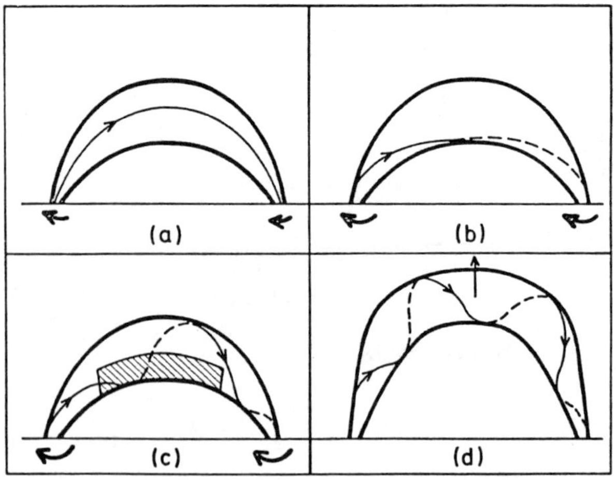

Fig.4: Sequence of twisting of a flux tube. From a) to c) the twist increases, and the prominence can form as indicated, d) prominence eruption when the twist reaches a critical value [Priest et al., 1989].

$$(\underline{B}\nabla)B_z < 0 \tag{14}$$

If we describe the flux tube by a part of a torus with major radius R and minor radius a and field components B_ϑ and B_ϕ then at the top of the loop condition (14) leads to

$$\frac{B_\phi^2}{a} > \frac{B_\vartheta^2}{R} \quad . \tag{15}$$

Let us now discuss the orientation of the flux rope with respect to the boundary of photospheric polarities. The observations indicate that the crossing should occur at very small angles as shown in Fig. 5. Such a stretching of the tube can be achieved by the differential rotation of the sun. For the twisting-up of the flux rope we suggest that the convective flow in supergranules is affected by Coriolis forces and this then will lead to a rotational motion. The sense of this rotation (relative to the stretching by differential rotation) is such that it produces preferentially configurations with $I-$ polarity.

In the context of this flux rope model it is important to realise that this giant magnetic flux rope is not identical

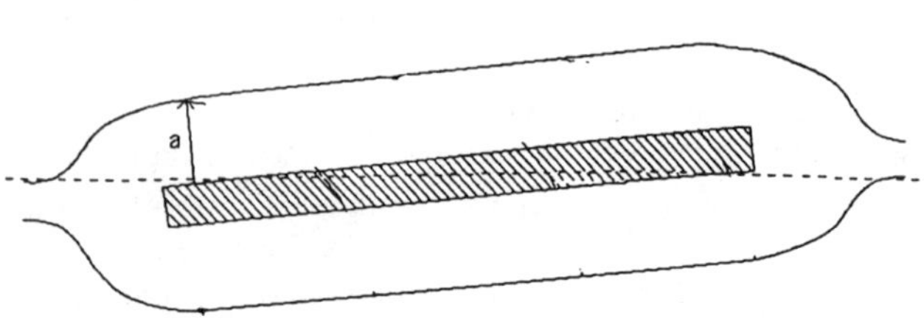

Fig.5: Top view of the twisted flux rope and of the prominence location [Priest et al., 1989].

with the prominence. The twisted flux tube only provides the dips in the field which are necessary for prominence formation. There are strong arguments against the interpretation that the entire twisted flux rope represents the prominence because prominences are thin, sheet-like structures whereas flux tubes are of circular cross section; magnetic field measurements do not show the helical geometry of a twisted flux tube; the pressure scale height along magnetic field lines is extremely small therefore it is impossible to fill an entire flux tube uniformly with cold prominence material. We want to emphasise that in our model the prominence has to be visualised as a sheet lying inside such a flux rope. Helical configurations for the prominence field have also been suggested by Anzer and Tandberg-Hanssen [1970].

4. Stability analysis

Since large prominences are in general stationary over time scales of up to several months a basic requirement for theoretical models is their stability. But because of the complexity of any theoretical configuration which is sufficiently close to reality such a stability analysis represents a difficult task. Kippenhahn and Schlüter [1957] in their early work addressed this problem and investigated the stability of sheet structures. But they considered only displacements which move the prominence as a whole, either horizontally or vertically. They found that for stability against horizontal displacements the supporting field has to be curved upward locally, i.e. it has to have a dip. The sheet configurations are always stable against uniform vertical displacements. Anzer [1969] investigated the stability of sheets against arbitrary perturbations. The following two necessary and sufficient stability conditions were obtained from a linearised stability analysis which was based on the energy principle:

$$[B_z]\,\frac{dB_x}{dz} \geq 0 \tag{16}$$

and

$$B_x\,\frac{d\,[B_z]}{dz} \leq 0 \quad . \tag{17}$$

The first condition can be interpreted as a generalisation of the finding by Kippenhahn and Schüter that dips are required. The second condition describes the stability against Rayleigh-Taylor modes. This condition requires that for stability the current density has to decrease with height. This can obviously not be fulfilled near the lower edge of the prominence where the current has to go to zero, but there also the idealisations under which condition (17) was derived break down. Therefore at present one cannot say what the appropriate stability condition at the lower edge would be. One important point which needs mentionning in this context is the fact that the above stability analysis was done be using potential fields outside the prominence. That means that the stabilising effects which result from line-tying in the photosphere could not be taken into account and prominences could in reality be more stable than conditions (16) and (17) would indicate.

Nakagawa and Malville [1969] investigated the stability of the lower boundary of prominences. They suggested that the quasi-periodic arch structures which are frequently seen in prominences are the result of an instability of the lower boundary. But the equilibrium model on which they based their stability analysis is very crude: they took a uniform horizontal vacuum field overlayed by a uniform dense plasma and separated by a planar horizontal interface. It seems that this configuration is quite different from solar prominences and therefore the interpretation that the arch structure results from the instability found by the authors is not very convincing.

The stability of models based on line currents has also been discussed [e.g. van Tend and Kuperus, 1978, and Démoulin and Priest, 1988]. In these investigations the prominence is treated as if it were a rigid, current-carrying wire. Démoulin and Priest take a model with a line current which is embedded in a special force free field of constant α. They calculate for a given prominence mass the equilibrium current as a function of height. Their results are shown in Fig. 6. They conclude that those parts of the curves where $dI/dh < 0$ holds are unstable. This conclusion is based on the finding that above each equilibrium curve the net force is upward, below it downward. This means that if the equilibrium is perturbed in such a way that $I =$ const. holds then those parts of the curve $I(h)$ with positive slope are stable, those with negative slope unstable. But the assumption that during the displacement $I =$ const. is somewhat arbitrary. It would be more realistic to conserve the magnetic flux which crosses the z-axis between the photosphere and the prominence. This would then require a change in the current, δI, for any given displacement δz and the quantity $(dI/dh - \delta I/\delta z)$ would determine the stability. The major advantage of using such line current models is that one can follow their nonlinear development which could describe the eruption of a prominence. But this kind of investigation ignores pinch instabilities and also kink instabilities, both of them are very important in many laboratory plasma configurations. Therefore the results obtained by this approach are not entirely conclusive.

5. Conclusions and outlook

The question of the energy balance of prominences has been left out completely in this review, because of lack of time. But this of course is another major field of prominence research where many problems are still unsolved,

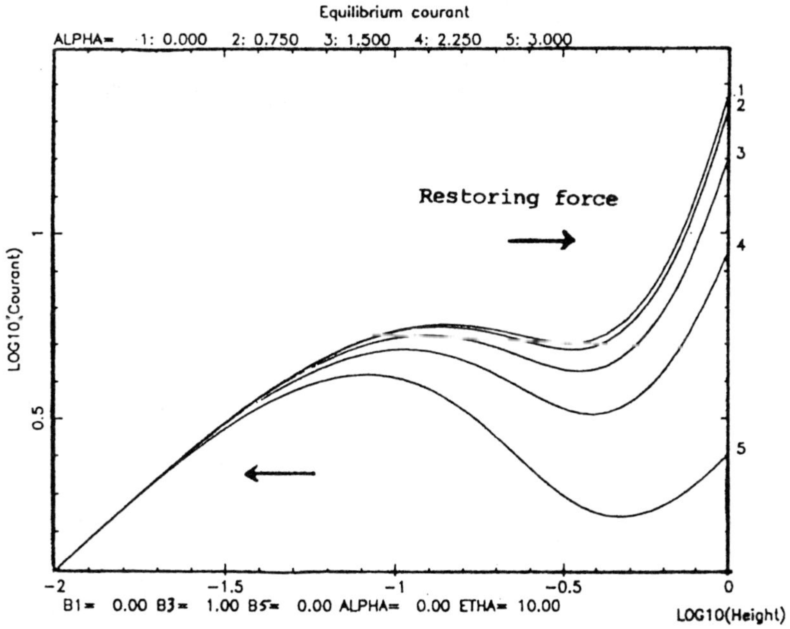

Fig.6: Equilibrium current as a function of height for line current models. The curves are for different values of α [Démoulin and Priest, 1988].

and it is my impression that at present the thermal equilibrium of prominences is even poorer understood than the mechanical equilibrium.

As far as the statics and the dynamics of prominences are concerned the following points require further investigations: There is the question of stationarity. The observations on this subject are somewhat conflicting: some of them report upward motions, others systematic downflows; and the magnitudes of the velocities are quite uncertain. This means that at present it is not clear from the observational standpoint whether static models are a reasonable approximation of the true situation or whether eventually dynamic models are needed. In this context it is also interesting to note that the frequently quoted problem of mass continuous supply might acutally not exist.

Another open field for future research is the modelling of the 3-d structure of prominences. On a large scale many prominences show a periodic arch-like sharp lower boundary with feet which extend to the solar surface and a typical separation of about 30 000 km. One explanation for the existence of the feet is that at their location the field is so weak that the material sags down very far before equilibrium can be reached. But then the question arises why there is no condensed material in the region between the feet where the field is strong and could therefore support much larger quantitities of matter. A different suggestion is that the field topology is such that the feet are thermally shielded by a field which is almost parallel to the prominence and therefore condensation is facilitated. But these are at present only vague ideas and no detailed modelling has been done. Prominences also show definite fine structures on a much smaller scale. Most prominences seem to consist of many thin vertical threads. In others cases inclined structures are observed which sometimes give the illusion that the prominence is of helical nature. A theoretical understanding of how these fine structures are produced is still lacking.

In this review I tried to line out what basic ideas exist about the equilibrium and stability of solar prominences. But one should also be aware that there are still many open questions. Even in the most simple case of 2-d modelling the situation is still far from being clear. In particular we still do not have 2-d equilibrium models for configurations with inverse magnetic polarity. Once one has found satisfactory 2-d models one could try to extend them to describe the 3-d structure. Another possibility is to start out with configurations which have already some aspects of three-dimensionality in them, as the recently developed flux rope model. But this kind of model still needs to be worked out mathematically in more detail. Once one has constructed detailed equilibrium models the next step would be to investigate their stability. This certainly is a demanding task which still lies ahead of us.

Questions and Answers

Chen: In a curved (toroidal structure), there are toroidal forces. What role can these forces play in prominence equilibria?

Anzer: In our twisted flux tube model, the field configuration which can lead to the formation of a prominence is assumed to be force-free and therefore there are no net toroidal forces.

Haerendel: Prominences seem to consist of a multitude of plasma tubes whose convex side is pointing upward. In the first model (Priest, Hood and Anzer, 1989), the part of the twisted flux tubes supporting matter against gravity seem to have their convex side pointing downward. How do you reconcile this conflict?

Anzer: The cold prominence plasma has a vertical pressure scale height of only around 100 km and, therefore, it cannot be used as a tracer of extended magnetic field structures. The observed prominence threads, therefore, have to be produced by some other mechanism which is still unknown.

Zirin: I cannot understand how you can support material with a field that is concave upwards. I should note that regions of filament footpoints often correspond to anomalies in the underlying photosphere, i.e., small dipoles that emerge in the horizontal field region.

Anzer: The flux tube model which I have described here starts out with a twisted force-free field which is stable for low enough amounts of twist. When it is loaded with cool material, a sheet configuration will develop. For the case that this sheet can be represented by a 2-D configuration, conditions (16) and (17) given in the paper guarantee the stability. For the general 3-D case, no stability conditions have been found so far.

Acknowledgement. I would like to thank the Deutsche Forschungsgemeinschaft and the American Geophysical Union for travel support.

References

Amari, T. and Aly, J. J., Interaction between a line current and a two-dimensional constant-α force-free field: an analytical model for quiescent prominences, Astron. Astrophys., 208, 261-270, 1989a.

Amari, T., and Aly, J.J., Structure of two-dimensional magnetostatic equilibria in the presence of gravity, these proceedings, 1989b.

Anzer, U., Stability analysis of the Kippenhahn-Schlüter model of solar filaments, Solar Phys., 8, 37-52, 1969.

Anzer, U., The global structure of magnetic fields which support quiescent prominences, Measurements of Solar Vector Magnetic Fields, (ed. M J Hagyard) NASA Conf. Pub. 2374, 101-106, 1984.

Anzer, U., and Priest, E. R., Remarks on the magnetic support of quiescent prominences, Solar Phys., 95, 263-268, 1985.

Anzer, U., and Tandberg-Hanssen, E., A model for quiescent prominences with helical structure, Solar Phys., 11, 61-67, 1970.

Brown,A., On the stability of a hydromagnetic prominence model, Astrophys. J., 128, 646-663, 1958.

Démoulin, P., and Priest, E. R., Non equilibrium of a prominence current in a linear force free field, Proc. Mallorca Workshop on "Dynamics and Structure of Solar Prominences", (eds. J. L. Ballester, and E. R. Priest), 45-48, 1988.

Kippenhahn, R., and Schlüter, A., Eine Theorie der solaren Filamente, Zs. Ap., 43, 36-62, 1957.

Kuperus, M., and Raadu, M., The support of prominences formed in neutral sheets, Astron. Astrophys., 31, 189-193, 1974.

Leroy, J. L., Chapter 4 in "Dynamics and Structure of Quiescent Solar Prominences" (ed. E. R. Priest), Kluwer Academic Publishers, Dordrecht, 1989.

Malherbe, J. M., and Priest, E. R., Current sheet models for solar prominences, Astron. Astrophys., 123, 80-88, 1983.

Nakagawa, Y., and Malville, J., Periodic structures in quiescent prominences, Solar Phys., 9, 102-115, 1969.

Priest, E.R. (ed.), Dynamics and structure of quiescent solar prominences, Kluwer Academic Publishers, Dordrecht, 1989.

Priest, E. R., Hood, A. W. and Anzer, U., A twisted flux tube model for solar prominences, I General properties, Astrophys. J., in press, 1989.

Poland, A.I. (ed.), Coronal and prominence plasmas, NASA Conf. Pub. 2442, 1986.

Tandberg-Hanssen, E., Solar prominences, D. Reidel, Dordrecht, 1974.

van Tend, W., and Kuperus, M., The development of coronal electric current systems in active regions and their relation to filaments and flares, Solar Phys., 59, 115-127, 1978.

Wu, F. and Low, B.C., Static current-sheet models of quiescent prominences, Astrophys. J., 312, 413-433, 1987.

FILAMENT COOLING AND CONDENSATION IN A SHEARED MAGNETIC FIELD

Gerard Van Hoven

Department of Physics, University of California, Irvine, California 92717

Abstract. Thermal instability driven by optically thin radiation in the corona is believed to initiate the formation of solar filaments [Parker, 1953]. The fact that filaments are observed generally to separate regions of opposite, line-of-sight, magnetic polarity in the differentially rotating photosphere suggests that filament formation requires the presence of a highly sheared magnetic field. In this paper we discuss the coupled energetics and dynamics of the most important condensation modes, those due to perpendicular thermal conduction at short wavelengths. We describe their linear structure in the sheared field and their growth rates. We have also performed two-dimensional, nonlinear, magnetohydrodynamic simulations of the evolution of these modes in a force-free field. To clarify the essential physics of the nonlinear behavior, we have traced the evolution of generic perturbations possessing broad spatial profiles. The simulations achieve the fine thermal structures, minimum temperatures and maximum densities characteristic of observed solar filaments.

The anisotropic influence of magnetic shear on the linear dynamics of the condensation instability has been examined extensively in a series of studies [Chiuderi and Van Hoven, 1979; Van Hoven and Mok, 1984; Van Hoven et al., 1986; and Sparks and Van Hoven, 1988]. Several conclusions have emerged from these studies:

1. A condensation will form preferentially in regions in which $k_{\|} \equiv \mathbf{k} \cdot \mathbf{B}/B \approx 0$ where $\mathbf{k}$ represents the (local) wave vector of the perturbation.
2. A condensation may be classified according to whether its spatial structure is determined primarily by considerations of force balance or energy balance. The former is identified as a dynamic condensation, and the latter a kinematic condensation.
3. The modes that exhibit the fastest growth exist only in the presence of anisotropic heat flow.
4. Plasma mass flow within kinematic condensations is directed parallel to the magnetic field.
5. Kinematic condensations are most compressible when sound waves traveling parallel to the magnetic field can maintain pressure balance.

Geophysical Monograph 58

To study the evolution of solar filaments, we adopt a one-fluid, transport model [Van Hoven et al., 1987]. The dynamical behavior of the plasma is governed by the nonlinear equations of resistive magnetohydrodynamics, specifically including the continuity equation, the momentum equation, Maxwell's equations, and Ohm's law. This set of equations is closed by including an energy equation of the form

$$\frac{dp}{dt} = \gamma \frac{p}{\rho}\frac{d\rho}{dt} + (\gamma - 1)[\nabla \cdot \boldsymbol{\kappa} \cdot \nabla T - C + H] \qquad (1)$$

where $\boldsymbol{\kappa}$ is the thermal conductivity tensor, $C = R\rho^2 T^{\tau}$ denotes cooling due to radiative loss [Hildner, 1974], and H is a heating function. We specify the magnetic field by the model form $\mathbf{B}_0 = B_0$ [sech (y/a) $\hat{\mathbf{e}}_x$ + tanh (y/a) $\hat{\mathbf{e}}_z$] where a is the shear scale, which is in dynamic equilibrium with a uniform temperature and density, with an accompanying energetic balance of uniform heating and radiation. The MHD system can be reduced to a simultaneous set of seven equations relating the mass density ρ, the temperature T, the three fluid-velocity components u_x, u_y, and u_z, the magnetic field component B_x, and the magnetic flux function $\Psi(y, z)$.

To understand the character of the initial plasma excitations that lead to significant cooling and condensation, we will consider a simplified model of the small-amplitude behavior of this system. To do so, we first linearize the MHD equations around the force-free, isobaric equilibrium just described by using $T = T_0 + T_1(y)$ exp $(\nu t + ikz)$

where $T_1 << T_0$. In two simple limits one can simplify the mass-conservation law

$$\nu \rho_1 = -\rho_0 \nabla \cdot \mathbf{v}_1 \rightarrow \begin{cases} -ik_{\parallel}\rho_0 v_{1\parallel} & k_{\parallel}v_A >> \nu \\ 0 & k_{\parallel}v_A >> \nu >> k_{\parallel}v_s \end{cases}$$

where $k_{\parallel} = k \tanh(y/a)$, $v_A = B_0/(\mu_0\rho_0)^{1/2}$ and $v_s = (p_0/\rho_0)^{1/2}$. [These limits are described in Sparks and Van Hoven (1988); the first eliminates the nearly singular dynamic modes of Chiuderi and Van Hoven (1979) which depend on cross-field motions, and the second states that parallel motions respond very weakly to pressure perturbations.]

In both of these $k_{\parallel}$ limits, the perpendicular (to **B**) flow perturbations disappear and we need only consider the parallel component of the force equation [Drake et al., 1988]

$$\nu \rho_0 v_{1\parallel} = -ik_{\parallel} p_1 = -ik_{\parallel} p_o(\hat{\rho}_1 + \hat{T}_1)$$

where $\hat{T}_1 = T_1/T_0$. Finally, we linearize (1) to obtain

$$\nu \hat{p}_1 = \gamma \nu \hat{\rho}_1 - (\gamma - 1)p_0^{-1} \left[\begin{array}{l} k_{\parallel}^2 \kappa_{\parallel} \hat{T}_1 + \frac{\partial C}{\partial \rho}\hat{\rho}_1 + \frac{\partial C}{\partial p}\hat{p}_1 \\ + \left(k^2 - k_{\parallel}^2\right) \kappa_{\perp} \hat{T}_1 - \kappa_{\perp} \hat{T}_1'' \end{array} \right]$$

where $\hat{T}_1'' = d^2\hat{T}_1/dy^2$. By eliminating $\hat{\rho}_1$ and $\hat{p}_1$, we can obtain a 1-D eigenvalue form of the energy equation

$$0 = \Omega_{\perp} T_1'' - \left[\begin{array}{c} k_{\parallel}^2 a^2 \Omega_{\parallel} + (k^2 - k_{\parallel}^2) a^2 \Omega_{\perp} + (\nu - \Omega_{\rho}) \\ + \frac{k_{\parallel}^2 v_s^2 [(\gamma - 1)\nu - \Omega_T]}{\nu^2 + k_{\parallel}^2 v_s^2} \end{array} \right] \hat{T}_1$$

where typical rates [Sparks and Van Hoven, 1988] include $\Omega_{\parallel} \equiv (\gamma - 1)\, \kappa_{\parallel} T_0/a^2 p_0$ and $\Omega_{\rho} \equiv -(\gamma - 1)[\partial C/\partial p]_{\rho}$. The last (fractional) term in the square bracket disappears when $\rho_1 \approx 0$.

One can have an idea of the transverse (shear-direction) character of these modes by looking at $\hat{T}_1''/\hat{T}_1$ which is shown in Fig.1(a), along with a schematic solution $\hat{T}_1$ (and $-\hat{\rho}_1$) for the (third) eigenvalue ν. The critical wavenumbers $k_1^2 \approx \nu^2(\nu - \Omega_{\rho})/v_s^2(\Omega_p - \nu)$ and $k_2^2 \approx \gamma(\Omega_p - \nu)/a^2\Omega_{\parallel}$ give the approximate location of the zeros of the square bracket [equivalent to α_1 and α_p of Chiuderi and Van Hoven, 1979].

Fig. 1(b) interprets the turning points (approximately) in terms of the essential local rates of the problem. At large y the parallel thermal conduction suppresses the temperature perturbation. For $k_{\parallel}(y) < k_2$, radiative losses lead to cooling, which can then lead to parallel-to-**B** flow and significant condensation in those locations $k_{\parallel}(y) > k_1$ where $k_{\parallel}v_s$ is the fastest rate in the problem.

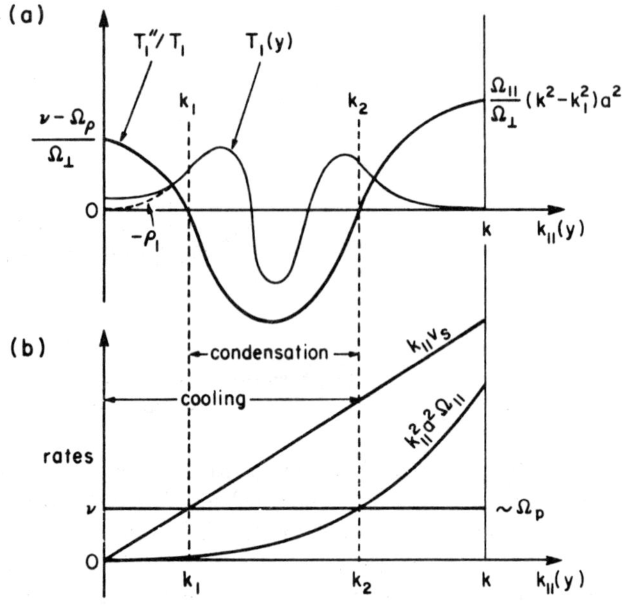

Fig. 1. (a) The effective binding potential for $\Omega_\rho < \nu < \Omega_p$ and $ka > (\gamma\Omega_p/\Omega_{\parallel})^{1/2}$, along with a typical T_1 (and $-\rho_1$) eigenfunction; (b) an interpretation of the turning points in terms of the fundamental local rates of the problem.

The growth rates and structure of these kinematic modes have been discussed by Drake et al. [1988], and numerically detailed and interpreted by Van Hoven et al. [1986; Sparks and Van Hoven, 1988]. Excitations composed of these short-wavelength modes provide the key to the attainment of significant nonlinear cooling and condensation in a sheared magnetic field.

In order to demonstrate this fact, and to understand the physical influences that guide the development of the filamentary condensations, it proves useful to consider the evolution of "generic" perturbations – smooth, isobaric excitations with broad spatial profiles. The spatial dependence of a generic temperature perturbation is specified according to the equation
$\hat{T}_1 = \epsilon \exp(-|y|) \sin kz$.

We initiate a simulation by specifying the equilibrium number density $n_0 = 10^{9.8}$, the field strength $B_0 = 10^{0.9}$, the shear scale $a = 10^{8.5}$ (in cgs units), and the temperature T_0. For illustrative purposes we choose $T_0 = 10^{5.7} K$, somewhat below that of the corona. We emphasize that, with this exception, values appropriate to the solar atmosphere are chosen for all equilibrium parameters and energy-transport coefficients.

We choose the length of the grid in z to be given by $k = 10/a$, and impose fixed-temperature boundary conditions in this direction. From our linear studies [Sparks and Van Hoven, 1988], we know that this length is sufficiently small for coronal conditions that short-wavelength, kinematic perturbations are the only thermally unstable, linear modes that can be excited. When we fix T at the boundary, we must also fix ρ, u_y, B_x, and Ψ, while the derivatives of u_x and u_z with respect to z must vanish. [The equations of motion determine these phase relationships among the dependent variables.] In this case there may be a net flow of mass across the z boundary over the course of a simulation.

Two aspects of the nonlinear evolution of this generic (multi-mode) excitation are shown in the following figures. The surfaces (and contours below) plotted in Fig. 2(a) represent the magnitude of the temperature $T(y, z)$ at $t = 293$ sec. Here the z axis is linear, but the y axis utilizes a coordinate transformation $y \rightarrow \sinh y$. We show this variation implicitly by plotting the direction of the equilibrium magnetic field as a function of y. Corresponding surfaces displaying the magnitude of the mass density $\rho(y, z)$ are plotted in Fig. 2(b). Vectors have been superimposed on the density contours to indicate the magnitude and direction of flows in the $y - z$ plane.

Fig. 3 displays the *d*ensity, *t*emperature, and *p*ressure as functions of time at a probe located at a point near the density maximum. At this point, $k_1 < k_{\parallel} < k_2$ and the temperature and density perturbations initially grow isobarically, as predicted by Fig. 1. A drop in temperature and a rise in density both produce increasing radiative energy losses that, in turn, cause further cooling and additional mass inflow. The local radiative-loss rate matches the constant-pressure value.

By $t \sim 200$ sec, a temperature well has formed in the *center* of the shear layer where the plasma has cooled from its equilibrium value by almost an order of magnitude, nearly to the point at which local nonlinear saturation occurs when the radiative cooling rate comes back into balance with the heating rate. Meanwhile, two peaks in the density arise off-center as a consequence of plasma flow that is directed primarily parallel to the magnetic field. By $t \approx 180$ sec, these peaks represent a 50% local increase in mass density, similar to that seen at $k_{\parallel} = 0$ by Van Hoven et al. [1987].

As the system continues to evolve, the temperature well expands toward the probe. By the time the probe temperature has cooled to $3 \times 10^5 K$, the local radiative-loss rate has climbed to the point where it exceeds the local dynamic-response rate. Sound waves traveling par-

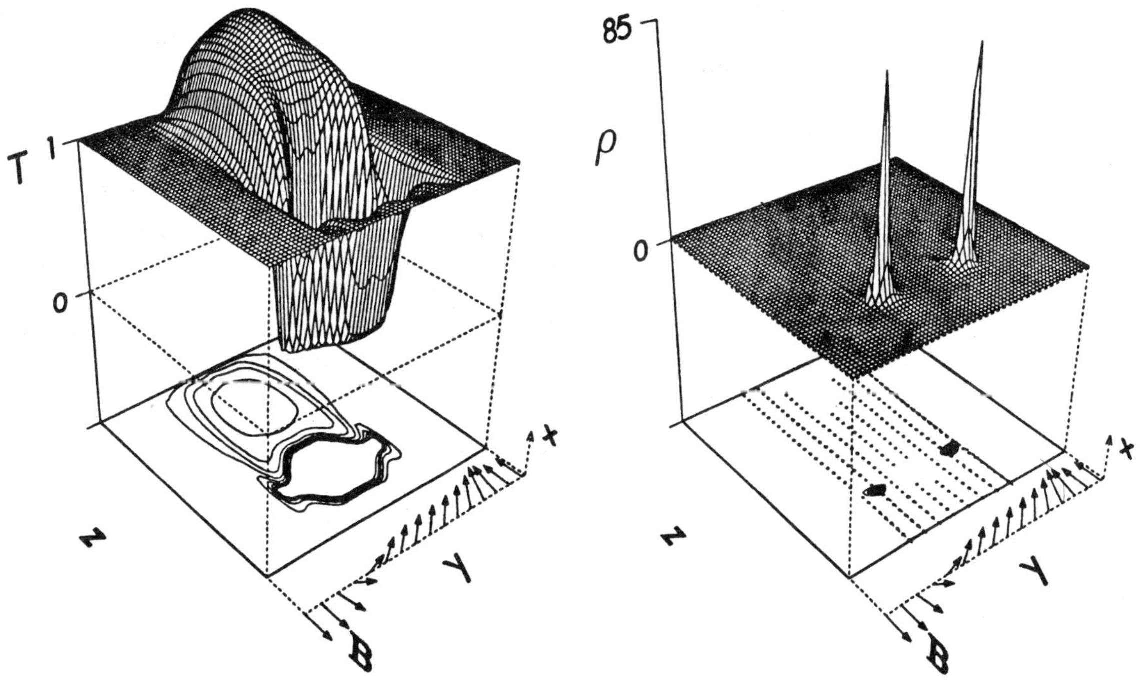

Fig. 2. The temperature $T(y, z)$ and density $\rho(y, z)$ profiles at 293 sec in the simulation, along with the sheared magnetic field $\mathbf{B}(y)$; the y axis is scaled as $\sinh y$, so that the central shear layer is expanded while the range $(-\infty, \infty)$ is covered.

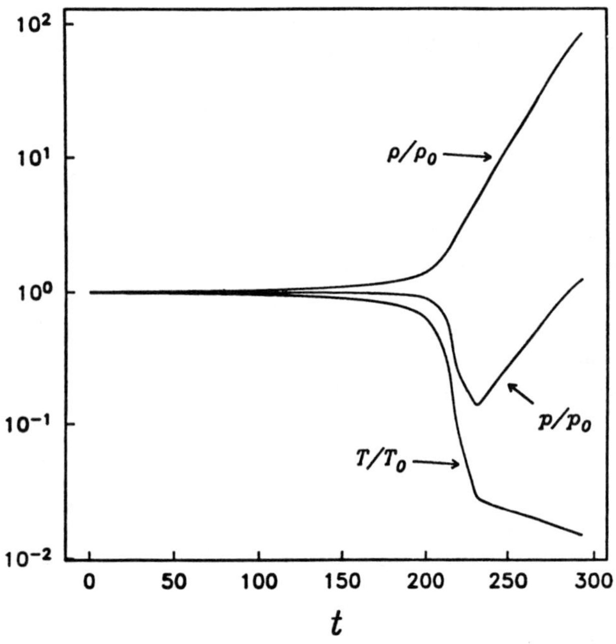

Fig. 3. Values of the thermodynamic variables *vs* t/τ_h taken from a diagnostic probe located at the position of the density peak, where the angle between the filament axis and the local magnetic field is 14°.

allel to **B** can no longer maintain constant pressure at the probe and, thus, the pressure begins to fall (Fig. 3).

The cooling rate ($dT/Tdt \propto C/p$) at the probe continues to accelerate until it achieves its maximum local value at $t \approx 220$ sec when $T = 8 \times 10^4 K$. The dependence of the cooling function on T is such that additional cooling now lowers the instantaneous radiative-loss rate. Some 10 seconds later the temperature has dropped to $1.5 \times 10^4 K$, and the pressure has attained its minimum value (Fig. 3). As the plasma continues to cool below $1.5 \times 10^4 K$, the radiation rate falls rapidly, and parallel flow can now respond to the pressure deficit, leading to a dramatic growth in the local density (and pressure).

At t = 293 sec the maximum number density has increased to $10^{11.7}$, nearly two orders of magnitude above its initial equilibrium value (Fig. 3). Wings have developed on the temperature contours that reflect the additional radiative cooling that ensues as the local density rises (Fig. 2a). At this position, the magnetic field forms an angle of approximately 14° with respect to the filament (x) axis, a value within the range of those observed [Leroy, 1989]. Nonlinear saturation of the condensation process should occur when temperature gradients become sufficiently large that thermal conduction can restore energy balance. To achieve such saturation in a simulation would require utilizing a grid with extremely fine resolution, which would require excessive computational expense.

By performing nonlinear magnetohydrodynamic simulations, we have demonstrated that the radiative condensational instability in a sheared magnetic field is capable of generating small-scale, filamentary, plasma structures with densities and temperatures, and angular orientations with respect to the local field, characteristic of solar prominences. Insight into the nonlinear, physical behavior of the instability has been achieved by studying the evolution of spatially broad, generic perturbations.

We have found that the most rapid growth of the condensation results from mass flow parallel to the magnetic field in regions of the plasma where the field inhibits thermal conduction (i.e., in the shear layer). Like the linear kinematic modes, nonlinear condensations possess density maxima located away from the center of the shear layer where the parallel sound-speed rate is finite. The growth that we observe appears to consist of the superposition of many linear kinematic modes with nearly degenerate growth rates such that the spatial scales characteristic of a nonlinear condensation are much broader than those of the individual sharply localized kinematic modes.

A fuller report of these nonlinear simulations, including the results of random perturbations, will appear in the *Astrophysical Journal* [Sparks, Van Hoven and Schnack, 1989].

Questions and Answers

Forbes: What role does ohmic heating play in the energetics of your calculation?

Van Hoven: ***During the early stages, the Lundquist number is too high for resistivity to have any effect. Later, the Coulomb resistivity does increase (by ~ 10^3 times), but, without reconnection or some other current-concentration mechanism, its ohmic effects usually remain insignificant.***

Acknowledgments. This research was performed in collaboration with J.F. Drake, D.D. Schnack and L. Sparks, and was supported by NSF and NASA, with computations provided by DOE and NSF.

REFERENCES

Chiuderi, C., and G. Van Hoven, The dynamics of filament formation: the thermal instability in a sheared magnetic field, *Astrophys. J. (Letts.)*, **232**, L69, 1979.

Drake, J.F., L. Sparks, and G. Van Hoven, Radiative instabilities in a sheared magnetic field, *Phys. Fluids*, **31**, 813, 1988.

Hildner, E., The formation of solar quiescent prominences by condensation, *Solar Phys.*, **35**, 123, 1974.

Leroy, J.L., Observation of prominence magnetic fields, in *Dynamics and Structure of Quiescent Solar Prominences*, ed. E.R. Priest (Dordrecht: Kluwer), pp. 77-113, 1989.

Parker, E.N., Instability of thermal fields, *Astrophys. J.*, **117**, 431, 1953.

Sparks, L., and G. Van Hoven, Thermal instability of a radiative and resistive coronal plasma, *Astrophys. J.*, **333**, 953, 1988.

Van Hoven, G., and Y. Mok, The thermal instability in a sheared magnetic field: filament condensation with anisotropic heat conduction, *Astrophys. J.*, **282**, 267, 1984.

Van Hoven, G., L. Sparks, and T. Tachi, Ideal condensations due to perpendicular thermal conduction in a sheared magnetic field, *Astrophys. J.*, **300**, 249, 1986.

Van Hoven, G., L. Sparks, and D.D. Schnack, Nonlinear radiative condensation in a sheared magnetic field, *Astrophys. J. (Letts)*, **317**, L91, 1987.

FIBRIL STRUCTURE OF SOLAR PROMINENCES

J. L. Ballester
Department de Fisica, Universitat de les Illes Balears
E-07071 Palma de Mallorca, Spain
E. R. Priest
Mathematical Sciences Department, University of St. Andrews
St. Andrews KY16 9SS, Scotland, UK

Abstract. Limb observations of quiescent solar prominences have revealed them to be composed of many fine structures. Also observations in Hα and UV lines suggest that quiescent prominences are made up of many clusters of small-scale loops at different temperatures inclined to the filament axis, with the CIV structures more extended than the Hα ones.

Active-region prominences have been modelled previously as single cool loops along the prominence. However, other observations suggest that they could be interpreted in terms of many loops of plasma inclined to the filament. Our aim has been to take into account such observations and to construct a model for the fibril structure of solar prominences in terms of slender flux tubes, in which the main observed parameters of both quiescent and active-region prominences are reproduced.

Introduction

Prominences are cool structures (10^4 K) imbedded in the solar corona. They appear as bright features at the limb and dark features on the disk and have been classified morphologically in several ways, but there appear to be two basic long-lived types: Quiescent and Active-region prominences. Reviews of the observations and theories can be found in Tandberg-Hanssen [1974], Jensen et al. [1979], Priest [1982], Hirayama [1985], Poland [1986], Ballester and Priest [1988], Priest [1989]. Some observations of quiescent prominences show very fine structures in the body of the filament, suggesting they are composed of small scale structures.

Simon et al. [1986] studied velocities in a quiescent prominence in Hα and CIV. They concluded that the filament is composed of many small-scale loops anchored at many different footpoints that are not aligned along the filament axis. The observed vertical velocities and the perspective suggest material falling down to the observed cool prominence.

Demoulin et al [1987] deduced statistical sizes of 10^3 km and 10^4 km for the thickness and length of the threads which form a filament.

Engvold et al. [1987] studied a quiescent prominence seen in projection against the disk. From the study of the prominence-corona interface, they deduced that the fine structure of the cool core of the prominence indicates that the region may consist of thin magnetic flux ropes, oriented at an angle of 20 degrees with the long axis.

Active-region prominences have been modelled as cool loops along the prominence [Hood and Priest, 1980; Schmieder et al., 1985].

However, Malherbe et al. [1983] studied an active region filament and performed a statistical analysis of radial velocities. They found that the observations could be interpreted in terms of loops of moving plasma crossing the filament.

Bearing in mind the observed fine structure of prominences and the suggestion that they may be composed of small scale loops, our aim here has been to try to model them in terms of slender flux tubes.

Model and Basic Equations

We consider the fibril structure of the prominence as composed of flux tubes containing hot plasma ($T_i \approx T_c$) over most of their lengths and cool parts ($T_i \ll T_c$) near their summit representing the cool region of the prominence. We start with a hot flux tube and assume that a cool condensation appears near the top of the flux tube with a temperature profile shown schematically in Figure 1. This produces a downward anti-buoyancy force that must be balanced by other forces acting on the flux tube and the general shape of the structure therefore be as shown in Figure 2. Thus,

Geophysical Monograph 58

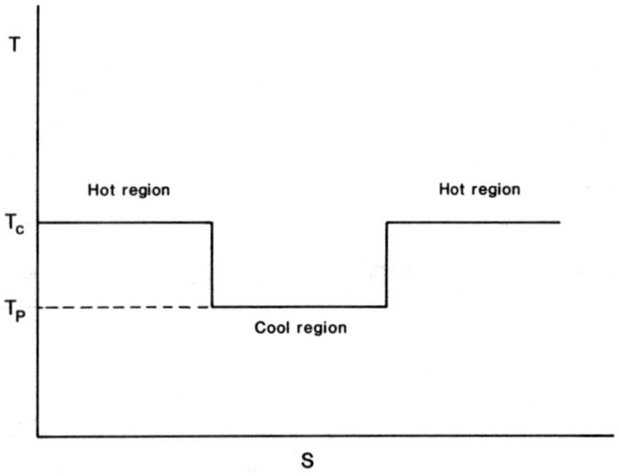

Fig. 1. Schematic of the temperature profile along the flux tube from one end to the other. T_c and T_p represent the coronal and prominence temperatures and s is the coordinate measured along the flux tube.

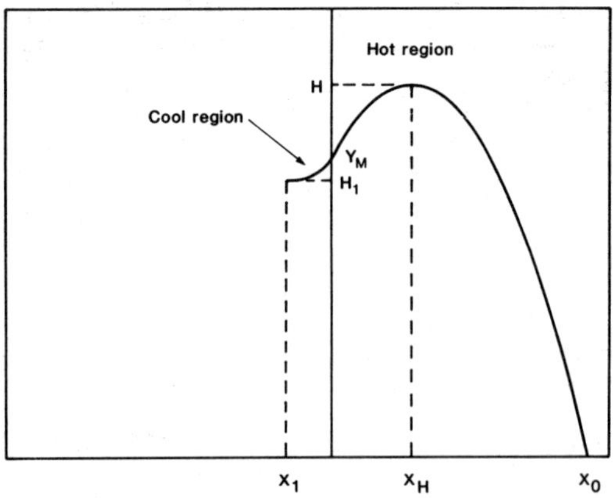

Fig. 2. Sketch of the fibril structure where (H, X_H) are the coordinates of the highest part of the hot flux tube, (X_0, 0) the coordinates of one of the feet, and (X_1, H_1) the coordinates of the lowest part of the cool flux tube. Y_m is the connection height between the hot and cool parts of the flux tube.

our first problem is to find the general equation that will allow us to determine the path of the flux tube.

We shall assume that the magnetic field is in equilibrium and that the plasma is in a state of hydrostatic equilibrium along the magnetic field lines. We are considering a scenario in which a hot coronal loop has been stretched out and has formed a dip near its summit and then, either the summit region has gone thermally unstable, or plasma has been ejected up from the footpoints and has settled down to a new equilibrium with cool plasma accumulated in the dip and the rest of the loop still being hot. Moreover, if the plasma velocities are subsonic in the final state then hydrostatic equilibrium is a reasonable approximation. Indeed the observed velocities in prominences [e.g., Schmieder et al., 1988] are highly subsonic. At the interface, in our model, between hot and cool parts of a loop the plasma is in equilibrium because of pressure balance and is Rayleigh-Taylor stable within the loop since the light plasma is resting above the denser plasma.

Starting from the general equation describing the equilibrium of a magnetic field in a plasma, with gravitational forces included, assuming that the global system is in equilibrium, so, total pressure must be continuous across the tube surface, and using as external field a potential magnetic field, we obtain:

$$\frac{1}{2}(1+X')\left(X'\frac{\partial B_E^2}{\partial y}-\frac{\partial B_E^2}{\partial x}\right)+\left(B_E^2+2\mu_0(P_E-P_I)\right)X'' \quad (1)$$
$$+\mu_0\left(\frac{P_I}{\Lambda_I}-\frac{P_E}{\Lambda_E}\right)X'(1+X^2)=0$$

where the first term is the external magnetic pressure, the second term is the loop tension and the third term is the buoyancy. For $T_e = T_i$, the equation (1) reduces to the general equation of Browning and Priest [1986]. For $B_e = 0$, reduces to that of parker [1975] and Spruit [1981], and with B_e = constant, it reduces to that of Browning and Priest [1984].

We relate the hot and cool parts of the flux tube imposing:

$$\rho^c_{im} = \rho^h_{im}\frac{T_i^h}{T_i^c}$$

where $\rho_{im}{}^h$ is the internal density of the hot flux tube at the connection point, $T_i{}^h$ is the internal temperature in the hot part of the flux tube and $T_i{}^c$ is the internal temperature in the cool part of the flux tube. Furthermore, we assume that the variation of the density inside the cool region is determined by hydrostatic equilibrium:
where Λ_p is the scale height in the cool region.

$$\rho^c = \rho^h_{im}\frac{T_i^h}{T_i^c}\exp\frac{(y_m-y)}{\lambda_p}$$

Fibril Structure with No External Field

If in the equation (1) we set $B_E = 0$, then it becomes:

$$2R(\rho_e T_e-\rho_i T_i)X'' + (\rho_i-\rho_e)gX'(1+X'^2) = 0$$

and for the particular case $T_i/T_e = 0.5$, we are able to obtain an analytical solution, which gives us the path of the hot part of the flux tube.

For the cool part, the general equation (1) becomes:

$$2R\left(\rho_{im}\frac{T_i^h}{T_p}\exp\left(\frac{(y_m-y)}{\lambda_p}\right)T_p-\rho_{eo}\exp\left(\frac{-y}{\lambda_e}\right)\right)\frac{d\theta}{dy}$$
$$=\left(\rho_{eo}\exp\left(\frac{-y}{\lambda_e}\right)-\rho_{im}\frac{T_i^h}{T_p}\exp\left(\frac{(y_m-y)}{\lambda_e}\right)\right)g\cot\theta$$

where T_p means prominence temperature.

Fibril Structure with an External Field

The two-dimensional external magnetic field we use here is a potential arcade having $B_x = B_0 \cos kx \exp(-ky)$, $B_y = B_0 \sin kx \exp(-ky)$. Then, the equation (1) becomes:

$$\frac{1}{2}(1+X'^2)X'[-2kB_0^2\exp(-2ky)$$
$$+2\mu_0 g(\rho_{io}\exp(-y/\lambda_i))-\rho_{eo}\exp(-y/\lambda_e)]$$
$$+[B_0^2\exp(-2ky)+2\mu_0 R(\rho_{eo}T_e\exp(-y/\lambda_e)$$
$$-\rho_{io}T_i\exp(-y/\lambda_i)]X''$$
$$=0$$

for the hot part. For the cool part it becomes:

$$2\tan\theta\frac{d\theta}{dy}[B_0^2\exp(-2ky)$$
$$+2\mu_0 RT_e(\rho_{eo}\exp(-y/\lambda_e)-\rho_{im}\exp\left(\frac{(y_m-y)}{\lambda_p}\right)]$$
$$=[-2kB_0^2\exp(-2ky)+2\mu_0 g\left(\frac{\rho_{im}T_i}{T_p}\right)$$
$$\exp\left(\frac{(y_m-y)}{\lambda_p}\right)-\rho_{eo}\exp(-y/\lambda_e)$$

In the two cases, we match the hot and cool solution imposing: a) $y = y_m$ at $x = 0$ and b) equality of slopes at $y = y_m$.

Supported Mass

The supported mass inside the cool part of the flux tube is given by:

$$\frac{M}{S} = \rho_{im}\frac{T_i^h}{T_p}\int_{H_1}^{y_m}\exp\left(\frac{(y_m-y)}{\lambda_p}\right)\sqrt{1+X'^2}\,dy$$

which must be performed numerically and furnishes us with an estimate of M/S for each cool flux tube. Assuming a thickness for the flux tubes and a length for the prominence we can then calculate the total mass in it.

Results

Our aim is to find the depression of the cool region calculated as the difference between y_m and H_1, while others are to obtain the width of this cool region and the mass contained in the cool flux tubes.

<u>Without an external field</u>

In this case, we have performed the calculations for two different values of x_0/Λ_e, three different values for the ratio ρ_{i0}/ρ_{e0} and two different values of T_e and T_i. Fixing a set of parameters we change the value of H (height of the hot part), which produces different values of y_m, and so we obtain different values for the depression, the width and the mass (Figures 3a-3c). A density of 10^{-13} kg/m^3, typical of quiet regions is adopted.

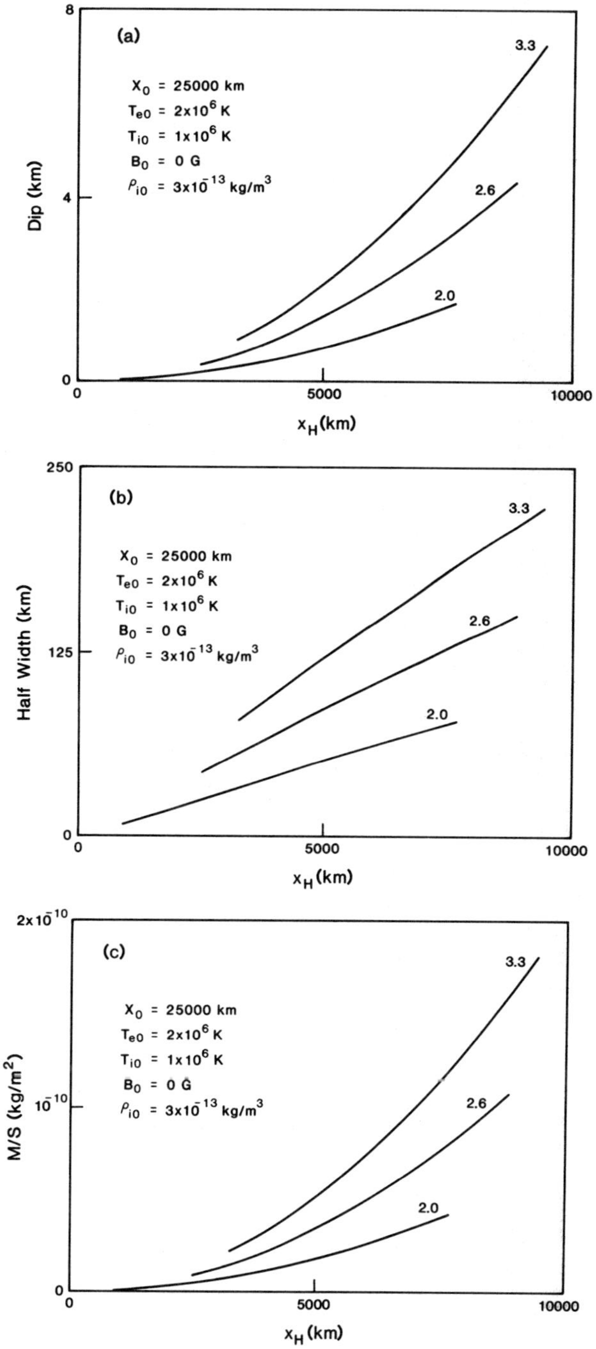

Fig. 3. (a) Dip, (b) half-width and (c) M/S ratio of cool region versus X_H with an external field. The curves are plotted for three different values of the density ratio.

For different T_e and T_i, the qualitative behavior of the solutions is the same. For given x_0, increasing ρ_{e0}/ρ_{i0} increases the dip, the half width, and the value M/S.

Lowering the external temperature, while keeping the same height of the hot part, x_h greater, which means that the depression and the width are larger too.

With an external field

In this case we have made a distinction between quiet and active regions, trying to model prominences that appear in those different situations.

We have performed the calculations for three different values of x_0/Λ_e and for ρ_{e0} and ρ_{i0} typical of quiet and active regions. In each of those cases, we use values of B_0 typical of quiet and active regions (Figures 4a-4c, 5a-5c). For the same x_0 and densities, increasing B_0 means an increase in the depression and the supported mass.

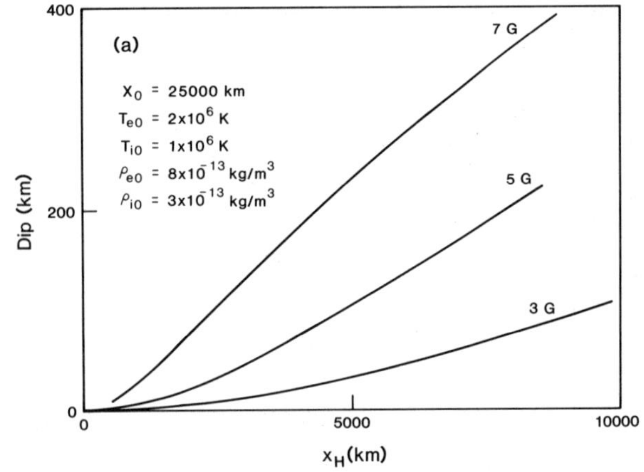

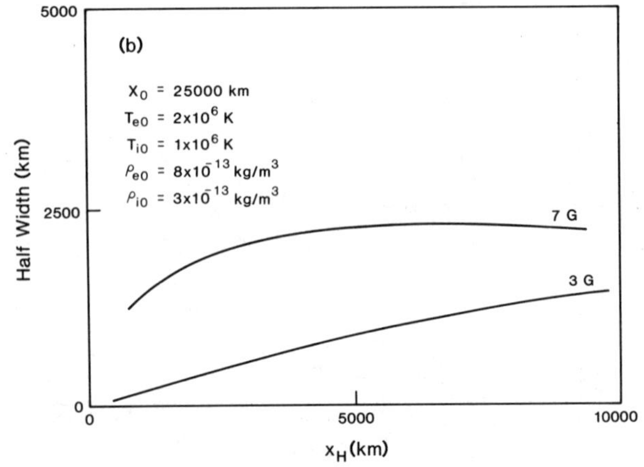

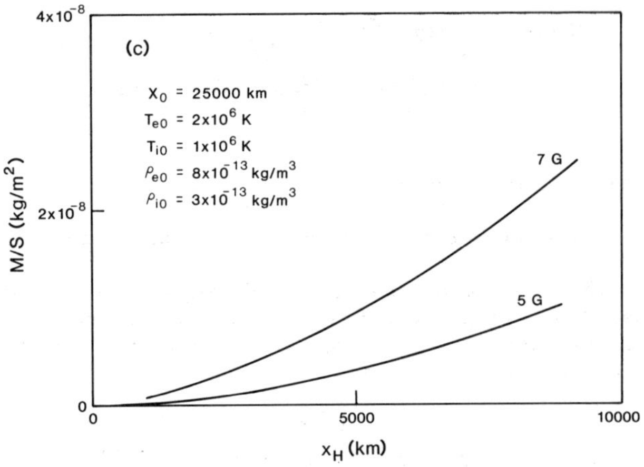

Fig. 4. (a) Dip, (b) half-width and (c) M/S ratio of cool region versus X_H with an external field. The densities are typical of a quiet region and the curves are labelled with the value of B_0.

The comparisons between cases with same densities, same field, same x_0 but different T_e and T_i, suggest that when T_e decreases we obtain a bigger depression, larger width and greater mass (Figures 6a-6c).

The variation of the half width of the cool region is rather complex as can be seen. For some values of B_0 below a critical one the half width decreases as x_H decreases, but above this critical value the behavior changes, raising the width up to a maximum and then decreasing. What happens is that the increase in B_0 produces a flattening in the shape of the cool part and the width increases. When x_H decreases, the value of y_m rises and at certain height, as the external field decreases with altitude, the behavior is reversed.

The mass of the cool region can be calculated assuming a length for the prominence. For example, taking a quiescent prominence in a quiet region, H = 20,000 km, depression = 300 km, M/S = 10^{-8}kg/m^2, assuming a thickness for the flux tube of 10^3 km and a length for the structure of

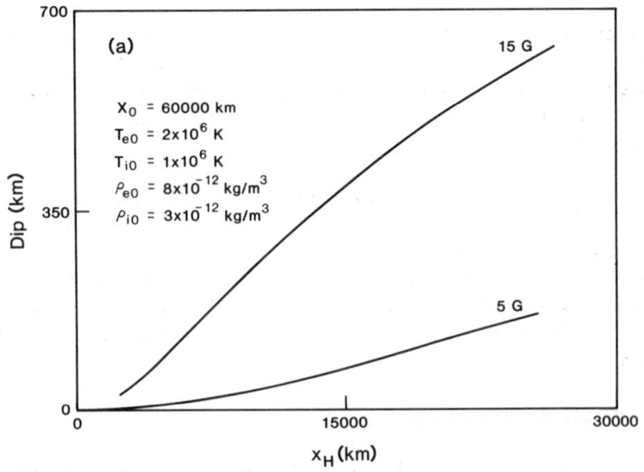

Fig. 5a-c. Flux tube properties for densities typical of active regions and including an external field.

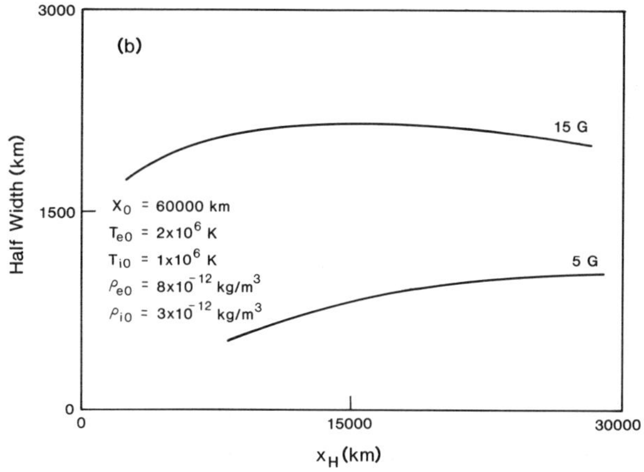

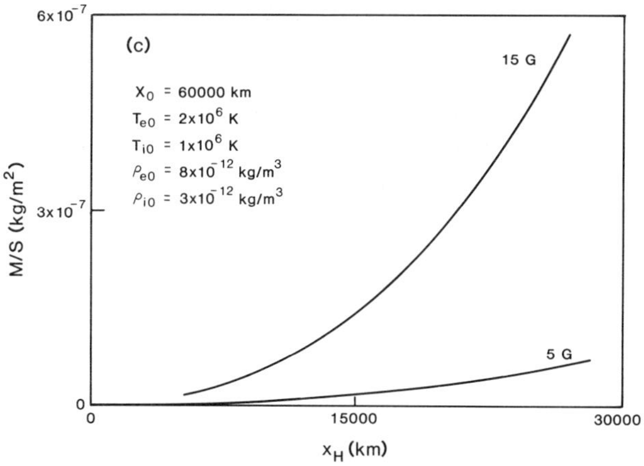

50,000 km, and B_0 = 5 G, the value of the mass is around 4 x 10^9 gm. For an active region, H = 10^4 km, depression = 590 km, M/S = 4 x 10^{-7} kg/m^2, length = 10,000 km, width of the cool region = 4,000 km, and B_0 = 20 G, the estimated mass is around 7 x 10^9 gm.

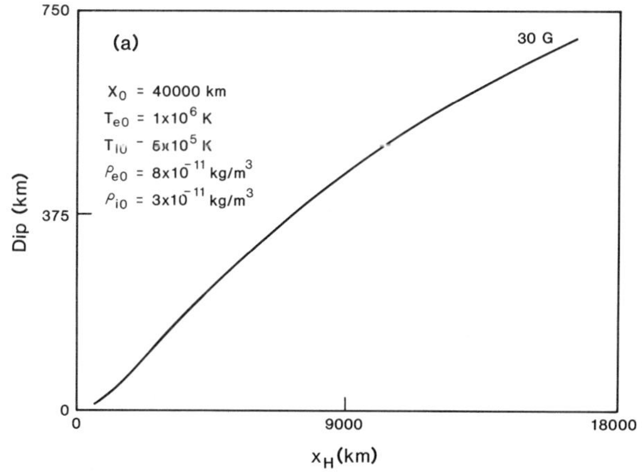

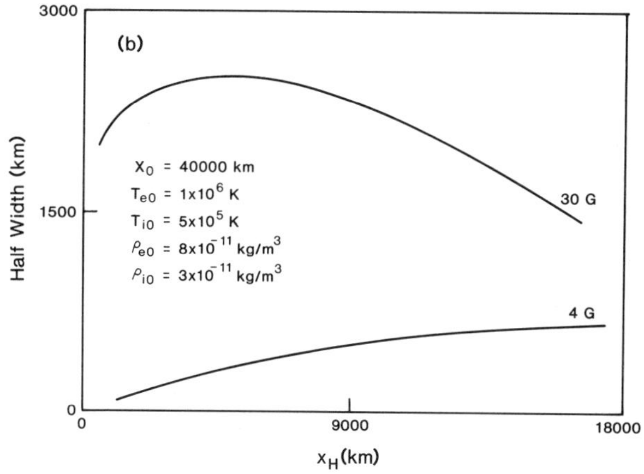

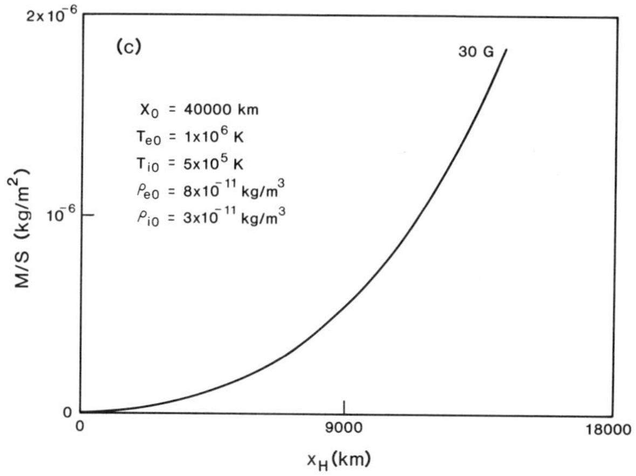

Fig. 6a-c. Flux tube properties for densities typical of active regions and including an external field.

Conclusions

Using the approximation of slender flux tubes, we have been able to construct a model for a structure consisting of a hot part and a depressed cool part in order to try to explain the observations of solar prominences that suggest they are composed of many small scale loops traversing the polarity inversion line. The most realistic results are obtained by including the effect of an external magnetic field in the corona. Then, for different values of internal and external densities and external magnetic field, representative of both quiet and active regions, we are able to reproduce realistic values for the width of the cool region and the mass contained in both quiescent and active-region prominences.

References

Ballester, J. L. and E. R. Priest, Dynamics and structure of solar prominences, Proceedings of Mallorca Workshop, Universitat de les Illes Balears Publications Service, 1988.

Browning, P. and E. R. Priest, The magnetic non-equilibrium of buoyant flux tubes in the solar corona, Solar Physics, 92, 173-188, 1984.

Browning, P. and E. R. Priest, The shape of buoyant coronal loops in a magnetic field and the eruption of coronal transients and prominences, Solar Physics, 106, 335-351, 1986.

Demoulin, P., M. A. Raadu, J. M. Malherbe and B. Schmieder, Fine structures in solar filaments I. Observations and thermal stability, Astron. Astrophys., 183, 142-150, 1987.

Engvold, O., O. Kjeldseth-Moe, J. D. F. Bartoe and G. Brueckner, Observations and modelling of the prominence/corona transition region, Proceedings of 21st ESLAB Symposium, ESA SP-275, 21-25, 1987.

Hirayama, T., Modern observations of solar prominences, Solar Physics, 100, 415-434, 1985.

Hood, A. W. and E. R. Priest, Are solar coronal loops in thermal equilibrium?, Astron. Astrophys., 87, 126, 1980.

Jensen, E., P. Maltby and F. Q. Orrall (eds.), Physics of Solar Prominences, IAU Colloq. 44, 1979.

Malherbe, J. M., B. Schmieder, E. Ribes and P. Mein, Dynamics of solar filaments II. Mass motions in an active region filament from Hα center to limb observations, Astron. Astrophys., 119, 197-206, 1983.

Parker, E., X-ray bright spots on the Sun and the non-equilibrium of a twisted flux rope in a stratified atmosphere, Astrophys. J., 201, 494-501, 1975.

Poland, A. I. (ed.), Coronal and Prominence Plasmas Workshop, NASA Conference Publication 2442, 1986.

Priest, E. R., Solar Magnetohydrodynamics, D. Reidel, 1982.

Priest, E. R. (ed.), Dynamics and Structure of Quiescent Solar Prominences, Kluwer, 1989.

Schmieder, B., J. M. Malherbe, A. I. Poland and G. Simon, Astron. Astrophys., 153, 64, 1985.

Schmieder, B., A. I. Poland, B. Thompson and P. Demoulin, Some dynamical aspects of a quiescent filament, Astron. Astrophys., 197, 281-288, 1988.

Simon, G., B. Schmieder, P. Demoulin and A. I. Poland, Dynamics of solar filaments VI. Center to limb study of Hα and CIV velocities in a quiescent filament, Astron. Astrophys., 166, 319-325, 1986.

Spruit, H. C., Motion of magnetic flux tubes in the solar convection zone and chromosphere, Astron. Astrophys., 98, 155-160, 1981.

Tandberg-Hanssen, E., Solar Prominences, D. Reidel, 1974.

STRUCTURE OF TWO-DIMENSIONAL MAGNETOSTATIC EQUILIBRIA IN THE PRESENCE OF GRAVITY

T. Amari* and J. J. Aly**

*Applied Mathematics Division, The University, St Andrews, Fift, Scotland
**Service d'Astrophysique, Centre d'etudes Nucléaires de Saclay, 91191 Gif-sur-Yvette Cedex, France

Abstract. We report the results of an analytical study of several 2D (x-invariant) models describing the equilibrium of a plasma occupying the half-space {z>0} and submitted to a magnetic field and to a constant vertical gravitational field. These models differ from each other by the assumptions which are made on the way matter is spatially distributed. In models 1 and 2 (directly aimed to describe the magnetic support of solar prominences), the plasma is taken to be concentrated in a vertical sheet and in a filament, respectively, the external supporting field being a constant-α force-free field, while in model 3, the plasma is taken to occupy all the half-space and to have a uniform temperature. For each of these models, we establish general results concerning in particular the existence and the uniqueness of solutions, thus bringing about several interesting "non-equilibrium" phenomena.

1. Introduction

Constructing a complete model of solar prominences appears to be a formidable task, as one has to deal with a system in which the equations for the mechanical equilibrium are strongly coupled to the equations for energy transport. Therefore, most authors up to now have tried to solve only one half of the problem, studying either the thermodynamics of the plasma in a given magnetic field, or the equilibrium of a plasma in a magnetic field under a simplifying assumption which is substituted for the correct energy equation. The work reported in this paper belongs to this second category. What we have done indeed is to build up several 2D models describing the support of a plasma against gravity by a magnetic field, these models differing from each other by an a priori assumption about the way matter is distributed over space.

In all our models, the corona is represented by the half-space {z>0} and is assumed to be invariant under translations along the x-axis ($\partial/\partial x = 0$). Therefore, the magnetic field may be written as

$$\underline{B}(y,z) = \nabla A(y,z) \times \hat{\underline{x}} + B_x(y,z)\hat{\underline{x}} \qquad (1)$$

where A is a scalar potential. On the other hand, we take the gravitational field to be vertical and constant:

$$\underline{g} = -g\hat{\underline{z}} \quad (g > 0). \qquad (2)$$

In our models 1 and 2, the plasma is assumed to be cold and concentrated in a thin sheet and a filament, respectively, while in our model 3, matter is distributed over all the corona with a uniform temperature (although this is certainly not a good assumption as far as prominences are concerned, it leads to a model which is interesting for giving insight into the mathematics).

2. Model 1: equilibrium of a massive current sheet in a constant-α force-free field

Geophysical Monograph 58

In the first model, we consider a massive sheet extending in the plane $\{y=0\}$ between the heights $z = a$ and $z = b = a+h$; a surface current $\underline{j} = (6\varepsilon\, I/h^3)(b-z)(z-a)\hat{\underline{x}}$ $(I > 0,\ \varepsilon = \pm 1)$ flows through the sheet, and is subjected to a constant-α force-free field $\underline{B}$ $(B_x = \alpha A)$ occupying the part of $\{z>0\}$ between the two planes $\{y = \pm L/2\}$; the potential A of $\underline{B}$ satisfies

$$-\Delta A = \alpha^2 A + (4\pi/c)\, j\delta(y), \qquad (3)$$

where $0 \le \alpha < \pi/L$, and the boundary conditions: $A(y,0) = B_0 L \cos(\pi y/L)$ and $A(\pm L/2, z) = 0$. The field $\underline{B}(y,z)$ is first determined as a function of the parameters $(a,h,\varepsilon,I,\alpha)$ by solving the previous problem for A; thus one gives to the sheet a mass density $\sigma(z) = (2\pi g)^{-1}(B_y B_z)(0^+,z)$ and one obtains a physically satisfying model of equilibrium if the condition $\sigma(z) > 0$ holds. It turns out that (Amari and Aly, 1989a):

i) if $\varepsilon = +1$, the necessary condition for the equilibrium at $(0,b)$ is never satisfied, no sheet can be in equilibrium;

ii) if $\varepsilon = -1$, an equilibrium exists only if $I \le I_c\ (a,h,\alpha) \le I_c{}^*$; correspondingly, the total mass per unit of x-length m of the sheet satisfies $m(a,h,\alpha,I) \le m_c\ (a,h,\alpha) \le m_c{}^*$; thus, a too large amount of condensed mass cannot find an equilibrium (i.e. adequate values of a and h) in a given field. On the other hand, the topology of the field lines is always of the "normal" type (Figure 1).

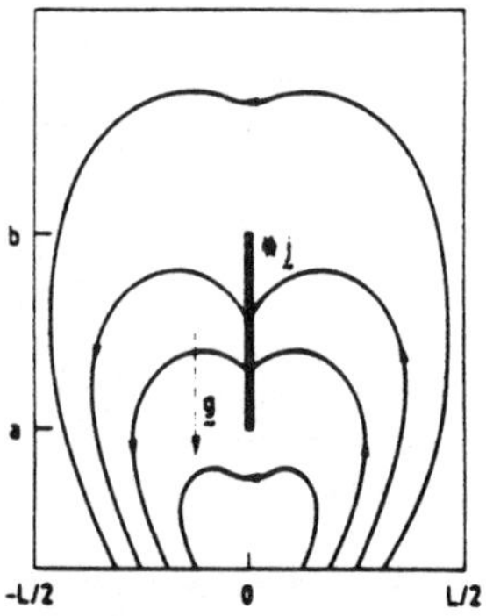

Figure 1. Topology of the lines of a field supporting a massive current sheet.

3. Model 2: equilibrium of a line current in a constant-α force-free field

The second model is obtained by making h=0 in the previous one. Considering now m as a parameter along with ε, I and α, one gets the following results (Amari and Aly (1989b):

i) there is a unique equilibrium position a $(\varepsilon, I, \alpha, m)$ for any values of the parameters (no loss of equilibrium); a is obtained by solving the equation

$$\frac{I}{c}\left\{\frac{4\pi I}{Lc}\sum_{0}^{\infty} e^{-2\gamma_{2p+1}a} - \varepsilon B_0 L \gamma_1 e^{-\gamma_1 a}\right\} = mg \qquad (4)$$

where $\gamma_n{}^2 = n^2\pi^2 L^{-2} - \alpha^2$; it must be noted, however, that a non-equilibrium phenomenon of the Kuperus-Van Tend type (Kuperus and Van Tend, 1981) may be recovered if a more complex expression for $A(y,0)$ is assumed (Demoulin and Priest, 1988);

ii) $\partial a/\partial I > 0$; $\partial a/\partial m < 0$; for $\varepsilon = +1$, $\partial a/\partial\alpha > 0$ and $\lim_{\alpha\to\pi/L} a = \infty$; for $\varepsilon = -1$, $\partial a/\partial\alpha)(m_c(I,\alpha)-m) \ge 0$, where m_c is a computable critical value;

iii) The topology of the lines is as shown on Figure 2.

As for the interpretation of this model:

i) for $\varepsilon = -1$, we may consider the filament to represent a sheet of small vertical extent $(h << a)$ when I is not too large;

ii) we may also $(\varepsilon = \pm 1$, m not too large) consider the filament to just represent a concentration of force- free currents, the matter being actually concentrated in a sheet located under it and locally perturbing very little the field (the "desingularization" procedure, which allows us to consider the singular field of the filament as the limit of a sequence of regular non-linear force-free fields, is studied in detail in a forthcoming paper).

4. Model 3: equilibrium of an isothermal plasma

In the general case where matter is continuously distributed over space, we need to add to the equation of equilibrium an equation of energy. As a first step towards solving the resulting system of coupled equations, we have considered in detail the problem of the equilibrium of a plasma held at a constant temperature T. Then one is led to consider the equation

$$-\Delta A = \frac{d}{dA}\left(\frac{B_x^2(A)}{2} + 4\pi p_0(A)\, e^{-\beta z}\right) \qquad (5)$$

where $\beta = kT/m_p g$ (m_p = proton mass) and $p_0(A)\, e^{-\beta z}$ represents the pressure of the plasma at a point located at height z on a field line labelled by the value A of the potential. In our study, we set: $4\pi\, dp_0/dA = \lambda\, f_1(A)$ and $dB_x^2/dA = 2\, \mu\, f_2(A)$, where $\lambda \geq 0$ and $\mu \geq 0$ are parameters, and the f_i are given functions such that $0 < f_i$ for $0 < A < A_i$ and $f_i = 0$ for $A \leq 0$ or $A_i \leq A$. On the other hand, we impose the boundary condition $A(y,0) = g(y)$ with $g \geq 0$ and $\lim_{|y|\to\infty} g(y) = 0$,

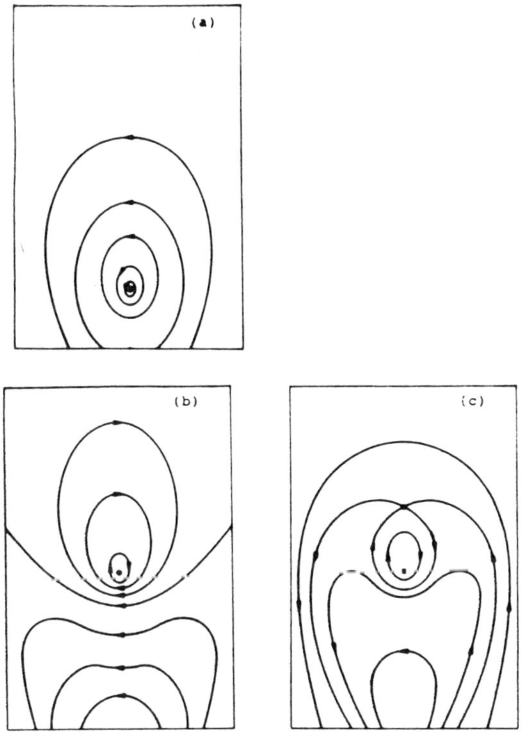

Figures 2. Topology of the lines of a field supporting a filament with: (a) $\varepsilon = +1$; (b) $\varepsilon = -1$; $m < m_c'(\alpha, I)$; (c) $\varepsilon = -1$, $m > m_c'(\alpha, I)$ (m_c' is a computable critical value).

and look only for bounded solutions $A(y,z) \leq M$.

The case $\lambda = 0$ has been studied by Heyvaerts et al. (1983). We simply report here the results we have obtained for $\mu=0$ - i.e. for an unsheared magnetic field (see Amari and Aly, 1989c for more details); the case $\lambda > 0$ and $\mu > 0$ is still

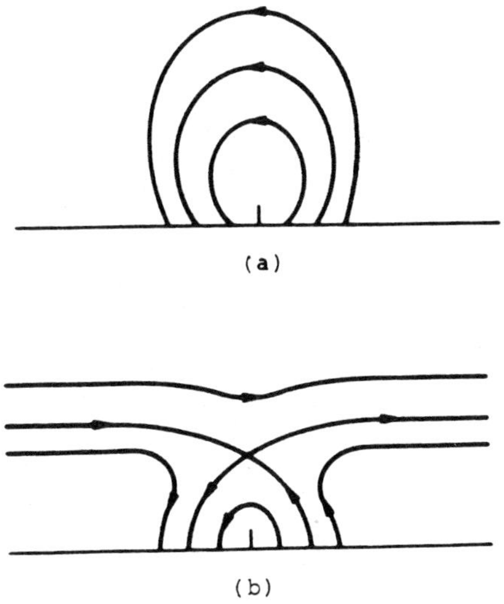

Figure 3. Topology of the lines of a field holding an isothermal plasma in equilibrium.

(a) lines of A_λ^- for $\lambda < \lambda_2^*$; (b) lines of A_λ for $\lambda > \lambda_2^*$ (the potential field A_0 - obtained for $\lambda = 0$ - is assumed to have an arcade-like structure).

under study, but we hope to present our results very soon:

i) $\forall\lambda$, there is always a maximal solution A_λ^+ and a minimal one A_λ^- (which actually may coincide);

ii) for $0 \leq \lambda < \lambda_1^*$, the solution is unique ($A_\lambda^- = A_\lambda^+$);

iii) there is a "closed topology" solution tending to zero at infinity for $0 \leq \lambda < \lambda_2^*$, but no such solution for $\lambda > \lambda_2^*$ (non-equilibrium behaviour); for $\lambda > \lambda_2^*$, the solutions have an open topology (see the example shown on Figure 3);

iv) for λ small enough, the total energy of A_λ^- is finite and a strictly increasing function of λ; on the other hand, for any other finite energy solution A_λ, one has $C[A_\lambda^-] < C[A_\lambda]$.

We have also investigated the more complicated case when T is only assumed to be constant along a given field line - then $T = T(A)$ and $\beta = \beta(A)$ - and proved results similar to those above.

5. Conclusion

We have reported here some of the results we have recently obtained in our study of 2D x-invariant magnetohydrostatic equilibria in the presence of gravity. In order to approach a more realistic representation of the support of prominences, we are currently trying to generalize the models we have just described. We would like in particular to introduce a component of the field along x with the functional form $B_x(A)$ not being given a priori, but being determined by shearing motions `imposed on {z=0}. We have already obtained some preliminary results in this direction and hope to report them soon.

References

Amari, T., and Aly, J.J., Extended massive current sheet in a two-dimensional constant-α force-free field: a model for quiescent prominences. I - Theory, Astron. Astrophys., in press, 1989a.

Amari, T., and Aly, J.J., Interaction between a line current and a two-dimensional constant-a force-free field: an analytical model for quiescent prominences, Astron. Astrophys., 208, 261, 1989b.

Amari, T., and Aly, J.J., Two-dimensional isothermal magnetostatic equilibria in a gravitational field. I - Unsheared equilibria, Astron. Astrophys., 208, 361, 1989c.

Démoulin, P., and Priest, E.R., Instability of a prominence supported in a linear force-free field, Astron. Astrophys., 206, 336, 1988.

Heyvaerts, J., Lasry, J.M., Schatzman, M., and Witomsky, P., Blowing up of a two-dimensional magnetohydrostatic equilibria by an increase of electric current on pressure, Astron. Astrophys., 111, 104, 1982.

Kuperus, M., and Van Tend, W., Solar Phys. 71, 125, 1981.

ON DRIVING THE ERUPTION OF A SOLAR FILAMENT

V. Gaizauskas

Herzberg Institute of Astrophysics, National Research Council of Canada
Ottawa, Canada K1A OR6

Abstract. The evolution of an active-region filament has been followed over its lifetime of 5 days. Its activation on one of those days was observed in detail by a spacecraft and ground-based telescopes. Impulsive axial flows along the filament, its untwisting and rapid expulsion all precede the eruption of a two-ribbon flare directly beneath its rest position. Local magnetic changes are ruled out by the observations as the origin of this dynamism. The evolution of the magnetic flux cells adjacent to either side of the disrupted filament shows prominent, steady changes remote from the filament for days. The filament disruption and subsequent flare can be reasonably explained by a gradual increase beyond a critical threshold of field-aligned currents generated by the expansion, shifting, and contraction of bipolar regions at the separator between adjacent flux cells.

Introduction

A characteristic chromospheric feature of mature solar active regions is a narrow dark filament threaded over the polarity inversion line between patches of opposite magnetic polarities. The exterior plasma of a stable active-region filament can sometimes be resolved into fine strands executing a twist along the length of the structure, giving it the appearance of a magnetic rope. Superposition of high-resolution filtergrams on photospheric magnetograms shows that footpoints at opposite ends of active region filaments are embedded in magnetic fields of opposite polarity. It is commonly supposed, therefore, that the braided rope is created from an arcade of loops which are originally transverse to the polarity inversion line but are then pulled into their elongated configuration by a shearing motion as adjacent patches of oppositely directed fields slide past each other. Vortical motions at the footpoints are assumed to braid the strands into a rope.

The support and stability of filaments are subjects of great theoretical interest owing to the frequent association of erupting filaments with two-ribbon flares. It has been proposed that reconnection leading to a flare might be initiated spontaneously in a sheared, twisted, magnetic arcade when the shear, twist, height, or length of the arcade exceed critical values [Priest, 1985, and references therein]. Or reconnection might be triggered by interaction between the filament and a separate flux system which either emerges from beneath the photosphere or moves horizontally and pushes against the filament [Heyvaerts et al., 1977].

During the past solar maximum, attempts to test these theories against observations of changes in fine-scale magnetic structures have met with limited success [Gaizauskas, 1989]. The purpose of this paper is to draw attention to large-scale aspects of active-region growth and their likely influence on the formation and stability of filamentary ropes.

Bipolar active regions rarely exist in isolation; they commonly cluster together in activity complexes. A complex is maintained by repeated injections of flux in the form of many active regions which form and disappear in rapid succession as closely spaced but distinct bipolar units [Gaizauskas et al., 1983]. During the active lifetime of a large complex, its total magnetic flux remains steady. Since most of the magnetic flux remains within the complex, this flux must emerge and be removed locally at the same rate. Individual bipolar units within an activity complex can be recognized provided that the history of the cluster can be traced continuously. What is often loosely called an 'active bipolar region' is likely to be a cluster, perhaps tightly packed, formed by at least one older with a newer bipolar region, i.e. a quadrupole. A large cluster of sunspots is a multipolar structure consisting of many interlocked bipolar units. The polarity inversion lines separating pairs of bipoles are sites where 'active-region' filaments can form.

Geophysical Monograph 58

It is observationally confirmed that filaments form between active regions as well as inside them [Martin, 1973]. A survey of two years of prominence data by Tang [1987] shows, in fact, that substantially more filaments are formed on polarity inversion lines between bipolar regions than on inversion lines inside bipolar regions. The position taken in this paper is that filaments form inside single bipolar regions when the region reaches an advanced stage of decay. Otherwise, filaments said to be 'inside' active regions are usually between individual pairs of bipolar regions which comprise a single cluster. This distinction is important because misconceptions could arise about the way the magnetic topology at internal boundaries of the cluster changes if, for example, the cluster is treated as a single bipolar entity instead of an assembly of individual bipolar units.

Some association thus seems likely between active-region filaments and the separators which must necessarily form between the interpenetrating cells of magnetic flux that comprise an activity complex [Baum and Bratenahl, 1980]. An interesting property of separators is that field-aligned currents can be driven along them by changing the flux distribution in the adjacent magnetic cells, i.e., by the addition or removal of flux, or by the rearrangement of sunspot positions. Flare onset can occur when an increasing current driven by such changes crosses a stability threshold [Bratenahl and Baum, 1976; Hénoux and Somov, 1987; Low and Wolfson, 1988]. No local magnetic changes need be invoked at the site of breakdown; the entire process can be driven by remote changes. Below we explore a specific case of the formation and eruption of an active-region filament in part of a large activity complex to see if the above concepts are applicable.

Specific Case of Filament Formation and Eruption

A two-ribbon flare with impulsive phase at 15:51 UT on 25 June 1980 was preceded a few minutes earlier by the disruption of a short filament lying over a segment of the irregular polarity boundary between active regions 2522 and 2530 (Boulder numbering system). This activity was jointly observed by the VLA, the SMM spacecraft, and ground-based observatories [Kundu et al., 1985]. The flare occurred in the middle of 7 hours of continuous Hα observations at the Ottawa River Solar Observatory (ORSO). The development of the activity complex surrounding the filament can be traced from ORSO data for 5 consecutive days ending on the day of the flare.

Figure 1 schematizes the growth and interaction of three bipolar regions. They were located at the SE end of a major activity complex which had 17 new bipolar regions form inside it during its disk passage in June 1980 [Martin et al., 1982]. Regions designated by I, II, III are, respectively NOAA 2519, 2522, and 2530. Day D, on which region III emerges, is 21 June. Region I undergoes little change during the period represented here, and is therefore used to register the field of view on successive days.

On D-2, region II is a mature EFR with a large leading spot of negative polarity. There are still arch filaments linking this -ve spot with the highly fragmented following (+) spot; the arches of the AFS traverse the polarity inversion line inside region II orthogonally. An existing large prominence lies eastward of regions I and II, close to the solar limb.

On day D, region III emerges swiftly on the eastern side of the pre-existing large prominence which is now labelled F_1. Region II has expanded, and what is left of AFS structure now links the +ve spot with small spots of -ve polarity in the middle of region II.

By D+1, region III has expanded greatly. Its leading -ve spot continues to advance westward at 0.1 to 0.2 km/s throughout the period depicted in Figure 1. As this spot approaches region II, filament F_1 is drawn along like a shield to isolate the two regions from each other. On this same day smaller amounts of flux of mixed polarity keep appearing in the mid-sections of regions II and III. Numerous small flares erupt around these small-scale emergences on succeeding days. For clarity, the bigger of the small spots are shown for region II only. The magnetic alignment of these small spots is reversed from the normal polarity order in this hemisphere of the sun. They move past each other at high speed, at times reaching the exceptional value of 0.6 km/s [Dezso et al.,1984]. Filament F_2 forms on D+1. There is no suggestion in the ORSO data that its growth results from the shearing of an arcade of loops initially oriented transversely to the final trajectory of F_2. Instead, F_2 develops from a thickening, darkening, and slight distortion of one of several pre-existing dark fibrils between the +ve trailer of region II and the -ve leader of III.

On D+2 and succeeding days, the polarity inversion line between the moving small spots inside region II becomes highly distorted. This complexity is related to the appearance of additional small concentrations of -ve polarity in the middle of region II.

By D+4 magnetic flux begins to diminish in the middle of region II. The strong relative motions of the small spots remaining there also ceases. On the region II side of F_2, therefore, flux emerging adjacent to the filament is no longer a factor for its disruption. Substantial flux continues however to emerge on the penumbral rims of the -ve leader spot of region II and of the +ve trailer spot of region III on D+4 and D+5 (not shown).

The eruption of F_2 on D+4 is viewed on the disk with sufficient perspective that horizontal motions can be distinguished from radial ones. The first stage of the eruption is the slow outward displacement for 2 hours of the filament at uniform velocity; Doppler shifts at the ends of the filament during this period suggest that

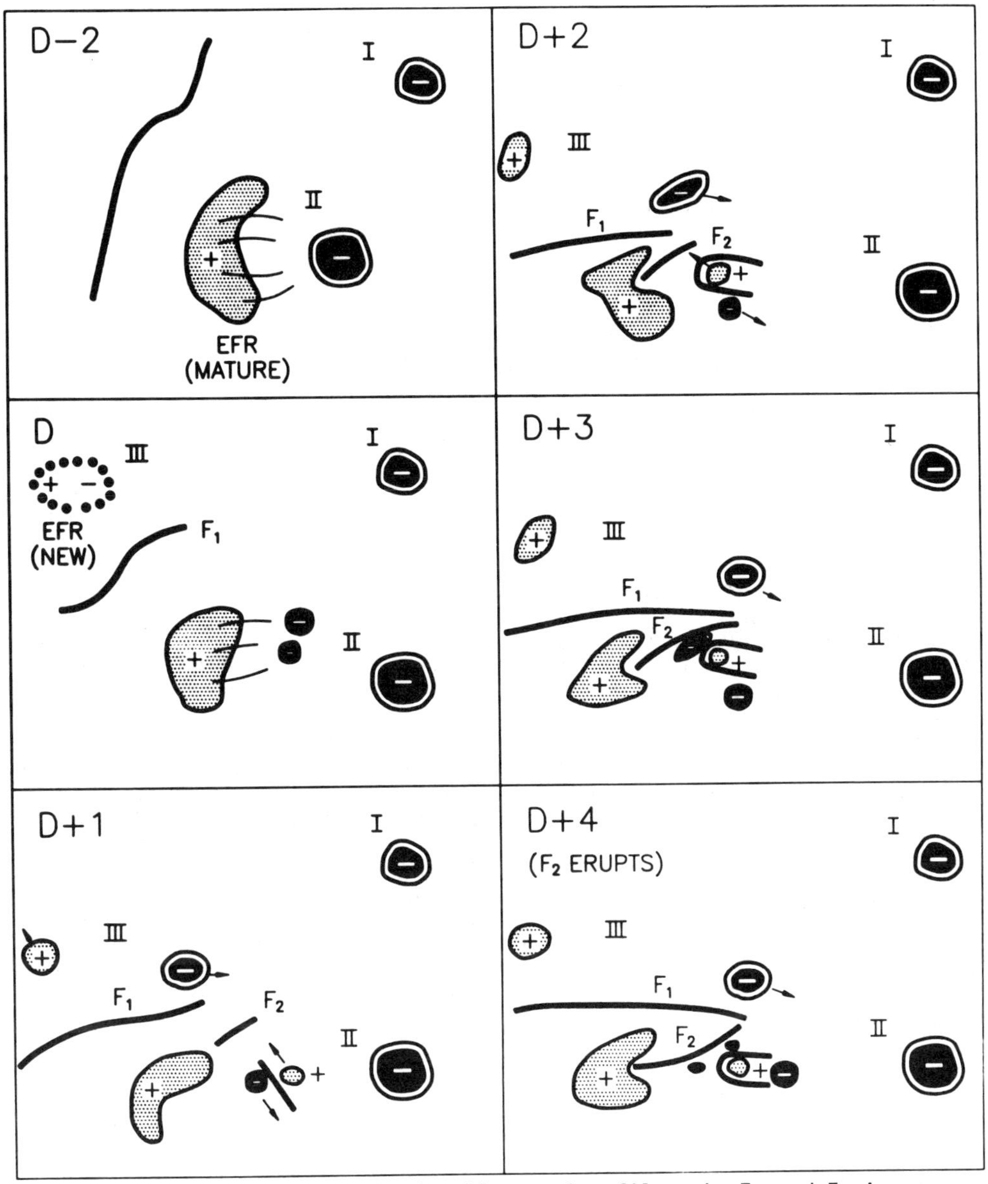

Fig. 1. Daily development of active region filaments F_1 and F_2 in an activity complex comprising large bipolar regions I, II, and III. Chromospheric filaments discussed in the text are indicated by heavy solid lines. Spots of leading (-ve) magnetic polarity are black; trailing polarities (+ve) are stippled (the fragmented trailer spots in region II are enclosed by a single contour). Features are traced from photographs and registered on successive days against region I. Each panel represents an area 0.13 x 0.2 Mm^2 and is oriented with North at the top, West to the right.

matter is draining from its midpoint toward both ends. Beginning 20 mins preflare, there are three episodes of enhanced axial flow of several mins duration each along F_2. During the second episode, the outward motion stops while the filament partially untwists and kinks. At the same time bright Hα kernels, not directly related to the subsequent flare, appear for a couple of minutes at the rest position of F_2, presumably beneath the rising and untwisting structure. The outward motion resumes within a few minutes but at much higher velocity. The distorted filament disrupts completely 2 to 3 minutes before the onset of a two-ribbon flare.

The restoration of the filament is surprisingly swift. Within 20 mins of the impulsive phase, the filament is completely reconstituted at the position it occupied before it began its slow rise more than 2 hrs earlier. It is significant that the restored filament immediately shows more signs, not less, of internal braiding than its earlier incarnation did in 3½ hours preceding the flare. Changes in the shape of F_2 proceed slowly thereafter until, about 5 hrs later, its western half fades away. The remainder merges with fibrils just south of the vanished portion. F_1 is involved in a minor disruption and subflare during this period. On the next day, several subflares erupt along the new trajectory adopted by the former F_2.

Discussion

In this example, the flare cannot cause a filament eruption which precedes it. The progression of events indicates that that the rising filament passed a critical threshold for stability, kinked, and untwisted in a way that allowed initial reconnection beneath it. We can speculate that the preflare Hα kernels of short duration signify that an early stage of rapid reconnection of magnetic field beneath the kinking filament precedes by a few minutes a later more energetic stage when the bulk of the filament is ejected. The rapid reappearance of the filament exactly in its original preflare location but in a braided form is unexpected. The stressed and twisted structure did not erupt again. It slowly faded away even though its outward appearance suggested a more complex magnetic configuration than its preflare manifestation. We can infer that the excess energy stored in this filament prior to the flare did not arise primarily from a twisting action at the footpoints.

It is impossible to account for the original activation of the filament, two hours before the main event, by means of a highly localized transient event. Furthermore, the very rapid reformation of F_2 after the impulsive phase of the flare, and its exact superposition on its original trajectory, argue in favour of a remote driver for the disruption rather than a local change which would alter that trajectory. The birth of the filament contains a direct clue about its activity. F_2 forms almost in situ as though F_1 is no longer an adequate barrier between II and III. It lies not on the main boundary between two growing bipolar active regions but on a secondary boundary between a small, oppositely oriented bipolar region and the two larger ones.

A physical connection between active-region filaments and separators is still a matter of conjecture. But the development of F_1 and F_2 make it reasonable to suppose that both are located near enough to separators that some parts of the magnetic structures supporting these filaments may intercept separators. Variations in the magnetic flux of the cells interacting at a separator will drive reconnection along it which can in turn perturb the equilibrium of the nearby filaments. The existence of a large-scale magnetic cell enveloping regions II and III is already implied in the observations of Kundu et al. [1984] of large X-ray structures propagating between the two regions across F_1 and F_2 in association with small flares on 24 and 25 June. In the case of F_2, a small magnetic cell interposes itself just where two much larger cells are already interpenetrating. The small cell loses flux on the day of the flare and the day after, while both of the large cells still gain new flux at locations remote from the activated filament.

The steady drainage of matter out of F_2 prior to its disruption leads us to speculate that slow reconnection across a boundary containing F_2 is pinching the plasma as well as heating it so that the filament slowly rises. Once the height of the filament exceeds a critical threshold, the instabilities discussed in Priest [1985] come into play and result in the observed enhanced axial flows along F_2 as well as its disruption.

Conclusions

A specific example of a filament activation preceding a flare, for which no local energy source could be observed, is consistent with the impulsive flux transfer model for flares developed by Bratenahl and Baum [1976]. The natural tendency for bipolar active regions to grow in clusters creates multiple boundaries between interlocked magnetic cells. Reconnection in field-aligned current sheets at these boundaries can be driven by the growth and collapse as observed of adjacent bipolar members of the cluster. Continuity in the observations for days is essential for tracking the interaction between members and for identifying the boundaries most sensitive to magnetic merging. In the case discussed above, the flare is secondary to a filament activation with a gradual onset indicative of slow reconnection.

Questions and Answers

Wolfson: You've emphasized the importance of the polarity reversal line in filaments. Yet, you show a region of positive polarity growing around a neutral line. What's going on here?

Gaizauskas: In simplifying my sketches, I have deliberately omitted many small islands of both magnetic polarities which are adjacent to filament F2 and which evolve in complicated ways from day to day. The general rule that filaments thread their way precisely between opposing polarities is not broken, even though these crude sketches may suggest anomalies such as you have detected.

Priest: Comments:

(1) The apparently highly twisted structure in the erupting prominence known as "grandpa" may not represent field lines with such a large twist.

Indeed, near one leg there is a suggestion of a structure (perhaps indicating the magnetic field) which has a much smaller twist.

(2) In the numerical experiment of Mikic, the field lines near the axis were extremely twisted, much more so than the observed twisting in filaments, and so one may at first sight think that the experiment is not relevant to reality. However, a plot of field lines further from the axis would have shown much less twist, similar to that seen in observed prominences.

(3) Your lack of an observed trigger for the flare is not too surprising. From a theoretical viewpoint, a magnetic arcade may evolve through a series of equilibria towards a point of nonequilibrium or instability. Thus, if the environment is quiet, the system may reach the threshold point before erupting spontaneously. But, if there is an external disturbance due to, for example, emerging flux, the system may be triggered to erupt just before it reaches the threshold.

Acknowledgments. Daily magnetograms from the National Solar Observatory used in this study were kindly provided by Harrison Jones. The extension of the study to the late post-flare phase was made possible with Hα filtergrams from Caltech supplied by Sara Martin.

References

Bratenhal, A. and P.J. Baum, Impulsive flux transfer events and solar flares, Geophys. J.R. Astr. Soc., 46, 259-293, 1976.

Baum, P.J. and A. Bratenhal, Flux linkages of bipolar sunspot groups: a computer study, Solar Phys. 67, 245-258, 1980.

Dezso, L., G. Csepura, O. Gerlei, A. Kovacs, and I. Nagy, Sunspot motions and magnetic shears as precursor of flares, Adv. Space Res., 4, No.7, 57-60, 1984.

Gaizauskas, V., K.L. Harvey, J.W. Harvey, and C. Zwaan, Large-scale patterns formed by solar active regions during the ascending phase of cycle 21, Astrophys. J., 265, 1056-1065, 1983.

Gaizauskas, V., Preflare activity, Solar Phys.,(in press), 1989.

Hénoux, J.-C. and B.V. Somov, Generation and structure of the electric currents in a flaring activity complex, Astron. Astrophys., 185, 306-314, 1987.

Heyvaerts, J., E.R. Priest, and D.M. Rust, An emerging flux model for the solar flare phenomenon, Astrophys. J., 216, 123-137, 1977.

Kundu, M.R., V. Gaizauskas, B.E. Woodgate, E.J. Schmahl, R. Shine and H.P. Jones, A study of flare buildup from simultaneous observations in microwave, H-alpha, and UV wavelengths, Astrophys J. Suppl., 57, 621-630, 1985.

Kundu, M.R., M.E. Machado, F.T. Erskine, M.G. Rovira, and E.J. Schmahl, Microwave, soft and hard X-ray imaging observations of two solar flares, Astron. Astrophys., 132, 241-252, 1984.

Low, B.C. and R. Wolfson, Spontaneous formation of electric current sheets and the origin of solar flares, Astrophys. J., 324, 574-581, 1988

Martin, S.F., The evolution of prominences and their relationship to active centers, Solar Phys., 31, 3-21, 1973.

Martin, S.F., L. Dezso, A. Antalova, A. Kucera, and K.L. Harvey, Emerging magnetic flux, flares and filaments - FBS Interval 16-23 June 1980, Adv. Space Res., 2, No.11, 39-51, 1982.

Priest. E.R., The magnetohydrodynamics of current sheets, Rep. Prog. Phys., 48, 955-1090,1985.

Tang, F., Quiescent prominences - where are they formed?, Solar Phys., 107, 233-237, 1987.

HELICAL FLUX ROPES IN SOLAR PROMINENCES

P. C. H. Martens and A. A. van Ballegooijen
Harvard-Smithsonian Center for Astrophysics

Abstract. We show that flux cancellation at the neutral line in a sheared magnetic arcade can lead to the formation of helical field lines which are capable of supporting prominence plasma. We have developed a numerical method for the computation of force-free, cancelling magnetic structures. We find a Kuperus-Raadu type of configuration for both an analytical, axisymmetric model, and a particular numerical model with prescribed photospheric footpoint motions. As more and more flux cancels, the axis of the helical field moves to larger heights at an increasing rate in the numerical solution, suggestive of a prominence eruption. For polar crown prominences the observed orientation of the magnetic field component transverse to the neutral line is consistent with Kuperus-Raadu type models. However, the axial component of the prominence magnetic field is systemically opposite to the one expected from the magnetic shear induced by the solar differential rotation. We propose two alternative scenarios for the formation of polar crown prominences that give the correct sign for the axial magnetic field: both models retain as their key component the formation of helical flux tubes through flux cancellation.

1. Introduction

Solar prominences consist of relatively cool ($\approx$ 8000 K) plasma which is elevated above the solar chromosphere and embedded in the much hotter corona (Tandberg-Hanssen 1974). When observed in Hα on the solar disk, prominences appear as dark, elongated structures (filaments) overlying the neutral lines that separate regions of opposite magnetic polarity in the solar photosphere. Prominences are found in both active and quiet regions, with some differences in morphology and typical size.

Magnetic field measurements using the Hanle effect provide information on the orientation of the supporting vector magnetic field at the height of the prominence (Leroy 1985, 1988).

Geophysical Monograph 58

Simultaneous measurements in two spectral lines indicate that the field is nearly horizontal (Athay et al. 1983; Querfeld et al. 1985; Bonnier et al 1986). Leroy et al. (1984) present a statistical analysis of a large number of measurements using the He D_3 line. They find that for quiescent prominences with maximum height larger than 30,000 km the component transverse to the filament axis generally points from the region of negative polarity in the photosphere to the region of positive polarity, i.e. opposite the direction of the potential field (also see Bommier et al. 1985). For prominences with maximum height less than 30,000 km, the transverse field is in the same sense as the potential field. The large prominences of the polar crown are all of the non-potential type (Leroy et al. 1983).

Recently it has been recognized that the disappearance of photospheric magnetic flux may play an important role in flare and filament activity. Martin et al. (1985) describe videomagnetograph observations of a decaying active region in the period 1984 August 3-8. They conclude that the 22 flares observed during this period were initiated at sites where photospheric magnetic fields were cancelling (here 'cancellation' is defined as 'the apparent mutual loss of magnetic flux in closely spaced features of opposite polarity'). Hermans and Martin (1986) studied small-scale eruptive filaments in the quiet Sun, and found that the majority of these structures were related to cancelling magnetic features in videomagnetograms. Zwaan (1987) and Priest (1987) point out the important role of reconnection processes in the cancellation of photospheric magnetic features.

2. Flux Cancellation in Sheared Fields

Disappearance of magnetic flux from the solar photosphere may occur in a variety of ways (e.g., Zwaan 1987). In the simplest possible scenario a preexisting loop is pulled down below the solar surface (submergence). The submergence of a magnetic loop through the solar atmosphere requires that the loop overcomes its magnetic buoyancy (Parker 1955). Because the kinetic energy of photospheric flows is small compared with the magnetic energy of the loop, the only mechanism by which a flux loop can be pulled down through the atmosphere and submerge is magnetic tension. To submerge, the downward curvature force must be larger than the upward force due to magnetic buoyance. From these considerations Parker (1979 pp. 136-141) shows that the distance between the photospheric footpoints of the loops cannot surpass about 1000 km (several photospheric scale heights) for submergence to occur. Frequently, however, the footpoints of observed coronal loops are much farther apart than 1000 km.

In Figure 1 we consider the response of a sheared arcade of loops to flux cancellation. In order for the flux to submerge, short loops have to be produced at the polarity inversion line through reconnection: only then is the above criterion satisfied. Figure 1 demonstrates that in this process a helical flux tube is formed above the polarity inversion line.

3. Models for the Field Evolution

We have developed a numerical method for the computation of force-free cancelling magnetic structures. We assume that the plasma β is low, and that the photospheric motions are very slow compared to the Alfven speed; then the magnetic field will evolve through a series of force-free equilibria. We further assume symmetry along the filament axis and let reconnection take place only at the polarity inversion line, not in the corona. Thus the toroidal and poloidal components of the coronal magnetic flux are conserved, but as photospheric line-of-sight magnetic flux cancels, flux is transferred from the arcade field to the helical field. In addition we have also derived a set of axially symmetric analytical models for cancelling magnetic structures.

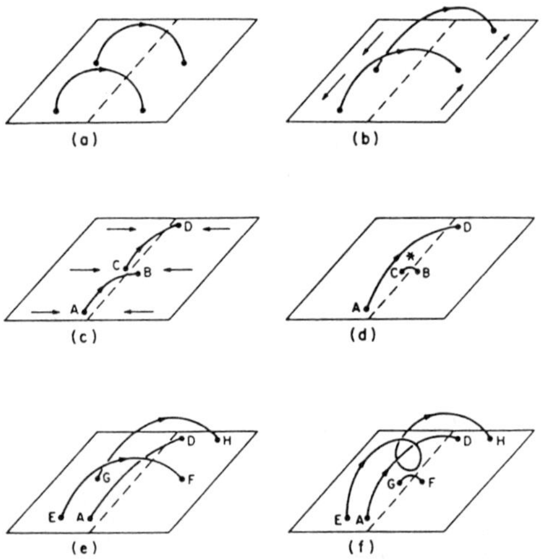

Fig. 1. Flux cancellation in a sheared magnetic field. The rectangle represents the solar photosphere, and the dashed line is the polarity inversion line separating two regions of opposite magnetic polarity. (a) Initial potential field; (b) Sheared magnetic field produced by flows along the polarity inversion line; (c) Magnetic shear is increased further due to flows toward the polarity inversion line; (d) Reconnection produces long loop AD and shorter loop CB which subsequently submerges; (e) Overlying loops EF and GH are pushed towards the polarity inversion line; (f) Reconnection produces the helical loop EH and a shorter loop GF which again submerges.

We prescribe the horizontal footpoint motions of the field lines at the boundary representing the photosphere-corona interface, and follow the evolutionary response of the field to these flows. The numerical and analytical methods have been described in detail elsewhere (van Ballegooijen and Martens 1989a); here we will focus on the results and their physical significance.

In Figure 2 we show the evolution of an axisymmetric field with constant shear displacement of the field lines. The motions towards the polarity inversion line are prescribed in such a manner that axial symmetry is preserved. This analytical result demonstrates clearly the formation of a flux tube with helical field lines, and its subsequent rise as a result of flux cancellation in the photosphere.

In the numerical calculations axisymmetry is no longer necessary. In the following x is the coordinate in the horizontal plane, perpendicular to the polarity inversion line, y is the coordinate along the inversion line, and z is the height with respect to the photosphere. We start with a potential, shearless, magnetic field, whose normal component at the photospheric boundary ($z = 0$) is given by

$$B_z(x,0) = \frac{2x}{(1 + x^2)^2} \qquad (1)$$

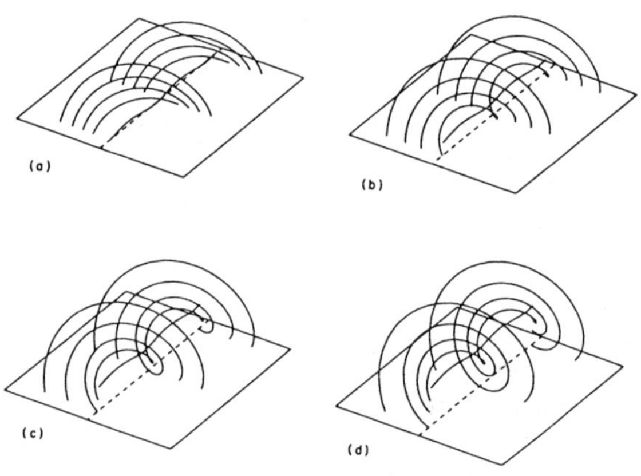

Fig. 2. The time sequence of an axisymmetric force-free field in which magnetic flux is cancelling at the polarity inversion line.

The prescribed shear velocity at the photosphere is

$$v_y(x,0) = v_0 \frac{|x|x}{c^2 + x^2} \qquad (2)$$

with the parameters $v_0 = 2$ and $c = 0.1$, corresponding to an abrupt change in v_y near the polarity inversion line. The velocity component towards the polarity inversion line is given by

$$v_x(x,0) = -\frac{1}{x} + \frac{x}{x_m^2}, \qquad (3)$$

with $x_m = 5$ defining the boundary of the numerical grid in the actual calculation.

In kthe numerical code a Lagrangian grid is used that is tied to the magnetic field lines. The shapes of the magnetic field lines are varied— conserving the field topology and the boundary connections— until an energy minimum is achieved. This energy minimum corresponds to a force-free state (Roberts 1967, pp. 110-113), that is stable for motions preserving the imposed symmetry. Here we have symmetry along the filament axis, so instabilities violating this symmetry, e.g. the kink instability, are still possible, and not much can be safely said about the overall stability of the fields we find. The minimization of the energy is achieved by iteration, using the 'conjugate gradient' method.

For numerical reasons the height h of the axis of the helical field is held fixed in our calculations. For each value of h we find a solution that is force-free everywhere in the corona, except at the boundary of a thin flux tube surrounding the axis. By varying h for given boundary conditions at the photosphere, we find the absolute energy minimum that corresponds to the true force-free equilibrium. Figure 3 shows the energy as a function of h at different times. Clearly the equilibrium height increases at an accelerating rate as time progresses and more and more flux is cancelled, which suggests an eruption of the filament.

We note that it appears that no further equilibrium exists after t = 0.75, and it would be tempting to identify this with a filament eruption. However, it has been shown by Aly (1984) that in a magnetic field with an ignorable coordinate the maximum energy situation is that of an open field. In our model that corresponds to the limit $h \to \infty$. Thus the energy W in Figure 3 has to go up again as h becomes large enough, and an equilibrium will always exist. However, in physical reality three-dimensional effects come into play when the height of the filament becomes comparable to its length. Then Aly's theorem is no longer applicable and loss of equilibrium may still occur. A significant new insight derived from the series of equilibrium solutions found in our calculations is that even without loss of equilibrium or MHD instability, there can still be a type of evolution that is eruptive in appearance.

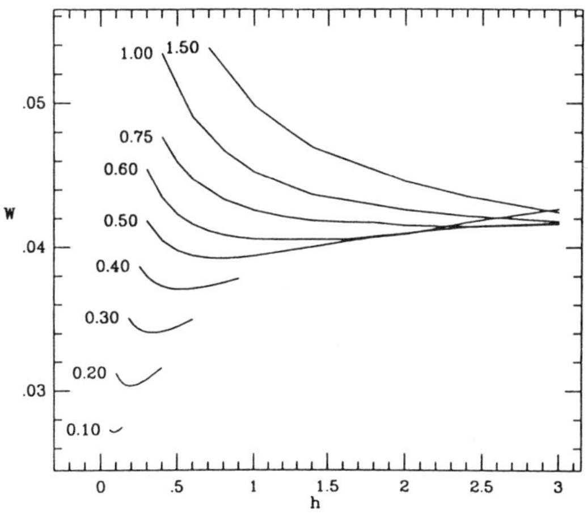

Fig. 3. Magnetic energy as a function of the height h of the axis of the helical field, at different times t. The force-free equilibrium corresponds to locations of the minimum energy in each curve. As t increases, more and more photospheric flux is cancelled, and the equilibrium height increases at an accelerating rate.

In summary, our numerical results show the formation, growth, and eruption of a helical magnetic flux tube in response to cancellation of line-of-sight magnetic flux in the photosphere. This Kuperus-Raadu (1974) type of structure is capable of supporting the observed cool prominence matter, and consistent with the observed transversal magnetic fields (Leroy et al. 1984) for most high altitude (> 30,000 km) and all polar crown prominences.

4. Polar Crown Prominences

The present model, as well as all other Kuperus-Raadu type models, predicts the correct orientation of the magnetic field component transverse to the polarity inversion line for all polar crown prominences (Leroy et al. 1983, 1984). However, as was already pointed out by Hyder (1965) and Rust (1967), the axial component of the prominence magnetic field at high latitudes is systematically opposite to the one expected from the magnetic shear induced by photospheric differential rotation. This is depicted in Figure 4. Because of the faster rotation at lower latitudes one would expect the lower latitude footpoints of field lines bridging the nearly East-West inversion lines, to move ahead relative to the higher latitude footpoints. In reality the lower latitude footpoints are actually behind, as can be inferred from the magnetic connections that follow from the field measurements in Figure 4a! (In this respect it makes no difference whether one is dealing with

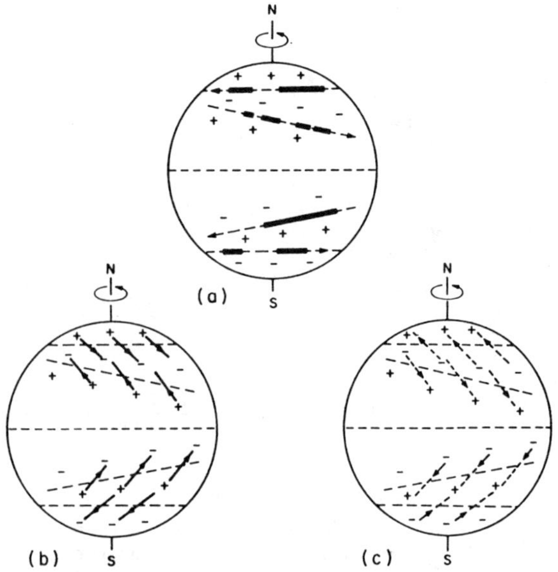

Fig. 4. Magnetic fields in quiescent prominences and their relation with magnetic fields in the surrounding photosphere. (a) Observed axial fields derived from Zeeman and Hanle measurements (after Leroy et al. 1983). (b) Axial fields predicted for differential rotation acting on line-tied magnetic fields above the solar surface. (c) Axial field predicted for differential rotation acting below the solar surface.

Kuperus-Raadu or Kippenhahn-Schluter (1957) prominences).

The intriguing aspect of this orientation is that it is systematic: if the orientation of the axial field component in prominences had nothing to do with differential rotation, then one would expect both orientations to be present. In the remainder of this section we will propose two scenarios for the formation of polar crown prominences which are based on this observation. Both scenarios are described in more detail in van Ballegooijen and Martens (1989b, submitted).

The first scenario is most easily understood by the following basic example. Consider an originally North-South oriented polarity inversion line on the Northern hemisphere, with the positive polarity to the East of it, and the magnetic field perpendicular to it, in the East-West direction. Under the influence of differential rotation the polarity inversion line will slowly rotate towards an East-West orientation. Meanwhile, the East-West directed field lines remain in that direction, and thus become sheared along the polarity inversion line. Flux cancellation at that point will lead to the emergence of prominences with the correct transverse and axial field. As is to be expected from this example, the detailed calculations demonstrate that prominences with the correct axial field orientation can only be formed when the polarity inversion line starts out close to the Solar meridian. Figure 5 illustrates this scenario and demonstrates that the axial field is identical to that of the North polar crown prominences of Figure 4.

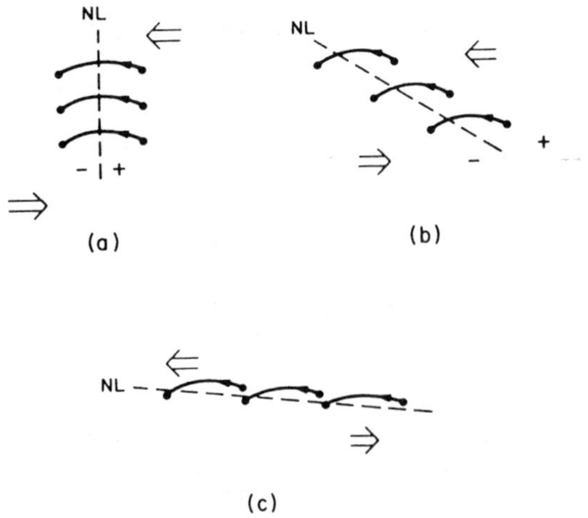

Fig. 5. Rotation of the polarity inversion line on the Northern hemisphere as a result of differential rotation. Time sequence (a), (b), (c). Flux cancellation in (c) generates a prominence with the correct transverse and axial magnetic field.

In the second scenario differential rotation plays a more constructive role. We assume that the subphotospheric field lines are of the general structure depicted in Figure 4c. Their shear component, generated by the differential rotation, is systematically opposite to that of the field lines above the surface. Hence below the surface axial fields may exist with the correct orientation. We propose that the subphotospheric fields emerge in the form of small bipoles, erupting in the 'neutral' zone between the regions with opposite magnetic polarity. We suggest that these bipoles emerge with a preferred orientation, defined by the direction of the underlying magnetic field in the convection zone. Once the bipoles have emerged, the poles of these bipoles move apart and cancel with elements of the background field, producing helical field lines in the corona.

Figure 6 shows three phases in the evolution of the magnetic structure. We consider an East-West polarity inversion line on the Northern hemisphere; the North pole is towards the back in Figure 6. Field lines above the photosphere are drawn as full curves, and field lines below the photosphere are drawn as dashed curves. Figure 6a shows an unsheared coronal arcade with field lines going from North to South, and an underlying field in the convection zone which

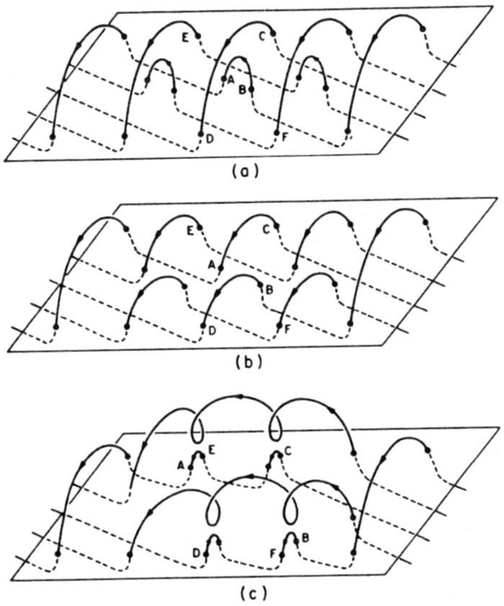

Fig. 6. Model of prominence magnetic structure. (a) Bipoles emerge within an overlying arcade. (b) First phase of reconnection: newly emerged flux elements become connected to elements of the background field. (c) Second phase of reconnection produces helical fields in the corona.

has been sheared by the solar differential rotation. We also show three bipoles emerging through the solar surface; the orientation of these bipoles is determined by the underlying magnetic field. As the photospheric footpoints of these bipoles move apart the bipoles interact with the overlying arcade and magnetic reconnection occurs in the corona. For example, the bipole AB in Figure 6a interacts with the loop CD of the arcade, forming two smaller loops AC and BD (See Figure 6b). We assume that footpoints A and B continue to move apart and eventually run into footpoints E and F. Then another round of reconnection occurs, and new loops AE and BF are formed (see Figure 6c). In the next phase of evolution the loops AE and BF will submerge below the photosphere, causing flux cancellation. During the latter phase of reconnection helical magnetic fields are formed above the solar surface through the process discussed in the previous section. Since flux cancellation occurs on both sides of the neutral zone, two strands of helical field are produced. Overlying both strands is a coronal arcade (not shown in Figure 6c).

As new sets of bipoles emerge and flux cancels at the boundaries of the 'neutral zone', more and more strands of helical field are produced. Each set of flux cancellations produces two new tubes of helical field, one on either side of the neutral zone (only the first pair is shown in Figure 6c). This is a difference with the model presented in the previous section, where the elements of the arcade field cancel with each other, and only a single helical tube is produced. The different strands may retain their identity, or they may coalesce into larger strands. Note that the different strands have the same sense of twist (left-handed in the Northern hemisphere, right-handed in the Southern hemisphere). Obviously, the present model (with the background arcade field cancelling against an element of a bipole) produces a more complicated magnetic structure with several parallel and perhaps intertwined helical tubes. We suggest that the observed fine-scale structure of quiescent prominences is related to the presence of multiple helical tubes.

5. Discussion

If the axial fields of mid- and high-latitude prominences indeed originate from below the photosphere, then prominences provide an indirect probe of the subsurface magnetic field. The stretching of subsurface fields by the differential rotation is one of the key processes of the dynamo, and it would be very interesting if the axial fields of prominences are a manifestation of the resulting 'toroidal' fields. We conclude that studies of prominence magnetic structure may have important implications for our understanding of the solar dynamo.

Testing the scenarios presented above requires magnetograph observations with very high sensitivity, since the high latitude magnetic fields are very weak. Moreover, to correctly identify the origin of magnetic elements and to follow their motion, the observations must be made continuously for long time spans. The videomagnetographs currently operated at Big Bear and Beijing Solar Observatory offer the best chance to observe the phenomena predicted here.

Further progress in the numerical modeling of the formation and eruption of solar prominences requires fully three-dimensional models which take the finite length of the magnetic structures into account. Although it is possible to extend our present Lagrangian method to the three-dimensional case, one disadvantage of the method is that resistive effects cannot easily be included. Therefore in the eruptive phase, when resistive energy release probably plays an important role, a more suitable approach would be to use a Eulerian grid (e.g., Mikic et al. 1988). Further work with both Lagrangian and Eulerian codes is needed to determine their relative merits.

Questions and Answers

Antiochos: Is it consistent to argue that reconnection forms the helical field? If so, shouldn't reconnection also relax it to a non-helical state?

Martens: *The reconnection we are talking about is* ***driven*** *by photospheric (or sub-photospheric) motions, and we have explicitly shown that it necessarily leads to the*

formation of helical fields, when there exists a shear displacement of the field lines with respect to the neutral line. The observations of Martin and co-workers show that the time-scale for this process is of the order of hours. So, in answer to your first question: Yes, it is consistent.

In answer to your second question: Yes, the dissipation of the electrical currents will indeed lead to reconnection to a non-helical state. However, the time-scale for this process is inversely proportional to the photospheric resistivity, which is very low. Estimates for this time-scale range from several days to months. Therefore, this process is not important for the much faster driven reconnection that we consider.

Acknowledgments. This work was supported by the National Aeronautics and Space Administration via grant NAGW-249 (AvB) and grant NAGW-112 (PM).

References

Aly, J. J., On some properties of force-free magnetic fields in infinite regions of space, Ap.J. 283, 349-362, 1984.

Athay, R. G., C. W. Querfeld, R. N. Smartt, E. Landi-Delg'Innocenti, and V. Bommier, Vector magnetic fields in prominences. III. He I D_3 Stokes profile analysis for quiescent and eruptive prominences, Solar Phys. 89, 3-30, 1983.

Bommier, V., J. L. Lelroy, and S. Sahal-Brechot, The linear polarization of hydrogen Hβ radiation and the joint diagnostic of magnetic field vector and electron density in quiescent prominences, Astron. Astrophys. 156, 79-89, 1986.

Hermans, L. M. and S. F. Martin, Small-scale eruptive filaments on the quiet Sun, in Coronal and Prominence Plasmas, NASA CP 2442, 369-375. 19986.

Hyder, C. L., The polarization of emission lines in astronomy: II. Prominence emission-line polarization and prominence magnetic fields, Ap.J. 141, 1374-1381, 1965.

Kippenhahn, R. and A. Schluter, Eine Theorie der solaren Filamente, Z. Astrophys. 43, 36-62, 1957.

Kuperus, M. and M. A. Raadu, The support of prominences formed in neutral sheets, Astron. Astrophys. 31, 189-193, 1974.

Leroy, J. L., The Hanle effect applied to magnetic field measurements, in Measurements of Solar Vector Magnetic Fields, NASA CP-2374, 121-140, 1985.

Leroy, J. L., The fine structure of prominence magnetic fields: Can we provide useful observational constraints?, in Solar and Stellar Coronal Structure and Dynamics, National Solar Observatory, 1988.

Leroy, J. L., V. Bommier, and S. Sahal Brechot, The magnetic field in the prominences of the polar crown, Solar Phys. 83, 135-142, 1983.

Leroy, J. L., New data on the magnetic structure of quiescent prominences, Astron. Astrophys. 131, 33-44, 1984.

Martin, S. F., S. H. B. Livi, and J. Wang, The cancellation of magnetic flux. II. In a decaying active region, Australian J. Phys. 38, 929-959, 1985.

Mikic, Z., D. C. Barnes, and D. D. Schnack, Dynamical evolution of a solar coronal magnetic field arcade, Ap.J., 328, 830-847, 1988.

Parker, E. N., The formation of sunspots from the solar toroidal field, Ap.J. 121, 491-507, 1953.

Parker, E. N., Cosmical Magnetic Fields, (pp. 841) Oxford University Press, England, 1979.

Priest, E. R., Appearance and disappearance of magnetic flux at the solar surface, in The Role of Fine-Scale Magnetic Fields on the Structure of the Solar Atmosphere, Proc. Tenerife Workshop, 297-313, 1987.

Querfeld, C. W., R. N. Smart, V. Bommier, E. Landi-Degl'Innocenti, and L. House, Vector magnetic fields in prominences. II. He I D, Stokes profiles analysis for two quiescent prominences, Sol. Phys. 96, 277-292, 1985.

Roberts, P. H., An Introduction to Magneto--hydrodynamics, (264 pp.) Longmans, London, 1967.

Rust, D., Magnetic fields in quiescent solar prominences. I. Observations, Ap.J. 150, 313-326, 1967.

Tanberg-Hanssen, E. Solar Prominences (148 pp.) Reidel, Dordrecht, 1974.

van Ballegooijen, A. A. and P. C. H. Martens, Formation and eruption of solar prominences, Ap.J., 343, 971-984, 1989a.

Zwaan, C., Elements and patterns in the solar magnetic field, in Ann. Rev. Astr. & Astrophys. 25, 83-111, 1987.

CORONAL MASS EJECTIONS AND MAGNETIC FLUX ROPES IN INTERPLANETARY SPACE

J. T. Gosling

Los Alamos National Laboratory
Los Alamos, New Mexico 87545

Abstract. Coronal mass ejections (CMEs) are important occasional sources of plasma and magnetic field in the solar wind at 1 AU. Formed in the corona by the ejection of solar material from closed field regions that were not previously participating in the solar wind expansion, CMEs at 1 AU generally have distinct plasma and field signatures by which they can be distinguished from the ordinary solar wind. Perhaps the most common of these signatures is a counterstreaming (along the magnetic field) flux of suprathermal electrons with energies $\gtrsim$ 80 eV. This signature indicates that CMEs at 1 AU typically are closed field structures either rooted at both ends in the Sun or entirely disconnected from it. A subset, perhaps 30% of all CME events at 1 AU, exhibit the large and coherent internal field rotations characteristic of magnetic flux ropes. Often, but not always, these special flux rope events have the low proton temperatures and stronger than average fields characteristic of "magnetic clouds." Although equilibrium flux rope models, in which the internal field pressure is balanced by the curvature stress of the field, generally are adequate for explaining the observed field rotations in these events, interplanetary flux ropes need not be equilibrium structures. We suggest that interplanetary magnetic flux ropes form as a result of reconnection within rising, previously sheared, coronal magnetic loops. Such reconnection probably occurs relatively rarely in CME events, since the majority of CME events in the solar wind at 1 AU do not appear to be flux ropes.

Introduction

Coronal mass ejection events, CMEs, are spectacular manifestations of the evolution of the solar magnetic field and occur frequently in the Sun's outer atmosphere (see Hundhausen [1988] and Kahler [1987, 1988] for recent reviews). During coronal mass ejection events 10^{15} - 10^{16} gms of solar material are propelled outward into interplanetary space from closed field regions in the corona and chromosphere that were not previously participating directly in the solar wind expansion. Ejection speeds range from less than 50 km/s in some of the slower events to greater than 1000 km/s in some of the faster ones, and frozen within the expelled material is a remnant of the solar magnetic field. The detection and identification of CMEs in the solar wind has been a subject of intense interest since the first observations of CMEs with satellite coronagraphs in the early 1970's. In fact, this interest actually preceded the coronagraph observations of CMEs [e.g., Hirshberg et al., 1972; Gosling et al., 1973; Montgomery et al., 1974], since it had previously been established that transient shock wave disturbances in the solar wind typically contain an excess amount of solar material [Hundhausen et al., 1970].

It is this author's personal opinion that we can now usually identify both slow and fast CMEs in the solar wind at 1 AU with a reasonable degree of confidence. Of the CMEs identified in solar wind data, a subset appear to have the magnetic topology of large, coherent flux ropes, characterized by relatively strong axial fields near their centers and increasingly strong poloidal fields near their outer edges. The purpose of the present paper is to review the evidence for interplanetary magnetic flux ropes, to place the observations of flux ropes into the overall context of CME events in interplanetary space, and finally to suggest how the flux rope magnetic topology might originate in reconnection near the Sun.

Fast Coronal Mass Ejection Events and Shock Wave Disturbances in the Solar Wind

The leading edges of the faster CMEs observed with coronagraphs have outward speeds considerably greater than that associated with the normal solar wind expansion [e.g., Gosling et al., 1976; Howard et al., 1985] Thus,

Geophysical Monograph 58

fast CMEs should and, in fact, usually do drive shock wave disturbances in the solar wind [e.g., Sheeley et al., 1985]. Indeed, we now believe that virtually all transient (as opposed to corotating) shocks at 1 AU are driven by CMEs. Figure 1 shows ecliptic and meridional cuts through a hypothetical interplanetary shock wave disturbance driven by a fast CME. For illustration purposes the CME has been drawn as a plasmoid magnetically disconnected from the Sun; however, such disconnection has not been definitely established by either coronal or solar wind measurements and is not essential to the discussion which follows. We will return later to the question of magnetic disconnection.

In Figure 1 the shock, which serves to initiate the deflection of the ambient solar wind around the CME, has propagated well ahead of it. Between the CME and the shock is a region of compressed ambient solar wind plasma and field similar in nature to the Earth's magnetosheath. Because of the relative speed between the CME and the ambient solar wind and because of the high electrical conductivity of the plasma (which prevents interpenetration of the ambient and CME fields), the magnetic field within the "sheath" drapes around the CME. Field rotations associated with this draping can mimic those associated with the CME itself. Thus in attempting to determine the internal magnetic field topology of the CME it is essential to be able to distinguish the compressed ambient plasma and draped magnetic field from that of the actual CME.

Not only is the ambient plasma and field ahead of the CME compressed as it is overtaken by the CME, but, as sketched in the bottom panel, the leading edge of the CME is also compressed as a result of the interaction [e.g., Gosling et al., 1987, 1988]. Such compression plays an important role in enhancing the field strength within CMEs which drive shocks and is also evident in some events where the relative speed between the CME and the ambient solar wind is insufficient to produce a shock. Finally, the sketches drawn in Figure 1 help illustrate that a CME in the solar wind at 1 AU is generally a large, 3-dimensional object whose internal structure is sampled only along a single (or, at best, several) radial trace(s). The overall plasma and field topology must therefore be inferred from a spatially limited measurement. Nevertheless, as we shall see, the inference derived from interplanetary measurements that a subset of all CMEs have the magnetic topology of flux ropes is quite strong.

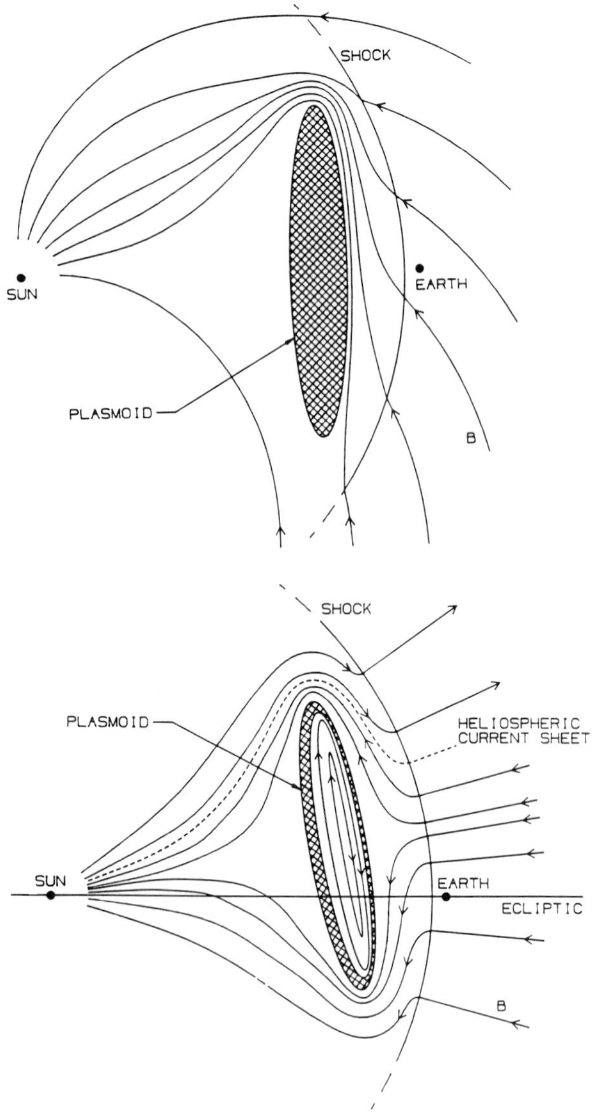

Fig. 1. Idealized sketches of an interplanetary shock wave disturbance driven by a fast-moving coronal mass ejection (CME), here drawn as a detached plasmoid, shortly before arrival at Earth. The upper panel shows a cut in the solar equatorial plane, while the lower panel shows a cut in a solar meridional plane. The ambient plasma ahead of the CME is compressed and deflected from its path, and the ambient magnetic field drapes around the front of the CME. The front portion of the CME is strongly compressed by the interaction in much the same way as the front of Earth's magnetosphere is compressed by its interaction with the solar wind. Adapted from Gosling and McComas [1987].

Signatures of Coronal Mass Ejection Events in Interplanetary Space

A number of studies have used interplanetary shocks as fiducials for searching for plasma and/or field signatures that might identify fast CMEs in the solar wind at 1 AU (see, for example, the review by Schwenn [1986]). As noted above with respect to Figure 1, the CMEs should be found not immediately behind the shocks, but rather a number of hours thereafter (typically the CME is first detected 8–12 hrs after shock passage, although the delay can be shorter or longer depending upon circumstances [e.g., Gosling et al., 1987]). Table 1 lists a variety of

TABLE 1. Common Signatures of Coronal Mass Ejection Events in the Solar Wind.

Signature	Representative References
1. Helium abundance enhancement	a,b
2. Ion and electron temperature depressions	c,d
3. Unusual ionization states	e,f,g,h,i
4. Strong magnetic field	j,k
5. Low magnetic field variance	l,m,n
6. Anomalous field rotations	o,p,q
7. Counterstreaming energetic protons	n,r,s
8. Counterstreaming suprathermal electrons	d,m,t,u

a. Hirshberg et al., 1972
b. Borrini et al., 1982
c. Gosling et al., 1973
d. Montgomery et al., 1974
e. Bame et al., 1979
f. Fenimore, 1980
g. Schwenn et al., 1980
h. Gosling et al., 1980
i. Zwickl et al., 1982
j. Hirshberg and Colburn, 1969
k. Burlaga and King, 1979
l. Pudovkin et al., 1979
m. Gosling et al., 1987
n. Marsden et al., 1987
o. Burlaga et al., 1981
p. Klein and Burlaga., 1982
q. Burlaga, 1989
r. Palmer et al., 1978
s. Kutchko et al., 1982
t. Temnyi and Vaisberg, 1979
u. Bame et al., 1981

plasma and field signatures which qualify as unusual compared to the normal solar wind but which are commonly observed a number of hours after shock passage. From the foregoing discussion it follows that fast CMEs in the solar wind often can be identified by these signatures; direct correlations of particular solar wind and coronal events confirms this expectation (see, for example, references in Table 1). Presumably the only thing unique about a fast CME as compared to a slow one is its speed; thus these same plasma and field signatures should serve to identify slow CMEs as well. Indeed, plasma and field signatures nominally similar to those frequently observed behind interplanetary shocks are often observed in the slow solar wind as well [e.g., Borrini et al., 1982; Burlaga and King, 1979; Klein and Burlaga, 1982; Gosling et al., 1987; Marsden et al., 1987], although the magnetic field tends to be weaker and more uniform in the generally slower, non-shock events since these CMEs do not interact strongly with the ambient solar wind.

It is worth emphasizing that relatively few CMEs at 1 AU exhibit all of the characteristics listed in Table 1 [e.g., Zwickl et al, 1983]. Further, some of these features are more commonly observed than are others. This is not surprising since each CME in the corona is a unique object, and the interplanetary evolution of a CME depends significantly on both its initial state and the ambient solar wind into which it propagates. Recent work indicates that a counterstreaming flux of suprathermal electrons is one of the more common signatures of a CME in the solar wind at 1 AU [Gosling et al., 1987]. Interestingly, as outlined below, this signature is one which also allows us to infer something about the overall global magnetic topology of CMEs in interplanetary space. Throughout the remainder of this paper we explicitly assume that for intervals when a spacecraft is not magnetically connected to a planetary bow shock, the onset (or end) of counterstreaming solar wind electron fluxes at energies above $\sim$ 80 eV signals entry into (or exit from) a CME. We recognize, of course, the possibility that some CMEs in the solar wind may lack the counterstreaming electron signature and thus will not be identified by this technique.

Origins of Counterstreaming Suprathermal Electron Events in Interplanetary Space

Figure 2 is a sketch illustrating possible origins of counterstreaming suprathermal electrons in interplanetary space. Since the solar corona is much hotter than interstellar space, on open field lines there commonly exists a unidirectional electron heat flux directed outward from the Sun along the interplanetary magnetic field, IMF (top of figure). This heat flux is carried by electrons with energies greater than about 80 eV [e.g., Feldman et al., 1975; Rosenbauer et al., 1977]; observations indicate that such electrons usually propagate nearly collision-free beyond several tenths of an AU from the Sun. As the Earth's bow shock is also a source of hot electrons, counterstreaming fluxes of suprathermal electrons are commonly observed on IMF field lines which connect an upstream spacecraft to the bow shock (bottom of figure) [e.g., Feldman et al., 1973]. For a spacecraft far upstream from Earth such as ISEE 3, bow shock connections occur only for a restricted set of IMF orientations, and the counterstreaming events associated with such connections can usually be distin-

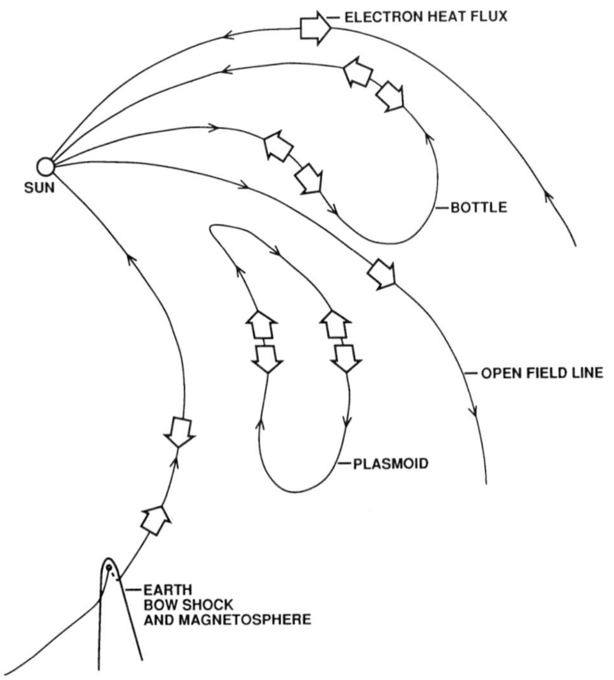

Fig. 2. A sketch illustrating possible origins of counterstreaming, suprathermal electrons in interplanetary space. Since they originate in closed field regions in the solar corona, coronal mass ejection events might be expected to have the magnetic structure of either a "bottle" or a detached plasmoid.

guished from events of solar origin both by the narrow field angle range over which they are observed and by the sporadic nature of their appearance [e.g., Stansberry et al., 1988]. By way of contrast, interplanetary shocks are generally much weaker sources of hot electrons and thus do not usually produce substantial fluxes of counterstreaming electrons in the solar wind.

Counterstreaming fluxes of suprathermal electrons are also expected on field lines which form extended magnetic loops (bottles) in interplanetary space. The counterstreaming flux arises because both ends of field lines within a magnetic bottle are rooted in the hot corona. As a typical travel time of a suprathermal electron along the IMF from the Sun to 1 AU is of the order of a couple of hours, magnetic mirroring near the Sun should play a role in maintaining the counterstreaming flux. If the field lines within the bottle become disconnected from the Sun to form a plasmoid, then the counterstreaming flux of suprathermal electrons is trapped on the closed field lines within the plasmoid. In view of the fact that a counterstreaming suprathermal electron flux (in common parlance, a bidirectional electron heat flux) is a common signature of a CME in the solar wind at 1 AU, CMEs must often be closed field structures, either magnetic bottles rooted at both ends in the Sun or disconnected plasmoids. Such magnetic topologies are consistent with the observation that CMEs generally originate in closed field regions in the solar corona [e.g., Hundhausen, 1988].

Coronal Mass Ejection Events in the Solar Wind Lacking Large Internal Field Rotations

Figure 3, which is a color-coded plot of suprathermal electron angular distributions obtained by the Los Alamos electron experiment on ISEE 3 on May 29–31, 1979, illustrates the counterstreaming electron signature characteristic of many CME events [e.g., Gosling et al., 1987]. Prior to about 2030 UT on May 29 and after about 0930 UT on May 31 the distributions were predominantly unidirectional as is usually the case in the solar wind, with short intermittent intervals of bidirectionality (such as in the time interval from about 1000 to 1300 UT on May 31) caused by brief magnetic connections with Earth's bow shock. But from about 2030 UT on May 29 until about 0930 UT on May 31 the distributions were persistently bidirectional along the IMF, indicating the passage of a CME.

Figure 4 displays more complete plasma and field data for the above time interval. From top to bottom the quantities plotted are the bulk flow speed, the proton density, the log of the proton temperature, the log of the electron temperature, the log of the total static pressure (ions plus electrons plus field), the magnetic field strength, and the field azimuth and polar angles in GSE coordinates. The CME, as identified by the counterstreaming event (delineated by vertical solid lines), was preceded by a shock which passed ISEE 3 at 1818 UT on May 29 (vertical dashed line). The delay from shock passage to CME was shorter than is typical of such events, but is compatible with the observed speed of the leading edge of the CME at 1 AU. The depressions in proton and electron temperature, the relatively strong field with low variance, and the sharp rotations in the field at the beginning and end of the CME are all characteristic, but not universal, signatures of these events, as noted previously. Also characteristic of CME events is the low plasma beta, although that quantity is not plotted directly in the figure. Finally, the declining speed, pressure, and field strength across the CME is a common observation, particularly in shock-associated events [e.g., Gosling et al., 1987]. The large gradient in speed from front to rear suggests that the CME was expanding as it passed over the spacecraft. This assumes that the front end of the CME, which was well beyond 1 AU when the rear end passed ISEE 3, did not slow substantially after it passed 1 AU. Although some slowing would be expected because of a continual transfer of momentum from the CME to the ambient plasma ahead, it seems unlikely that the front end of the CME would have slowed from ~ 650 to ~ 450 km/s within 36 hours of Earth-passage.

Of particular interest to the present paper is the lack of

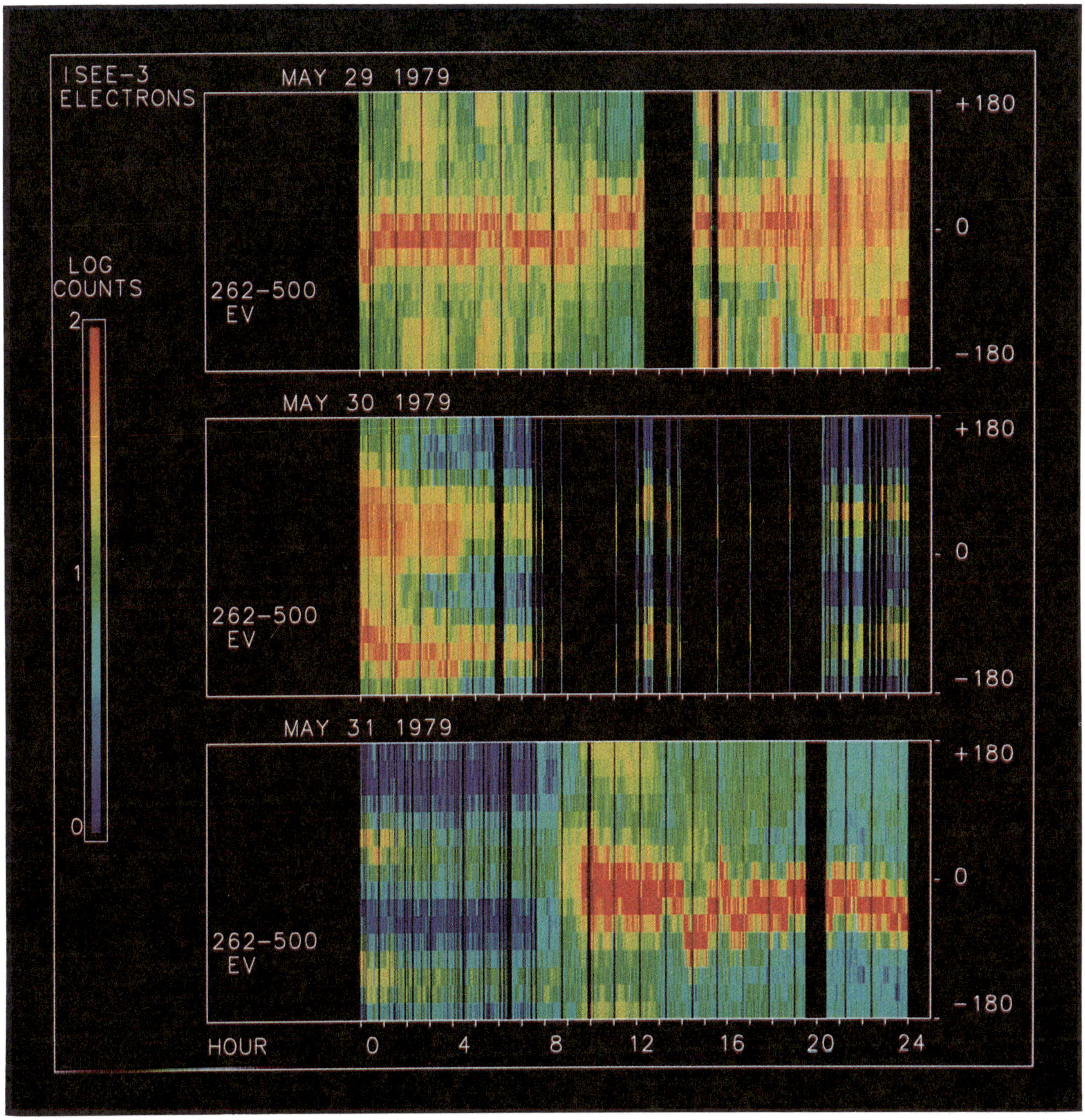

Fig. 3. A color-coded representation of ISEE 3 measured solar wind electron angular distributions within an energy passband extending from 262 to 500 eV. As indicated by the color bar on the left, the color coding is related to the log of the measured counts within ± 67.5 deg of the ecliptic plane. Numbers at the right-hand edge of the panels refer to the azimuthal look angle of the measurement, with 0 deg corresponding to the solar direction. The energy passband shown is dominated by the solar wind halo population which carries the electron heat flux. Note the intense counterstreaming electron event beginning about 2000 UT on May 29 and persisting until about 0900 UT on May 31. Counterstreaming events such as this are believed to signal passage of coronal mass ejection events. Intermittent intervals of counterstreaming electrons at other times in this plate (for example, between about 1000 and 1300 UT on May 31) are caused by magnetic connections to Earth's bow shock.

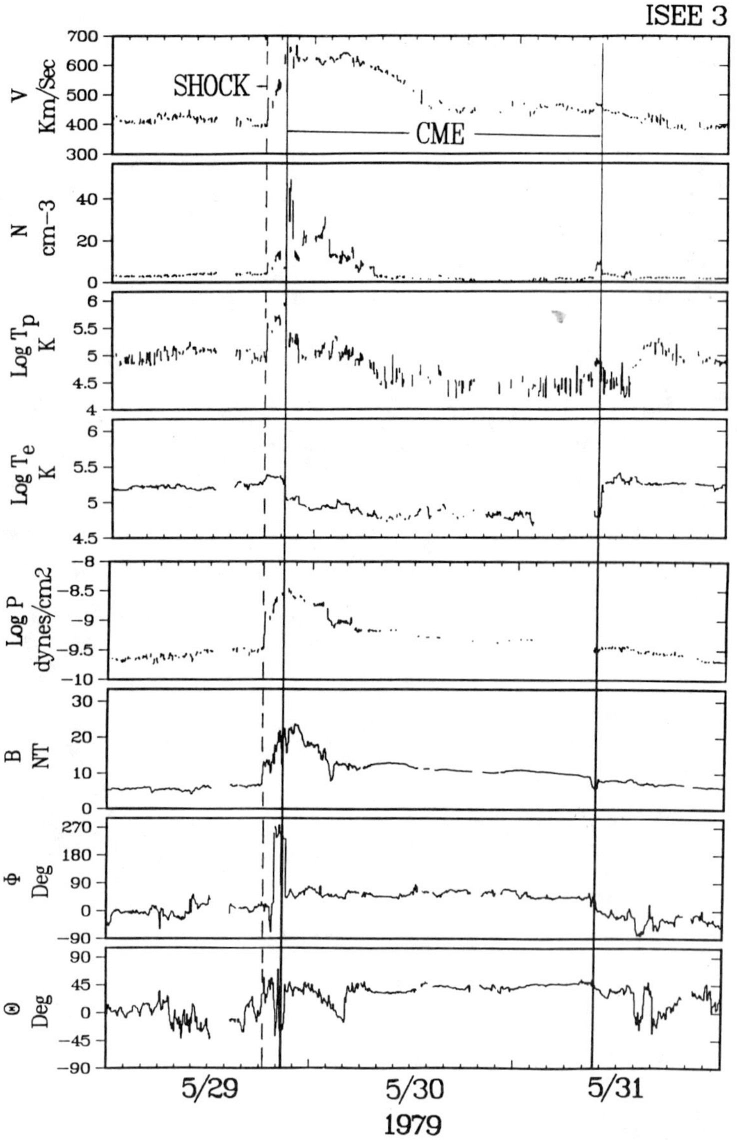

Fig. 4. Solar wind parameters surrounding the May 29-31, 1979, counterstreaming electron event. Note particularly the relative lack of reorientation of the magnetic field within the counterstreaming electron event (CME).

any substantial, coherent field rotation within the CME itself. This is characteristic of many, perhaps most, CME events as identified by the counterstreaming electron signature. Nevertheless, as we shall see, a subset of all CMEs do exhibit the coherent internal field rotations characteristic of magnetic flux ropes.

Approximately half of all observed counterstreaming suprathermal electron events do not drive interplanetary shocks [Gosling et al., 1987]. Figure 5 shows plasma and field data from ISEE 3 for a counterstreaming event (bracketed by the solid vertical lines) which was not preceded by a shock. Examination of the speed profile (top panel) reveals why this event did not produce a shock - its outward speed was less than that of the ambient solar wind ahead. Nevertheless, the counterstreaming event was a distinct plasma and field entity characterized by low proton and electron temperatures, a moderately strong field of low variance, and sharp field rotations at either end. Although not demonstrated explicitly in the figure, the average density, low temperature, and moderately strong field make this CME a low beta structure. The more modest internal field strength for this event as compared to the previous one can be at least partially attributed to the lack of any substantial interaction (com-

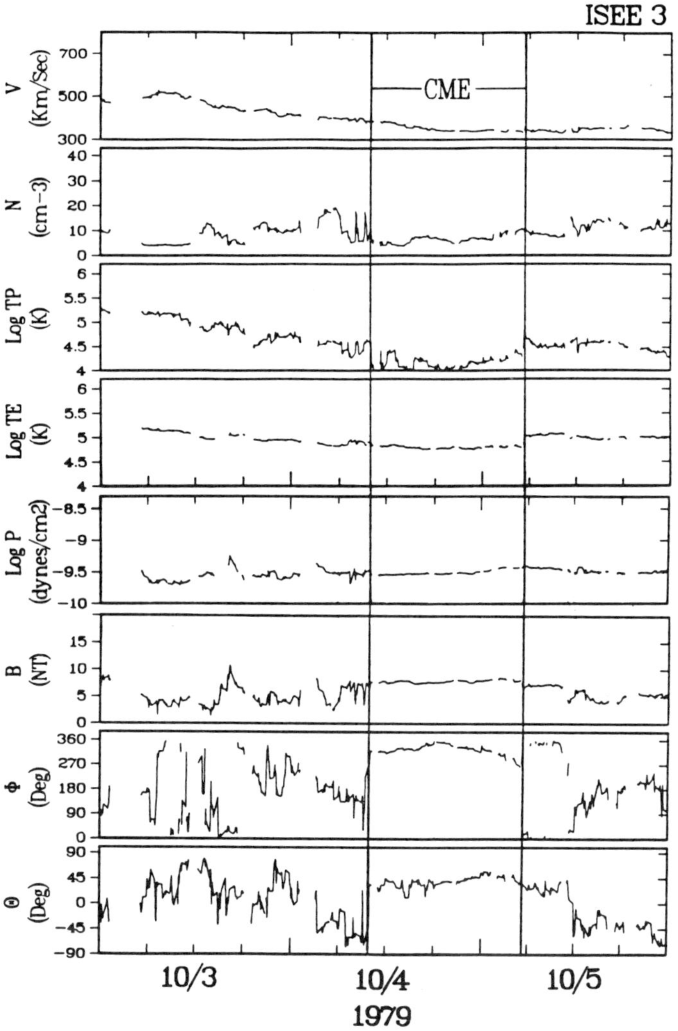

Fig. 5. Solar wind parameters surrounding a counterstreaming electron event (CME) on October 4 and 5, 1979. This CME did not have a sufficiently high outward speed to produce a shock wave disturbance. Note again the relative lack of reorientation of the magnetic field within the event. Adapted from Gosling et al. [1987].

pression) with the ambient solar wind ahead. As in the previous example, the field orientation remained essentially constant throughout the event.

Magnetic Clouds

To this point we have focussed on CME events lacking substantial internal field rotations. Such events are common; however, we will henceforth concentrate on a subset of all CME events characterized by large, coherent internal field rotations. When such events also have internal magnetic fields greater than ~ 10 nT (at 1 AU) and lower than average proton temperatures, they are commonly called magnetic clouds [e.g., Klein and Burlaga, 1982; Burlaga, 1989]. Figure 6 shows plasma and field data for a shock disturbance driven by a magnetic cloud and observed by Helios 1 at 0.54 AU. From top to bottom the parameters plotted are the magnetic field strength, the latitude (polar) and azimuth field angles, the bulk flow speed, the proton density, and the proton temperature. The cloud is identified by the low proton temperature, the strong, declining field, and steady, near monotonic north-south rotation in the field in the interval from about 0230 to about 2030 UT on June 20. Similar to the event shown in Figure 4, interplanetary compression was largely responsible for the very strong field observed within the leading portion of the cloud. As for many CMEs which drive shocks in the solar wind, the speed declined across the cloud. Because of the average density, depressed temperature, and strong field, the cloud was a low beta structure.

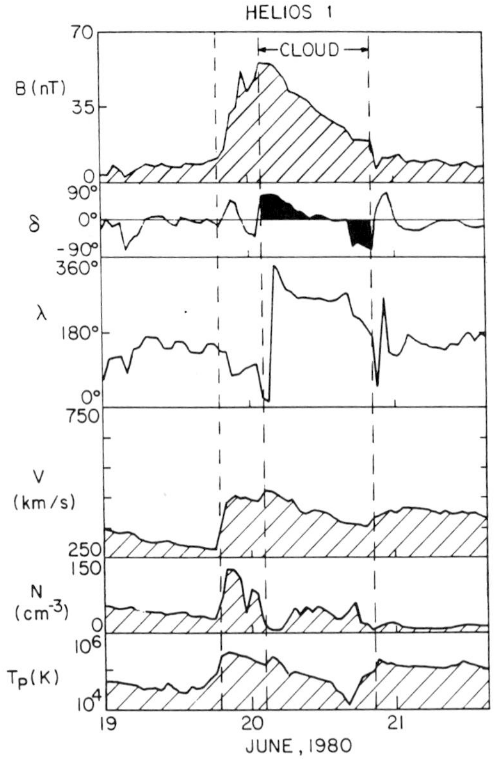

Fig. 6. Solar wind parameters for a shock disturbance driven by a magnetic cloud. The angles δ and λ are the polar (latitude) and azimuthal angles of the field, indicated by θ and ϕ elsewhere in this paper. In contrast to the events shown in Figures 4 and 5, this magnetic cloud event is characterized by a large internal rotation in the magnetic field. From Burlaga et al. [1982].

The original papers on magnetic clouds concentrated on events where the rotation in the field occurred in a plane perpendicular to the ecliptic (that is, they concentrated on events such as that shown in Figure 6 and characterized by large, coherent changes in latitude angle) [e.g., Burlaga et al., 1981; Klein and Burlaga, 1982; Burlaga and Behannon, 1982; Burlaga et al., 1982]. However, field rotations within CMEs are not necessarily restricted to planes which are perpendicular to the ecliptic. Figure 7 shows color-coded plots of suprathermal electron angular distributions measured by the solar wind electron experiment on ISEE 3 for a 4-day interval in February 1981 in which two field-aligned, counterstreaming electron events were detected. The first of these began about 1800 UT on February 6 and lasted until about 1900 UT on February 8, while the second began about 2200 UT on February 8 and lasted until about 0500 UT on February 9. The first of these events drove a shock which passed ISEE 3 about 0800 UT on February 6, while the second drove a shock which propagated into the rear end of the first event and passed ISEE 3 about 1330 UT on February 8. Of particular interest here is the steady drift in azimuth experienced by the counterstreaming electron beams in the first of these events; from beginning to end this drift was of the order of 180 degrees.

Figure 8 shows the corresponding electron moments (above) and magnetic field data (below) for the Feb. 6–9, 1981 interval. (No solar wind ion data are available from ISEE 3 for this event.) Again the vertical dashed lines indicate the shocks, and the vertical solid lines bracket the CMEs as distinguished by the suprathermal electron data. As noted above, the second of these shocks was propagating into the rearward portion of the first CME producing a field enhancement there. Typical of shock-associated events, the field was strong within the CMEs and the flow speed declined from front to rear. On the other hand, the elevated electron temperatures throughout a large portion of the first of the CMEs and average electron temperature across the second one help emphasize the point that one can not always distinguish CME events on the basis of electron temperature alone.

Keying on the rotation in polar angle (bottom panel), Zhang and Burlaga [1988] identified the interval from 1900 UT on Feb. 6 until 0900 UT on Feb. 7 as a magnetic cloud. Clearly, however, the suprathermal electron measurements indicate that the event started at 1800 UT on Feb. 6 and persisted for many hours after 0900 UT on Feb. 7. This event thus demonstrates the utility of using suprathermal electron measurements to determine the extent and boundaries of magnetic clouds. Further, the rotation in polar angle was a relatively minor portion of the total field rotation in the event; the major field rotation for this event was in the azimuthal component. Here then is an excellent example of a magnetic cloud where the internal field rotation was not confined to a plane perpendicular to the ecliptic, and other similar events have been documented in the literature recently [e.g., Burlaga, 1988, 1989]. Finally, the second CME shown in these figures, which was of much shorter duration than the first, did not exhibit any substantial coherent internal rotation, and was clearly not a magnetic cloud.

Occurrence Rates of Coronal Mass Ejection Events and Magnetic Clouds

It is instructive to compare the relative occurrence rates of events which have been identified as magnetic clouds and CMEs in general as identified by counterstreaming electron events. From an examination of the ISEE 3 electron data for the interval from launch in mid-August 1978 until entry into the geomagnetic tail in mid-October 1982, we have identified 194 interplanetary counterstreaming events similar to those displayed in this paper. Interpreting these events as CMEs, we find that CMEs were detected upstream from the Earth at an average rate of about 4 events per month. For this same time interval, Zhang and Burlaga [1988] have identified 15 magnetic clouds in the data (14 of which are counterstreaming electron events), corresponding to an occurrence rate of approximately 0.3 events per month, more than a factor of 10 lower than the occurrence rate of counterstreaming electron events. We believe the actual occurrence rate of CMEs with large and coherent internal field rotations is somewhat higher (by perhaps a factor of 2 or 3) than can be inferred from the magnetic cloud statistics of Zhang and Burlaga because many CMEs at 1 AU with large and coherent internal field rotations do not have internal field strengths greater than 10 nT (see below). Nevertheless, the message is clear: magnetic clouds as presently defined form a small subset of all CMEs in the solar wind at 1 AU. From the observation that 14 out of 15 magnetic cloud events identified by Zhang and Burlaga are also counterstreaming, suprathermal electron events, we can infer that magnetic clouds are generally closed field structures.

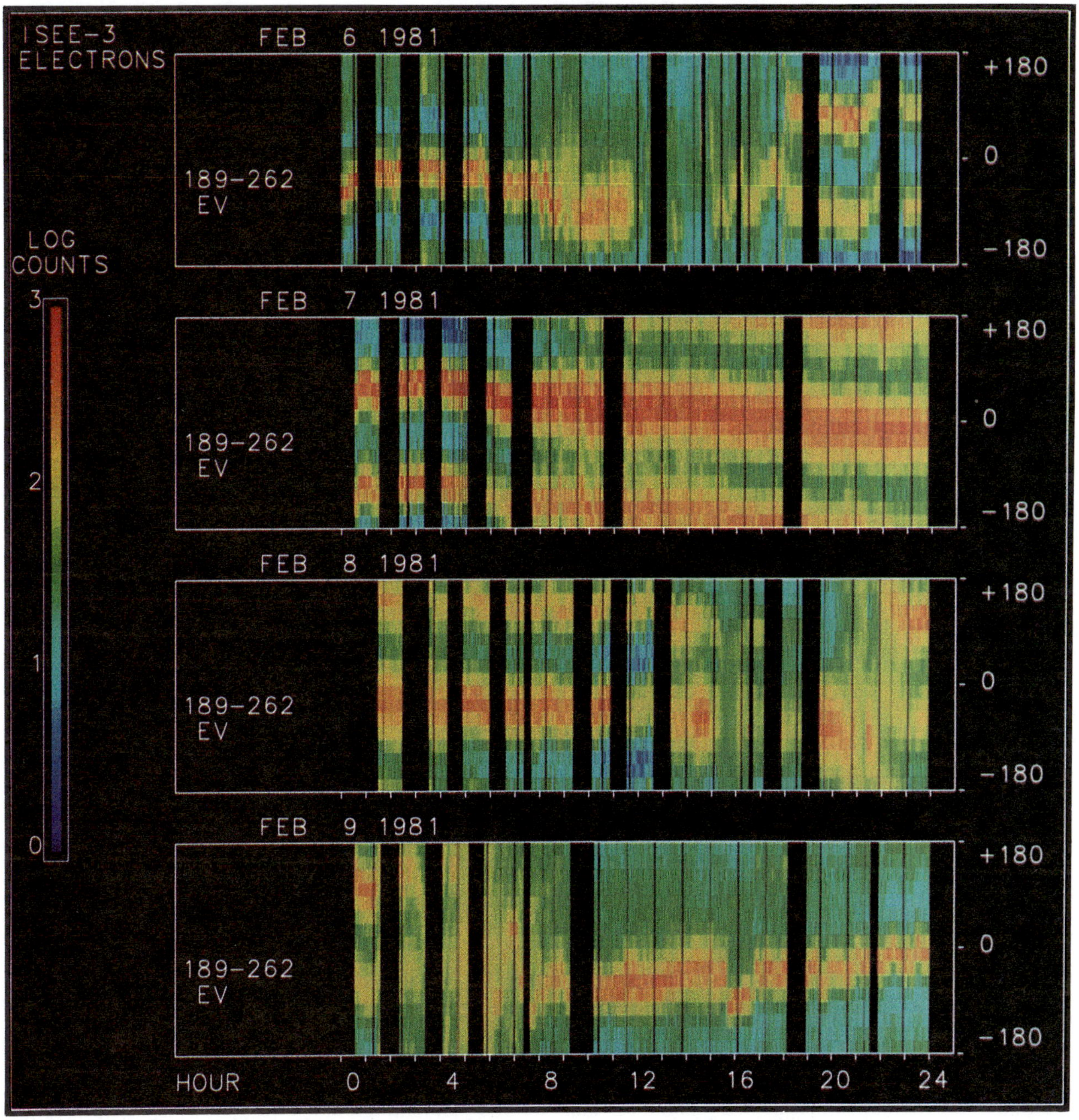

Fig. 7. Color-coded electron angular distributions from ISEE 3 in an energy passband extending from 189 to 262 eV for February 6–9, 1981. This interval contained two separate counterstreaming electron events (CMEs), each of which drove a shock disturbance. Note the steady drift in azimuth of the counterstreaming flux during the first of these events.

Other Coronal Mass Ejection Events With Large Internal Field Rotations

We have already noted that compression in interplanetary space is at least partly responsible for the elevated field strengths observed within the leading portions of CMEs which drive shocks in the solar wind at 1 AU. Lacking the benefit of substantial compression in interplanetary space, slow CMEs generally have weaker internal fields at 1 AU than do shock-associated CMEs, although as illustrated by the event shown in Figure 5, even slow, non-interacting CMEs generally have moderately strong internal fields [e.g., Gosling et al., 1987]. However, the internal fields may not be strong enough for slow CMEs to be identified as magnetic clouds. Figure 9 shows color-coded plots of electron angular distributions from ISEE 3 for a counterstreaming suprathermal electron event observed on December 31, 1979 and January 1, 1980. The event began just prior to 2140 UT on Dec. 31 and appeared to end at about 1530 UT on Jan. 1. We strongly suspect, however, that the event actually persisted until almost 2100 UT since, for the most part, the suprathermal electron flux from all azimuths nearly disappeared during the 1530–2100 UT interval. The acceptance fan of

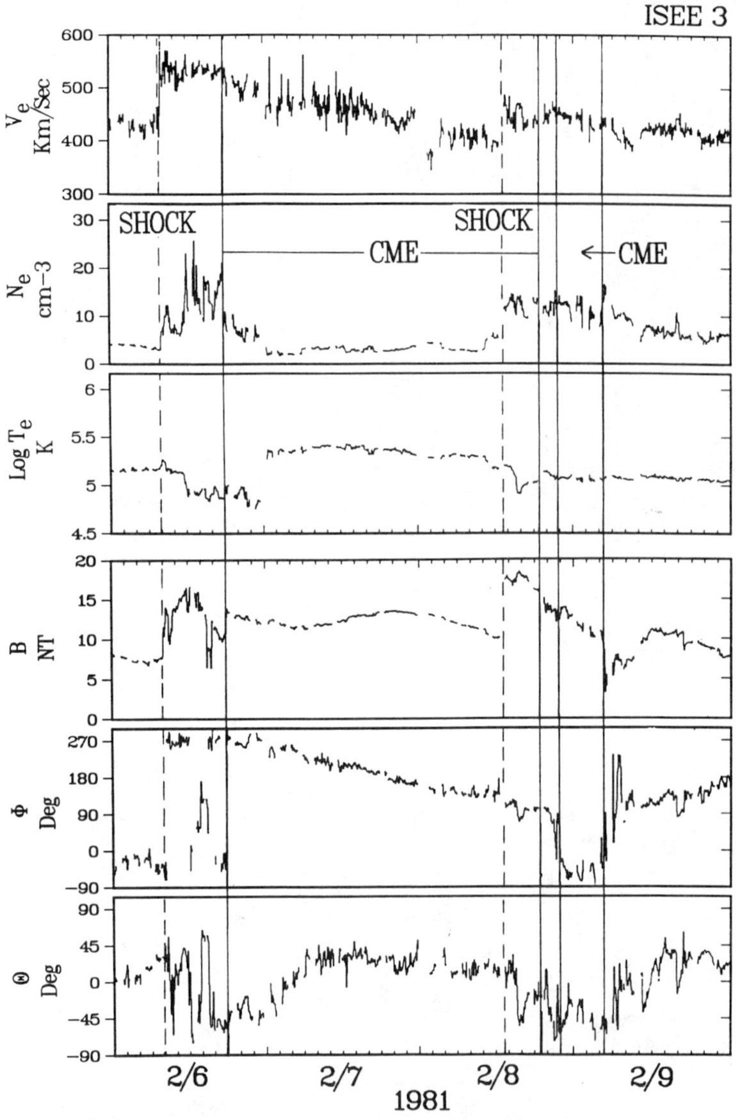

Fig. 8. Solar wind electron and field parameters from ISEE 3 for the February 6-9, 1981 interval. Note the smooth drift in field azimuth during the first counterstreaming electron event (CME), which is also a magnetic cloud.

the ISEE 3 experiment extends only ± 67.5 deg above and below the spin plane of the spacecraft (nominally the ecliptic); thus when the IMF is inclined more that ± 67.5 deg to the ecliptic, the field direction falls outside the acceptance fan of the electron experiment. Figure 10 shows that the IMF was nearly perpendicular to the ecliptic from about 1530 UT until about 2100 UT on January 1 so that any counterstreaming flux of suprathermal electrons present then would not have been detected.

The top panel of Figure 10 demonstrates that the flow within the Dec. 31/Jan. 1 event, which we identify as a CME, was about the same as that of the ambient solar wind ahead but was slower than that of the trailing solar wind. Consequently, the rear of the CME was moder-

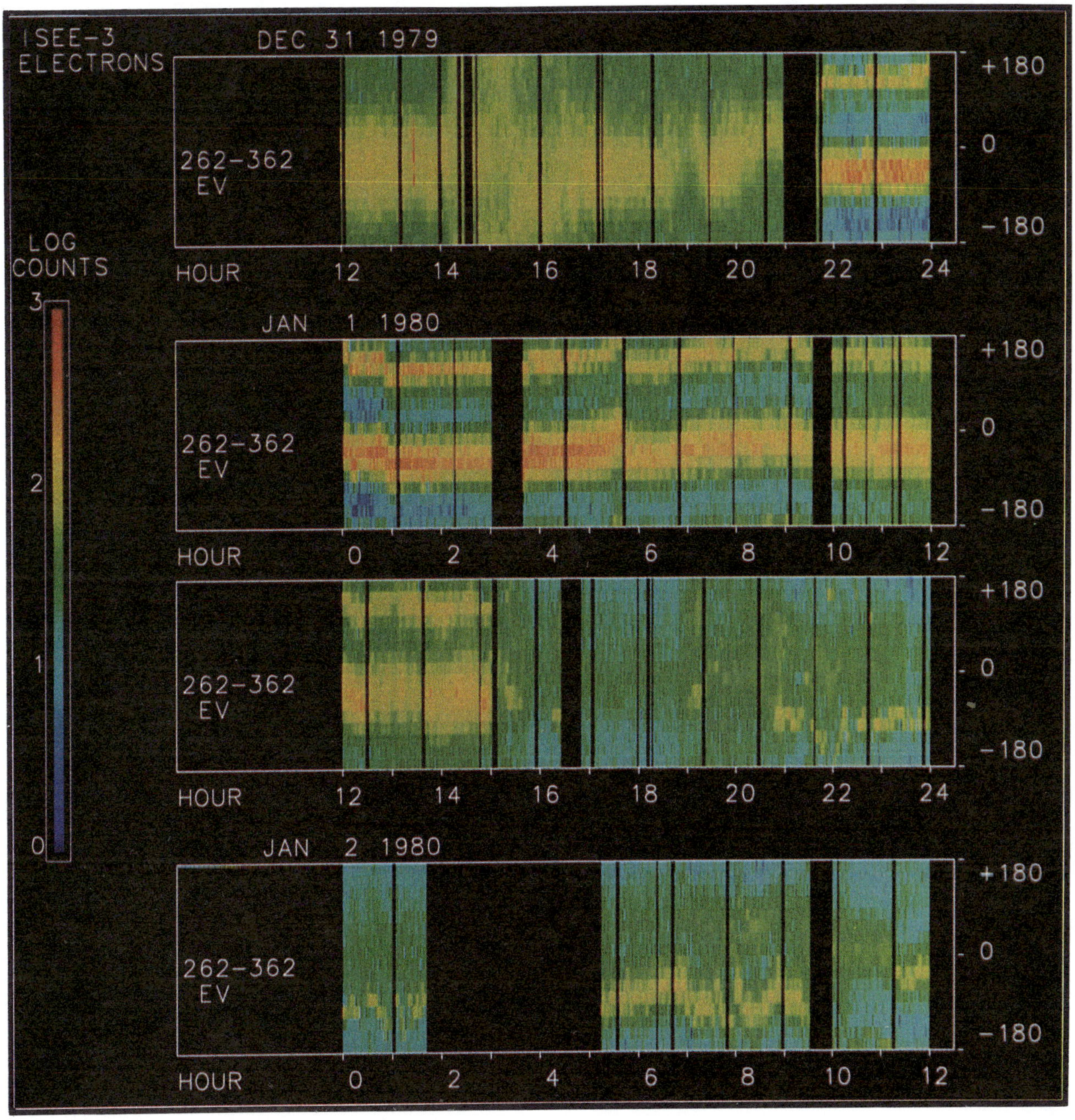

Fig. 9. Color-coded electron angular distributions from ISEE 3 in an energy passband extending from 262 to 362 eV for a 48-hr interval in late 1979 and early 1980. The counterstreaming electron event beginning about 2140 UT on December 31 was associated with a large internal field rotation, primarily in polar angle, but this event has not previously been identified as a magnetic cloud.

ately compressed, resulting in a somewhat stronger magnetic field at the back of the event as compared to the front. As for many events, the field variance was relatively low within the CME and the proton temperature was depressed; on the other hand, the electron temperature was not depressed. Of prime interest here is the fact that within the CME the field rotated from an orientation approximately 45 deg north of the ecliptic to one almost 90 deg south of the ecliptic. Note also that the field azimuth shifted abruptly by nearly 180 deg at about 1530 UT on January 1. Such rotations in polar angle and azimuth are characteristic of magnetic clouds (as is the low proton temperature), but this event has not been identified as a magnetic cloud because of the relatively modest strength of the internal field.

A number of other events similar to this exist in the ISEE 3 data although we have not yet fully catalogued them. For example, in the August 1978–December 1979 interval at least six more counterstreaming electron events have large coherent internal field rotations but were not identified as magnetic clouds by Zhang and Burlaga. All of these events were characterized by only moderately strong

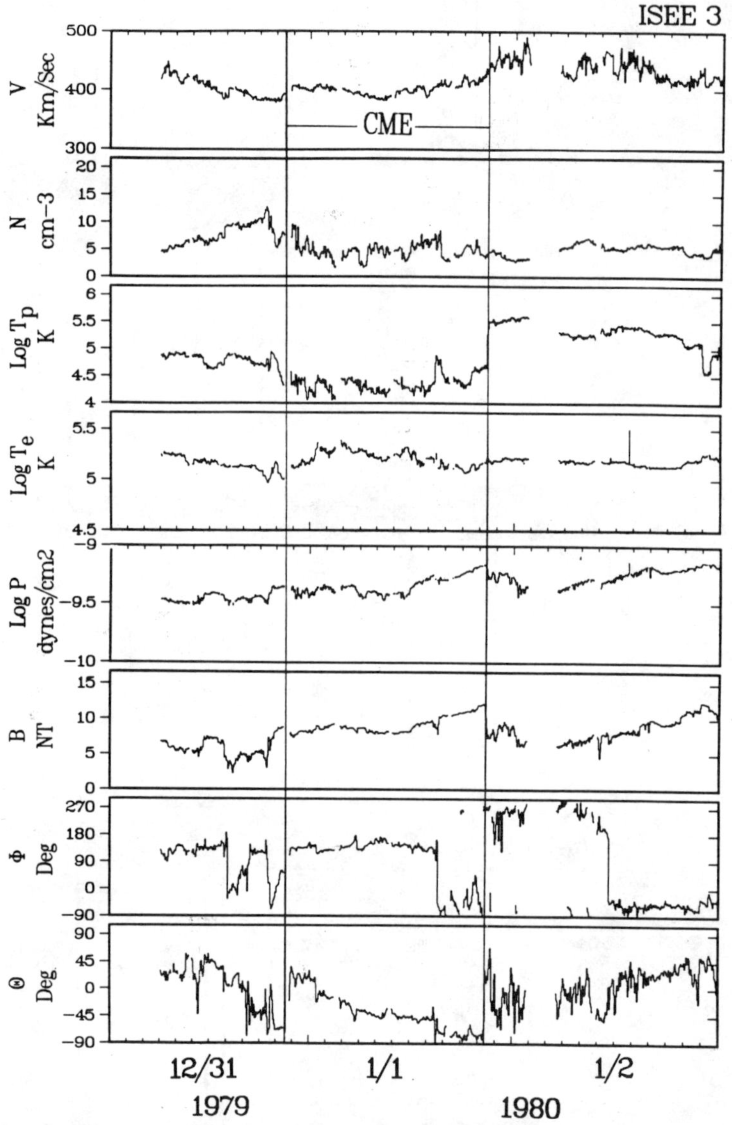

Fig. 10. Solar wind parameters from ISEE 3 surrounding the counterstreaming electron event (CME) shown in Figure 9. Note the substantial rotation in polar angle (bottom panel) during the event. Although enhanced, the magnetic field within this particular event was not sufficiently strong for the event to be classified as a magnetic cloud.

internal fields, and thus did not meet their selection criteria. The important point is that the number of CMEs having substantial, coherent internal field rotations is larger than would be inferred from statistics of magnetic clouds alone, but is still considerably smaller than the total number of CMEs in the solar wind at 1 AU, constituting perhaps 30% of the total.

Magnetic Clouds as Magnetic Flux Ropes

Various attempts have been made to visualize the global magnetic field topology responsible for the coherent internal field rotations measured as magnetic clouds and certain other CMEs pass over a spacecraft. Goldstein [1983] was apparently the first to suggest that the magnetic cloud observations could be consistently interpreted in terms of cylindrical magnetic flux ropes, characterized by axial fields near their centers and increasingly poloidal fields near their outer edges. At intermediate distances from the axis the field lines are helices with increasingly steeper pitch away from the axis of the rope. Taking note of the fact that the field minimum variance direction for magnetic clouds is commonly roughly parallel to the radial (from the sun) direction [e.g., Klein and Burlaga, 1982; Burlaga and Behannon 1982], he proposed that magnetic clouds are cylindrical flux ropes oriented with their axes roughly perpendicular to the radial direction such as shown, for example, in the lower portion of Figure 11. Ignoring the plasma pressure both internal and external to the flux rope and assuming that magnetic clouds

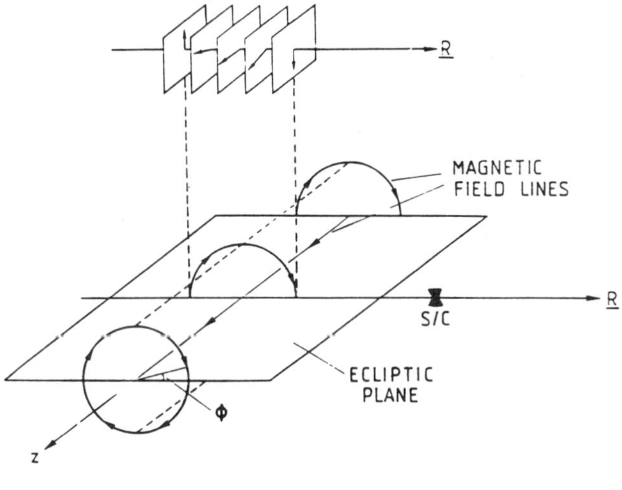

Fig. 11. Sketch of a cylindrical magnetic flux rope in the solar wind with its axis in the equatorial plane perpendicular to the radial (from the Sun) direction. The magnetic field is parallel to the axis of the rope at the center and is increasingly poloidal near the outer edge. When the flux rope passes over a spacecraft (S/C) a rotation of the field vector in a plane perpendicular to the radius vector is observed. This is illustrated in the upper part of the figure. From Goldstein [1983].

are equilibrium structures in which the internal field pressure is balanced by the curvature stress of the field lines, Goldstein showed that the field strength within such structures would be strongest near the center and would be weakest near the outside edge, consistent with the field variations observed for some magnetic clouds. The upper portion of the figure shows the characteristic internal rotation in the field that would be observed as such a cylindrical flux rope convects over a spacecraft. Note that the minimum variance direction is in the radial direction, as is commonly observed. Other orientations of the axis of the cylinder would lead to field rotations in different planes from that illustrated.

Goldstein did not attempt to make a detailed comparison between the flux rope model and observations of magnetic clouds. Subsequently Marubashi [1986] (K. Marubashi, Interplanetary magnetic clouds and solar filaments, extended abstract, Colorado Springs Meeting on Solar Events and their Influence on the Interplanetary Medium, March 1988) and Burlaga [1988] have done so for a number of events, and Suess (1988) has examined a variety of equilibrium pinch-type models to explain the observations. Following Goldstein, both Marubashi and Burlaga have assumed that magnetic clouds are equilibrium, force-free ($\nabla\times\ \mathbf{B} = \alpha\mathbf{B}$) structures in which the plasma pressure is negligible both internal and external to the flux rope and in which the internal field pressure is balanced by magnetic tension. They both allow the axis of the flux rope to have an arbitrary orientation, assume that the flux rope is intercepted at a distance, d, from the axis, and then vary the spatial scale of the flux rope, its orientation, and the interception distance until reasonable agreement is obtained with the field observations. Marubashi has assumed further that the pitch of the helical field relative to the rope axis, ω, varies with d as $\omega \sim d^2$, while Burlaga has assumed constant α throughout the flux rope.

Figure 12, adapted from Marubashi's work, provides a sample comparison between observations (upper) and model calculations (lower) for a magnetic cloud event observed in March and April 1973. The agreement between the model prediction of the field rotation within the cloud with that actually observed as the cloud passed over the spacecraft is quite good. On the other hand, the model does less well in predicting the observed variation in field strength within the cloud. In particular, typical of all equilibrium flux rope models, the calculated field strength peaks at the center of the flux rope, whereas the observed field magnitude actually peaked near the rear edge. The enhancement in field strength within the rearward portion of the cloud was almost certainly a result of interplanetary compression associated with the overtaking of the cloud by the faster solar wind behind it.

Figure 13, taken from Burlaga's work, provides a similar comparison between observations (left panel) and model calculations (right panel). Again the model does a better job of matching the observed changes in field orientation

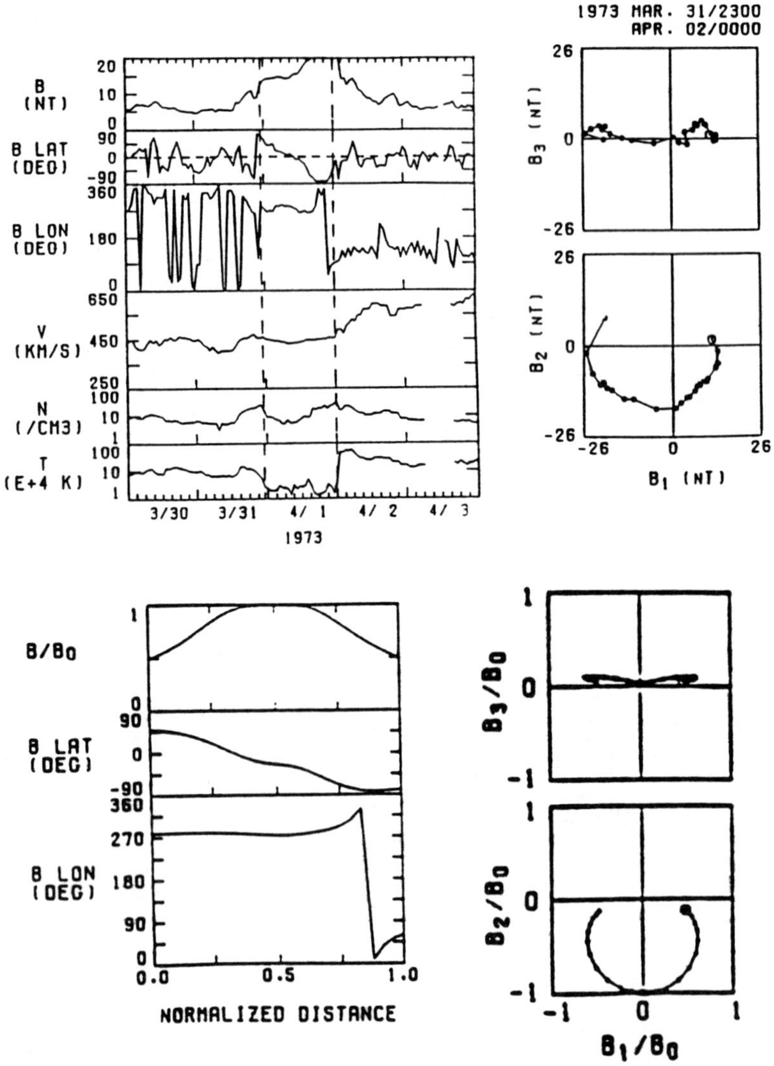

Fig. 12. A representative comparison between observations (above) and equilibrium flux rope model calculations (below) for a magnetic cloud event (indicated by the dashed lines in the upper left panel) observed in 1973. In the right side of the figure the observed and calculated field is presented in a hodogram format, with B1 and B3 being the maximum and minimum variance components respectively. The axis of the model flux rope is inclined 25 deg below the ecliptic plane at an azimuth of 270 deg. Note that the model does an excellent job of reproducing the observed change in field orientation within the cloud, but does less well in reproducing the observed change in field magnitude. Adapted from Marubashi (unpublished, extended abstract, 1988).

within the cloud than in matching the observed changes in field strength. In this case the field strength peaked near the front of the cloud rather than at the center, a consequence of the fact that this CME was overtaking ambient plasma and driving a shock disturbance.

The Marubashi and Burlaga equilibrium flux rope models are not identical, yet each does a creditable job of reproducing observations. The comparisons shown in Figures 12 and 13 are representative, and in general both models do a better job of matching observed rotations in the field than changes in field magnitude. The discrepancies between the model calculations and the observations,

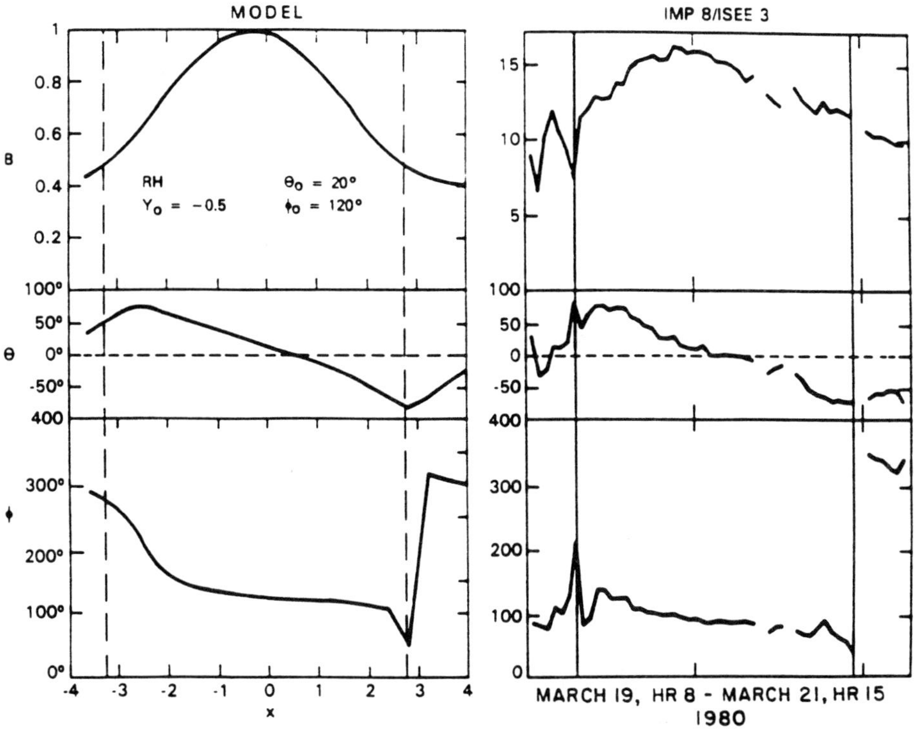

Fig. 13. A representative comparison between observations (right) and equilibrium flux rope model calculations (left) for a magnetic cloud event observed March 19-21, 1980. The axis of the model flux rope is inclined 20 deg above the ecliptic plane at an azimuth of 120 deg. Again, the model does a better job of reproducing the observed change in field orientation than the observed change in field magnitude. From Burlaga [1988].

which suggest that magnetic clouds generally are not truly force-free structures, can be at least partially attributed to dynamic processes occurring in interplanetary space. In addition, for some events it is probably not reasonable to assume that the plasma pressure is negligible, particularly in the ambient plasma surrounding the flux rope.

One characteristic predicted by all equilibrium flux-rope models is that the total static pressure, P, (ions, electrons, plus field) should peak in the center of the flux rope and be a minimum at the outer edges. Figure 14, which shows plots of log P versus time, provides a test of this prediction. Eight events are included in the figure: these include the six magnetic cloud events identified by Zhang and Burlaga [1988] for which simultaneous ion, electron, and magnetic field data are available from ISEE 3, plus 2 additional flux rope-like events (the 5th and 8th events from the top in the figure) identified using the combined magnetic field and suprathermal electron measurements. The first 6 of the events in the figure from top to bottom were shock-associated events where, as we have already seen, one expects the equilibrium models to break down. In addition, we have already shown (in Figure 10) that the last event in the figure was compressed in the rear by its interaction with a trailing high speed flow. Given all of these qualifications, it is perhaps not surprising that the predicted equilibrium pressure variation is not readily discernible for most of the events plotted in Figure 14. Indeed, variations associated with dynamic processes in interplanetary space generally dominate the total pressure profiles. We note here that the lack of equilibrium does not negate the interpretation of these structures as flux ropes.

The foregoing suggests that CMEs for which large, coherent internal field rotations are measured can reasonably be interpreted in terms of magnetic flux ropes. Nevertheless, it is well to remember that the flux rope field topology has been inferred from single spacecraft passages through what are undoubtedly large, 3-dimensional objects. Thus, although the flux rope interpretation is reasonably consistent with the observations, it is not necessarily a unique interpretation. Indeed, the possibility remains that some CMEs exhibiting large internal field rotations may be better explained in terms of other magnetic topologies.

Speed Gradients and the Radial Expansion of CMEs and Flux Ropes

We have already noted that the bulk flow speed within shock-associated CMEs at 1 AU (and magnetic clouds) usually declines steadily from the leading edge to the trailing edge, and that this front-to-rear gradient suggests that the CMEs are expanding radially as they pass 1 AU. Observations of magnetic clouds with the Voyager 1 and 2 experiments indicate that this is the case and that expansion continues unabated at larger heliocentric distances [e.g., Burlaga and Behannon, 1982]. Clear examples of the negative front-to-rear speed gradient can be seen in Figures 4, 6, 8, and 12. Radial expansion is consistent with the observation [e.g., Borrini et al., 1982; Klein and Burlaga, 1982; Gosling et al., 1987] that CMEs at 1 AU commonly have radial thicknesses of about 0.2 AU (about 45 solar radii) whereas near the sun they appear to have dimensions of the order of several solar radii. It has often been suggested that this expansion is a consequence of a higher intrinsic internal pressure within the CME than in the surrounding plasma [Klein and Burlaga, 1982; Burlaga and Behannon, 1982; Burlaga, 1989]. If true, then, as noted for example by Suess [1988], the observed expansion is a problem for equilibrium flux rope models since the higher internal pressure in a force-free flux rope should be balanced by the field curvature stress and there would be no expansion. Of course, we have already indicated that

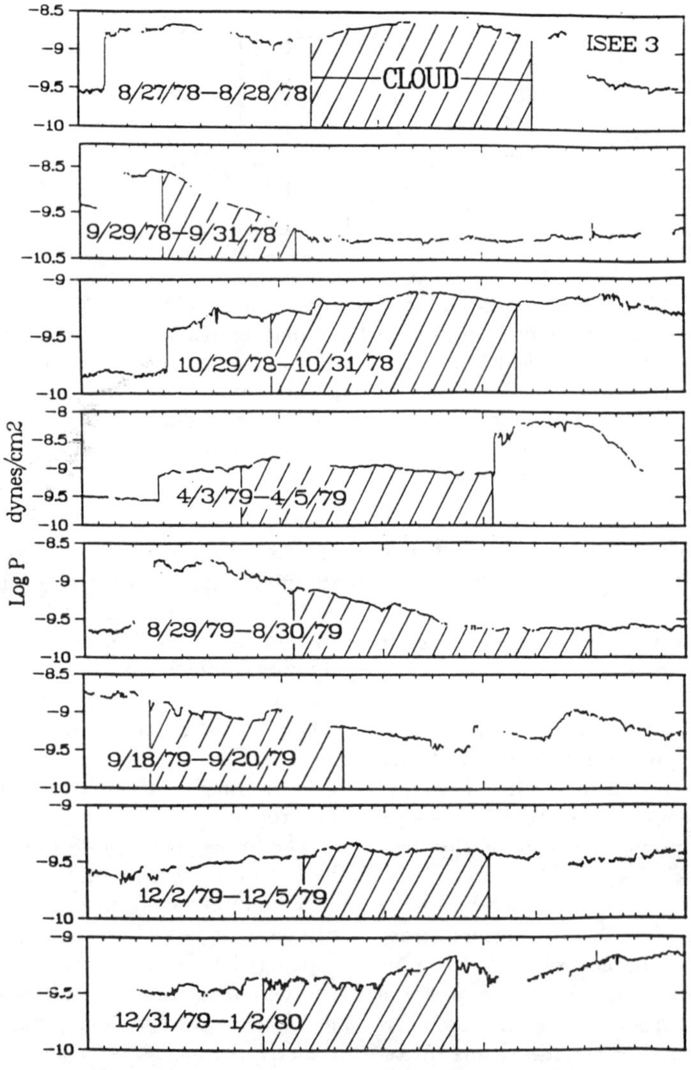

Fig. 14. Total static pressure versus time for 8 flux rope events (indicated by the cross hatching) observed by ISEE 3. By and large the pressure variations are dominated by dynamic processes operating in interplanetary space and the flux ropes are therefore often not equilibrium structures.

shock-associated magnetic clouds are almost certainly not force-free; more importantly, we wish to point out the radial expansion can be driven by forces other than an intrinsically high internal pressure. For example, CMEs may originate in the corona with a substantial front/back speed differential to begin with [e.g., Hildner, 1977; Webb and Jackson, 1981; Low and Hundhausen, 1987]. Further, as we show below, even if a fast CME originates without a higher internal pressure and without a substantial front/back speed differential, such a differential will be created in interplanetary space as the disturbance evolves with increasing distance from the sun.

Hundhausen [1985] has reviewed how transient disturbances in the solar wind should evolve with increasing heliocentric distance. Figure 15 has been adapted from that review and shows calculated radial speed and pressure profiles of a shock disturbance that has just arrived at 1 AU. As indicated at the top of the figure, the disturbance was initiated at 0.14 AU in the 1-d model calculation by abruptly raising the flow speed from 275 to 980 km/s, sustaining it at this level for 6 hours, and then returning it to its original value of 275 km/s. The temperature and density at the inner boundary were left unchanged. (There was no magnetic field in the calculation; inclusion of the field would produce quantitative differences in the calculated result but the qualitative result would be the same.) The initial disturbance thus mimics a uniformly fast, spatially limited CME with an internal pressure equal to that of the surrounding plasma.

A region of high pressure does, however, develop on the leading edge of the disturbance as the faster CME overtakes the ambient solar wind. This region of higher pressure is bounded by a forward shock on its leading edge which propagates into the ambient solar wind ahead and a reverse shock on its trailing edge which propagates backward into the CME. Both shocks are, however, carried outward away from the sun by the convective flow of the solar wind. By the time the leading edge of the disturbance reaches 1 AU, the reverse shock in this example has propagated entirely through the CME and is much weaker than the forward shock. Both simulations [D'Uston et al., 1981] and observations [Gosling et al., 1988] indicate that the reverse shock should ordinarily be observed only near the center line of the disturbance.

During the outward transit of the disturbance the speed and pressure profiles evolve into a characteristic sawtooth form and the overall CME slows considerably as a result of momentum transfer to the ambient medium. A rarefaction wave, produced as the CME pulls away from the slower, trailing solar wind, propagates forward toward the front of the disturbance, causing an erosion of the rearward portion of the region of high pressure. This erosion and the resulting sawtoothlike pressure and speed profiles are characteristic of disturbances initiated by speed enhancements where the input time (in this case, 6 hours) is short compared to the transit time to 1 AU (in this case, 44 hours) [e.g., Hundhausen and Gentry, 1969]. Thus,

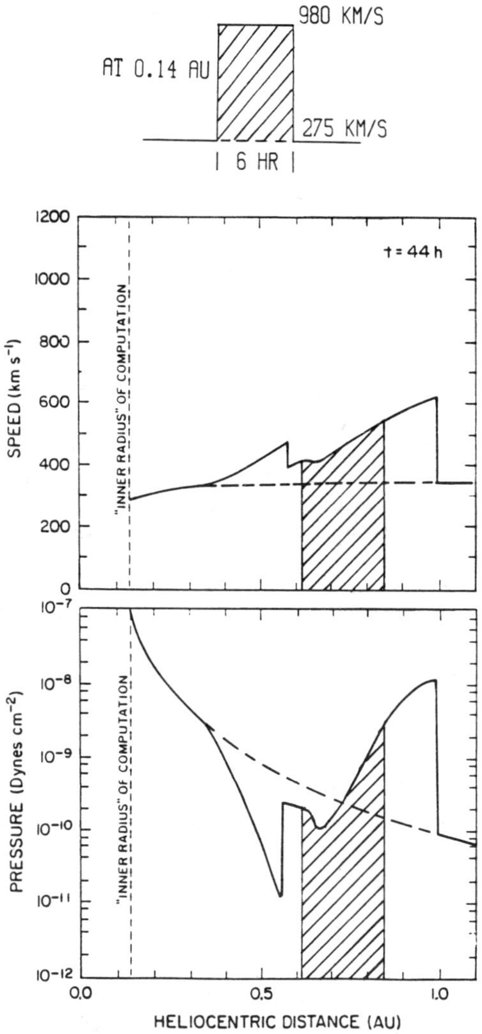

Fig. 15. Solar wind speed and pressure as functions of heliocentric distance for a numerical simulation of a weakly driven disturbance. The temporal variation in flow speed imposed at the inner boundary of 0.14 AU is shown at the top of the figure. The hatching identifies material that was introduced with a speed of 980 km/s at the inner boundary, and therefore identifies the CME in the simulation. Adapted from Hundhausen [1985] and reproduced in Gosling et al. [1988].

the radial speed gradient across a CME and the resulting expansion of the CME can result solely from dynamical effects in interplanetary space and may have nothing at all to do with a higher intrinsic internal pressure. (In the particular case illustrated the CME has expanded by almost a factor of two in travelling from 0.1 to 0.85 AU even though there was no initial front-to-rear speed gradient to begin with, and even though the internal pressure initially was the same as the external pressure.) We suspect, however, that both a higher intrinsic internal pressure and an initial front-to-rear speed differential do also contribute substantially to the expansion of many CMEs, especially close to the Sun.

Observational Summary – Magnetic Flux Ropes in Interplanetary Space

From the foregoing it is apparent that a strong case can be made for the existence of large, coherent magnetic flux ropes which propagate outward from the Sun through interplanetary space. These flux ropes at 1 AU have typical radial dimensions of the order of 0.1–0.2 AU and form a subset constituting about 30% of all coronal mass ejection events in the solar wind. The flux ropes are closed magnetic structures with field lines being either rooted at both ends in the Sun or completely disconnected from it (plasmoids). Often, but not always, the flux ropes can be recognized as "magnetic clouds," characterized by low proton temperatures, strong internal magnetic fields ($\gtrsim$ 10 nT), relatively large and coherent internal field rotations, and low beta. Compression in interplanetary space contributes substantially to the particularly strong fields observed in many shock-associated magnetic clouds (and fast CMEs in general); however, CMEs do tend to have somewhat stronger internal magnetic fields than the surrounding solar wind even without the added boost from interplanetary compression. Nevertheless, some CME events with large, coherent internal field rotations do not have sufficiently strong internal fields to be identified as magnetic clouds according to the usual definition. These previously unrecognized flux rope-like events appear to be somewhat more numerous than events which qualify as magnetic clouds.

The axes of interplanetary flux ropes usually lie in a plane which is oriented roughly perpendicular to the radial direction. The flux ropes need not be equilibrium, force-free structures, although force-free models often do an excellent job of modeling the internal field rotations observed as a flux rope passes over a satellite. The flux ropes expand radially (and probably in the transverse direction as well) as they propagate outward from the Sun. This expansion can arise from an initial spread in outward speed close to the sun or from an imbalance between the internal plasma and field pressure, the curvature stress of the flux rope field lines, and the external plasma and field pressure. For shock-associated events this imbalance can result solely from the dynamic interaction of the flux ropes with the ambient solar wind.

Flux rope boundary determinations are an imprecise art when based solely on magnetic field and/or plasma moment determinations. These boundary determinations can be made more precise with the aid of suprathermal electron measurements. In particular, whenever there is relative motion between a flux rope and the surrounding solar wind, the electron measurements help one to distinguish the closed magnetic field lines of the flux rope from the compressed, draped magnetic field lines of the ambient plasma.

Discussion: A Suggested Origin for Interplanetary Flux Ropes

One of the remarkable aspects of interplanetary magnetic flux ropes is the large spatial scale over which these structures are coherent entities. For example, the plasma and field data shown in Figure 7 and Figure 8 indicate that the transverse dimension of a flux rope can extend over a radial distance as large as 0.4 AU at 1 AU, and even larger events can be found in the data. It is interesting that CMEs at 1 AU never seem to be comprised of more than a single flux rope; that is, CME events consisting of multiple flux ropes do not seem to exist, or at least have not yet been identified in the data.

It is also interesting that optical observations of CMEs in the outer corona with satellite coronagraphs do not lead one to suspect that CMEs in interplanetary space should have a flux rope structure. To the contrary, as reviewed by Hundhausen at this meeting, CMEs in the outer corona generally appear to be comprised largely of untwisted magnetic field lines. (One must remember, or course, that the coronagraph observations do not provide a direct measurement of the magnetic field structure within CMEs; rather, the magnetic structure must be inferred from the visible appearance of events.) On the other hand, eruptive solar prominences, which are often embedded within CMEs but which typically occupy only a small fraction of their total volume, usually do have the optical appearance of twisted flux ropes. However, we believe it is unlikely that the interplanetary structures which we have been calling CMEs are instead nothing more than solar prominences.

Coronal observations suggest strongly that CMEs usually are 3-dimensional structures comprised largely of a series of closed coronal loops, which become extended outward into interplanetary space [e.g., Hundhausen, 1988; Kahler, 1988]. Although the physical processes responsible for extending the loops outward are not entirely understood, there is some suggestion that CME events are often initiated by a gradual readjustment of the global coronal field leading to an instability in which the coronal loops comprising the helmet portion of coronal streamers are propelled outward by the buoyancy of the underlying coronal cavity (see, for example, the review by Hundhausen, [1988]). In contrast to some early models of CME initiation [e.g., Anzer and Pneuman 1982], magnetic reconnection plays no central role in this scenario; however, reconnection may still occur as a result of the outward extension of the coronal loops [Kopp and Pneuman, 1976].

Figure 16, which was inspired by conversations with N. Crooker, J. Birn, and M. Hesse, provides a series of three sketches illustrating the possible temporal evolution of a set of large rising coronal loops, such as might originally overlay a solar prominence and whose footpoints in the photosphere are sheared relative to one another. If reconnection occurs as the loops rise, adjacent magnetic loops interconnect (the individual magnetic loops do

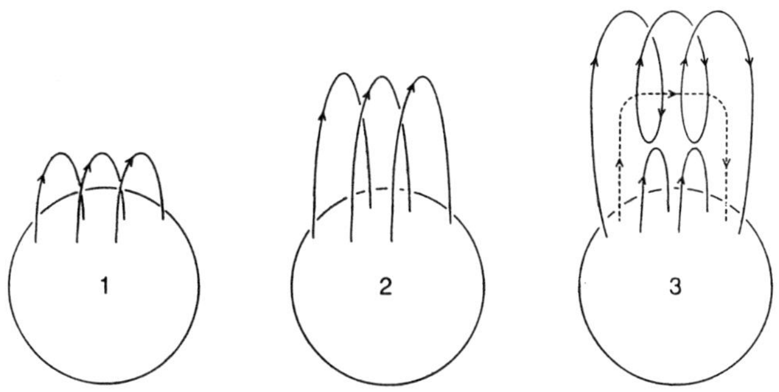

Fig. 16. A series of sketches illustrating the possible temporal evolution of a set of large rising magnetic loops in the solar corona such as might be associated with a coronal mass ejection event arising from within a large coronal streamer. Magnetic reconnection causes a flux rope to form when the footpoints of the loops are sheared relative to one another.

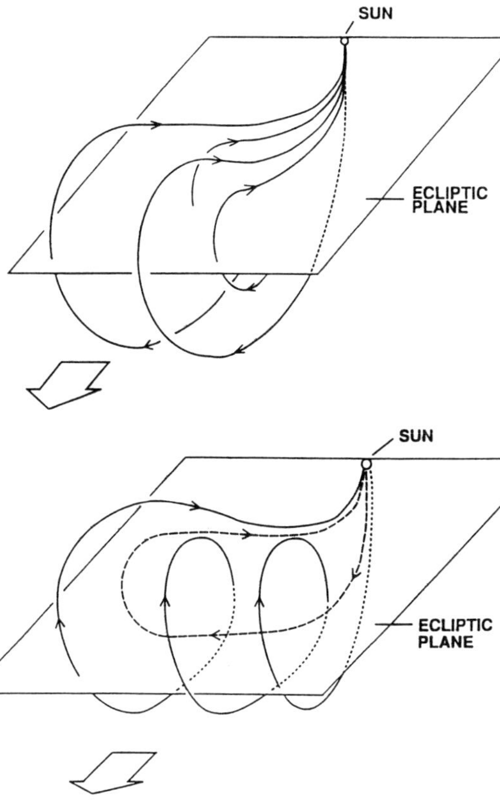

Fig. 17. The interplanetary extensions of rising coronal loops which have and have not undergone reconnection (lower and upper panels, respectively). For simplicity the loops are drawn roughly perpendicular to the ecliptic, although any orientation relative to the ecliptic is possible in principle.

not reconnect with themselves) and a flux rope is formed. (See also, Crooker et al., [1989].) The situation is similar to what is believed to occur within the plasma sheet of Earth's magnetotail during substorms in the presence of a finite cross-tail field component. Such reconnection has been modeled by Birn and Hesse [1989a,b] whose simulations demonstrate formation of helical field lines similar to that sketched in Figure 16. The pitch of these field lines typically decreases toward the center of the rope, eventually becoming nearly parallel to the axis of the rope (dashed line in the figure) at the center. Thus, the net effect of reconnection in rising coronal loops should be the formation of a rising flux rope whose ends are still rooted in the Sun, plus the formation of a new series of closed magnetic loops low in the corona. Note that reconnection need not necessarily occur when the leading edges of the coronal loops are still relatively close to the sun. On the contrary, reconnection may preferentially occur when the leading edges of the loops are well out into interplanetary space. One might possibly identify the ends of the flux rope with the "legs" of a CME, which are known to persist rooted in the photosphere for days [e.g., Gosling et al., 1974], and the newly formed closed loops with "post-flare" loops. Interestingly, reconnection as described here seems to provide a natural way of relating the relatively large scale size of the outer envelope of a CME with the much smaller scale size of the underlying "post-flare" (actually post-CME is a better word choice) structure in the chromosphere and low corona.

Figure 17 provides two sketches illustrating the interplanetary extension of a set of rising magnetic loops from the corona. In the upper portion of the figure the loops have been extended outward under the assumption that reconnection does not occur as the loops rise. In the case shown the loops have been drawn nearly perpendicular to the ecliptic for ease of illustration; however, virtually any other orientation is possible, depending upon the original orientation of the loops in the corona. The loops will be bent by solar rotation somewhat as drawn. The passage of such a set of loops over a spacecraft at (say) 1 AU would produce a counterstreaming suprathermal electron event, field rotations at entry and exit (the magnitude of which would depend on the relative orientation between the ambient field and the loops), and a relatively steady field orientation internal to the event. Of course, the loops need not necessarily lie in parallel planes so some fluctuation in field direction might be expected, as is often observed. The sketch in the lower portion of Figure 17, on the other hand, shows the interplanetary extension of a series of loops which have reconnected to form a flux rope as in Figure 16. In this case a coherent rotation in the field would be observed as the flux rope passes over a spacecraft.

The foregoing suggests that interplanetary magnetic flux ropes are a consequence of reconnection of rising, previously sheared, coronal loops. It is of interest that this reconnection should produce an interplanetary structure that is something of a mixture between a fully disconnected plasmoid and a magnetic bottle rooted at both ends in the solar photosphere. Note that the type of reconnection envisioned produces a single large flux rope in interplanetary space, as is observed, rather than multiple flux ropes.

On the other hand, we are aware that there is little direct optical evidence, other than the persistence of CME "legs" and the formation of "post-flare" loops, for the type of structure we have sketched. (However, see Illing and Hundhausen [1983].) Further, the fact that flux ropes form a relatively small subset of all coronal mass ejection events in the solar wind at 1 AU indicates that this type of reconnection probably occurs in only a small fraction (~ 0.3) of all CME events. For most other events reconnection must be occurring elsewhere on the sun in order to avoid a net long-term build up of magnetic flux in interplanetary space [e.g., Gosling, 1975; Newkirk et al., 1981]. It has recently been suggested [McComas et al., 1989] that the reconnection required to maintain a relatively constant magnetic flux in interplanetary space may commonly occur at the heliospheric current sheet, producing "U-shaped" magnetic structures in interplanetary space which are disconnected from the solar photosphere and which are relatively devoid of the hot electrons which normally carry the solar wind electron heat flux.

Questions and Answers

Akasofu: How can you observe the central part of CMEs by earth-bound satellites? How can you be sure what you observe are CMEs? You have identified a distinct decrease of the temperature as an important indication of CME (=driver gas). Yet, at least three cases you presented, the temperature showed a clear increase during the passage of what you identified to be CMEs.

Gosling: There is no universal signature of a CME that I am aware of. However, as I outlined in my talk, CMEs in the solar wind at 1AU commonly are characterized by one or more of a variety of signatures, the most common of which appears to be a bidirectional electron heat flux. The solar wind structures which we identify as CMEs are often (~50%) found behind interplanetary shocks and are separated from the shocks by disturbed sheaths of shocked plasma. There is very good reason to believe that these structures are fast CMEs since they are the objects driving the shocks. When observed in the absence of shocks, these structures have essentially the same type of signature except that their velocities are lower than or the same as that of the ambient plasma ahead. Further, a number of events which we recognize as CMEs in the solar wind have been correlated directly with CMEs observed with coronagraphs.

I see no problem with identifying the central part of a CME since the entire structure is convected past a spacecraft by its outward motion.

Statistically speaking, the structures we identify as CMEs in the solar wind often have anomalously low kinetic temperatures. However, there are exceptions to the rule, and the events you refer to are such exceptions. Indeed, that is one of the reasons I chose to show them. It helps emphasize why the bidirectional electron heat flux is a more reliable signature. Observed kinetic temperatures will depend on the initial kinetic temperatures (which can vary from event to event), the amount of expansion experienced en route to 1AU (which also can vary), and factors such as density, velocity, and magnetic field orientation(which also very and which all affect temperature evolution with distance from the sun).

Forbes: Do you see any special magnetic field features in those CMEs which contain He in low energy ionization states?

Gosling: Not that I am aware of. However, I personally have never looked closely at the field variations within He+ events and I suspect nobody else has either.

Luhmann: Jack, Rich Elphic pointed out that there is an ambiguity in the interpretation of magnetic structures where the external draping geometry can "mimic" the magnetic signature of a flux rope. Have any of the studies where clouds or CMEs have been fitted with flux rope models taken into account that there is also this exterior draping?

Gosling: Draping should only occur for CMEs where there is relative motion between the CME and the ambient solar wind. In several of our papers we have considered the draping problem - indeed, draping can provide external field rotations which can mimic part of the signature of a flux rope. To the best of my knowledge, none of the magnetic cloud studies in the literature have taken draping into account specifically. However, my examination of magnetic cloud events reported by Burlaga and colleagues suggests that they have not, in general, confused draping for flux ropes. The models have assumed that the magnetic clouds are equilibrium structures riding along with the solar wind. With no relative motion and no expansion there is no draping. On the other hand, some of the model comparisons with actual observations have been for events driving shocks where the internal pressure is not balanced by magnetic tension and where there is significant relative motion between the CMEs (clouds) and the ambient plasma ahead. Nevertheless, examination of the bidirectional electron signature for these events suggests they have not been misled by draping effects.

Acknowledgments. I wish to thank J. Birn, M. Hesse, and A. J. Hundhausen for conversations on various aspects of this paper, T. Onsager for reviewing a draft of this manuscript, J. L. Phillips for assistance in extending the survey of counterstreaming electron events through 1982, E. J. Smith for use of the ISEE 3 magnetic field data, and L. F. Burlaga and both referees for their comments. This work was performed under the auspices of the U.S. Department of Energy with support from NASA under S-04039-D.

References

Anzer, U. and G. W. Pneuman, Magnetic reconnection and coronal transients, *Solar Phys.*, *79*, 129, 1982.

Bame, S. J., J. R. Asbridge, W. C. Feldman, E. E. Fenimore, and J. T. Gosling, Solar wind heavy ions from flare heated coronal plasma, *Sol. Phys.*, *62*, 179, 1979.

Bame, S. J., J. R. Asbridge, W. C. Feldman, J. T. Gosling, and R. D. Zwickl, Bi-directional streaming of solar wind electrons > 80 eV: Evidence for a closed field structure within the driver gas of an interplanetary shock, *Geophys. Res. Lett.*, *8*, 173, 1981.

Birn, J., and M. Hesse, MHD simulations of magnetic reconnection in a skewed three-dimensional tail configuration, *J. Geophys. Res.*, in press, 1989a.

Birn, J., and M. Hesse, The magnetic topology of the plasmoid flux rope in a MHD simulation of magnetic reconnection in the magnetotail, this volume, 1989b.

Borrini, G., J. T. Gosling, S. J. Bame, and W. C. Feldman, Helium abundance enhancements in the solar wind, *J. Geophys. Res.*, *87*, 7370, 1982.

Burlaga, L. F., Magnetic clouds and force-free fields with constant alpha, *Geophys. Res.*, *93*, 7217, 1988.

Burlaga, L. F., Magnetic clouds, chapter 5, in *Physics of the Inner Heliosphere*, edited by L. Lanzerotti, R. Schwenn, and E. Marsch, Springer-Verlag, in press, 1989.

Burlaga, L. F., and J. H. King, Intense interplanetary magnetic fields observed by geocentric spacecraft during 1963-1975, *J. Geophys. Res.*, *84*, 6633, 1979.

Burlaga, L. F., and K. W. Behannon, Magnetic clouds: Voyager observations between 2 and 4 AU, *Sol. Phys.*, *81*, 181, 1982.

Burlaga, L. F., E. Sittler, F. Mariani, and R. Schwenn, Magnetic loop behind an interplanetary shock: Voyager, Helios, and IMP 8 observations, *J. Geophys, Res.*, *86*, 6673, 1981.

Burlaga, L. F., L. Klein, N. R. Sheeley, D. J. Michels, R. A. Howard, M. J. Koomen, R. Schwenn, and H. Rosenbauer, A magnetic cloud and a coronal mass ejection, *Geophys. Res. Lett.*, *9*, 1317, 1982.

Crooker, N. U., J. T. Gosling, E. J. Smith, and C. T. Russell, A bubblelike coronal mass ejection flux rope in the solar wind, this volume, 1989.

D'Uston, C., M. Dryer, S. M. Han, and S. T. Wu, Spatial structure of flare-associated perturbations in the solar wind simulated by a two-dimensional numerical MHD model, *J. Geophys. Res.*, *86*, 525, 1981.

Feldman, W. C., J. R. Asbridge, S. J. Bame, and M. D. Montgomery, Solar wind heat transport in the vicinity of the earth's bow shock, *J. Geophys Res.*, *78*, 3697, 1973.

Feldman, W. C., J. R. Asbridge, S. J. Bame, M. D. Montgomery, and S. P. Gary, Solar wind electrons, *J. Geophys. Res.*, *80*, 4181, 1975.

Fenimore, E. E., Solar wind flows associated with hot heavy ions, *Astrophys. J.*, *235*, 245, 1980.

Goldstein, H., On the field configuration in magnetic clouds, in *Solar Wind Five*, NASA Conference Publ. 2280, edited by M. Neugebauer, pp. 731–733, 1983.

Gosling, J. T., Large scale inhomogeneities in the solar wind of solar origin, *Rev. Geophys. and Space Phys.*, *13*, 1053, 1975.

Gosling, J. T., and D. J. McComas, Field line draping about fast coronal mass ejecta: A source of strong out-of-the-ecliptic interplanetary magnetic fields, *Geophys. Res. Lett.*, *14*, 355, 1987.

Gosling, J. T., V. Pizzo, and S. J. Bame, Anomalously low proton temperatures in the solar wind following interplanetary shock waves: Evidence for magnetic bottles?, *J. Geophys. Res.*, *78*, 2001, 1973.

Gosling J. T., E. Hildner, R. M. MacQueen, R. H. Munro, A. I. Poland, and C. L. Ross, Mass ejections from the sun: A view from Skylab, *J. Geophys. Res.*, *79*, 4581, 1974.

Gosling, J. T., E. Hildner, R. M. MacQueen, R. H. Munro, A. I. Poland, and C. L. Ross, The speeds of coronal mass ejection events, *Solar Phys.*, *48*, 389, 1976.

Gosling, J. T., J. R. Asbridge, S. J. Bame, W. C. Feldman, and R. D. Zwickl, Observations of large fluxes of He^+ in the solar wind following an interplanetary shock, *J. Geophys. Res.*, *85*, 3431, 1980.

Gosling, J. T., D. N. Baker, S. J. Bame, W. C. Feldman, R. D. Zwickl, and E. J. Smith, Bidirectional solar wind electron heat flux events, *J. Geophys. Res.*, *92*, 8519, 1987.

Gosling, J. T., S. J. Bame, E. J. Smith, and M. E. Burton, Forward-reverse shock pairs associated with transient disturbances in the solar wind at 1 AU, *J. Geophys. Res.*, *93*, 8741, 1988.

Hildner, E., Mass ejections from the solar corona into interplanetary space, in *Study of Travelling Interplanetary Phenomena*, edited by M. A. Shea, D. Smart, and S. T. Wu, p. 3, D. Reidel, Hingham, Mass., 1977.

Hirshberg, J., and D. S. Colburn, Interplanetary field and geomagnetic variations: A unified view, *Planet. Space Sci.*, *17*, 1183, 1969.

Hirshberg, J., S. J. Bame, and D. E. Robbins, Solar flares and solar wind helium enrichments: July 1965-July 1967, *Sol. Phys.*, *23*, 467, 1972.

Howard, R., N. R. Sheeley, M. J. Koomen, and D. J. Michels, Coronal mass ejections: 1979–1981, *J. Geophys. Res.*, *90*, 8173, 1985.

Hundhausen, A. J., Some Macroscopic properties of shock waves in the heliosphere, in *Collisionless Shocks in the Heliosphere: A Tutorial Review*, Geophys. Monogr. Ser., vol. 34, edited by R. G. Stone and B. T. Tsurutani, pp. 37–58, AGU, Washington D.C., 1985.

Hundhausen, A. J., The origin and propagation of coronal mass ejections, in *Proceedings of the Sixth International Solar Wind Conference*, edited by V. Pizzo, T. E. Holzer, and D. G. Sime, NCAR/TN-306+Proc, Boulder, Colo., pp. 181–214, 1988.

Hundhausen, A. J., and R. A. Gentry, Effects of solar flare duration on a double shock pair at 1 AU, *J. Geophys. Res.*, *74*, 6229, 1969.

Hundhausen, A. J., S. J. Bame, and M. D. Montgomery, The large scale characteristics of flare-associated solar wind disturbances, *J. Geophys. Res.*, *75*, 4631, 1970.

Illing, R. M. E., and A. J. Hundhausen, Possible observation of a disconnected magnetic structure in a coronal transient, *J. Geophys. Res.*, *88*, 10,210, 1983.

Kahler, S., Coronal mass ejections, *Revs. Geophys.*, *25*, 663, 1987.

Kahler, S., Observations of coronal mass ejections near the sun, in *Proceedings of the Sixth International Solar Wind Conference*, edited by V. Pizzo, T. E. Holzer, and D. G. Sime, NCAR/TN-306+Proc, Boulder, Colo., pp. 215–231, 1988.

Klein, L. W., and L. F. Burlaga, Interplanetary magnetic clouds at 1 AU, *J. Geophys. Res.*, *87*, 613, 1982.

Kutchko, F. J., P. R. Briggs, and T. P. Armstrong, The bidirectional particle event of October 12, 1977, possibly associated with a magnetic loop, *J. Geophys. Res.*, *87*, 1419, 1982.

Kopp, R. A., and G. W. Pneuman, Magnetic reconnection in the corona and loop prominence phenomenon, *Sol. Phys.*, *50*, 85, 1976.

Low, B. C., and A. J. Hundhausen, The velocity field of a coronal mass ejection: The event of September 1, 1980, *J. Geophys. Res.*, *92*, 2221, 1987.

Marsden, R. G., T. R. Sanderson, C. Tranquille, K.-P. Wenzel, and E. J. Smith, ISEE 3 observations of low-energy proton bidirectional events and their relation to isolated interplanetary magnetic structures, *J. Geophys. Res.*, *92*, 11,009, 1987.

Marubashi, K., Structure of the interplanetary magnetic clouds and their solar origins, *Adv. Space Res.*, *6*, 1, 1986.

McComas, D. J., J. T. Gosling, J. L. Phillips, S. J. Bame, J. G. Luhmann, and E. J. Smith, Electron heat flux dropouts in the solar wind: Evidence for interplanetary magnetic field reconnection?, *J. Geophys. Res.*, *94*, 6907, 1989.

Montgomery, M. D., J. R. Asbridge, S. J. Bame, and W. C. Feldman, Solar wind electron temperature depressions following some interplanetary shock waves: Evidence for magnetic merging?, *J. Geophys. Res.*, *79*, 3103, 1974.

Newkirk, G., A. J. Hundhausen, and V. Pizzo, Solar cycle modulation of galactic cosmic rays: Speculation on the role of coronal transients, *J. Geophys. Res.*, *86*, 5387, 1981.

Palmer, I. D., F. R. Allum, and S. Singer, "Bidirectional anisotropies in solar cosmic ray events: Evidence for magnetic bottles," *J. Geophys. Res.*, *83*, 75, 1978.

Pudovkin, M. I., S. A. Zaitseva, and E. E. Benevslenska, The structure and parameters of flare streams, *J. Geophys. Res.*, *84*, 6649, 1979

Rosenbauer, H., R. Schwenn, E. Marsch, B. Meyer, H. Miggenrieder, M. D. Montgomery, K.-H. Muhlhauser, W. Pilipp, W. Voges, and S. M. Zink, A survey of initial results of the Helios plasma experiment, *J. Geophys.*, *42*, 561, 1977.

Schwenn, R., Relationship of coronal transients to interplanetary shocks: 3D aspects, *Space Sci. Rev.*, *44*, 139, 1986.

Schwenn, R., H. Rosenbauer, and K.-H. Muhlhauser, Singly ionized helium in the driver gas of an interplanetary shock wave, *Geophys. Res. Lett.*, *7*, 201, 1980.

Sheeley, N. R., R. A. Howard, M. J. Koomen, D. J. Michels, R. Schwenn, K.-H. Muhlhauser, and H. Rosenbauer, Coronal mass ejections and interplanetary shocks, *J. Geophys. Res.*, *90*, 163, 1985.

Stansberry, J. A., J. T. Gosling, M. F. Thomsen, S. J. Bame, and E. J. Smith, Interplanetary magnetic field orientations associated with bidirectional electron heat fluxes detected at ISEE-3, *J. Geophys. Res.*, *93*, 1975, 1988.

Suess, S. T., Magnetic clouds and the pinch effect, *J. Geophys. Res.*, *93*, 5437, 1988.

Temnyi, V. V., and O. L. Vaisberg, A dumbbell distribution of epithermal electrons in the solar wind based on observations on the Prognoz 7 satellite, *Kosm. Issled.*, *17*, 580, 1979. (Cosmic Res., Engl. Transl., 17, 476, 1979).

Webb, D. F., and B. V. Jackson, Kinematical aspects of flare spray ejecta observed in the corona, *Solar Phys.*, *73*, 341, 1981.

Zhang, G., and L. F. Burlaga, Magnetic clouds, geomagnetic disturbances, and cosmic ray decreases, *J. Geophys. Res.*, *93*, 2511, 1988.

Zwickl, R. D., J. R. Asbridge, S. J. Bame, W. C. Feldman, and J. T. Gosling, He^+ and other unusual ions in the solar wind: A systematic search covering 1972-1980, *J. Geophys. Res.*, *87*, 7379, 1982.

Zwickl, R. D., J. R. Asbridge, S. J. Bame, W. C. Feldman, J. T. Gosling, and E. J. Smith, Plasma properties of driver gas following interplanetary shocks observed by ISEE 3, in *Solar Wind 5*, NASA Conf. Publ., CP-2280, edited by M. Neugebauer, pp. 711–717, 1983.

A BUBBLELIKE CORONAL MASS EJECTION FLUX ROPE IN THE SOLAR WIND

N. U. Crooker

Department of Atmospheric Sciences, University of California
Los Angeles, CA 90024–1565

J. T. Gosling

University of California, Los Alamos National Laboratory
Los Alamos, NM 87545

E. J. Smith

Jet Propulsion Laboratory, California Institute of Technology
Pasadena, CA 91109

C. T. Russell

Institute of Geophysics and Planetary Physics, University of California
Los Angeles, CA 90024–1567

Abstract. A resolution to the question of whether coronal mass ejections are loops or bubbles is proposed and applied to the geometrical analysis of a solar wind event detected at 1 AU by ISEE 1 and 3. The discontinuity orientations, the size determined by time of passage, and the magnetic cloud signature are fit into the topology of a flux rope loop distorted by expansion into a thick rope with comparable dimensions in both the ecliptic and meridional planes. The looped rope fills a bubblelike cavity, thus preserving both types of proposed coronal mass ejection geometries.

Other interesting features of the data include an apparent separation by the rope core of bidirectionally streaming protons in the leading section from electrons in the trailing section, possible vortical flow within the magnetic cloud, and a well-defined filamentary structure behind the shock.

Introduction

The geometry of coronal mass ejections has been the subject of considerable discussion [e.g., Schwenn, 1986; Webb, 1988]. A central question concerns whether ejections are loops or bubbles. Loops are treated as two-dimensional, with a thickness that is small compared to the dimensions in the plane of the loop, whereas bubbles are three-dimensional objects. Schwenn and others argue that purported coronal mass ejection signatures in the solar wind appear to be bubblelike plasmoids, and Webb presents evidence that looplike ejections observed in white light coronagraphs have a bubblelike rather than a planar geometry. Thus the observations favor the three-dimensional geometry, although no magnetic topology has been proposed for bubbles.

On the other hand, as discussed by Webb [1988], the close association between erupting prominences and many coronal mass ejection events suggests that ejections have a looplike magnetic topology based on the known topology of prominences. In this view, as a prominence erupts to form a simple magnetic loop, the overlying, nearly orthogonal, coronal arcade magnetic field reconnects, forming a helical shell around the prominence loop. The net topology is a magnetic flux rope loop rather than a simple magnetic loop, but models of its geometry remain basically planar [e.g., Mouschovias and Poland, 1978].

Roughly 10–30% of solar wind events that have been classified as coronal mass ejections by the presence of bidirectionally streaming electrons include signatures of magnetic clouds [Gosling, this volume]. We restrict

Geophysical Monograph 58

our discussion to this subclass of events. Analysis of magnetic cloud data supports flux rope topology. The characteristic magnetic signature of a magnetic cloud was immediately recognized as a possible closed magnetic configuration of a coronal mass ejection [e.g., Klein and Burlaga, 1982] and has been modeled as a flux rope [Goldstein, 1983; Marubashi, 1986; Burlaga, 1988; Suess, 1988]. Global models of how these flux ropes close in space have not yet been developed, but rooting both ends of the rope on the sun, in an expanded version of the erupting prominence topology described above, is an obvious closure choice [e.g., Marubashi, 1989].

One reason for the lack of global models for coronal mass ejections in the solar wind concerns the question of whether or not they are magnetically disconnected from the sun. In studies that treat ejections as simple magnetic loops, it has been argued that without disconnection, an unrealistic, long-term magnetic flux increase would occur in interplanetary space [e.g., Gosling, 1975]. Thus, ejections are sometimes pictured as disconnected magnetic bottles or plasmoids, spherical near the sun but distorted into bubblelike shells with sausage-shaped cross-sections (in any plane containing the central, flow-aligned axis) by advection with the spherically expanding solar wind [e.g., Newkirk et al., 1981; Gosling and McComas, 1987; Gosling et al., 1987a]. The magnetic field in the bubble, if drawn at all, closes upon itself, consistent with assumed pinching off of a two-dimensional loop in the plane of the figure and not inconsistent with magnetic cloud data; but no magnetic topology has been proposed for the third dimension of a completely disconnected ejection. Nevertheless, its existence has been postulated because it is a likely candidate for the three-dimensional bubble consistent with the many other observations reviewed by Schwenn [1986] and Webb [1988], as mentioned above.

These questions of coronal mass ejection geometry and topology—loop or bubble, disconnected or not—can be reconciled for magnetic cloud events with a model that is a looped, thick, flux rope, distorted by expansion into a bubblelike shape. Sliced in half in the plane perpendicular to the rope axis, the cross-section appears as a disconnected plasmoid, as described above, except that only the projection of the magnetic field closes upon itself. The actual field coils through the cross-section and connects to the sun. The problem of increasing flux in interplanetary space is diminished (at least for this subclass of events) by the amount of coronal arcade flux that reconnects to form the rope helices. Thus, the model is a bubblelike loop, with both connected and disconnected magnetic field, together forming the larger structure of a rope. It maintains the topology of the erupting prominence while satisfying the desired, three-dimensional form.

The distorted flux rope model suggested itself as a solution to the set of observational constraints put on constructing a global picture of a magnetic cloud detected at the ISEE 1 and 3 spacecraft in the fall of 1978. This particular cloud is a prime candidate for beginning to unravel the complexities of coronal mass ejection geometry. It contains most of the plasma signatures associated with ejection passage—a preceding shock wave, low temperature within the cloud, and bidirectional electron and proton events [Gosling et al., 1987b; Marsden et al., 1987]—and the signatures are well-ordered by the magnetic field.

In the next section, we present the data from this notable event and demonstrate how they constrain the geometry and topology of a coronal mass ejection interpretation.

Data and Interpretation

Figure 1 shows magnetic field and plasma data from ISEE 3 when the spacecraft was located about 200 R_E upstream of Earth. The time variations cover the period from October 29 to November 1, 1978. In the top panel, two magnetic features stand out. First, pronounced, directional discontinuities occurred during the three-day interval. They are marked with numbered, vertical, dashed lines. Second, between discontinuities 5 and 8 lie the large-scale variations in B_y and B_z and the relatively smooth, strong field strength that characterize a magnetic cloud. This cloud was identified by Zhang and Burlaga [1988] with nearly the same start and stop times (2300 on the 29th and 1200 on the 31st, respectively).

The event began with a somewhat weak shock, marked as discontinuity 1. The magnetosonic Mach number of ~ 1.4 and the vector normal to the shock surface were determined by Russell et al. [1983]. The GSE normal components are listed on the first line of Table 1. The components of the vectors normal to five other discontinuity surfaces are also listed in Table 1. Discontinuities 2, 4, 5, and 7 were determined to be tangential by minimum variance analysis, since the magnetic field had no significant component normal to the discontinuity surfaces. The eigenvalues in Table 1 are well-separated, indicating unambiguous results, except for discontinuity 7. In this case, however, the pressure on each side of the discontinuity was calculated (with use of both electron and proton temperature components perpendicular to the magnetic field) and found to be essentially equal, consistent with the properties of tangential discontinuities. As a check, the normal vectors for all four of these discontinuities were recalculated by cross-product analysis and found to be within $\sim 10°$ of the minimum variance vectors, except for number 4, where the fields on each side of the discontinuity were

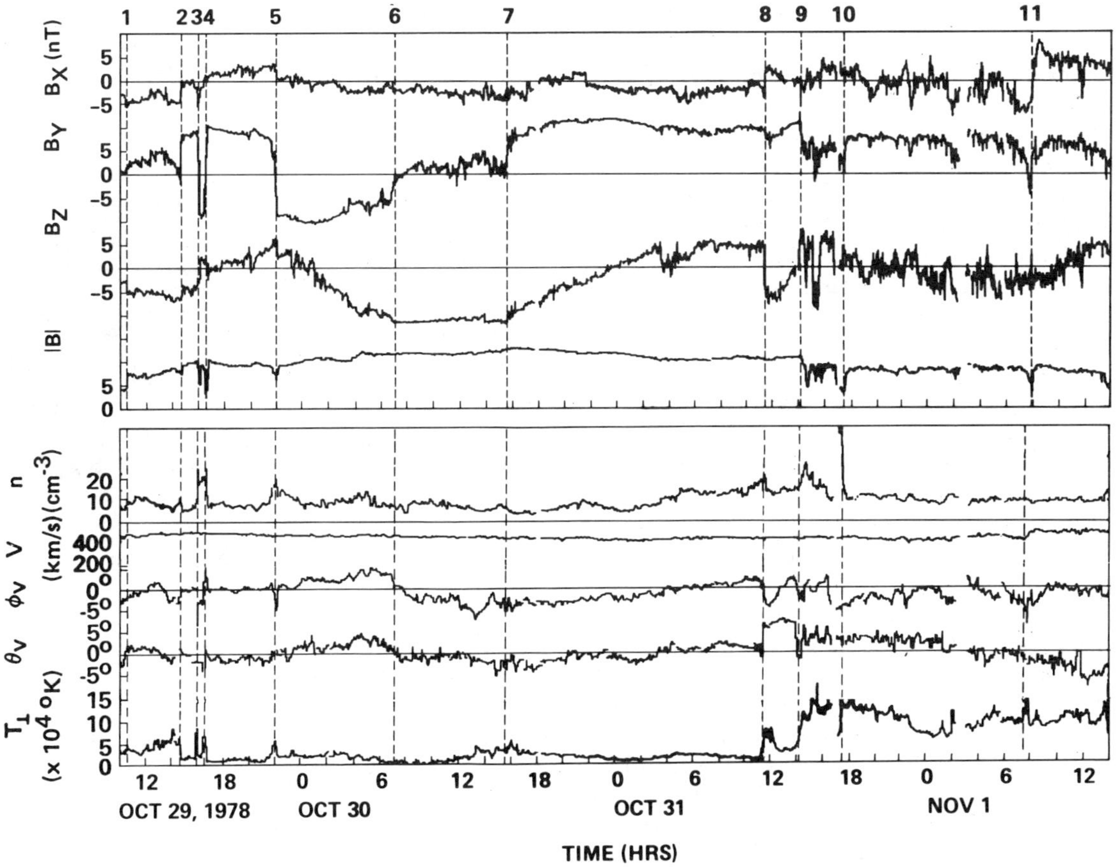

Fig. 1. Time variations of ISEE 3 magnetic field components B_x, B_y, B_z in GSE coordinates, and magnitude B (top panel) and plasma density n, speed V, westward and northward flow angles ϕ_V and θ_V, and temperature $T_\perp$ perpendicular to B (bottom panel). The dashed, numbered vertical lines mark discontinuities pictured in Figure 2.

TABLE 1. GSE components and eigenvalues of discontinuity normals.

Discontinuity	n_x	n_y	n_z	λ_1	λ_2	λ_3
1	−.89	.05	.45	–	–	–
2	−.79	.19	.58	19	.49	.07
4	−.97	.10	.24	27	1.1	.05
5	−.87	.07	.49	25	1.0	.17
6	−.99	.04	.11	–	–	–
7	−.91	.20	.36	11	.18	.13

nearly antiparallel, thus invalidating the cross-product calculation. Discontinuities 2 and 4 also appear in the ISEE 1 data (not shown). Minimum variance analysis yields normal vectors that are 13° and 10°, respectively, from the ISEE 3 normals in Table 1.

Minimum variance analysis of discontinuity 6 yielded an unreasonably large magnetic field component normal to the surface. Although the eigenvalues were well-separated, it appears that small-scale fluctuations in one component had the effect of rotating the surface normal by 90°. Since the calculated pressure across discontinuity 6 is constant, we assume the discontinuity is tangential and list normal components obtained from cross-product analysis in Table 1.

Analysis of the remaining discontinuities 3, 8, 9, 10, and 11 identified in Figure 1 yielded ambiguous results. It is likely that much of the structure is evolving and should be treated not as a set of steady-state features convecting with the solar wind but rather as a dynamically changing object. However, for the purpose of constructing the geometry of an ejection, we use the orientations of the well-defined discontinuities and leave the analysis of the remainder for future dynamical studies.

The main feature of the normal components in Table 1 is that they are roughly the same for each discontinuity.

Thus the discontinuity surfaces are almost parallel. The normal vectors point mostly in the minus x direction, with an upward tilt in the positive z direction (GSE coordinates).

Figure 2 illustrates how placing the ecliptic measurements in the context of the geometry of a model coronal mass ejection structure organizes a large amount of information. The discontinuities identified in Figure 1 are drawn as surfaces of various features, labeled by number along the x axis in the ecliptic plane. The discontinuity surfaces are concentric and tilted back at the x axis, consistent with the results in Table 1.

The magnetic cloud signature in Figure 1 is interpreted as a flux rope in Figure 2, in accord with the successful modeling efforts mentioned in the introduction. The rope center has a finite-width core of zero pitch field, between discontinuities 6 and 7. The core field points mostly in the z direction, with a small component in the x direction, consistent with a rope axis aligned in a meridional plane, parallel to the discontinuity surfaces. For clarity, the helical field outside the core is illustrated separately in Figure 3. It is shown distorted by expansion into the shape drawn in the rope cross-section in Figure 4, described below. The shaded area in Figure 3 matches the meridional cross-section of the rope in Figure 2, and the solid and dashed curves indicate portions of the distorted helix in front of and behind that plane, respectively.

Under the assumption that the flux rope ends connect back to the sun, the midplane of the rope must lie below the ecliptic plane, as indicated in Figure 2, consistent with the observed axis and discontinuity orientations. The size of the flux rope is scaled to the time it took to pass ISEE 3, multiplied by the solar wind speed. It is about twice as large as the average magnetic cloud size of .25 AU determined by Klein and Burlaga [1982].

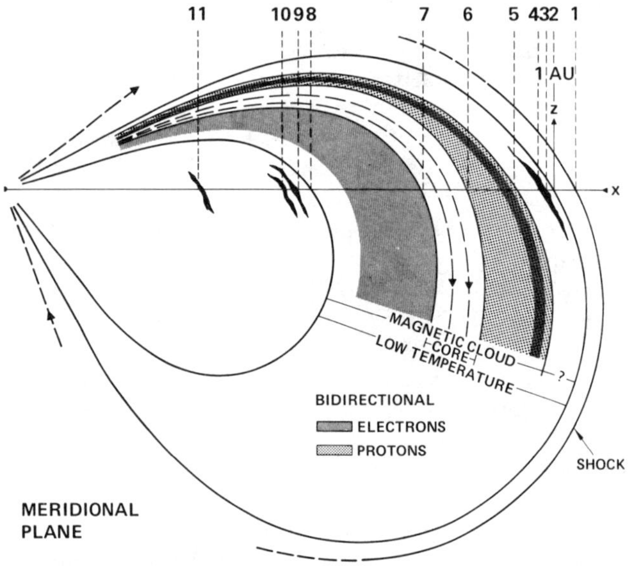

Fig. 2. Coronal mass ejection structure based on observations and assumed flux rope topology.

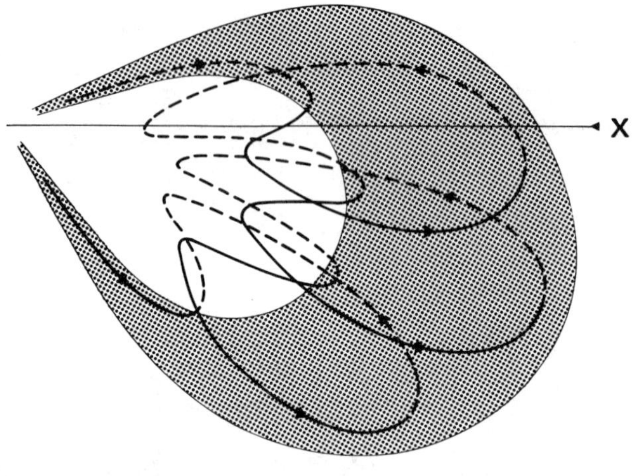

Fig. 3. Helical portion of the flux rope magnetic field, drawn separately as if it were an overlay for Figure 2, for clarity. The shaded area is the meridional cross-section of the rope and fits between discontinuities 5 and 8. The solid (dashed) curves trace the distorted helix in front (back) of the shaded plane. The helix is distorted according to the ecliptic cross-sectional shape in Figure 4.

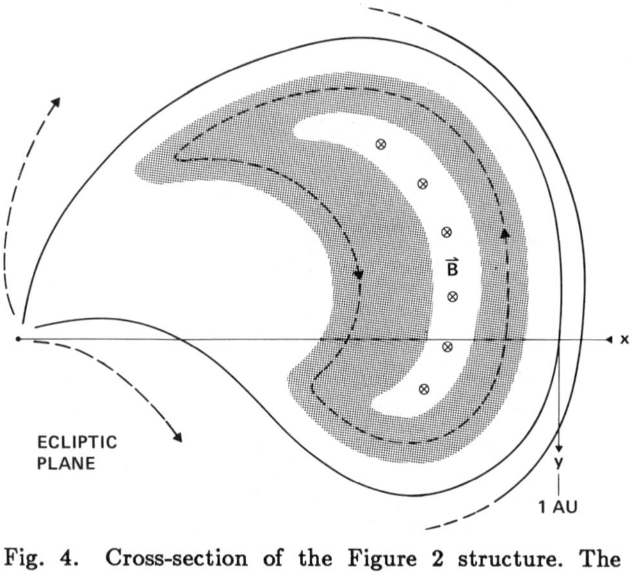

Fig. 4. Cross-section of the Figure 2 structure. The shaded region is the helical portion of the flux rope, between discontinuities 5 and 8. The core of the rope, between discontinuities 6 and 7, is unshaded, and the southward direction of the magnetic field is indicated there. The dashed contour inside the shaded region is a projection of one turn in the helical magnetic field. The dashed curves spiraling out from the sun indicate the IMF polarity.

The shaded regions in Figure 2 mark the locations of the bidirectional streaming particles reported by Gosling et al. [1987b] and Marsden et al. [1987]. They are nearly coincident with the helical portions of the rope field. The protons in the front half of the helix are separated by the core from the electrons in the back half. The separation seems curious, since there is no obvious barrier to prevent the particles from circulating around the helical field to the other side of the rope. It is possible that the separation, once established, is maintained by particle mirroring, not back near the sun at the feet of the flux rope, as is usually assumed, but at the distorted extremities of each turn in the helical coils. These extremities are clearest in the dashed contour inside the shaded region of the ecliptic view in Figure 4.

On the other hand, separation of the bidirectional electrons and protons probably is not as clear as the figure suggests. There is some evidence of proton bidirectionality in the region of bidirectional electrons, although the flow in one direction is much stronger than in the other (T. Sanderson, private communication), and a careful analysis of the electron data reveals possible bidirectional electrons throughout about half of the region of bidirectional protons. (All of the bidirectional electron data in this event show a strong flow asymmetry, not uncommon to other events as well.) The lack of bidirectionality in the core is consistent with the arguments of Marsden et al. [1987] that mirroring at the sun may be ineffective in producing the observed signatures, or it may reflect the detectors' inability to measure bidirectional streaming at large angles from the ecliptic plane. Clearly, the relation between bidirectional electrons and protons is complicated. Nevertheless, the first order pattern illustrated in Figure 2 is intriguing and invites further study of additional cases.

The plasma data in the lower panel in Figure 1 show east-west flow deflections within the magnetic cloud consistent with a vortex circulating in the direction of the coiled magnetic field. Similar evidence for vortical flow was observed in one of the cases discussed by Klein and Burlaga [1982], except that the deflections were north-south in a cloud lying at right angles to the one analyzed here. Eastward flow deflections within bidirectional electron events preceded by westward deflections were found by Gosling et al. [1987a] in a statistical study and interpreted as eastward deflected mass ejections preceded by westward deflected sheath flow. This particular case, however, does not fit that interpretation. Although the eastward flow occurred in the region of bidirectional electrons, the westward deflection occurred not in the sheath but in the region of bidirectional protons, within the body of the magnetic cloud itself, which, presumably, is the mass ejection. The interpretation of vortical flow seems better suited to this case, although its significance is not yet understood.

Although the magnetic cloud appears to extend only between discontinuities 5 and 8, according to the variations in magnetic field direction, other characteristics of the driver gas commonly associated with shocks and coronal mass ejection passage extend over slightly larger regions. Low temperature extends forward to discontinuity 2, and high field magnitude covers the entire region between discontinuities 2 and 9, with the exception of the sharp decreases at 3, 4, and 5. If the driver gas begins at discontinuity 2, as drawn in Figure 2, consistent with the temperature and field magnitude observations, then the sheath between the shock and the driver gas is narrower than would be expected. For example, scaling to Earth's magnetosheath with an extrapolation from the Spreiter et al. [1966] model, the calculated Mach number of 1.4 for this shock would give a minimum sheath thickness of about half the radius of curvature of the driver gas volume. Thus, the region between discontinuities 2 and 5 is labeled with a question mark in Figure 2, since it could also be part of the sheath. Even with this addition, however, it appears that the sheath would still be thinner than half the radius of curvature of the driver gas volume as drawn in Figure 2, perhaps because the scaling yields an overestimate.

The solid black, filamentary shapes in Figure 2 correspond to features in Figure 1 that are bounded by pronounced directional discontinuities and have sharp decreases in field magnitude and increases in density, especially at their edges. The feature between discontinuities 3 and 4, near the front of the cloud, is particularly pronounced, with systematic variations in the flow angles. It was also observed clearly at ISEE 1 in abbreviated form. Burlaga et al. [1981] found the same type of feature associated with a different cloud, but on the trailing edge. These features may be remnants of the bright, filamentary material observed in coronograph images of ejections.

The flux rope structure in Figure 2 is drawn in cross section in Figure 4. The shaded cylinder of the flux rope is shown distorted into the crescent shape often used to represent a disconnected plasmoid coronal mass ejection structure in the ecliptic plane [e.g., Newkirk et al., 1981; Gosling et al., 1987b]. However, here the shape encloses a flux rope that is connected to the sun above and below the ecliptic plane. The distortion is anticipated as a result of both kinematic spherical expansion [e.g., Suess, 1988] with a rotating source and dynamic expansion of the exploding ejection itself. Consequently, the back side of the helical magnetic field has the same curvature as the front side, resulting in sharp field line bends and possible particle mirroring sites where the front and back sides join, as mentioned above. The observed forward displacement of the unshaded rope core is consistent with the proposed distortion by expansion. The Figure 4 shape differs considerably from the circular cross sections assumed in the flux rope modeling of magnetic clouds cited in the introduction and may provide some guidance for future modeling.

The Figure 4 rope cross-section is enclosed in a large tongue of solar material extending from the sun. This shape was drawn by Hundhausen [1972] for nonrecurrent, shock-wave disturbances in the solar wind and remains appropriate for coronal mass ejections with shocks. The position of the x axis through the east side of the structure is consistent with the small, positive y components of the discontinuity normals in Table 1. For simplicity, the longitudinal displacement of the structure during the two-day magnetic cloud passage interval has not been taken into account.

The observed away polarity of the IMF before and after the magnetic cloud passage, indicated in Figure 4 by the dashed curves emanating from the sun, is consistent with explosion of the looped flux rope from a helmut streamer spanning the solar magnetic equator below the ecliptic plane, with the heliospheric current sheet extending outward from the rope's midplane (in the plane of the solar magnetic equator). The solar polarity of 1978 had away fields at the north pole, implying that the heliospheric current sheet was below Earth throughout the event, consistent with the deduced passage through the northern portion of the rope in Figure 2 and with the polarity of the dashed IMF lines drawn there.

It is clear from the above analysis that Figures 2 and 4 are not merely sketches of a coronal mass ejection but rather constructions that are relatively tightly constrained by the observed discontinuity and magnetic field orientations. Although the observations are essentially single-point measurements and the model flux rope loop is assumed, placing the observations in the context of the model roughly fixes where the spacecraft passed through that loop and, consequently, its size and orientation. The data are consistent with a flux rope that has similar dimensions in the meridional and ecliptic planes, thus filling a three-dimensional, bubblelike cavity.

Conclusions

1. The magnetic topology of a three-dimensional, bubblelike coronal mass ejection can be a flux rope, distorted by expansion into the solar wind. Flux ropes need not be two-dimensional loops. Thus, the question of whether ejections are loops or bubbles can be resolved by answering that they are both, i.e., a bubble-shaped loop.

2. In view of the predicted distortion of the flux rope loop, there may be substantial uncertainty introduced in present flux rope modeling of magnetic clouds by the required assumption of cylindrical symmetry.

3. In the past, coronal mass ejections in the solar wind were often pictured as distorted, magnetically disconnected plasmoids. In the context of the bubblelike flux rope model, these plasmoids become the rope cross-section perpendicular to its axis. Thus, the question of whether a coronal mass ejection is magnetically connected or disconnected can be resolved for the flux rope model by answering that they are both connected, at the rope ends, and disconnected, to form the rope's helices.

4. A detailed analysis of a prime candidate for a coronal mass ejection in the solar wind yields geometrical constraints consistent with a bubblelike flux rope model.

5. The bidirectionally streaming protons in the event appear to be confined to the front of the rope, and the electrons to the rear. This separation might possibly be maintained by particle mirroring at the distorted extremities of each turn in the helical fields.

Acknowledgements. Preliminary results of this research were presented at the Workshop on Solar Events and their Influence on the Interplanetary Medium, Colorado Springs, March, 1988, and completion of the project benefitted from collaborations begun there. This work was supported by the National Aeronautics and Space Administration under subcontracts 955376 (JPL to UCLA) and UC884C27 (LANL to UCLA) and grant NAG5-1067. The work at Los Alamos was performed under the auspices of the Department of Energy with NASA support under S-04039-D.

References

Burlaga, L. F., Magnetic clouds and force-free fields with constant alpha, *J. Geophys. Res., 93,* 7217-7224, 1988.

Burlaga, L., E. Sittler, F. Mariani, and R. Schwenn, Magnetic loop behind an interplanetary shock: Voyager, Helios, and IMP8 observations, *J. Geophys. Res., 86,* 6673-6684, 1981.

Goldstein, H., On the field configuration in magnetic clouds, in *Solar Wind Five*, edited by M. Neugebauer, NASA CP-2280, pp. 731-733, 1983.

Gosling, J. T., Large scale inhomogeneities in the solar wind of solar origin, *Rev. Geophys. Space Phys., 13,* 1053-1058, 1975.

Gosling, J. T., and D. J. McComas, Field line draping about fast coronal mass ejecta: A source of strong out-of-the-ecliptic interplanetary magnetic fields, *Geophys. Res. Lett., 14,* 355-358, 1987.

Gosling, J. T., M. F. Thomsen, S. J. Bame, and R. D. Zwickl, The eastward deflection of fast coronal mass ejecta in interplanetary space, *J. Geophys. Res., 92,* 12399-12406, 1987a.

Gosling, J. T., D. N. Baker, S. J. Bame, W. C. Feldman, and R. D. Zwickl, Bidirectional solar wind electron heat flux events, *J. Geophys. Res., 92,* 8519-8535, 1987b.

Hundhausen, A. J., *Coronal Expansion and Solar Wind*, p. 192, Springer-Verlag, New York, 1972.

Klein, L. W., and L. F. Burlaga, Interplanetary magnetic clouds at 1 AU, *J. Geophys. Res., 87,* 613-624, 1982.

Marsden, R. G., T. R. Sanderson, C. Tranquille, K.-P. Wenzel, and E. J. Smith, ISEE 3 observations of low-energy proton bidirectional events and their relation to isolated interplanetary magnetic structures, *J. Geophys. Res., 92,* 11009-11019, 1987.

Marubashi, K., Structure of the interplanetary magnetic clouds and their solar origins, *Adv. Space Res., 6,* 335-338, 1986.

Marubashi, K., The space weather forecast program, *Space Sci. Rev.,* in press, 1989.

Mouschovias, T. Ch., and A. I. Poland, Expansion and broadening of coronal loop transients: A theoretical explanation, *Astrophys. J., 220,* 675-682, 1978.

Newkirk, G., Jr., A. J. Hundhausen, and V. Pizzo, Solar cycle modulation of galactic cosmic rays: Speculation on the role of coronal transients, *J. Geophys. Res., 86,* 5387-5396, 1981.

Russell, C. T., E. J. Smith, B. T. Tsurutani, J. T. Gosling, and S. J. Bame, Multiple spacecraft observations of interplanetary shocks: Characteristics of the upstream ULF turbulence, in *Solar Wind Five*, edited by M. Neugebauer, NASA CP-2280, pp. 385-400, 1983.

Schwenn, R., Relationship of coronal transients to interplanetary shocks: 3D aspects, *Space Sci. Rev., 44,* 139-168, 1986.

Suess, S. T., Magnetic clouds and the pinch effect, *J. Geophys. Res., 93,* 5437-5445, 1988.

Webb, D. F., Erupting prominences and the geometry of coronal mass ejections, *J. Geophys. Res., 93,* 1749-1758, 1988.

GLOBAL CONFIGURATION OF A MAGNETIC CLOUD

L. F. Burlaga and R. P. Lepping

Laboratory for Extraterrestrial Physics
NASA Goddard Space Flight Center
Greenbelt, MD 20771

J. A. Jones

STX, 4400 Forbes Blvd., Lanham, MD 20706

Abstract. A magnetic cloud associated with a 2N flare on January 1, 1978 was observed by IMP-8, Helios A, Helios B, and Voyager 2. The variation of the magnetic field observed at each spacecraft is represented to good approximation by Lundquist's solution for a cylindrically symmetric force-free magnetic field with constant α. A least squares fit of Lundquist's solution to the data from each spacecraft gives the local orientation of the axis of the magnetic cloud. The times of the estimated boundaries of the magnetic cloud at each spacecraft together with the speeds of the boundaries and the spacecraft position give the positions of the boundaries at a given time. From these results we determine that the magnetic cloud resembles a flux rope whose minor radius is approximately 0.15 AU at 1 AU and whose radius of curvature at 1 AU is approximately 1/3 AU.

1. Introduction

A "magnetic cloud" following a solar flare was identified in the interplanetary medium near 1 AU by Burlaga *et al.* [1981] on the basis of the following characteristics: 1) a smooth rotation of the magnetic field direction during an interval of the order of one day; 2) low values of $\beta = P_T/P_B$, where P_T is the thermal pressure and P_B is the magnetic pressure; and 3) relatively high magnetic field strengths. Magnetic clouds are observed at 1 AU at the rate of at least one every three months (Klein and Burlaga, 1982). Magnetic clouds are a subset of the regions with very low temperature (Geranios, 1987) and a subset of the interplanetary ejecta. A magnetic cloud observed by Helios A over the west limb of the sun was directly associated with a coronal mass ejection (Burlaga *et al.*, 1982), and statistical studies (Wilson and Hildner, 1984) provide further support for the association of magnetic clouds with coronal mass ejections. Magnetic clouds are also associated with major geomagnetic storms, since the magnetic field strength in a magnetic cloud is usually relatively high and the field direction is usually directed southward for several hours during the motion of the magnetic cloud past the earth (Burlaga *et al.*, 1981; Wilson 1987; Zhang and Burlaga, 1988; Lepping *et al.*, 1989a). For a review of the literature on magnetic clouds see Burlaga [1989].

Goldstein [1983] suggested that magnetic clouds are force-free configurations, and Marubashi [1986] showed by means of two examples that magnetic clouds can be described by a force-free field with variable α, where curl $\mathbf{B} = \alpha$ x $\mathbf{B}$. Burlaga [1988] found that the essential features of magnetic clouds observed locally by a single spacecraft at 1 AU can be modeled as a cylindrically symmetric force-free field with constant alpha, for which an analytical solution is known (Lundquist, 1950). This suggests that the time-scale of the process that brings the structure back to equilibrium is much smaller than the expansion time-scale which tends to destroy the cylindrical symmetry. We also assume that the interactions of the cloud with the neighboring flows do not change the magnetic field line geometry substantially, although the interactions may produce distortions in the magnetic field strength profile. In short, we assume that magnetic clouds are stable equilibrium structures to first approximation. Ivanov (1989) fit magnetic clouds to the solution for a force-free torus. Simulations by Suess [1988] also support the idea that magnetic clouds may be represented as constant α force-free field configurations. Lepping *et al.* [1989b] developed a method to fit Lundquist's solution to observations of a magnetic cloud in an optimal way, which gives the local orientation of the axis of a magnetic cloud. Little is known about the large-scale configuration and topology of magnetic clouds. It is unlikely that they are toroidal, since magnetic cloud signatures are not observed in pairs. Nevertheless, a magnetic cloud might be better described by a toroidal solution than a cylindrical solution if the local curvature of the axis of a magnetic cloud is comparable to the radius of the magnetic cloud. Chen [1989] presented a model of an expanding current loop which might be applicable to magnetic clouds, and the stability of an expanding magnetic cloud was examined by Yang [1989].

Geophysical Monograph 58

The aim of this work is to use observations from four spacecraft (IMP-8, Helios A, Helios B, and Voyager 2) together with the method of Lepping *et al.* [1989b] to determine the large-scale configuration of a magnetic cloud. The magnetic cloud was associated with a 2N solar flare at 2153 UT on January 1, 1978 located at E06°, S21°. This flare was accompanied by Type II, III, IV bursts as well as microwaves and X-rays.

2. Observations

The event that we shall analyze is the magnetic cloud discussed by Burlaga *et al.* [1981]. The subsequent discovery that a magnetic cloud can be approximately represented as a force-free magnetic field and the development of a technique for fitting observations to the constant alpha force-free model provide a means to determine the local orientation of the axis of the magnetic cloud at several positions, thereby providing important new constraints on the global configuration of the magnetic cloud.

The magnetic cloud as seen at Helios B is shown in Figure 1, which gives the magnitude of the magnetic field B, the elevation δ and azimuth λ of the magnetic field in heliographic coordinates, the bulk speed V, the sum of the magnetic and thermal pressure P_T, and the ratio of the gas pressure to the magnetic pressure β. The thermal pressure is the sum of the proton pressure NkT_p and the electron pressure P_e. P_e is estimated as 9.09 $N^{1.185}$ (10^{-12} dyn/cm^2) in accordance with the results of Sittler and Scudder [1980]. It is not always possible to objectively identify the front and rear boundaries of a magnetic cloud. It is not even clear that a magnetic cloud generally has a unique thin boundary; for example, it is conceivable that the boundary of a magnetic cloud is subject to the ballooning instability (Strauss, 1989), which is driven by a gradient in β. The rear boundary of the magnetic cloud is chosen as the time when the monotonic decrease in B and P_T ends, when the onset of a new stream is occurring. The front of the magnetic cloud probably occurred in the data gap; it is indicated by the line drawn at hour 13 on January 4 in Figure 1, but of course it might be drawn anywhere in the data gap.

Using Lundquist's solution in the way described by Burlaga [1988] and the method of Lepping *et al.*[1989b], we obtained the fit to the magnetic cloud observed by Helios B shown as the dashed curve in Figure 2. The fit was made only to the data between the front and rear boundary times given in Table 1. There are seven parameters in the fitting procedure, as described by Lepping *et al.* [1989b], viz., two angles giving the attitude of the magnetic cloud axis, the maximum magnetic field strength, the size of the magnetic cloud, the sign of the helicity, the time of closest approach to the axis of the magnetic cloud, and the distance of closest approach to the magnetic cloud axis. The fitting procedure works best when the spacecraft goes through the axis of the magnetic cloud, but it also works well whenever the distance between the spacecraft and the axis is less than 2/3 the radius of the magnetic cloud. Note that the magnetic field directions in Figure 2 are given in solar ecliptic coordinates, with $\theta \sim \delta$ and $\phi \sim \lambda$ - 180°. Lundquist's solution for a constant α force-free magnetic field is $B_z = B_o J_o(ar)$, $B_T = B_o J_1(ar)$, $B_r = 0$, where J_o and J_1 are Bessel functions, and we assume that the boundary is

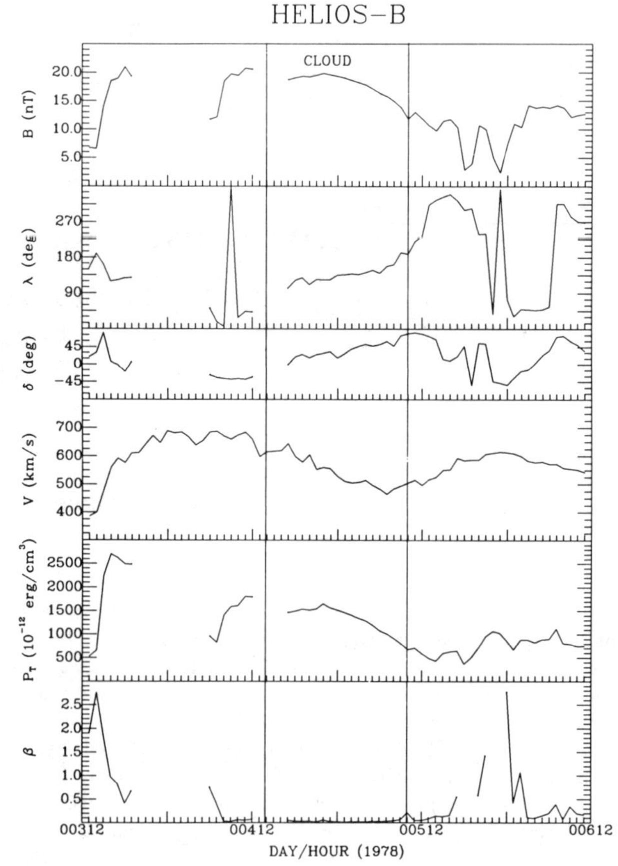

Fig. 1.
Observation of the magnetic field and plasma parameters for a magnetic cloud. The magnetic cloud is between the two vertical lines. (See Table 1).

where $B_z = 0$. Lundquist's solution (the dotted curve in Figure 1) provides a very good fit to the data in the magnetic cloud, between the two vertical lines in Figure 2.

Formally, one can extend the solution beyond the magnetic cloud boundaries, and this extended solution is shown in Figure 2, but we attribute no physical significance to the extended solution beyond the vertical lines in Figure 2. The cylindrically symmetric constant alpha force-free magnetic field solution is a flux rope, i.e., the magnetic field line on the axis of the cylinder is a straight line and the other lines of force are helices, the helices farthest from the axis having the largest pitch angles. The agreement between the observations of the magnetic field in the magnetic cloud and the solution for a force-free field is the basis for our conclusion that the magnetic cloud is a flux rope.

From the fit of the data between the boundaries of the magnetic cloud (Table 1) to Lundquist's solution, one obtains the direction of the axis of the magnetic cloud at the position of Helios B, which is given in Table 1 and shown as an arrow in Figure 3. Choosing the start time at the beginning of the data gap (4-11) instead of after it (4-17) as shown in the Table

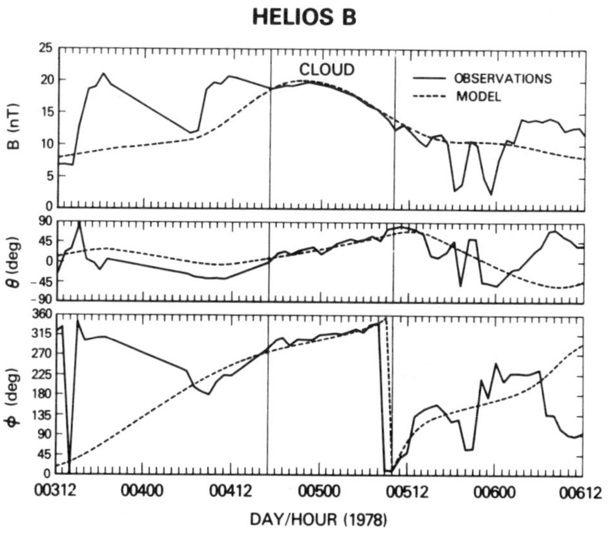

Fig. 2.
Magnetic field observations of a magnetic cloud (solid curves) are shown together with the fit to Lundquist's solution (dashed curves). Only the dashed curves between the two vertical lines have meaning in the context of our model for the magnetic cloud.

TABLE 1

S/C	BOUNDARY TIME (DAY - HR)		BOUNDARY SPEED (KM/S)		S/C DISTANCE (AU)	CLOUD AXIS DIRECTION	
	FRONT	REAR	FRONT	REAR		ϑ_a	ϕ_a
IMP-8	4-17	5-14	662	474	1.0	27°	334°
HELIOS-B	(4-17)	5-10	644	505	0.94	-12°	332°
VOYAGER-2	7-13	8-8	582	553	1.96	35°	271°
HELIOS-A	4-18	5-19	615	506	0.95	-5°	173°

gives $\theta_a = 31^o$ and $\phi_a = 337^o$. From the estimated times of the front and rear boundaries of the magnetic cloud observed by Helios B and from the observed speeds at these times, one can compute the positions of these boundaries at some arbitrary instant which we take to be 1400 UT on January 5, 1978. The position of the rear boundary is shown by the "o" in Figure 3 and the beginning and end positions of the data gap containing the front boundary are shown by the "x's" in Figure 3.

The magnetic cloud was also observed by IMP-8, Helios A, and Voyager 2, whose positions on January 5, 1978 are shown in Figure 3. The radial scale is given by the fact that the rear boundary of the magnetic cloud at IMP-8 was at 1 AU at the time represented by the figure. Note that Voyager 2 was at 2 AU, which lies off the figure. The magnetic cloud was also observed by Voyager 1, but we could not analyze it because of the large data gaps. The estimated times of the boundaries of the magnetic clouds at these spacecraft are given in Table 1. Note that these times differ somewhat from the times of the boundaries selected by Burlaga et al. [1981]. The times of the boundaries of the magnetic cloud at Helios A are particularly difficult to identify. An important task for future studies is the determination of an objective criterion for identifying the boundaries of magnetic clouds. Using the times of the boundaries and the speeds measured at the boundaries (Table 1) and assuming that the plasma moves radially at constant speed, we compute the positions of the boundaries at 1400 UT on January 5, 1978 which are shown in Figure 3. Choosing the boundaries given in Table 1 and fitting the data between these boundaries to the constant alpha force-free solution of Lundquist using the method of Lepping et al. gives the local directions of the axis of the magnetic cloud (θ_a, ϕ_a) shown in Table 1.

We estimate the uncertainty in θ_a and ϕ_a as approximately $\pm$ 15°, based on both the error given by the fitting procedure and on fits obtained using different but plausible boundaries. Smaller uncertainties could be achieved if the boundaries could be determined more accurately. The average inclination of the axis of the magnetic cloud is $11^o \pm 15^o$, and the local azimuthal directions ϕ_a are shown as the arrows in Figure 3. If the positions of the boundaries and the directions of the axis of the magnetic cloud are as shown in Figure 3, and given that the magnetic field profile in the magnetic cloud is described locally by the cylindrically symmetric solution of Lundquist, we infer that in the region near 1 AU the magnetic cloud has the form of a flux rope whose axis has a radius of curvature of approximately 1/3 AU and whose minor radius is approximately 0.15 AU, as illustrated in Figure 3.

Since the magnetic cloud was associated with a flare at E06°, S21°, and since the plasma in the magnetic cloud moves nearly radially away from the sun, one expects that to first approximation the outer part of the magnetic cloud was symmetric with respect to the radial line extending from the flare site at the time of the flare, as shown in Figure 3. The structure of the magnetic cloud close to the sun is not known, so we show an extrapolation of the magnetic cloud boundaries by dashed lines in Figure 3. The magnetic field lines in the magnetic cloud might extend to the sun as suggested by the dashed lines in Figure 3, or they might terminate at neutral points in the solar wind. There is no evidence that the magnetic cloud reconnected to form a torus (which would imply that each spacecraft would have seen two magnetic cloud signatures, since the axis of the magnetic cloud was close to the ecliptic). If the magnetic

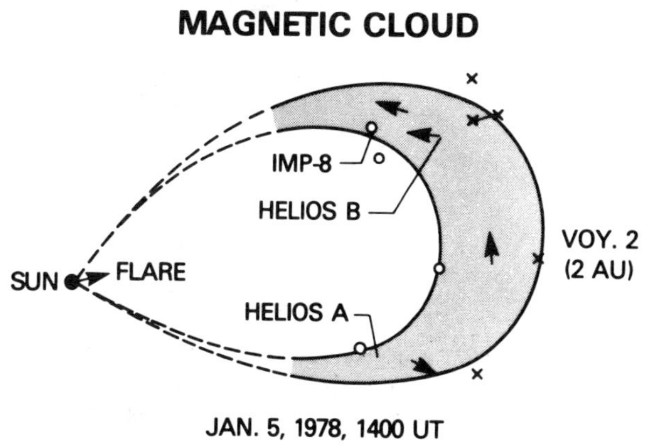

Fig. 3. Global configuration of the magnetic cloud.

field lines remained attached to the sun, then they would be curved near the sun as a result of the solar rotation, since the flare site moved westward from the location shown in Figure 3.

3. Summary and Discussion

Using observations from four spacecraft and the constant alpha force-free magnetic model for a magnetic cloud we have shown that the magnetic cloud which moved past earth on January 4, 1978 was a flux rope. The radius of curvature of the axis of the magnetic cloud near 1 AU was approximately 1/3 AU and the minor radius of the magnetic cloud near 1 AU was approximately 0.15 AU. Thus the configuration of the magnetic cloud in the interplanetary medium is like that illustrated schematically in Figure 4. The projections of the helical magnetic field lines are not shown exactly to scale.

Consider the magnetic field lines on a cylinder of radius R, and let L be the distance that the field line extends along the cylinder when it wraps around the cylinder once. Lundquist's solution implies that the ratio L/R is approximately 4.5 for the field lines at a radius of 0.4 in units of the radius R of the magnetic cloud; this ratio is 1.4 for field lines at 0.8 R. The outer helices are more tightly wound than the inner helices. Figure 4 shows the magnetic field lines connecting back to the sun, which is certainly one possibility, but the magnetic field line configuration in the corona was not determined in this study. For example, it is conceivable that the magnetic field lines end on two neutral points in the interplanetary medium so that the magnetic cloud has the form of a gigantic banana. It is unlikely that the magnetic field lines reconnect to form a torus, because that would imply two passages through the cloud, which is not generally observed. The shape of the magnetic cloud may also be somewhat different in detail from that shown here. It may be either narrower or broader; it may be either thinner or thicker; it may be pinched here and bulge there; the pitch of a magnetic field line may be different from that illustrated here, it may bend owing to rotation of the sun, etc. The essential point is that the interplanetary configuration of the magnetic field lines in a magnetic cloud resembles a flux rope, insofar as the magnetic field lines form nested helices with approximate cylindrical symmetry locally.

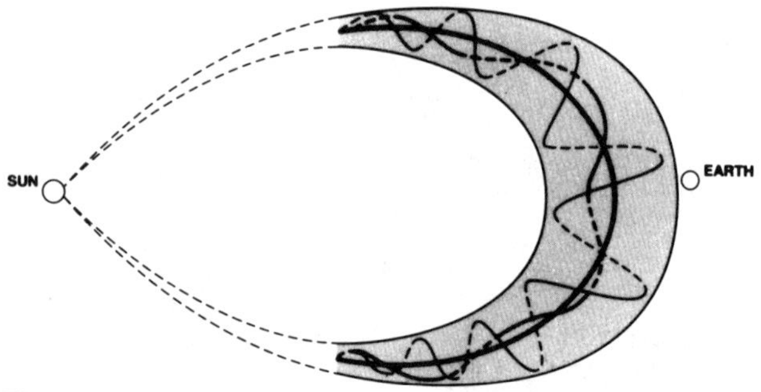

Fig. 4.
A schematic view of a magnetic cloud illustrating the global shape and the flux rope geometry of the magnetic field lines in the magnetic cloud. No attempt is made to include the effect of solar rotation, because the form of the magnetic cloud close to the sun (dashed lines) is not known.

The results presented above suggest opportunities for collaborative studies among solar physicists, interplanetary physicists and magnetospheric physicists. Solar and coronal observations are needed to determine whether or not the magnetic field lines in a magnetic cloud remain attached to the sun. Observations of suprathermal electrons and energetic protons will also aid in answering this question. There is much to be learned about the magnetic clouds in the solar wind, such as the nature of the boundaries, the fine structure and the composition. A magnetic cloud provides a simple "polarized" input signal for driving geomagnetic activity, auroral phenomena and ionospheric responses, and one might expect to observe a variety of east-west effects in these phenomena depending on the orientation of the magnetic cloud. For example, the unusual auroral features reported by Akasofu and Tsurutani [1984] were probably associated with a magnetic cloud, although the authors did not identify the cause as such.

Finally, the magnetic cloud configuration illustrated in Figures 3 and 4 implies that the dynamics of magnetic clouds is more akin to that of solar flux ropes than to the traditional models of interplanetary transient ejecta, and it suggests the appropriateness of a fresh approach to the theory of interplanetary ejecta.

Questions and Answers

Wolfson: Gene Parker reminded us that all those flux ropes are immersed in an ambient field. Since your flux rope increases in pitch as you move toward the edge, shouldn't you see a current layer at the boundary?

Burlaga: Magnetic clouds are created in explosive events and move through a turbulent, highly structured medium, so that it is not surprising that a single current layer is generally not seen at the boundary of a magnetic cloud. A single current layer is seen in some cases.

Forbes: It looks like to me that there is a contradiction between the ratio of the strength of longitudinal to transverse (poloidal) component in your flux rope model and the ratio observed in prominences which is about 10 to 1. Is there really such a contradiction?

Burlaga: Since a magnetic cloud is described by the solution for a cylindrically symmetric force-free field with constant alpha, the ratio of the maximum field to that on the boundary (longitudinal to transverse component) is approximately 2 to 1. A relation between magnetic clouds and stationary prominences has not been established.

References

Akasofu, S.-I., and B. Tsurutani, Unusual auroral features observed on January 10-11, 1983 and their possible relationships to the interplanetary magnetic field, *Geophys. Res. Lett.*, **11**, 1086, 1984.

Burlaga, L. F., Magnetic clouds and force free fields with constant alpha, *J. Geophys. Res.*, **93**, 7217, 1988.

Burlaga, L. F., Magnetic clouds, *Chapter 6 in Physics of the inner Heliosphere*, edited by R. Schwenn and E. Marsch, Springer- Verlag, 1989, To appear.

Burlaga, L. F., E. Sittler, F. Mariani, and R. Schwenn, Magnetic loop behind an interplanetary shock: Voyager, Helios and IMP-8 observations, *J. Geophys. Res.*, **86**, 6673, 1981.

Burlaga, L. F., L. W. Klein, N. R. Sheeley, Jr., D. J. Michels, R. A. Howard, M. J. Koomen, R. Schwenn and H. Rosenbauer, A magnetic cloud and a coronal mass ejection, *Geophys. Res. Lett.*, **9**, 1317, 1982.

Burlaga, L. F., K. W. Behannon and L. W. Klein, Compound streams, magnetic clouds and major magnetic storms, *J. Geophys. Res.*, **92**, 5725, 1987.

Chen, James, Effects of toroidal forces in current loops embedded in a background plasma, *Ap. J.*, **453**, 1989.

Geranios, A., Statistical analysis of magnetically closed structures, *Planetary Space Science*, **35**, 727, 1987.

Goldstein, H., On the field configuration in magnetic clouds, *Solar Wind Five*, NASA Conference Publ. 2280, p. 731 1983.

Ivanov, K. G., A. F. Harshiladze, E. G. Eroshenko, and V. A. Styazhkin, Configuration, structure and dynamics of magnetic clouds from solar flares in light of measurements on board VEGA 1 and VEGA2 in January-February, 1986, *Solar Physics*, **120**, 407, 1989.

Klein, L. W., and L. F. Burlaga, Interplanetary magnetic clouds at 1 AU, *J. Geophys. Res.*, **87**, 613, 1982.

Lepping, R. P., F. M. Ipavich, and L. F. Burlaga, A flare-associated shock pair at 1 AU and related magnetic cloud, *J. Geophys. Res.*, submitted, 1989a.

Lepping, R. P., J. A. Jones, and L. F. Burlaga, Fitting of the field in a magnetic cloud using least-squares, to appear, 1989b.

Lundquist, S., Magnetohydrostatic fields, *Arkiv Fysik*, **2**, 361, 1950.

Marubashi, K., Structure of the interplanetary magnetic clouds and their solar origins, *Adv. Space Sci.*, **6**, 335, 1986.

Sittler, E. C., Jr., and J. D. Scudder, An empirical polytrope law for solar wind thermal electrons between 0.45 AU and 4.67 AU: Voyager 2 and Mariner 10, *J. Geophys. Res.*, **85**, 5131, 1980.

Strauss, H. R., The effect of ballooning modes on thermal transport and magnetic field diffusion in the solar corona, *Geophys. Res. Lett.*, **16**, 219, 1989.

Suess, S. T. Magnetic clouds and the pinch effect, *J. Geophys. Res.*, **93**, 5437, 1988.

Wilson, R. M. and E. Hildner, Are interplanetary magnetic clouds manifestations of coronal transients at 1 AU?, *Solar Physics*, **91**, 168, 1984.

Wilson, R. M., Geomagnetic response to magnetic clouds, *Planet. Space Sci.*, **35**, 329, 1987.

Yang, W.-H., Expansion of solar-terrestrial low-β plasmoid, *Ap. J.*, 1989, in press.

Zhang, G., and L. F. Burlaga, Magnetic clouds, geomagnetic disturbances, and cosmic ray decreases, *J. Geophys. Res.*, **93**. 2511, 1988.

EFFECTS OF THE DRIVING MECHANISM IN MHD SIMULATIONS OF CORONAL MASS EJECTIONS

J. A. Linker and G. Van Hoven
Department of Physics, University of California, Irvine
D. D. Schnack
Science Applications International Corporation

Abstract: We present results of time-dependent MHD simulations of mass ejections in the solar corona. Previous authors have shown that results from simulations using a thermal driving mechanism are consistent with the observations only if an elaborate model of the initial corona is used. Our first simulation effort, using a simple model of a plasmoid as the driving mechanism and a simple model of the initial corona, produces results that are also consistent with many observational features, suggesting that the nature of the driving mechanism plays an important role in determining the subsequent evolution of mass ejections. Our first simulations are based on the assumption that mass ejections are driven by magnetic forces; we are now developing simulations where the initial corona is perturbed magnetically by introducing a "plasmoid-like" current perturbation. The preliminary results from these simulations show some features that are consistent with the observations, others that are not. The discrepancies may be caused by the lack of internal force balance in the initial plasmoid structure. In the future, we plan to perform simulations where plasmoid formation occurs self-consistently.

Introduction

Coronal mass ejections (CMEs) have been observed in space-based white light coronagraphs since the early 1970s. Because of the complex nature of CMEs, their initiation and propagation in the solar corona is still not well understood. An important class of CME events are the looplike CMEs [Wagner, 1984]. Sime et al. [1984] identified three general characteristics of the looplike CMEs observed with the Skylab coronagraph: (1) the sides of the loop (the "legs" of the mass ejection) have a greater density enhancement than the top of the loop; (2) a depletion of density occurs within the loop; (3) the legs exhibit very little lateral motion throughout the time evolution of the mass ejection.

One approach for investigating CMEs is the time-dependent magnetohydrodynamic (MHD) simulation. The first MHD simulations of CMEs were based on the assumption that mass ejections are initiated by the rapid injection of thermal energy at the base of the corona from a solar flare [e.g., Dryer et al., 1979; Wu et al., 1982]. In these simulations, the initial corona was modeled hydrostatically with a current-free magnetic field. A pressure pulse was introduced to model the thermal energy released in the flare, and the subsequent time evolution was followed. The resulting expanding fast-mode shock wave was identified as the CME. Although these simulations did give a coherent explanation for the initiation and propagation of CMEs, Sime et al. [1984] pointed out that these simulations failed to replicate the observed characteristics of looplike ejections. Because of these discrepancies, Sime et al. questioned the idea that CMEs could be regarded as compressional disturbances initiated by thermal energy release.

In order to obtain simulation results that are compatible with the observations, Steinolfson and Hundhausen [1988] examined the role of the initial corona. They first verified that simulations with a thermal driving mechanism and an initially hydrostatic, current-free corona with a closed magnetic topology could not reproduce any of the observed features of mass ejections. They then showed that if one used a specially tailored initial condition and a thermal driving mechanism, many of the observed features of CMEs could be reproduced. Their model assumes that the evolution of CMEs in the solar corona is primarily influenced by the pre-event corona, and that the nature of the driving mechanism does not greatly affect the dynamics.

Geophysical Monograph 58

We are investigating the possiblility that the driving mechanism that initiates coronal mass ejections may play an important role in their subsequent evolution. We have performed MHD simulations that start with a simple model for the initial corona (hydrostatic equilibrium and a potential magnetic field) and nonthermal driving mechanisms. Anzer and Pneumann [1982] suggested that mass ejections could be caused by plasmoids formed from magnetic reconnection, and a simulation study by Mikić et al. [1988] indicates that shearing photospheric flows can lead to unstable magnetic field configurations, reconnection, and the subsequent ejection of plasmoids. In our simulations, we proceed from the assumption that the onset of magnetic instability in the corona is the driving mechanism for CMEs.

Linker et al. [1989] give a description of the simulation code and parameters. Here we note that the code solves the nonlinear, compressible MHD equations in spherical coordinates by using a two-step Lax-Wendroff finite-difference scheme. Thus far we have performed two-dimensional simulations (azimuthal symmetry assumed). In our first set of simulations, we use a simple model of a plasmoid as the driving mechanism for the mass ejection: a cold, dense, parcel of plasma is propelled upward into the initially static corona. The perturbation of the initial corona is accomplished by placing the plasma parcel (density $\rho = 5\rho_0$, temperature $T = T_0/5$) near the base of the corona and including a force term of the form

$$F = \begin{cases} (\rho - \rho_b)A, & \text{if} \rho - \rho_b \geq 0, \\ 0, & \text{if} \rho - \rho_b < 0, \end{cases}$$

in the momentum equation. The unperturbed density is ρ_b and A is chosen to achieve the desired acceleration. The acceleration is performed for about 15 minutes, at which point the velocity of the parcel is close to the Alfvén speed at the base of the corona. We then turn off the force term and allow the parcel to "coast" through the rest of the corona. For these simulations, the magnetic Reynolds number R_m (the ratio of the characteristic time for diffusion of the magnetic field into the plasma parcel to the characteristic advection time for the parcel) is 1000. The ratio of plasma to magnetic pressure, β, was 0.05 the equatorial base of the initial corona.

Results for Simulations of an Advected Plasma Parcel

In order to assess the effect of a nonthermal driving mechanism, we compare our simulation results with observations of looplike CMEs. Sime et al. [1984] showed plots of the time evolution of D, the fractional change in brightness of the K corona:

$$D(r, \theta, t) = \frac{I(r, \theta, t) - I_0(r, \theta)}{I_0(r, \theta)}.$$

Here $I(r, \theta, t)$ is the observed scattered white-light intensity of the corona at a given spatial location and time, and $I_0(r, \theta)$ is the background intensity observed at the same spatial location prior to the start of the event. D is proportional to the fractional change in the plasma density in the corona. Sime et al. [1984] compared the observed value of D with a quantity analogous to D but computed from the fractional density change

$$\rho_f = \frac{\rho - \rho_0}{\rho_0}$$

in the results of Dryer et al. [1979], where $\rho_0 = \rho_0(r, \theta)$ was the initial density used in the simulation. We perform a similar comparison for our simulation results; we compare ρ_f in our simulations with plots of D from the Sime et al. observations. Qualitative comparisons between the observations and the simulation are useful; however, even when one compares the fractional density in the simulation to the observations, there is a fundamental difference between the form of the observed data and that of the simulations. The simulation data are the result of a two-dimensional calculation, whereas the observations represent line-of-sight measurements of a three-dimensional object.

Figure 1 shows a comparison of the fractional density obtained from our simulation and the observations reported by Sime et al. [1984]. Distances shown in this and all succeeding figures are in solar radii (R_s). The left-hand frames show the observed evolution of the fractional brightness D and the right-hand frames show the evolution of ρ_f as found in our simulation. The position of the occulting disk (at about $2R_s$) is marked on both sets of frames as a circle. We see that, in both the observations and the simulation, the major characteristics of mass-ejection evolution noted by Sime et al. [1984] appear. In the early evolution a dense loop forms. As the loop evolves, the fractional density becomes larger in the legs of the loop, and a fractional density depression occurs in the center of the loop.

Figure 2 shows a surface representation of the plasma density for a simulation with slightly different parameters than Figure 1, but the same qualitative results. The time evolution shown is for the period when the plasma parcel first evolves into a looplike structure. Figure 2 shows that a break-up of the plasma parcel in the center of the simulation region is responsible for the legs of the loop having a larger density than the top of the loop. We have found that both the advective and compressional terms in the mass equation are important in the loop formation.

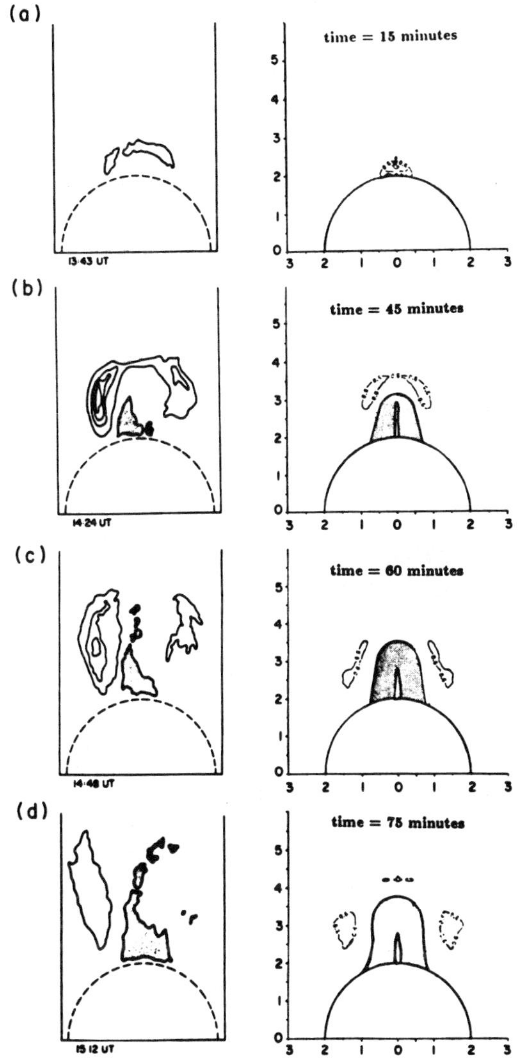

Fig. 1. On the left, a sequence showing the temporal evolution of the contours of the fractional brightness for a coronal transient that occurred on August 10, 1973, as measured by the white light coronagraph aboard Skylab [adapted from Sime et al., 1984]. On the right, a sequence showing the temporal evolution of contours of the fractional density $(\frac{\rho-\rho_0}{\rho_0})$ in the simulation of the advected plasma parcel. The contour levels for the fractional brightness are at 0.5. The contour levels for the fractional density in the simulation were chosen at 0.6 to show the difference in density between the loop sides and loop top more clearly. Regions where the fractional density is negative are shaded on both figures. The plotted circles in both sets of frames mark the position of the occulting disk (adapted from Linker et al. [1989]).

Although our simulation results are qualitatively similar to the observations, we note that quantitative differences do appear. For example, the density enhancement of the legs of the mass ejection relative to the top is generally not as large in our simulations as in the observations. Also, the density loop in our simulations decelerates throughout the time evolution, whereas observations of CMEs show evidence of mild acceleration [Gosling et al., 1976; Rust et al., 1980]. This deceleration appears to be an artifact of the way that we initially propel the plasma parcel. Despite these quantitative differences, Figure 1 shows that by using a nonthermal driving mechanism, simulations using a simple model for the initial corona can produce results qualitatively consistent with the observations.

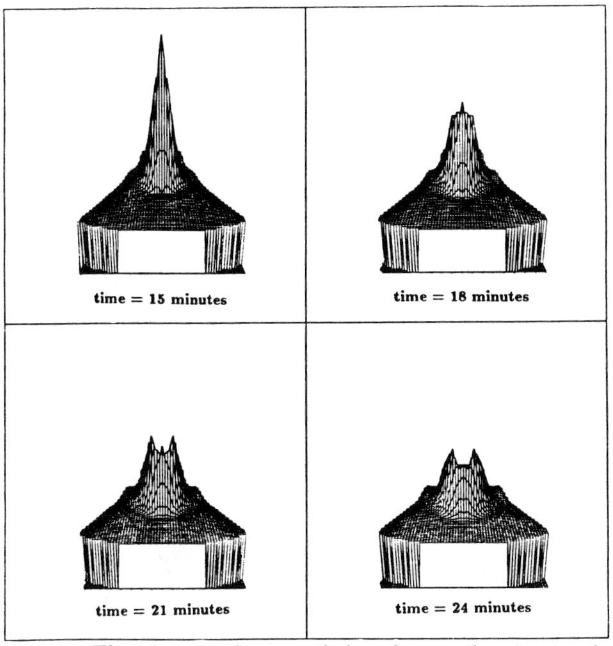

Fig. 2. The time evolution of the plasma density in the simulation. The plasma density in each frame is represented as a topographic surface and viewed looking towards the sun. The plasma density first becomes greater on the sides of the loop as the result of a break-up of the plasma in the center of the loop.

Figure 3 shows the time evolution of the magnetic field in our simulation. The first frame shows the initial current-free magnetic field, before acceleration is applied to the parcel. Figure 3 shows that the upward motion of the plasma parcel stretches out the magnetic field.

Simulations of Magnetically Driven CMEs

The results from our first simulations suggest that the form of the driving mechanism is important in determining the evolutions of CMEs. The basic assumption underlying these simulations is that CMEs are driven by the onset of magnetic instability in the corona; however,

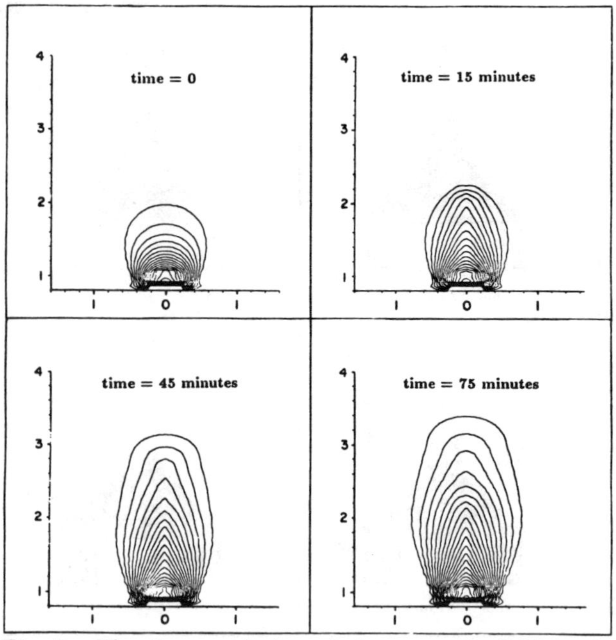

Fig. 3. The time evolution of the magnetic field in the simulation. The uppermost frame on the left shows the initial magnetic field, before acceleration is applied to the plasma parcel. The sequence of frames shows that the upward motion of the parcel stretches the magnetic field (adapted from Linker et al. [1989]).

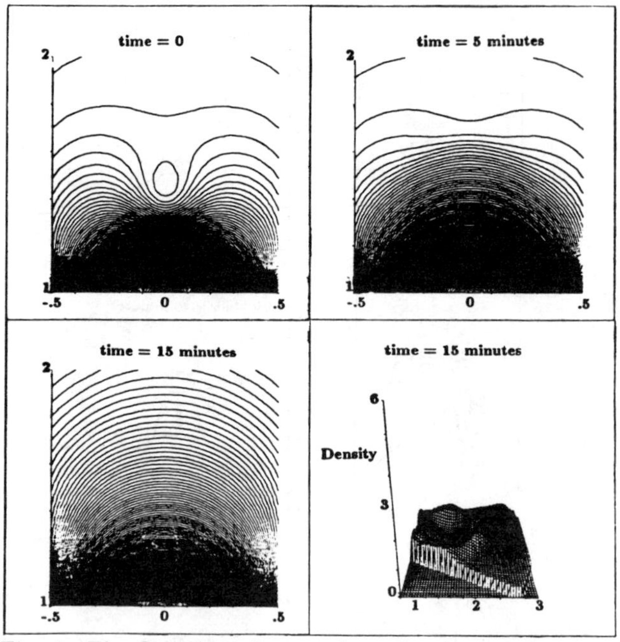

Fig. 4. The first three frames show the time evolution of the magnetic field for a simulation where a current perturbation is applied to the initial corona. The applied current results in an initially plasmoid-like magnetic field structure, and an upward force on the plasma. However, the applied forces also tend to pull the initial current distribution apart. The last frame shows the plasma density. A looplike density distribution with higher density on the sides has started to form; however there has also been large lateral movement.

the actual perturbation is accomplished using an artifical force term. We are now developing simulations that perturb the initial corona magnetically.

To apply a magnetic force to the initial corona, we need our "plasmoid" model to have a magnetic topology that gives an outward force on the plasma, i.e., we need a perturbation magnetic field that yields a current out of the page in Figure 3. To obtain a perturbation magnetic field of this form, we solve the equation $\nabla^2\vec{A} = -\vec{J}$ (normalized units), where we specify the desired current perturbation $\vec{J}$. Because we are concerned with two-dimensional simulations, both $\vec{A}$ and $\vec{J}$ are in the ϕ direction and the above equation reduces to a scalar Poisson equation. As we are solving this equation to develop our initial condition and not at every time step, a simple iteration method (successive-over-relaxation, e.g., Potter, 1973) suffices. We then calculate the perturbation magnetic field $\vec{B}_p = \nabla \times \vec{A}$ and add this field to the initial magnetic field $\vec{B}_0$. Thus far we have only considered a current perturbation; no excess density perturbation was added.

Figure 4 shows the time evolution of the magnetic field ($\beta = .005$ at the equatorial base for this simulation). The first frame shows that the addition of a current perturbation results in a magnetic field structure that is reminiscent of a plasmoid. In the subsequent evolution, the stressed magnetic field "snaps back" and applies an outward force on the plasma. However, the initial current distribution breaks up, and the plasmoid-like structure of the magnetic field is not maintained. The reason for this break-up is that, in addition to the outward radial force introduced as result of the perturbation current acting on the background B_θ, there are also oppositely directed forces (in the $\pm\hat{\theta}$ direction) imposed from the perturbation current acting on the background B_r. The last frame in figure 4 shows the plasma density 15 minutes after the current perturbation was introduced. An upward-moving density perturbation has formed that is looplike and is stronger on the sides than on the top, and we have already noted that this behavior is seen in observations of CMEs. However, in contrast to the results shown in Figures 1 and 2 (and to CME observations), the "legs" of the density perturbation have expanded laterally and have in fact impacted on the side boundaries of the simulation.

The lateral motion appears to be caused by the break-up of the current structure. We are now proceeding with simulations where the current perturbation is somewhat larger and of different form than what we have used here, and we are also considering adding a B_ϕ component to the perturbation field to change the internal stability of the perturbation.

The forces that break-up the initial current perturbation structure occur because the structure was not formed self-consistently. In the future, we plan to perform simulations where we stress the magnetic field by introducing shearing flows at the coronal base, as in the work by Mikić et al. [1988], and cause a plasmoid to form in a self-consistent manner. Such simulations will require the incorporation of implicit or semi-implicit finite-differencing.

Summary

Steinolfson and Hundhausen [1988] have recently shown that, for simulations of mass ejections using a thermal driving mechanism, consistency with the observations could be obtained only if an elaborately tailored initial corona was used. Our initial simulations used a simple plasmoid model as the driver, and a model of the corona that is initially hydrostatic with current- free fields to obtain results that are also qualitatively consistent with the main observational features of mass ejections, as noted by Sime et al. [1984]. We do not claim that these results represent a complete modeling of CMEs; rather, they suggest that the nature of the driving mechanism may play an important role in the evolution of CMEs in the solar corona. Our initial simulations proceeded from the assumption that the onset of magnetic instability is the driving mechanism for CMEs; we are now developing simulations that actually perturb the initial corona with magnetic forces. Our initial results show looplike density perturbations with some features that are consistent with the observations; however, the lateral movement of the density loop seen in these simulations is not observed in actual CMEs. We are now investigating the effects of different forms of the current perturbation. In the future, we plan to perform simulations where the onset of magnetic instability occurs as a result of shearing flows imposed at the coronal base.

Questions and Answers

Steinolfson: The Steinolfson and Hundhansen work you referenced did not claim or infer that the driving mechanism was not important. We put the same driver in different initial atmospheres and, since the CME signature was significantly different ion all cases, we concluded that the initial atmosphere was important. With your solid wall boundary conditions at 30°, are you not just seeing material pileup at the boundaries? Consequently, the legs have the appearance of not moving laterally.

Linker: We have examined the time evolution of the plasma density, and it is quite clear that the formation of denser "legs" occurs because the parcel breaks up in the center of the simulation region, far from the boundaries. For the first hour to one-and-a-half hours of the simulation (which is comparable with the time-scale of the main dynamics in the event we are simulating), the outer boundaries are not having any significant effect on the results. We do not show results for time-scales longer than this because then we are no longer sure that the boundaries do not affect the calculation.

Wolfson: I have a philosophical objection to this focus on "drivers" for CMEs. As we've moved away from the notion of flare-driven CMEs, the new view is that the CME results from a gradual evolution of the global magnetic field to the point where further equilibrium is not possible. The resulting dynamic event is the CME. Clearly, your first calculation (with plasmoid driver) is not consistent with this view. What about your magnetically driven CME? Is the current perturbation really small, or is it a large effect imposed to drive the CME? If the latter, it, too, is not consistent with the view of global nonequilibrium put forward by Low (1988) and Priest (1988).

Linker: I think that your objection may be related to terminology. As I stated in the beginning of the talk, we would agree that CMEs are not driven by flares, but instead are caused by magnetic instability. We refer to the onset of this magnetic instability as "the driver."

Our first simulations clearly use a model of this driver. However, the question that we are investigating is whether or not changing the initiation mechanism in a simulation from a thermal pulse (as was used in previous studies) to something else makes a difference for the subsequent evolution of the simulated event. We have found that the way we initiate the CME in the simulation does change the evolution, suggesting that the driving mechanism is, in fact, an important aspect of CME.

Our most recent simulations use a perturbation magnetic field that is smaller than the background, but not insignificant compared to the background. However, it is the current in the perturbation field pushing against the background field that drives the magnetic structure outward.

Ultimately, we would like to have the initial stage of our simulation be like that of Mikic et al (1988), and slowly build up energy in the magnetic field until instability occurs. However, developing a code that can follow the fast advection of structures over flare length-scales and also efficiently compute during the slow energy buildup stage is a difficult computational problem.

Acknowledgements: This research was supported by the Solar Terrestrial Theory Program of NASA and the Atmospheric Sciences Section of NSF. Computational resources were provided by NSF at the San Diego Supercomputer Center.

References

Anzer, U., and G. W. Pneuman, Magnetic reconnection and coronal transients, *Sol. Phys., 79,* 129, 1982.

Dryer, M., S. T. Wu, R. S. Steinolfson, and R. M. Wilson, Magnetohydrodynamic models of coronal transients in the meridional plane, II, simulation of the coronal transient of 1973, August 21, *Astrophys. J., 227,* 1059, 1979.

Gosling, J. T., E. Hildner, R. M. MacQueen, R. H. Munro, A. I. Poland, and C. L. Ross, The speeds of cornal mass ejection events, *Sol. Phys., 48*, 389, 1976.

Linker, J. A., G. Van Hoven, and D. D. Schnack, MHD simulations of coronal mass ejections: importance of the driving mechanism, in press, *J. Geophys. Res.*, 1989.

Mikić Z., D. C. Barnes, and D. D. Schnack, Dynamical evolution of a solar coronal magnetic field arcade, *Astrophys. J., 328*, 830, 1988.

Potter, D., *Computational Physics*, John Wiley and Sons, Chichester, England, 1973.

Rust, D. M., et al., Mass ejections, in *Solar Flares* edited by P. A. Sturrock, 273, Colorado Associated University Press, Boulder, Colorado, 1980.

Sime, D. G., R. M. MacQueen, and A. J. Hundhausen, Density distribution in looplike coronal transients: a comparison of observations and a theoretical model, *J. Geophys. Res.*, 89, 2113, 1984.

Steinolfson, R. S., and A. J. Hundhausen, Density and white light brightness in looplike coronal mass ejections: temporal evolution, *J. Geophys. Res., 93,* 14269, 1988.

Wagner, W. J., Coronal mass ejections, *Ann. Rev. Astron. Astrophys., 22*, 267, 1984.

Wu, S. T., Y. Nakagawa, S. M. Han, and M. Dryer, Magnetohydrodynamics of atmospheric transients, IV, Nonplane two-dimensional analysis of energy conversion and magnetic field evolution, *Astrophys. J., 262*, 369, 1982.

ENERGETIC ION AND COSMIC RAY CHARACTERISTICS OF A MAGNETIC CLOUD

T. R. Sanderson, J. Beeck, R. G. Marsden, C. Tranquille, K.-P. Wenzel

Space Science Dept. of ESA, Estec, Noordwijk, The Netherlands

R. B. McKibben

LASR, Enrico Fermi Institute, University of Chicago, Chicago

E. J. Smith

Jet Propulsion Laboratory, California Institute of Technology, Pasadena

Abstract. We present energetic ion and magnetic field observations from ISEE-3 and ground based cosmic ray observations from selected neutron monitors, of the magnetic cloud associated with the large interplanetary shock event of February 11, 1982. This shock event had distinct features which included an upstream foreshock, a turbulent downstream region lasting for several hours, and a large magnetic cloud containing bidirectional energetic ion fluxes lasting for several days. Associated with the event was a Forbush decrease observed at the earth. The timing of the onset of the Forbush decrease coincides with the arrival at the earth of the magnetic cloud, and the duration of the decrease corresponds to the duration of the passage of the cloud past the earth. In this event the magnetic cloud appears to be the major cause of the Forbush decrease, whilst the shock and the post-shock turbulent region do not appear to play a major role. Within the cloud we observe highly anisotropic bidirectional energetic ion fluxes and a well ordered magnetic field, both of which signify a large scattering mean free path. The large scattering mean free path suggests that the ions within the magnetic cloud can easily travel along the magnetic field lines, but not across, and therefore, once inside the cloud cannot get out. In the same way, the well ordered magnetic field lines act as a barrier to the cosmic rays crossing the field lines, thereby preventing them from entering into the cloud. The onset of the Forbush decrease occured when the asymptotic arrival direction of the cosmic ray particles for the Deep River, Climax and Huancayo neutron monitors was from an anti-sunward direction. Therefore when the magnetic cloud arrived at the earth, the cosmic rays were immediately prevented from reaching the earth. We conclude that the magnetic cloud is the major cause of the Forbush decrease, whilst the post-shock region plays only a minor role, if any, in this decrease.

Geophysical Monograph 58

Introduction

Detailed studies of large shock events, by for example, Borrini et al. [1982], van Nes et al. [1984], Sanderson et al. [1985], and Tranquille et al. [1987], have shown that many large shock events have a foreshock extending several earth radii upstream from the shock, and a turbulent region lasting for several hours immediately after the shock, usually terminated by a tangential discontinuity. This is often followed by a region containing bidirectional fluxes of energetic particles [Sanderson et al., 1983, Tranquille et al. 1987] and which is magnetically quiet [Smith 1983]. This region, where the variance of the magnetic field and the magnetic field power spectral density are lower than average [Tranquille et al, 1987, Zhang and Burlaga, 1988], can last for several days, and is often identified with driver gas from the flare [Zwickl et al., 1983]. Large scale magnetic loop-shaped regions, identified by the rotation of the magnetic field direction in a plane perpendicular to the ecliptic plane, were called magnetic clouds by Burlaga et al. [1981] and Klein and Burlaga [1982]. These regions are often considered to be the signature at 1 AU of a coronal mass ejection (CME) [Wilson and Hildner, 1984]. Large scale structures in the solar wind, some of which are magnetic clouds, were identified by Palmer et al. [1978], Kutchko et al. [1982], Sanderson et al. [1983] and Marsden et al. [1987] using the bidirectional signature of energetic protons, and by Montgomery et al. [1974], Bame et al. [1981], Pilipp et al. [1987] and Gosling et al. [1987] using the bidirectional signatures of electron heat flux.

There have been many attempts to identify the causes of Forbush decreases. Both transient events (see for instance Barouch and Burlaga [1975] and references therein), and long lasting events [Duggal et al., 1981] have been proposed for the source of this modulation of the cosmic rays. Barouch and Burlaga [1975], and Sarris et al. [1989] suggested that gradient drift could be responsible for the decreases. Nishida [1982] studied the magnetic structures reported by Bame et al. [1981] and Burlaga et al. [1981] and concluded that the post-shock region is responsible for Forbush decreases. Thomas and Gall [1984] suggested that the post-shock region acted as a barrier to cosmic rays, and that the decrease is caused by adiabatic cooling in the region behind. Badruddin et al. [1986] used a superposed epoch analysis of neutron monitor data, and concluded that the onset of the decreases started before the arrival of the magnetic cloud. Zhang and Burlaga [1988] using a superposed epoch analysis of shock events, showed that although magnetic clouds could produce Forbush decreases, magnetic clouds preceded by a shock could produce larger decreases than clouds not preceded by a shock, and therefore that the turbulent magnetic fields associated with the sheath between a shock and a magnetic cloud are more effective in modulating the cosmic rays than the strong ordered fields in the magnetic cloud itself.

In this paper we examine in detail the timing of a Forbush decrease as measured by selected neutron monitors, and its association with the passage of a magnetic cloud as identified with the magnetic field and the bidirectional signature of the energetic ions measured on ISEE-3 at the sunward libration point. We conclude that this Forbush decrease is caused by the presence of the magnetic cloud and not the turbulent region between the shock and the magnetic cloud.

Instrumentation

The proton observations presented here are taken from the low energy proton experiment (DFH) on ISEE-3. This instrument measures the 3-dimensional distribution of protons in the energy range 35-1600 keV, using three telescopes mounted at 30°, 60°, and 135° with respect to the spacecraft spin axis. The spin axis is oriented along the north ecliptic pole. A description of this instrument and the methods used in analysing the 3-dimensional distributions can be found in Sanderson et al. [1985]. The magnetic field measurements used here are from the Vector Helium magnetometer [Frandsen et al., 1978], also on the same spacecraft. One-minute values of the magnetic field and its variance have been used in this study.

The ground based observations used in this paper are taken from the Deep River, Climax, Huancayo, and the Kiel, Jungfraujoch, and Rome neutron monitors [Solar Geophysical Data, 1982]. For this study, we have grouped the neutron monitors into two groups on opposite sides of the earth relative to the dipole axis, according to their geomagnetic longitude. Table 1 lists the characteristics of the two groups of neutron monitors.

Observations

In the left hand panel of Figure 1 we show the relative counting rates for the Deep River, Climax and Huancayo neutron monitors for the ten day period beginning at 0000 UT on February 9, 1982. At the time of arrival of the magnetic cloud at the earth, these neutron monitors were monitoring particles coming from the antisunward direction. A Forbush decrease, lasting for several days, commenced at around 0000 UT on February 12, after a period of relatively constant cosmic ray intensity. The decrease was associated with an interplanetary shock which passed the ISEE-3 spacecraft at around 1315 UT on February 11, approximately 11 hours before the onset, as shown by the first vertical dashed line. However, the onset of the decrease was at the time of passage at ISEE-3 of the discontinuity marking the boundary between the post-shock turbulent region and the magnetic loop, as shown by the second vertical dashed line. Only a small change in the counting rate is observed at the time of passage of the shock over the earth, and virtually no change is seen during the period between the passage of the shock and the end of the post-shock turbulent region.

The right hand panel of Figure 1 shows neutron monitor counting rates for three stations which sample cosmic ray

TABLE 1. Neutron monitor characteristics

Station	Geomag.Lat., N.	Geomag. Long. E.	Cutoff, GV
Deep River	57.4	-10.0	1.02
Climax	48.2	-43.5	3.03
Huancayo	-3.4	-40.6	13.45
Kiel	54.7	96.0	2.28
Jungfraujoch	47.8	90.1	4.53
Rome	42.4	92.6	6.32

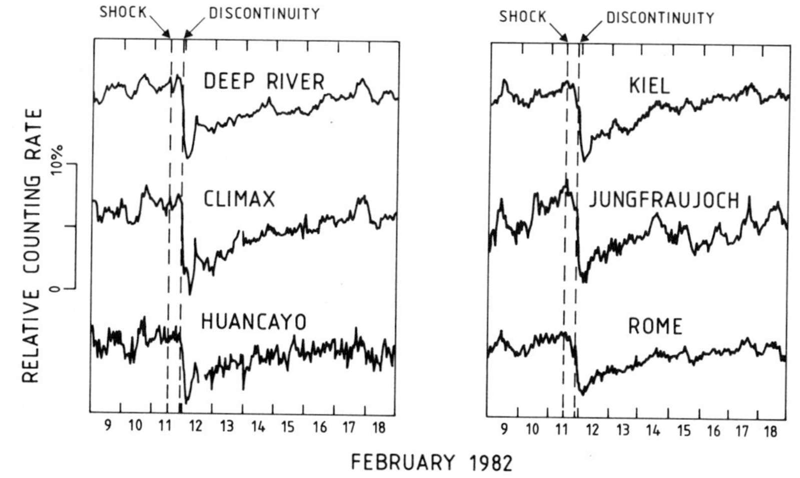

Fig. 1. Left hand side: Counting rate of the Deep River, Climax and Huancayo neutron monitors for the ten-day period commencing on February 9, 1982. Right hand side: Counting rate of the Kiel, Jungfraujoch and Rome neutron monitors for the same ten-day period. The vertical lines show the arrival times of the shock and magnetic cloud at ISEE-3.

particles coming from a sunward direction. Again, only a small change in counting rate is seen at the time of arrival of the shock at the earth, and a small decrease is seen a few hours before the arrival of the magnetic cloud. The main decrease however, also commenced at around the time of arrival of the magnetic cloud at the earth. All six neutron monitors observed the maximum decrease a few hours after the arrival of the magnetic cloud at the earth.

Figure 2 shows the low-energy proton intensity and anisotropy parameters and magnetic field data measured at ISEE-3 for the 5-day period commencing on February 10, 1982. The top panel shows the 35-56 keV spin averaged intensity. The anisotropy amplitudes, shown in the next two panels, have been derived by first transforming the proton data into a frame of reference moving with the instantaneous value of the solar wind velocity, and then fitting the data to a 3-dimensional spherical harmonic series [Sanderson et al., 1985].

The first harmonic can be described by three coefficients, A'_{10}, A'_{11}, and B'_{11}, each coefficient representing the flow along one of the axes of the coordinate system. In our coordinate system, with the z-axis aligned along the instantaneous magnetic field direction, A'_{10}/A'_{00} represents the flow along the magnetic field direction, as shown in the second panel.

The second harmonic can be described by five coefficients. If the flow is axially symmetric along the magnetic field, then only the A'_{20} coefficient is significant. Positive values of A'_{20}, when the second harmonic amplitude exceeds the first harmonic amplitude, denote bidirectional fluxes, whilst negative values of A'_{20} denote intensities peaked at 90 degree pitch angles. A'_{20}/A'_{00} is shown in the

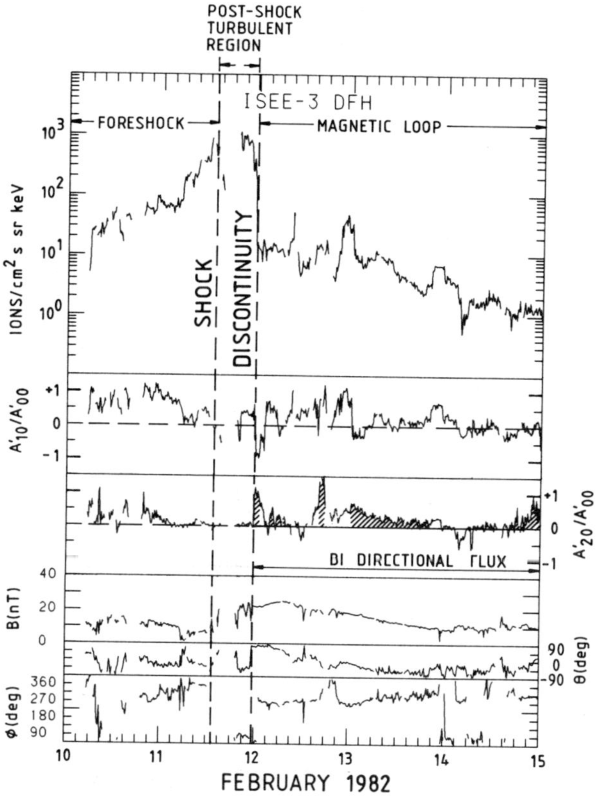

Fig. 2. From top to bottom: 35-56 keV proton intensity, 1st-order field aligned anisotropy amplitude in the solar wind frame of reference, A'_{10}/A'_{00}, 2nd-order field aligned anisotropy, A'_{20}/A'_{00}, magnetic field amplitude B, elevation θ, and azimuth ϕ.

third panel. The second harmonic amplitude exceeds the first harmonic amplitude at various times during the period February 11 to February 14, signifying bidirectional fluxes, as shown by the shading in Figure 2.

The bottom three panels show the magnetic field amplitude and directions. A region of turbulent magnetic field was observed immediately after the shock, lasting for around 10 hrs. This turbulent post-shock region was bounded by a discontinuity at 2320UT on February 11, denoted by the second vertical dashed line in Figure 2. The discontinuity also marked the arrival of a large magnetic cloud, containing the bidirectional proton fluxes mentioned above [Marsden et al., 1987].

We have been unable to find an event on the Sun which we can positively identify as the source of the event. A large event (class 3B) was observed at E54 at 0056 UT on February 10, but the location is unusual for a magnetic cloud to be observed at the earth, and the implied shock speed is very high. We would expect however, the source to be located either centrally, or west of center.

Figure 3 shows the counting rate of the Climax neutron monitor plotted together with the 35-56 keV proton intensity, the second harmonic amplitude, A'_{20}/A'_{00}, the magnetic field strength B, elevation θ, azimuth ϕ, and the variance of the magnetic field, for the 2-day period commencing at 0900 UT on February 11. The arrival of the shock, the passage of the post-shock turbulent region, the onset of the Forbush decrease and the arrival of the magnetic cloud at the spacecraft can clearly be seen from this figure. The precise timing of the onset of the decrease

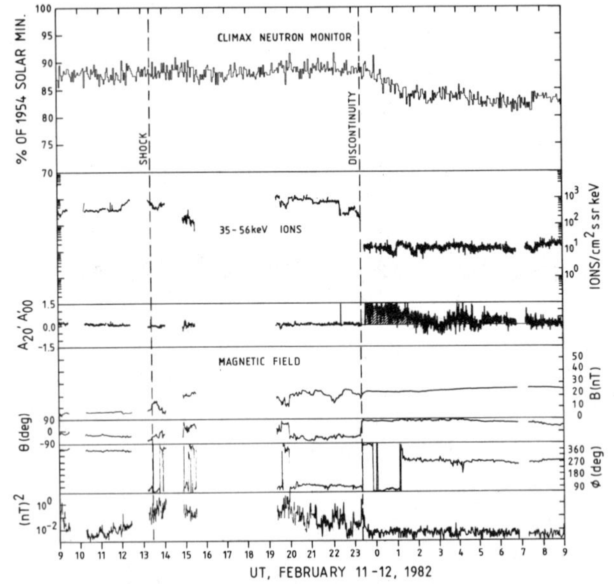

Fig. 3. From top to bottom: Climax neutron monitor counting rate, 35-56 keV proton intensity, 2nd-order field aligned anisotropy, A'_{20}/A'_{00}, magnetic field magnitude B, elevation θ, azimuth ϕ, and magnetic field variance.

and the effects observed at the earth must take into account the finite upstream distance of ISEE-3. The magnetic discontinuity was observed at a time when the solar wind velocity was around 600 km/s, so that the transit time from ISEE-3 to the earth was around 40 minutes, if the discontinuity was travelling at the solar wind speed. This delay means that the cloud must have arrived at the earth at around 0000 UT on February 12.

It is clear that the onset of the Forbush decrease does not coincide either with the passage of the shock at the earth, or with the passage of the turbulent region following the shock, but with the time when the discontinuity at the edge of the magnetic cloud arrives at the earth. No change is seen in the neutron monitor counting rate at the time of passage of the shock. In principle, the turbulent region following the shock could act as a barrier to high energy cosmic rays. However, due to the turbulent nature of the region, cosmic rays can propagate across the field lines because of the presence of scattering centers, and so cross the region, albeit with difficulty.

The magnetic field in the post shock turbulent region has a high variance. Comparing this event with, for example the November 12, 1978 event, which has been widely quoted in the literature as one of the largest events observed on ISEE-3, we find that variance of both the magnetic field magnitude and the sum of the variances of the components measured during the post-shock turbulent region are comparable with that observed in the post-shock region of the November 12, 1982 event. We are unable to make a thorough comparison of the two events, due to a data gap in the middle of the post shock region of the February 1982 event. Despite this high level of turbulence, the post shock turbulent region does not significantly affect the cosmic rays.

Unlike the post-shock region, the magnetic cloud has a magnetic field with an extremely low variance, in fact one of the lowest values observed in interplanetary space on ISEE-3. This means that the mean free path is extremely high, possibly as high as 1 AU, such as was found for another cloud by Tranquille et al. [1987]. In such a region, protons can freely propagate along the field lines without encountering scattering centers. This also implies that the protons cannot cross the field lines within the cloud, so that the cloud can act as a closed-off volume into which cosmic rays cannot easily enter. The leading edge of the cloud acts as a barrier, sweeping away cosmic rays as it propagates outwards from the Sun, and causing the onset of the Forbush decrease as the edge passes over the observer. The minimum of the decrease occurs only a few hours after the leading edge passes over the observer.

The leading edge of the cloud is well defined. On the upstream side of the cloud there is a population of isotropic energetic ions with high intensity, separated by a tangential discontinuity from a much lower intensity population of ions with a bidirectional anisotropy inside the cloud. The onset of these bidirectional anisotropies clearly identifies the boundary of the magnetic cloud. The time at which the maximum cosmic ray decrease was observed was only a few hours after the cloud passed over the earth. This further strengthens the argument that the arrival of the cloud causes the decrease. (If the post shock turbulent region were responsible for the decrease, we should expect instead to see the cosmic ray intensity start to decrease when the shock arrives, and have maximum decrease at the time of passage of the end of the post shock turbulent region).

The recovery continues for approximately the duration of the passage of the magnetic cloud. The cloud is observed for at least 2.5 days, when the solar wind velocity was around 400 km/s, implying a radial extent of at least 0.6 AU. The duration of the cloud corresponds approximately to the duration of the Forbush decrease. This can also be seen in the example shown in Figure 1 of Zhang and Burlaga [1989]. The exact end of the Forbush decrease and the end of the cloud however, are not so clearly distinguishable as the onset of the decrease and the leading edge of the cloud. The same features which are often used to identify the onset of the cloud do not give a clear indication of the end of the cloud. There is no discontinuity marking the end of the cloud, the magnetic field gradually decays to its background level, the rotation of the magnetic field vector slows down, and there is no sudden switch-off of the bidirectional fluxes.

The timing of the onset of the Forbush decrease will depend on many factors. Each neutron monitor responds to cosmic rays which are coming from a direction which is quite different to the pointing direction of the axis of the instrument, due to the effect of the earth magnetic field on the trajectories of the particles. This direction is the asymptotic direction for the neutron monitor [Shea and Smart, 1975]. (For example, for the Deep River neutron monitor, located at 46.10° N, and 282.50 ° W geographic, the asymptotic arrival direction for 1.5 GV particles is 14° N, 73° E). If the asymptotic arrival direction of the neutron monitor is oriented towards the Sun, then the approaching barrier caused by the cloud could already cause a decrease in the cosmic ray intensity when it is a few cosmic-ray gyroradii away from the earth. At the time of the onset of the Forbush decrease described here, at 0000 UT on February 12, the Kiel, Jungfraujoch and Rome neutron monitors were sampling cosmic rays from the sunward direction and exhibited a small decrease a few hours before the arrival of the cloud. However, if the asymptotic arrival direction is from the antisunward direction, as is the case for the Deep River, Climax and Huancayo neutron monitors, the approaching cloud will not cause a decrease until the barrier passes over the earth. At the time of onset, the above three neutron monitors were sampling cosmic rays from an antisunward direction, and observed the decrease when the barrier passed over the earth.

In Figure 4, we sketch, in 2-dimensions, the magnetic cloud at the time of the passage of the shock, at the onset of the Forbush decrease, and during the Forbush decrease, assuming a source on the central meridian. Also shown is the longitude of the Deep River, Climax, and Huancayo neutron monitors, together with an arrow which shows the approximate arrival direction of the cosmic ray particles. On the left hand side is shown the situation when the shock arrives at the earth. At this time cosmic ray particles are able to penetrate through the post shock region, and can reach the earth with little or no attenuation. The central panel shows the situation at the time of the arrival of the cloud at the earth. The Deep River, Climax and Huancayo neutron monitors are now in a different position, due to the rotation of the earth, and the asymptotic arrival direction of the neutron monitors means that the monitors are observing cosmic ray particles coming from an anti-sunward direction. The leading edge of the cloud now acts as a barrier, attenuating the cosmic rays, and causing the onset of the decrease. The right hand panel shows the situation when the earth is inside the cloud. At this time, cosmic rays have difficulty in reaching the earth whatever the arrival direction, and so the minimum cosmic ray intensity is observed.

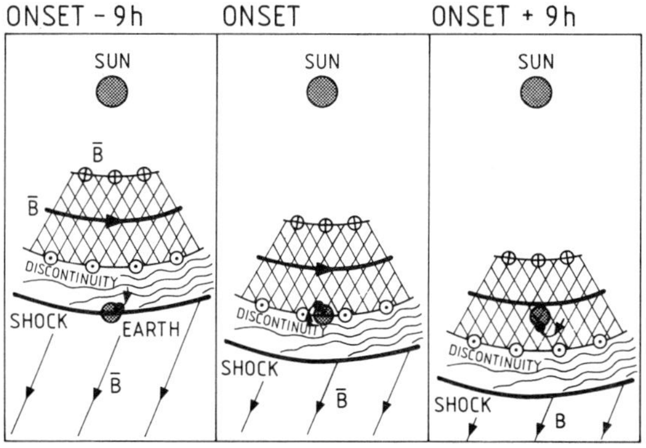

Fig. 4. Configuration of the shock, post-shock region, and magnetic loop at the time of the passage of the shock, at the onset of the Forbush decrease, and during the Forbush decrease.

Finally, in Figure 5 we show a possible 3-dimensional representation of the magnetic cloud, again, assuming a source on the central meridian. This configuration has been suggested by several authors, e.g. Goldstein [1983], Marabushi [1986], Burlaga et al. [1988], Burlaga et al. [1989], Crooker et al. [1989], Gosling [1989]. This shows the situation soon after the cloud has reached the earth, when the shock and the post-shock turbulent region and the discontinuity have already passed the earth. At ISEE-3, a northward magnetic field is observed. Later, around the centre of the cloud, a westward magnetic field is observed, whilst at the trailing edge of the cloud, a southward pointing magnetic field is observed. The cloud can be interpreted as a twisted magnetic flux rope, with a

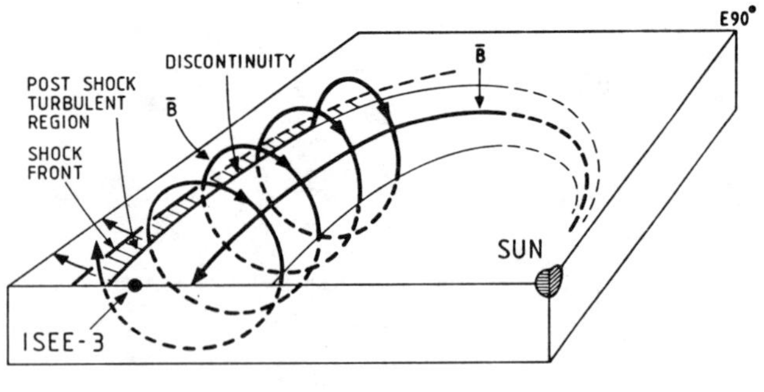

Fig. 5. Three dimensional representation of the mag netic cloud soon after the onset of the Forbush decrease. The cloud has the appearance of a twisted magnetic flux rope, with a northward magnetic field at the leading edge of the cloud, a westerly magnetic field in the centre, and a southward magnetic field at the trailing edge of the cloud.

central field oriented along a westerly direction, and at the edges a strong tangential component. The flux rope may still be connected to the the Sun, although from our observations, it is not possible to deduce whether the flux rope is connected or not.

This event is one of many studied in a statistical survey of Forbush decreases and bidirectional events observed during the time when ISEE-3 was in its orbit around the upstream Libration point. Several other events of this type have been observed. A paper is in preparation to discuss this.

Conclusions

Having examined the timing of a Forbush decrease, and compared it with the interplanetary conditions as observed at ISEE-3, we find

1. The onset of the Forbush decrease coincides with the arrival at the earth of the discontinuity at the leading edge of the magnetic cloud.

2. The duration of the main part of the recovery of the decrease corresponds approximately to the duration of the passage of the magnetic cloud.

3. The magnetic cloud has a quiet magnetic field with very low magnetic field variance, and contains anisotropic, bidirectional distributions of 35-1600 keV protons, consistent with a large scattering mean free path. In such a region, particles can propagate easily along the field lines, but not across them.

We conclude that the magnetic cloud is responsible for the Forbush decrease and that the post-shock turbulent region does not play a significant role in this Forbush decrease.

Questions and Answers

Priest: Your cartoon of a highly twisted and curved tube would have a higher magnetic field strength on the inside of the curve, so how do you explain the decrease of field strength in the event as it passes over the spacecraft?

Sanderson: The magnetic field actually peaks around 12-hours after entering the flux rope. This is probably the center of the tube. The field then decays due to the expansion of the cloud, which explains the weaker magnetic field on the inside of the curve.

Russell: Since the axis of the cloud, if it is a rope, could be at an angle to the ecliptic plane and not necessarily in it, could this not explain the appearance of the shock of Helios without a following magnetic structure? The shock should be more spherically symmetric.

Sanderson: Yes, quite possibly. We would expect the structure to be symmetric about the flare site, but so far we have not found evidence for that either in the Helios magnetic field or the plasma data.

Kahler: You showed that the Forbush decrease of this event clearly began with the cloud rather than the preceding shock. However, the analysis of Zhang and Burlaga showed that most Forbush decreases began with the shock rather than the cloud.

Sanderson: The event presented was part of a statistical survey that we are conducting. A preliminary result of this survey shows that this type of event constitutes approximately one-quarter of the events that we have studied.

Gosling: You mentioned that the bidirectional proton signature is an equally good signature of a CME in interplanetary space as a bidirectional electron heat flux signature. I strongly disagree. I have compared our bidirectional electron events with yours on an event by event basis. Only 7 out of 60 of your events in a 4-year period are definitely not bidirectional electron events. On the other hand, more than a hundred of our events, including a number of spectacular ones, apparently are not bidirectional proton events in this same ~4-year period. I am not sure why such a discrepancy in numbers of events (by a factor of ~3) should exist, but there can be little doubt that it does. Since I strongly believe the bidirectional electron events we report are a signature of CMEs in interplanetary space, I must conclude that for some reason many CMEs do not also have a bidirectional proton signature.

Sanderson: I do not claim that we can identify CMEs on the basis of our bidirectional signatures. As yet, we have not studied the relation between CMEs and our bidirectional ions. I agree that there is a discrepancy between your events and ours. The identification of those bidirectional events is very subjective, and perhaps the best way to resolve this

discrepancy is to compare both data sets, and if necessary, revise the criteria for the identification of the events.

Acknowledgements. We thank S. T. Ho and M. Szumlas for assistance in the processing of the low energy proton experiment data. We thank Dr. S. J. Bame for permission to use ISEE-3 solar wind velocity measurements for transforming the proton data into the solar wind frame of reference. This work was supported in part by NSF Grant ATM-86-20160.

References

Badruddin, R. S. Yadav, and N. R. Yadav, Influence of magnetic clouds on cosmic ray intensity variations, *Solar Phys.*, *105*, 413, 1986.

Bame, S.J., J.R. Asbridge, W.C. Feldman, J.T. Gosling, and R.D. Zwickl, Bidirectional streaming of solar wind electrons 80eV: ISEE evidence for a closed field structure within the driver gas of an interplanetary shock, *Geophys. Res. Lett.*, *8*, 173, 1981.

Barouch, E., and L. F. Burlaga, Causes of Forbush decreases and other cosmic ray variations, *J. Geophys. Res.*, *80*, 449, 1975.

Borrini, G., J.T.Gosling, S.J. Bame, and W.C. Feldman, An analysis of shock wave disturbances observed at 1 AU from 1971 through 1978, *J. Geophys. Res.*, *87*, 65, 1982.

Burlaga, L.F., E. Sittler, F. Mariani, and R. Schwenn, Magnetic loop behind an interplanetary shock : Voyager, Helios, and IMP 8 observations, *J. Geophys. Res.*, *86*, 6673, 1981.

Burlaga, L. F., Magnetic clouds and force-free fields with constant alpha, *J. Geophys. Res.*, *93*, 7217, 1988.

Burlaga, L. F., R. P. Lepping, and J. A. Jones, Global configurations of magnetic clouds, *This volume*, 1989.

Crooker, N. U., J. T. Gosling, E. J. Smith, and C. T. Russell, A bubblelike coronal mass ejection flux rope in the solar wind, *This volume*, 1989.

Duggal, S. P., M. A. Pomerantz, C. H. Tsao, B. T. Tsurutani, and E. J. Smith, *J. Geophys. Res.*, *86*, 7473, 1981.

Frandsen, A. M. A., B. V. Connor, J. Van Amersfoort, and E. J. Smith, The ISEE-C vector helium magnetometer, *IEEE Trans. Geosci. Elect.*, *GE-16*, 195, 1978.

Goldstein, H., On the field configuration in magnetic clouds, *NASA conference publication 2280*, 731, 1983.

Gosling, J. T., D. N. Baker, S. J. Bame, W. C. Feldman, and R. D. Zwickl, Bidirectional solar wind electron heat flux events, *J. Geophys. Res.*, *92*, 8519, 1987.

Gosling, J. T., Coronal mass ejections and magnetic flux ropes in interplanetary space, *This volume*, 1989.

Klein, L.W., and L. F. Burlaga, Interplanetary magnetic clouds at 1 AU, *J. Geophys. Res.*, *87*, 613, 1982.

Kutchko, F.J., P.R. Briggs, and T.P. Armstrong, The bidirectional particle event of October 12, 1977, possibly associated with a magnetic loop, *J. Geophys. Res.*, *87*, 1419, 1982.

Marabushi, K., Structure of the interplanetary magnetic clouds and their solar origin, *Adv. Space Res.*, *6*,1, 1986.

Marsden, R. G., T. R. Sanderson, C. Tranquille, K.-P. Wenzel, and E. J. Smith, ISEE-3 observations of low energy proton bidirectional events and their relation to isolated interplanetary magnetic structures, *J. Geophys. Res.*, *92*, 11009, 1987

Montgomery, M. D., J. R. Ashbridge, S. J. Bame, and W. C. Feldman, Solar wind electron temperature depressions following some interplanetary shock waves: Evidence for magnetic merging, *J. Geophys. Res.* *79*, 3103, 1974.

Nishida, A, Numerical evaluation of the precursory increase to the Forbush decrease expected from the diffusion convection model, *J. Geophys. Res.*, *87*, 6003, 1982.

Palmer, I.D., F.R. Allum, and S. Singer, Bidirectional anisotropies in solar cosmic ray events: evidence for magnetic bottles, *J. Geophys. Res.*, *83*, 75, 1978.

Pilipp, W. G., H. Miggenrieder, M. D. Montgomery, K.-H. Muhlhauser, H. Rosenbauer, and R. Schwenn, Unusual electron distribution functions in the solar wind derived from the Helios plasma experiment: Double strahl distributions with an extremely anisotropic core, *J. geophys Res.*, *92*, 1093, 1987.

Sanderson, T.R., R. G. Marsden, R. Reinhard, K.-P. Wenzel, and E.J. Smith, Correlated particle and magnetic field observations of a large scale magnetic loop structure behind an interplanetary shock, *Geophys. Res. Lett.*, *10*, 916, 1983.

Sanderson, T.R., R. Reinhard, P. van Nes, and K.-P. Wenzel, Observations of three-dimensional anisotropies of 35- to 1000-keV protons associated with interplanetary shocks, *J. Geophys. Res.*, *90*, 19, 1985.

Sarris, E. T., C. A. Dodopoulos, and D. Venkatesan, On the E-W asymmetry of Forbush decreases, *Solar Phys.* *120*, 153, 1989.

Shea, M. A., and D. F. Smart, Asymptotic directions and vertical cut-off rigidities for selected cosmic-ray stations as calculated using the international geomagnetic reference field model appropriate for epoch 1975, *AFCRL-TR-75-0247*, May, 1975.

Smith, E. J., Observations of interplanetary shocks: Recent progress, *Space Sci. Rev.*, *34*, 101, 1983.

Thomas, B. T., and R. Gall, Solar-flare induced Forbush decreases: Dependence on shock geometry, *J. Geophys. Res.*, *89*, 2991, 1984.

Tranquille, C., T.R. Sanderson, R.G. Marsden, K-P. Wenzel, and E.J. Smith, Properties of a large-scale interplanetary loop structure as deduced from low-energy proton anisotropy and magnetic field measurements, *J. Geophys. Res.*, *92*, 6, 1987 van Nes, P., R. Reinhard, T.R. Sanderson, K-P. Wenzel, and R.D. Zwickl, The energy spectrum of 35- to 1600 keV protons associated with interplanetary shocks, *J. Geophys. Res., 89,* 2122, 1984.

Wilson, R.M., and E. Hildner, Are interplanetary magnetic clouds manifestations of coronal transients at 1 AU ?, *Solar Phys., 91,* 169, 1984.

Zhang, G., and L. F. Burlaga, Magnetic clouds, geomagnetic disturbances, and cosmic ray decreases, *J. Geophys. Res., 93,* 2511, 1988.

Zwickl, R. D., J. R. Ashbridge, S. J. Bame, W. C. Feldman, J. T. Gosling, and E. J. Smith, Plasma properties of driver gas following interplanetary shocks observed by ISEE-3, Solar Wind Five, *Nasa Conference Publication 2280,* 711, 1983.

FORMATION OF SLOW SHOCK PAIRS ASSOCIATED WITH CORONAL MASS EJECTIONS

Y. V. Whang

Department of Mechanical Engineering, Catholic University of America, Washington, D.C.

Abstract. An MHD model is used to simulate the formation of a forward-reverse slow shock pair in a solar coronal environment. The model uses the Rankine-Hugoniot solution to calculate the jumps in flow properties at all shock crossings. The shocks divide the domain of solution into several continuous flow regions. In each continuous region, the governing equations are solved by the method of characteristics. The initial condition represents the impact of a high-speed (500 km/s) mass ejecta on a low-speed (100 km/s) ambient solar wind. The momentum impact compresses the plasma near the front of the coronal mass ejection (CME). Large pressure disturbances propagate in both the forward direction relative to the ambient solar wind and the reverse direction relative to the ejecta flow. The pressure fronts steepen to form forward and reverse MHD shocks. In a $\beta = 0.1$ plasma, the resulting shocks consist of a forward slow shock and a reverse slow shock. As the CME associated slow shock pair moves outwards in interplanetary space, it evolves into a pair of fast shocks. The CME and its associated shock pair eventually manifest in interplanetary space as a magnetic cloud accompanied by a fast shock pair: a forward shock precedes the cloud and a reverse shock either within or behind the cloud.

Introduction

This paper presents a simulation study for the formation of slow shocks in a solar coronal environment. Several papers published in the past four years deal with CME associated slow shocks. The possible existence of forward slow shocks preceding some coronal mass ejections (CMEs) was suggested by Whang [1986, 1987] and by Hundhausen et al. [1987]. Whang also showed that transition from a traveling forward slow shock to a forward fast shock can take place in interplanetary space. In a recent paper, Whang [1988] predicted that for some CMEs, their interaction with the ambient solar wind can produce a forward-reverse shock pair.

Geophysical Monograph 58

When a coronal mass ejecta exerts a direct impact on the ambient solar wind, the high-speed mass ejecta compresses the plasma near the top of the CME on both sides of the interface which separates the CME plasma from the ambient solar wind plasma. Large pressure disturbances propagate in both the forward direction relative to the ambient solar wind and the reverse direction relative to the ejecta flow. The pressure fronts steepen to form forward and reverse MHD shocks. In a low-β plasma in the coronal environment, the resulting shocks consist of a forward slow shock and a reverse slow shock.

In order to understand the basic physics involved in the shock formation process, we use an unsteady, one-dimensional model. Similar unsteady 1-D models have been used to study shock formations in the coronal and interplanetary environment by Hundhausen and Gentry [1969], Steinolfson et al. [1975] and Whang [1984]. The time-dependent MHD flow is governed by a system of hyperbolic, nonlinear equations. We use the method of characteristics to carry out the numerical solution.

We treat the shocks as surfaces of discontinuity with zero thickness. The shock surfaces divide the domain of solutions into several continuous flow regions. The model uses the Rankine-Hugoniot solution to calculate the jumps in flow properties at all shock crossings between neighboring continuous flow regions. In each flow region the variation of the thermal pressure and the plasma density follows the isentropic relation that p/ρ^γ remains constant following the motion of each fluid element where γ is the ratio of specific heats. The path of fluid element is defined by $dx/dt = \mathbf{U}$ where $\mathbf{U}$ is the fluid velocity. Across the shock surfaces the entropy increases according to the Rankine-Hugoniot solution. This method has been proved to be very successful in studying the formation and interaction of shocks in the heliosphere [Whang, 1984; Whang and Burlaga, 1985a,b; 1986; 1988]

Magnetohydrodynamic Waves

The model assumes that flow properties are functions of x and t only. Throughout this paper, we use an x,y,z coordinate system in such a way that the x-axis is normal to the wave front pointing the forward direction of the solar wind flow and the magnetic field vector is parallel to the xy coordinate plane.

Small disturbances propagate in the solar wind as three modes of magnetohydrodynamic waves (fast mode, slow mode and Alfven mode) in both the forward and the reverse directions relative to the solar wind flow. We shall first review the theory of linearized one-dimensional MHD waves with special attention given to the directions of the small perturbation vectors for each mode of wave propagation.

The governing equations for linearized one-dimensional MHD waves may be decoupled into two subsystems: one for the compressive mode of wave propagation, the other for the Alfven mode. Figure 1 illustrates that the small perturbations generated at a source point O propagate as three modes of MHD waves. The two compressive modes of magnetoacoustic waves propagate along $dx/dt = U \pm C_i$ with $C_i = C_f$ or C_s. Here U is the x component of the solar wind velocity. For fast mode magnetoacoustic waves, the normal speed

$$C_f^2 = (a^2 + c^2)/2 + \{(a^2 + c^2)^2 - 4a^2c^2\cos^2\theta\}^{1/2}/2 \quad (1)$$

and for slow mode magnetoacoustic waves, the normal speed

$$C_s^2 = (a^2 + c^2)/2 - \{(a^2 + c^2)^2 - 4a^2c^2\cos^2\theta\}^{1/2}/2 \quad (2)$$

Here $c = (\gamma p/\rho)^{1/2}$ is the gasdynamic speed of sound, $a = B_o/(4\pi\rho)^{1/2}$ is the Alfven speed, and θ is the angle between the wave normal and the direction of the background magnetic field. The upper and the lower signs respectively represent the forward and the reverse family of wave propagation. Alfven waves propagate along $dx/dt = U \pm a \cos\theta$. Both C_s and a are less than C_f. Throughout this paper, calculations are carried out by using $\gamma = 5/3$ for fully ionized plasma.

We can write the solutions for various small perturbation quantities in terms of wave functions [Whang et al., 1987]. Fast and slow mode magnetoacoustic waves carry the x and y components of the perturbation velocity u and v, the y component of the perturbation magnetic field b_y, and the perturbation pressure p′,

$$\begin{aligned} u &= \pm (1 - a^2\cos^2\theta/C_i^2)(C_i/a \sin\theta)g \\ v &= \mp (a\cos\theta/C_i)g \\ b_y &= (4\pi\rho)^{1/2}g \\ p' &= (\rho/a\sin\theta)(C_i^2 - a^2)g \end{aligned} \quad (3)$$

where $g \equiv g\{x - (U \pm C_i)t\}$ is a wave function with $C_i = C_f$ for fast waves and $C_i = C_s$ for slow waves. Alfven waves carry the z component of the perturbation velocity w and the z component of the perturbation magnetic field b_z,

$$\begin{aligned} b_z &= (4\pi\rho)^{1/2}h \\ w &= \mp h \end{aligned} \quad (4)$$

where $h \equiv h\{x - (U \pm a\cos\theta)t\}$. Equations (3) and (4) mean that the perturbations lying on the plane defined by the wave normal and the background magnetic field propagate as magnetoacoustic waves and the perturbation velocity and field perpendicular to both the wave normal vector and the background magnetic field vector propagate as Alfven waves. If the small perturbations in velocity and magnetic field generated at the source point O are parallel to the xy-plane defined by **n** and **B**, then the wave motion consists of fast waves and slow waves only. This means no Alfven waves are excited.

Next, we consider the propagation of large disturbances. Again, if the initial disturbances in velocity and magnetic field are parallel to the xy-plane defined by the wave normal and the background magnetic field, then only fast and slow waves are generated by the large-amplitude disturbances. Alfven waves are still absent in the domain of solution. In this paper, we use the method of characteristics to calculate the evolution of the flow field to form a pair of

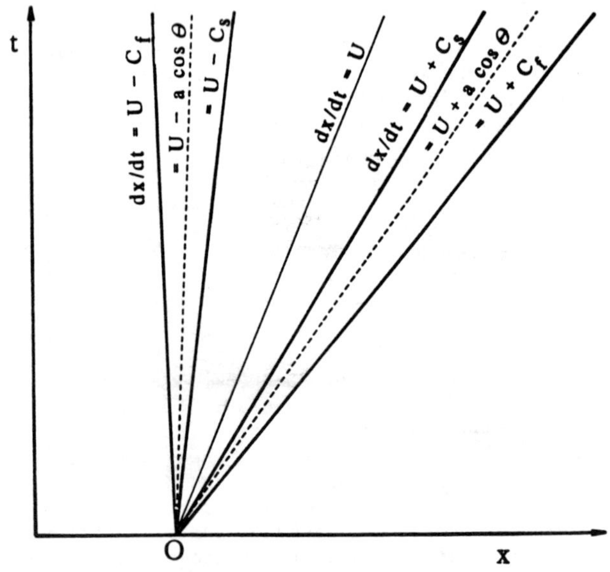

Fig. 1. Small perturbations generated at a source point O propagate as Alfven waves along $dx/dt = U \pm a\cos\theta$ and as compressive magnetoacoustic waves along $dx/dt = U \pm C_i$ with $C_i = C_f$ for fast waves and $C_i = C_s$ for slow waves. The upper and the lower sign respectively represent the forward and the reverse family of the wave propagation.

forward-reverse slow shocks. We consider that the initial large disturbances are generated by the impact of a high-speed ejecta on the ambient solar wind. Our model assumes that throughout the solution domain the magnetic field vectors and the flow velocity vectors are parallel to the xy coordinate plane. The assumption that $B_z = V_z = 0$ can substantially simplify the interaction relationships among the three modes of MHD waves. Because Alfven waves carry the propagation of disturbances for the z-components of the velocity and magnetic field vectors, no disturbances are carried by the Alfven-mode characteristic curves in the solution domain of this model. The MHD equations are now reduced to a system which governs the variations of flow properties along the compressive mode characteristics. Those lines with slopes

$$dx/dt = U \pm C_f$$

and

$$dx/dt = U \pm C_s$$

are fast and slow mode characteristic curves. The upper signs represent the families of forward characteristics, and the lower signs the reverse characteristics. This method of solution integrate the governing equations along the four families of characteristic curves.

Formation of Slow Shock Pairs Associated With CMEs

The simulation demonstrates that the difference in momentum flux is the driving force for the formation of shock pairs. This simulation uses an idealized initial condition which represents the impact of a high-speed mass ejecta on a low-speed ambient solar wind. Figure 2 shows the initial condition and some simulation results for U, p, n and T. At t = 0 the momentum flux changes smoothly over a distance of about 2 x 10^5 km near x = 0. On the upstream (mass ejecta) side

$$U_1 = 500 \text{ km/s}$$

and $n_1 = 20.98 \times 10^6$ protons/cm^3.

On the downstream (ambient solar wind) side

$$U_o = 100 \text{ km/s}$$

and $n_o = 7.0 \times 10^6$ protons/cm^3.

Throughout the flow field,

$$\beta = 0.1, \quad B_x = 0.8 \text{ Gauss}, \quad \theta = 15°.$$

At t = 0, the flow is field aligned in the rest frame of reference and p = constant.

The simulation result shows that the dynamic interaction between the coronal mass ejecta and the ambient solar wind produces a pair of forward and reverse slow shocks. The shocks have reached a fully developed state at t = 25 minutes. The reverse slow shock located at x = 3.71 x 10^5 km propagates upstream at a speed of 248 km/s relative to the ejecta flow. The forward slow shock at x = 7.49 x 10^5 km propagates downstream at a speed of 427 km/s relative to the ambient solar wind. The strengths of the two shocks are almost equal. The relative shock speeds are 2.31 times the slow speed on the front side. The shocks have the same density ratio of 2.55 and the same pressure ratio of 6.40.

The two families of slow mode characteristic curves merge in the region where the shock formation takes place. The merging of characteristic curves means that strong disturbances generated by the momentum impact pile up to form the shock fronts. The forward slow mode characteristic curves merge to form the forward slow shock, and the reverse slow mode characteristic curves merge to form the reverse slow shock.

At t = 0, the number density and the temperature vary smoothly over a transition layer with a thickness of about 2 x 10^5 km. This transition layer evolves to form an interface instead of a single contact surface. At t = 25 minutes, the fully developed interface between the ejecta and the ambient solar wind has a thickness of about 0.5 x 10^5 km. The forward slow shock is at a standoff distance of about 2.2 x 10^5 km in front of the interface. During the formation process, a shock grows stronger steadily until it reaches a fully developed state. The jump in entropy across a shock varies during the formation process. Thus, the entropy also changes smoothly in the interaction region bounded by the shock pair.

Figure 3 shows the change in magnetic field calculated from the simulation. The strength of the magnetic field decreases across each slow shock. The perturbations generated by the momentum impact also disturb the plasma and the magnetic field outside the interaction region bounded by the forward-reverse slow shock pair. But their influence is limited to the region bounded by a forward fast mode characteristic and a reverse fast mode characteristic. Note that the perturbations in magnetic field generated by the momentum impact is less than 3% of the initial field, but the perturbation in pressure is greater than 500% of the initial pressure. The disturbances propagate as fast mode waves are negligibly small compared with the slow shocks.

Interplanetary Manifestation of CME-Associated Shock Pairs

CMEs supply significant momentum to interplanetary space extending over times that can be greater than one day. Figure 4 depicts that a CME pushes its way through preceding slower solar wind. The impact due to the large momentum of the high-speed CME plasma generates a narrow high pressure interaction region near the

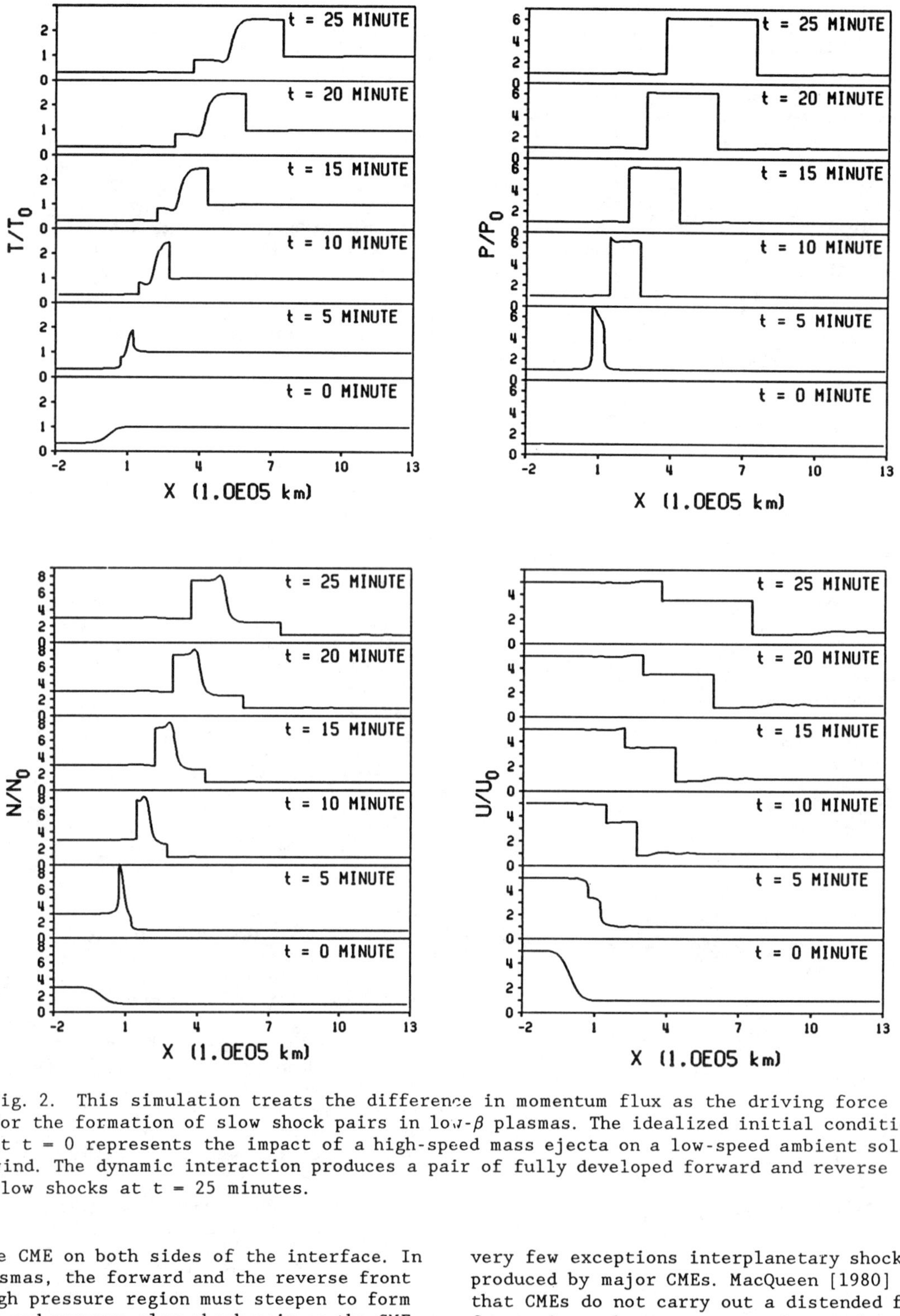

Fig. 2. This simulation treats the difference in momentum flux as the driving force for the formation of slow shock pairs in low-β plasmas. The idealized initial condition at t = 0 represents the impact of a high-speed mass ejecta on a low-speed ambient solar wind. The dynamic interaction produces a pair of fully developed forward and reverse slow shocks at t = 25 minutes.

top of the CME on both sides of the interface. In low-β plasmas, the forward and the reverse front of the high pressure region must steepen to form a forward and reverse slow shock pair as the CME moves outward from the Sun.

Sheeley et al. [1985] studied interplanetary shocks detected at Helios 1 and CMEs observed from Solwind coronagraph and inferred that with very few exceptions interplanetary shocks are produced by major CMEs. MacQueen [1980] suggested that CMEs do not carry out a distended field very far to interplanetary space, but rather are subject to a reconnection process. CME loops must magnetically disconnect from the Sun and form a closed magnetic bubble which continues outward into interplanetary space. The disconnected

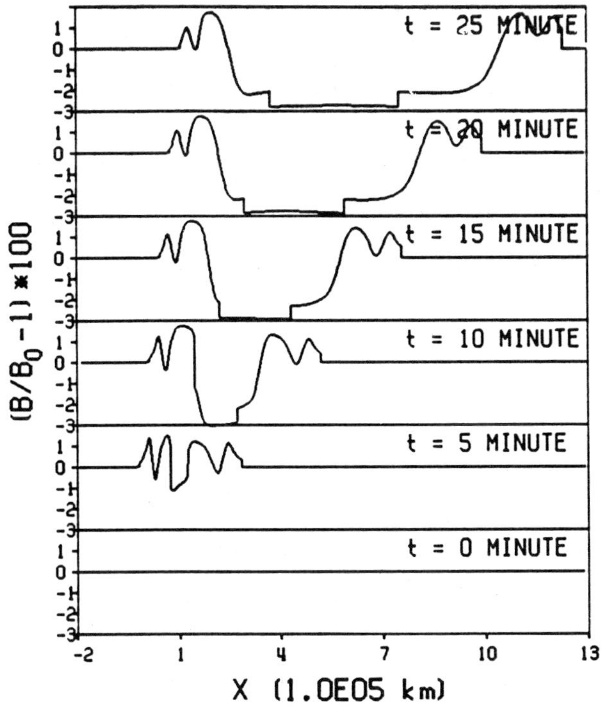

Fig. 3. This figure shows the change in magnetic field calculated from the simulation. The strength of the magnetic field decreases across each slow shock. The perturbations generated by the momentum impact disturb the plasma and the magnetic field in the region bounded by a forward fast mode characteristic and a reverse fast mode characteristic.

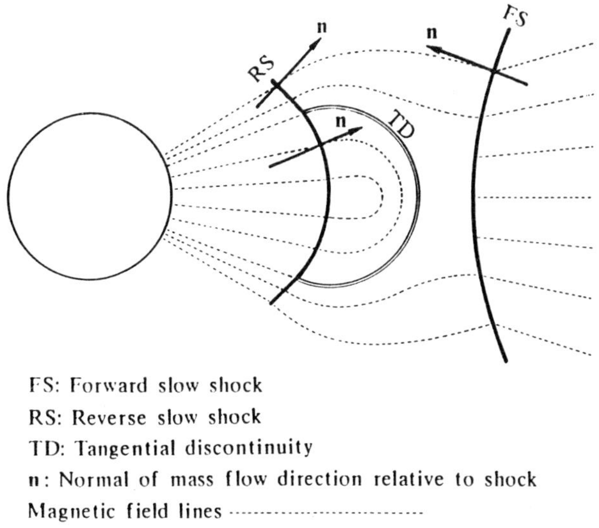

Fig. 4. The direct impact of high-speed mass ejecta on the ambient solar wind compress the plasma near the top of CME on both sides of the interface. This interaction produces a forward slow shock preceding the CME and a reverse slow shock closely behind the leading edge.

bubble is believed to manifest as a magnetic cloud.

Burlaga et al. [1981] analyzed the magnetic field configuration behind three shocks. They found a systematic configuration of the magnetic field and called it the magnetic cloud. Klein and Burlaga [1982] identified 45 clouds in interplanetary data obtained near 1 AU between 1967 and 1978. About one third of them are preceded by a shock. They compared the physical properties of magnetic clouds with those of CMEs and suggested an association of some clouds with disconnected magnetic structures of CMEs. A good case of an association between a CME observed by Solwind and a magnetic cloud observed at 0.54 AU by Helios 1 two days later was presented by Burlaga et al. [1982].

As a slow shock pair travels outward from the Sun, the decrease in the Alfven speed with increasing heliocentric distance causes the slow shocks to evolve into fast shocks [Whang, 1987]. As the forward (reverse) slow shock moves outwards from the sun, the decrease in the Alfven speed causes the shock Alfven number to reach 1 near the nose of the shock surface, the transition from a forward (reverse) slow shock to a forward (reverse) fast shock begins to take place at a point of the shock surface where the shock normal is aligned with the magnetic field. The onset of transition must occur in the region of interplanetary space where $\beta < 1.2$. At the onset of the transition the forward (reverse) slow shock smoothly converts to a system consisting of a forward (reverse) slow shock, a very weak forward (reverse) rotational discontinuity, and a very weak forward (reverse) fast shock.

During the transition, the system consists of a slow shock, a fast shock, and a rotational discontinuity. The three surfaces of discontinuity intersect along a closed transition line. As the system moves outward from the sun the transition line moves laterally across the field lines, the area enclosed by the closed loop of the transition line expands continuously, the fast shock grows stronger, and the slow shock becomes weaker. Eventually, the slow shock diminishes and the entire system evolves into a forward fast shock. A simulation study of this transition process will be presented in a future paper.

In summary, two important evolutions of the shock pair and the CME may occur in interplanetary space. First, a CME loop eventually disconnects from the sun to form a closed magnetic structure. The disconnected bubble manifests as a magnetic cloud in interplanetary space. Second, as the CME associated forward-reverse shock pair moves outwards in interplanetary space, they evolve into a pair of fast shocks. The reverse shock first propagates within the magnetic cloud, but eventually the reverse shock must exit the cloud to become detached from the magnetic cloud. Then both the forward and the reverse shocks propagate in the ambient solar wind as shown in Figure 5. Therefore, as a CME and its associated slow shock pair move outwards in interplanetary space, the system may evolve into a magnetic cloud accompanied by a fast shock pair: a forward shock precedes the cloud and a reverse shock either within or behind the cloud [Whang, 1988].

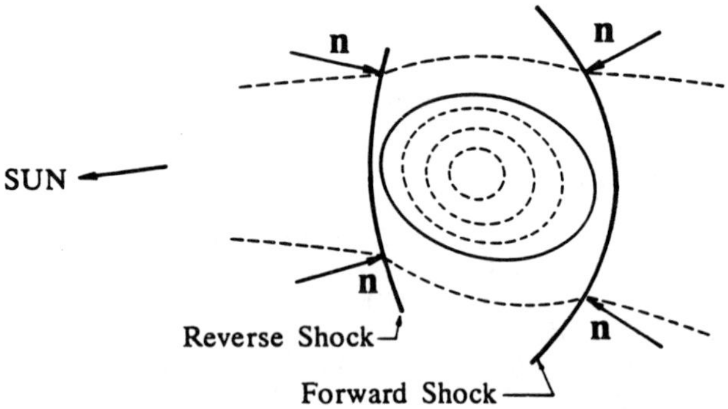

Fig. 5. As the cloud moves outward from the Sun, the slow shocks must convert to fast shocks. The reverse shock propagates within the cloud and eventually the reverse shock exits the cloud to become detached from the magnetic cloud. The cloud is now wrapped around by an interaction region which is itself bound by a pair of forward-reverse fast shocks.

Klein and Burlaga [1982] reported that about one third of clouds identified near 1 AU between 1967 and 1978 are preceded by a shock. There are several observations of a shock pair associated with a CME or magnetic cloud. Lepping et al. [1988] reported that a large magnetic cloud accompanied by a shock pair was observed on October 31 and November 1 of 1972 from IMP 7. The event was associated with an unusually longlasting solar flare which occurred some 49 hours earlier. Gosling et al. [1988] have identified two CME associated shock pairs at 1 AU from ISEE 3. Another event involving a shock pair and a magnetic cloud was observed on August, 1982, from Voyager 2 at 10.3 AU [Burlaga et al., 1985]. This observation shows that for a long-lived cloud, the reverse shock had enough time to propagate through the cloud, exit the rear of the cloud, and appear closely behind the cloud.

Questions and Answers

Steinolfson: Are slow shocks seen ahead of magnetic clouds near 1AU?

Whang: Two important evolutions of the shock pair and the CME may occur in interplanetary space. First, a CME loop manifests as a magnetic cloud in interplanetary space. Second, as the CME associated forward-reverse shock pair moves outwards in interplanetary space, they evolve into a pair of fast shocks (Whang, JGR, 92, 4349, 1987). Therefore, the interplanetary consequence of some CMEs may consist of a magnetic cloud accompanied by a fast shock pair. The forward fast shock propagates ahead of the cloud and the reverse fast shock is either within or behind the cloud.

Acknowledgments. This work was supported by the National Aeronautics and Space Administration under grant NAGW-1323.

References

Burlaga, L. F., E. Sittler, F. Mariani, and R. Schwenn, Magnetic loop behind an interplanetary shock: Voyager, Helios and IMP 8 observations, J. Geophys. Res., 86, 6673-6684, 1981.
Burlaga, L. F., L. Klein, N. R. Sheeley, Jr., D. J. Michels, R. A. Howard, M. J. Koomen, R. Schwenn, and H. Rosenbauer, A magnetic cloud and a coronal mass ejection, Geophys. Res. Lett., 9, 1317-1320, 1982.
Burlaga, L. F., F. B. McDonald, M. L. Goldstein, and A. J. Lazarus, Cosmic ray modulation and turbulent interaction regions near 11 AU, J. Geophys. Res., 90, 12,027-12,039, 1985.
Gosling, J. T., S. J. Bame, E. J. Smith, and M. E. Burton, Forward-reverse shock pairs associated with transient disturbances in the solar wind at 1 AU, J. Geophys. Res., 93, 8741-8748, 1988.
Hundhausen, A. J., and R. A. Gentry, Numerical simulation of flare-generated disturbances in the solar wind, J. Geophys. Res., 74, 2908-2918, 1969.
Hundhausen, A. J., T. E. Holzer, and B. C. Low, Do slow shocks precede some coronal mass ejections?, J. Geophys. Res., 92, 11,173-11,178, 1987.
Klein, L. W., and L. F. Burlaga, Interplanetary magnetic clouds at 1 au, J. Geophys. Res., 87, 613-624, 1982.
Lepping, R. P., F. M. Ipavich, and L. F. Burlaga, A flare-associated shock pair at 1 AU and related phenomena, EOS, 69, 458, 1988.
MacQueen, R. M., Coronal transients: A summary, Philos. Trans. R. Soc. London., Sec. A, 297, 605-620, 1980.
Sheeley, N. R. Jr., R. A. Howard, M. J. Koomen, D. J. Michels, R. Schwenn, K. H. Muhlhauser, and H. Rosenbauer, Coronal mass ejections and interplanetary shocks, J. Geophys. Res., 90, 163-175, 1985.
Steinolfson, R. S., M. Dryer, and Y. Nakagawa, Numerical MHD simulation of interplanetary shock pairs, J. Geophys. Res., 80, 1223-1231, 1975.
Whang, Y. C., The forward-reverse shock pair at

large heliocentric distances, J. Geophys. Res., 89, 7367-7379, 1984.

Whang, Y. C., Transition of traveling interplanetary slow shock to fast shock, Eos Trans. AGU, 67, 327, 1986.

Whang, Y. C., Slow shocks and their transition to fast shocks in the inner solar wind, J. Geophys. Res., 92, 4349-4356, 1987.

Whang, Y. C., Forward-Reverse Shock Pairs Associated with Coronal Mass Ejections, J. Geophys. Res., 93, 5897-5902, 1988.

Whang, Y. C., and L. F. Burlaga, Coalescence of two pressure waves associated with stream interactions, J. Geophys. Res., 90, 221-232, 1985a.

Whang, Y. C., and L. F. Burlaga, Evolution and interaction of interplanetary shocks, J. Geophys. Res., 90, 10,765-10,778, 1985b.

Whang, Y. C., and L. F. Burlaga, The coalescence of two merged interaction regions between 6.2 and 9.5 AU: September 1979 event, J. Geophys. Res., 91, 13,341-13,348, 1986.

Whang, Y. C., F. S. Wei and H. Du, Critical angles of incidence for transmission of magnetohydrodynamic waves across shock surfaces, J. Geophys. Res., 92, 12,036-12,044, 1987.

Whang, Y. C., and L. F. Burlaga, Evolution of recurrent solar wind structures between 14 AU and the termination shock, J. Geophys. Res., 93, 5446-5460, 1988.

THE SOLAR WIND INTERACTION WITH UNMAGNETIZED PLANETS: A TUTORIAL

J. G. Luhmann

Institute of Geophysics and Planetary Physics
University of California, Los Angeles

Abstract. Venus serves as the prototype of the solar wind interaction with unmagnetized planets because we have a fairly complete observational picture relative to that of Mars. To a first approximation, Venus appears as a highly electrically conducting obstable in the supermagnetosonic flow of the solar wind. The "surface" of the obstacle, or ionopause, is located near the locus of points where the component of solar wind pressure normal to the surface is balanced by the thermal pressure of the ionospheric plasma. The subsolar location of this obstacle is typically around 300 km at solar maximum. Upstream of the obstacle, a bow shock forms to slow and divert the flow around it. Any planetary ions produced in the intervening region or magnetosheath between the shock and obstacle surface are picked up by the solar wind convection electric field. Some of these ions are swept away, while others are energized and returned to the planet. The magnetic field of the solar wind piles up and drapes around the obstacle. Inside the ionopause, the ionospheric plasma is found to be affected by the solar wind interaction in several ways. There is usually a sharp gradient in plasma density at the ionopause, and the ionospheric electron and ion temperatures indicate that there are heating processes other than those due to solar photon radiation absorption. There are also ionospheric magnetic fields that are related to the solar wind interaction.

In the limit where the ionopause is depressed below ~240 km altitude due to an increase in the magnitude of the solar wind dynamic pressure, the ionospheric magnetic field is a large scale horizontal field with a fairly well defined altitude profile. When the ionopause is near or above its typical solar maximum position, the ubiquitous "flux rope" features are seen. The connection between these two types of ionospheric fields is uncertain. However, both phenomena require a consideration of how magnetic flux of solar wind origin is transported through the ionopause and distributed within the ionosphere. The production and evolution of the large scale magnetic field is relatively well understood in terms of diffusion and downward convection, but the behavior of flux ropes continues to present a challenge for both theorists and observationalists. The relative strength of the solar wind and ionospheric pressures at Mars suggest that if the Martian intrinsic field is insignificant, the ionosphere of Mars is likely to exhibit a large scale field instead of flux ropes.

Geophysical Monograph 58

I. Introduction

Venus, and perhaps at times Mars, together represent a class of direct planetary atmosphere-solar wind interaction which somehow gives rise to ionospheric flux ropes. In order to appreciate the circumstances surrounding their generation, it is useful to review the general features of the solar wind interaction and the related problem of the large scale ionospheric magnetic field. In this spirit, this paper starts with the basic concept of how a planet without an intrinsic magnetic field perturbs the solar wind flow. It then moves on to describe how magnetic fields of interplanetary origin can produce large scale ionospheric fields with the observed characteristics if the draped interplanetary field in the magnetosheath is convected into the ionosphere by ionospheric convection and then decays at low altitudes (below ~160 km) due to collisional dissipation of the associated currents. It is pointed out that although the presence of the large scale field precludes the formation of flux ropes, some of the same factors that enter into the modeling of the large field must also be considered in theories of the small scale field structures.

II. The Solar Wind Interaction with Venus at Solar Maximum

Spacecraft measurements have shown us that Venus, at solar maximum, stands in clear contrast to the earth in that it forms an obstacle to the solar wind purely by virtue of its ionosphere [see Russell and Vaisberg, 1983, or Luhmann, 1986, for reviews of the observations]. Solar radiation incident on a neutral atmosphere of substantial density appears to be the essential requirement (although some ionization of neutral atmosphere by charge exchange with solar protons and impact ionization by solar wind or energetic interplanetary particles can also play a role). When the ionospheric plasma pressure exceeds the incident solar wind pressure, the potential for an obstacle interaction might be expected to exist. Yet, due to the collisionless nature of space plasmas, if the solar wind were not magnetized, the solar wind plasma would essentially flow straight into the atmosphere where it would be absorbed. It is the presence of an interplanetary magnetic field that makes Venus an obstacle.

To lowest order, one can envision Venus in the interplanetary magnetic field as a conducting sphere in a uniform vacuum magnetic field as shown in Figure 1. Currents flow on the surface of the sphere to produce a perturbation field that has the form of an opposing dipole which cancels the field in the interior of the sphere and distorts the exterior field so that it remains outside of it; the obstacle of the conducting sphere thus forms a cavity in the surrounding environment. This same magnetic field configuration would result if the

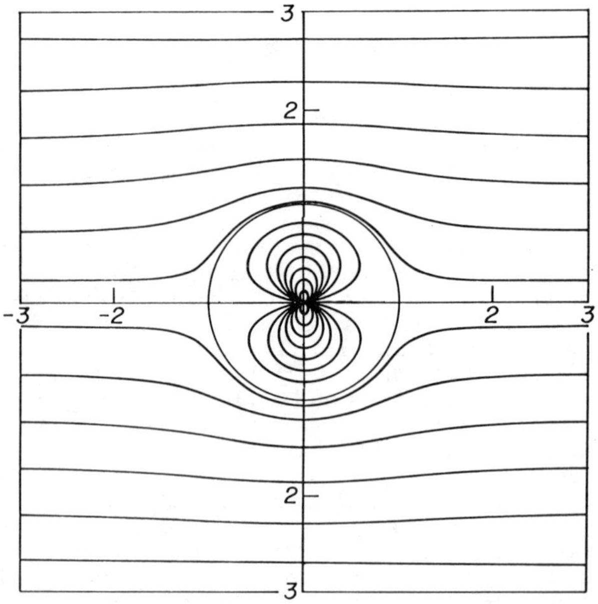

Fig. 1. Vacuum superposition of a uniform magnetic field and opposing dipole field, illustrating how currents induced on the surface of a conducting sphere shield out of the external field. The internal dipole field is only "virtual" in the sense that currents on the surface of the sphere produce the extension of such a dipole field external to the sphere only.

magnetic field was embedded in plasma flowing slowly and incompressibly around the sphere. The convection electric field associated with the flow would be excluded since magnetic field lines are equipotentials and a single field line effectively "paints" the surface of the conducting sphere.

To next order, one needs to take into account the fact that the magnetic field is actually embedded in the compressible supermagnetosonically flowing plasma of the solar wind. This distorts the way that the magnetic field "drapes" around the sphere. The flowing plasma, in which the field is almost frozen-in, is shocked and compressed against the dayside and a cavity forms in the obstacle's wake. The perturbation field that arises in this case can be calculated by subtracting the interplanetary field from a model constructed by Spreiter and Stahara [1980] to describe a convected field frozen in a compressible hypersonic gas dynamic flow around a blunt obstacle. This separation is illustrated in Figure 2, where one can see that the perturbation field is no longer a dipole, but is instead described by closed loops of field lines. The perturbation field is caused by shielding currents which flow in the ionosphere.

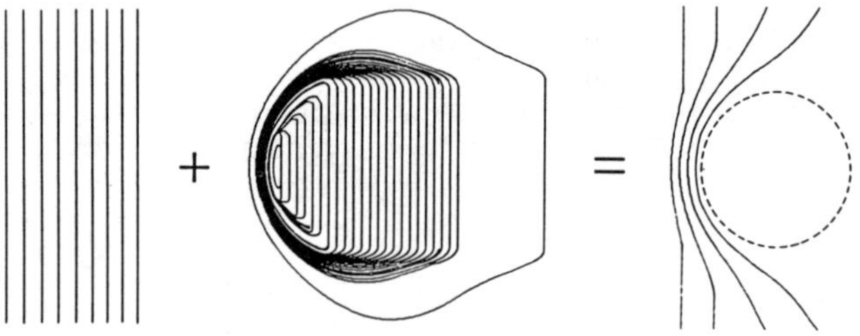

Fig. 2. Decomposition of the gas dynamic magnetosheath field into uniform (interplanetary) field and shielding current field parts.

The nomenclature used to identify the various features caused by the perturbation is given in Figure 3.

In cases where flux ropes are present, the shielding current is virtually confined to a boundary of sharp ionospheric plasma density gradient called the ionopause, which effectively separates the planetary and solar wind plasmas. Planetary ions produced above this boundary are "picked up" by the convection electric field $\mathbf{E} = -\mathbf{V}\times\mathbf{B}$ (where $\mathbf{V}$ = plasma velocity and $\mathbf{B}$ = magnetic field) in the magnetosheath and nearby solar wind as illustrated in Figure 4 [see also Cloutier et al., 1974]. As will be discussed below, the position of this boundary is determined by pressure balance between the incident solar wind dynamic pressure and the pressure of the ionospheric plasma. The nature and origin of the bow shock and magnetosheath is basically the same for magnetospheric obstacles except that the perturbing currents flow on a magnetopause instead of in an ionopause. (Because of this difference, the Venus magnetosheath is sometimes called an "ionosheath.")

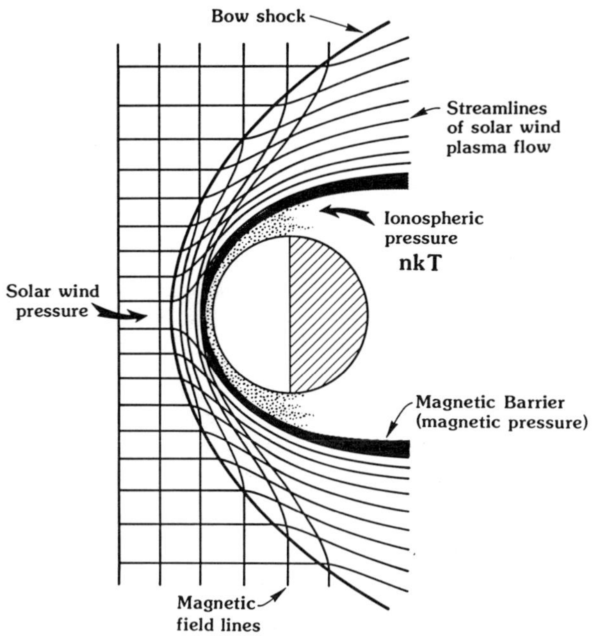

Fig. 3. Schematic illustration of the major features of the solar wind interaction with Venus.

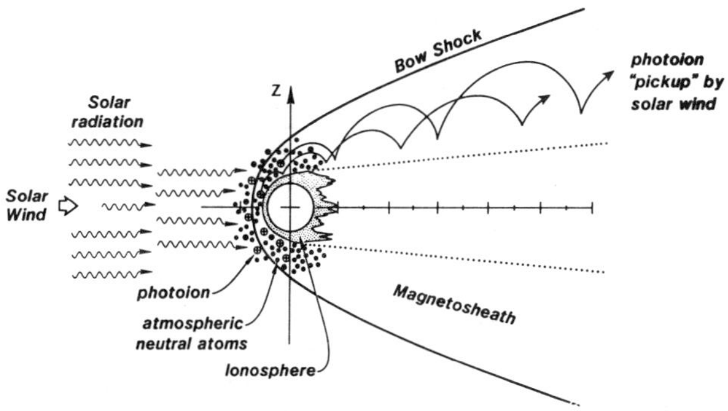

Fig. 4. Schematic illustration of planetary ion pickup by the solar wind outside of the ionopause.

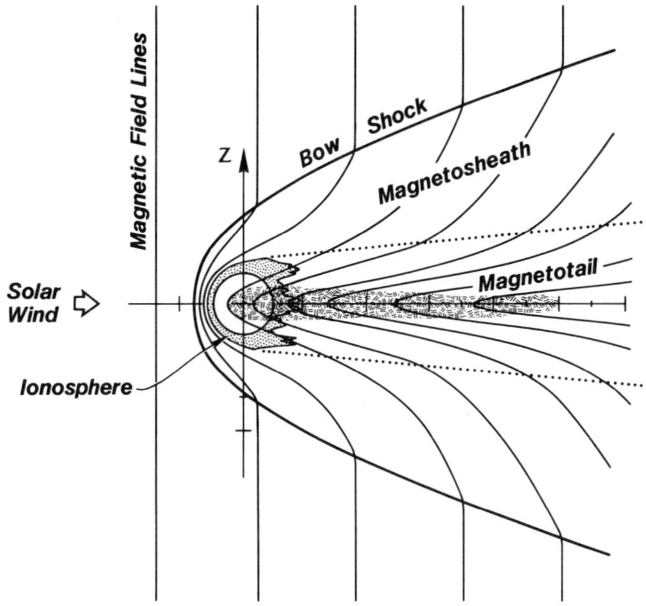

Fig. 5. Schematic illustration of tail formation by mass-loaded, draped interplanetary field lines.

The "magnetic barrier" in Figure 3 is simply the inner region of the magnetosheath where solar wind plasma is excluded because the compressed magnetic field virtually squeezes it out of the stagnation region flux tubes [cf. Zwan and Wolf, 1979].

An additional current system is also present in the wake of the planet. This system is in the form of a current sheet that separates two "lobes" of magnetic field which follow the draping pattern of the magnetosheath field as suggested by Figure 5 [see also Saunders and Russell, 1986]. This current sheet is thought to be at least partially attributable to the picked-up planetary ions. Given the significant curvature of the field lines, the JxB = $1/\mu_0(\nabla \times B \times B)$ force is also thought to be important in accelerating ions in this "magnetotail" [cf. McComas et al., 1986]. The field in the tail is connected to the downstream solar wind. Thus its general origin and configuration are somewhat reminiscent of those of the tail of a comet.

It is worth noting that solar wind absorption is not needed to make any of these currents flow. They flow simply because the dayside upper ionosphere is a very good conductor exposed to an external magnetic field and because planetary ions are picked up by the solar wind. Ionospheric magnetic flux ropes are one of the interesting phenomena connected with the dayside shielding currents. However, the problem of the origin and evolution of these structures cannot be addressed without first taking a closer look at the solar-wind ionosphere boundary and other features of the ionospheric magnetic field.

The Concept of Pressure Balance and the Ionopause

As was mentioned above, observations obtained with experiments on the Pioneer Venus Orbiter spacecraft indicate that the solar wind-ionosphere interface in its simplest form is the boundary where the incident dynamic pressure of the solar wind is balanced by the thermal pressure of the ionospheric plasma [cf. Brace et al., 1983, Phillips et al., 1984]. This "pressure balance" occurs through the intermediary of the piled-up magnetic field in the magnetosheath in such a way that, between the upstream solar wind and the top of the ionosphere, the condition

$$\rho V^2 + B^2/2\mu_0 + P \approx \text{constant}$$

(where ρV^2 is dynamic pressure, $B^2/2\mu_0$ is magnetic pressure and P is thermal pressure nk $(T_i + T_e)$) is satisfied. The dominant component of pressure simply changes from almost purely dynamic pressure upstream to plasma pressure in the ionosphere. The scales in Figure 3 have been distorted to show detail. The boundary where ionospheric plasma pressure takes over actually occurs near 300 km altitude at the subsolar point and near 1000 km at the terminator as shown by the observations in Figure 6. On one side of the boundary one finds the magnetosheath with its diverted solar wind plasma flowing around the ionospheric obstacle, and in the process compressing and draping the interplanetary magnetic field around it. Planetary ions undergoing pick-up and removal as described in Figure 4 are also present here with a density that increases with decreasing altitude. Because of the MHD effects noted above, the piled-up field forms an almost purely magnetic "barrier" between the solar wind plasma and the cold ionospheric plasma that is found on the inside. The transition between the field of the magnetic barrier and the ionospheric plasma occurs in a layer that may be as thin as a few times the gyroradius of an ionospheric ion (~10 km) which is O^+ at the altitudes where this simple balance occurs [cf. Elphic et al., 1981]. It is in this layer that the shielding current evidently flows to produce the perturbation magnetic field that cancels the interplanetary field on the inside of the ionosphere and forms the magnetic barrier. Since the ionospheric ions produced in the atmosphere on the outside of this boundary are accelerated by the solar wind to energies well beyond the energies of the thermal ionospheric ions, the name "ionopause" is certainly appropriate. Essentially all of the cold, ionospheric plasma is found on the inside. It is important to appreciate that because solar wind plasma is largely excluded from the overlying magnetic barrier, this magnetic layer forms the effective obstacle to the solar wind flow rather than the ionopause. Thus, in this simplest scenario, the ionopause is actually the boundary between the magnetic barrier and the ionosphere rather than a boundary between solar wind plasma and ionospheric plasma.

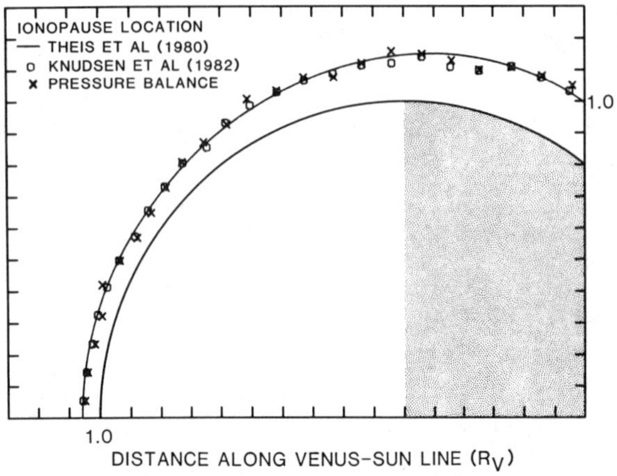

Fig. 6. Observed average location of the ionopause [from Phillips et al., 1984].

Ionosphere Magnetization

<u>Large Scale Fields</u>. As the altitude profiles of magnetic field and electron density in Figure 7 illustrate, the Venus ionopause is not entirely impermeable to magnetic field. Because there are both Coulomb collisions between ions and electrons, and collisions between ions or electrons and neutrals in the weakly ionized ionosphere, the nearly horizontal magnetic field in the barrier region diffuses downward to diminish the field gradient. Once the field is inside the ionosphere, it is also subject to redistribution by ionospheric convection. The appropriate mathematical description for this diffusion/convection scenario in the subsolar region is the familiar dynamo

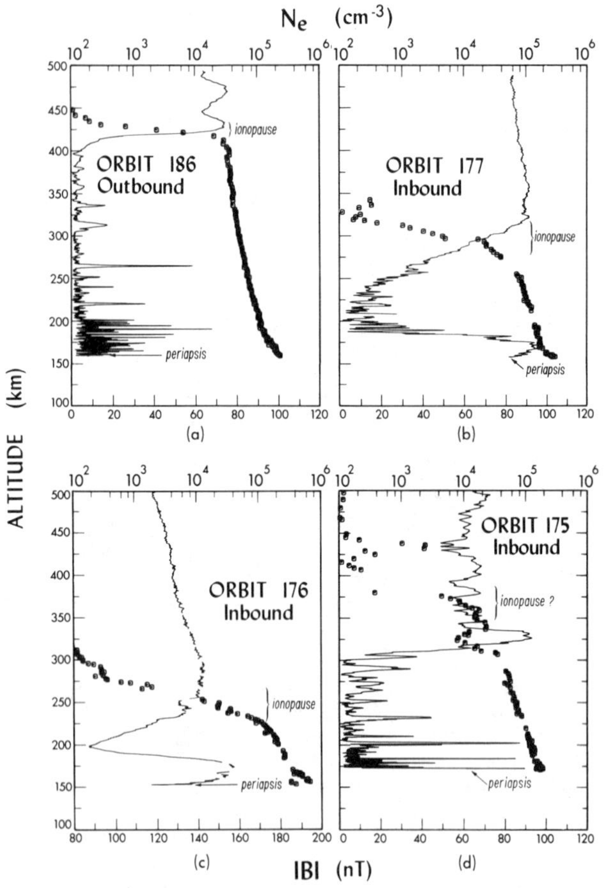

Fig. 7. Altitude profiles of magnetic field magnitude (B) and ionospheric electron density (N_e) as measured by instruments on the Pioneer Venus Orbiter [from Elphic et al., 1980]. The ionopause is located in the altitude range of sharpest density gradient on the topside. Its thickness grows as the ionopause altitude decreases.

equation, which in one dimension (altitude = z) is given by

$$\frac{\partial B}{\partial t} = \frac{\partial}{\partial z} D(z) \frac{\partial B}{\partial z} - \frac{\partial}{\partial z} (W(z)B)$$

where, in this case

$$D = \frac{m_e (\nu_{en} + \nu_{ei})}{ne^2 \mu_0}$$

(ν_{ab} frequency of collisions between species a and b, m_e = electron mass, e = electron charge, n = electron density) and

$$W(z) = \frac{1}{nm_i \nu_{in}} \{ - \frac{\partial}{\partial z} (nk (T_i+T_e) + \frac{B^2}{2\mu_0}) + nm_i g \}$$

(g = gravitational acceleration, m_i = ion mass).

The altitude dependent diffusion coefficient D(z), shown in Figure 8, is easily related to the background plasma and neutral densities through the collision frequencies which are also shown in Figure 8, but the bulk vertical velocity W(z) is somewhat more difficult to determine. In a steady situation where the solar wind pressure is not changing, it can be obtained to first order from the observations. Measurements of the ionospheric properties at Venus have been used to infer that, at least in the subsolar region, the ionospheric plasma below the ionopause is generally drifting downward in response to thermal pressure gradients [Cravens et al., 1984]. The altitude profile of that downward drift in Figure 9 indicates that its magnitude ranges from a few m s^{-1} to ~60 m s^{-1} under normal conditions when the effect of the ionospheric field is negligible. The downward drift implies that magnetic field that diffuses across the pressure balance boundary is transported to lower altitudes. The shape of the velocity profile is reflected in the characteristic ionospheric magnetic field profiles in Figure 7 which have a minimum near 190 km where the velocity is highest. Below ~170 km the growing diffusion coefficient from the increasing neutral density causes the field to dissipate.

Some profiles that have been modeled by numerically solving the above dynamo equation are shown in Figure 10. The altitude of the upper boundary and the boundary magnetic field strength were varied according to the observed relationship between the solar wind pressure, the magnetic barrier field, and the ionopause height [cf. Phillips et al., 1984]. As Figure 10 illustrates, the large scale ionospheric field has its greatest magnitudes (~150 nT) when the ionopause is at its lowest altitudes of ~220-240 km where the diffusion coefficient and velocity are both larger, and is weakest when the ionopause is high where both of these quantities are much smaller. Since the

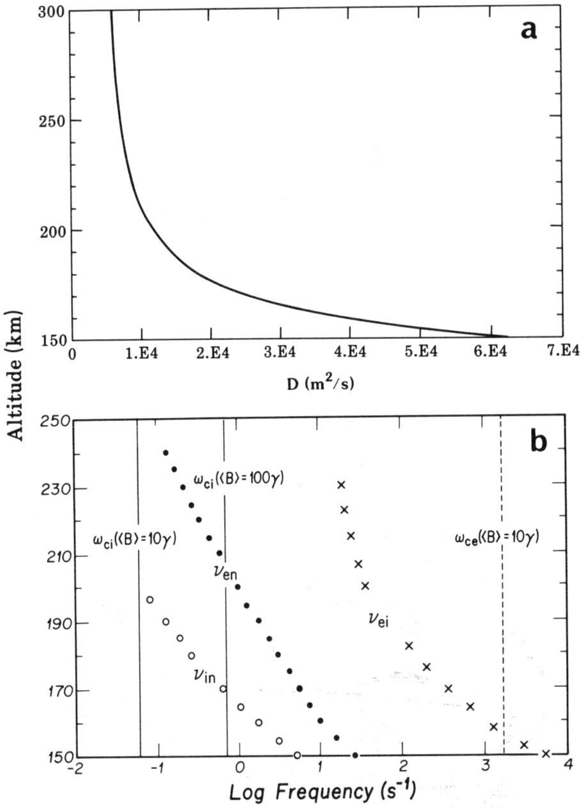

Fig. 8. (a) Diffusion coefficient in the Venus ionosphere calculated from models of the electron and neutral densities. (b) Collision frequencies used to calculate the diffusion coefficient: ν_{ie} = coulomb collision frequency, ν_{in}, ν_{en} = ion-neutral and electron-neutral collision frequencies, respectively. Gyrofrequencies ω_g for a few field magnitudes are indicated for comparison.

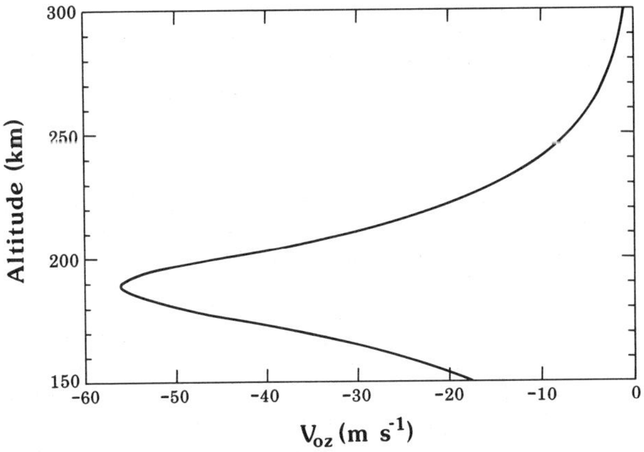

Fig. 9. Vertical ionospheric velocity profile calculated by Cravens et al. [1984] from measured plasma parameters, assuming no effect of an ionospheric magnetic field.

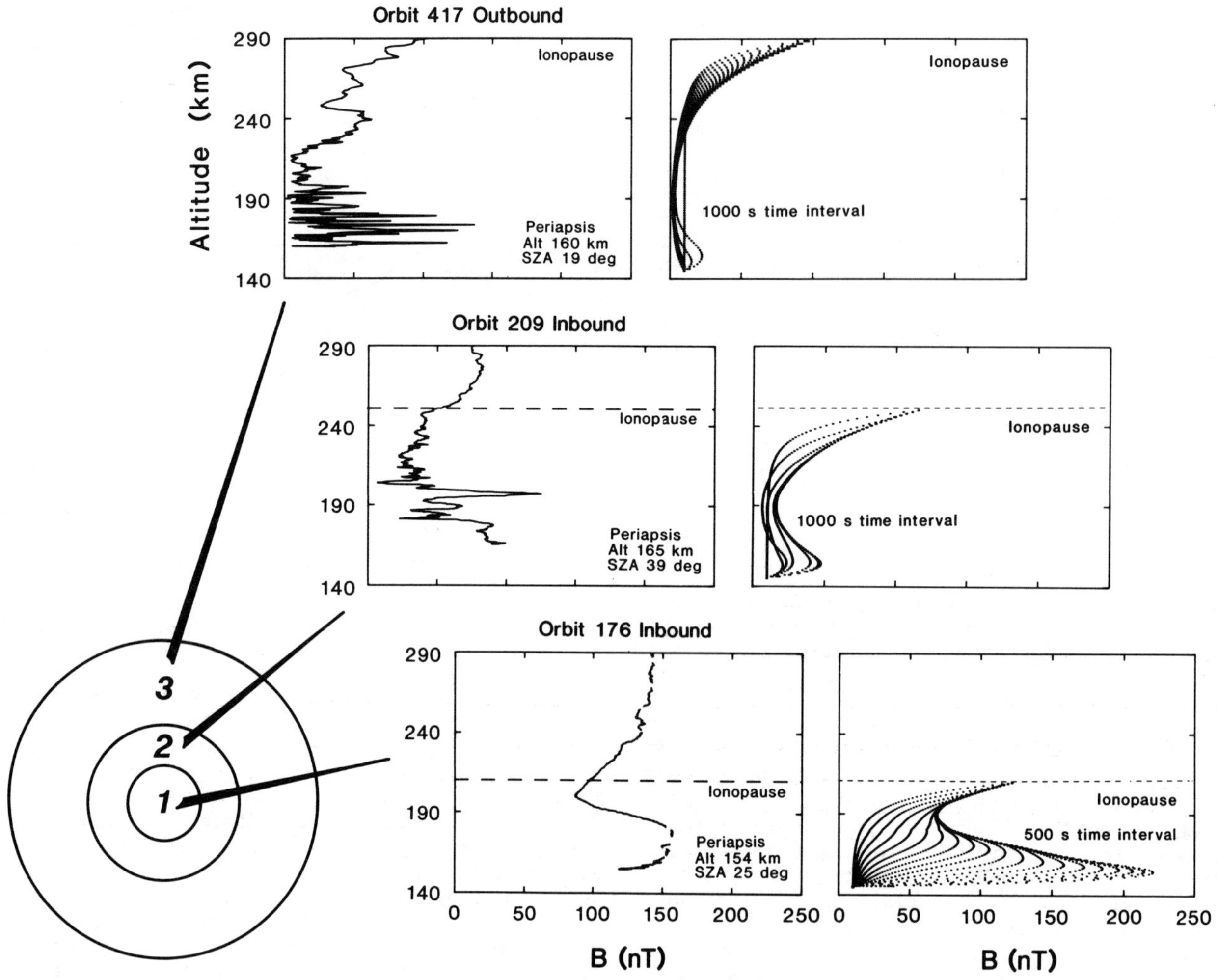

Fig. 10. Comparison of some observed altitude profiles of the magnetic field magnitude (left) with models (right) obtained by numerically solving the one-dimensional dynamo equation [from Phillips et al., 1984]. Dotted lines are drawn for equal time intervals to illustrate the rate of growth of the field from an initial small value. The "bullseye" is a schematic illustration of the dayside locations of different strength large scale ionospheric fields consistent with the ionopause shape. High fields appear most frequently near the subsolar point where the ionopause is lowest.

ionopause is lowest in the subsolar region due to the geometric reduction of the normal (to the ionopause) incident solar wind pressure with distance to the terminator, substantial large scale fields most frequently appear at solar zenith angles $\leq 45°$ as schematically illustrated by Figure 10. A high solar wind pressure will reduce the ionopause altitude everywhere and cause a more widespread large scale magnetic field, but one must be somewhat cautious in spreading the "bulls-eye" too far. When the ionopause is very low, the magnetic pressure gradients in the ionosphere alter the velocity so that both the profile shown in Figure 6b, and the kinematic dynamo description of the field, are no longer appropriate [cf. Shinagawa and Cravens, 1988].

Away from the subsolar region, the ionosphere is not only drifting downward, but is convecting in the antisolar direction in response to day-to-night plasma pressure gradients. Figure 11 shows some measurements and model altitude profiles of this velocity for several solar zenith angles. This additional motion complicates the dynamo description because the magnetic field will be convected horizontally as well as vertically; but observationally, the general features of the large scale field do not seem to be greatly affected. This may be because the ionopause has to be fairly low before the large scale field appears, and most of the transterminator flow occurs near the terminator at the higher altitudes where the ionopause is high. However, solutions of the

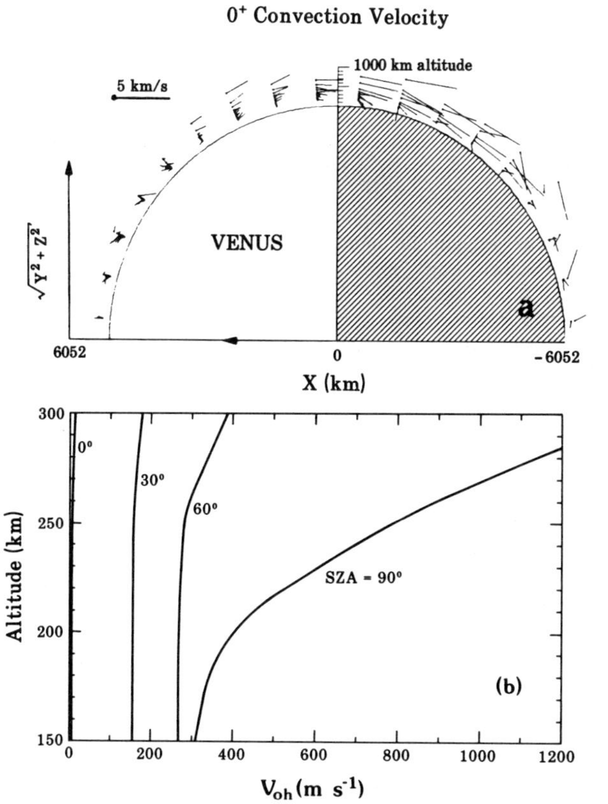

Fig. 11. Horizontal (antisolar) velocities of the ionospheric plasma (a) from the observations of Knudsen et al. [1980] and (b) from a model of Theis et al. [1984] based on observed plasma pressure gradients.

three-dimensional dynamo equation suggest that some distortion of the large scale field away from the direction of the overlying magnetosheath field may result as illustrated schematically in Figure 12.

An additional point of considerable interest is that the downward drift of the ionosphere is in fact related to the effects of the solar wind interaction. If one constructed a model of the ionosphere purely from a model of the neutral atmosphere and the ion production from solar radiation, it would have much lower temperature gradients and a different density profile. Beneath the ionopause, this ionosphere would still have a net drift, but the vertical velocity would have a different altitude profile. Thus the alteration of the ionosphere by the solar wind is in part what causes the ionopause to be at its observed location and the large scale magnetic field of the ionosphere to have its particular appearance. The nature of the solar wind-related heating is still a subject of debate. One proposed heat source for the electrons is the Landau damping of plasma waves of undetermined origin observed in the ionopause current layer [cf. Taylor et al., 1979]. These plasma waves are probably also related in some way to the solar wind interaction.

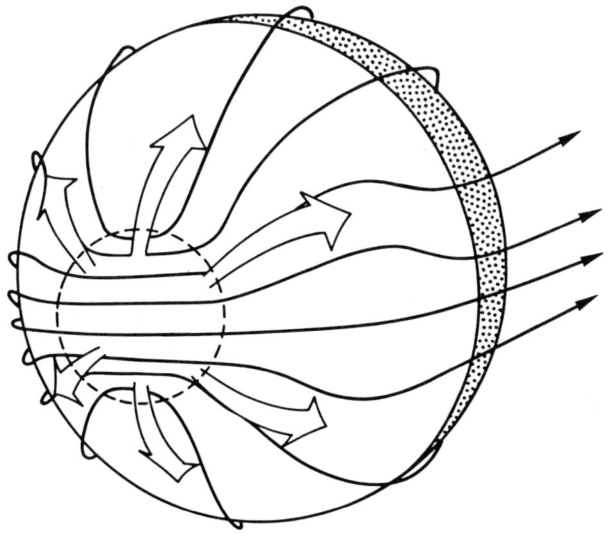

Fig. 12. Geometry of the distortion of the ionospheric field resulting from the horizontal ionospheric plasma flow [based on a model of Luhmann, 1988].

A Note on Small Scale Fields. Small scale ($\leq$ 10 km) magnetic structures are also observed in the dayside ionosphere of Venus. A subset of these structures has been shown to be modelable in terms of magnetic flux ropes which have strong, straight axial fields surrounded by gradually weaker, more helical fields. It is an important observation that when the large scale ionospheric field appears, the small scale field structures that are apparent in its absence disappear. This exclusivity, which may be telling us something about their origin, is illustrated in Figure 7 where fine structure is seen in the altitude profile without significant large scale fields. The phenomenon of ionospheric flux ropes continues to present a challenge for both theorists and observationalists. However, since another paper in this volume deals specifically with the details of the observations and models proposed for the generation of the small scale magnetic fields in the ionosphere of Venus, the subject will not be elaborated upon here. Nevertheless, for the sake of perspective, it is worth noting that these fields, like the large scale fields discussed above, are subject to modification by convection and collisional dissipation within the ionosphere.

III. The Solar Wind Interaction with Mars and with Venus at Solar Minimum

Whereas at solar maximum at Venus the situation depicted in Figure 3 prevails, at solar minimum the ionosphere of Venus generally has a thermal plasma pressure that is everywhere less than the incident solar wind pressure, a condition that is "normal" at Mars [cf. Slavin and Holzer, 1982]. However, at both planets the observed bow shock position indicates that under such conditions there is still an obstacle larger than the planet around which the solar wind flows. Observations at Venus during solar maximum for the occasional periods when a similar pressure imbalance occurred due to extraordinary solar wind conditions showed us that the ionospheric magnetic field can "supplement" the obstacle strength since the magnetic field pressure can contribute significantly to the ionosphere pressure. In fact, the ionospheric magnetic field pressure at Venus can exceed the peak ionosphere thermal pressure by a factor of ~3 [cf. Luhmann et al., 1987]. Thus, the situation at Mars and at Venus at solar minimum may look more like that sketched in Figure 13a than that shown in Figure 3. As before, the planetary ions are removed from the upper atmosphere by the solar wind, and the magnetic barrier continues to provide an effective obstacle, but the ionopause is no longer a boundary where the thermal pressure of the ionosphere becomes dominant. Rather, we know from the Venus observations that the ionospheric density profile attains a minimum state which is insensitive to solar wind pressure. In this minimum state, the cold ionospheric plasma is largely confined to altitudes below ~220 km and is permeated by the piled-up horizontal magnetic field that is no longer confined to the overlying magnetosheath. The pressure of this magnetic field is then what reflects the incident solar wind pressure.

Of course, there must be a limit to the amount of magnetic field that the ionosphere can accommodate. We know that in the case of a very weak ionosphere, like that of the moon, there is no deflection of the solar wind. The solar plasma flows directly into the surface where it is absorbed. The interplanetary magnetic field, on the other hand, suffers negligible distortion. If Venus and Mars had no ionospheres at all, a similar absorption of solar wind plasma would occur except that unlike the moon, the planets have a conducting core. Thus, although the solar wind plasma would be absorbed in the neutral atmosphere, the magnetic field that continues to move unimpeded into the solid body would be diverted around the core (as for the conducting sphere in a vacuum field). The addition of a weak ionosphere (in fact, the absorption of solar wind plasma should produce some weak impact ionization) would presumably introduce partial shielding currents and thereby some weak draping of the interplanetary field prior to its entry into the solid body. The field structure that might prevail on such occasions is suggested by Figure 13b. Since absorption of the solar wind plasma is a part of this scenario, it follows that one should detect a reduced deflection of the solar wind flow in the neighborhood of the planet. Thus far, the solar minimum data from Venus and the available data from Mars show no clear evidence for incomplete deflection of the solar wind. This implies that the ionospheric magnetic field is practically always sufficient to keep these obstacles "impenetrable" and that the situation depicted in Figure 13b is rare. Nevertheless, we can continue to search for evidence of this

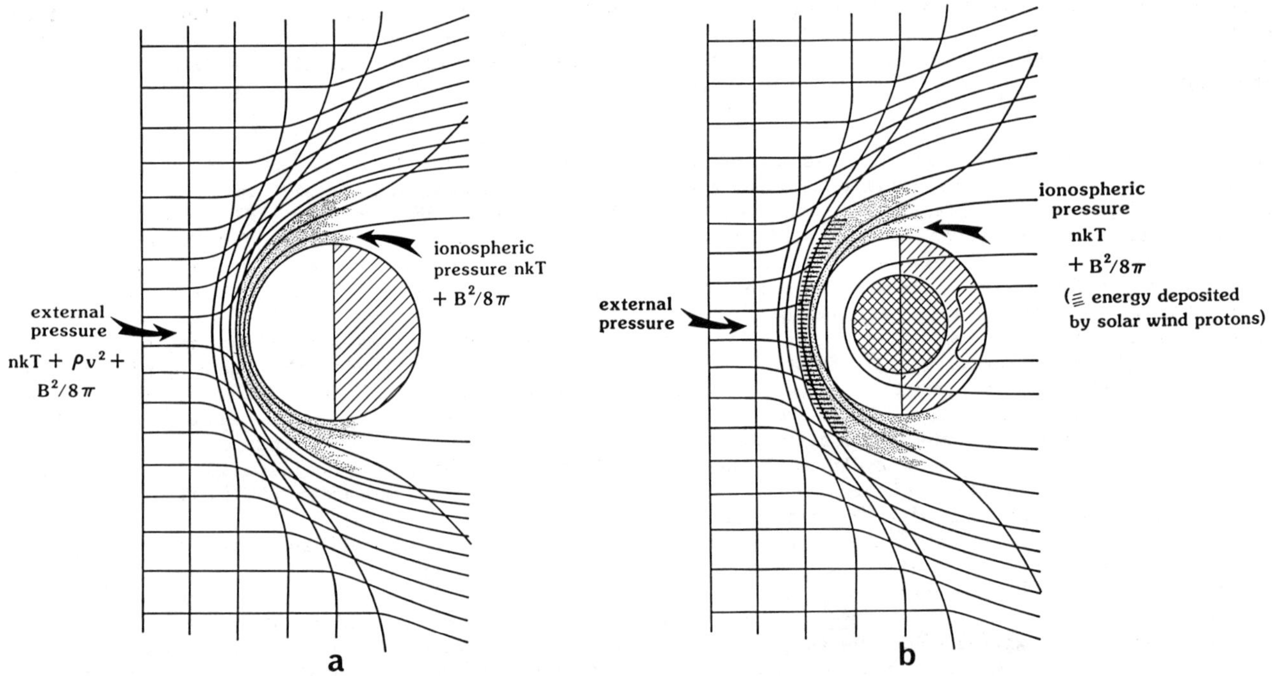

Fig. 13. (a) Schematic illustration of the penetration of interplanetary field into the ionosphere and resulting alteration of the solar wind interaction. (b) Suggested nature of the planet-solar wind interaction on occasions when solar wind pressure is so strong that the ionosphere is an ineffective obstacle.

"breakdown" of the obstacle, especially at Mars where a weak ionosphere is the rule.

IV. Concluding Remarks

The solar wind interaction with the ionosphere of Venus induces currents in the planetary ionosphere which can, for the most part, exclude the solar wind and interplanetary magnetic field from the dayside ionosphere beneath the "ionopause" where ionosphere thermal pressure is equal to the incident solar wind dynamic pressure. However, the field diffuses through the ionopause with increasing speed as the ionopause lowers. It is normally lowest in the subsolar region due to the geometrical reduction of incident solar wind pressure, and lowers everywhere in response to increasing solar wind pressure. Once inside the ionopause, the magnetic field is redistributed by ionospheric convection and decays at low altitudes by collisional dissipation of the associated currents. Since the convection is generally downward, the large scale magnetic field is horizontal and has an orientation similar to that of the overlying magnetosheath/magnetic barrier field, although an additional horizontal velocity component distorts this draping configuration to some extent. A maximum in the downward velocity near 190 km altitude produces characteristic minima in the altitude profiles of field magnitude, and dissipation reduces the field to small values below ~150 km. The maximum ionospheric field magnitudes observed, of ~150 nT, provide magnetic pressures that exceed the ionosphere thermal pressure by a factor of ~3. These values occur after the ionopause reaches its limiting low altitude of ~220 km.

Interestingly, the small scale field structure ($\leq$ 10 km scale size) associated with flux ropes appears only when and where the large scale field is weak ($\leq$ 10 nT), or similarly, when and where the local ionopause is high. It is not at present clear what causes these small structures, but from what we know about the large scale field we can surmise that Mars, which has a significantly weaker ionosphere than Venus does at solar maximum, generally possesses only large scale fields. The Venus ionosphere at solar minimum is probably in a similarly "permanently" magnetized state. We can also surmise that there is some connection between the diffusion and convection processes that cause the large scale field, and those responsible for the generation and evolution of Venus' ionospheric flux ropes.

Questions and Answers

Priest: Are the observed flow speeds much smaller than the sound and Alfvén speeds so that the large-scale and flux rope structures are in equilibrium? Is the magnetic Reynolds number of order unity so that advection balances diffusion? In the low ionosphere is the large-scale magnetic field increasing with depth due to the stagnation point flow? Is this field unstable to magnetic buoyancy instability? Is the large-scale magnetic radius of curvature much larger than the scale-height so that magnetic tension cannot inhibit magnetic buoyancy?

Luhmann: The observed flow speeds are smaller than the sound speed except perhaps at high altitudes in the ionosphere at the terminator where transonic flow is sometimes observed. The Alfvén speed is low enough compared to the spacecraft velocity (~10km/s) that one can consider the structures as "static" structures that the spacecraft flies through. I never thought of the magnetic Reynolds number as an index in this problem, but when the ionospheric field structure reaches its equilibrium state, according to the dynamo equation description, diffusion and convection balance to make $\partial B/\partial t = 0$. In the model I've described the magnetic field convection "stagnates" at low altitude. I think one has to be careful about thinking of magnetic buoyancy when these ionospheric flux tubes are in a region where local ion production and recombination is very important. The radius of curvature is much larger than the scale height. The former is the order of a planetary radius while the latter is only tens of km like the atmospheric or ionospheric scale height. Therefore, I agree that magnetic tension is not important, but once again I feel magnetic buoyancy in the presence of so much local ion production and loss is a tricky concept.

Haerendel: Did you consider the competitive model promoted by Cloutier, according to which the main force opposing the penetration of solar wind magnetic field is ion-neutral friction the ions being dragged by the magnetic field? The plasma pressure just causes a diamagnetic suppression of |B|. The downward transport of B that you derive must be continued by a horizontal transport mostly below the level of PVO-Observations.

Luhmann: In our picture, neutrals are also important. In particular, collision frequencies affect both the altitude variation of the vertical plasma velocity that convects the field downward and the diffusion coefficient which dissipates the field at the bottom of the structure. Our pictures of the large-scale field in the Venus ionosphere were originally quite different, with Cloutier's picture based on solar wind absorption and latitude (instead of altitude) gradients. I believe that now our models are beginning to converge, albeit slowly.

Fedder: Does the vertical profile of horizontal velocity appear to be a boundary layer flow? If it does, how thick is the boundary layer?

Luhmann: No, the bulk of the antisolar horizontal flow seems to be explainable in terms of a simple pressure gradient driven flow within the ionosphere as a result of day to night differences in plasma densities. There may be some kind of boundary layer at the highest altitudes within the ionopause (especially in the flanks near the terminator), but it's hard to tell observationally with available data.

Zelenyi: (1) Are the horizontal convection velocities you have shown sufficient to maintain the nightside ionosphere? (2) Do you see any evidence in experimental data that the venusian magnetotail is more developed and lengthy when the solar wind pressure is high? (This seems to be the direct consequence of your model.)

Luhmann:(1) Yes, the nightside ionosphere can be supplied by transport from the dayside except when the ionopause is depressed (and the ionosphere magnetized). Since most of the flow occurs at high altitudes, a lowered ionopause "shuts off" the source for the nightside ionosphere. At these times, the weak nightside ionosphere that is seen is probably caused by particle impact ionization. (2) No, in fact the tail seems to be weaker at these times. This may be because the transterminator flow of ions plays a role in the tail formation.

Dubinin: What is the role of picked-up ions in the formation of magnetopause (its position)? (2) What kind of plasma flows do you observe in plasma-sheet? Is the flow pattern regular in plasma-sheet or not?

Luhmann:(1) The ionopause position is to first order insensitive to picked-up ions. It is well described as a balance between solar wind dynamic pressure and ionospheric thermal pressure. However, it is true that the picked-up ion current seems to alter the magnetic field strength in the magnetic barrier such that the barrier magnetic pressure pushes harder on the side of the planet where the picked-up ions go. Still, this is a relatively small perturbation. (2) We really don't have the instrumentation to observe the plasma flows within the tail.

Atkinson: Where is the lowest conductivity layer, lower atmosphere or upper crust?

Luhmann: Lower atmosphere.

Drake: Why aren't the magnetic fields in the lower Venus ionosphere buoyant? Because of local pressure balance, the density inside flux-ropes must be lower than the ambient density.

Luhmann: Since the thermal pressure is considerably higher than the magnetic pressure, there should be only a small dip in thermal pressure inside flux-ropes. We sometimes see this, but sometimes the structures also appear to be force-free. Again, I think that considerations of buoyancy (or lack thereof) must take into account the fact that there is a lot of local plasma production and loss within ionospheric flux tubes, that gravity is significant, and that ambient ionospheric dynamics and collisional dissipation can largely explain the appearance of the field.

Acknowledgement. The author acknowledges support by Grant NAG2-501 from the Planetary Exploration Division of the National Aeronautics and Space Administration.

References

Brace, L. H., H. A. Taylor, Jr., T. I. Gombosi, A. J. Kliore, W. C. Knudsen and A. F. Nagy, The ionosphere of Venus: Observations and their interpretations, p. 779 in Venus, eds. D. M. Hunten, L. Colin, T. M. Donahue, V. I. Moroz, University of Arizona Press, 1983.

Cloutier, P. A., R. E. Daniell, Jr. and D. M. Butler, Atmospheric ion wakes of Venus and Mars in the solar wind, Planet. Space Sci., 22, 967, 1974.

Cravens, T. E., H. Shinagawa and A. F. Nagy, The evolution of large-scale magnetic fields in the ionosphere of Venus, Geophys. Res. Lett., 11, 267, 1984.

Elphic, R. C., C. T. Russell, J. A. Slavin and L. H. Brace, Observations of the dayside ionopause and ionosphere of Venus, J. Geophys. Res., 85, 7679, 1980.

Elphic, R. C., C. T. Russell, J. G. Luhmann, F. L. Scarf and L. H. Brace, The Venus ionopause current sheet: thickness, length scale and controlling factors, J. Geophys. Res., 86, 11430, 1981.

Knudsen, W. C., K. Spenner, K. L. Miller and V. Novak, Transport of ionospheric O^+ ions across the Venus terminator and implications, J. Geophys. Res., 85, 7803, 1980.

Luhmann, J. G., The solar wind interaction with Venus, Space Sci. Rev., 44, 241, 1986.

Luhmann, J. G., C. T. Russell, F. L. Scarf, L. H. Brace and W. C. Knudsen, Characteristics of the Marslike limit of the Venus-solar wind interaction, J. Geophys. Res., 92, 8545, 1987.

Luhmann, J. G., A three-dimensional diffusion/convection model of the large scale magnetic field in the Venus ionosphere, J. Geophys. Res., 93, 5909, 1988.

McComas, D. J., H. E. Spence, C. T. Russell and M. A. Saunders, The average magnetic field draping and consistent plasma properties of the Venus magnetotail, J. Geophys. Res., 91, 7939, 1986.

Phillips, J. L., J. G. Luhmann and C. T. Russell, Growth and maintenance of large-scale magnetic fields in the dayside Venus ionosphere, J. Geophys. Res., 89, 10676, 1984.

Russell, C. T. and O. Vaisberg, The interaction of the solar wind with Venus, p. 873 in Venus, eds. D. M. Hunten, L. Colin, T. M. Donahue, V. I. Moroz, University of Arizona Press, 1983.

Saunders, M. A. and C. T. Russell, Average dimension and magnetic structure of the distant Venus magnetotail, J. Geophys. Res., 91, 5589, 1986.

Shinagawa, H. and T. E. Cravens, A one-dimensional multispecies magnetohydrodynamic model of the dayside ionosphere of Venus, J. Geophys. Res., 93, 11263, 1988.

Slavin, J. A. and R. E. Holzer, The solar wind interaction with Mars revisited, J. Geophys. Res., 87, 10285, 1982.

Spreiter, J. R. and S. S. Stahara, Solar wind flow past Venus: Theory and comparisons, J. Geophys. Res., 85, 7715, 1980.

Taylor, W. W. L., F. L. Scarf, C. T. Russell and L. H. Brace, Absorption of whistler mode waves in the ionosphere of Venus, Science, 205, 112, 1979.
Theis, R. F., L. H. Brace, R. C. Elphic and H. G. Mayr, New empirical models of the electron temperature and density in the Venus ionosphere with application to transterminator flow, J. Geophys. Res., 89, 1477, 1984.
Zwan, B. J. and R. A. Wolf, Depletion of solar wind plasma near a planetary boundary, J. Geophys. Res., 81, 1636, 1976.

MAGNETIC FLUX ROPES IN THE IONOSPHERE OF VENUS

C. T. Russell

Department of Earth and Space Sciences and Institute of Geophysics and Planetary Physics, University of California, Los Angeles, California 90024-1567

Abstract. The Pioneer Venus Orbiter has shown that the ionosphere of Venus is often filled with filamentary magnetic fields. These magnetic flux tubes apparently are created at the ionopause in the subsolar region and sink to low altitudes. As they sink they twist and compress, maintaining about the same magnetic flux, 2 to 3 Webers, until they sink below 175 km. Ropes also appear to become kink unstable at about 200 km generating spatial helices. These 3-dimensional structures can lead to increased interaction between flux ropes.

Introduction

The Pioneer Venus Orbiter was injected into orbit about Venus on December 4, 1978 and its periapsis gradually lowered on succeeding days so that it could perform in-situ studies of the upper atmosphere and ionosphere. The first deep penetration of the ionosphere provided a great surprise for the magnetic fields investigation team. The ionosphere was magnetized in filaments separated by nearly field-free plasma. Study of these filaments revealed them to be magnetic ropes or twisted tubes of magnetic flux. The left-hand panel of Figure 1 shows an example of such ropes in an altitude profile of the magnetic field and electron density. Under normal solar wind conditions the altitude of the ionopause in the subsolar region is about 300 km. However, when the solar wind dynamic pressure increases, the ionosphere begins to become magnetized and the flux rope formation process appears to be suppressed as shown in the middle panel. When the solar wind becomes very strong the ionosphere becomes fully magnetized. Since the solar wind is usually stronger than the maximum ionospheric pressure at Mars, we expect the Martian ionosphere to be magnetized (Luhmann et al., 1987). Thus the Venus ionosphere may provide us with the only opportunity to make in-situ measurements of such structures.

Geophysical Monograph 58

In this review we will examine first the evidence that these structures are indeed magnetic ropes. Then we will review the modeling efforts which have been used to deduce the structure of the fields and currents within the ropes. We will examine the evolution of magnetic flux ropes in the Venus ionosphere and finally discuss several of the various theories for the evolution of flux ropes.

Modeling Flux Ropes

Often the best way to attempt to understand a magnetic variation is to attempt to express it in 2 dimensions. Usually the magnetic variations associated with wave phenomena or plasma boundaries are confined to planes. To find the plane of maximum variance or equivalently the direction of minimum variance we use the principal axis technique pioneered by Sonnerup and Cahill (1967). It is useful also for the analysis of magnetic flux ropes also but as we shall see below the directions returned by this analysis are not always simply interpretable as they are with a wave. We should note that in analyzing structures in the Venus ionosphere we can assume that the observed temporal profile is in fact a spatial profile converted into a temporal profile by the velocity of the spacecraft. Even when the magnetic field in a flux rope reaches 50 nT the Alfven velocity is still less than a 1 km/sec in the Venus ionosphere compared with the satellite velocity of about 7 km/sec.

The left-hand panel of Figure 2 shows the magnetic profile through one of the earliest measured Venus flux ropes (Russell and Elphic, 1979). Here principal axis analysis has been used in an attempt to reduce the variation to a single plane but clearly the variation in the field is 3 dimensional. On the right is a model field which reproduces the observed field. This model field was constructed by the use of the following equations:

$$B_\phi = B(\rho) \sin (\alpha(\rho)) \qquad (1)$$

$$B_z = B(\rho) \cos (\alpha(\rho)) \qquad (2)$$

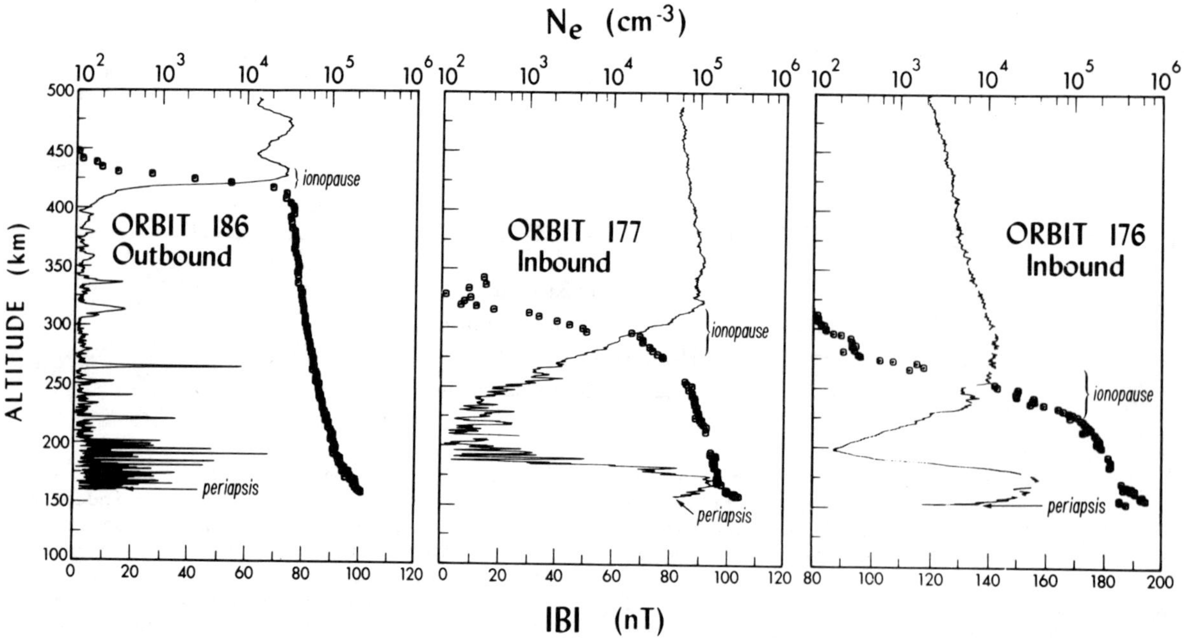

Fig. 1. Altitude profiles of the magnetic field and the electron density observed by the Pioneer Venus Orbiter on three different orbits of increasing solar wind dynamic pressure and decreasing ionopause altitude.

where $B(\rho) = B_0 \exp(-\rho^2/b^2)$ (3)

$\alpha(\rho) = \pi/2\ (1 - \exp(-\rho^2/a^2))$ (4)

The Z-direction is along the axis of the rope, the ϕ-direction, the azimuthal direction and ρ is the distance from the center of the rope. This model represents a structure whose magnetic field at the center of the rope is parallel to its axis. The magnetic field strength, $B(\rho)$, weakens with distance from the center of the rope and the field becomes more azimuthal. The Gaussian form for $\alpha(\rho)$ was chosen because it was evident from the data that the direction of the field varied more slowly as the center of the structure was reached. For the model in Figure 2 a = 3.2 km and b = 4.5 km and the spacecraft trajectory was chosen such that the distance of closest approach to the axis of the rope, the impact parameter was 2 km.

Another way of looking at the magnetic structure is shown in Figure 3. This shows a hodogram of the magnetic field variation in 2 orthogonal planes. Here the end of the magnetic field projections at successive times are plotted and joined by straight-line segments. The panels on the top show the observations and the panel on the bottom shows the model. The left-hand panels show the plane of maximum variance and the right-hand panels show the orthogonal plane containing the direction of minimum variance.

If the magnetic flux tube were completely untwisted then the hodogram on the left would be a narrow vertical line. The width of the hodogram is a measure of how twisted is the flux rope. The angle at which the hodogram approaches the origin

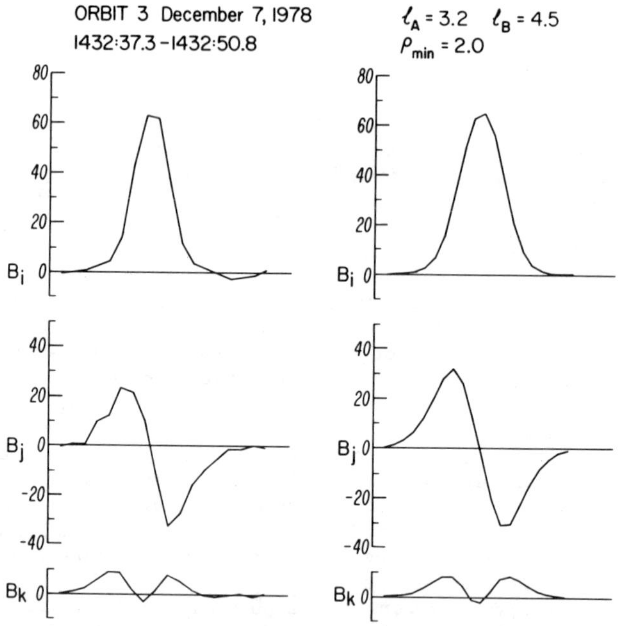

Fig. 2. Comparison of the magnetic structure of a Venus flux rope as displayed in the principal axis coordinate (left) system with a model rope (right). The data segment lasts 13.5 sec (Russell and Elphic, 1979).

in the plane of maximum variance is a measure of the twist at the edge of the flux rope. The curvature of the right-hand panels, resulting from the overall potato-chip shape of the hodogram is an

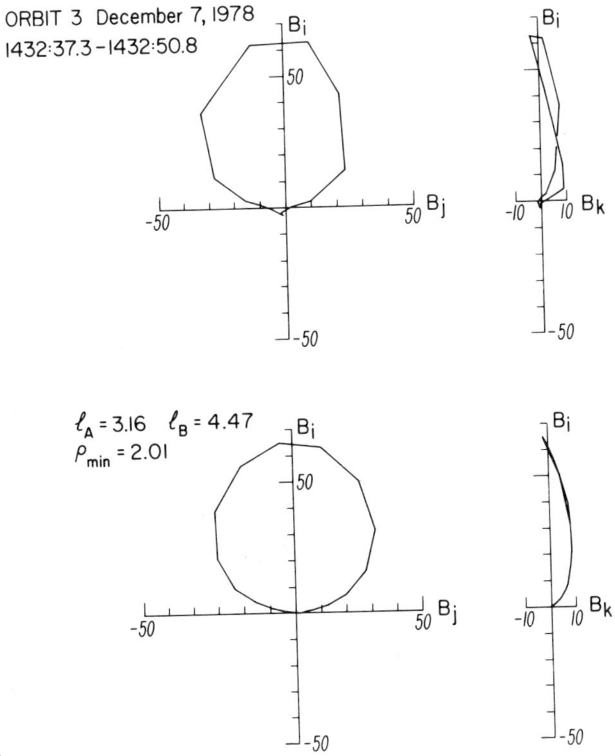

Fig. 3. Hodograms of the magnetic field of a Venus flux rope (top) with the model field (bottom). These data correspond to those shown in Figure 2 (Russell and Elphic, 1979). Unless the impact parameter is zero none of the principal axis directions i, j, k will coincide with the Z-direction, the axis of the flux rope.

indication that the spacecraft crossed the rope off axis.

The magnetic structure of our model magnetic rope is shown in a cut-away diagram in Figure 4. In the center of the rope the magnetic field is strong and along the axis of the rope. With increasing distance from the center of the rope the magnetic field strength decreases and the pitch of the magnetic field, i.e., the angle between the magnetic field direction and the axis of the rope, increases.

The key indicator that we are observing a magnetic flux rope and not some planar magnetic structure lies in the behavior of the magnetic field in the minimum variance direction, that is the component that gives the hodogram its "potato chip" shape in 3 dimensions. The way this component arises is illustrated in Figure 5 which shows schematically the variation of the magnetic field along the trajectory of a spacecraft flying through a rope, together with two orthogonal projections of the hodogram of the field variations in the principal axis system. The principal axes of this system are not along the axis of the rope nor is the direction of the peak magnetic field. Only when the satellite passes through the exact center of the rope is the peak magnetic field necessarily along the rope axis. At this time the direction of maximum variance or the direction of intermediate variance will also be along the rope axis. Which of the two is parallel depends on the radial structure of the twist and magnetic field strength. In other words, in some ropes the long

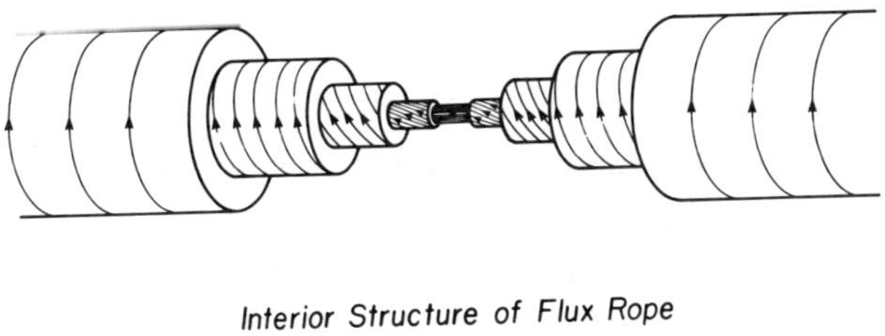

Fig. 4. Inferred magnetic structure of a Venus flux rope. The field is weak and azimuthal in the outer regions becoming much stronger and more axial near the center of the rope (Russell and Elphic, 1979).

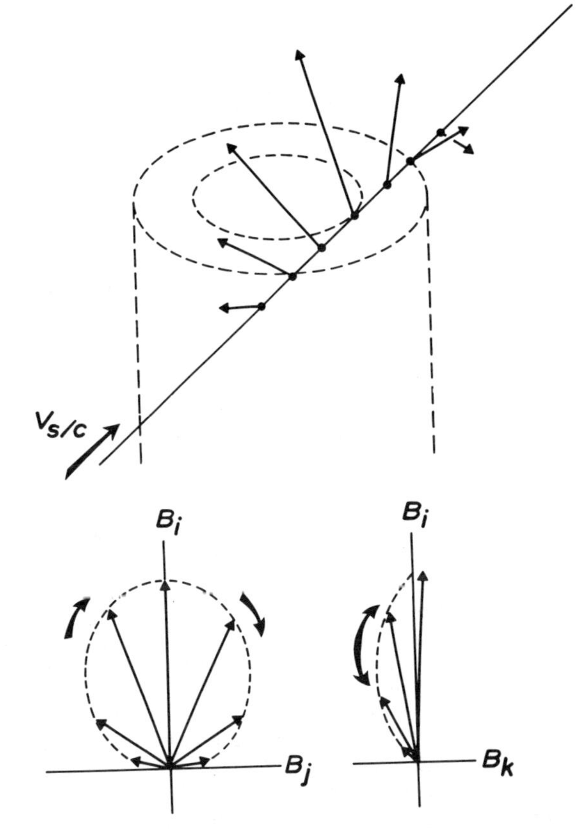

Fig. 5. Schematic diagram of the flight of Pioneer Venus through a magnetic flux rope and the resulting hodograms of the magnetic field variation in the principal axis coordinate system (Elphic et al., 1980).

axis of the hodogram is not vertical as it is in Figure 3 but rather horizontal.

Several different functional forms have been used to model these flux ropes. In their second paper on flux ropes Elphic et al. (1980) changed equation (4) to:

$$\alpha = \alpha_s \; (1 - \exp\,(-\rho^2/a^2)) \quad \rho \geq a \qquad (4')$$

$$\text{and } \alpha = C\rho^2 + D\rho \quad \rho < a \qquad (5)$$

This adjustment allowed the modeled rope to have an asymptotic pitch angle different than 90° and gave a parabolic behavior of the pitch of the field near the center of the rope. Both these adjustments were necessary to more nearly approximate the observed structure of the ropes.

In their subsequent work Elphic and Russell (1983a) deemed it necessaary to increase the number of free parameters in the flux rope. The model (on which the properties of the ropes reported in this review are based) is as follows:

$$B(\rho) = B_o \exp\,(-\rho^2/a^2) \quad \rho \leq b \qquad (6)$$

$$B(\rho) = B_o \; (C(a/\rho) + D(a/\rho)^2 + E(a/\rho)^3) \quad \rho > b \qquad (7)$$

$$\alpha(\rho) = \alpha_o \; (F(\rho/a) + G(\rho/a)^2) \quad \rho \leq c \qquad (8)$$

$$\alpha(\rho) = \alpha_o \; (1 - \exp(-\rho^2/d^2)) \quad \rho > c \qquad (9)$$

Equations (6) and (7) define the field strength as a Gaussian in ρ for $\rho \leq b$ and a third degree polynomial for $\rho > b$. The former is a compact form easily fit to the data yielding an exponential scale length, a, to which all other lengths in the problem may be normalized. The latter equation provides an asymptotic $(1/\rho)$ fall-off of B at large (ρ/a) while preserving the continuity of B, $(\partial B/\partial \rho)$ and $(\partial^2 B/\partial \rho^2)$ across $\rho = b$. This is equivalent to maintaining the continuity of the magnetic field strength, the current, and the gradient in the current across the point $\rho = b$. The location of b is chosen to best fit the data while C, D and E are obtained from the continuity requirements.

Equations (8) and (9) provide a quadratic increase in the pitch angle α from zero at the center of the rope to some finite value at $\rho = c$. In general it is not the same distance as b. Beyond c, α asymptotically approaches α_o. The coefficients F, and G, the length, d, and the asymptotic angle, α_o, are chosen as before to provide continuity in α and its first and second derivative and to provide a good fit to the data. Further details of the modeling procedure are given by Elphic and Russell (1983a).

Figure 6 shows the hodograms associated with the crossing of a number of flux ropes displayed in the principal axis system. Each of them shows the characteristic warp of the minimum variance hodogram. However, the orbit 7 crossing and the middle event on the left (orbit 185) show a minimal warp.

We can use this analysis to determine if the spacecraft has passed close to the center of the magnetic flux rope by examining the eigenvalues returned by the analysis. Unless the radial behavior of the magnetic field and pitch angle are linked by a very specific functional relationship, the minimum variance will approach zero as the spacecraft "impact parameter" approaches zero. It will be finite otherwise. Figure 7 shows a set of such flux ropes (Elphic and Russell, 1983a). The line shows the model fit to the spatial variation in field strength and pitch. The cases shown for orbit 7 and 185 are those shown in Figure 6. We see that there is variation among the ropes, even these small impact parameter ones. This can be seen especially in the variation of the pitch of the field near the center of the rope. Sometimes it is sharp and other times very broad.

Even more dramatic is the variation in current through the ropes which we can obtain by taking the curl of the field model. These are shown in Figure 8. The rope of orbit 7 for example has a strong perpendicular current quite in contrast to the rope on orbit 204 which appears to be almost all parallel current. Parallel current structures are force-free, i.e., the twist in the field completely balances the outward pressure gradient of the magnetic field so it exerts no force on the plasma. Table 1 lists a number of the model parameters for these ropes. The parameter, r_i, is the modeled thermal ion gyro radius; β^* is the ratio of the ionospheric thermal pressure to the peak rope magnetic pressure; and u^* is the peak ion/electron field-aligned relative drift velocity.

Global Properties of Flux Ropes

Flux ropes are observed everywhere in the dayside Venus ionosphere (Elphic and Russell, 1983b). However, their properties abruptly change near the terminators, i.e., the dawn-dusk plane (Luhmann, this volume) at which time the properties evolve into those of terminator waves (cf. Brace et al., 1983). Large scale fields are more often observed in the subsolar region (Luhmann et al., 1980). Hence by definition flux ropes must be less often observed there. Nevertheless, these two phenomena may still be linked since the large-scale fields are thought to be due to the transport of magnetic flux (Luhmann et al., 1984; Cravens et al., 1984) and flux ropes do transport magnetic flux.

In order to study the global properties of flux ropes one must identify a homogeneous set of data for which the properties of the ropes are well determined. To do this Elphic et al. (1980) and Elphic and Russell (1983b) identified the small impact parameter subset for which the ratio of the intermediate to the minimum eigenvalue was greater than 100. Figure 9 shows the altitude dependence of the maximum magnetic field for these flux ropes. The bimodal structure at low altitudes is due to the fact that grazing passes also have a large eigenvalue ratio. Thus, to determine which were

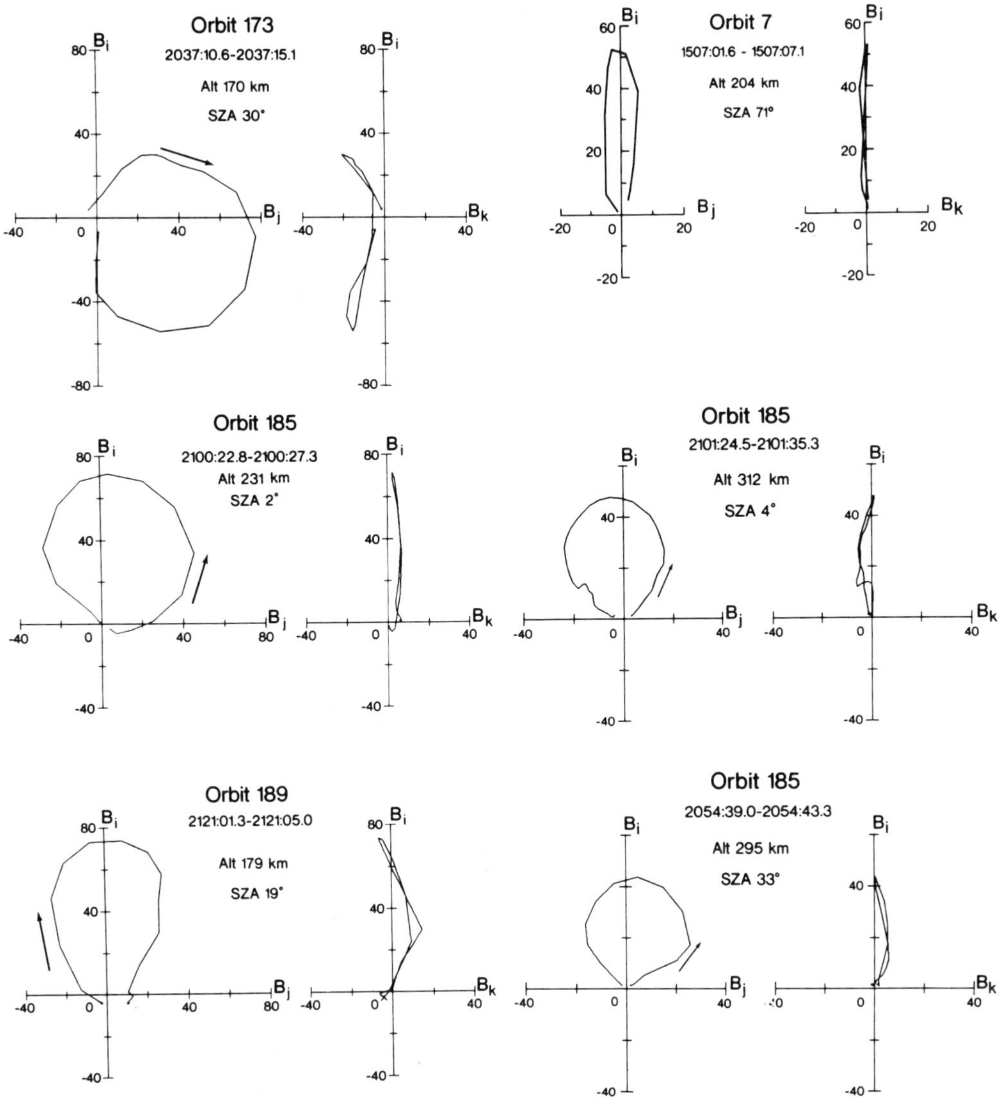

Fig. 6. Typical hodograms of the field variation through magnetic flux ropes as expressed in principal axis coordinates (Elphic and Russell, 1983a).

true low impact parameter cases only values to the right of the line in Figure 9 were considered to be actual low impact parameter crossings. At highest altitudes peak flux rope field strengths are about 20 nT increasing to about 70 nT at lowest altitudes. Flux ropes at higher solar zenith angles are slightly weaker than those at lower solar zenith angles (Elphic and Russell, 1983b).

Flux ropes occur everywhere in the dayside ionosphere. In particular they are observed right at the subsolar ionopause (Russell et al., 1987). This fact argues against the Kelvin-Helmholtz instability as being responsible for flux rope formation because there is no sheared flow at the subsolar point. Flux ropes increase in occurrence rate with decreasing altitude as shown in Figure 10 (Elphic and Russell, 1983b). High in the ionosphere, flux ropes seem to be more prevalent near the terminators than near the subsolar point but at low altitudes quite the reverse is true. We attribute this distribution to the flow properties of the Venus ionosphere which generally flows from high altitudes to low where it recombines and "deposits" any magnetic field which it has transported. Flow also proceeds from the subsolar region across the terminators for the same reason. The relative paucity of flux ropes at low altitudes in the near terminator region might be due to the fact that the streamlines are more antisunward than vertically downward in this region. Finally, at the very lowest altitudes as shown in Figure 11, the flux-rope occurrence rate decreases markedly.

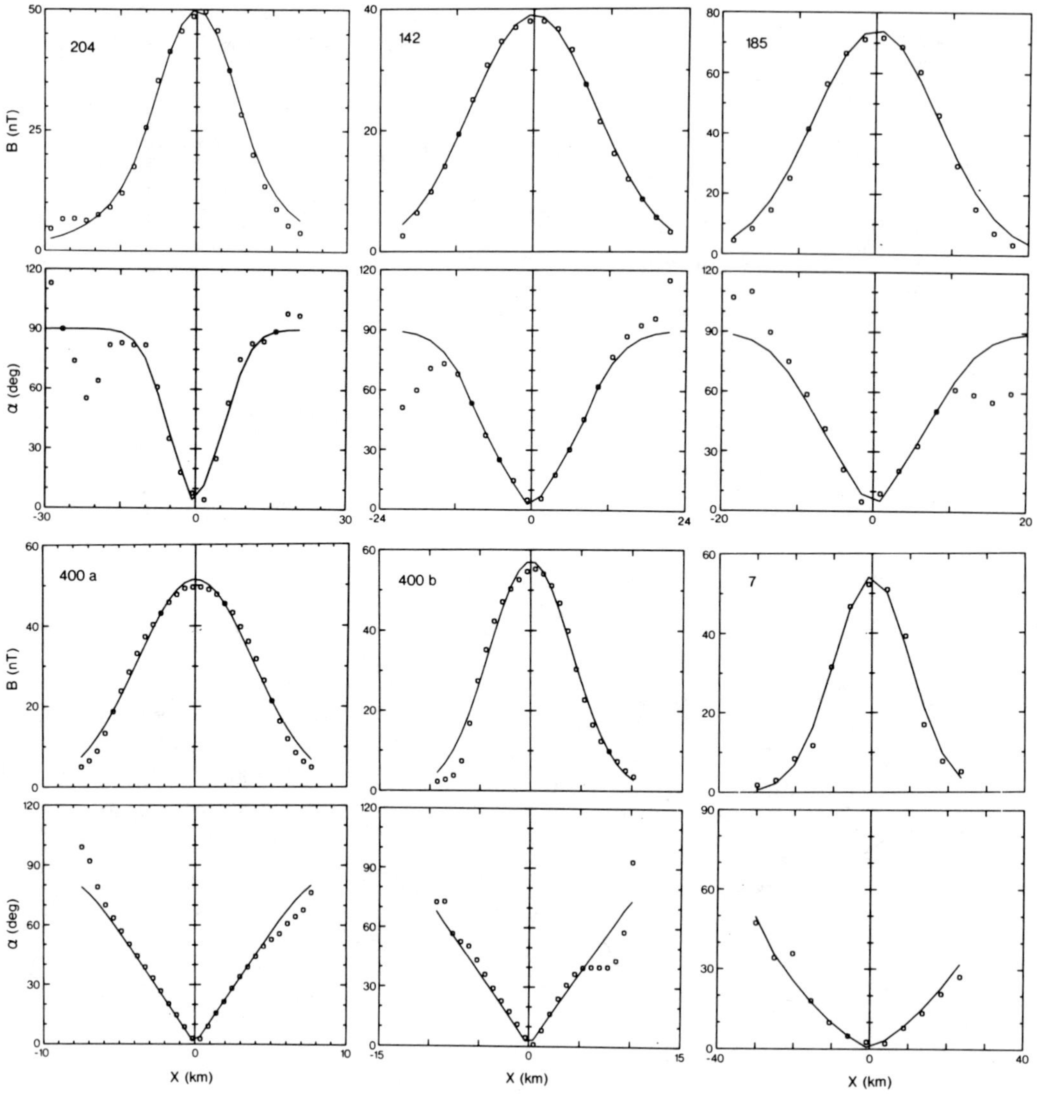

Fig. 7. Observed magnetic field variations and the model fields expressed as a function of distance from the axis of the rope for six small impact parameter rope traversals. The cases for orbits 7 and 185 are the same as shown in Figure 6 (Elphic and Russell, 1983a).

We attibute this drop off to the increase in collisional resistivity which dissipates the energy stored in the flux rope.

The size of flux ropes is also a function of position. As illustrated in Figure 12 the diameter of a flux rope is about 12 km at high altitudes in the subsolar region and about 6 km at lowest altitudes. Nearer the terminator the flux ropes are about 16 km in diameter at high altitudes and about 7.5 km in diameter at low altitudes. Within the accuracy of our statistics this variation implies that the net magnetic flux in a 'rope remains constant with altitude being about 2 Webers in the subsolar region and about 3 Webers in the near terminator region. In other words, the decrease in cross section of flux ropes with decreasing altitude seems to be due to an increased compression of the rope, presumably due to increased twist of the ropes with decreasing altitude. The increase of ionospheric thermal pressure external to the rope is only a factor of 2 between 300 and 150 km. As measured by the ratio of the eigenvalues in the principal axis analysis, flux ropes are more twisted at low altitudes than at high (Elphic and Russell, 1983b). There also seems to be no control of the twist or helicity of the flux ropes by the orientation of the IMF or position (Elphic and Russell, 1983; Luhmann, this

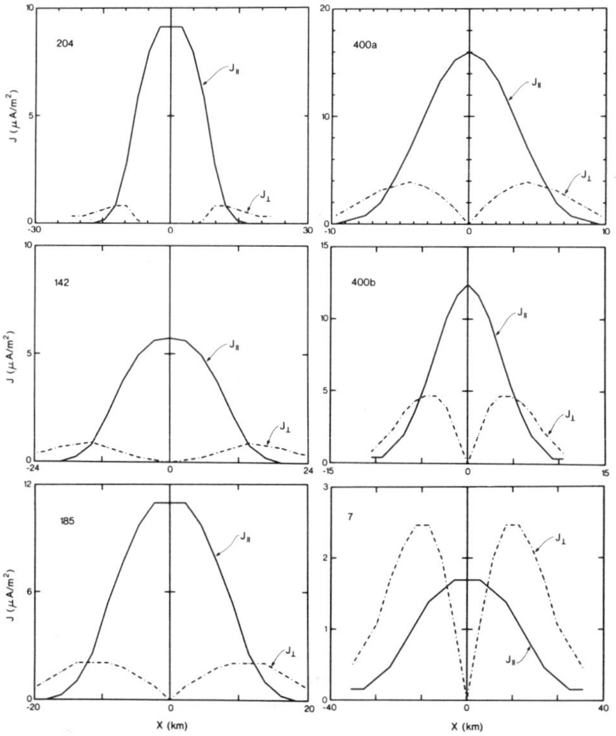

Fig. 8. Electric current parallel (solid line) and perpendicular (broken line to the local magnetic field for the small impact parameter ropes shown in Figure 7 (Elphic and Russell, 1983a).

volume). Thus either flux ropes are long-lived entities and the measured IMF's have little correspondence to the observed ropes or the flux rope formation process creates ropes with no net helicity such as would occur if they were formed in pairs with opposite helicity.

The Helical Kink Instability

A band of rubber is basically a 2-dimensional object with a width and length. If one twists the ends of such an object at first the rubber band remains basically a 2-dimensional object, albeit twisted around its axis. However, if the twisting continues long enough, the rubber band develops a helical kink. The axis of the rubber band is no longer a straight line but is a helix in 3-dimensional space. The rubber band has become a 3-dimensional object. As such, it will now begin to interact with other rubber bands, especially if they also have become kink unstable and have formed 3-dimensional structures. Thus in the Venus ionosphere the weakly twisted flux ropes may not interact with each other. However, once they become strongly twisted, they might become intertwined to form complex rope-like structures. One way to check this is to study the orientation of the axes of the ropes.

Figure 13 shows the inclination of the axes of the low impact flux ropes at high and low solar zenith angles and high and low altitudes (Elphic and Russell, 1983c). The angle 'i' is the angle between the flux rope axis and the local horizontal direction. If the flux ropes were mainly horizontal the distributions of Figure 13 would be peaked near zero but if they were mainly vertical they would be peaked near 90°. By plotting occurrence rate as a function of 'sin i', we normalize the distribution so that a random orientation of the axes will result in a constant occurrence rate.

The top panels show the distributions for low solar zenith angles and high solar zenith angles in separate panels summed over all altitudes. This

TABLE 1. Characteristics of Modeled Flux Ropes

Case	a, km	a/r_i	β^*	J^* μA/m²	E^* nV/m	P^* W/m³	u^*/c_s	h km	χ deg,	B_0 nT
7	14.1	7.0	1.9	2	33	7×10^{-14}	0.22	204	71	54
136	8.4	3.9	2.2	12	220	3×10^{-12}	1.34	199	78	51
142	13.9	3.2	1.6	6	26	2×10^{-13}	1.88	455	65	39
185	11.5	8.8	2.3	11	110	1×10^{-12}	1.53	232	2	74
204	12.1	4.8	2.9	9	71	6×10^{-13}	0.81	271	38	50
400a	5.4	2.2	2.8	16	127	2×10^{-12}	1.47	270	37	51
400b	5.8	3.4	3.1	12	123	2×10^{-12}	0.71	236	34	57

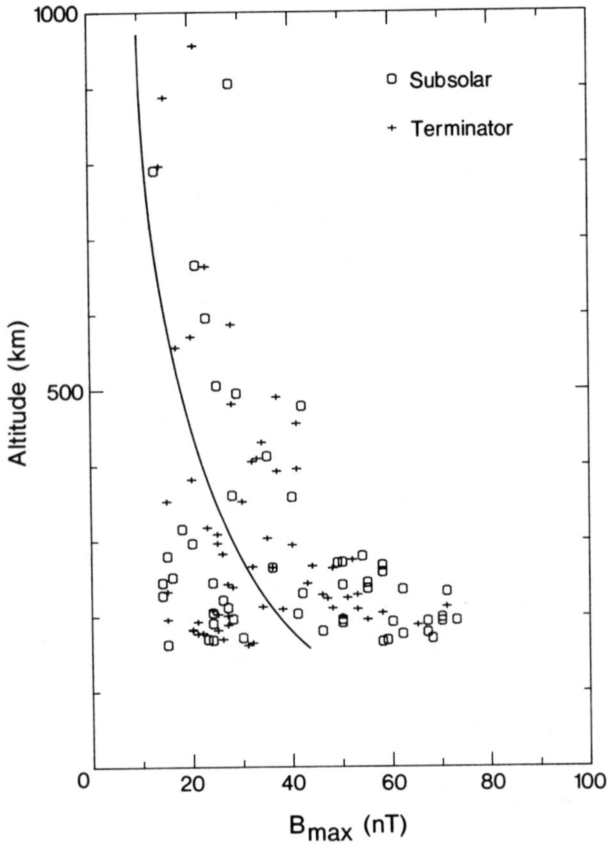

Fig. 9. Maximum observed flux rope field strengths as a function of altitude for cases with a low minimum variance ratio. Circles correspond to subsolar cases. The solid line separates two populations. The high field cases are felt to be true small impact parameter rope traversals while the low field cases are probably grazing encounters (Elphic and Russell, 1983a).

diagram shows that the orientations are fairly random in the subsolar region but much more horizontal in the near terminator region. When divided into low and high altitudes around the 200 km level as shown in the lower panels, a difference with altitude appears. In the subsolar region, the high altitude ropes are more horizontal than vertical and at low altitudes more vertical than horizontal. At high solar zenith angles, the high altitude ropes are nearly exclusively horizontal while the low altitude ropes are more random in orientation. This pattern suggests that at high altitudes ropes near the subsolar point are formed in a principally horizontal orientation but as they sink they become more twisted and become kink unstable. The axis of the rope then adopts a 3-dimensional helical form of its own as illustrated in Figure 14. Such ropes can interact and lead to even more complex magnetic structures.

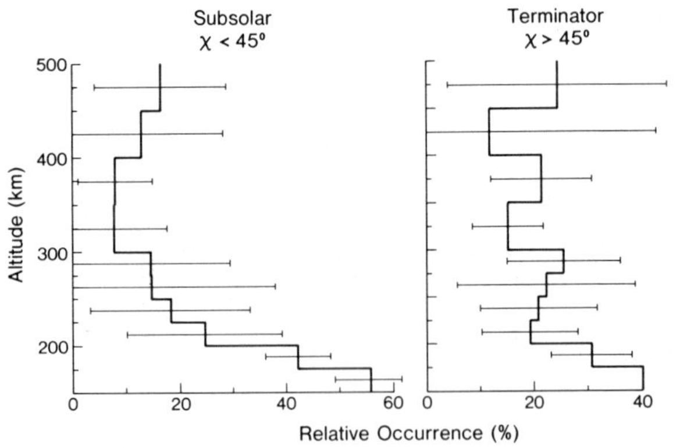

Fig. 10. Relative occurrence of flux ropes as a function of altitude in the near subsolar region (left) and the near terminator (right) regions. The relative occurrence rate is the ratio of the time spent within flux ropes at a particular altitude to the total time spent at that altitude (Elphic and Russell, 1983b).

The helical kink instability also answers another question about the behavior of flux ropes at low altitudes. As shown in the left-hand panel of Figure 1, at high altitudes there is much space between flux ropes relative to the space occupied by flux ropes. At low altitudes the reverse is true. If flux tubes remained straight as they sank to low altitudes this relative volume would remain constant. However, the formation of 3-dimensional helices allows the low altitude region to become filled with flux tubes (Elphic and Russell, 1983c).

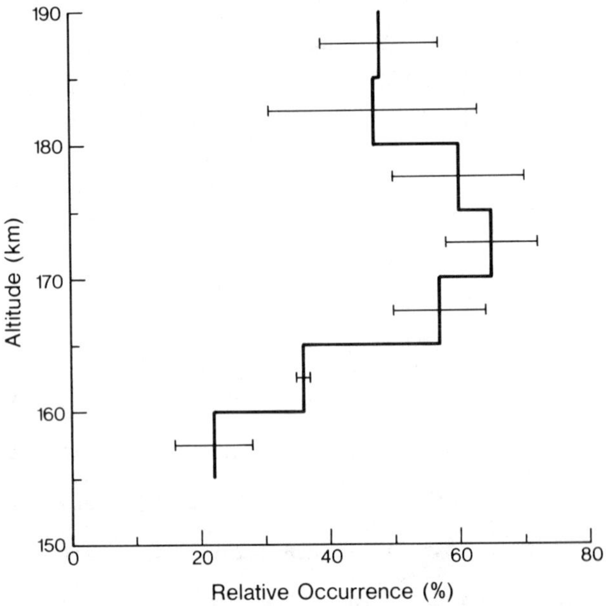

Fig. 11. Relative occurrence rate of flux ropes between 155 and 190 km (Elphic and Russell, 1983b).

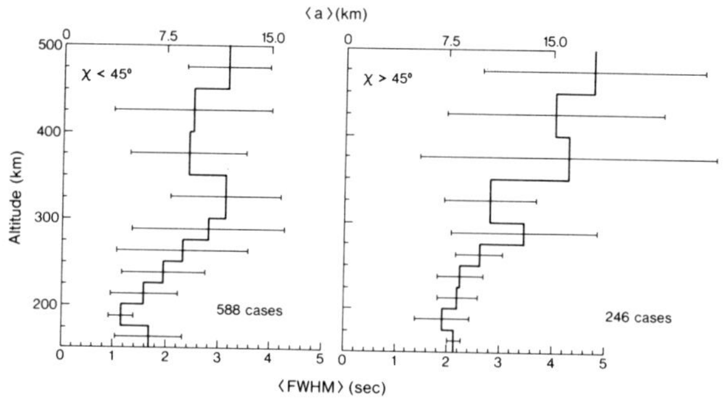

Fig. 12. Average full width at half maximum of flux ropes as a function of altitude for near subsolar and near terminator regions. Error bars denote the standard deviation of the samples (Elphic and Russell, 1983b).

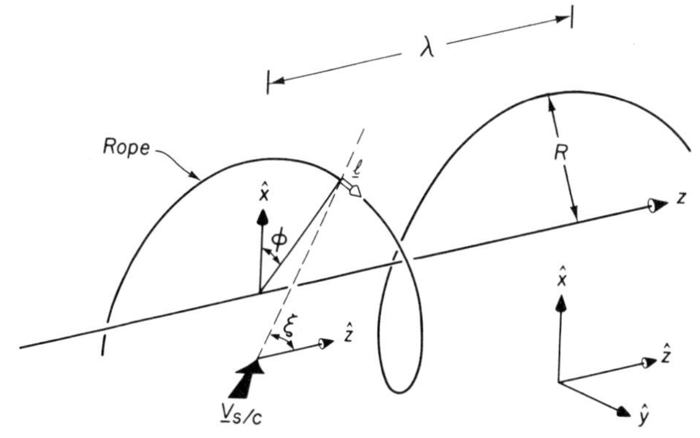

Fig. 14. Schematic diagram showing the traversal of a Venus flux rope that has become helical kink unstable. The axis of the helix is along the Z-direction (Elphic and Russell, 1983c).

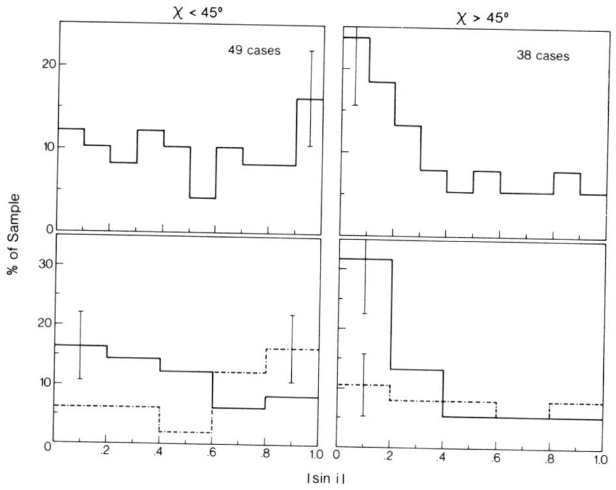

Fig. 13. The upper panels show the number of cases of flux rope inclinations with respect to the local horizontal as a function of the sine of the inclination for all altitudes for the near subsolar region (left) and the near terminator region (right). In the lower panels these distributions have been separated by altitude. The distributions above 200 km are shown by a solid curve and below 200 km by a broken curve (Elphic and Russell, 1983b).

The Source Region of Venus Flux Ropes

The increasing field strength and helicity of the magnetic flux ropes with decreasing altitude in the subsolar region and the transition from horizontal to vertical orientation of the ropes at low altitudes suggests that the source region for Venus flux ropes is at high altitudes and the sink at low. The Kelvin-Helmholtz instability is not expected to be an effective source for these ropes because the magnetic barrier shields the ionosphere from the solar wind and there is little velocity shear at the ionopause. Moreover, these ropes are observed in the subsolar region where the shear would be low even without the magnetic barrier.

To determine if flux ropes were indeed forming at the subsolar point we examined those orbits of the Pioneer Venus spacecraft that passed closest to the subsolar point (Russell et al., 1987). Figure 15 shows the magnetic field on orbit 187 when the spacecraft moved outward across the ionopause less than 3° from the expected sub-flow point. If there were no twist of the magnetic field then the magnetic field would have no radial component.

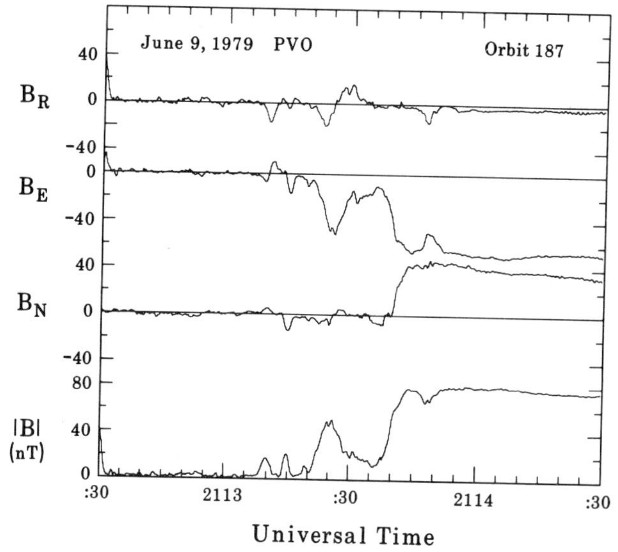

Fig. 15. Time series of the vector magnetic field in radial east and north components for 2 minutes on the outbound portion of orbit 187 when the Pioneer Venus orbiter was within 3° of the expected subflow point (Russell et al., 1987).

Clearly, near the ionopause both inside the ionosphere and in the magnetic barrier it does have a radial component. Moreover, the east-west component in the "ropes" is different from that in the magnetic barrier. Thus these structures at the subsolar ionopause are not fluctuations in the location of the ionopause but in fact are due to the formation of twisted ropes.

How this might occur is illustrated in Figure 16. Near the subsolar point the flux tubes in the magnetic barrier float on the heavy ionospheric plasma. However, these tubes become heavier with time as the neutral oxygen exosphere is photoionized in them. In the ionosphere, such newly created ions will just be incorporated in the ionosphere and slowly drift down to eventually recombine at low altitudes. However, the ions in the magnetic barrier will continue to build up in density unless there is some convective flow to carry them away. In the magnetic barrier, especially at the subsolar point, this flow should be very weak. Thus flux tubes can become heavy enough to sink in the ionosphere, but even before they do so the curvature stresses in the field will assist the tube in overcoming the buoyancy in the ionosphere. Once the tube enters the ionosphere as illustrated in Figure 16 then it will continue to become twisted and more compressed as it sinks. We have drawn a single tube in Figure 16 but tubes could form in pairs in a manner as to produce no net helicity in the totality of ropes. This seems to be implied by the data.

One problem with this mechanism is that if the newly created plasma in the magnetic barrier is as hot as the plasma in the ionosphere, it will reduce the field strength as its density increases, weaking the flux rope and the curvature force. We need a quantitative model of this process before we can claim that it is fully understood.

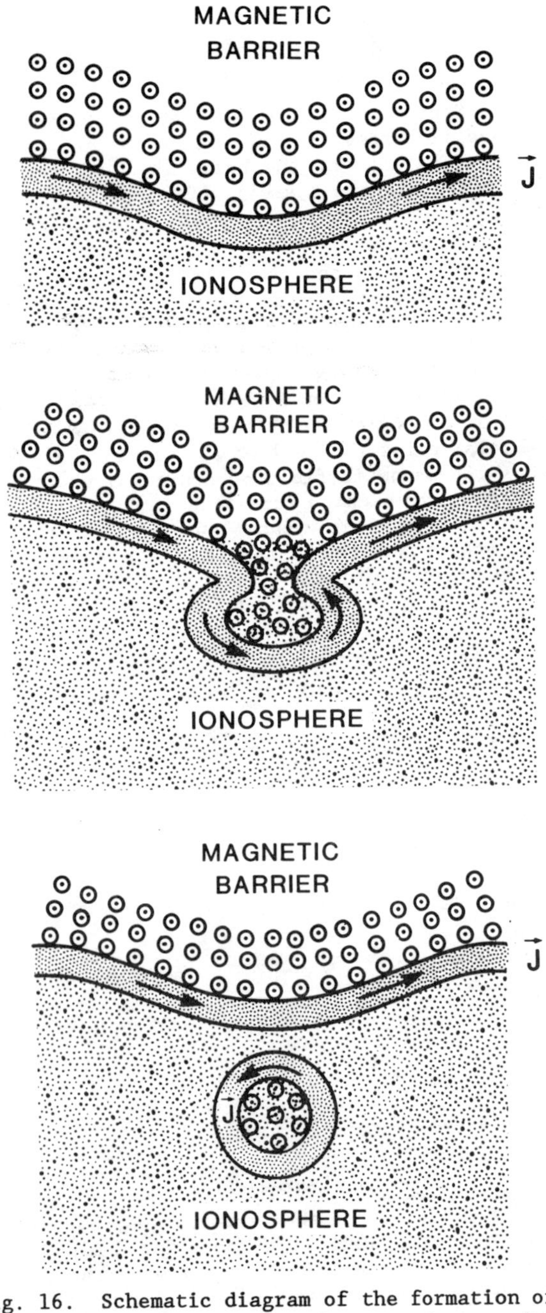

Fig. 16. Schematic diagram of the formation of a flux rope at the interface between the magnetic barrier and the ionopause.

Conclusions

Our observations clearly lead to a paradigm of Venus flux ropes in which the flux ropes are created at the base of the magnetic barrier where it is in contact with the Venus ionosphere. The data further suggest that mass loading is responsible for making these otherwise buoyant flux tubes sink. When the tubes sink in the ionosphere they become further twisted and shrunk in radius maintaining roughly the same total flux about 2-3 Webers. Below 200 km in the subsolar region these ropes appear to become kink unstable and develop into 3-dimensional helices. Below 175 km these structures reach the more resistive portions of the ionosphere and begin to dissipate.

Questions and Answers

Priest: Can you measure the pressure gradient across flux tubes and so compare with $j_\perp B$ from your models? In general, since the plasma beta is greater than of order unity, I would not have expected the flux tubes to be force-free. As an alternative to kink instability as an explanation for the tendency of flux tubes to become more vertical as they sink downwards, is it possible that, if part of a tube is inclined to the horizontal, plasma flows downwards along that part of the tube making it heavier and so more vertical, i.e., magnetic buoyancy instability?

Russell: The time-resolution of the Pioneer Venus thermal electron instrument is barely adequate to measure the electron pressure (the dominant contributor to the thermal pressure) in the larger flux ropes which occur predominantly at high altitudes. There are changes in density at high altitudes but we have not yet compared the pressure change observed with our estimates of perpendicular current. We have not considered the magnetic buoyancy instability. However, at low altitudes, the plasma is close to being in photo-chemical equilibrium with the neutrals and transport may not be very important for the neutrals.

Haerendel: Did you make a statistical analysis of occurrence probability of $j_{\perp}/j_{\parallel}$? This may give you another clue as to the regions where the twist is applied.

Russell: We did not model enough ropes to do statistical analyses of the model parameters. Instead we used parameters from our principal axis analyses that gave insight into the properties of the ropes. For example, we used the orientation of the maximum field as the axis of the ropes for those ropes we thought were crossed through the center of the rope. For twistedness we used the ratio of the intermediate to maximum eigen value.

Schindler: What is the reason for not considering resistive instabilities as a cause of flux ropes? Is the energetic shear near the ionopause coming from changes of the IMF-direction too small?

Russell: Often there is shear in the magnetic barrier caused by the varying IMF direction. However, when I examine cases in which flux ropes appear to be forming at the ionopause there is little shear except that which occurs within the rope.

van Ballegooijen: What about the possibility that the ropes are formed in pairs, with opposite handedness in the two pairs?

Russell: I think that such a mechanism would be consistent with our data. The problem is to create such a pair of helices. L. C. Lee has such a mechanism based on the Kelvin Helmholtz instability. However, I feel that the subsolar ionopause where these structures appear to be formed is not Kelvin-Helmholtz unstable.

Acknowledgments. The author wishes to thank R. C. Elphic and J. G. Luhmann with whom most of the work reported herein was performed. This work was supported by the National Aeronautics and Space Administration under research grant NAG2-501.

References

Brace, L. H., R. C. Elphic, S. A. Curtis and C. T. Russell, Wave structure in the Venus ionosphere downstream of the terminator, Geophys. Res. Lett., 10, 1116-1119, 1983.

Cravens, T. E., H. Shinagawa and A. F. Nagy, The evolution of large-scale magnetic fields in the ionosphere of Venus, Geophys. Res. Lett., 11, 267, 1984.

Elphic, R. C. and C. T. Russell, Magnetic flux ropes in the Venus ionosphere: Observations and models, J. Geophys. Res., 88, 58-72, 1983a.

Elphic, R. C. and C. T. Russell, Global characteristics of magnetic flux ropes in the Venus ionosphere, J. Geophys. Res., 88, 2993-3004, 1983b.

Elphic, R. C. and C. T. Russell, Evidence for helical kink instability in Venus magnetic flux ropes, Geophys. Res. Lett., 10, 459-462, 1983c.

Elphic, R. C., C. T. Russell, J. A. Slavin and L. H. Brace, Observations of the dayside ionopause and ionosphere of Venus, J. Geophys. Res., 85, 7679-96, 1980b.

Luhmann, J. G., "Wave" analysis of Venus ionospheric flux ropes, in Physics of Ionospheric Flux Ropes, this volume, 1989.

Luhmann, J. G., R. C. Elphic, C. T. Russell, J. D. Mihalov and J. H. Wolfe, Observations of large-scale steady magnetic fields in the dayside Venus ionosphere, Geophys. Res. Lett., 7, 917-920, 1980.

Luhmann, J. G., C. T. Russell and R. C. Elphic, Time scales for the decay of induced large-scale magnetic fields in the Venus ionosphere, J. Geophys. Res., 89, 362-368, 1984.

Luhmann, J. G., C. T. Russell, F. L. Scarf, L. H. Brace and W. C. Knudsen, Characteristics of the Mars-like limit of the Venus-solar wind interaction, J. Geophys. Res., 92, 8545-8557, 1987.

Russell, C. T. and R. C. Elphic, Observation of magnetic flux ropes in the Venus ionosphere, Nature, 279, 616-618, 1979.

Russell, C. T., R. N. Singh, J. G. Luhmann, R. C. Elphic and L. H. Brace, Waves on the subsolar ionopause of Venus, Adv. Space Res., 7(12), 115-118, 1987.

Sonnerup, B. U. O. and L. J. Cahill, Jr., Explorer 12 observations of the magnetopause current layer, J. Geophys. Res., 73, 1757, 1968.

"WAVE" ANALYSIS OF VENUS IONOSPHERIC FLUX ROPES

J. G. Luhmann

Institute of Geophysics and Planetary Physics
University of California, Los Angeles, CA 90024

Abstract. The details of the formation and evolution of filamentary magnetic structures observed in the Venus ionosphere by the Pioneer Venus Orbiter (PVO) remain subjects of debate. It is understood that magnetic flux from the overlying magnetosheath can be convected deep into the dayside ionosphere, but it is not known how this flux is intensified in localized regions or how the field is reoriented so that a nearly isotropic distribution of vectors is seen in the subsolar ionosphere. The initial analyses of these structures in terms of "flux rope" models concentrated on a special class which was designated the "small impact parameter" subset. However, the most frequently observed features do not fit into this (small impact parameter) category. To investigate the behavior of the general field fluctuations observed by the PVO magnetometer, wave analysis techniques were applied to study their compressional amplitude, ellipticity and sense of polarization. The most remarkable change in these properties occurrs within ~15° of the terminator. The ellipticity and compressional power drop markedly, implying that flux ropes either drastically change their character there or are simply confined to the dayside hemisphere within ~75° of the subsolar point. A heuristic model which seeks to unify the interpretations of the dayside flux ropes and the "terminator waves" is suggested.

Introduction

The phenomenon of ionospheric flux ropes appears to occur only under conditions of weak induced magnetization in the ionosphere of a planet without an appreciable intrinsic field. So far, Venus at solar maximum has provided the only example; it is not yet known whether the Martian ionosphere sometimes also exhibits these structures.

Geophysical Monograph 58

Although flux ropes were found in the ionosphere of Venus early in the Pioneer Venus mission [cf. Russell and Elphic, 1979], they still present a puzzle in that no satisfying picture of their origin and evolution has since emerged. Several scenarios have been proposed which invoke Kelvin-Helmholtz instabilities either in the shear layer of the ionopause [cf. Wolff et al., 1980] or within the ionosphere itself [cf. Cloutier et al., 1983]. Others have considered how a flux tube of magnetosheath origin might evolve after being entrained into the ionosphere at the subsolar ionopause by some unspecified instability [cf. Elphic and Ershkovich, 1984, Russell et al., 1987], with the aim of ordering the observed handedness or "twist" of the field lines [cf. Elphic and Russell, 1983a]. Luhmann and Elphic [1986] considered that turbulence in the ionospheric plasma could create the observed small scale structures via a kinematic dynamo action on a weak "seed" field.

Implicit in the above "models" are predictions of the global properties of flux ropes which can be used to test the various hypotheses concerning their formation. In particular, the Kelvin-Helmholtz mechanisms illustrated by Figure 1 could lead to a spatial distribution of flux ropes that has them concentrated at high solar zenith angles where the largest velocity shears are found. On the other hand, the subsolar instabilities could produce subsolar flux ropes with the characteristic distribution of handedness suggested by Figure 2 [adapted from Elphic, 1982].

To date, the only global surveys of flux rope properties that have been used to test these heuristic models are those of Elphic and Russell [1983a,b]. Their surveys included only the restricted subset of several hundreds of flux ropes which they considered were traversed by the Pioneer Venus Orbiter at small "impact parameters." They found that globally, flux rope axis inclinations were generally random and that their handedness was not well organized by the draped interplanetary field direction. The average field in the ionosphere was practically zero, a finding not consistent with either of the pictures in Figures 1 and 2. They also found that near the terminators

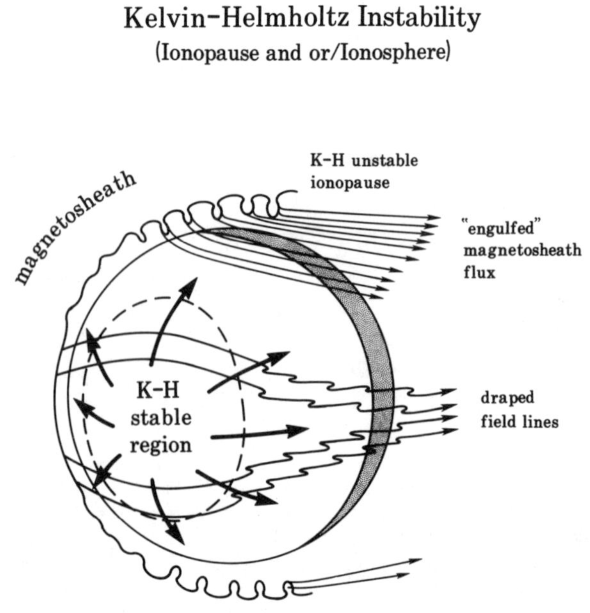

Fig. 1. Schematic illustration of the characteristics expected for ionospheric fields resulting from Kelvin-Helmholtz instabilities. The operation of the instability in the velocity shear layer of the ionopause should give rise to flux tubes "entrained" from the overlying magnetosheath at the flanks where the shear is greatest. This mechanism should work preferentially where the flow is most perpendicular to the field. Both at the ionopause and within the ionosphere there should be a stable region surrounding the stagnation point where the velocity shear is insufficient to drive the instability.

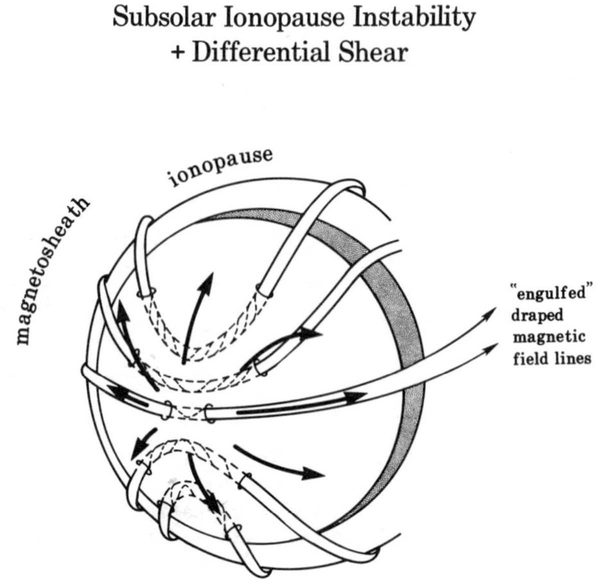

Fig. 2. Schematic illustration of the possible global ordering of the "polarity" of flux tubes created at the subsolar ionopause. The antisunward flow of the magnetosheath ends of the flux tubes "winds up" the field as the tube is pulled into the ionosphere and over the "poles."

and at the highest altitudes, the small scale field structures become less twisted and the fields are more nearly horizontal. The terminator structures were in fact given special attention by Brace et al. [1986], who found them to be unlike the dayside flux ropes in that they generally had fields that oscillated in direction and also had associated wavelike variations in the plasma density or pressure.

Since much of the small scale ionospheric field structure did not qualify for inclusion in Elphic and Russell's [1983a,b] studies, it was decided that a less selective approach should be used to carry out an investigation of the structure in its entirety. For this purpose, and based on the appearance of the ionospheric magnetic field data, Fourier analysis of the time series of the magnetic vectors was chosen. Although this approach treats the spatial distribution of the field structure as temporal structure, effectively as waves, it allows one to examine the full range of spatial field fluctuations for characteristics such as polarization (handedness), wavelength (scale size), ellipticity (circularity) and degree of compression (field line distortion versus bunching). One only needs to keep in mind that these time series analyses are really wavenumber analyses and not frequency analyses. As Elphic and Russell [1983a,b] have pointed out, the Alfvén velocity in the ionosphere is small compared to the spacecraft velocity of ~10 km/s ($V_A \approx .5$ km/s) so that the magnetic structure is practically stationary while the spacecraft passes through. The spacecraft moving through the ionosphere thus provides a spatial to temporal "transformation."

Observations

The best ionospheric data were obtained during the first two years of the Pioneer Venus Orbiter mission when the spacecraft periapsis was maintained at its lowest altitude of ~150 km. During this period the spacecraft periapsis traversed the dayside ionosphere from the dawn to the dusk terminator twice. These data are found within ~5-15 minutes of periapsis on orbits 120-250 and 350-480. Magnetic field vectors were obtained every 1/4 s, permitting an effective spatial resolution of a few kilometers for the detection of small scale field structure.

Some examples of time series of the ionospheric field from orbits in the subsolar and intermediate solar zenith angle region are shown in Figure 3a, while examples from the near-terminator ionosphere are shown in Figure 3b. The examples in Figure 3a are typical "flux rope" orbits. The time series in

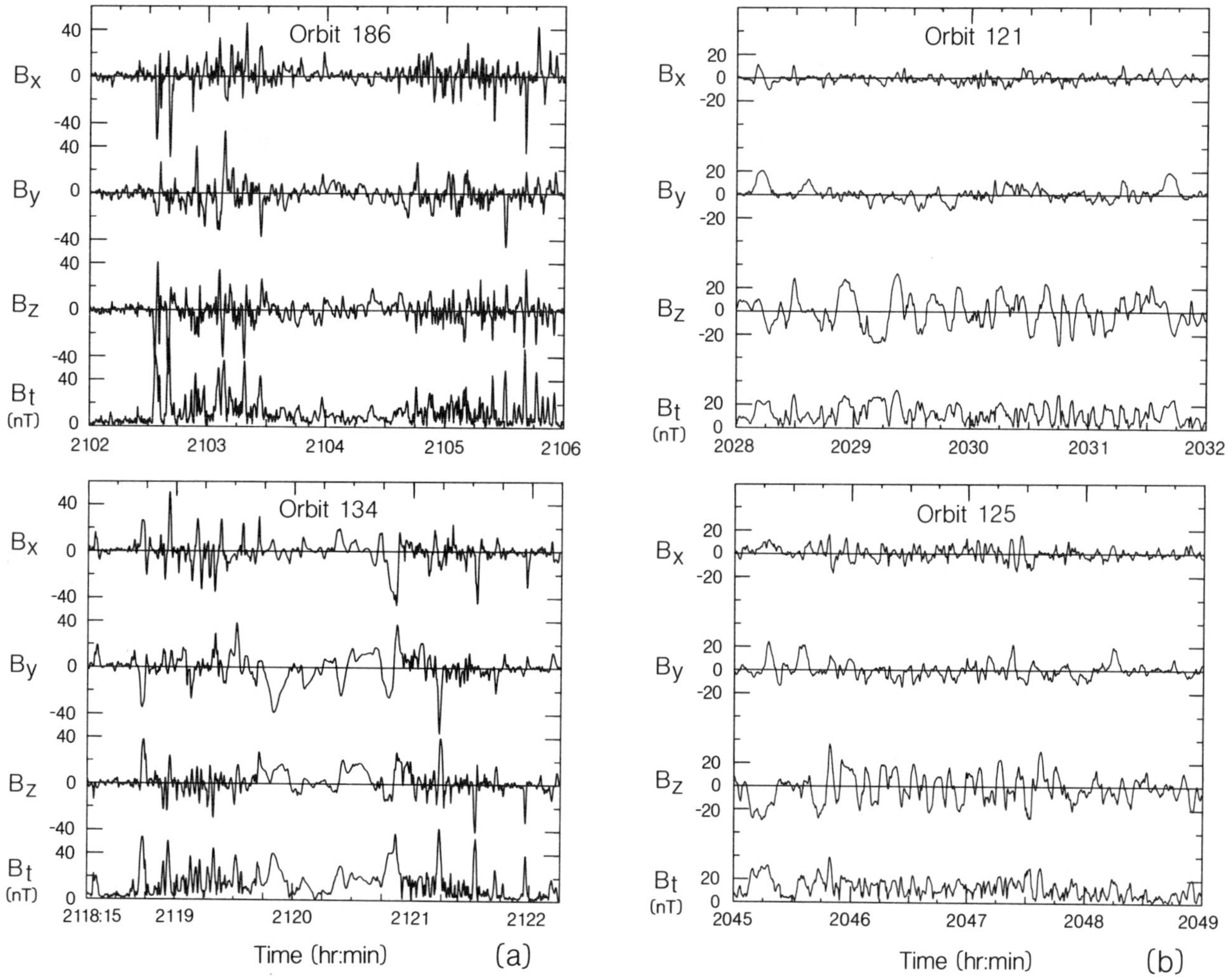

Fig. 3. Examples of times series of Pioneer Venus Orbiter magnetometer vector components obtained within the dayside ionosphere of Venus during orbits when (a) flux ropes were present and (b) terminator waves were present. The coordinates are defined such that x points toward the sun, z points north from the planetary orbit plane and y is the direction opposite planetary orbital motion. B_T is the field magnitude.

Figure 3b, on the other hand, are considered examples of "terminator waves." Note the distinctive difference in the appearance of these structures. Hodogram or minimum variance analysis of isolated flux ropes spikes often show circular traces in one plane and quasi-linear traces in the other, while terminator waves are characterized by flat traces in all principal axis projections.

Wave Analysis

To obtain information about the statistical behavior of the magnetic fluctuations in the time series, standard Fast Fourier Transforms (FFTs) of each of the vector components and the total field were calculated. The intervals analyzed were always taken between the highest altitude in the ionosphere where the small scale field structures regularly appeared and the lowest altitude where they either diminished in strength or merged with a remnant large scale field structure. Typical durations of these intervals, for both inbound and outbound legs, were ~1 minute each. Wave analyses were performed over the entire spectral range of each spectrum obtained from the FFTs: the compressional power is derived from the integrated power of the spectrum of the total field; the percent polarization is defined as

$$R = [1 - 4D/(P_{ii} + P_{jj})^2]^{1/2} \times 100$$

where P_{ii} and P_{jj} are the powers in the maximum and intermediate variance directions of the rotated time series and D is the determinant of the 2x2 matrix corresponding to the maximum and intermediate eigenvalues; the ellipticity is defined as $\epsilon = \tan \beta$ where

$$\sin (2\beta) = \frac{2Q_{ij}}{[(P_{ii} + P_{jj})^2 - 4|D|]^{1/2}}$$

where Q_{ij} is the non-zero element of the imaginary part of the spectral matrix. The ellipticity is given a sign which is positive for right-handed waves and negative for left-handed waves. A value of zero corresponds to linear polarization and unity to circular polarization. One hundred percent polarization means that the wave vector is always oscillating in the same direction. A low value of percent polarization is found if there is equal power in all 3 components of the vector time series.

Figure 4 shows power spectra obtained from a time series analysis of one leg (outbound) of an orbit during which flux ropes were observed. These spectra, one for each component of the field in the standard Venus-solar-ecliptic coordinates and one for the field magnitude, are typically undistinguished. In general, the spectra from both terminator wave and flux rope orbits are broad and featureless except for a few randomly located maxima and minima. At the low end the spectra fall off at frequencies of ~.1 Hz, representing 100 km, while at high frequencies the power drops several orders of magnitude by ~3 Hz, representing ~3 km or the resolution of the measurement. The peak power generally lies between ~.1 Hz and 1 Hz, indicating

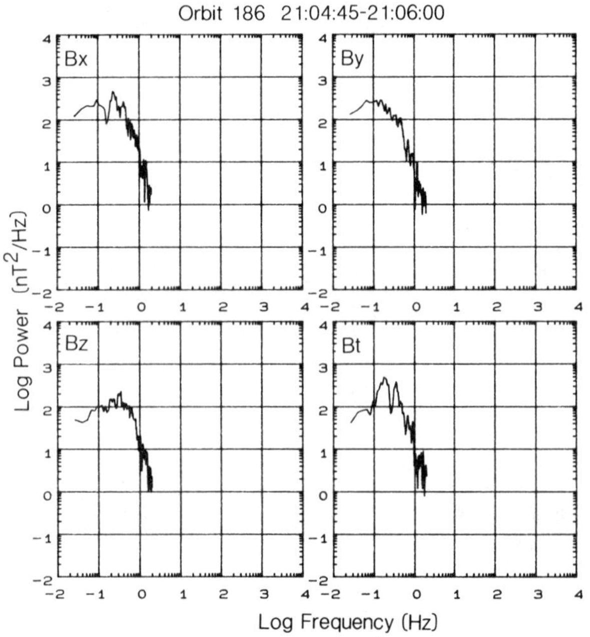

Fig. 4. Typical power spectra obtained by Fourier analysis of the inbound or outbound segments of time series such as those shown in Fig. 3.

that most of the structure is between 10 and 100 km in apparent size.

Properties of spectra such as those shown in Figure 4 and of the time series from which they are derived can be used to gain a "global" picture if they are examined for a full dayside pass of the spacecraft. Hence, various spectral parameters obtained from the two dayside passes described above have been plotted versus orbit number to illustrate the local time and/or solar zenith angle variations of a particular property of the small scale field structure. Properties obtained from inbound and outbound legs were treated as separate data points since flux ropes sometimes disappear in the few minutes around periapsis [cf. Elphic and Russell, 1983b], but since inbound and outbound data did not show any average differences in behavior, they are usually not identified separately in the plots. The exception is Figure 5, wherein a plot of the peak spectral power (from the spectra of the field magnitudes) is plotted against the ionopause altitude as identified by Phillips et al. [1984].

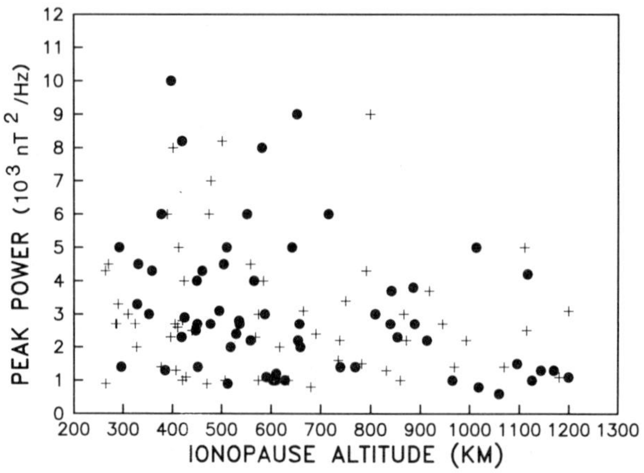

Fig. 5. The peak power in the spectra from data obtained from inbound (dots) and outbound (crosses) legs of dayside ionosphere transits versus the associated ionopause height as defined by Phillips et al. [1984].

Figure 5 examines the question of whether flux rope activity (e.g., power) is related to ionopause altitude as is the large scale ionospheric magnetic field [cf. Phillips et al., 1984]. Although it appears that for the occasions where the highest ionopauses were observed, the flux ropes were never at their highest level of activity, there was also low activity when the ionopause was at its typical lower altitude of ~300 km. Thus it seems unnecessary to include ionopause height as a discriminator in the present analysis. In fact, since flux rope activity always seems to maximize below 200 km, regardless of ionopause height, one can simply take spectra from all of the time series during the interval of greatest activity and be

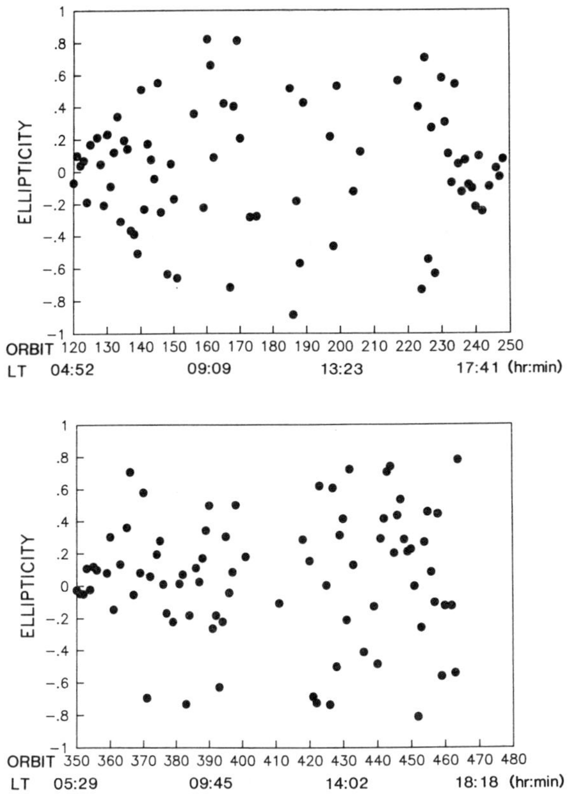

Fig. 6. Plots of ellipticity versus orbit number for the first two dayside passes. The spacecraft periapsis moves from the dawn to the dusk terminator (noon is at the center). Each point is obtained from the spectrum of an inbound or outbound leg time series.

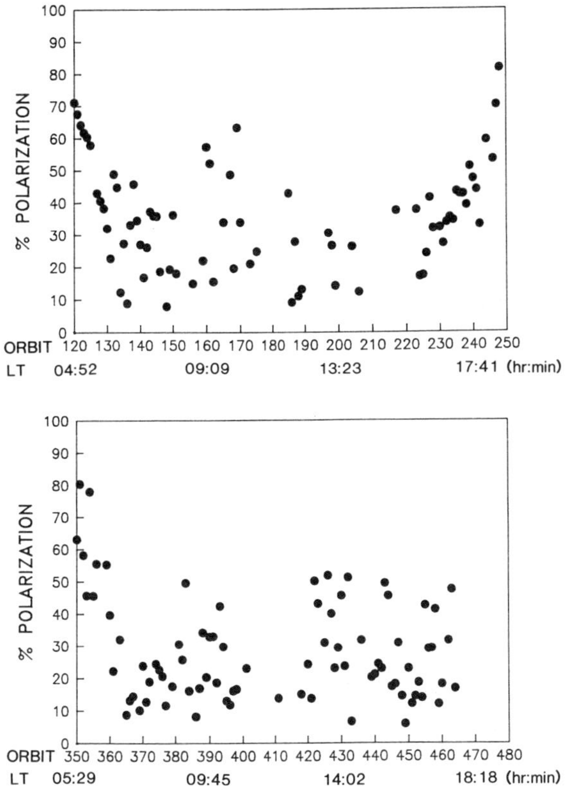

Fig. 7. Same as Fig. 6 but for percent polarization.

assured that the samples come from essentially the same altitude range.

Figures 6-8 contain the spectral parameters of most interest. The left and right-hand sides of these plots can be thought of as the dusk and dawn terminators, respectively. Figure 6 shows the progression of "ellipticity" in the magnetic field time series as periapsis moved across the dayside ionosphere. Ellipticity can be thought of here as indicating the degree of circularity in hodograms obtained from the time series. Large values of the ellipticity indicate that very circular traces are typical, while small values imply that linear traces are the rule. As mentioned earlier, a positive ellipticity means the net rotation in the maximum variance hodogram of the complete time series under consideration would be right-handed, while a negative ellipticity denotes a left-handed net rotation or twist. While it was found that, away from the terminators, the ellipticity could vary greatly depending on where the time series was started and stopped, the absolute value of the ellipticity was always small within a ~10°-20° solar zenith angle range of the terminator. This is most clearly seen in the envelope of the points from the first dayside pass (orbits 120-250). Similarly, "percent polarization," plotted in Figure 7 shows that if the spacecraft is near the terminator, the "waves" are highly linear [e.g., highly polarized) and oscillate in the same general direction, while at more subsolar locations they are more elliptically polarized with no preferred orientation of the polarization ellipse. The "compressional power" in Figure 8 adds the information that the magnitude of the magnetic field fluctuations has a fairly strong dependence on solar zenith angle, with the largest fields occurring in the subsolar region.

It is significant that the sign, as well as the magnitude, of the ellipticity is highly sensitive to the starting and stopping points of the time series. This lack of consistency suggests that there is no overriding order in the sense of twist either in a single data point or across the dayside ionosphere. Thus this study provides no observational support for a simple pattern of prevailing handedness such as that illustrated in Figure 2. Of course, one could argue that Figure 2 only describes the initial state of polarization which evolves somewhat chaotically as the flux tube moves into the ionosphere [e.g., see Elphic and Russell, 1983c].

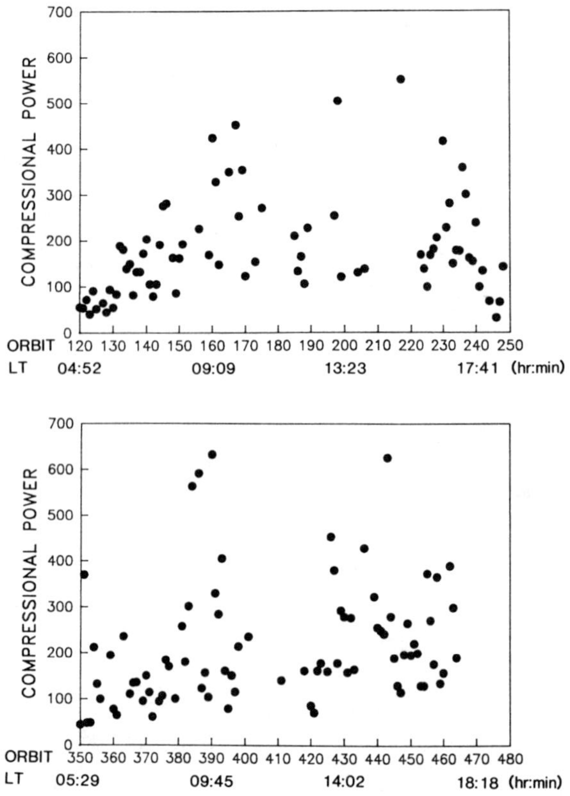

Fig. 8. Same as Fig. 6 but for compressional power.

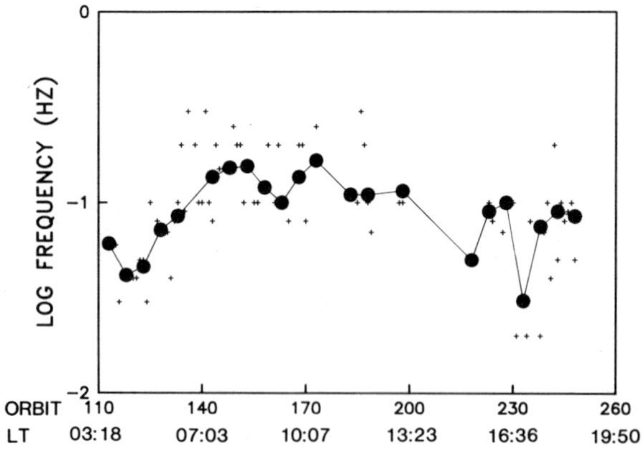

Fig. 9. Progression of the frequency at the peak power in each spectrum for the first dayside pass. The dots are median values.

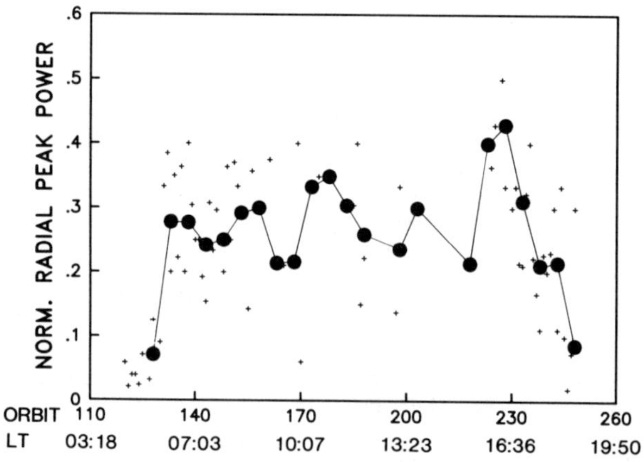

Fig. 10. Progression of the fraction of the total power in the vertical or radial direction across the first dayside pass. The dots are median values.

Figures 6-8 do establish the existence of a fairly sudden change in the field structure near the terminator. To further investigate this change, spectra from the better-behaved first dayside pass were examined for the frequency at the maximum power and for the fraction of the power in the vertical (radial) direction. These data and their medians are displayed in Figures 9 and 10. The maximum power frequency in Figure 9 is seen to be slightly smaller near the terminators, suggesting an increase in average "wavelength" or scale size of the structure there to ~100-200 km from tens of kilometers in the subsolar region. For Figure 10, the time series were first rotated into radial-east-north coordinates, after which spectral analysis gave the fraction of the total power in the radial component. This display emphasizes the sudden decrease in the radial power that accompanies the appearance of the terminator waves. The field vectors are practically randomly oriented in flux rope orbits (all components contain about the same power spectral density) but nearly horizontal in terminator wave orbits. It is also notable that the horizontal terminator wave fields generally point alternately toward and away from the sun (east-west) as if they are in some way connected with the magnetosheath field draping geometry.

Discussion

One possible scenario for the formation of the terminator waves is suggested by their east-west field oscillations (see Figure 3b). This behavior is most easily produced if a wavy current sheet is present in the ionosphere as illustrated in Figure 11. If some large scale interplanetary field penetrates the ionosphere as it is known to do [e.g., see the tutorial by Luhmann in this volume], and then the interplanetary magnetic field orientation reverses, layers of different field orientation will be present. Any disturbances of the constant plasma pressure surfaces, such as the gravity waves that are observed in the terminator atmosphere by Kasprzak et al. [1988], can corrugate the resulting ionospheric current sheet and cause the appearance of waves in the draped field along the spacecraft trajectory.

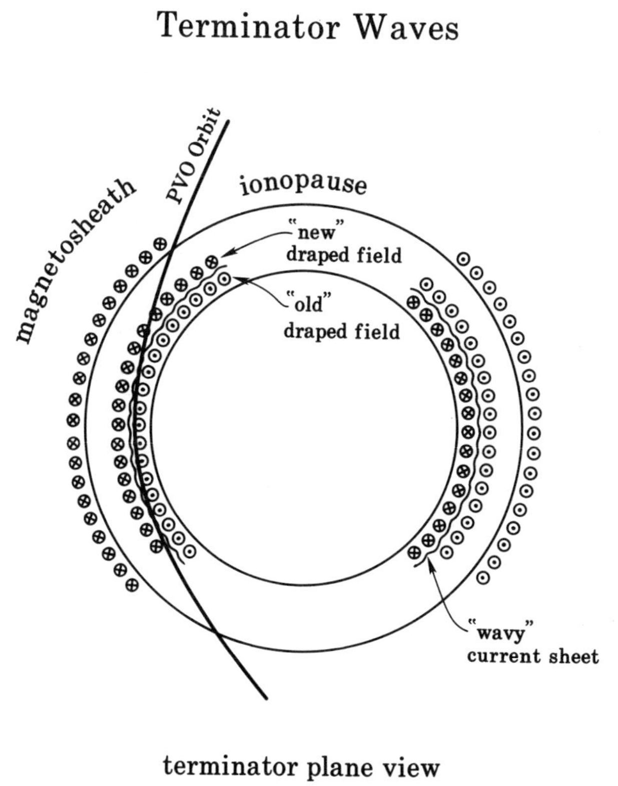

Fig. 11. Schematic suggesting a possible interpretation of terminator waves in terms of passage of the spacecraft through corrugated current sheets in the terminator ionosphere.

A consequence of the above picture that relates to the more subsolar fields is that current sheets are present in a highly dissipative medium. This means that non-explosive merging of the draped fields in the dayside ionosphere can occur as illustrated in Figure 12a. The result may be the detachment of subsolar ionospheric flux tubes from the terminator region where ionospheric conductivity is low and the distortion and eventual dissipation of these detached fields as illustrated in Figure 12b. Of course this model is highly speculative and difficult to prove, but it does have an attractive aspect in that it attempts to explain the observed features of both terminator waves and flux ropes in concert. One can also work in some of the other mechanisms mentioned earlier, like the "dynamo" mechanism of Luhmann and Elphic [1986] as a means of distorting the fields in the detached ionospheric flux tubes. However, they could also twist up purely by virtue of the nature of the dissipative merging process.

Concluding Remarks

While the study described here has not been conclusive, it does bring out some features of the small scale structure of the dayside ionospheric

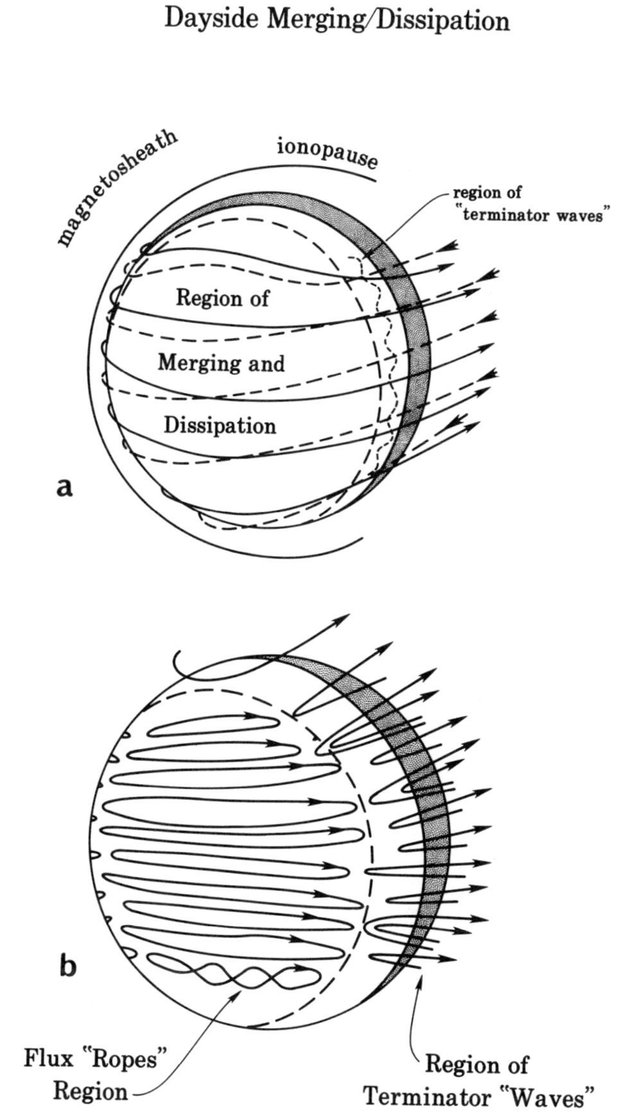

Fig. 12. (a) Schematic suggesting a possible extension of the terminator wave concept in Fig. 11. The current sheets extend into the subsolar ionosphere where the opposing fields are subject to merging processes, leading to the formation of transient flux rope structures. (b) A suggestion of how magnetic merging within the ionosphere can lead to a change in the character of the ionospheric field near the terminator where conductivity drops.

fields of Venus which were not obvious from the earlier analyses. In particular, it shows how rapidly the transformation between fluctuations with properties more like those of the "classic" flux ropes and terminator waves occurs. It also reinforces the earlier finding that the polarity or sense of twist does not appear to be organized in any particular global pattern. Future theories of flux rope formation will need to account for these properties. Even if the flux ropes and terminator waves are in fact independent entities, one will need to explain both why the fields in flux rope orbits are randomly oriented and why the field structure either disappears or drastically changes character within ~15° of the terminator.

Acknowledgement. This work was supported by NASA Grant NAG 2-501 from the Solar System Exploration Division. The author benefitted from discussions with C. T. Russell.

References

Brace, L. H., R. C. Elphic, S. A. Curtis, and C. T. Russell, Wave structure in the Venus ionosphere downstream of the terminator, Geophys. Res. Lett., 10, 1116, 1983.

Cloutier, P. A., T. F. Tascione, R. E. Daniell, Jr., H. A. Taylor, Jr. and R. S. Wolff, Physics of the interaction of the solar wind with the ionosphere of Venus: Flow/field models, in Venus, eds. D. M. Hunten, L. Colin, T. M. Donahue and V. I. Moroz, University of Arizona Press, p. 941, 1983.

Elphic, R. C., A study of magnetic flux ropes in the Venus ionosphere, Ph.D. Thesis, University of California at Los Angeles, 1982.

Elphic, R. C. and C. T. Russell, Magnetic flux ropes in the Venus ionosphere: Observations and models, J. Geophys. Res., 88, 58, 1983a.

Elphic, R. C. and C. T. Russell,Global characteristics of magnetic flux ropes in the Venus ionosphere, J. Geophys. Res., 88, 2993, 1983b.

Elphic, R. C. and C. T. Russell, Evidence for helical kink instability in the Venus magnetic flux ropes, Geophys. Res. Lett., 10, 459, 1983c.

Elphic, R. C. and A. I. Ershkovich, On the stability of the ionopause of Venus, J. Geophys. Res., 89, 997, 1984.

Luhmann, J. G. and R. C. Elphic, On the dynamo generation of flux ropes in the Venus ionosphere, J. Geophys. Res., 90, 12047, 1985.

Phillips, J. L., J. G. Luhmann and C. T. Russell, Growth and maintenance of large scale magnetic fields in the dayside Venus ionosphere, J. Geophys. Res., 89, 10676, 1979.

Russell, C. T. and R. C. Elphic, Observations of flux ropes in the Venus ionosphere, Nature, 279, 616, 1979.

Russell, C. T., R. N. Singh, J. G. Luhmann, R. C. Elphic, and L. H. Brace, Waves on the subsolar ionopause of Venus, Adv. Space Res., 7, 115, 1987.

Wolff, R. S., B. E. Goldstein and C. M. Yeates, The onset and development of Kelvin-Helmholtz instability at the Venus ionopause, J. Geophys. Res., 85, 7697, 1980.

THE MODEL OF THE VELOCITY SHEAR INSTABILITIES AT VENUSIAN IONOPAUSE AND THE PROBLEM OF MAGNETIC FLUX ROPES FORMATION

E. V. Belova and L. M. Zelenyi

Space Research Institute, Academy of Sciences, USSR, Moscow 117810

Abstract. Shear velocity instability in the Venusian ionopause is studied as a possible mechanism of the formation of magnetic flux ropes . All characteristic spatial scales related to processes in ionopause, such as its thickness, ion Larmor radius and magnetic flux ropes diameters are of comparable values. It complicates analysis of the instability. In addition the charge exchange and photoionization effects, density and magnetic field gradients may also be important. Here we try to study the linear theory of instability taking into account all this effects. We find that for the realistic ionopause conditions the growth of small scale velocity shear driven instability can give the mechanism of formation of magnetic flux ropes with typical scales $d \sim 10$–20 km.

Introduction

The instabilities of the ionopause caused by the velocity shear between the solar wind flowing around the Venus and the quiet plasma within the ionosphere (Kelvin–Helmholz instability in particular) have been widely discussed before as a possible mechanism of the destruction of the regular ionopause structure and formation of magnetic flux ropes (MFR) within the Venusian ionosphere [Wolff et al., 1980; Cloutier et al., 1981]. A lot of important simplifications have been made including the assumption about the infinitely small thickness of the ionopause, incompressibility of the plasma, zero Larmor radius of ions. However, not all of them are justified according to observations: all characteristic spatial scales, related with the ionopause processes – its thickness L, ion Larmor radius ρ_i, diameters of MFR d , have the comparable values. Moreover, charge exchange and photoionization effects not included before in stability analysis can also influence the development of ionopause instability. We consider here the realistic model of development of the velocity shear driven mode and take into account all the factors mentioned above.

We consider here the small scale local mode of instability with $\lambda \sim \rho_i < L$. The large scale mode, although possible will have the typical wavelengths of the order of few hundreds kilometers ($\lambda \geq 4\pi L \sim 300$–$600$ km) and it is reasonable to discuss it not in relation with such small structures as MFR ($d \approx 10$–30 km), but for global phenomena like the magnetic belt.

Local Instability of Solar Wind Stream Flowing Around the Ionopause

We consider the slab geometry of interaction where the stream with sheared velocity profile $V_0(x)$ is moving along the magnetic field $\mathbf{B}_0 = (0, 0, B_0(x))$. The corresponding density profile $n_0(x)$ has the characteristic thickness L and satisfies the pressure balance condition. The value of L corresponds to the observed thickness of the ionopause $L \sim 100$ km $> \rho_i$, where ρ_i is the ion Larmor radius. Such configuration is unstable and we will study here the growth of short wavelength modes with $\lambda < L$. Gravity appears to be a negligible effect for the wavelengths being considered here ($k\rho_i \sim 1$). Only if $k\rho_i < \sqrt{L/R_V} \ll 1$ becomes growth rate of an interchange mode comparable to that of the shear velocity instability (R_V is Venus' radius). We start with the study of the influence of the finiteness of the value of $k\rho_i$, where $k = 2\pi/\lambda$ is the wave vector of perturbation. The mass–loading effects will be neglected in this section, but we take them into account later.

Drift instabilities caused by velocity shear in an inhomogeneous plasma have been considered by a number of authors [Dobrovolny, 1972; Mikhailovskii, 1982; Mikhailovskii and Klimenko, 1980] but mainly in the limit of small Larmor radius $z_i = (k_\perp \rho_i)^2/2 \ll 1$. The existence of two unstable modes – magnetosonic and Alfven ones have been found in these papers. Local dispersion equations for the slow magnetosonic (SMS) and Alfven (A) modes have respectively the following forms:

$$(1 + \beta/2)(\omega' - \omega_{ne})(\omega' + \Omega_i) - k_z^2 \frac{T_i + T_e}{m_i} -$$

$$-\alpha_e k_z^2 \frac{T_e}{m_e}\left(1 - \frac{\beta}{2}\frac{T_i}{T_e}\right) = 0 \qquad (1)$$

$$\omega'^2 - \omega'(\omega_{ni} + \Omega_i) - k_z^2 v_A^2 + \alpha_i k_z^2 \frac{T_i}{m_i} = 0. \qquad (2)$$

Geophysical Monograph 58

We assume here that $\omega \ll k_z v_{Te}$, $k_\perp \gg k_z$, $\omega' = \omega - k_z V_0$, $\alpha_j = k_\perp / k_z (dV_0/dx)\omega_{Bj}^{-1}$; β – ratio of thermal and magnetic pressure; ω_{nj}– diamagnetic drift frequency; $\Omega_j = -\beta\omega_{nj}/2$ – the frequency of magnetic drift , provided $T_{i,e}(x)$ =const ; ω_{Bj} – cyclotron frequency; v_A – Alfven velocity.

After solving dispersion equations (1) and (2), and maximizing the expressions for the growth rates of instability γ over k_z value we find in the limit $z_i \ll 1$:

$$\max_{k_z} \gamma^2_{SMS} = z_i \left(\frac{dV_0}{dx}\right)^2 \frac{\left(1 - \frac{\beta}{2}\frac{T_i}{T_e}\right)^2 \frac{T_e}{T_i}}{4(1+\beta/2)(1+T_i/T_e)} - \frac{(\omega_{ne} + \Omega_i)^2}{4} \quad (3)$$

$$\max_{k_z} \gamma^2_{A} = z_i \frac{\beta}{8}\left(\frac{dV_0}{dx}\right)^2 - \frac{(\omega_{ni} - \Omega_i)^2}{4} \quad (4)$$

Here the effects of density, velocity and magnetic field gradients are taken into consideration, but for simplicity we neglect the gradients of temperature. One can see from Eqs. (3) and (4) that the density gradient as well as magnetic field gradient (term proportional to Ω_i) have the stabilizing influence on instability growth.

The threshold conditions for the instability of SMS and A modes can be easily obtained from Eqs. (3) and (4), provided that velocity and density profiles have the same characteristic thickness — L:

$$V_0 > \sqrt{\frac{1+\beta}{2}}\, c_s \frac{1 + \frac{\beta}{2}\frac{T_i}{T_e}}{1 - \frac{\beta}{2}\frac{T_i}{T_e}}, \qquad c_s = \sqrt{\frac{T_i + T_e}{m_i}} \quad (5)$$

$$V_0 > 2v_A(1 + \beta/2) \quad (6)$$

As the drift frequency depends on z_i : $\omega^2_{nj} \sim z_i$, the value of $\max \gamma^2$ grows linearly with the increase of z_i and to find the limit of such growth one should consider the opposite approximation $z_i \gg 1$.

The dispersion equation for arbitrary z_i could be obtained from the standard expressions for the dielectric permeability tensor of inhomogeneous plasma [Mikhailovskii, 1977]. With the only assumption $\beta \ll 1$, we have:

$$\varepsilon_{00} = \frac{1}{(kd_e)^2}\left[1 + \frac{T_e}{T_i} - I_0 exp(-z_i)\left(1 - \frac{\omega_{ni}}{\omega} - \frac{A_i}{\omega^2}\right)\frac{T_e}{T_i}\right]$$

$$\varepsilon_{03} = \varepsilon_{30} = \frac{1}{k_z k d_e^2}\left[1 - \frac{\omega_{ne}}{\omega} + I_0 exp(-z_i)\frac{A_i}{\omega^2}\frac{T_e}{T_i}\right]$$

$$\varepsilon_{33} = \frac{1}{(k_z d_e)^2}\left[1 - \frac{\omega_{ne}}{\omega} - \frac{A_e}{\omega^2}\right],$$

where d_e – electron Debye length; $A_j = \alpha_j k_z^2 T_j / m_j$; $I_0 = I_0(z_i)$ – is the modified Bessel function of the zero order.

Dispersion equation has the form:

$$\varepsilon_{00}(\varepsilon_{33} - k^2c^2/\omega^2) - \varepsilon_{03}\varepsilon_{30} = 0. \quad (7)$$

The calculations in this case appears to be lengthy but straightforward and their results are shown on Figure 1. To understand the physics of unstable modes better we discuss the simple case $\nabla n_0 = 0$ in more detail. The solution of Eq. (7) in such a case is:

$$\gamma^2_{SMS,A} = -G \mp \sqrt{G^2 + (1-\Delta)A_e\left(A_i - \frac{z_i k_z^2 v_A^2}{\Delta}\right)}, \quad (8)$$

where

$$G = \frac{1}{2}\left(A_e - A_i + \frac{z_i k_z^2 v_A^2}{\Delta}\left(1 + \Delta\frac{T_e}{T_i}\right)\right),$$

$$\Delta = 1 - I_0 exp(-z_i).$$

If $z_i \ll 1$ in Eq.(8), the solution with minus before square root corresponds to the SMS mode (3), with plus – to the A mode (4). The solutions of Eq.(8) are shown on Figure 1a, 1b

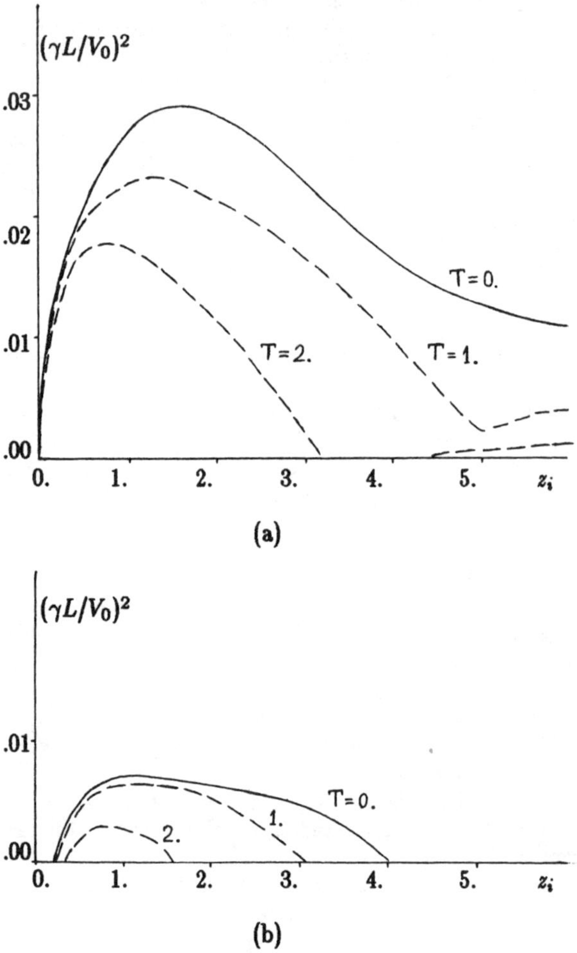

Fig. 1. The dependence of the square of normalized growth rate $(\gamma L/V_0)^2$ from z_i for different $T = |d(\ln n_0)/d(\ln V_0)| = 0; 1; 2$ for slow magnetosonic mode – (a) and for Alfven mode – (b).

by solid lines (case with $\nabla n_0 = 0$) as the dependencies of the growth rate on parameter z_i. The numerical values entering the Eq.(8) are $T_e/T_i = 1$, $\beta = 0.4$, $V_0/v_A = 4.$, $k_z L = 0.5$. Figure 1a shows the results for SMS mode, Figure 1b for the A mode . One can see that the growth rates for both modes reach their maximums at $z_i \approx 1$. For large z_i, A mode becomes stable and the growth rate of SMS mode approaches to the constant value independent on z_i.

Numerical results obtained by solving dispersion equation (7) for the case $\nabla n_0 \neq 0$ are shown on Figure 1a,b by the dashed lines. In Figure 1 the value of $T = |d(\ln n_0)/d(\ln V_0)| = L_V/L_n$ determines the ratio of the characteristic thickness of velocity profile $L_V = (d \ln V_0/dx)^{-1}$ to that of density profile $L_n = (d \ln n_0/dx)^{-1}$. We do not assume the specific value of this parameter here, assuming only that at the ionopause it is of the order of unity and make our calculations for $T = 0, 1, 2$, to understand the respective roles of L_n and L_V. Both the numerical and analytical results (from Eqs. (3) and (4)) confirm that the density gradient has stabilizing influence on A as well on SMS mode.

So, in local approximation when the thickness of ionopause is assumed to be large in comparison with λ and ρ_i the most unstable modes have the wavelengths $\lambda \approx 2\pi\rho_i$, scaled with the ion Larmor radius. Two types of modes can be exited in the system : slow magnetosonic and Alfvenic, both propagating at the large angle with respect to magnetic field ($k_z \ll k_\perp$). The magnetosonic mode is of primary importance for the ionopause processes because of larger growth rate and lower threshold of excitation (see Eqs. (5) and (6)).

The Influence of the Mass–Loading of Solar Wind Flow by Ionospheric Ions

The presence of extended neutral atmosphere surrounding Venus can significantly influence the dynamics of solar wind interaction with planetary ionosphere via the processes of photoionization of neutral atoms of planetary origin and charge exchange of solar wind ions with them.

The major part in mass–loading processes at Venus plays the hot oxygen corona [Nagy et al., 1987, Gombosi et al., 1981], so we will take into account below only photoionization and charge exchange with oxygen atoms. Also we will assume in calculations below that at the ionopause heights protons of solar wind are already completely replaced by ions of planetary origin (O^+), so that the plasma stream with sheared velocity profile consists from oxygen ions with density $n_0(x)$.

As will be shown below effective loading frequency is $\nu_{load} = \nu_{ph} N_n/n_0 + \nu$, ν_{ph} and ν here are the frequencies of photoionization and charge exchange respectively; ν_{ph} for Venus is $3 \cdot 10^{-7} s^{-1}$, N_n – is the density of neutral component ; $\nu = \sigma N_n V_0$, where $\sigma = 3 \cdot 10^{-15}$ cm^2 — is the cross–section of the charge exchange of O^+ with O. After the numerical estimate of ν_{load} it appears that it can be of the same order as the instabilities growth rates (3) and (4), so for proper study of the ionopause stability problem the effects related with the charge exchange and "birth" of new photoions should be taken into account. We will do that including the model "collisional" operator of BGK type into kinetic equation.

We will consider two groups of ions: oxygen ions appearing due to photoionization and charge exchange with average velocity $< \mathbf{v} >= 0$, and streaming ions with $< \mathbf{v} >= V_0$ ($V_0 \gg v_{Ti}$) that does not lose yet their velocities after the charge exchange process. To take the charge exchange effects into account is essential only for the "moving" ions because the momentum of the stream will be lost in this case. So three particle species should be considered: ions O^+ moving with the stream, newly born O^+ ions and electrons, and the charge exchange effects will be taken into account only for the first of them. Let the f_i^m, f_i^T, f_e — be the corresponding distribution functions. After that we have the following "collisional" terms for the considered processes

$$\frac{df_i^m}{dt} = -\nu f_i^m, \tag{9}$$

$$\frac{df_i^T}{dt} = \frac{1}{\sqrt{2\pi}}\left(\frac{m_i}{T_i}\right)^{3/2}(\nu_{ph}N_n + \nu n_0)\exp\left(-\frac{m_i v^2}{2T_i}\right), \tag{10}$$

$$\frac{df_e}{dt} = \frac{1}{\sqrt{2\pi}}\left(\frac{m_e}{T_e}\right)^{3/2}\nu_{ph}N_n\exp\left(-\frac{m_e v^2}{2T_e}\right), \tag{11}$$

Eq. (9) describes the disappearance of "moving" ions due to charge exchange. Eq. (10) – birth of photoions due to photoionization and charge exchanged thermal ions, Eq. (11) – birth of photoelectrons after photoionization. It was assumed for simplicity that the particles appearing due to charge exchange and photoionization effects have the Maxwellian velocity distribution with the equal temperatures of the newly born ions and ions of the stream.

The linearized kinetic equations for the perturbed distribution functions with the collisional terms given by Eqs. (9)–(11) acquire the forms

$$\begin{aligned}\frac{\partial f_1^\alpha}{\partial t} + \mathbf{v}\frac{\partial f_1^\alpha}{\partial \mathbf{r}} + [\mathbf{v}\times\omega_{B\alpha}]\frac{\partial f_1^\alpha}{\partial \mathbf{v}} = \\ = -\frac{e_\alpha}{m_\alpha}\left(\mathbf{E} + \frac{\mathbf{v}\times\mathbf{B}}{c}\right)\frac{\partial f_0^\alpha}{\partial \mathbf{v}} + I^\alpha,\end{aligned} \tag{12}$$

where I^α – collisonal terms for the particle of the α–species obtained after the linearization of Eqs. (9)–(11) :

$$I_i^m = -\nu f_{1i}^m; \quad I_e = 0;$$

$$I_i^T = \frac{1}{\sqrt{2\pi}}\left(\frac{m_i}{T_i}\right)^{3/2}\nu n_{1m}\exp\left(-\frac{m_i v^2}{2T_i}\right)$$

Here $n_{1m} = \int f_{1i}^m \, d\mathbf{v}$ is the perturbed density of "moving" ions, $\mathbf{E}$ and $\mathbf{B}$ are perturbed electric and magnetic fields.

The solution of Eq. (12) can be obtained after the usual integration over the unperturbed particle trajectories. In the previous section we have considered the case of $\nabla n_0 \neq 0$ for arbitrary z_i to clarify the influence of drift effects on mode growth. Here we concentrate on mass–loading effects and consider the more simple case with $\nabla n_0 = 0$ subjected to analytical investigation. The dispersion equations in this limit with the inclusion of the mass–loading terms are given by:

$$(\omega'^2+\gamma_{\rm SMS}^2)(\omega'^2+\gamma_{\rm A}^2) =$$

$$= -i\frac{\nu}{\omega} I_0 exp(-z_i)\left[\omega' k_z V_0 - A_i + \frac{z_i k_z^2 v_A^2}{\Delta}\right]\omega'^2, \qquad (13)$$

$$(\omega'^2+\gamma_{\rm SMS}^2)(\omega'^2+\gamma_{\rm A}^2) = -i\nu_{\rm ph}\frac{N_n}{n_0}\omega'^3, \qquad (14)$$

respectively for charge exchange and photoionization effects; the values of $\gamma_{\rm SMS,A}$ are determined by Eq. (8).

As one can see from Eqs. (13) and (14) the SMS and A modes are coupled already due to the inclusion of mass–loading effects. As we will see after the numerical estimates the charge exchange plays a main role in mass–loading influence on the mode growth. Figure 2 shows the dependencies of dimensionless γ^2 on z_i, obtained by solving Eq. (13) for different values of the loading term $\nu_0 = \nu L/V_0$: $\nu_0 = 0.0$ (solid lines), $\nu_0 = 0.05$ and 0.1 (dashed lines). Figure 2a refers to SMS mode, Figure 2b to the A mode. The value of $\gamma_{\rm SMS}$ is usually larger then $\gamma_{\rm A}$ and mass–loading effects as one can see from Figure 2 slightly increase the growth rate of slower growing A mode and has some stabilizing influence on the SMS mode which we consider here to be of primary importance for the flux rope formation problem.

Let us make now some quantitative estimates using the results of "Pioneer Venus Orbiter" observations [Elphic et al., 1980]. Parameters of plasma in the near ionopause region depend appreciably on the solar wind dynamic pressure ($p_{\rm SW}$) and are varied with the zenith angle θ.

For the average values of $p_{\rm SW}$ typical parameters are: $B_0 \sim$ 70 nT, $n_0 \sim 10^3{\rm cm}^{-3}$, $L \approx 50$ km, $T_i \sim 1$ eV, so we have $v_A \approx$ 15 km/s, $\omega_{Bi} \approx 0.4{\rm s}^{-1}$, $\rho_i \approx 10$ km.

The value of tangential velocity component V_0 depends on θ. It is small near the subsolar point $V_0 < 10$–30 km/s, but it grows with increase of θ and reaches $V_0 \approx 100$ km/s near the terminator. The instability thresholds (5) and (6) require V_0 >10km/s for the development of SMS mode and $V_0 >$ 30km/s for the development of A mode. So both modes can in principle be unstable at the ionopause with the possible exception of the small vicinity of the subsolar point. The dominant mode with $k\rho_i \sim 1$ has wavelength $\lambda \sim 2\pi\rho_i \approx 60$ km.

Assuming $V_0 \sim 30$ km/s we find from Eqs. (3) and(4): $\gamma_{\rm SMS} \approx 0.2{\rm s}^{-1}$, $\gamma_{\rm A} \approx 0.1{\rm s}^{-1}$. If $N_n \sim 10^6{\rm cm}^{-3}$ the effective frequencies of charge exchange and photoionization correspondingly are $\nu \approx 10^{-2}{\rm s}^{-1}$, $\nu_{\rm ph}N_n/n_0 \approx 3\cdot10^{-4}{\rm s}^{-1}$. So $\gamma_{\rm SMS,A} \gg \nu$, $\nu_{ph}N_n/n_0$ and there is at least one order of magnitude margin to neglect by the mass–loading influence of instability growth for the small and moderate solar wind pressures $p_{\rm SW}$.

With the increase of $p_{\rm SW}$ the amplitude of the magnetic field inside the magnetic barrier increases and the height of the ionopause lowers down [Phillips et al., 1985]. The density of neutrals is appreciably higher there and the influence of mass–loading becomes significant. For $p_{\rm SW} > 3\cdot10^{-8}{\rm din}\cdot{\rm cm}^{-2}$ — $H_{\rm ionop} \approx 250$ km, $B \sim 140$ nT, $N_n \approx 5\cdot10^6{\rm cm}^{-3}$. The velocity threshold (6) at this height is V_0 >60 km/s and mass–loading (charge exchange) appreciably influence already the development of instability: $\nu = 0.1{\rm s}^{-1}$, $\nu_{\rm ph}N_n/n_0 = 0.001{\rm s}^{-1}$. The growth rate of SMS mode diminishes, of A mode increases (especially for $z_i < 1$). So with the mass–loading effects taken into account we have for the lowest ionopause positions $\gamma_{\rm SMS} \approx \gamma_{\rm A} \approx 0.15{\rm s}^{-1}$. In any case even for the worst condition mass–loading effects does not stabilize completely the velocity shear driven modes at the Venus ionopause.

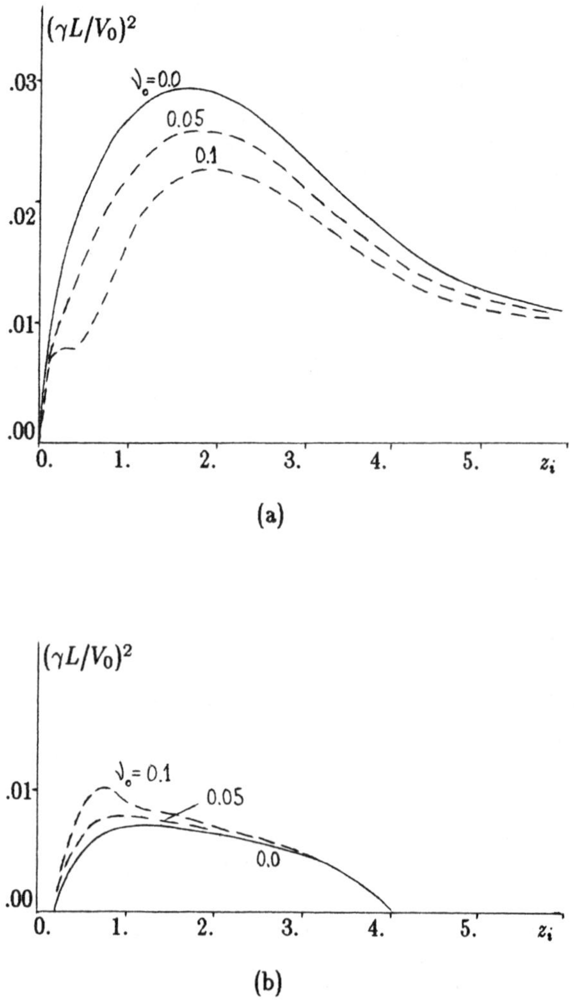

Fig. 2. The dependence of the square of normalized growth rate $(\gamma L/V_0)^2$ from z_i for different values of normalized charge exchange frequency $\nu_0 = \nu L/V_0 = 0$; 0.05; 0.1 for slow magnetosonic mode – (a) and for Alfven mode – (b).

Summary and Conclusions

The growth of the velocity shear driven modes for realistic conditions at Venus ionopause have been studied in this paper in local approximation. Such approach can be applied to the magnetic field aligned plasma streams formed due to the squeezing of the plasma along the magnetic flux tubes from the regions with high plasma pressure near the subsolar point.

We find that slow and Alfven modes can be destabilized by the velocity shear. The perturbations propagate almost perpendicularly to the ambient magnetic field ($k_z \ll k_\perp$) and have the characteristic wavelength $\lambda \approx 2\pi\rho_i \approx 60$ km. Slow MHD mode have the lower threshold and larger growth rate than the Alfven mode. Mass–loading effects have been taken into consideration by means of model "collisional" operator (similar to BGK integral) included into the kinetic equation. This results in coupling of Alfven and slow modes, and leads to the decrease of the growth rate for the slow mode and increase for the Alfven mode. Numerical estimate shows that for lower and moderate values of solar wind pressure p_{SW} the loading effects can be disregarded ($\nu_{load} \ll \gamma$ where growth rate $\gamma \approx 0.1 \div 0.2s^{-1}$). For large p_{SW} when the ionopause lowers to smaller heights with large density of neutral particles, charge exchange influence becomes important ($\nu_{load} \approx 0.1s^{-1}$). However even in this case mode will not be completely stabilized and remains unstable up to $H_{ionop} \approx 250$ km .

The theoretical value of λ for the local mode conforms rather well with the observed MFR diameters. This enables to consider this small scale instability as a possible mechanism of MFR formation. The slow mode development accompanied by the growth of field–aligned component of the perturbed magnetic field seems to be the more probable candidate for supporting filamentary MFR structure. Instability growth can maintain also the field line twisting because of the appreciable azimuthal component k_y of the wave vector the perturbations possess. The most favorable conditions for the instability growth are accomplished in some distance from the subsolar point. This conform well with the occurrence frequency pattern of MFR observations over the ionopause [Elphic and Russell, 1983].

The model discussed here differs from that studied by Wolff et al. [1980] who considered the ionopause as a tangential discontinuity ($L = 0$). Our point is that for understanding of such small structures as MFR with $\lambda < L$ you should get "inside" the ionopause and consider the problem not in global but in local approximation. Cloutier et al. [1981] studied the shear velocity instability deep inside the ionosphere itself as being a source of flux ropes. We have serious doubts about excitation of this instability inside the ionosphere. Even if flow velocity will be sufficient (really it is small: $V_0 \sim 1$km/s), mass–loading effects will suppress the instability growth. In any case the model proposed here meets serious problems when one analyses the nonlinear stage. It is necessary to explain how such regular (coherent) structures like MFR can be created by small scale turbulence developed inside relatively thick ionopause . The mechanism of MFR transport from the ionopause to the ionospheric heights 160-200 km (where their observational occurrence frequency has a peak) should also be clarified.

References

Cloutier, P. A., et al., An electrodynamic model of electric currents and magnetic fields in the dayside ionopause of Venus, *Planet. Space Sci.*, *28*, 635–652, 1981.

Dobrowolny, M., Kelvin–Helmholtz instability in a high–β collisionless plasma, *Phys. Fluids*, *15*, 2263–2270, 1972.

Elphic, R. C., and C. T. Russell, Global characteristics of magnetic flux ropes in the Venus ionosphere, *J. Geophys. Res.*, *88*, 2993–3003, 1983.

Elphic, R. C., C. T. Russell and J. A. Slavin, Observations of the dayside ionopause and ionosphere of Venus, *J. Geophys. Res.*, *85*, 7679–7696, 1980.

Gombosi, T. I., et al., The role of charge exchange in the solar wind absorption by Venus, *Geophys. Res. Lett.*, *8*, 1265–1268, 1981.

Mikhailovskii, A. B., Hydrodynamic theory of drift Kelvin–Helmholtz instabilities, *J. Plasma Phys.*, *28* ,1-11, 1982.

Mikhailovskii, A. B., and V. A. Klimenko, The microinstabilities of a high–β plasma flow with a non–uniform velocity profile, *J. Plasma Phys.*, *24*, 387–407, 1980.

Mikhailovskii, A. B., *Theory of plasma instabilities*, *2*, 360 pp, Consultants Bureau, New York, 1974.

Nagy, A. F., et al., Hot Oxygen atoms in the upper atmosphere of Venus, *Geophys. Res. Lett.*, *8*, 629–632, 1981.

Phillips, J. L., et al., Dependence of Venus ionopause altitude and ionosphere magnetic field on solar wind dynamic pressure, *Adv. Space Res.*, *5*, 173–176, 1985.

Wolff, R. S., B. E. Goldstein and C. M. Yeates, The onset and development of Kelvin–Helmholtz instability at the Venus ionopause, *J. Geophys. Res.*, *85*, 7697–7707, 1980.

THE MAGNETOPAUSE

C. T. Russell
Earth and Space Sciences Department and Institute of Geophysics
and Planetary Physics, University of California,
Los Angeles, California 90024–1567

Abstract. The magnetopause is the interface between the shocked solar wind in the magnetosheath and the geomagnetic field and plasma in the magnetosphere. This interface is far from simple because both sides of the interface contain magnetized plasma. As a result there are boundary layers on both sides of the interface so the resulting structure is many ion gyro radii thick. There is also substructure which may be much less than an ion gyro radius in thickness. The structure of the magnetopause is also sensitive to the Mach number and beta of the plasma. When the beta is very high the magnetopause resembles a slow mode wave. When the IMF is southward and the Mach number and/or beta is low the plasma is accelerated much as Dungey predicted. However, at other times reconnection seems to be less steady and perhaps patchy. Rope-like structures are seen which may be connected to the magnetosphere. These structures which have been called FTE's are still not fully understood.

Introduction

While the dipolar magnetic field of the Earth has been known for at least four centuries, the fact that it is bounded on the sunward side by a magnetopause is but a recent concept. The magnetopause was invented to explain the phenomena surrounding geomagnetic storms. Chapman and Ferraro [1931a,b; 1933; 1940] proposed that a plasma, consisting of equal numbers of ions and electrons was emitted by the sun at active times and disturbed the Earth's magnetic field. When this stream hit the Earth's magnetic field it was deflected. Figure 1 shows the situation for a half-space filled with this highly electrically conducting gas. The magnetic field lines are confined to the half-space in which there is a vacuum. Mathematically the field lines in this case are equivalent to those distorted by the presence of an image dipole parallel to the first dipole and an equal distance on the other side of the plasma boundary.

Geophysical Monograph 58

In reality the plasma surface is not planar but curves around the magnetic field as shown in Figure 2. The magnetic cavity is leaky and solar wind plasma enters the interior of the magnetosphere to cause the late or main phase of the storm. The early or compressional phase is caused by the sudden increase in solar wind pressure that pushes the boundary closer to the Earth.

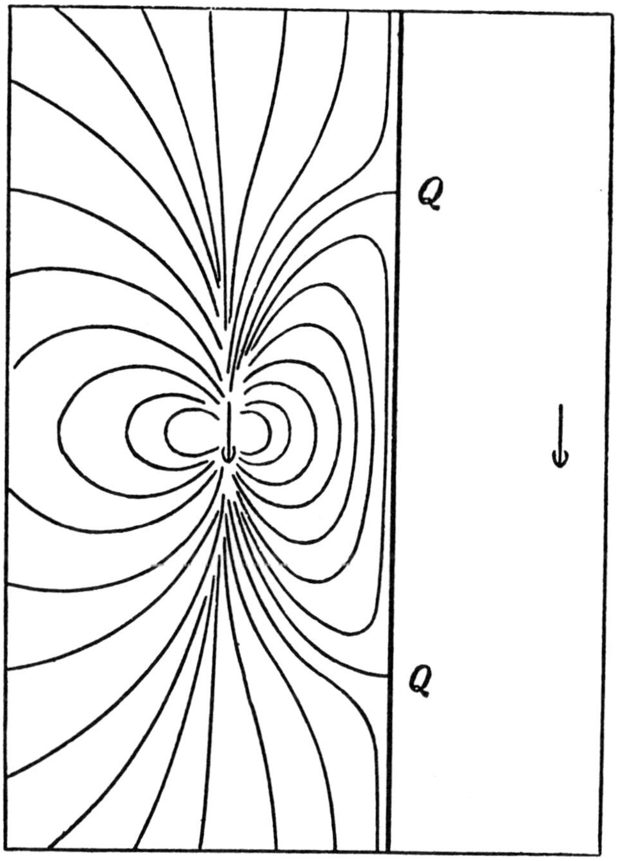

Fig.1. Magnetic field of a dipole in the presence of an infinitely conducting plane. The strength of the magnetic field goes to zero at the points labelled Q (Chapman and Bartels, 1940).

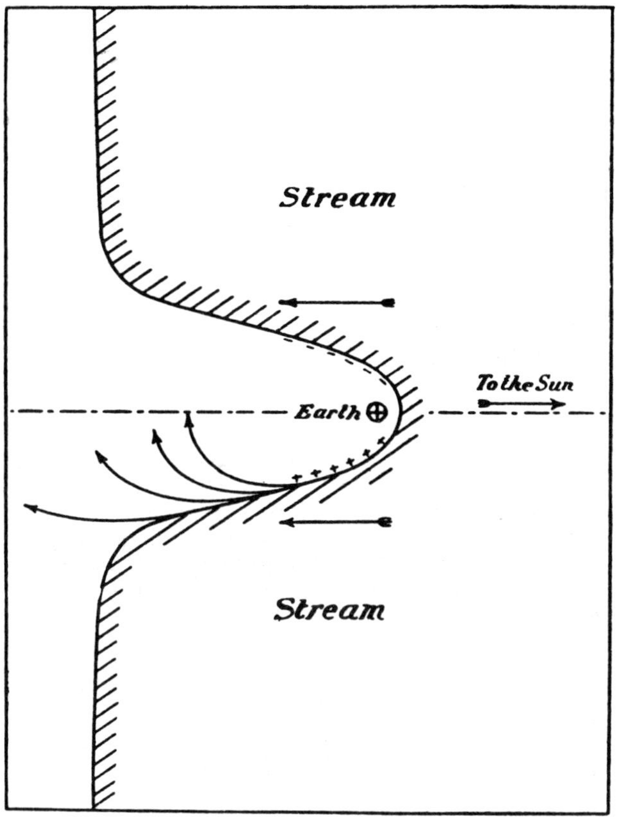

Fig. 2. Distortion of the advancing front of solar plasma (ion-electron gas) by the Earth's magnetic field. This model was proposed by Chapman and Ferraro (1931a,b, 1933) to explain the sudden compression of the Earth's magnetic field followed by a more gradual depression of the field during a geomagnetic storm.

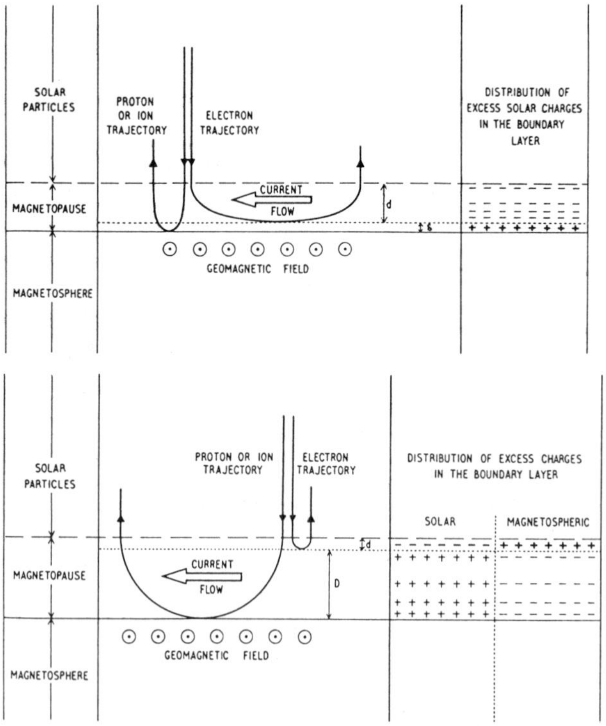

Fig. 3. The motion of charged particles flowing from a cold unmagnetized region against a slab of magnetic field. The magnetic field is out of the page. The ion trajectories are shortened and the electron trajectorieslengthened by a charge separation electric field in the top panel. In the bottom panel this charge has been neutralized and the ion and electron gyrate with the radii expected for their incoming energy (Willis, 1971).

Originally, it was thought that the streams were intermittent. However, we now know that the solar wind blows all the time but with varying intensity. Thus there is always a magnetopause. However, as we will see herein the properties of the magnetopause are quite variable.

It was not long before speculation arose about the nature and structure of the magnetopause despite the fact that the evidence of such a boundary was indirect. Figure 3 shows the original simple, perhaps overly simple, paradigm of the structure of the magnetopause [Ferraro, 1952; Willis, 1971]. In this model the flow of cold, unmagnetized plasma comes in from the top and blows against the terrestrial magnetic field. Ions are deflected to the right and electrons to the left. If a charge separation electric field is set up by the differing gyro radii of the ions and electrons then the thickness of the boundary will be the geometric mean of the electron and ion gyro radii. However, if the electric field is neutralized (say by ionospheric electrons) then the thickness will be that of the ions.

The magnetopause is not so simple. As we will see in this review the boundary is several ion gyro radii thick. Nevertheless there is substructure that could be the Ferraro boundary. The reason for this greater than expected thickness is at least in part due to the magnetic field carried by the solar wind. The boundary is not simply a current layer between a cold flowing unmagnetized plasma and a magnetic field. It is the boundary between two hot magnetized plasmas and these two plasmas can interact at the interface.

One way these plasmas can interact is through the process called reconnection. Reconnection affects the plasmas at the magnetopause in many ways. The most important effect is to join magnetic field lines which were originally not joined. This allows the plasmas to mix along the magnetic field lines. The magnetic geometry of the reconnected field lines accelerates plasma initially but later, downstream, decelerates it. This deceleration takes energy out of the flow and stores it in the magnetic field built up in the tail. Thus the magnetopause is a region of both acceleration and deceleration, as well as one of mixing and heating.

Early History

The year 1961 was a very important year for the magnetopause both observationally and theoretically. Explorer 10 was launched on March 25 of that year into a very eccentric orbit in the antisunward direction. Only 52 hours of data were obtained on this battery-powered mission which provided measurements out to 43 Earth radii (R_e). Part of the orbit skimmed what we know now to be the tail magnetopause, the interface between the geomagnetic tail and the magnetosheath. These first data on the magnetopause are shown in Figure 4 which were obtained at 15 R_e by Heppner et al. [1963]. While these data provided little resolution on the structure of the magnetopause they did reveal that the magnetopause .was constantly in motion. This motion which is much more rapid than the velocity of the spacecraft complicated the study of the magnetopause because it is difficult to convert the temporal profile of a boundary crossing into a spatial profile, and compare the thickness of the boundary with that expected theoretically.

Later this same year, on August 16, Explorer 12 was launched into an elliptical orbit with an apogee of 14 R_e initially near noon. As the Earth moved around the sun, apogee which was approximately fixed in inertial space moved around the magnetosphere to dawn. Thus when transmissions ceased on this solar-panel powered spacecraft nearly 4 months later on December 6, Explorer 12 had sampled the entire dawnside near equatorial magnetopause [Cahill and Amazeen, 1963; Cahill and Patel, 1967]. Figure 5 shows the earliest published magneto-pause crossing from this mission. The analysis of the magnetic field data from Explorer 12 provided us with the foundation for our understanding of the magnetopause for the

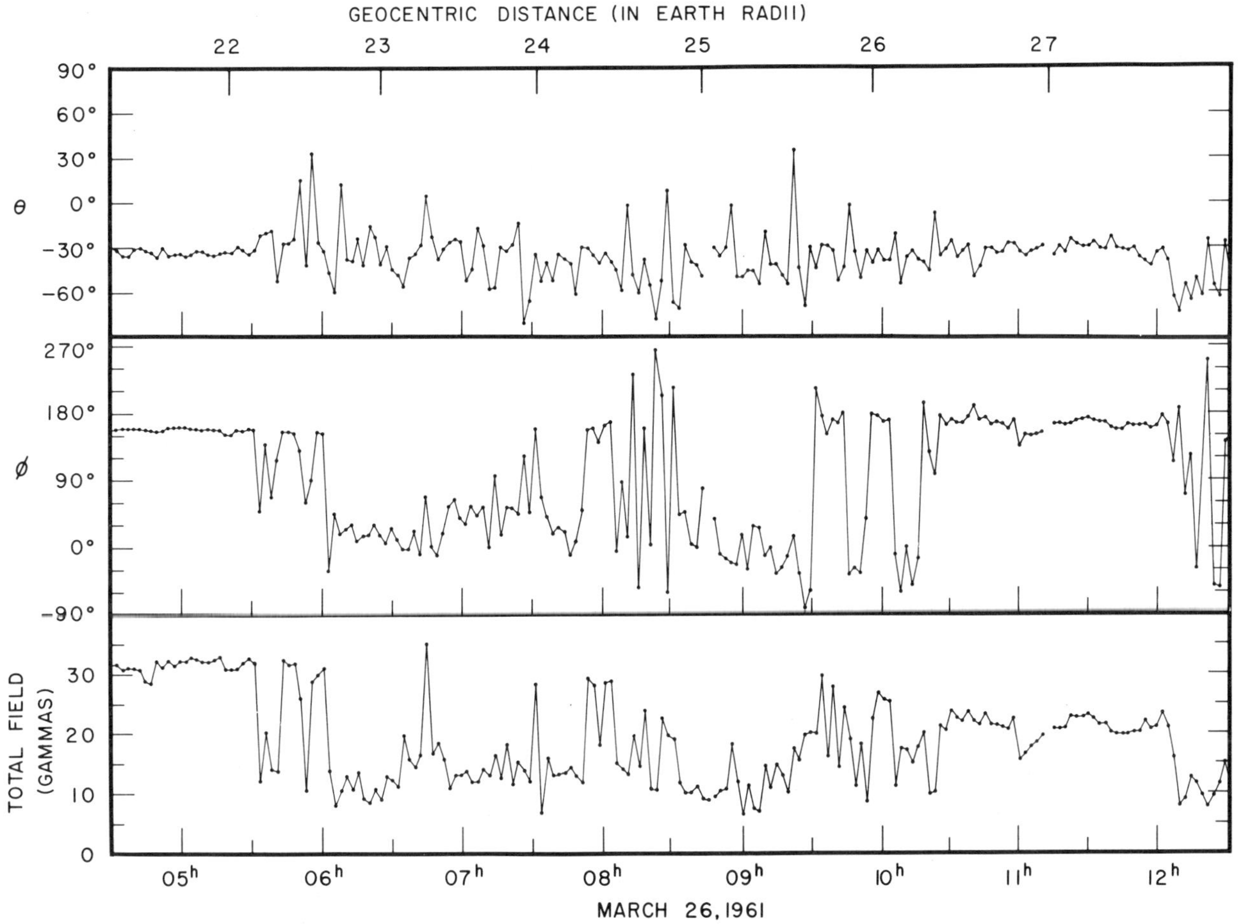

Fig. 4. Magnetic field measurements made by Explorer 10 on 3/26/61, 15 Re behind the Earth. These were the first measurements taken across the magnetopause and the first measurements of the geomagnetic tail (Heppner et al., 1963).

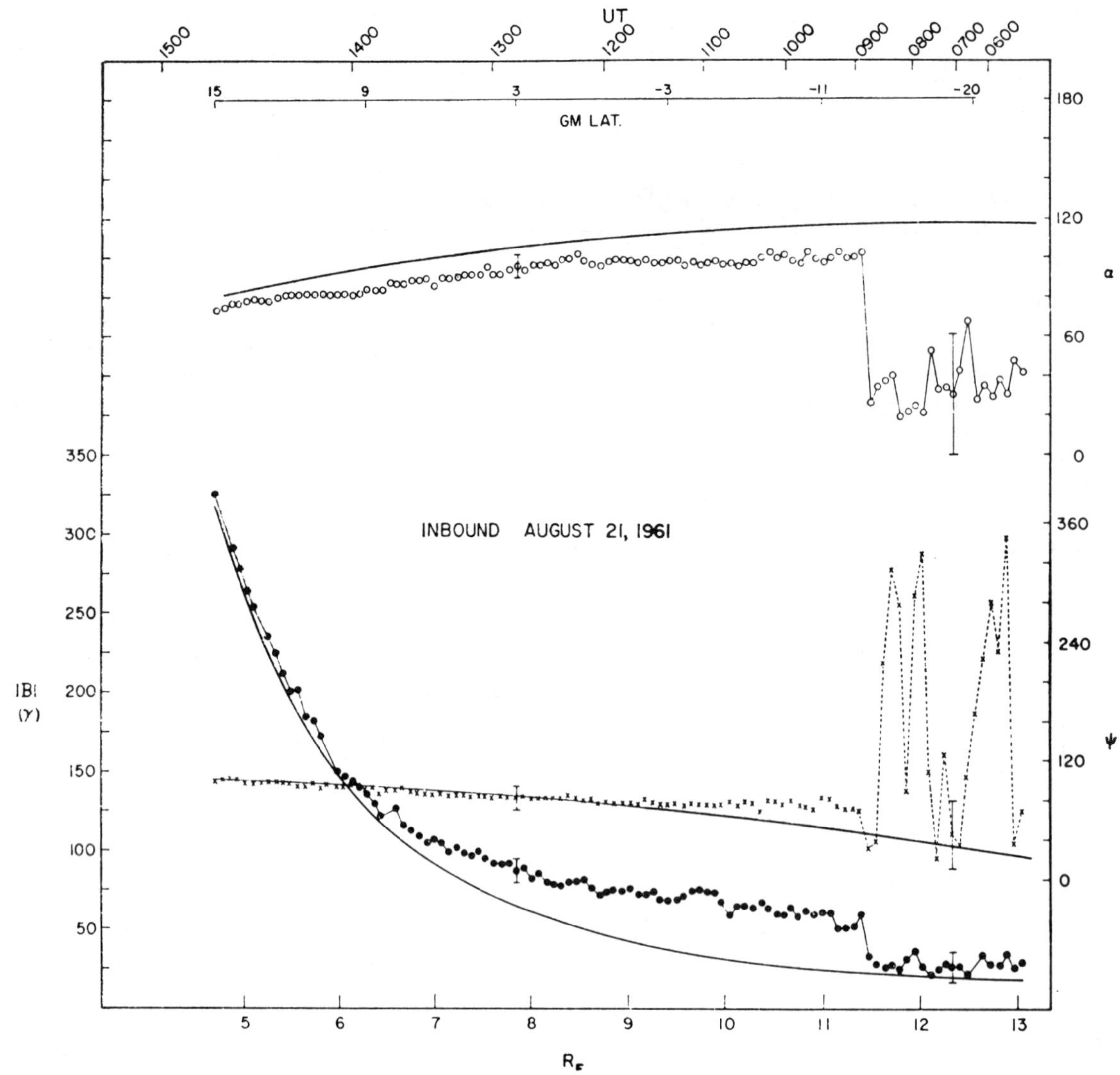

Fig. 5. Magnetic field measurements made by Explorer 12 on 8/21/61. These measurements are the earliest published measurements taken across the subsolar magnetopause (Cahill and Amazeen, 1963; Cahill and Patel, 1967).

next decade until the launch of the ISEE spacecraft. Fairfield and Cahill [1966] used these data to provide the first real evidence for the reconnecting magnetopause, for example. However, it was data from another spacecraft, OGO-5, which showed that reconnection eroded the magnetic flux from the dayside magnetopause and transported it to the tail [Aubry et al., 1970; Russell et al., 1971].

The year 1961 also marked the publication of J. W. Dungey's classic model of the reconnecting magnetosphere in which magnetic field lines carried by the solar wind link up with the terrestrial field where the magnetic field lines are antiparallel [Dungey, 1961]. Field lines in the noon-midnight meridian are sketched in Figure 6 for the two cases when the interplanetary magnetic field is antiparallel to the magnetic field near the nose of the magnetosphere (top) and when it is parallel (bottom). In the former case a circulation pattern is set up which can be a steady state pattern but which in practice is quite time varying because the direction of the interplanetary magnetic field varies. In fact, it is the imbalance of the reconnection on the front side of the magnetosphere with the return of magnetic flux from the nightside which causes the erosion of the magnetopause associated with a southward turning of the interplanetary magnetic field. When the interplanetary magnetic field is northward, reconnection still occurs but this time on open magnetic field lines which transports no

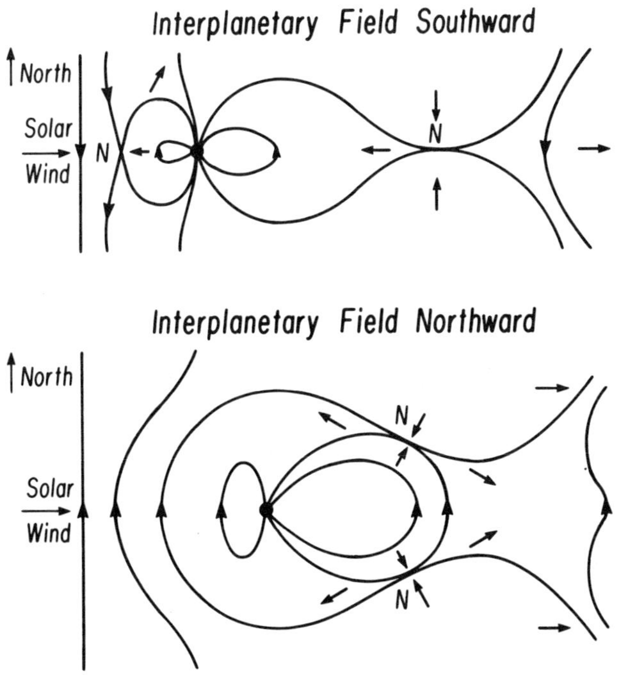

Fig. 6. Magnetic topology of the reconnecting magnetosphere in the noon midnight meridian for southward interplanetary magnetic fields (top) and northward fields (bottom) (Dungey, 1963).

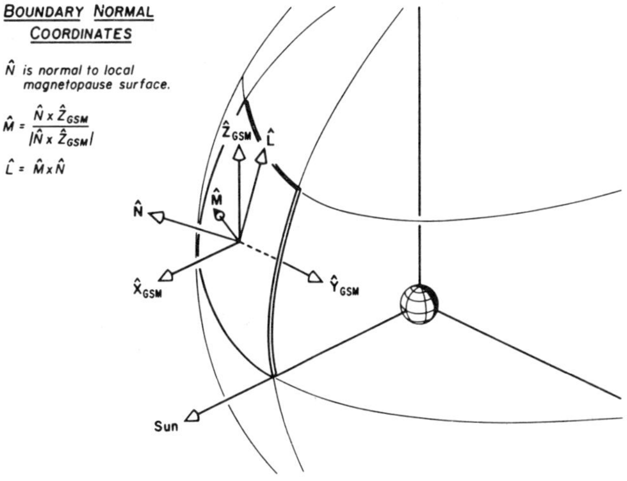

Fig. 7. Boundary normal coordinate system (Elphic and Russell, 1979).

flux or as sketched here which creates a dayside field line out of a nightside one. While the evidence for this model was strong it did not gain wide acceptance until the 1980's when the results of the ISEE mission became available. The direct observation of the predicted accelerated plasma was the decisive piece of evidence [Paschmann et al., 1979; Sonnerup et al., 1981].

Magnetopause Motion and Thickness

In 1977 the co-orbiting International Sun Earth Explorer 1 and 2 spacecraft were launched into the same highly elliptical orbit extending out to 23 Earth radii. Gas carried on ISEE-2 allowed the separation of ISEE-1 and -2 to be varied so that time delays between boundary crossings could be used to measure the boundary velocities. In order to understand the observed magnetic structure and to measure spacecraft separation, boundary normal coordinates were used [Russell and Elphic, 1978]. These are illustrated in Figure 7. The N direction is chosen to be along the magnetopause normal. The L direction is northward along the magnetospheric field and the M direction is tangential to the boundary toward dawn. The difficulty in using these coordinates is in determining the direction of the normal. Sometimes when the magnetopause is quiet and one may assume that there is no reconnection across the interface one can simply assume that the magnetopause is a tangential discontinuity and take N along the vector cross product of the magnetospheric and magnetosheath magnetic fields. At other times there is sufficient rotational structure that one can find the normal to be along the direction of most constant field or the minimum variance direction [Sonnerup and Cahill, 1968]. At other times one simply has to use the average geometry of the boundary.

The magnetopause is constantly in motion and the velocity of this motion is variable. Figure 8 shows the L-component of the magnetic field in

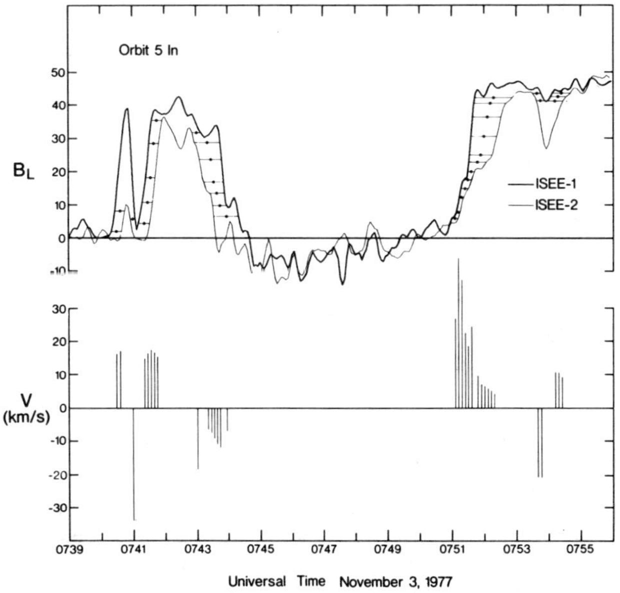

Fig. 8. The velocity of the magnetopause on November 3, 1977, (Russell and Elphic, 1978).

boundary normal components measured by ISEE-1 (heavy line) and ISEE-2 (light line) as the spacecraft made multiple crossings of the magnetopause [Russell and Elphic, 1978]. The L-component increases as the magnetopause is entered. The first entry by ISEE-1 is only partially followed by ISEE-2. After retreating into the magnetosheath again both spacecraft then enter the magnetosphere solidly from about 0742 to 0743 and then return to the magnetosheath. Then after 0751 they again enter the magnetosphere. The horizontal lines joining the magnetic measurements made by the two spacecraft show the time delay between the arrival of the two spacecraft at equivalent locations in the magnetopause. If we divide these time delays into the separation along the normal we obtain the velocities shown on the bottom of the plot. The boundary velocity is quite variable ranging from about 3 to over 40 km/s and typically being about 20 km/s much faster than the velocity of the spacecraft.

Similar analyses have been carried out on a large number of magnetopause crossings by Berchem and Russell [1982a]. The results of these analyses are shown in Figure 9. Velocities over 300 km/s have been recorded. Typically the thickness of the magnetopause as seen in the change in magnetic field strength and direction is between 400 and 900 km, which is equivalent to many thermal ion gyro radii. Since the velocity and thickness measurements are coupled an error in one would affect the other. Figure 10 shows a plot of the measured thickness versus the velocity. They appear to be uncorrelated. Thus we feel these measurements are in fact accurate representations of the velocity and thickness of the magnetopause defined to be the region in which the magnetic field changes from its magnetosheath orientation and strength to those in the magnetosphere.

The motion of the magnetopause seems to be driven by pressure fluctuations in the solar wind and by reconnection but not by the Kelvin-Helmholtz instability at least over the dayside of the magnetopause. Figure 11 shows the amplitude of motion of the magnetopause as measured by ISEE-1 and -2 from the subsolar point past the terminator when the interplanetary magnetic field (IMF) was strongly southward (top) and strongly northward (bottom) [Song et al., 1988]. For southward IMF the boundary motion increases with distance from the subsolar point. For northward IMF the amplitude remains constant and perhaps decreases. Examination of the amplitude of pressure fluctuations on the magnetopause show that they are large enough to cause the observed boundary motions for northward fields. However, when the IMF is southward the amplitude of motion is much greater than can be explained by solar wind pressure variations. Reconnection must be the cause of these variations. The Kelvin-Helmholtz instability could cause an increase in amplitude with distance

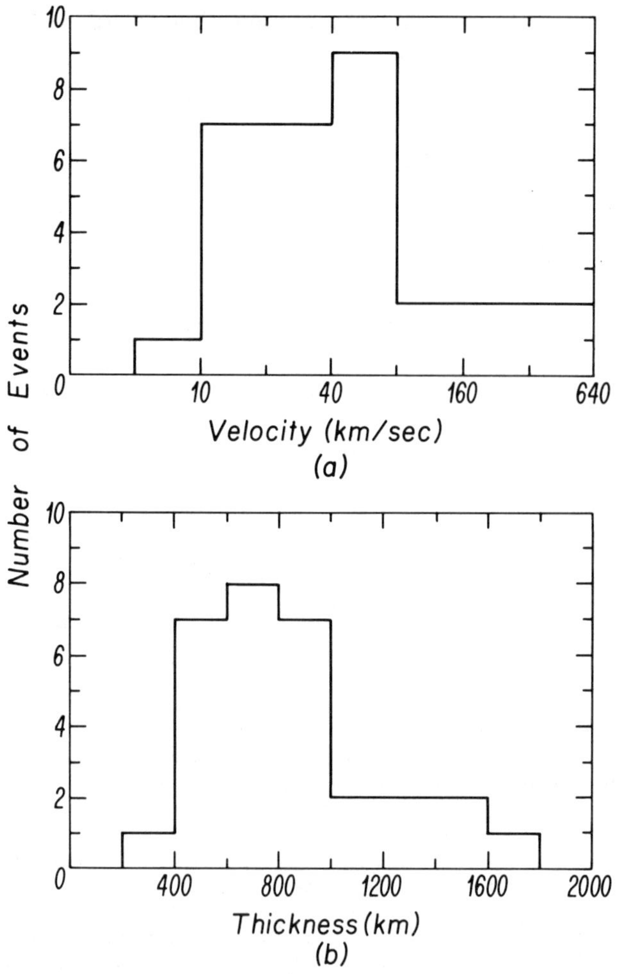

Fig. 9. Velocity and thickness of the magnetopause (Berchem and Russell, 1982a).

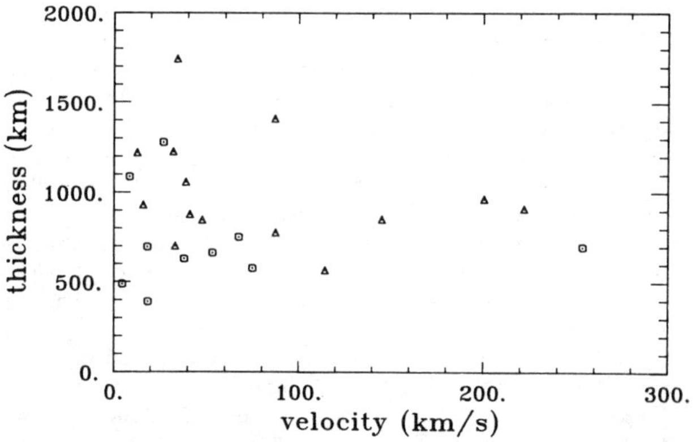

Fig. 10. The thickness of the magnetopause versus the velocity. These quantities would be expected to be independent unless there was some bias in their determinations.

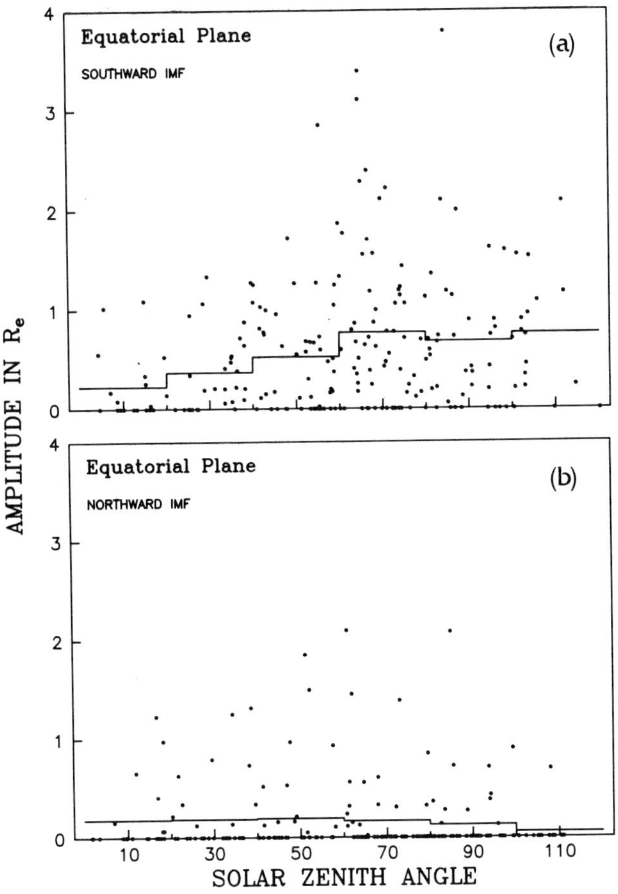

Fig. 11. The radial distance between the first and last magnetopause crossing as a function of the angle from the subsolar point for southward interplanetary magnetic fields (top) and northward interplanetary fields (bottom). Only observations obtained within 30 deg of the equatorial plane are used. Southward magnetic fields had to have latitudes less than -27 deg and northward fields more than 27 deg in this study. The solid line give the average in each bin (Song et al., 1988).

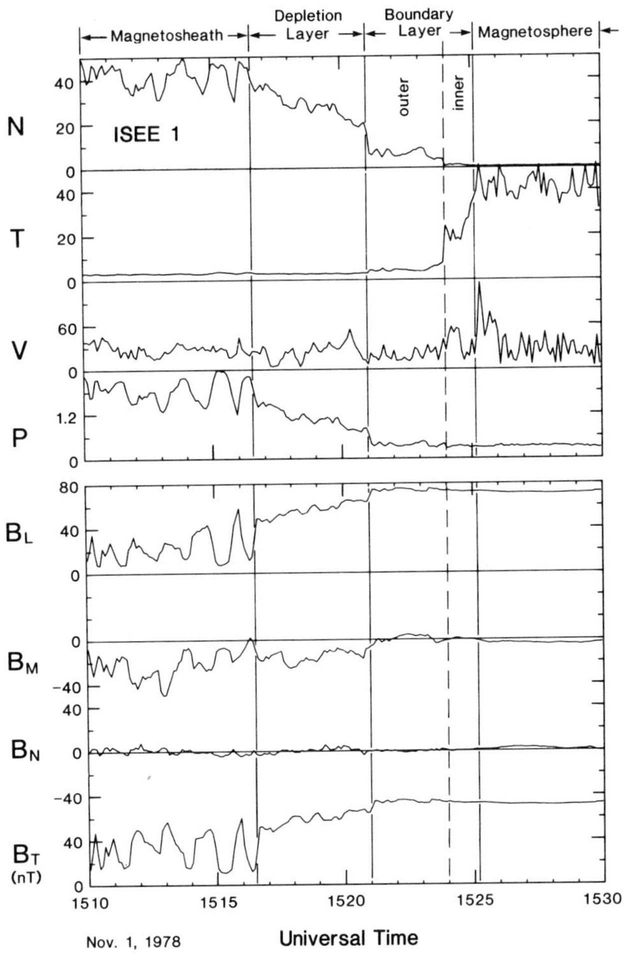

Fig. 12. Fast plasma data and magnetic field measurements across the magneto-pause at the subsolar point when the IMF was strongly northward. The ion density is given in ions cm^{-3}, the ion temperature in milllions of degrees, K, the velocity in km/s; the pressure in 10^{-8} dyne cm^{-2}, and the magnetic field in nT. The magnetic field is in boundary normal coordinates (Song et al., 1989).

from noon but it cannot be operative here because it would also cause such an increase for northward IMF and it does not.

Magnetopause Structure

The magnetopause is a complicated plasma boundary even under the best conditions. Figure 12 shows plasma data and magnetic field data across the subsolar magnetopause when the interplanetary magnetic field was strongly northward [Song et al., 1989]. There are five regions of different plasma conditions on this plot. On the left is the magnetosheath where the ion density is high about 40 cm^{-3} and the temperature relatively cool. The magnetic field and the plasma pressure oscillate out of phase so that the total pressure is fairly constant. These are probably mirror mode waves. As the spacecraft moves closer to the Earth it encounters a region of decreasing density and increasing field strength. Again, the total pressure is fairly constant. This sheath transition layer may be just a boundary layer in the magnetosheath formed as the magnetosheath flows along the magnetopause. Streamlines closest to the magnetopause move the slowest and hence have the oldest plasma. These tubes then have longer time to empty (along the field line) via pitch angle diffusion and thermal motion. Hence the tubes near the magnetopause are less dense. This may correspond to the Zwan-Wolf

[1976] depletion layer. We note that this entire region would be included in our previous determination of magnetopause thickness.

The magnetic field and plasma then undergo a small but abrupt change and enter an outer boundary layer. This abrupt change may in fact correspond to the classical Ferraro current layer with a thickness of less than an ion gyro radius. In the outer boundary layer the density drops to a roughly constant level and the temperature rises. As the spacecraft proceeds inwards it enters the inner boundary layer again with a discrete density and temperature. Finally, the spacecraft enters the magnetosphere. The fact that the boundaries between the layers are sharp and the regions between the boundaries moderately uniform indicates that there is little diffusion present at this magnetopause crossing.

The distribution functions of electrons and ions for this magnetopause traversal are shown in Figure 13. The interesting feature of these distributions is that they all cross at a single point and that they are bounded by the magnetosheath and magnetospheric distributions. This means that all interior distributions can be made from the 2 limiting distributions by simple mixing with no acceleration or heating. This raises the question as to how the magnetosheath and magnetospheric plasmas mix without diffusion. Perhaps there is some transitory reconnection, possibly at high latitudes or at all latitudes as proposed by Nishida [1989].

The magnetopause is somewhat different when the IMF is southward. Figure 14 shows the plasma and magnetic field data for such a crossing when ISEE was near the subsolar point [P. Song et al., unpublished manuscript, 1989]. The plasma density is more nearly constant in the transition layer but there is heating and perhaps some acceleration of the bulk motion. As before there is a small but abrupt jump in the magnetic field and plasma at what may be the Ferraro current. In the model of Heyn and Rijnbeek [Heyn et al., 1988; Rijnbeek et al., 1989] the region of depressed field is bounded by two slow shocks one of which is this current layer. However since we do not have full 3D plasma data here we cannot further check this hypothesis. The boundary layers here are quite

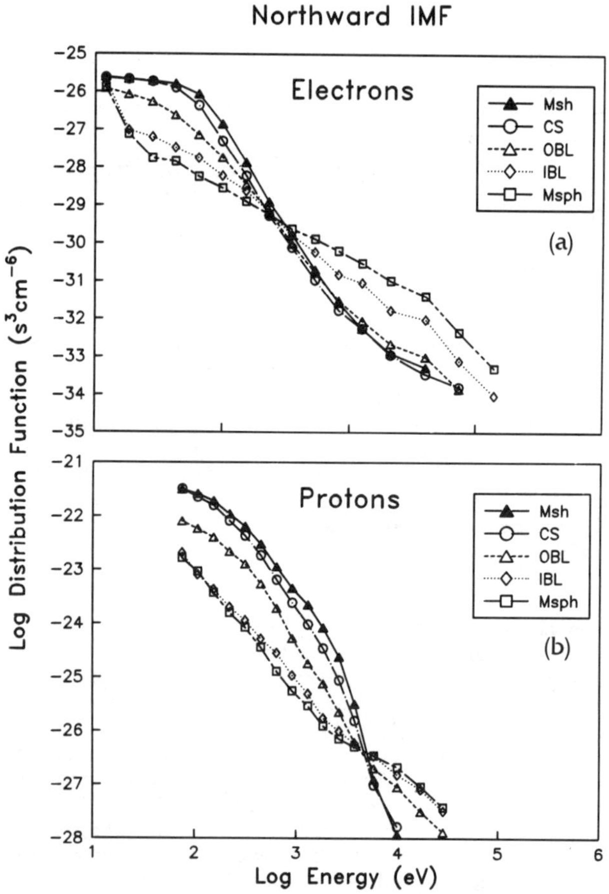

Fig. 13. Distribution functions of the electrons and ions obtained during the interval shown in Figure 12 (Song et al., 1989).

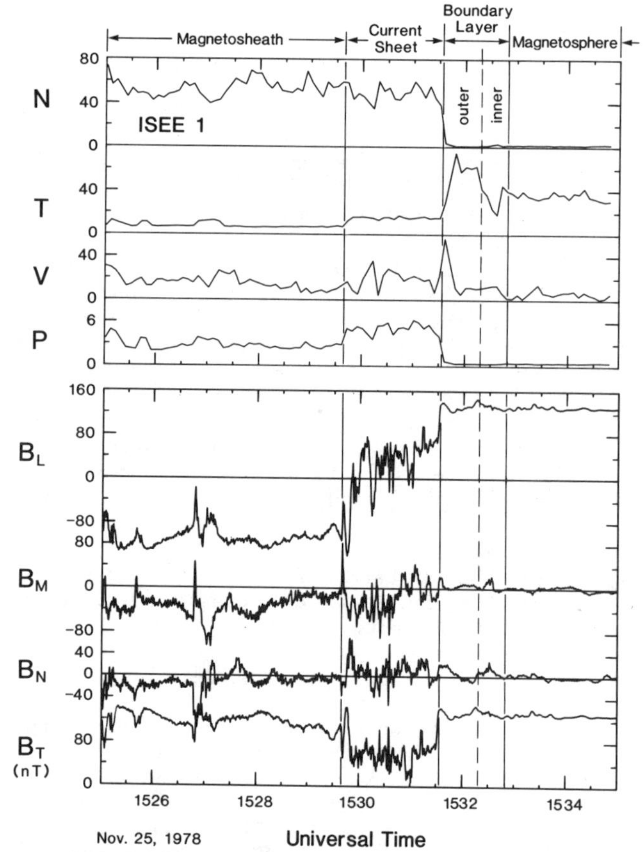

Fig. 14. Plasma and magnetic field measurements obtained near the subsolar point when the interplanetary magnetic field was strongly southward. See figure caption of Figure 12.

rarefied and consequently the temperatures are probably in error. The important point to note is that the boundary layers are so rarefied that they are probably open to the magnetosheath or tail and not trapping any plasma. Figure 15 shows

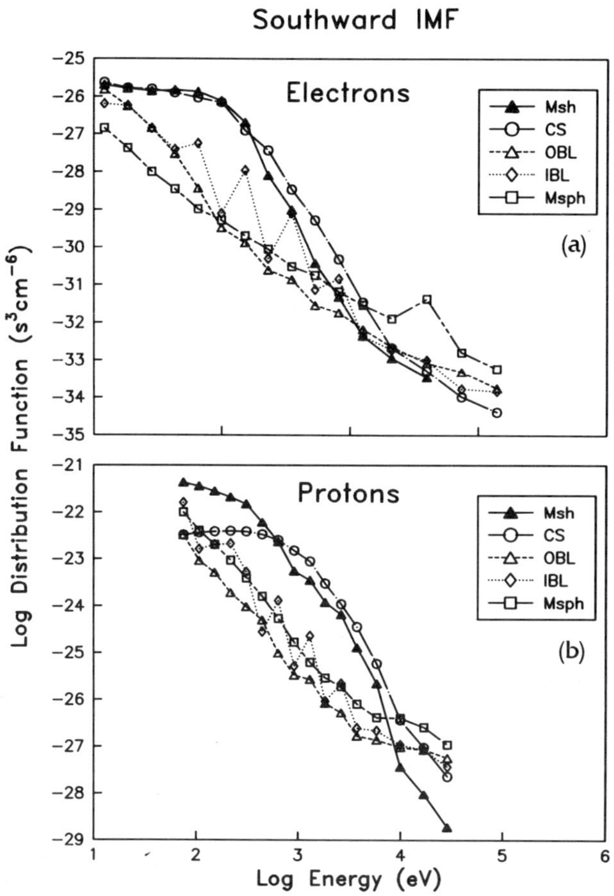

Fig. 15. Distribution functions of the electrons and ions obtained during the interval shown in Figure 14.

the electron and ion distribution functions across the magnetopause. The distributions do not cross at a single point and the magnetosheath and magnetospheric distributions do not bound the other curves. There is acceleration heating and particle loss occurring at this magnetopause.

The acceleration of plasma at the magnetopause depends on more than just direction of the IMF. Figure 16 shows the change in plasma flow velocity seen when the magnetosheath magnetic field was strongly southward as a function of the plasma beta [Paschmann et al., 1986]. It is apparent from this plot that the magnetic field has to be strong relative to the plasma pressure for reconnection to accelerate plasma to high velocities.

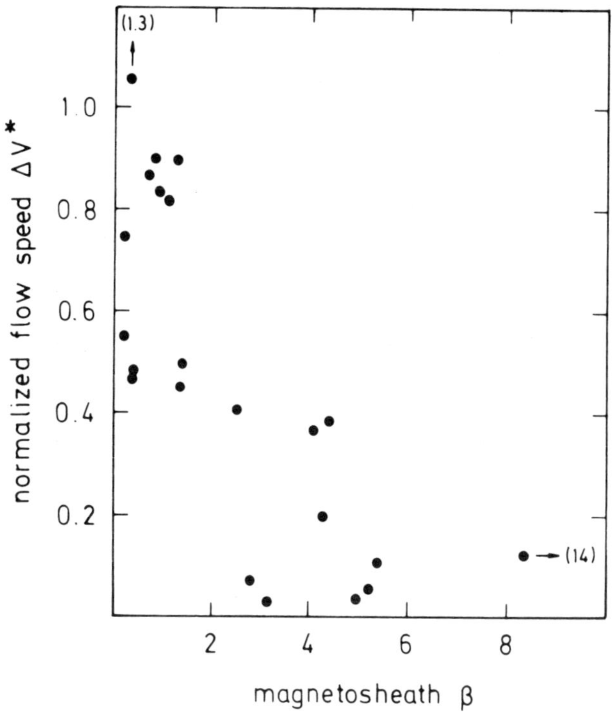

Fig. 16. Normalized flow speed of accelerated ions at the magnetopause as a function of the beta value of the magnetosheath. The normalization is by the Alfven speed (Paschmann et al., 1986).

Another behavior of the magnetopause that seems to be dependent on the beta of the plasma is the rotation of the magnetic field in the plane of the magnetopause. Berchem and Russell [1982b] showed that the magnetopause usually rotated the magnetic field vector through the shortest rotational path as shown in Figure 17. While generally true this is not universally true. Figure 18 shows the magnetic variation through the Uranian magnetopause which occurred under very high plasma conditions. Here the field does rotate through more than 180°. This behavior is also seen in some terrestrial magnetopause crossings when β is very high [C. T. Russell et al., unpublished manuscript, 1989]. The behavior is similar to that of a slow mode magnetohydrodynamic wave, and appears to be analogous to the whistler standing wave often observed on the leading edge of a fast collisionless shock.

Flux Transfer Events

Early in the ISEE-1 and -2 mission it was recognized that what once had been thought to be multiple crossings of the magnetopause were in fact something quite different, the possible formation of magnetic ropes on the boundary. Figure 19 shows an example of these structures

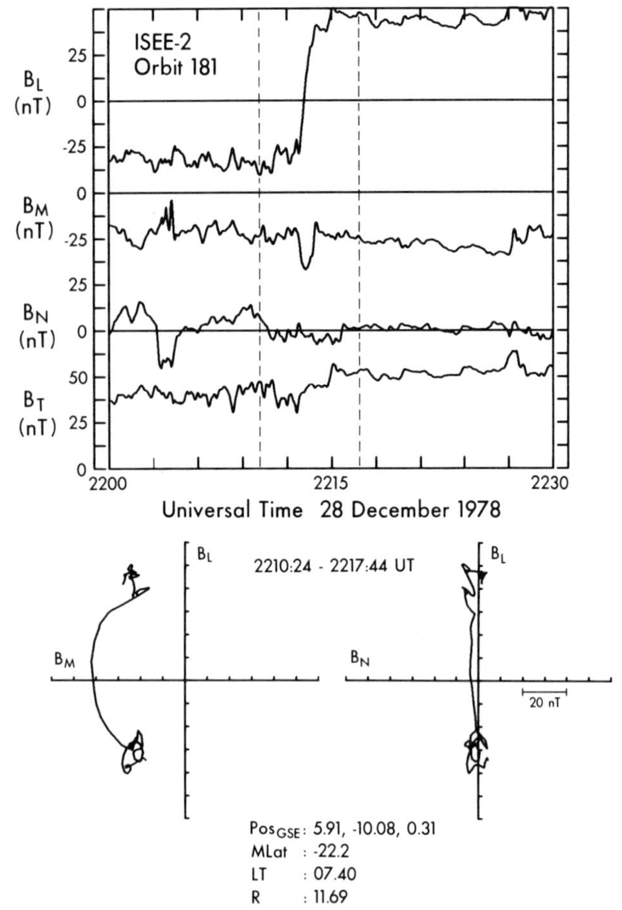

Fig. 17. Magnetic field observations during a magnetopause crossing on 28 December 1978 and corresponding hodogram of the magnetic field variation. This shows the usual less than 180 deg rotation in direction of the magnetic field vector. The data are in boundary normal coordinates.

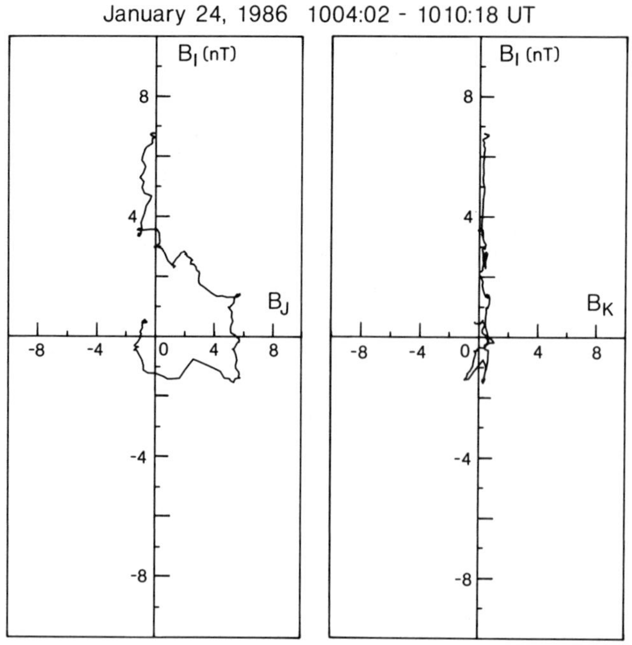

Fig. 18. Hodogram of the magnetic field variation as Voyager 2 crossed the Uranian magnetopause. The directional change following the path of the magnetic field vector is greater than 180°.

which have been called Flux Transfer Events [Russell and Elphic, 1978; 1979]. The measurements are in boundary normal coordinates. The most noticeable signature of a Flux Transfer Event (FTE) is its positive then negative signature in the normal component. The magnetic field is first strongly out of the magnetopause and then into the magnetopause. Both spacecraft see the same normal component although they see slightly different L-components. If we use the characteristic positive then negative normal component then large FTE's are seen at 0213 and 0235 UT with smaller ones at 0241, 0255 and 0300 UT, the latter two in the magnetosphere. It is clear that these events are not multiple crossings of the magnetopause from the behavior of the M-component. The M-component becomes significantly stronger in the magnetosheath FTE's and stronger in the direction of the surrounding magnetosheath field. The plasma within the magnetosheath FTE's was flowing at or above the surrounding magnetosheath flow velocity and consisted of a mixture of magnetospheric and magnetosheath plasma. The name Flux Transfer Event was chosen because magnetic flux was being transported and the phenomenon had a beginning and an end. The presence of the magnetospheric electrons implied that FTE's were connected to the magnetosphere [Russell and Elphic, 1978, 1979]. The amount of flux in a typical FTE is about 5 M Weber. The rate of flux transport depends on the interval between FTE's and the number simultaneously present at the same time on the magnetopause. For example if each FTE contained 5 M Webers and was connected to the polar cap and if there were 3 independent sites creating 10 FTE's per hour these structures could contribute up to 40 kV of potential drop across the polar cap.

It is difficult to determine the geometry of such 3-dimensional structures with two spacecraft. Thus there is still some ambiguity. Simulations both in the laboratory and in the computer can provide some guidance here, even though the exact relative scale lengths cannot be maintained in the laboratory and the resistivity of computer simulations may be artificial. Terrella experiments in which a magnetized plasma is fired down a tube at a magnetic dipole have shown that when the Alfven Mach number is low the reconnection pattern at the magnetopause strongly resembles the Dungey picture sketched in Figure

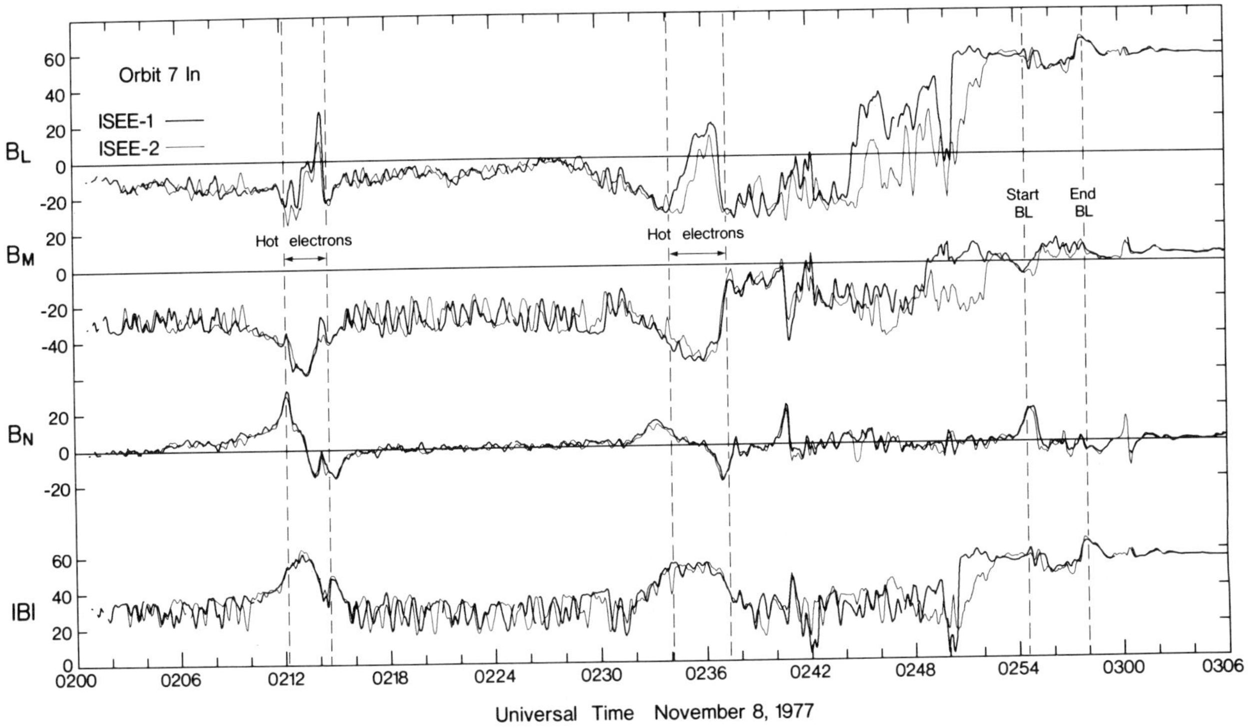

Fig. 19. ISEE-1 (heavy line) and ISEE-2 (light line) observations of flux transfer events on November 8, 1977. The measurements are in boundary normal coordinates.

6. However at high Mach numbers when the IMF is strongly southward reconnection appears to take place at high latitudes in the south and north forming a visor or curl over the front of the magnetosphere [Dubinin et al., 1977; 1980]. The visor itself is subject to the tearing mode and then forms small tearing islands as shown in Figure 20. We note that FTE's do not look like tearing islands because they usually are well isolated from each other and they do not represent the passage of x-points because the field is strong when normal component of the field reverses, not weak.

When there is an east-west component of the IMF in addition to the southward component, rather than tearing islands, a rope appears [Dubinin et al., 1980]. The formation of this structure is shown in Figure 21. The rope is not connected by field lines to the magnetosphere. This rope cannot build up indefinitely. Eventually it must slide off to the north or to the south. Such a structure has many attractive properties. It is twisted so that it remains a discrete entity. Moreover, not only do FTE's appear to be twisted in directions different from the magnetosphere or magnetosheath but they also appear to require a twisted field configuration to balance their internal pressure [Paschmann et al., 1982]. It also provides connection to the magnetospheric plasma but since field lines are open to the solar wind on both ends the plasma can escape rapidly.

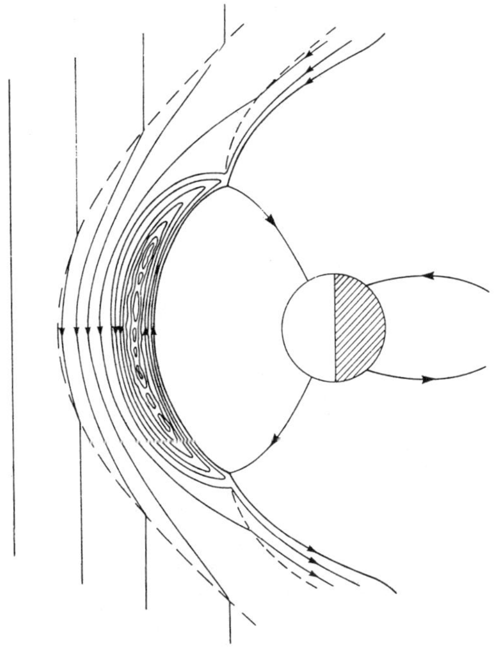

Fig. 20. Schematic representation of the magnetic configuration observed in the noon-midnight meridian in laboratory terrella experiments for high beta when the IMF was due South. Tearing islands are observed on the magnetopause (Dubinin et al., 1977).

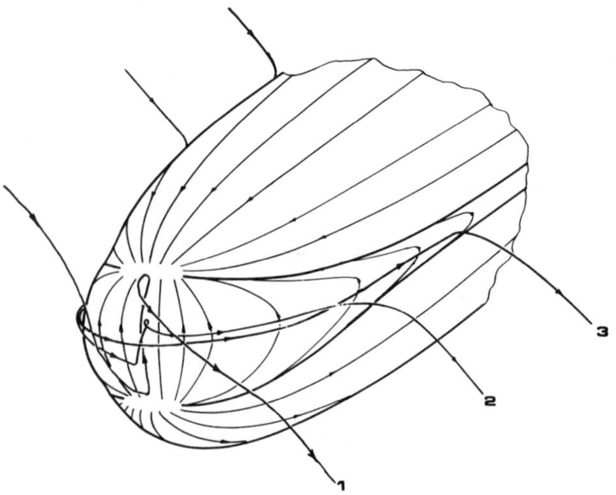

Fig. 21. Formation of a magnetic rope across the dayside magnetopause as observed in laboratory terrella experiments for high beta when there was a significant east-west component of the IMF (Dubinin et al., 1980).

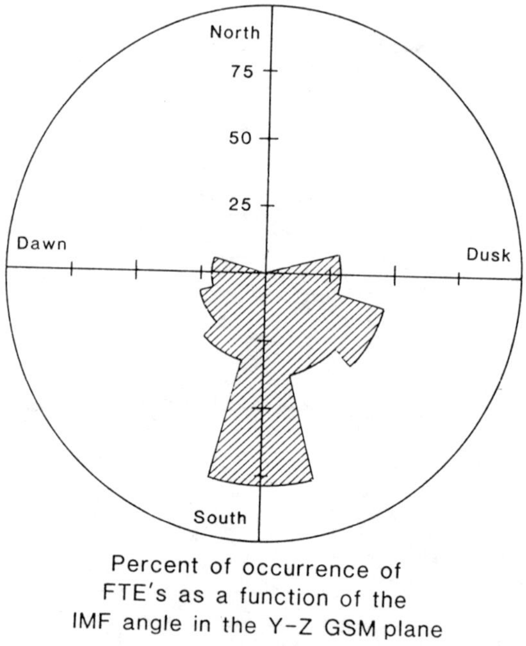

Fig. 22. The normalized occurrence rate of FTE's as a function of the direction of the IMF projected on the Y-Z plane perpendicular to the solar wind flow (Berchem and Russell, 1984).

Moreover, this configuration provides only a single rope across the magnetosphere that can only flow or roll one way north or south at a time while Elphic and Southwood [1987] feel that the

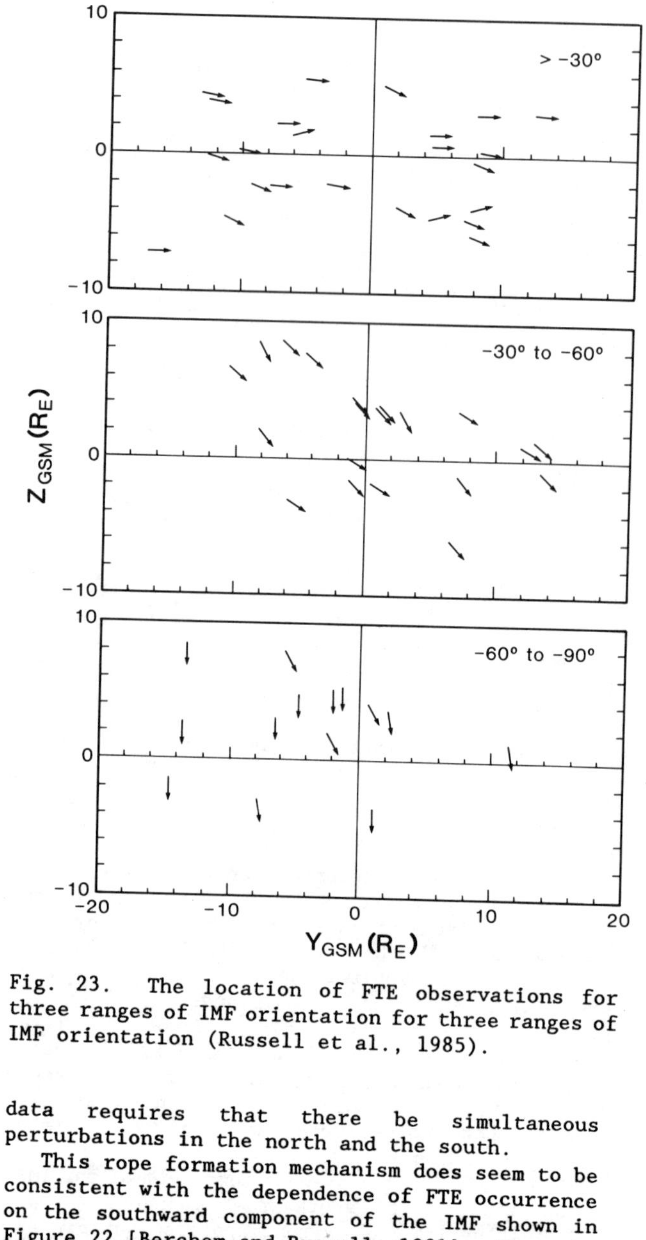

Fig. 23. The location of FTE observations for three ranges of IMF orientation for three ranges of IMF orientation (Russell et al., 1985).

data requires that there be simultaneous perturbations in the north and the south.

This rope formation mechanism does seem to be consistent with the dependence of FTE occurrence on the southward component of the IMF shown in Figure 22 [Berchem and Russell, 1981]. When the IMF is more than slightly northward FTE's do not occur. In the Dubinin et al. model reconnection switches from field lines connected to the high latitude dayside magnetosphere to those connected to the polar cap thus shutting off rope formation. Another constraint is where FTE's are observed as a function of the IMF direction. This is shown in Figure 23 [Russell et al., 1985]. When the IMF is horizontal, FTE's are seen in a horizontal band along the equator. When the IMF is at 45° to the horizontal, the FTE occurrence band swings around the subsolar point following the magnetic field. When the IMF is southward, there seems to be no clearly defined band of occurrence. This behavior at first glance is consistent with the Dubinin et al. picture, because the band where the ropes are formed should follow the IMF direction. However, since in this model the rope has to roll or slip around the magnetosphere it has to be detected everywhere on the boundary, not just in a band.

The only model that limits the FTE occurrence to a band is the transient, patchy reconnection model originally proposed by Russell and Elphic [1978, 1979] and shown in Figure 24. In this model magnetic field lines in the magnetosheath become connected to the magnetospheric field in the subsolar region and then create 2 connected tubes one convecting over the north and one the south. Simultaneous observations above and below the equator by ISEE and AMPTE seem to require two simultaneously oppositely moving tubes [Elphic and Southwood, 1987].

Many computer models have produced flux ropes in two dimensions and in three. Some of these result from subsolar multiple x-line reconnection [cf. Lee and Fu, 1985] and others are the result of high latitude reconnection [cf. Ogino et al., 1989]. However, the relevance of these structures to the observed properties of FTE's is still not clear. The exact geometrical structure of FTE's is still somewhat of a mystery.

Finally we note that Flux Transfer Events occur on the magnetopause of both Mercury [Russell and Walker, 1985] and Jupiter [Walker and Russell, 1985]. Figure 25 shows an example of a Mercury FTE. At Mercury the FTE's are brief, lasting only a second or two and more frequent than on Earth. At Jupiter the FTE's are similar in size and frequency to their terrestrial counterparts. This has led to a suggestion that both the magnetopause thickness and the dimensions of the system play a role in determining the size of FTE's [Kuznetsova and Zeleny, 1986].

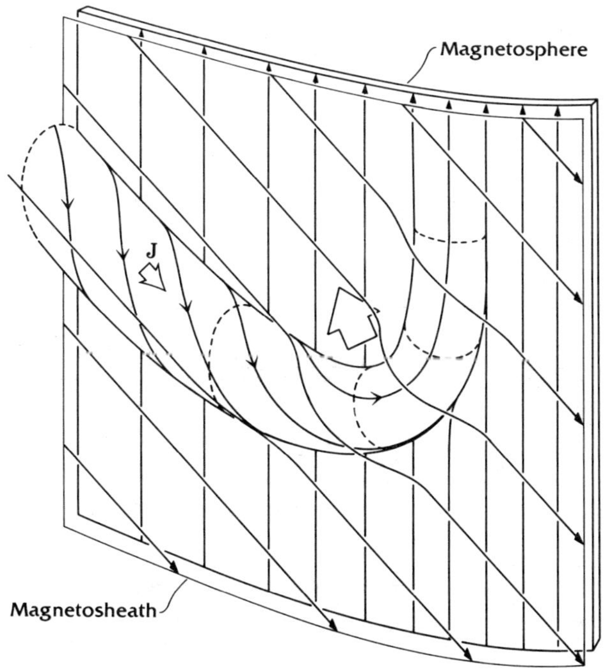

Fig. 24. Schematic of the Russell and Elphic model of the top half of an FTE with a twisted magnetic field (Russell, 1984).

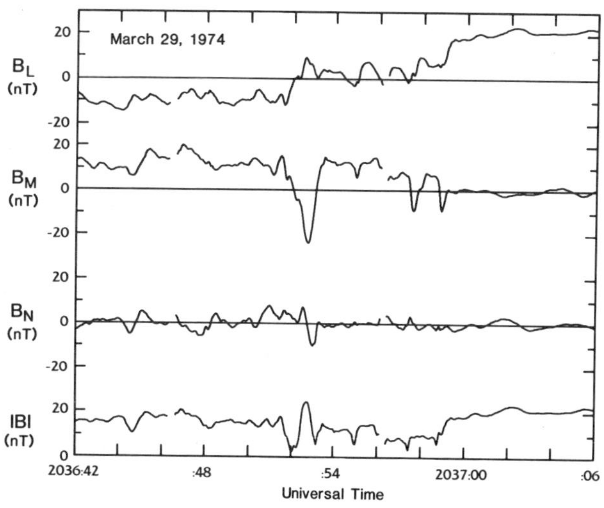

Fig. 25. A flux transfer event at the magnetopause of Mercury (Russell and Walker, 1985).

Summary

The magnetopause is not a simple boundary even at the subsolar point when the IMF is northward. There are boundary layers in the magnetosheath and the magnetosphere. While the overall structure is several ion gyro radio thick there is substructure with thickness less than an ion gyro radius. When the IMF is southward reconnection takes place which accelerates and heats the plasma. The reconnection process seems not to be steady because the boundary oscillates and magnetic flux rope-like structures appear. The geometry of such structures has not unambiguously been determined as yet.

Questions and Answers

Kivelson: Comment regarding the Kelvin-Helmholtz (KH) instability: The data were 1-minute time resolution, which limits analysis to waves with periods > 2 minutes. As waves are propagating away from noon at, say, hundreds of km/s, there is probably only about one consecutive growth-period on the dayside. Your work has effectively ruled out KH as important on the dayside, but leaves open the possibility that it is important on the nightside (where a few growth-periods are possible).

***Russell:** The data we used to measure the boundary amplitudes were 12-second averages every 4 seconds. Thus, we could have detected <2-minute* *oscillations at the boundary. These periods were not observed. Typical periods of oscillation are about 5 to 15 minutes. There is little power in waves at the shorter periods. In fact, from the observed thickness of the magnetopause and the velocity shear across it, we would not expect the dayside magnetopause to be unstable to the Kelvin-Helmholtz instability. On the other hand, in the Earth's tail, Hones and co-workers have reported vortices in the plasma flow. These may be, in fact, due to the Kelvin-Helmholtz instability which would have the opportunity to grow to detectable amplitudes well behind the terminators.*

Zelenyi: Can you comment whether the normal component during magnetopause crossings can be turbulent?

Russell: Yes, the normal component is more often turbulent than coherent. Since we have concentrated our analyses on features in the normal component that repeat, there is some bias in published examples. I see no contradiction between the observations and your percolation model.

Sonnerup: Is there a β dependence in the occurrence rate of FTEs?

Russell: I expect that there is and have several times suggested that the dependence be examined. Thus far, I have not had the time to examine this problem myself. However, I note that many of the FTEs reported occur when the magnetosheath magnetic field is less than half of that of the magnetosphere field. In such situations, the magnetosheath field pressure must be less than $\frac{1}{3}$ of the thermal pressure. Hence, FTEs are a "high" beta phenomenon.

Hollweg: You suggested that you can get rid of the KH by making the discontinuity thicker, but in my talk I showed how a nondiscontinuous surface can lead to a resonant instability below the KH threshold. So, making things thicker can make things more easily destabilized. [The resonance occurs on the field-line where $(\omega/k)_{\text{surface wave}} - V_{\text{flow}} = \pm V_A$.] Do you have any thoughts?

Russell: Since the Kelvin-Helmholtz instability is driven by velocity shear, it seems counter-intuitive that a smaller shear would lead to greater instability. In any event, it is clear that waves do not grow significantly from noon to the terminators.

Forbes: How far down the magnetotail do the flux-transfer events exist as observable phenomena on the surface of the magnetopause?

Russell: D. Sibeck examined this question in his Ph.D. thesis using IMP-8 data. At about 30 R_E, he found a variable orientation of the magnetopause normal in the plane perpendicular to the solar wind flow as if the boundary were corrugated perhaps caused by flux ropes lying on the surface of the tail.

Acknowledgments. The author thanks, R. C. Elphic and P. Song with whom most of the work described here was performed. This work was supported by the National Aeronautics and Space Administration under research grant NAG5-1067.

References

Aubry, M. P., C. T. Russell and M. G. Kivelson, On inward motions of the magneto- pause before a substorm, J. Geophys. Res., 75, 7018, 1970.

Berchem, J. and C. T. Russell, The thickness of the magnetopause current layer: ISEE-1 and -2 observations, J. Geophys. Res., 87, 2108-2114, 1982a.

Berchem, J. and C. T. Russell, Magnetic field rotation through the magnetopause: ISEE-1 and -2 observations, J. Geophys. Res., 87, 8139-8148, 1982b.

Berchem, J. and C. T. Russell, Flux transfer events on the magnetopause: Spatial distribution and controlling factors, J. Geophys. Res., 89, 6689-6703, 1984.

Cahill, L. J. and P. G. Amazeen, The boundary of the geomagnetic field, J. Geophys. Res., 68, 1835-1843, 1963.

Cahill, L. J. and V. L. Patel, The boundary of the geomagnetic field, August to November 1961, Planet. and Space Sci., 15, 997-1033, 1967.

Chapman, S. and J. Bartels, Geomagnetism, Vol. II, Oxford University, 1940.

Chapman, S. and V. C. A. Ferraro, A new theory of magnetic storms, Terr. Mag., 36, 77-97, 1931a.

Chapman, S. and V. C. A. Ferraro, A new theory of magnetic storms, Terr. Mag., 36, 171-186, 1931b.

Chapman, S. and V. C. A. Ferraro, A new theory of magnetic storms, II. The main phase, Terr. Mag., 38, 79, 1933.

Chapman, S. and V. C. A. Ferraro, The theory of the first phase of the geomagnetic storm, Terr. Mag., 45, 245, 1940.

Dubinin, E. M., I. M. Podgorny and Yu. N. Potanin, Experimental proof of the existence of open and closed models of the magnetosphere, Kosmich. Issled., 15, 866, 1977.

Dubinin, E. M., I. M. Podgorny and Yu. N. Potanin, Structure of the magnetic field at the boundary of the magnetosphere (Analysis of a simulation experiment), Kosmich. Issled., 18, 99, 1980.

Dungey, J. W., Interplanetary magnetic fields and the auroral zones, Phys. Rev. Lett., 6, 47-48, 1961.

Dungey, J. W., The structure of the exosphere or adventures in velocity space, in Geophysics The Earth's Environment, (C. De Witt, J. Hieblot and A. Lebeau, Eds.), Gordon Breach, New York, 1963.

Elphic, R. C. and C. T. Russell, ISEE-1 and -2 magnetometer observations of the magnetopause, in Magnetospheric Boundary Layers, (B. Battrick, Ed.), 43-50, ESA SP-148, Paris, 1979.

Elphic, R. C. and D. J. Southwood, Simultaneous measurements of the magnetopause and flux transfer events at widely separated sites by AMPTE UKS and ISEE-1 and -2, J. Geophys. Res., 92, 13666-13672, 1987.

Fairfield, D. H. and L. J. Cahill, Jr., Transition region magnetic field and polar magnetic disturbances, J. Geophys. Res., 71, 155 169, 1966.

Ferraro, V. C. A., On the theory of the first phase of a geomagnetic storm: A new illustrative calculation based on an idealized (plane, not cylindrical) model field distribution, J. Geophys. Res., 57, 15, 1952.

Heyn, M., R. P. Rijnbeek, H. K. Biernat, V. S. Semenov, C. J. Farrugia, D. J. Southwood, G. Paschmann, N. Sckopke and C. T. Russell, Energy flow inside a reconnection layer containing slow shocks, Adv. Space Res., 8, (9)239-(9)244, 1988.

Heppner, J. P., N. F. Ness, C. S. Scearce and T. L. Skillman, Explorer 10 magnetic field measurements, J. Geophys. Res., 68, 1, 1963.

Kuznetsova, M. M. and L. M. Zeleny, Spontaneous reconnection at the boundaries of planetary magnetospheres, in Proceedings of Workshop on Plasma Astrophysics, 1-10, ESA Publication NSP-251, Noordwijk, Netherlands, 1986.

Lee, L. C. and Z. F. Fu, A theory of magnetic flux transfer at the earth's magnetopause, Geophys. Res. Lett., 12, 105-108, 1985.

Nishida, A., Can random reconnection on the magnetopause produce the low latitude boundary layer, Geophys. Res. Lett., 16, 227-230, 1989.

Ogino, T., R. J. Walker and M. Ashour-Abdalla, A magnetohydrodynamic simulation of the formation of magnetic flux tubes at the Earth's dayside magnetopause, Geophys. Res. Lett., 16, 155-158, 1989.

Paschmann, G., B. U. O. Sonnerup, I. Papamastorakis, N. Sckopke, G. Haerendel, S. J. Bame, J. R. Asbridge, J. T. Gosling, C. T. Russell and R. C. Elphic, Plasma acceleration at the earth's magnetopause: Evidence for reconnection, Nature, 282, 243-246, 1979.

Paschmann, G., G. Haerendel, I. Papamastorakis, N. Sckopke, S. J. Bame, J. T.Gosling and C. T. Russell, Plasma and magnetic flux transfer events, J. Geophys. Res., 87, 2159-2168, 1982.

Paschmann, G., I. Papamastorakis, W. Baumjohann, N. Sckopke, C. W. Carlson, B. U. O. Sonnerup and H. Luhr, The magnetopause for large magnetic shear: AMPTE/IRM observations, J. Geophys. Res., 91, 11099-11115, 1986.

Rijnbeek, R. P., H. K. Biernat, M. F. Heyn, V. S. Semenov, C. J. Farrugia, D. J. Southwood, G. Paschmann, N. Sckopke and C. T. Russell, The structure of the reconnection layer observed by ISEE-1 on 8 September 1978, Annal. Geophys., 7, 297-310, 1989.

Russell, C. T. and R. C. Elphic, Initial ISEE magnetometer results: Magnetopause observations, Space Sci. Rev., 22, 681-715, 1978.

Russell, C. T. and R. J. Walker, Flux transfer events at Mercury, J. Geophys. Res., 90, 11067-11074, 1985.

Russell, C. T., R. L. McPherron and P. J. Coleman, Jr., Magnetic field variations in the near geomagnetic tail associated with weak substorm activity, J. Geophys. Res., 76, 1823, 1971.

Russell, C. T. and R. C. Elphic, ISEE observations of flux transfer events at the dayside magnetopause, Geophys. Res. Lett., 6, 33-36, 1979.

Russell, C. T., J. Berchem and J. G. On the source region of flux transfer events, Adv., Space Res., 5, 363-368, 1985.

Song, P., R. C. Elphic and C. T. Russell, ISEE-1 and -2 observations of the oscillating magnetopause, Geophys. Res. Lett., 15, 744-747, 1988.

Song, P., R. C. Elphic, C. T. Russell, J. T. Gosling and C. A. Cattell, Structure and properties of the magnetopause for northward IMF: ISEE observations, J. Geophys. Res., submitted, 1989.

Sonnerup, B. U. O. and L. J. Cahill, Jr., Explorer 12 observations of the magnetopause current layer, J. Geophys. Res., 73, 1757, 1968.

Sonnerup, B. U. O., G. Paschmann, I. Papamostorakis, N. Sckopke, G. Haerendel, S. J. Bame, J. R. Asbridge, J. T. Gosling and C. T. Russell, Evidence for magnetic field line reconnection at the earth's magnetopause, J. Geophys. Res., 86, 10,049-10,067, 1981.

Walker, R. J. and C. T. Russell, Flux transfer events at the Jovian magnetopause, J. Geophys. Res., 90, 7397-7404, 1985.

Willis, D. M., Structure of the magnetopause, Rev. Geophys. Space Phys., 9, 953, 1971.

Zwan, B. J. and R. A. Wolf, Depletion of solar wind plasma near a planetary boundary, J. Geophys. Res., 81, 1636-1648, 1976.

OBSERVATIONS OF FLUX TRANSFER EVENTS: ARE FTES FLUX ROPES, ISLANDS, OR SURFACE WAVES?

R. C. Elphic

Los Alamos National Laboratory, Los Alamos, New Mexico, USA

Abstract. Flux transfer events (FTEs) are widely regarded as a signature of transient reconnection between the magnetic fields in the solar wind and magnetospheric plasmas. Until recently the prevailing model held that the structure of FTEs is that of a magnetic flux rope, with a strong central axial core field surrounded by a weaker twisted azimuthal field. Lately the advent of multiple x-line reconnection models has raised the question of whether FTEs are tearing islands, or time-varying single x-line reconnection. If the former, FTEs may indeed have a rope-like structure; if the latter, then FTEs ought not to be rope-like. We reexamine the evidence that first led to the suggestion that FTEs are related to a non-time-stationary reconnection process. In particular we discuss how the combination of field and plasma variations suggest that FTEs are magnetic flux ropes. Both time-varying single x-line and multiple x-line reconnection can disturb the surrounding plasma and produce a signature which 'mimics' that of a flux rope without flux rope topology. In fact any localized perturbation of the magnetopause surface can produce such a signature. However, the evidence is against a surface wave explanation of FTEs: their occurrence during southward IMF, mixture of solar wind and magnetospheric plasmas, leakage of energetic particles, accelerated plasma flows and peculiarities of the magnetic signature all point to a reconnection-related phenomenon.

Introduction

The terrestrial magnetopause is the boundary of the Earth's magnetosphere. To lowest order it is a current sheet separating the plasmas and topologically-distinct magnetic fields of the solar wind and magnetosphere. Consequently the magnetopause is potentially the prime site for instabilities leading to mass, momentum and energy transfer from the solar wind to the magnetosphere. It has long been known that geomagnetic activity is linked to solar wind conditions. When the interplanetary magnetic field (IMF) has a southward orientation, opposite to that of the northward terrestrial magnetic field, there is enhanced energy input from the solar wind to the magnetosphere. When the IMF has a northward orientation, parallel to that of the Earth's field, little energy is transferred. Dungey [1961] was the first to suggest that magnetic reconnection at the magnetopause is the principal mechanism for energy transfer when the IMF is southward.

Geophysical Monograph 58

While there was at the time indirect evidence of the validity of Dungey's 1961 hypothesis, the first convincing in situ measurements of reconnection at the magnetopause came much later, after the launch of the ISEE spacecraft in 1977. ISEE plasma and magnetic field measurements revealed that the time-independent tangential stress balance expected in a rotational discontinuity sometimes appears to hold across the magnetopause, consistent with quasi-steady state reconnection both on the dayside [Paschmann et al., 1979; Sonnerup et al., 1981] and along the flanks [Gosling et al., 1986]. But even under favorable conditions these steady reconnection signatures were not always observed; evidently reconnection occurs sporadically.

Haerendel et al. [1978] noted evidence in the HEOS 2 data that reconnection can occur at the high latitude magnetopause in a spatially and temporally limited manner. Russell and Elphic [1978] suggested that impulsive reconnection can occur at the low-latitude magnetopause, based on ISEE observations of isolated but large scale disturbances of the magnetic field, plasma and energetic particle environment at and near the magnetopause. These disturbances are brief (1 - 2 minutes) and separated by a longer period of quiet (typically 6 - 9 minutes). What Russell and Elphic [1978] suggested, in effect, was a magnetopause analog to solar flares. Because of their episodic nature and their association with enhanced convective plasma flow, they were dubbed flux transfer events (FTEs).

Russell and Elphic [1978] found that the magnetic field of FTEs was highly distorted from that found in the magnetosphere, magnetosheath or even in the quiescent magnetopause boundary itself. The details of the observations are discussed in the next section, but in brief Russell and Elphic [1978] found that the field executed a peculiar bipolar oscillation in the component normal to the magnetopause surface. In addition the orientation of the peak field in an FTE was usually part way between the magnetosheath and magnetosphere directions. These magnetic features together with the observed convective plasma flow and mixed magnetospheric and magnetosheath plasma populations led to the schematic shown in Figure 1.

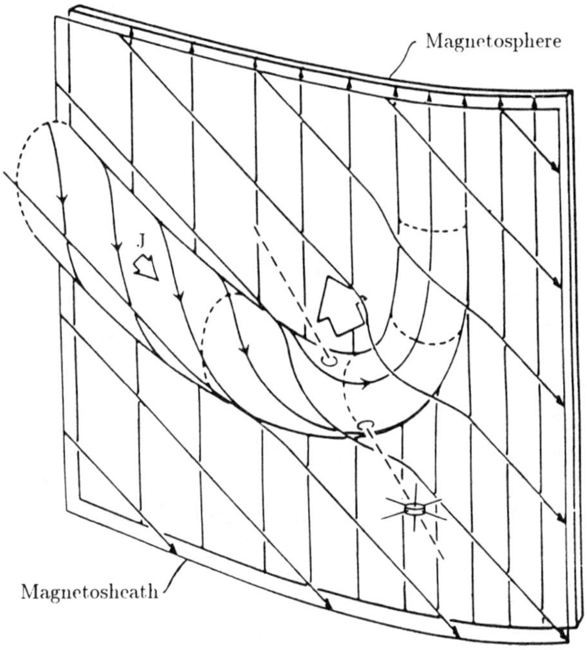

Fig. 1. Schematic diagram of an early interpretation of FTE structure. A bundle of flux tubes is shown having reconnected between the magnetosheath (foreground) and the magnetosphere (background). This isolated reconnected flux tube is shown accelerating upward and to the left, as shown by the arrow. It is disturbing the surrounding environment, as can be seen in the magnetosheath field lines draped over the flux tube. There is also a field-aligned current flowing in the flux tube, shown by the smaller arrow. This current produces the twisted field structure inferred from magnetic field and plasma measurements. Not shown here is the downward-going counterpart FTE below and to the right of the diagram. (Adapted from Paschmann et al. [1982]).

The figure portrays one of a pair of bundles of magnetic field lines formed by spatially-limited transient reconnection between the magnetic fields of the magnetosheath (foreground) and the magnetosphere (background). This isolated, topologically-distinct reconnected bundle (or flux tube) accelerates upward and to the left due to magnetic tension, shown by the arrow. As it progresses the bundle disturbs the surrounding environment, visible as the draping of field lines over the flux tube. There is also a twist to the field lines in he bundle, corresponding to a field-aligned current flowing in the flux tube. A spacecraft sitting essentially stationary in the magnetosheath would, in the frame of the flux tube, trace out the trajectory shown by dashed lines. This model was used to explain the ISEE field and plasma observations of FTEs.

Magnetic flux is transported in FTEs; that this flux transport is related to reconnection is suggested by both plasma and energetic particle data. Surveys by Berchem and Russell [1984] and Rijnbeek et al. [1984] showed that FTEs are observed all across the dayside magnetopause with highest occurrence when the IMF is strongly southward, whereas FTEs are never seen when the IMF has a significant northward component. They are often observed near magnetopause crossings where quasi-steady reconnection is found. FTEs may thus be an important form of reconnection at the magnetopause.

Farrugia et al. [1988] and Sonnerup [1988] have recently reviewed, respectively, the observations and theoretical interpretation of FTEs. Our goal here is to present observations of FTEs and to compare them with the various hypotheses put forward to explain those observations. In particular we wish to test if FTEs are actually magnetic flux ropes, as certain models predict. In the process we shall visit the question of whether or not FTEs have anything to do with reconnection at all – can they be explained as mere surface waves on the magnetopause? We shall briefly review magnetic, plasma and energetic particle behavior in FTEs, and then discuss some of the hypotheses put forward to explain that behavior.

FTE Observations

Magnetic Field Signature

As mentioned above, FTEs were first identified by Russell and Elphic [1978] as a particular signature in the magnetic field of the magnetosheath near the magnetopause. Examples of this signature can be seen in Figure 2, magnetic field data near a magnetopause crossing on 21 October, 1980. The data shown span the period 1200 to 1340 UT, during which the ISEE spacecraft passed from the magnetosphere into the magnetosheath. The magnetopause crossing is at about 1247 UT. The data have been cast into a coordinate system ordered with respect to the magnetopause surface: the N component is normal

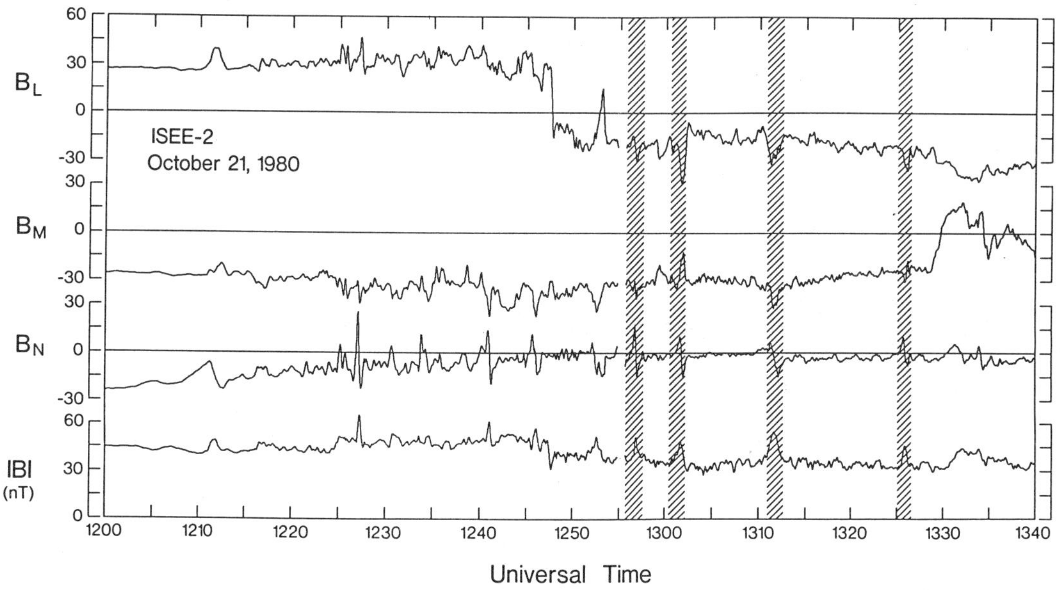

Fig. 2. ISEE 2 magnetic field data for a crossing of the magnetopause on October 21, 1980. The data are shown in boundary normal coordinates, where N is the inferred direction normal to the magnetopause surface, and L is the projection of the Earth's dipole on that surface. M then completes the right-handed LMN system. The magnetopause crossing can be seen as the sudden transition at 1247 UT from $B_L > 0$ (magnetosphere) to $B_L < 0$ (magnetosheath). Throughout the interval there are noticeable disturbances in the B_N component, each lasting 1 - 2 minutes and separated by about 6 minutes. Four of these signatures can be seen in the magnetosheath, indicated by hatching. Most of the B_N variations are bipolar, a positive followed by a negative excursion, and many are associated with local enhancements of the field strength. (From Berchem and Russell [1984]).

to that surface, while the L component lies in the plane of the magnetopause surface and is directed northward; the M component completes the right-handed LMN set. If the magnetopause were a simple one-dimensional current sheet, the field would change only in the plane of the current sheet, i.e., only in the L and M components.

Figure 2 shows that field variations are not confined to the magnetopause plane. Indeed, large variations in the N component indicate that the boundary is disturbed, or filamentary currents are present, or both. Several of the B_N variations observed in the magnetosheath are demarcated by hatching; however, there are similar signatures seen in the magnetosphere as well. The most striking feature is the tendency for these B_N variations to be bipolar and to occur at rather regular intervals of about every six minutes. The bipolarity is almost without exception a positive followed by negative variation. These B_N variations are often accompanied by an increase in the field strength, and by changes in B_L and B_M away from either the magnetospheric or magnetosheath orientation. These three-dimensional magnetic field variations indicate a localized breakdown in the planar structure of the magnetopause.

The persistent $+/-$ bipolar B_N signature in both the magnetospheric and magnetosheath FTEs in Figure 2 indicates that the field in both locations bends first away from the plane of the magnetopause, then toward it. Thus the disturbance on the magnetopause must both push outward into the magnetosheath and inward into the magnetosphere. Moreover, the magnitude of the B_M component is often seen to maximize at the center of the B_N signature, indicating a field strength and orientation unlike that in the magnetosphere, magnetosheath or in the magnetopause. This combination of field variations led Russell and Elphic [1978] to reject surface waves as an explanation of the FTE phenomenon (but this topic arises again later). The bipolar signatures in B_N have the same $+/-$ sense whether they are seen inside the magnetosphere, or in the magnetosheath. In order for a surface wave to produce both signatures, it must selectively push the magnetopause only inward for magnetospheric FTEs and only outward for magnetosheath FTEs. The simpler explanation is that the FTE is a disturbance like a blister on the boundary, distorting both the southward magnetosheath and northward magnetospheric fields to produce the characteristic $+/-$ B_N signature if the disturbance moves northward, and a $-/+$ signature if it moves to the south. Russell and Elphic [1978] proposed that the disturbance is an isolated reconnected flux tube, topologically distinct from the surrounding magnetosheath and magnetospheric plasmas. Such tubes would originate as pairs at the reconnection site, one traveling north and the other south.

Comprehensive surveys of the ISEE 1 and 2 magnetic field data by Berchem and Russell [1984] and Rijnbeek et al. [1984] showed that FTE signatures are seen with nearly equal frequency inside and outside of the magnetopause, and that FTEs are observed almost exclusively when the magnetosheath magnetic field is strongly southward. They also found that the bipolar B_N component variation tends to be $+/-$ (termed 'direct') north of the magnetic equator, while the opposite sense $-/+$ ('reverse') is found to the south. Figure 3 shows the distribution of FTE direct and reverse signatures on the dayside magnetopause together with a histogram showing FTE occurrence as a function of IMF orientation, based on the above surveys. Dailey et al. [1985] examined the measured electric field in FTEs and found that those with direct signatures have northward $\mathbf{E}\times\mathbf{B}/B^2$ drift velocities

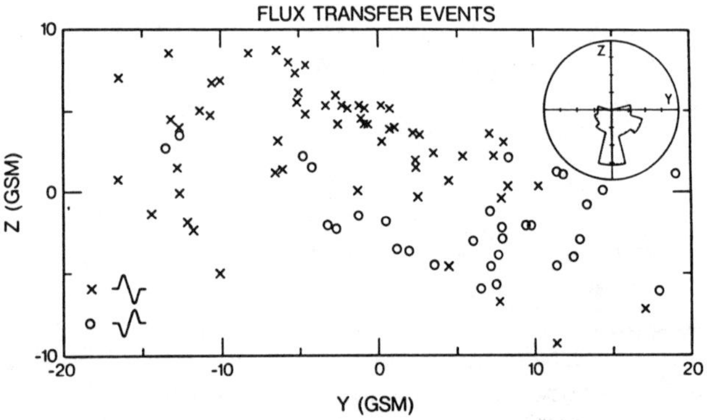

Fig. 3. The distribution of FTEs across the dayside magnetopause, shown as position in Y and Z GSM. The different symbols denote the two observed senses of B_N variation: the direct sense is $+/-$ (denoted by $\times$), the reverse sense $-/+$ is (denoted by $\circ$). There is a very clear tendency to find direct cases to the north and dawnward, while the reverse cases are found to the south and duskward. This result implies that direct FTEs represent northward-going reconnected flux tubes and reverse FTEs southward-going flux tubes. Moreover, the distribution of direct and reverse signatures is the same whether the FTE is observed in the magnetosphere or in the magnetosheath. The inset shows an angular histogram of FTE occurrence as a function of IMF orientation in GSM coordinates. FTEs are observed preferentially when the IMF is southward. (From Baumjohann and Paschmann [1987], adapted from Berchem and Russell [1984]).

and those with reverse signatures are traveling southward.

Elphic and Southwood [1987] examined a case where widely separated spacecraft, AMPTE UKS and ISEE 1 and 2, were passing through the magnetopause at roughly the same time. UKS, near noon LT and a few degrees above the equator, observed FTE signatures at the same time as ISEE 1 and 2, near the same local time but several R_E to the south. The implication is that FTEs truly originate as pairs near the equator and convect away, north and south.

One important, but not widely appreciated ramification of FTE observations is that not all encounters by spacecraft with FTEs are equal. Some are grazing passages, and the B_N signature then reflects merely the local plasma response to a passing disturbance on the magnetopause. Some, however, must be passages through the true reconnected magnetic flux; there we must expect the plasma and field signatures to be different from the grazing cases. In order to distinguish between exterior and interior structure, we study first the grazing cases.

Farrugia et al. [1987a] investigated the magnetic signatures of grazing FTEs and showed that they are approximately consistent with incompressible plasma flow about an impenetrable cylinder lying in the plane of the magnetopause. The $+/-$ B_N variation corresponds to the deflection of plasma out of the way of the approaching cylinder, while the B_L and B_M variations follow from the draping of field over the cylinder and the plasma flow around it. This process is illustrated in Figure 4, adapted from Farrugia et al. [1987a]. For oblique cylinder orientations there is a peak in B_L. Consequently the field strength maxima so often seen in FTEs do not necessarily correspond to the core field of a twisted flux rope structure, but rather to the draping field around the FTE core region, the impenetrable cylinder. Strictly speaking, these results do not address the issue of what is inside the core, nor what produced it.

It can be argued that the incompressible model, with its assumption of potential flow and magnetic fields, is not strictly applicable to magnetospheric plasmas. The model is, however, a tractable analytic basis for understanding the fundamentals of the field and flow signatures. A comparison of predicted and observed magnetic field and convective plasma flow around an FTE is shown in Figure 5. The prediction is based on the incompressible flow/field assumption discussed by Farrugia et al. [1987a]. The data, in boundary normal coordinates, are from an FTE observed by ISEE 1 and 2 while in the magnetosphere at about 0700 UT on September 3, 1978. The convective plasma velocity $\mathbf{V}_c = \mathbf{E}\times\mathbf{B}/B^2$ is derived from electric and magnetic field measurements (see Dailey et al. [1985]). We have scaled the abscissa to be a measure of spacecraft trajectory distance in arbitrary units assuming the FTE convects at a steady speed. Solid lines denote the best fit incompressible flow and field solution, while dotted lines denote the data.

The magnetic field variations appear to fit the model at least qualitatively, with the B_N signature due to the northward passage of the FTE cylinder past the spacecraft. The negative/positive/negative variations in B_M correspond to a cylinder inclination of 51° relative to the magnetospheric field, canted in the direction of the magnetosheath field. As expected for a northward-going FTE as seen in the magnetosphere, the convective flow component V_{cN} is in antiphase to the B_N component. Likewise, the observed $+V_{cM}$ flow is consistent with the expected reconnection flow based on the external magnetosheath field orientation. However, the details of the data are not in agreement with the prediction, suggesting the breakdown of one or more of our assumptions. Nevertheless, the qualitative agreement of field and flow with the incompressible model suggests that the notion of an FTE obstacle as a convecting cylinder is not completely unreasonable. Very recently, Papamastorakis et al., [1989] have found FTE plasma flow and magnetic field variations in a grazing incidence case agree with the Farrugia draping model.

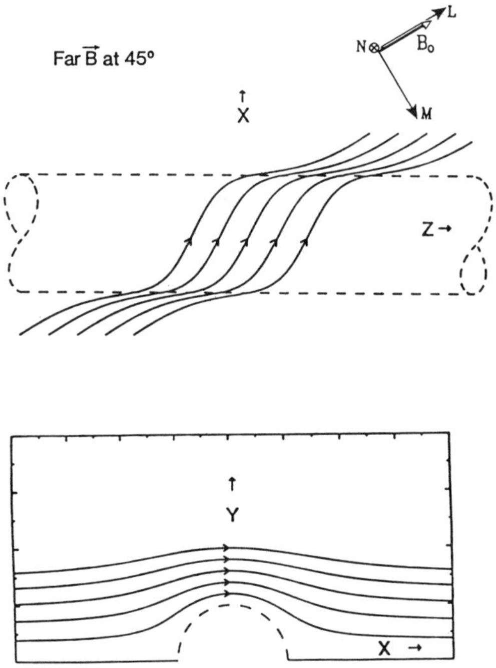

Fig. 4. Much of the observed magnetic field variation associated with an FTE can be qualitatively explained as the disturbance field around an impenetrable obstacle, shown here as a cylinder. Farrugia et al. [1987a] compared the expected signatures of grazing impacts with such an obstacle to observed magnetic signatures of FTEs. Shown here is the field disturbance discussed by Farrugia et al.. The top panel is a view down at the plane of the magnetopause, illustrating how the ambient field drapes about the cylinder traveling in the X direction. The lower panel is an orthogonal view, looking along the cylinder axis and illustrating the perturbations which give rise to the $+/-$ B_N component and the peak in the B_L component.

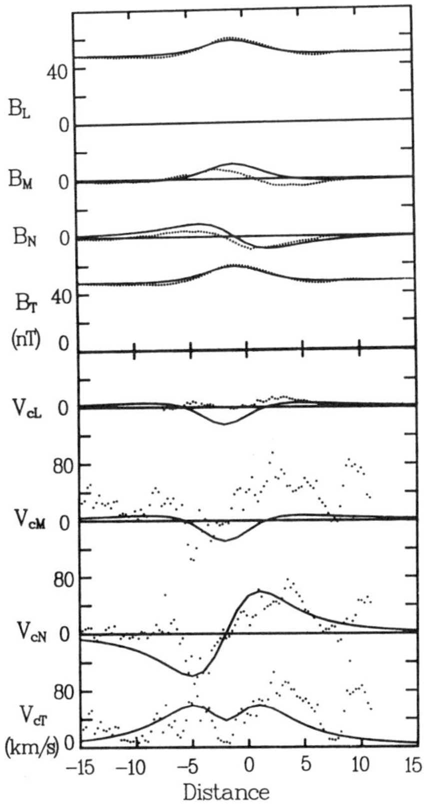

Fig. 5. Comparison of best-fit model and measured field and convective $(\mathbf{E}\times\mathbf{B}/B^2)$ plasma flow perturbations due to the grazing passage of an FTE. The data are from a magnetospheric FTE observed by ISEE 1, cast into boundary normal coordinates. The model magnetic field predicts an enhancement of the B_L component at closest approach, a tripolar $-/+/-$ variation in the B_M component, and the usual $+/-$ bipolar B_N variation. The data generally follow this behavior, but the phases and amplitudes are in slight disagreement. The model convective flow field predicts primarily a $-/+$ V_{cN} signature as expected for a northward-going cylinder; the data also show this. The other flow components show less agreement, however. While the predicted V_{cM} variation is a tripolar $+/-/+$ variation, the observed signature shows largely a $+V_{cM}$ flow increase commencing at about the FTE midpoint. The departures from the incompressible plasma prediction probably reflect both a breakdown in the incompressible assumption and in the geometry of the obstacle, which is not expected to be a cylinder.

Another, almost brute-force demonstration of the FTE morphology comes from simultaneous observations of the magnetic fields on either side of the magnetopause. This was possible when ISEE 1 and 2 were at their greatest separations on the dayside in 1979, with one spacecraft in the magnetosphere, the other in the magnetosheath. Farrugia et al. [1987b] discussed these observations in detail. Figure 6 shows the observed variations of the magnetic field in the magnetosheath (ISEE 1) and in the magnetosphere (ISEE 2) for an FTE at 1358 - 1400 UT on November 9 1979. The field vectors are shown in the L-N plane along trajectories parallel to the magnetopause surface, assuming that the FTE is convecting as a whole at 150 km/s from north to south (the spacecraft are south of the equator). The spacecraft are separated by 3300 km normal to the magnetopause, with ISEE 1 some 3800 km ($\approx 0.6\ R_E$) further north along the L direction.

ISEE 1, the more northerly spacecraft, observes the first disturbance of the oncoming FTE: a negative B_N perturbation. At the time of maximum negative B_N on ISEE 1, ISEE 2 has begun to sense a negative excursion as well. When the ISEE 1 B_N signature passes through zero, at the midpoint of the FTE, ISEE 2 is approaching its maximum negative excursion. As the ISEE 2 B_N signature swings through zero, ISEE 1 is just past its maximum positive B_N excursion. Finally, ISEE 1 is sensing the last vestige of its positive B_N signature when ISEE 2 passes through its positive B_N maximum. The combination of spacecraft separation normal to the boundary, the separation along the boundary in the north-south direction, and the phase relationship between the two sets of B_N signatures indicates that a structure much like the impenetrable cylinder discussed above is passing on the magnetopause.

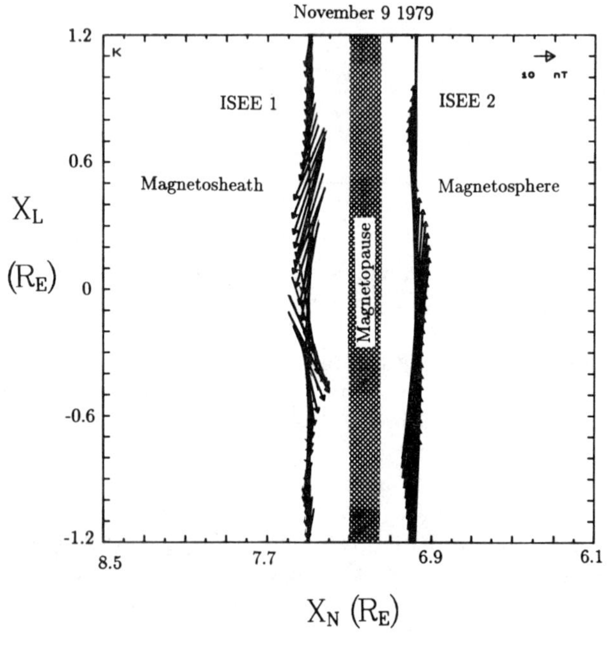

Fig. 6. FTE magnetic field variations in the B_L - B_N plane, as observed by ISEE 1 (magnetosheath) and ISEE 2 (magnetosphere) while the two spacecraft were separated by more than 0.5 R_E. The vertical trajectories of the spacecraft represent an assumed 150 km/s advection speed from north to south. The separation in L (pointing from ISEE 2 to 1) is 3800 km, and in N it is 3300 km.

Plasma and Particle Observations

The observed FTE magnetic field variation is not in itself necessarily evidence for reconnection. However, the observed plasma and particle signatures within FTEs are consistent with reconnection. Daly et al. [1981] found that energetic magnetospheric ions stream along the local magnetic field in FTEs, away from the Earth. Paschmann et al. [1982] found that the bulk properties of the FTE plasma are a mixture of tenuous, hot magnetospheric plasma, and the dense, thermal magnetosheath plasma. They also found that the FTE plasma is sometimes accelerated above the local magnetosheath speed. Scudder et al. [1984] have reported that a net parallel electron heat flux is found on one or both edges of magnetosheath FTEs; they ascribe the signature to ongoing reconnection. More recently, Thomsen et al. [1987] reaffirmed the view that FTE plasma is a mixture but also found significant modifications to the magnetosheath electron component, particularly heating preferentially parallel to the field. They found that the electron heat flux could be as readily explained by leakage of a hot magnetospheric component as by heating through reconnection. Sibeck et al. [1987] echoed this assessment for the energetic ions. Figure 7 shows the ion anisotropies as organized (a) by the merging model of ion escape and (b) by simple leakage through the boundary. In the merging model, the data have been organized by folding the -B_y cases about noon, so that the distribution of events corresponds to a duskward and southward IMF. For the leakage model, the positions are just as they were observed; Sibeck et al. argued that the observed sense of anisotropy corresponds to escape of magnetospheric particles onto magnetosheath field lines at roughly 1500 LT, where the Parker spiral field lines drape closely about the magnetopause. The observations include magnetopause crossings where no FTEs were observed.

It is instructive to view FTE plasma and magnetic field data together. Figure 8 shows a pass through the magnetopause near the nose by AMPTE UKS, the same crossing examined by Elphic and Southwood [1987]. The top panels contain ion density, temperature, thermal pressure, and vector flow velocity, respectively. The bottom panels show magnetic field; both field and flow are in boundary normal coordinates. UKS is initially in the magnetosphere

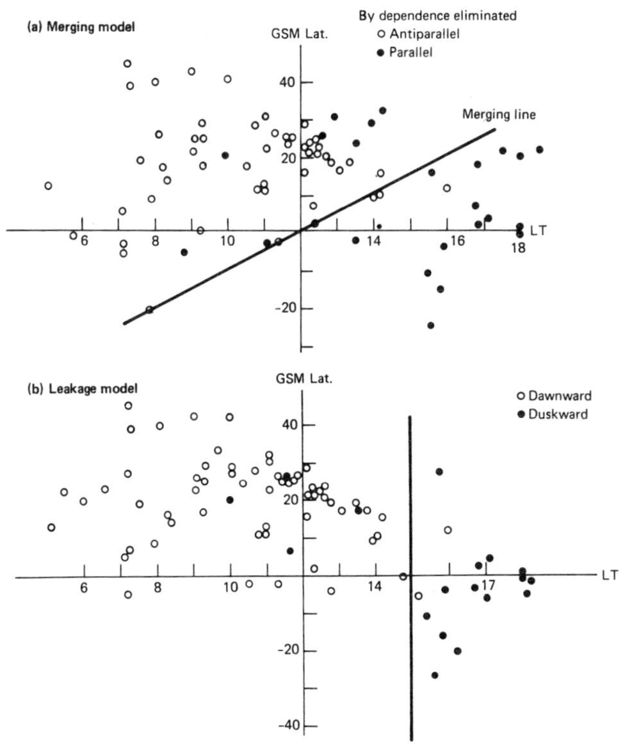

Fig. 7. A comparison of the distribution of energetic ion anisotropies parallel (dots) and anti-parallel (open circles) to the local magnetic field, (a) assuming the merging model and (b) based on leakage. Positions of the observations are shown on the dayside magnetopause as a function of local time and GSM latitude. (From Sibeck et al. [1987]).

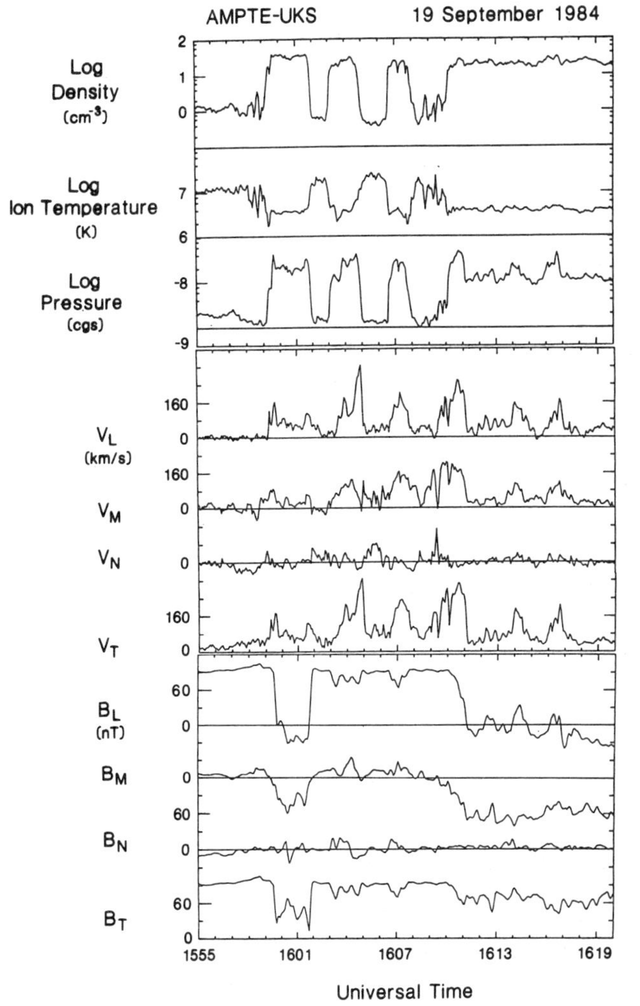

Fig. 8. Plasma ion and magnetic field data from AMPTE UKS for magnetopause crossings on September 19, 1984. The velocity and field are in boundary normal coordinates. As can be seen in the B_L component, the crossings occur at 1559:30, 1602:00 and 1611:00 UT, and accelerated flows are observed at these times. Rapid flows are also seen associated with the FTEs at 1604, 1607, 1614 and 1617 UT. The accelerated plasma flows appear to occur in a quasi-periodic manner, every 2 to 3 minutes. Each flow burst is associated with a maximum in ion thermal pressure.

as evidenced by the low plasma density and high temperature; there is a brief exit to the magnetosheath between 1559 and 1602 UT, characterized by high densities and low temperatures. Thereafter UKS returns to the magnetosphere but has two encounters with boundary layer-like plasma at 1604 and 1608 UT; the satellite does not exit the magnetosphere completely until 1610 UT.

There are flow bursts throughout the pass, with center times of 1559:30, 1601:45, 1604:30, 1607:15, 1610:30, 1614:00 and 1616:30 UT. The first two and the event at 1610:30 UT are associated with magnetopause current sheet crossings, and may reflect quasi-steady reconnection. The others appear to be associated with bipolar variations in the B_N component, a signature of FTEs. The magnitude of the plasma flow in each burst is considerably larger than the background magnetosheath flow speed of 40 – 80 km/s. All flow bursts are associated with ion thermal pressure maxima. Two possible explanations of this behavior are (1) Motions of the magnetopause, including undulations or small wavelength surface waves, carry a relatively steady-state fast flow layer over the spacecraft, giving the illusion of temporal burstiness; (2) Rapid reconnection flows occur over a variety of time scales, from quasi-steady to impulsive. In the former only quasi-steady reconnection is required, along with surface motion of the magnetopause; however, the arguments advanced in the last section put this explanation in doubt. The second picture explains why there should be a B_N signature in some events and not in others.

Figure 9 shows observations of an FTE by AMPTE UKS, adapted from Rijnbeek et al. [1987], and illustrating some of the plasma features of an FTE. Here selected moments of the ion distribution function, electron flux intensities and magnetic field variations are shown. This is one of a class of FTE signatures dubbed "crater FTEs", so named because of the local magnetic field minimum in the center of the event (cf. Lühr and Klöcker [1987] and LaBelle et al. [1987]). The data begin at 1043 UT in the magnetosphere where the ion plasma has a roughly 1 keV temperature, and a density of about 4 cm^{-3}. The intensities of 0.2 and 1 keV electrons is comparable within a factor of 2, characterizing a warm electron distribution in the magnetosphere. In region (1) the magnetic field gradually changes, the B_N component making a positive excursion. At 1045 UT the plasma density begins to increase, the temperature to drop, and the flow speed to rise. In the region marked (2) the 12 eV electron intensity rises and the 1 keV intensity falls, indicating the presence of a much cooler electron population. The 205 keV electron intensity has a local maximum in this region. The magnetic field magnitude rises. In region (3) the plasma has the characteristic density and temperature of the magnetosheath, and the magnetic field is skewed by nearly 30° from the magnetospheric orientation. The second half of the FTE shows this sequence of events in reverse, except that the B_N perturbation is negative.

Taken together, the particle and field characteristics are consistent with reconnection. Rijnbeek et al. have argued that Region (1) represents the disturbance region outside the reconnected flux tube, and Region (2) corresponds to the separatrix layer, and the enhanced 205 eV intensity is a signature of the electron heat flux out of the reconnection site, similar to that reported by Scudder et al. [1984]. It is a possible indication that reconnection is still occurring somewhere along field lines connected to the observation point. Region (3) is near the core of the reconnected flux tube, with accelerated convective plasma flow away from the reconnection site, and the magnetic field in an orientation intermediate between that of the magnetosphere and the magnetosheath. Not shown here is the streaming of energetic ions out of the magnetosphere.

There may be reasons to doubt that the Rijnbeek et al. [1987] FTE is in fact a reconnection event. It may be nothing more than a brief exit from the magnetosphere into the magnetosheath while the IMF is northward [D. Sibeck, private communication, 1989]. ISEE 1 and 2 were in the solar wind at the time and observed a quite variable IMF with a strong radial component. These conditions lead to the development of upstream waves which, when processed by the bow shock and convected to the magnetopause, may cause pressure fluctuations and boundary motions.

Paschmann et al. [1982] noted that the plasma and

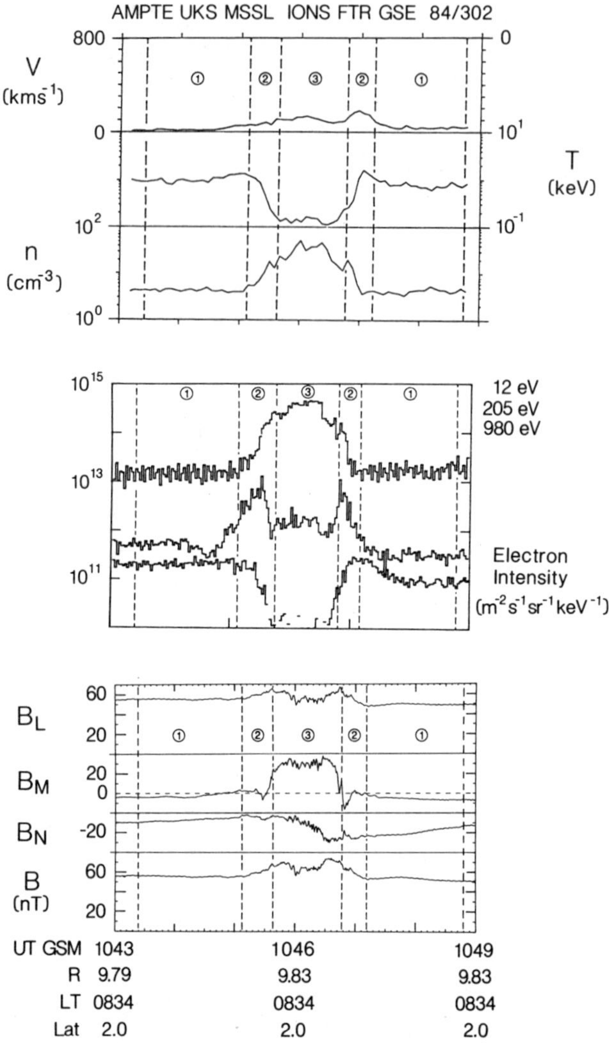

Fig. 9. AMPTE UKS observations of an FTE on October 28, 1984, adapted from Rijnbeek et al. [1987]. Top panels show the ion bulk flow, temperature and density moments. Middle panels show electron intensities at 12, 205 and 980 eV energies, characteristic of magnetosheath, transition and magnetospheric populations respectively. The bottom panels show the magnetic field in boundary normal coordinates. Region 1 corresponds to the draping or disturbance region outside the FTE proper, while Region 2 marks the transition from magnetosphere to magnetosheath-like plasma. Note the dramatic peak in the 205 eV intensity in Region 2. Finally, Region 3 is characterized by magnetosheath-like plasma, and a nonmagnetospheric field orientation. There is also an absence of the more energetic electrons. It has been suggested (Sibeck, private communication) that this is actually a brief exit from the magnetosphere during northward IMF due to a solar wind pressure pulse.

magnetic pressures in an FTE often maximize at the center of the event. They found that this overpressure could be balanced by the tension associated with the Maxwell stress of field lines wrapped around the core region. Their results are shown in Figure 10. They approximated the tension force of the wrapped field lines using the observed B_N variations and the background field. This relationship between tension and internal pressure implies that the FTE is a self-balancing entity, and led to the supposition that at least part of the B_N signature corresponds to a helically twisted outer field in FTEs [Cowley, 1982; Paschmann et al., 1982]. Interestingly, Paschmann et al. [1982] did not distinguish between grazing and penetrating passages through FTEs; both cases appear to support the tension/pressure balance picture.

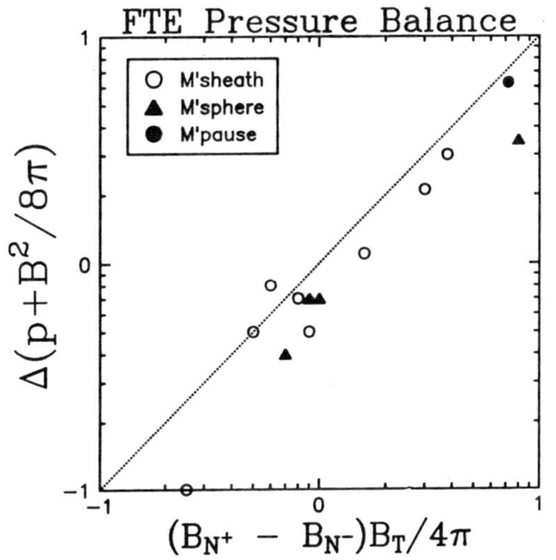

Fig. 10. Comparison of the total overpressure in FTEs versus the inferred magnetic tension due to the presence of an azimuthal field component, represented by the observed B_N signature. These results, from Paschmann et al. [1982], suggested that FTEs have a pinch-like structure reminiscent of a magnetic flux rope.

FTE Occurrence

FTEs have been interpreted as highly time-dependent reconnection events. Sometimes quasi-steady reconnection and FTEs are observed on the same magnetopause pass. Are the two seemingly distinct forms of reconnection related? Is there an intrinsic time scale associated with the growth and decay of reconnection? The quasi-periodic occurrence of FTEs suggests that the process has some intrinsic time scale for the buildup and release of free energy in the magnetopause current sheet. If so, there should be a relationship between the energy released in an FTE and the free-energy buildup time: the longer the buildup time, the greater the energy available for release, and hence the greater the energy in the FTE. The quasi-periodic buildup and release of energy is analogous to proposed substorm mechanisms in the magnetotail, to the unsteady flow of water drops from a leaky faucet, and even to the occurrence of earthquakes. It is, in short, a characteristic of certain states of a highly nonlinear dynamical system.

So we wish to explore the relationship between FTE (released) energy and the accumulated free energy since the last release. Because it is impossible to determine the total energy content of an FTE, we must use a measurable quantity which is in some way related to energy content. One possible parameter is simply FTE size, as characterized by the duration of the event. For this quantity to be a valid size parameter, we must assume that all FTEs travel at the same speed. To characterize the free energy accumulated at the magnetopause, we use the time since the last FTE. For this quantity to be a valid parameter we must assume that the last FTE released all free energy from the boundary, and moreover that the free energy accumulation rates are always the same.

We have measured FTE durations and inter-FTE times for a subset of events observed by AMPTE UKS or IRM, and ISEE 1 and 2; these are shown in Figure 11. There are 53 cases from ISEE 1 or 2, 24 from AMPTE UKS or IRM. Two points emerge: (1) Most of the FTEs observed by AMPTE, sampled closer to the equator, have shorter durations; (2) Larger FTEs tend to be observed for longer inter-FTE times. There is considerable scatter in

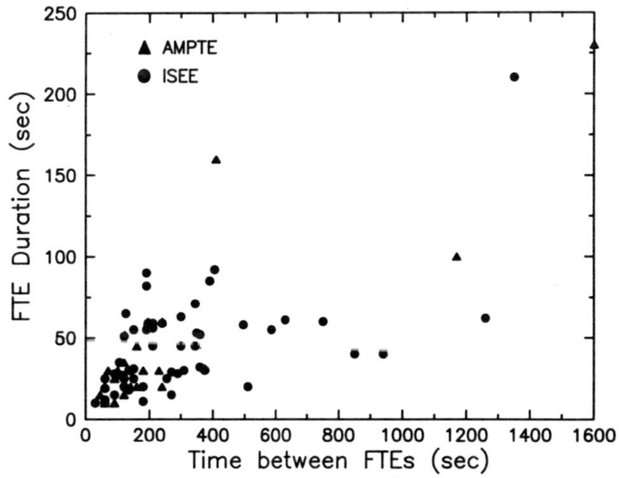

Fig. 11. FTE durations versus inter-FTE time. FTE duration is defined as the time between the extrema of the bipolar B_N signature; inter-FTE time is simply the time elapsed since the last FTE. If duration is an indication of FTE size, hence energy content, and inter-FTE time is an indicator of magnetopause free-energy accumulation time, then longer accumulation times lead to larger FTEs. Most of the AMPTE FTEs, sampled at lower latitudes, are smaller than those at ISEE.

the data, suggesting that our assumptions are not entirely good. Another parameter relating to FTE energy content would be an estimate of FTE cross-section. A measure of the FTE extent normal to the boundary is the ratio of the bipolar B_N excursion to the background field. The larger the size of the FTE normal to the boundary compared to its extent along the boundary, the larger the value of $\Delta B_N/\langle B\rangle$ (for a given impact parameter). When multiplied by FTE duration, the quantity becomes an estimate of FTE size in the boundary normal direction. Figure 12 shows how this quantity varies with inter-FTE time. Once again there is a trend suggesting that "larger" FTEs are found after longer energy accumulation times, and most of the AMPTE FTEs are smaller than the ISEE FTEs.

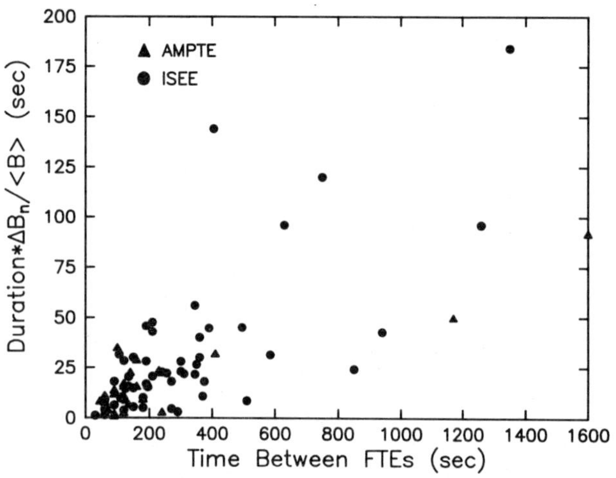

Fig. 12. Another FTE "size" measure versus inter-FTE time. The product of duration and $\Delta B_N/\langle B\rangle$ is a measure of FTE size normal to the magnetopause surface. Once again, larger FTEs appear to be associated with longer inter-FTE times.

The correlation coefficients for the above plots are approximately 0.5. There are many reasons for the scatter in Figures 11 and 12. Our simple measures of FTE size or energy content, and of the boundary's free energy accumulation time are crude. It is unlikely that all FTEs convect past the spacecraft at the same speed; thus duration is not an ideal measure of FTE size. Similarly, the observed amplitude of B_N depends on the spacecraft impact parameter, so it too is an imperfect measure of FTE extent normal to the boundary. The free-energy accumulation rate varies with IMF and solar wind dynamic pressure changes. Moreover, the quantity "Time Between FTEs" hides the fact that, if the last FTE was a small one, little free energy was removed from the boundary. Thus, a large FTE could follow a small one by a very short time.

Like unsteady water drops, there should be a relationship between the time since the last FTE and the time since the last one before that. This relationship, a kind of FTE strange attractor, would not be obvious until hundreds or thousands of FTEs had been observed, and then only under absolutely constant external conditions. In practice the external conditions are constantly changing. Thus, an intrinsically endogenic process (the quasi-periodic accumulation and shedding of free energy in the boundary) could be triggered irregularly by exogenic processes (solar wind pressure pulses, or changes in IMF orientation). A solar wind pressure pulse, for example, could pinch an initially stable magnetopause current sheet, drive it unstable, and produce a burst of reconnection.

FTE Models

Transient Reconnection

Russell and Elphic's [1978] original model for FTEs involved spatially and temporally limited reconnection. If a patch of the magnetopause were to become unstable to reconnection for a limited time the result would be a bundle of reconnected flux lines threading the magnetopause surface. The bipolar B_N variation is the signature of draping of the surrounding fields around the reconnected flux tube, and the intermediate field orientation within the FTE is the core field of the flux tube. A mixture of magnetosheath and magnetospheric plasma would be seen in the open flux tube, and that mixed plasma would be flowing at a velocity different from the background magnetosheath flow. The sense of the FTE plasma velocity would be in approximate agreement with quasi-steady reconnection stress balance. On the other hand, a grazing encounter with the reconnected flux tube would not necessarily show such an agreement.

Because of the relationship between FTE internal pressure and tension forces discussed above, Paschmann et al. [1982] and Cowley [1982] argued that the FTE magnetic field must have a twist, like a magnetic flux rope. In this view a field-aligned current flowing in the reconnected tube produces a net azimuthal field which helps to pinch the FTE tube. Such a configuration was shown in Figure 1. Saunders et al. [1984] examined FTE magnetic and flow field variations and found they obey the Walén relation, $\Delta\mathbf{V} = \pm\Delta\mathbf{V}_A = \pm\Delta\mathbf{B}(1-\alpha)^{1/2}(\mu_o n m_p)^{-1/2}$, where $\alpha = (p_{\parallel} - p_{\perp})\mu_o/B^2$, $p_{\parallel}$ and $p_{\perp}$ are plasma thermal pressure parallel and perpendicular to the local magnetic field, n is number density, and m_p is the proton mass. In other words, the flow perturbations are Alfvénic. This would be expected if the field twist were an Alfvén wave propagating along the FTE flux tube. However the sense of wave propagation on the magnetospheric side of an FTE is found to be different from that on the magnetosheath side. Consequently the field twist could not simply arise from shear in the plasma flow on the magnetosheath 'end' of the reconnected flux tube.

It now appears that Saunders et al. analyzed the disturbance flow and field discussed above, namely the perturbed flow of plasma about the onrushing FTE obstacle. As Farrugia et al. [1987a] point out, this incompressible flow/field perturbation obeys the relation $\Delta \mathbf{V} = c\Delta \mathbf{B}$, which resembles the Walén relation given an appropriate value of c. This disturbance flow about the FTE obstacle would appear to produce a field twist in one sense on the magnetosheath side and the opposite twist on the magnetospheric side. Nevertheless, Sonnerup [1987] argues, there should still be a field twist within the reconnected FTE tube. The field-aligned current producing the twist is related to the orientation of the magnetosheath field, and to the $\mathbf{J}\times\mathbf{B}$ forces acting on the kinked reconnected flux tube. Wright [1987] has also pointed out that such a twist is intrinsic to localized, or patchy reconnection. (See also the paper by Song and Lysak, this volume).

Scholer [1988] and Southwood et al. [1988] have introduced a revised view of FTEs as transient reconnection at a single x-line, drawing from work by Biernat et al. [1987] and Semenov et al. [1984] on time-dependent Petschek-type reconnection. Figure 13, taken from Scholer [1988], illustrates this transient reconnection process. The figure shows a snapshot from a 2-D MHD simulation of field and plasma flow at a field-reversing current sheet after the onset of reconnection at Z = 0. The plane of the figure is defined by the magnetic field vectors and the normal to the current sheet. Near the reconnection site the field and plasma have established the inflow and outflow characteristics of steady Petschek reconnection, while far from the reconnection site (near Z = 160) the initial Harris equilibrium current sheet is undisturbed. In the intermediate region the field and plasma undergo a transition from static equilibrium to the dynamic expansion fan geometry. The transition field is bubble-like, and contains the accelerated plasma. Outside the bubble the ambient plasma and magnetic field are forced out of the way in a fashion that is qualitatively very similar to that described by Farrugia et al. [1987a] for incompressible flow about a cylindrical obstacle.

This transient single x-line reconnection scenario contains most of the observed attributes of FTEs, including the proper sense of the B_N variation, accelerated plasma flow, energetic particle escape, and electron heat flux on field lines mapping to the reconnection site. Moreover, the statistically determined results for the sense of energetic ion anisotropy, the sense of the bipolar B_N signature, and the convective flow direction that were successfully explained by the Russell and Elphic [1978] picture are still consistent with the updated model.

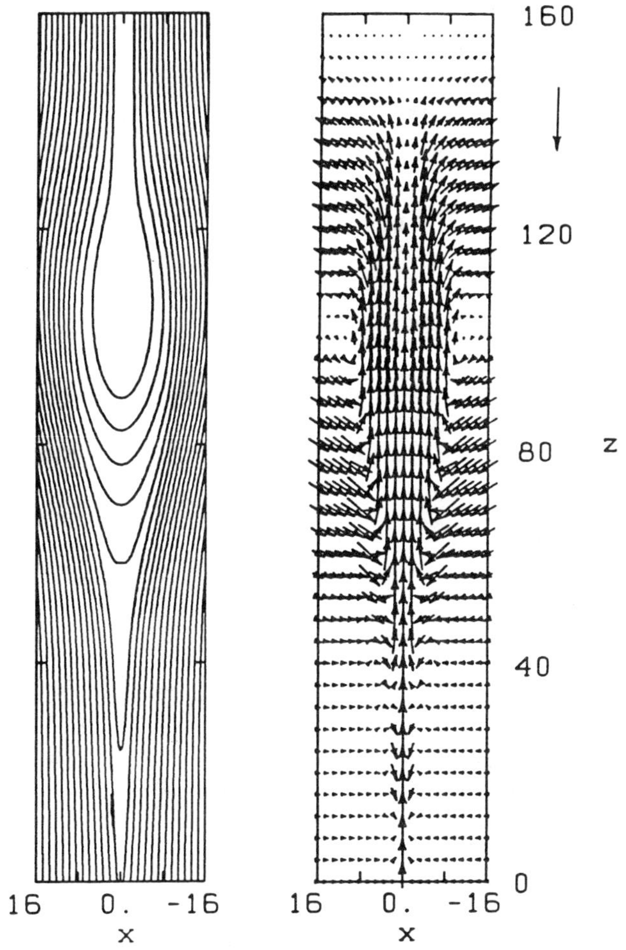

Fig. 13. Snapshot of magnetic field lines and plasma flow vectors from one time step of a two-dimensional MHD simulation of transient reconnection by Scholer et al. [1988]. The configuration began as a stable neutral sheet. Finite and spatially-limited resistivity was then imposed near the origin. Near the reconnection site the field and plasma have established the inflow and outflow characteristics of steady Petschek reconnection, while far from the reconnection site (beyond Z = 140) the initial equilibrium current sheet is undisturbed. In the intermediate region the field and plasma undergo a transition from static equilibrium to a standing expansion fan geometry. The transition field is bubble-like, and contains the accelerated plasma.

Multiple X-line Reconnection

Lee and Fu [1985] advanced an alternative mechanism for creating FTEs at the magnetopause. As Podgorny et al. [1980] had done, they suggested that multiple x-lines due to tearing mode reconnection could produce magnetic islands at the magnetopause current sheet. But Lee and Fu pointed out that, if the magnetosheath and magnetospheric fields are not precisely antiparallel, the islands will contain a field component along the island's axis. The resulting overall field configuration is that of a magnetic flux rope, as illustrated in Figure 14. As the tearing mode saturates, the islands grow to some maximum size and are eventually convected away down the flanks of the magnetosphere, and the process begins anew.

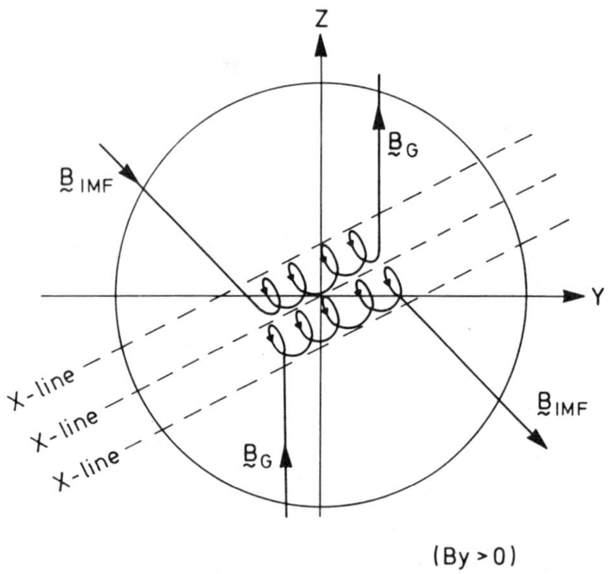

Fig. 14. Lee and Fu's [1985] picture of multiple x-line reconnection at the magnetopause when there is a finite field component along the x-lines. As the tearing mode develops the tearing islands contain an axial field component, and the overall structure of the islands is that of a magnetic flux rope. Eventually the islands develop nonlinearly and are convected away, then the process repeats. Poleward of the islands the signature of quasi-steady reconnection would be observed.

Variations on the multiple x-line reconnection have been advanced. Crooker [1986] invoked antiparallel merging to produce a large single island-like structure at low latitudes. There are also questions concerning the role of the Kelvin-Helmholtz instability in reconnection; Labelle-Hamer et al. [1988] have suggested that the tearing mode may feed off the K-H instability. There is also the possibility that the reverse takes place, that strongly sheared reconnection flows could drive the K-H instability.

The multiple x-line scenario explains the observed magnetic field variations, energetic particle anisotropy, plasma acceleration, the global distribution of B_N signatures and the episodic occurrence of FTEs. However the observed electron heat flux signatures of Scudder et al. [1984] may not be consistent with the tearing island picture. Multiple x-lines should produce multiple heat flux signatures, some nested within others. As an FTE passes across a spacecraft, electron heat fluxes both parallel and antiparallel to the field should be seen consecutively. FTE electron observations do not appear to support this. In fact, Scudder [1989] has recently shown preliminary evidence that, on a particular magnetopause pass the magnetosheath FTE electron heat flux is parallel to the local field, while for the magnetospheric FTEs it is antiparallel. If confirmed, this observation cannot be explained by simple leakage, and it is not consistent with multiple-x-line reconnection. The former would produce a single sense of heat flux parallel or antiparallel to **B** whether seen in the magnetosheath or in the magnetosphere. As noted above, multiple x-lines should produce a nested series of alternating heat flux senses. The Scudder [1989] observation is consistent with single x-line reconnection as observed from a location north of the reconnection site.

Kan [1988] has described a synthesis of the multiple and single x-line pictures. He suggests that any one x-line will be limited in longitudinal extent, so that along a given meridional cut through the magnetopause there is but one merging site. This is shown schematically in Figure 15. Each x-line may turn on and off intermittently, producing an FTE signature. Along any given meridional cut the observed plasma and field signature would be that of single x-line reconnection.

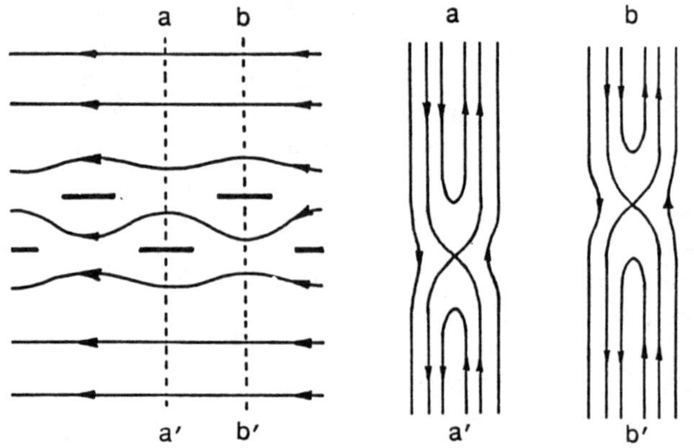

Fig. 15. Kan's [1988] synthesis of single and multiple x-line formation. The view is looking from the sun down on the magnetopause plane. X-lines (bold lines without arrows) are limited in longitude and turn on intermittently. Along any meridional cut (lines a - a' and b - b') there is at most one x-line. Lines with arrow heads are current flow lines.

FTEs and Solar Wind Pressure Pulses

Over the last decade there has arisen a minority view of solar wind/magnetosphere interactions based on nonsteady solar wind conditions. This view holds that parcels of solar wind plasma may impulsively penetrate the geomagnetic field [Lemaire et al., 1979; Heikkila, 1982], and cause the FTE signature. Lundin [1987] has recently reviewed the impulsive penetration picture; we shall not discuss it further. However, another side of the nonsteady solar wind picture has received increasing attention.

Recently Sibeck et al. [1989] have published observations of upstream solar wind dynamic pressure pulses that cause localized compressions of the magnetopause. These brief ($\approx$ 1 min.) pulses lead to signatures resembling the FTE $|\mathbf{B}|$ and B_N variations for the same reasons that Farrugia et al. [1987a] described; the ambient magnetospheric plasma and field are forced to circulate about any traveling magnetopause disturbance. A convecting tearing island, a single-x-line "bubble", or a surface wave all distort the boundary and force the surrounding medium into motion. This process sends Alfvén waves to the ionosphere and produces the vortices observed there [Friis-Christiansen et al., 1988; Elphic, 1988]. But as discussed earlier, in order to produce FTE signatures in both the magnetosheath and magnetosphere, the surface disturbance must protrude both into the magnetosphere and into the magnetosheath.

In a further development of the pressure pulse idea, Sibeck [1989] has argued how such a double-peaked protrusion may occur. The pressure pulse, traveling along the magnetopause, sends a fast mode wave ahead of it. The increased total pressure in the fast mode wave sends the local magnetopause out of equilibrium: the boundary must move outward in response. Consequently the whole train of magnetopause disturbances is first an outward movement of the boundary associated with the leading fast mode wave, followed by an inward movement associated with the external pressure pulse. A spacecraft in the magnetosheath would sense the $+/-$ B_N disturbance of the outward magnetopause protruberence, while a spacecraft in the magnetosphere would sense the disturbance associated with the external pressure pulse. The boundary motion and perturbed magnetic fields are shown schematically in Figure 16.

There are a number of open questions in the pressure pulse model. One is whether or not there is evidence for a fast mode wave ahead of the magnetosheath pressure pulse signature. In the ISEE data shown in Figure 6, and discussed by Farrugia et al. [1988] there was no indication of a field enhancement in the magnetosphere ahead of the magnetosheath B_N signature. ISEE 2 was ideally placed to observe such an enhancement. Instead, the magnetospheric field maximum was inferred to coincide spatially with the one in the magnetosheath. Another question centers on how the pressure pulse can cause accelerated plasma flow, as is observed. FTEs are observed to occur almost exclusively during southward IMF; pressure pulses have no preference for southward IMF orientation.

Discussion

Elements of each of the above scenarios may occur at the magnetopause at one time or another. The question is whether any one of them can explain the observed suite of FTE attributes. If one such model can be singled out, the question of FTE topology may be addressed. There

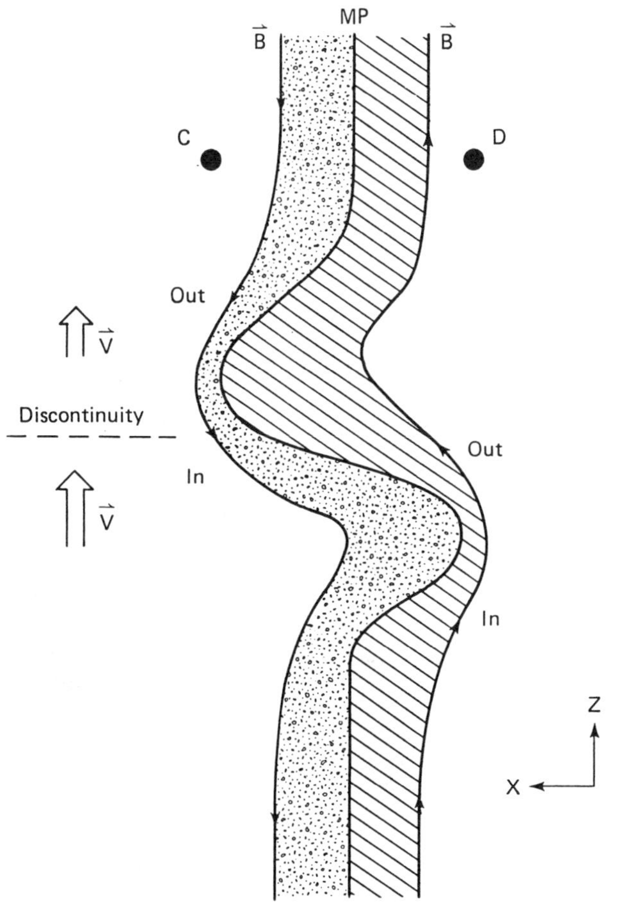

Fig. 16. Schematic diagram of the magnetopause response to an isolated solar wind pressure pulse. The local magnetosheath pressure maximum forces a localized inward displacement of the magnetopause. A fast-mode wave can travel more rapidly in the magnetosphere, and so runs ahead of the magnetosheath pressure pulse. However, the higher pressure in the fast mode wave pushes the magnetopause outward. In this way a single pressure pulse can produce an outward deflection on the magnetosheath side and an inward deflection on the magnetospheric side. (From Sibeck [1989]).

is, however, the possibility that a combination of these models may apply.

The transient single x-line picture appears to be able to reproduce the detailed characteristics and global statistics of FTEs. In addition it offers an explanation for the observed association between quasi-steady and impulsive reconnection: FTEs can be regarded as a very large modulation of the reconnection rate. The multiple x-line scenario also explains the observed FTE characteristics, with the possible exception of electron heat flux. However, since there is some question concerning the origin of the electron signature, it cannot be used as a discriminator between the two reconnection scenarios.

Finally, the pressure pulse scenario offers an explanation for the apparently impulsive nature of FTEs, but does not invoke reconnection. In this case there is some question of how accelerated plasma flow arises, as well as whether or not the required fast mode wave is observed in the magnetosphere ahead of the magnetosheath pressure pulse. On the other hand, there may be times when a pressure pulse can mimic an FTE signature. Any boundary indentation, if in the canonical few-minute FTE time scale, could be identified as an FTE.

As stated above, while a single model explaining FTEs is desirable, it may be a combination of the above scenarios that relates to the formation of FTEs. The simplest explanation for many of the observed plasma and energetic particle characteristics of FTEs is a form of reconnection. Solar wind pressure variations may trigger the reconnection process in the first place by forcing the magnetopause current sheet to thin and destabilize. Initially, at small scales, it may be tearing mode island formation that is the embryonic state of FTEs. As islands form, they may coalesce leaving a single localized active reconnection site. The reconnection rate at this site could grow and diminish, providing the observed FTE signature as well as that of quasi-steady reconnection. During northward IMF there is less total current (and lower current density) flowing in the magnetopause, and the boundary may be stable to reconnection even when pressure pulses force it to thin.

Finally, the question of whether FTEs are magnetic flux ropes has not been answered. If the multiple x-line scenario obtains, then FTEs must have a rope-like structure. In the transient single x-line picture, the field topology is not that of a rope. However, it has certain rope-mimicking qualities. The pressure pulse/surface wave model does not alter the magnetopause field topology.

Summary and Conclusions

In order to address the question of FTE origin, structure and topology, we have reviewed their salient observational features and statistical attributes. In particular we have focused on their magnetic, plasma and energetic particle signatures. We have also discussed the occurrence of FTEs, their quasi-periodic nature and how that might be a clue to their origin. Finally we have discussed three basic hypotheses put forward to explain the physics of FTEs.

The magnetic signature of FTEs includes a distinctive bipolar signature in the component normal to the magnetopause surface. Roughly, the direct (+/−) signature is found north of the equator, while the reverse (−/+) is found to the south. The same sense is found inside as outside the magnetosphere, and the direct signature corresponds to northward-going plasma, the reverse to southward. There are other field variations which suggest that at least some FTE signatures are due to grazing encounters with the FTE obstacle and do not correspond to passage through the reconnected flux tube at all. Consequently the field strength maxima often observed in FTEs can be ascribed (at least in some cases) to the draping or disturbance field around the FTE obstacle, not to the 'core field' at all. Observations of FTE fields when two spacecraft are on either side of the magnetopause strongly suggest that the FTE obstacle is a bubble-like structure (as opposed to a single indentation of the boundary). FTEs are found to occur preferentially when the local magnetosheath field has a southward orientation.

Plasma and particle observations tend to support the reconnection picture of FTEs. The combination of accelerated plasma flows, energetic particle anisotropies consistent with the emptying of a newly opened magnetic flux tube, and the mixture of magnetosheath and magnetospheric plasmas all point to a reconnection-related phenomenon. The observation of electron heat flux at the outer edge of magnetosheath FTEs may also support this interpretation, though the heat flux may just be magnetospheric leakage. Cases where no plasma acceleration is observed may correspond to grazing encounters with the reconnected flux tube. Finally, the observational agreement between the total (thermal and magnetic) pressure maxima within FTEs and the inferred external magnetic tension in the B_N field component points to a localized, self-pressure-balancing entity.

FTEs were first interpreted as a form of spatially and temporally-limited, each FTE a single flux tube connecting magnetosheath and magnetospheric fields. More recently it has become clear that FTE signatures may arise simply from the sudden and extreme modulation of the reconnection rate at a single x-line; there is no reason why they should be limited in longitudinal extent. Multiple reconnection sites can also produce a flux rope form of tearing island having the attributes of an FTE. There are few convincing ways to discriminate between the two forms of reconnection. At present the best hope is to use electron heat flux observations; but even this is possibly ambiguous. Though surface waves were initially rejected as an explanation for FTEs, pressure pulse-driven boundary motion has recently been advanced as a means of producing an FTE-like signature. At a minimum, this view implies that not all B_N bipolar signatures are due to transient reconnection.

Taken together, FTE plasma, particle and field characteristics strongly suggest a reconnection-related origin. However, since FTE identification is based largely on the B_N magnetic signature, it is possible to find many cases lacking one or more of the other FTE characteristics. For example, in a grazing encounter little or no plasma acceleration would be observed. By the same token, a B_N signature caused by a solar wind pressure pulse will probably not be accompanied by other characteristics of transient reconnection; it is possible to misidentify FTEs.

Nevertheless, in the case of FTEs possessing the requisite reconnection behavior the evidence points toward single x-line reconnection modulated at some characteristic (1 – 2 minute) time scale, with a recurrence interval of several times the modulation time scale (5 – 10 minutes). What remains unknown is the reason for these two time scales, which are far longer than the Alfvén travel time across the magnetopause current sheet. Rather, the modulation timescale is more like the time required for a fast mode wave to transit the dayside magnetosphere, while the recurrence interval is approximately the time for an Alfvén wave to propagate along the geomagnetic field to the ionosphere, and return.

Possibly solar wind pressure pulses and transient reconnection are related. If upstream measurements reveal a pressure pulse associated with every FTE at the magnetopause, this does not necessarily mean that FTEs are pressure pulses. Instead the pulse may thin the magnetopause current sheet, destabilizing it to reconnection. The reconnection event may be short-lived or long, depending on the free-energy available in the current sheet. In any case only the transient part of the event causes the FTE signature. The recurrence rate of reconnection bursts may be externally-driven or a function of the coupled magnetopause/ionosphere system. Future work on FTEs may reveal that the local magnetopause reconnection rate is governed both by local plasma conditions and by the response of the global magnetospheric system to those conditions.

Questions and Answers

Priest: What is the location and frequency of FTEs on the magnetosphere and how do they vary with the direction of the IMF? How much of the magnetosphere is covered with FTE holes? What proportion of the flux is reconnected in FTEs and in steady reconnection?

Elphic: Chris Russell showed a diagram that suggests FTEs are seen almost everywhere on the dayside magnetopause when the interplanetary field is southward. Their occurrence is greatest when the IMF is due southward, and falls off rapidly as the IMF orientation passes through "horizontal." According to two independent FTE surveys, FTE signatures are not seen for northward IMF.

How much of the magnetopause is "open" is impossible to answer based on local spacecraft measurements. However, either the single-x-line model or the multiple-x-line can extend across most of the dayside magnetopause.

One of the possibilities I mentioned is that the bursty reconnection picture gives the FTE signature. Once reconnection turns on, it may continue for some time in a quasisteady manner. In this case, much more flux may be reconnected in the quasi-steady manner than in the FTE portion itself. In some cases, quasisteady reconnection has been observed for 30 minutes before the spacecraft left the vicinity of the magnetopause. For a typical FTE duration of, say, three minutes, the proportion of flux reconnected would be 10% of that in quasi-steady reconnection, provided reconnection rates are the same in either case. However, if FTEs are simply the signature of the onset or dramatic change in the role of reconnection, the question of the proportion of reconnected flux becomes meaningless.

Linker: If variations in the solar wind dynamic pressure are responsible for FTEs, why are FTEs correlated with southward IMF?

Elphic: Apologists for the pressure pulse explanation of FTEs argue that conditions at the magnetopause may be different for northward IMF. Specifically, the wave that makes the bulge on the magnetopause may propagate differently for northward IMF, and may not give a noticeable B_N signature. I know of no published observations of plasma conditions being different for northward or southward IMF.

This is not to say that pressure pulses have nothing to do with FTEs: indeed, pulses may cause FTEs. A solar wind pressure pulse could compress and destabilize the magnetopause current sheet, leading to a brief episode of reconnection. For northward IMF, the free energy available even in a compressed current sheet is insufficient to produce energetically important reconnection.

Sibeck: Why do FTEs occur primarily for southward IMF if they are driven by solar wind pressure variations? Solar wind pressure variations do not occur primarily for southward IMF.

Elphic: I believe that solar wind dynamic pressure variations drive magnetopause motion, that the amplitude of this motion depends (indirectly) on the IMF orientation, and that the plasma, magnetic field, and particle signatures of this motion resemble those reported for FTEs.

Following Kaufmann and Konradi (1969), I believe that an increase in the solar wind dynamic pressure launches a fast-mode compressional wave in the magnetosphere. If the fast-mode wave speed exceeds that of the solar wind discontinuity in the magnetosheath, the magnetopause moves outward in advance of the discontinuity. The disturbances which the advancing magnetopause bulge creates in the magnetosheath plasma, magnetic field, and energetic particles are similar to those previously associated with FTEs (Sibeck, 1989).

The speed at which the fast-mode compressional wave advances depends upon the density and temperature of the low-latitude boundary layer (LLBL), a region of magnetosheath-like plasma just inside the equatorial magnetopause. I would argue

that the density of this layer increases, and the temperature decreases, during periods of northward IMF (Haerendel et al., 1978; Mitchell et al., 1987). If so, the fast-mode wave would be less likely to advance ahead of the solar wind discontinuity, and there would be less magnetopause motion (FTEs?), during periods of northward IMF.

Finally, I note that many (perhaps all?) of the "crater" FTEs recently reported by LaBelle et al. (1987) and Farugia et al. (1988) occurred during periods of northward IMF according to simultaneous IMP-8/ISEE measurements. The IMF dependence of FTEs may require further work.

Van Ballegooijen: Do the flux ropes associated with FTE's ever move away from the magnetopause?

Elphic: All evidence I am aware of points to the conclusion that the FTE "tube" or "island" is centered on the magnetopause current sheet.

Zelenyi: Islands are just the two-dimensional cross-sections of the ropes (or tubes). Why do you mention both assumptions as contradicting one to another?

Elphic: The tearing islands I showed are indeed flux ropes. So are the FTEs pictured in the original Russell and Elphic model. The topology of the bursty reconnection picture of Scholer (1988) and Southwood et al. (1988) is not that of a flux rope. Field lines do not wrap completely around an axial core.

Acknowledgments. The author benefited from discussions with D. Sibeck, V. Sergeev, H. Biernat, M. Thomsen, J. Gosling and many other colleagues. He is grateful to Mark Smith for AMPTE UKS plasma data, and to Cindy Cattell for ISEE electric field data. The work at UCLA was supported by NASA grants NAG5-536 and NAGW-1663; the work at LANL was done under the auspices of the U.S. Department of Energy.

References

Baumjohann, W. and G. Paschmann, Solar wind – magnetosphere coupling: Processes and observations, *Phys. Scripta, T18,* 73, 1987.

Berchem, J., and C. T. Russell, Flux transfer events on the magnetopause: Spatial distribution and controlling factors, *J. Geophys. Res., 89,* 6689, 1984.

Biernat, H., M. F. Heyn, and V. S. Semenov, Unsteady Petschek reconnection, *J. Geophys. Res., 92,* 3392, 1987.

Cowley, S. W. H., The causes of convection in the Earth's magnetosphere, *Rev. Geophys. Space Phys., 20,* 531, 1982.

Crooker, N. U., An evolution of antiparallel merging, *Geophys. Res. Lett., 13,* 1063, 1986.

Dailey, R., C. A. Cattell, F. S. Mozer and J. Berchem, Electric fields and convection velocities associated with flux transfer events, *Geophys. Res. Lett., 12,* 843, 1985.

Daly, P. W., D. J. Williams, C. T. Russell and E. Keppler, Particle signature of magnetic flux transfer events at the magnetopause, *J. Geophys. Res., 86,* 1628, 1981.

Daly, P. W., M. A. Saunders, R. P. Rijnbeek, N. Sckopke and C. T. Russell, The distribution of reconnection geometry in flux transfer events using energetic ion, plasma and magnetic data, *J. Geophys. Res, 89,* 3843, 1984.

Dungey, J. W., Interplanetary magnetic field and the auroral zones, *Phys. Rev. Lett., 6,* 47, 1961.

Elphic, R. C., Multipoint observations of the magnetopause: Results from ISEE and AMPTE, *Adv. Space Res., 8,* 223, 1988.

Elphic, R. C., and D. J. Southwood, Simultaneous measurements of the magnetopause and flux transfer events at widely separated sites by AMPTE UKS and ISEE 1 and 2, *J. Geophys. Res., 92,* 13,666, 1987.

Farrugia, C. J., D. J. Southwood and S. W. H. Cowley, Observations of flux transfer events, *Adv. Space Res., 8,* 249, 1988.

Farrugia, C. J., R. C. Elphic, D. J. Southwood and S. W. H. Cowley, Field and flow perturbations outside the reconnected field line region in flux transfer events: Theory, *Planet. Space Sci., 35,* 227, 1987a.

Farrugia, C. J., D. J. Southwood, S. W. H. Cowley, R. P. Rijnbeek, and P. W. Daly, Two-regime flux transfer events, *Planet. Space Sci., 35,* 737, 1987b.

Friis-Christiansen, E., M. A. McHenry, C. R. Clauer and S. Vennerstrom, Ionospheric traveling convection vortices observed near the polar cleft: A triggered response to sudden changes in the solar wind, *Geophys. Res. Lett., 15,* 253, 1988.

Gosling , J. T., M. F. Thomsen, S. J. Bame and C. T. Russell, Accelerated plasma flows at the near-tail magnetopause, *J. Geophys. Res., 91,* 3029, 1986.

Haerendel, G., G. Paschmann, N. Sckopke, H. Rosenbauer and P. C. Hedgecock, The frontside boundary layer of the magnetosphere and the problem of reconnection, *J. Geophys. Res., 83,* 3195, 1978.

Heikkila, W. J., Impulsive plasma transport through the magnetopause, *Geophys. Res. Lett., 9,* 159, 1982.

Kan, J. R., A theory of patchy and intermittent reconnections for magnetospheric flux transfer events, *J. Geophys. Res., 93,* 5613, 1988.

LaBelle, J., R. A. Treumann, G. Haerendel, O. H. Bauer, G. Paschmann, W. Baumjohann, H. Lühr, R. R. Anderson, H. C. Koons and R. H. Holzworth, AMPTE IRM observations of waves associated with flux transfer events in the magnetosphere, *J. Geophys. Res., 92,* 5827, 1987.

Labelle-Hamer, A. L., Z. F. Fu, and L. C. Lee, A mechanism for patchy reconnection at the dayside magnetopause, *Geophys. Res. Lett., 15,* 152, 1988.

Lee, L. C. and Z. F. Fu, A theory of magnetic flux transfer at the earth's magnetopause, *Geophys. Res. Lett.*, *12*, 105, 1985.

Lemaire, J., M. J. Rycroft and M. Roth, Control of impulsive penetration of solar wind irregularities into the magnetosphere by the interplanetary magnetic field direction, *Planet. Space Sci.*, *27*, 47, 1979.

Lühr, H. and N. Klöcker, AMPTE IRM observations of magnetic cavities near the magnetopause, *Geophys. Res. Lett.*, *14*, 186, 1987.

Lundin, R., Processes in the magnetospheric boundary layer, *Physica Scripta*, *T18*, 85, 1987.

Papamastorakis, I., G. Paschmann, W. Baumjohann, B. U. Ö. Sonnerup and H. Lühr, Orientation, motion and other properties of flux transfer event structures on September 4, 1984, *J. Geophys. Res.*, *94*, 8852, 1989.

Paschmann, G., B. U. Ö. Sonnerup, I. Papamastorakis, N. Sckopke, G. Haerendel, S. J. Bame, J. R. Asbridge, J. T. Gosling, C. T. Russell and R. C. Elphic, Plasma acceleration at the earth's magnetopause: Evidence for reconnection, *Nature*, *282*, 243, 1979.

Paschmann, G., G. Haerendel, I. Papamastorakis, N. Sckopke, S. J. Bame, J. T. Gosling and C. T. Russell, Plasma and magnetic field characteristics of magnetic flux transfer events, *J. Geophys. Res.*, *87*, 2159, 1982.

Podgorny, I. M., E. M. Dubinin, and Yu. N. Potanin, On magnetic curl in front of the magnetosphere boundary, *Geophys. Res. Lett.*, *7*, 247, 1980.

Rijnbeek, R. P., S. W. H. Cowley, D. J. Southwood and C. T. Russell, A survey of dayside flux transfer events observed by ISEE 1 and 2 magnetometers, *J. Geophys. Res.*, *89*, 786, 1984.

Rijnbeek, R. P., C. J. Farrugia, D. J. Southwood, M. W. Dunlop, W. A. C. Mier-Jedrzejowicz, C. P. Chaloner, D. S. Hall and M. F. Smith, A magnetic boundary signature within flux transfer events, *Planet. Space Sci.*, *35*, 871, 1987.

Russell, C. T., and R. C. Elphic, Initial ISEE magnetometer results: Magnetopause observations, *Space Sci. Rev.*, *22*, 681, 1978.

Saunders, M. A., C. T. Russell and N. Sckopke, Flux transfer events: Scale size and interior structure, *Geophys. Res. Lett.*, *11*, 131, 1984.

Scholer, M., Magnetic flux transfer at the magnetopause based on single X line bursty reconnection, *Geophys. Res. Lett.*, *15*, 291, 1988.

Scudder, J. D., K. W. Ogilvie and C. T. Russell, The relation of flux transfer events to magnetic reconnection, in *Magnetic Reconnection in Space and Laboratory Plasmas, Geophys. Monogr. Ser.*, vol. 30, edited by E. W. Hones, Jr., pp. 153-154, Washington, D. C., 1984.

Scudder, J. D., Experimental differentiation of magnetic topologies near "canonical" sites of "reconnection", (abstract) *EOS Trans. AGU*, *70*, 426, 1989.

Semenov, V. S., I. V. Kubyshkin, M. F. Heyn, and H. K. Biernat, Temporal evolution of the convective plasma flow during a reconnection process, *Adv. Space Res.*, *4*, 471, 1984.

Sibeck, D. G., The ionospheric response to solar wind dynamic pressure variations, (abstract) *EOS Trans. AGU*, *70*, 408, 1989.

Sibeck, D. G., R. W. McEntire, A. T. Y. Lui, R. E. Lopez, S. M. Krimigis, R. B. Decker, L. J. Zanetti and T. A. Potemra, Energetic magnetospheric ions at the dayside magnetopause: leakage or merging?, *J. Geophys. Res.*, *92*, 12,097, 1987.

Sibeck, D. G., W. Baumjohann, R. C. Elphic, D. H. Fairfield, J. F. Fennell, W. B. Gail, L. J. Lanzerotti, R. E. Lopez, H. Lühr, A. T. Y. Lui, C. G. Maclennan, R. W. McEntire, T. A. Potemra, T. J. Rosenberg and K. Takahashi, The magnetospheric response to 8-minute period strong-amplitude upstream pressure variations, *J. Geophys. Res.*, *94*, 2505, 1989.

Sonnerup, B. U. Ö., On the stress balance in flux transfer events, *J. Geophys Res.*, *92*, 8613, 1987.

Sonnerup, B. U. Ö., Experimental tests of FTE theories, *Adv. Space Res.*, *8*, 263, 1988.

Sonnerup, B. U. Ö., G. Paschmann, I. Papamastorakis, N. Sckopke, G. Haerendel, S. J. Bame, J. R. Asbridge, J. T. Gosling and C. T. Russell, Evidence for magnetic field reconnection at the Earth's magnetopause, *J. Geophys. Res.*, *86*, 10,049, 1981.

Southwood, D. J., C. J. Farrugia and M. A. Saunders, What are flux transfer events?, *Planet. Space Sci.*, *36*, 503, 1988.

Thomsen, M. F., J. A. Stansberry, S. J. Bame, S. A. Fuselier and J. T. Gosling, Ion and electron velocity distributions within flux transfer events, *J. Geophys. Res.*, *92*, 12,127, 1987.

Wright, A. N., The evolution of an isolated reconnected flux tube, *Planet. Space Sci.*, *35*, 813, 1987.

THE THEORY OF FTE STOCHASTIC PERCOLATION MODEL

M. M. Kuznetsova and L. M. Zelenyi

Space Research Institute, Academy of Science, U.S.S.R.

Abstract. Stochactic percolation model of spontaneous localized reconnection of magnetic field lines through the magnetopause current layer (MCL) due to the growth of multiple collisionless tearing-mode within it is proposed. The suggested mechanism of magnetic reconnection for plasma without collisions or noise is based on an intrinsic property of MCL — the presence of magnetic shear there. Reconnection appears to be a complex irregular multiscale process associated with the diffusion of magnetic field on self-consistently generated magnetic turbulence. We call this process magnetic percolation to emphasize its stochastic turbulent nature and finally it results in establishing of a topological connection of field lines on both sides of the MCL. There are two bounds on the thickness of the MCL L_0 for the formation of reconnection "patchy" with characteristic spatial scales along magnetopause $\lambda_z \times \lambda_y$. One is related to the conditions of linear destabilization the tearing perturbation with wave length λ_z at all magnetic surfaces within the MCL. The other is associated with the diffusion of magnetic field lines and is governed by the width w^* of nonlinear saturation of the magnetic island growth — the length of magnetic field line that accomplished the diffusion $s_0 \sim L_0^2/w^{*3}k_\parallel'$ should not exceed λ_y. Further behavior of such percolated field lines depends on the specific global magnetic field and plasma flow pattern and may be coupled with some macroscopic models of FTE formation through the diffusion term. For MCL with thicknesses below the critical value the diffusion of a single "elementary" magnetic filament is rather fast process. So during the time which the reconnection "patchy" spend in the stagnation area the whole bunch of percolated field lines can gather to form the FTE magnetic tube. The average angle at which this FTE tube of percolated magnetic lines "transects" the magnetopause surface depends on the level of magnetic turbulence within the MCL. Two possible geometries of the FTE tube (elbow-shaped and extended along the magnetopause) are discussed.

1. Introduction.

The discovery of flux transfer events — FTEs has emphasized the importance of the localized impulsive forms of reconnection and, after the original papers by Russell and Elphic [1978, 1979] the unsteady reconnection models have begun to be intensively explored. In the meantime the internal structure and the occurrence statistics of FTEs have been studied in more detail (for example, [Paschmann et al., 1982; Berchem and Russell, 1984; Daly et al.; Saunders et al.,1984]). Since the initial observational papers a number of models for FTE [Lee and Fu, 1985; Pudovkin and Semenov, 1985; Galeev et al., 1986; Scholer, 1988; Southwood et al.,1988] have been proposed. Significant progress has been achieved in 2D, 3D and time dependent MHD simulations of magnetic driven reconnection [Fu and Lee, 1985; Sato et al., 1986; La Belle-Hamer et al., 1988; Shi et al., 1988; Walker and Ogino, 1988]. Most of this models imply the formation of an X-line (single of multiple) with an associated diffusion region (region where the finite resistivity is present) having a thickness δ of the order of ion inertial length (>100 km) and even grater latitude width. For example, the multiple X-line reconnection model proposed by Lee and Fu [1985], which is widely discussed now in literature and has an original geometry different from that suggested by Russell and Elphic, assumes the existence of the diffusion regions extended over the half of the dayside magnetopause with thickness $\delta \approx (0.1\text{–}0.2)a$, length $l_y > 10a$ and resistivity $\eta \sim (0.00075\text{–}0.015)4\pi v_A a/c^2$, where v_A is Alfven speed and a is the simulation length in the Earth-Sun direction which is comparable with the FTE tube diameter (>1000 km).

Such an approach implies an important role for microscopic plasma waves in collisionless plasma in the magnetopause region. Experimental investigations of plasma waves at the magnetopause, including recent results from the AMPTE/IRM satellite are represented in a review paper by LaBelle and Treumann [1988]. Wave instruments have failed to see the strong wave turbulence which might characterize the diffusion region. Even the maximum observed wave amplitude (few mV/m) does not correspond to the anomalous resistivity required by the MHD reconnection model. Analyzing the correlation between observed FTEs and wave morphology LaBelle et al. [1987] come to the conclusion that "either all of the FTEs are observed at some distance from the diffusion region or the observable waves play no significant role in the diffusion process, or "traditional" reconnection models are not applicable to reconnection in FTEs". Thus, the important question concerning the origin

Geophysical Monograph 58

of X–line where, in fact, tearing of magnetic field line occurs, remains unclear in the "forced" MHD reconnection models. In addition to a global picture we need to investigate the physical mechanism for tearing of the magnetic field lines just inside the local regions which have much smaller scales than the global region that is generally described by MHD models, i.e., to study the fine dynamics of the thin magnetopause current layer. The most difficult problem is to couple these local processes with global magnetic field and plasma flow pattern.

The experimental observations mentioned above give arguments logically justifying the application of the collisionless tearing–mode mechanism for the study of magnetopause reconnection. The tearing instability driving source is the free energy of the sheared magnetopause magnetic field itself, without any ad hoc assumptions. Initially the theory of the tearing–mode has been developed for the problems of plasma confinement in tokamaks [Laval et al., 1966]. Since the end of the 1970th the importance of the tearing–mode process for the dayside magnetopause has been emphasized by Galeev and Zelenyi [1977].

The main problem one meets in studying the reconnection is to understand the bursty, sporadic character of these phenomena. It seems as if the system possesses matastable properties with respect to the reconnection process. If one studies these phenomena with the help of plasma instability formalism it is clear that some linear or nonlinear thresholds should exist for the growth of tearing perturbations.

The present report is devoted to the discussion of the fine dynamics and stability of the magnetopause current layer (MCL). The generalized equilibrium Harris–type model used in the paper may, of course, be refined in the future. Nevertheless, it contains the main shear property of the magnetopause — the rotation of a magnetic field within the MCL. This slab equilibrium model has plane undestroyed magnetic surfaces, so one can imagine the spontaneous establishing of topological connection between interplanetary and magnetospheric magnetic field as a result of the magnetic field line diffusion across the MCL in some spatially limited region of it (the appearance of a reconnection "patchy"). Further behavior of such reconnected field lines (common for magnetospheric and solar wind plasma) depends on the specific global magnetic field and plasma flow pattern and may be coupled with some macroscopic MHD models of FTE formation through the diffusion term. If the spatial scale of a reconnection patchy is sufficiently large, reconnected magnetic field lines can simply gather into tubes. The motion of such a reconnected flux tube (FTE) is determined by its convection with the solar wind plasma flow in the magnetosheath to the high–latitude regions or to the flanks of the magnetosphere. Flux tubes can experience twisting in the course of this convection, which results in the generation of field–aligned currents within them [Saunders et al., 1984; Sonnerup, 1987; Wright, 1987]. However, we shell focus our attention here primarily on the nontrivial subject of formation of the reconnection "patchy" at the magnetopause.

2. The Dynamics of the Drift Tearing Perturbations at the Magnetic Surfaces within the Magnetopause Boundary Layer.

The model of magnetic field configuration at the magnetopause is shown in Figure 1a. Such a layer with a characteristic thickness $L_0 = 2L$ $(-L < X < L)$ where the magnetic field rotates by an arbitrary angle θ_0 from a value $\vec B_{IMF}$ at one side to a value $\vec B_M$ at the other side, can be modelled by the well–known generalized Harris equilibrium

$$\vec B = B_0 \mathrm{th}(X/L)\vec e_Z + B_y \vec e_Y, \qquad n = n_0 \mathrm{ch}^{-2}(X/L). \qquad (1)$$

$$\vec B_{IMF,M} = B_y \vec e_Y \pm B_0 \vec e_Z, \qquad b_y = B_y/B_0.$$

The self-consistent distribution functions $f_{0j}(j = e, i)$ are, in fact, usual Maxwellian functions shifted by the diamagnetic drift velocity $u_j = 2cT_j/e_jB_0L$. The parameter b_y may be considered as a measure of the magnetic shear (i.e. a measure of the magnetic field rotation through the layer $\theta_0 = 2\mathrm{arctg} b_y^{-1}$). The parameter

$$e_j = u_j/v_{Tj} = \frac{\varrho}{L}\sqrt{(1+b_y^2)\frac{m_jT_j}{m_iT_i}} = v_{Tj}m_jc/eB_0L,$$

where v_{Tj} is the thermal velocity of the j–species, $\varrho = v_{Ti}m_ic/e\sqrt{B_0^2+B_y^2}$ is the ion gyroradius in an asymptotic $(X \gg L)$ magnetic field, characterizes both the plasma anisotropy and the relative thickness of the layer.

The investigation of the instability at a given magnetic surface within the MCL (microscopic reconnection) may be considered as a first necessary step in establishing threshold conditions for global macroscopic reconnection. The magnetic surface $X = X_S$ within the layer is initially smooth, but becomes

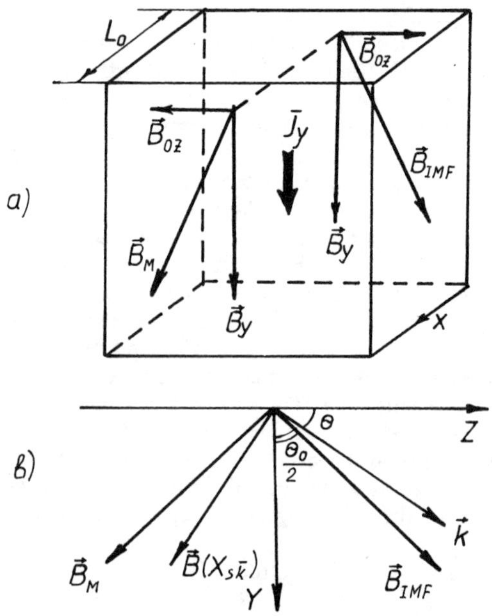

Fig. 1. (a) Model of magnetic field configuration at the dayside magnetopause. (b) The geometry of the singular surface. The view along the X–axis. The frozen–in condition breaks at the magnetic surface $X_{s\vec k}$ due to the growth of tearing–mode with $\vec k \perp \vec B(X_{s\vec k})$.

unstable with respect to the excitation of low-frequency electromagnetic perturbations (standardly described by the vector and scalar potentials $(\vec{A}, \varphi \sim \exp(-\imath\omega t + \imath\vec{k}\vec{r}))$ with the wave vector $\vec{k}$ perpendicular to the local direction of the equilibrium magnetic field at $X = X_S$: $k_\parallel(X_S) = \vec{k}\vec{B}/|B| = 0$. The subscript "$\parallel$" refers to the components parallel to the local direction of the equilibrium magnetic field. For such perturbations the inductive and potential parts of the electric field $E_\parallel = (\imath\omega/c)A_\parallel - \imath k_\parallel\varphi$ cannot compensate each other in some δ_φ-vicinity of the $X = X_S$, so $E_\parallel$ has a finite value in this region. This results in a strong nonadiabatic interaction of electromagnetic perturbations with particles and in a formation of a chain of magnetic islands. The plane $X = X_S$ is usually called the singular surface for the $\vec{k}$-mode. The region $|x| = |X - X_S| < \delta_\varphi$ around $X = X_S$, where $E_\parallel \neq 0$ will be called the interaction region. The geometry of the singular surface is shown in Figure 1b.

The dynamics of the drift tearing-mode has been thoroughly investigated by Galeev et al. [1986], and Kuznetsova and Zelenyi [1985, 1989]. The structure of a parallel electric field and its influence on ion dynamics as well as the influence of "integral" effects dealing with finiteness of ion Larmor radius ϱ_i have been incorporated into the analysis. The primary effect in the study of the nonlinear dynamics of magnetic perturbations $\tilde{B}$ is the modification of the orbits of resonant particles caused by the chain of magnetic islands with halfwidth $w = 2\sqrt{\tilde{B}/k'_\parallel B}$ ($k'_\parallel = \partial k_\parallel/\partial x$), moving along the magnetic surface due to drift effects. Here we represent the main results of these studies.

The general dispersion relation for the tearing-mode development acquire the form (see, for example, [Galeev and Zelenyi, 1977])

$$\Delta' = \frac{1}{2}\sum_j \int_{-\infty}^{\infty} \hat{V}_j\, dx = \sum_j U_j. \tag{2}$$

U_j is a total nonadiabatic response of the j-species of particles to perturbations. Δ' is the jump in logarithmic derivative of $A_\parallel$ on a largest scale L, which is proportional the value of the free energy of the growing mode. The dependencies of Δ' on X_S and wave number $m = kL$ are calculated in detail by Kuznetsova and Zelenyi [1985]. The operator $\hat{V}_j$, which is proportional to the perturbed current density $j_\parallel \sim \sigma_{\parallel j}E_\parallel$, includes terms resulting from taking into account both the inductive and electrostatic parts of the perturbed electric field $E_\parallel$ integrated along the particle's orbit.

$$\hat{V}_j A_\parallel = -\imath\frac{4\pi e^2}{c^2 T_j}(\omega - \omega_j)\int_{-\infty}^{\infty} v_\parallel\, dv_\parallel$$

$$\int_{-\infty}^{0} f_{0j}(v_\parallel)(v_\parallel A_\parallel - c\hat{\Gamma}_j\varphi(x))\exp(-\imath\omega\tau + \imath\vec{k}\vec{r}(\tau))d\tau, \tag{3}$$

where $\omega_j = -c\vec{k}[T_j\nabla n\vec{B}]/e_j B^2 n(X)$ is drift frequency of j-species of particles. The nonlinear operator $\hat{\Gamma}_j$ acting on φ comes from averaging the electrostatic field over the particle motion along the Larmor orbit and takes the form

$$\hat{\Gamma}_j = \exp(-\frac{1}{2}\hat{k}_x^2\varrho_j^2)I_0(\frac{1}{2}\hat{k}_x^2\varrho_j^2),$$

where $\hat{k}_x = -\imath\partial/\partial x$, and I_0 is a modified Bessel function. It is the average scalar potential $\hat{\Gamma}_i\varphi$ that determines the cutting distance of the parallel electric field

$$\delta_\varphi = \begin{cases} \varrho_i, & \varrho_i > \delta_i \\ \sqrt[4]{(\delta_i\varrho_i)^4 + w^4/32}, & \varrho_i < \delta_i \end{cases}, \tag{4}$$

where $\delta_j = |\omega|/k'_\parallel v_{Tj}$ is the thicknesses of electron $j = e$ and ion $j = i$ resonant regions.

Integration in equation (3) is carried out along the perturbed guiding center orbits. The orbits of linear resonant (R) particles with $\delta_j > w$ are practically not disturbed by the moving magnetic island. When $\delta_j < w$ the main part of resonant particles are trapped (TR) in region of closed magnetic surfaces. The period with which the TR particles circumnavigate the island τ_j is less than the wave period $2\pi/\omega_e$ in this case.

$$\tau_j \approx 2/k'_\parallel v_{Tj} w. \tag{5}$$

The trajectories of TR particles can be explicitly expressed through the elliptic functions [Zelenyi and Taktakishvily, 1984]. The orbits of untrapped (UT) particles far from the island are slightly disturbed by the island and in a first order of perturbation theory can be considered as linear.

The explicit form for operator $\hat{V}_j$ in dispersion relation (2) can be obtained after inserting $\hat{\Gamma}_i\varphi(x)$ expressed through the $A_\parallel$ with the help of the quasineutrality equation, the form of which depends on conditions $\delta_j \gtrless w$ and $\varrho_i \gtrless \delta_i$ (see the paper [Kuznetsova and Zelenyi, 1989]).

Figure 2 illustrates the solution of the dispersion equation (2) for the drift tearing mode ($\mathrm{Re}\omega \approx \omega_e \gg \mathrm{Im}\omega = \gamma$). When the halfwidth of the magnetic island w is less than the thickness of electron resonant region δ_e the growth rate γ takes the form

$$\gamma(w \to 0) = \gamma_L(1 - a), \quad \gamma_L = \omega_e\frac{c^2\Delta'}{\omega_{pe}^2\delta_e\sqrt{\pi}}. \tag{6}$$

Parameter a characterizes how far the mode is from its linear stability threshold. The stabilization of the mode when parameters are below threshold is related to its coupling with the field-aligned ion sound waves [Coppi et al., 1979; Galeev et al., 1986].

$$a = \frac{\omega_{pi}^2}{c^2\Delta'}\frac{T_e + T_i}{T_e}\frac{\pi\Gamma(3/4)}{\sqrt{2}\Gamma(1/4)}\sqrt{\varrho_i\delta_i}\min(1, (\delta_i/\varrho_i)^{3/2}). \tag{7}$$

To understand these results physically let's consider the energy balance equation

$$\Delta W_F \downarrow + \Delta W_i \uparrow + \Delta W_e^R \uparrow = 0, \quad w < \delta_e < \delta_i, \tag{8}$$

Arrows indicate whether the relevant term increases (↑) or decreases (↓) in the course of instability growth. $\Delta W_F = \Delta W_B + \Delta W^{NR}$ is the power of the free energy source. $\Delta W_B > 0$ (↑) is the power needed for the support of the growth of the perturbed magnetic field, $\Delta W^{NR} < 0$ (↓) is the energy change in a unit of time of nonresonant particles adiabatically interacting with perturbations ("reactive" contribution). The value $\Delta W_F < 0$ (↓) decreases, which related with the thermodynamical nonequilibrium state of the system (anisotropy of the distribution function), i.e. the tearing instability is the wave of negative energy resulted in increasing the energy of resonant particles. The other terms in equation (8) corresponds to nonadiabatic particle–perturbation interaction — electric field work upon the electric current ("active" contribution) and at the linear stage describes the nonreversible increase of resonant particle energy ($\Delta W_e^R > 0 \uparrow$, $\Delta W_i > 0 \uparrow$). The energy expenditures for the excitation of the field–aligned ion oscillations $\Delta W_i \uparrow$ reduce the free energy gain $\Delta W_F \downarrow$ resulted from the formation of magnetic islands in a sheared magnetic field and consequently decrease the part of energy absorbed by resonant electrons $\Delta W_e^R \uparrow$. That can lead to the complete stabilization of the instability when $\Delta W_i > \Delta W_F$, i.e. when

$$a(X_S, L, k, \theta_0) > 1. \qquad (9)$$

Figure 2 shows the dependencies of the growth rate $\gamma(w)/\gamma_L$ on the halfwidth of the magnetic island w in two limiting cases $\varrho_i > \delta_i$ and $\varrho_i < \delta_i$ for different values of the parameter a. When $w > \delta_e$ the nonlinear growth rate is independent on the initial margin of the free energy Δ' as if the instability "forgets" at nonlinear stage about its linear history. The energy balance equation at nonlinear stage takes the form

$$\Delta W_F \downarrow + \Delta W_i^{UT} \uparrow + \Delta W_e^{TR} \downarrow = 0, \quad \delta_e < w < \delta_i. \qquad (10)$$

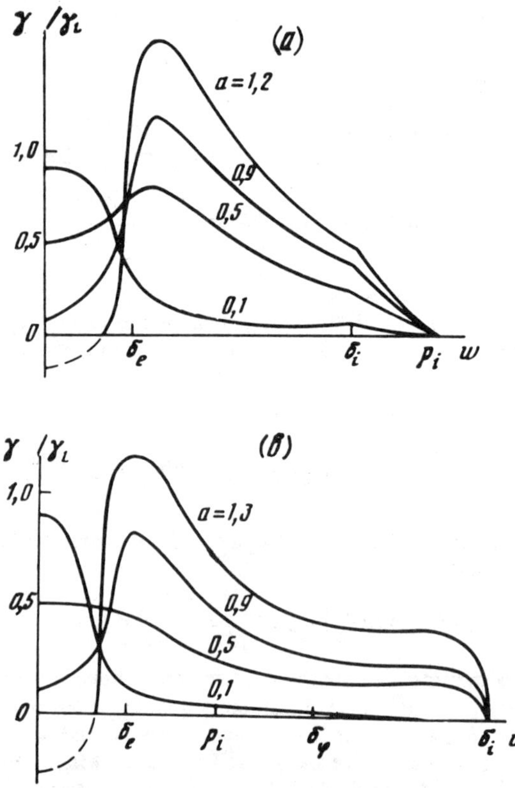

Fig. 2. The dependencies of the normalized growth rate γ/γ_L on the halfwidth of the magnetic island w:
(a) $\delta_i < \varrho_i$; (b) $\delta_i > \varrho_i$.

At the nonlinear stage particles change their roles in interaction with perturbations. Trapped electrons become the particles "feeding" the development of instability ($\Delta W_e^{TR} \downarrow$), instead of particles adiabatically oscillating outside the resonant region at the linear stage ($\Delta W^{NR} \downarrow$), and ions begin to play the role of the single "sink" of the released energy (since the resonant electrons are trapped), which is required for the maintenance of the instability energy balance. When $w > \max(\delta_\varphi, \delta_i)$ resonant ions are trapped and UT ions are squeezed out of the interaction region by the growing magnetic islands. The disappearance of UT and R ions absorbing the energy of developing perturbations results in the cessation of instability growth.

Thus, the width of the nonlinear saturation of the islands growth in the absence of collisional or effective resistivity reaches a "finite" level

$$w^* = \max(\delta_i, \varrho_i), \qquad (11)$$

which is much less than the thickness of the layer $2L$.

The principle conclusion of this section is that the destruction of a single magnetic surface cannot result in the macroscopic reconnection of the magnetic fields on both sides of the layer. This result forces us to reject the models considering reconnection at the dayside magnetopause as a regular single–mode process (like the reconnection within the magnetotail).

3. Stochastic Percolation Model

If the number of the growing modes is sufficiently large the nearby growing islands can overlap and magnetic field lines begin to wander stochastically between them (see Figure 3). One can easily imagine this stochastic process as a random Brownian walk across the MCL with the spatial step $\Delta x \sim 2w$ and some time step $\tau_w = s_w/v_A$. The length of the magnetic field line s_w corresponding to one Brownian step can be easily evaluated from equation (5)

$$s_w = \tau_j v_{Tj} \approx 2/k'_\parallel w. \qquad (12)$$

Finally, the magnetic field line can diffuse in such a manner from one side of the MCL to another with the diffusion coefficient

$$D = D_F v_A, \qquad D_F = \frac{(2w)^2}{s_w} \approx 2w^3 k'_\parallel. \qquad (13)$$

It is reasonable to relate D with the general level of magnetic turbulence within the layer $b_0^2 = \sum_k (\tilde{B}_k^2/B^2)$. For simplicity one can assume that the magnetic island have approximately equal thicknesses w through the layer. Then the number of modes required to cover the entire MCL ($-L < X < L$) at the margin of their overlapping may be estimated as $N = L/w$. Accordingly, we have the simple approximate expression

$$b_0^2 \simeq N \frac{\tilde{B}_k^2}{B^2} \simeq w^3 L k_\parallel'^2. \qquad (14)$$

The quantity b_0^2 is proportional to the energy released during the development of the instability. It is easy to see from equation (14) that it is more favorable energetically to "pave" the MCL by magnetic islands with maximum possible thickness w^*,

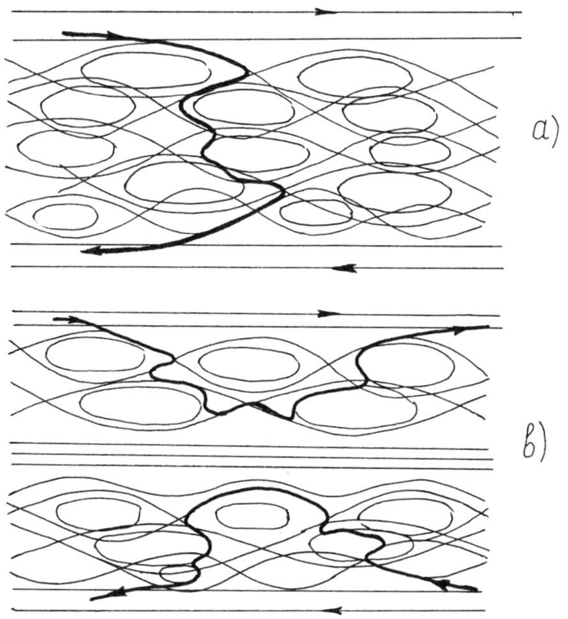

Fig. 3. Diffusion of magnetic field lines through the MCL with (a) and without (b) the magnetic percolation. The view is from the Y-axis. Topological connection of the magnetic field on both sides of the MCL is absent in case (b).

but without significant overlapping. This simple consideration indicates the importance of the single mode effects discussed in Section 2 for the study of magnetopause processes. Note that the magnetic field line diffusion coefficient given by equation (13) expressed through b_0^2

$$D_F \simeq b_0^2 \frac{1}{k_\parallel' L} \tag{15}$$

coincides with the well-known quasilinear expression obtained by Rosenbluth et al. [1966]. The nonlinear quenching of the growth of magnetic islands occurs, as was shown in Section 2, at the value $w^* = \max(\delta_i, \varrho_i)$, which is slightly changes across the layer and takes the form

$$w^*(X_S) = \varrho \sqrt{\frac{1+b_y^2}{b_y^2 + \mathrm{th}^2 \frac{X_S}{L}}} \max\left(1, \frac{T_e}{T_i} \mathrm{th}\frac{X_S}{L} \sqrt{b_y^2 + \mathrm{th}^2 \frac{X_S}{L}}\right). \tag{16}$$

In order to estimate the dependency of D_F on b_y we substitute into equation (13) the values $\overline{w^{*2}}$ and $\bar{s}_w$ averaged across the MCL

$$\overline{w^{*2}} = \frac{1}{L} \int w^{*2}\, dX_S, \quad \bar{s}_w = \frac{1}{L} \int 2\, dX_S / k_\parallel' w^*(X_S).$$

The final expression for the magnetic field line diffusion coefficient takes the form:

$$D_F = \frac{4\varrho^3}{L^2} \frac{1}{b}, \qquad b = \frac{3b_y^2 + 1}{\sqrt{1+b_y^2}} \left(1 + \frac{\theta_0}{2b_y}\right)^{-2}. \tag{17}$$

For $\theta_0 = \pi/2$ the coefficient $b \approx b_y \approx 1$. For small angles ($\theta_0 < \pi/2$, $b_y < 1$) $b \approx 3b_y$. For large angles ($\theta_0 > 2\pi/3$) $b \approx b_y^2$. It is clear that the larger the angle θ_0 is the faster the diffusion process within the MCL will be accomplished.

The magnetic field line diffusion coefficient enables us to evaluate the length of magnetic field line $s(X)$ which has diffused across the layer in a distance X: $s(X) = 2X^2/D_F$. When $X = L = L_0/2$ the length $s(X)$ is equal to

$$s_0 = 2L^2/D_F = \frac{b}{4} L_0 \left(\frac{L_0}{2\varrho}\right)^3. \tag{18}$$

The dependencies of the dimensionless length

$$s_0/\varrho \approx s_y/\varrho = \bar{s}_y$$

on θ_0 for different values of dimensionless magnetopause thickness $\bar{L}_0 = 2L/\varrho$ are shown in Figure 4.

An attempt to visualize the process of magnetic field stochastic diffusion across the MCL is shown in Figure 5 (in projections on the XY, XZ and YZ planes). If all magnetic surfaces within the MCL become "shaggy" due to the growth of magnetic

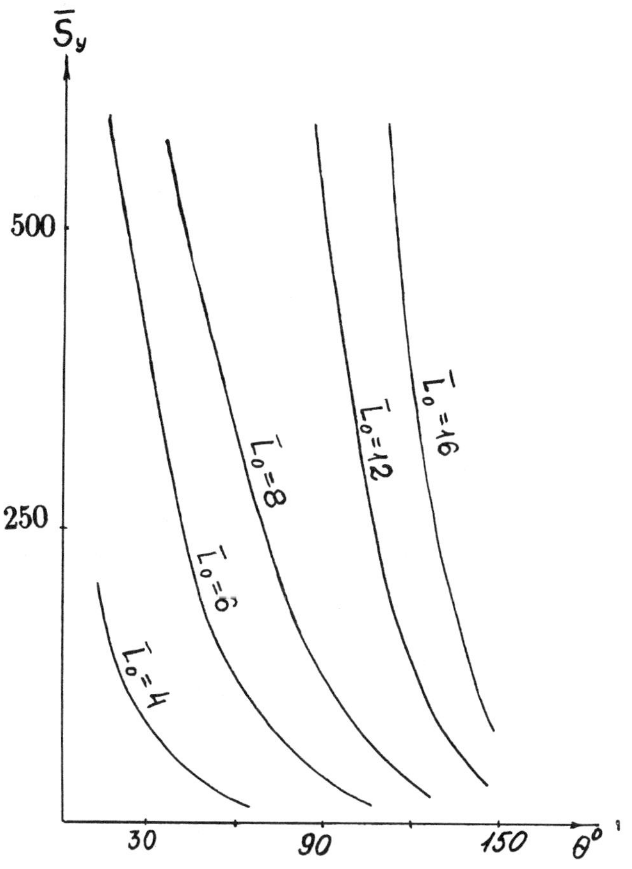

Fig. 4. The dependencies of the dimensionless shift of the magnetic field line along the magnetopause in the Y-direction in the course of its percolation through the MCL on the angle θ_0 for different MCL dimensionless thicknesses $\bar{L}_0 = L_0/\varrho$.

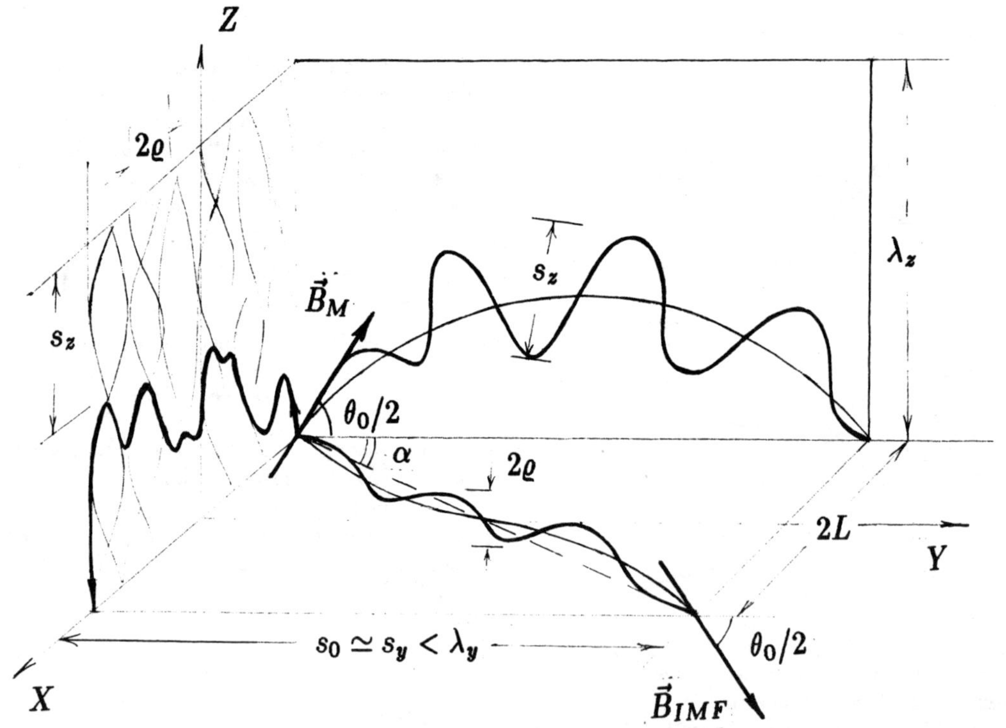

Fig. 5a. The picture of magnetic percolation through the MCL with thickness $L_0 = 2L$ and characteristic spatial scales along the magnetopause λ_y ("length") and λ_z ("width"). s_z is the maximum wavelength of perturbations. s_y is the shift of the magnetic field line along the magnetopause in the course of diffusion. Bold solid lines — projections of magnetic field line diffusing across the magnetopause on the XY, XZ and YZ planes. Thin solid curves on planes XY and ZY — projections of the trajectory of field line averaged over its random motion across the MCL with a diffusion spatial step $\sim 2\varrho$ and a "spatial smearing" s_z on the magnetopause surface. If all magnetic surfaces within the MCL are destroyed ($s_z < \lambda_z$) percolation can be accomplished. The magnetic field line penetrates from the magnetosphere to the magnetosheath if $s_y < \lambda_y$. α — is the average angle of transecting the MCL.

islands and the characteristic length of MCL along the Y-axis is larger than $s_y \approx s_0$, the magnetic field line can accomplish the diffusion from one side of MCL to another (see Figures 3a and 5a). This means the establishment of the topological connection between interplanetary and geomagnetic field lines. So the macroscopic reconnection in our model consists microscopically in the diffusion of magnetic field lines through the layer due to the small scale magnetic field inhomogeneities developing within it. We call this process magnetic percolation. This term, to our mind, describes better the physical nature of the process than the "magnetic migration" suggested later by LaBelle and Treumann [1988].

It is worth mentioning that the detailed analysis of 18 ISEE-1 magnetopause crossings [Eastman et al., 1985], based on the three-dimensional plasma measurements demonstrates that the plasma transport across the MCL is governed by the process of diffusive type and no evidence of plasma acceleration usually attributed to reconnection have been observed. We want to emphasize that although the final result of percolation is the macroscopic reconnection of magnetic field lines, the process is of a stochastic diffusive nature and results in the deceleration and isotropization of the plasma flow along the reconnected field lines, rather than its acceleration. This is the principal difference of the percolation model of magnetic reconnection from the well-known "driven" MHD models.

The growth of the magnetic island saturates rather quickly at finite island width $2w^*$ which is much less than the thickness of the layer $2L$. So if even a very narrow region (but wider than w^*) with stable magnetic surfaces exists within the layer it cannot be overlapped by the nearby growing magnetic islands and thus it appears to be impenetrable for the diffusing field lines. In this case shown in Figure 3b and 5b there is no topologiocal connection of field lines on both sides of the MCL, and so the reconnection (in global macroscopic sense) is absent (despite the possible observation of the developed magnetic turbulence within the magnetopause $\sim b_0^2$). Therefore the first necessary condition for the magnetic percolation through the MCL will be the destruction of all magnetic surfaces within it. Possible appearance of the undestroyed domain within MCL can be attributed to the effect of tearing-mode stabilization due to its coupling with the field-aligned plasma motions of the ion-sound wave type. It is reasonable to assume that the excitation of per-

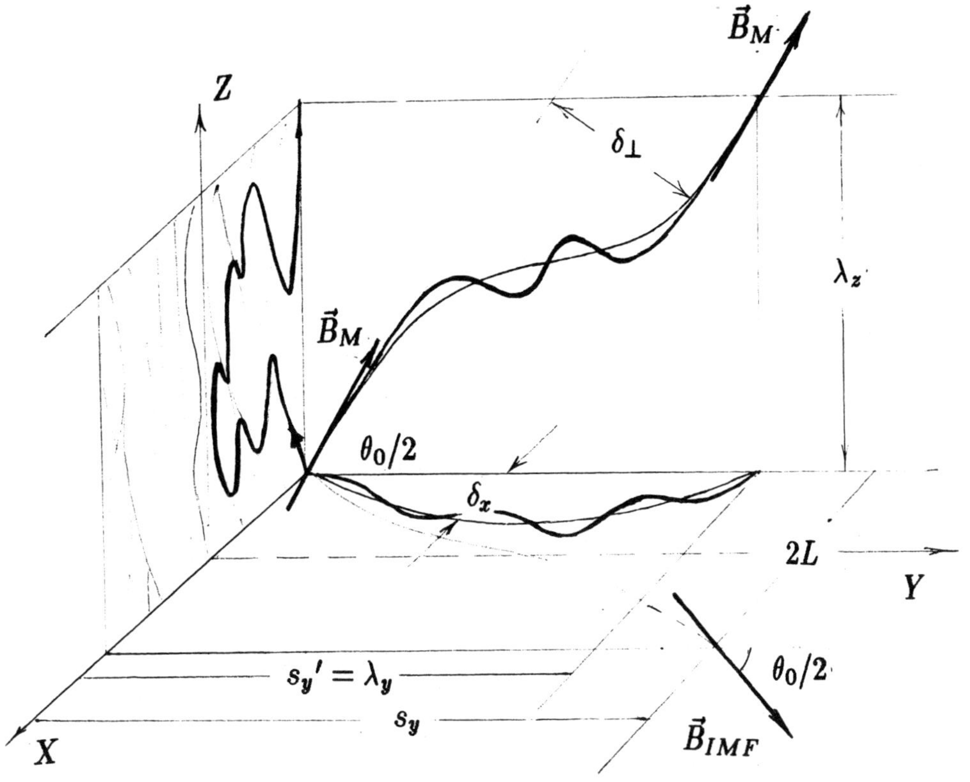

Fig. 5b. If $s_y > \lambda_y$ and/or a region wider than 2ϱ with stable magnetic surfaces exists within the MCL, only partial immersion of the field line into the MCL on the depth $\delta_x < L$ can be achieved. Magnetic field line returns to the same side of the MCL being shifted at a distance $\delta_\perp$.

turbations is possible in the layer only if it has the characteristic width λ_z along the Z-axis (i.e. perpendicular to the equilibrium electric current J_{0Y}) larger or at least of the same order as the maximum wavelength of the perturbations s_z. The marginal size s_z can be obtained [see Galeev et al., 1986] by the comparison of the stability thresholds (9) for all magnetic surfaces X_S within the MCL and by the selection of the destabilization condition for some domain within the MCL with the magnetic surfaces most stable against reconnection. Figure 6 shows the dependencies of the dimensionless marginal "latitudinal" width of the MCL $\overline{\lambda}_z \geq \overline{s}_z = (s_z/\varrho)(4\beta T_e/T_i)$ on the angle of magnetic field rotation θ_0 for different values of the dimensionless magnetopause thickness $\overline{L}_0 = (2L/\varrho)(4\beta T_e/T_i)$. Below we will omit the numerical factor $4\beta T_e/T_i$ of the order of unity.

As the magnetic surfaces are destroyed by growing islands a magnetic field line should circumnavigate each island and, viewing from the X-axis (i.e. in projection on the YZ plane in Figure 5), it appears to be smeared around its average position (thin solid curve). As the island maximum length in Z-direction (i.e. in the direction perpendicular to the equilibrium current J_{0Y}) is s_z, the value of s_z may be considered as a measure of spatial smearing of a field line of the magnetopause surface.

If the characteristic length of the MCL λ_y is less than s_0 the magnetic field line diffusion will not result in establishing of a topological connection of field lines of both sides of the MCL. Thus, the other necessary condition for the magnetic percolation

$$s_y \approx s_0 < \lambda_y \tag{19}$$

imposes restrictions on the thickness of the magnetopause L_0.

One can assume that the characteristic spatial scales along the magnetopause ("length" λ_y and "width" λ_z) may be determined by the external conditions (the size of the magnetopause, the convection pattern in the magnetosheath). The dependencies of the dimensionless magnetopause thicknesses $\overline{L}_0 = 2L/\varrho$ on θ_0 for several values of dimensionless length $\overline{\lambda}_y = \lambda_y/R_E$ and latitudinal width $\overline{\lambda}_z = \lambda_z/R_E$ (R_E=6000 km) are shown in Figure 7. For a given value $\overline{\lambda}_z = \lambda_z/R_E$ all magnetic surfaces within the layers with a thickness less than the marginal one $\overline{L}_0(\overline{\lambda}_z)$ will be destroyed. The thickness of the MCL with characteristic spatial scales along magnetopause $\lambda_z \times \lambda_y$ subjected to percolation should lie below both curves ($\overline{L}_0(\lambda_y)$ and $\overline{L}_0(\lambda_z)$).

Thus, there are two bounds on the thickness L_0 of the MCL for the formation of reconnection "patchy" $\lambda_z \times \lambda_y$. One is related to the conditions of the tearing-mode linear destabilization and growth ($\overline{L}_0(\lambda_z)$) and has been discussed in detail in our previous paper [Galeev et al., 1986], the other considered here for the first time is associated with the diffusion of magnetic field lines and is governed by the width of nonlinear saturation of the magnetic island growth ($\overline{L}_0(\lambda_y)$). As it can be seen

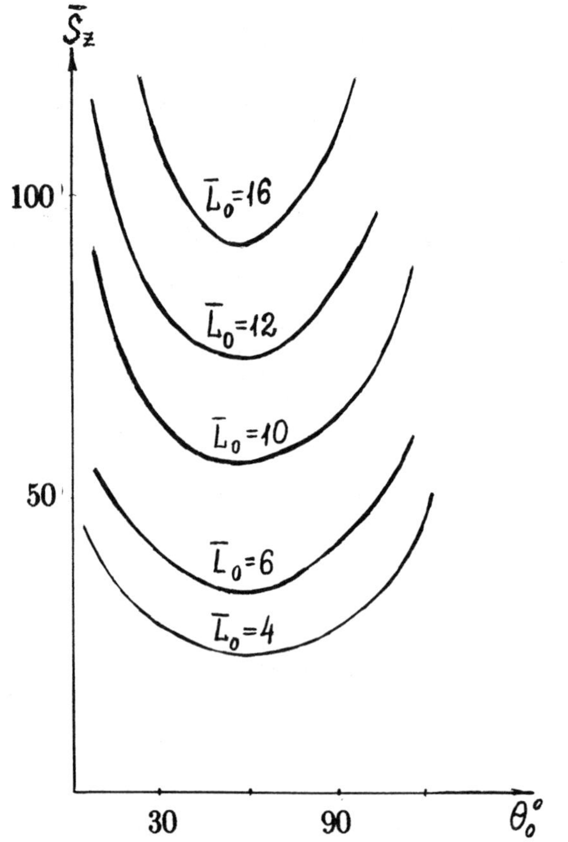

Fig. 6. The dependencies of dimensionless perturbation wavelength $\bar{s}_z = s_z/\varrho$ on angle θ_0 for different values of MCL dimensionless thicknesses $\bar{L}_0 = L_0/\varrho$.

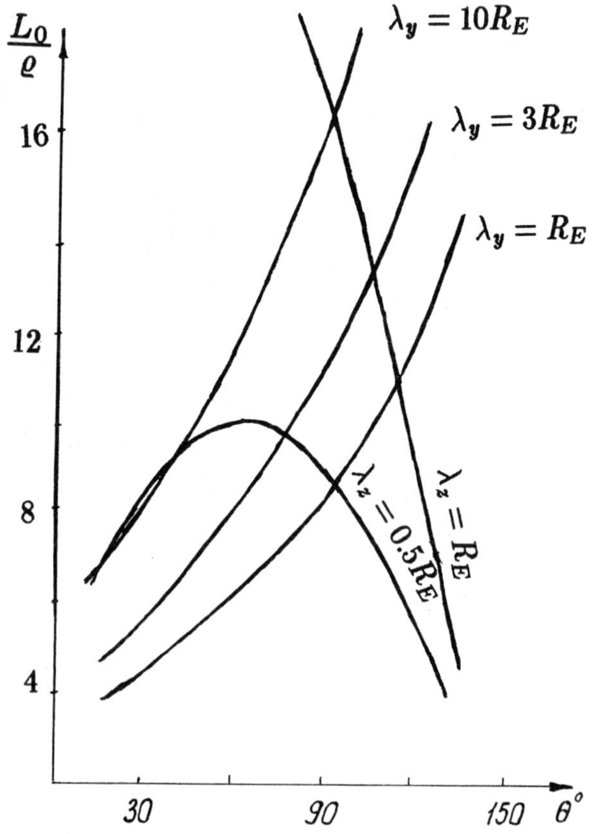

Fig. 7. The threshold curves $\bar{L}_0^{cr}(\theta_0)$ of magnetic field percolation for different scales of the reconnection patchy $\lambda_z \cdot \lambda_y$ at magnetopause surface. For the MCL with angle of magnetic field rotation θ_0 percolation can be accomplished only if dimensionless MCL thickness $L_0 < \left\{\bar{L}_0^{cr}(\theta_0, \lambda_y), \bar{L}_0^{cr}(\theta_0, \lambda_z)\right\}$.

from Figure 7, the restriction imposed by the nonlinear theory is more stringent for small angles $\theta_0 < \pi/2$, but for large angles $\theta_0 > 2\pi/3$ the linear stability threshold $(\bar{L}_0(\lambda_z)$ lies below the curve $(\bar{L}_0(\lambda_y)$ and therefore more restricting.

As one can see from Figure 7, the MCL with $L_0 < 4 \div 6\varrho$ (L_0 <200–400 km) will be unstable against magnetic percolation for a wide range of parameters λ_y, λ_z. MCLs with thicknesses greater than $L_0 \sim 10 \div 16\varrho$ are stable practically in all cases. This value L_0 ~500–800 km conforms well with the typical thicknesses of magnetopause observed in ISEE–1,2 measurements [Berchem and Russell, 1982].

We see that the percolation process has different spatial scales $w^* \ll L_0 < \lambda_y, \lambda_z$, so it proceeds in many steps and has a diffusive, nonregular multiscale character.

4. Macroscopic Pictures of Magnetopause Reconnection

We have discussed above the microscopic picture of reconnection on a fast time scale associated with internal MCL processes. The next time scale is related already with the macroscopic plasma dynamics governed by plasma convection and we want in this section to speculate on coupling of these phenomena both on temporal and spatial scales [see Galeev et al., 1986 and Kuznetsova and Zelenyi, 1986].

If the force driving the magnetic field lines to the reconnection site is absent (purely spontaneous case) only field lines located at the magnetic barrier not farther than $2\varrho_i \leq 100$ km (the spatial step of diffusion) will be involved in the diffusion process. The group of magnetic field lines with cross-section $\lambda_z \times \lambda_z \approx 2\varrho \times R_E \approx 100$ km$\times 6 \cdot 10^3$ km constitute the magnetic filament containing the flux $\Delta\Phi \sim 4 \cdot 10^4$ Wb, which is two orders of magnitude smaller than the fluxes observed within FTE tubes in measurements near the Earth's magnetopause. Such small filaments can be twisted only slightly and therefore could not be observed in experiments near the Earth. However, these considerations will be important for giant planets, Jupiter for example, where puzzling small scale magnetic filaments

$$\Delta\Phi_J \leq 10^5 \text{ Wb} < \Delta\Phi_E \sim 10^7 \text{ Wb}$$

have been identified [Walker and Russell, 1985].

To explain the larger fluxes in FTE tubes corresponding to the real experiment data some driving force pushing magnetic field lines to the reconnection site should be taken into account. Following this conclusion few possible ways to develop theory further are possible.

The first one is to couple our microscopic theory with some of the MHD driven reconnection models widely discussed in literature [e.g. Lee and Fu, 1986; Scholer, 1988]. MHD models usually do not consider the internal magnetopause processes, but use instead of it some phenomenological transport coefficient of unspecified nature. This is an "ad hoc" assumption in fact, which unfortunately deprives these models of self-consistency. The necessary value of resistivity η used for example by Lee and Fu [1986] in their numerical computations is in dimensionless units $\bar{\eta} = (c^2/4\pi v_A a)\eta = 0.0008$–$0.015$ (a is the simulation length in X-direction). The appropriate diffusion coefficient is $D = \eta c^2/4\pi = \bar{\eta} v_A a > (3 \cdot 10^8 \div 4 \cdot 10^9) \mathrm{m^2 s^{-1}}$ even for rather small $a \sim 1000$ km. The usual belief was that this diffusion can be maintained by the plasma microturbulence within the magnetopause.

Let us discuss this hypothesis in more detail. Microturbulence really is a permanent feature of magnetopause crossings. It was observed by Prognoz-8 [Vaisberg et al., 1983], Imp 6 [Gary and Eastman, 1979] ISEE 1,2 [Anderson et al., 1982], and AMPTE/IRM [LaBelle and Treumann, 1988] spacecraft in the form of small scale, low-frequency (1–100 Hz) oscillations of the electric field. A comparison of the experimental data with the theory shows that the lower hybrid drift instability is the most probable candidate to explain these oscillations. The amplitude of the electric field measured by Prognoz-8 satellite in this range of frequencies is 0.1–0.5 mV/m. LaBelle and Treumann [1988] give for the amplitude of the most intense turbulence observed in their measurements a value of the order of few mV/m.

The theoretical estimates for the nonlinear saturation level of such turbulence obtained independently by different authors [see review by Galeev et al., 1986] conform well with each other. The corresponding expression up to the numerical coefficients of the order of unity can be represented in the following form

$$\frac{(\delta E)^2}{8\pi n_0 T_i} \approx \left(\frac{m_e \omega_{Be}^2}{m_i \omega_{pe}^2}\right)\left(\frac{\varrho}{L}\right)^2. \tag{20}$$

The numerical estimate obtained with the help of equation (20) agrees with observations only for rather thick layers $L/\varrho > 10$. For layers with smaller thicknesses it gives an overestimated value. This is because the theory applied to the magnetopause has not taken into account up to now the strong stabilizing influence of magnetic shear [Krall, 1977; Huba et al., 1982; Gladd et al., 1985], which could be very important for this problem.

The interaction of electrons with lower hybrid turbulence results in an anomalous loss of electron momentum. This corresponds with the appearance of some anomalous collision frequency with a value [Huba et al., 1977]

$$\nu_{\text{eff}} = \omega_{LH} \frac{(\delta E)^2}{8\pi n_0 T_i}\left(\frac{m_e \omega_{Be}^2}{m_i \omega_{pe}^2}\right)\sqrt{\frac{\pi}{4}}, \quad \omega_{LH} = \sqrt{\omega_{Be}\omega_{Bi}}. \tag{21}$$

More general analysis for other microturbulence (ion acoustic drift, electron-cyclotron) modes has been done recently by LaBelle and Treumann [1988]. Using the measurements done recently by AMPTE/IRM satellite these authors find an upper possible estimate for the diffusion coefficient $D < 10^7 \mathrm{m^2 s^{-1}}$. This is at least an order of magnitude smaller than is required for the driven MHD models. So the microturbulence in realistic conditions failed to provide the necessary dissipation. This agrees with the general reconsideration of the role of turbulent dissipation in space plasma initiated by Coroniti [1985].

To overcome this difficulty we suggest here the new mechanism of magnetic diffusion maintained by larger scale magnetic turbulence generated by the tearing-mode instability within the magnetopause. It operates in purely collisionless plasma, does not use any "ad hoc" assumptions and utilizes only free energy intrinsic to magnetically sheared configurations. The associated diffusion coefficient already obtained above is

$$D \approx \frac{4\varrho^2}{L^2}\frac{1}{b} v_{Ti}\varrho. \tag{22}$$

Comparing it even with the obviously overestimated theoretical estimates for diffusion due to lower hybrid drift microturbulence based on equations (20) and (21) and given by the expression

$$D_{LH} \sim \sqrt{\frac{\pi}{4}\frac{m_e}{m_i}}\left(\frac{\varrho}{L}\right)^2 v_{Ti}\varrho \tag{23}$$

we find that D is $\sqrt{m_i/m_e}$ larger than D_{LH}. The numerical estimate of D for the magnetopause can be obtained assuming that $\varrho \simeq 50$ km, $v_{Ti} \simeq 300$ km/s, $T_i \approx 900$ eV, $b \sim 1$. It gives

$$D \sim 2.4 \cdot 10^{11} (\varrho/L)^2, \text{ i.e.}$$

$$D \sim \begin{cases} 1.5 \cdot 10^{10} \,\mathrm{m^2 s^{-1}} & \text{for } L_0/\varrho \sim 4 \\ 2.4 \cdot 10^{9} \,\mathrm{m^2 s^{-1}} & \text{for } L_0/\varrho \sim 10 \end{cases}. \tag{24}$$

Now we want to proceed to the discussion of global pictures of driven reconnection where the external boundary condition play an essential role together with local conditions determined by the value of D. A variety of magnetopause reconnection theories, for example, the well-known Petchek model, the multiple X-line reconnection model later suggested by Lee and Fu [1985], or the elbow-shaped configuration proposed by Russell and Elphic [1979], all can be formulated in terms of different boundary conditions. The same refers to the nonstationary models by Syrovatskii [1981], Pudovkin and Semenov [1985], and Scholer [1988].

The structure of plasma flows near the stagnation region in the vicinity of the subsolar point is of principal importance for reconnection theories. There are two principally different approaches to this problem. The first suggests the presence of a small stagnation region (SR) near the magnetopause surface [e.g. Spreiter et al., 1966; Russell et al., 1980], the other implies the presence of a stagnation line (SL) extended over comparatively large distances (fractions of magnetospheric scale $R_m \simeq 15 R_E$) in the direction of the equilibrium current [Pudovkin et al., 1977; Sonnerup, 1980]. The observations summarized in the paper by Crooker et al. [1984] do not support the suggestion about the stagnation line. However, we do not feel that this dilemma is entirely solved and consider below both global geometries trying to explain the FTE formation using the above obtained results about the appearance of small scale magnetic filaments in the course of tearing-modes development.

We assume that the stagnation area scaled as $D_Y \times D_Z$ exists at the dayside magnetopause and consider two options:

(A) SL-configuration $D_S = D_Z \sim R_E \ll D_Y \sim R_m$.

(B) SR-configuration $D_S \approx D_Z \approx D_Y \sim 1 \div 3 R_E \ll R_m$.

To make the rough estimate of the convection time τ_{conv}, which flux tubes spend in the stagnation area one can use the expansion of the flow velocity within it proposed by Zwan and Wolf [1976] $V(\xi) \simeq V_{SW}\xi$, where $V_{SW} \sim 400$ km/s is the characteristic velocity of the magnetosheath flow close to the terminator, ξ is the dimensionless distance measured along the magnetopause. In case (A) $\xi = z/R_m$ (asymmetric flow of plasma), for case (B) $\xi = r/R_m$ (symmetric plasma flow). Here z and r correspondingly the distance in the Z-direction and the radial distance from the subsolar point. After that we have in a very qualitative manner

$$\tau_{conv} = R_m \int_{D_S} \frac{d\xi}{V(\xi)} \simeq \frac{R_m}{V_{SW}} \ln \frac{R_m}{D_S}. \quad (25)$$

The process of magnetic field line percolation through the magnetopause takes some time $\tau = \tau_{tear} + \tau_{diff}$ needed for the growth of magnetic islands, their overlapping (τ_{tear}) and corresponding magnetic field line diffusion (τ_{diff}). The diffusion time τ_{diff} given by s_0/v_A is a rather quick process. It is reasonable to estimate τ_{tear} as the inverse linear growth rate γ_L^{-1} of the tearing-mode (see equation (6)). Assuming that $T_i = 4T_e$, $\beta = 1$, B =60 nT, we find

$$\tau_{tear} \simeq b_y(L_0/\varrho)^3 \text{ s}. \quad (26)$$

The critical MCL thickness L_0^* obtained from the condition

$$\tau_{tear} < \tau_{conv} \quad (27)$$

assuming that $D_S \sim R_E$ is shown in Figure 8 by dashed line. We see that for a MCL with $L_0 < 8\varrho$ the tearing-mode growth is a fast process with respect to the convection. The diffusion of a single "elementary" magnetic filament (with characteristic thickness 2ϱ) occurs even faster $\tau_{diff} \sim s_0/v_A$. So, for $L_0 < L_0^*$ not only the single filament could be reconnected for the time τ_{conv}, but the whole bunch of percolated field lines can gather at the stagnation region to form the FTE magnetic tube.

The characteristic diffusion velocity across the layer can be easily estimated, because the field line shifts across the layer (distance L_0) over the time $\tau_{diff} = L_0^2/2D$.

$$V_{Dz} = \frac{L_0}{\tau_{diff}} = v_A \sin\alpha, \qquad \sin\alpha = \frac{4}{b}\left(\frac{2\varrho}{L_0}\right)^3, \quad (28)$$

where α is the average angle of "incidence" of percolated magnetic line to the magnetopause surface. The value of α is small, because the field line "transects" the magnetopause at an oblique angle. According to equation (14) α depends on the level of magnetic turbulence b_0^2 within the MCL. So the time of macroscopic reconnection of flux tube with transverse dimension λ_z is

$$\tau_{Rec} = \lambda_z/V_{Dz} = \lambda_z/v_A \sin\alpha. \quad (29)$$

To accomplish the reconnection on such a scale, τ_{Rec} should not exceed τ_{conv}. This condition limits the possible values of

$$\lambda_z < (v_A/V_{SW}) \sin\alpha\, R_m. \quad (30)$$

For the scales $\lambda_z \leq R_E$ and $L_0 < L_0^*$ the condition (30) is easily satisfied.

If the FTE magnetic tube with flux $\Delta\Phi \sim B \cdot \lambda_z \cdot \lambda_z \sim B_{IMF} R_E^2$ (B_{IMF} ~60 nT is the compressed interplanetary field in magnetic barrier) obliquely transects the MCL, the corresponding reconnection "patchy" can be identified at magnetopause surface. It will have the area

$$\lambda_s^2 = R_E^2 q^2, \qquad q^2 = 1/\sin\alpha. \quad (31)$$

Inserting expression (28) for $\sin\alpha$ into equation (31) we find

$$\frac{L_0}{\varrho} = 2\left(\frac{4}{b} q^2\right)^{1/3}. \quad (32)$$

Figure 8 illustrates the dependencies of the MCL thickness L_0 on θ_0 for different squares of reconnection patchy characterizing by the factor $q = 1/\sqrt{\sin\alpha}$. Since the reconnection patchy on the magnetopause should be smaller than the stagnation area on it, q^2 should be smaller than 10–20. For such q^2 the condition (32) is more restricting than the already discussed limitation (27) (see Figure 8).

The attempt to visualize the process of reconnection of magnetic tube $B \times R_E^2$ at the MCL for the SL (case (B) above) geometry of convection is shown in Figure 9. The reconnection patchy is rather elongated in Y-direction in this case $\lambda_z \sim \lambda_z \sim R_E \ll \lambda_y = \lambda_z q^2 \leq R_m$. The resulting geometry resembles the

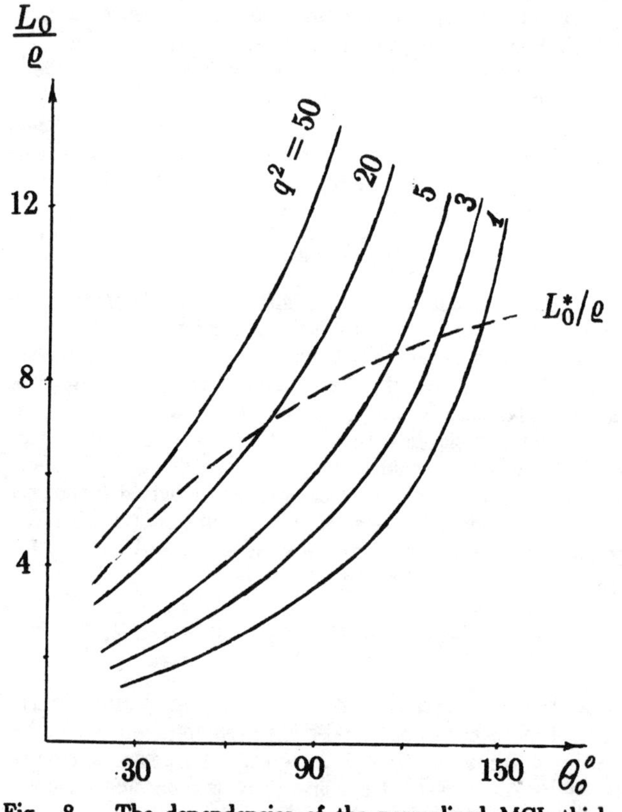

Fig. 8. The dependencies of the normalized MCL thickness L_0/ϱ on θ_0 for different squares of reconnection patchy $\lambda_s^2 = q^2 R_E^2$. The dashed line shows the critical MCL thickness L_0^* obtained from the condition $\tau_{tear} = \tau_{conv}$.

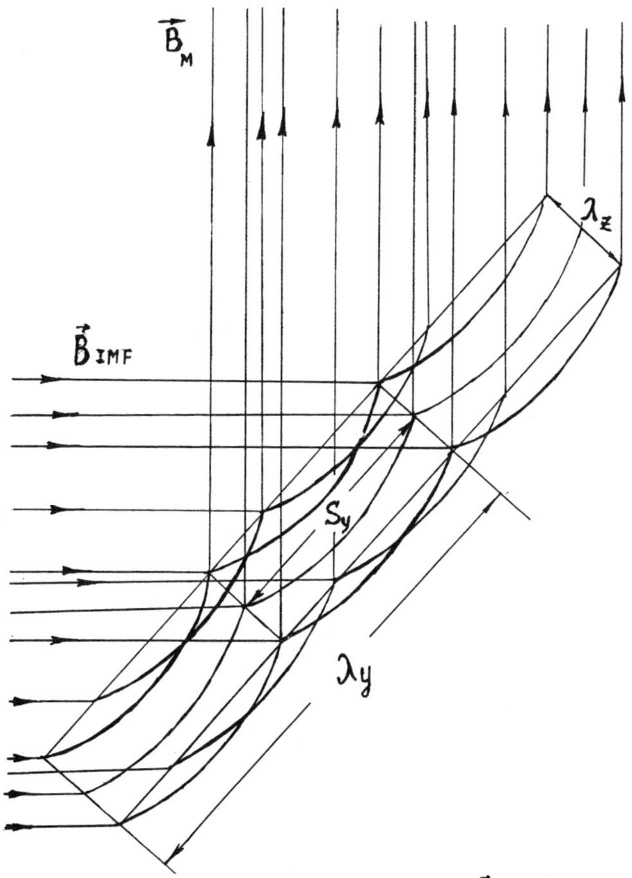

Fig. 9. The reconnection of magnetic tube $\vec{B} \times R_E^2$ at the MCL for the SL (stagnation line) geometry of convection ($s_x, s_y \ll \lambda_y = \lambda_x q^2 \le R_m, \lambda_x \sim \lambda_z \sim R_E$).

picture proposed by Lee and Fu [1985] in their multiple X-line reconnection model. The principle difference is that they suggested the coherent structure of reconnection region and in our model it is irregular and turbulent.

For the SR model of plasma flow (case (A) above) the reconnection patchy should be more or less symmetric $\lambda_x \simeq \lambda_y \simeq \lambda_s = R_E q$, $\lambda_z \sim \lambda_s \sin\alpha = R_E/q$. Appropriate geometry of reconnection is illustrated in Figure 10, which resembles the pioneering elbow-shaped picture by Russell and Elphic [1979]. The percolation processes discussed in this paper occur in a thin current layer separating interplanetary and magnetosphere plasmas.

If the threshold conditions discussed above (see Figures 7 and 8) are satisfied, the reconnection between interplanetary and geomagnetic flux tubes occurring via magnetic percolation will be accomplished and as a result a macroscopic hole through which occurs the exchange between two plasmas will be formed. This is the complicated process proceeded in several stages through multiple scales

$$w^* < L < s_x < \{\lambda_x, s_y\} < \lambda_y \quad \text{SL case,}$$

$$w^* < L < \{s_x, s_y\} < \lambda_s \quad \text{SR case.}$$

After the time $\tau_{conv} \sim$200–400 s estimated above (equation (25)) the newly formed FTEs will be pulled away from the stagnation area and will be draped away to the flanks of the magnetosphere. If the external condition are not too variable on the time scale of the FTE formation (a few minutes), the percolation process would be repeated with new flux tubes brought to the magnetopause by the magnetosheath flow after the previously reconnected flux tubes were carried away from the site of their formation. The estimates discussed above give a minimum repetition time-scale of the order of a few minutes. This also, in principle, agrees with the multiple FTE observations by ISEE satellites (4–5 events on one pass) with an average time interval between them of the order of 400 s [Rijnbeek et al., 1984].

The model of the FTE formation in general is shown in Figure 11. FTEs originate at the equatorial magnetosphere near the stagnation area. For simplicity only the north half of the FTE pair is shown. Despite the internal complexity of the process the final picture seems to be remarkably similar to the famous Dungey reconnection model. The principal generalization is that the flux transfer occurs here not in a steady state, but in an impulsive regime by means of the finite magnetic flux "quanta", $\Delta\Phi_E \sim 10^7$ Wb. The convection of the reconnected flux tubes in northern and southern hemispheres, and its characteristic features at the dayside, were discussed in a review by Cowley [1982]. Flux tubes can experience twisting in the course of convection, which results in the generation of field-aligned currents within them [Saunders et al., 1984; Sonnerup, 1987; Wright, 1987]. Finally, the magnetosheath plasma motion brings the FTE tubes to the nightside thus increasing the magnetotail magnetic content. This is also illustrated schematically in Figure 11.

Let us discuss in conclusion one more interesting consequence of the developed percolation theory. What will be the fate of field lines starting the percolation process, but not finishing it? (See Figure 5b). Such lines will enter into the MCL and reenter back to the same region of space, but being shifted in transverse direction with respect to their initial one (see Figure 5b, the projection on the ZY plane) on a distance

$$\delta_\perp = \frac{\delta_x^3 L_0}{4\varrho^3} \frac{b_y b}{\sqrt{(1+b_y^2)\left(b_y^2 + \mathrm{th}^2(1-\frac{\delta_x}{L})\right)}}, \qquad (33)$$

where δ_x is the depth of immersion of field line into the MCL (see Figure 5b, the projection on the XY plane). Assuming for simple estimates $\delta_x = L_0/2$ and $b \sim b_y \sim 1$ we will have $\delta_\perp = s_0/\sqrt{2}$. So such lines will be appreciably bent after such excursion into the MCL. For magnetosheath field lines the consequences of such incomplete penetration will not be dramatic because of the fast plasma flow there, but for the slowly convecting magnetospheric field lines the bent magnetic flux tubes can be formed (see Figure 12). So even if the macroscopic reconnection is not accomplished, the magnetic diffusion practically always occurring within MCL (in a larger or lesser extent) can have the influence on global magnetic topology. Looking carefully at the recently published results of the numerical simulations of three dimensional global MHD reconnection by Walker and Ogino [1988], we see some confirmation of this ideas. In ad-

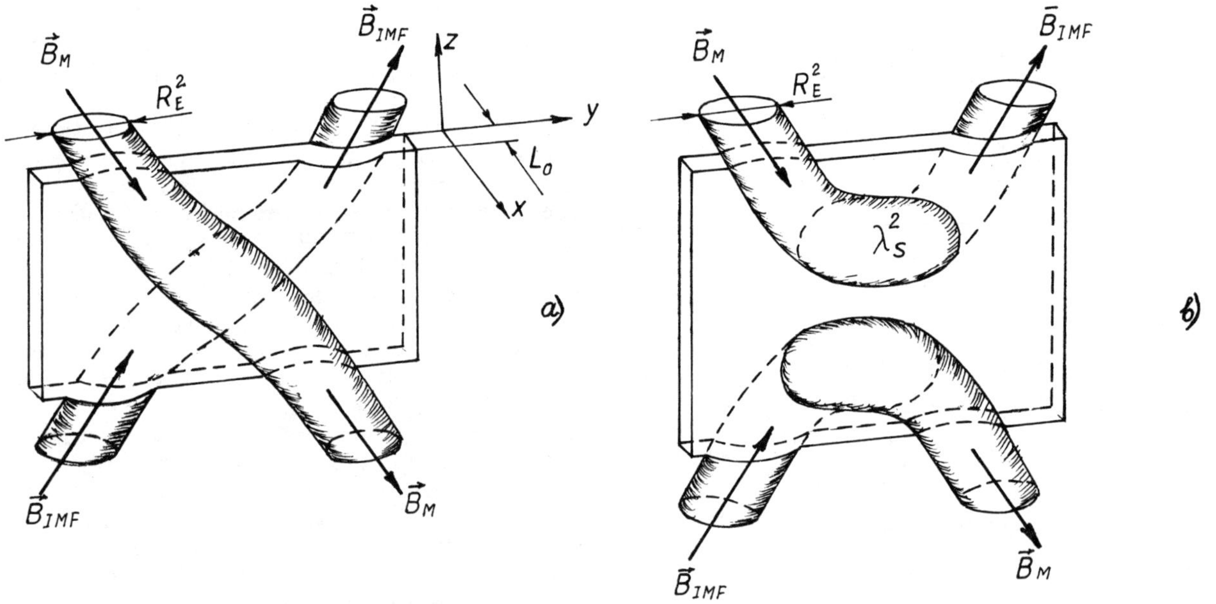

Fig. 10. (a) The scenario of FTE formation for SR (stagnation region) plasma flow pattern. The reconnection patchy with the scale $\lambda_s > \lambda \sim R_E$ appears on the magnetopause surface. Reconnection of flux tubes is accomplished via magnetic percolation through the region with (x, y, z) scales — $(L_0, \lambda_s, \lambda_s)$. (b) The convection of newly formed FTE tube pair from the site of its formation with the magnetosheath plasma flow.

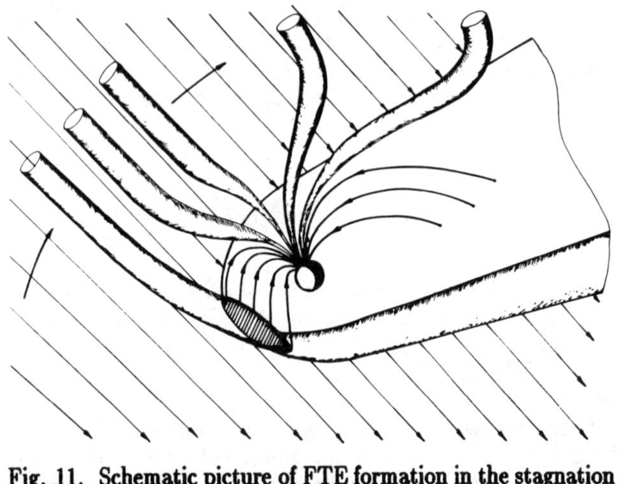

Fig. 11. Schematic picture of FTE formation in the stagnation equatorial area. The figure illustrates the convection of newly reconnected flux tubes and shows the downstream extension of the FTE to the magnetotail. FTE tubes are topologically connected with the auroral ionosphere which should result in a number of low-altitude signatures of these phenomenon. Multiple subsequent formation of FTE pairs can also be accomplished in many cases.

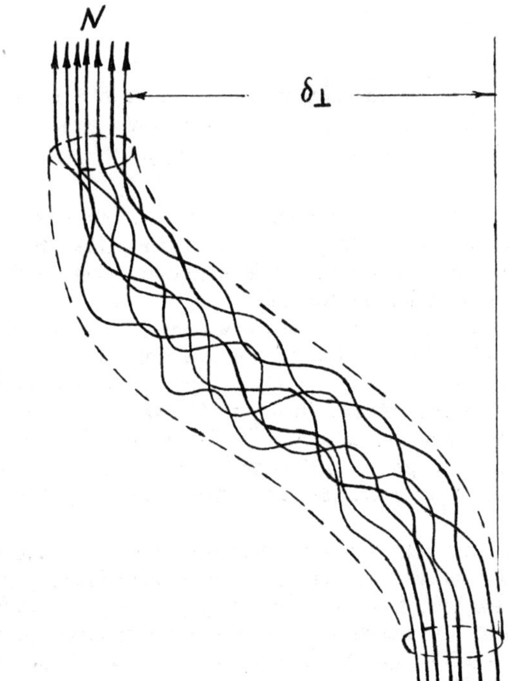

Fig. 12. The formation of the bent magnetic flux tube of geomagnetic field lines due to its partial immersion into MCL during time period $\tau \sim \tau_{conv}$.

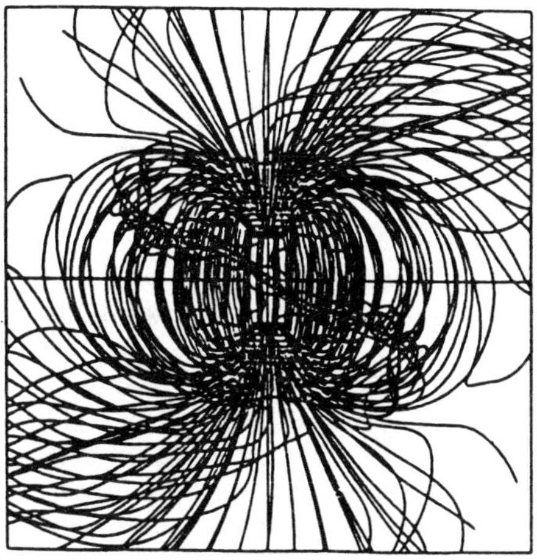

Fig. 13. Results of the numerical simulation of three dimensional global MHD reconnection by Walker and Ogino [1988]. The closed bent flux tubes at this figure can be explained in terms of the partial magnetic percolation (see Fig. 12).

dition to usual FTEs corresponding to the open twin flux tubes Walker and Ogino [1085] find the twisted and bent flux tubes on closed lines (see Figure 13).

Conclusion

The main aim of this study is to propose a mechanism of magnetic reconnection for plasma without collisions or noise. The suggested mechanism is based on an intrinsic property of the magnetopause — the presence of magnetic shear there. Reconnection appears to be a complex irregular multiscale process associated with the diffusion of magnetic fields on self-consistently generated magnetic turbulence. We refer to this process as magnetic percolation to emphasize its stochastic turbulent nature.

It is very interesting that the complicated magnetic topology of the weak magnetic field near Mars can also be discussed in terms of magnetic percolation models [Kennel et al., 1989, in press].

New measurements, of course, should be done to have a deeper insight into the nature of FTEs. The shortcoming of the rather abundant experimental material obtained by ISEE-1,2 spacecraft is the insufficient information about the processes in the equatorial regions of the magnetopause where, according to Russell et al. [1985], is located the most probable site of FTE formation. The recently begun activity to organize within the framework of the GGS program the joint USA-USSR Equator mission supports the hopes of specialists for resolving this important and challenging problem of space plasma physics.

Questions and Answers

Schindler: In the magnetotail there is the nontrivial problem arising from electrostatic coupling between electrons and ions. How do you treat this problem in the Percolation model? Are space-charge fields reduced by electron streaming along the sheared magnetic field? Are the corresponding currents important?

Zelenyi: Yes, these effects are really very important for the problem under consideration. Field-aligned ion motion is coupled with electron motion via a potential electric field. This results in linear stabilization of the drift tearing mode in many cases. In our calculations, these effects have been self-consistently taken into account, both in linear and nonlinear regimes (where they are also important). The solution of the coupled system of equations for electrostatic and vector potentials in the magnetopause case appears to be simpler than in the case of magnetotail geometry. This is because in the magnetpause case we assume the presence of a guiding magnetic field B_y, which magnetizes the particle trajectories. So, instead of complicated magnetotail trajectories, we have simple guiding center drift orbits and the problem can be solved analytically.

Sonnerup: I can see how your percolation model generates an effective diffusion coefficient, but how does it generate organized flux tubes of the size of FTEs ($1R_E$ diameter or more)?

Zelenyi: I agree with this comment. In my talk, I really have discussed the diffusion of a single field line. In the written paper, some speculations as to how these field lines could be gathered into flux bundles forming the FTE tube will also be given. Anyway, it is a complicated problem for this model because to explain such bunching, one should describe both the small-scale internal processes and large-scale plasma flow around the magnetopause.

Russell: I just want to emphasize a point I mentioned in my paper. The magnetopause current does not vary smoothly. It does have substructure that is in agreement with your model.

Zelenyi: That is very important. Theory predicts that substructures should have the scale $W^* = max(\delta_i, \rho_i)$, which for realistic conditions gives $W^* \sim \rho_i$. If this theory is valid the changes in the B_n should be observed sometimes on this scale.

Birn: The percolation model would predict that reconnection occurs preferably in magnetopause configurations in which the field rotates from one direction to another, rather than changes its magnitude and sign only. Is there any observational support for this?

Zelenyi: It is not exactly the case because we have considered just a symmetric configuration, so at this moment we can't make definite predictions of what will happen if $|B_M|, \neq |B_{IMF}|$. But, if $|B_M|, = |B_{IMF}|$ and even changes the sign (southward polarity of FTE) both theory and experiment predict preferable conditions for FTE formation.

Acknoledgments. The authors are vary grateful to Professor A. Galeev for useful discussions and suggestions.

References

Anderson R.R., Harvey C.C.., Hoppe M.M., Tsurutani B.T., Eastman T.E., and Etcheto J., Plasma waves near the magnetopause, *J. Geophys. Res.*, *87*, 2087, 1982.

Berchem J., and Russell C.T., The thickness of the magnetopause current layer: ISEE 1 and 2 observations, *J. Geophys. Res.*, *87*, 2108, 1982a.

Berchem J., and Russell C.T., Magnetic field rotation through the magnetopause: ISEE 1 and 2 observations, *J. Geophys. Res.*, *87*, 8139, 1982b.

Berchem J., and Russell C.T., Flux transfer events on the magnetopause. Spatial distributions and controlling factors, *J. Geophus. Res*, *89*, 6689, 1984.

Coppi B., Mark J.W.-K., Sugiyama L., and Bertin G., Reconnecting modes in collisionless plasma, *Phys. Rev. Letters*, *42*, 1058, 1979.

Coroniti F.V., Space plasma turbulent dissipation: reality or myth? *Space Sci. Rev.*, *45*, 399-410, 1985.

Crooker N.U., Siscoe G.L., Eastman T.E., Frank L.A., and Zwickl R.D., Large-scale flow in the dayside magnetosheath, *J. Geophys. Res.*, *A89*, 9711-9719, 1984.

Daly P.W., Saunders M.A., Rijnbeek R.P., Sckopke N., and Russell C.T., The distribution of reconnection geometry in flux transfer events using energetic ion, plasma, and magnetic data, *J. Geophys. Res.*, *89*, 3843, 1984.

Eastman T.E., Popielawska B., and Frank L.A., Three-dimensional plasma observations near the outer magnetospheric boundary, *J. Geophys. Res.*, *A90*, 9519 1985.

Elphic R.C., and Russell C.T., ISEE 1 and 2 magnetometer observations of the magnetopause, in *Magnetospheric boundary layers*, Proceedings of a Sydney Chapman Conference, Alpbach, 11-13 June, p. 43, ESA publication, 1979.

Fu Z.F., and Lee L.C., Simulation of multiple X-line reconnection at the dayside magnetopause, *Geophys. Res. Lett.*, *12*, 291, 1985.

Galeev A.A., and Zelenyi L.M., The model of magnetic field reconnection in a slab collisionless plasma sheath, *Pisma Zh. Eksp. Teor. Fiz.*, *25*, 407, 1977.

Galeev A.A., Kuznetsova M.M., and Zelenyi L.M., Magnetopause stability threshold for patchy reconnection, *Space Sci. Rev.*, *44*, 1, 1986.

Gary S.P., and Eastman T.E., The lower hybrid drift instability at the magnetopause, *J. Geophys. Res.*, *84*, 7378, 1979.

Gladd N.T., Sgro A.G., and Hewett D.W., Microstability properties of the sheath region of a field-reversed configuration, *Phys. Fluids*, *28(7)*, 2222-2234, 1985.

Huba J.B., Gladd N.T., and Papadopoulos K., The lower hybrid drift instability as a source of anomalous resistivity for magnetic field line reconnection, *Geophys. Res. Letters*, *4*, 125, 1977.

Huba J.B., Gladd N.T., and Drake J.F., The lower hybrid drift instability in non-antiparallel reversed field plasmas, *J. Geophys. Res.*, *87*, 1697, 1982.

Kennel C., Coroniti F., Moses S. et al., Unsteady dynamics of Mars magnetic field, *Nature*, in press, 1989.

Kuznetsova M.M., and Zelenyi L.M., Structure and stability of perturbations of magnetic surfaces in transitional layers, *Plasma Physics Contr. Fusion*, *27*, 363, 1985.

Kuznetsova M.M., and Zelenyi L.M., Spontaneous reconnection at the boundaries of planetary magnetospheres, *Proceedings ESA SP-251*, pp. 137-146, 1986.

Kuznetsova M.M., and Zelenyi L.M., Nonlinear dynamics of the drift tearing-mode, *Proceedings ESA SP-285*, Vol. 11, pp. 29-34, 1989.

Krall N.A. Shear stabilization of lower hybrid drift instabilities, *Physic. Fluids*, *20*, 311, 1977.

LaBelle J., Treumann R.A., Haerendel G., Bauer O.H., Paschmann G., Baumjohann W., Luhr H., Anderson R.R., Koons H.C., and Holzworth R.H., *J. Geophys. Res.*, *92*, 5827, 1987.

LaBelle J., and Treumann R.A., 1988, Plasma waves at the dayside magnetopause, *Space Science Reviews*, *47*, 175-202, 1988.

La Belle-Hamer, A.L. Fu Z.F., and Lee L.C., A mechanism for patchy reconnection at the dayside magnetopause, *Geophys. Res. Letters*, *15*, 152-155, 1988.

Laval G., Pellat R., and Vuillemin M., Instabilities electromagnetiques des plasmas sans collisions, *Plasma Phys. and Contr. Fusion Res.*, *2*, p. 259 IAEA, Vienna, 1966.

Lee L.C., and Fu Z.F., A theory of magnetic flux transfer at the Earth's magnetopause, *Geophys. Res. Lett.*, *12*, 105, 1985.

Lee L.C., and Fu Z.F., Multiple X-Line reconnection. 1. A criterion for the transition from a single X-line to a multiple X-line reconnection, *J. Geophus. Res.*, *91*, 6807-6815, 1986.

Paschmann G., Haerendel G., Papamastorakis I., Sckopke N., Bame S.G., Gosling J.T., and Russell C.T., Plasma and magnetic field characteristics of magnetic flux transfer events, *J. Geophys. Res.*, *87*, 2159, 1982.

Pudovkin M.I., and Semenov V.S., Stationary frozen-in coordinate system, *Ann. Geophys.*, *33*, 429-433, 1977.

Pudovkin M.I., and Semenov V.S. Magnetic field line reconnection theory and the solar wind magnetosphere interaction. A review, *Space Sci. Rev.*, *41*, 1, 1985.

Rijnbeek R.P., Cowley S.W.H., Southwood D.J., and Russell C.T., A survey of dayside flux transfer events observed by ISSE 1 and 2 magnetometers, *J. Geophys. Res.*, *89*, 786, 1984.

Rosenbluth M.N., Sagdeev R.Z., Taylor J.B., and Zaslavsky G.M., The destruction of magnetic surfaces, *Nucl. Fusion*, *6*, 297, 1966.

Russell C.T., and Elphic R.C., Initial ISEE magnetometer results: Magnetopause observations, *Space Sci. Rev.*, *22*, 681, 1978.

Russell C.T., and Elphic R.C., ISEE observations of flux transfer events at the dayside magnetopause, *Geophys. Res. Lett.*, *6*, 33, 1979.

Russell C.T., Zhuang H.C., Walker R.J., and Crooker N.U., A note on the location of the stagnation point in the magnetosheath flow. *Geophys. Res. Lett.*, *8*, 984, 1981.

Russell C.T., Berchem J., and Luhmann J.G., On the source region of flux transfer events, *Adv. Space Res.*, *5*, 363, 1985.

Sato T., Shimada T., Tanaka M., Hayashi T., and Watanabe K., Formation of field twisting flux tubes on the magnetopause and solar wind particle entry into the magnetosphere, *Geophys. Res. Letters*, *13*, 801, 1986.

Saunders M.A., Russell C.T., and Sckopke N., Flux Transfer Events: Scale size and interior structure, *Geophys. Res. Letters*, *11*, 131, 1984.

Scholer M., Magnetic flux transfer at the magnetopause based on single X-line bursty reconnection, *Geophys. Res. Letters*, *15*, 295, 1988.

Shi Y., Wu C.C., and Lee L.C., A study of multiple X-line reconnection at the dayside magnetopause, *Geophys. Res. Letters*, *15*, 295, 1988.

Sonnerup B.U.O., Transport mechanisms at the magnetopause, in: *Dynamics of the Magnetosphere*, ed. by S.-I.Akasofu, pp. 77-100, D. Reidel, Hingham, Mass, 1980.

Sonnerup B.U.O., On the stress Balance in Flux Transfer events, *J. Geophys. Res.*, *92*, 8613-8620, 1987.

Southwood D.J., Farrugia C.J., and Saunders M.A., What are flux transfer events?, *Planet. Space Sci.*, *36*, 503-508, 1988.

Spreiter J.R., Summers A.L., and Alksne A.Y., Hydromagnetic flow around the magnetosphere, *Planet. Space Sci.*, *14*, 223-253, 1966.

Syrovatskii S.I., Pinch sheets and reconnection in astrophysics, *Ann. Rev. Astron. Astrophys.*, *19*, 163-229, 1981.

Vaisberg O.L., Galeev A.A., Zelenyi L.M., Zastenker G.N., Omeltchenko A.N., Klimov S.I., Savin S.P., Ermolaev V.I., Smirnov V.N., and Nozdratchev M.N., The fine structure of the magnetopause: Prognoz 7, Prognoz 8 measurements, *Kosmich. Issled.*, *21*, 57, 1983.

Walker R.J., and Russell C.T., Flux transfer events at the Jovian magnetopause, *J. Geophus. Res.*, *A90*, 7397-7404, 1985.

Walker R.J., and Ogino T., The formation of isolated magnetic flux tubes on the day side magnetopause, *Preprint-PPG-1210, IGPP-3227*, University of California, Los Angeles, 1988.

Wright A.N., The evolution of an isolated reconnected flux tube, *Planet. Space Sci.*, *35*, 813-819, 1987.

Zelenyi L.M., and Taktakishvili A.L., Nonlinear theory of magnetic islands growth in sheared magnetic field, *Fisika Plasmy*, *10*, 50, 1984.

Zwan R.J., and Wolf R.A., Depletion of solar-wind plasma near a planetary boundary, *J. Geophys. Res.*, *86*, 1635, 1976.

IMBEDDED OPEN FLUX TUBES AND "VISCOUS INTERACTION" IN THE LOW LATITUDE BOUNDARY LAYER

N. U. Crooker

Department of Atmospheric Sciences
University of California, Los Angeles, CA 90024

Abstract. In a magnetosphere that is closed except for open tail lobes extending to infinity, the addition of small patches of open flux on the magnetopause along a streamline leading from the dayside to the flanks produces a series of isolated flux tubes that form elongated ionospheric footprints fanning outward from the cusp. The cusp is the only point of contact between the footprints and the polar cap. If the patches convect with the magnetosheath flow, as proposed by Sonnerup [1987], then the footprints rotate about the cusp and feed into the polar cap at dawn and dusk. The closed flux sandwiched between the rotating footprints must also move with the flow. The net flow pattern is a large-scale convection cell with a mixture of open and closed flux flowing antisunward on the poleward edge, which maps out to the low latitude boundary layer. Formerly, the convection cell for the low latitude boundary layer was thought to be driven by a viscous interaction between the magnetosheath and closed magnetospheric field lines, a mechanism separate from reconnection, which presumably drives the major convection through the polar cap. However, in the proposed model, no mechanism other than reconnection is needed to drive both polar cap and low latitude boundary layer convection.

Ionospheric footprints of flux transfer events are often shown as rounded extrusions on the polar cap boundary, as in Figure 1a, with tailward motion [e.g., Cowley, 1984; Southwood, 1987; Lockwood et al., 1988]. This representation is based on the idea that isolated flux tubes with round cross-section, freshly merged on the dayside magnetopause, accelerate poleward. Their ionospheric footprints, also round, are topologically required to be contiguous with the open flux in the polar cap, and the poleward motion feeds them directly into it. However, there are two problems with this picture.

Geophysical Monograph 58

First, not all freshly merged flux tubes will convect over the polar cap. Sonnerup [1987] has proposed that small holes of open flux in the magnetopause will convect with the external flow rather than accelerate poleward, owing to balancing magnetic forces exerted by the field within the magnetopause layer. Consequently, small flux tubes formed at low latitudes on streamlines that lead to the flanks will convect around the sides of the magnetosphere rather than over the poles.

Second, isolated tubes of open flux with round cross-section at the magnetopause do not map to round ionospheric footprints. Crooker and Siscoe [1989] have shown that round patches of open flux distort to form elongated footprints that fan outward from the cusp. The farther down the flank the patch is located, the longer and narrower is its footprint. If an open polar cap is added to the Crooker-Siscoe mapping (work in progress), as in Figure 1b, the cusp will be the only point of contiguity between the footprints and the cap. The assumed convective motion of the magnetopause patches will produce a rotation of the footprints about the cusp, as indicated by the arrows. Eventually the footprints will coalesce with the polar cap, forming layers of fresh flux hugging the dawn and dusk edges. This pattern of feeding flux into the polar cap is considerably different from the rounded extrusion pattern in Figure 1a.

A feature of particular interest in Figure 1b is the implied motion of closed flux between the open footprints. As is the case for ionospheric convection in general, the motion of one flux tube necessarily causes motion of neighboring flux tubes, whether they are open or closed. If only one open flux tube moves, the neighboring tubes will simply move out of the way, forming two localized vortices [e.g., Southwood, 1987]. However, an array of open tubes rotating tailward, as in Figure 1b,

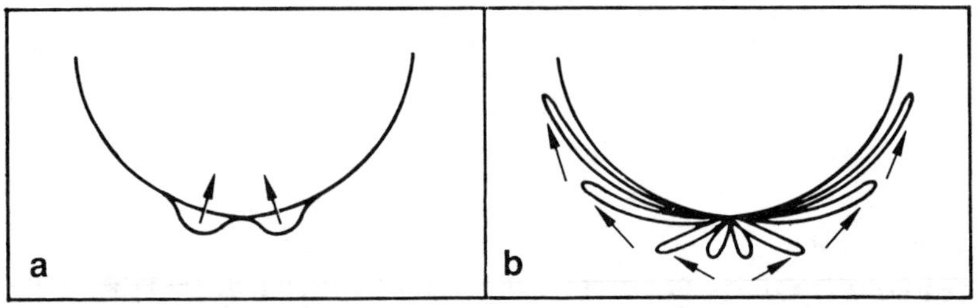

Fig. 1. Patterns of open flux tube footprints in the ionosphere on the polar cap boundary, with noon at the bottom. a) Rounded tubes move poleward to coalesce directly into the polar cap, as suggested previously by other authors. b) In the proposed model, tubes distorted by boundary currents fan outward from the cusp and rotate into the polar cap at dawn and dusk.

will "paddle" the intervening closed flux into organized antisunward flow, setting up a large-scale convection cell.

The antisunward flowing, open flux tube footprints and closed flux between them presumably map out together to the low latitude boundary layer on the magnetopause flanks. Figure 2 shows a schematic view of the flux tubes there. Open tubes formed periodically near the equator on the dayside alternate with closed tubes as they convect together around to the flanks. The elbows or bends in the open flux tubes where the fields merged upstream remain fairly sharp, consistent with Sonnerup's [1987] proposed stress balance within the hole for small tubes. That is, the tension in the tube which would straighten the bend is balanced by tension in the magnetopause field lines swept up by the moving tube. Farther down the tail, the open flux tubes should stretch and coalesce into the tail lobes at the cleft, while the closed field lines pass into the plasma sheet.

Low latitude boundary layer data are consistent with Figure 2. For example, Lundin and Evans [1985] found magnetosheath injection structures with excess momentum imbedded in the layer. They assume that the structures are on closed field lines and that the source of the excess momentum is the initial magnetosheath momentum, but an open field line interpretation is equally viable and, perhaps, preferable in view of the continual supply of momentum available to flux tubes connected to the solar wind. Another example of observations consistent with the model is the energetic particle results of Mitchell et al. [1987]. They conclude that for southward IMF, the low latitude boundary layer consists of a combination of open and closed field lines, although the open field lines are generally limited to the outer portion of the layer.

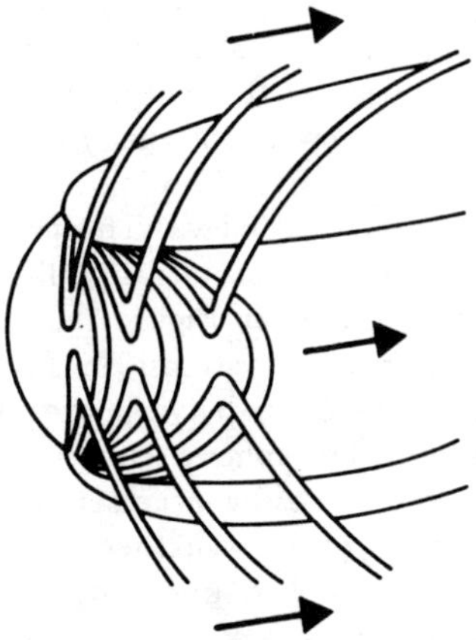

Fig. 2. Sketch of open flux tubes imbedded in low latitude boundary layer.

Observed antisunward convection on presumably closed field lines in both the low latitude boundary layer and the ionosphere has long been attributed to a viscouslike interaction of unknown origin [e.g., Axford, 1969; Eastman et al., 1976; Crooker, 1977; Reiff, 1979; Burke et al., 1979] or, more specifically, to the impulsive penetration of plasma filaments onto closed field lines by an unknown mechanism [e.g., Lemaire et al., 1979; Lundin and Evans, 1985]. On the other hand, a large body of evidence suggests that the main driver of magnetospheric convection is magnetic merging [e.g., Hill, 1979; Cowley, 1982]. As voiced by Axford twenty years ago, it seems unlikely that we would be so unlucky as to have two competing mechanisms for driving convection. This paper offers a way to drive antisunward convection on closed field lines in the low latitude boundary layer by magnetic merging, thus unifying the driving mechanisms for all convection. (See, also, Nishida [1989].)

Although a southward IMF has been assumed in the illustrations given here, it seems likely that the proposed closed field line convection will also occur for all other IMF orientations. As the IMF rotates northward, the merging sites presumably move off the equator and away from the subsolar region. However, as long as small patches of open flux form on streamlines covering extensive regions of otherwise closed magnetopause, antisunward convection on closed flux will be driven in the ionosphere by the open patch footprints. Studies of the magnetospheric topology and geometry for open patches during periods of northward IMF are currently in progress.

We note that larger open patches on the magnetopause, such as the longitudinal extensions in the flux transfer event models of Southwood et al. [1988] and Scholer [1988], would tend to accelerate poleward and feed large amounts of flux directly into the polar cap, generating a large potential drop across it. In comparison, the proposed feeding of small flux patches at dawn and dusk would be a minor contributor to the total potential generated by merging, of the same order as the potential generated on the closed field lines.

Magnetosheath plasma in the low latitude boundary layer lies on the open flux tubes in the proposed model (cf., e.g., Cowley [1982] and Saunders [1983]). As suggested by Cowley [1982], this plasma may also populate the surrounding closed flux tubes by diffusion. Tsurutani and Thorne [1982] point out that diffusion should be an inevitable consequence of the ever-present wave emissions observed at the magnetopause. They demonstrate that diffusion by wave-particle interactions could populate a low latitude boundary layer entirely on closed field lines, but the process would seem even more efficient for a boundary layer with imbedded open flux tubes. LaBelle et al. [1987] find peak wave activity at the boundaries of flux transfer events, consistent with strong particle diffusion on the surface of freshly merged flux tubes. Alternatively, Nishida [1989] has suggested that magnetosheath plasma on closed field lines in the low latitude boundary layer results from reclosure of the open flux tubes by random reconnection, especially during periods of northward IMF.

In summary, if the dayside magnetopause has small patches of open flux on streamlines that lead to the flanks, then their ionospheric footprints do not feed directly into the polar cap but pivot about the cusp and rotate into the boundary along the dawn and dusk sides. The rotating footprints "paddle" the adjacent closed flux into organized antisunward convection. The convecting open and closed flux tubes map out to the low latitude boundary layer. Thus, indirectly, magnetic merging becomes the viscouslike interaction that drives closed field line convection on the flanks of the magnetosphere, and no second driving force is necessary to account for the whole of magnetospheric convection.

Questions and Answers

Gosling: (1) Is there any direct evidence that FTEs experience the asymmetric convection required to explain the Svaalgard-Manaurov effect? (2) What inhibits the reconnected field lines from being pulled into the polar cap in your model? (3) If FTEs move like you suggest, they probably cannot contribute to polar cap convection. Are you suggesting that FTEs do not contribute directly to polar cap convection?

Crooker: (1) None that I know of. (2) I base my model on the hypothesis that the flux tubes convect with the flow rather accelerate directly into the polar cap. This hypothesis is the result of Bengt Sonnerup's theory, in which the flux tube is inhibited by the magnetic tension in the magnetopause field that links the elbow in the flux tube formed by merging at the x-line. I am not completely comfortable with this reason for force balance and plan to explore the theory further. (3) No. Only small FTEs formed on low-latitude streamlines leading to the flanks will feed into the dawn and dusk sides of the polar cap. There they will still contribute to polar cap convection, but not strongly. However, small FTEs formed on streamlines leading over the polar cap will feed directly into it and contribute to convection in the usual way. Further, larger FTEs will be accelerated poleward and contribute strongly to polar cap convection.

Kivelson: How do you rule out the development of vortical flows around the tailward-moving "paddles" producing equatorward-directed flows at latitudes below the polar cap boundary?

Crooker: I do not rule out vortical flow. Each "paddle" will push closed flux out of its way, forming a small vortex. However, when the vortices of all the "paddles" are superposed, they will produce a global convection cell with antisunward flow at high latitudes and sunward flow at lower latitudes.

Forbes: I think it would be very nice if only one convection mechanism were needed in stead of separate reconnection and viscous (Axford-Hines) mechanisms, but wouldn't the magnetotail collapse completely for a northward magnetic field if there were no separate viscous mechanism?

Crooker: The lobes of the magnetotail are generally always present, even when the IMF is northward, presumably because of past merging with southward IMF. The IMF rarely remains in a given orientation for long. However, there are occasions when observers claim that the open polar cap disappears. The plasma sheet seems to expand and cool at these times, forming the full body of the magnetotail. The presence of plasma filling the magnetotail at these times probably prevents its collapse.

Atkinson: How small does the flux tube have to be for it to go around the dawn and dusk rather than poleward?

Crooker: I refer you to Bengt Sonnerup.

Sonnerup: The theory has a number of free parameters about which we have little information to fix the critical size.

Acknowledgement. This work was supported by the National Science Foundation under grant ATM87-22962.

REFERENCES

Axford, W. I., Magnetospheric convection, *Rev. Geophys.*, *7*, 421-459, 1969.

Burke, W. J., M. C. Kelley, R. C. Sagalyn, M. Smiddy, and S. T. Lai, Polar cap electric field structures with a northward interplanetary magnetic field, *Geophys. Res. Lett., 6,* 21-24, 1979.

Cowley, S. W. H., The causes of convection in the Earth's magnetosphere–A review of developments during the IMS, *Rev. Geophys. Space Phys., 20,* 531-565, 1982.

Cowley, S. W. H., Evidence for the occurrence and importance of reconnection between the Earth's magnetic field and the interplanetary magnetic field, in *Magnetic Reconnection in Space and Laboratory Plasmas,* edited by E. W. Hones, Jr., pp. 375-378, American Geophysical Union, Washington, D. C., 1984.

Crooker, N. U., The magnetospheric boundary layers: A geometrically explicit model, *J. Geophys. Res., 82,* 3629-3633, 1977.

Crooker, N. U., and G. L. Siscoe, On mapping flux transfer events to the ionosphere, *J. Geophys. Res.,* submitted, 1989.

Eastman, T. E., E. W. Hones, Jr., S. J. Bame, and J. R. Asbridge, The magnetospheric boundary layer: Site of plasma, momentum and energy transfer from the magnetosheath into the magnetosphere, *Geophys. Res. Lett., 3,* 685-688, 1976.

Hill, T. W., Rates of mass, momentum, and energy transfer at the magnetopause, in *Magnetospheric Boundary Layers,* ESA SP-148, edited by B. Battrick, pp. 325-332, European Space Agency, Paris, 1979.

LaBelle, J., R. A. Treumann, G. Haerendel, O. H. Bauer, G. Paschmann, W. Baumjohann, H. Lühr, R. R. Anderson, H. C. Coons, and R. H. Holzworth, AMPTE IRM observations of waves associated with flux transfer events in the magnetosphere, *J. Geophys. Res., 92,* 5827-5843, 1987.

Lemaire, J., M. J. Rycroft, and M. Roth, Control of impulsive penetration of solar wind irregularities into the magnetosphere by the interplanetary magnetic field direction, *Planet. Space Sci., 27,* 47-57, 1979.

Lockwood, M., M. F. Smith, C. J. Farrugia, and G. L. Siscoe, Ionospheric ion upwelling in the wake of flux transfer events at the dayside magnetopause, *J. Geophys. Res., 93,* 5641-5654, 1988.

Lundin, R., and D. S. Evans, Boundary layer plasmas as a source for high-latitude, early afternoon, auroral arcs, *Planet. Space Sci., 32,* 1389-1406, 1985.

Mitchell, D. G., F. Kutchko, D. J. Williams, T. E. Eastman, L. A. Frank, and C. T. Russell, An extended study of the low-latitude boundary layer on the dawn and dusk flanks of the magnetosphere, *J. Geophys. Res., 92,* 7394-7404, 1987.

Nishida, A., Can random reconnection on the magnetopause produce the low latitude boundary layer?, *Geophys. Res. Lett., 16,* 227-230, 1989.

Reiff, P. H., Low altitude signatures of the boundary layers, in *Magnetospheric Boundary Layers,* ESA SP-148, edited by B. Battrick, pp. 167-173, European Space Agency, Paris, 1979.

Saunders, M. A., Recent ISEE observations of the magnetopause and low latitude boundary layer: A review, *J. Geophys., 52,* 190-198, 1983.

Scholer, M., Magnetic flux transfer at the magnetopause based on single X line bursty reconnection, *Geophys. Res. Lett., 15,* 291-294, 1988.

Sonnerup, B. U. Ö., On the stress balance in flux transfer events, *J. Geophys. Res., 92,* 8613-8620, 1987.

Southwood, D. J., The ionospheric signature of flux transfer events, *J. Geophys. Res., 92,* 3207-3213, 1987.

Southwood, D. J., C. J. Farrugia, and M. A. Saunders, What are flux transfer events?, *Planet. Space Sci., 36,* 503-508, 1988.

Tsurutani, B. T., and R. M. Thorne, Diffusion processes in the magnetopause boundary layer, *Geophys. Res. Lett., 9,* 1247-1250, 1982.

COUPLING OF THE TEARING MODE INSTABILITY WITH K-H INSTABILITY AT THE MAGNETOPAUSE

Z. Y. Pu
Department of Geophysics, Peking University, Beijing, China
M. Yei
Center for Space Research and Applications, Chinese Academy of Sciences
Beijing, China

Abstract. The coupling of the tearing mode instability with Kelvin-Helmholtz instability (K-H) at the magnetopause is studied by performing a 2-D MHD simulation. It is found that in the case that both Reynolds and magnetic Reynolds numbers are fixed, the Alfven Mach number, M_A plays an essential role in determining the linear properties and nonlinear evolution of the coupled instability. If $M_A < 0.4$, the spontaneous tearing mode is dominated in the system. When $0.4 \leq M_A \leq 1$, the tearing mode is apparently modified by K-H. As $1 < M_A$, the coupled instability, called the vortex induced tearing mode instability (VITM), appears to be intrinsically different from the conventional tearing mode instability. The long time asymptotic quasi-static state for VITM is characterized by a large scale fluid vortex together with a concurrent magnetic island.

I. Introduction

In the past decade considerable attention has been given to the tearing mode instability and Kelvin-Helmholtz instability (K-H) instability at the magnetospheric boundary [Sonnerup and Ledley, 1979; Quest and Coroniti, 1981; Sonnerup, 1984; Paschmann et al., 1986; Russell, 1984; Fu and Lee, 1985; Southwood, 1979; Pu and Kivelson, 1983a,b; Kivelson and Pu, 1984; Miura, 1984, 1986]. Much has been achieved. For example, the multiple reconnection model based on the forced tearing mode instability has presented an interesting picture for the FTEs at the dayside magnetopause [Lee and Fu, 1985]. The transport effects associated with the K-H instability have been found to contribute significantly to the magnetospheric boundary dynamics [Miura, 1984, 1986; Pu and Kivelson, 1983b]. It is well known that the tearing mode instability occurs when (at least one component of) the magnetic field $\vec{B}$ is reversed within a short distance, while the K-H instability appears in a sheared flow field. Almost all previous work treated these two instabilities separately. However, when the interplanetary magnetic field (IMF) has a southward orientation, a sheared velocity field and a sheared magnetic field both exist simultaneously near the magnetospheric boundary. What will happen in this case if the tearing mode and K-H are coupled with each other? The problem is obviously of great interest to the solar terrestrial physics.

Recently, Liu and Hu, La Belle-Hamer et al. and Hu et al. suggested that local magnetic field reconnection can be caused by vortices in a flow field [Liu and Hu, 1988; Hu et al., 1988; La Belle-Hamer et al., 1988]. They pointed out that if there is a strong velocity shear in the current sheet, the K-H instability is excited to produce large-scale fluid vortices; the magnetic field response to the vortices causes the magnetic field lines to twist and so to generate local reconnection. Apparently, in this case the coupling of K-H with the tearing mode has occurred, leading to the change of features of instability in the transition region. How do the features of instability vary with the flow speed? Under what condition can the vortex induced reconnection take place? How is the vortex induced reconnection different from the conventional reconnection? In this paper a two dimensional MHD simulation is performed to investigate the coupling of the tearing mode with K-H in an attempt to the answer the above questions.

It is found that the Alfven mach number, M_A, plays an essential role in determining the properties of the coupled instability. In the case of $B_0 = 1.0$, $R_m^{-1} = 2.0$ and $R^{-1} = 0.1$, if $M_A < 0.4$, spontaneous tearing mode is dominated in the system. When $0.4 \leq M_A \leq 1$, the tearing mode is apparently modified by K-H. As $1 < M_A$, the coupled instability, called the vortex induced tearing mode instability (VITM) in this paper, appears to be

Geophysical Monograph 58

intrinsically different from the conventional tearing mode. The vortex induced magnetic reconnection (VIMR) discussed by Liu et al. appears in this circumstance. The long time asymptotic quasi-static state (AQSS) for VITM is found to be composed of a large scale vortex together with a co-located magnetic island.

II. Simulation Model

Basic Assumptions

For simplicity, we assume that: (a) in the initial state a one dimensional current sheet exists together with a velocity shear: both $\vec{B}$ and $\vec{V}$ vary with z only and direct in the x direction; (b) all perturbation quantities are invariant in the y direction, i.e., $\partial/\partial y = 0$; (c) plasmas can be regarded as incompressible and the number density is uniform everywhere.

Dimensionless Equations

The 2-D dimensionless MHD equations can be rewritten as

$$\partial\Omega/\partial t = -\vec{V}\cdot\nabla\Omega + \vec{B}\cdot\nabla J + R^{-1}\nabla^2\Omega, \quad (1)$$

$$\partial A/\partial t = -\vec{V}\cdot\nabla A + R_m^{-1}\nabla^2 A, \quad (2)$$

$$\vec{V} = -\nabla\times(\phi\vec{e}_y) \quad (3)$$

and

$$\vec{B} = \nabla\times(A\vec{e}_y) \quad (4)$$

with $\vec{V} = V_x\vec{e}_x + V_z\vec{e}_z$, $\vec{B} = B_x\vec{e}_x + B_z\vec{e}_z$, $\nabla^2\phi = \Omega$, $\nabla^2 A = -J$, where A, Ω and J represent the y component of the magnetic potential vector, fluid vorticity and current density, respectively. $R = \ell_{00}^2/\mu t_0$ and $R_m = \ell_{00}^2/\mu_m t_0$ refer to the Reynolds and magnetic Reynolds number, $t_0 = \ell_{00}/v_A$, ℓ_{00} indicates a unit length used to measure all distances in the system, $v_A = B_{00}/(\mu_0\rho_{00})^{1/2}$, B_{00} is the characteristic magnitude for the sheared component of $\vec{B}$, ρ_{00} denotes the mass density, μ and μ_m are the fluid and magnetic viscosity, respectively.

Initial and Boundary Conditions

The simulation region is taken to be a square of $-L/2 \leq x \leq L/2$, $-L/2 \leq z \leq L/2$ on the (xz) plane. The boundary conditions imposed at $x = \pm L/2$ are periodic in the x direction, while at the upper and lower boundaries, $z = \pm L/2$, perturbations all tend to zero. The initial profiles of $\vec{V}$ and $\vec{B}$ are assumed as

$$\vec{V} = V_0 \tanh(z/\ell_0)\ \vec{e}_x \quad (5)$$

and

$$\vec{B} = -B_0 \tanh(z/\ell_0)\ \vec{e}_x \quad (6)$$

with ℓ_0 indicating the scale length of the shear region.

Numerical Methods

A mesh system of (32x32) grid points is used. The differential derivatives in space are replaced by the central difference formulas, while for the time evolution we use the 4th-order Runge-Kutta scheme. The following parameters are chosen in the simulations: $B_0 = 1$, $\ell_0 = 6.4$, $R^{-1} = 0.1$, $R_m^{-1} = 2.0$.

III. Simulation Results

Suppose that small perturbations appeared in both Ω and A at t = 0. Figure 1 shows how the maximum values of $|\Omega|$ and $|A|$ evolve with time afterwards. It can be seen clearly that the linear growth rates for Ω and A are the same. During the nonlinear stage, the system will be saturated and approaches an asymptotic quasi-static state (AQSS).

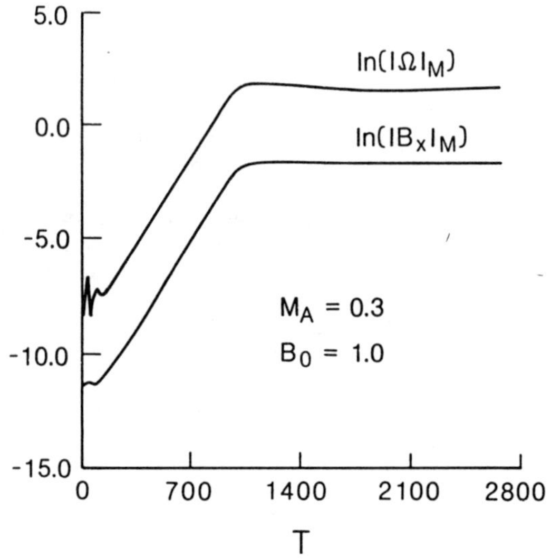

Fig. 1 Time evolutions of maximum $|\Omega|$ and $|A|$ for $M_A = 0.3$.

Figure 2 plots how $\alpha = \gamma/\gamma_{TM}$ varies with $M_A = V_0/B_0$, where γ is the linear growth rate of the simulation system and γ_{TM} denotes the linear growth rate for the spontaneous tearing mode instability ($M_A = 0$) under the same condition. If $M_A < 0.4$, $\alpha \to 1$. This growth rate is very similar to that obtained by La Belle-Hamer et al. [1988] in their Figure 3b. The tearing mode is dominant in this case. When $0.4 \leq M_A \leq 1$, the growth rate gradually increases. This means that the tearing mode is enhanced and modified by the K-H. As $M_A > 1$, α rises very rapidly, implying that the instability property has been changed and the system is controlled by the K-H instability in this circumstance.

The saturation level of the instability, which is defined as

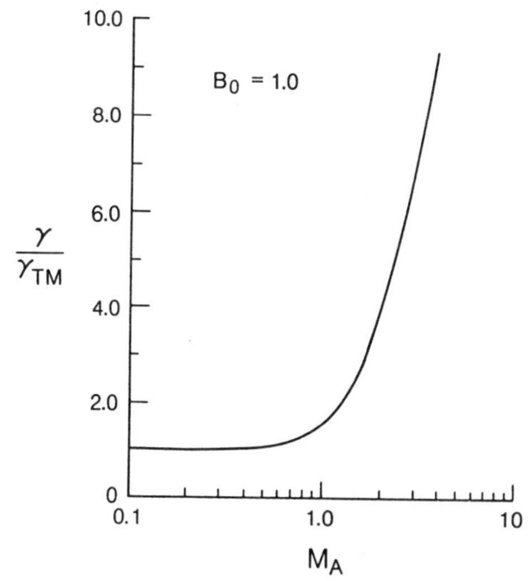

Fig. 2 The variation of γ/γ_{TM} with different M_A, where γ is the linear growth rate and γ_{TM} represents the linear growth rate of the spontaneous tearing mode instability.

$$S = \left. \frac{<(B_x - B_{0x})^2 + (B_z - B_{0z})^2>}{B_0^2} \right|_{max}$$

and used to describe the maximum deviation of $\vec{B}$ in saturated states from that of the initial state, is dependent upon M_A as well. Figure 3 shows how S varies with M_A. It can be seen that if $M_A < 0.4$, S is close to its minimum value at $M_A = 0$. When $0.4 \leq M_A \leq 1$, S apparently increases. As $M_A > 1$, S gradually tends to an asymptotic value.

Simulation results indicate that as $M_A > 0.2$, the difference between the time of occurrence for the magnetic island and fluid vortex reflects, to some extent, the relative importance of the magnetic field shear over the flow shear in determining the instability development of the system. It is found that if $M_A < 0.8$, magnetic islands appear earlier than vortices, while for $M_A > 0.8$, vortices are formed earlier than magnetic islands. Figure 4, 5 and 6 present the time of occurrence for the large scale magnetic islands and vortices as $M_A = 1.0$, 0.7 and 0.8, respectively.

Time evolutions of the instability and patterns of magnetic islands and vortices developed for distinct ranges of M_A are more easily found to be intrinsically different from each other. If $M_A < 0.4$, plasmas in the shear region are driven to the center line (z = 0) and then move to the flank edges, leading to the appearance of two pairs of

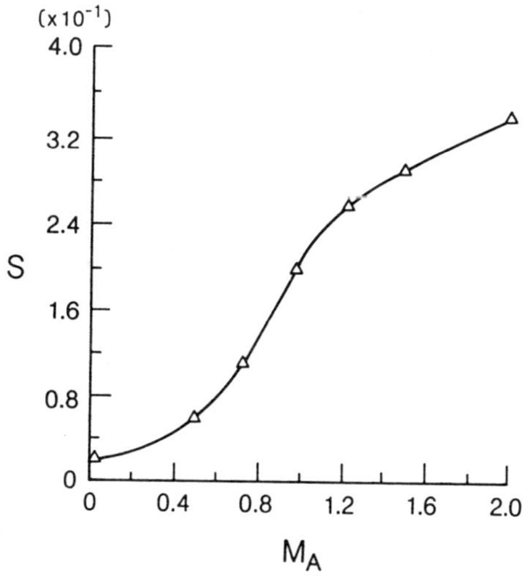

Fig. 3 The saturation level, S, varying as a function of time with the time unit being 12.8 t_0.

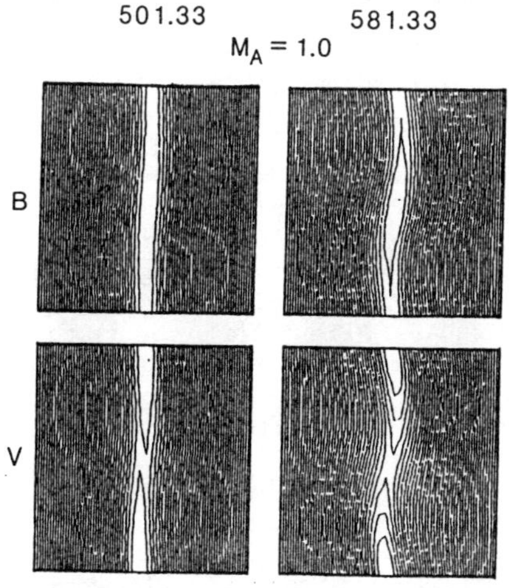

Fig. 4 The time of occurrence of the magnetic island and fluid vortex for $M_A = 1.0$. The time unit in the figure equals 12.8 t_0.

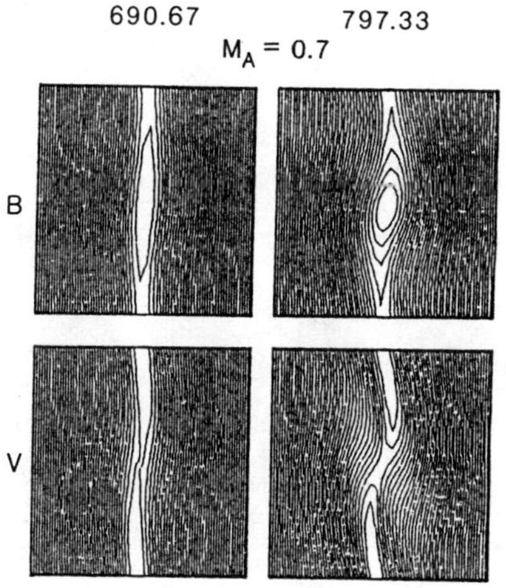

Fig. 5 The time of occurrence of the magnetic island and fluid vortex for $M_A = 0.7$.

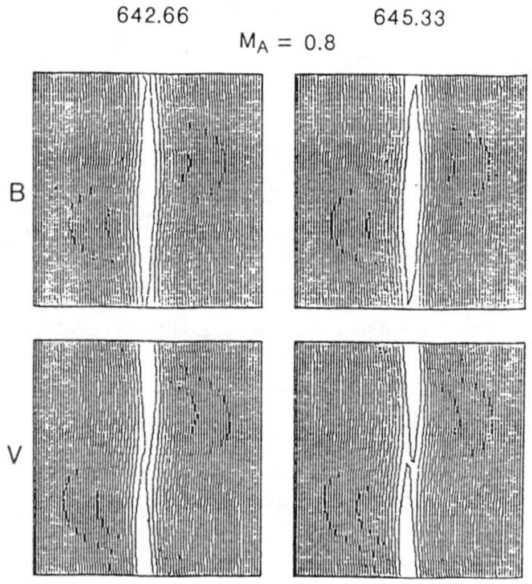

Fig. 6 The time of occurrence of the magnetic island and fluid vortex for $M_A = 0.8$.

vortices with the scale length in the z direction smaller than that in the x-direction. Two X-points occur at z = 0 where field lines with opposite directions are contacted. A magnetic island is thus formed with the O-point in the center of the simulation square. Figure 7a presents the AQSS the system finally achieves for $M_A = 0.3$, which looks quite similar to that of spontaneous tearing mode. As $0.4 \leq M_A \leq 1$, the vortex and magnetic island develop independently at first. The vortex is of K-H type with its center being L/2 apart from that of the magnetic island. However, as time goes on, a new vortex concentric with the island appears, which gradually grows and finally dominates the flow system. Figure 7b plots the field line and

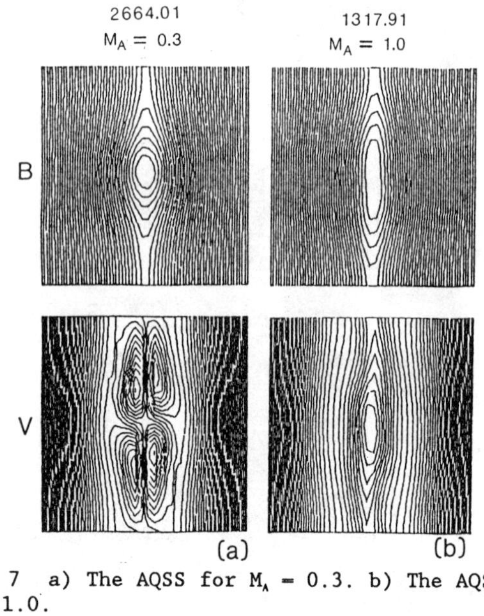

Fig. 7 a) The AQSS for $M_A = 0.3$. b) The AQSS for $M_A = 1.0$.

flow patterns at t = 1318 as $M_A = 1.0$, showing the final state for this medium range of M_A. Furthermore, when $M_A > 1$, as we can see in Figure 8 which plots the stream lines and magnetic field lines for $M_A = 2.0$ at t = 234, 251, 320 and 338, respectively, that a large scale fluid vortex occurs soon as the K-H instability has developed. The vortical motion in this case is strong enough to twist significantly the magnetic field lines and drive the field lines with opposite directions on each side of the vortex center to close to each other. Magnetic reconnection then takes place with two X-points occurring near, respectively, the left and right edges of the vortex. Two magnetic islands are thus formed with one being concentric with the vortex, and the other being limited in the center region of the simulation square. The former grows and expands as reconnection of field lines continues, while the latter is getting smaller and smaller, and finally disappears. The system then approaches the asymptotic quasi-static state, which is composed of a large scale fluid vortex together with a concentric magnetic island. This merging process is the so-called vortex induced magnetic reconnection (VIMR) first recognized by Liu and Hu, and Hu et al. [Liu and Hu, 1988; Hu et al., 1988].

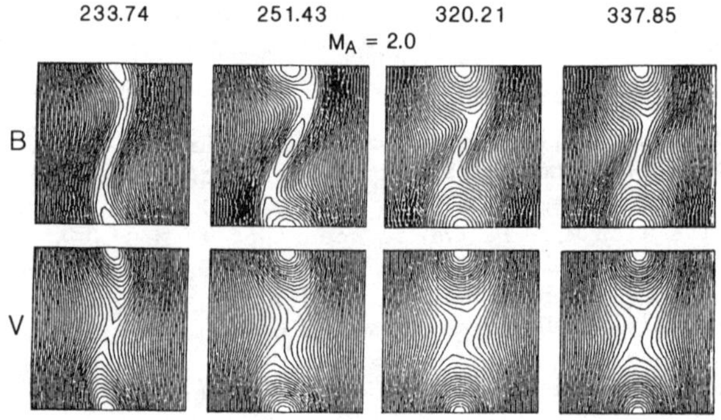

Fig. 8 Time evolution of the system for $M_A = 2.0$.

IV. Discussion

Our simulation shows that when a sheared magnetic field and a sheared velocity field both exist at the magnetospheric boundary, the time development of the instability at the boundary depends strongly upon the Alfven Mach number. Under the conditions taken in our calculations, we find that the instability properties for distinct ranges of M_A are intrinsically different from each other. If $M_A < 0.4$, the spontaneous tearing mode is dominant. For $0.4 \leq M_A \leq 1$, the tearing mode and K-H are 'comparable' to each other. The coupling of these two leads to a modified type of tearing mode instability. As $M_A > 1$, the K-H instability governs the evolution of the system. The vortical motion generated by the instability drives lines of force with opposite directions to meet each other, causing the merging and reconnection of the field lines. The size, shape and position of the magnetic island are all related to those of the fluid vortex. Time development in this case is much faster than that of spontaneous tearing mode. The reconnection rate is much stronger as well. Apparently, the instability appearing in this situation is essentially different from the conventional tearing mode. From the viewpoint of the magnetic field morphology, we call it the vortex induced tearing mode instability (VITM).

It is anticipated that the Alfven Mach number plays an essential role in determining the vortex pattern and instability features. The ratio of the first term on the right hand side of Equation 1 over the second term gives the order of M_A^2, which measures the relative importance of the inertia force exerted on a plasma element over the electromagnetic force. In the early stage of the spontaneous tearing mode instability, plasmas are driven mainly by the electromagnetic force. On the contrary, in the case of the pure K-H, the vortical motion is generated basically by the inertia force associated with the velocity shear. Then it is natural to see in our simulation that the less M_A is, the more the coupled instability appears tearing mode-like; the larger M_A becomes, the more the K-H controls the system.

In our simulation, a frame of reference is chosen in which $\vec{V}_1 = -\vec{V}_2 \| \vec{e}_x$, where 1 and 2 indicate the magnetosheath and magnetosphere, respectively. The plasmas are assumed to be incompressible with $N_1 = N_2$ and $\vec{B}_1 = -\vec{B}_2$ for simplicity [Fu and Lee, 1985; Liu and Hu, 1988; La Belle-Hamer et al., 1988]. It can easily be proved that in two dimensional cases $(\partial/\partial y = 0)$, B_y and V_y are decoupled from other quantities, and hence the presence of $B_y \neq 0$ does not make any change of Equations 1-4. Thus, in our model, B_0 (see Equation 5) may be regarded as the "reconnected component" of the initial magnetic field at the magnetopause, and M_A and v_A, in practice, are only a fraction of the real Alfven Mach number and Alfven speed, respectively. Taking $B_{00} = 20$ nT and $N = 18/cm^3$ from both ISEE 1, 2 and Pioneer 6 observations and theoretical prediction [Haerendel and Paschmann, 1982; Spreiter et al., 1967], we obtain $v_A \sim 100$ km/s. This means that as long as the magnetosheath flow passing over the magnetosphere at $v > 200$ km/s, VITM and hence VIMR will then take place, provided plasma parameters at the boundary are close to those used in this paper. Besides, the time to form the final state as $M_A = 2$ is estimated to be ~100 s. Therefore, one may expect the VITM to provide a possible mechanism for the formation of FTEs at the dayside magnetopause away from the stagnation point.

An important conclusion obtained in our simulation lies in the fact that the AQSS for both K-H modified tearing mode and VITM are composed of a large scale magnetic island and a concentric fluid vortex. A 2-D magnetic island corresponds to a magnetic vortical flux tube in the 3-D situation in the case of $B_y \neq 0$. Thus if an FTE is formed through the VITM at the dayside magnetopause, its flux tube must be a current tube and a vortex tube as well. We have investigated the properties of the AQSS for VITM in detail and presented the results in a separate paper [Pu et al., 1989].

R_m^{-1} and R^{-1} have been taken as 2.0 and 0.1, respectively, in this work. The following way can be used to examine whether the choice is permissible. The anomalous resistivity may be expressed as $\sigma_a^{-1} = \mu_0 R_m^{-1} \ell_{00} v_A$, where μ_0 is the vacuum magnetic permeability. Assuming $v_A \sim 100$ km/s, we find $\sigma_a^{-1} \sim 8.2 \times 10^3$ mΩ, which can be shown to be an order higher than that obtained by using the quasi-linear theory of the lower hybrid drift turbulence [Guan et al., 1984], while an order lower than that required for fast reconnection at the dayside magnetopause [Coroniti, 1985]. Meanwhile, the cross-field anomalous diffusivity can be written as $D_\perp = R^{-1} \ell_{00} v_A$ and found to be $\sim 3.3 \times 10^8$ m^2/s in our case, which is an order less than that needed for the viscous model of the solar wind-magnetosphere coupling [Haerendel et al., 1982; Pu et al., 1986]. Therefore, both values of R^{-1} and R_m^{-1} used in this work seem to be acceptable. The details of simulation results depend, of course, on the dissipation parameters used. We will discuss how instability properties change with different R_m^{-1} and R^{-1} in a separate paper.

Acknowledgements. We are grateful to Dr. Z. F. Fu for his helpful assistance. The authors also thank Prof. Z. X. Liu for useful discussions. This work is supported by Chinese Science Foundation.

References

Coroniti, F. V., Space plasma turbulent dissipation: Reality or myth, Space Sci. Rev., 42, 399, 1985.

Fu, Z. F. and L. C. Lee, Simulation of multiple X lines in reconnection at the dayside magnetopause, Geophys. Res. Lett., 12, 291, 1985.

Guan, J., R. J. Yin, K. H. Zhao and C. Y. Tu, A simplified model of the magnetopause and the lower hybrid drift instability, Chinese J. Space Res., 4, 112, 1984.

Haerendel, G. and G. Paschmann, in Magnetospheric Plasma Physics, ed. by A. Nishida, Center for Academic Publications, Tokyo, Japan & D. Reidel Publishing Company/Dordrecht, Boston, London, p. 49, 1982.

Hu, Y. D., Z. X. Liu and Z. Y. Pu, Response of the magnetic field to the flow vortex field and reconnection in the magnetopause boundary region, Scientia Sinica (A), 10, 1100, 1988.

La Belle-Hamer, A. L., Z. F. Fu, and L. C. Lee, A mechanism for patchy reconnection at the dayside magnetopause, Geophys. Res. Lett., 15, 152, 1988.

Lee, L. C. and Z. F. Fu, A theory of magnetic flux transfer at the Earth's dayside magnetopause, Geophys. Res. Lett., 12, 105, 1985.

Liu, Z. X. and Y. D. Hu, Local magnetic reconnection caused by vortices in the flow field, Geophys. Res. Lett., 15, 752, 1988.

Miura, A., Anomalous transport by magnetohydrodynamic Kelvin-Helmholz instability in the solar wind-magnetosphere interaction, J. Geophys. Res., 89, 801, 1984.

Miura, A., Simulation of Kelvin-Helmholtz instability at the magnetospheric boundary, J. Geophys. Res., 92, 3195, 1987.

Paschmann, G., I. Papamastorakis, W. Baumjohann, N. Sckopke, C. W. Carlson, B. U. O. Sonnerup and H. J. Luhr, The magnetopause for large magnetic shear: AMPTE/IRM observation, J. Geophys. Res., 91, 11099, 1986.

Pu, Z. Y. and M. G. Kivelson, Kelvin-Helmholtz instability at the magnetopause: energy flux into the magnetosphere, J. Geophys. Res., 88, 853, 1983a.

Pu, Z. Y. and M. G. Kivelson, Kelvin-Helmholtz instability at the magnetopause: Energy flux into the magnetosphere, J. Geophys. Res., 88, 853, 1983b.

Pu, Z. Y., C. Q. Wei and Z. X. Liu, Kinetic drift instabilities on Earth's magnetopause, in Proceedings of the International Symposium on Space Physics, Beijing, China, Chinese Society of Space Research, 1986.

Pu, Z. Y., P. T. Hou and Z. X. Liu, A two dimensional MHD simulation study on vortex induced tearing mode instability at the magnetopause: long time asymptotic quasi-static state, submitted to J. Geophys. Res., 1989.

Quest, K. B. and F. V. Coroniti, Tearing at the dayside magnetopause, J. Geophys. Res., 86, 3289, 1981.

Russell, C. T., Magnetic reconnection, in Magnetic Reconnection in Space and Laboratory Plasmas, ed. by E. W. Hones, Jr., American Geophys. Union, p. 124, 1984.

Sonnerup, B. U. O. and B. G. Ledley, OGO 5 magnetopause structure and classical reconnection, J. Geophys. Res., 84, 399, 1979.

Sonnerup, B. U. O., Magnetic field reconnection at the magnetopause: an overview, in Magnetic Reconnection in Space and Laboratory Plasmas, ed. by E. W. Hones, Jr., American Geophys. Union, p. 92, 1984.

Southwood, D. J., Magnetopause Kelvin-Helmholtz instability, in Proceedings of Magnetospheric Boundary Layers Conference, ESA Scientific and Technical Publications Branch, Noordwijk, the Netherlands, p. 357, 1979.

THE ASYMPTOTIC QUASI-STATIC STATE OF THE VORTEX INDUCED TEARING MODE INSTABILITY AT THE MAGNETOPAUSE

Z. Y. Pu and P. T. Hou

Department of Geophysics, Peking University, Beijing, China

Z. X. Liu

Center for Space Research and Applications, Chinese Academy of Science
Beijing, China

Abstract. A 2-D MHD simulation is performed to investigate the asymptotic quasi-static state (AQSS) of the vortex induced tearing mode instability (VITM) at the magnetopause. The AQSS is composed of a large scale magnetic island together with a co-located vortex tube. The properties of the AQSS and the detailed structures of the island and vortex tube are studied.

I. Introduction

A number of investigations have shown that the flux transfer events (FTEs) [Russell and Elphic, 1978; Haerendel et al., 1978] are a common feature of the dayside magnetospheric boundary when the interplanetary magnetic field (IMF) has a southward orientation [Cowley, 1982; Russell, 1984; Rijnbeek et al., 1984; Berchem and Russell, 1984]. It is likely that FTEs may contribute significantly to the solar wind-magnetosphere coupling [Russell, 1984].

Russell and Elphic [1979] first suggested that FTEs result from patchy reconnection near the equator and are convected to higher latitudes. Lee and Fu [1985] considered an alternate model in which flux tubes giving rise to the FTEs signature are magnetic islands created by the forced non-linear tearing mode instability. Liu and Hu, La Belle-Hamer et al. and Hu et al. presented a new model suggesting that the vortices generated by the Kelvin-Helmholtz instability can lead to reconnection in the presence of a sheared magnetic field [Liu and Hu, 1988; Hu et al., 1988; La Belle-Hamer et al., 1988]. Recently, Pu et al. [1989a] have pointed out that the formation of islands in this case is the consequence of the nonlinear vortex induced tearing mode instability (VITM). The VITM occurs as M_A, the Alfven Mach number, exceeds a threshold determined by both the Reynolds and magnetic Reynolds number, where $M_A = V_0/B_0$, V_0 and B_0 representing the characteristic value of the sheared velocity field and the 'reconnected component' of the sheared magnetic field. This critical condition may be satisfied at the dayside solar wind-magnetosphere boundary away from the stagnation point. The patterns of vortices and magnetic islands of the non-linear VITM and their generation mechanism are completely different from those of conventional tearing mode instability.

In order to make it possible to compare the structure of the magnetic island formed through the VITM with observations, we have performed several 2-dimensional MHD simulations to investigate the evolution and the final state of the VITM. It is found that after being saturated, the system will arrive at an asymptotic quasi-static state (AQSS) composed of a large scale magnetic island and a concurrent vortex tube. The detailed structures of the island and vortex tube have been studied. A part of our results are presented in this paper. It will be seen that the VITM model may also provide an interesting picture to explain the formation of the flux tubes of FTEs.

II. Simulation Model

Assumptions

To simplify the problem, we make the following assumptions:

(a) In the initial state, a sheared magnetic field and a sheared velocity field exist at the magnetospheric boundary:

$$\vec{B}(z) = -B_0 \tanh (z/l_0) \vec{e}_x \qquad (1)$$

and

$$\vec{V}(z) = V_0 \tanh (z/l_0) \vec{e}_x, \qquad (2)$$

together with a number density profile described as

$$N = (N_1 + N_2)/2 + (N_1 + N_2) \tanh (z/l_0)/2, \qquad (3)$$

Geophysical Monograph 58

where B_0, V_0 and $N_{1,2}$ are all dimensionless and expressed in units of B_{00}, $v_A = B_{00}/(\mu_0\rho_{00})^{1/2}$ and N_{00}, respectively, subscript 1 (2) refers to the magnetosphere (magnetosheath), and l_0 represents the scale length of the shear region, $\rho_{00} = N_{00}m_p$.

(b) All quantities are invariant in the y direction, i.e., $\partial/\partial y = 0$.

(c) The plasma can be considered incompressible, i.e., $\nabla\cdot\vec{v} = 0$.

Dimensionless equations

The 2-D dimensionless MHD equations can be rewritten as:

$$\partial\Omega/\partial t = -\vec{V}\cdot\nabla\Omega + \vec{B}\cdot\nabla J/N - J(\vec{B}\cdot\nabla N)/N^2 + R^{-1}\nabla^2\Omega, \quad (4)$$

$$\partial\vec{B}/\partial t = -(\vec{V}\cdot\nabla)\vec{B} + (\vec{B}\cdot\nabla)\vec{V} + R_m^{-1}\nabla^2\vec{B}, \quad (5)$$

$$\partial N/\partial t = -\vec{V}\cdot\nabla N, \quad (6)$$

$$\nabla^2\phi = \Omega, \quad (7)$$

$$\vec{V} = e_y \times \nabla\phi \quad (8)$$

and

$$J = (\nabla \times \vec{B})_y, \quad (9)$$

where R and R_m denote the Reynolds and magnetic Reynolds numbers, $\vec{V} = V_x\vec{e}_x + V_z\vec{e}_z$, $\vec{B} = B_x\vec{e}_x + B_z\vec{e}_z$, J and Ω represent the y component of the current density and vorticity, respectively.

Simulation model

Simulations are performed in a square region of $-L/2 \leq x \leq L/2$, $-L/2 \leq z \leq L/2$ in the (x,z) plane. The boundary conditions imposed at $x = \pm L/2$ are periodic in the x direction, while at upper and lower boundaries, $\vec{B}\ (x, \pm L/2) = \mp B_0\vec{e}_x$, $\vec{V}\ (x, \pm L/2) = \pm V_0\vec{e}_x$, $N\ (x, L/2) = N_1$ and $N\ (x, -L/2) = N_2$, representing the steady state of the magnetosphere and magnetosheath, respectively. A mesh system of (32x32) grid points is used. The differential derivatives in space are replaced by the central difference formulas, while for the time evolution we use the 4th-order Runge-Kutta scheme. Following parameters are chosen in the simulations: $V_0 = 2$, $B_0 = 1$, $l_0 = 5$, $R^{-1} = 0.08$, $R_m^{-1} = 0.19$, $N_1 = 9$ and $N_2 = 1$.

III. Simulation Results

Figure 1 plots how the normalized perturbed magnetic energy density, averaged over (32x32) grid points, develops as a function of time. It is clear from this figure that after saturation the system gradually approaches an asymptotic quasi-static state (AQSS). Figures 2(a-d) display the AQSS evolved from an initial disturbance containing, in the x-direction, a sinusoid perturbation of $\lambda \sim L$. Several features are found immediately:

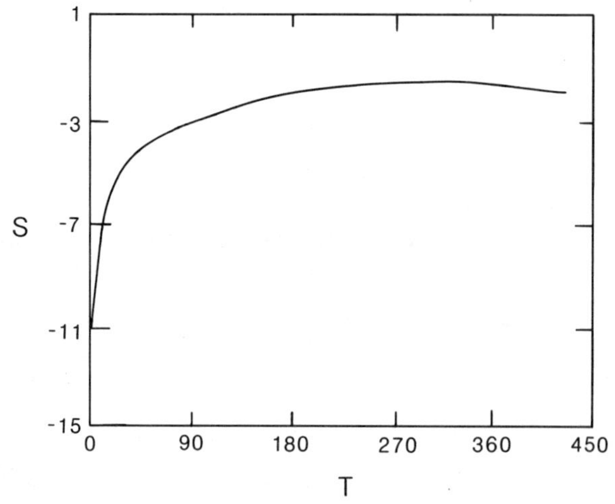

Fig. 1. The time development of the averaged, normalized, perturbed magnetic energy density.

(a) The vortex, magnetic island and contour lines of Ω are concentric. The vortex pattern is completely different from that of conventional tearing mode instability. In the latter case, four smaller vortices are formed around the O-point of the magnetic island, while for the VITM, only one vortex appears in the whole simulation region, which drives the lines of force with opposite directions to meet each other, leading to the

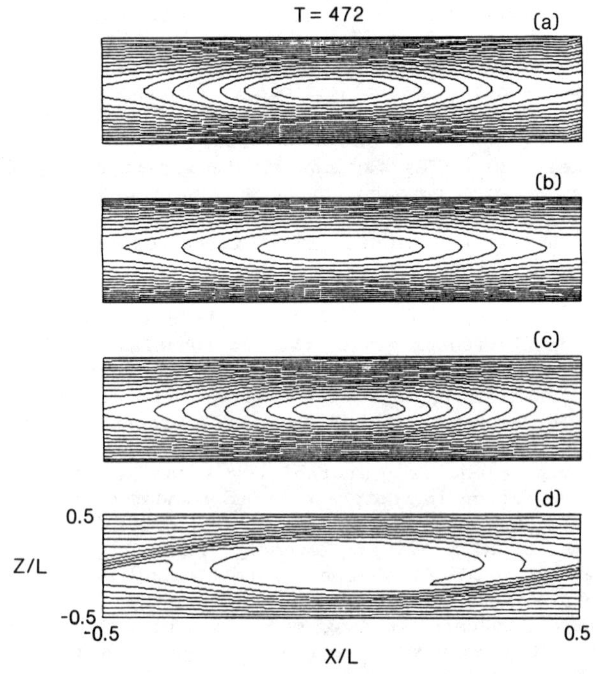

Fig. 2. The asymptotic quasi-static state of the non-linear vortex induced tearing mode instability. (a) The stream lines. (b) The magnetic field lines. (c) The contour lines of the vorticity. (d) The contour lines of the number density.

formation of the co-located magnetic island [Pu et al., 1989a].

(b) Since a magnetic island corresponds to a magnetic vortex tube (or, a magnetic flux tube if $B_y \neq 0$) when viewed 3-dimensionally, and a region in which $|\Omega|$ peaks is referred to as a vortex tube, we can see in Figures 2(b,c) that the magnetic vortex tube coincides, almost exactly, with the fluid vortex tube.

(c) Inside the magnetic island, the mass density tends to be uniform everywhere. This asymptotic uniform state is obtained through strong mixing of the fluids with different densities.

In the case that N is constant, the shape of the magnetic island, the pattern of the fluid vortex and the form of the contour lines of Ω are all almost the same as those in Figures 2(a-c).

To reveal the spatial structure of the magnetic island and vortex tube, we have plotted the variations of various quantities along the x and z axes. Figure 3 shows how B_z (x, z = 0) varies with x. Obviously, B_z component is positive as x > 0 and negative as x < 0. The same property holds for B_z (x, z = ±L/4). Figure 4 draws the curve of V_z (x, z = 0). The bipolar feature appears as well. The variation of Ω along the central abscissa is drawn in Figure 5. Ω increases towards the center of the tube. Figures 6 and 7 represent the diagrams for J (x, z = 0) and N (x = 0, z). Evidently, inside the tube, $N \simeq (N_1 + N_2)/2$, and J also tends to be approximately constant. It is worth noting here that for the case of constant N, the variations of B_z, V_z, Ω and J inside the tube are almost the same as those illustrated in Figures 3-6.

We have also investigated the behavior of the magnetic field and the velocity. Figure 8 shows how the convection electric field, $\vec{E}'$, varies along the central abscissa. Clearly it vanishes inside the tube. A similar behavior is observed along the x = 0 axis. Since $\vec{E}' = \vec{V} \times \vec{B}$, it is expected

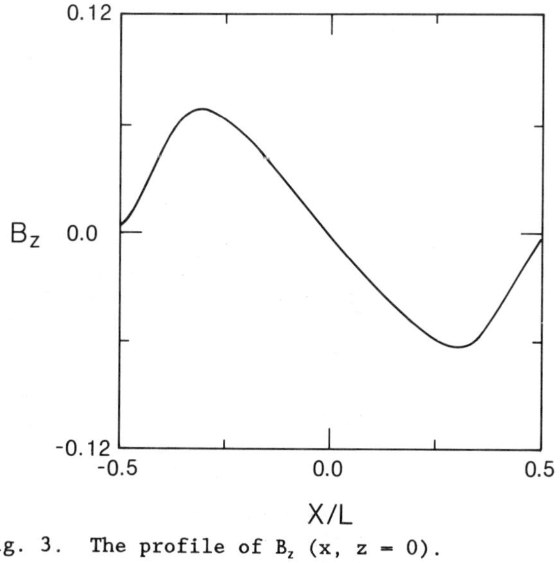

Fig. 3. The profile of B_z (x, z = 0).

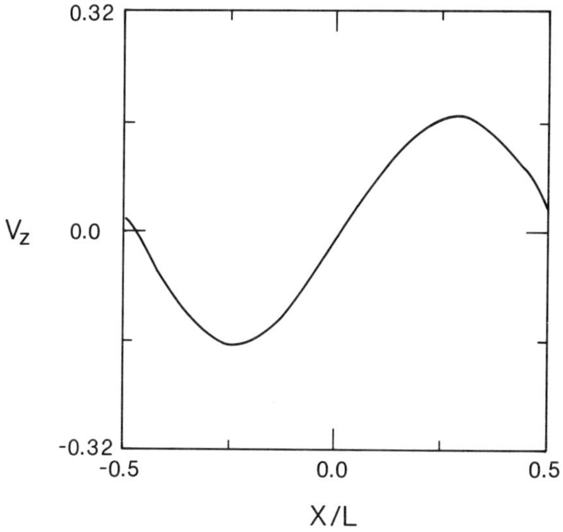

Fig. 4. The profile of V_z (x, z = 0).

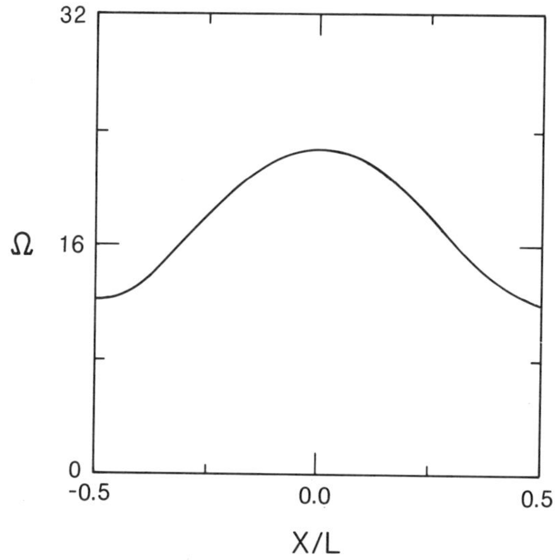

Fig. 5. The variation of Ω along the central abscissa.

that the flow is approximately field-aligned inside the magnetic vortex. Figures 9 and 10 plot $\vec{V}\cdot\nabla\Omega$ and $\vec{B}\cdot\nabla J$ along the z = 0 axis. The former vanishes everywhere and the latter is zero within the tube. Similar behavior appears along the x = 0 axis. Therefore, $\vec{V}\perp\nabla\Omega$ in the entire simulation region, while $\vec{B}\perp\nabla J$ for the internal fields. Calculations also show that similar behavior holds for $\vec{B}$, $\vec{V}$, Ω and J in the case of N = constant.

IV. Discussion

The spatial properties of both magnetic island and fluid vortex tubes in the AQSS can be understood easily. Rewriting Eq. (6) as

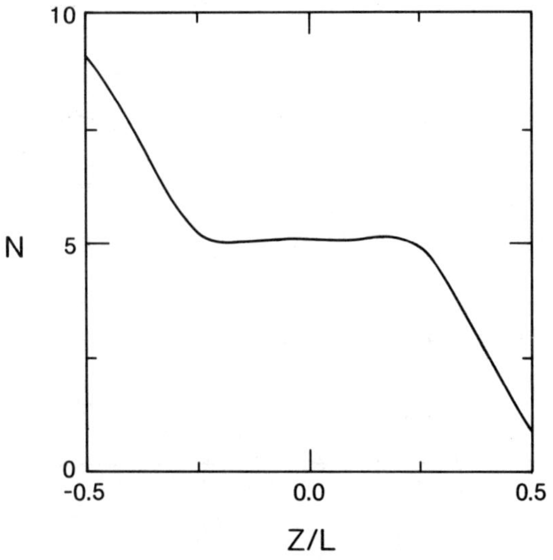

Fig. 6. The profile of N (x = 0, z).

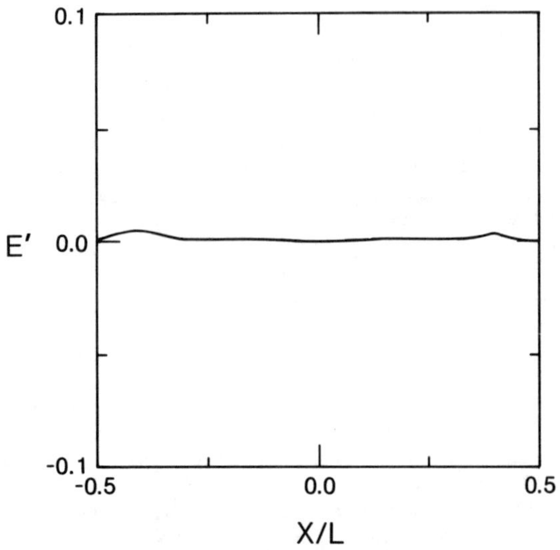

Fig. 8. The variation of $\vec{E}' = \vec{V} \times \vec{B}$ along the central abscissa.

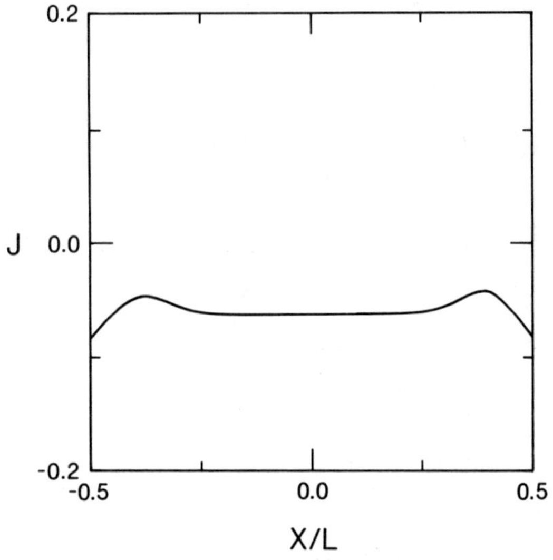

Fig. 7. The profile of J (x, z = 0).

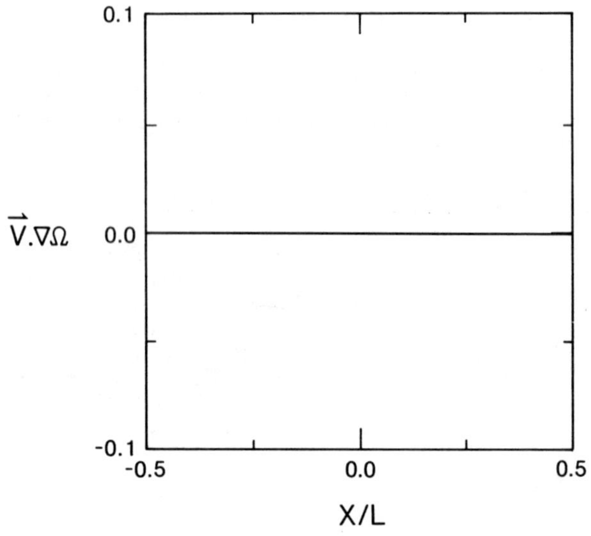

Fig. 9. The variation of $\vec{V}\cdot\nabla\Omega$ along the central abscissa.

$$\partial A/\partial t = -\vec{V}\cdot\nabla A + R_m^{-1}\nabla^2 A \tag{10}$$

and

$$\vec{B} = \nabla \times (A\vec{e}_y), \tag{11}$$

we find that, since R_m^{-1} is a small quantity, $\vec{V}\|\vec{B}$ gives $\partial A/\partial t \to 0$. Furthermore, as $R^{-1} \ll 1$, the fact that $\vec{V}\perp\nabla\Omega$ and $\vec{B}\perp\nabla J$ together with N are approximately constant then leads to $\partial\vec{V}/\partial t \to 0$ and $\partial\Omega/\partial t \to 0$. Finally, $\nabla N \to 0$ yields $\partial N/\partial t \to 0$.

It can be demonstrated that the asymptotic quasi-static states (AQSSs) achieved from a series of initial perturbations containing sinusoidal perturbations of $\lambda \sim L$ in the x direction are almost the same as that illustrated in Figures 2(a-d). Futhermore, so long as the VITM is generated, the AQSSs and the spatial structures of the magnetic islands and vortex tubes obtained for various M_A and for $B_y \neq 0$, are also similar to those plotted in section III. Therefore, the results presented in this paper seem to reveal some common features of VITM relevant to the simulation model used.

As we pointed out in the previous paper that B_0 in Equation 1 may be regarded as the 'reconnected component' of the sheared magnetic field, and the critical condition for driving the VITM ($M_A > 1$)

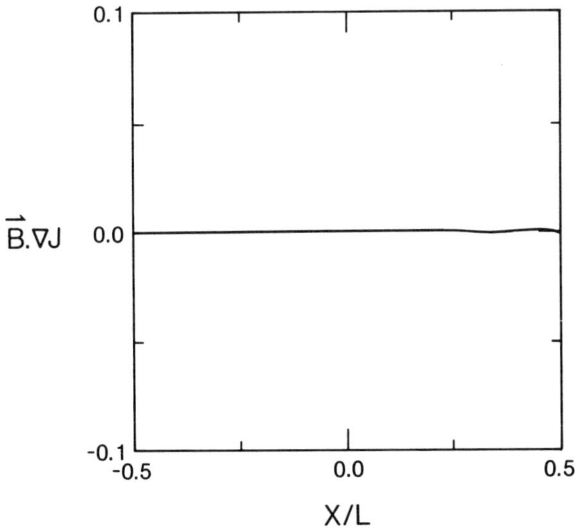

Fig. 10. The variation of $\vec{B}\cdot\nabla J$ along the central abscissa.

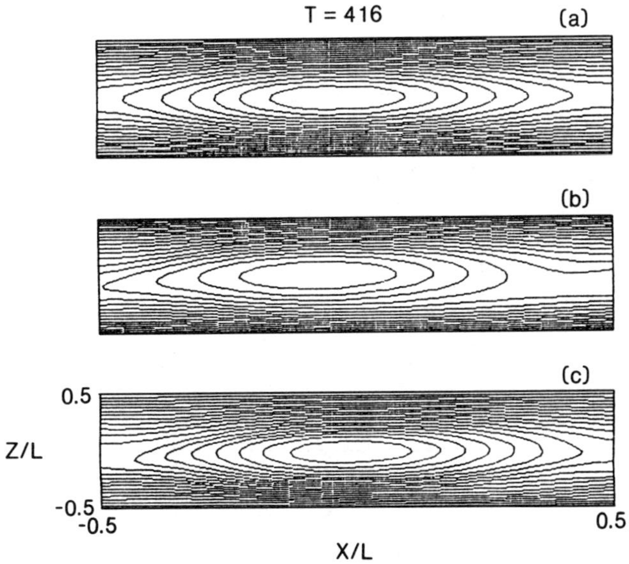

Fig. 11. The AQSS in the two step simulation study for VITM. (a) Streamlines. (b) Field lines. (c) Vorticity contours.

can be satisfied at the magnetopause away from the stagnation point [Pu et al., 1989a]. In addition, if we choose l_0 = 500 km, B_{00} = 20 nT and N_{00} = 4, the time for approaching AQSS, T, is found to be ~60s. Therefore, one might conjecture that the magnetic flux tubes associated with FTEs away from the stagnation point may probably be formed through, or strongly influenced by, the VITM, although at low latitudes, the conventional tearing mode and time-dependent reconnection are believed to play a major role [Lee and Fu, 1985; Scholer, 1986; Southwood et al., 1988]. We have done a two-step calculation: first to run a code of spontaneous tearing mode by setting $N_1 = N_2 = 1$ and $V_0 = 0$, and then taking the obtained island and vortex as initial perturbations to run the VITM code. Figure 11 shows what AQSS thus found. Apparently it is quite similar to that illustrated in Figures 2a-c. Besides, the time needed for the system to change from the AQSS of spontaneous tearing mode to the AQSS of the VITM is ~T also. Thus, the flux tubes at middle and high latitudes may indeed possibly be formed through, or strongly influenced by, the VITM.

It is then very interesting to compare the structures of the magnetic island obtained in our simulation with the observed spatial properties of FTEs. FTEs are identified from the bipolar behavior of B_N, the magnetic component normal to the magnetopause, which corresponds to B_z in our notation. The variation of B_z (x, z = 0) plotted in Figure 3 is qualitatively similar to the characteristic signature of standard FTEs. This behavior is also observed for $z \neq 0$. If we set $B_z(x, \pm L/2) = \pm B_0\vec{e}_x$ to model the southern hemisphere, we then get a negative-positive pattern of B_z (x, z = 0), similar to the signature corresponding to reverse FTEs [Rijnbeek et al., 1984]. Interest also lies in the fact that, on the magnetosphere side, the total plasma density is found to be higher inside the tube than outside, while on the magnetosheath side, the density varies oppositively [Baumjohann and Paschmann, 1987; Saunders et al., 1984]. The variations of N (x, z = ±L/4) of the AQSS plotted in Figure 12a,b are qualitatively in agreement with the ISEE and AMPTE/IRM observations. Plasma mixing along open field lines may, of course, lead to the increase of the density in the magnetosphere. However, mass mixing does not necessarily lead to a reduction of the number density in the magnetosheath part of the

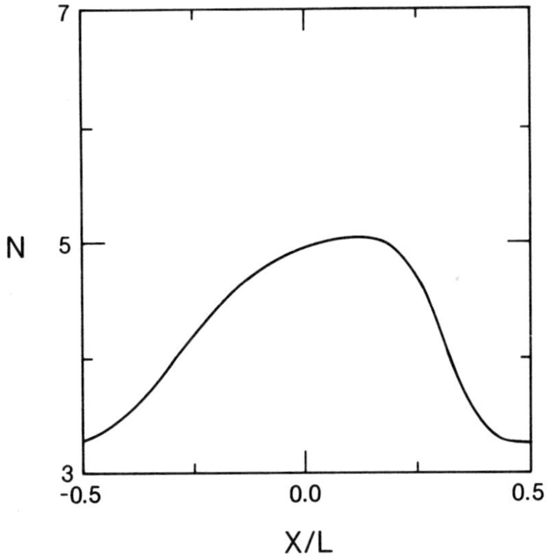

Fig. 12(a). The number density profile N (x, z = L/4).

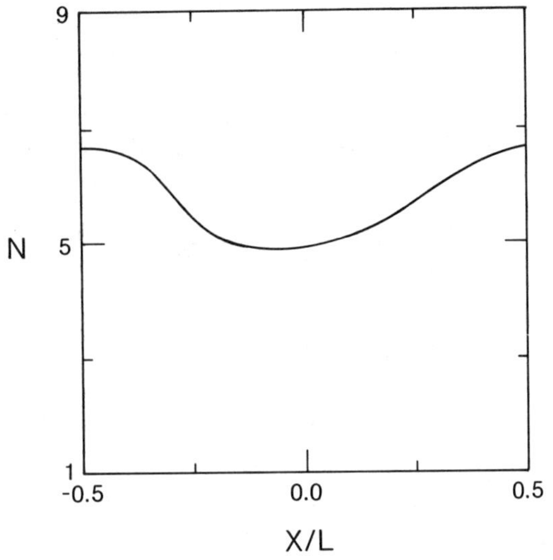

Fig. 12(b). The number density profile N (x, z = -L/4).

open flux tube, since the abundance of ions and electrons in the magnetosheath can quickly compensate for the loss of plasma. In our opinion, the mixing of plasmas with different density by the strong vortex motion provides an alternate possibility.

It has also been shown that FTEs typically have a scale size of d ~ 1-2 R_E [Saunders et al., 1984]. Figure 13 presents the time evolution of $\ln(|B_x|_{max})$ obtained in our simulations for different ratios of L/R_E for the case of l_0 = 400 km. Apparently, the magnetic islands with the diameter d < 0.5 R_E do not reach quasi-steady states because of strong dissipation, while those with d > 2.5 R_E grow too slowly to reach their shape in the time interval of interest. Only the magnetic island-flux tube with

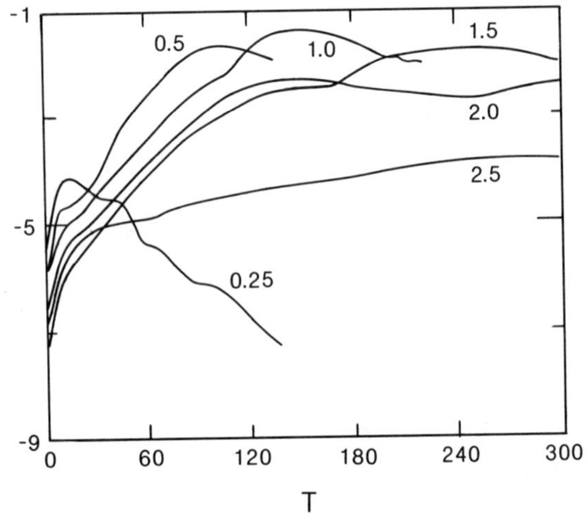

Fig. 13. The time development of $\ln(|B_x|_{max})$ for different ratio of L/R_E in the case of l_0 = 400 km.

scale size of (0.5-2) R_E grows at the appropriate rate to be observed. In a separate paper we have shown that magnetic islands of small scales may coalesce to form a large scale one [Pu et al., 1989b].

Acknowledgments. We are grateful to Dr. Z. F. Fu for his helpful assistance. This work is supported by Chinese Science Foundation.

References

Baumjohann, W. and G. Paschmann, Solar wind-magnetosphere coupling: processes and observations, Physica Scripta, T18, 61, 1987.

Berchem, J. and C. T. Russell, Flux transfer events on the magnetopause: Spatial distribution and controlling factors, J. Geophys. Res., 89, 6689, 1984.

Cowley, S. W. H., The causes of convection in the Earth's magnetosphere - A review of developments during the IMS, Revs. Geophys. Space Physics, 20, 531, 1982.

Haerendel, G., G. Paschmann, N. Sckopke, H. Rosenbauer and P. C. Hedgecock, The frontside boundary layer of the magnetosphere and the problem of reconnection, J. Geophys. Res., 83, 3195, 1978.

Hu, Y. D., Z. X. Liu and Z. Y. Pu, Response of the magnetic field to the flow vortex field and reconnection in the magnetopause boundary region, Scientia Sinica (A), 10, 1100, 1988.

La Belle-Hamer, A. L., Z. F. Fu, and L. C. Lee, A mechanism for patchy reconnection at the dayside magnetopause, Geophys. Res. Lett., 15, 152, 1988.

Lee, L. C. and Z. F. Fu, A theory of magnetic flux transfer at the Earth's dayside magnetopause, Geophys. Res. Lett., 12, 105, 1985.

Liu, Z. X. and Y. D. Hu, Local magnetic reconnection caused by vortices in the flow field, Geophys. Res. Lett., 15, 752, 1988.

Pu, Z. Y., M. Yein and Z. F. Fu, Coupling of the tearing mode with the Kelvin-Helmholtz instability and the vortex induced tearing mode instability, to be published in Acta Geophysica Sinica, 1989a.

Pu, Z. Y., P. T. Hou, Y. Ming and Z. X. Liu, A 2-D MHD simulation study on vortex induced tearing mode instability at the magnetopause: The coalescence of vortex tubes and magnetic islands, submitted to A monthly Journal of Science, Academia Sinica, 1989b.

Rijnbeek, R. P., S. W. H. Cowley, D. J. Southwood and C. T. Russell, A survey of dayside flux transfer events observed by ISEE-1 and -2 magnetometers, J. Geophys. Res., 89, 786, 1984.

Russell, C. T. and R. C. Elphic, Initial ISEE magnetometer results: Magnetopause observations, Space Sci. Rev., 22, 681, 1978.

Russell, C. T. and R. C. Elphic, ISEE observations of flux transfer events at the dayside magnetopause, Geophys. Res. Lett., 6, 33, 1979.

Russell, C. T., Magnetic reconnection, in Magnetic Reconnection in Space and Laboratory Plasmas, ed. by E. W. Hones, Jr., American Geophysical Union, 124, 1984.

Saunders, M. A., C. T. Russell and N. Sckopke, Flux transfer events: Scale size and interior structure, Geophys. Res. Lett., 11, 131, 1984.

Scholer, M., Magnetic flux transfer at the magnetopause based on single x line bursty reconnection, Geophys. Res. Lett., 15, 291, 1988.

Southwood, D. J., C. J. Farrugia, and M. A. Saunders, What are flux transfer events?, Planet Space Sci., 36, 503, 1988.

A SIMULATION STUDY OF PARTICLE HEAT FLUX AND PLASMA WAVES ASSOCIATED WITH MAGNETIC RECONNECTIONS AT THE DAYSIDE MAGNETOPAUSE

D. Q. Ding and L. C. Lee

Geophysical Institute and Department of Physics, University of Alaska, Fairbanks, AK 99775–0800

Abstract. Flux transfer events (FTEs) observed at the dayside magnetopause are believed to be caused by magnetic reconnection. Satellite observations of FTEs have shown the presence of particle heat flux and intense plasma waves. The particle heat flux and plasma waves associated with the magnetic reconnection processes at the dayside magnetopause are studied by using $2\frac{1}{2}$-dimensional electromagnetic particle simulations. It is observed in the simulations that superthermal particles are generated, leading to the presence of particle heat flux. Intense plasma waves are also observed. Typical power spectra of fluctuating magnetic and electric fields are found to be $P_{B_x} \sim f^{-3.8}$ and $P_E \sim f^{-1.8}$, respectively, where f is the frequency. For a symmetric magnetic field configuration, both multiple X line reconnection (MXR) and single X line reconnection (SXR) processes lead to a simultaneous bipolar FTE signature on both sides of the current sheet; whereas for an asymmetric magnetic field configuration, MXR again generates the simultaneous FTE signature on both sides of the current sheet, however, SXR can produce the FTE signature only on the weak magnetic field side.

1. Introduction

The concept of magnetic field line reconnection was first proposed in 1940s by solar physicists to explain solar flares observed in the solar atmosphere [e.g., Giovanelli, 1947]. In the 1960s, Dungey [1961] proposed that magnetic reconnections may also occur in the Earth's magnetosphere, both at the dayside magnetopause and in the nightside magnetotail. The ISEE data provided some observational evidence of magnetic reconnection at the dayside magnetopause in late 1970s [e.g., Paschmann et al., 1979]. However, the detailed picture of the dayside magnetic reconnection process remains a debatable issue, and hence an active research topic [see the reviews in this volume by Russell, 1989, and Lee and Ding, 1989].

Geophysical Monograph 58

Flux transfer events (FTEs) were first reported by Russell and Elphic [1979] based on ISEE satellite data. Observations of FTEs indicate that the dayside magnetic reconnection is an intermittent and sporadic process rather than a steady-state one. Various observational features of FTEs have been reported by many authors. Particle heat flux, which is formed by the energetic particles streaming along magnetic field lines, has been observed to be associated with FTEs at the dayside magnetopause [e.g., Scudder et al., 1984; Thomsen et al., 1987; Klumpar et al., 1989]. The heat flux associated with FTEs may be caused by the particle acceleration during magnetic reconnection process and/or by the leakage of hot magnetospheric plasma along the reconnected open magnetic field lines [e.g., Daly et al., 1981; Scholer et al., 1981; Sonnerup et al., 1981; Sibeck et al., 1987]. Various plasma waves have been observed at the dayside magnetopause [e.g., Gurnett et al., 1979; Anderson et al., 1982; LaBelle and Treumann 1988]. The observed feature that wave intensities increase with the increase of the negative IMF B_z indicates a possible correlation between these waves and the magnetic reconnection process [Tsurutani et al., 1989]. Intense plasma waves associated with FTEs have also been reported by LaBelle et al. [1987].

To explain the occurrence and other observed features of FTEs, a multiple X line reconnection (MXR) model was proposed by Lee and Fu [1985] and simulated by Fu and Lee [1985], Ding et al. [1986], and Shi et al. [1988]. In the particle simulations of MXR, energetic particles are found to be generated during the reconnection process [Ding et al., 1986]. However, Ding et al. [1986] used a magnetoinductive code, in which the electron dynamics and electrostatic interaction were not included. Particle simulations of dayside magnetic reconnection have also been reported by Swift [1986], Hoshino [1987], Francis et al. [1988], and Allen and Swift [1989].

In the present paper, the particle heat flux and plasma waves generated during the magnetic reconnection process is studied based on $2\frac{1}{2}$-dimensional particle simulations. The results are compared with satellite observations of FTEs.

2. Simulation Model

The $2\frac{1}{2}$-dimensional electromagnetic particle code used for the present study includes ion dynamics, electron dynamics, electrostatic interaction, and the self-consistently generated magnetic field component perpendicular to the simulation plane. Darwin approximation is used in the present code, in which the transverse part of the displacement current is neglected. Basic equations used in the simulations are

$$\nabla^2 \mathbf{A} = -(\frac{4\pi}{c})\mathbf{J} + \frac{1}{c}\nabla\chi \tag{1}$$

$$\nabla^2 \chi = 4\pi\nabla\cdot\mathbf{J} \tag{2}$$

$$\nabla^2 \phi = -4\pi\rho \tag{3}$$

$$\frac{d\mathbf{p}_j}{dt} = \pm\frac{e}{m_j c}(\nabla\mathbf{A})\cdot(\mathbf{p}_j \mp \frac{e}{c}\mathbf{A}) \mp e\nabla\phi \tag{4}$$

$$\frac{d\mathbf{x}_j}{dt} = \frac{1}{m_j}(\mathbf{p}_j \mp \frac{e}{c}\mathbf{A}) = \mathbf{v}_j \tag{5}$$

where $\mathbf{A}$ is the vector potential, ϕ is the electrostatic potential, $\chi = \partial\phi/\partial t$ is the time derivative of ϕ, $\mathbf{J}$ is the current density, ρ is the charge density, and c is the speed of light, e is the magnitude of an electron charge, m_j, $\mathbf{x}_j$, $\mathbf{p}_j$, and $\mathbf{v}_j$ are, respectively, the mass, position, canonical momentum and velocity of a particle. The upper parts of $\pm$ and $\mp$ signs are for ions ($j = i$) while the lower parts are for electrons ($j = e$). Our formulation uses the Coulomb gauge, $\nabla\cdot\mathbf{A} = 0$. The charge density and current density are calculated from

$$\rho = (n_i - n_e)e \tag{6}$$

$$\mathbf{J} = \left[(n\mathbf{u})_i - (n\mathbf{u})_e\right]e \tag{7}$$

where

$$n = \sum_k S\left[\mathbf{x} - \mathbf{x}_k(t)\right] \tag{8}$$

$$(n\mathbf{u}) = \sum_k [\mathbf{p}_k \mp \frac{e}{c}\mathbf{A}(\mathbf{x}_k)]S\left[\mathbf{x} - \mathbf{x}_k(t)\right] \tag{9}$$

$$\mathbf{A}(\mathbf{x}_k) = \int S\left[\mathbf{x} - \mathbf{x}_k(t)\right]\mathbf{A}(\mathbf{x})d^2x \tag{10}$$

where $S\left[\mathbf{x} - \mathbf{x}_k(t)\right]$, the weighting factor, is determined by the first-order PIC bilinear interpolation method.

The simulation domain can be considered as a rectangular box near the subsolar region of the dayside magnetopause. The origin of the coordinate system is located at the center of the magnetopause current sheet, with the x-axis pointing to the Sun, the z-axis to the north, and the y-axis to the dusk side. The simulation is performed in the $x - z$ plane with $-L_x \le x \le L_x$ and $-L_z \le z \le L_z$. The variations of all physical quantities in y-direction are neglected, i.e. $\partial/\partial y = 0$. The simulation is initialized with a current sheet, which is located in the $x = 0$ plane with antiparallel magnetic field on the two sides. Both symmetric and asymmetric magnetic field profiles are used. For the symmetric initial configuration, a symmetric boundary condition is imposed in the $x = 0$ plane and the simulation is performed only in the region with $x \ge 0$. The initial plasma distribution is a drift-Maxwellian distribution, in which the local drift speed is determined by dividing the local current density with the local plasma density. Since ions are loaded on the top of electrons, the initial electric field is set to zero. A driven boundary condition with a constant particle influx is imposed at $x = \pm L_x$, while an open boundary condition applied at $z = \pm L_z$ allows the plasma to flow out of the simulation domain. A detailed discussion of the boundary conditions used in the present simulation can be found in Ding et al. [1989]

3. Simulation Results

A series of simulations have been carried out to study the driven magnetic reconnection process. Various length ratios (L_z/L_x) and plasma parameters have been used in the simulations. In this section, we present examples of our simulation results with emphasis on the particle heat flux and power spectrum of the fluctuating magnetic and electric fields associated with the driven magnetic reconnection.

First, we present results obtained from a symmetric magnetic field configuration. The initial magnetic field is given by

$$\mathbf{B}(x) = -B_0\tanh(\frac{x}{a})\mathbf{e}_z. \tag{11}$$

The initial plasma distribution is decided by balancing the magnetic pressure with the plasma pressure. The simulation is run from $t = 0$ to $t = 1500\Omega_e^{-1}$ with a time step of $0.2\Omega_e^{-1}$, where $\Omega_e = eB_0/m_e c$ is the electron gyrofrequency based on the asymptotic magnetic field outside the current sheet. The grid size is chosen to be $\Delta/\rho_e = 1.5$, where ρ_e is electron gyroradius based on the electron thermal speed v_{the} and Ω_e. The lengths of simulation domain are $L_z = 96\rho_e$ and $L_x = 48\rho_e$, respectively. The current sheet thickness is $a/\rho_e = 5$. The temperature ratio of ions to electrons is $T_i/T_e = 6.25$. The mass ratio of ions to electrons is assumed to be $m_i/m_e = 25$. Therefore, we have $\rho_i/\rho_e = 12.5$, $\Omega_i/\Omega_e = 0.04$, $v_{thi}/v_{the} = 0.5$, where ρ_i, Ω_i, and v_{thi} are the ion gyroradius, gyrofrequency, and thermal speed, respectively. At the inflow boundary, the Alfvén speed is $v_A = 1.52 v_{thi}$, the plasma $\beta = 1.0$, the imposed electric field $E_1 = 0.3 v_A B_0$. Initially, the simulation is started with 40,000 ions and electrons.

The intermittent and sporadic multiple X line reconnection (MXR) is observed during the simulation. The driving electric field applied at the boundary causes a pile-up of the magnetic field and an enhancement of the current density, and thereby enhances the growth rate of the tearing mode. The formation and convection of magnetic islands are observed intermittently. However, during the course of the simulation, the quasi-steady single X line reconnections (SXR) are also observed. The generation of superthermal particles is found to be coincident with the period of enhanced magnetic reconnections, which are indicated by the peaks in the time profile of inductive electric field. The energetic electrons and ions are found to have an energy spectrum of $f_e(E) \sim E^{-3.8}$ and $f_i(E) \sim E^{-4.4}$, respectively, where E is the particle energy. Bursty high speed plasma flows and particle heat fluxes along the magnetic field lines are also observed during reconnections. As an example, magnetic field lines, contours of electrostatic potential, ion flow pattern, and ion heat flux pattern at simulation time $t = 1020\Omega_e^{-1}$ are plotted in Figure 1.

Figure 1a shows the magnetic field pattern, in which two reconnection points (X points) are present. The center of a large magnetic island has moved out of the simulation domain from the left boundary, while a smaller magnetic islands is being formed and convected towards the right boundary. The sizes of both islands grow because the magnetic reconnections continue to take place at both X points. The magnetic tension force associated with the bending reconnected field lines pulls the smaller magnetic island towards the right.

Figure 1b shows contours of the electrostatic potential. Outside the current sheet region, the contours are more or less field-aligned and the electric field is basically perpendicular to the magnetic field lines. Inside the current sheet, contours of electrostatic potential are not aligned with the magnetic field lines, leading to the presence of parallel electric fields. The presence of parallel electric fields can be explained by the different behaviors of the ion and electron motion near the current sheet. Near the current sheet, ions are unmagnetized while electrons are still magnetized by the week magnetic field. These magnetized electrons tend to be trapped in the closed field line region (magnetic islands), leading to an excess negative charge accumulation in that region. On the other hand, in the open field line region, an excess positive charge accumulation is observed due to the escape of electrons along the open field lines. Therefore, complicated electric contour pattern can exist near the current sheet region, which leads to the presence of parallel electric field.

Figure 1c, the ion flow pattern is plotted. The plasma flows into the simulation domain from the top and bottom boundaries under the influence of the driving force

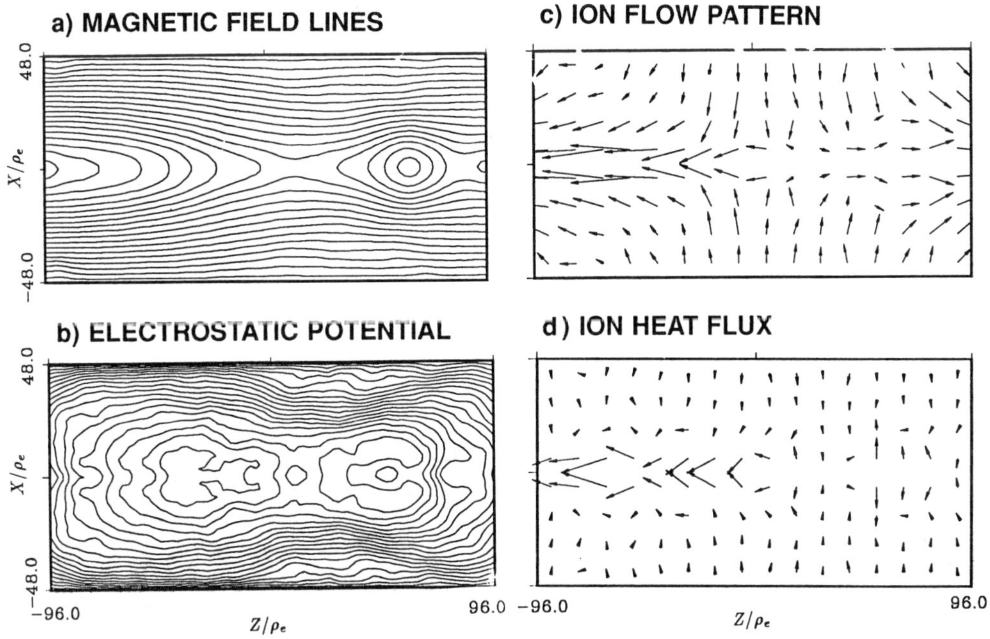

Fig. 1. (a) Magnetic field lines, (b) contours of electrostatic potential ϕ, (c) ion flow pattern, and (d) ion heat flux pattern at simulation time $t = 1020\Omega_e^{-1}$.

applied at the boundaries. The plasma is then accelerated by the reconnection process and flows outwards towards the left and right boundaries. The maximum speed in Figure 1c is estimated to be $1.3v_A$. Apart from the high speed plasma flow, particle heat flux is also generated during the simulation. Figure 1d shows the ion heat flux pattern. It is found that the heat flux generated during the reconnection is mainly confined to the current sheet and its direction is mainly along the magnetic field lines.

To further demonstrate the presence of particle heat flux during the reconnection process, the ion distribution as a function of $V_{\|}$, the velocity component along magnetic field lines, at simulation time $t = 1020\Omega_e^{-1}$ is plotted in Figure 2. The distribution function has a high energy component with $-2.4 \leq V_{\|}/v_{the} \leq -1.5$. The number of energetic particles is estimated to be 3% of the number of total particles.

To examine the electric and magnetic waves associated with the driven magnetic reconnection process, the magnetic field **B** and electric field **E** are recorded at a time interval of $1\Omega_e^{-1}$ at selected points in the simulation domain. The power spectra of magnetic field **B** and electric field **E** are computed. An example of the power spectrum of the electromagnetic waves associated with driven magnetic reconnection process is plotted in Figure 3.

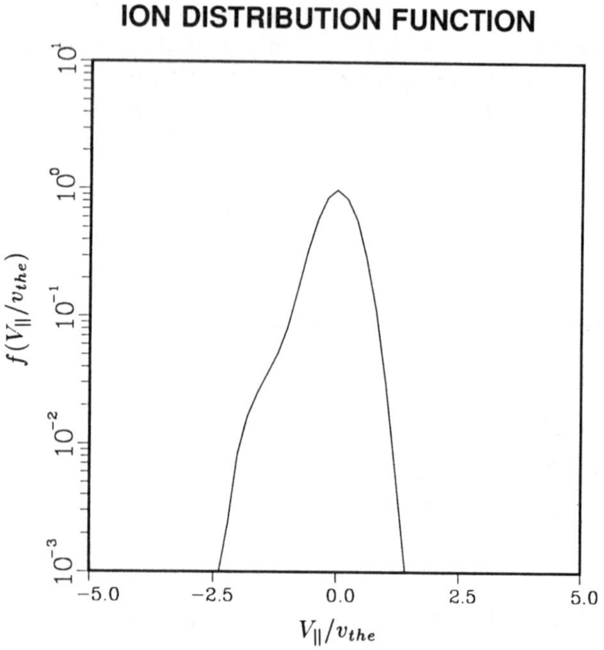

Fig. 2. Ion distribution as a function of $V_{\|}$, the component of velocity parallel to the magnetic field at simulation time $t = 1020\Omega_e^{-1}$. The distribution shows the presence of a high energy component with $-2.4 \leq V_{\|}/v_{the} \leq -1.5$, corresponding to the presence of a particle heat flux.

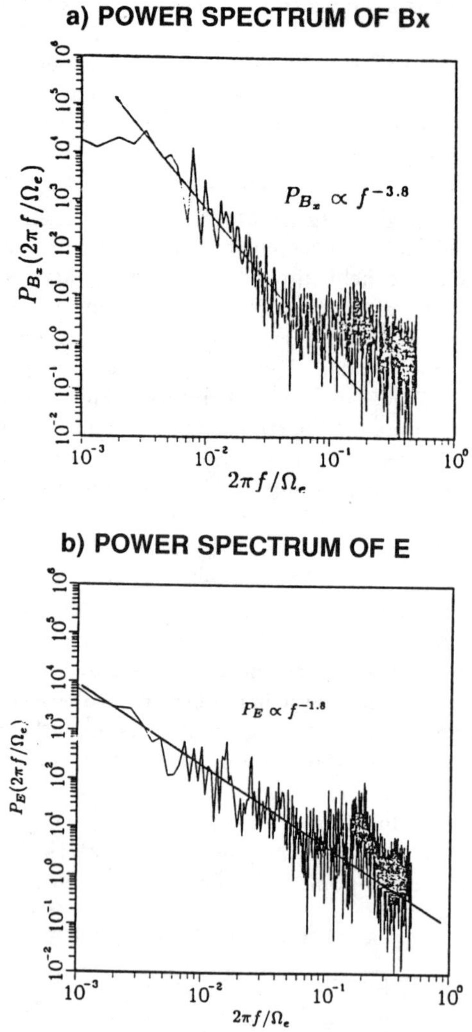

Fig. 3. (a) Power spectrum (P_{B_x}) of magnetic field component B_x and (b) power spectrum (P_E) of electrostatic field $E(=\sqrt{E_x^2+E_z^2})$.

In Figure 3a, the power spectrum of magnetic component B_x is shown. The B_x is measured at ($x/\rho_e = 12, z/\rho_e = 72$). The lower frequency part of the spectrum can be fitted by a power law, $P_B(f) \sim f^{-3.8}$. The power spectrum of the electrostatic field **E**, which is measured at ($x/\rho_e = 24, z/\rho_e = 0$), is plotted in Figure 3b. The approximate power law fit for the electrostatic field **E** is found to be $P_E(f) \sim f^{-1.8}$. Figure 3a and 3b show the typical spectra for the magnetic component B_x and electric field $E = \sqrt{E_x^2 + E_z^2}$, although the power spectra are found to vary a little at different points. Further investigation of the electromagnetic waves associated with the driven reconnection process is undertaken.

Second, we have simulated several cases with an asymmetric magnetic configuration. For the asymmetric magnetic field configuration, the initial magnetic field and temperature profiles are given by

$$\mathbf{B}(x) = -B_0[(\frac{1+R_1}{2}) + (\frac{1-R_1}{2})\tanh(\frac{x}{a})]\mathbf{e}_z \quad (12)$$

$$T(x) = T_0[(\frac{1+R_2}{2}) + (\frac{1-R_2}{2})\tanh(\frac{x}{a})] \quad (13)$$

where $B_0 = B_z(x=\infty)$, $T_0 = T(x=\infty)$, $R_1 = -B_z(x = -\infty)/B_z(x=\infty)$, and $R_2 = T(x=-\infty)/T(x=\infty)$. The plasma density profile is determined by balancing the magnetic pressure with the plasma pressure. The ratio of magnetic fields is chosen to be $R_1 = 2.0$ and the temperatures ratio is chosen to be $R_2 = 4.0$. The parameters used are the same as the symmetric case in Figure 1, except $T_i/T_e = 4.0$.

Figure 4a shows the magnetic field configuration and Figure 4b shows the positions of electrons at simulation time $t = 150\Omega_e^{-1}$ and $1080\Omega_e^{-1}$. At $t = 150\Omega_e^{-1}$, two X points are present and one large magnetic island is formed near the center of the simulation domain. The normal component ($B_n = B_x$) of the perturbed magnetic field associated with the presence of magnetic island can be observed in both sides of the current sheet. However, the disturbance in the weaker magnetic field side ($x > 0$) is larger than that in the strong magnetic field side ($x < 0$). The cold dense plasma, which is originated from the weaker magnetic field side and trapped by the magnetic island, is found to bulge into the strong magnetic field side. At $t = 1080\Omega_e^{-1}$, a single X line quasi-steady reconnection is observed in the simulation. The perturbed normal magnetic field can only be observed in the weak magnetic field side and the bulge of cold dense plasma in the magnetic field side disappears. However, for the symmetric case ($R_1 = 1$), the perturbed normal magnetic field can be observed on both sides of the current sheet, for both MXR and SXR processes. Similar result has also been obtained in the MHD simulation of bursty single X line reconnection by Scholer [1989].

The magnetic field configuration is asymmetric at the dayside magnetopause. The magnetic field strength in the magnetosheath is weaker than that in the magnetosphere. Our present result provides a potential explanation why the B_n signature of FTEs can sometimes be observed on both sides of the magnetopause and sometimes only on the magnetosheath side.

4. Discussion

Observations of FTEs have modified our understanding of the magnetic reconnection process at the dayside magnetopause. The observed ion and electron velocity distributions during FTEs have been reported by Thomsen et al., [1987] with the ISEE data and by Klumpar et al., [1989] with the AMPTE/CCE data. Both ion and electron distributions show the presence of heat flux along magnetic field lines. In the simulation shown in section

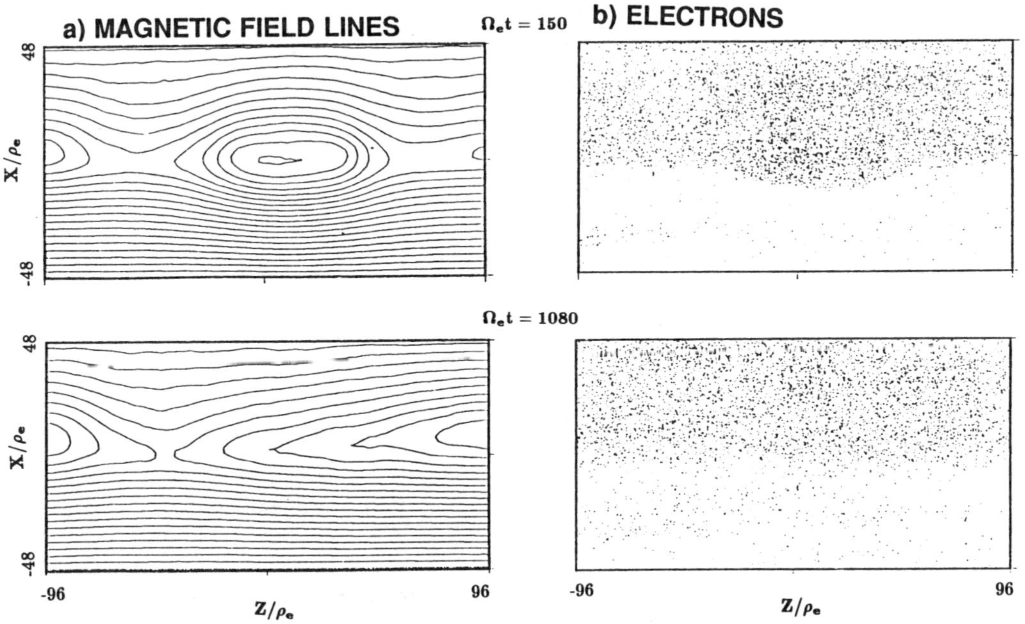

Fig. 4. An example of driven magnetic reconnection with an asymmetric initial magnetic field configuration. (a) Magnetic field pattern and (b) electron scatter plots in the simulation domain at $t = 150\Omega_e^{-1}$ and $1080\Omega_e^{-1}$.

3, the particle heat flux is mainly generated by the particle acceleration during the magnetic reconnection process. The leakage of hot magnetospheric particles along the reconnected open magnetic field lines may also contribute to the particle heat flux observed during the FTEs. To examine the leakage effect, new simulations are being undertaken and the results will be reported in the future.

Observations of plasma waves at the dayside magnetopause have been reported by Gurnett et al. [1979], Anderson et al. [1982], LaBelle et al., [1987], LaBelle and Treumann [1988], and Tsurutani et al. [1989] based on ISEE and AMPTE/IRM satellite data. Enhanced plasma waves observed during the period with a negative IMF B_z and during FTEs indicate that the magnetic reconnection may be responsible for the generation of these waves. The power spectrum for the electric field is observed to be $P_E \sim f^{-2}$ and the power spectrum for the magnetic field is $P_B \sim f^{-3.3 \text{ to } -4.7}$. The results in Figure 3 show that our simulation results of P_E and P_B are consistent with the observed spectra at the dayside magnetopause.

Observations of FTEs have shown that sometimes the bipolar B_n signature can be detected simultaneously on both sides of magnetopause and sometimes the B_n signature can be identified only on the magnetosheath side [e.g., Farrugia et al., 1987]. In the present simulations, it is shown that for the case with a symmetric magnetic field both MXR process and SXR process can produce a simultaneous B_n signature on both sides of the current sheet. On the other hand, in the asymmetric field case the simultaneous B_n signatures can only be generated by the MXR process. At the dayside magnetopause, the magnetosheath field is usually weaker than the magnetospheric field, corresponding to the asymmetric case shown in Figure 4. Therefore, we suggest tentatively that the observation of a simultaneous FTE signature on both sides of magnetopause would indicate an MXR process at the dayside magnetopause, whereas the observation of FTE signature only on the magnetosheath side would indicate an SXR process. Further statistical study of FTE signatures on both sides of the magnetopause may provide an experimental verification for the above suggestion.

Acknowledgements. This work is supported by Department of Energy grant DE-FG06-86ER 13530 and National Science Foundation grant ATM88-20992 to the University of Alaska, Fairbanks. The computation is supported by the San Diego Supercomputer Center.

References

Allen, C. W., and D. W. Swift, A particle simulation of the tearing mode instability at the dayside magnetopause, *J. Geophys. Res., 94*, 6925, 1989.

Anderson, R. R., C. C. Harvey, M. M. Hoppe, B. T. Tsurutani, T. E. Eastman, and J. Etcheto, Plasma waves near the magnetopause, *J. Geophys. Res., 87*, 2087, 1982.

Daly, P. W., D. J. Williams, C. T. Russell, and E. Keppler, Particle signatures of magnetic flux transfer events at the magnetopause, *J. Geophys. Res., 86*, 1628, 1981.

Ding, D. Q., L. C. Lee, and Z. F. Fu, Multiple X line reconnection, 3: A particle simulation of flux transfer events, *J. Geophys. Res., 91*, 13384, 1986.

Ding, D. Q., L. C. Lee, and D. W. Swift, A particle simulation of reconnection processes at the dayside magnetopause including electron dynamics, to be submitted to *J. Geophys. Res.*, 1989.

Dungey, J. W., Interplanetary field and auroral zones, *Phys., Rev., Lett., 6*, 47, 1961.

Farrugia, C. J., D. J. Southwood, S. W. H. Cowley, R. P. Rijnbeek, and P. W. Daly, Two-regime flux transfer events, *Planet. Space Sci., 35*, 737, 1987.

Francis, G. E., D. W. Hewett, and C. E. Max, Kinetic simulation of magnetic reconnection in the presence of shear, in *Proceedings of an International Workshop on Reconnection in Space Plasma, ESA SP-285*, edited by T. D. Guyenne and J. J. Hunt, p. 61, Netherland, 1988.

Fu, Z. F., and L. C. Lee, Simulation of multiple X line reconnection at the dayside magnetopause, *Geophys. Res. Lett., 12, 291*, 1985.

Giovanelli, R. G., Magnetic and electric phenomena in the Sun's atmosphere associated with sunspots, *Mon. Notices Roy. Astron. Soc., 107*, 338, 1917.

Gurnett, D. A., R. R. Anderson, B. T. Tsurutani, E. J. Smith, G. Paschmann, G. Haerendel, S. J. Bame, and C. T. Russell, Plasma wave turbulence at the magnetopause: Observations from ISEE 1 and 2, *J. Geophys. Res., 84*, 7043, 1979.

Hoshino, M., The electrostatic effect for the collisionless tearing mode, *J. Geophys. Res., 92*, 7368, 1987.

Klumpar, D. M., S. A. Fuselier, T. A. Potemra, and K. Takahashi, AMPTE/CCE observations of electron distributions in flux transfer events, *EOS Trans. AGU, 70*, 437, 1989.

LaBelle, J., and R. A. Treumann, Plasma waves at the dayside magnetopause, *Space Sci. Rev., 47*, 175, 1988.

LaBelle, J., R. A. Treumann, G. Haerendel, O. H. Bauer, G. Paschmann, W. Baumjohann, H. Lühr, R. R. Anderson, H. C. Koons, and R. H. Holzworth, AMPTE IRM observations of waves associated with flux transfer events in the magnetopause, *J. Geophys. Res., 92*, 5827, 1987.

Lee, L. C., and D. Q. Ding, Theoretical and simulation models of flux transfer events at the dayside magnetopause, *in this volume*, 1989.

Lee, L. C., and Z. F. Fu, A theory of magnetic flux transfer at the Earth's magnetopause, *Geophys. Res. Lett., 12*, 105, 1985.

Paschmann, G., B. U. Ö. Sonnerup, I. Papamastorakis, N. Sckopke, G. Haerendel, S. J. Bame, J. R. Asbridge, J. T. Gosling, C. T. Russell, and R. C. Elphic, Plasma acceleration at the earth's magnetopause: Evidence for reconnection, *Nature, 282*, 243, 1979.

Russell, C. T., and R. C. Elphic, ISEE observations of flux transfer events at the dayside magnetopause, *Geophys. Res. Lett., 6*, 33, 1979.

Russell, C. T., The magnetopause, *in this volume*, 1989.

Scholer, M., Asymmetric time-dependent and stationary magnetic reconnection at the dayside magnetopause, manuscript, 1989.

Scholer, M., F. M. Ipavich, G. Gloeckler, D. Hovestadt, and B Klecker, Leakage of magnetospheric ions into the magnetosheath along reconnected field lines at the dayside magnetopause, *J. Geophys. Res., 86*, 1299, 1981.

Scudder, J. D., K. W. Ogilvie, C. T. Russell, The relation of flux transfer events to the magnetic reconnection, in *Magnetic Reconnection in Space and Laboratory Plasmas*, edited by E. W. Hones, Jr., p153, AGU, Washington, D.C., 1984.

Shi, Y., C. C. Wu, and L. C. Lee, A study of multiple X line reconnection at the dayside magnetopause, *Geophys. Res. Lett., 15*, 295, 1988.

Sibeck, D. G., R. W. McEntire, A. T. Y. Lui, R. E. Lopez, S. M. Krimigis, R. B. Decker, L. J. Zanetti, and T. A. Potemra, Energetic magnetic ions at the dayside magnetopause: Leakage or merging, *J. Geophys. Res., 92*, 12097, 1987.

Sonnerup, B. U. Ö., G. Paschmann, I. Papamastorakis, N. Sckopke, G. Haerendel, S. J. Bame, J. R. Asbridge, J. T. Gosling, and C. T. Russell, Evidence for magnetic field reconnection at the Earth's magnetopause, *J. Geophys. Res., 86*, 10049, 1981.

Swift, D. W., Numerical simulation of tearing mode instabilities, *J. Geophys. Res., 91*, 219, 1986.

Thomsen, M. F., J. A. Stansberry, S. J. Bame, S. A. Fuselier, and J. T. Gosling, Ion and electron velocity distributions within flux transfer events, *J. Geophys. Res., 92*, 12127, 1987.

Tsurutani, B. T., A. L. Brinca, E. J. Smith, R. T. Okida, R. R. Anderson, T. E. Eastman, A statistical study of ELF-VLF plasma waves at the magnetopause, *J. Geophys. Res., 94*, 1989.

A THREE-DIMENSIONAL MHD SIMULATION OF THE MULTIPLE X LINE RECONNECTION PROCESS

Z. F. Fu[1], L. C. Lee[2] and Y. Shi[2]

[1]Center for Space Science and Applied Research, Academia Sinica, P.O. Box 8701, Beijing, China
[2]Geophysical Institute and Department of Physics, University of Alaska, Fairbanks, AK 99775

Abstract. The multiple X line reconnection (MXR) process is studied through a 3-dimensional incompressible MHD simulation. The simulation is conducted in a local rectangular box for which a constant plasma flow and magnetic inflow flux are maintained on on the two plasma incoming boundaries. Reconnection takes place under such driven boundary conditions.

Magnetic reconnections at multiple sites are observed. As the reconnections proceed, several magnetic flux tubes are formed and convected outward after growing to a large size. Tube-aligned plasma flows are observed inside the moving magnetic flux tubes. As reconnection proceeds, the active reconnection regions shift towards the ends of the reconnection lines. Of the most importance is the field line topology revealed at the two ends of the flux tube. The simulation shows that the magnetic ropes usually have "frayed" ends.

Introduction

The study of magnetic reconnection was initiated by the attempt to explain the occurrence of solar flares. The emphasis in the past has been mainly on the 2-dimensional and steady state aspects of magnetic reconnection [Petschek, 1964; Sonnerup, 1980; Vasyliunas, 1975; Priest and Forbes, 1986].

The discovery of flux transfer events (FTEs) at the earth's dayside magnetopause [Russell and Elphic, 1978] has inspired some workers to propose time-dependent reconnection models. Motivated by the observed FTEs, Lee and Fu [1985] proposed a multiple X line reconnection (MXR) model which is time-dependent and intrinsically 3-dimensional. The model predicts the formation of helical magnetic flux tubes at the dayside magnetopause and various reconnection geometries at the ends of the flux tubes. In the earth's magnetotail, the traditional 2-dimensional plasmoid picture is also being considered as insufficient to describe the reconnection process in the magnetotail [Hones et al., 1982]. Birn et al. [1988] have studied the field topology of the magnetotail plasmoid in a 3-dimensional geometry. A 3-dimensional global simulation of dayside reconnection was carried out by Sato et al. [1986] and Ogino et al. [1988]. Various properties of the 3-D reconnection have been discussed by a number of authors [e.g., Schindler et al., 1988; Hesse and K. Schindler, 1988; Song and Lysak, 1989; Priest and Forbes, 1989].

Our present work aims to study the basic properties of the 3-dimensional multiple X line reconnection process and the resultant magnetic flux tubes. The results can be applied to magnetic reconnection processes in planetary as well as in laboratory plasmas.

Simulation Model

A 3-D incompressible MHD code is used in the simulation. The normalized governing equations are (hereinafter all the physical quantities and equations are expressed in the normalized form unless otherwise noted)

$$\frac{\partial \mathbf{\Omega}}{\partial t} = \nabla \times (\mathbf{v} \times \mathbf{\Omega}) + \nabla \times (\mathbf{j} \times \mathbf{B}) + \nu \nabla^2 \mathbf{\Omega} \quad (1)$$

$$\frac{\partial \mathbf{B}}{\partial t} = \nabla \times (\mathbf{v} \times \mathbf{B}) + \nabla \times (\eta \nabla \times \mathbf{B}) \quad (2)$$

where $\mathbf{v}$ is flow velocity, $\mathbf{\Omega}$ is vorticity defined as $\mathbf{\Omega} = \nabla \times \mathbf{v}$, $\mathbf{B}$ is magnetic field, $\mathbf{j}$ is the current related to $\mathbf{B}$ by $\mathbf{j} = \nabla \times \mathbf{B}$, ν is the viscosity, and η is the magnetic diffusivity. Eq. (1) is obtained by taking the curl of momentum equation in the standard MHD formulation, and

Geophysical Monograph 58

the density ρ has been assumed to be a constant normalized to one.

In (1) and (2) and simulation results presented hereinafter, the magnetic field, length, velocity, current, resistivity, viscosity, and time are expressed in units of the asymptotic magnetic field value B_0, the initial central current sheet thickness a_0, the corresponding Alfvén speed $V_A = B_0/\sqrt{\rho_0\mu_0}$, $B_0/\mu_0 a_0$, $\mu_0 a_0 V_A$, $\rho_0 a_0 V_A$, and the Alfvén transit time $t_A = a_0/V_A$, respectively, and μ_0 is the permeability in free space.

The magnetic diffusivity η in (2) is assumed to be nonlinear:

$$\eta(j) = \begin{cases} \lambda(j - j_0)^2 + \eta_0, & \text{if } j > j_0, \\ \eta_0, & \text{otherwise,} \end{cases} \tag{3}$$

where λ is a constant, j_0 is the critical current for the enhancement of the local resistivity, j is the total current, and η_0 is a constant background resistivity.

The simulation is carried out in a rectangular box, which extends from $-L_x$ to L_x, $-L_y$ to L_y, and $-L_z$ to L_z, in x, y, and z directions, respectively. The initial velocity field of plasmas is specified by a curl-free 3-D flow field satisfying the imposed boundary conditions (see below). The initial magnetic field configuration is

$$\mathbf{B}(x, y, z) = B_{0z}\tanh(x/a(y))\hat{\mathbf{z}} + B_{0y}\hat{\mathbf{y}}, \tag{4}$$

where $\hat{\mathbf{z}}$ and $\hat{\mathbf{y}}$ are the unit vectors in z and y directions, respectively, B_{0z} and B_{0y} are related to the normalization unit, B_0, for magnetic field by $B_0 = (B_{0z}^2 + B_{0y}^2)^{1/2}$. Note that the initial current sheet thickness $a(y)$ varies with y as $a(y) = 1 + (y/L_y)^2$.

The actual simulation calculates only one quadrant of the simulation box which extends from $-L_x$ to L_x, 0 to L_y, and 0 to L_z, in x, y, and z directions, respectively. A set of symmetry operations is specified on the two symmetric planes (at $y = 0$ and $z = 0$). For the above setup, there are six boundary planes: two imposed ones (inflow boundaries), two free (outgoing) ones, and two symmetric (or antisymmetric) ones. Boundary conditions on these six planes are prescribed as follows.

(1) On the inflow boundaries at $x = L_x$ and $-L_x$, a constant inward plasma velocity $\mathbf{v} = -v_0\hat{\mathbf{x}}$ and $\mathbf{v} = v_0\hat{\mathbf{x}}$, respectively, is maintained and magnetic field is also fixed to initial magnetic fields at these two boundaries.

(2) Boundaries at $y = 0$ and $z = 0$ are symmetric ones.

(3) Boundaries at $z = L_z$ and $y = L_y$ are "free" outgoing boundary planes. By "free" we mean that these boundaries are so set up that the plasmas can flow freely out of the simulation box.

A finite differencing technique is used in the present simulation. In particular, all the derivative terms with respect to spatial coordinates in the governing equations (1) and (2) are approximated by the standard central differencing scheme of second-order accuracy while the equations are advanced in time by a fourth-order Runge-Kutta scheme.

Simulation Results

In the present study, magnetic reconnection is simulated under the driven boundary condition in a three-dimensional incompressible fluid. As predicted in Lee and Fu's earlier theoretical model [1985], helical magnetic flux tubes are formed as a result of reconnection at multiple sites and the reconnected field lines assume a complicated geometry, especially at the ends of the flux tubes.

One 3-D simulation case will be presented in this paper. The parameters used are: $B_{0z}/B_{0y} = 4$, $v_0 = 0.2$, $\eta_0 = 0.01$, $\lambda = 6 \times 10^{-4}$, $j_0 = 4$, $\nu = 0.025$, $L_x = 6$, $L_y = 76$, and $L_z = 48$. The number of grid points used for the simulated quadrant are $65 \times 33 \times 33$.

1. Formation of Magnetic Flux Tubes

Under the driven boundary conditions described above, magnetic reconnections start in the current sheet without the use of any artificial resistivity trigger. At time $t = 58t_A$, a pair of magnetic flux tubes is observed to be formed as a result of magnetic reconnections along three X lines approximately in the y direction. Figure 1 shows the lower ($z < 0$) one of the two magnetic flux tubes formed at $t = 58t_A$. Eight field lines are traced and plotted. The

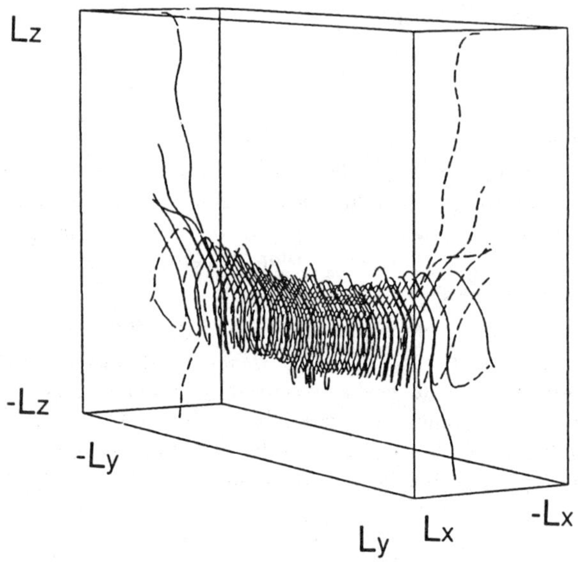

Fig. 1. Magnetic field lines showing the flux tube structure with "frayed" ends.

solid lines represent the magnetic field lines on Side 1 (the $x > 0$ region of the simulation box) and dashed lines the magnetic field lines on Side 2 ($x < 0$ region). It is seen that at both ends of the flux tube, magnetic field lines are connected either to Side 1 or Side 2 in a seemingly unpredictable way, forming the two "frayed" ends of the flux tubes. The geometry revealed at the ends of the flux tube shown is quite different from that in single X line reconnection.

As reconnection goes on, a second pair of magnetic flux tubes is formed at $t = 70t_A$ between the first pair. At $t = 78t_A$, one more flux tube is formed between the second pair of flux tubes. In the process of reconnection, one interesting feature is that reconnections first start in the current sheet near the central $y = 0$ plane and then "propagate" sideways in the positive and negative y direction. As a result, the y-direction extent, or the length, of the flux tubes increase in the course of reconnection and convection of flux tubes. The length of flux tubes formed later is therefore shorter than the length of those formed earlier.

2. Flux Tube-Aligned Flow

The response of plasma flows to the formation of magnetic islands in the 2-D cases has been studied relatively well [e.g., Fu and Lee, 1986]. We next show the plasma flows associated with the 3-D magnetic flux tubes. Shown in Figure 2 are the velocity vectors in y-z plane at $x = 0$ for $t = 40t_A$ and $t = 65t_A$, respectively. Figure 2(a) shows the flow pattern just before the onset of reconnections ($t = 40t_A$). By $t = 64t_A$, two magnetic flux tubes are formed. In Figure 2(b), the plasma flow pattern in x-y plane at $t = 65t_A$ is shown. Two magnetic field lines outlining the formed flux tubes located in the upper ($z > 0$) and lower ($z < 0$) parts of the box are projected onto the same plane by dashed lines. From Figure 2(b), one can identify two apparent flow features: (1) that the plasmas trapped within the flux tubes due to magnetic reconnection leak out from the ends of the flux tubes, forming enhanced flux tube-aligned plasma flows, and (2) that the central segment of the flux tube is convected outward (positive z direction for the upper tube and negative z direction for the lower one) at a faster speed than the two end segments, resulting in bow-shaped tubes.

3. Topology of Magnetic Field Lines

As shown in Figure 1, the topology of the magnetic flux tube is complicated (see also Birn et al., 1989). It is, however, possible for individual field lines to be classified into distinct categories based on their topology. Our simulation shows that there are four types of reconnection topology for individual field lines. In Figure 3, these four reconnection topologies are displayed. Shown in each box of Figure 3 are two symmetric magnetic field lines in each of the two magnetic flux tubes formed at $t = 58t_A$ (recall that at $t = 58t_A$, only two magnetic flux tubes are formed) and the dashed and solid line representations of magnetic field lines follows the same conventions as in Figure 1. We will first concentrate on the lower flux tube ($z < 0$). It is seen that the reconnected field lines in the lower flux tube can go from side 2 to side 2: Type 1 field lines (Figure 3a); from side 1 to side 1: Type 2 field lines (Figure 3b); from side 1 to side 2: Type 3 field lines (Figure 3c); from side 2 to side 1: Type 4 field lines (Figure 3d). The corresponding field lines in the upper flux tube also show the four types of reconnection geometry but in a symmetric way.

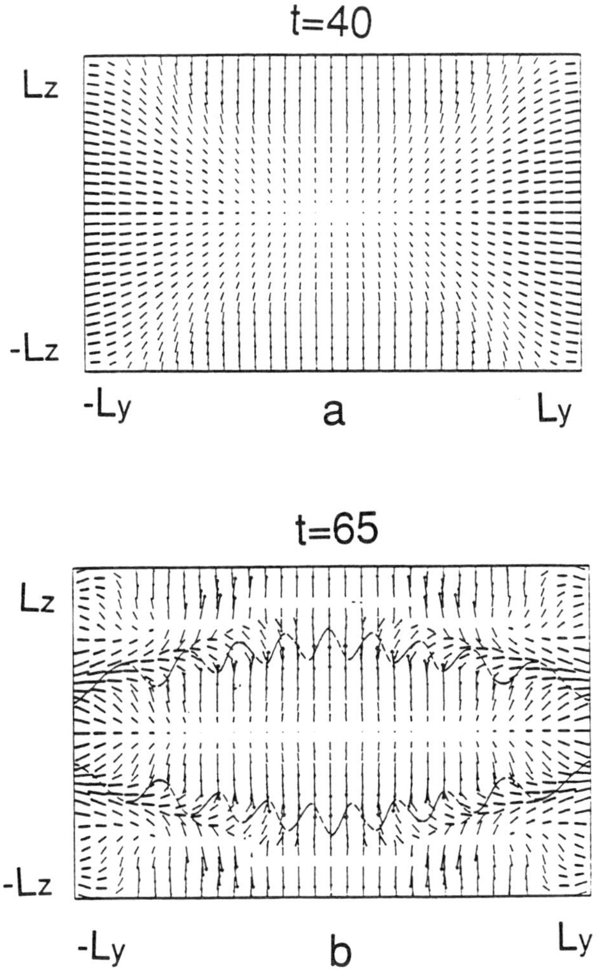

Fig. 2. Flow patterns of plasma in the y-z plane at $x = 0$: (a) $t = 40t_A$ (before reconnection) and (b) $t = 65t_A$ (after reconnection). Dashed lines outline the positions of the flux tubes in this plane.

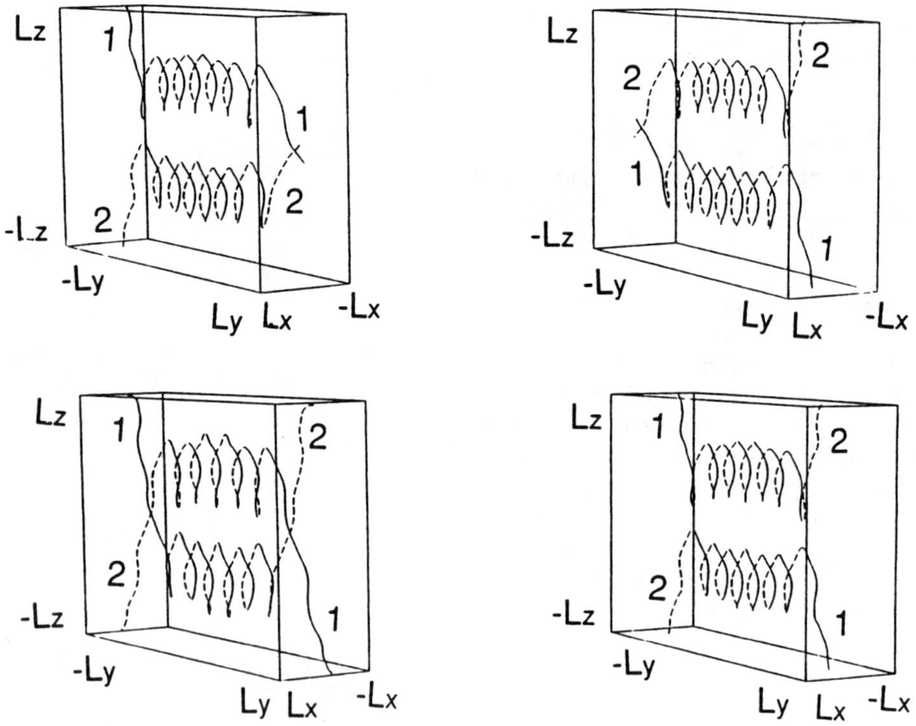

Fig. 3. Four types of field line topology in the magnetic flux tubes.

Discussions and Conclusions

Our simulation results confirms some aspects of the theoretical 3-D multiple X line reconnection model by Lee and Fu [1985]. Figure 4 shows the topology of magnetic reconnections at the dayside magnetopause in Lee-Fu's model. The model predicts that magnetic reconnection can occur along multiple X lines and that in the presence of a finite B_y component in the interplanetary magnetic fields, magnetic flux tubes aligned in the y direction can be formed. Our simulation further suggests that the magnetic reconnection geometry for these flux tubes will be complicated.

The flux tubes in the Lee-Fu model is also assumed to be of finite length. This finiteness can be caused by many processes. For example, the sheared flow at the magnetopause can lead to patchy reconnection which may produce flux tubes of finite length in 3-D situation [see discussion in La Belle-Hamer et al., 1988]. In our simulation, it is seen that the finite length of flux tubes is a natural consequence of a non-uniform current sheet.

In conclusion, we have simulated the 3-D magnetic reconnection in an incompressible MHD fluid. Multiple X line reconnections are observed to occur under driven boundary conditions. The multiple X line reconnections can further lead to the formation of magnetic flux tubes of finite length. The magnetic flux tubes are found to have "frayed" ends. In addition to the convection motion along with the magnetic flux tubes, the flow velocity of plasmas has a strong tube-aligned component.

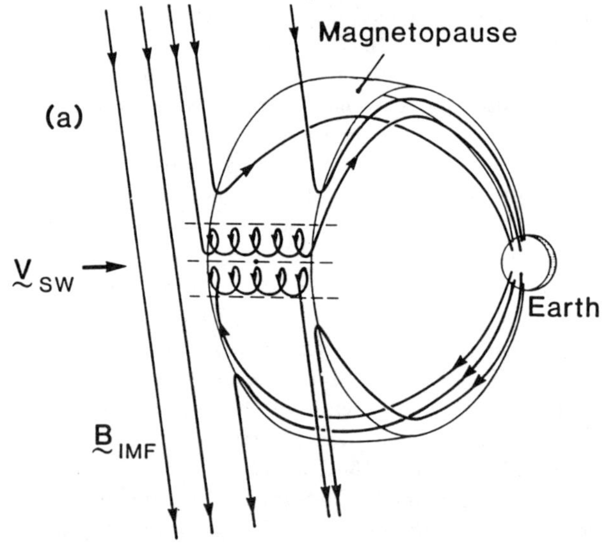

Fig. 4. A perspective of magnetic reconnection at the dayside magnetopause in the multiple X line reconnection model.

Questions and Answers

Priest: What keeps most of the magnetopause stable? What determines the time between bursts? If you add an extra field component normal to the plane of your 2-D experiments, why should tearing take place preferentially when the fields are nearly antiparallel (in order to agree with observations of FTEs)?

Lee: The magnetopause can be unstable everywhere in the presence of a southward IMF. However, reconnection may take place at those sites that are heavily pinched. The time t_B *between bursts can be estimated as* $t_B \simeq B_I S_x / 2V_I B_0$, *where* B_0 *is the external field,* B_I *is the magnetic field inside the magnetic island,* S_x *is the half-width of the island, and* V_I *is the inflow speed.* S_x *can be related to the wavelength* λ *of the most unstable mode by* $S_x \simeq 0.4\lambda$. *If we set* $S_x \simeq 1R_E$, $V_I \simeq 0.1V_A \simeq 20 km/s$, $B_I \simeq 2B_0$, *we have* $t_B \simeq 300s \simeq 5$ *min. The subsolar reconnection may be caused by the fact that the magnetopause current sheet is heavily pinched near the subsolar region by the solar wind pressure.*

Sonnerup: In his simulation of the reconnection process, Biskamp finds a long current sheet, the thickness of which is independent of conductivity and is proportional to reconnection rate (evaluated at the inflow edges of the simulation box). Do you find this behavior in your simulations?

Lee: As presented in Lee and Fu [1986], the thickness of the diffusion region δ *is related to the the resistivity* η *and the imposed reconnection rate* R_0 *by* $\delta \sim \eta^{0.85} R_0^{-0.1}$. *This result is different from Biskamp's, probably due to the difference in the imposed boundary conditions between the two simulations.*

Atkinson: The Russell and Elphic model for FTEs implies conjugate ionospheric footprints. Does your model imply nonconjugate footprints and can this be used as a test of the theories?

Lee: In the Lee-Fu model, the open geomagnetic field lines consist of two parts: (a) the helical field lines associated with the magnetic ropes, and (b) the open field lines wrapping around the magnetic islands. The open field lines in the part (a) have nonconjugate footprints, while the open field lines in the part (b) have conjugate footprints. In the Russell-Elphic model, the nonconjugate part does not exist. This difference, of course, can be used as a test of the theories.

Acknowledements. This work is supported by Department of Energy grant DE-FG06-86ER 13530 and National Science Foundation grant ATM88-20992 to the University of Alaska, Fairbanks. And one of the authors, Z. F. Fu, is also partially supported by National Science Foundation of China grant 4860174 to the Center for Space Science and Applied Research, Academia Sinica. The computation part is done at the San Diego Supercomputer Center.

References

Birn, J., M. Hesse, and K. Schindler, Filamentary structure of a three-dimensional plasmoid, *J. Geophys. Res., 94*, 241, 1989.

Fu, Z. F, and L. C. Lee, Multiple X line reconnection 1. A criterion for the transition from a single X line to a multiple X line reconnection, *J. Geophys. Res., 91*, 6807, 1986.

Hesse, M., and K. Schindler, A theoretical foundation of general magnetic reconnection, *J. Geophys. Res., 93*, 5559, 1988.

Hones, E. W., Jr., J. Birn, S. J. Bame, G. Paschmann, and C. T. Russell, On the three-dimensional magnetic structure of the plasmoid created in the magnetotail at substorm onset, *Geophys. Res. Lett., 99*, 203, 1982.

La Belle-Hamer, A. L., Z. F. Fu, and L. C. Lee, A mechanism for the patchy reconnection at the dayside magnetopause, *Geophys. Res. Lett., 15*, 152, 1988.

Lee, L. C., and Z. F. Fu, A theory of magnetic flux transfer at the earth's magnetopause, *Geophys. Res. Lett., 12*, 105, 1985.

Ogino, T., R. J. Walker, and M. Ashour-Abdalla, A magnetohydrodynamic simulation of the formation of magnetic flux tubes at the earth's dayside magnetopause, *Geophys. Res. Lett., 16*, 155, 1989.

Petschek, H. E., Magnetic field annihilation, in AAS-NASA Symposium on the Physics of Solar Flares, *NASA Spec. Publ., SP-50*, 425-439, 1964.

Priest, E. R., and T. G. Forbes, New models for fast steady state magnetic reconnection, *J. Geophys. Res., 91*, 5579, 1986.

Priest, E. R., and T. G. Forbes, Steady magnetic reconnection in three dimensions, *Solar Phys., 119*, 211, 1989.

Russell, C. T., and R. C. Elphic, ISEE observations of flux transfer events at the dayside magnetopause, *Geophys. Res. Lett., 6*, 33, 1979.

Sato, T., T. Shimada, M. Tanaka, T. Hayashi, and K. Watanabe, Formation of field twisting flux tubes on the magnetopause and solar wind particle entry into the magnetosphere, *Geophys. Res. Lett., 13*, 801, 1986.

Schindler, K., M. Hesse, and J. Birn, General Magnetic Reconnection, parallel electric field, and helicity, *J. Geophys. Res., 93*, 5547, 1988.

Song, Y., and R. L. Lysak, Evaluation of twist helicity of flux transfer event flux tubes, *J. Geophys. Res., 94*, 5273, 1989.

Vasyliunas, V. M., Theoretical models of magnetic field line merging, 1, *Rev. Geophys., 13,*, 303, 1975.

THE GENERATION OF TWISTED FLUX ROPES DURING MAGNETIC RECONNECTION

Mitchell A. Berger[1]

Department of Mathematical Sciences, University of St Andrews, Fife, Scotland

Andrew N. Wright

School of Mathematical Sciences, Queen Mary College, London

Abstract. The reconnection of bundles of magnetic flux can lead to the creation of twisted magnetic flux ropes. Magnetic helicity conservation provides an important constraint on this process. This paper reviews recent calculations of the net twist generated during the reconnection of flux tubes. Twist acquired due to shearing motions will also be discussed.

1. Introduction

Magnetic reconnection is an extremely complicated process – it is three dimensional, non-linear, and time-dependent. It can also involve non-MHD processes on very small length scales. Fortunately, some questions about reconnection do not depend on the extremely difficult fine details. Wright [1987] showed that reconnection of flux tubes would generally lead to twisted tubes by considering the field topology and the torques near the reconnection region. Here we describe some more recent work on how flux ropes become twisted during reconnection [Song and Lysak, 1989; Wright and Berger, 1989]. This work employs magnetic helicity conservation to calculate the twist gained by reconnecting flux. The helicity calculations depend only on which bundles of flux reconnect with each other, and the number of reconnection regions.

Magnetic helicity is essentially a volume integral of $\vec{A} \cdot \vec{B}$, involving the magnetic field $\vec{B}$ and its vector potential $\vec{A}$. This integral can be made gauge-invariant; various (equivalent!) expressions can be found in Berger and Field [1984], Jensen and Chu [1984], and Finn and Antonsen [1985].

[1]Present Address: Department of Mathematics, University College London.

Geophysical Monograph 58

The helicity integral measures the twisting and linking of magnetic lines of force [Moffatt, 1969; Berger and Field, 1984]. The total helicity of a set of flux tubes can be expressed as a sum of *mutual helicities* due to linking of different flux tubes with one another, and *self helicities* due to twisting or kinking of individual tubes. Total magnetic helicity is approximately conserved on reconnection timescales [Berger, 1984]. However, reconnection can redistribute the total helicity between self and mutual helicities. Thus if reconnection were to unlink two flux tubes then the new flux tubes would be more twisted; increasing the self helicity compensates for a reduced mutual helicity. In the context of flux transfer events, flux from the interplanetary field and the magnetospheric field can reconnect to form flux ropes. This generation of twisted magnetic structures from the interactions of very large-scale background fields has been examined in terms of a dynamo process [Song and Lysak, 1989, *these proceedings*].

Let us look at a few simple examples of how helicity relates to field structure. First consider a single flux tube extending between the two planes $z = 0$ and $z = L$. Suppose this tube is uniformly twisted; i.e. every field line twists about the central axis through an angle $2\pi T$. In this case the magnetic helicity is simply $H = T\Phi^2$ where Φ is the flux within the tube.

Next consider a field consisting of two thin flux tubes extending between the two planes (see figure 1a). These tubes have fluxes Φ_1, Φ_2, and twists T_1, T_2. At the lower plane $z = 0$ imagine a line segment drawn between the centers of the two tubes. Now go upwards in z, keeping the ends of the line segment tied to the center field lines of the tubes. The line segment will rotate by some angle $2\pi\Theta_{12}$ going from $z = 0$ to $z = L$. Thus Θ_{12} increases by 1 each time the tubes twine about each other through one turn. The angle Θ_{12} is called the *winding number* of the two tubes. The total helicity is

$$H = T_1\Phi_1^2 + T_2\Phi_2^2 + 2\Theta_{12}\Phi_1\Phi_2 \qquad (1)$$

where $H_1 = T_1\Phi_1^2$, $H_2 = T_2\Phi_2^2$ are self helicities, and the Θ_{12} term is the mutual helicity.

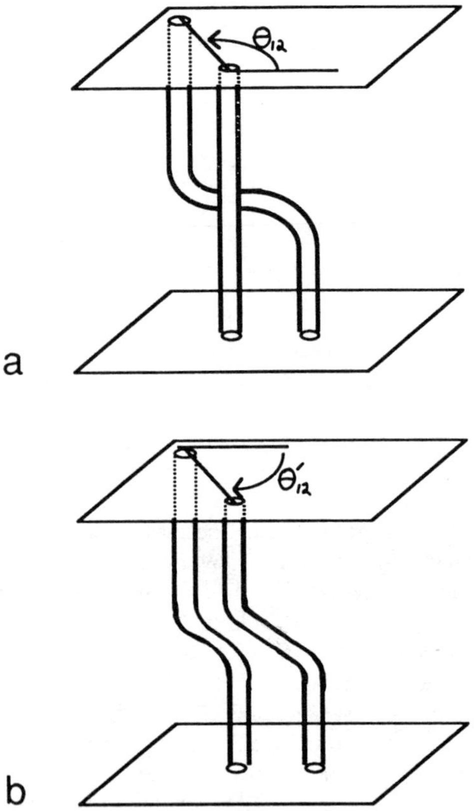

Fig. 1. The winding number Θ_{12}. (a) Before reconnection. (b) after reconnection. Θ_{12} has decreased by $1/2$.

2. Reconnection of Two Tubes

Assume that $\Phi_1 = \Phi_2 = \Phi$. Now suppose that the two tubes reconnect. Let T_1', T_2', and Θ_{12}' be the twist and winding numbers after reconnection. By conservation of the total helicity H,

$$T_1' + T_2' + 2\Theta_{12}' = T_1 + T_2 + 2\Theta_{12}. \quad (2)$$

During reconnection the winding number changes by $\pm\frac{1}{2}$ (see figure 1b): $\Theta_{12}' = \Theta_{12} \pm \frac{1}{2}$. Thus

$$T_1' + T_2' = T_1 + T_2 \pm 1. \quad (3)$$

This equation confirms earlier results obtained by topological considerations and by estimating the torques felt by flux tubes undergoing reconnection [Wright, 1987]. Under fairly general conditions [Wright and Berger, 1989] the extra self helicity will be shared equally between the two flux tubes. Both T_1 and T_2 increase (or decrease) by 1/2 (half a complete uniform twist) as a result of reconnection. It should be emphasized that the twist produced by a single reconnection line is indeed uniform inside the flux tubes. Sometimes theorists study uniformly twisted tubes because of their relative simplicity. However, uniform twist can arise naturally as a consequence of reconnection.

3. Multiple Reconnection Lines

Can reconnection create flux tubes with more than one-half a twist? Let us look at what happens when there are two or three reconnection lines. Multiple reconnection line models have been presented by Lee and Fu [1985] to explain flux rope signatures at the Earth's magnetopause. In figure 2 tubes A and D (with flux Φ_1) reconnect once; tubes B and C (with flux Φ_2) may reconnect at several places. In figure 2b these tubes enter the region between two reconnection lines from the left. After reconnection the field lines 'bounce' one or more times (twice for the figure) between the reconnection lines before finally emerging to the right to again form two tubes. If

$$\Phi_1 = n\Phi_2 \quad (4)$$

(with n an integer), then each field line in tubes B and C reconnects (bounces) n times. Furthermore, if and only if n is an integer will all the flux within tube B (or C) map to a single tube on the right. The incoming flux in tube B maps to the outgoing flux in the upper/lower bundle for n odd/even. For $\Phi_1 \neq n\Phi_2$ some of the incoming flux from tube B still maps to the lower bundle on the right.

Let us look at the special case where $\Phi_1 = n\Phi_2$. The total helicity before reconnection (assuming initially untwisted tubes) is all stored in mutual helicity:

$$H = (\Phi_1 + \Phi_2)^2 = (n+1)^2\Phi_2^2. \quad (5)$$

After reconnection, A and D each have half a twist; these tubes each contribute $n^2/2\ \Phi_2^2$ to the total helicity. A field line of B bounces n times; so B has $n/2$ twists. The self helicity of B is then $n/2\ \Phi_2^2$ (similarly for C). This implies a mutual helicity between B and C of

$$H - 2(n^2/2\ \Phi_2^2) - 2(n/2\ \Phi_2^2) = (n+1)\Phi_2^2. \quad (6)$$

Tubes B and C are now braided about each other! The number of crossovers is $n+1$ (i.e. the tubes wind about each other by $\Theta_{BC} = (n+1)/2$ turns).

With more reconnection lines we could again achieve the situation where all the helicity resides in self helicity. For example in figure 2d a third reconnection line midway between the two shown would prevent tubes B and C from crossing over each other. The twists of A and D are still 1/2, whereas B and C acquire $(n+1/2)$ twists. (In Song and Lysak [1989] this result is stated using the parameter $M = n+1 = (\Phi_1 + \Phi_2)/\Phi_2$.)

4. Boundary Motions

Twisting and shearing motions, as well as reconnection, can generate twist inside flux ropes. Fluid motions at a boundary

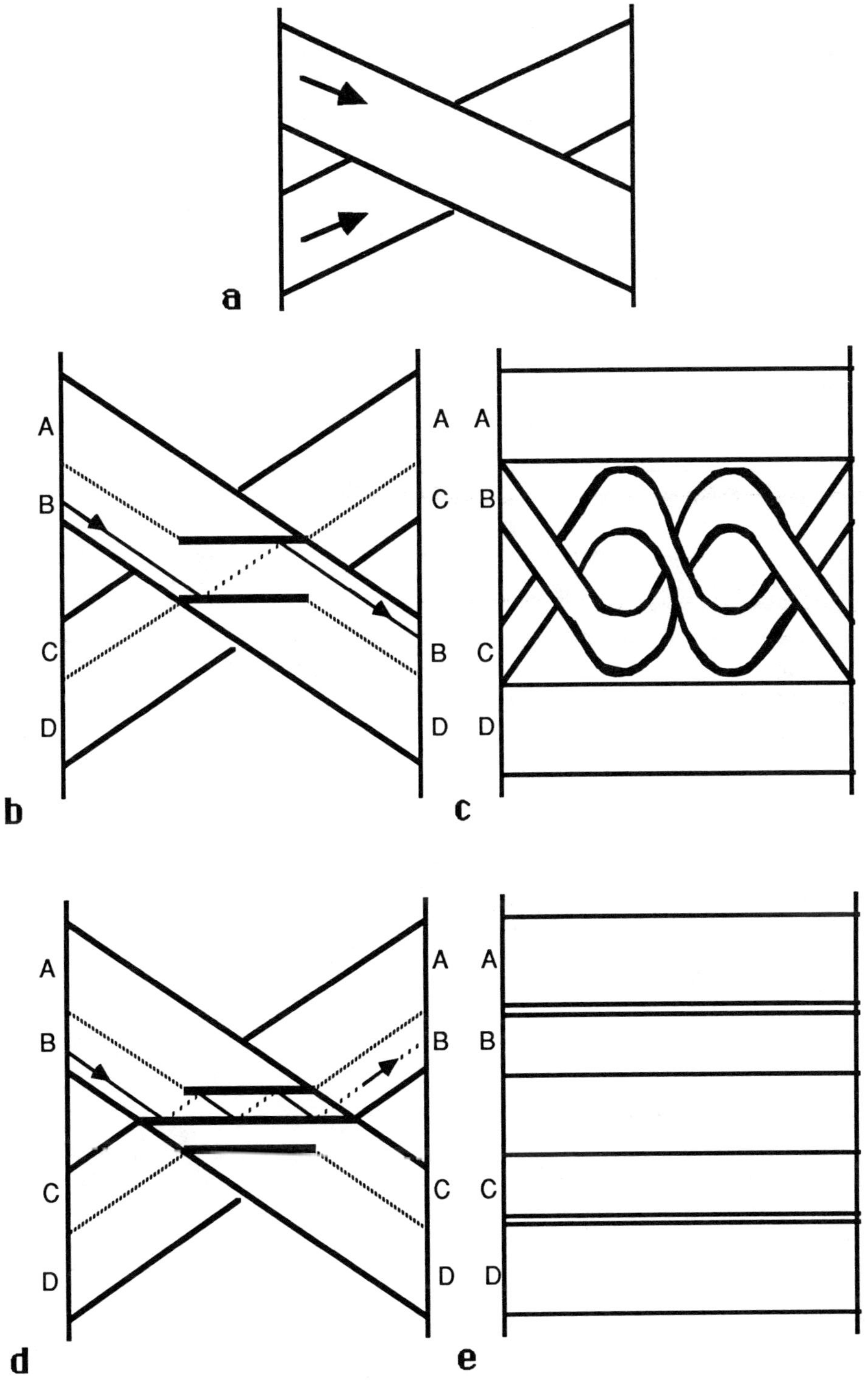

Fig. 2. (a) Before reconnection. (b) and (c) Reconnection at two reconnection lines. A typical field line of B is shown. Flux tubes B and C wind about each other through 3/2 turns. A and D have twist 1/2, while B and C have twist 1. (d) and (e) Three reconnection lines. Both B and C now have twist 5/2.

may give rise to a flow of helicity from one side to the other. For example, the Northern and Southern hemispheres of interplanetary space have substantial helicities (of opposite sign); the source is ultimately the rotation of field lines at the solar surface [W. H. Matthaeus, private communication]. (rotating open coronal field lines twist to form the Parker spiral). Similarly, the fields inside astrophysical jets can become twisted if they connect to a central rotating driver.

As a second example, shearing motions at the photosphere can increase the helicity of the coronal field above. A sheared coronal arcade contains a net helicity; if the arcade reconnects to form a prominence filament [Pneuman, 1983; House and Berger, 1987] then the helicity can be transfered to the filament. A similar mechanism can create a helical plasmoid structure [Schindler, Hesse, and Birn, 1988].

The twist generated during FTE reconnection may propagate away from the reconnection site via torsional Alfvèn waves [Wright, 1987; Southwood et al., 1988]. Since reconnection is often a dynamic process, the effects of both reconnection and large scale motions must be considered in order to predict the magnitude of these waves. Further details will be published elsewhere; here we review the basic equations.

Divide space into a volume $\mathcal{V}$ and its exterior $\mathcal{V}'$. In general $\mathcal{V}$ will not be bounded by a magnetic surface; field lines may pass through the boundary. In this case, the helicity inside $\mathcal{V}$ can change due to boundary motions. If the field within $\mathcal{V}$ is divided into a collection of flux tubes, then reconnection redistributes mutual and self helicities, keeping the total helicity of $\mathcal{V}$ approximately unchanged. Boundary motions, on the other hand, redistribute helicity between $\mathcal{V}$ and $\mathcal{V}'$.

Consider the case where $\mathcal{V}$ is the upper half space $\{z = 0\}$. Let $\vec{a}_1$ and $\vec{a}_2$ be the endpoints of two field lines at $z = 0$, with separation $\vec{r}_{12} = \vec{a}_1 - \vec{a}_2$. Also let θ_{12} be the orientation of $\vec{r}_{12}$ with respect to the x-axis. Then the helicity change due to motions of the field line endpoints $\vec{a}_1$ and $\vec{a}_2$ is [Berger, 1988]

$$\frac{dH}{dt} = \int\int \left(\frac{-1}{2\pi} B_z(\vec{a}_1) B_z(\vec{a}_2) \frac{d\theta_{12}}{dt} \right) d^2 a_1 d^2 a_2 . \quad (7)$$

Thus dH/dt measures the net angular rotation of flux at the boundary.

Suppose we have two tubes A and B which form arches in $\mathcal{V}$, each with two footpoints at $z = 0$. From equation we can easily derive the evolution of self helicity H_A and mutual helicity $2\,H_{AB}$ (total $H = H_A + H_B + 2\,H_{AB}$). If the two footpoints of A are A_+ and A_- then

$$\frac{dH_A}{dt} = \int_{A_+}\int_{A_+} + \int_{A_-}\int_{A_-} + 2\int_{A_+}\int_{A_-} \quad (8)$$

(with the integrand as in equation 7). Similarly, dH_{AB}/dt would be integrated over all combinations of footpoints, one from A and one from B.

Acknowledgments The authors are grateful to E. Priest, S. Schwartz, and Y. Song for helpful discussions.

References

Berger, M. A., Rigorous new limits on magnetic helicity dissipation in the solar corona,, *Geophysical and Astrophysical Fluid Dynamics* **30**, 79, 1984.

Berger, M. A., An energy formula for nonlinear force-free magnetic fields,, *Astronomy and Astrophysics* **201**, 355, 1988.

Berger, M. A. & G. B. Field, The topological properties of magnetic helicity,, *J. Fluid Mechanics* **147**, 133, 1984.

Finn, J. H. & T. M. Antonsen, Magnetic Helicity: What is it and What is it Good For?,, *Comments on Plasma Physics and Controlled Fusion* **9**, 111, 1985.

House, L. L. & M. A. Berger, Ejection of helical field structures through the outer corona,, *Astrophysical J.* **323**, 406, 1987.

Jensen, T. H. & M. S. Chu, Current Drive and Helicity Injection,, *Physics of Fluids* **27**, 281, 1984.

Moffatt, H. K., The Degree of Knottedness of Tangled Vortex Lines,, *J. Fluid Mechanics* **35**, 117, 1969.

Pneuman, G. W., The formation of solar prominences by magnetic reconnection and condensation,, *Solar Physics* **88**, 219, 1983.

Schindler, K., M. Hesse, & J. Birn, General magnetic reconnection, parallel electric fields and helicity,, *J. Geophys. Res.* **93**, 5547, 1988.

Song, Y. & R. L. Lysak, Evaluation of twist helicity of FTE flux tubes,, *J. Geophys. Res.* **94**, 5273, 1989.

Wright, A. N., The evolution of an isolated reconnected flux tube,, *Planet. Space Sci.* **35**, 813, 1987.

Wright, A. N. & M. A. Berger, The effect of reconnection upon the linkage and interior structure of magnetic flux tubes,, *J. Geophys. Res.* **94**, 1295, 1989.

M. A. Berger, Department of Mathematics, University College London, Gower Street London WC1E 6BT, England.

A. N. Wright, Queen Mary College, University of London, Astronomy Unit, Mile End Road, London E1 4NS, England.

FORMATION OF FLUX ROPES BY TURBULENT RECONNECTION

Robert L. Lysak and Yan Song

School of Physics and Astronomy
University of Minnesota, Minneapolis, MN 55455

Abstract. Although traditional reconnection models have assumed a two-dimensional geometry and steady-state conditions, reconnection at the earth's magnetopause and in other applications is likely to be time-dependent, turbulent, and three-dimensional in nature. At the high magnetic Reynolds numbers present in most reconnection sites, the evolution of the plasma is expected to follow nearly ideal MHD conditions. Under these conditions, the invariants of three-dimensional MHD, namely the energy, magnetic helicity, and the cross helicity are expected to be conserved. When ideality is violated in a localized region, the energy is dissipated at a higher rate than the other two invariants, so that the magnetic helicity is essentially conserved during the reconnection process. In addition, during the three-dimensional evolution of an MHD fluid, magnetic helicity undergoes an inverse cascade to large scales, indicating that a region in which many small-scale reconnections take place will self-organize into a large scale flux rope.

A magnetic flux rope intrinsically contains magnetic helicity which exhibits itself topologically in the twisting of the rope. Under the above scenario, this helicity can be produced by the transformation of helicity contained in the topology of the flux tubes which existed before reconnection. Such a model can satisfactorily account for the amount of twist observed in flux transfer event (FTE) flux ropes. In addition, observations of the small-scale structure of FTE flux tubes indicates that the velocity and magnetic field perturbations are aligned. Such a situation is expected to develop since the cross helicity, i.e., the correlation between the velocity and magnetic field, also decays less rapidly than the energy. Thus, a number of features of FTEs indicate that they were formed by a turbulent reconnection process. By extension, it appears likely that turbulent, three-dimensional reconnection may be an effective mechanism to produce flux ropes in general.

Geophysical Monograph 58

Introduction

Although traditional reconnection models have assumed a two-dimensional geometry and steady-state conditions, reconnection at the earth's magnetopause and in other applications is likely to be time-dependent, turbulent, and three-dimensional in nature. Minor extensions of the usual two-dimensional reconnection picture, e.g., simply adding a uniform magnetic field component along the separator line, can not account for the richness of possibilities inherent in three dimensions. Simple results from the two-dimensional model, such as the presence of high speed plasma flow along the separatrices, and the monotonic decrease of magnetic energy, do not necessarily occur in a three dimensional geometry. On the other hand, new phenomena such as the possibility of dynamo action can occur in three dimensions although the symmetries of the two-dimensional model prevent their occurrence in two dimensions. The purpose of this paper is to suggest that the formation of flux ropes is such a phenomenon which naturally occurs in three-dimensional reconnection.

Magnetic reconnection is commonly thought to be an important, if not the dominant, process by which mass, momentum, and energy are transported from the solar wind into the magnetosphere. While the basic process of reconnection has been studied in two dimensions for nearly 30 years [e.g., Dungey, 1961; Petschek, 1964], recent observations of the reconnection region [e.g., Russell and Elphic, 1978; Saunders et al., 1984; Farrugia et al., 1988] have indicated that the simple two-dimensional models are not adequate to describe reconnection at the dayside magnetopause, where the result of reconnection is often a twisted flux tube, which has been termed a flux transfer event. Such structures are inherently three dimensional, and thus cannot be discussed in terms of the classic two-dimensional models. In addition, such flux transfer events are often quite turbulent and contain a great deal of structure [e.g., LaBelle et al., 1987; Farrugia et al., 1988]. Most models and simulations of reconnection suppress such turbulence and thus may miss crucial features of the reconnection process.

The magnetopause region, particularly during southward interplanetary magnetic field, is known to contain turbulence on a wide range of spatial scales. Observations of plasma waves from satellites show [see, for example, LaBelle and Treumann, 1988] that the turbulent spectra of the fluctuating electric and magnetic fields in the dayside magnetopause extend from large to small scales. Estimates of microscopic turbulent transport under different assumptions for the wave mode are always too small to explain the required magnetic diffusivity. The macroscopic diffusive process caused by stochastic **E**×**B** scattering has been suggested by LaBelle and Treumann [1988]. In addition, LaBelle et al. [1987] have shown that the region around flux transfer events is characterized by enhanced levels of turbulence. Thus, it appears likely that turbulence may play an important role in reconnection at the magnetopause.

The formation of flux ropes during reconnection will be addressed in two ways: in the following section, we will consider the quantities which are conserved during the evolution of a nearly ideal MHD fluid; then we will discuss the dynamics of a simple model of the magnetopause reconnection region. During reconnection, the relevant conservation law is the conservation of magnetic helicity, which remains roughly constant despite the presence of a localized resistive region in which reconnection takes place. The magnetic helicity may be interpreted in terms of the topology of field lines, and when reconnection takes place, this topology is altered in such a way that the magnetic helicity remains constant. In the initial configuration, magnetic helicity is present due to the crossing of flux tubes on either side of the reconnection region. After reconnection, the net result is that the helicity released by reconnection manifests itself in the twisting of the reconnected field lines, forming a flux rope.

Although the helicity conservation argument above suggests that reconnection should produce twisted flux tubes, it would be preferable to understand the dynamics of the reconnection region. As a simple example to illustrate the physical principles involved, we will consider the consequences of a localized region of resistivity present in a current sheet which separates topologically distinct regions of magnetic field, such as is present at the magnetopause. Such a region may result from the excitation of the tearing mode or another microinstability which would locally increase the effective resistivity. It will be shown that the presence of a localized region of resistivity leads to the formation of a component of the magnetic field normal to the current sheet which can connect flux tubes on either side of the current sheet. In addition, a normal component of the current will also be generated, consistent with the presence of a twist in the newly connected tubes. This process constitutes a current dynamo effect, which will be analyzed in terms of a statistical description of the turbulent reconnection region in a companion paper [Song and Lysak, this meeting].

Magnetic helicity conservation and reconnection

At the high magnetic Reynolds numbers present in most reconnection sites, the evolution of the plasma is expected to follow nearly ideal MHD conditions. In incompressible ideal magnetofluids, the total energy $E=1/2\int(B^2+U^2)dV$, cross helicity $K_c=1/2\int\mathbf{U}\cdot\mathbf{B}dV$ and magnetic helicity $K=\int\mathbf{A}\cdot\mathbf{B}dV$ are exact invariants of the MHD equations, where **A** is the vector potential of the magnetic field. These quantities are strictly invariant in periodic systems, and in addition when the volume of integration has no normal magnetic or velocity component, and no tangential electric field in the fluid rest frame. In addition, the magnetic helicity is gauge invariant when there is no normal magnetic field, and differences in helicity between two states are gauge invariant when they satisfy the same boundary condition on the normal magnetic field [Berger and Field, 1984]. Although all of the ideal MHD invariants are dissipated by resistivity and/or viscosity, the dissipation rates differ greatly. In particular, the energy can be dissipated much faster than the magnetic helicity. The approximate conservation of the magnetic helicity can be explained in terms of the selective decay process [Matthaeus and Montgomery, 1984]. An upper limit on helicity dissipation has been derived by Berger [1984], who showed that the ratio between the helicity dissipation and the energy dissipation is proportional to the resistivity averaged over the volume. Thus, if the resistivity is small and/or localized, the helicity may be treated as approximately constant. Since the magnetopause has a very low resistivity, except perhaps in localized patches where reconnection takes place, the magnetic helicity will be approximately conserved.

The magnetic helicity in a region of space may be interpreted in terms of the topology of magnetic field lines in this region. For example, if two isolated flux tubes are interlinked with each other, the total helicity of the structure is $K=\pm2\Phi_1\Phi_2$, where Φ_1 and Φ_2 are the magnetic fluxes contained in the two tubes and the sign is given by the sense of linkage [Moffatt, 1978]. In general, a helicity of $\Phi_1\Phi_2$ is associated with each crossover of two flux tubes; thus, an untwisted, figure 8 shaped flux tube with flux Φ contains a helicity of $\pm\Phi^2$. In addition, a single uniformly twisted closed flux tube contains a helicity $K=T\Phi^2$, where Φ is the flux in the tube and T is the number of twists [Berger and Field, 1984]. In ideal MHD, the mutual helicity due to the linkage of different flux tubes and the self helicity caused by the twisting of a single flux tube are conserved separately. During reconnection, the total helicity remains nearly constant as seen above; however, the reorganization of the topology of the magnetic field allows mutual and self helicity to be converted to each other.

In general, when two regions of approximately uniform magnetic field are separated by a current sheet, there will be a finite angle between the two fields and thus a helicity of $\pm\Phi^2$, where Φ is the reconnected flux and the sign depends on the sign of B_y, where y is the direction of the magnetic component which does not change across the current sheet. Note that what the flux tube does as it closes somewhere outside the reconnection region is not important, since, if surfaces are defined on which the normal component of the magnetic field is specified, the relative helicity of the magnetic field subject to this boundary condition does not change [Berger and Field, 1984]. Reconnection effectively removes the crossover between the two flux tubes, and the helicity released manifests itself in the twist of the two newly formed flux tubes. Wright and Berger [1989] have shown that this helicity will be divided evenly between the two flux tubes, and so each new flux tube will contain a helicity of $\pm\Phi^2/2$, corresponding to a half twist. The twisting of the newly reconnected flux tubes depends on the geometry of the reconnection region. If the reconnection region is extended in either latitude or longitude (see Figure 1) and a flux tube undergoes reconnection at multiple sites, as in the multiple X-line model of Lee and Fu [1985], additional twist will be produced. On the other hand, if an FTE is formed by reconnection along a single line, as in the models of Scholer [1988] and Southwood et al. [1988], each flux tube formed will have only a half twist (Figure 1c).

As a reconnected flux tube propagates, the ambient magnetic field drapes around the flux tube, and the resulting tension tends to resist the motion of the flux tube. Such a flux tube can continue to propagate only if additional reconnection occurs in the wake of the flux tube [Sonnerup,

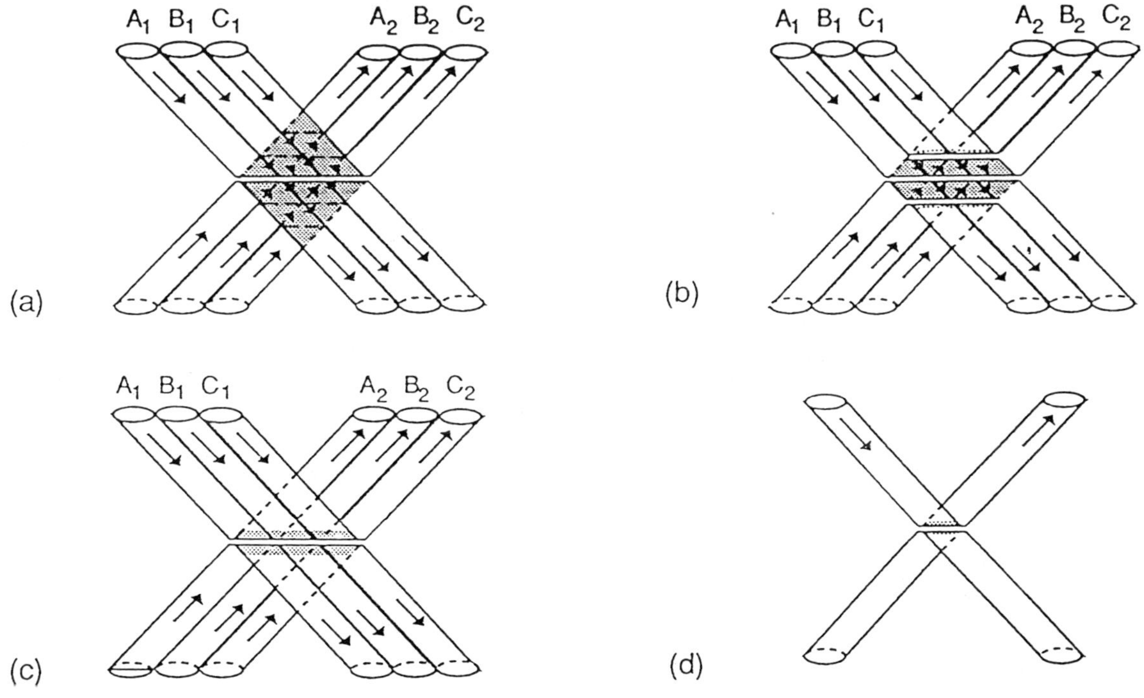

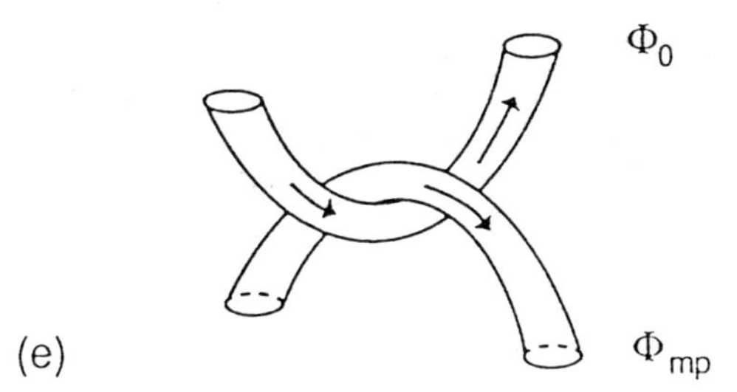

Fig. 1. Illustration of helicity conversion in various reconnection geometries. (a) Crossing of three flux tubes in a reconnection region extended in both latitude and longitude. Two tubes with 2 1/2 twists, two with 1 1/2 twists and two with 1/2 twist are formed. (b) A reconnection region extended in longitude [cf. Lee and Fu, 1985]. Two tubes with 2 1/2 twists and 4 with 1/2 twist are formed. (c) A single reconnection line model [Scholer, 1988; Southwood et al., 1988]. Six flux tubes with 1/2 twist are formed. (d) Reconnection of a single pair of tubes, producing two tubes with 1/2 twist. (e) Interaction of a flux tube with flux Φ_0 with the magnetopause magnetic field with flux Φ_{mp}. Amount of twist after reconnection depends on amount of flux reconnected.

1987]. Figure 1e shows that this configuration contains a helicity of $\pm 2\Phi_0\Phi_{mp}$, where Φ_0 is the flux in the initial flux tube and Φ_{mp} is the amount of flux swept up by the flux tube. This leads to an additional twist of the flux tube and additional field-aligned current as described by Sonnerup [1987], although he did not discuss this situation in terms of magnetic helicity. Similar arguments may be made in the case where an impulsive penetration of magnetosheath plasma occurs [Lemaire, 1977; Heikkila, 1982; Lundin, 1988]. In this case, the penetrating plasma creates an indentation of the magnetopause. As the plasma propagates further into the magnetosphere, the magnetopause field lines must eventually close behind the penetrating plasma and reconnect much as in the Sonnerup [1987] picture. The amount of magnetic helicity converted to twist in this case is again given by $\pm 2\Phi_0\Phi_{mp}$, where now Φ_0 is the flux contained in the penetrating plasma. The amount of helicity produced in these models of reconnection at the magnetopause has been summarized by Song and Lysak [1989].

The twist produced at the magnetopause reconnection site will propagate away from this region in the form of Alfven waves propagating along the flux tube. Thus, the notion of helicity conservation is only applicable on time scales small to the Alfven transit time to the ionosphere, a time of a few minutes. This time scale may be not too much greater than the growth of tearing modes; e.g., Quest and Coroniti [1981] estimate the collisionless tearing mode growth time to be 30 seconds. On the other hand, effective collisions due to lower hybrid drift turbulence could enhance tearing mode growth. Galeev et al. [1986] estimate that growth times in the semi-collisional case could be as short as 3 seconds. In this case, reconnection could occur before coupling to the ionosphere becomes important. In any case, the results from the helicity conservation argument should be taken as indicative of the tendency of flux tubes to twist during reconnection. The net twist of the flux tube will be the result of the twisting of the flux tube during reconnection and during propagation along the magnetopause, balanced by the radiation of Alfven waves away from the magnetopause. In order to evaluate this balance, a detailed model of the reconnection region is necessary, but such a model is not yet available. A model which can begin to describe the dynamics and time scales in this region is discussed in the following section.

Basic dynamics of the magnetopause reconnection region

Although much work has gone into the theory of two-dimensional reconnection and into simulation of two and three-dimensional reconnection, less thought has been given to the physical processes which occur in the reconnection region. We will consider these processes on the intermediate scale level, considering processes which occur on scales larger than the ion gyroradius and comparable to the magnetopause thickness. On this scale, the onset of reconnection is caused by a violation of ideal MHD by the formation of a resistive region of finite spatial extent. Such a region could form, for example, due to localized excitation of an instability such as the tearing mode or lower hybrid drift instabilities, and may be associated with the increase in the magnetopause current due to the arrival at the magnetopause of a solar wind pressure pulse [Sibeck et al., 1989] which compresses the magnetopause current sheet.

Consider the effect of such a localized region of resistivity which develops in the current sheet. The initial configuration can be specified as:

$$\mathbf{B}_0 = B_{y0}\hat{\mathbf{y}} + B_{z0}(x)\hat{\mathbf{z}} \tag{1}$$

indicating the initial current sheet is in the $\hat{\mathbf{y}}$ direction. At the magnetopause, the $\hat{\mathbf{x}}$ direction is normal to the magnetopause and points outward, the $\hat{\mathbf{z}}$ direction points generally northward, and $\hat{\mathbf{y}}$ is directed from dawn to dusk. (These coordinates reduce to the usual GSE coordinate system in the case where the solar wind magnetic field points directly southward, in which case B_y is zero.) For simplicity, let the initial velocity be zero (appropriate, for example, to the stagnation point which exists at the subsolar point of the magnetopause). Thus the $\mathbf{j}_0\times\mathbf{B}_0$ force which is in the $\hat{\mathbf{x}}$ direction is balanced by a zero order pressure gradient. Let the resistivity $\eta(\mathbf{x})$ be introduced at t=0. (It is assumed that η is maximum at $\mathbf{x}$=0 and decreases monotonically in all directions.) The magnetic induction equation can be written:

$$\frac{\partial \mathbf{B}}{\partial t} = \nabla\times(\mathbf{v}\times\mathbf{B}) + \eta\nabla^2\mathbf{B} - \nabla\eta\times(\nabla\times\mathbf{B}) \tag{2}$$

where we have explicitly separated a term proportional to the gradient of the resistivity. This term has a normal component directed into the magnetosphere on the northern edge of the resistive region and out of the magnetosphere on the southern edge. The other term involving the resistivity leads to a diffusion of the B_z component. If we expand the magnetic field and the velocity in a Taylor series in time:

$$\mathbf{B}(\mathbf{x},t) = \mathbf{B}_0(\mathbf{x}) + t\,\mathbf{b}_1(\mathbf{x}) + (t^2/2)\mathbf{b}_2(\mathbf{x}) + \cdots \tag{3}$$

and similarly for $\mathbf{v}$, we can solve the induction equation and the Navier-Stokes equation order by order in time, resulting in

$$\mathbf{b}_1 = \eta\nabla^2\mathbf{B}_0 - \nabla\eta\times(\nabla\times\mathbf{B}_0) \tag{4}$$

$$\mathbf{v}_1 = 0 \tag{5}$$

where the first term in equation (4) gives the diffusion in B_z and the second term gives the normal component B_x. The combined magnetic perturbation rotates in a solenoidal fashion around the resistive region, with the result being to divert the current around the resistive obstacle (see Figure 2). These normal components, superposed on the initial sheared magnetic field, are of the correct sense to link the magnetosheath and magnetospheric magnetic fields.

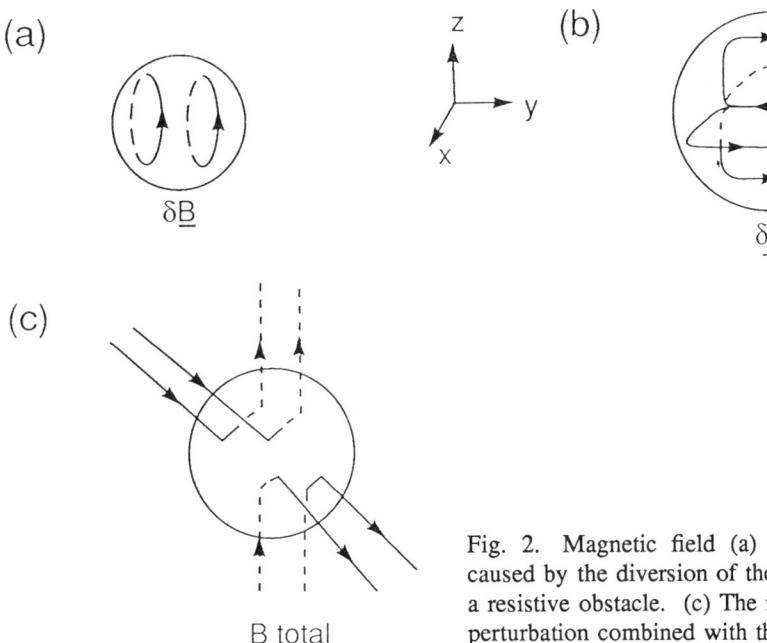

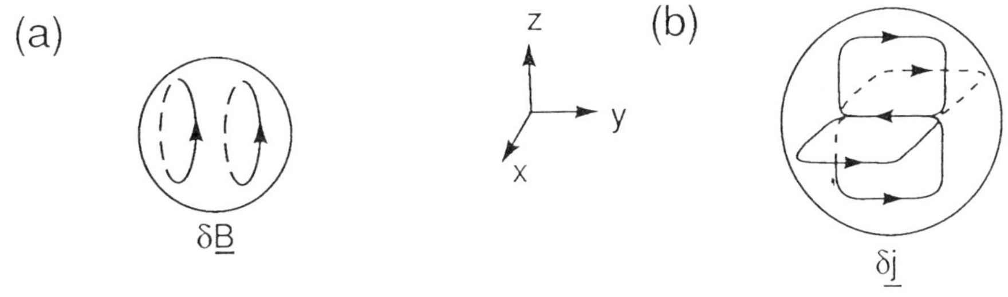

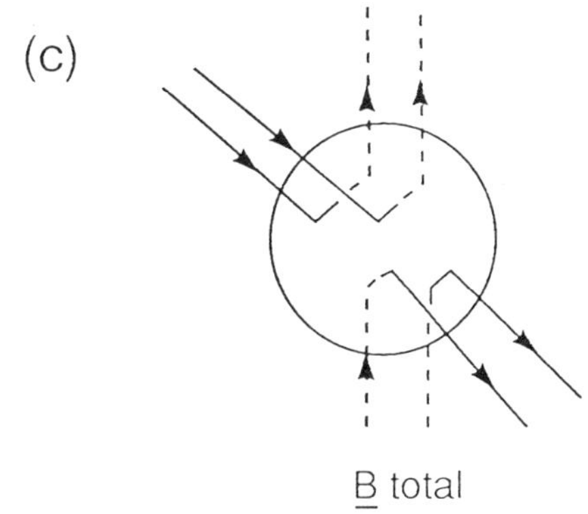

Fig. 2. Magnetic field (a) and current (b) perturbations caused by the diversion of the magnetopause current around a resistive obstacle. (c) The net effect of the magnetic field perturbation combined with the initial field.

At the next order, the perturbed current and magnetic field combine to give a **j**×**B** force which produces a second order velocity perturbation $\mathbf{v}_2$. Since this **j**×**B** force is nonuniform, the torques resulting from the curl of this force will induce vorticity in the fluid, which, by twisting up the nearly frozen-in magnetic field, will generate additional currents. In the case of positive B_y, it is instructive to consider the effect of these torques on an incoming magnetosheath field line (see Figure 3). In this situation, there is a northward component of **j**×**B** on the dawn side of the resistive region and a southward component on the dusk side. An incoming magnetosheath field line with southward B_z and duskward B_y will then be deflected toward the north on the dawn side of the field line and toward the south on the dusk side, implying a field-aligned current into the magnetosphere on the dawn and out of the magnetosphere on the dusk side, consistent with the sense of the Region 1 currents, the poleward ring of field-aligned currents in the auroral zone [e.g., Iijima and Potemra, 1977], as well as observed FTEs. For dawnward B_y, the sense of these torques is reversed, but so is the background field, so that the currents produced are in the same direction (Figure 3c,d). Note that we do not mean to imply that this process is the only source of the Region 1 currents; however, this mechanism does produce currents that are in the same direction as this current system. Additional torques on both magnetosheath and magnetospheric field lines in the equatorial plane tend to lead to currents parallel to the magnetic field on the dawn side and antiparallel at dusk, again consistent with the observed sense of these currents.

In the general three-dimensional geometry, these torques have components in all three directions, and will lead to a complicated pattern of flow, which can be treated by a statistical turbulence model which will be described in an accompanying paper [Song and Lysak, this meeting]. In addition, the current sheet will tend to be squeezed by the incoming flow outside of the resistive region, increasing the current density there and possibly initiating new regions of resistivity. The interactions between plasma accelerated in these various resistive sites will lead to additional vorticity in the plasma. In this turbulent, three-dimensional MHD fluid, turbulence theory and simulations predict that the small scale magnetic perturbations will coalesce to form large twisted magnetic flux tubes [Pouquet et al., 1976; Meneguzzi et al., 1981; Horiuchi and Sato, 1986]. This so-called inverse cascade of the magnetic helicity is associated with the tendency of parallel current filaments to attract one another, forming large scale magnetic islands. Since the initial evolution leads to magnetic perturbations normal to the magnetopause pointing both into and out of the magnetosphere, it is expected that two flux ropes with core magnetic fields in opposite directions will be generated. In addition, the solar wind often contains cross-helicity, with the velocity and magnetic field tending to align with each other [Belcher and Davis, 1971; Barnes, 1979]. In this situation, the turbulent evolution of the plasma leads to a state of dynamic alignment, in which the cross-helicity maximizes relative to the energy [Matthaeus and Montgomery, 1984; Ting et al., 1986; Pouquet et al., 1986]. Such an aligned state is observed in the outer regions of FTE flux tubes [Farrugia et al., 1988].

The nonlinear **j**×**B** forces play an important role in the subsequent evolution of the newly formed flux rope. The presence of azimuthal magnetic flux around the rope

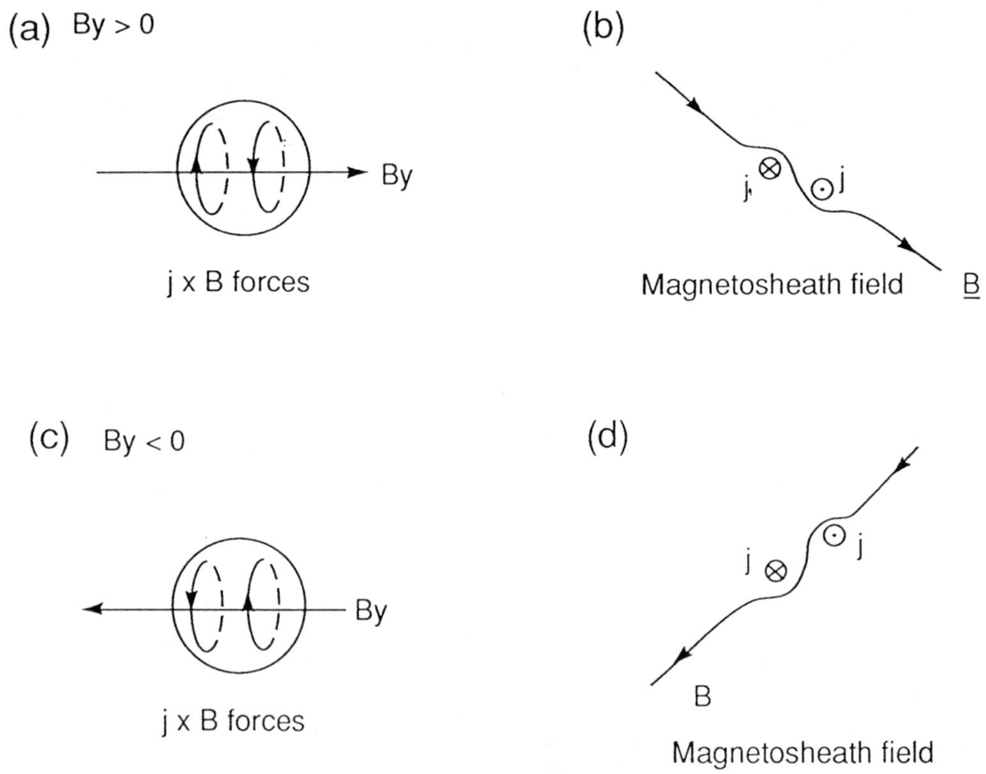

Fig. 3. (a) **j**×**B** forces produced by the perturbation currents of Figure 2 interacting with the background B_y field, for the case B_y>0. (b) Resultant perturbations of an incoming magnetosheath field line subjected to these torques. (c) Torques and (d) field line perturbations for B_y<0.

creates a current pinch geometry, which tends to reduce the size of the rope and enhance the axial magnetic field [Shi and Lee, this meeting]. This enhanced axial magnetic field must be supported by an azimuthal current, which balances the pinch and leads to a force-free state. In addition, the newly formed flux rope will in general be kinked, which leads to an additional **j**×**B** torque which can generate additional current [Sonnerup, 1987; Wright, 1987]. This process can be described as a current dynamo process in which current is generated along the magnetic field. In this process an inductive electric field directed along the axis of the flux rope is generated. This dynamo process will be described more completely in the companion paper [Song and Lysak, these proceedings].

Discussion

The preceding model suggests a scenario for reconnection at the magnetopause which requires only that a localized resistive region occur in the magnetopause current sheet. This resistivity may occur due to the excitation of some current driven instability, or may be associated with the onset of magnetic percolation due to the tearing mode instability [Galeev et al., 1986; Kuznetsova and Zelenyi, this meeting]. Whatever the mechanism, a region in which the resistivity is enhanced locally will induce magnetic and current perturbations which will lead to the formation of a normal magnetic field component which can link magnetosheath and magnetospheric fields. While a quantitative numerical model describing these processes is left for future work, it is clear from an analysis of the basic equations describing this system that a complicated and likely turbulent system will be produced. It should be emphasized that the three-dimensional nature of the reconnection region is an essential feature of this model, and leads to the conservation of magnetic helicity and the production of a current dynamo effect, both of which are absent in the classical two-dimensional reconnection picture.

Although developed primarily with the magnetopause application in mind, this model should be applicable to other reconnection sites. For example, in the geomagnetic tail, similar considerations will give rise to the formation of a flux rope geometry and the generation of field-aligned currents on newly reconnected flux tubes, both tubes which are connected to the earth and plasmoid flux tubes which are closed or extend into interplanetary space. This model would indicate that field-aligned currents on field lines which connect to the earth would be directed into the northern (southern) hemisphere when the dawn-dusk component of the tail magnetic is directed duskward (dawnward). A similar geometry for the newly reconnected tail field lines was drawn by Hughes and Sibeck [1987]. To the best of our knowledge, this correlation between the tail magnetic field and the current sense of the substorm currents has not been investigated observationally.

Questions and Answers

Birn: You said that magnetotail reconnection in a configuration that includes a net B_y component across the tail should generate a field-aligned current from one hemisphere to another. Actually, such a current is already present in the skewed initial configuration that exists before reconnection occurs. The simulations that I will discuss later in this meeting will, however, show that this current indeed increases through reconnection.

Lysak: This is certainly true. The current I refer to, however, is a localized line current such as may initiate substorm phenomena, such as the westward traveling surge.

Zelenyi: What are the dissipation effects taken into account in your model: ionospheric resistance or electron inertia (inertial conductivity)?

Lysak: *Electron inertial effects are in fact included in the calculation leading to the scale length given, but the coupling to the ionospheric conductivity must also be included. In addition, it is really the electron inertial length in the acceleration region, and not the ionosphere which enters the model. This length is the order of a few kilometers.*

Acknowledgements. This work has been supported by National Science Foundation grants ATM-8451168 and ATM-8810208.

References

Barnes, A., Hydromagnetic waves and turbulence in the solar wind, in *Solar System Plasma Physics*, E. N. Parker, C. F. Kennel, and L. J. Lanzerotti (eds.), North Holland, Amsterdam, 1979, p.249.

Belcher, J. W., and L. Davis, Jr., Large-amplitude Alfven waves in the interplanetary medium, *J. Geophys. Res., 76,* 3534, 1971.

Berger, M. A., Rigorous new limits on magnetic helicity dissipation in the solar corona, *Geophys. Astrophys. Fluid Dynamics, 30,* 79, 1984.

Berger, M. A., and G. B. Field, The topological properties of magnetic helicity, *J. Fluid Mech., 147,* 133, 1984.

Farrugia, C. J., R. P. Rijnbeek, M. A. Saunders, D. J. Southwood, D. J. Rodgers, M. F. Smith, C. P. Chaloner, D. S. Hall, P. J. Christiansen, and L. J. C. Woolliscroft, A multi-instrument study and flux transfer event structure, *J. Geophys. Res., 93,* 14465, 1988.

Galeev, A. A., M. M. Kuznetsova, and L. M. Zeleny, Magnetopause stability threshold for patchy reconnection, *Space Sci. Rev., 44,* 1, 1986.

Heikkila, W. J., Impulsive plasma transport through the magnetopause, *Geophys. Res. Lett., 9,* 159, 1982.

Horiuchi, R., and T. Sato, Self-organization and energy relaxation in a three-dimensional magnetohydromagnetic plasma, *Phys. Fluids, 29,* 1161, 1986.

Hughes, W. J., and D. G. Sibeck, On the 3-dimensional structure of plasmoids, *Geophys. Res. Lett., 14,* 636, 1987.

Iijima, T. and T. A. Potemra, The amplitude distribution of field-aligned currents at northern high latitudes observed by TRIAD, *J. Geophys. Res., 81,* 2165, 1976.

Lee, L. C., and Z. F. Fu, A theory of magnetic flux transfer at the earth's magnetopause, *Geophys. Res. Lett., 12,* 105, 1985.

Lemaire, J., Impulsive penetration of filamentary plasma elements into the magnetospheres of Earth and Jupiter, *Planet. Space Sci., 26,* 889, 1977.

Lundin, R., On the magnetospheric boundary layer and solar wind energy transfer into the magnetosphere, *Space Sci. Rev., 48,* 263, 1988.

Matthaeus, W. H., and D. Montgomery, Dynamic alignment and selective decay, in *Statistical Physics and Chaos in Fusion Plasmas,* C. W. Horton, and L. E. Reicel (eds.), Wiley, New York, 1984.

Meneguzzi, M., U. Frisch, and A. Pouquet, Helical and nonhelical turbulent dynamos, *Phys. Rev. Lett., 47,* 1060, 1981.

Moffatt, H. K., *Magnetic Field Generation in Electrically Conducting Fluids,* Cambridge University Press, Cambridge, 1978.

Pouquet, A., U. Frisch, and J. Leurat, Strong MHD helical turbulence and the nonlinear dynamo effect, *J. Fluid Mech., 77,* 321, 1976.

Pouquet, A., M. Meneguzzi, and U. Frisch, Growth of correlations in magnetohydrodynamic turbulence, *Phys. Rev. A, 33,* 4277, 1986.

Quest, K. B., and F. V. Coroniti, Tearing at the dayside magnetopause, *J. Geophys. Res., 86,* 3289, 1981.

Scholer, M., Magnetic flux transfer at the magnetopause based on single X-line bursty reconnection, *Geophys. Res. Lett., 15,* 291, 1988.

Sibeck, D. G., W. Baumjohann, R. C. Elphic, D. H. Fairfield, J. F. Fennell, W. B. Gail, L. J. Lanzerotti, R. E. Lopez, H. Luehr, A. T. Y. Lui, C. G. Maclennan, R. W. McEntire, T. A. Potemra, T. J. Rosenberg, and K. Takahashi, The magnetospheric response to 8-minute period strong-amplitude upstream pressure variations, *J. Geophys. Res., 94,* 2505, 1989.

Song, Y. and R. L. Lysak, Evaluation of twist helicity in FTE flux tubes, *J. Geophys. Res., 94,* 5273, 1989.

Sonnerup, B. U. O., On the stress balance in flux transfer events, *J. Geophys. Res., 92,* 8613, 1987.

Southwood, D. J., C. J. Farrugia, and M. A. Saunders, What are flux transfer events? *Planet. Space Sci., 36,* 503, 1988.

Ting, A. C., W. H. Matthaeus, and D. Montgomery, Turbulent relaxation processes in magnetohydrodynamics, *Phys. Fluids, 29,* 3261, 1986.

Wright, A. N., The evolution of an isolated reconnected flux tube, *Planet. Space Sci., 35,* 813, 1987.

Wright, A. N., and M. A. Berger, The effect of reconnection upon the linkage and interior structure of magnetic flux tubes, *J. Geophys. Res., 94,* 1295, 1989.

THE CURRENT DYNAMO EFFECT AND ITS STATISTICAL DESCRIPTION DURING 3-D TIME-DEPENDENT RECONNECTION

Yan Song and Robert L. Lysak

School of Physics & Astronomy, University of Minnesota, Minneapolis, MN 55455

Abstract. Time-dependent magnetic reconnection is not only a dissipation process ($\underline{J} \cdot \underline{E} > 0$), but also a dynamo process ($\underline{J} \cdot \underline{E} < 0$). The current dynamo effect during (and after) 3-D magnetic reconnection corresponds to the formation of the field-aligned current for the geomagnetic flux tubes, which can be described as result of injection of twist helicity. The self-organization process of the local turbulent magnetofluids is used to explain the initial formation of the twisted flux tubes (i.e. the formation of the holes at the magntopause and the diversion and disruption of the magnetopause current). The theory of the conversion and conservation of the magnetic helicity during reconnection is used to estimate the average dynamo effect. The torques which come from the volume integral of the curl of Lorentz force and/or the drag force by the magnetosheath flow can also provide the source of twist helicity during (and after) reconnection.

The average input rate of the electromagnetic energy during reconnection is formulated in terms of the input rate of the total twist helicity or an equivalent inductance. For a turbulent reconnection region, the induced electric field is formulated in terms of the α and β effect, where α corresponds to the current dynamo effect and β provides the MHD turbulent diffusivity required for reconnection.

1 Introduction

Magnetic reconnection is an intrinsic 3-D time-dependent physical process, which can be considered to be a self-organized evolution process of a non-equilibrium dissipative system. The 2-D reconnection model with an extended separatrix is structurally unstable [Schindler et al., 1988]. The 2-D reconnection theory, which is based on a structurally unstable state, loses some important physical content, such as the conversion of the magnetic helicity and the dynamo effect of the field-aligned current during magnetic reconnection. It is not surprising that the physical phenomena predicted by the over-simplified 2-D steady state model such as energized particles, high speed plasma jets and a tangential electric field with $\underline{J} \cdot \underline{E} > 0$ are not frequently observed [e.g., Heikkila, 1975; Paschmann et al., 1976; Mozer et al., 1979].

Observations of electric fields suggest that the magnetofluids at the magnetopause and boundary layer are directly involved in a dynamo effect [e.g., Lundin, 1988]. The possible dynamo effect during reconnection was first suggested by Heikkila [1975] to be the result of the irregular entry of the plasma clouds, so called impulsive penetration, which is independent of magnetic field merging [Lemaire and Roth, 1978; Heikkila, 1982]. However, the large field-aligned voltages along the open field lines on either side of the magnetopause [Heikkila, 1984, figure 1b] are not observed in the electric field data [e.g., Mozer et al., 1979].

Theoretical effort has been made on 3-D reconnection [e.g., Schindler et al., 1989; Schindler and Hesse, 1989; Greene, 1988] and the dynamic evolution of reconnection [e.g., Matthaeus and Lamkin, 1986; Horiuchi and Sato, 1985]. Lee and Fu [1985] first proposed a multiple X-line reconnection model. In fact, reconnection may often occur in the multiple sites in an irregular time-dependent and localized form [e.g., Matthaeus and Lamkin, 1986]. It was suggested [Song and Lysak, 1989c] that the turbulent magnetofluid in a reconnection site can experience a self-organization process toward a force- and torque-free state, which may play an important role on the current diversion and disruption during reconnection, i.e., FTE formation. However, the relationship between reconnection and the generation of the field-aligned current has not been considered. The theoretical basis of the formation of flux ropes and the important microscopic dynamics of the

Geophysical Monograph 58

current dynamo of reconnection was discussed in the previous paper [Lysak and Song, 1989].

The purpose of the present paper is to describe the mechanism of the current dynamo effect of magnetic reconnection in terms of the injection of twist helicity (section 2). A statistical description for the energy transfer during time-dependent, multiple-site reconnection is given in section 3. The meaning of the statistical description is two-fold. First of all, the average input rate of the electromagnetic energy is formulated in terms of the average area of the reconnection region and an average reconnection rate. Secondly, the turbulent statistically averaged induced electric field is given in terms of the α and β effect.

2 The Dynamo Effect of 3-D Magnetic Reconnection

In 2-D steady state reconnection, the magnetic energy can only be annihilated and is in general converted into bulk kinetic energy and heat ($\underline{J}\cdot\underline{E}>0$) [e.g., Cowley, 1985] (Figure 1a). In 3-D time-dependent reconnection, when magnetic flux approaches the reconnection region, instead of a complete annihilation, part of the magnetic flux can be converted into the azimuthal component of a twisted or writhed flux tube, which corresponds to the generation of field-aligned currents (Figure 1b).

In general, twisting a cylindrical magnetic flux tube is a dynamo effect. The dynamo effect comes mainly from the addition of energy of azimuthal magnetic field, since the change of the magnetic energy of the axial magnetic field is usually small. The dynamo effect can be expressed as $\int\underline{J}\cdot\underline{E}\ dV = -(\partial/\partial t)\int(B_\theta^2/8\pi)dV<0$. The equation for energy conservation in the reconnection region is

$$-\frac{c}{4\pi}\int(\underline{E}\times\underline{B})\cdot d\underline{S}=\int\left(\frac{J^2}{\sigma}+\frac{1}{C}(\underline{J}\times\underline{B})\cdot\underline{V}\right)dV+\Sigma\left<\frac{\partial}{\partial t}\left(\int\frac{B_\theta^2}{8\pi}dV\right)\right> \quad (1)$$

where B_θ represents the azimuthal magnetic field and Σ here is a sum over the multiple site reconnection regions. The Poynting flux of electromagnetic energy flowing into the reconnection region gives not only the energy dissipation, which includes plasma heating and the kinetic energy of plasma flow, but also the time averaged increase of the energy of the azimuthal magnetic field summed over the whole magnetic reconnection region, i.e., a dynamo effect.

In the cylindrical coordinates, the increase of the twist helicity K_T with the adding of azimuthal magnetic flux is

$$\frac{dK_T}{dt}=\frac{d}{dt}\left(\int 2\Phi_\theta(r)\ d\Phi_z(r)\right) \quad (2)$$

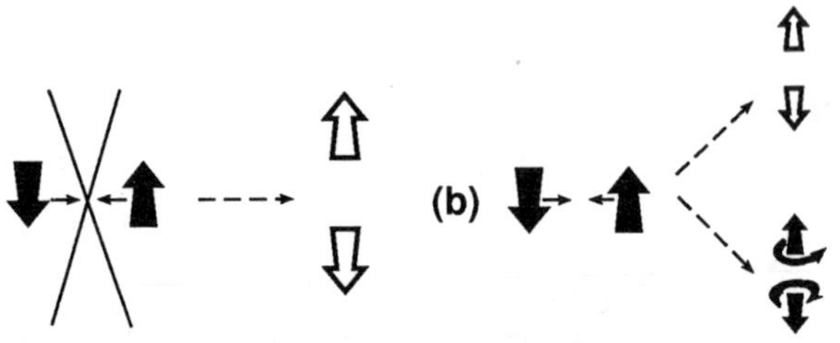

Fig. 1. A schematic diagram for energy conversion during reconnection. In the 2-D model (1a), when the magnetic flux (solid arrow) with anti-parallel magnetic fields approaches the X-line, magnetic energy is annihilated and the plasma flow leaves the reconnection region with increased thermal and non-thermal energy (open arrow). In the 3-D case (1b), part of the magnetic flux becomes the azimuthal flux of the flux rope.

where Φ_θ is the azimuthal magnetic flux and Φ_z is the axial magnetic flux through a circle of radius r. For a uniformly twisted flux tube with N turns and length L_ℓ, the twist helicity is $K_T=\pm N\Phi^2$ [Berger and Field, 1984] and the magnetic energy is $W_B=N^2\Phi_z^2/(4L_\ell)$. Then, we have $|(dW_B/dt)/(dK_T/dt)| = N/(2L_\ell)$.

If we consider a portion of a flux tube confined between two planes on which a vortical motion is imposed, the derivative of the twist helicity $K_T(t)$ [cf. Berger and Field, 1984] is

$$\frac{dK_T}{dt}=-\frac{dK_K}{dt}-\frac{dK_{ex}}{dt}+2c\int_S(\underline{A}_\theta\times\underline{E})\cdot\hat{\underline{n}}\ dS-2c\int\underline{E}\cdot\underline{B}dV \quad (3)$$

where $\underline{A}_\theta$ is the vector potential and $\nabla\cdot\underline{A}_\theta=0$. The first two terms of the RHS are the changes of kink helicity $K_K(t)$ and the mutual helicity $K_{ex}(t)$ respectively. The third term on the RHS is the flow of relative helicity across the boundary of the open ends of the flux tubes (S). The last term on the RHS gives the internal helicity dissipation, i.e., $2c\int\underline{E}\cdot\underline{B}\ dV = 2(c/\sigma(t))\int\underline{J}\cdot\underline{B}\ dV$, where $\sigma(t)$ is time-dependent conductivity. When twist propagates as an Alfven wave and arrives at the ionosphere, $\sigma(t)$ is nearly equal to the ionospheric conductivity.

An evaluation of twist helicity by releasing the internal helicity due to kinking of the flux tube ($-dK_K/dt$) and by transforming the external (mutual) helicity due to magnetic field linkage ($-dK_{ex}/dt$) has been given [Song and Lysak, 1989a; Wright and Berger, 1989] and used to estimate the generation of twist for previous flux transfer event models [Song and Lysak, 1989a]. In the evaluation, helicity conservation during reconnection is assumed [Berger, 1982, 1984], i.e., the dissipation and the flow of relative helicity through the open boundary are assumed to be zero. It was noted [Song and Lysak, 1989c] that if reconnection occurs slowly in a narrow reconnection region, for which the time for the dissipation of helicity is shorter than the time for reconnection, the kinked flux tubes (mentioned above by Wright, 1987) are very thin, and most of the magnetic energy within the tubes is dissipated before twisting. Even if current filaments on small twisted tubes are formed, the magnetic energy can dissipate in a short time. In this situation, reconnection can be approximately described by a 2-D model. It has been suggested [Song and Lysak, 1987; 1989c; Lysak and Song, 1989] that large twisted flux tubes can be formed by a self-organization process (Figure 2). These large flux tubes have longer dissipation times and thus can survive dissipation during the reconnection process.

The dynamo effect caused by releasing kink helicity during reconnection can be described as an "overlap-reconnect-twist" process [Song and Lysak, 1989b, figure 4]. The flow of relative helicity across S can be obtained from the vertical motion of the plasma flow, i.e., $2c\int(\underline{A}_\theta\times\underline{E})\cdot\hat{\underline{n}}dS = -(1/2\pi)\int\omega_z(r)\Phi(r)d\Phi(r)$ (in the paper by Berger and Field [1984], ω_z is the angular velocity, ω_z here represents the vorticity). The equation of the conservation of vorticity $\underline{\omega}$, i.e.,

$$\frac{\partial\underline{\omega}}{\partial t} = \nabla\times(\underline{u}\times\underline{\omega}) + \frac{1}{c\rho}\nabla\times(\underline{J}\times\underline{B}) + \nabla\times\underline{F}_D + \nu\nabla^2\underline{\omega} \quad (4)$$

shows that the torques come from the volume integral of $(1/c\rho)\nabla\times(\underline{J}\times\underline{B})$ and $\nabla\times\underline{F}_D$ terms, where ν is the viscosity, ρ is the plasma density and $\underline{F}_D$ is the external force. In cylindrical coordinates, the axial vorticity ω_z can be produced by the azimuthal component of the Lorentz force and the external force $F_\theta = 1/c(\underline{J}\times\underline{B})_\theta + F_{D\theta}$, i.e., $\rho(d\omega_z/dt) = (1/r)(\partial/\partial r(rF_\theta))$. Then, the non-zero volume integral of the $(1/c\rho)\nabla\times(\underline{J}\times\underline{B})$ and $\nabla\times\underline{F}_D$ can provide twisting for a newly formed flux tube (or sheet). The mechanism of twisting of a kinked flux tube was discussed as a result of the non-zero volume integral of $\nabla\times(\underline{J}\times\underline{B})$ [Wright, 1987]. Recently, Saunders [1989] showed that the twisting can be caused by the tilting of the magnetic field when the flux tube (or sheet) moves from an east/west orientation to an antisunward orientation.

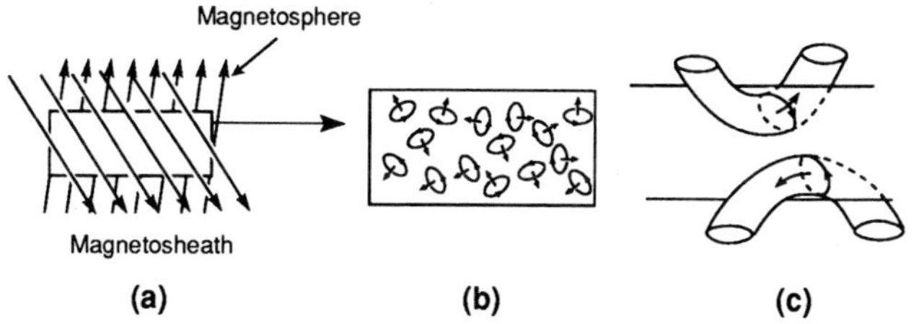

Fig. 2. The formation of the large twisted flux tubes.
(a) before reconnection.
(b) the formation of the small twisted tubes by the torques of $\nabla\times(\underline{J}\times\underline{B})$.
(c) condensation to two large twisted tubes by the self-organization process.

In summary, there are two kind of source for the injection of twist helicity, i.e., (1) releasing kink and mutual helicity through reconnection; (2) imposing the external torques on the flux tube. A quantitative calculation of the twist helicity change is based on a careful analysis of the topological change of the flux tubes during and after reconnection.

3 Statistical Description

Average Input of Electromagnetic Energy (the Dynamo Effect)

From equation (1), the free energy available for dissipation (W_f) per unit time can be determined by the difference of the total input of the magnetic energy (W_{in}) and the magnetic energy stored in the final state summed over the multiple reconnection sites (ΣW_b) per unit time, i.e., $W_f/\Delta t = W_{in}/\Delta t - \Sigma W_b/\Delta t$. For a slow and steady reconnection in a smooth magnetic structure, $\Sigma W_b/\Delta t \sim 0$, implying $W_f/\Delta t \sim W_{in}/\Delta t$, which basically gives a quasi-2D description. If the time scale for helicity decay is much greater than the Alfven wave travel time, ΣW_b term represents the input of electromagnetic energy into the magnetosphere and restricts the amount of energy dissipated.

The average dynamo effect can be described in terms of the average rate of increase of twist helicity, which can be estimated from the area, the distribution of reconnection patches and the reconnection rate. For M_1 uniform twisted tubes with length ℓ_1 and flux ϕ_1 (i.e. FTE flux tubes), and M_2 uniform twisted tubes with length ℓ_2 and flux ϕ_2 (i.e., current filaments of mini-FTEs or impulsive penetration), the total magnetic energy $W_B = \Sigma W_b$ is

$$W_B = (N^2/4\ell)(M_1\phi_1^2 + M_2\phi_2^2) = (N/4\ell)\,K_T \quad (5)$$

where $\ell_1=\ell_2=\ell$ and N is the number of twists. The total twist helicity is $K_T=N(M_1\phi_1^2 + M_2\phi_2^2)$. If the time for the formation of these twisted tubes is Δt, the average energy input rate is $W_B/\Delta t$. An alternative description is that energy is stored in a set of current filaments, in which case the energy can be expressed in terms of inductance:

$$W_B=(1/2c^2)\sum_i L_i I_i^2 + (1/c^2)\sum_{i\neq k} M_{ik} I_i I_k \qquad (6)$$

where L_i is the self-inductance of the i-th current filament and M_{ik} are the mutual inductances. The equation for energy conservation during reconnection becomes $IV=I_1^2R_t + \underline{F}\cdot\underline{v} + (L_t/2c^2)(dI_\ell^2/dt)$, where IV is the energy input rate and $I_1^2R_t$ and $\underline{F}\cdot\underline{v}$ give plasma heating and the kinetic energy of plasma flow, respectively. R_t and L_t are the equivalent resistivity and inductance of the whole circuit, respectively. Time constant for releasing of twisting can be determined by L(t)/R(t). The last term on the r.h.s. of the energy conservation equation shows the current generation from the dynamo effect.

A mean-field description of turbulence induced electric field

For a high Reynolds number case, many twisted flux tubes are formed in a turbulent reconnection patch. The large twisted tube can be formed by an inverse cascade process and the turbulent diffusivity can be provided by the interaction of the small twisted tubes. The magnetic field, vector potential and velocity field are split into the mean and fluctuating part, i.e., $\underline{B}=\underline{B}_0+\delta\underline{B}$, $\underline{A}=\underline{A}_0+\delta\underline{A}$ and $\underline{u}=\delta\underline{u}$. The Ohm's law for the mean current $\underline{J}_0$ is $\underline{J}_0=\sigma(\underline{E}_0+1/c\langle\delta\underline{u}\times\delta\underline{B}\rangle)$. The turbulent mean electric field, i.e., $\langle\underline{E}_T\rangle=(1/c)\langle\delta\underline{u}\times\delta\underline{B}\rangle$, has been mentioned by Matthaeus and Lamkin [1986]. However, the mechanism of the generation of $\langle\underline{E}_T\rangle$ is unclear.

Three time scales are set: (a) T_{tear}, the formation time for each twisted flux tube; (b) T_0, the time scale for $\delta\underline{u}$ and $\delta\underline{B}$, (c) T_{diffu}, the time scale for large scale magnetic field diffusion. These three time scales correspond to small, medium and large spatial scale phenomena, respectively. The condition $T_{tear}\ll T_0\ll T_{diffu}$ is assumed, which gives the time period T_0 during which the interaction among the small tubes occurs. A "frozen in" condition can be used to study the interaction of small twisted tubes, since $T_0\ll T_{diffu}$. Also, small tubes can penetrate magnetic field lines and coalesce, since $T_0\gg T_{tear}$.

The injection of twist helicity for each small tube is $d\delta K_T/dt \sim 2\int(\partial\delta\Phi_{\theta'}/\partial t)d\delta\Phi_{\ell'}$, where ℓ' and θ' are the direction along the magnetic flux tube and the poloidal direction for each small tube. $\delta\Phi_{\theta'}$ and $\delta\Phi_{\ell'}$ are the magnetic flux in the θ' and ℓ' directions for each small tube. The rate of change of the total mean helicity can be expressed by

$$\left\langle\frac{dK_T}{dt}\right\rangle = -2c\langle\delta E_{\ell'}\cdot\delta B_{\ell'}\rangle V_\Delta \qquad (7)$$

where V_Δ is the mean volume of the reconnection region, which is a product of the mean length and the mean area of the reconnection region ($V_\Delta=\langle\ell\rangle\langle S\rangle$). During time scale T_0, the "frozen in" condition is valid and the fluctuation of the electric field by twisting is equivalent to the fluctuation of the velocity field, i.e., $\delta\underline{u}=c(\delta\underline{E}\times\underline{B}_0/B_0^2)$. The twist helicity injection by the torques of the Lorentz force provides the source of velocity fluctuation on a small scale [also see Lysak and Song 1989].

The turbulent average electric field $\langle\underline{E}_T\rangle$ can be determined by the average change of the total helicity ΔK_T during time Δt. Considering helicity conservation, ΔK_T can be estimated from the released kink helicity.

$$\langle\underline{E}_T\rangle \sim -\frac{\Delta K_T}{2cB_0V_\Delta\Delta t} \qquad (8)$$

where V_Δ is the volume of the reconnection region. From turbulent dynamo theory (e.g., Moffatt 1978), the turbulent electric field can be expressed in terms of the α and β effect, i.e., $\langle\underline{E}_T\rangle=\alpha\underline{B}_0-\beta(\nabla\times\underline{B}_0)$. α represents the field-aligned current dynamo effect and the average α is $\langle\alpha\rangle \sim -\Delta K_T/(2cB_0^2V_\Delta\Delta t)$. The Ohm's law becomes $\underline{J}_0=\sigma^*(\underline{E}_0+\alpha\underline{B}_0)$, where the anomalous conductivity is $\sigma^*\sim c^2/(4\pi\beta)$. The anomalous magnetic diffusivity β can provide the required diffusivity for the coalesence of the small tubes. β can be determined by the energy and helicity spectrum.

4 Discussion

The current dynamo effect is a natural result from 3-D reconnection. The magnetic energy stored in the solar wind-magnetosphere configuration can directly enter the magnetosphere in the form of the electric currents through many time-dependent and moveable reconnection patches.

The effect of the turbulence on the magnetic reconnection is (1) to create the condition for self-organization (α effect) and (2) to provide the turbulent diffusivity (β effect) required by the coalesence. It is noticed that the total helicity and net current for the whole reconnection region can not be produced by the turbulence of the magnetofluids in the reconnection region. However, the turbulence creates the condition for the conversion of the magnetic helicity (the formation of the twisted magnetic flux tubes) and the diversion and disruption of the current. The degree of openness of the magnetosphere depends on the reconnection rate and the level of the fluctuations, irregularities and non-linearity.

The concept of injection and transport of helicity is useful for considering the current dynamo effect during and after reconnection. The electric field during reconnection consists of the convection electric field and the time-dependent induced electric field.

Questions and Answers

Berger: In the overlap-reconnection-twist dynamo, magnetic energy is converted into kinetic energy (and vorticity), then back into magnetic energy of a different form. I am guessing that the flux tubes shrink, i.e., their length decreases as they drive the vorticity. This may be one way of looking at the energetics - long untwisted tubes are converted into shorter twisted tubes. The energy of the axial magnetic field decreases because of the reduced length; some of this energy goes into creating the azimuthal field inside the twisted tubes.

Song and Lysak: The mechanism you suggest represents a part of the energy conversion. On the other hand, the ultimate energy for this process comes from the work done by the solar wind in overcoming the J × B force which would keep geomagnetic and interplanetary field lines apart.

Acknowledgments. This work has been supported by National Science Foundation Grants No. ATM-8451168, ATM-8508949, and ATM-8810208.

References

Berger, M. A., Rapid reconnection and the conservation of magnetic helicity, (Abstract), Bull. Am. Astr. Soc., 14, 978, 1982.

Berger, M. A., and G. B. Field, The topological properties of magnetic helicity, J. Fluid Mech., 147, 133, 1984.

Berger, M. A., Rigorous new limits on magnetic helicity dissipation in the solar corona, Geophys. Astrophys. Fluid Dynamics, 30, 79, 1984.

Cowley, S. W. H., Magnetic reconnection, in Solar System Magnetic Fields, ed. by E. R. Priest, D. Reidel Publishing Co., 1985.

Greene, J. M., Geometrical properties of three-dimensional reconnecting magnetic field with nulls, J. Geophys. Res., 8, 8583, 1988.

Heikkila, W. J., Is there an electrostatic field tangential to the dayside magnetopause and neutral line? Geophys. Res. Lett. 2, 954, 1975.

Heikkila, W. J., Impulsive plasma transport through the magnetopause, Geophys. Res. Lett., 9, 159, 1982.

Heikkila, W. J., The electromagnetic field for an open magnetosphere, in Magnetic Reconnection In Space and Laboratory Plasmas, Ed. E. W. Hones, Jr., AGU, Washington, D. C., 92, 1984.

Horiuchi, R., and T. Sato, Three-dimensional self-organization of a magnetohydrodynamic plasma, Phys. Rev. Lett., 55, 211, 1985.

Lee, L. C., and Z. F. Fu, A theory of magnetic flux transfer at the Earth's magnetopause, Geophys. Res. Lett., 12, 105, 1985.

Lemaire, J., and M. Roth, Penetrating of solar wind plasma elements into the magnetosphere, J. Atmos. Terr. Phys., 40, 331, 1978.

Lundin, R., On the magnetospheric boundary layer and solar wind energy transfer into the magnetosphere, Space Sci. Rev., 48, 263, 1988.

Lysak, R. L., and Y. Song, Formation of flux ropes by turbulent reconnection, in this proceedings, 1989.

Matthaeus, W. H., and S. L. Lamkin, Turbulent magnetic reconnection, Phys. Fluids, 29, 2513, 1986.

Moffatt, H. K., Magnetic Field Generation in Electrically Conducting Fluids, Cambridge Uni. Press, 1978.

Mozer, F. S., et al., Direct observations of a tangential electric field component at the magnetopause, Geophys. Res. Lett., 6, 305, 1979.

Paschmann, G., G. Haerendel, N. Sckopke, and H. Rosenbauer, Plasma and magnetic field characteristics of the distant polar cusp near local noon: the entry layer, J. Geophys. Res., 81, 2883, 1976.

Saunders, M. A., Origin of the cusp Birkeland currents, Geophys. Res. Lett., 16, 151, 1989.

Schindler, K., M. Hesse and J. Birn, General magnetic reconnection, parallel electric fields and helicity, J. Geophys. Res., 93, 5547, 1988.

Schindler, K., and M. Hesse, General magnetic reconnection, parallel electric fields and helicity, J. Geophys. Res., 93, 5547, 1988.

Song, Y., and R. L. Lysak, Fine structure of FTE flux tube. A 2-D spectral MHD turbulent simulation, (Abstract), Book of Abstract, The Third International School for Space Simulation. ISEE-3, Beaulieau, France, June 22-27, 1987.

Song, Y., and R. L. Lysak, Evaluation of twist helicity of FTE flux tubes, J. Geophys. Res., 94, 5273, 1989(a).

Song, Y., and R. L. Lysak, Flux transfer events-the result of three dimensional time-dependent magnetic reconnection, Turbulence and Nonlinear Dynamics in MHD Flows, M. Meneguzzi, A. Pouquet and P. L. Sulem (eds.), Elsevier Science Publishing B. V. (North-Holland), 75, 1989(b).

Song, Y., and R. L. Lysak, Current dynamo effect of 3-D time-dependent reconnection in the dayside magnetopause, in press, Geophys. Res. Lett., 1989(c).

Wright, A. N., The evolution of an isolated reconnected flux tube, Planet. Space Sci., 35, 813, 1987.

Wright, A. N., and M. A. Berger, The effect of reconnection upon the linkage and interior structure of magnetic flux tubes, J. Geophys. Res., 94, 1295, 1989.

FIELD-ALIGNED CURRENTS IN THE EARTH'S MAGNETOSPHERE

Gerhard Haerendel

Max-Planck-Institut für Physik und Astrophysik
Institut für extraterrestrische Physik, 8046 Garching, FRG

Abstract. Field-aligned currents are generated when momentum is transferred from one plasma regime to another via magnetic shear stresses. Therefore, they play a key role in establishing the force balance between the hot magnetospheric plasma and the cold, dense plasma of the ionosphere. This interaction has many aspects. Six of them are addressed in this paper, the *source*, *propagation*, *reflection* and *closure* of field-aligned currents and the possibility of finite field-aligned voltages for narrow transverse scales (*kinetic corrections*) and/or high current densities, which in some situations may lead to current *instability* and anomalous resistivity. The paper emphasizes the dynamical aspects of field-aligned currents, i.e. their relation to the overall momentum balance as well as propagation and collisionless dissipation effects. In the first part, the development of concepts of magnetospheric current systems based on the existence of field-aligned currents is briefly described.

Introduction

The dynamical role of field-aligned currents is to transport transverse momentum along magnetic lines of force. Practically all magnetized stars are in one way or another interacting with the plasma environment via their magnetic field in the sense that momentum is being interchanged between the star and its atmosphere, on the one hand, and an external plasma, on the other. Hence field-aligned currents should exist in practically all stellar environments in which an appreciable magnetic field is present.

The Earth's magnetosphere presents a most convenient opportunity to study the role of field-aligned currents in a strong magnetic field, i.e. in an environment in which the plasma beta ($\beta = 8\pi p/B^2$, p = pressure, B = magnetic field strength) is smaller or equal to one in the outer regions (plasma sheet, boundary layers) and much smaller than one in most of the inner regions (ring current). Field-aligned currents are most noticeably present at high magnetic latitudes where they transport magnetic shear stresses from the magnetospheric boundary layers and the tail towards the ionosphere. Less intense currents also exist where the reverse happens, i.e. where atmospheric (tidal) forces are communicated to the higher altitude plasma (e.g., interaction of equatorial E and F regions).

The aim of this paper is neither to give an overview of the existing body of knowledge of magnetospheric/ionospheric field-aligned currents, nor to present a history of the recognition of their importance. This has been done at many previous occasions. The most comprehensive recent review of both aspects can be found in the Geophysical Monograph 28 of the American Geophysical Union [Potemra, 1984]. My aim is rather to discuss principal physical aspects of the existence of field-aligned currents. This will be done in the second and main part of this paper. The first part is a very brief outline of the development of basic concepts of global current systems, of which field-aligned currents only constitute a part. Nowhere in this paper will I make an attempt of being nearly complete in acknowledging previous work. The examples and references chosen to support or illustrate the discussed topics reflect personal preferences and the desire to keep the paper short.

1. Concepts of Magnetospheric Current Systems

Almost six decades passed following the initial postulate of Kristian Birkeland [1908] that field-aligned currents feed horizontal ionospheric currents, before clear-cut experimental evidence of their existence was begun to be collected. Much of this period was filled with a controversy between the schools of Birkeland and Alfvén, on the one hand, and that of Chapman, on the other, about existence, significance and need for such currents. Their proponents were mostly concerned with the question of current closure and the

Geophysical Monograph 58

relation between current and auroral particle precipitation. Féjer [1961] was one of the first to consider the dynamical role of the currents. In 1964, Boström [1964] published his famous paper on two competing configurations of the magnetospheric current system connected to the auroral electrojet (AEJ). Figure 1 is taken from this work. With minor adjustments these two configurations have not only remained valid until today, but also essentially comprise all more complex suggestions made later on. In the first configuration (Type I), the auroral (horizontal) electrojet is thought to be mainly fed by a downward current on its eastern end and upward current on its western end. Since upward currents can to a large extent be carried by precipitating hot electrons, the western end tends to be optically more active. A great spiral feature, called the "Westward Traveling Surge" (WTS) [Akasofu, 1968], is connected with it [see also Baumjohann et al., 1981]. The second configuration (Type II), with sheet currents on the poleward and equatorward sides of the AEJ, closes via Pedersen currents in the ionosphere. In addition, it drives a Hall current which may largely close within the ionosphere and which gives rise to the magnetic perturbations measured on the ground. Superpositions of both systems may exist. The electrojet model of Baumjohann et al. [1981] is such an example, with the refinement that the sense of the sheet currents is opposite to that in Boström's Type II. The implications of field-aligned current closure will be addressed in Section 2.4.

After Boström's [1964] work the next major step in development of concepts of global current systems was made by McPherron et al. [1973]. The intervening years had brought the first identification of $j_{\parallel}$ at low altitudes by Zmuda et al. [1966], by Coleman and McPherron [1970] at geosynchronous altitudes and by Haerendel et al. [1971] in the tail. The proposal of McPherron et al. [1973], as shown in Figure 2a, was that the auroral electrojet during the break-up phase of a substorm is connected via $j_{\parallel}$ of the Boström Type I to a westward current at the inner edge of the tail due to a disruption of unspecified nature of the latter current. Some authors, like Akasofu et al. [1983], have pointed out the equivalence of McPherron et al.'s current system and Boström's Type I, if one subtracts the westward tail current from the first model. However, this equivalence is just mathematical. In terms of driving forces they are distinctly different. In Boström's model, the generator is located *inside* the current wedge; in McPherron et al.'s model, it must be located on the *outside*. The latter model has appealed greatly to the magnetospheric community, and many

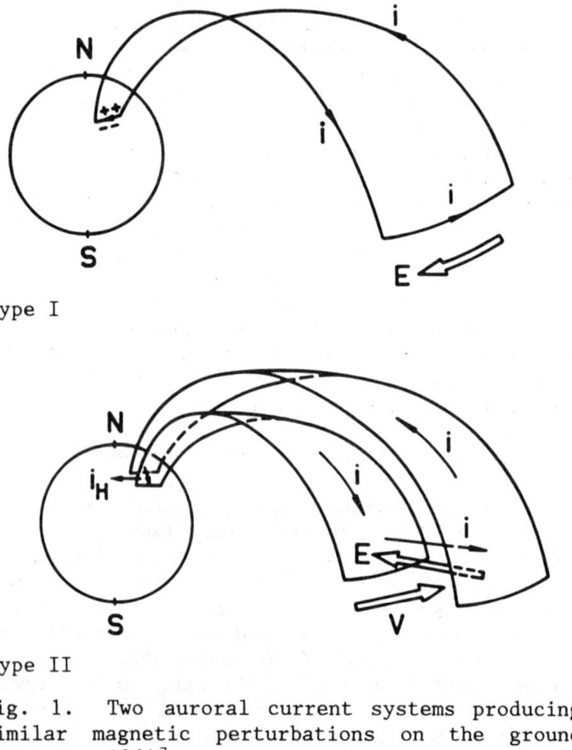

Fig. 1. Two auroral current systems producing similar magnetic perturbations on the ground [Boström, 1964].

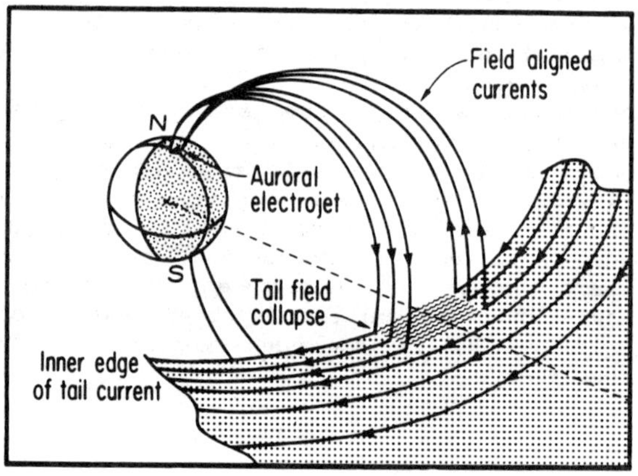

(a)

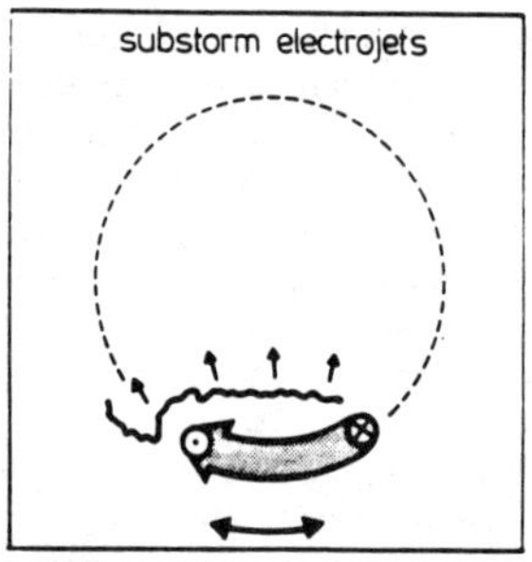

(b)

Fig. 2. (a) Disruption of part of the cross-tail current and diversion via field-aligned currents through the ionosphere [McPherron et al., 1973]. (b) Blow-up of the ionospheric part of the substorm current system during the break-up phase showing the growing E-W extent and poleward expansion (from [Baumjohann, 1983]).

interpretations of substorm related data were based on it, although no convincing model of the disruption process has ever been presented. Figure 2b, taken from Baumjohann [1983], has been added in order to highlight the low altitude part of this current system. The figure tries to express the temporal development of the closure current, its growing E-W extent and poleward expansion.

The third major step in assessing the role of field-aligned currents in the magnetosphere came from the derivation of a quasi-permanent global pattern of $j_{\parallel}$ from the measurements of the TRIAD satellite by Iijima and Potemra [1976]. Figure 3 shows the regions of inward and outward directed currents above the high latitude ionosphere. One can distinguish between a poleward located Region 1 system and an equatorward located Region 2 system. They differ by overall strength and response to variations of the interplanetary conditions. The pattern of $j_{\parallel}$ agrees very well with theoretical conclusions drawn by Vasyliunas [1970, 1972] and model calculations by the Rice University group [Wolf, 1972; Harel et al., 1981], who considered the dynamics of the hot ring current plasma under a prescribed electric potential across the polar cap as boundary condition and a variable ionospheric conductivity, which is enhanced along the auroral oval by particle precipitation. It clearly appears that the ionosphere occupied by the Region 1 and 2 systems of $j_{\parallel}$ is the preferred area where energy

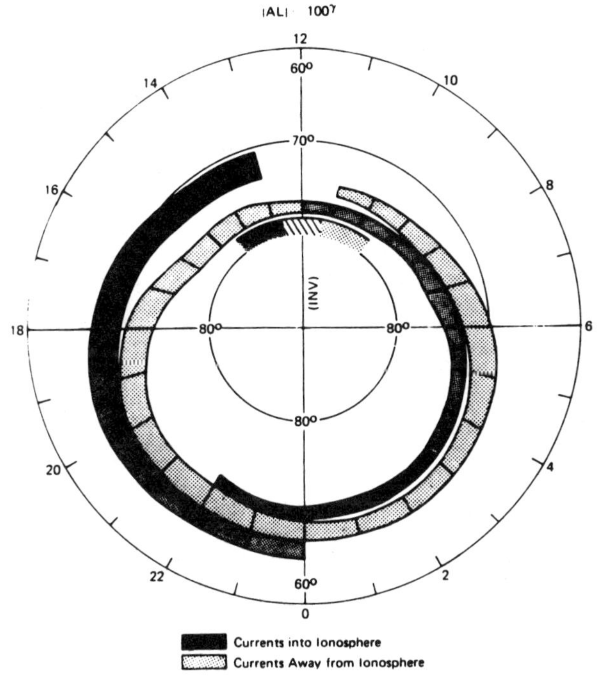

Fig. 3. The distribution of field-aligned currents above the ionosphere in invariant latitudes and magnetic local time coordinates [Iijima and Potemra, 1976].

and momentum contained in the hot plasma of the tail's plasma sheet are being dumped. A more or less strongly expressed convection pattern from midnight (Harang discontinuity) towards daytime on the evening as well as on the morning sides is a witness thereof. Numerous measurements of electric fields and convection flows by satellites and incoherent scatter radars have established their consistency with the $j_{\parallel}$-pattern of Iijima and Potemra [1976]. Current closure through the ionosphere follows largely the pattern of Boström's type II system, however not necessarily with perfect balance of the sheet currents in any chosen local time sector.

From studies near the geosynchronous orbit [Robert et al., 1984], in the tail [Frank et al., 1981; Sibeck et al., 1984] and near the magnetopause [Paschmann et al., 1982], it has become evident that field-aligned currents like to be concentrated in tubes or filaments. This has led to the concept of flux-ropes, which gave the name to this symposium. A most important feature of highly concentrated field-aligned current is that they can be subject to resistive forces which are commonly absent in more diffuse currents. The resistivity can be either a consequence of inertial or magnetic mirror forces or of current-driven instabilities. These topics will be addressed in Sections 2.5 and 2.6.

2. Physical Properties of Field-Aligned Currents

Field-aligned currents are found where transverse momentum is transferred along magnetic lines of force from one plasma regime to another. As such they are force-free, but they are part of a current system on whose ends forces are exerted. In general, they connect a *source* or dynamo region with a region of *closure* in which electromagnetic energy is either dissipated or at least converted into another form of energy. In the prevailing situation where steady-state descriptions are inappropriate we have to consider the *propagation* of $j_{\parallel}$, as well as *reflection* at sharp transitions that separate one plasma regime from another as, for example, the collision dominated lower ionosphere. As mentioned above, concentrated field-aligned currents may be subject to *kinetic effects* and/or *instability*. The six topics highlighted by script letters are those that we will subsequently address. This cannot be exhaustive, neither with respect to referencing previous work (see Introduction), nor - which is worse - with respect to the physics.

2.1. Sources

Field-aligned currents arise from the divergence of transverse currents. Transverse currents determine the Lorentz force on a quasi-neutral plasma and can thus be derived from the momentum balance equation. In the MHD description and for isotropic pressure this is:

$$\rho \frac{d\underline{v}}{dt} = - \nabla p + \frac{1}{c} \underline{j} \times \underline{B} \qquad (1)$$

With Ampère's law Equation 1 can be rewritten as

$$\rho \frac{d\underline{v}}{dt} = - \nabla \left(p + \frac{B^2}{8\pi}\right) + \frac{1}{4\pi} (\underline{B}\cdot\nabla)\underline{B} \qquad (2)$$

Equation 2 expresses the balance of four forces, inertial, $\underline{I}$, gas pressure, $\underline{P}$, magnetic pressure, $\underline{M}$, and magnetic tension, $\underline{T}$:

$$\underline{I} + \underline{P} + \underline{M} + \underline{T} = 0 \qquad (3)$$

In any large-scale dynamical equilibrium of a collision-free plasma, it is easy to visualize the role of these four forces. Here, we are interested in their action as drivers of field-aligned currents. From Equation 1 one can derive a simple expression for the divergence of the transverse current, $\underline{j}_\perp$, which is just the negative divergence of the field-aligned current, $j_\|$, since $\nabla\cdot\underline{j} = 0$ [Vasyliunas, 1984]:

$$B \frac{\partial}{\partial s_\|} \left(\frac{j_\|}{B}\right) = - \nabla\cdot\underline{j}_\perp = 2\, \underline{j}_\perp \cdot \frac{\nabla B}{B} \quad \frac{c}{B} \underline{e}_B \cdot (\nabla x \underline{I}) \qquad (4)$$

($\underline{e}_B = \underline{B}/B$). The expression for $\nabla_\| \cdot \underline{j}_\|$ on the ℓ.h.s. of Equation 4 takes into account the varying cross-section of a flux-tube, which is inversely proportional to B. In static equilibrium, $\underline{I} = 0$, the sources of $j_\|$ result from transverse currents across surfaces of constant $|\underline{B}|$. There are many ways of transforming Equation 4 without rendering it necessarily more transparent. One such example is the decomposition of the inertial term on the r.h.s. of Equation 4 by Hasegawa and Sato [1980]:

$$\frac{c}{B} \underline{e}_B \cdot (\nabla \times \underline{I}) = \rho \frac{d}{dt} \left(\frac{\Omega}{B}\right) + \frac{\nabla n}{nB}\cdot(\underline{e}_B \times \underline{I}), \qquad (5)$$

$$\text{where} \quad \Omega = \underline{B} \cdot \left(\nabla \times \frac{\underline{v}}{B}\right) . \qquad (6)$$

If $\underline{I} = 0$ (static equilibrium), the source term of $j_\|$ is most easily derived from Equations 3 and 4:

$$B \frac{\partial}{\partial s_\|} \left(\frac{j_\|}{B}\right) = \frac{8\pi c}{B^3} (\underline{T} \times \underline{P}) \cdot \underline{e}_B \qquad (7a)$$

$$\text{or:} \quad B \frac{\partial}{\partial s_\|} \left(\frac{j_\|}{B}\right) = \frac{8\pi\ c}{B^3} (\underline{M} \times \underline{T}) \cdot \underline{e}_B \qquad (7b)$$

Vasyliunas [1984] has treated equivalently the case of anisotropic pressure.

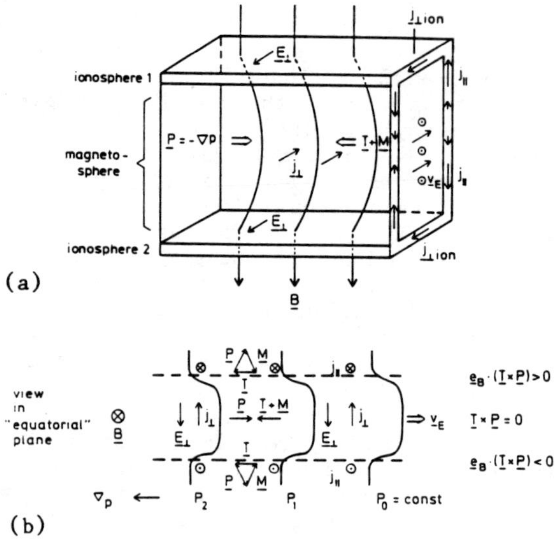

Fig. 4. (a) Box model of the magnetosphere-ionosphere current system at auroral latitudes with identification of the driving forces ($\underline{P} = -\nabla p$). This corresponds to the Boström Type II model (Figure 1b) and is consistent with the current system shown in Figure 3. (b) Force balance, perpendicular current, and current divergence below the mid-plane of the model current system.

Figure 4a serves as an illustration of the typical force balance in the magnetosphere. For sake of simplicity, I choose a nearly homogeneous magnetic field pervading a hot plasma (plasma sheet or ring current) inside a "box" with two end plates (ionosphere) in which an Ohm's law holds. A pressure force $\underline{P} = -\nabla p$ is balanced by the sum of magnetic shear and normal stresses, $\underline{T} + \underline{M}$. A current $\underline{j}_\perp$ flows at right angles to $\underline{B}$ and these forces. The box is limited in the direction of $\underline{j}_\perp$. This represents the radial dependence of magnetospheric quantities. Hence there are current divergences. A view of forces and currents in the ("equatorial") midplane of the box as given in Figure 4b demonstrates the relation between the sources and sinks of $j_\|$ and the forces as expressed by Equation 7a. If $\underline{P}$, $\underline{M}$, and $\underline{T}$ are not collinear, the source term of $j_\|$ has a finite value.

Figure 4a contains already the aspect of current closure which we will address in more detail in Section 2.4. If a simple Ohm's law characterizes the closure regions (ionosphere), an electric field, $\underline{E}_\perp$, proportional to resistivity and $\underline{j}_\perp$ is set up. Since in an MHD situation $E_\| = 0$, this field exists all along the magnetic lines of force. Inside the hot plasma region, it thus opposes $\underline{j}_\perp$; we have the case of a generator. It is also clearly seen, how the energy dissipated in the end plates is supplied. There is a Poynting flux:

$$\underline{S} = \frac{c}{4\pi} \underline{E} \times \underline{B} \quad , \tag{8}$$

whose component due to the bending of the field lines, $c\underline{E} \times \Delta\underline{B}_{\perp}/4\pi$, $(\Delta\underline{B}_{\perp} = \underline{B} - \underline{B}_o)$ is directed into the end plates where it has a negative divergence (sink).

For applications to a low beta plasma (much of the inner magnetosphere), it is often less convenient to derive $j_{\parallel}$ from the balance of forces. One prefers to refer explicitly to the motion of the hot plasma constituent in a model magnetic field. The equivalence of both views was demonstrated by Parker [1957]. Since only the gradient (G), curvature (C) and inertial drifts (I) of the ions (i) and electrons (e) can have finite divergences (in contrast to the diamagnetic drift), and since flux-tube integrated quantities are needed in order to derive the total field-aligned current entering and leaving the ionosphere, $j_{\parallel,o}$, the following expression (see also [Vasyliunas, 1972]) is most useful:

$$j_{\parallel,o} = -\nabla_{\perp} \cdot \{eN_H[(\hat{\underline{v}}_{Gi}+\hat{\underline{v}}_{Ci}+\hat{\underline{v}}_{Ii})-(\hat{\underline{v}}_{Ge}+\hat{\underline{v}}_{Ce})]\} \ , \tag{9}$$

where

$$N_H = B_o \int \frac{n_H}{B} ds_{\parallel} \tag{10}$$

is the flux-tube content of hot magnetospheric plasma above the ionosphere and

$$\hat{\underline{v}}_j = \frac{B_o}{N_H} \int \underline{v}_j \cdot \frac{n_H}{B} ds_{\parallel} \tag{11}$$

is a weighted average of the gradient (G), curvature (C) and inertial (I) drifts of ions and electrons. Vasyliunas [1972] pointed out that since the flux-tube content of ions changes very slowly, i.e.

$$\nabla_{\perp} \cdot \{N_H [\hat{\underline{v}}_E + \hat{\underline{v}}_{Gi} + \hat{\underline{v}}_{Ci} + \hat{\underline{v}}_{Ii}]\} \simeq 0 \ , \tag{12}$$

$j_{\parallel,o}$ can be solely derived from the electron component (their inertia being neglected):

$$j_{\parallel,o} \simeq e \cdot \nabla_{\perp} \cdot \{ N_H (\hat{\underline{v}}_E + \hat{\underline{v}}_{Ge} + \hat{\underline{v}}_{Ce})\} \tag{13}$$

Since in the absence of appreciable induction fields magnetic field lines are electric equipotentials, the weighted electric drift, $\hat{\underline{v}}_E$, is nothing else than the well-known $\underline{E} \times \underline{B}$-drift referred to the ionosphere:

$$\hat{\underline{v}}_E = \underline{v}_{Eo} = c \left. \frac{\underline{E} \times \underline{B}}{B^2} \right|_{\text{ionosphere}} \tag{14}$$

Physical insight into the origin of field-aligned currents in the magnetosphere can be gained only by considerations whose mathematical expressions have been given in Equations 4, 7a, or 9(13). "Mystical" notions like "current disruption" would have to be cast into this language before their meaning could be appreciated. The field-aligned current pattern derived by Iijima and Potemra [1976], for instance, has a very simple interpretation. In general and in particular during magnetospheric substorms, hot plasma piles up in the midnight sector of the inner edge of the tail (plasma sheet). The resulting pressure maximum is radially confined between inward directed magnetic shear and normal stresses from the tail (plasma sheet and lobes) and an outward directed magnetic pressure force from the dipole field of the inner magnetosphere. In longitude the situation is different. The pressure force acts against magnetic stresses which are in part communicated (via $j_{\parallel}$) to the ionosphere and taken up by ion friction with the neutral atmosphere. The convection pattern thus established is a necessary consequence of momentum balance, in which frictional forces balance magnetic shear stresses at high altitude. At this point it becomes amply clear that the sources of $j_{\parallel}$ in the hot plasma regime cannot be derived without due account of current *closure* since without knowledge of the frictional forces on one end of the system the shear stresses on the other end can not be defined. The existence of field-aligned currents is just an expression of a *non-local* force balance. We shall return to this subject in Section 2.4.

2.2. Propagation

Before addressing the subject of current closure we have to deal with the consequences of temporal dependence. Forces may suddenly be set up in a limited region of the hot plasma, for instance by some type of enhanced collisionless dissipation (e.g., tearing instability, reconnection) or plasma injection or penetration. If these forces have a component normal to $\underline{B}_o$, we may find essentially a situation as sketched in Figures 5a and b. A plasma blob is injected transverse to $\underline{B}_o$ with velocity $v_{\perp}$. The magnetic field convects with the blob and, on large scales, behaves essentially as if it were "frozen" in the plasma. The magnetic perturbation arising therefrom propagates with Alfvén speed, v_A , along the lines of force in both directions. Figure 5b shows the current circuit set up if the blob has finite lateral extent. Closure of the current occurs mainly in the propagating fronts of the "Alfvén wings" attached to the blob. The resulting magnetic perturbation field, $\Delta B_{\perp}$ obeys Walén's relation:

$$\frac{\Delta B_{\perp}}{B_o} = \frac{v_{\perp}}{v_A} \ . \tag{15}$$

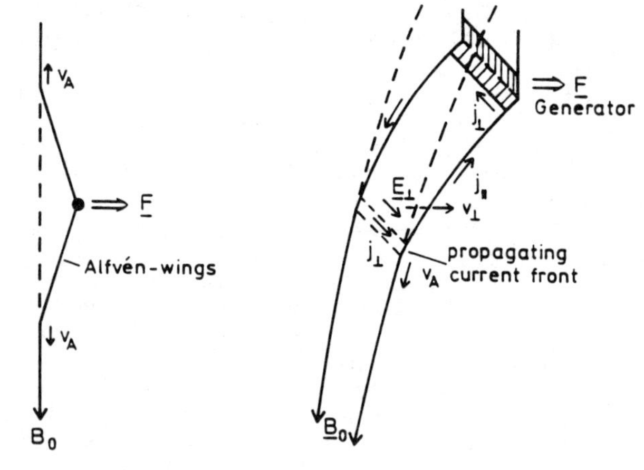

(a) (b)

Fig. 5. (a) Generation of "Alfvén-wings" by impulsive injection or sudden acceleration of a plasma blob. (b) A 3D-view of the current circuit within an Alfvén-wing of finite extent.

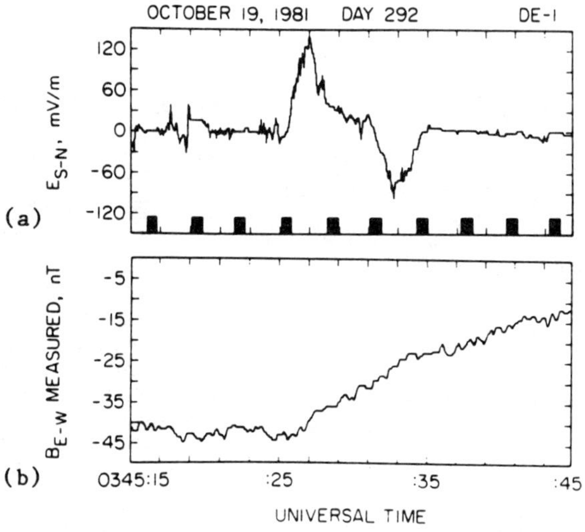

Fig. 6. Electric (a) and transverse (E-W) magnetic perturbation (b) fields measured with the DE-1 satellite at 10520 km altitude and 20:31 MLT [Weimer et al., 1987]. The total width of the event corresponds to ~ 50 km (N-S).

Another way of expressing this is to say that the integrated inertial current, $\underline{J}_\perp$, in the propagating fronts is subject to a wave impedance, R_w:

$$\underline{E}_\perp = R_w \underline{J}_\perp \quad , \tag{16}$$

with

$$R_w = \frac{4\pi v_A}{c^2} \quad . \tag{17}$$

Since $$\underline{E}_\perp = -\frac{1}{c} \underline{v}_\perp \times \underline{B} \tag{18}$$

we recover Walén's relation from Equation 16 by using Ampère's law. Another feature of the wave impedance is an *equipartition of energy* between magnetic and kinetic energy of the ambient plasma:

$$\frac{(\Delta B_\perp)^2}{8\pi} = \frac{\rho_a}{2} v_\perp^2 \quad . \tag{19}$$

A spacecraft equipped with a magnetometer and electric field or plasma drift sensors, when traversing such a region of confined field-aligned currents, can provide an easy check whether the current circuit is quasi-static and closes in the ionosphere, or whether it is freely propagating. In the latter case the ratio of the measured values of $E_\perp$ and $\Delta B_\perp$ should yield:

$$c \frac{E_\perp}{\Delta B_\perp} = v_A \tag{20}$$

This check, of course, requires the knowledge of the local plasma density, ρ_a, and of B_o. Figure 6a,b taken from Weimer et al. [1987] shows a measurement of Dynamics Explorer 1 of a confined bipolar electric field of large amplitude at an altitude of ~ 10 000 km (B_o = 0.03 G). Relation 20 would be fulfilled for an ambient density of about 10 hydrogen atoms per cm^3. This is a plausible value. Proper interpretation of the observed bipolar electric field structure would involve more than we are able to address in the context of this paper. However, one specific model is to be discussed in Section 2.6 (Figure 14).

2.3. Reflection

The magnetospheric plasma is bounded by the ionosphere. Transient magnetic perturbations as considered in the last section will eventually (after a few tens of seconds) strike the ionosphere. Here they encounter a region of completely different relation between integrated current, $J_\perp$, and electric field, $E_\perp$:

$$\underline{J}_\perp = \Sigma_p \underline{E}_\perp \quad . \tag{21}$$

Σ_p is the integrated Pedersen conductivity. Scholer [1970] studied what happens in this case. In general, the Alfvén wave is reflected unless there is perfect matching of wave impedance, R_w, and Σ_p^{-1}. The reflection coefficient, $\mathcal{R}$, of the wave electric field is given by:

$$\mathcal{R} = \frac{1 - R_w \Sigma_p}{1 + R_w \Sigma_p} \quad . \tag{22}$$

In most cases, $R_w \Sigma_p >> 1$ and $R \simeq -1$. This leads to a reversal of the incoming current $J_{\perp o}$ and, to a large degree, to a cancellation of the incoming electric field, E_o, by the reflected one, E_r, resulting in a very small value of the field E_a actually applied to the ionosphere. This is sketched in Figure 7. With $E_r = R \cdot E_o$

$$E_a = E_o + E_r = \frac{2 E_o}{1 + \Sigma_p R_w} \quad . \qquad (23)$$

The current flowing inside the reflected front, J_r, is related to E_r by Equation 16 and the applied current to E_a by Equation 21. This yields for $\Sigma_p R_w >> 1$:

$$J_a = \Sigma_p E_a \simeq \frac{2 E_o}{R_w} = 2 J_{\perp o} \qquad (24)$$

Thus the current applied to the ionosphere is approximately twice as large as the front current, $J_{\perp o}$, of the incoming wave. As the reflected front returns to the plasma blob from which it originated, the latter will suddenly find itself subject to a doubled wave drag. It will respond accordingly by a sudden decrease of $v_\perp$. The play can repeat itself several times, until eventually the increasing E_a becomes adjusted to the decreasing E_o. Each step is connected with a downward jump of $E_\perp$ and an upward jump of $J_{\perp o}$, $J_\parallel$ and $\Delta B_\perp$. Hence the relation between $E_\perp$ and $\Delta B_\perp$ deviates rapidly from the one for free propagation as given in Equation 20.

Adjustment of the hot plasma dynamics to the frictional control by the ionosphere thus necessarily involves flux-tube oscillations with periods of the Alfvén travel time from generator to ionosphere and back. Such oscillations are well-known under the names of geomagnetic pulsations of type Pc5 or Pi2 (e.g. [Jacobs, 1970]). Of course, reality is much more complex than the simple picture outlined above, mainly because of mode coupling of the shear wave and additional damping arising therefrom. However, the qualitative picture of the reflection remains.

Nishida [1979] considered situations different from our example of an injected plasma blob, namely constant electric field sources on the one hand (case a) and constant current sources on the other (case b). The results are shown in Figures 8a and b. In the first case, the electric field applied to the ionosphere builds up to the constant value at the generator and is accompanied by a current increase. In the second case, the magnetospheric field adjusts to the ionospheric value. Our example of Figure 7 lies in between, since it is an initial value problem with neither constant voltage nor current.

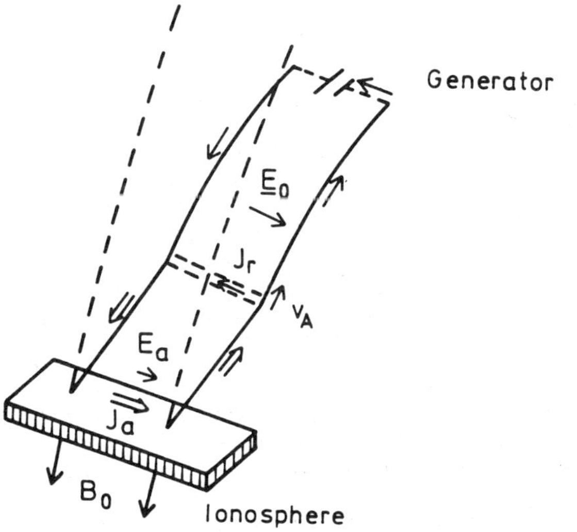

Fig. 7. The current pulse of Figure 5(b) after one reflection from the ionosphere.

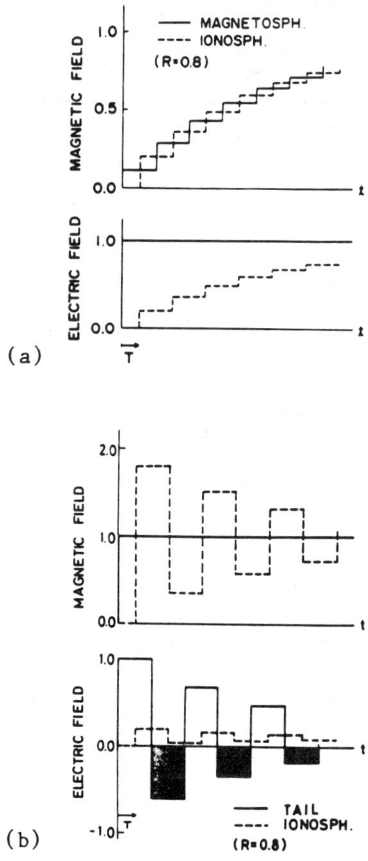

Fig. 8. Development of magnetic and electric fields in the magnetospheric tail and ionosphere after (a) a constant electric field source and (b) a constant current source have been switched on in the tail [Nishida, 1979].

2.4. Closure

The sources and sinks of field-aligned current above the ionosphere are matched to the sources and sinks of the horizontal ionospheric current. We are only interested in the total height integrated current and neglect any contribution from an atmospheric dynamo, because our focus is on the high-latitude magnetosphere-ionosphere system. The ionospheric current is composed of two constituents, the Pedersen and Hall currents (considered to be approximately horizontal):

$$\underline{J}_{ion} = \Sigma_p \underline{E}_\perp + \Sigma_H (\underline{e}_B \times \underline{E}_\perp) \quad . \quad (25)$$

If we take the divergence of $\underline{J}_{ion}$ and equate it to $j_{\parallel,o}$ of Equation 9, we get a very general relation which allows derivation of $\underline{E}_\perp$ (or $-\nabla_\perp\phi$) for given magnetic field and hot plasma distribution in the magnetosphere and known conductivity distribution in the ionosphere. In reality, these are not independent quantities. The true magnetic field is modified by the presence of the internal transverse and parallel currents, and the conductivity is modified by accumulations and depletions of plasma at the locations of current entry into and exit from the ionosphere plus additional ionization by impact of energetic electrons as current carriers. A wider set of relations including transport equations for the hot magnetospheric and the cold ionospheric plasmas would have to be solved with the current continuity equation, and all of this in a magnetic field that is made consistent with the currents. This full program has not even been attempted.

Partial solutions of the problem of magnetosphere-ionosphere interaction for the sake of deriving a realistic convection pattern have been given by Vasyliunas [1970, 1972], Wolf [1974], Harel et al. [1981] and others. Here we will not deal with this most impressive work, since our focus is on field-aligned currents. Only one aspect will be mentioned because of its elegant nature [Vasyliunas, 1972]. If we assume slow convection (no appreciable inertial currents) and conservation of the flux-tube content of hot ions, we can equate $\nabla\cdot\underline{J}_{ion}$ with $j_{\parallel,o}$ of Equation 13 to obtain:

$$\nabla_\perp\cdot(\Sigma_p\underline{E}_\perp)+(\underline{e}_B\times\underline{E})\cdot\nabla_\perp(\Sigma_H + \Sigma^*) = \Sigma_\perp(eN_H\,\hat{\underline{v}}_{Be}) \quad (26)$$

Here $\hat{\underline{v}}_{Be} = \hat{\underline{v}}_{Ge} + \hat{\underline{v}}_{Ce}$, and Σ^* is defined by:

$$\Sigma^* = \frac{eN_H c}{B} \quad . \quad (27)$$

Equation 25 makes use of the fact that Hall current as well as electron flow due to the $\underline{E}\times\underline{B}$-drift in the magnetosphere are divergence-free, unless there are gradients in Σ_H or Σ^*. The latter quantity behaves like a Hall conductivity. Its numerical value is, however, typically one order of magnitude greater than high-latitude values of Σ_p and Σ_H. Current continuity then requires that $\underline{E}_\perp$ is essentially normal to the surfaces of constant Σ^* or hot plasma density, N_H. Otherwise, the second term in Equation 25 could not be balanced. This is basically the reason why the sunward convection has a relatively sharp equatorward boundary which coincides with the inner edge of the ring current [Vasyliunas, 1972]. The Region 2 current system of Iijima and Potemra [1976] (see Figure 3) arises from the associated divergence of $\underline{J}_{ion}$.

A typical value for $\underline{E}_\perp$ can be found easily from the current continuity equation, if one uses Equation 9 for the magnetospheric current sources. With $\underline{I} = 0$, this equation becomes:

$$\nabla\cdot(\Sigma_p\underline{E}_\perp) + (\underline{e}_B\times\underline{E})\cdot\nabla\Sigma_H = -\nabla\cdot\{eN_H(\hat{\underline{v}}_{Bi}-\hat{v}_{Be})\} \quad (28)$$

In the plasma sheet and ring current one can mostly neglect $|\hat{v}_{Be}|$ against $|\hat{v}_{Bi}|$. It is essentially the ionic pressure force that is balanced by the magnetic shear stresses which are transferred to the ionosphere via $j_\parallel$. Equation 27 then leads to the estimate:

$$|\underline{E}_\perp| \simeq \frac{\Sigma^*}{\Sigma_p} \frac{\langle W_{iH}\rangle}{e\,\ell_p} \quad , \quad (29)$$

where $\langle W_{iH}\rangle$ is the mean kinetic energy of the hot magnetospheric ions and ℓ_p the e-folding length of the ion pressure in E-W direction, projected into the ionosphere. The strength of $j_\parallel$ turns out to be of the order:

$$|j_\parallel| \sim \Sigma^* \cdot \frac{\langle W_{iH}\rangle}{e\ell_p w_A} \quad , \quad (30)$$

where w_A is the N-S width of the auroral current system. Setting $\langle W_{iH}\rangle = 10$ keV, $\ell_p = 3300$ km, $w_A = 50$ km, one obtains $\langle j_\parallel\rangle \doteq 2\ \mu A/m^2$. This is in good agreement with typical TRIAD values. Equation 29 is a very condensed expression of the fact that the E-W convection in the polar ionosphere is essentially driven by the "internal heat engine" constituted by the hot magnetopsheric plasma in the near-Earth tail. The current system of Iijima and Potemra [1976] (Figure 3) can be seen as the "gear" by which the engine is coupled to the load.

In the preceding discussions, the ionosphere was considered to be the load region of the magnetospheric-ionospheric current circuit. This is not necessarily always the case. Besides the dynamo action of thermospheric winds and gravity waves, which even at high latitudes can at times be significant, the ionosphere can act as a *transformer*. This situation can exist inside the auroral electrojet when a primary westward electric field, $\underline{E}_o$, exists parallel to the long

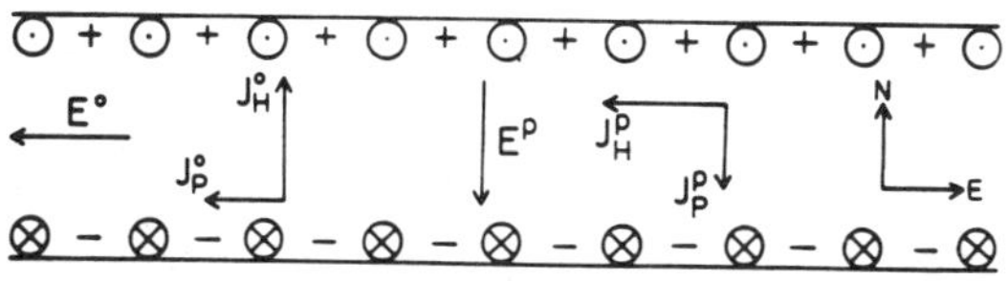

Fig. 9. Primary (index "o") and secondary (index "p") electric fields and Pedersen (J_p) and Hall (J_H) currents in an elongated region of enhanced ionization (Cowling channel). Finite divergence of the N-S currents in the ionosphere and field-aligned currents at the edges can exist. The ionosphere acts as a dynamo [Boström, 1975].

axis of a region of enhanced conductivity, a so-called Cowling channel [Boström, 1975]. Figure 9 shows the primary Hall and Pedersen currents, J_H^o and J_p^o, a secondary poleward electric field, E^p, whose role is to balance partially the primary Hall current by a secondary Pedersen current, J_p^p, and field-aligned currents at the edges of the channel. Allowing for finite divergence of $J_H^o - J_p^p$, the total current along the channel is of the order of $\Sigma_c \cdot E^o$, where the Cowling conductivity, Σ_c, is given by

$$\Sigma_c = \Sigma_p + \frac{\Sigma_H^2}{\Sigma_p} \quad . \qquad (31)$$

The value of the effective conductivity depends, however, on the degree of depolarization (reduction of E^p) by current diversion into the magnetosphere. What happens is that the plasma in the magnetosphere above the Cowling channel experiences a time-varying electric field and thus sustains an inertial current:

$$\underline{j}_I = \frac{\rho c^2}{B^2} \frac{d\underline{E}_\perp}{dt} \quad . \qquad (32)$$

The divergence of this current and resulting $j_\parallel$ above the ionosphere can be expressed as:

$$j_{\parallel,o} = - \frac{\Sigma^*}{\bar{\Omega}} \frac{\partial}{\partial x} \left(\frac{dE_x}{dt}\right) \quad , \qquad (33)$$

where $\bar{\Omega}$ is a weighted average of the gyrofrequency along the field lines with an effective value somewhat above the minimum at high altitudes.

The magnetospheric plasma now constitutes the load in the current circuit, whereas with respect to the poleward current (transverse to the Cowling channel), the ionosphere is the dynamo: $(J_H^o + J_p^p) \cdot E^p < 0$. Of course, the energy is supplied from the magnetosphere, and $E^o \cdot (J_p^o + J_H^p) > 0$. Recently, Rothwell et al. [1988] have included this inertial current in a model of the current system at auroral break-up.

2.5. Kinetic Effects

So far we have regarded the magnetospheric plasma as infinitely conducting. The magnetic field lines are electric equipotentials, as long as temporal variations are slow in comparison with the travel time of an Alfvén wave. However, this convenient property (frozen-in magnetic fields), which seems to apply to the normal state of most cosmical plasmas, can be lost under certain circumstances. They can be classified into two classes, kinetic corrections and current instability. In this section we address the first class. Here we can again subdivide into corrections due to electron inertia and due to the magnetic mirror force.

The first type of effects occurs only for current structures of sufficiently short transverse scale-lengths which appear in the context of so-called *kinetic Alfvén waves* [Hasegawa and Mima, 1978; Goertz and Boswell, 1979; Goertz, 1984]. For the low-altitude magnetosphere, in which $\beta \sim 0\ (10^{-6})$, the phase velocity of such kinetic Alfvén waves becomes:

$$\frac{\omega}{k_z} = v_A \left[1 + \left(k_\perp \frac{c}{\omega_{pe}}\right)^2\right]^{-1/2} \qquad (34)$$

Thus kinetic effects arise for $k_\perp^{-1}$ of the order of c/ω_{pe} which ranges around 1 km above the auroral ionosphere. In such waves, a parallel potential drop exists with $E_\parallel$ of the order of $(k_\parallel/k_\perp)\cdot E_\perp$ and $k_\parallel = \omega/v_A$. This leads to parallel electron heating and wave damping.

More important for the properties of large-scale current systems is the finite resistance parallel to $\underline{B}$ that can exist for $j_\parallel$ flowing along diverging magnetic field lines, i.e. away from the ionosphere. Necessary conditions are high temperature of the current carrying electrons and sufficiently low density [Knight, 1973; Fridman and Lemaire, 1980]. The sketch of Figure 10 tries to demonstrate the origin of what could be called "mirror resistivity". Electrons moving into a converging field are subject to the magnetic mirror force and will eventually be reflected. If the field-aligned current solely carried by these electrons, is so strong that the directed velocity (opposite to $j_\parallel$) is comparable to the thermal velocity, continuity of this current can necessitate a lowering of the mirror points by means of an upward directed potential drop (Figure 10). The energy thus fed into the electrons is supplied by a Poynting flux from inside the current circuit. Thus the field-aligned current can become "dissipative" ($E_\parallel \cdot j_\parallel > 0$), in the sense of conversion of electromagnetic into kinetic energy. No noise is needed, the electron motion is strictly adiabatic. This process is thought to contribute greatly to the generation of primary electron beams [Lyons, 1980; Chiu and Cornwall, 1980].

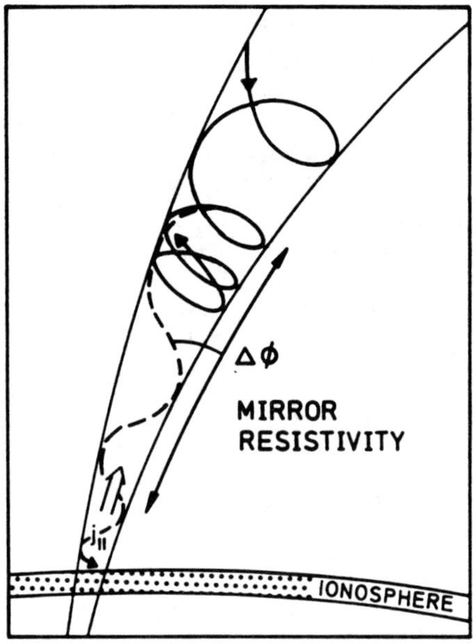

Fig. 10. Lowering of the mirror points of energetic electrons by an upward potential drop ($\Delta\phi$). This way parallel current continuity can be maintained.

Figure 11 taken from Burke [1984] shows the current/voltage characteristics for electrons of initial temperature (at high altitudes) T and varying mirror ratios. $j_{\parallel}$ is normalized to j_0 = e n v_T ($v_T = \sqrt{kT/2\pi m_e}$) and $e\phi_{\parallel}$ to the electron temperature at the source. For large mirror ratios, B_i/B_v, and $1 << eV_{\parallel}/kT << B_i/B_v$, there is a linear relation between $j_{\parallel}$ and $V_{\parallel}$ (not $E_{\parallel}$!). It is customary to write this relation as:

$$j_{\parallel} = K V_{\parallel} \quad (35)$$

with

$$K = \frac{e^2 n}{(2\pi m_e kT_e)^{1/2}} \quad (36)$$

This is in a way not quite appropriate, since the current is to be considered as primary. Its need for continuity sets up the voltage, $V_{\parallel}$, and the necessary energy is supplied from the magnetic energy constant of the overall current circuit. The nature of a resistivity (or overall field-aligned resistance) becomes more apparent if one writes Equation 34 as

$$V_{\parallel} = K^{-1} j_{\parallel} \quad (37)$$

The quasi-Ohmic relations (35, 37) yield a very simple energy conversion rate (electromagnetic into kinetic energy of parallel beams) per unit cross-section of a flux-tube:

$$\dot{\varepsilon}_{\parallel} = j_{\parallel} \cdot V_{\parallel} = K^{-1} j_{\parallel}^2 = K \cdot V_{\parallel}^2 \quad (38)$$

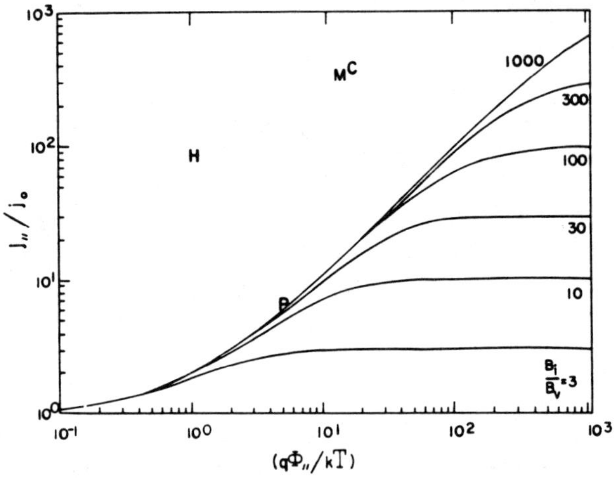

Fig. 11. Current (density)-voltage relation in a diverging magnetic field for various values of the mirror ratio (B_i/B_v). Current density and potential are normalized to thermal velocity and energy of the electrons at the source [Burke, 1984].

The applicability of this relation to the energy flux of auroral electron beams (whereby the value of $V_{\parallel}$ can be deduced from the shape of the energy spectrum) has been convincingly demonstrated by Lyons et al. [1979].

Continuation of $j_{\parallel}$ by an ionospheric Pedersen current yields:

$$K V_{\parallel} + \frac{\partial}{\partial x} (\Sigma_p \frac{\partial \phi_i}{\partial x}) = 0 \quad , \quad (39)$$

with $V_{\parallel} = \phi_m - \phi_i$ (magnetospheric and ionospheric potentials). This equation shows immediately that a current circuit characterized by these two resistive elements must have a typical transverse spatial scale, x_0, of

$$x_0 = \sqrt{\frac{\Sigma_p}{K}} \quad (40)$$

[Lyons, 1980; Chiu and Cornwall, 1980], which is of the order of a few 100 km. Hence, it has been suggested that the mirror resistivity effect is responsible for the more prominent auroral precipitation regions called "inverted V" events [Frank and Ackerson, 1971]. Figure 12 provides a sketch of such a current system with an upward current subject to mirror resistivity and the corresponding spectrogram of electrons accelerated by $V_{\perp}$ and precipitating into the ionosphere. In principle, we have the same situation as in Boström's Type II current circuit (Figure 1b), here driven by forces out of the plane. The new element is that dissipation and

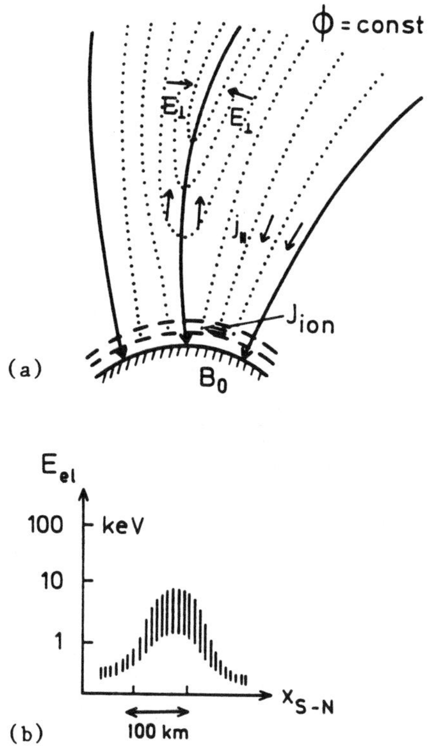

Fig. 12. (a) Cross-section of an auroral current system with electric equipotential lines for the case of "mirror resistivity". (b) Resulting spectral distribution of accelerated electrons above the ionosphere ("inverted V" signature).

potential drop are also found in the magnetosphere. This enhances the rate of current decay. Little potential drop should exist where $j_{\parallel}$ is downward, since current continuity does not pose a problem there.

As an alternative to the static situation with ionospheric current closure, one can consider a continuation of $j_{\parallel}$ by transverse inertial currents to which one can attach the wave impedance R_w (Equation 17 [Lysak, 1981]. This defines a much smaller scale:

$$d_{\perp} = (K R_w)^{-1/2} \tag{41}$$

of the order of 1 km above the polar ionosphere.

2.6. Instability

The last aspect of field-aligned currents to be covered in this paper is that of current-driven instabilities, which may yield anomalous resistivity. This process is completely different from the "mirror resistivity" discussed in the previous section. It involves noise, true Ohmic dissipation, albeit accompanied by a powerful run-away component. Since the condition for instability involves some threshold of the electron-ion drift velocity, which ranges somewhere between the ion and electron thermal velocities (depending on T_e/T_i), this process occurs preferentially in a dilute cold plasma background as found in the topside ionosphere at several 1000 km height. Kindel and Kennel [1971] have postulated the existence of regions of instability of $j_{\parallel}$ above the auroral ionosphere, once $j_{\parallel}$ is of the order of 10^{-5} A/m^2. Such current densities do indeed exist. Production of primary auroral electrons may well be related to field-aligned potential drops as sketched in Figure 13 [after Gurnett, 1972].

Some type of wave must exist to scatter particles and randomize the energy received from $E_{\parallel} \neq 0$. Candidates of such waves have been experimentally identified by Temerin et al. [1982] and Boström et al. [1987]. The waves appear to be ion-acoustic solitons of several Debye-lengths parallel extent associated with potential jumps of the order of kT_e (T_e = temperature of the cold background electrons), i.e. very small in comparison with the overall voltage over the height range of current instability (several kV over several 1000 km). The auroral electrons and upward accelerated ions, which are a common and well explored phenomenon, are probably only a run-away

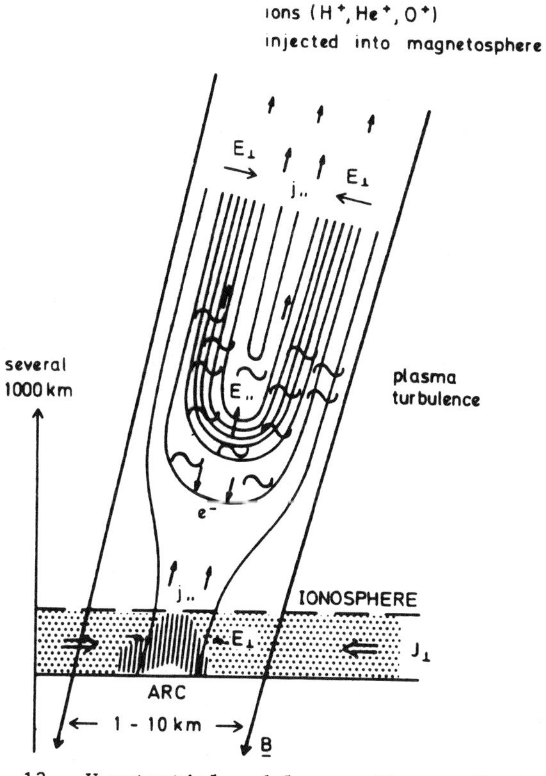

Fig. 13. U-potential model according to Gurnett [1972] with indication of current flow, ion and electron beam formation, plasma turbulence sustaining $E_{\parallel}$, and auroral arc formation. The ratio of width to height is grossly exaggerated.

population, small by number density, great by energy flux and carrying a substantial fraction of the current.

Besides the particle signature, also the existence of high transverse electric fields, which are an important ingredient of the U-shaped potentials shown in Figure 13, has been verified [Mozer et al., 1977]. They follow from Equation 16 with $J_\perp = j_\parallel d$ (d being the width of the unstable current), if the current is not strictly field-aligned, but has a finite transverse component constituted by an ion inertial current. Quasi-stationary solutions with such high transverse electric fields are only possible, if the latter are shorted out above the ionosphere as shown in a qualitative way in Figure 13. The corresponding parallel potential drop can be sustained only if

$$j_\parallel > j_{crit} = e\, n\, c_{crit} \tag{42}$$

For instance in a plasma density of 10^2 cm^{-3} with $kT_e = 1$ eV and $c_{crit} \cong 10\, c_{si}$ (c_{si} = ion-acoustic speed), we find $j_{crit} = 1.6 \cdot 10^{-6}$ A/m^2. This would appear at low altitudes as a current more intense by almost one order of magnitude. Such current densities have been found in association with thin auroral arcs (e.g., Spiger and Anderson [1975]). Also the event shown in Figure 6 yields $j_\parallel = 3 \cdot 10^{-6}$ A/m^2 which could well be unstable in an environment of $n = 10$ cm^{-3} (see Section 2.2.). (At this point it should be mentioned that the authors of Figure 6 [Weimer et al., 1987] used the data contained in this figure for a check of the mirror resistivity concept (Equation 35) and found good agreement, as well.)

Much theoretical work has been devoted to the study of the current instability process and its role in the formation of structured auroral arcs (e.g. [Lysak and Dum, 1983; Haerendel, 1983; 1989]). This author compares the macroeffect of current instability with a fracture process [Haerendel, 1980; 1989], i.e. a self-sustained propagation of the sheet of unstable current into the current circuit, which constitutes the reservoir of magnetic energy, thereby reducing the overall energy content. Figure 14 is taken from this work. It shows one half of the auroral current circuit in a straight field-line geometry. The distinguishing feature of this model is the standing oblique Alfvén waves attached to the region of current instability (acceleration region). Knowledge of j_{crit} (Equation 42) and of the thickness 2d of the total system yields a simple expression for the field-aligned voltage (with help of Equation 16):

$$V_{\parallel,max} = V_\perp = E_\perp \cdot \frac{d}{2} = R_w\, j_{crit} \frac{d^2}{4} \,. \tag{43}$$

If, for instance, $j_{crit} = 2 \cdot 10^{-6}$ A/m^2, it needs a half width of only ~ 15 km in order to

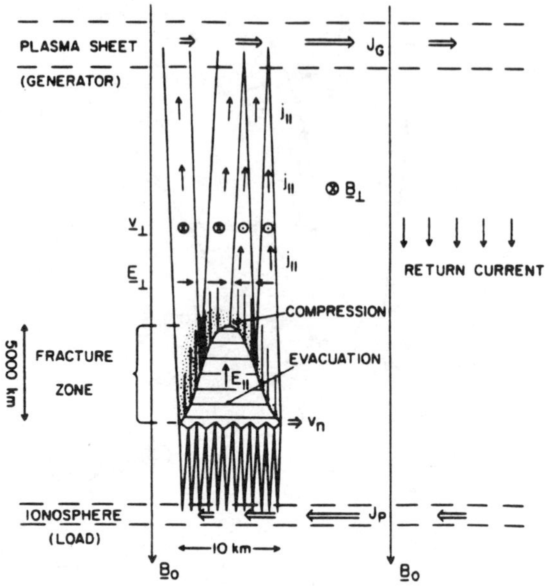

Fig. 14. Model of the auroral acceleration region or "fracture zone" according to Haerendel [1989]. Standing oblique Alfvén waves are attached to the fracture zone and communicate with the generator, on the one hand, and the ionosphere, on the other. The latter leads to a narrow interference pattern of kinetic Alfvén waves because of the short travel time. The fracture zone ($E_\parallel \neq 0$) is subject to an evacuation process and propagates spontaneously with velocity v_n into the current circuit.

produce $V_{\parallel,max} = 5$ kV, a typical energy of primary auroral electrons. These figures apply to a height of about 6000 km. Projected into the ionosphere, d becomes about 5 km. This is a frequently observed width of structured auroral arcs. Also the energy flux carried by the primary electrons agrees well with $j_{crit} \cdot V_{\parallel,max}$, since most of the energy is transferred to the run-away electron component. The normal propagation speed of the arcs is very small; it is of the order of d/τ_A (tens of m/s), where τ_A is the travel time of an Alfvén wave.

It should be noted that the current circuit shown in Figure 14 is of the same type as those in Figures 12 and 1b and driven the same way by forces out of the plane. The new feature here is that the sheets of additional current dissipation (generating the auroral particle beams) propagate spontaneously into the energy reservoir, i.e. the current circuit, and thus reduces its energy content. This can explain the motion of auroral arcs relative to the background plasma.

An interesting feature of the current instability model (Figure 14) is the process of evacuation of the acceleration region ($E_\parallel \neq 0$). This results from the combination of two effects, anomalous dissipation and strongly reduced thermal conductivity $\parallel$B. Thus heat accumulates in this region and the pressure increases. It leads to parallel plasma expansion and lowers the threshold for current instability (Equation 42), because the reduction in density is stronger than the increase of $\sqrt{T_e}$. This can explain the persistence of acceleration regions as witnessed by the persistence of auroral arcs. The existence of very low-density regions above the auroral ionosphere has been established by measurements of plasma wave excitation [Persoon et al., 1988].

Final Remark.

A brief survey of a wide field of phenomena and processes must always remain unsatisfactory. I tried to confine this review of field-aligned currents in the magnetosphere to the principal physical aspects. This was only possible by refraining largely from describing the great zoo of geophyisical manifestations, as well as the many elaborate theoretical models involving the flow of field-aligned currents. The main aim of this paper is to remind the reader that an understanding of the role of currents can only come from an analysis of the forces acting on a magnetized plasma. Much of this can be visualized in the framework of an MHD description with infinite conductivity, i.e. in the "frozen field" concept. Exceptions, where some form of finite resistivity destroys the latter concept, have been pointed out. As so often, these exceptions contain some of the most dramatic manifestations of $j_\parallel$ and, therefore, receive particular attention.

Questions and Answers

Atkinson: Are these ideas relevant to the Sun, too?

Priest: On the Sun we have instead an MHD situation in which currents are not driven along field lines by electric fields as in wires in the laboratory. Instead, since the magnetic Reynolds number is so large, the electric fields are determined by field line motion and the currents are determined by force balance in a static state and by the equation of motion in a dynamic one.

Atkinson: Comment: If regions of $E_{//}$ exist on the sun, then $E_\perp$ is discontinuous along a flux tube. This would imply that helicity was not conserved. I wonder if double layers, mirror resistivity, etc., exist on the sun.

Haerendel: I agree with you that narrow regions of strong and strongly variable $E_\perp$ must exist in conjunction with $E_{//}$. This implies shear flows in the sense of untwisting a severely twisted flux-tube, a flux-rope. The main question with respect to the dense plasma of the solar atmosphere is whether macroscopic configurations can remain stable with growing concentration of $J_{//}$ until the threshold of microinstability is reached. If so, the maintenance of $E_{//}$ would be no problem.

Forbes: Is there really any evidence showing that the current in the tail is diverted to the ionosphere, rather than simply dissipated *in situ* in the tail? Either way the tail current is disrupted.

Haerendel: No, I believe there is no such evidence. It is a model. We see the signatures of the field-aligned currents connecting to the auroral electroject, but how they are closed in the tail is not really known. Nor is it known what current "disruption" means. I personally favor a model in which $j_\parallel$ does not connect to currents originating from the magnetopause, but rather to an eastward current at the outer edge of the dipolarization region.

Martens: I have some difficulty understanding the concept of mirror-resistivity. First, don't you need cross-field currents in the converging magnetic field region to preserve $\nabla \cdot \mathbf{J} = 0$? Second, where does the energy go?

Haerendel: Yes, you have to close $j_{//}$ eventually. In the context which I discussed the current closed in the ionosphere by a Pederson current and at high altitudes inside the generator plasma. The force balance established between that plasma and the ionosphere dictates how much $j_{//}$ has to flow. It is needed to maintain the upward $j_{//}$ in a low density hot plasma environment which creates the field-aligned voltages, $V_{//}$. By lowering the mirror points of the hot electrons, $j_{//}$ is kept continuous. The energy is flowing as Poynting flux into the region of $V_{//}$ where it is converted into kinetic energy of accelerated electrons.

Nordlund: What is the reason the double layers have typical potential drops of no more than kT/e and a large number are created, rather than a few with larger potential drops. Do they merge? Is there, in fact, a distribution of potential drops?

Haerendel: I cannot give a simple answer to that. Not much is known on the origin and decay of such solitary waves. There is always a balance of momentum between particles falling through the potential drop and background trapped and scattered particles. The temperature probably limits the potential jump. Why the double layers do not merge to larger ones, I really do not know, but it may be related to the background plasma temperature. One has also to consider the runaway particles which experience only the potential jump without scattering. They impart momentum to the background plasma that is carrying the double layer. This interaction may be very important for the creation and maintenance of the solitary structures.

Zelenyi: I want to comment that at lower altitudes where a lot of cold plasma is present, another scale can determine the fine structure of FAC

instead of ρ_i. I ,mean the electron inertial length c/ω_{pe} which can result in a small scale structures of the order of hundreds of meters. Such things have been observed in ARCAD-3 measurements.

Haerendel: Yes, I agree. The cutoff scale $(R_\omega K)^{-\frac{1}{2}}$ I was referring to applies to the high temperature, low density plasma at high altitudes where the mirror effect acts on the energetic electrons as dominant charge carriers.

Acknowledgement. I am grateful to Dr. W. Baumjohann for valuable suggestions during the preparation of the lecture.

References

Akasofu, S.I., *Polar and Magnetospheric Substorms*, Reidel Publ. Co., Dordrecht-Holland, 1968.

Akasofu, S.-I., B.-H. Ahn, and G.J. Romick, A study of the polar current systems using the IMS meridian chains of magnetometers, *Space Sci. Rev.* 36, 337-413, 1983.

Baumjohann, W., Ionospheric and field-aligned current systems in the auroral zone: A concise review, *Adv. Space Res.*, 2(10), 55-62, 1983.

Baumjohann, W., R.J. Pellinen, H.J. Opgenoorth, and E. Nielsen, Joint two-dimensional observations of ground magnetic and ionospheric electric fields associated with auroral zone currents: Current systems associated with local auroral break-ups, *Planet. Space Sci.*, 29, 431-447, 1981.

Birkeland, K., *The Norwegian aurora polaris expedition 1902-1903, Volume I*, H. Aschelong and Co., Christiania, 1-315, 1908.

Boström, R., A model of the auroral electrojets, *J. Geophys. Res.* 69, 4983-4999, 1964.

Boström, R., Mechanisms for driving Birkeland currents in *Physics of the Hot Plasma in the Magnetosphere*, ed. by B. Hultqvist and L. Stenflo, 431-447, Plenum Press, New York, 1975.

Burke, W.I., Electric fields and currents observed by S3-2 in the vicinity of discrete arcs, in *Magnetospheric Currents*, Geophysical Monograph 28, American Geophys. Union, 294-303, 1984.

Chiu, Y.T., and J.M. Cornwall, Electrostatic model of a quiet auroral arc, *J. Geophys. Res.*, 85, 543-556, 1980.

Coleman, P.J., Jr., and R.L. McPherron, Fluctuations in the distant geomagnetic field during substorms: ATS 1, in *Particles and Fields in the Magnetosphere*, ed. by B.M. McCormac, Reidel Publ. Co., Dordrecht-Holland, 171-194, 1970.

Féjer, J.A., The effects of energetic trapped particles on magnetospheric motions and ionospheric currents, *Can. J. Phys.* 39, 1409, 1961.

Fridman, M. and J. Lemaire, Relationship between auroral electron fluxes and field-aligned electric potential difference, *J. Geophys. Res.*, 85, 664-670, 1980.

Frank, L.A., and K.L. Ackerson, Observations of charged particle precipitation into the auroral zone, *J. Geophys. Res.*, 76, 3612-3643, 1971.

Frank, L.A., R.L. McPherron, R.J. de Coster, B.G. Burek, K.L. Ackerson, and C.T. Russell, Field-aligned currents in the Earth's magnetotail, *J. Geophys. Res.*, 86, 687-700, 1981.

Goertz, C.K. Kinetic Alfvén waves on auroral field lines, *Planet. Space Sci.*, 32, 1387-1392, 1984.

Goertz, C.K., and R.W. Boswell, Magnetosphere-ionosphere coupling, *J. Geophys. Res.*, 84, 7239-7246, 1979.

Gurnett, D.A., Electric field and plasma observations in the magnetosphere, in *Critical Problems of Magnetospheric Physics*, ed. by E.R. Dyer, 123-138, Nat. Academy of Sciences, Washington, D.C., 1972.

Haerendel, G., Auroral particle acceleration - an example of a universal plasma process, *ESA Journal*, 4, 197-210, 1980.

Haerendel, G., An Alfvén wave model of auroral arcs, in *High-Latitude Space Plasma Physics*, ed. by B. Hultqvist and T. Hagfors, 515-535, Plenum, New York, 1983.

Haerendel, G., Cosmic linear accelerators, in *Plasma Astrophysics*, ESA SP-285, ESA Publications Division, 37-44, 1989.

Haerendel, G., P.C. Hedgecock, and S.-I. Akasofu, Evidence for magnetic field-aligned currents during the substorms of March 18, 1969, *J. Geophys. Res.*, 76, 2382-2395, 1971.

Harel, M., R.A. Wolf, P.H. Reiff, R.W. Spiro, W.J. Burke, F.J. Rich, and M. Smiddy, Quantitative simulation of a magnetospheric substorm, 1, Model logic and overview, *J. Geophys. Res.*, 86, 2217-2241, 1981.

Hasegawa, A., and K. Mima, Anomalous transport by kinetic Alfvén wave turbulence, *J. Geophys. Res.*, 83, 117-1123, 1978.

Hasegawa, A., and T. Sato, Generation of field-aligned current during substorm, in *Dynamics of the Magnetosphere*, ed. by S.-I. Akasofu, Reidel Publ. Co., Dordrecht-Holland, 529-542, 1979.

Iijima, T., and T.A. Potemra, The amplitude of field-aligned currents at northern high latitudes observed by Triad, *J. Geophys. Res.* 81, 5971-5979.

Kindel, J.M., and C.F. Kennel, Topside current instabilities, *J. Geophys. Res.*, 76, 3055, 1971.

Knight, S., Parallel electric fields, *Planet. Space Sci.*, 21, 741-750, 1973.

Lyons, L.R. Generation of large-scale regions of auroral currents, electric potentials, and precipitation by the divergence of the convection electric field, *J. Geophys. Res.*, 85, 17-24, 1980.

Lyons, L.R., D.S. Evans, and R. Lundin, An observed relation between magnetic field-aligned electric fields and downward electron energy fluxes in the vicinity of auroral forms, *J. Geophys. Res.*, 84, 457-461, 1979.

Lysak. R.L. Electron and ion acceleration by

strong electrostatic turbulence, in *Physics of Auroral Arc Formation*, ed by S.-I. Akasofu and J.R. Kan, 444-450, Geophysical Monograph 25, AGU, Washington, DC, 1981.

Lysak, R.L., and C.T. Dum, Dynamics of magnetospheric-ionospheric coupling including turbulent transport, *J. Geophys. Res.*, *88*, 365-380, 1983.

McPherron, R.L., C.T. Russell, and M.P. Aubry, Satellite studies of magnetospheric substorms on August 15, 1968: 9. Phenomenological model for substorms. *J. Geophys. Res.*, *78*, 3131-3149, 1973.

Mozer, F.S., C.W. Carlson, M.K. Hudson, R.B. Torbert, B. Parady, J. Yatteau, and M.C. Kelley, Observations of paired electrostatic shocks in the polar magnetosphere, *Phys. Rev. Lett.*, *38*, 292-295, 1977.

Nishida, A., Possible origin of transient dusk-to-dawn electric field in the nightside magnetosphere, *J. Geophys. Res.*, *84*, 3409-3412, 1979.

Parker, E.N., Newtonian development of the dynamic properties of ionized gases of low density, *Phys. Rev.* *107*, 924-933, 1957.

Paschmann, G., G. Haerendel, I. Papamastorakis, N. Sckopke, S.J. Bame, J.T. Gosling, and C.T. Russell, Plasma and magnetic field characteristiscs of magnetic flux transfer events, *J. Geophys. Res.*, *87*, 2159-2168, 1982.

Persoon, A.M., D.A. Gurnett, W.K. Peterson, J.H. Waite, Jr., J.L. Burch, and J.L. Green, Electron density depletion in the nightside auroral zone, *J. Geophys. Res.* 93, 1871-1895, 1988.

Potemra, T.A., ed., *Magnetospheric Currents*, Geophysical Monograph, American Geophys. Union, Washington, DC, 1984.

Robert, P., R. Gendrin, S. Perraut, and A. Roux, GEOS 2 identification of rapidly moving current structures in the equatorial outer magnetosphere during substorms, *J. Geophys. Res.* *89*, 819-840, 1984.

Rothwell, P.L., L.P. Block, M.B. Silevitch, and C.-G. Fälthammar, A new model for substorm onsets: the pre-breakup and triggering regimes, *Geophys. Res. Lett.* *15*, 1279-1282, 1988.

Scholer, M., On the motion of artificial ion clouds in the magnetosphere, *Planet. Space Sci.*, *18*, 977-1004, 1970.

Sibeck, D.G., G.L. Siscoe, J.A. Slavin, E.J. Smith, S.J. Bame, and F.L. Scarf, Magnetotail flux ropes, *Geophys. Res. Lett.*, *11*, 1090-1093, 1984.

Spiger, R.J., and H.R. Anderson, Electron currents associated with an auroral band, *J. Geophys. Res.*, *80*, 2161-2164, 1975.

Temerin, M., K. Cerny, W. Lotko, and F.S. Mozer, Observations of double layers and solitary waves in the auroral plasma, *Phys. Rev. Lett.*, *48*, 1175-1179, 1982.

Vasyliunas, V.M., Mathematical models of magnetospheric convection and its coupling to the ionosphere, in *Particles and Fields in the Magnetosphere*, ed. by B.M. McCormac, D. Reidel Publ. Comp., Dordrecht-Holland, 60-71, 1970.

Vasyliunas, V.M., The interrelationship of magnetospheric processes, in *Earth's Magnetospheric Processes*, ed. by B.M. McCormac, D. Reidel Publ. Comp., Dordrecht-Holland, 29-38, 1972.

Vasyliunas, V.M., Fundamentals of current description, in *Magnetospheric Currents*, ed. by T.A. Potemra, Geophys. Monograph 28, American Geophysical Union, Washington, DC, 63-66, 1984.

Weimer, D.R., D.A. Gurnett, C.K. Goertz, J.D. Menietti, J.L. Burch, and M. Sugiura, The current-voltage relationship in auroral current sheets, *J. Geophys. Res.*, *92*, 187-194, 1987.

Wolf, R.A., Calculations of magnetospheric electric fields, in *Magnetospheric Physics*, ed. by B.M. McCormac, D. Reidel Publ. Comp., Dordrecht-Holland, 167-177, 1974.

Zmuda, A.J., J.H. Martin, and F.T. Heuring, Transverse magnetic perturbations at 1100 km in the auroral zone, *J. Geophys. Res* *71*, 5033-5045, 1966.

SATELLITE OBSERVATIONS OF FINE SCALE STRUCTURE IN AURORAL FIELD-ALIGNED CURRENT SYSTEM

E. M. Dubinin

Space Research Institute, Academy of Sciences, USSR

Abstract. Strong disturbances ($\Delta E_\perp$ ~100 mV/m and $\Delta B_\perp$ ~100 nT) are often observed on a background of large-scale auroral quasi-stationary field-aligned current structures. The small scales have been difficult to understand due to the lack of such fine-scale current structures in the outer magnetosphere. The ratio between the amplitudes of the mutually perpendicular components of the electric and magnetic fields for small-scale structures ($\ell_\perp \leq 20$ km) at altitudes ~1000 km turns out to be V_A/c. This amplitude corresponds to the nonstationary shear Alfven waves. Bursts of precipitating electrons accompany these waves. The fine scale structure of the fields is analyzed. The structure of the fields resembles convective plasma vortices and peculiar magnetic islands that correspond to filamentary field-aligned current tubes. Phase shifts between perturbations of the electric and magnetic fields are often observed. Interference effects can be responsible for these phase shifts. Among small-scale events the localized disturbances, Alfven vortices are elucidated. Beside them wide zones (100-1000 km) of Alfven wave turbulence are often observed. The properties of this turbulence resemble the properties of the usual two-dimensional hydrodynamic turbulence (power law spectra; vortex structure of the disturbances at different spectral intervals). It is shown that the effects of the nonlinear interaction of shear Alfven waves with finite amplitudes can be responsible for some of the observed peculiarities.

Introduction

There are numerous satellite observations of transverse magnetic field disturbances with large amplitudes and with characteristic scales at ionospheric altitudes from hundreds of meters up to some hundreds of kilometers. Since Zmuda et al.'s [1966] pioneering measurements these observations have attracted attention since transverse perturbations could be considered manifestations of Birkeland field-aligned currents. The existence of such currents was widely disputed in the past.

Birkeland currents can be separated into two categories [Potemra, 1988]: 'large scale' currents with spatial scales larger than 50 km (> 0.5° in latitude) and 'small scale' structures ($\ell_\perp \leq 50$ km). The former typically have a stable distribution and a close association with large-scale electric field. These current systems are the result of the large-scale solar wind-magnetosphere interaction. The generation of region 1 currents is related to dynamo-processes in boundary layers. The so-called NBZ-system is a result of changes in the magnetospheric configuration for northward component of the IMF. The region 2 current is supplied by the pressure-gradient current in the internal magnetosphere and so on.

On the other hand the Earth's magnetosphere is a very dynamical object. So it is natural to expect the existence of 'small-scale' disturbances on the background of large-scale field-aligned currents.

These fine field-aligned currents reveal a close association with visible auroral forms, micropulsations and other irregular phenomena of the auroral magnetosphere.

With an assumption of plane and stationary current sheets the observations of small scale magnetic disturbances can be interpreted as due to small scale and very high density Birkeland currents. The current densities are up to some tens of $\mu A/m^2$. The measurements carried out on DE-2 [Maynard et al., 1982] and ICB-1300 [Dubinin et al., 1983] have shown that intense electric field disturbances are associated with fine-scale field-aligned currents. The amplitudes of the electric fields correspond to drift velocities up to several km/s.

The acceleration processes in auroral field tubes are also associated with high density field aligned currents. Indeed there is maximum current for upward currents carried by precipitating electrons: $j_{\parallel} \sim 1\mu A/m^2$ ($j_{\parallel} = neV_e$ where $n \sim 1$ cm^{-3}, $T_e \sim 100$ eV). This value is comparable

Geophysical Monograph 58

to that typical for large-scale field-aligned auroral currents. The presence of acceleration processes in small-scale and high density upward field aligned current regions can increase this limiting current. Measurements carried out on satellites have shown that small-scale magnetic disturbances are accompanied by spikes of precipitating electrons with typical energy of 1 KeV. Figure 1 shows an example of simultaneous measurements on the ICB-1300 satellite of transverse magnetic disturbances and of fluxes of precipitating electrons with an energy of 1 KeV. The spikes in the electron flux correspond well to upward (positive gradient of magnetic variation) field-aligned currents. The other interesting phenomenon in the auroral magnetosphere that is associated with the intense small-scale upward field-aligned currents is the large upgoing thermal ion flux in excess of 6×10^{10} $cm^{-2}s^{-1}$ [Heelis et al., 1984]. The latter is especially important for understanding the presence of large fluxes of O^+ ions in the Earth's magnetosphere.

Acceleration of electrons is generally associated with parallel electric fields. The latter can appear due to anomalous resistivity. The existence of anomalous resistivity along magnetic field lines in a collisionless plasma is widely disputed. This phenomenon has a great interest for fundamental plasma physics. The measurements carried out on the S3-3 and Viking satellites [Temerin et al., 1982; Koskinen et al., 1988; Boström et al., 1988] have shown that parallel fields were often localized in double layers. According to theory such strong plasma turbulence can be excited when the drift velocity of the electrons in field-aligned currents is about

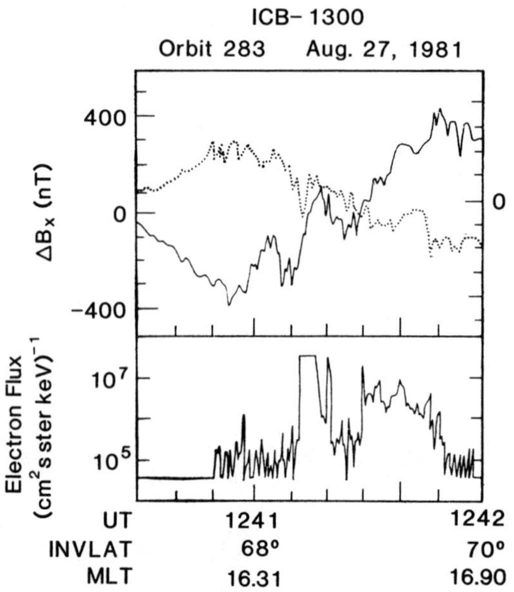

Fig. 1. Transverse magnetic disturbances and fluxes of precipitating electrons with E = 1 KeV (ICB-1300; orbit 283, Aug. 27, 1981).

their thermal velocity. Such drift velocities can be expected for small-scale field-aligned current with transverse scales $\sim c/\omega_{pe}$

$$(V_{dr} = j_{\parallel}/ne = cK_\perp B_\perp/4\pi ne = V_{the}\,\frac{cK_\perp}{\omega_{pe}}\,\frac{B_\perp}{B_0}\,\frac{1}{\beta_e^{1/2}},$$

$$\text{where } \beta_e = \frac{4\pi n T_e}{B_0^2} \simeq 2\cdot 10^{-6},\quad B_\perp/B_0 \sim 3\cdot 10^{-3}).$$

This corresponds to scales of several kilometers at altitudes of 1 to 2 R_E, where parallel electric fields were observed.

Are the fine scale structures waves or static?

Localized narrow field-aligned currents and broad zones (with widths up to some hundreds of km), filled by small-scale structures are observed. Figure 2a shows an example of an isolated event recorded on ICB-1300 satellite. Variations of the E_x and B_y components are shown. The satellite had a three-axis stabilized orientation. The X-axis was along the satellite velocity, the Z-axis was along the radius-vector from the Earth, and the Y-axis completed the right hand coordinate system. The satellite had a circular polar orbit ($h \simeq 900$ km).

A rather intense electromagnetic disturbance with amplitudes of 150 mV/m and 120 nT was recorded at $19^h\ 08^m\ 30^s$ UT. Figure 2b shows an example of a lot of small-scale field-aligned currents embedded in the large-scale upward current. The measurements were made in the premidnight sector during a disturbed time (2 March, 1982).

Figure 3 shows the fine structure of a localized event recorded in the morning sector. From top to bottom the variations of ΔB_y, ΔE_x, downward fluxes of electrons with energy of 1 KeV, and variations of the ion density measured by the flat plasma trap are shown. The vectors of the drift velocity and the structure of the field-aligned currents are shown on the right side of Figure 3. The vortex-like character of the plasma motion is distinctly seen. The scale of the vortices is ~10 km. The current density of the field-aligned currents is up to 30 $\mu A/m^2$. Variations in the ion density reach ~50%.

The two-dimensional structure of the electric and magnetic fields corresponding to a small time interval (~12 s) of the region filled by small-scale disturbances (Figure 2b) is shown in Figure 4. The data are represented in hodogram-form. Sampling time is 0.08 sec. This corresponds to 0.6 km. It is seen that $\vec{E}_\perp$ and $\vec{B}_\perp$ vectors experience a chaotic rotation. With the assumption of stationarity each rotational structure can be interpreted as a convective vortex motion of the plasma and filamentary field-aligned current. The picture resembles MHD-turbulence.

There are two different approaches to understanding the origin of small-scale disturbances. According to one of them transverse

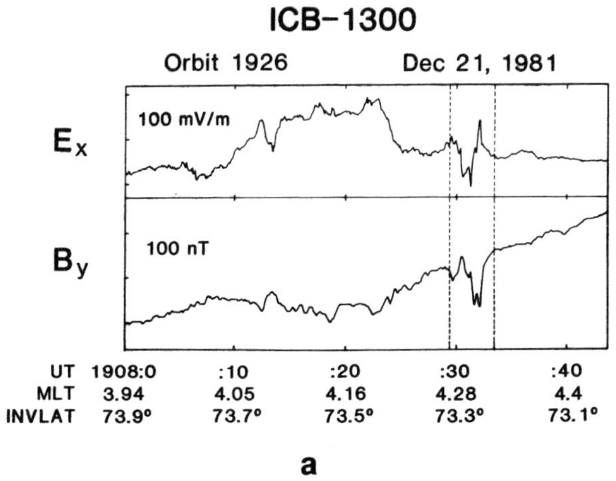

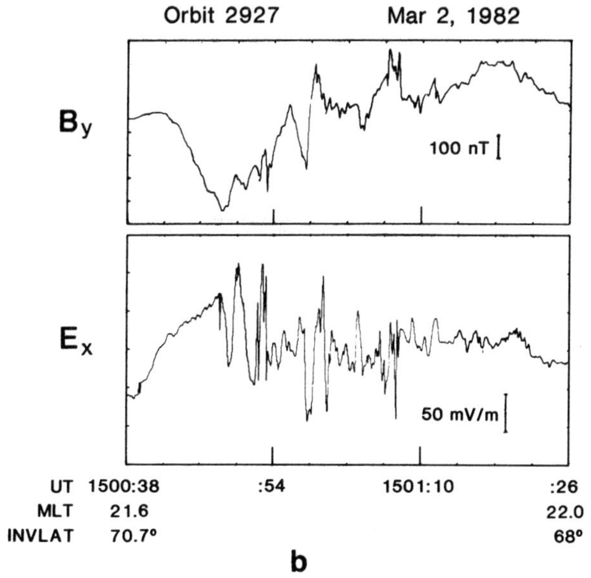

Fig. 2. a) An isolated disturbance ($19^h\ 08^m\ 30^s$) with amplitudes 150 mV/m and 120 nT (orbit 1926, 21.12.1981). b) Many small-scale field-aligned currents and narrow plasma "jets" (orbit 2927 02.03.1982) imbedded in large-scale upward current.

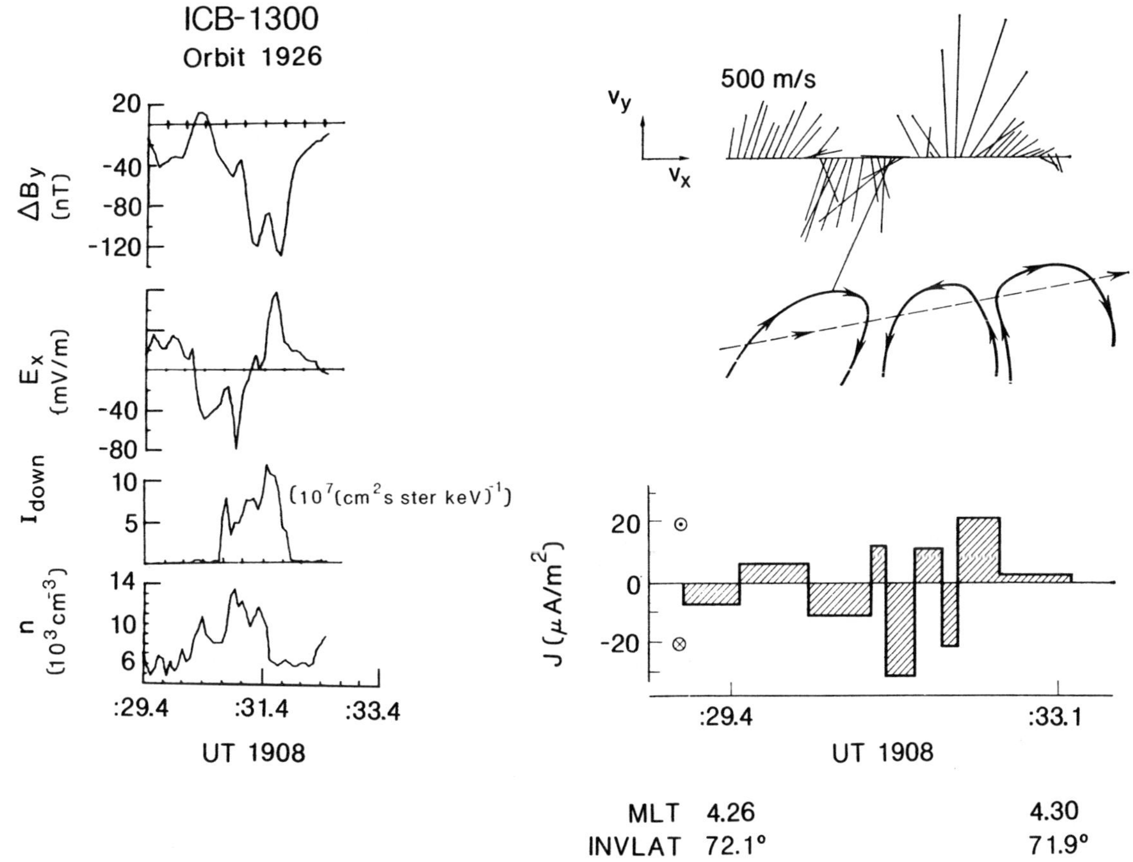

Fig. 3. From top to bottom are shown variations of ΔB_y, ΔE_x downward fluxes of electrons (E = 1 KeV), and variations of the ion density. The vectors of the drift velocity and the structure of field-aligned currents are also shown (orbit 1926).

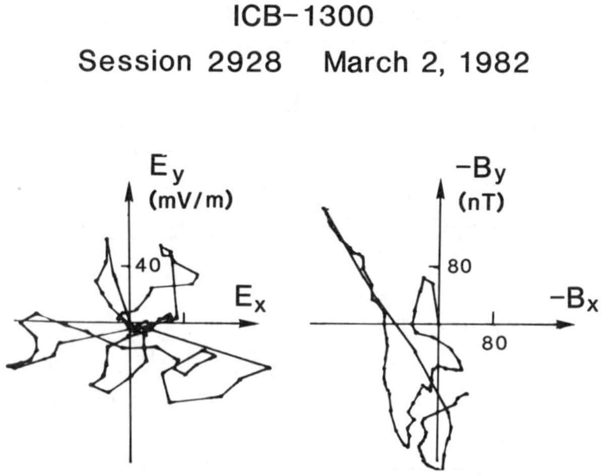

Fig. 4. Two-dimensional structure of fields, corresponding to a 12 s interval (Fig. 2b - orbit 2927).

disturbances of the magnetic field correspond to narrow current sheets embedded in large-scale Birkeland currents. Closing of these currents in the ionosphere by meridional Pedersen currents gives a simple functional relationship between the magnetic and electric fields: $j_{\parallel} = \nabla I_{ionosp.} = \Sigma_p \nabla E_{\perp}$ (it is assumed that Σ_p = const), or $\Delta B_{\perp} \simeq \Sigma_p E_{\perp}$. Sugiura et al. [1982] examined the relationship between the electric field and the perturbation magnetic field on the DE-2 satellite for large-scale (>100 km) disturbances. They showed that extremely good correlations existed between them. The ratio of amplitudes corresponds to the stationary pattern: $\Delta E_{\perp}/\Delta B_{\perp} = c/4\pi\Sigma_p$. On the other hand the irregular character of small-scale disturbances allows us to propose their nonstationarity and their electromagnetic origin. The idea that the observed phenomena are sheared Alfven waves seems most fruitful. Alfven wave can be generated, for example, by an inhomogeneity in the magnetospheric convection that supplies charge separation across the magnetic field. To produce the polarization current that maintains the quasi-neutrality of Alfven wave, field-aligned currents flow. Propagating along magnetic field lines Alfven waves transport magnetospheric perturbation energy from the edge of the magnetosphere to the ionosphere. However the transverse scales of inhomogeneities of magnetospheric convective flows should be at least some Larmor ion radii. So the wave model should be able to explain the existence of fine scale current structures.

The simultaneous measurements of electric and magnetic fields carried out at altitudes ~900 km on the ICB-1300 satellite have shown that these small-scale disturbances indeed had an electromagnetic origin. The ratio of the amplitudes of the mutually-perpendicular components of the magnetic and electric fields in small-scale structures ($\ell_{\perp} \leq 10$ km) were proportional not to the Pedersen ionospheric conductivity as in the stationary field-aligned current structures, but to c/V_A, where c/V_A is the ratio of speed of light to Alfven speed. This relationship between electric and magnetic fields corresponds to nonstationary events-sheared Alfven waves. The order of magnitude difference between the Pedersen ionospheric conductivity ($\Sigma_p \sim 5$ Mho) and Alfven wave conductivity at altitudes of ~1000 km ($\Sigma_A = c^2/4\pi V_A \sim 0.4$ Mho) makes the identification of Alfven waves possible [Dubinin et al., 1985; 1988; 1989]. The measurements of field fluctuations at higher altitudes have shown that the effective conductivity (coefficient of proportionality between $B_{\perp}$ and $E_{\perp}$) appears to decrease with increasing radial distance. This also supports the Alfven wave model [Gurnett et al., 1984]. Figure 5 shows an example of small-scale structures on the background of two large-scale (~80 km) field-aligned current sheets ($j_{\parallel} \sim 5$ μ_A/m^2). The measurements were made on ICB-1300 satellite. The result of a filtration procedure (low-frequency 0-0.2 Hz (Figure 5b) and higher frequency > 0.2 Hz (Figure 5c) components) is shown too. The rather high degree of correlation between the electric and magnetic components in the large-scale (low-frequency) structures and the small value of ratio of amplitudes $E_{\perp}/B_{\perp}$ in comparison with V_A/c is consistent with a stationary structure. From the ratio of amplitudes it is easy to estimate $\Sigma_p \sim 5$ mhos. This is in good agreement with typical values of auroral ionospheric conductivity. On the

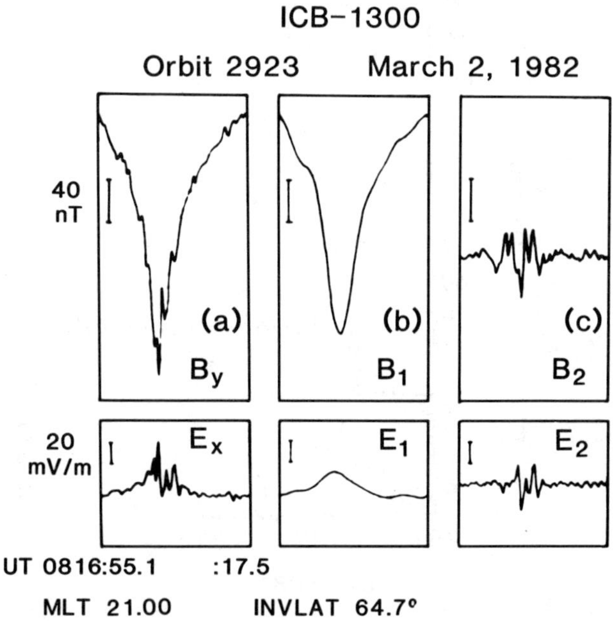

Fig. 5. a) Small-scale structures on a background of two large scale field-aligned currents (orbit 2923, March 2, 1982). b) The low frequency (0-0.1 Hz) components of the disturbance. c) Higher frequency (>0.2 Hz) components are shear Alfven waves.

other hand the ratio of $E_\perp/B_\perp$ for the smaller-scale (high-frequency) structures is greater and it is about V_A/c, i.e., Alfven waves are observed on the background of stationary field-aligned currents [Dubinin et al., 1988].

The simultaneous measurements of the two transverse components of the electric and magnetic fields on ICB-1300 satellite have shown that the observed Alfven waves had an elliptical rather than a linear polarization. A typical example of field distribution is shown on Figure 6. Although we observe here some structures, their elliptical polarization is clearly seen. Since measurements on only one satellite do not allow spatial and temporal variations to be distinguished there is uncertainty in the interpretation. Indeed two-dimensional structures of electric and magnetic field perturbations resemble convective plasma vortices and peculiar magnetic islands that correspond to filamentary field-aligned current tubes. On the other hand they can be considered Alfven waves. In the latter the convective flow changes its direction after distance $\sim V_E/\omega$ (V_E is the drift velocity in electric field $E_\perp$, ω is frequency of Alfven wave). Direction of field-aligned current changes also after $t \sim 1/\omega$. Measurements on one satellite do not allow a convective cell flow (stationary closed plasma motion) to be distinguished from a plasma flow in an Alfven wave. Volokitin and Dubinin [1989] have shown that convective cells and magnetostatic modes can be excited as a result of the nonlinear interaction of shear Alfven waves. Convective cells are collective modes with $K_\perp \neq 0$ and $\omega \simeq 0$ resembling hydrodynamical vortices. Similar modes are excited as the result of the parametric interaction of drift waves in plasmas [Sagdeev et al., 1978] and they are characteristic element of intense drift-wave turbulence. The picture shown in Figure 4 indeed resembles the results of numerical simulations of drift wave turbulence [Cheng and Okuda, 1977]. Recently similar structures were reproduced during numerical simulations of shear Alfven waves in a cold ($\beta \leq$ m/M) plasma [Seyler, 1988].

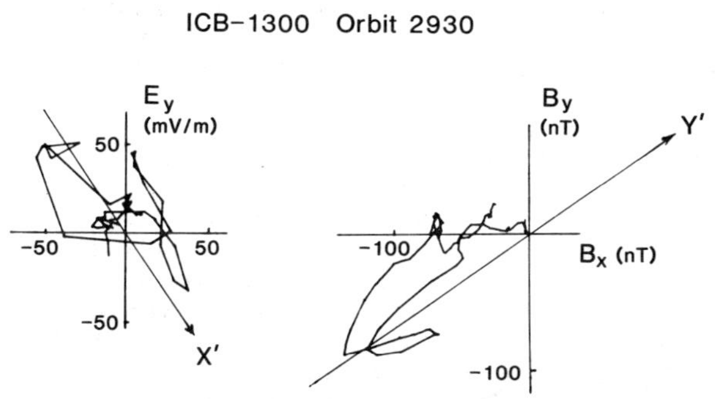

Fig. 6. The two transverse components of the electric and magnetic field for an intense disturbance. The data are represented in hodogram form (orbit 2930, March 2, 1982; 2008.5 UT).

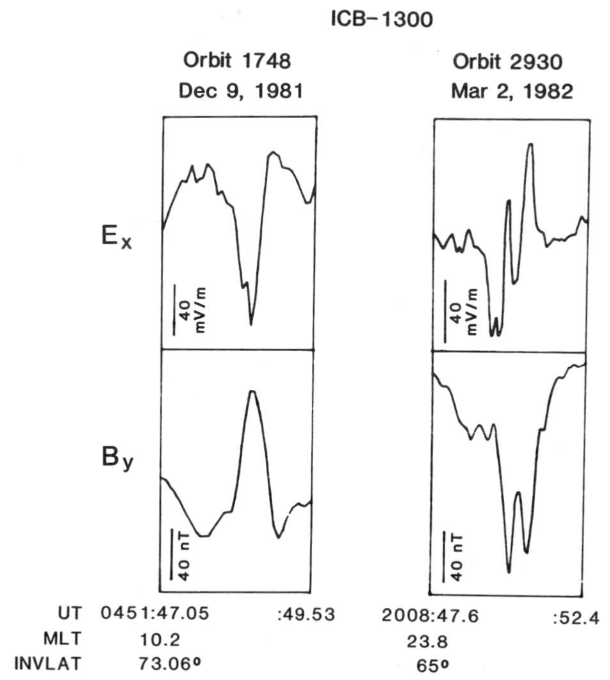

Fig. 7. Two small-scale structures observed in the noon (orbit 1748) and midnight (orbit 2930) sectors.

It should be noted that wave forms of electric and magnetic field components at low altitudes are often not similar to each other. The absence of such similarity makes the interpretation of data more complicated [Dubinin et al., 1985, 1989]. Figure 7 shows examples of two small-scale structures, observed in the noon (orbit 1748) and midnight (orbit 2930) sectors. In the first example the similarity of the wave forms of ΔE_x and ΔB_y is typical. The ratio of amplitudes $\Delta E_x/\Delta B_y$ is $\sim V_A/c$. In the second example the ratio of amplitudes is also about V_A/c, but the absence of similarity between the components is clearly seen. The immediate proximity of the ionosphere with its good reflecting properties can result in the interference of the incident and reflected waves. Dubinin et al. [1985, 1989] tried to separate the spatial and temporal effects for a certain group of localized disturbances. Figure 8 shows three such structures. It should be noted that the electric field structures resemble those that have been observed on the S3-3 satellite - so-called "electrostatic shocks" [Mozer et al., 1980]. However the analysis of the electric and magnetic measurements at altitudes of 1000 km reveals their electromagnetic origin [Dubinin et al., 1985, 1988]. Dubinin et al. [1985, 1989] used the model of quasi-monochromatic standing shear Alfven waves that arise due to the superposition of shear Alfven wave incident and reflected from the ionosphere. The typical frequency and transverse scale are about ~1 Hz and ~10 km.

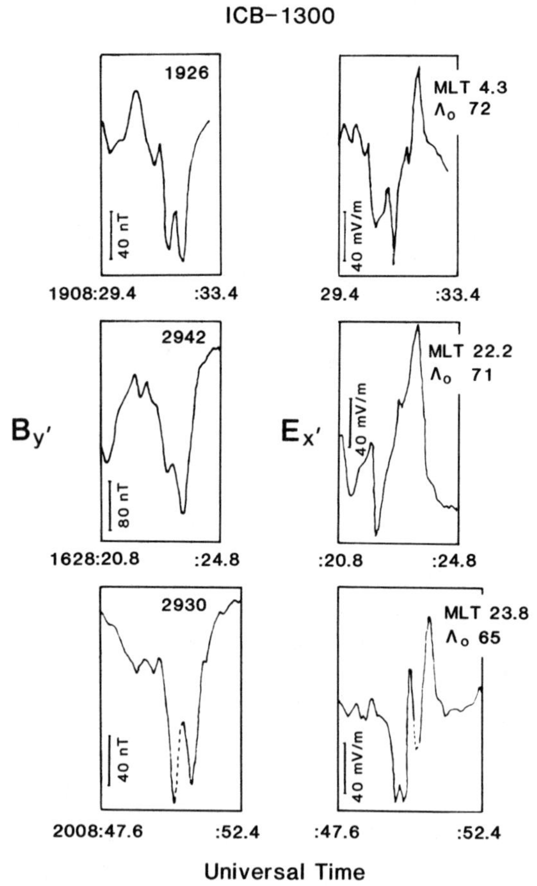

Fig. 8. A group of events with similar wave forms is shown.

The physical mechanism of the generation of quasi-monochromatic standing Alfven waves with f ~ 1 Hz has been discussed by Dubinin et al. [1985, 1986]. It can be related to the formation of a peculiar resonator consisting of the ionosphere and an anomalous resistive layer at ~1 R_E altitude [Haerendel, 1983; Lysak and Dum, 1983]. Such a resistive layer may be the result of the excitation of plasma instabilities by the currents that can lead to the formation of anomalous resistivity. Lysak and Dum [1983] discussed the case of "classic" anomalous resistivity that results in a smooth distribution of the parallel electric field. Kan et al. [1988] analyzed the interaction between Alfven waves and double layers that can appear when strong plasma turbulence is excited. In both cases such a resistive slab reflects Alfven waves. These reflections are of an opposite sense from the ionospheric reflections. They lead to an intensification of the perpendicular electric field. In reality the situation is more complicated due to the spatial inhomogeneity of the plasma parameters (n, B) along the magnetic field tube. There inhomogeneities also result in partial reflections on Alfven waves. The minimum in the Alfven conductivity is at approximately the same altitude (~1 R_E) as the resistive layer. The reflections of waves from these altitudes gives the same values of characteristic frequencies (f ~ 1 Hz).

The simple relationship between $E_\perp$ and $B_\perp$ can also break down when the source region for the Alfven waves is moving with respect to the plasma and the incident and reflected waves interfere with each other [Lysak, 1985]. All these factors confuse the observed pattern.

More detailed analysis of the relationship between $E_\perp$ and $B_\perp$ at low altitudes is given by Dubinin et al. [1989]. It is characterized by phase shifts between them. It was shown that in some cases there is so-called reversed Hilbert transformation $E_\perp \sim H\{ B_\perp \}$, $B_\perp \sim H\{ E_\perp \}$, where

$$H\{ f(t) \} = \frac{1}{\pi} \int_{-\infty}^{t} \frac{f(T)}{T-t} dT$$

It takes into account some kind of cause-effect relations between $E_\perp$ and $B_\perp$: electric (magnetic) field at a certain moment of time is determined not only by the magnetic (electric) field at this moment but also by their values at earlier times [see also Mallinckrodt and Carlson, 1978]. This relationship can merely be a result of interference effects (field structure at some point will be determined not only by the field structure in the incident wave but also by the field structure in reflected wave, i.e., by the field structure at earlier moments of time). Interference will produce the phase shifts between $E_\perp$ and $B_\perp$. The reversed Hilbert transformation is reduced to the $\pi/2$ phase shift of all harmonics (without change of their amplitudes).

However the events with distinctly pronounced temporal effects are observed sometimes. Figure 9 shows the variations of the magnetic field in a region of large-scale downward field-aligned current. The fluxes of precipitating electrons with energy 1 KeV are also shown. The periodicity of magnetic field disturbances with f ~ 1 Hz is clearly seen. The amplitude of them increase gradually up to 250 nT. A similar periodicity is seen in the electron fluxes. So it is proposed that the observed Alfven waves are responsible for acceleration of the electrons.

Contrary to the measurements at low altitudes, the measurements of low frequency electric and magnetic fields that have been carried out on the DE-1 satellite are well interpreted in the framework of a model of static small-scale field aligned currents. Comparisons of low-altitude (800-900 km) electric field spectra measured by the DE-2 satellite with high-altitude (h ~ 2 R_E) magnetic field spectra measured by DE-1 satellite, when both satellites were nearly on the same magnetic field line, showed good agreement [Weimer et al., 1985]. The ratio of corresponding spectral components (B_k/E_k) mapped to a common wave number scale gives a Pedersen conductivity. This means

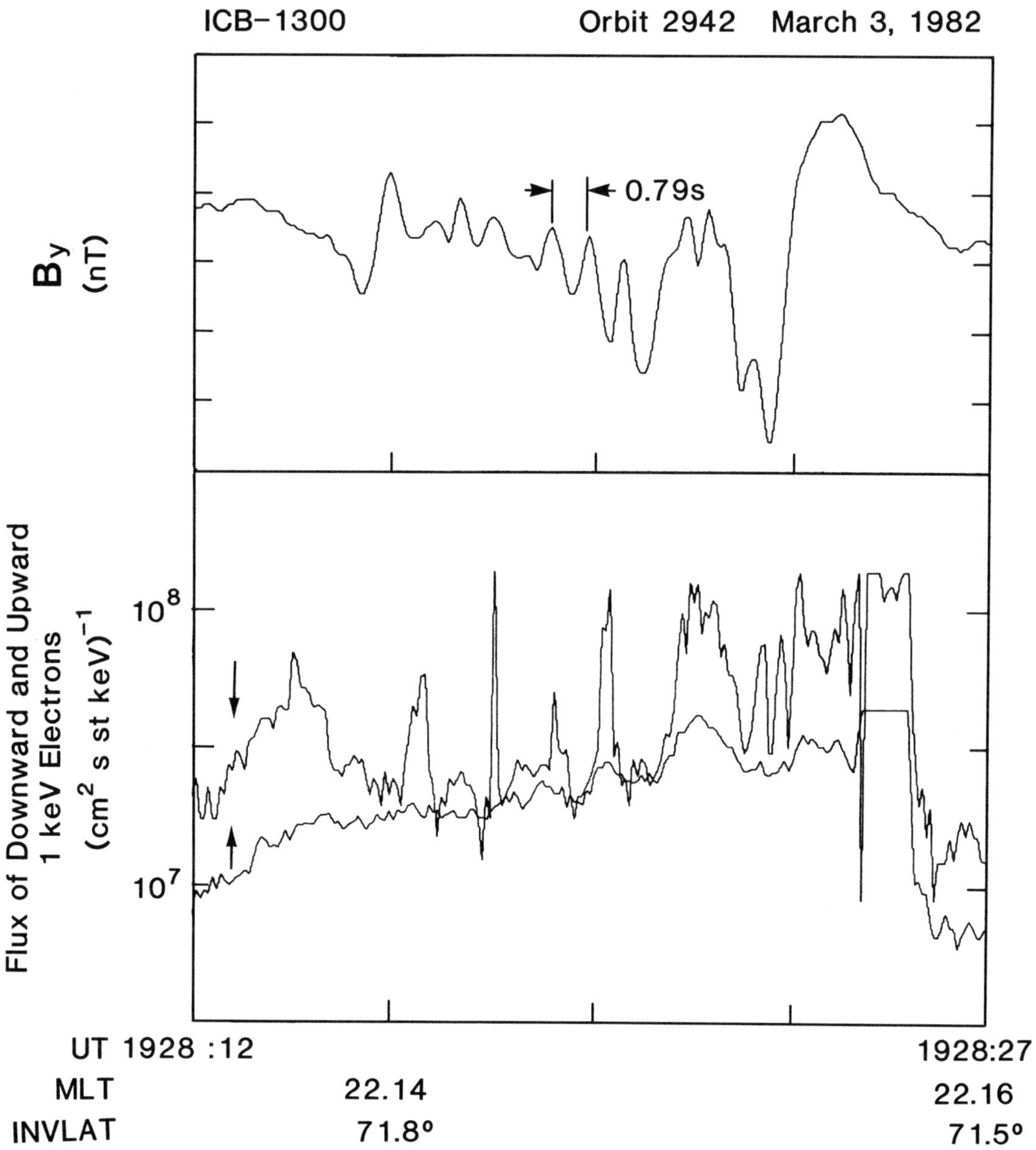

Fig. 9. Oscillating behavior of the magnetic field on the background of large-scale downward field-aligned currents. Periodicity of precipitating electrons is observed too.

that low-altitude electric fields are generated by static small-scale field-aligned currents. Additionally, however, Weimer et al. [1985] found that small-scale ($\ell_\perp$ < 100 km) electric fields are strongly attenuated at low altitudes. This was interpreted by Weimer et al. [1985] as evidence of parallel finite resistivity. Weimer et al. [1987] have also shown that for the "smallest scales" ($\ell_\perp$ < 20 km) the ionospheric conductivity is no longer important in the relationship between the transverse components of the magnetic and the electric fields at high altitudes. The relationship is determined only by parallel resistivity.

Thus the measurements of small-scale field-aligned currents at different altitudes are interpreted differently -- in the framework of static and wave models. However, it turns out that some "static" peculiarities can be qualitatively understood while analyzing shear Alfven waves with finite amplitudes [Volokitin and Dubinin, 1989; Dubinin et al., 1988]. The dispersion of shear Alfven waves taking place for small transverse scales results in the possibility of decay

TABLE 1. Shear Alfven Wave Comparisons at Two Altitudes

Low altitudes ($h < 3$-$4\ R_E$)
(cold plasma $\beta < m/M$)

When the wavelength perpendicular to the geomagnetic field becomes small the dispersion effects due to the inertia of the electrons take place:

$$\omega^2 = K_z^2 V_A^2/(1 + k_\perp^2 a^2)$$

where $a = c/\omega_{pe}$ is the electron inertial length. Shear Alfven wave propagating in cold plasma decay to two Alfven waves [Dubinin et al., 1988; Volokitin and Dubinin, 1989].

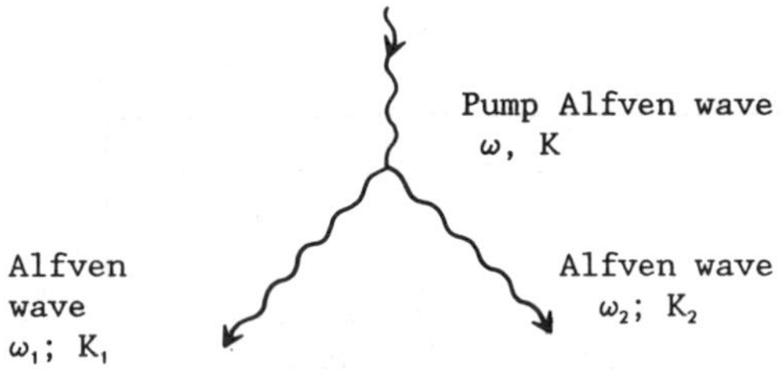

The growth rate of the decay instability is $\gamma \sim K_\perp V_A B_\perp / B_0 (aK_\perp)^2$. The instability arising at low altitudes should result in the appearance of elliptical structures in the electric and magnetic field disturbances (the growth rate of decay instability has maximum for $\vec{K}_1 \perp \vec{K}_2$) and in the appearance of beating that is manifested in an alternating direction of rotation of the polarization vector (superposition of two Alfven waves polarized in mutually perpendicular directions that have similar frequencies $\omega \sim \omega_1$). The measurements carried out at low altitudes (ICB-1300) often observed such elliptical structures. The decay processes can form a spectrum of Alfven waves. The dynamics of its formation can be described in the framework of the theory of weak plasma turbulence when the frequencies of the interacting modes are much higher than γ. A power-law spectrum is expected.

Another interesting effect - the excitation of convective cells and filamentary currents - can be manifested at low altitudes too. Such forced static modes ($K_z \simeq 0$) can arise due to beatings of Alfven waves if there is a developed spectrum of weak Alfven wave turbulence [Dubinin et al., 1988]. Amplitudes of electric fields should be higher than those that occur in Alfven modes ($E_\perp/B_\perp \sim 1$). This means that the electric field should be increasingly more electrostatic with increasing $K_\perp$ [Dubinin et al., 1985].

Intermediate altitudes ($h \geq 3$-$4\ R_E$)
(warm plasma $1 \gg \beta \geq m/M$)

In warm plasmas kinetic effects due to the finite ion Larmor radius become essential:

$$\omega^2 = K_z^2 V_A^2 (1 + K_\perp^2 \rho^2)$$

where ρ is the ion gyroradius. The modified decay instability exists at intermediate altitudes. It results in the appearance of electrostatic and magnetostatic modes.

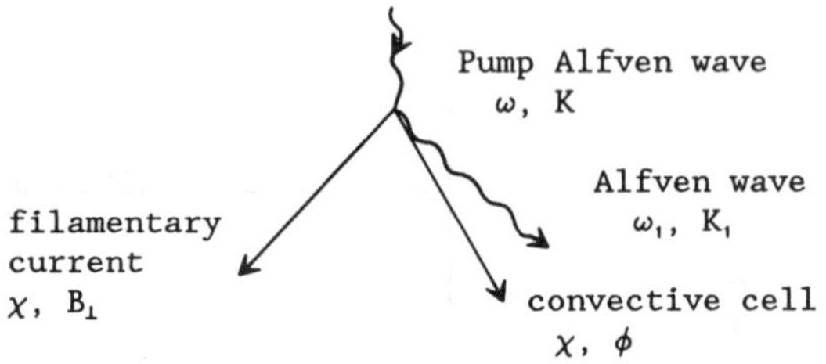

The growth rate is $\gamma_{m.d.} \sim \omega_k(\delta B_k/B)^2 K_\perp^2/K_z^2$. The threshold for amplitude of the pump Alfven wave is $(\delta B/B_0)^2 > 1/8 (K_z/K_\perp)^2 \rho^2 \chi^2$ (χ^{-1} is the transverse scale of the convective cell and of the filamentary field-aligned current).

For typical disturbances $\delta B/B_0 \sim 2.5 \cdot 10^{-3}$ for transverse scales of ~20 km at altitudes ~900 km (this corresponds to ~150 km at h ~~$3R_E$) we obtain the characteristic time of instability: (1-2)s ($K_\perp/K_{||} \sim 10^2$). Thus an Alfven wave propagating to the ionosphere can decay into another Alfven wave, a convective cell and a magnetostatic mode. As a result of the instability the amplitudes of the static modes increase and become comparable with the amplitudes of the Alfven waves. The process cannot be described in the framework of the theory of weak plasma turbulence. The dispersion effect is nonessential and the equations describing the dynamics of the Alfven waves are reduced to the equations describing MHD-turbulence. The turbulence becomes strong. It consists of an ensemble of shear Alfven waves, convective cells (vortices) and magnetostatic fluctuations.

The spectra of disturbances should have a power law with energy cascading to small transverse scales. Cascading mechanisms can be responsible for the excitation of small-scale field-aligned currents. The observations of power-law spectra and of a vortex structure in the field distribution in different spectral intervals are in good agreement with these notions [Gurnett et al., 1984; Weimer et al., 1985; Dubinin et al., 1988].

processes (in the MHD approximation Alfven waves even with large amplitudes don't interact between themselves) that can be responsible for the excitation of fine static structures. Other peculiarities ("V"-shaped potential structures) are reproduced in numerical simulations of time-dependent models [Lysak and Dum, 1983; Lysak, 1985; Seyler, 1988].

Two altitude scales should be distinguished while analyzing shear Alfven waves in the polar magnetosphere as shown in Table 1.

Summary

The following qualitative model can be proposed for shear Alfven waves with large amplitudes propagating from the deep magnetosphere to the Earth. At intermediate altitudes ($h \geq 3\ R_E$) where the warm plasma approximation ($m/M \leq \beta << 1$) is valid, Alfven wave can decay to Alfven and static modes. The sources of free energy are shears in the plasma flow and in the magnetic field. The latter are caused by field-aligned currents. Magnetic field bendings are untied from drift plasma motions for scales of $K_\perp \rho \sim 0.1$. The structures are then mainly potential. At altitudes 3-4 R_E, where the ion Larmor radius is several kilometers, the typical scales of such structures are ~100 km (~5 km at ionospheric altitudes). Strong Alfven turbulence with the cascade of energy to smaller scales is excited. The slope of the magnetic spectrum should increase with increasing of $K_\perp$ due to the excitation of anomalous resistivity and the dissipation of magnetic field energy ($K_\perp^2 a^2 \leq \omega/\nu_{eff} << 1$). For $\nu_{eff} \sim 0.4\ \Omega_{Hi} \sim 25\ s^{-1}$ [Lysak and Dum, 1983] and $\omega \sim 1\ s^{-1}$ we obtain $K_\perp a \sim 1/5$ and $\lambda_\perp \sim 30\ a \sim 50$-$150$ km ($n \sim 10 \div 1\ cm^{-3}$). So the characteristic scale of the smallest field-aligned currents is determined by the value of the ion Larmor radius and (or) by the diffusion scale. This gives scales ~1 km at ionospheric altitudes. The closing of electric field equipotentials at altitudes of (1-2) R_E produces "V"-shaped structures that should be responsible for electron acceleration.

At lower altitudes where the dispersion is determined by electron inertia, the structures generated by these "static" small-scale field-aligned currents reveal their electromagnetic origin more distinctly. The observations on ICB-1300 satellite [Dubinin et al., 1985, 1986, 1988, 1989] and the similarity of magnetic spectra on DE-1 with electric field spectra on DE-2 [Weimer et al., 1985] support this point of view. The excitation of static structures (convective cells; magnetostatic modes) can take place at low altitudes too. The characteristic smallest scale for vortices is c/ω_{pe}.

Questions and Answers

Haerendel: What is the main effect of the small-scale Alfvén wave turbulence? Do they produce low energy electrons which may produce radio aurora and could they also create ion conics by transverse ion heating?

Dubinin: Indeed, the main effect of small-scale Alfvén waves at low altitudes may be the production of ion conics due to stochastic transverse ion acceleration in random electric fields. The dissipation of field-aligned currents can apparently result in electron heating, but we have no observational evidence of it.

References

Bostrom, R., G. Gustafsson, B. Holback, G. Holmgren, H. Koskinen, and P. Kintner, Characteristics of solitary waves and weak double layers in the magnetospheric plasma, IRF preprint 105, 1988.

Cheng, C. and H. Okuda, Formation of convective cells, anomalous diffusion and strong plasma turbulence due to drift instabilities, *Phys. Rev. Lett.*, *38*, 708, 1977.

Dubinin, E. M., N. S. Nikolaeva, I. M. Podgorny, V. M. Balebanov, L. Bankov, N. Bankov, I. Kutiev, P. Marinov, K. Serafimov, and L. Todorieva, Large and small scale plasma motions in topside ionosphere ICB-1300, *Kosmicheskie Issledovanija*, *21*, 697, 1983.

Dubinin, E. M., I. M. Podgorny, V. M. Balebanov, I. Bankov, N. Bankov, G. L. Gdalevich, Z. Dachev, L. I. Zhuzgov, I. Kutiev, V. I. Lazarev, N. S. Nikolaeva, K. Serafimov, G. Stanev and D. Teodosiev, The intense localized disturbances of the auroral ionosphere, *Kosmicheskie Issledovanija*, *23*, 449, 1985a.

Dubinin, E. M., P. L. Israelevich, I. Kutiev, N. S. Nikolaeva and I. M. Podgorny, Localized auroral disturbance in the morning sector of topside ionosphere as a standing electromagnetic wave, *Planet. Space Sci.*, *33*, 597, 1985b.

Dubinin, E. M., P. L. Israelevich, N. S. Nikolaeva, I. M. Podgorny, N. Bankov, and L. Todoreva, The electromagnetic structures at the auroral altitudes, *Kosmicheskie Issledovanija*, *24*, 434, 1986.

Dubinin, E. M., A. A. Volokitin, P. L. Israelevich, and N. S. Nikolaeva, Auroral electromagnetic disturbances at altitudes 900 km: Alfven wave turbulence, *Planet. Space Sci.*, *36*, 949, 1988.

Dubinin, E. M., P. L. Israelevich, and N. S. Nikolaeva, Auroral electromagnetic disturbances at altitudes 900 km: The relationship between electric and magnetic fields, *Planet. Space Sci.*, In press, 1989.

Gurnett, D. A., R. L. Huff, J. D. Menietti, J. D. Winningham, I. L. Burch, and S. D. Shawhan, Correlated low frequency electric and magnetic noise along the auroral field lines, *J. Geophys. Res.*, *89*, 8971, 1984.

Haerendel, G., An Alfven wave model of auroral arcs, in *High Latitude Space Plasma Physics*, edited by B. Hultquist and T. Hadfords, Plenum Publishing Corporation, NY, p. 515, 1983.

Heelis, R. A., J. D. Winningham, M. Sugiura, and N. C. Maynard, Particle acceleration parallel and perpendicular to the magnetic field observed by DE-2, *J. Geophys. Res.*, *89*, 3893, 1984.

Kan J. R., L. H. Lyu, and V. M. Vasyliunas, Interaction between Alfven waves and field-aligned potentials, Biennial Report 1985-1986 Geophysical Institute, University of Alaska, Fairbanks, p. 30, 1988.

Koskinen, H., R. Bostrom, and B. Holback, Viking observations of solitary waves and weak double layers on auroral field lines, in *Ionosphere-Magnetosphere-Solar Wind Coupling Processes*, edited by T. Chang, G. B. Crew, and J. R. Jasperse, Scientific, Cambridge, MA, 1988.

Lysak, R. L. and C. T. Dum, Dynamics of magnetosphere-ionosphere coupling including turbulent transport, *J. Geophys. Res.*, *88*, 365, 1983.

Lysak, R. L., Auroral electrodynamics with current and voltage generators, *J. Geophys. Res.*, *90*, 4178, 1985.

Lysak, R. L., Coupling of the dynamic ionosphere to auroral flux tubes, *J. Geophys. Res.*, *91*, 7047, 1986.

Mallinckrodt, A. J. and C. W. Carlson, Relations between transverse electric fields and field-aligned currents, *J. Geophys. Res.*, *83*, 1426, 1978.

Maynard, N. C., J. P. Heppner, and A. Egeland, Intense, variable electric fields at ionospheric altitudes in high latitude regions as observed by DE-2, *Geophys. Res. Lett.*, *9*, 981, 1982.

Mozer, F. S., C. A. Cattell, M. K. Hudson, R. L. Lysak, M. Temerin, and R. B. Torbert, Satellite measurements and theories of auroral particle acceleration, *Space Sci. Rev.*, *27*, 155, 1980.

Sagdeev, R. Z., V. D. Shapiro, and V. I. Shevchenko, Convective cells and anomalous diffusion of plasma, *Physics Plasma*, *4*, 451, 1978 (in Russian).

Seyler, C. E., Nonlinear 3-D evolution of bounded kinetic Alfven waves due to shear flow and collisionless tearing instability, *Geophys. Res. Lett.*, *15*, 756, 1988.

Sugiura, M., N. C. Maynard, W. H. Farthing, J. P. Heppner, B. G. Ledley, and L. J. Cahill, Initial results on the correlation between the electric and magnetic fields observed from the DE 2 satellite in the field-aligned current regions, *Geophys. Res. Lett.*, *9*, 985, 1982.

Temerin, M., K. Cerny, W. Lotko, and F. S. Mozer, Observations of double layers and solitary waves on auroral zone field lines, *Phys. Rev. Lett.*, *48*, 1175, 1982.

Volokitin, A. S. and E. M. Dubinin, The turbulence of Alfven waves in the polar magnetosphere of the Earth, *Planet. Space Sci.*, in press, 1989.

Weimer, D. R., C. K. Goertz, D. A. Gurnett, N. C. Maynard, J. L. Burd, Auroral zone electric fields from DE-1 and 2 and magnetic conjunctions, *J. Geophys. Res.*, *90*, 7479, 1985.

Weimer, D. R., D. A. Gurnett, C. K. Goertz, J. D. Menietti, J. L. Burch, and M. Sugiura, The current-voltage relationship in Auroral current sheets, *J. Geophys. Res.*, *92*, 187, 1987.

Zmuda, A. J., J. H. Martin, and F. T. Heuring, Transverse magnetic disturbances at 1100 km in the auroral region, *J. Geophys. Res.*, *71*, 5033, 1966.

OBSERVATIONS OF FILAMENTARY FIELD-ALIGNED CURRENT COUPLING BETWEEN THE MAGNETOSPHERIC BOUNDARY LAYER AND THE IONOSPHERE

C. R. Clauer, M. A. McHenry[1]

STAR Laboratory, Stanford University, Stanford, California, 94305

E. Friis-Christensen

Division of Geophysics, Danish Meteorological Institute, Lyngbyvej 100
DK-2100, Copenhagen, Denmark

Abstract. A distinct class of dayside high-latitude magnetic pulsations can be identified from the spatial characteristics of the disturbance field. These pulsations exhibit traveling radial patterns such as would result from moving filaments of field-aligned current interacting with the ionosphere to produce cells of Hall current and vortex-like plasma flow. Time intervals containing a series of continuous multiple vortices are investigated here. We find that the vortices occur on the boundary between sunward and anti-sunward ionospheric plasma convection. Low altitude DMSP satellite particle measurements indicate that the vortices are on magnetic field lines which map to the inner edge of the magnetospheric low latitude boundary layer. No repetitive solar wind disturbance (eg. pressure variations) appears to be associated with the events suggesting that the vortices are related to a local magnetospheric instability. No strong correlation between interplanetary field conditions and the detection of vortices is found, but conditions which slightly favor the detection of vortices are IMF $B_x < 0$, $B_y > 0$, high solar wind density and slow solar wind speed.

[1] now at Remote Measurements Laboratory, SRI International, 333 Ravenswood Ave. Menlo Park, California, 94025

Geophysical Monograph 58

Introduction

Ground magnetic observations made in the high latitude dayside region have recently become the focus for a variety of investigations. This focus has resulted from the desire to observe phenomena associated with the momentum coupling occurring between the solar wind and the Earth's magnetic field. Magnetic field lines at invariant latitudes around 75° on the dayside map to the vicinity of the magnetospheric boundary where the solar wind first encounters the geomagnetic field. Because the ionosphere is one of the electrically conducting boundaries of the system, phenomena measured at the ionospheric intersection of these field lines are likely to show manifestations of the coupling processes occurring between the solar wind and the magnetosphere.

One indication of coupling between the solar wind, magnetosphere and ionosphere is in the strong relationship observed between the large scale dayside high latitude F-region ionospheric plasma convection and the strength and orientation of the interplanetary magnetic field (IMF) observed upstream near the magnetopause [Clauer et al., 1984; Rishbeth et al., 1985; Clauer and Banks, 1986; Clauer and Friis-Christensen, 1988; Etemadi et al., 1988; Lockwood and Freeman, 1988]. F-region plasma convection is the result of $\mathbf{E} \times \mathbf{B}$ plasma drift and changes in the convection are the result of changes in the applied electric field and associated field-aligned current distributions. The response of the dayside high latitude ionosphere to changes in the solar wind is observed to occur within a few minutes of the IMF change encountering the magnetopause. Since the observed response time is on the order of the time for an Alfven wave to travel from the magnetopause to the ionosphere, this is strong evidence for a direct electrical connection between the solar wind and the dayside high latitude ionosphere [Clauer and Friis-Christensen, 1988; Etemadi et al., 1988].

A variety of ground based, high latitude, irregular magnetic pulsations have been identified which are also considered to be indications of momentum coupling between the solar wind and the Earth's magnetosphere and ionosphere [Bolshakova et al. 1975; Kleymenova et al. 1982, 1985; Bolshakova and Troitskaya, 1982]. Some of these pulsations are particularly interesting because of the spatial characteristics of the magnetic disturbance. For these pulsations, the horizontal magnetic perturbation vectors appears to point radially toward or away from a single point [Friis-Christensen et al., 1988a; McHenry et al., 1988]. Such a signature would be expected from the circular Hall current in the ionosphere associated with a filamentary field-aligned current [McHenry and Clauer, 1987]. The radial perturbation pattern is observed to move tailward along lines of invariant latitude.

Two classes of these traveling ionospheric current vortices have been observed. The first class of phenomena consists of transient, large amplitude, single cycle pulsations observed in the magnetic records [Friis-Christensen et al., 1988a,b]. These events appear to be associated with sudden changes in the solar wind dynamic pressure and/or interplanetary magnetic field (IMF) orientation. This class is discussed in greater length in the paper by Friis-Christensen [1989], in this volume. The second class which will be discussed in more detail in this paper consists of multiple vortices that are aligned in anti-sunward traveling 'trains'. These pulsations may exist for several hours and it is found that neighboring vortices in these events have alternating rotation direction. This is consistent with trains of alternately directed filamentary field-aligned current pairs moving tailward.

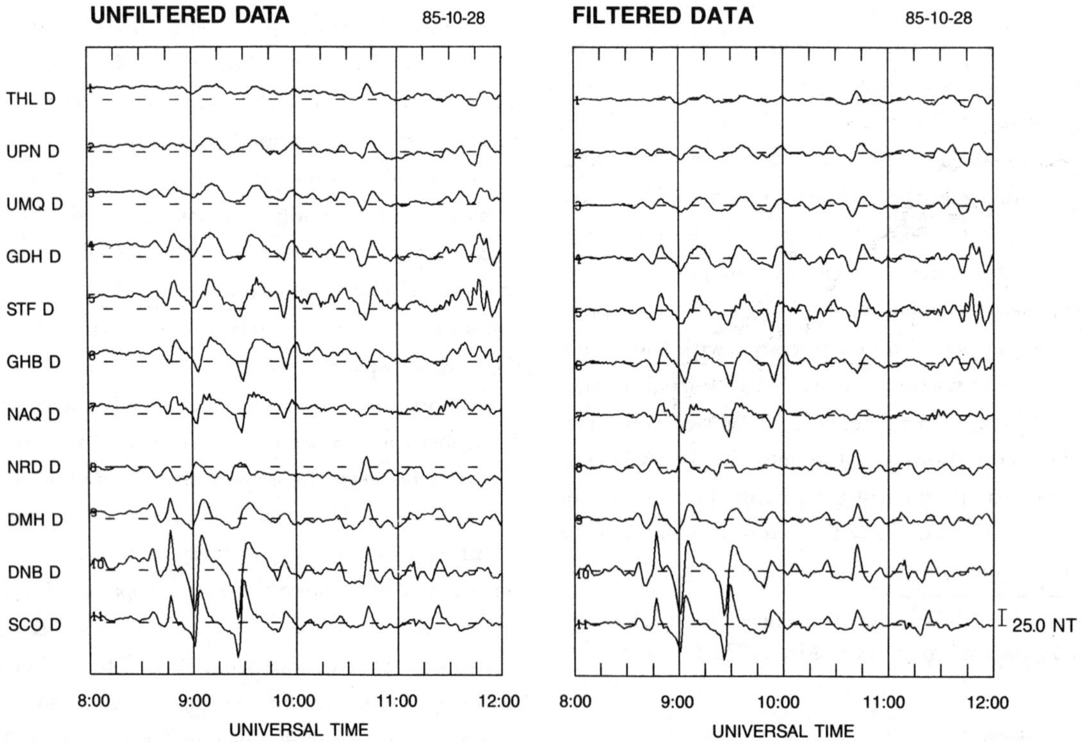

Fig. 1. D component of the magnetic field from the Greenland magnetometer chains on October 28, 1985. Top 7 traces are from the west coast stations, bottom 4 traces are from the east coast stations. Unfiltered data are shown in the left panel and the results of applying a 30 minute high pass filter to the data are shown in the right panel.

Observations

The magnetic data presented here are from the Greenland magnetometer chains located on the east and west coasts of Greenland (see Figure 1 of Friis-Christensen et al., 1988a and Figure 3 of the present paper). A 30 minute high pass filter is applied to the data. This removes the Earth's background field, errors due to equipment drift, and diurnal variations. The periods of the variations of interest here are on the order of 10 to 20 minutes.

Figure 1 shows the unfiltered (left) and filtered (right) D component from the Greenland magnetometers for the interval 08 to 12 UT on October 28, 1985. At 08:30 UT (06:30 LT) continuous irregular pulsations begin at both east coast (top 7 curves) and west coast (bottom 4 curves) stations. The pulsations last for several hours but are greatly reduced in amplitude by 12 UT.

It is well known that the F-region plasma velocity is the result of $\mathbf{E} \times \mathbf{B}$ drift and is directed roughly 90° CCW (viewed from above) to the observed ground magnetic perturbations produced by E-region Hall currents. Using Ohms law, $J_H = \Sigma_H E$ where J_H and Σ_H are the Hall current and integrated Hall conductivity respectively, the F-region velocity can be related to the strength of the Hall currents by

$$v = E/B = \frac{J_H}{\Sigma_H B_0}.$$

If the ionospheric current dimension scale is larger than the distance to the ground, roughly 100 km, then it produces a uniform magnetic field on the ground. Thus,

$$B = \mu H = \frac{\mu J_H}{2} = \frac{\mu \Sigma_H B_0 v}{2}$$

$$v = \frac{2B}{B_0 \mu \Sigma_H}$$

where B is the magnetic perturbation due to the Hall current and B_0 is the background geomagnetic field. This approach is commonly used to estimate ionospheric flows from ground magnetometer data. The derivation ignores ground induction effects which would tend to increase the horizontal field and reduce the vertical magnetic field for the moving vortex-like currents systems which will be discussed below. Currents moving at 1 to 6 km s^{-1} should produce negligible induction effects. [McHenry, 1989].

In Figure 2 we show the ionospheric velocity perturba-

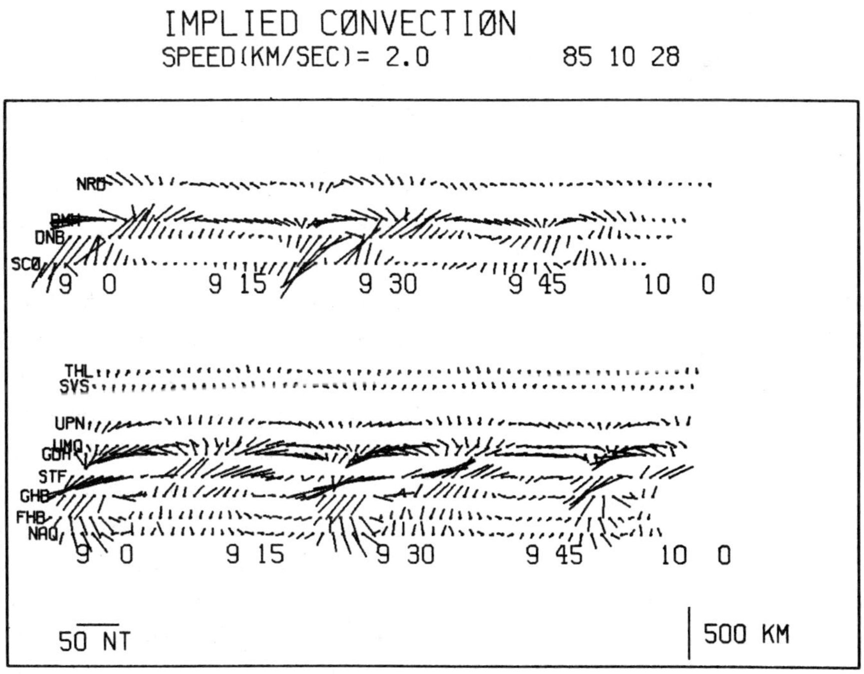

Fig. 2. Inferred F-region plasma flow perturbations from 9:00 to 10:00 UT on October 28, 1985. A westward (antisunward) velocity of 2 km s^{-1} is assumed in this standard plot format. The top portion of the plot shows observations from the east coast stations, the bottom portion shows observations from the west coast stations.

tions from 9:00 to 10:00 UT, October 28, 1985 inferred from the magnetic data shown in Figure 1. Each horizontal magnetic perturbation is rotated 90° CCW and the station location for sequential observations is offset a distance corresponding to an ionospheric speed of 2 km s^{-1} which is our standard plot format. While the speed of the current systems may vary from this assumed speed, this format shows the qualitative characteristics of the vortex patterns associated with these systems. In contrast to the single cycle event discussed by Friis-Christensen et al. [1988a,b], this event shows a succession of vortices lasting for many cycles.

Since these pulsations are apparent at stations on both coasts of Greenland, it is possible to identify the same features in the vortices on both coasts and thereby determine the velocity. Vortices are observed in the east coast data about 3.5 to 4 minutes earlier than in the west coast data. Thus the velocity is estimated to be about 6 km s^{-1} westward (anti-sunward). The center of the vortices (and hence the inferred filamentary field-aligned currents) appears to be near station STF (invariant latitude 74°.

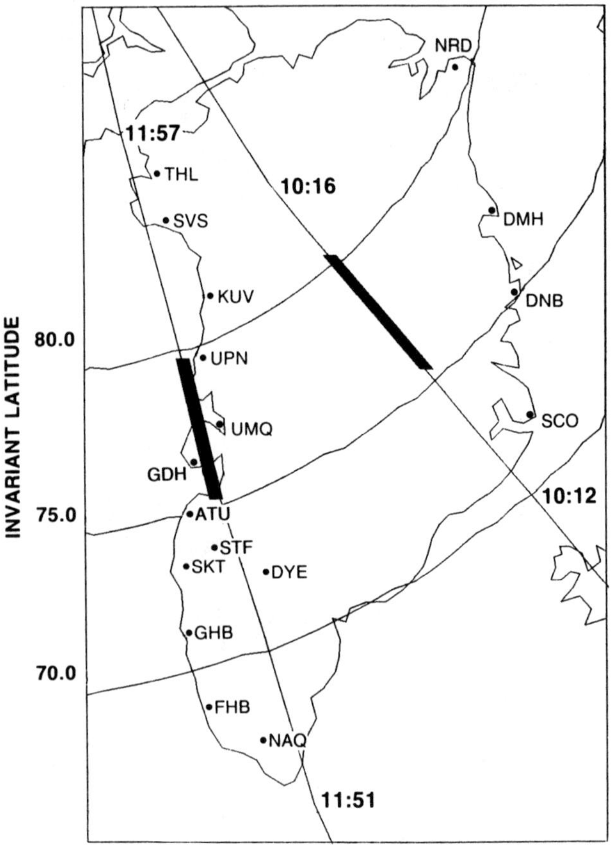

Fig. 3. DMSP satellite trajectories at 10:16 and 11:57 UT on October 28, 1985. Invariant latitude lines are drawn across the plot. Portions of the orbit where boundary layer plasma are observed are indicated by wider trajectory lines.

Figure 3 shows the trajectory of two consecutive passes by the DMSP satellite over Greenland on October 28, 1985 during these pulsations. The wider trajectory lines show the location of the magnetospheric boundary layer identified using the plasma data [Newell and Meng, 1988]. The identification of the boundary layer is based upon examination of the average electron and ion energies of the precipitating particles. From the boundary layer identification shown in Figure 3, it appears that the source field-aligned currents for the vortices appear to originate near the inner edge of the magnetospheric boundary layer. Indeed, this appears to be typical for most events examined where DMSP data are available [McHenry et al., 1989].

The location of the vortices also appears to be very near the ionospheric convection reversal boundary. Figure 4 shows a series of afternoon vortices observed from 15 to 16 UT on October 29, 1989. The center of the vortices appears to be near the station UMQ or GDH. Simultaneous measurements of the high latitude plasma convection and reversal boundary location were obtained using the Sondre Stromfjord incoherent scatter radar using fixed elevation azimuth scans. In Figure 5, the line of sight (LOS) velocity data (either away from or toward the radar) are shown for several scans using a vector plot. Each scan took approximately 5 minutes and covered between 45 and 60 degrees of azimuth. The data were integrated for 10 second intervals. The view in the figure is from above looking down on the Earth and the location of nearby Greenland magnetometer stations are shown with their 3-letter names. The shear in the ionospheric flow, called the convection reversal boundary, is located by identifying the points where the LOS velocity changes sign. These are marked by lines in the plots. Successive scans show that the boundary moves or undulates with time and is near the location of UMQ and GDH. Thus, the vortices appear to be centered on the reversal boundary and the vortex flow perturbations may well be associated with the observed deflections of the boundary in the radar measurements.

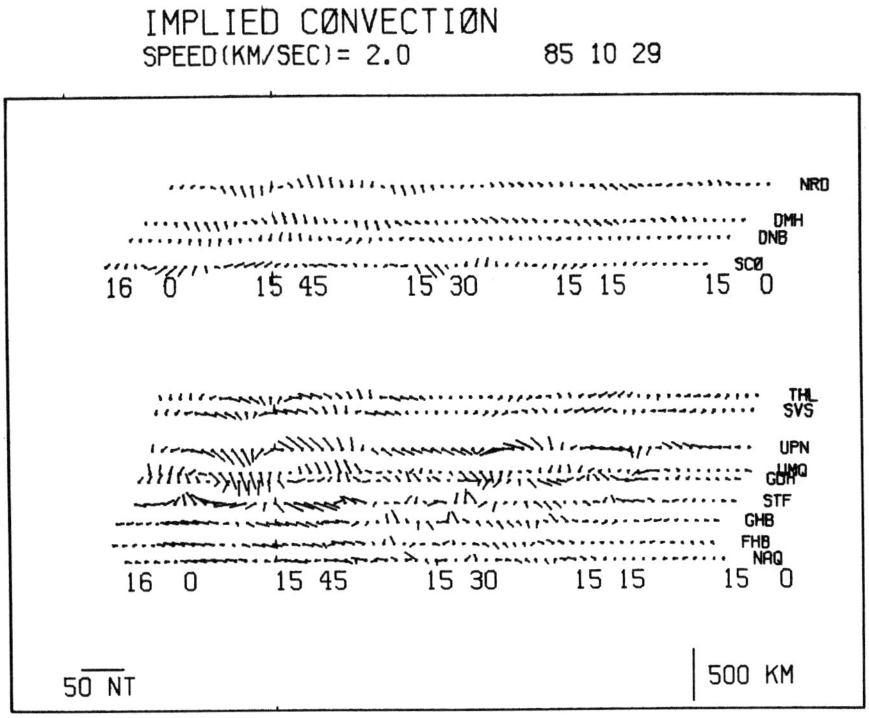

Fig. 4. Inferred F-region plasma flow perturbations from 15:00 to 16:00 UT on October 29, 1985 shown in the same format as Figure 2. An eastward (anti-sunward) velocity of 2 km s^{-1} is assumed.

In order to obtain a statistical description of these dayside multiple vortex intervals, we performed a computer search on 50 days of Greenland magnetometer data. Universal time hours between 09 and 21 were searched (local geographic noon is at 15 UT). The search utilized a least squares error pattern match to the modeled ground magnetic field produced by a filamentary field-aligned current interacting with a uniform ionosphere [McHenry and Clauer, 1987; McHenry et al., 1989]. The modeled field is fit to the field perturbations measured by the magnetometer array at each measurement time. The free parameters in the fit are the current in the filament and the north-south and east-west location where the filament intersects the ionosphere relative to the magnetometer stations.

Figure 6 is a horizontal bar chart showing the time distribution when vortices were found using the pattern matching technique. The horizontal axis show universal time (local magnetic noon is at 14 UT) and each bar indicates an interval when vortices were detected. The width of each bar indicates the average amount of current in the fit. There is a clear preference for vortices to occur in the post noon period.

Figure 7 shows the relationship between upstream solar wind measurements and the occurrence of vortices. The histograms have been normalized by the corresponding solar wind parameter probability distribution which is also shown on the plot. No strong correlation appears, but conditions which perhaps slightly favor vortices appear to be $B_x < 0$, $B_y > 0$, high solar wind density and slow solar wind speed.

Discussion

We find that under quiet conditions, characterized by small amplitude variations in the Greenland magnetometer data, a significant fraction of days have dayside, high latitude, continuous, irregular, magnetic pulsations which are caused by trains of steady, alternately directed, field-aligned currents producing ionospheric vortices which tra vel anti-sunward. The vortices are located at the convection reversal boundary in the ionosphere and the field lines associated with the vortices appear to map near to the inner edge of the low latitude boundary layer.

Vortex intervals show a weak preference for solar wind

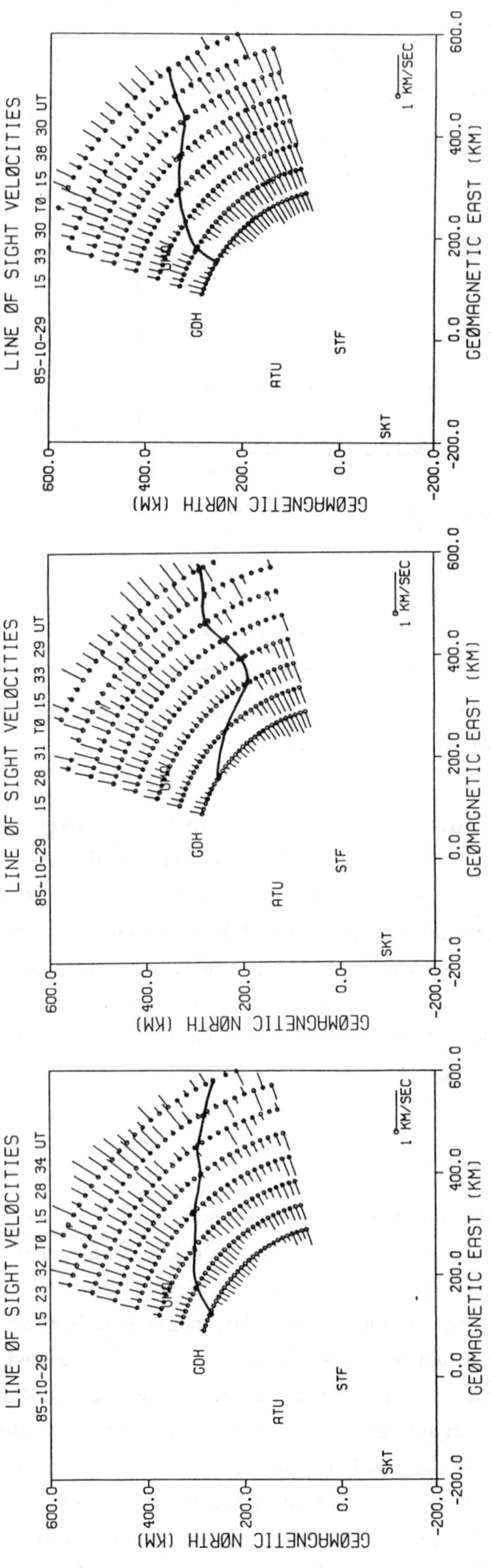

conditions with slow speed, high density, negative B_x, positive B_y and large values of $|B_z|$. Quiet days which have no pulsations are characterized also by slow solar wind speed and high density, however B_x and B_z are near zero or positive.

We believe that the vortices are the result of wave instability at the inner edge of the low latitude boundary layer electrically coupling to the ionosphere. A likely instability is the Kelvin Helmholtz shear instability. Higher than average solar wind density is consistent with this.

We believe that the slow solar wind speed dependence is related to the vortex identification process. The clearest vortex patterns are found at low solar wind speeds.

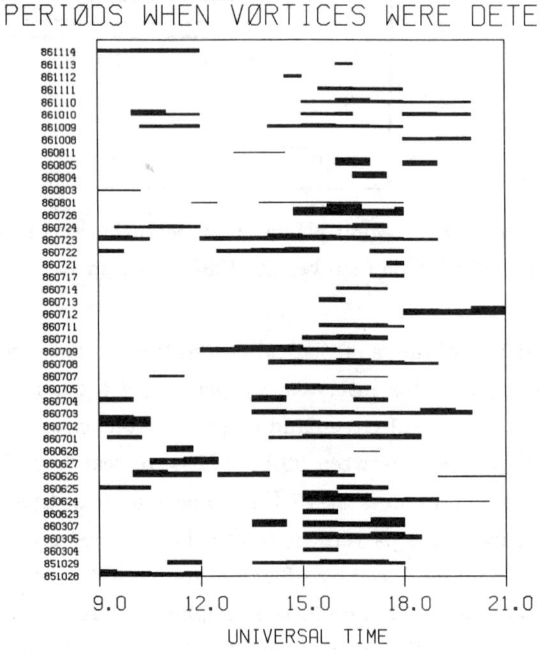

Fig. 6. Local time dependence of vortex detection. Local magnetic noon is at 14:00 UT, local geographic solar noon is at 15:00 UT.

Fig. 5. Line of sight ionospheric flow vectors measured by the Sondre Strømfjord, Greenland incoherent scatter radar on October 29, 1985. The line drawn through the vectors divides toward and away flow and marks the location of the ionospheric convection reversal boundary. Three sequential scans are shown, 15:23:32 to 15:28:34 UT (left), 15:28:31 to 15:33:29 UT (middle), and 15:33:30 to 15:38:30 UT (right) .

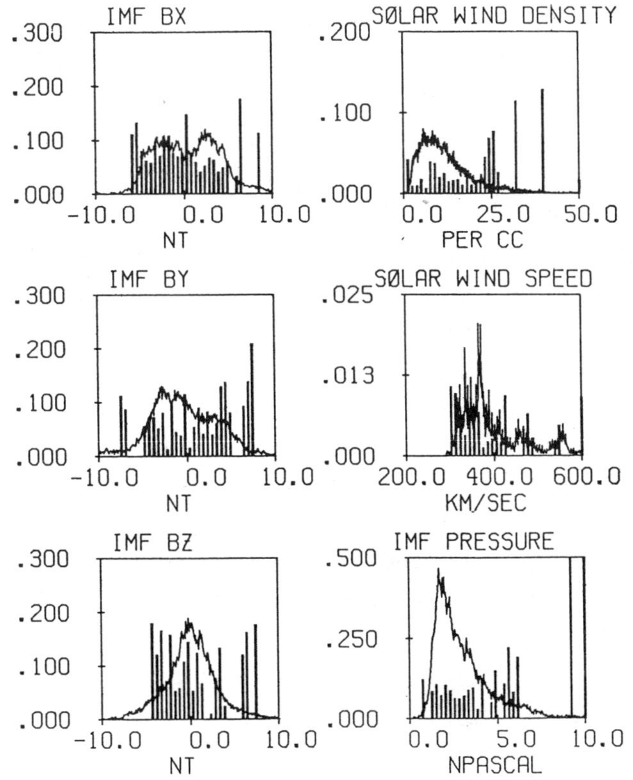

Fig. 7. Histograms showing periods when vortices were detected vs. solar wind parameters. The probability distribution for each parameter is shown on each plot and the histograms are normalized to this distribution.

Assuming approximately 10 samples per second are required to determine the two dimensional patterns of a traveling vortex, the 1 minute data available for this investigation will not be adequate to identify fast moving patterns. Thus, we have potentially missed events which may be associated with faster solar wind speeds. Examination of a small amount of 20-second data obtained from an temporary array of magnetometers indicate that higher frequency pulsations are associated with higher solar wind speeds, and this would be consistent with the Kelvin Helmholtz mechanism

Occurrence frequency also appears to favor afternoon hours. Other studies have reported local time asymmetries in the occurrence of pulsations at high latitudes. Olson and Rostoker [1978] showed that high latitude morning pulsations tend to be more sinusoidal and afternoon pulsations more irregular. Gupta [1975] found than diurnal variations in the numbers of large Pc5 events observed depends upon invariant latitude. A polar cap station ($\Lambda = 83.1°$) observed more Pc5 events post-noon while at lower latitudes ($\Lambda = 73.9°$) more events occur pre-noon. Lee and Olson [1980] attribute diurnal asymmetries to differences in the dawn and dusk magnetosheath field strength and suggest that the dawn boundary should be more Kelvin-Helmholtz unstable than the dusk boundary. Our observations appear to either contradict this or to be inconsistent with the Kelvin-Helmholtz mechanism. However, it is possible that the vortices can only be identified when the boundary layer is not overly turbulent. The dawn boundary being more unstable could produce smaller, broken and/or unsteady vortices which are more difficult to identify on the ground. Further investigation of the spatial characteristics and local time asymmetries of the vortex pulsations is required.

The IMF conditions required for Kelvin Helmholtz instability are not well known. Recent work suggests that the thickness of the boundary layer [Miura, 1984; Miura 1985] and the electrical coupling of the boundary layer to the ionosphere [Keskinen et al., 1988] are important parameters of the instability. They also find that the wavelengths which have the maximum growth rate are similar to the boundary layer thickness. Our observations, however, indicate that the wavelengths are much larger than the boundary layer thickness. We estimate that the phase velocity at the boundary layer is on the order of 200 km s^{-1} with a 10 to 20 minute period. The wavelength is, therefore, roughly equal to $19R_e$ and would cover the entire dayside magnetopause. With such large disturbances, the analysis must include the magnetosphere's three dimensional aspects.

Since the discovery of isolated magnetic reconnection events on the surface of the magnetopause called flux transfer events (FTEs) by Russell and Elphic [1978; 1979], there has been a strong linkage of vortex-like pulsations and FTEs. Models of the ground magnetic field of FTEs have suggested that they would produce a radial or a dipolar ground magnetic field [McHenry and Clauer, 1987]. Lanzerotti et al., [1986] noted that several pulsation events had a moving radial horizontal field similar to FTE models. Friis-Christensen et al. [1988], however, noted that these traveling vortices move in the direction of the line connecting the vortex centers which is not in accordance with the FTE models. Also, even though vortices and FTEs share an azimuthal ionospheric signature, a key feature of FTEs is the increased probability of detection when IMF $B_z < 0$ [Berchem and Russell, 1984; Rijnbeek et al., 1984]. No such dependence was found in this investigation. Thus, it is our opinion that the multiple vortices are not the result of FTEs nor directly caused by reconnection phenomena.

Since we also find that the occurrence frequency of vortices shows a slight B_x dependence, magnetosheath turbulence generated by a quasi-parallel bow shock must be considered. Perhaps pressure variations on the magnetopause surface due to this turbulence are the source of the waves associated with the ionospheric vortices. In this mechanism, however, turbulence is generated when the IMF B_x cone angle is near 0° or 180° [Greenstadt and Olson, 1976]. We find, however, a dependence on the sign of B_x rather than cone angle.

Finally, we should state that we have surveyed the upstream solar wind density and velocity data and find no periodic variations which might be driving the vortices. Thus, we feel that the continuous traveling ionospheric vortices are related to a local magnetospheric instability.

Acknowledgements. Support for this research at Stanford University has been provided by National Science Foundation grant ATM-8800327. M. A. M. was partially supported by the NASA Student Researchers Program through grant NGT 50016. The authors would like to thank Drs. R. Lepping and J. King for the IMP-8 magnetic field data, and the solar wind group at MIT, supported in part by NASA Goddard under grant NAG5-584, for the IMP-8 solar wind plasma data and Patrick Newell for the DMSP plasma data identification of magnetospheric boundary layer field lines.

References

Berchem, J. and C. T. Russell, Flux transfer events on the magnetopause: Spatial distribution and controlling factors, *J. Geophys. Res., 89*, 6689, 1984.

Bolshakova, O. V., V. A. Troitskaya and V. P. Hessler: Determination of the position of the polar boundary of the day side cusp from the intensity of high-latitude pulsations, *Geomagnetism and Aeronomy, 15*, 569, 1975.

Bolshakova, O. V. and V. A. Troitskaya: Pulsed reconnection as a possible source of ipcl type pulsations, *Geomagnetism and Aeronomy, 22*, 723, 1982.

Clauer, C. R. et al., Observations of interplanetary magnetic field and of ionospheric plasma convection in the vicinity of the dayside polar cleft, *Geophys. Res. Lett., 11*, 891, 1984.

Clauer, C. R., and P. M. Banks, Relationship of the interplanetary electric field to the high-latitude ionospheric electric field and currents: Observations and model simulation, *J. Geophys. Res., 91*, 6959, 1986.

Clauer C. R. and E. Friis-Christensen, High latitude dayside electric fields and currents during strong northward IMF: Observations and model simulation, *J. Geophys. Res.,93*, 2749, 1988.

Etemadi, A., S. W. H. Cowley, M. Lockwood, B. J. I. Bromage, D. M. Willis and H. Lühr, The dependence of high-latitude dayside ionospheric flows on the north-south component of the IMF: A high time resolution correlation analysis using EISCAT "Polar" and AMPTE UKS and IRM data, *Planet. Space Sci., 36*, 471, 1988.

Friis-Christensen, Eigil, Terrestrial ionospheric signatures of field-aligned currents, this volume, 1989.

Friis-Christensen, E., M. A. McHenry, C. R. Clauer, and S. Vennerstrøm, Ionospheric traveling convection vortices observed near the polar cleft, *Geophys. Res. Lett., 15*, 253, 1988a.

Friis-Christensen, E., S. Vennerstrøm , C. R. Clauer, and M. A. McHenry, Irregular magnetic pulsations in the polar cleft caused by traveling ionospheric current vortices, *Adv. Space Res.*, in press, 1988b.

Greenstadt, E. W. and J. V. Olson, Pc3, 4 activity and interplanetary field orientation, *J. Geophys. Res., 81*, 5911, 1976.

Gupta, J. C., Some characteristics of large amplitude Pc5 pulsations, *Aust. J. Phys., 29*, 67, 1975.

Keskinen, M. J., H. G. Mitchell, J. A. Fedder, P. Satyanarayana, S. T. Zalesak, and J. D. Heba, Nonlinear evolution of the Kelvin-Helmholtz instability in the high-latitude ionosphere, *J. Geophys. Res., 93*, 137, 1988.

Kleymenova, N.G., O.V. Bolshakova, V.A. Troitskaya , and E.

Friis-Christensen: Long period geomagnetic fluctuations and the polar chorus at latitudes corresponding to the daytime polar cusp, *Geomagnetism and Aeronomy, 22*, 580, 1982.

Kleymenova, N.G., O V. Bolshakova, V.A. Troitskaya , and E. Friis-Christensen: Two forms of long-period geomagnetic pulsations near the equatorial border of the dayside polar cusp, *Geomagnetism and Aeronomy, 25*, 139, 1985.

Lanzerotti, J.L. et al., Possible evidence of flux transfer events in the polar ionosphere, *Geophys. Res. Lett. 13*, 1089, 1986.

Lee, L. C. and J. V. Olson, Kelvin-Helmholtz instability and the variation of geomagnetic pulsation activity, *Geophys. Res. Lett., 7*, 777, 1980.

Lockwood, M. and M. P. Freeman, Recent ionospheric observations relating to solar wind - magnetosphere coupling, *Phil. Trans. R. Soc. (London)*, **A**, 1988 (in press).

McHenry, Mark A., and C. Robert Clauer, Modeled ground magnetic signatures of flux transfer events, *J. Geophys. Res., 92*, 11231, 1987.

McHenry, M. A., C. R. Clauer, E. Friis-Christensen, and J. D. Kelly, Observations of ionospheric convection vortices: Signatures of momentum transfer, *Adv. Space Res., 8*, (9)315, 1988.

McHenry, M. A., C. R. Clauer, and E. Friis-Christensen, Relationship of solar wind parameters to continuous, dayside, high latitude traveling ionospheric vortices, *J. Geophys. Res.*, 1989, (submitted).

McHenry, Mark A., *Ground Signatures of Dayside Magnetospheric Boundary Layer Phenomena*, Ph. D. dissertation, Stanford University, Stanford, California, 1989.

Miura, A., Anomalous transport of magnetohydrodynamic Kelvin-Helmholtz instabilities in the solar wind - magnetosphere interaction, *J. Geophys. Res., 89*, 801, 1984.

Miura, A., Kelvin-Helmholtz instability at the magnetopause boundary, *Geophys. Res. Lett., 12*, 635, 1985.

Newell, P. T., and C. Meng, The cusp and the cleft/LLBL: low altitude identification and statistical local time variation, *J. Geophys. Res., 93*, 14,549, 1988.

Olson, J. V. and G. Rostoker, Longitudinal phase variations of Pc 4-5 micropulsations, *J. Geophys. Res., 83*, 2481, 1978.

Rijnbeek, R. P., S. W. H. Cowley, D. J. Southwood, and C. T. Russell, A survey of dayside flux transfer events observed by ISEE 1 and 2 magnetometers, *J. Geophys. Res., 89*, 786, 1984.

Risbeth, H. et al., Ionospheric response to changes in the interplanetary magnetic field observed by EISCAT and AMPTE-UKS, *Nature, 318*, 451, 1985.

Russell, C. T. and R. C. Elphic, Initial ISEE magnetometer results: Magnetopause observations, *Space Sci. Rev., 22*, 681, 1978.

Russell, C. T. and R. C. Elphic, ISEE observations of flux transfer events at the dayside magnetopause, *Geophys. Res. Lett., 6*, 33, 1979.

MEASUREMENT OF FIELD-ALIGNED CURRENTS BY THE SABRE COHERENT SCATTER RADAR

M. P. Freeman[1], D. J. Southwood[1], M. Lester[2], and J. A. Waldock[3].

Abstract. We outline, and discuss the calibration of, an experimental method for the measurement of magnetospheric field-aligned currents using coherent scatter radar data. The technique is applied to identify and analyse a localised upward field-aligned current sheet observed in the late afternoon mid-latitude ionosphere. We argue that this current arises from the earthward transport of hot plasma due to ongoing magnetospheric erosion.

Introduction.

Field-aligned currents are of central importance in understanding the coupling within the Earth's magnetic system. They are the agents by which momentum may be transferred from the magnetosphere to the ionosphere. Here we outline and discuss an experimental method for their detailed study using ionospheric radar data.

Most measurements of field-aligned currents have used spacecraft magnetometers. Armstrong and Zmuda [1973] identified the magnetic signature characteristic of the large-scale magnetospheric current systems: as a polar-orbiting satellite traversed the afternoon auroral zone an eastward magnetic perturbation was measured over ~ 2-3 ^{o} latitude. The observation was interpreted as showing the presence of two oppositely directed current sheets, each aligned along a latitude circle with the poleward current system directed outwards from the ionosphere. In the morning sector the sense of the solenoidal current system was reversed. The global distribution of these large-scale currents was subsequently confirmed by statistical studies [Iijima and Potemra, 1978]. The currents were found to intensify and move equatorward with increasing magnetic activity.

The current systems are intimately related to the global magnetospheric convection and arise from the divergent electric fields associated with latitudinal velocity gradients at the pole ward and equatorward edges of the auroral zone flow cells [Vasyliunas, 1975]. Though the spacecraft observations have been successful in identifying the large-scale magnetospheric current systems, the single point sequential measurements preclude detailed knowledge of their structure and rely upon assumptions about their invariance transverse to the satellite trajectory. In this paper we show how observations of the ionospheric convection field at high temporal and spatial resolution can be used to measure field-aligned currents and determine their spatial structure. As an example we present observations of a localised current sheet which we interpret as being generated by the ring current in response to enhanced reconnection-induced flow.

Analysis Technique.

The Sweden and Britain auroral radar experiment (SABRE) was a bistatic radar facility which transmitted at 140 MHz and measured the amplitude and Doppler velocity of the received signal backscattered from naturally occuring E-region density irregularities. Each of the two antennae arrays received signals from various positions along eight narrow beams which view a common area of the ionosphere of approximate dimension 500 km x 500 km, located in geographic coordinates between 63.6 and 68.6 ^{o}N and -0.5 and 12 ^{o}E (Fig. 1). This allows the calculation of the two dimensional velocity vector for the irregularities at approximately 400 locations with a spatial resolution of 20 km. The temporal resolution is 20 s. From this the two dimensional electron drift velocity field can be inferred.

The irregularities arise from the gradient-drift and the two-stream instabilities which depend upon a relative electron-ion velocity [see e.g. Fejer and Kelley, 1980]. Thus they are generated by plasma convection in a medium where the ion population suffer collisions (e.g. with neutrals) on a time scale much shorter than the electron collision time scale. A number of assumptions are employed in deriving the electron drift velocity from the measured irregularity velocity and the validity of these is still the subject of research [e.g. Robinson, 1986].

We shall assume, a priori, that the inferred velocity field is the true electron drift velocity field. By looking at quantities that we expect to be conserved we might then be able to identify, a posteriori, regimes where the assumptions used in the measurement process are invalid. We find that one quantity that we expect to be conserved in the steady state is div $\mathbf{v}_e$, the divergence of the electron drift velocity field. We can also identify a quantity, involving the measured electron drift velocity, which is a direct measure of a magnetospheric physical parameter that we wish to observe. We find that in the steady state the quantity curl $\mathbf{v}_e$, the curl of the electron drift velocity field, is directly proportional to the field-aligned current. We can therefore use this quantity to remotely sense and understand the physics of the magnetospheric region.

[1]Blackett Lab., Imperial College, London, UK.
[2]Leicester University, Leicester, UK.
[3]Sheffield City Polytechnic, Sheffield, UK.

Geophysical Monograph 58

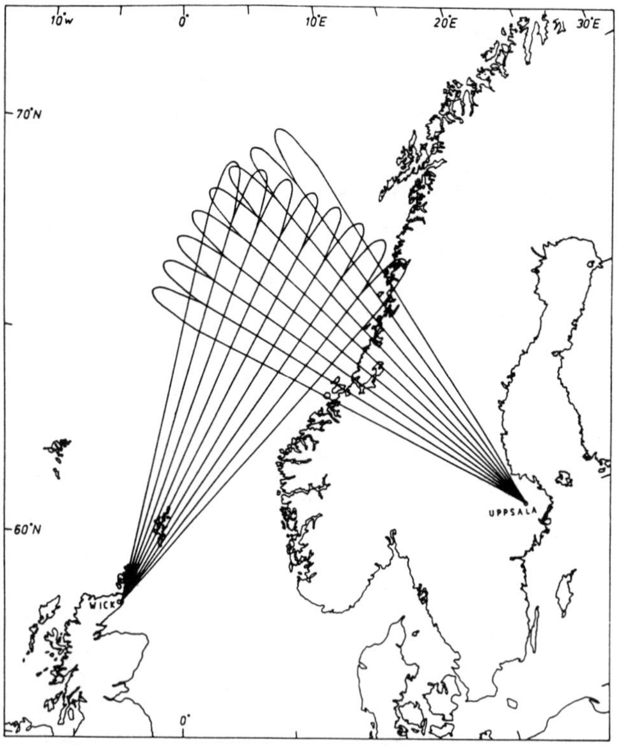

Fig. 1. Illustrating the location and field of view of the SABRE radars at Wick in Scotland and at Uppsala in Sweden.

The electron-neutral collisional frequency is negligible in the E-region ionosphere and so magnetic field lines are "frozen in" to the electron flow:

$$\mathbf{E} = -\, \mathbf{v_e} \times \mathbf{B} \qquad (1)$$

We shall assume that in the absence of an electric field the Earth's magnetic field is purely dipolar and shall also ignore the inclination of the field to the vertical. When an electric field is imposed the magnetic field is perturbed:

$$\mathbf{B} = \mathbf{B_0} + \mathbf{b} \qquad (2)$$

Thus $\mathbf{E}$, $\mathbf{v_e}$, and $\mathbf{b}$ are first order terms with respect to $\mathbf{B_0}$.

The divergence field.

Resolving the curl of equation (1) along the magnetic field we have:

$$\mathbf{B}.\mathrm{curl}\ \mathbf{E} = \mathbf{B}.\mathrm{curl}\ (\mathbf{B} \times \mathbf{v_e})$$

$$= B^2\ \mathrm{div}\ \mathbf{v_e} - \mathbf{B}.\mathbf{v_e}\ \mathrm{div}\ \mathbf{B}$$

$$= B^2\ \mathrm{div}\ \mathbf{v_e}$$

In steady state, curl $\mathbf{E} = 0$, and thus the electron motion is incompressible.

When the derived value of div $\mathbf{v_e}$ is non-zero then we might expect it to be due to one of two possible causes. Firstly, the steady state approximation may no longer be valid. However, the ionosphere remains essentially incompressible even under time-dependent conditions because of the large ambient geomagnetic field; an order of magnitude argument shows that fluctuations in the magnetic field of order 1000 nT on time scales of 1 min generate divergences of less than 1 mHz. Alternatively, the electron drift velocity or its gradient inferred from the backscattered signals may be in error, as we discuss later.

The vorticity field.

Taking the divergence of the frozen-in field condition given by equation (1) yields:

$$\mathbf{B}\ .\ (\mathrm{curl}\ \mathbf{v_e}) = -\ \mathrm{div}\ \mathbf{E} + \mathbf{v_e}\ .\ (\mathrm{curl}\ \mathbf{B}) \qquad (3)$$

Substituting for $\mathbf{B}$ using equation (2) and neglecting terms of second order or higher, equation (3) simplifies to:

$$\mathrm{div}\ \mathbf{E} = -\ \mathbf{B_0}\ .\ (\mathrm{curl}\ \mathbf{v_e}) \qquad (4)$$

By definition, curl $\mathbf{B_0} = 0$.
If the Pedersen conductivity is locally uniform in the ionosphere then:

$$j_{par} = \mathrm{div}\ \mathbf{I} = -\ \Sigma_P\ \mathbf{B_0}\ .\ (\mathrm{curl}\ \mathbf{v_e}) \qquad (5)$$

where $\mathbf{I}$ and Σ_P are the height integrated Pedersen current density and conductivity respectively, and j_{par} is the field-aligned current density. If the vorticity is calculated by a difference equation using discrete velocity measurements on a spatial grid, then equation (5) relates the vorticity to the local field-aligned current density if the Pedersen conductivity varies on a length scale much longer than the distance between neighbouring grid points (~ 20 km). In the absence of any field-aligned current source within the field of view we expect the vorticity field to be given by the lower order terms. Putting div $\mathbf{E} = 0$ in equation (3) yields:

$$\mathbf{B}\ .\ (\mathrm{curl}\ \mathbf{v_e}) = \mathbf{v_e}\ .\ (\mathrm{curl}\ \mathbf{B})$$

$$= \mu_0\ \mathbf{v_e}\ .\ (\mathbf{j_P} + \mathbf{j_H}) = -\ \mu_0\ v_e\ j_H$$

where j_P and j_H are the Pedersen and Hall current densities respectively, and μ_0 is the permeability of free space. Writing $j_H = \sigma_H E = \sigma_H v_e B$, we have:

$$(\mathrm{curl}\ \mathbf{v_e})_{par} = \mu_0\ \sigma_H\ v_e^2 \qquad (6)$$

Thus for an electron drift speed of 500 m s^{-1} and a typical Hall conductivity (~ 10^{-4} Siemens) we find that the expected value of the vorticity in the absence of local field-aligned currents is ~ 0.03 mHz. Values of the vorticity in excess of this figure can arise from one of two sources. Firstly, as indicated above, it may be due to a source of field-aligned current at this point. The structure of the current sheet could be seen over several neighbouring data points if it is of dimensions greater than the range cell. Alternatively, vorticity values above 0.03 mHz could arise from errors in the inferred electron drift velocity or its gradient.

Method for the evaluation of the divergence and vorticity fields.

Fig. 2 shows an example of electron drift velocity vectors inferred from the SABRE. A vector is plotted wherever the two radars view a common ionospheric volume and where the

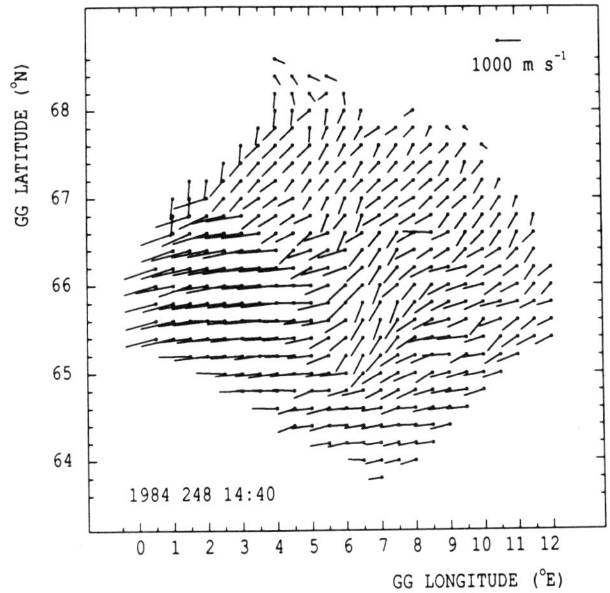

Fig. 2. A "snapshot" of the ionospheric electron drift velocity field at 14:40 UT over the SABRE field of view.

backscatter signal power at both stations is above a set threshold. The major scattering instability occurs only when the electric field magnitude is more than ~ 15 mV m^{-1}, and thus there is a natural cut-off in the backscatter power when the electron drift velocity is below ~ 300 m s^{-1}. In the figure backscatter is above threshold at nearly all common viewing sites, so that the plotted vectors delineate the maximum possible viewing area of the SABRE shown earlier in Fig. 1. The viewing area is diamond shaped within the larger square on which the axes lie. There are about 350 independent velocity measurements in this snapshot with a spatial separation of 20 km. Each measurement in fact represents the electron drift velocity averaged over the range cell area. The electron drift field is stored on magnetic tape as a spatial array of geographic North-South and East-West velocity components at each sample time. The spatially discrete nature of the data means that we have to rewrite the expressions used to calculate quantities such as the divergence and vorticity as difference equations.

To find the divergence or curl of some vector we need to know the spatial partial derivatives of its components. To estimate these derivatives using discrete data we first consider the Taylor expansion of a scalar quantity, f, which is known at a point ($x \pm a$), where a is a small displacement from the position, x, at which the value of the spatial derivative is to be found:

$$f(x \pm a) = f(x) \pm f'(x).a + {}^1\!/_2 f''(x).a^2 + ... \qquad (7)$$

where the dash (′) indicates the spatial derivative operator, d/dx. Now from equation (7) we can see that an expression for the spatial derivative, $f'(x)$, can be obtained:

$$f'(x) = \{f(x+a) - f(x-a)\}/2a + O(a^2) \qquad (8)$$

This expression is accurate to second order in a.

The result is generalised to two dimensions by replacing the total derivative by the appropriate partial derivative. Consider a grid where Δn and Δe are the distances between neighbouring data points in the geographic northward and eastward directions respectively. Let $n.\Delta n$ and $e.\Delta e$ be the north and east coordinates of a point within the field of view (n,e integer). Then the expressions for curl $\mathbf{v_e}$ and div $\mathbf{v_e}$ to second order are:

$$\begin{aligned} \text{curl } \mathbf{v_e} &= \partial v_N / \partial e - \partial v_E / \partial n \\ &= \{v_N(n, e+1) - v_N(n, e-1)\} / 2\Delta e \\ &\quad - \{v_E(n+1, e) - v_E(n-1, e)\} / 2\Delta n \end{aligned} \qquad (9)$$

$$\begin{aligned} \text{div } \mathbf{v_e} &= \partial v_N / \partial n + \partial v_E / \partial e \\ &= \{v_N(n+1, e) - v_N(n-1, e)\} / 2\Delta n \\ &\quad + \{v_E(n, e+1) - v_E(n, e-1)\} / 2\Delta e \end{aligned} \qquad (10)$$

where v_N and v_E are the northward and eastward components of the electron drift velocity. $v_N(n+1, e)$ is the northward flow component at position $((n+1).\Delta n, e.\Delta e)$.

Thus to calculate the divergence or vorticity at a given point we use data from the four nearest neighbours. To minimise errors arising from this technique we check that all of the data points used have a backscatter power above a set threshold.

Results.

As an illustration of how the divergence and vorticity fields can be used as a diagnostic tool with SABRE we present some results from a case study on September 4, 1984. AMPTE-IRM measurements from the magnetopause boundary layer showed that the dayside magnetosphere was undergoing erosion, at least from 14:42 UT to 15:01 UT. Concurrent measurements on the ground support this view by the observation of an expanding polar cap between ~ 13:50 UT and ~ 15:00 UT [M. P. Freeman, unpublished data, 1989].

Fig. 3 shows the variation of the electron drift velocity field with time near 16:00 MLT. The data have been averaged over three areas within the SABRE viewing area. Each area is two degrees of longitude by one degree of latitude, centred on 5 °E and 65.0, 66.0, and 67.0 °N in geographic coordinates (MLT ~ UT + 1.5 h, L ~ 5 in geomagnetic coordinates). The latitudinal bins are denoted by the different line types as indi-

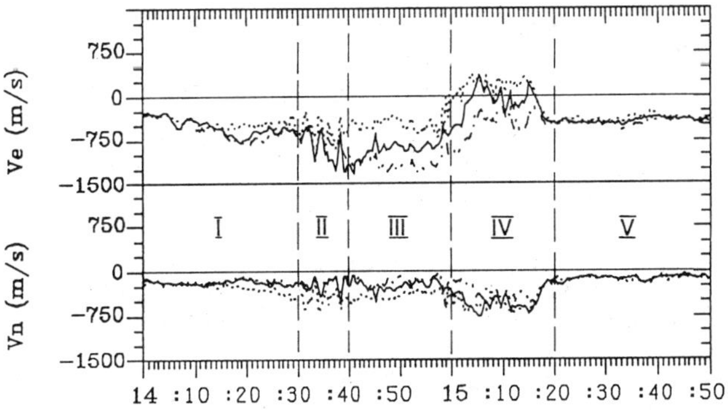

Fig. 3. Measured electron drift velocity versus Universal Time on September 4, 1984. The data are averaged over the longitudinal range 4-6 °E and over three 1° latitudinal ranges between 64.5 and 67.5 °N (geographic coordinates).

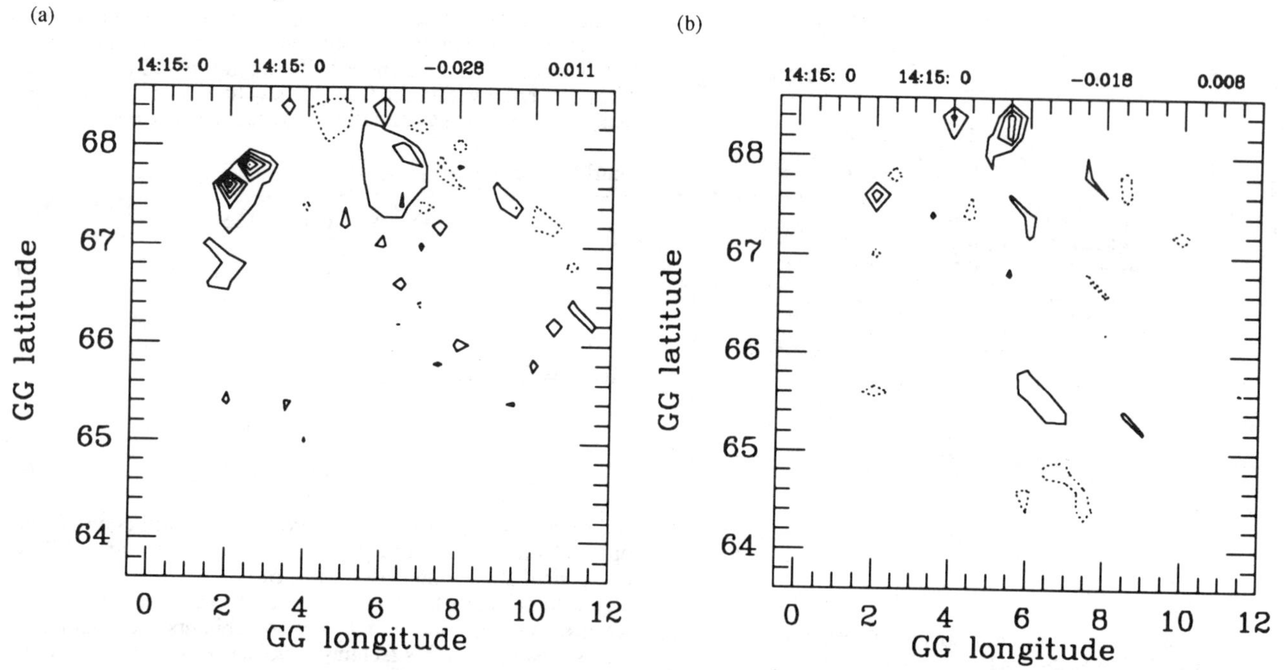

Fig. 4. The divergence (a) and vorticity (b) fields at 14:15 UT over the SABRE field of view, calculated using the discrete electron drift velocity measurements (see text).

cated. The upper and lower panels show the geographic eastward and northward electron drift flow speeds, respectively (geographic North is ~ 20 $^{\circ}$ clockwise of geomagnetic North). The time sequence has been divided into 5 intervals, indicated by the dashed vertical lines in the figure.

In intervals I and V the flows were similar in all three bins indicating laminar flow, and the flow magnitude was fairly steady. In interval I the equatorward component to the flow, increasing with latitude, is thought to be due to the polar cap expansion. The divergence and vorticity fields at 14:15 UT in interval I are shown in Fig. 4. The fields are presented in the form of contour plots with the interval between contour levels set at 5 mHz. Positive values of the divergence and curl are denoted by a dashed linetype, negative values by a solid linetype. It can be seen that the values of both the divergence and vorticity were below 5 mHz almost everywhere. The divergence and vorticity fields during interval V (not shown) yield similarly low values. The low values of divergence were measured at times when the flow field was steady and demonstrate the incompressibility of the ionospheric electron plasma under such conditions. During these intervals there appeared to be no local field-aligned current sheet, as evidenced by the small value of vorticity everywhere within the field of view. There were, however, small regions where the values of the curl and divergence were significant. We note that these are of only one pixel in size, requiring only one of the four data points used in the calculations to be in error. The small dimensions of these features thus suggest that they arise due to erroneous measurements of the electron drift velocity at individual points. Their location at the northern edge of the field of view means that they use data from the furthest range gates of the radar beams. It may be that the radar measurement technique is unreliable at these outer limits.

At 14:50 UT, in interval III, the flow components were again fairly steady, but there existed a large latitudinal velocity gradient. The divergence and vorticity fields at this time are shown in Fig. 5. The former is again very small with only isolated data points having a divergence above 5 mHz. The flow speeds at this time range from ~ 600 m s^{-1} in the highest latitude bin to in excess of 1200 m s^{-1} at low latitude. Even in these exceptional conditions the incompressible flow approximation holds good. The vorticity field however shows some very interesting features. The most striking is the long, linear plateau of high vorticity (> 10 mHz). This we interpret as a field-aligned current sheet (cf. equation (5)). The feature persists over the whole of interval III (cf the sharp rotation of the flow at 14:40 UT in Fig. 2), unlike the smaller scale structures in the divergence field which come and go in a random manner. The source of the steady state current sheet we shall argue to be due to hot plasma effects in the magnetosphere which severely modifies the imposed electric field and hence the convection pattern. To assess whether the pixel sized regions of high divergence and curl were associated with random errors in the measurement technique, we have also temporally averaged the divergence and vorticity fields over the central 10 min interval of interval III. The divergence field (not shown) was indeed reduced further, whilst the vorticity sheet persisted.

At 14:35 UT, in interval II, the central and high latitude bins experienced large amplitude pulsations. The divergence and vorticity fields (not shown) were highly disturbed. In the vorticity field could be seen the beginnings of the current sheet that extends right across the field of view by interval III. The sources of the large divergences are thought to be threefold. Firstly, the measured Doppler velocity of the irregularities can, at times, depart from the actual line of sight electron drift velocity [Robinson, 1986]. The effect is thought to lead to systematic

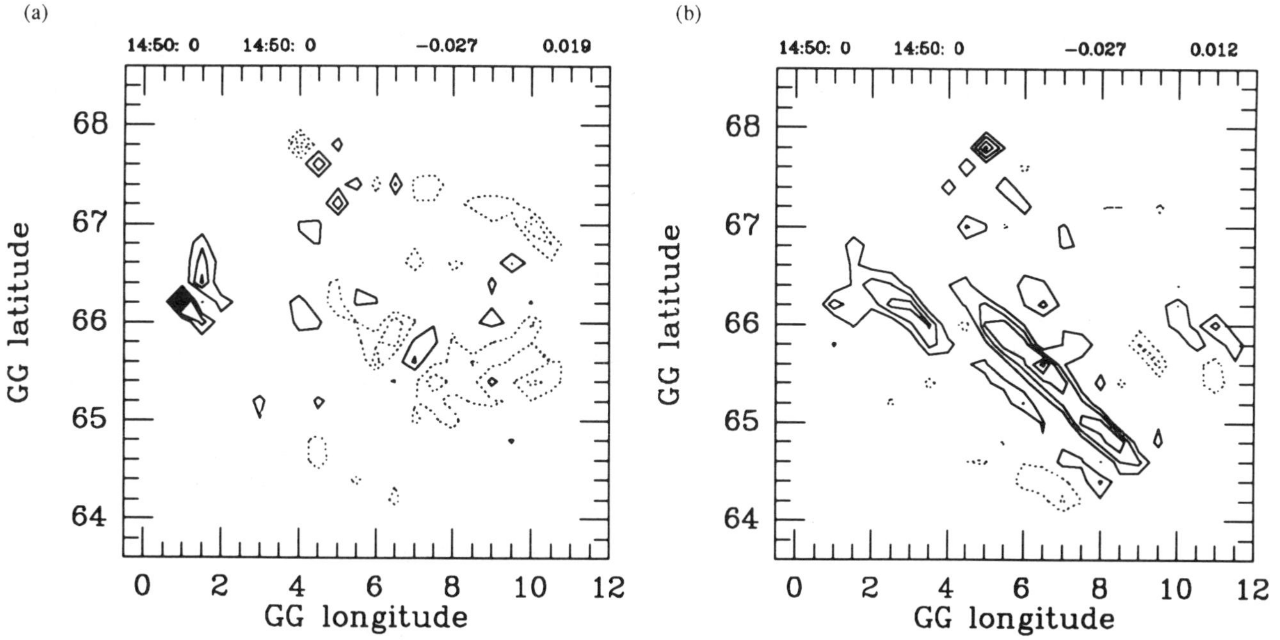

Fig. 5. As for Fig. 4, but using measurements at 14:50 UT.

under-estimates of flows in excess of ~ 1 km s^{-1}, as mentioned earlier. Secondly, the finite integration time for the complete velocity field measurement may lead to inaccuracies during intervals of highly time-varying flow, such as during interval II. Thirdly, when the spatial scale of a flow shear is comparable to the grid scale for the measurements, the discretisation can lead to erroneous relative weightings of the individual velocity measurements used to calculate the differential fields. Thus the presence of a shear in the flow which is not aligned with an axis of symmetry of the grid results in the calculation of erroneous divergences. The effect of these errors is the subject of further detailed analysis.

The situation at 15:10 UT, in interval IV, was similar to that at 14:35 UT with large temporal variations in the flow. Large flow divergences were measured and the vorticity field (not shown) gives evidence for the break-up of the current sheet that was present in interval III, though the situation is highly complicated. The decay of the current sheet during interval IV coincides with a rapid poleward contraction of the polar cap boundary [M. P. Freeman, unpublished data, 1989].

Discussion and Conclusion.

In summary, we have investigated some properties of the ionospheric plasma by looking at quantities which we expect to be conserved (div $\mathbf{v}_e = 0$ in the steady state), or which are a direct measure of some interesting physical entity (curl $\mathbf{v}_e$ is directly proportional to the local field-aligned current). Our expectation that the electron plasma is incompressible in the steady state has been supported by our analysis, even during periods of high electric field strength and strong velocity gradients.

We have shown the presence of a strong, large-scale vorticity of the electron flow following pulsation activity. Waldock et al. [1988] demonstrated a close correspondence of the pulsation behaviour to the vorticity sheet structure. Referring to equation (5), the sense and strength of the vorticity sheet implies the presence of a co-located upward field-aligned current of ~ few μA m^{-2} across the SABRE field of view, unless there was a precisely matched change in the ionospheric conductivity across the sheet. The vorticity sheet stands in the flow for ~ 20 min before its decay suggesting that it is a current sheet of magnetospheric origin.

We shall now consider how such a current sheet might arise in the middle magnetosphere. Let us employ a simple fluid model where the plasma moves in the presence of electric and magnetic fields and the plasma pressure is isotropic. Now, neglecting inertia, the momentum equation can be written:

$$\mathbf{j} \times \mathbf{B} = \nabla p$$

Integrating along field lines this may be re-expressed in the form:

$$B_{eq}^2 \nabla . \mathbf{I} = B_I \nabla p . (\mathbf{B}_{eq} \times \nabla V)$$

where $\mathbf{I}$ is the height-integrated ionospheric current, B_I and B_{eq} are the magnetic field strengths in the ionosphere and magnetospheric equatorial plane respectively, and V is the flux tube volume [Southwood, 1977].

In Fig. 6 we sketch a scenario showing a typical fluid streamline in the equatorial plane of the magnetosphere, arising from a high latitude boundary condition appropriate to magnetospheric erosion. In the late afternoon magnetosphere plasma is carried onto flux tubes of smaller volume distorting the ambient pressure gradient and giving it a component along the flux tube volume contour. Thus, if the local pressure gradient is distorted from a generally radial (anti-radial) orientation, we expect a strong upward (downward) field-aligned current to be driven at the conjugate point in the ionosphere. The sense of this current depends upon the direction of the ambient pressure gradient

BIRKELAND CURRENT SYSTEMS ARISING FROM MAGNETOSPHERIC EROSION

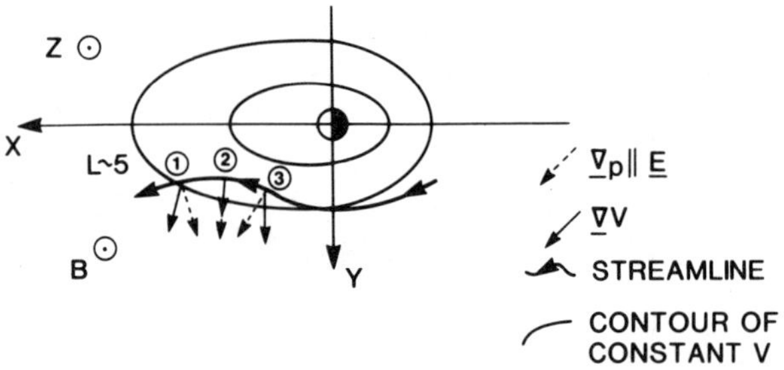

$$B^2_{EQ}\,\underline{\nabla}.\underline{I} = B_I\,\underline{\nabla}p.(\underline{B}_{EQ}\wedge\underline{\nabla}V)$$

REGION	$\underline{\nabla}.\underline{I}$	BIRKELAND CURRENT SENSE
①	<0	UPWARD
②	0	-
③	>0	DOWNWARD

⇒UPWARD BIRKELAND CURRENT AT ~1600 MLT, L~5

Fig. 6. The Birkeland current systems in the middle magnetosphere arising from magnetospheric erosion.

which would be generally anti-radial in the middle and outer magnetosphere, outside the steep radial gradient at the ring current inner edge. However, measurements of the particle pressure in the mid-afternoon sector just prior to the event studied here indicate that enhancements to the plasma pressure in the outer magnetosphere could have generated a radial pressure gradient outside the ring current inner edge in a localised region located at L ~ 4-6 R_E [Lui et al., 1987], which is precisely the region covered by the SABRE field of view. In addition, Lui et al. show evidence that, though the pressure is not isotropic as we have assumed, the dominant contribution to the ring current is the pressure gradient effect, rather than the field curvature effect which involves the pressure anisotropy. Thus, during the interval when we observed the upward current sheet at SABRE, the radar was ideally located to image the current system of the same polarity arising from an equatorial motion of plasma due to the prevailing polar cap expansion. We note that when magnetospheric erosion ceased after ~ 15:00 UT the current system immediately decayed.

We conclude that the concept of flow vorticity and its measurement using ionospheric radars can be a useful tool in the identification of field-aligned current structure in the magnetosphere. In this study it has been successfully shown how field-aligned currents are generated by hot plasma effects which can modify the imposed convection pattern and prevent its penetration to low latitudes [Southwood, 1977]. We encourage the further use of the technique in identifying other magnetospheric current systems e.g. in the cusp, where an analysis might shed further light on the nature of the solar - terrestrial coupling process.

Acknowledgments. The SABRE radar was operated jointly by the University of Leicester, UK, and the Max-Planck Institut für Aeronomie, Lindau, FRG, in cooperation with the Uppsala Ionospheric Observatory, Sweden. M. P. Freeman is currently a UK Science and Engineering Research Council postgraduate research student.

References.

Armstrong, J. C., and A. J. Zmuda, Triaxial magnetic measurements of field-aligned currents at 800 kilometers in the auroral region: initial results, *J. Geophys. Res.*, *78*, 6802, 1973.

Fejer, B. G., and M. C. Kelley, Ionospheric Irregularities, *Rev. Geophys.*, *18*, 401, 1980.

Iijima, T., and T. A. Potemra, Large-scale characteristics of field-aligned currents associated with substorms, *J. Geophys. Res.*, *83*, 599, 1978.

Lui, A. T. Y., R. W. McEntire, and S. M. Krimigis, Evolution of the ring current during two geomagnetic storms, *J. Geophys. Res.*, *92*, 7459, 1987.

Robinson, T. R., Towards a self-consistent non-linear theory of radar auroral backscatter, *J. Atmos. Terr. Phys.*, *48*, 417, 1986.

Southwood, D. J., The role of hot plasma in magnetospheric convection, *J. Geophys. Res.*, *82*, 5512, 1977.

Vasyliunas, V., M., Concepts of magnetospheric convection, *The Magnetospheres of the Earth and Jupiter* (ed. V. Formisano), D. Reidel, 179, 1975.

Waldock, J. A., Southwood, D. J., Freeman, M. P., and M. Lester, Pulsations observed during high-speed flow in the ionosphere, *J. Geophys. Res.*, *93*, 12883, 1988.

OBSERVATIONS OF IONOSPHERIC FLUX ROPES ABOVE SOUTH POLE

Z. M. Lin, J. R. Benbrook, E. A. Bering, G. J. Byrne

[1]Physics Department, University of Houston
Houston, Texas 77204–5504

E. Friis-Christensen

[2]Division of Geophysics, Danish Meteorological Institute
Lyngbyvej 100, DK 2100, Copenhagen, Denmark

D. Liang, B. Liao, J. Theall

[1]Physics Department, University of Houston
Houston, Texas 77204–5504

Abstract. Two different models of the current flow in ionospheric flux ropes have been studied by numerical techniques. In each case, we assume that the ionosphere is flat and parallel to the Earth's surface, and that the height-integrated conductivity is uniform, and we solve the problem of mapping the electric field from the ionosphere down to balloon altitude. The observed electric signals at balloon altitudes and magnetic signals on the ground may be a superposition of signals of the towing system and the twisting system. We are particularly interested in an isolated event that occurred at ~1625 UT on 3 January 1986. It exhibits many features predicted by these two models. We present a fit of the models to the electric and magnetic field data, which were acquired at South Pole Station during the 1985-1986 South Pole Balloon Campaign. We also have examined Greenland magnetometer chain data and Iqaluit magnetometer data. All these data imply that this event was localized in longitude and global in the sense of occurring at conjugate points. IMP 8 solar wind plasma and interplanetary magnetic field (IMF) data show no evidence of a significant pressure perturbation associated with the event. They do show that the event occurred just as a strong southward turning of the IMF, seen slightly earlier at IMP 8, passed the earth.

Introduction

The transfer of energy from solar plasma to the Earth's magnetosphere has attracted much attention from physicists. It has long been recognized that there is a correlation between geomagnetic activity and solar wind properties, especially southward directed interplanetary field. Observations of these correlations led to the development of the idea of reconnection at the magnetopause [Dungey, 1961]. Signatures of episodic magnetic field reconnection processes at the magnetopause were among the first to be found. Russell and Elphic [1978, 1979] used observations from ISEE-1 and ISEE-2 to show evidence for a patchy impulsive reconnection process termed the flux transfer event (FTE), and they suggested that FTE's provide much of the flux transport associated with magnetospheric dynamics. Many studies have concentrated on searching for the ionospheric or ground signature of flux transfer events. Based on ground radar data, Goertz *et al.* [1985] found that sporadic and spatially isolated flow across the dayside ionospheric convection boundary occurs on a time scale of a few minutes and in regions which, when mapped down to the ionosphere, range in size from 50 km to 300 km. They interpret this as a result of a flux tube connecting with the magnetosheath field and believe that the boundary layer flux is transported from within the magnetopause by relatively large $\mathbf{E} \times \mathbf{B}$ flows across the inner boundary. Lanzerotti *et al.* [1986] presented magnetic field data from the cusp-latitude South Pole station that exhibit the signature expected in the ionosphere from a flux transfer event at the magnetopause. Southwood [1985] pointed out that the convection of the flux tube in the ionosphere will generate two convection vortices around a flux tube and a $\mathbf{J} \times \mathbf{B}$ force will drive the flux tube through the ionosphere. Lee and Fu [1985] suggested that the multiple X-line reconnection process produces a helical magnetic flux tube at the magnetopause.

Geophysical Monograph 58

Arrays of detectors are very useful in observing electric and magnetic field patterns on the ground since their data contain information about the motion of the magnetic flux rope. Friis-Christensen *et al.* [1988], and Glaßmeier *et al.* [1989] reported observations of transient magnetic variations similar to those reported by Lanzerotti *et al.* [1986] and others. Friis-Christensen *et al.* [1988] analyzed magnetometer data from an array of Greenland stations. They presented an isolated event that consisted of a twin vortex pattern moving westward, parallel to the line of centers between the vortices. The event was associated with relatively quiet-time ionospheric convection. It occurred during an interval of northward IMF and was associated with a large sudden decrease in the solar wind number density. The equivalent current system observed by Glaßmeier *et al.* [1989] also showed two current vortices, rotating clockwise in the west and counterclockwise in the east. Glaßmeier [1988] classified observed current systems and those observed by Friis-Christensen *et al.* [1988] as examples of a parallel dipole model.

Most studies of high latitude impulsive field perturbations in recent years have used ground based magnetic field data and only a few have been based on balloon or radar electric field data. Actually, the electric field is as important as the magnetic field in revealing the nature of the correlation between geomagnetic activity and the solar wind. The electric field is also useful in understanding the current configuration, convection patterns, and group velocity of ionospheric flux ropes. Bering *et al.* [1988] pointed out that the electric field of the towing system attenuates less than the electric field of the twisting system as the field is mapped down to balloon altitude, so it should govern the sub-ionospheric electric field traces.

Calculation of the stratospheric signature of ionospheric electric fields is a well understood boundary value problem that has been treated extensively in the literature. On August 5 and 6, 1968, two balloons were launched from Fort Churchill, in an attempt to measure magnetospheric dc electric fields by observing their extensions into the upper atmosphere [Mozer and Serlin, 1969]. In interpreting the balloon data, the ionospheric electric field was mapped to balloon altitude by assuming that the source potential in the ionosphere varied sinusoidally with horizontal distance. Boström *et al.* [1973] considered models where the electric field was a standing or travelling sinusoidal wave. It was concluded that ionospheric field variations give rise to substantial disturbances in the atmospheric electric field, but at low altitude they were hard to distinguish from disturbances of meteorological origin. The mapping of thundercloud electric fields at middle and subauroral latitudes was investigated analytically as a three-dimensional boundary value problem by Park and Dejnakarintra [1973]. Their results suggested that giant thunderclouds might be an important source of localized electric fields that could form field-aligned electron density irregularities in the ionosphere and the magnetosphere. Numerical techniques were adopted by Park [1976] that allow direct calculations of electric fields and currents between 0 and 150 km altitude for any arbitrary potential distribution at the upper boundary. Hughes and Southwood [1976a, b] and Southwood [1975] have done studies regarding mapping of micropulsations to the Earth's surface. The basic feature of Southwood's [1975] model is that the ambient magnetic field is straight, uniform, and in the $\mathbf{z}$ direction. An inhomogeneous Alfvén speed $A(x)$ is introduced by allowing the cold plasma density to vary in the $\mathbf{x}$ direction. It was found that the ionosphere-atmosphere system screens out rapid horizontal variations [Hughes and Southwood, 1976].

The purpose of this paper is to give the ionospheric electric field boundary conditions of two models of the small size current in ionospheric flux ropes (~400km scale size or less), and to determine the electric field at balloon altitiude (30 km) for new ionospheric boundary conditions specific to the flux rope problem. We also present electric and magnetic field data acquired from South Pole, and magnetometer data acquired in Greenland and Iqaluit. The features of these data are clearly consistent with the models. The IMP 8 solar wind data show no significant dynamic pressure perturbations, but do show a strong southward turning of the IMF associated with the event.

Theory

Southwood [1985, 1987] and Lockwood *et al.* [1988] suggested that a reconnected flux rope produces a 2-dimensional electric dipole that convects poleward from the noon sector auroral oval. In their model, which we call the towing system, the dividing line between two oppositely directed field-aligned currents is parallel to the direction of motion of the magnetic flux rope. This current system has the same basic ionospheric footprint as that used by Friis-Christensen *et al.* [1988] to interpret their magnetometer data from the Greenland array, but the assumed direction of motion of the structure is rotated 90°. For the twisting system, Lanzerotti *et al.* [1986] presented a cylindrically symmetrical system, which is composed of a central current core with a co-axial return current.

Figure 1 shows two different current models. The twisting system we have used is a little different from that suggested by

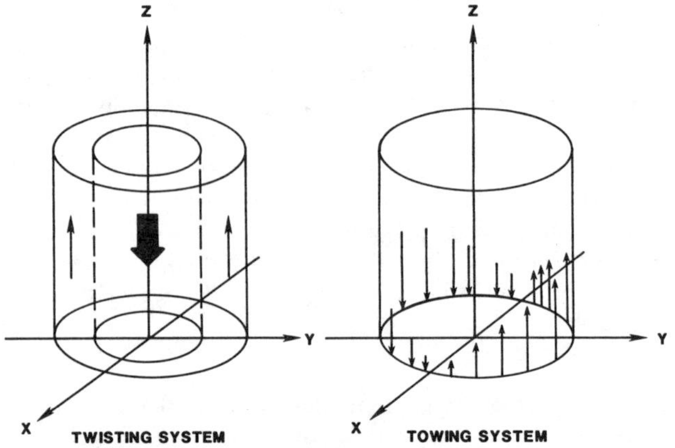

Fig. 1. Postulated twisting system composed of a central current core with a co-axial return current. Postulated towing system composed of two oppositely directed field-aligned currents on opposing edges of a flux rope.

Lanzerotti *et al.* [1986] in that the field-aligned current is volume distributed. We assume the ionosphere is flat and parallel to the Earth's surface, and the conductivity of the atmosphere is a scalar $\sigma(z)$, that depends only on the altitude z. We also assume the Earth's surface is a perfect conductor. Thus the horizontal E-field perturbations on the ground caused from the flux rope are supposed to be zero. Since stratospheric balloons are above most of the column resistance of the atmosphere, the solutions found are insensitive to the exact details of the lower boundary condition. Under the steady-state condition, the basic equations to describe the electric phenomena in the atmosphere are

$$\nabla \bullet \mathbf{j} = 0 \tag{1}$$

$$\mathbf{j} = \Sigma \bullet \mathbf{E} \tag{2}$$

$$\mathbf{E} = -\nabla V \tag{3}$$

The mapping of the electric field of the ionospheric flux ropes down to the balloon becomes a boundary value problem. We should first calculate the ionospheric boundary conditions of these two models. We choose the origin to be on the lower boundary of the ionosphere. The z direction upward, coincident with the center axis of these two cylindrical models. The x direction is parallel to the dividing line between upward and downward field-aligned currents of the towing system, and the y direction completes the right hand orthogonal system. For the towing system we assume the direction of current flow in the ionosphere is parallel to the y direction. Under the steady-state condition (1) is also satisfied in the ionosphere. From equation (1) we have

$$\frac{1}{\rho}\frac{\partial}{\partial\rho}(\rho j_\rho) + \frac{1}{\rho}\frac{\partial}{\partial\phi}(j_\phi) + \frac{\partial}{\partial z}(j_z) = 0 \tag{4}$$

Integrating along the z direction gives

$$\frac{1}{\rho}\frac{\partial}{\partial\rho}(\rho J_\rho) + \frac{1}{\rho}\frac{\partial}{\partial\phi}(J_\phi) + j_z + c = 0 \tag{5}$$

where J_ρ and J_ϕ are height-integrated currents and c is a constant. We assume the height-integrated conductivity is uniform in a region around the models. The ionosphere is then represented by a two-dimensional layer with height-integrated conductivities.

For the twisting system the field-aligned current j_z is constant, equal to $2I/\pi R_0^2$ and changes direction for $\rho > R_0/\sqrt{2}$. The height-integrated J_ϕ is zero. E_ρ at the edge ($\rho = R_0$) is zero. We can integrate (5) from the origin along any radius on the ionosphere , and use (2) to get the ionospheric boundary condition of the twisting system as

$$E_\rho = -\frac{I}{\pi R_0^2 \Sigma_p}\rho, \qquad \left(0 \le \rho \le \frac{R_0}{\sqrt{2}}\right) \tag{6}$$

$$E_\rho = \frac{I}{\pi R_0^2 \Sigma_p}\left(\frac{\rho^2 - R_0^2}{\rho}\right), \qquad \left(\frac{R_0}{\sqrt{2}} \le \rho \le R_0\right) \tag{7}$$

$$E_\rho = 0, \qquad (R_0 \le \rho) \tag{8}$$

where I is the integral of the upward current in the core, R_0 is the radius of the current model, and ϕ is the azimuthal angle ($0° \equiv +x$). Σ_p is the height-integrated Pedersen conductivity in the ionosphere. From $E_\rho(R_0) = 0$, we know that c is equal to zero. The boundary condition at the ground of the twisting system is assumed to be

$$E_\rho = 0 \tag{9}$$

For the towing system we assume a two dimensional physical dipole of radius R_0. The height-integrated horizontal currents are

$$J_\rho = \frac{I}{2R_0}\sin\phi \tag{10}$$

$$J_\phi = \frac{I}{2R_0}\cos\phi \tag{11}$$

and the field-aligned current density in the ionosphere is

$$j_z = \frac{I}{2R_0}\delta(\rho - R_0)\sin\phi \tag{12}$$

This means that the horizontal electric field in the ionosphere within the towing system is uniform. This term comes directly from the multipole expansion of solutions to Laplace's equation in cylindrical coordinates and therefore is a necessary consequence of a 2-D dipole assumption. The ionospheric boundary condition of towing system within the model is

$$V = -\frac{I}{2R_0\Sigma_p}\rho\sin\phi, \qquad (\rho \le R_0) \tag{13}$$

The electric potential of any point outside towing system can be obtained from the solution of the two-dimensional Laplace equation

$$\frac{\partial^2 V}{\partial\rho^2} + \frac{1}{\rho}\frac{\partial V}{\partial\rho} + \frac{1}{\rho^2}\frac{\partial^2 V}{\partial\phi^2} = 0 \tag{14}$$

with the boundary condition

$$V(R_0, \phi) = -\frac{I}{2\Sigma_p}\sin\phi \tag{15}$$

and is given by

$$V = -\frac{I}{2\Sigma_p}\frac{R_0}{\rho}\sin\phi, \qquad (R_0 \le \rho) \tag{16}$$

It is obvious that for the towing system the electric field inside the flux rope is uniform and is a dipole field outside the flux rope. The boundary condition at the ground of the towing system is assumed to be

$$V = 0 \tag{17}$$

For these two models the electric field should be finite on the Z axis, and the electric field should vanish at infinity. Empirically the E-field perturbations of magnetic flux ropes are observed to drop off quickly with distance. Thus, we can choose the outer boundary ρ_0 to be between $20R_0$. At $\rho = \rho_0$, the

E-field perturbations of two models are assumed to be zero. For a flux with a radius > 100 km, $\rho_0 = 20R_0$ is of the order of the size of the earth. Since we are using a flat earth approximation, letting ρ_0 go to infinity and using an integral Hankel transform solution does not produce any meaningful improvement in the accuracy of the solutions. Having set the boundary conditions we now solve the problem of mapping the electric fields of these two current structures down to balloon altitude. We choose the XY plane to be on the ground, and X, Y, Z axes to be parallel separately to the x, y, z axes that we have defined above. As a convenient large-scale model, the conductivity profile in the atmosphere can be approximated by

$$\sigma(z) = \sigma_0 e^{z/H}, \qquad (H = 6km) \tag{18}$$

with the conductivity at the ground $\sigma_0 = 9\times10^{-14}$ ohm^{-1}m^{-1}. The equation (1) in the atmosphere can be written as

$$\frac{\partial^2 V}{\partial \rho^2} + \frac{1}{\rho}\frac{\partial V}{\partial \rho} + \frac{1}{\rho^2}\frac{\partial^2 V}{\partial \phi^2} + \frac{\partial^2 V}{\partial z^2} + \frac{1}{H}\frac{\partial V}{\partial z} = 0 \tag{19}$$

The separation of the variables is accomplished by the substitution:

$$V(\rho,\phi,z) = R(\rho)\Phi(\phi)Z(z) \tag{20}$$

Then equation (19) becomes

$$\frac{1}{R}\frac{d^2R}{d\rho^2} + \frac{1}{\rho R}\frac{dR}{d\rho} + \frac{1}{\rho^2\Phi}\frac{d^2\Phi}{d\phi^2} + \frac{1}{Z}\frac{d^2Z}{dz^2} + \frac{1}{HZ}\frac{dZ}{dz} = 0 \tag{21}$$

For the towing system the analytical solution of equation (21) is then

$$V(\rho,\phi,z) = \sum_n C_n(e^{a_{n+}z} - e^{a_{n-}z})J_1(k_n\rho)\sin\phi \tag{22}$$

From the boundary conditions we know

$$k_n = \frac{J_1^n}{\rho_0} \tag{23}$$

where J_1^n is the nth root of $J_1(x)$, and ρ_0 is the outer boundary. And

$$a_{n\pm} = \frac{-1 \pm \sqrt{1+4k_n^2H^2}}{2H} \tag{24}$$

Equation (22) is Bessel series in ρ. At the ionospheric boundary $z = z_0$, equation (22) gives

$$V(\rho,\phi,z_0) = \sum_n C_n(e^{a_{n+}z_0} - e^{a_{n-}z_0})J_1(k_n\rho)\sin\phi \tag{25}$$

From the orthogonality condition of the the Bessel function and the ionospheric boundary conditions (13) and (16), it can be found that

$$C_n = \frac{2E_{0to}R_0^2}{k_n\rho_0^2J_0^2(k_n\rho_0)(e^{a_{n+}z_0} - e^{a_{n-}z_0})}\left(J_0(k_n\rho_0) - \frac{2J_1(k_nR_0)}{k_nR_0}\right) \tag{26}$$

where z_0 is the altitude of the ionosphere, assumed to be $z_0 =$ 100 km, and

$$E_{0to} = \frac{I}{2R_0\Sigma_p} \tag{27}$$

The electric field of the towing system is then

$$\mathbf{E}(\rho,\phi,z) =$$

$$-\mathbf{e}_\rho \sum_n C_n(e^{a_{n+}z} - e^{a_{n-}z})\left(k_nJ_0(k_n\rho) - \frac{J_1(k_n\rho)}{\rho}\right)\sin\phi$$

$$-\frac{\mathbf{e}_\phi}{\rho}\sum_n C_n(e^{a_{n+}z} - e^{a_{n-}z})J_1(k_n\rho)\cos\phi$$

$$-\mathbf{e}_z\sum_n C_n(a_{n+}e^{a_{n+}z} - a_{n-}e^{a_{n-}z})J_1(k_n\rho)\sin\phi \tag{28}$$

For the twisting system, we assume azimuthal symmetry, so that

$$\Phi(\phi) = \text{const.} \tag{29}$$

and equation (21) becomes

$$\frac{1}{R}\frac{d^2R}{d\rho^2} + \frac{1}{\rho R}\frac{dR}{d\rho} + \frac{1}{Z}\frac{d^2Z}{dz^2} + \frac{1}{HZ}\frac{dZ}{dz} = 0. \tag{30}$$

With all the boundary conditions that we have calculated, the analytical solution is found to be

$$V(\rho,z) = \sum_n C_nJ_0(k_n\rho)(e^{a_{n+}z} - e^{a_{n-}z}) \tag{31}$$

where k_n, and $a_{n\pm}$ are the same as for the towing system. The electric field of the twisting system is then

$$\mathbf{E}(\rho,z) = \mathbf{e}_\rho\sum_n C_nk_n(e^{a_{n+}z} - e^{a_{n-}z})J_1(k_n\rho)$$

$$-\mathbf{e}_z\sum_n C_n(a_{n+}e^{a_{n+}z} - a_{n-}e^{a_{n-}z})J_0(k_n\rho) \tag{32}$$

At the ionospheric boundary $z = z_0$, the radial electric field is

$$E_\rho(\rho,z_0) = \sum_n C_nk_n(e^{a_{n+}z_0} - e^{a_{n-}z_0})J_1(k_n\rho) \tag{33}$$

From the orthogonality of Bessel function and the ionospheric boundary conditions (6), (7), and (8) we have

$$C_n = \frac{2E_{0tw}(J_1(k_nR_0) - \sqrt{2}J_1(k_nR_0/\sqrt{2}))}{k_n^3\rho_0^2J_0^2(k_n\rho_0)(e^{a_{n+}z_0} - e^{a_{n-}z_0})} \tag{34}$$

where

$$E_{0tw} = \frac{2I}{\pi R_0\Sigma_p} \tag{35}$$

The equations we have developed above show that the mapping of the electric fields down to the balloon is quite different for the two models. The electric potential of the twisting system in the atmosphere is proportional to the zeroth order

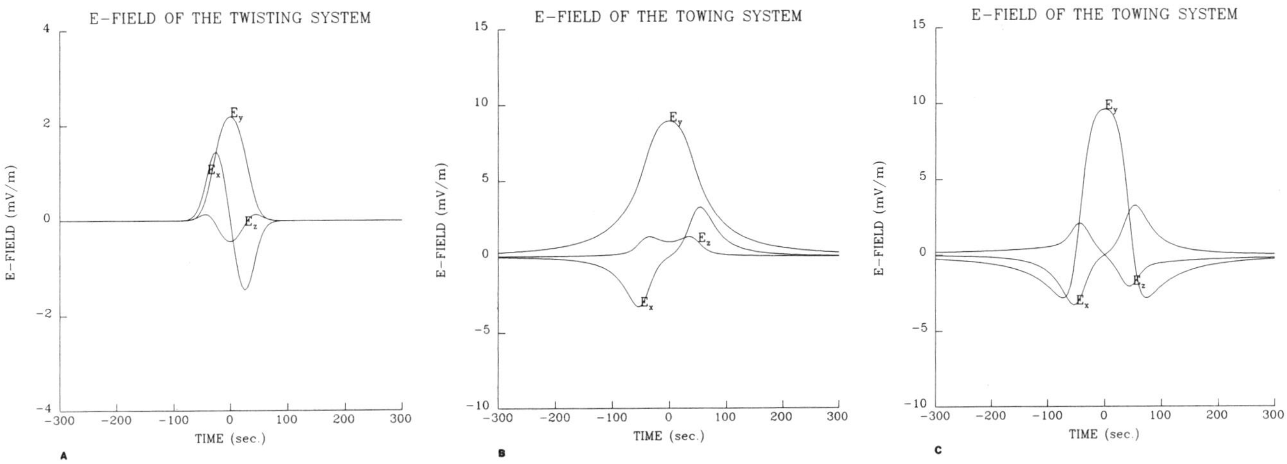

Fig. 2. Electric field at balloon altitude (30 km) versus time. The flux rope is assumed to move at a rate of 2 km/s and to pass 50 km away from the center of the balloon. The radius of the flux rope is 100 km. The height-integrated Pedersen conductivity is 5 mhos. The total current is 1×10^4 A.

Bessel function. This symmetry is a necessary consequence of the assumed cylindrically symmetrical current configuration for the twisting system. The electric potential of the towing system in the atmosphere is proportional to the first order Bessel function, and exhibits the features of a dipole for a similar reason. These differences are important in explaining the different electric field signatures that were acquired in the balloon experiment. Figure 2 shows the electric field of these two models at balloon altitudes (30 km). We assume that the radius of the magnetic flux rope is 100 km, and the total current I is 1×10^4A. In all three panels, the flux tube is assumed to move in a straight line across the ionosphere at the rate of 2 km/s and to pass at a distance of 50 km from the center of the magnetic flux rope to the balloon. The height-integrated Pedersen conductivity is assumed to be 5 mhos. In Fig. 2(a), the electric signature of the symmetrical twisting system is shown. In Fig. 2(b), the signature from the towing system moving parallel to the X-axis, as suggested by Southwood [1985,1987] and Lockwood [1988], is shown. Fig. 2(c) shows the signature expected for motion along the Y-axis, the model used by Friis-Christensen *et. al.* [1988]. In each case, a very characteristic signature is predicted at balloon altitude. The electric field at balloon altitude is dependent on the current structure of the models, and is also dependent on the location of the balloon beneath the flux rope models. Attenuations of the electric fields at the balloon altitude are different everywhere. Figure 2 shows that towing system produces bigger signals at the balloon than the other system when these two systems have the same amount of total current and the same size. However it should be noticed that the electric field of the towing system on the ionosphere under this situation is ~ 2.2 times the maximum electric field of the twisting system on the ionosphere. Figure 2 also shows that the electric field perturbations of the twisting system drop off with distance from the tube more quickly than those of the towing system.

At high latitude, where the magnetic field is roughly perpendicular to the ionosphere, the magnetic field of a field-aligned current cancels that of the Pedersen current [Fukushima, 1969]. The Hall current generates the magnetic signature measured on the ground. McHenry and Clauer [1987] have already calculated the ground magnetic signature of these two models.

$$\mathbf{B} = \mathbf{B}_{hall} = -\frac{\mu_0 \Sigma_h}{4\pi\Sigma_p} \int\int \left(\frac{1+\cos\alpha}{\sin\alpha}\rho + \mathbf{z}\right)\frac{j(\mathbf{r})}{r'}dA \qquad (36)$$

where Σ_h and Σ_p are height-integrated Hall and Pedersen conductivities respectively, $j(\mathbf{r})$ is the field-aligned current density, α is the vertical angle of the line connecting the source and observer, and r' is the distance from the source point to the observer. The unit vector ρ is along the horizontal projection of $\mathbf{r}'$ pointing from the source to the observer.

Observations

As the magnetic flux rope passes overhead, the electric and magnetic fields projecting beneath it leave characteristic traces in electrometer and magnetometer records. As our models predict, the signature of the magnetic flux rope is relatively isolated and localized to a certain path. This isolation is enhanced in the case of the twisting system, for which the signature attenuates quickly when the distance between observer and magnetic flux rope increases. If the magnetic flux rope is the result of magnetic reconnection occurring on the magnetopause, then there is a possibility to find a similar signature at the conjugate point, especially when the event has just occurred or the Y component of the IMF vanishes.

We are interested in the event that occurred at $\sim$ 1625 UT on 3 January 1986 (magnetic local time was $\sim$ 1300). This isolated event seems to have the typical features of a magnetic flux rope. Figure 3 shows six panels of data for a one hour period beginning at 1600 UT on 3 January 1986. The

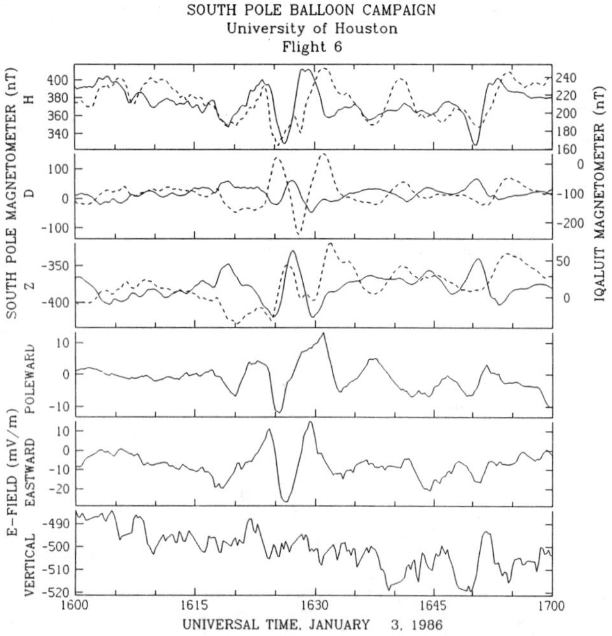

Fig. 3. The top three panels show 15 s averages of the geomagnetic field observed at South Pole Station (solid line) and at Iqaluit (dashed line). The next three panels show 15 s averages of the electric field measured by the payload on balloon flight 6.

bottom three panels show 15 s average electric field data that were acquired at the cusp latitude during the 1985-1986 balloon campaign. The noise level of the double probe electric field detector was ~0.4 mV/m and the digitization increment was 0.1 mV/m. The top three panels show 15 s averages of the geomagnetic data observed at South Pole Station and at Iqaluit. The solid line is the signature at the South Pole, and the dashed line is the signature at Iqaluit. The noise level of the instrument was about 0.2 nT, and the digitization increment was about 0.06 nT. An isolated electric and magnetic pulsation was recorded at ~ 1625 UT both at the South Pole and at Iqaluit. At that time the invariant longitude of the balloon was ~ 10.87°, and the invariant latitude of the balloon was ~ −73.88°, using the Baker and Wing [1989] coordinate system. The invariant longitude of South Pole was ~ 18.26°, and the invariant latitude of South Pole was ~ −74.42°. The invariant longitude of Iqaluit was ~ 14.07°, and the invariant latitude of Iqaluit was ~ 74.31°. The magnetic component amplitudes were ~50-100 nT. The poleward and eastward electric component amplitudes were ~ 20-40 mV/m, but the vertical electric field perturbation was lost in the background noise. The time dependence of electric and magnetic signatures and the relationship between their different components are similar to those predicted by the models.

The signatures that occurred at the same time in the northern hemisphere were also quite interesting. The conjugate area ground magnetic field data at Iqaluit are shown in Figure 3 (dashed line). Figure 4 presents the Greenland magnetometer chain data at the stations Thule(THL, $\Lambda = 86.3°$, $\Phi = 37.3°$), Godhavn(GDH, $\Lambda = 76.8°$, $\Phi = 42.1°$), Sondre Stomfjord(STF, $\Lambda = 74.2$, $\Phi = 43.1°$), Narssarssuaq(NAQ, $\Lambda = 67.3°$, $\Phi = 45.0°$) and Daneborg(DNB, $\Lambda = 75.3°$, $\Phi = 83.0°$) sampled at one minute intervals. The locations of Greenland Stations, Iqaluit(FRB), and the conjugate points of South Pole (PLE) and the balloon (BAL) are shown in Figure 5. The isolated magnetic signatures with a width of ~ 5-10 minutes are quite obvious. The peak value of the z component at the STF station is ~ 70 nT. Most features predicted by the two models appear in the data. It was only visible at GDH, STF, and NAQ in western Greenland, but was very weak at THL (also western Greenland) and DNB in eastern Greenland, Leirvogur(LVR, $\Lambda = 65.4°$, $\Phi = 69.0°$) and Post de la Baleine(PBQ, $\Lambda = 66.6°$, $\Phi = 347.3°$). Figure 6 shows Greenland chain magnetic perturbations plotted as equivalent convection. The observed perturbation vectors have been rotated by 90° counterclockwise to be consistent in direction with an assumed convection pattern. Figure 6 and the phase rela-

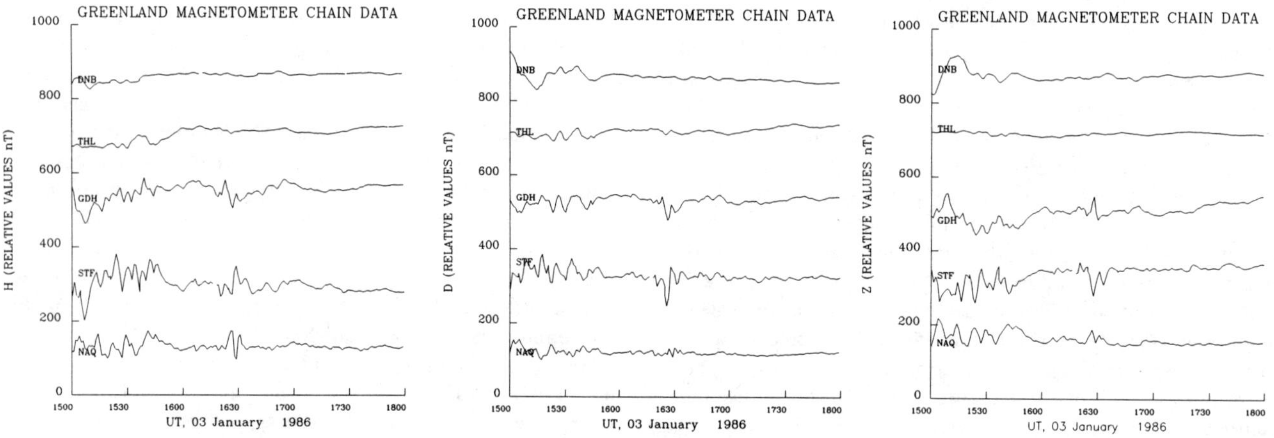

Fig. 4. Greenland magnetometer chain data at the stations THL, GDH, STF, NAQ, and DNB, sampled at one minute intervals.

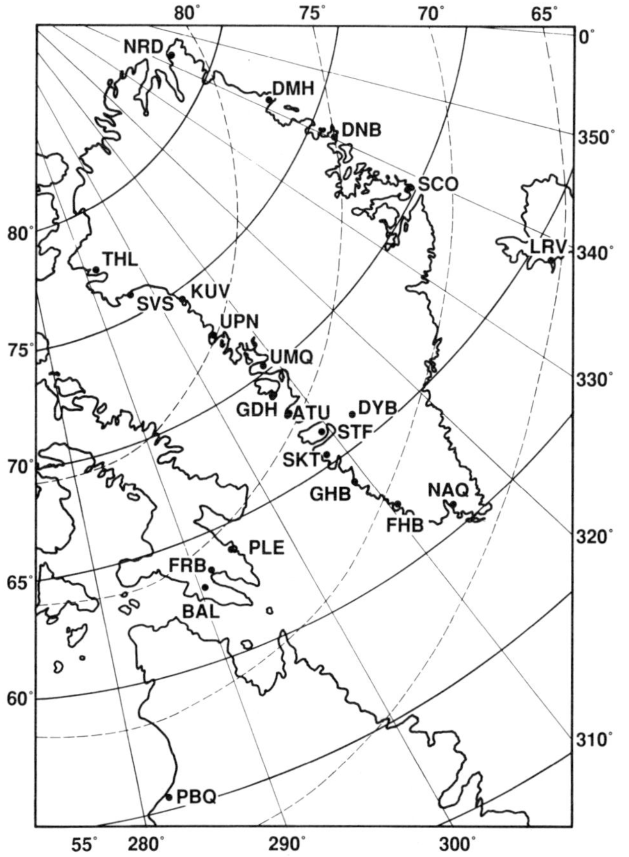

Fig. 5. The locations of Greenland stations, Iqaluit (FRB), and the conjugate points of South Pole (PLE) and balloon (BAL).

tion in Figure 5 strongly suggest that the magnetic flux rope to have moved geomagnetically eastward.

From equations (26), (27), (34), (35), and (36), there are nine parameters to choose when we try to fit the electric field data acquired at the balloon, and the magnetic field data acquired at South Pole under the same condition. These nine parameters are the radius of the magnetic flux rope R_0, equivalent electric fields in the ionosphere E_{0to} and E_{0tw}, the velocity of the magnetic flux rope, the distance from the observer to the line that the center of the magnetic flux rope moved along, the times that the center of magnetic flux rope passed the balloon and South Pole, and the height-integrated conductivities Σ_p and Σ_h. The velocity and the direction of the magnetic flux rope and the distance from the observer to the line that the center of the magnetic flux rope move along are also important in determining the shape and the amplitude of the signals. Fortunately in our case, the Greenland magnetometer data strongly suggested that the magnetic flux rope in the conjugate point at that time moved eastward and almost parallel to the geomagnetic latitude.

We assume the magnetic flux rope moved with a constant velocity roughly geomagnetically eastward, parallel to the line which connected the centers of two vortices (Y direction), similar to the model used by Friis-Christensen *et al.* [1988]. We also assume the current configuration of the magnetic flux rope did not change during the interval of the fit. The distance along the geomagnetic latitude between the balloon and South Pole was $\sim$ 224 km. The velocity of the flux rope can be calculated from the distance and the phase shift between the balloon and South Pole. The center of the magnetic flux rope passed the balloon at $\sim$1625:15UT, and passed South Pole at $\sim$1626:13UT. The time difference was $\sim$ 58 sec. The calculated velocity of the flux rope was $\sim$ 3.9 km/s. The radius could be estimated from the velocity and the width of the signal of the event. Σ_h/Σ_p was choosen to be 2 [McHenry *et al.*, 1987]. The other six parameters were found by fitting to all six components of the electric field data which were acquired at the balloon and the magnetic field data which were acquired at South Pole. A fit of the electric and magnetic field data between 1620 UT and 1630 UT is shown in Figure 7. We made a superposition of the twisting system and the towing system. The solid line is the experimental data and the dashed line is calculated. The radius of the magnetic flux rope was $\sim$ 350 km. In order to get a better fit, we found that the height-integrated Pedersen conductivity should be $\sim$5 mhos. The total current of the twisting system was $\sim 2.7 \times 10^5$ A with the center core current downward and the outside return current upward. The total current of the towing system was $\sim 1.7 \times 10^4$ A with the height-integrated horizontal current antiparallel to the Y direction. It seems this event was dominated by the twisting system. In Fig. 6, we can roughly see a strong single-vortex pattern in the center, and a weak twin-vortex pattern superimposed on it. Figure 6 suggested that in the northern hemisphere the event was also a superposition of these two systems and was also dominated by the twisting system. This conclusion is consistent with the result derived from electric and magnetic field fits in the south hemisphere. All these data imply that this event was localized in longitude, and global in the sense of occurring at conjugate points.

Discussion

Magnetospheric activity is strongly related to the interplanetary magnetic field. The magnetopause is more oscillatory for southward IMF than for northward IMF [Song *et al.*, 1988]. The southward interplanetary magnetic field plays the crucial role in magnetopause reconnection. The data base from the balloon campaign has been searched for events similar to this one. We found that more than 70 percent of the impulsive events were related to southward B_z [Bering *et al.*, 1989]. IMP 8 magnetic field data on 3 January 1986 between 1550-1650 UT, delayed by the calculated 616 s transit time from IMP 8 to the ionosphere of South Pole [Formisano, 1979; Formisano *et al.*, 1979; Etemadi *et al.*, 1988] in Geocentric Solar Magnetospheric coordinates (GSM) were show in Figure 8. Figure 9 shows the solar wind dynamic pressure and IMF orientation, where Θ is the cone angle of the magnetic field vector about the X_{GSM} axis and Φ is the azimuth angle in the $Y-Z$ plane ($0° \equiv +Y$). The IMP 8 position at that time was

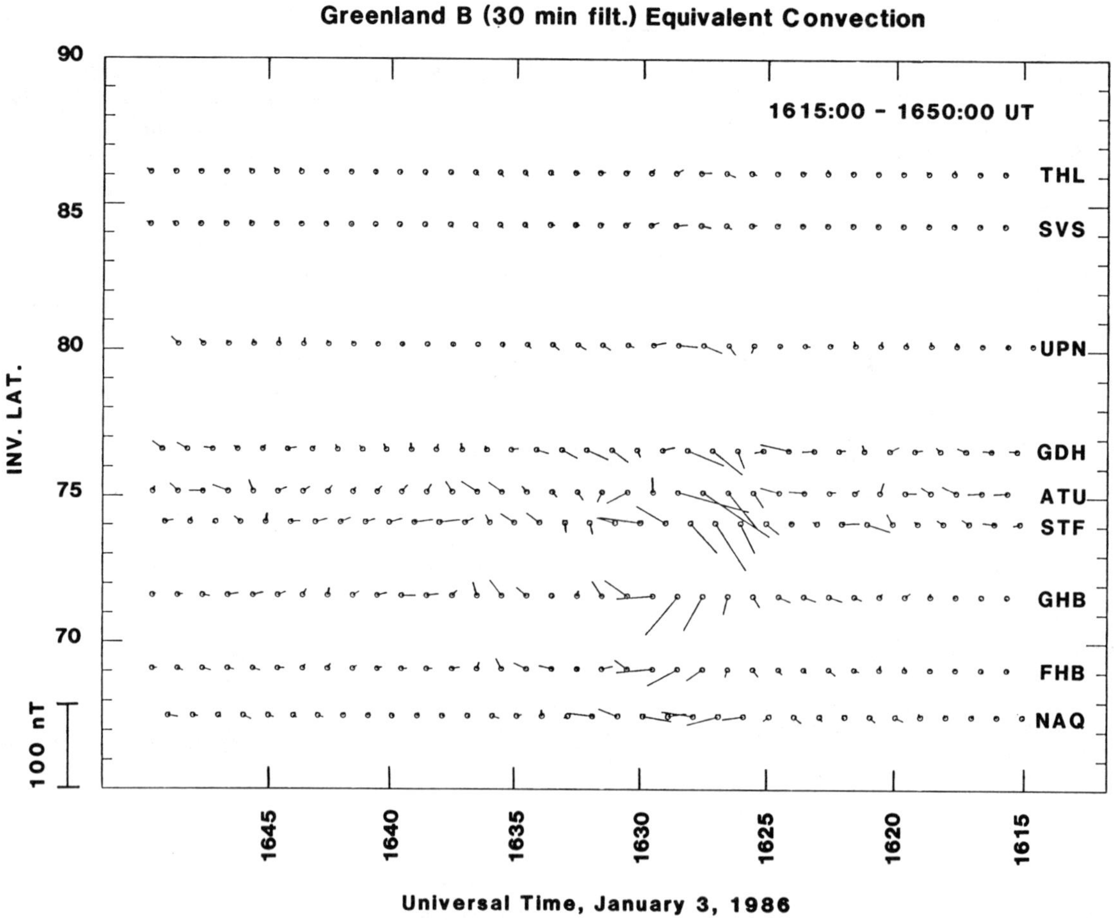

Fig. 6. Greenland Chain magnetic perturbations plotted as equivalent convection between 1615 UT and 1650 UT on 3 January 1986. This figure was provided by Dr. E. Friis-Christensen.

$X_{GSM} =\sim 21.9R_E$, $Y_{GSM} =\sim 21.6R_E$, $Z_{GSM} =\sim 16.8R_E$. At $\sim$ 1616 UT, B_z experienced a strong southward turning, from $\sim$ 3.1 nT to $\sim$ -2.4 nT. The azimuth angle Φ in the Y-Z plane changed from $\sim 90°$ to $\sim -50°$, but the the cone angle Θ was $\sim 108°$ and almost constant. This was a typical example of southward IMF as seen by IMP 8. The fluctuations of solar wind dynamic pressure are also a cause of magnetic activity in the magnetosphere. However, during the event we report here, the fluctuations of solar wind dynamic pressure were relatively small. There was also a B_y change just before this event. However, the amplitude of the B_y change was smaller than the amplitude of the B_z change, and the rate of the B_y change was less than the rate of the B_z change. Thus in this case, the strong southward turning of IMF might play a crucial role.

Conclusion

Irregular pulsations are common phenomena in the magnetosphere and it is easy for small events to be lost in the ULF background noise. Figure 2 shows that even a small sized magnetic flux rope is sufficient to contribute significantly to the signature at balloon altitude. Nonetheless, there are a number of possible sources of the signatures that we detected. It will be very helpful to acquire more data about these events, including ground array magnetometer data, electric field data, solar wind data, and DC background data. The motion of the magnetic flux rope is very important and plays an important part in determining the shape of the signatures.

The event that we report occurred at $\sim$ 1625 UT on 3 January 1986. For this event, we present South Pole balloon electric field data, South Pole ground magnetometer data and ground magnetometer data which were acquired in the conjugate area, such as in the Iqaluit and Greenland. We also present IMP 8 magnetic field and dynamic pressure data. All these data are consistent with each other and can be identified as a magnetic flux rope caused by a strong southward turning of the interplanetary magnetic field. It exhibited the features that are predicted by the two models. The signature of the magnetic flux rope appears as an isolated pulsation with a width $\sim$ 4-10 minutes. It was only visible in the area near

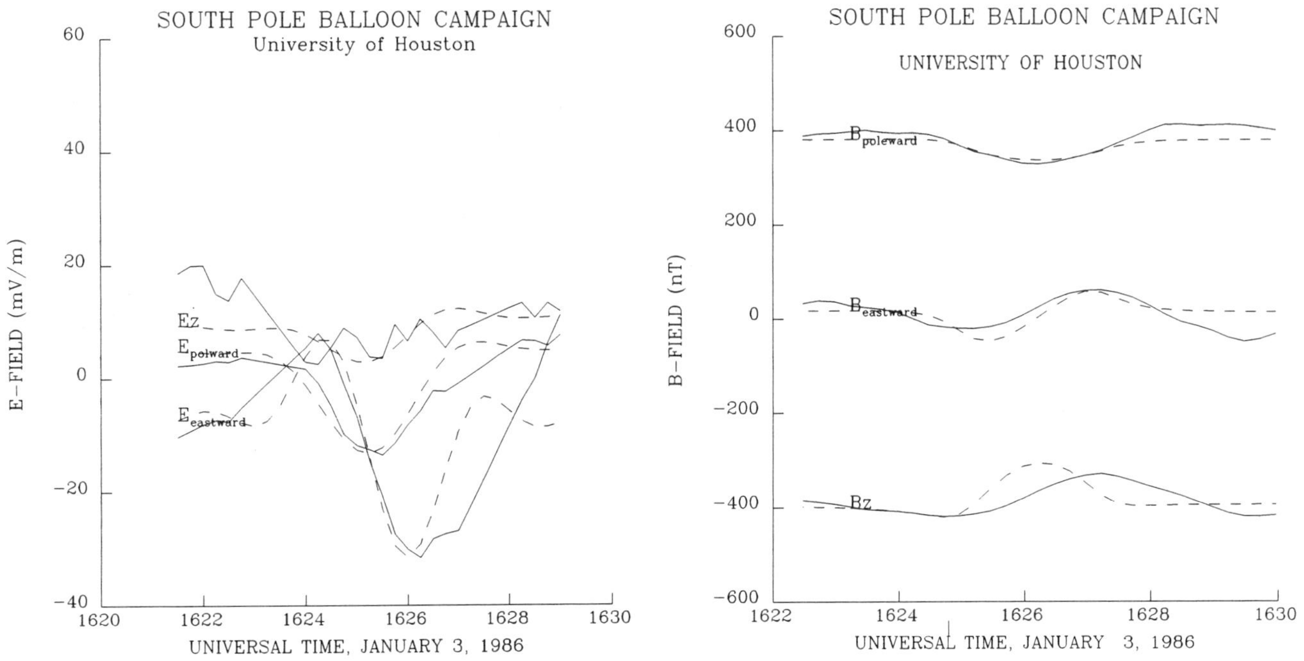

Fig. 7. A fit of the measured electric and magnetic field between 1620 and 1630 UT on 3 January 1986, and the prediction of the models. The solid line is experimental data and the dashed line is calculated. The radius of the flux rope is 350 km, moving geomagnetically eastward at a rate of 3.9 km/s, parallel to the line which connected the centers of two vortices. The total current of the towing system is $\sim 1.7 \times 10^4$ A, and the total current of the twisting system is $\sim 2.7 \times 10^5$ A. Since our interest here is only in perturbations, in order to plot three components together, we added a constant value of 508 mV/m to the experiment E_z component.

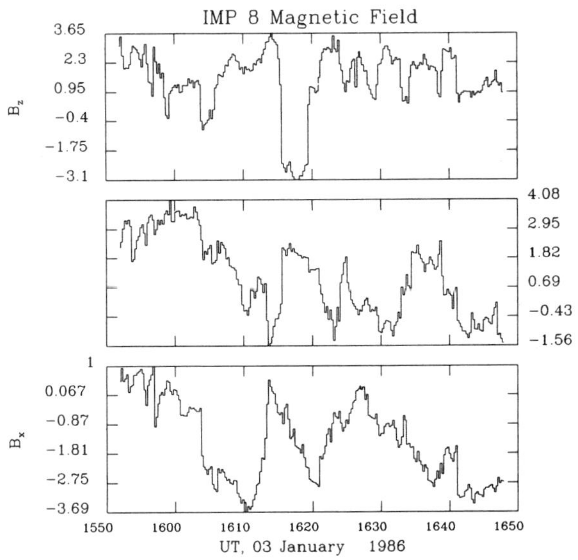

Fig. 8. The IMP magnetic field data between 1550-1650 UT on 3 January 1986. We use geocentric solar magnetospheric coordinate system. At $\sim$ 1616 UT B_z experienced a strong southward turning, and at $\sim$ 1625 UT an isolated electric pulsation was recorded at the balloon.

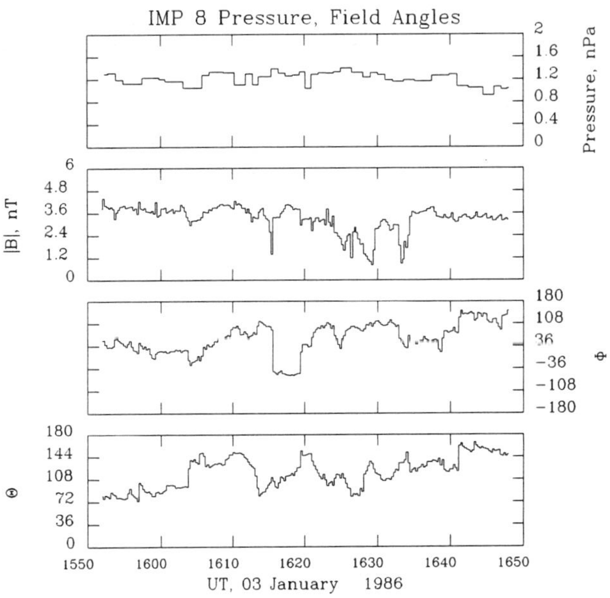

Fig. 9. Solar wind dynamic pressure and IMF orientation, where Θ is the cone angle of the magnetic field about X_{GSM} and Φ is the azimuth angle in the $Y-Z$ plane ($0° \equiv +Y$).

the magnetic flux rope. The radius of the magnetic flux rope at the South Pole was $\sim$350 km and it was moving at a rate of 3.9 km/s geomagnetically eastward. The total current of the towing system was $\sim 1.7 \times 10^4$ A and the total current of the twisting system was $\sim 2.7 \times 10^5$ A. The sub-ionospheric signature is determined by the current structure of the flux rope and its motion.

Acknowledgments. We thank Dr. L. J. Lanzerotti, Dr. C. G. Maclennan, and Dr. A. Wolfe for the South Pole and Iqaluit magnetometer data and many discussions. We also thank Dr. D. G. Sibeck and Dr. P. Song for useful discussion. This work was supported by NSF grants DPP-8415203 and DPP-8614091. IMP 8 IMF data of Dr. N. Ness were provided by Dr. R. Lepping. The IMP 8 plasma data were provided by Dr. A. Lazarus who was supported by NASA grant NAG5-584.

References

Baker, K. B., and S. Wing, A new magnetic coordinate system for conjugate studies at high latitudes, J. Geophys. Res., 94, 9139, 1989.

Bering, E. A., J. R. Benbrook, G. J. Byrne, B. Liao, J. R. Theall, L. J. Lanzerotti, C. G. Maclennan, A. Wolfe, and G. L. Siscoe, Impulsive electric and magnetic field perturbations observed over South Pole: flux transfer events? *Geophys. Res. Lett., 15*, 1545, 1988.

Bering, E. A., L. J. Lanzerotti, J. R. Benbrook, Z.-M. Lin, C. G. Maclenna, A. Wolfe, R. E. Lopez and E. Friis-Christensen, Solar wind properties observed during high-latitude impulsive perturbation events, paper presented at the 6th Scientific Assembly, International Association of Geomagnetism and Aeronomy, Exeter, United Kingdom, 24 July - 4 August, 1989.

Boström, R., U. Fahleson, L. Olausson, and G. Hallendal, Theory of time varying atmospheric electric fields and some applications to fields of ionospheric origin, TRITA-EPP-73-02, January 1973.

Dungey, J. W., Interplanetary magnetic field and the auroral zones, *Phys. Rev. Lett., 6*, 47, 1961.

Formisano, V., Orientation and shape of the Earth's bow shock in three dimensions, *Planet. Space Sci. 27*, 1151, 1979.

Formisano, V., V. Domingo, and K. -P. Wenzel, The three dimensional shape of magnetopause, *Planet. Space Sci. 27*, 1137, 1979.

Etemadi, A., S. W. H. Cowley, M. Lockwood, B. J. I. Bromage, D. M. Willis, and H. Lühr, The dependence of high-latitude dayside ionospheric flows on the north-south component of the IMF: a high time resolution correlation analysis using eiscat "polar" and *ampte UKS* and *IRM* data, *Planet. Space Sci., 36*, 471, 1988.

Friis-Christensen, E., M. A. McHenry, C. R. Clauer, and S. Vennerstrom, Ionospheric traveling convection vortices observed near the polar cleft: a triggered response to sudden changes in the solar wind, *Geophys. Res. Lett., 15*, 253, 1988.

Fukushima, N., Equivalence in ground geomagnetic effect of Chapman-Vestine's and Birkeland-Alfven's electric current-systems for polar magnetic stroms, *Rep. Ionos. Space Res. Jpn., 23(3)*, 1969.

Glaßmeier, K. H., M. Hönisch, and J. Untiedt, Ground-based and satellite observations of traveling magnetospheric convection twin-vortices, *J. Geophys. Res., 94*, 2520, 1989.

Glaßmeier, K. H., ULF pulsation in the polar cusp and cap, invited talk, presented at the NATO Advanced Research Workshop, Lillehammer, September 20-24, 1988.

Goertz, C. K., E. Nielsen, A. Korth, K. H. Glaßmeier, C. Haldoupis, P. Hoeg, and D. Hayward, Observations of a possible ground signature of flux transfer events, *J. Geophys. Res., 90*, 4069, 1985.

Hughes, W. J., and D. J. Southwood, The screening of micropulsation signals by the atmosphere and ionosphere, *J. Geophys. Res., 81*, 3234, 1976a.

Hughes, W. J., and D. J. Southwood, An illustration of modification of geomagnetic pulsation structure by the ionosphere, *J. Geophys. Res., 81*, 3241, 1976b.

Lanzerotti, L. J., L. C. Lee, C. G. Maclennan, A. Wolfe and L. V. Medford, Possible evidence of flux transfer events in the polar ionosphere, *Geophys. Res. Lett., 13*, 1089, 1986.

Lee, L. C., and Z. F. Fu, A theory of magnetic flux transfer at the earth's magnetopause, *Geophys. Res. Lett., 12*, 105, 1985.

Lockwood, M., M. Smith, C. J. Farrugia, and G. S. Siscoe, Ionospheric ion upwelling in the wake of flux transfer events at the dayside magnetopause, *J. Geophys. Res., 93*, 5641, 1988.

McHenry, M. A., and C. R. Clauer, Modeled ground magnetic signatures of flux transfer events, *J. Geophys. Res., 92*, 11231, 1987

Mozer, F. S., and R. Serlin, Magnetospheric electric field measurements with balloon, *J. Geophys. Res., 74*, 4739, 1969.

Park, C. G., and M. Dejnakarintra, Penetration of thundercloud electric fields into the ionosphere and magnetosphere, 1, Middle and subauroral latitudes, *J. Geophys. Res., 78*, 6623, 1973.

Park, C. G., Downward mapping of high-latitude ionospheric electric fields to the ground, *J. Geophys. Res., 81*, 168,1976.

Russell, C. T., and R. C. Elphic, Initial ISEE magnetometer results: magnetopause observations, *Space Science Rev., 22*, 681, 1978.

Russell, C. T., and R. C. Elphic, ISEE observations of flux transfer events at the dayside magnetopause, *Geophys. Res. Lett., 6*, 33, 1979.

Song, P., R. C. Elphic, and C. T. Russell, ISEE 1 and 2 observations of the oscillating magnetopause, *Geophys. Res. Lett., 15*, 744, 1988.

Southwood, D. J., Comments on field line resonances and micropulsations, *Geophys. J. Roy. Astron. Soc., 41*, 425, 1975.

Southwood, D. J., Theoretical aspects of ionospheric-magnetosphere-solar wind coupling, *Adv. Space Res., 5*, 7, 1985.

Southwood, D. J., The ionospheric signature of flux transfer events, *J. Geophys. Res., 92*, 3207, 1987.

DE-2 OBSERVATIONS OF FILAMENTARY CURRENTS AT IONOSPHERIC ALTITUDES

M. F. Smith and J. D. Winningham

Department of Space Sciences
Southwest Research Institute, San Antonio, TX 78228

J. A. Slavin

Laboratory for Extraterrestrial Physics
Goddard Space Flight Center, Greenbelt, MD 20771

M. Lockwood

Rutherford Appleton Laboratory
Chilton, Didcot, UK

Abstract. Conjunctive measurements made by the Dynamics Explorer 1 and 2 spacecraft on October 22, 1981, under conditions of southward IMF, suggest the existence of a cusp ion injection from a region at the magnetopause with a scale size of $\sim 1/2$ to 1 R_E. Current signatures observed by the LAPI and MAGB instruments on board DE-2 indicate the existence of a rotation in the magnetic field that is consistent with a filamentary current system. The observed current structure can be interpreted as the ionospheric signature of a flux transfer event (FTE). In addition to this large-scale current structure there exist three small-scale filamentary current pairs. These current pairs close locally and thus, if our interpretation of this event as an FTE is correct, represent the first reported observations of FTE interior structure at low-altitudes.

Introduction

The cusp is a region where there is significant mixing of solar wind and ionospheric plasma. Solar wind plasma appears to be injected onto cusp field lines and then dispersed by the convective electric field. This effect has been observed on many occasions (e.g., Heikkila and Winningham [1971] and Shelley et al. [1976]). Burch et al. [1982] and Menietti and Burch [1988] have shown data from the high-altitude DE-1 spacecraft that demonstrate the existence of a narrow region of solar wind plasma injection on cusp field lines. Velocity dispersion causes a reduction in the observed energy of the injected ions at higher latitudes for a southward directed IMF [Shelley et al., 1976; Reiff et al., 1977]. At DE-1 altitudes the travel time is a strong function of pitch angle, leading to characteristic V-shaped signatures in the energy/pitch angle spectrograms of injected ions (see, for example, Burch et al. [1982]). Menietti and Burch [1988] have shown that the data can be interpreted as a limited spatial injection region with a scale length of about 1/2 R_E in the direction perpendicular to the magnetic field. They have speculated on the link between these injection events and flux transfer events (FTEs).

At lower altitudes the injected ions no longer show a pitch angle variation [Reiff et al., 1977], and a simple ion dispersion signature is observed. Smith et al. [1988] have presented data which strongly suggest a connection between the low-altitude plasma signatures observed by DE-2 in the cusp and FTEs. More recently, Lockwood and Smith [1989] have shown the DE-2 data to be consistent with the expected ionospheric signature of a growing FTE. Specifically, they interpret the ion dispersion signature to be the result of particle injection on newly opened field lines.

In addition to the injected solar wind ions the cusp is also a region where significant ion outflows, particularly of O^+ ions [Lockwood et al., 1985], are found. These are associated with ion heating perpendicular to the magnetic

Geophysical Monograph 58

field, bipolar field aligned currents, convective electric shear flows, and intense plasma wave activity [Moore et al., 1985]. Lockwood et al. [1988] noted that the twin vortex pattern predicted by Southwood [1987] as the ionospheric signature of FTEs would produce non-thermal plasma distributions. These ions have been shown to bunch in the wake of the FTE [Lockwood et al., 1988] and thus are important in identifying the existence of FTEs. We show the presence of large fluxes of upwelling ions in the cusp during the ion injection. The upwelling ion fluxes are consistent with the observations of Lockwood et al [1988].

In this paper we focus on the low-altitude current signatures observed by DE-2 on October 22, 1981, and show that the large-scale current structure is consistent with that predicted for the ionospheric signature of an FTE. In addition, small-scale features such as electron bursts, small-scale magnetic field rotations, and flow enhancements have also been observed and may be interpreted as structure within the FTE. These small-scale current structures close locally and are not part of the global current system.

DE-1/2 Observations

The DE-1 plasma data from the HAPI instrument for October 22 between 09:42 and 09:45 are shown in Figure 1. The top panel is an energy-time spectrogram for electrons with energies between 5 eV and 17 keV; the bottom panel is an energy-time spectrogram for the combined ion population with E/q from 5 eV to 17 keV. The middle panel gives the pitch angle for both electron and ion sensors. Prior to 09:42:42 UT the electron data are typical of the cleft/low-latitude boundary layer and include narrow upward electron beams [Burch et al., 1983]. After 09:42:20 and for the rest of the data shown here the electrons are typical of those found in the cusp. The incoming solar wind ions show a general decrease in energy, along with distinctive V-shaped signatures, as the spacecraft moves more poleward. This is typical of injected solar wind ions [Burch et al., 1982]. In addition to the ion V's a single electron V can be seen on the equatorward edge of the cusp at 09:42:00. As the electron velocity is high, the electrons are not dispersed in latitude by the poleward electric field. Thus the field-line

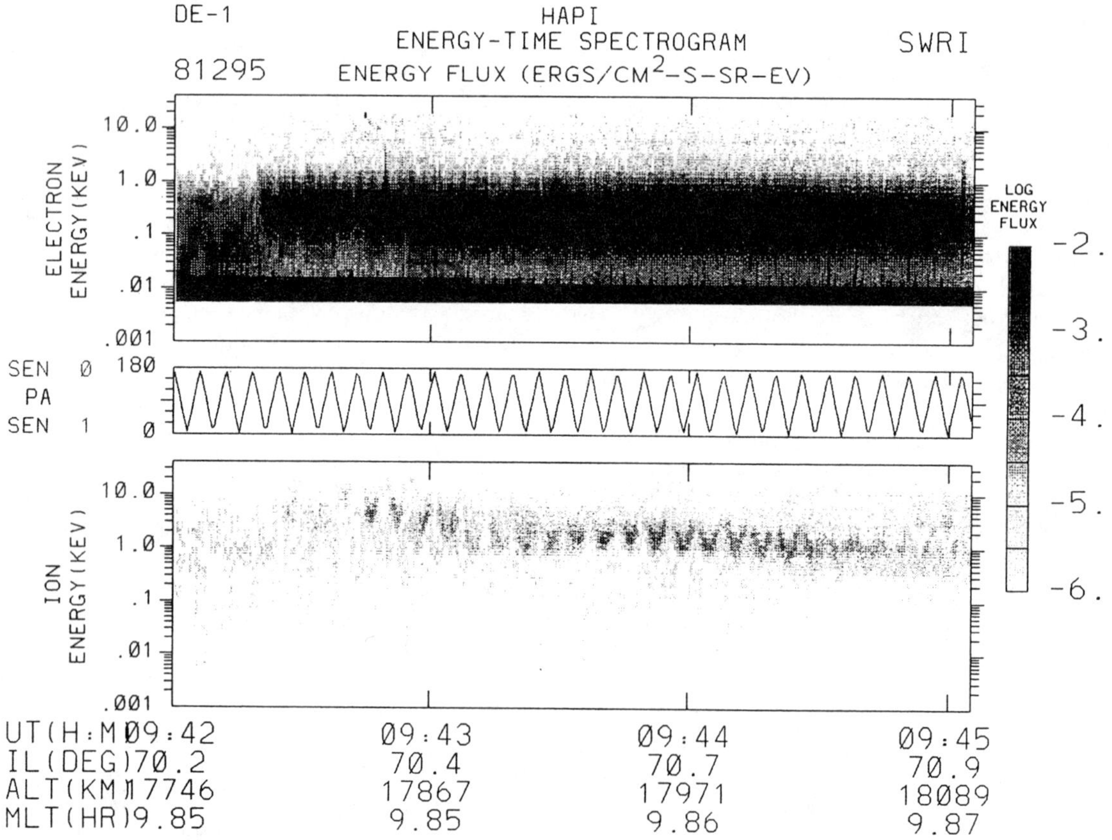

Fig. 1. Data from the HAPI instrument on DE-1 for the period between 09:42 and 09:45. The top and bottom panels are energy-time spectrograms for electrons and ions respectively, between 5 eV and 17 keV. The middle panel gives the pitch angle of the sensors.

on which the electron V occurs probably signifies the injection region. By using the ion signature and the method of Menietti and Burch [1988], the injection region size at the magnetopause is estimated to be 0.75 R_E in the direction perpendicular to B. The ion and electron densities track each other throughout the cusp region due to ambipolar diffusion at the magnetopause [Burch, 1985].

The LAPI particle data from the DE-2 spacecraft, during a pass in close conjunction with the above DE-1 pass, are shown in Figure 2. The top panel shows the fluxes of electrons with energies greater than 35 keV for 0° (blue) and 90° (red) pitch angles. The next two panels are energy-time spectrograms for up-going (upper panel) and down-going (lower panel) ions. The bottom two panels present electron data in the same format as the ion spectrograms. Prior to 09:55:20 UT there are anisotropic fluxes of > 35 keV electrons normally identified with a trapped magnetospheric population. The ion and electron plasma data are also consistent with a trapped population until 09:55:20 UT. After 09:55:20 UT, however, the > 35 keV electron fluxes become isotropic which is normally understood to indicate open field lines. During the period 09:55:20 to 09:55:50 the ions show an injection signature with the energy decreasing from about 1 keV to 100 eV as the spacecraft moves poleward. This region is the cusp. In addition to the ion dispersion signature, there are small, but significant fluxes of > 10 keV ions in the downgoing ion spectrogram throughout the period from 09:55:20 to 09:55:50. The usual interpretation of these fluxes would suggest that this region is on closed field lines. However, noted above, the > 35 keV electron fluxes are isotropic, suggesting open field lines. The apparent inconsistency between the 10 keV

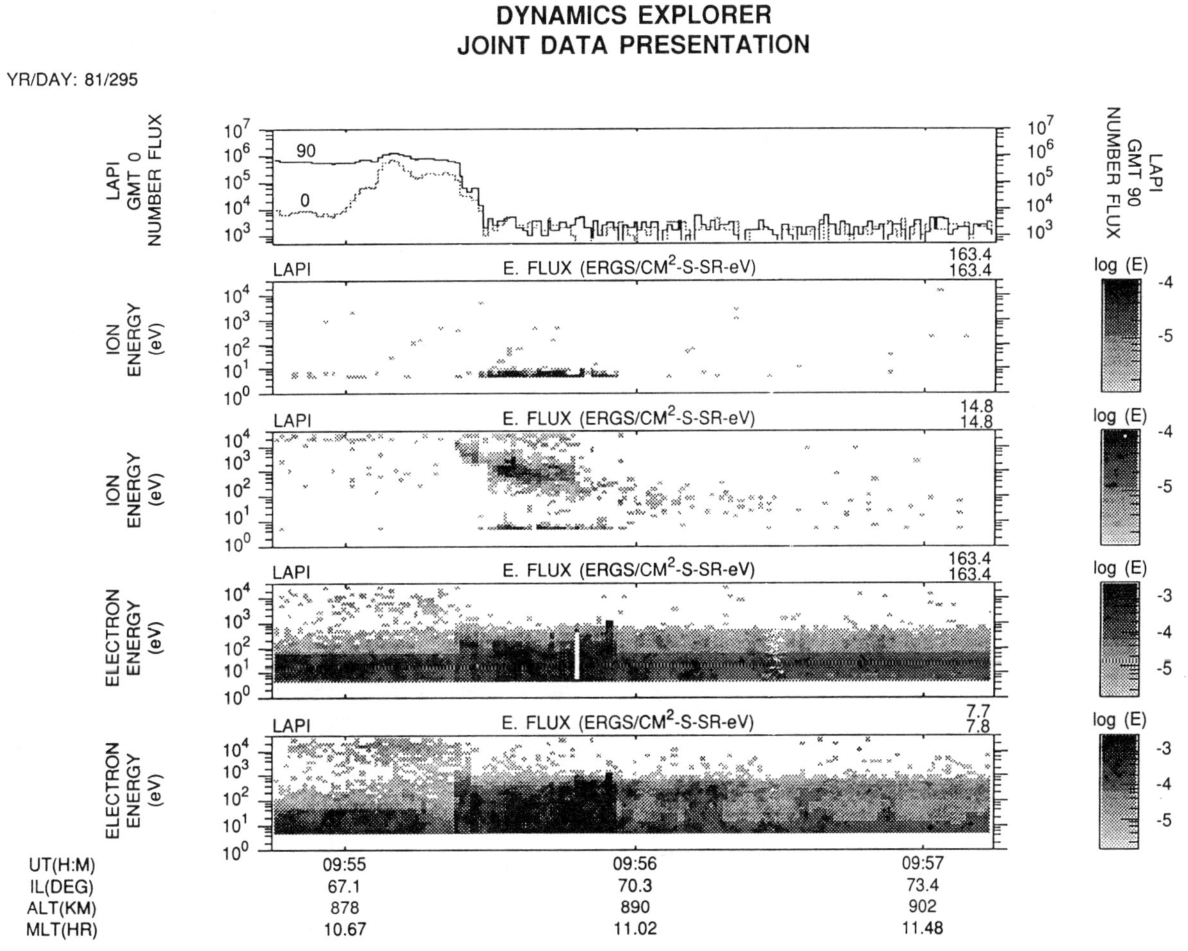

Fig. 2. Data from the DE-2 LAPI instrument between 09:54:45 and 09:57:15. The top panel shows the intensities of > 35 keV electrons for 0° and 90° pitch angles. The next two panels are energy-time spectrograms for upgoing (upper) and downgoing (lower) ions. The bottom two panels are similar to the ion spectrograms except for electrons.

ions and 35 keV electrons may be resolved if it is postulated that the field lines in this region are newly opened, i.e., the structure may be time-dependent. Alternatively, the 10 keV ions may have diffused from closed field lines onto open field lines.

The tail of an upwelling ion distribution can be seen at low energies in the up-going ion spectrogram of Figure 2. These upwelling low-energy ions are thought to be generated by frictional heating [Lockwood and Fuller-Rowell 1987]. The electron distributions detected between 09:55:20 and 09:55:50 UT are typical of a cusp population with peak fluxes in the hundreds of eV range. However, superimposed on the cusp population are two large bursts of low-energy (< 200 eV) electrons at 09:55:30 and 09:55:47. After 09:55:50 both the ion and electron populations are typical of the polar cap. Figure 3 shows the current calculated by using moments of the LAPI particle distribution for the period 09:55:20 to 09:56:00 (i.e., the cusp). The two bursts seen in the electron spectrogram show up as upward currents at 09:55:30 (~ 23 mA m^{-2}) and 09:55:47 (~ 25 mA m^{-2}). It should be noted that the low-energy cut-off of the detector will affect the magnitude of the current, but the direction will be correct. At this time the ion currents were an order of magnitude lower than the electron currents.

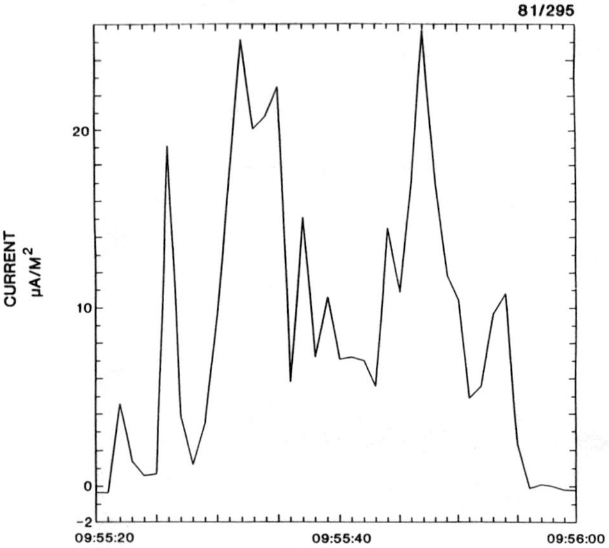

Fig. 3. Calculated current density from LAPI using moments of the distributions function.

The magnetic field for these events shows three large shears in the ΔB_ϕ component at 09:55:20-:40, 09:55:40-:50 and 09:55:50-09:56:00 (Figure 4). However, examination of the horizontal magnetic field perturbation shows that in these intervals the north/south (ΔB_θ positive south) and east/west components (ΔB_ϕ positive east) are not linearly

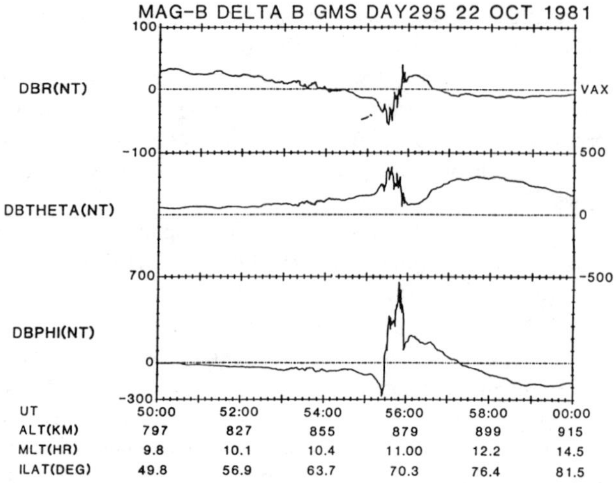

Fig. 4. Data from the magnetometer on DE-2. The panels show the three components after removal of a model field. The top panel is the ΔB_R component (positive radially outward), the middle panel is the ΔB_θ component (positive southward), while the bottom panel is the ΔB_ϕ component (positive eastwards).

correlated. This indicates that the magnetic perturbations are not associated with the traversal of infinite current sheets. In fact, the magnetic field shows a rotation in the horizontal plane (Figure 5), the plane perpendicular to B, from a southwesterly direction to a southeasterly direction between 09:55:20 and 09:55:50 as DE-2 moved northward. This is consistent with a net downward current, in disagreement with the direction of current flow measured by the LAPI instrument. This disagreement is expected, because low-altitude measurements are generally unable to detect downward currents [Burch et al., 1983]. Between 09:55:50 and 09:56:00 the rotation continues in a clockwise direction, leading to a total rotation of about 270°. The rotation implies that the spacecraft is passing to the right of a pair of oppositely directed line currents. The fact that the rotation is less than 360° means that the spacecraft passed further away from the second (upward) current than from the first (downward) current. This inferred spacecraft trajectory agrees with the calculations of Lockwood and Smith [1989] which are based on a simple model of an ionospheric FTE footprint. It is important to note that the large-scale currents lie at the beginning and end of the ion dispersion signature.

In addition to the large-scale clockwise rotation seen from 09:55:20 until 09:56:00, three smaller rotations occur. These rotations are labelled AA′ (09:55:31–09:55:35), BB′ (09:55:40–09:55:45) and CC′ (09:55:50–09:55:55) in Figure 5 and are the three shears seen in the ΔB_ϕ component of Figure 4. These rotations suggest that there are small-scale filamentary currents within the overall large-scale current struc-

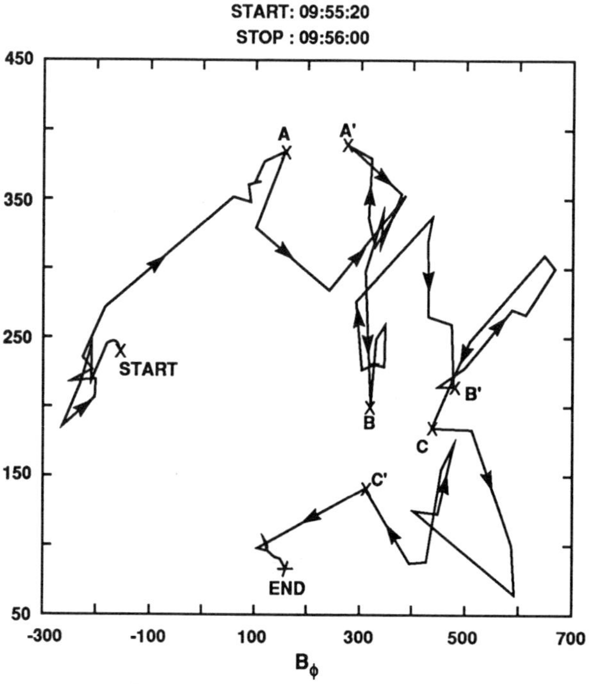

Fig. 5. Hodogram of the horizontal magnetic field from 09:55:20 until 09:56:00. The points AA′, BB′ and CC′ are the start and end points of the three small-scale rotations.

ture which may be interpreted as up/down current pairs. By this interpretation, all three small-scale current pairs are aligned approximately along the spacecraft track with the spacecraft passing either on the poleward or equatorward side of the pairs. Using the direction of rotation, one can calculate which of the current pairs was passed first and to which side of the pair the spacecraft passed. The first of these rotations, AA′, is anti-clockwise, meaning that the spacecraft passed poleward of an upward current followed by a downward current. This current pair, AA′, coincides with the first electron burst at 09:55:30 and thus the precipitating electrons in this first burst are likely to be the source of the upward current portion of AA′. For the rotation BB′ the spacecraft passed equatorward of a down-up pair, while for CC′ it passed equatorward of an up-down pair. Neither of the rotations, BB′ or CC′, are coincident with the second electron burst at 09:55:47 and indeed they appear to straddle the electron burst. However, the upward filament currents in both BB′ and CC′ lie next to one another, and hence one would expect the precipitating electrons for each current to be close to one another. Thus, the downward precipitaing electrons appear to be part of two filamentary pairs.

Figure 6 shows data from the Ion Drift Meter on DE-2 during the period 09:54:00 to 09:58:00. The top panel is the cross-drift; for this particular set of data positive is approximately southeastwards. The next panel is the ram drift (positive antiparallel to the spacecraft velocity vector), while the bottom panel is the vertical drift (positive upwards). Throughout the whole cusp region the convection velocity is enhanced. Using the calculated electric field from the IDM and an estimate for the Pedersen Conductivity from the LAPI plasma moments, we estimate the horizontal current to be ~ 250 mA m^{-2}. This current is almost certainly closed by the large vertical currents seen by the magnetometer and far exceeds the currents calculated from the LAPI moments. This intense current structure is the main source of momentum transfer and will dominate magnetic records. However, the small-scale current pairs will also effect the convection velocity. As all three current pairs lie roughly along the spacecraft track, the changes in flow velocity will be most apparent in the component perpendicular to the spacecraft track, i.e., the cross-drift component. According to the analysis of the current pairs, the flow due to the pair AA′ should be to the left of the track, in other words, negative cross-drift. This enhancement can be seen in the IDM data as a spike in the cross-drift at about 09:55:33. The pair BB′ should produce the opposite effect, i.e., a positive cross-drift component. Study of Figure 6 in fact suggests a slight negative enhancement in the cross-drift component. However, there is much variation in the ram component, and if the current pair is not aligned along the spacecraft track, the flow may have a significant ram component. Finally, the pair CC′ should generate a westward flow. However, the strong westward flow spike at 09:55:47 lies between the BB′ and CC′ current pairs. This spike is probably due to the enhanced precipitation which occurs in the region between BB′ and CC′. The IDM also measures a large flux of upwelling ions that reach fluxes exceeding 5×10^9 cm^{-2} s^{-1}. The high-energy tail of this distribution is seen in the upgoing ion spectrogram in Figure 2.

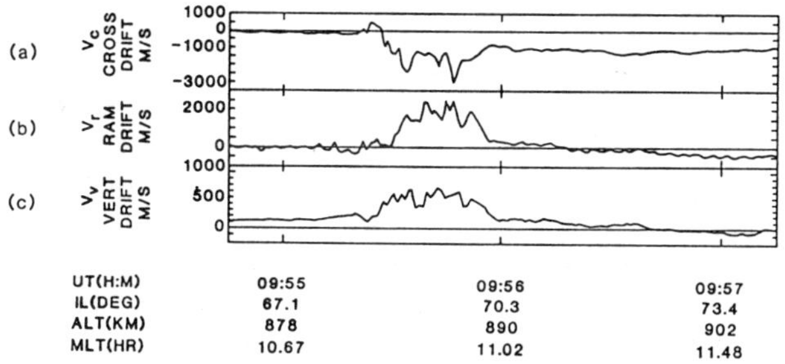

Fig. 6. The three components of ion drift as measured by the IDM along with N_i (bottom panel). The drifts are the cross-drift (positive to the right of the spacecraft track), ram drift, and vertical drift (positive upwards).

Discussion

The magnetometer and particle data can be reconciled if it is realized that the magnetometer signal is integrated over a large region. As we have shown above, the current measured by the magnetometer is much greater than that from the particle moments, suggesting that the particle currents are a part of a smaller-scale current system superimposed on the large system. The large upward and downward currents measured by the magnetometer are closed by the horizontal current, which can be seen in the IDM data as the large horizontal drifts in the cusp. The existence of spikes in the drift data at the time of the electron bursts and small-scale current rotations again suggests that these smaller magnitude currents form part of smaller spatial structures superimposed on the large current system. It is important to note that these current systems are closed locally and are not part of the global convection pattern.

An interpretation of the complete current system is shown in Figure 7a. The spacecraft is moving in a northward (poleward) direction. It first encounters a large downward current (as measured by the magnetometer) followed by an upward current of similar magnitude, poleward of the first current. The motion of this current system is in a southwesterly direction, which gives an electric field (from $V = \mathbf{E} \times \mathbf{B}$) pointing to slightly west of the spacecraft track. This direction is consistent with the ion drift measured by the IDM. The large-scale rotation in the hodogram suggests that the spacecraft is moving to the right of a vertical field-aligned current pair. We can thus assume that the spacecraft has passed through the eastward edge of a westward moving flux tube. The data presented above then show a striking similarity to the ionospheric signature of an FTE (Figure 7b) [Lockwood et al., 1988]. However, the observations presented here show more detail than just the large-scale pattern. On the edge of the open flux tube regions lie current pairs which are closed locally. The upward currents in these small filamentary currents pairs are carried by intense bursts of low energy electrons. These electron bursts are qualitatively similar, in terms of energy range, to those observed on the edges of FTEs at the magnetopause [Scudder et al., 1984; Farrugia et al., 1988]. The electrons observed at the magnetopause are thought to be generated by a continuing reconnection process as the flux tube propagates [Sonnerup, 1987; Southwood et al., 1988; Scholer, 1988] and thus must lie on the edge of the open flux tube. As we have shown above, the bursts observed by LAPI would lie on the edge of the open flux tube, if our interpretation of this event as an FTE is correct. If we use the electron bursts as the signature of the open flux tube region, then we obtain a scale size at the magnetopause of 1/2–1 R_E for this event. This size corresponds to the open flux tube size and is not the size measured by a magnetometer at the magnetopause, where field line draping over the open flux tube is included.

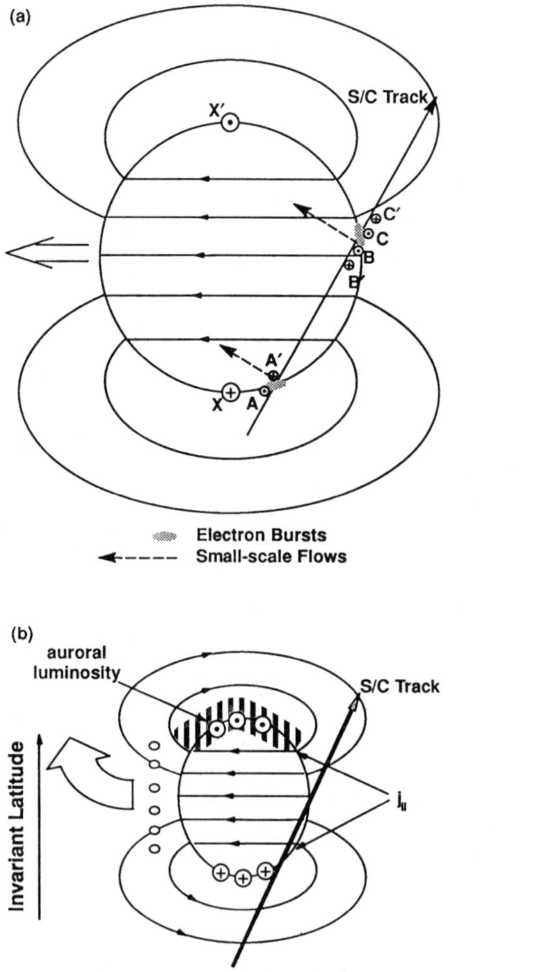

Fig. 7. Panel (a) is a schematic of the data presented in this paper showing both the large- and small-scale features, while (b) shows an FTE footprint from Lockwood et al. [1989]. A spacecraft track has been drawn through the event showing the direction of traversal of the DE-2 spacecraft.

The observed flux of upwelling ions, seen during the event, is considerable ($\sim 5 \times 10^9$ cm^{-2} s^{-1}) and exceeds the classical polar wind flow. These upwelling fluxes also exceed those measured by the AMPTE UKS at the magnetopause [Lockwood et al., 1988] by two orders of magnitude. However, as stated by Lockwood et al. [1988], the outflow estimated by the AMPTE ion experiment is probably low by several orders of magnitude, thus the AMPTE UKS and DE-2 results are qualitatively consistent. Lockwood et al. [1988] have argued that the upwelling ion fluxes observed by AMPTE UKS agree with the predicted upwelling ion flux from an FTE. This suggests that the DE-2 fluxes observed during the event are also due to an FTE.

As we have shown, there is strong evidence to suggest that this event is indeed the ionospheric footprint of an FTE. In Lockwood and Smith [1989] we show how the observed electric and magnetic field signatures can be produced by a simple ionospheric FTE model. These observations are of importance in understanding the relationship between the cusp and FTEs. For the particular case studied here and in Lockwood and Smith [1989], the FTE signature, as defined by the large-scale current system, covers almost the whole region normally called the cusp. The high-energy ion data presented here also suggest that the cusp field-lines are newly opened and that the observed effects are time-dependent. This time-dependency is supported by the interpretation of this event by Lockwood and Smith [1989] as being from a growing FTE. At present it is impossible to tell whether or not these observations are typical of the cusp. The data also suggests the structure we have interpreted as an FTE is coextensive with the entire region normally considered the cusp, raising the possibility that the two phenomena may, in some cases, be synonymous.

Summary

We have presented data from the DE-1 spacecraft typical of a pass through the cusp during times of southward IMF. Data from a conjunctive pass of the low-altitude DE-2 spacecraft show the existence of a filamentary current structure which is similar to that predicted and reported for the ionospheric signature of FTEs within the cusp region. We have presented the first observations of a small-scale filamentary current structure existing within the putative FTE. A detailed study of the data shows that the solar wind ion injection signature is best interpreted as being due to time-dependent ion injection as opposed to convection. In addition, we have demonstrated that the cusp and FTE signature occupy roughly the same region of space, leading us to speculate on the possible synonymity of the two phenomena.

Acknowledgments. The authors thank R. A. Heelis for use of the IDM data and for useful discussions. We also wish to thank the referees for many useful and illuminating comments. This work at SwRI was supported by NASA grants and contracts NAGW-1638, NAS5-28711, NAS5-28712 and by SwRI internal research grant 15-9557.

References

Burch, J. L., P. H. Reiff, R. A. Heelis, J. D. Winningham, W. B. Hanson, C. Gurgiolo, J. D. Menietti, R. A. Hoffman, and J. N. Barfield, Plasma injection and transport in the mid-altitude polar cusp, *Geophys. Res. Lett., 9*, 921–924, 1982.

Burch, J. L., P. H. Reiff, and M. Sugiura, Upward electron beams measured by DE-1: A primary source of dayside region-1 birkeland currents, *Geophys. Res. Lett., 10*, 753–757, 1983.

Burch, J. L., Quasi-neutrality in the polar cusp, *Geophys. Res. Lett., 12*, 469–472, 1985.

Farrugia, C. J., R. P. Rijnbeek, M. A. Saunders, D. J. Southwood, D. J. Rodgers, M. F. Smith, C. P. Chaloner, D. S. Hall, P. J. Christiansen, and L. J. C. Woolliscroft, A multi-instrument study of flux transfer event structure, *J. Geophys. Res., 93*, 14465–14477, 1988.

Heikkila, W. J., and J. D. Winningham, Penetration of magnetosheath plasma to low altitudes through the dayside magnetospheric cusps, *J. Geophys. Res. 76*, 883–892, 1971.

Lockwood, M., M. O. Chandler, J. L. Horwitz, J. H. Waite, Jr., T. E. Moore, and C. R. Chappell, The cleft fountain, *J. Geophys. Res., 10*, 9736–9749, 1985.

Lockwood, M., and T. J. Fuller-Rowell, The modelled occurrence of non-thermal plasma in the ionospheric F region and the possible consequences for ion outflows into the magnetosphere, *Geophys. Res. Lett., 14*, 371–374, 1987.

Lockwood, M., M. F. Smith, C. J. Farrugia, and G. L. Siscoe, Ionospheric ion upwelling in the wake of flux transfer events at the dayside magnetopause, *J. Geophys. Res., 93*, 5641–5654, 1988.

Lockwood, M., and M. F. Smith, Low-altitude signatures of the cusp and flux transfer events, *Geophys. Res. Lett., 16*, 879–882, 1989.

Lockwood, M., P. E. Sandholt, and S. W. H. Cowley, Dayside auroral activity and magnetic flux transfer from the solar wind, *Geophys. Res. Lett., 16*, 33–36, 1989.

Menietti, J. D., and J. L. Burch, Spatial extent of the plasma injection region in the cusp-magnetosheath interface, *J. Geophys. Res., 93*, 105–115, 1988.

Moore, T. E., C. R. Chappell, M. Lockwood, and J. H. Waite, Jr., Superthermal ion signatures of auroral acceleration processes, *J. Geophys. Res., 90*, 1611–1619, 1985.

Reiff, P. H., T. W. Hill, and J. L. Burch, Solar wind plasma injection at the dayside magnetospheric cusp, *J. Geophys. Res., 82*, 479–492, 1977.

Scholer, M. Magnetic flux transfer at the magnetopause based on single X line bursty reconnection, *Geophys. Res. Lett., 11*, 291–295, 1988.

Scudder, J. D., K. W. Ogilvie and C. T. Russell, The relation of flux transfer events to magnetic reconnection, in *Magnetic Reconnection in Space and Laboratory Plasmas*, edited by E. W. Hones, pp. 153–154, AGU, Washington, D.C., 1984.

Shelley, E. G., R. D. Sharp, and R. G. Johnson, He^{++} and H^{+} flux measurements in the dayside magnetospheric cusp, *J. Geophys. Res., 81*, 2363–2376, 1976.

Smith, M. F., J. D. Winningham, and J. A. Slavin, A filamentary current structure at ionospheric altitudes, in press, SPI Proceedings Series, MIT Press, 1988.

Sonnerup, B. U. O., On the stress balance in flux transfer events, *J. Geophys. Res., 92*, 8613–8621, 1987.

Southwood, D. J., The ionospheric signature of flux transfer events, *J. Geophys. Res., 92*, 3207–3214, 1987.

Southwood, D. J., C. J. Farrugia and M. A. Saunders, What are flux transfer events?, *Planet. Space Sci., 36*, 503–508, 1988.

A MODEL OF FTE FOOTPRINTS IN THE POLAR CAP

F. R. Toffoletto, T. W. Hill and P. H. Reiff

Department of Space Physics and Astronomy
Rice University, Houston, TK 77251–1892

Abstract. We have investigated the mapping of FTE's onto the polar cap ionosphere using the closed magnetic field model of Voigt [1981] and the perturbed (open) version of that model by Toffoletto and Hill [1989]. We have assumed that the magnetic flux associated with the FTE's crosses the magnetopause through small regions of large normal component, and that these small flux tubes map down to the polar cap ionosphere. This procedure essentially follows that of Crooker and Siscoe [1988], using a more realistic magnetopause and internal field model. In agreement with Crooker and Siscoe, we find that the footprint of a circular hole in the magnetopause of an otherwise closed magnetosphere becomes increasingly distorted as the hole moves from the day side to the night side. However, a similar region of enhanced open flux in an otherwise open magnetosphere maintains an approximately circular footprint. The first case, representing the hypothesis that magnetopause merging is an intrinsically sporadic process, results in a polar cap constructed from a scattered distribution of FTE's at the magnetopause. The second case, representing the hypothesis that FTE's are formed by sudden localized enhancements of merging along an already active quasi-steady-state merging line, results in an open polar cap dotted with FTE footprints. Such regions moving through the ionosphere at a different rate than the background, would produce the dipolar electric-field and current structures predicted by Southwood [1987] and observed by Bering et al. [1988] and Lockwood et al. [1989].

Introduction

The process of magnetic merging changes the topology of magnetic field lines: closed field lines (with both ends connected to the Earth) become open (with one end still connected to the Earth and the other end extending into the the solar wind), or vice versa, by localized violation of the frozen-in-flux condition. Observations of flux transfer events (FTE's) near the dayside magnetopause [Russell and Elphic, 1978] were interpreted in terms of isolated flux tubes containing open field lines, implying that the merging process at the magnetopause is an intrinsically sporadic, time dependent process. In this picture, FTE's are presumed to be the major mechanism for transferring open magnetic flux from the dayside to the nightside tail (i.e., most of the open flux in the magnetosphere is in the form of FTE's). This is in contrast to the classic Dungey [1961] picture of the open magnetosphere where, for constant IMF conditions (favorable for merging), merging is a steady, continuous process whereby closed field lines are converted to open field lines and convected tailward. The observed FTE's may then result from a temporary local enhancement of an otherwise steady merging process, producing intermittent patches of enhanced flux across an already open magnetopause [Southwood, 1987].

It is possible that the IMF direction controls which of the above two merging scenarios occurs. For example, during times of northward IMF, it is possible that merging is patchy, resulting in the former picture. Small fluctuations in the IMF direction and/or the magnetopause shape (in response to a pressure pulse) could make conditions favorable for merging for a short period and/or over a spatially confined region. The scarcity of observations of FTE's during periods of northward IMF [Rijnbeek et al., 1984; Berchem and Russell, 1984] may indicate that they occur lee often during northward versus southward IMF, or that they occur only poleward of the cusp during northward IMF, as implied by the antiparallel merging hypothesis of Crooker [1979], i.e., in regions not observed by the ISEE spacecraft. When the IMF turns southward, quasi-steady merging of the Dungey type with intermittent enhancements in the form of FTE's may be a more appropriate description.

Even if we assume that merging occurs along a single x-line on the dayside magnetopause, spatial and/or temporal variations of the merging rate can produce a wide variety of open field-line configurations. For example, if the merging rate were enhanced in a localized region comprising a small fraction of the length of the dayside merging line, for a time interval that is long compared to the magnetosheath flow time around the magnetosphere, the result would be a long narrow flow-aligned "window" of open flux crossing the magnetopause (which maps to a narrow window in the solar wind, as suggested by Stern [1973]). On the other hand, if the merging rate were enhanced over the length of the merging line for a relatively brief time interval (for example, as the result of a sudden and short-lived change in the solar wind that makes conditions favorable for merging over a large portion of the magnetopause), the open "window" would be elongated in the direction transverse to the flow, as

Geophysical Monograph 58

proposed by Scholer [1988] and Southwood et al. [1988]. Of the various configurations that have been proposed [e.g., Russell and Elphic 1978; Lee and Fu; 1985, Southwood 1987; Scholer, 1988], we concentrate here on the simple configuration of Russell and Elphic [1978], which results from a merging-rate enhancement is localized both in position and in time. This case is investigated by imposing a short-lived, highly localized enhancement of the merging rate along a given merging line. This condition is imposed on two background magnetospheric configurations, one that is closed and one that is already open, thus simulating the two scenarios described above. We model this situation by adding a perturbation field consisting of cylinders of flux across the magnetopause at various locations and mapping the cross section of these cylinders from the magnetopause down to the polar-cap ionosphere. The other cases mentioned above can be modeled by varying the geometry, location, and orientation of the cylinders. For example, the case in which the enhanced merging rate is extended in duration but confined spatially can be modeled by imposing a cylinder that is elongated in the direction of the flow but constricted in its width.

Model Outline

We have adapted the approach of Crooker [this volume, also Crooker and Siscoe, 1988] to a more realistic magnetic field and magnetopause model. The magnetic field configuration is constructed by a superposition of a model of a cylindrical perturbation magnetic field onto a background magnetospheric field model. This magnetic field model is then used to map the cylindrical boundary from the magnetopause down to the polar cap ionosphere. The magnetospheric field model is the Voigt [1981] closed model, as modified by Toffoletto and Hill [1989] to reproduce a realistic open magnetospheric magnetic field configuration. The open model varies depending on the boundary conditions specified at the magnetopause, which are determined by the orientation of the interplanetary magnetic field (IMF). For the cases presented here, we have assumed that when dayside merging occurs in a steady state, the merging line passes through the subsolar point at an orientation depending on the direction of the IMF. For southward IMF, it lies in the equatorial plane; if the IMF is rotated away from southward, the merging line is rotated in the same direction, but by an angle equal to half the IMF rotation angle [Toffoletto and Hill, 1989]. In this paper the merging line is coincident with the equatorial plane. The normal component is finite everywhere on the magnetopause, and is antisymmetric in sign about the merging line (i.e., the normal component is negative/positive to the north/south of the merging line). In contrast, the closed model (by definition) does not allow any flux to cross the magnetopause.

The magnetic field for the FTE model is illustrated in Figure 1a which shows a cross section of a cylinder in the noon-midnight plane. This is the perturbation magnetic field that is added to each of the magnetospheric field models discussed above. The configuration is designed to simulate the penetration of the magnetopause by the FTE; its effect on the mapping near the Earth is negligibly small because the dipole field is dominant there. An example of the resultant field line configuration is shown in Figure 1b for two resultant flux tubes in the noon midnight plane. Note that the perturbation fields are directed radially toward and away from the Earth on the dayside, while on the nightside they are perpendicular to the plane of the merging line (in this paper the merging line is in the equatorial plane). This model has a discontinuous change in the perturbation field orientation as it moves from the dayside to the nightside (its configuration suddenly changes from pointing Earthward to pointing perpendicular to the merging line). The transition is continuous only if the cylinder resides in the noon-midnight plane. This field configuration, although it is somewhat arbitrary, was chosen for simplicity; other simple geometric variations were investigated with little change in the results and arguments presented here. The important property is that the magnetic field at and near the magnetopause is modeled reasonably accurately. The field inside each of the cylinders is assumed to be constant in magnitude (B_0) and directed parallel to the cylinder axis, but as the shape of the cylinder changes the magnitude is required to vary such that the total flux contained within the cylinder is unchanged.

As the cylinder flows downstream its cross sectional shape changes in a manner determined by the magnetosheath streamlines at the magnetopause; its shape will distort. The magnetosheath flow is specified by the gas-dynamic solution derived from the Spreiter model [Stahara et al., 1980], which

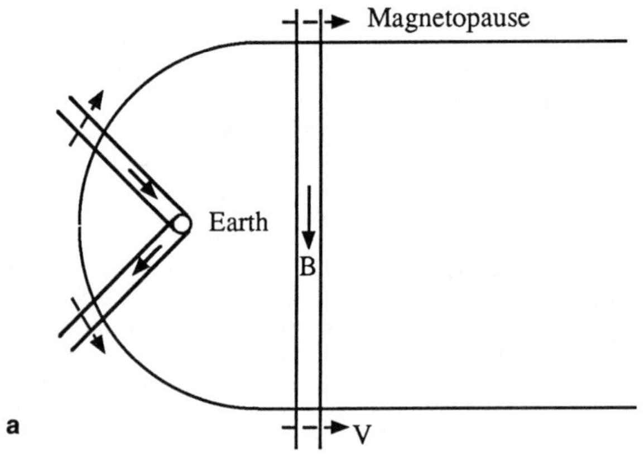

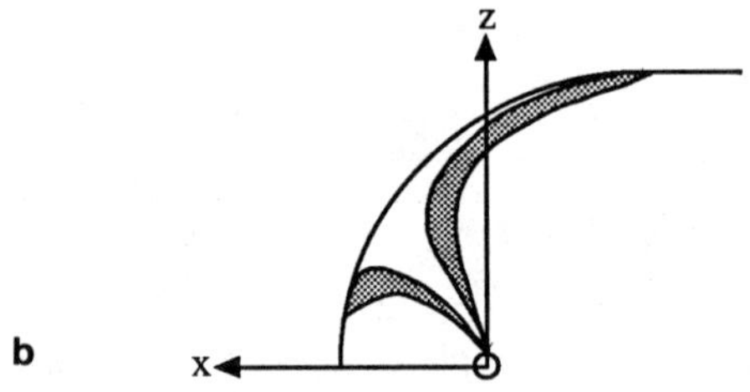

Fig. 1. (a) Sketch of the perturbation field configuration used to model FTE penetration of the magnetopause of a model magnetosphere. The configuration is unrealistic in regions close to the Earth, but this has little effect on the mapping, which is dominated by Earth's dipole field in this region. (b) Total field line plot of two flux tubes (shown shaded) that reside in the noon-midnight plane.

has been modified to incorporate the outflow from the merging process [Toffoletto and Hill, 1989]. The shape of the cylinder on the day side magnetopause that evolves to a circular shape in the tail is determined by following streamlines tangent to the edges of the cylinder, upstream from the tail back to the merging line. This allows the time evolution of the cylinder size and shape to be determined. As the cross sectional shape of the cylinder changes, the magnitude of the magnetic field inside the cylinder is determined by the requirement of flux conservation. The downstream perturbation field strength in the tail was chosen to be 10 nT, with a cross sectional radius 1 R_E. This choice of field strength is probably unrealistically large, but is necessary to ensure a footprint in the ionosphere that is large enough to be easily resolved by the model (i.e. there is sufficient flux crossing the magnetopause to produce a footprint larger than the numerical resolution of the model; smaller footprints would require excessive computational time).

We assume for simplicity that the cross sectional shape of the cylinder at the magnetopause is an ellipse, with semi-minor axis a perpendicular to the flow and semi-major axis b along the flow. If a and b are expressed in R_E and B_0 in nT, the flux tube maps to an area A in the polar cap of

$$A \approx 2000\ \text{km}^2\, a\, b\, B_0$$

For $B_0 = 10$ nT and $a = b = 1$ R_E, a circular footprint in the ionosphere would have a radius of ≈ 80 km (or 0.75° in latitude).

Model Results

FTE's in an Otherwise Closed Magnetosphere

We first investigate the mapping of an open flux tube as it moves downstream from the dayside merging line to the tail within an otherwise closed magnetosphere. The result (Fig. 2) is similar to that of Crooker [this volume], in which all the footprints appear to be anchored to a small region in the ionosphere (marked with an 'x'). Owing to small numerical errors, the background field model is not completely closed, so that a small amount of flux leaks out, forming a small open region centered around the cusp. Also, the small scale distortions of the shape of the footprints are the result of numerical errors in the field line tracing. This effect results from the convergence of magnetopause field lines in the vicinity of the cusp. In a closed magnetosphere, a single cusp field line defines the magnetopause. Therefore an open flux tube in the far tail has an ionospheric footprint that is elongated to satisfy the condition that the first (sunward-most) open field line corresponds to the single cusp field line.

If there are two such open flux tubes separated by a closed region of the magnetopause but connected by a magnetopause field line of the background closed model (for example, two open tubes centered on the noon-midnight plane, one downstream from the other), their footprints coalesce in the polar cap to form a single region of open flux. This configuration would result from merging that occurs sporadically in time but always in the same location. An example is shown in Figure 2 for five regions of open flux that reside simultaneously along the symmetry streamline (labelled 'a' to 'e'). Table 1 shows the location of the magnetopause crossing points of the centers of the flux tubes, along with the approximate downstream flow time from the

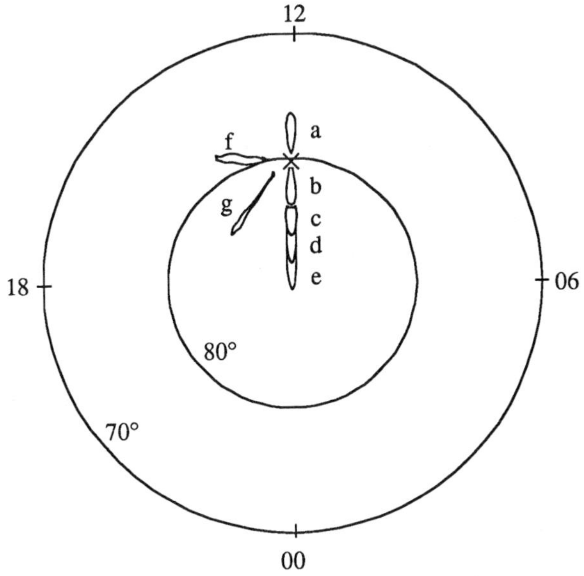

Fig. 2. Ionospheric footprints of FTE's penetrating an otherwise closed magnetopause. The five contiguous footprints on the noon-midnight axis cross the magnetopause at y = 0, while the other two cross at y = 9 R_E and 15 R_E (Table 1).

TABLE 1. Magnetopause parameters of FTE flux tubes whose ionospheric signatures are shown in Figures 2 and 3.

label[1]	x[2] (R_E)	y[2] (R_E)	z[2] (R_E)	Δt[3] (min)	B_0[4] (nT)
a	9.5	0.0	4.2	1	20.4
b	2.5	0.0	15.5	5	12.9
c	-10.0	0.0	20.0	8	10.0
d	-23.2	0.0	20.0	12	10.0
e	-35.6	0.0	20.0	16	10.0
f	7.6	8.9	3.3	1	14.6
g	-40.0	15.0	13.0	21	10.0
h	9.5	0.0	4.2	1	20.4
i	2.5	0.0	15.5	5	12.9
j	-13.2	0.0	20.0	9	10.0
k	-166.0	0.0	20.0	51	10.0
l	7.6	8.9	3.3	1	14.6
m	166.0	15.0	13.0	51	10.0

[1]Labels 'a' to 'g' refer to Fig. 1; labels 'h' to 'm', to Fig. 2.
[2]Locations (GSM) of perturbation magnetic field intersection with magnetopause, which is also the magnetopause end of the FTE flux tube.
[3]Downstream flow time from merging line.
[4]Magnetic field strength within the perturbation field.

dayside merging line, and the magnetic field intensity within the cylinder. The shape of the magnetopause ends of the flux tubes is approximately circular (with radius 1 R_E), except for the flux tube closest to the magnetopause (labelled 'a') which has a half-width of approximately ≈ 0.5 R_E (= a)

perpendicular to the flow and a half-length of $\approx 0.9\ R_E$ ($= b$) parallel to it.

Open flux tubes that cross the magnetopause off the noon-midnight symmetry plane map to fan-like open regions on the ionosphere, and as the flux tube moves downstream, the ionospheric footprint pivots about the cusp footprint. The motion of the footprint is constrained by this pivoting around the cusp, so that regardless of whether the flux tube, after reconnection, moves antisunward over the pole or around the flanks of the magnetosphere, the footprint is tied to the cusp in the polar cap. This result is similar to that of Crooker [this volume). Figure 2 illustrates this fan-like structure for footprints marked 'f ' and 'g'. Table 1 has the corresponding locations, flow times and field strengths. The magnetopause end of the flux tube corresponding to 'f ' has approximate dimensions a $\approx 0.8\ R_E$ and b $\approx 0.9\ R_E$, while flux tube 'g' is circular (1 R_E in radius).

FTE's in an Open Magnetosphere

Our second example is based on the assumption that the magnetosphere is already open and that FTE flux tubes are the result of localized and short-lived enhancements of an otherwise steady merging process. Figure 3 shows the ionospheric mapping of a series of flux tubes centered around the noon-midnight plane. In Figure 3 we have superimposed the perturbation magnetic field model onto the open magnetic field model of Toffoletto and Hill [1989]. The footprints labelled 'h' to 'k' are the ionospheric mappings of flux tubes that reside on the symmetry streamline in the noon-midnight plane. (Their corresponding locations on the magnetopause, etc. are shown in Table 1) The magnetopause cross sections of these flux tubes, as in Figure 2 are approximately circular with a radius of 1 R_E, except 'h' which is the same as 'a'. The first three of the four footprints shown

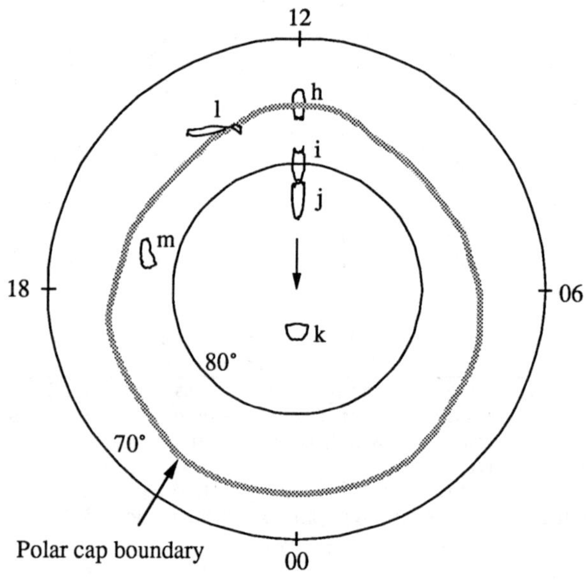

Fig. 3. Ionospheric footprints of FTE's in an already open magnetosphere, four of them crossing the magnetopause at y = 0 and the other two crossing at y = 9 R_E and 15 R_E (Table 1). The polar-cap boundary is the limit of open field lines before the addition of the FTE's.

for y=0 are separated at their magnetopause ends by 4 minutes in time; i.e., the downstream flow time in the magnetosheath from one to the next flux tube is 4 minutes. The first (sunward-most) footprint is mapped from a location approximately 1 minute downstream from the merging line. The fourth footprint is mapped from a flux tube crossing the magnetopause approximately 1 hour downstream from the merging line. The footprints retain an approximately circular footprint and, in contrast to the closed model, do not coalesce when the flux tubes are on the same magnetopause field line. These flux tubes were not present simultaneously but were mapped from four different magnetic field models corresponding to four locations of the flux tubes; a qualitatively similar result is obtained by using a model that has all four flux tubes present simultaneously. The polar cap boundary, also shown in Figure 3, is the boundary between open and closed field lines for the open model before the FTE flux tubes are added. Part of the footprint that maps from the FTE flux tube closest to the merging line on the magnetopause extends outside the previously open polar cap, thus causing an expansion of the polar cap in a form similar to that of Figure 2. The subsequent separation of the flux tube footprints results from the small but finite amount of open flux between the FTE tubes, allowing a detachment of the flux tubes at their ionospheric ends. In an open model, the FTE flux tubes are not required to remain tied to the cusp field line.

One can, however, obtain extended footprints in an open model. Figure 3 also shows the footprints of two flux tubes that cross the magnetopause at distances y=9 and 15 R_E from the noon-midnight plane (labelled 'l' and 'm' respectively). As in the previous case, part of the FTE flux tube initially maps to a region outside the previous polar cap (open field-line region), resulting in an extended footprint shape similar to that obtained in the closed model. As the flux tube moves further down the tail, the footprint moves into the polar cap and becomes more nearly circular. The footprint is always expected to move eventually into the polar cap because the movement of the flux tube, after merging, is always away from the merging line owing to the merging outflow.

Discussion and Conclusion

We have presented preliminary results of a model of FTE footprints as mapped to the polar cap ionosphere. The footprints in an otherwise closed model assume the elongated shape previously obtained by Crooker and Siscoe, and are tied at one point to the cusp, pivoting about that point as the flux tubes move downstream. In an otherwise open model, however, the presence of a small background normal component at the magnetopause allows the FTE footprints to disconnect from the cusp field line and proceed tailward with a more nearly circular shape. Only if the FTE flux tube penetrates an otherwise closed section of the magnetosphere does the cross section remain elongated (as shown in Figure 3, where the footprint 'l' fans out outside the polar cap).

These preliminary results indicate that the model may provide a useful tool for studying the geometry of ionospheric footprints of open flux tubes in the magnetosphere. The results suggest that the shape of the FTE footprint is determined by whether it is surrounded by closed or open field lines. The fan-like patterns obtained in Figure 2 would be expected for FTE's in an otherwise closed magnetosphere. If the FTE's result from localized enhancements of an otherwise steady merging process, then the footprints may initially have the elongated structure if they map to outside the previously open field line region, but eventually become rounded as they move into the open polar cap. Further work is required to investigate the effects on the footprint of a y-component of the IMF, of variations in the perturbation magnetic field strength and size, and of dipole tilt. The observed twisting of the magnetic field within FTE's has not been included in this model. One could accomplish this by the simple addition of an azimuthal component to the perturbation magnetic field, although it is unlikely that this would materially affect the mapping geometry.

Because the electric field within an FTE flux tube is expected to be larger than in the background ionosphere (owing to the larger magnetic-field normal component at the magnetopause) the flux tube footprint is expected to move through the ionosphere at a faster rate than the background. This will result in Birkeland currents, which in turn will effect the mapping geometry. Preliminary results suggest that field aligned currents of a Region 1 sense will make the flux tube geometry more tail like and thus the footprint will map further sunward. This is a subject of ongoing work: for the present paper we have assumed that the Birkeland currents are small enough to have a negligible effect on the mapping.

Variation of the merging-line geometry may also produce interesting effects. For example, if we impose the condition that merging occurs only where the internal magnetospheric field is antiparallel to the external magnetosheath field, then for a non-southward IMF the merging line splits into two segments, one attached to each of the (northern and southern) cusp regions of the magnetopause [Crooker, 1979]. Field lines near the subsolar point do not participate in the merging process, and the polar cap is bifurcated by a "convection gap" whose field lines are not attached to the solar wind, but which is surrounded on either side by open field lines [Crooker, 1979; Toffoletto and Hill, 1989]. This closed field line region near the subsolar point could be the site of sporadic, patchy merging of the type that is conducive to the formation of FTE's [Crooker, 1989]. The present model could be adapted to investigate the ionospheric footprints of FTE's that form in this convection gap, including a scenario in which steady merging occurs in the cusp region concurrently with transient FTE formation in the subsolar point region.

Thus, extensions of the present model may provide a useful tool in identifying ionospheric signatures of FTE's on the assumption that they are on open field lines, as well as providing important clues as to the nature of the merging process at the dayside magnetopause.

Acknowledgements. We have benefitted from conversations with N. U. Crooker and M. F. Smith. This work was supported in part by the Upper Atmosphere Research Section of the National Science Foundation under Grant ATM-8822662, and in part by the Solar Terrestrial Theory Program of the National Aeronautics and Space Administration under Grant NAGW-482. Some of the computing was performed on the Cray X-MP at NCSA (Champaign, Ill.) as well as the Cray Research X-MP at Mendota Heights, Ill.

References

Berchem, J., and C. T. Russell, Flux transfer events on the magnetopause: Spatial distribution and controlling factors, *J. Geophys. Res.*, 89, 6689, 1984.

Bering, E. A., J. R. Benbrook, G. J. Byrne, B. Liao, J. R. Theall, L. J. Lanzerotti, C. G. Maclennan, A. Wolfe, and G. L. Siscoe, Impulsive electric and magnetic field perturbations observed over South Pole: Flux Transfer events? *Geophys. Res. Lett.*, **15**, 1545, 1988.

Crooker, N. U., Dayside merging and cusp geometry, *J. Geophys. Res.*, **84**, 951, 1979.

Crooker, N. U., and G. L. Siscoe, On mapping flux transfer events to the ionosphere, *EOS*, **69**, 1383, 1988.

Crooker, N. U., and G. L. Siscoe, On mapping flux transfer events to the ionosphere, submitted *J. Geophys. Res.*, 1989.

Crooker, N. U., Imbedded open flux tubes and "viscous interaction" in the low latitude boundary layer, Submitted for publication in *Physics of Magnetic Flux Ropes*, 1989 (C. T. Russell, ed.).

Dungey, J. W., Interplanetary magnetic field and auroral zones, *Phys. Rev Lett.*, **6**, 47, 1961.

Lee, L. C., and Z. F. Fu, A theory of magnetic flux transfer at the Earth's magnetopause, *Geophys. Res. Lett.*, **12**, 105, 1985.

Lockwood, M., P. E. Sandholt, and S. W. H. Cowley, Dayside auroral activity and magnetic flux transfer events, *Geophys. Res. Lett.*, **16**, 33, 1989.

Rijnbeek, R. P., S. W. H. Cowley, D. J. Southwood, and C. T. Russell, A survey of dayside flux transfer events observed by ISEE 1 and 2 magnetometers, *J. Geophys. Res.*, **89**, 786, 1984.

Russell, C. T., and R. C. Elphic, Initial ISEE magnetometer results: Magnetopause observations, *Space Sci. Rev.*, **22**, 681, 1978.

Scholer, M., Magnetic flux transfer at the magnetopause based on single x line bursty reconnection, *Geophys. Res. Lett.*, **15**, 291, 1988.

Southwood, D. J., The ionospheric signature of flux transfer events, *J. Geophys. Res.*, **92**, 3207, 1987.

Southward, D. J. , C. J. Farrugia, and M. A. Saunders, What are flux transfer events?, *Planet. Space Sci.*, **36**, 1, 1988.

Stahara, S. S., D. Klenke, B. C. Truddinger, and J. R. Spreiter, Application of advanced computational procedures for modeling solar-wind interactions with Venus- Theory and computer code, *NASA Contract Rep. CR 3267*, 1980.

Stern, D. P., A study of the electric field in an open magnetospheric model, *J. Geophys. Res.*, **78**, 7292, 1973.

Toffoletto, F. R., and T. W. Hill, Mapping the solar wind electric field onto the Earth's polar caps, *J. Geophys. Res.*, **94**, 329, 1989.

Voigt, G.-H, A mathematical magnetospheric field model with independent physical parameters, *Planet. Space Sci.*, **29**, 1, 1981.

TERRESTRIAL IONOSPHERIC SIGNATURES OF FIELD-ALIGNED CURRENTS

E. Friis-Christensen

Division of Geophysics, Danish Meteorological Institute, Lyngbyvej 100
DK-2100, Copenhagen, Denmark

Abstract. Field-aligned currents play an important role in the magnetosphere and in particular in the physical processes involved in the coupling between the solar wind plasma and the Earth's magnetic field. The field-aligned currents consist of a large-scale part which is closely related to the solar wind magnetic fields and plasma. In addition to the large-scale part, a number of observations indicate the importance of small-scale field-aligned current filaments for the understanding of the magnetospheric plasma processes. Based on extensive magnetic field measurements collected by low altitude satellites, the large-scale field-aligned currents have been relatively well described. It is, however, considerably more difficult to deduce small-scale field-aligned currents from single satellite magnetic field observations. Since, however, the field-aligned current filaments close in the ionosphere, they will set up ionospheric currents, which may be observed by using ground-based magnetometers. With measurements of good spatial and temporal resolution it is possible to observe the ionospheric distribution and motion of the field-aligned current filaments. One type of field-aligned current filaments observed in this way corresponds to a pair of oppositely directed current filaments moving primarily tailward along the auroral oval in the ionospheric projection of the dayside polar cleft. Since the cleft is associated with the low-latitude boundary layer of the magnetosphere, the presence of these field-aligned current filaments is probably a manifestation of boundary layer processes associated with reconfigurations of the magnetosphere due to sudden changes in the solar wind.

Introduction

Large-scale field-aligned currents form current sheets which are extended more or less along the invariant latitude circles. These current sheets are permanent features of the magnetosphere although their location and intensity may vary according to the varying solar wind and magnetospheric conditions [Iijima and Potemra, 1976; Zanetti and Potemra, 1986]. A magnetic observatory on the ground will be affected by these large-scale currents, and in particular from the associated horizontal Hall currents in the ionosphere. An analysis of data from a distribution of magnetic observatories in the polar regions [Friis-Christensen, 1984] shows a pattern of field-aligned currents consistent with the general pattern observed from satellites.

In addition to the large-scale field-aligned currents, the magnetic data from satellites show the presence of small-scale perturbations which have been interpreted as due to very high density current filaments in the magnetosphere. In contrast to the large-scale field-aligned current sheets, it is not possible from single satellite data alone to unambiguously identify the location and intensity of the current filaments. With ground-based observations the possibility exists of having good temporal and spatial resolution which may allow a determination of the location and motion of field-aligned current filaments. Auroral observations display spectacular sudden illuminations and rapid movements of auroral forms created by precipitating particles which may be associated with intense field-aligned current filaments. Ground magnetic observations show the presence of small-scale fluctuations which probably could be related to such filamentary currents.

In particular in the cleft region, which maps to the magnetospheric boundary layers, spike-like magnetic variations are often observed. Some of these impulsive events have been proposed to be the ionospheric manifestations of flux transfer events (FTE's) which were observed on the dayside magnetopause by Russell and Elphic [1979]. According to a model by Saunders et al. [1984] and Lee [1986] the FTE is associated with a central core of field-aligned current which spreads radially in the ionosphere and closes through a local or distant sheath of field-aligned currents. Contrary to this model, Southwood [1985] proposed that the FTE consists of two oppositely directed field-aligned currents on the flanks of the FTE flux tube producing flux tube motion perpendicular to a line connecting the vortex centers.

A number of studies of possible ionospheric signatures of FTE's have been made using radars [Goertz et al., 1985]

Geophysical Monograph 58

and magnetometers [Lanzerotti et al., 1986], but without definitive conclusions. Recent observations from a dense array of magnetometers in Greenland around the cleft [Friis-Christensen et al., 1988a] show, however, that previous possible FTE ground magnetometer observations are not consistent with any of the proposed FTE models. Although the ionospheric convection pattern observed by the magnetometer station resembles the twin-vortex system proposed by Southwood [1985], its direction of motion is perpendicular to the motion predicted by the model. The relatively high velocity and a direction of motion parallel to the electric field impressed on the ionosphere by the two oppositely directed field-aligned currents means that the event does not correspond to a net transport of plasma from closed to open field-lines. Consequently, the velocity probably corresponds to the phase velocity corresponding to a tailward moving structure or wave in the magnetospheric boundary layer.

Ionospheric traveling convection vortices

Friis-Christensen et al. [1988a,b] and McHenry et al. [1988] examined different classes of cleft region magnetic perturbations and pulsations and found that a number of these could be explained by traveling small-scale ionospheric current systems which are associated with series of field-aligned current filaments moving rapidly along the cleft. The field-aligned current filaments normally occur in pairs, creating a basic twin-vortex convection pattern in the ionosphere with a size of order 1000 km.

A typical, nearly ideal, twin-vortex pattern of ionospheric convection is shown in Figure 1 taken from Friis-Christensen et al. [1988a]. This pattern has been obtained from a dense magnetometer chain in Greenland on June 28, 1986 around 1000 UT, approximately 4 hours prior to magnetic local noon. The data were recorded with a time resolution of 20 seconds and filtered using a high-pass filter with a corner frequency of .56 mHz corresponding to a period of about 30 minutes. The pattern shown in Figure 1 represents the equivalent convection pattern, i.e. the vectors correspond to the horizontal magnetic perturbation vectors rotated 90° counterclockwise according to the assumption that the magnetic perturbations are caused by Hall currents in the ionosphere.

Since the magnetometer chain is basically one-dimensional, situated nearly along a magnetic meridian along the west-coast of Greenland, it is not immediately possible to derive the east-west distribution of the convection pattern. One of the stations in the chain, however, is located on the Greenland ice cap further east, and this station shows a phase lag corresponding to a motion of a stationary current pattern across the line of stations. Using the east-west velocity estimated from this phase lag, consecutive sets of vectors may be drawn offset to the right corresponding to a motion of the pattern towards the left (west) of the plot. The westward motion corresponds to a tailward (antisunward) motion in the morning sector. The ground magnetic field of a single filamentary field-aligned current and its associated ionospheric currents has

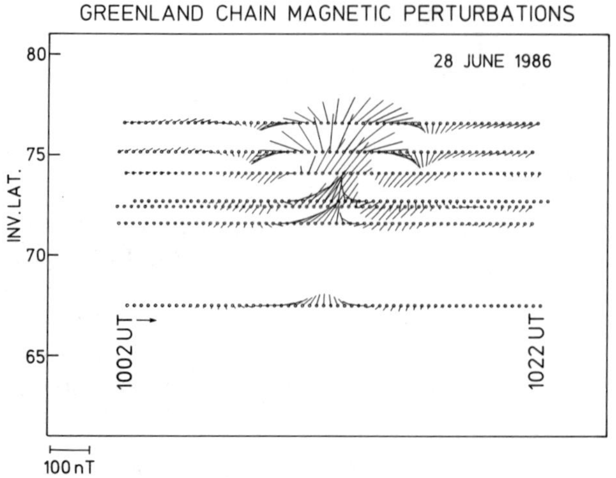

Fig. 1. Total horizontal magnetic perturbation vectors measured by the Greenland magnetometer chain on June 28, 1986 have been rotated by 90° counterclockwise and plotted every 20 seconds during the interval from 10:02:00 to 10:22:00 UT. For each time the position of the vectors have been off-set to the right by a distance corresponding to 80 km to account for an assumed 4 km/s westward motion (to the left) of the pattern.

been calculated by McHenry and Clauer [1987] for a uniform ionosphere. During these conditions and assuming vertical magnetic field-lines the magnetic field from the field-aligned current filament is completely cancelled by the magnetic effects from the ionospheric Pedersen currents. Two oppositely directed field-aligned current filaments would produce a horizontal pattern very close to what is being observed with the magnetometer chain.

The event shown in Figure 1 is located in the prenoon sector. A similar event, observed around 1600 UT (1400 MLT) in the postnoon sector is shown in Figure 2. In this case no data were available from the ice cap station, but assuming an eastward velocity of 4 km/s and plotting each set of vectors displaced to the left to account for this motion we observe that also this event corresponds to a twin-vortex pattern. The eastward motion in the postnoon sector represents a tailward motion consistent with the case of the prenoon event interpreted as a tailward moving structure in the low-latitude boundary layer. The observations of twin-vortex patterns in the afternoon differs from the results of Glassmeier et al. [1989] who found a pronounced peak of occurrence frequency around 0800 MLT and no observations at all in the afternoon sector.

Figure 3 shows an event observed at the same local time as the event of Figure 2. This pattern which was observed on July 21, 1986 around 1600 UT (1400 MLT) is, however, moving westward (sunward or anti-tailward) in contrast to the normal direction of motion in the afternoon, which is eastward (tailward). The direction of motion and the approximate speed of 4 km/s has been obtained by comparing the two stations at approximately the same

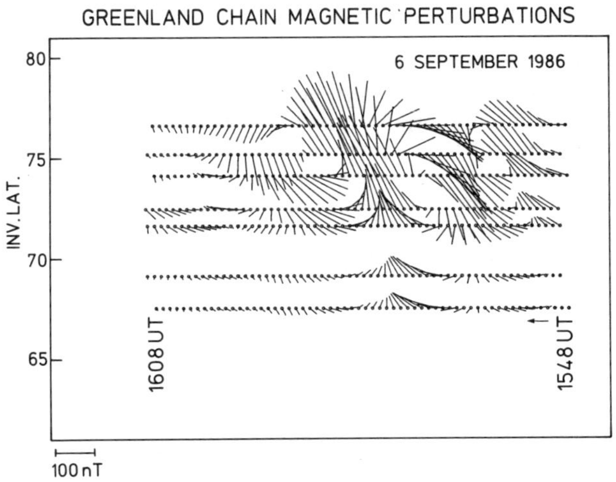

Fig. 2. Total horizontal magnetic perturbation vectors measured by the Greenland magnetometer chain on September 06, 1986 have been rotated by 90° counterclockwise and plotted every 20 seconds during the interval from 15:48:00 to 16:08:00 UT. For each time the position of the vectors have been off-set to the left by a distance corresponding to 80 km to account for an assumed 4 km/s eastward motion (to the right of the pattern).

invariant latitude but separated in longitude. The solar wind plasma data in Figure 4 shows some, but not exceptional, density variations whereas the interplanetary magnetic field data shows a very abrupt change in the B_y component, varying from positive to negative values and back again to positive, within an interval of about 15 minutes from 1603 UT. The B_z component is close to zero

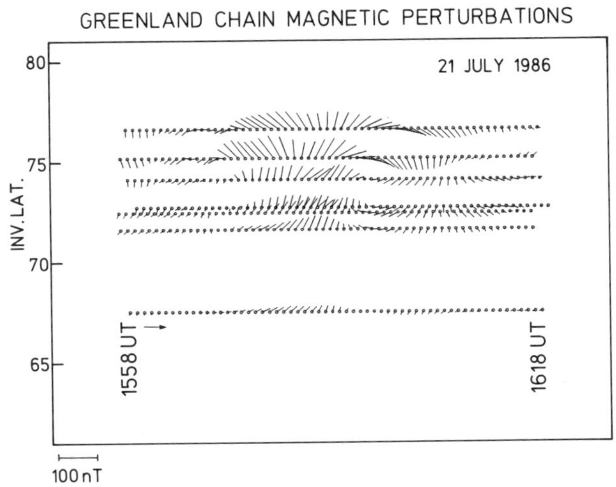

Fig. 3. Total horizontal magnetic perturbation vectors measured by the Greenland magnetometer chain on July 21, 1986 have been rotated by 90° counterclockwise and plotted every 20 seconds during the interval from 15:58:00 to 16:18:00 UT in a manner corresponding to Figure 1.

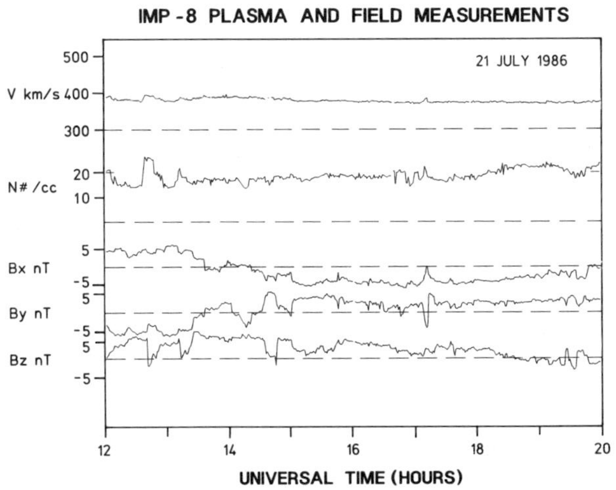

Fig. 4. Solar wind data for 1200 to 2000 UT on July 21, 1986 measured by IMP-8. From top to bottom is shown the solar wind velocity, the number density, and the interplanetary magnetic field components in GSM coordinates, B_x, B_y, B_z.

during these variations in the B_y component. The IMF variations occur prior to the observation of the magnetic effect on the ground. This is probably explained by the position of IMP-8 at $(X_{GSE}, Y_{GSE}) = (31R_E, -20R_E)$. This means that the interplanetary discontinuity will reach the Earth's magnetopause before it reaches the satellite. It is conceivable that the interaction of the discontinuity with the magnetosphere creates a disturbance of the magnetosphere which travels away from the point of interaction.

One might, however, also speculate that the anomalous direction of motion is only apparent and is caused by a change in the mapping from the magnetopause to the ionosphere related to a sudden reconfiguration of the magnetosphere in response to a sudden change in the solar wind-magnetosphere coupling.

Mapping to the magnetosphere

McHenry et al. [1988] compared the location of non-impulsive pulsations with particle data on the low altitude satellite DMSP F7 passing over Greenland close to the magnetometer chain. From the electron and ion fluxes they mapped the peak of the pulsations to the equatorward edge of the boundary layer and they concluded that the waves creating the pulsations were located at the inner edge of the magnetopause boundary layer.

In the pattern of the traveling twin-vortices in Figure 1 and Figure 2 as well as in the patterns of the multiple vortices corresponding to the non-impulsive pulsations, we observe a fundamental deviation from the expected symmetric twin-vortex pattern. A model consisting of two oppositely directed field-aligned current filaments will produce a twin-vortex convection pattern which is symmetric relative to a line perpendicular to the line connecting the current filaments. All the observations, however, indicate a systematic deviation from this model. For the westward moving vortices, the equivalent convection in the center has an eastward component (see Figure 1), and the eastward moving vortices have a westward component of convection in the center of the twin-vortex pattern (see Figure 2).

A possible explanation of this systematic feature is that it is due to the mapping along the geomagnetic field-lines from the boundary layer to the ionosphere. Crooker and Siscoe [1989] calculated the possible ionospheric footprints of FTE's on the magnetopause taking into account the draping of the field lines and found that round patches of open flux piercing the magnetopause map down to the ionosphere along flux tubes that form an elongated form. Although the vortices discussed here are not on open flux tubes, similar arguments concerning the influence of the draped field-lines may be valid. Assuming a radius in the ionosphere of the current filaments of 300 km corresponding to somewhat less than half the distance between the oppositely directed current filaments, this corresponds to about three degrees in latitude. The high-latitude part of the current filament maps to the magnetospheric boundary along field-lines which are considerably more draped than those connecting to the lower latitude part of the current filament. In Figure 5 is sketched a round structure in the low-latitude boundary layer moving tailward in the dawn sector. The field-lines connected to the innermost and lower latitude part of the structure are assumed to be nearly dipolar whereas the field-lines connected to the outermost and higher latitude part are considerably distorted. This corresponds to a motion of the ionospheric projection where the lower latitude part of the structure moves faster than the higher latitude part, implying an apparent 'delay' of the higher latitude part of the ionospheric convection pattern consistent with the skewing of the direction of the central convection.

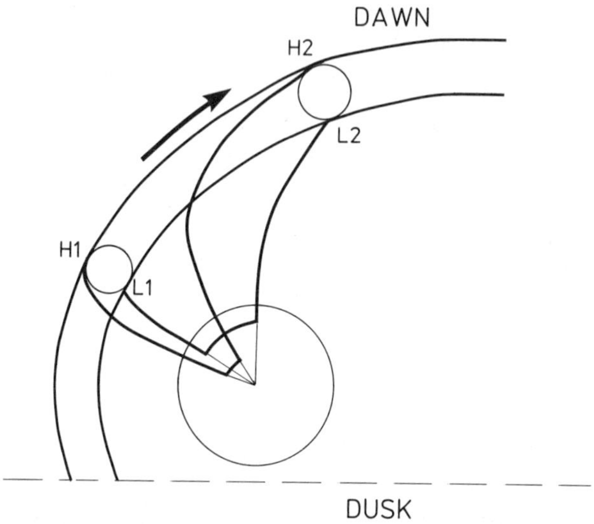

Fig. 5. Sketch showing an equatorial cross section of the magnetosphere. A circular structure is moving tailward in the boundary layer. The outermost part of the structure (H1/H2) is connected to the high-latitude polar ionosphere through highly draped magnetic field-lines whereas the innermost and lower latitude part of the structure (L1/L2) is assumed to be connected to nearly dipolar field-lines. The resulting effect of the draped field-lines is a motion in the ionosphere where the equatorward part of the ionospheric projection of the structure moves faster than the poleward part.

Summary and discussion

The occurrence of the traveling field-aligned current filaments and twin-vortex patterns seems to be related to sudden changes in the solar wind dynamic pressure and/or the interplanetary magnetic field. The event discussed by Friis-Christensen et al. [1988a], shown in Figure 1 was associated with a large and abrupt change in the solar wind dynamic pressure accompanied by large fluctuations in the interplanetary magnetic field. Potemra et al. [1989] used magnetic measurements on the ground as well as on two satellites in a study where they concluded that a possible cause was solar wind dynamic pressure changes which created surface waves on the magnetopause. Farrugia et al. [1989] studied events of a similar kind which could not be related to IMF variations but more probably to density variations in the solar wind. Sibeck et al. [1989] reported on magnetic field signatures corresponding to tailward-moving magnetopause surface wavelets apparently driven by pressure oscillations in the solar wind.

If solar wind dynamic pressure variations are causing the traveling current filaments they should be moving antisunward or tailward, i.e. westward in the morning and eastward in the afternoon sector. Most of the events found in the Greenland data are consistent with this view [Friis-Christensen et al., 1988b; McHenry et al., 1988]. There are, however, events which do not fit into this general pattern. An example of this kind has been presented in Figure 3. During this event, and during several other events, large IMF variations occurred which were associated with considerable ionospheric current changes in the polar cap.

The creation of the traveling convection vortices may therefore not only be caused by solar wind dynamic pressure variations as a number of observations indicate. It is possible that they could also be caused by a sudden change of the global magnetospheric convection pattern, related to a sudden reorientation of the IMF. In the solar wind the IMF variations and changes in the plasma properties are closely related, and it is therefore difficult to distinguish the effects from these two possible sources. A different explanation might be that the cases of anomalous direction of motion are related to abnormal mappings from the magnetopause to the ionosphere caused by a sudden reconfiguration of the magnetosphere in response to a sudden change in the interaction region of the solar wind-magnetosphere coupling.

The mapping of the ionospheric convection patterns seems to involve field-lines which are considerably distorted from the dipolar form, consistent with the fact, that the field-aligned current filaments originate in the boundary layer of the magnetosphere. The prenoon and postnoon events display systematic differences which may be explained by the different mapping effects on the two flanks of the magnetosphere.

The Greenland magnetometer observations of the cleft region magnetic perturbations associated with small-scale twin-vortex patterns show that the events occur on both sides of magnetic local noon. This is in disagreement with the results of Glassmeier et al. [1989] who performed a statistical analysis of 82 twin-vortex events found at sub-auroral latitudes using the Scandinavian Magnetometer Array during the time interval 1975–1979. They found a pronounced peak of occurrence frequency at 0800 MLT and did not find any afternoon events. The fact that Glassmeier et al. [1989] observe a maximum in the occurrence frequency around 0800 MLT could be related to the sub-auroral latitude of the Scandinavian Magnetometer Array which means that the stations are closer to the ionospheric projection of the cleft earlier in magnetic local time compared with the Greenland stations which are located across the cleft and observe the twin- vortices at all local times on the dayside. The latitudinal location of the Scandinavian magnetometer stations does not, however, explain that vortices are not observed during the afternoon.

Questions and Answers

Luhmann: We all know that the ionospheric convection pattern at high latitudes is different for different directions of the IMF and different solar wind dynamic pressures. Can't we regard what you see as simply transients in the convection pattern as the ionosphere adjusts to new external conditions (both IMF and solar wind pressure)?

Friis-Christensen: I certainly agree that this is a possibility, and a possibility which is supported by the fact that we do observe vortex patterns which seem unaccompanied by changes in the dynamic pressure, at least as far as IMP-8 measurements are concerned. I have never seen an event which was not accompanied by either a pressure change or an IMF change.

Sibeck: Is there any preferred IMF orientation for the transient auroral zone events? Also, what fraction move sunward, and is the sunward motion consistent with convection patterns dependent on the IMF B_y?

Friis-Christensen: We have searched for preferred IMF orientations, but not found any conclusive results yet. Sudden variations seem more important than the initial IMF orientation. We have until now only observed one sunward moving twin-vortex in the postnoon sector and none in the prenoon sector. During the event, B_y becomes suddenly negative, and thereafter, positive. So, the event could correspond to a motion of the merging region from the postnoon sector to the prenoon sector across local noon.

Lemaine: Comment: Kp is correlated with σ_B / |B| in the interplanetary medium (Svalgaard, 1978; Garret, 1981; Lemaine, 1989). These results support and complement the results presented by Friis-Christensen. They also support the idea of impulsive penetration of small-scale plasma irregularities always present in the solar wind.

Indeed, when the IMF has large fluctuations (σ_B /B) (i.e., when the solar wind plasma is more inhomogenous the level of geomagnetic activity (Kp, Ap, Am, Aα indices) are significantly higher indicates that the source of certain classes of perturbations of the magnetosphere (observed at ground level) is located deep in the solar wind, (i.e., external to the magnetosphere and magnetosheath), and not necessarily a consequence of locally excited instabilities at the magnetopause (like K-H instability, spontaneous reconnection) as invoked in FTE theories.

Friis-Christensen: When looking at magnetometer data, particularly in the cleft, irregular pulsations seem more prominent when the IMF is varying. I would not, however, on this ground, rule out the possibility that, for example, boundary motions could increase the probability of an instability occurring.

Dubinin: What is smallest scale of vortices can you observe? The existence of vortices with scales ~10-20km is followed from satellite observations?

Friis-Christensen: The scale size of what can be observed from the ground is determined by two factors. One is the height of the ionosphere, or rather the E-layer where the Hall currents are located around 100-120km. The other factor is the actual horizontal spacing of the observing stations which is around 200 km. So, vortices with scale sizes of 10-20km would be averaged out in the ground-based measurements.

Acknowledgements. The author would like to thank Drs. R. Lepping and J. King for the IMP-8 magnetic field data, and the solar wind group at MIT, supported in part by NASA Goddard under grant NAG5-584, for the IMP-8 solar wind plasma data.

References

Crooker N. U., and G. L. Siscoe, On mapping flux transfer events to the ionosphere, *J. Geophys. Res.*, submitted 1989.

Farrugia C. J., M. P. Freeman, S. W. H. Cowley, D. J. Southwood, M. Lockwood, and A. Etemadi, Pressure-driven magnetopause motions and attendant response on the ground, *Planet. Space Sci.*, submitted 1989.

Friis-Christensen, E., Polar cap current systems, in T. A. Potemra, editor, *Magnetospheric Currents,* American Geophysical Union, Wash., D.C., p. 86, 1984.

Friis-Christensen, E., M.A. McHenry, C.R. Clauer, and S. Vennerstrøm, Ionospheric traveling convection vortices observed near the polar cleft: A triggered response to sudden changes in the solar wind, *Geophys. Res. Lett., 15*, 253, 1988a.

Friis-Christensen, E., S. Vennerstrøm, C. R. Clauer, and M.

A. McHenry, Irregular magnetic pulsations in the polar cleft, caused by traveling ionospheric convection vortices, *Adv. Space Res.*, *8*, (9)311, 1988b.

Glassmeier K.-H., M. Hönisch, and J. Untiedt, Ground-based and satellite observations of traveling magnetospheric convection twin-vortices, *J. Geophys. Res.*, *94*, 2520, 1989.

Goertz, C. K., E. Nielsen, A. Korth, K. H. Glassmeier, C. Haldoupis, P. Høeg, and D. Hayward, Observations of a possible ground signature of flux transfer events, *J. Geophys. Res*, *90*, 4069, 1985.

Iijima, T. and T. A. Potemra, Field-aligned currents in the dayside cusp observed by TRIAD, *J. Geophys. Res.*, *81*, 5971, 1976.

Lanzerotti, J. L., L. C. Lee,C. G. Maclennan, A. Wolfe, and L. V. Medford, Possible evidence of flux transfer events in the polar ionosphere, *Geophys. Res. Lett.* *13*, 1089, 1986.

Lee L. C. , Magnetic flux transfer at the Earth's magnetopause, in Y. Kamide and J. A. Slavin, eds., *Solar Wind - Magnetosphere Coupling*, Terra, Tokyo, p. 297, 1986.

McHenry, M. A., and C. R. Clauer, Modeled ground magnetic signatures of flux transfer events, *J. Geophys. Res.*, *92*, 11231, 1987.

McHenry, M. A., C. R. Clauer, E. Friis-Christensen, and J. D. Kelly, Observations of ionospheric convection vortices: Signatures of momentum transfer, *Adv. Space Res.*, *8*, (9)315, 1988.

Potemra, T. A., L. J. Zanetti, K. Takahashi, R. E. Erlandsen, H. Lühr, G. T. Marklund, L. P. Block, and A. Lazarus, Multi satellite and surface observations of transient ULF waves, *J. Geophys. Res.*, *94* , 2543, 1989.

Russell, C. T. and R. C. Elphic, ISEE observations of flux transfer events at the dayside magnetopause, *Geophys. Res. Lett.*, *6*, 33, 1979.

Saunders, M. A., C. T. Russell, and N. Sckopke, Flux transfer events: Scale size and interior structure, *Geophys. Res. Lett.*, *11*, 131, 1984.

Sibeck, D. G., W. Baumjohann, R. C. Elphic, D. H. Fairfield, J. F. Fennell, W. B. Gail, L. J. Lanzerotti, R. E. Lopez, H. Lühr, A. T. Y. Lui, C. G. Maclennan, R. W. McEntire, T. A. Potemra, T. J. Rosenberg, and K. Takahashi, The magnetospheric response to 8 minute-period strong-amplitude upstream pressure variations, *J. Geophys. Res.*, *94*, 2505, 1989.

Southwood, D. J., Theoretical aspects of ionosphere - magnetosphere - solar wind coupling, *Adv. Space Res.*, *5*, 7, 1985.

Zanetti, L. J. and T. A. Potemra, The relationship of Birkeland and ionospheric current systems to the interplanetary magnetic field, in Y. Kamide and J. A. Slavin, eds., *Solar Wind - Magnetosphere Coupling*, Terra, Tokyo, p. 547, 1986.

THE RESPONSE OF THE MAGNETOSPHERE-IONOSPHERE SYSTEM TO SOLAR WIND DYNAMIC PRESSURE VARIATIONS

M. P. Freeman[1], C. J. Farrugia[1], S. W. H. Cowley[1], D. J. Southwood[1], M. Lockwood[2], and A. Etemadi[3]

Abstract. We study the causal chain of events by which variations in the solar wind dynamic pressure cause the magnetopause boundary to move and excite magnetic perturbations at the ground. The observation of large ground magnetic transients is argued to be due to the coupling of the magnetohydrodynamic compressional wave to the field-guided Alfvén wave, which carrying current, can thereby transfer momentum to the ionosphere. The study highlights the similarity of the ionospheric signatures at a single station arising from the response of the coupled magnetosphere-ionosphere system to disparate impulsive processes at the magnetopause.

Introduction.

The Flux Transfer Event [FTE; Russell and Elphic, 1978] is thought to be a central agent in the transfer of momentum from the solar wind to the magnetosphere and ionosphere [Cowley, 1984]. The dynamics and internal stucture of the FTE have thus been extensively studied. Initially, the bipolar variation in the component of the magnetic field normal to the magnetopause plane, characteristic of the FTE, was argued to be due to a draping of the ambient field over a moving, isolated tube of magnetic flux of approximately circular cross-section. This tube is formed by a burst of reconnection at the dayside magnetopause, localised in time and space. The reconnected field threading the magnetopause connects the shocked interplanetary magnetic field (IMF) to the geomagnetic field to form a kinked flux tube which moves in a generally anti-solar direction under the action of magnetic tension. Subsequently, the same bipolar variation was found internal to the flux tube and thus the reconnected field lines were argued to be twisted into a flux rope [Saunders et al., 1984].

These arguments led to the development of two models for the way in which a FTE couples to the ionosphere and imparts momentum there. Saunders et al. [1984] and Lee [1986] emphasised the twisted nature of the FTE field which must have associated with it a central core current. Southwood [1985; 1987] stressed the dynamics of the FTE, which gives rise to field-aligned current sheets of opposite polarity on the flanks of the isolated flux tube. The closure of these current systems at ionospheric heights drive plasma motions therein. It is tacitly assumed that a FTE that has a circular cross-section at the magnetopause maps to a circular ionospheric foot. In the Southwood model the resultant ionospheric flows within the ionospheric foot are matched to its velocity, and continuity demands a concommitant return flow of plasma outside to produce a twin vortical flow pattern. The strength of the return flow relative to that within the foot depends upon its geometry, the flows being approximately equal in the case of a circular foot. In the Saunders et al. - Lee model a single flow vortex is produced. In general, these flow systems may co-exist as it appears to us that a translating FTE must have a twin vortical flow component to it. The ionospheric plasma motions drive oppositely directed Hall currents due to ion-neutral collisions in the E-region ionosphere which generate a magnetic perturbation that is measurable on the ground. By assuming a velocity and scale size of the FTE ionospheric footprint, the models yield clear predictions for the FTE ground magnetic signature at an individual station [McHenry and Clauer, 1987].

The study we present here [see also Farrugia et al., 1989] arose from a search for impulsive ground signatures that could be attributed to FTEs at the magnetopause. Recently a number of studies have identified transient signatures in the ionosphere and have attributed them to a variety of causes [e.g. Lanzerotti et al., 1986; Friis-Christensen et al., 1988].

In Fig. 1 we present magnetic field perturbations recorded at two ground stations during an interval we shall study in detail. In addition we have superposed the perturbations that would result from the passage overhead of a circular FTE footprint of radius 100 km, as modelled by McHenry and Clauer [1987]. In Fig. 1a we use the Southwood model and assume a poleward motion at 2.3 km s^{-1} and a flank current of 5.8×10^5 A. In Fig. 1b we use the Saunders et al.-Lee FTE model and assume a poleward and eastward motion at 400 m s^{-1} with an axial current of 3.3×10^5 A. In both cases we have further assumed a reduction by a factor of two of the modelled perturbation in the vertical (Z) field component, due to finite ground conductivity effects [Boteler, 1978]. It is apparent that these model current systems describe the observations quite well. A comparable level of agreement has previously been claimed to satisfactorily identify the cause of similar ground signatures as being that of impulsive reconnection at the magnetopause. However, we shall show that

---

[1] Blackett Lab., Imperial College, London, SW7 2BZ, UK.
[2] Rutherford Appleton Lab., Chilton, Didcot, Oxon., OX11 0QX, UK.
[3] Mullard Space Science Lab., Holmbury St. Mary, Dorking, RH5 6NT, UK.

Geophysical Monograph 58

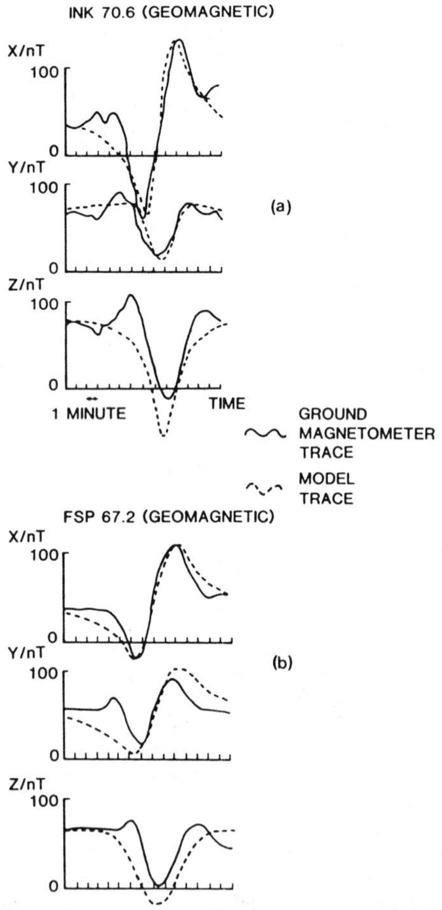

Fig. 1. Ground magnetic perturbations (solid line) recorded between 01:00 UT and 01:15 UT on September 10, 1978 at two stations in the mid-afternoon MLT sector. Station code and geomagnetic latitude are indicated at the top of each figure. Panels show from top to bottom the northward (X), eastward (Y), and downward (Z) field components. The postulated ground magnetic signature of a FTE (see text) is also shown (dashed line).

the perturbations presented in Fig. 1 were in fact due to a sudden enhancement in the solar wind dynamic pressure. Thus we argue that other impulsive processes at the magnetopause can mimic the postulated effect on the ground of FTEs and hence we urge caution in the interpretation of such signatures, particularly when using measurements from only a single station.

Data Overview and Analysis.

For the case study we present here we have a serendipitous arrangement of observing stations that is ideal for examining the solar-terrestrial interaction. Over the interval of interest the IMP 8 satellite monitored the solar wind conditions outside the dawn flank of the magnetosphere at a radial distance, R ~ 32 R_E. The spacecraft pair, ISEE-1 and -2, were just south of the magnetic equator in the vicinity of the mid-afternoon magnetopause (R ~ 9-12 R_E), which they crossed and re-crossed repeatedly. Fourteen ground magnetometers gave information about the ionospheric current distribution over the dayside, northern hemisphere, mid- and high-latitude region in the afternoon local time sector (see Farrugia et al. [1989] for a tabulation of magnetometer locations). In addition, we also use data from a near-equatorial ground magnetometer located near the noon meridian.

Firstly, we examine the solar wind behaviour in the period of interest. Fig. 2 shows, from top to bottom, the quantities: density, velocity, X, Y, Z GSM components of the IMF, and its magnitude, B. The measurements are 5 min averages.

Across midnight in the long data gap, IMF Bz changed from a southward to northward polarity and remained strongly northward throughout the interval we study in detail between midnight and 03:00 UT. We note that the northward field orientation is unpropitious for the generation of FTEs [Rijnbeek et al., 1984]. In fact, when the magnetosheath field was southward three FTEs at ~ 5 min intervals were observed by the spacecraft [see Fig. 2a, Farrugia et al., 1989], but no FTE signatures were seen after the field had turned northward.

Between 00:15 UT and 00:45 UT the field and plasma conditions recorded by IMP 8 were very steady, a fact that we shall use later. However, just before 01:00 UT there was a large sudden increase in solar wind density accompanied by a field compression. After this, plasma and field conditions exhibited much variability. We shall examine the effect of the corresponding solar wind dynamic pressure variations on the magnetospheric cavity.

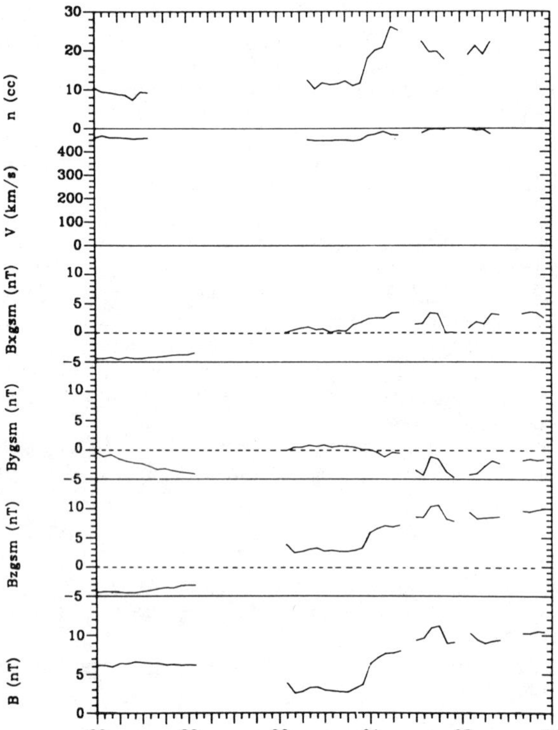

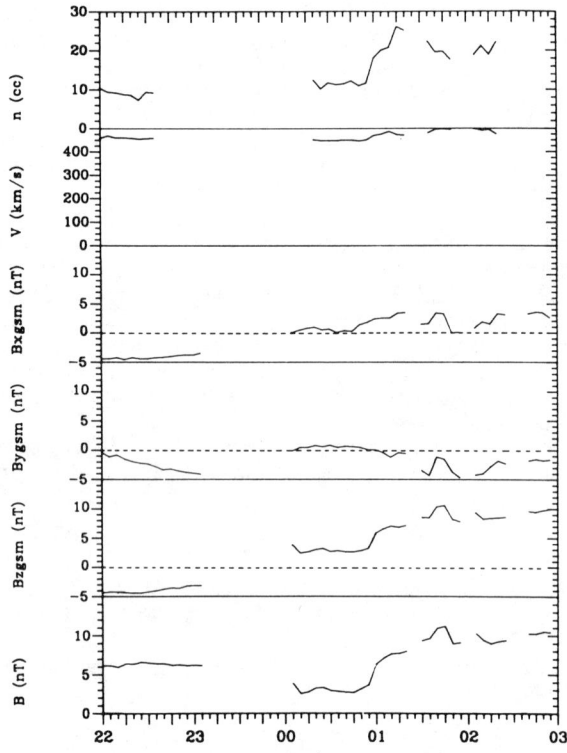

Fig. 2. IMP 8 data from the solar wind. Panels show from top to bottom: density (cm^{-3}), flow speed (km s^{-1}), X, Y, Z GSM components and magnitude of the IMF (nT).

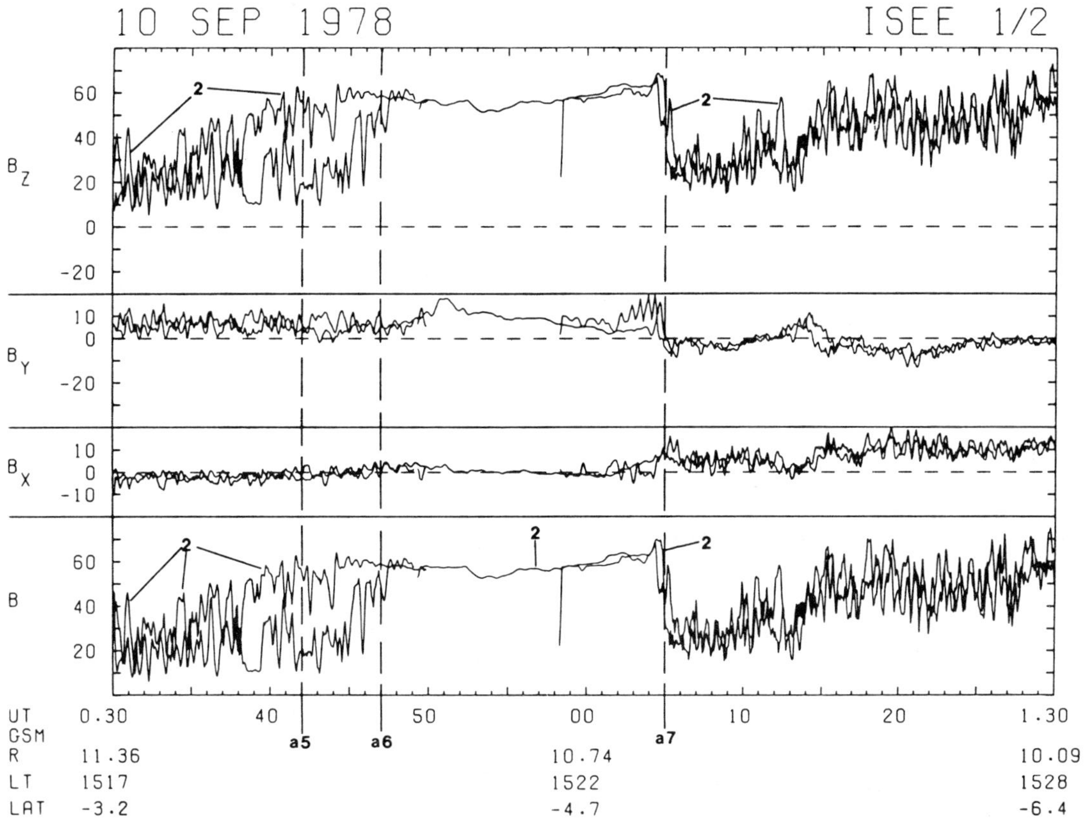

Fig. 3. ISEE 1 and 2 GSM magnetic field components and magnitude measured in the vicinity of the magnetopause.

Next we consider the ISEE 1 and 2 magnetic field data recorded in the vicinity of the magnetopause. As an example, we show in Fig. 3 the GSM field components and magnitude over the hour interval from 00:30 UT to 01:30 UT. The ISEE 2 spacecraft led ISEE 1 on an inbound trajectory, separated by ~ 2000 km.

Referring to Fig. 3, both spacecraft begin and end in the noisy and northward directed magnetosheath field. However, near 01:00 UT ISEE 1 and 2 record the quieter and stronger field of the magnetosphere. Plasma data (not shown) confirm the spacecraft to be in this region [N. Sckopke, private communication]. The individual spacecraft crossings of the magnetopause boundary are shown in the figure by the vertical guidelines [see also Farrugia et al., 1989]. It is apparent that the crossing into the magnetosphere was very different from the exit; the spacecraft entry crossings were widely separated in time, but the exits were almost coincident. Since the spacecraft separation was almost constant between entry and exit, the observations clearly show the magnetopause to have been in varying states of motion. In fact, the ISEE dual spacecraft arrangement allows us to calculate the radial speed of the boundary whenever it moves over both satellites. We find that on entry the magnetopause was stationary to within 1 km s^{-1}, but on exit it was moving earthward at ~ 90 km s^{-1}. In the following hour interval (not shown) the boundary moved inward and outward over the ISEE spacecraft a further five times, before the inbound satellites finally remained within the magnetosphere.

We now ask what is causing the magnetopause boundary to move. In particular, we wish to determine whether the boundary motions are due to the solar wind dynamic pressure variations recorded at IMP 8.

In the simplest approximation the subsolar magnetopause will be located at a radial distance, R, such that the magnetic pressure of the compressed magnetospheric dipole field is balanced by the stagnation pressure, P_s, at the subsolar point

$$(R/R_E)^6 = \alpha^2 B_{eq}^2/(2\mu_0 P_s)$$

where μ_0 is the permeability of free space, R_E is the radius of the Earth, B_{eq} is the dipole field strength on the Earth's surface at the equator, and α is a measure of the dipole field compression. Assuming the magnetopause to be an infinite plane current sheet we have $\alpha = 2$, though in reality the field curvature can increase this value by ~20 % [Mead and Beard, 1964]. For an off-subsolar location the pressure acting normal to the boundary is less and hence the dipole field stands off the flow more effectively. Thus the above expression may be generalised for any magnetopause position with α a function of local time and latitude.

For the high Mach number solar wind the stagnation pressure is ~88 % of the upstream solar wind dynamic pressure [Siscoe et al., 1968, and references therein]. Allowing for alpha particles to contribute ~4% to the measured solar wind proton number density, n_p, the stagnation pressure is given by:

$$P_s = 1.02\ m_p n_p V^2 = 1.02\ P_{sw}$$

where m_p is the proton mass, V is the solar wind bulk speed, and P_{sw} is the solar wind dynamic pressure assuming the particles to be protons. Thus the magnetopause position at a given local time and latitude is expected to be proportional to $(P_{sw})^{-1/6}$.

To test this we must know where the magnetopause is for a given steady solar wind dynamic pressure, and how long it will take for a change in the solar wind to be effective on the magnetopause at the local time of the ISEE spacecraft.

To determine the steady state proportionality constant (α) between the solar wind dynamic pressure and the magnetopause position at the spacecraft local time we isolate a period of very steady solar wind conditions when the magnetopause was crossed. This corresponds to the widely separated ISEE 1 and 2 crossings near 00:45 UT seen in Fig. 3 and referred to above. Using the measured solar wind dynamic pressure and the magnetopause location at this time we find the value of α to be 4.5.

To estimate the lag we assume information to travel through the solar wind and magnetosheath at the local convection speed and use the Spreiter and Stahara [1980] gas-dynamic model of the planetary interaction with the wind. The lag thus calculated is ~7.5 min.

Fig. 4 shows the predicted magnetopause location versus time using the IMP 8 data and the above assumptions. Also shown are the ISEE 1 and 2 trajectories. We have coded these to indicate which plasma region, magnetosheath or magnetosphere, the spacecraft was in. The large dots represent the magnetopause crossings made by the spacecraft. The agreement between the predictions and observations is generally good, showing that the estimated lag is appropriate, that the magnetospheric cavity responds rapidly to solar wind dynamic pressure changes, and that the magnetopause position scales with $(P_{sw})^{-1/6}$. The interval of little magnetopause motion is seen prior to a sudden large inward compression. Subsequently the magnetopause was observed to move in and out in an oscillatory manner, though the 5 min averaged IMP 8 data cannot predict these rapid variations so well.

To confirm the gross magnetopause motions at high time resolution we examine magnetic data from a ground station, WKE, which is located near to the geomagnetic equator (12.5° N; 233.0° E in geomagnetic coordinates) and very close to noon MLT. The north-south (X) component of the geomagnetic field is shown in Fig. 5. To the right of the panel is given the station code and geomagnetic latitude, whilst MLT is plotted on the ordinate, together with UT at the base of the figure. The panel height is 50 nT.

The dominant effects which cause magnetic perturbations at these low latitudes are thought to be the equatorial electrojet

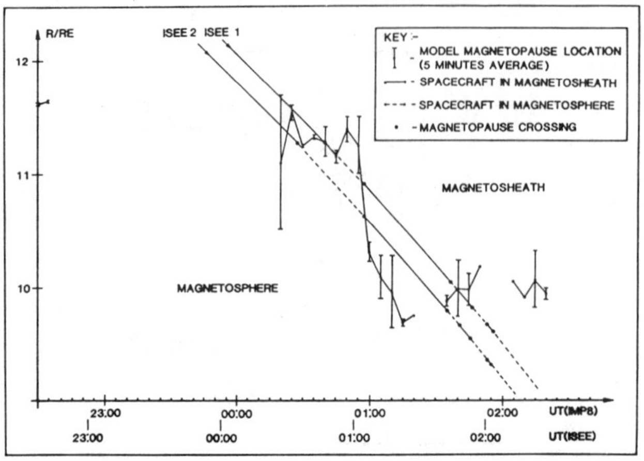

Fig. 4. The observed and expected location of the mid-afternoon magnetopause (see text).

[e.g. Sugiura and Cain, 1966], the ring current [Sugiura, 1964] and the magnetopause current [Chapman and Ferraro, 1930]. We consider here the latter effect. In the approximation of the magnetopause as an infinite plane current sheet, the magnetic perturbation is independent of distance from the current sheet and is equal to the dipole field strength at the subsolar magnetopause stand-off distance, or proportional to $(P_{sw})^{1/2}$. For a more realistic magnetopause geometry the constant of proportionality between an equatorial ground magnetic perturbation and a change in P_{sw} is found to be 24 nT $(nPa)^{-1/2}$ [Mead, 1964]. Thus, following the sudden increase in the solar wind dynamic pressure at IMP 8 from (4.0 ± 0.5) nPa at 00:55 UT to (6.7 ± 0.2) nPa at 01:00 UT which caused the large magnetospheric compression observed by the ISEE spacecraft at ~01:05 UT, we estimate the perturbation in the dipole field to be an increase of (14 ± 3) nT. The effect of currents induced within the Earth's surface is thought to enhance the ground magnetic perturbation due to a sudden increase in dynamic pressure by ~50% [see Siscoe et al., 1968].

Referring to the Fig. 5 a sudden large enhancement in the X component was indeed seen almost coincident with the earthward magnetopause motion. (The Y and Z components showed no change). The amplitude of the step-like rise was 18 nT, in good agreement with the expected perturbation using the measured solar wind dynamic pressure. Subsequently, between ~01:40 UT and ~02:10 UT further large variations in the near-equatorial ground magnetic field were recorded. We conclude

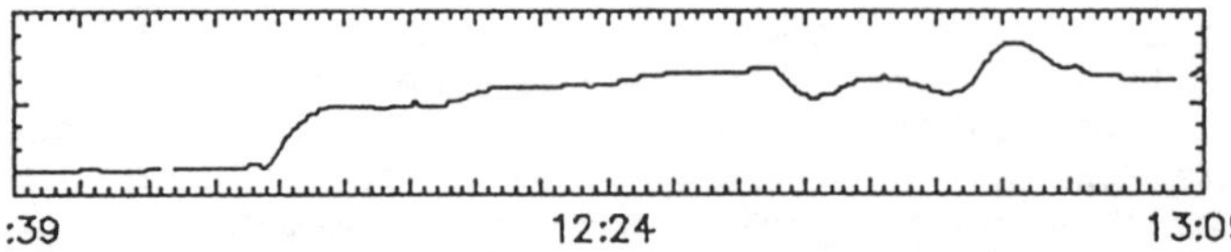

Fig. 5. The time variation of the north-south component of the ground magnetic field at a near-equatorial station (see text).

that these oscillations (period ~ 10 min) were associated with solar wind dynamic pressure fluctuations which could not be resolved by the 5 min averaged IMP 8 data, though they contributed to the variance of the averages (note the large error bars at this time in Fig. 4).

Now that we know and understand the gross magnetospheric behaviour in terms of the solar wind dynamic pressure changes, we investigate how the coupled magnetosphere-ionosphere system adapts in detail to these changes.

We have seen that the ground response at the equatorial station WKE is simply a compression of the dipole field. In Fig. 6 we show data from 8 mid- and high-latitude ground magnetometers. We present only the north-south (X) component of the geomagnetic field. Each panel in the figure presents data from one magnetometer station in the same format as for Fig. 5, except that the panel height is now 100 nT. The stations are arranged in order of increasing geomagnetic latitude.

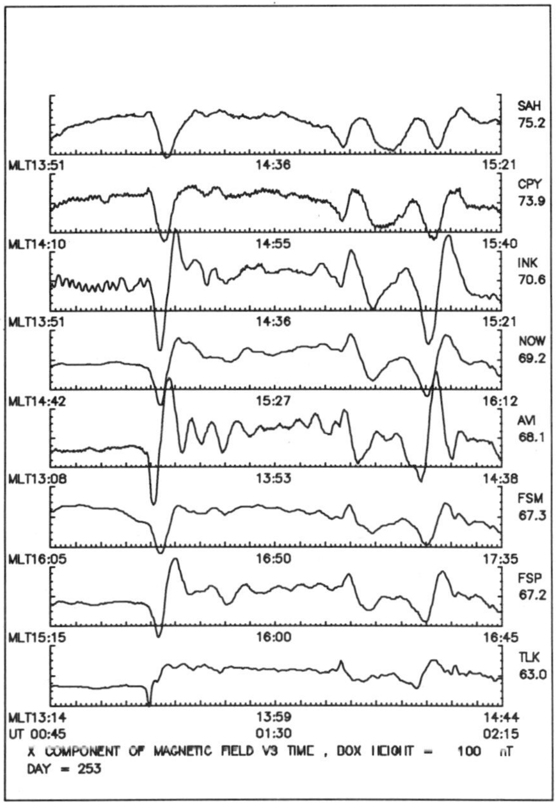

Fig. 6. Stackplot of the north-south component of the geomagnetic field versus time at eight mid- and high-latitude dayside ground stations (see text).

The data correlates with the solar wind and magnetopause observations in the following way: initially there was a quiet field at a time of similarly quiet solar wind and little magnetopause motion. Just after 01:00 UT all stations exhibited a large perturbation, though the nature of this changed with position. At station AVI a large, fairly continuous oscillation was observed whereas elsewhere the signal duration was more brief and generally smaller. The onset time of the signal closely matched the time of observation of the sudden increase in the equatorial ground geomagnetic field and the inward magnetopause motion, both previously correlated with the large rapid increase in solar wind dynamic pressure recorded earlier at IMP 8. The ground magnetic perturbations at stations INK and FSP shown in Fig. 1 were recorded at this time. Subsequently, after ~01:40 UT, a second packet of magnetic perturbations were observed. Oscillations of similar phase but different amplitudes were seen at all stations which also closely correlated with the phase of the equatorial geomagnetic field perturbations. At this time the magnetopause position and solar wind dynamic pressure are known to also have been varying as discussed above.

To investigate further the relationship between the ground observations and the spacecraft data we present in Fig. 7 a summary plot of the salient physical parameters against time. In the middle of the plot is shown the solar wind dynamic pressure calculated from the 5 min averaged IMP 8 plasma data. Above it is shown the X component of the geomagnetic field at AVI. The time axes are offset by the estimated lag of 7.5 min for information to propagate from IMP 8 to the ionosphere. Taking the lag into account, the first oscillation onset at AVI just after 01:00 UT correlates well with the sudden dynamic pressure increase recorded at IMP 8.

The bottom plot shows the near-instantaneous magnetopause speeds inferred from ISEE 1 and 2 data using two-spacecraft timing (positive values correspond to Earthward motion). The rapid inward magnetopause motion can be seen just after 01:00 UT coincident with the lagged solar wind dynamic pressure increase and the first oscillatory event at AVI. This followed the earlier, quiet solar wind interval when the magnetopause was observed to be stationary and the geomagnetic field was steady. The latter period was used for the determination of the magnetopause proportionality constant, α, above and is indicated by the bar marked 'Reference' in Fig. 7.

Later on, the aforementioned second packet of oscillations in the geomagnetic field occurred when the solar wind variations were quite large and the magnetopause was known to be moving. In particular, with the lag shown, the field at AVI was temporally stationary whenever the magnetopause was still, as shown by the vertical guidelines; and there was a non-zero $\partial B/\partial t$ whenever the magnetopause was in motion.

Discussion and Conclusion

We have shown that the magnetopause responds promptly and in a predictable way to a change in solar wind dynamic pressure. The equilibrium boundary position is determined by a pressure balance between the stagnation pressure of the solar wind flow and the magnetic pressure of the compressed dipole field.

Temporal changes in the dipole compression are detectable at equatorial locations on the Earth's surface. The variations in the field can be related quantitatively to the Chapman-Ferraro magnetopause current strength required for pressure balance at the boundary. Following a rapid increase in the solar wind dynamic pressure the observed near-equatorial surface field enhancement of ~ 18 nT was in agreement with simple theoretical predictions. Additionally, the time averaged field at most high-latitude ground stations increased by about 20-30 nT following the step-like change, although the effect was smaller at high latitudes and later local times (see Fig. 6). This global part of the response detected on the ground can also be attributed to the increase in Chapman-Ferraro currents on the magnetopause after the compression (cf Fig. 5).

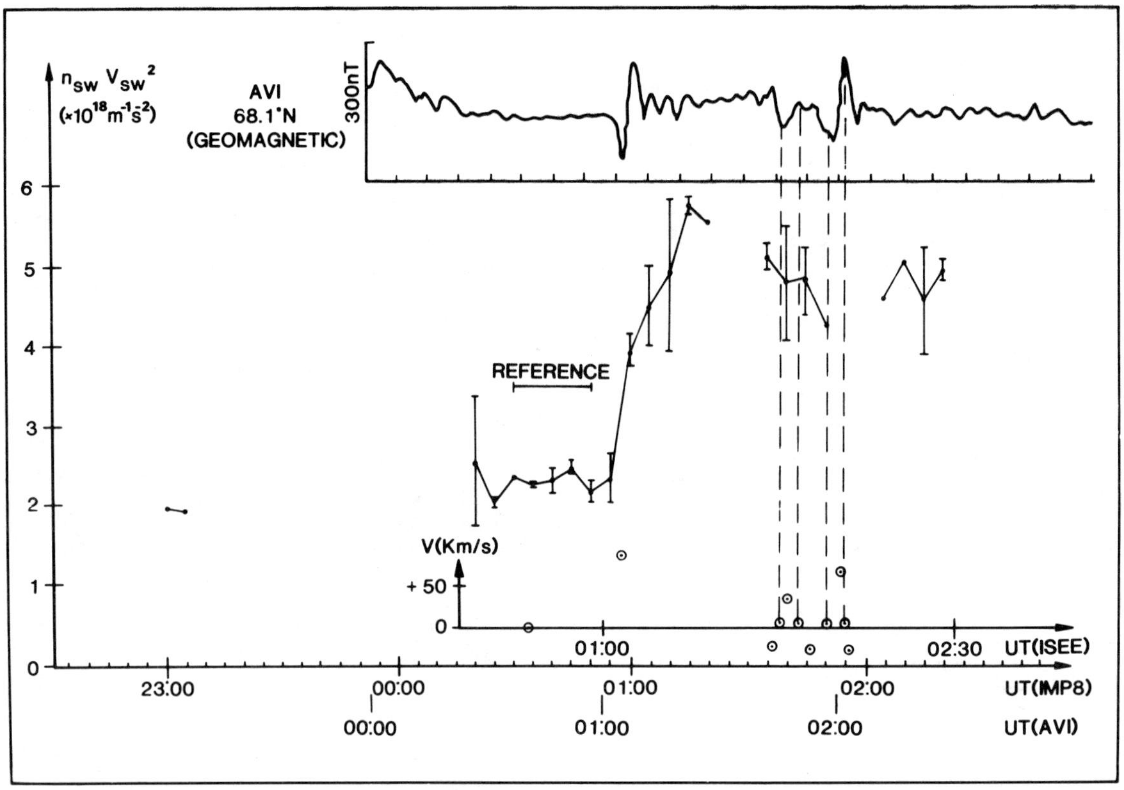

Fig. 7. Composite figure showing the relationship between the solar wind dynamic pressure (centre panel), the radial motion of the mid-afternoon magnetopause (lower panel), and the north-south component of the geomagnetic field at high latitude (upper panel) (see text).

However, there are also transients detected in association with the compression events. The nature of the transients differed from position to position and between the 01:00 UT and 01:40 UT events. At the equatorial station, WKE, oscillatory signals are detected in the latter event but the amplitude is of the same order as the background change in field strength during the previous compression event. In contrast, the amplitude of the transients and oscillations at the high latitude stations is substantially greater than at WKE and also greater than the time-averaged field increase at these stations. As all ground stations are at roughly the same distance from the magnetopause, it follows that the high latitude transient signals are not due to the Chapman-Ferraro currents alone. We conclude that the bulk of the transient signals seen at high latitude are due to current induced in the ionosphere above the station.

The transients can be understood by considering how the magnetosphere adapts to a compression. A pressure front that is planar in the solar wind is unlikely to impinge upon all regions of the magnetopause at the same time. On time scales short compared with that for the pressure front to propagate past the magnetospheric cavity the magnetopause will be deformed from its equilibrium shape. Behind and ahead of the front the magnetopause will be at its equilibrium position. However, in the region of the front the magnetopause boundary will be distorted. The length scale of the boundary deformation will be determined by the duration of the pressure variation and the propagation speed of the front.

The response of the cavity to compression is to launch a compressional (fast) MHD mode propagating isotropically into the cavity. At the same time, if the magnetopause boundary is distorted by the compression, Alfvén waves are set up. The non-uniform medium will allow compressional energy to be coupled into the shear Alfvén mode, which is field guided. In particular, the latter mode carries Birkeland current and is thus the only mode that can set up flows in the ionosphere at the feet of the field lines [Southwood and Kivelson, 1989]. In contrast, the fast mode signal is almost perfectly reflected from the ionosphere [Kivelson and Southwood, 1988].

Because the shear Alfvén mode is field guided and reflected at the ionosphere (albeit imperfectly), the magnetospheric portion of a field line forms a resonant cavity for the mode. It follows that if the source contains power at some integer multiple of the eigenfrequency of the field line, standing waves can be excited. The resonant frequencies vary with latitude and thus the oscillations will be localized in this dimension. We interpret the first packet of oscillations in this way. At 01:00 UT five large amplitude linearly polarized oscillations of ~5 min period were observed at AVI alone. Elsewhere, the response was of much shorter duration and elliptically polarized. Close examination of Fig. 7 reveals a poleward and anti-sunward phase motion to the ground response. We conclude that station AVI was close to resonance but that elsewhere the field lines were not in resonance and the observed brief perturbations were due to a shear flow excited in the local ionosphere directly driven by coupling into the Alfvén mode from the high altitude compression. The longitudinal variation of the ground magnetic perturbation depends upon the structure of the magnetopause deformation.

Oscillations are seen at all latitudes in the 01:40 UT event. One explanation could be that the solar wind pressure changes have excited compressional eigen-oscillations of the magnetospheric cavity as a whole, but the ten minute period observed is rather long. However, there is a stronger argument against the signals being compressional eigenoscillations; there are large vertical magnetic components detected at the high latitude stations (not shown here, but see Farrugia et al., [1989]) which would be inconsistent with the compressional hydromagnetic mode. Not only is the ionospheric electric field of the compressional mode small, but so also should be any vertical magnetic field in the atmosphere.

There is an alternative and more likely explanation. Referring to Fig. 6 the magnetopause appears to be in oscillation at this time and furthermore quite closely locked in phase with the oscillations on the ground. We thus propose that the very low frequency ten minute period magnetic perturbations commencing at ~ 01:40 UT are directly driven by the magnetopause motion which is itself being forced by external variations in pressure, presumably of the same frequency. Potemra et al. [1989] and Sibeck et al. [1989a] have reported observations where external pressure variations were associated with ground magnetic perturbations of the same frequency. We then deduce the large high latitude amplitudes to be due to coupling of the compressional motion into shear motions which produce ionospheric electric fields and currents which are the immediate source of the signals detected on the ground. Hence we expect the amplitude of the ground perturbation at a given station to be correlated with the size of the compressional wave (note the correlation of the magnetopause speed with perturbation amplitude at AVI in Fig. 6). From station to station the amplitude varies due to the differing ionospheric conductivities.

Finally we return to the initial hypothesis of this paper, namely the possible identification of FTEs at the magnetopause by their ground magnetic signature. The measured ground magnetic perturbations shown in Fig. 1 come from two of the stations (INK and FSP) shown in Fig. 6. They occurred just after 01:00 UT during the first of the two ground magnetic perturbation events, at the time when we have established that a solar wind dynamic pressure enhancement caused these and other large ground magnetic perturbations. The apparent agreement of the predictions of two different FTE models with single station observations illustrates an important point: impulsive pressure changes acting at the magnetopause can drive compressional motions within the magnetosphere which may couple to the Alfvén wave mode and excite plasma motions in the ionosphere. The similarity of the pressure-excited signatures with those predicted by FTE models reflects only the similarity of the time-varying field-aligned currents at a given observing point. Since any ionospheric flow must be driven by field-aligned currents, single station observations may not describe the current system sufficiently well to unambiguously identify its cause.

This concurs with the view of Sibeck et al. [1989b] who have shown that ionospheric events that have previously been reported as candidates for the ionospheric signature of a FTE were likely instead to have been caused by a variation in the solar wind dynamic pressure. The apparent absence of an ionospheric counterpart to the magnetopause FTE observations may indicate that the model of its ionospheric signature is incorrect. Recently the magnetopause FTE has been viewed as being due to a temporal fluctuation in the dayside reconnection rate along a neutral line that may extend over a considerable portion of the dayside magnetopause [Biernat et al., 1987; Southwood et al., 1988], rather than being due to sporadic, spatially localised reconnection as in the original Russell and Elphic picture. Thus the ionospheric foot of a FTE may be elliptical rather than circular, extending over a wide longitudinal range. In this case, if the event were to initially move zonally under magnetic tension, the associated flow pattern would form a strong flow channel aligned with the dayside polar cap boundary with only weak return flows outside it. The concommitant ground magnetic perturbations would then be quite distinct from those considered hitherto.

In conclusion, compressional changes in the high altitude magnetosphere have been seen to give rise to field aligned current systems coupling ionospheric motion to the magnetospheric change. Oscillations have been found to occur localized in latitude. We interpret these as resonant excitation of field lines by the compressional changes.

We have also presented an instance where oscillatory signals were detected over a large range of latitude. The ground signals at high latitude seem to originate from currents set up in the local ionosphere, a fact that precludes the signals being in the fast mode. We have proposed that the magnetosphere may be directly driven from the magnetopause.

Questions and Answers

Smith: When you look at upstream data for FTEs observed at the magnetopause, you always see some sort of variation in the pressure. Have you any qualitative idea for the range of pulse that produces a filamentary current signature?

Freeman: I am looking into this at present. It appears that you get the strongest signature when the dominant frequency in the pressure pulse matches the characteristic frequency of the L-shell. But more importantly, when the dynamic pressure variation is very rapid ($\simeq$1 min.) then the magnetopause cannot respond directly and the amplitude of the boundary motion is reduced. Also, the coupling of the compressible signal to the shear Alfvén wave which causes the ground signature wavelength of the magnetopause perturbation is small. This may be related to whether the dynamic variation is discontinuous or smooth.

Acknowledgments. The authors would like to thank N. Sckopke for the provision of ISEE plasma data, R. Lepping and M. Hapgood for IMP 8 data, and the World Data Center in Boulder, Colorado for ground magnetometer data. MPF is supported by a British SERC post-graduate award and CJF is supported by a SERC research assistantship.

References.

Biernat, H. K., M. F. Heyn, and V. S. Semenov, Unsteady Petschek reconnection, *J. Geophys. Res., 92,* 3392, 1987.

Boteler, D. H., The effect of induced currents in the sea on magnetic bays observed at a coastal observatory, *J. Atmos. Terr. Phys., 40,* 577, 1978.

Chapman, S., and V. C. A. Ferraro, A new theory of magnetic storms, *Nature*, *126*, 129, 1930.

Cowley, S. W. H., Solar wind control of magnetospheric convection, *Proc. Conf. Achievements of the IMS, ESA SP-217*, 483, 1984.

Farrugia, C. J., M. P. Freeman, S. W. H. Cowley, D. J. Southwood, M. Lockwood, and A. Etemadi, Pressure-driven magnetopause motions and attendant response on the ground, *Planet. and Space Sci.*, *37*, 589, 1989.

Friis-Christensen, E., M. A. McHenry, C. R. Clauer, and S. Vennerstrom, Ionospheric traveling convection vortices observed near the polar cleft: a triggered response to sudden changes in the solar wind, *Geophys. Res. Lett.*, *15*, 253, 1988.

Kivelson, M. G., and D. J. Southwood, Hydromagnetic waves and the ionosphere, *Geophys. Res. Lett.*, *15*, 1271, 1989.

Lanzerotti, L. J., L. C. Lee, C. G. MacLennan, A. S. Wolfe, and L. V. Medford, Possible evidence of flux transfer events in the polar ionosphere, *Geophys. Res. Lett.*, *13*, 1089, 1986.

Lee, L. C., Magnetic flux transfer at the Earth's magnetopause, *Solar Wind-Magnetosphere Coupling*, edited by Y. Kamide and J. Slavin, Terra, Tokyo, 1986.

McHenry, M. A., and C. R. Clauer, Modeled ground magnetic signatures of flux transfer events, *J. Geophys. Res.*, *92*, 11231, 1987.

Mead, G. D., and D. B. Beard, Shape of the geomagnetic field solar wind boundary, *J. Geophys. Res.*, *69*, 1169, 1964.

Mead, G. D., Deformation of the geomagnetic field by the solar wind, *J. Geophys. Res.*, *69*, 1181, 1964.

Potemra, T. A., H. Lühr, L. J. Zanetti, K. Takahashi, R. E. Erlandson, G. T. Marklund, L. P. Block, L. G. Blomberg, and R. P. Lepping, Multisatellite and ground-based observations of transient ULF waves, *J. Geophys. Res.*, *94*, 2543, 1989.

Rijnbeek, R. P., S. W. H. Cowley, D. J. Southwood, and C. T. Russell, A survey of dayside flux transfer events observed by ISEE 1 and 2 magnetometers, *J. Geophys. Res.*, *89*, 786, 1984.

Russell, C. T., and R. C. Elphic, Initial ISEE magnetometer results: magnetopause observations, *Space Sci. Rev.*, *22*, 681, 1978.

Saunders, M. A., C. T. Russell, and N. Sckopke, Flux transfer events: scale size and interior structure, *Geophys. Res. Lett.*, *11*, 131, 1984.

Sibeck, D. G., W. Baumjohann, R. C. Elphic, D. H. Fairfield, J. F. Fennell, W. B. Gail, L. J. Lanzerotti, R. E. Lopez, H. Lühr, A. T. Y. Lui, C. G. Maclennan, R. W. McEntire, T. A. Potemra, T. J. Rosenberg, and K. Takahashi, The magnetospheric response to 8-minute period strong-amplitude upstream pressure variations, *J. Geophys. Res.*, *94*, 2505, 1989a.

Sibeck, D. G., W. Baumjohann, and R. E. Lopez, Solar wind dynamic pressure variations and transient magnetospheric signatures, *Geophys. Res. Lett.*, *16*, 13, 1989b.

Siscoe, G. L., V. Formisano, and A. J. Lazurus, Relation between geomagnetic sudden impulses and solar wind pressure changes- an experimental investigation, *J. Geophys. Res.*, *73*, 4869, 1968.

Spreiter, J. R., and S. S. Stahara, A new predictive model for determining solar wind-terrestrial planetary interactions, *J. Geophys. Res.*, *85*, 6769, 1980.

Southwood, D. J., and W. J. Hughes, Theory of hydromagnetic waves in the magnetosphere, *Space Science Rev.*, *35*, 301, 1983.

Southwood, D. J., Theoretical aspects of ionospheric-magnetosphere-solar wind coupling, *Adv. Space Res.*, *5*, 7, 1985.

Southwood, D. J., The ionospheric signature of flux transfer events, *J. Geophys. Res.*, *92*, 3207, 1987.

Southwood, D. J., C. J. Farrugia, and M. A. Saunders, What are flux transfer events?, *Planet. Space Sci.*, *36*, 503, 1988.

Southwood, D. J., and M. G. Kivelson, The magnetohydrodynamic response of the magnetospheric cavity to changes in solar wind pressure, *J. Geophys. Res.*, in press, 1989.

Sugiura, M., Hourly values of equatorial Dst for the IGY, *Ann. Int. Geophys. Year*, *35*, 1964.

Sugiura, M., and J. C. Cain, A model equatorial electrojet, *J. Geophys. Res.*, *71*, 1869, 1966.

MAGNETOPAUSE PRESSURE PULSES AS A SOURCE OF LOCALIZED FIELD-ALIGNED CURRENTS IN THE MAGNETOSPHERE

Margaret G. Kivelson[1,2] and David J. Southwood[1,3]

Abstract. Spatial structures exhibiting twisted magnetic fields and associated vortical flows have been observed on the terrestrial magnetopause where they are called flux transfer events (FTE's) and are generally thought to be signatures of bursty reconnection. However, it has recently been argued that many of the events can be interpreted as magnetopause responses to impulsive changes of solar wind dynamic pressure, an interpretation that does not require any magnetic reconnection. One might expect that it is possible to distinguish the two mechanisms by examining the signatures produced within the magnetosphere and in the ionosphere. Within the ionosphere, impulsive magnetic disturbances and traveling vortical flows have been interpreted as ground signatures associated with isolated FTE's. In this paper we use a simple model magnetosphere to demonstrate the qualitative features of magnetospheric and ionospheric responses to FTE-like compressional disturbances on the magnetopause. We show that many aspects of the magnetospheric and ionospheric responses also mimic features that have been interpreted as FTE signatures.

Introduction

Among the earliest and most intriguing revelations of the ISEE 1 and 2 spacecraft was that the magnetopause contains magnetic flux ropes called flux transfer events or FTE's [Russell and Elphic, 1978]. Characteristic signatures of these flux ropes include a rotation of the magnetic field into the directions normal to the boundary and perpendicular to the background magnetic field and changes in the magnetic field intensity followed by a return to ambient conditions. Signatures of particle flows are also commonly associated with these structures [Farrugia et al., 1989]. The principal features of the FTE have been explained by various forms of intermittent reconnection on the magnetopause [Russell and Elphic, 1979; Lee and Fu, 1985, Southwood et al., 1988; Scholer, 1988].

It was soon recognized that the rotation of the magnetic field at the magnetopause implies the presence of field aligned currents that must close in the ionosphere. Several models of the expected ionospheric signature were proposed [Southwood, 1985 and 1987; Lanzerotti, et al., 1986; McHenry and Clauer, 1987]. The expected ground-based observations of impulsive magnetic perturbations [Lanzerotti and Maclennan, 1988; Lanzerotti et al., 1987], of electric and magnetic field perturbations [Bering et al., 1988] and of vortical disturbances in the ionosphere [Friis-Christensen et al., 1988; McHenry et al., 1989] were also reported.

Recently, Sibeck et al. [1989] and Elphic [1989] have proposed that on and near the magnetopause the perturbations produced by a pressure pulse traveling along the magnetopause can mimic the signature of an FTE. Our interest here is in investigating the ionospheric signature of such a traveling pulse [see also Southwood and Kivelson, 1989b]. Using a simple model to represent the magnetosphere, we are able to show that the form of the response within the magnetosphere and in the ionosphere depends on the form of the pressure impulse on the magnetopause. We will point out that even well away from the boundary, and in particular within the ionosphere, field disturbances that are caused by pressure pulses can be very similar to the perturbations produced by FTE's.

As an example, consider a pressure perturbation producing a fluted indentation of limited azimuthal scale that travels azimuthally, say from noon to night, along the magnetopause. Pressure balance across the magnetopause requires that the magnetic pressure just within the boundary change as the external signal moves by. The magnetic perturbation serves as a source of waves that propagate into the magnetospheric cavity. Compressional waves propagate inward across field lines, setting up disturbances throughout the system. Shear Alfven waves, coupled to the compressional disturbances, are guided along the field and drive field-aligned currents into the ionosphere. These currents set up vortical flows in the ionosphere on field lines that map to the boundary source. They also set up magnetic perturbations that may be impulsive with a wave form resembling that of the boundary perturbation or oscillatory at the frequency of field line resonance on certain field lines. The model provides an alternative interpretation of several recent space, ground based and ionospheric radar measurements.

1. Institute of Geophysics and Planetary Physics, University of California, Los Angeles, CA 90024-1567
2. Department of Earth and Space Sciences, University of California, Los Angeles, CA 90024-1567.
3. Department of Physics, Imperial College of Science and Technology, London SW7 2BZ, U.K.

Geophysical Monograph 58

The Model

Let us consider, then, how the magnetosphere and ionosphere respond to a pressure perturbation on the magnetopause that propagates around the boundary from the day side to the night side with the velocity of the magnetosheath flow. As our interest is in the qualitative nature of the response, we dispense with the mathematical complexity of a realistic magnetospheric geometry and instead introduce a representation of a bounded magnetized plasma that is mathematically more tractable.

A model magnetosphere that retains much of the interesting physics fills a rectangular cavity with the z axis parallel to the uniform background magnetic field, **B**, the x axis parallel to the gradient of the plasma density (or Alfven velocity) and $\hat{\mathbf{y}} = \hat{\mathbf{z}} \times \hat{\mathbf{x}}$. The relation between the model and a more realistic magnetospheric geometry is illustrated schematically in Figure 1. The magnetopause lies in a yz plane at x=a. The boundaries at $z = \pm c$ represent the ionospheres that impose boundary conditions related to their height integrated Pedersen and Hall conductivities. We assume that the magnetospheric plasma pressure is negligible and consider linear perturbations only.

Suppose that a pressure perturbation $\delta p(a+,y,z,t)$ is imposed on the boundary. Pressure balance requires that the magnetic pressure change, $\mathbf{b}\cdot\mathbf{B}/\mu_o$, just within the boundary satisfy

$$b_z(a-,y,z,t)B/\mu_o = \delta p(a+,y,z,t) \qquad (1)$$

where **b** is the perturbation magnetic field. Within the cavity, the response to a perturbation can be expressed in terms of the field aligned magnetic perturbation, $b_z(x,y,z,t)$ and the transverse (perpendicular to **B**) displacement, $\xi(x,y,z,t)$. The solutions in the cavity simplify if it is assumed that they depend on y and z as

$$\exp(ik_y y + ik_z z)$$

It is also convenient to express the time dependences in terms of Laplace transforms over frequency components $b_{\omega z}(x)$, $\xi_\omega(x)$. Then the perturbations within the cavity satisfy the relations

$$(\omega^2/A^2(x) - k_z^2)\xi_\omega(x) = -(1/B)\nabla_\perp b_{\omega z}(x)$$
$$b_{\omega z}/B = -\nabla_\perp \cdot \xi \qquad (2)$$

where A is the Alfven velocity. The subscript $\perp$ refers to the directions perpendicular to the unperturbed field. Note that the requirement that the pressure perturbation have structure in the azimuthal direction requires that the solution include contributions with k_y non zero. This implies that there must be azimuthal displacement or flow of the plasma within the boundary.

If the plasma is uniform throughout the cavity [A(x) = constant], the initial response is a propagating signal that preserves the form of the initial excitation. The x-dependence of signals that persist after a wave transit time can be represented in terms of $\exp(ik_x x)$ and are the normal modes of the cavity. These are a compressional mode (magnetic field polarized in the xz plane) that is spatially oscillatory in x and satisfies the fast mode dispersion relation,

$$\omega^2 = (k_x^2 + k_y^2 + k_z^2)A^2 > (k_y^2+k_z^2)A^2$$

and a transverse (Alfven) wave mode (magnetic perturbation polarized in the y direction) that is non oscillatory in x and satisfies $\omega^2 = k_z^2A^2 < (k_y^2+k_z^2)A^2$. Southwood and Kivelson [1989] show that the transverse wave produced by a boundary excitation decays inward, i.e., it has a surface wave structure. The presence of this wave is of considerable importance because in a uniform system only Alfven waves carry field aligned current that can couple energy into the ionosphere and produce ground signatures. In the uniform cavity, the field aligned currents flow only on the boundary across which a shear in field and flow occur.

For a monochromatic excitation of the inhomogeneous cavity, the transverse mode can occur not only on the boundary but also on isolated (resonant) field lines where the transverse mode dispersion relation [$\omega^2 = k_z^2A^2$] is satisfied. Pure fast mode waves cannot occur if their frequencies match the field line resonant frequency somewhere in the cavity. If the frequencies match at some field line, the two fundamental wave modes are coupled [Kivelson and Southwood, 1985, 1986; Allan et al., 1985 a and b, 1986]. In the vicinity of resonant field lines, the coupled waves are similar to shear Alfven waves. In particular, field aligned currents may be particularly strong in the vicinity of resonant field lines within the cavity and the wave energy can penetrate the ionosphere. Away from the boundary and the resonant field line, the coupled wave characteristics are similar to those of compressional waves and the ionosphere acts as an almost perfect reflector [Kivelson and Southwood, 1988].

Elsewhere we have described various types of perturbations that may be interesting [Southwood and Kivelson, 1989b]. Here we limit our attention to some common types of magnetopause disturbances produced when the pressure in the solar wind changes for a limited time interval and then either levels off at the new level or returns to its original level. The displacement of the magnetopause is then of limited spatial extent with a spatial scale size that we shall designate as L. The boundary perturbation occurs first near the subsolar point and sweeps around the magnetopause with a velocity U corresponding to the magnetosheath flow velocity. The characteristic time scale of the pulse at the boundary is evidently

$$T \sim L/U$$

and the scale length of the perturbation is important in determining how the magnetosphere responds. It is evident that in a realistic magnetosphere the perturbation travels along the boundary in directions both along and across the field. For the simplified treatment of this paper, we ignore the propagation of the perturbations along the field and at first restrict our consideration to a boundary displacement of the model magnetosphere produced by a pressure change of the form:

$$\delta p \sim b_z(a,y,z,t)$$
$$\sim [H(y-Ut) - H(y+\pi/k_y-Ut)] \sin k_y(y-Ut) \exp(ik_z z - i\omega t) \qquad (3)$$

If this form is thought of as one term in a spatial Fourier decomposition of a more general perturbation, the assumption is seen not to be very restrictive. Here H is the Heaviside step function and its form ensures that the pressure perturbation is everywhere negative. We shall assume that the scale length of the ripple, k_y^{-1}, is small compared to the system dimension but the scale along the field is relatively large (i.e., $k_y^{-1} \ll k_z^{-1}$). (The situation envisaged is thus more appropriate for modeling the manner in which a ripple sweeps around the flanks rather than its motion over the polar magnetosphere.)

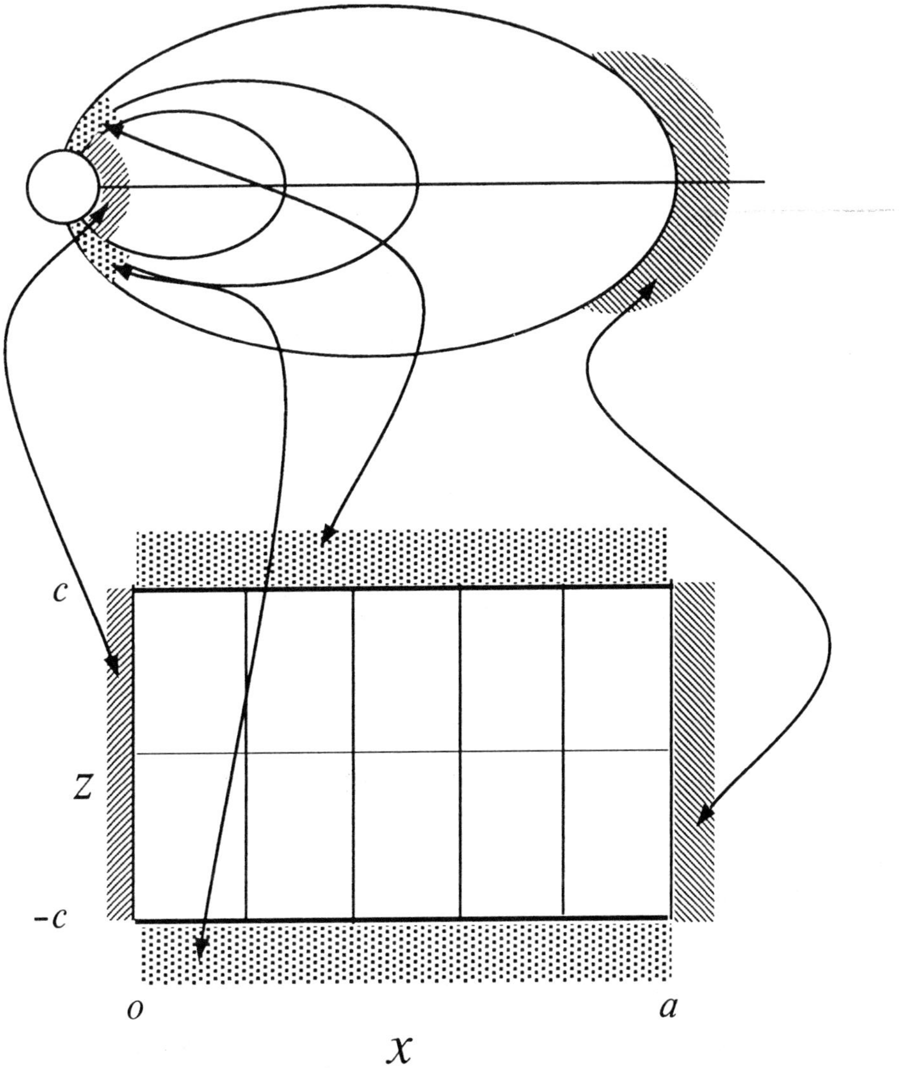

Fig. 1. Schematic of the mapping of the quasidipolar magnetosphere onto the cuboidal box used for the analysis of this paper. Boundaries corresponding to the magnetopause, the northern hemisphere and southern hemisphere ionospheres, and a low latitude near-equatorial boundary surface are illustrated. The magnetic field is in the z direction. Ionospheric boundaries are planes at z=0 and z=c. The magnetopause is a plane at x=a and a reflecting boundary is assumed at x=0. The y-direction represents the azimuthal direction in the magnetosphere. Plasma density is assumed to be either constant (the uniform model) or to vary with x (the non-uniform model).

A Laplace transform in time is convenient for revealing the response of the system to an initial perturbation. The transform of the source expression in equation (3) is

$$b_{\omega z}\ (x) = -\left[1 + e^{i\omega\pi/k_yU}\right] e^{i\omega y/U} \frac{k_yU}{\left[\omega^2 - (k_yU)^2\right]} \tag{4}$$

The denominator of (4) makes it apparent that significant power is present in the band near the central frequency, k_yU. The bulk of the power will be fed into transverse signals that evanesce away from the boundary as for the uniform case. An estimate of the amplitude in the interior of the cavity can be obtained by substituting the central frequency, $\omega = k_yU$, into the equations (2) and solving for b_z. Let us assume the corresponding solution is g(x). If the time scale of the compressional perturbation is long compared with the time for perturbations to cross the cavity in the x-direction, it follows that the compressional field within the cavity can be approximately written as

$$b_z(x,y,t) = g(x)[H(y-Ut) - H(y+\pi/k_y-Ut)]\sin(k_yy-Ut)\exp(-i\omega t) \tag{5}$$

and the frequency transform, $b_{z\omega}(x,y)$ in the cavity is g(x) times the form given in (4). The frequency transform of the associated field aligned current within the cavity is given by

$$j_{\|\omega} = \mathbf{B}\cdot(\nabla\times\mathbf{b}_\omega)/\mu_o B$$
$$= - ik_z\omega^2\,(dA^2/dx)\,(\partial b_{z\omega}/\partial y)\,(\omega^2 - A^2k_z^2)^{-2}(\mu_o)^{-1} \qquad (6)$$

As the source perturbation at the boundary varies in the y direction, i.e., in longitude, the term on the r.h.s. of equation (6) is non-zero. Thus a field aligned current flows within the magnetosphere. The field aligned currents in turn act as a source of ionospheric electric fields [see e.g. Southwood and Kivelson, 1989a].

The transform of the field aligned current [equation (6)] back to the time domain contains a pole of order one at $\omega = \pm k_yU$ and poles of order two at $\omega = k_zA(x)$ and at $\omega = - k_zA(x)$. Note that the first order poles are x-independent whereas the second order poles occur on localized shells where the resonance condition is satisfied. A pole of order two in a transform has the effect of producing an oscillatory signal at frequency $k_zA(x)$ whose amplitude on inversion (i.e. when the response in the time domain is calculated) grows secularly with time.

Evidently, the nature of the cavity response depends on whether the local Alfven resonance frequency, $k_zA(x)$, is near the central frequency of the source perturbation, k_yU. Where k_yU is well separated from $k_zA(x)$, the field-aligned current contains signatures of two sorts. In the interval immediately following the arrival of the perturbation, the field-aligned current reflects the source time structure as it did in the similar limit for the transverse perturbation of the uniform cavity described earlier. In addition, as in the uniform case, field-lines respond to the perturbation by oscillating at their resonant frequencies. Unlike the uniform case, the local resonant frequencies are x-dependent and each field line drives field-aligned current at its local resonant frequency. After the perturbation passes, only the oscillations at the local resonance frequencies persist and normally such oscillations are expected to have small amplitude. However, where $k_zA(x)$ is close to k_yU, resonance occurs and the amplitude of the resonant oscillations becomes large.

The boundary perturbation (3) requires motions both out and in. Possibly a unidirectional displacement behind a traveling front that moves the boundary to a new distance (in the x-direction) from the origin would appear to be a more elementary form of perturbation. That is not the case as both the traveling bulge and the traveling unidirectional displacement require a balanced current pair. Currents must flow in one sense to set the plasma into motion and in the opposite sense to terminate the motion. The field aligned currents that flow at different distances from the boundary in such circumstances are illustrated in Figure 2. In the figure, x_{res} represents the resonant magnetic shell.

Evidently, $j_{\|} = 0$ for $t < t_1 = y/U$, i.e., for times before the perturbation has moved across the boundary to the y location of interest. For $t > y/U$, a field-aligned current proportional to the derivative of the boundary displacement (i.e., positive, then negative, and then zero) develops and begins to propagate into the cavity. Within the cavity, field lines begin to oscillate at the local resonant frequency $k_zA(x)$. During the initial interval, the signal within the cavity is in phase with the driving signal. Near the field lines where $k_zA(x)$ is approximately k_yU, the amplitude is large. There the amplitude growth is quadratic in time while the ripple passes. Thereafter the vibrations at the local resonant frequencies continue until damped by whatever dissipation process is dominant in the system.

As noted previously, the field aligned currents couple into the ionosphere where they drive flows and magnetic perturbations that can be observed on the ground. In making connection between the model and the real magnetosphere, one needs to recall that x is analogous to radial distance in the magnetosphere and to latitude on the surface. The schematic perturbations of Figure 2 would produce signatures in the transverse magnetic field recorded at different latitudes by a chain of ground-based observatories. For a solution of the form (5), $j_{\|}$ and $\pm b_{y\omega}$ are in phase with each other and $\pm b_{z\omega}$ is in quadrature; $b_{\omega y}$ changes sign across the resonance and the oscillatory signal should be largest and longest lived near the location of the resonance. Freeman et al. [in this volume] and others [Farrugia et al., 1989] have reported magnetic perturbations following a solar wind pressure pulse that show many of the features described above.

Stations near the polar cap boundary map to the dayside magnetopause where, as noted above, the propagating disturbance sets up field aligned currents. Thus a traveling pressure pulse on the magnetopause would produce impulsive traveling magnetic perturbations on the polar cap boundary that could be hard to distinguish from the signatures of flux transfer events on the magnetopause [Goertz et al., 1985; Lanzerotti et al., 1986, 1987; Lanzerotti and Maclennan, 1988].

As previously noted, the traveling pressure pulses initiate vortical motion of the plasma just inside the cavity. The patterns would be those naturally associated with the current dipole corresponding to the balanced currents into and out of the ionosphere. The ionospheric flows corresponding to the elementary field aligned current dipole are illustrated schematically in Figure 3. Recalling that the ionospheric signature is mapped quite directly from the magnetopause boundary, one can see that the center of the pertubation flow corresponds to the inward or outward (radial) displacement that initiates the disturbance. As the ionosphere is close to incompressible, the surrounding fluid must be oppositely displaced by continuity. The perturbation moves azimuthally at the velocity U of the boundary plasma mapped into the polar cap boundary, setting up a rapidly moving vortex. Such traveling vortices on the polar cap boundary have been reported by Friis-Christensen [1988] and by McHenry et al. [1989]. The fluid signature is very similar to what Southwood [1985, 1987] anticipated as the ionospheric signature of a flux transfer event.

Summary

Using a simplified model of the magnetosphere, we have described how a compressional signal on the magnetopause can drive a signal with incompressible components elsewhere in the magnetosphere. The incompressible flow has field-aligned currents associated with it and these couple into the ionosphere to produce flows there. The ionospheric fields and flows determine the response observed on the ground.

The temporal variations within the cavity often will mimic the temporal variations of the source but they may be out of phase.

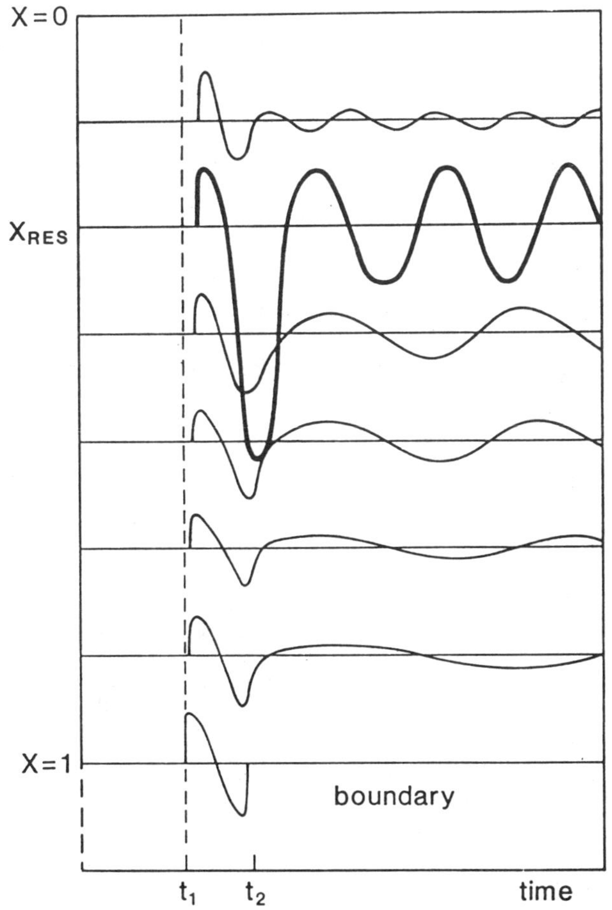

Fig. 2. Schematic of the temporal variation of the field aligned current at fixed y and different positions in x for a pressure pulse traveling at velocity U along the boundary. x_{res} is the location where $k_yU = k_zA(x)$.

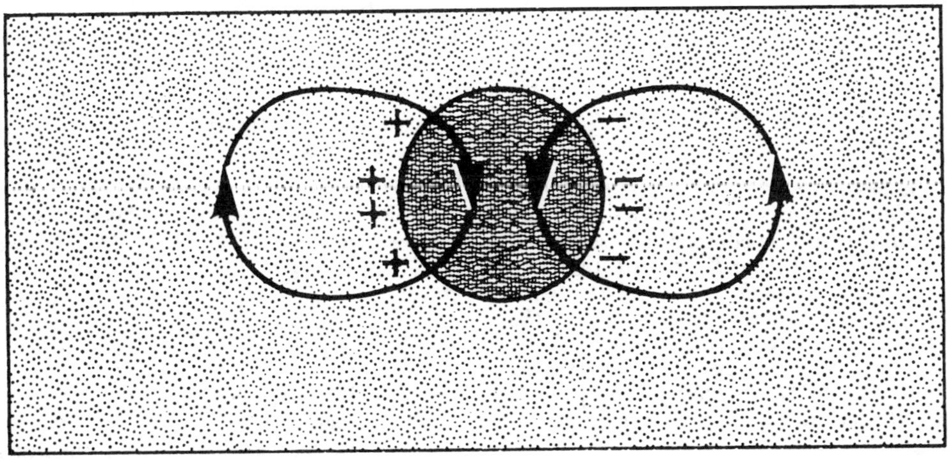

Fig. 3. Schematic of the flows (curves with arrows) in the ionosphere corresponding to the antiparallel currents driven near the boundary by a perturbation as in Figure 2. For pulses of the types described in this paper, the velocity in the center of the pattern is along the normal to the boundary. The pattern is at rest in a system moving at velocity U in the y-direction which is perpendicular to the flow in the center of the pattern. Plus and minus signs represent currents into and out of the ionosphere.

Relatively large amplitude persistent oscillations can be initiated by a traveling pulse on the boundary at interior locations where the dominant frequencies of the source spectrum match the local field line resonant frequency.

Finally, both the ground magnetic perturbations and the flows set up in the ionosphere can correspond to the expected signatures of flux transfer events.

Questions and Answers

Elphic: Your simple magnetopause dent produces a twin vortex in the ionosphere. Friis-Christensen and McHenry also observe single vortex signatures. What could produce that?

Friis-Christensen: To supplement Elphic's question: The basic mode seen in the ground magnetic response is a twin vortex, but often the trailing vortex is distorted so that the total picture looks more like a single vortex.

Kivelson: No further comment.

Drake: What is the perpendicular scale size of your current? You may have coupling to the kinetic Alfven wave if $\rho_L \nabla_\perp$ is not too small and this might lead to outward propagation of the wave energy.

Kivelson: The perpendicular scale size of the disturbance is determined by the duration of the pressure pulse at the boundary. We consider the typical time scales of solar wind disturbances to be sufficiently long that finite Larmor radius effects are irrelevant. This assumption is consistent with the scale sizes of greater than 1 R_E inferred from observations of ionospheric vortices and of FTE-like signatures on the magnetopause.

Benbrook: Margaret, would you care to speculate on what the E - field beneath the ionosphere will look like for your simple sinusoid Δp?

Kivelson: Yes. For a simple pressure step, the electric field is everywhere perpendicular to the flow lines illustrated in Figure 3. If you think of the positive and negative charges in the figure as the source of the perturbation electric field, you will have a general idea of what the two dimensional structure looks like but the current sources will be more distributed. The whole structure moves azimuthally (or left-right) in the diagram. For a sinusoidal impulse, the ionospheric signature would be composed of a pair of these dipolar patterns, with leading and trailing currents reversed, aligned roughly azimuthally. As the intensity of the ground signature depends on the current densities, the relative amplitudes of the ionospheric fields corresponding to upward and downward currents would be affected by the precise form of the temporal variation of the pressure pulse.

Acknowledgements. This work was supported by the Division of Atmospheric Sciences of the National Science Foundation under grant ATM 86–10858.

References

Allan, W., S. P. White, and E. M. Poulter, Magnetospheric coupling of hydromagnetic waves–Initial results, *Geophys. Res. Lett., 12,* 287, 1985.

Allan, W., S. P. White, and E. M. Poulter, Impulse-excited hydromagnetic cavity and field line resonances in the magnetosphere, *Planet. Space Sci., 34,* 371, 1986a.

Allan, W., S. P. White, and E. M. Poulter, Hydromagnetic wave coupling in the magnetosphere-plasmapause effects on impulse-excited resonances, *Planet. Space Sci., 34,* 371, 1986b.

Bering, E. A., III, J. R. Benbrook, G. J. Byrne, B. Liao, J. R. Theall, L. J. Lanzerotti, C. G. Maclennan, A. Wolfe, and G. L Siscoe, Impulsive electric and magnetic field perturbations observed over south pole: Flux transfer events?, *Geophys. Res. Lett., 15,* 1545, 1988.

Elphic, R. C., Multipoint observations of the magnetopause: Results from ISEE and Ampte, in *Multipoint Magnetospheric Measurements [Adv. Space Res., Volume 8 (#9–10)].*, (edited by C. T. Russell), p. 223, Pergamon, N.Y., 1988.

Farrugia, C. J., M. P. Freeman, S. W. H. Cowley, D. J. Southwood, M. Lockwood, and A. Etemadi, Pressure-driven magnetopause motions and attendant response on the ground, *Planet. Space Sci., 37,* 589, 1989.

Friis-Christensen, E., M. A. McHenry, C. R. Clauer, S. Vennerstrom, Ionospheric traveling convection vortices observed near the polar cleft: a triggered response to sudden changes in the solar wind, *Geophys. Res. Lett., 15,* 253, 1988.

Goertz, C. K., E. Nielsen, A. Korth, K. H. Glassmeier, C. Haldoupis, P. Hoeg, and D. Hayward, Observations of a possible ground signature of flux transfer events, *J. Geophys. Res., 90,* 4069, 1985.

Kivelson, M. G., and D. J. Southwood, Resonant ULF waves: A new interpretation, *Geophys. Res. Lett., 12,* 49, 1985.

Kivelson, M. G. and D. J. Southwood, Coupling of global magnetospheric MHD eigenmodes to field line resonances, *J. Geophys. Res., 91,* 4345, 1986.

Kivelson, M. G., and D. J. Southwood, Hydromagnetic waves and the ionosphere, *Geophys. Res. Lett., 15,*1271, 1988.

Lanzerotti, L.J., and C. G. Maclennan, Hydromagnetic waves associated with possible flux transfer events, *Astrophys. Space Sci., 144,* 279, 1988.

Lanzerotti, L.J., L.C. Lee, C.G. Maclennan, A. Wolfe, and L.V. Medford, Possible evidence of flux transfer events in the polar ionosphere, *Geophys. Res. Lett., 13,* 1089, 1986.

Lanzerotti, L.J., R.D. Hunsucker, D. Rice, L.C. Lee, A. Wolfe, C.G. Maclennan, and L.V. Medford, Ionosphere and ground-based response to field-aligned currents near the magnetospheric cusp regions, *J. Geophys. Res., 92,* 7739, 1987.

Lee, L.-C., and Z. F. Fu, A theory of magnetic flux transfer at the earth's magnetopause, *Geophys. Res. Lett., 12,* 105, 1985.

McHenry, M. A., and C. R. Clauer, Modeled ground magnetic signatures of flux transfer events, *J.Geophys. Res., 92,* 11231, 1987.

McHenry, M. A., C. R. Clauer, E. Friis-Christensen, J. D. Kelly, Observations of ionospheric convection vortices: Signatures of momentum transfer, in *Multipoint Magnetospheric Measurements [Adv. Space Res., Volume 8 (#9–10)].*, (edited by C. T. Russell), p. 315, Pergamon, N.Y., 1988.

Russell, C. T., and R. C. Elphic, Initial ISEE magnetometer results: magnetopause observations, *Space Science Reviews, 22,* 681, 1978.

Russell, C. T., and R. C. Elphic, ISEE observations of flux transfer events at the dayside magnetopause, *Geophys. Res. Lett., 6,* 33, 1979.

Scholer, M., Magnetic flux transfer at the magnetopause based on single X line bursty reconnection, *Geophys. Res. Lett., 15,* 291, 1988.

Sibeck D. G., W. Baumjohann, R. C. Elphic, D. H. Fairfield, J. F. Fennell, W. B. Gail, L. J. Lanzerotti, R. E. Lopez, H. Luhr, A. T. Y. Lui, C. G. Maclennan, R. W. McEntire, T. A. Potemra, T. J. Rosenberg, and K. Takahashi, The magnetospheric response to 8–minute period strong-amplitude upstream pressure variations, *J. Geophys. Res., 94,* 2505, 1989.

Southwood, D. J., Theoretical aspects of ionosphere -magnetosphere - solar wind coupling, in *Physics of Ionosphere - Magnetosphere, Adv. Space Res., 5,* 7, 1985.

Southwood, D.J., The ionospheric signature of flux transfer events, *J. Geophys. Res, 92,* 3207-3213, 1987.

Southwood, D. J., C. J. Farrugia, and M. A. Saunders, What are flux transfer events?, *Planet. Space Sci., 36* 503, 1988.

Southwood, D. J., and M. G. Kivelson, Magnetospheric interchange motions, *J. Geophys. Res., 94,* 299, 1989a.

Southwood, D. J., and M. G. Kivelson, The magnetohydrodynamic response of the magnetospheric cavity to changes in solar wind pressure, *J. Geophys. Res.*, in press, 1989b.

SUBSTORMS AND FLUX ROPE STRUCTURES

Wolfgang Baumjohann and Gerhard Haerendel

Max-Planck-Institut für extraterrestrische Physik, D-8046 Garching, FRG

Abstract. The magnetospheric substorm comprises directly driven dissipation as well as sudden unloading of energy. The global behavior of the geomagnetic tail during the substorm expansion phase can be understood in the framework of a revised near-Earth neutral line model. However, none of the presently available substorm models yet includes a coherent interpretation of some of the smaller scale features observed during substorms, namely current tubes observed in the auroral ionosphere and geostationary orbit as well as plasmoids and flux ropes detected in the near and distant tail.

1. Introduction

The term magnetospheric substorm comprises many different phenomena and processes as well as numerous models trying to explain these features or at least certain facets. It is virtually impossible to describe all these phenomena and models in a single review. Thus we shall focus on three points.

First, we shall define the magnetospheric substorm as a process which comprises both directly driven dissipation as well as storage and subsequent unloading of enhanced solar wind energy input.

Second, we shall describe the global behavior of the geomagnetic tail during the substorm expansion phase and its interpretation in the most-accepted and developed substorm model, the near-Earth neutral line model. The description of this model will include some recent modifications needed to explain the tail observations between 10 and 20 R_E.

Finally, we shall discuss some of the smaller scale features observed during substorms, namely those which resemble flux rope-like structures. These include current tubes observed in the auroral ionosphere and geostationary orbit as well as plasmoids and flux ropes detected in the near and distant tail.

The reader interested in a broader coverage of all substorm phenomena is referred to some recent reviews by Baker et al. [1985] Baumjohann et al. [1986,1988] and Rostoker et al. [1987]. A description of some other concurrent substorm models can be found in, for example, Roux [1985], Smith et al. [1986], Rostoker and Eastman [1987], and Rothwell et al. [1988].

Geophysical Monograph 58

2. Definitions

The Earth's magnetosphere is a cavity filled with hot dilute plasma embedded in the fast-flowing denser but colder solar wind plasma. Due to a couple of transfer processes operating in the vicinity of the dayside magnetopause, a small fraction of the kinetic energy of the solar wind impinging on the magnetosphere (about 2–5%) can enter this cavity [e.g., Baumjohann and Paschmann, 1987]. Since magnetic reconnection is the process which accounts for most, if not all, transfer of energy, momentum, and mass, the transfer rate depends on the polarity of the interplanetary north-south magnetic field component (IMF B_z) and is thus time-dependent.

The enhanced energy input must be compensated by enhanced dissipation somewhere in the magnetosphere-ionosphere system. While dissipation may occur in a quasi-steady manner when the magnetopause energy transfer rate is low (especially during periods of northward IMF B_z), the magnetosphere must react nonstationary when the energy input is high. The magnetospheric substorm comprises all phenomena by which the magnetosphere tries to adjust itself to enhanced solar wind input (during periods of southward IMF B_z).

Nowadays widespread agreement exists that the magnetospheric substorm comprises two basic concurrent processes, the driven process and the loading-unloading process [Baker et al., 1984; Rostoker et al., 1987]. As schematically displayed in Figure 1, some part of the enhanced solar wind energy input is directly dissipated by means of global convection, leading to Joule heating of the auroral ionospheres by the enhanced convection currents and deposition of particle energy in the auroral ionospheres as well as in the ring current. This energy dissipation is directly correlated with the solar wind energy input and constitutes a driven process, i.e., directly driven by the solar wind [e.g., Akasofu, 1981]. The remainder of the enhanced energy input is, at the same

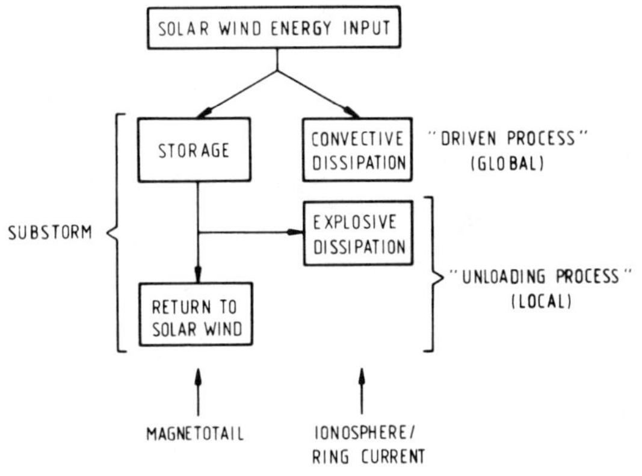

Fig. 1. Schematic illustration summarizing the storage and dissipation of solar wind energy by both driven and loading-unloading process during a magnetospheric substorm [adapted from Baker et al., 1984].

time, stored intermediately in the Earth's magnetotail and then, at substorm onset, it is rather explosively released via Joule and particle heating of localized regions of the auroral ionospheres, injection of particles into the ring current, and return to the solar wind by the downtail release of plasmoids. This second substorm process, which operates concurrently to the driven process, has been named loading-unloading process.

Figure 2 gives a schematic illustration on the concurrent operation of the driven and the loading-unloading process during and after a period of enhanced solar wind

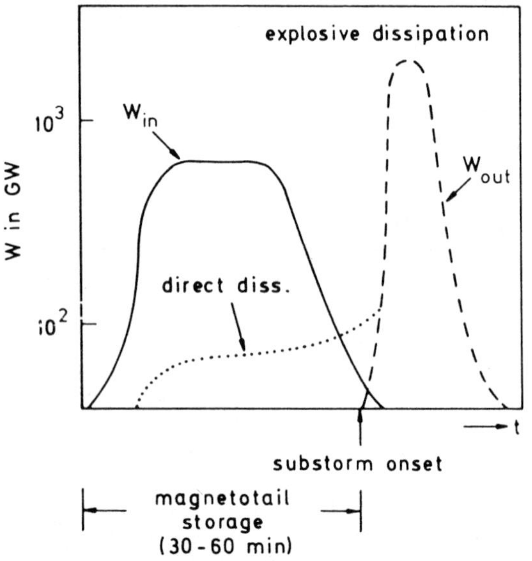

Fig. 2. Schematic diagram describing the directly driven dissipation and the concurrent magnetotail storage and explosive dissipation (loading-unloading) during a magnetospheric substorm [after Baker et al., 1985].

input [Baker et al., 1985]. Soon after the energy coupling between solar wind and magnetosphere is enhanced due to a southward IMF B_z component, energy is stored in the magnetotail. Concurrently, with a time delay of about 10–20 min due to the inductance of the system, energy is directly dissipated in the auroral ionosphere in the form of Joule heat. The stored energy is then, typically after about 30–60 min, explosively released during the so-called expansion phase, often but not necessarily after reducing the solar wind input.

It should, however, be noted that in contrast to the impression one gets from Figure 2, direct and explosive dissipation of energy may coexist during the expansion phase. Furthermore, the question which of these two processes dominates, i.e., dissipates more energy during a substorm, cannot be answered easily. In the case shown in Figure 2 explosive dissipation is clearly dominant, but other cases have been observed in which the dominant energy dissipation had to be attributed to the driven process [e.g., Pellinen et al., 1982].

3. Key Features of the Unloading Process

The key features of the unloading phase are most clear-cut in near-Earth observations (auroral ionosphere and tail magnetosphere up to distances of about $10\,R_E$) and in the distant tail measurements by the ISEE-3 satellite (beyond $100\,R_E$). Inbetween, especially around 20–$40\,R_E$, the observations are much more ambiguous and debated, even though, and perhaps because, this is the most likely region where the unloading is triggered.

The dynamic behaviour of the near-Earth tail during the expansion phase is governed by the substorm current wedge. During the growth or loading phase, energy which cannot be directly dissipated via enhanced convection in the auroral electrojets is stored in the tail magnetosphere, mainly in the form of magnetic energy by enhancing the neutral sheet-tail current circuit [e.g., McPherron, 1979]. The storage is accompanied by a distortion of the nightside magnetospheric field lines into a strongly tail-like configuration, even at radial distances where the field is usually rather dipolar, e.g., at geostationary distances of $6.6\,R_E$. According to a popular view sketched in Figure 3 [e.g., McPherron, 1979], the magnetic tension is suddenly released, after a 30–60 min phase of storage, at expansion phase onset by "disrupting" the neutral sheet current in a azimuthally limited region of the tail and diverting its current along magnetic field lines and through the midnight sector auroral ionosphere.

The current "disruption" has rather drastic effects in the near-Earth tail and auroral ionosphere. In the ionosphere, it is accompanied by the break-up of aurora and the formation of a westward travelling surge. It is inside this region covered by active aurora where the ionospheric part of the current wedge is closed by a strong westward Cowling current [e.g., Baumjohann et al., 1981]. Satellites in the near-Earth tail observe the sudden dipolarization of the stretched magnetic field, first in the midnight

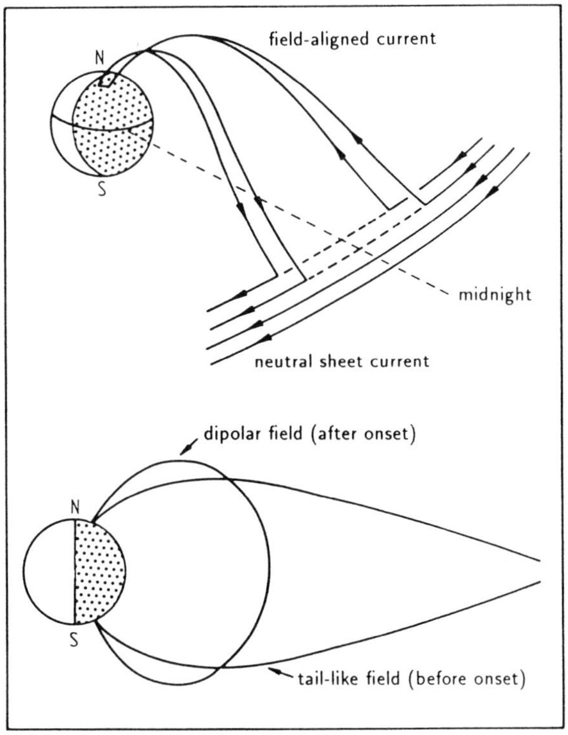

Fig. 3. Upper panel: the substorm current wedge short-circuits the neutral sheet current in the magnetotail. Lower panel: reconfiguration of the nightside magnetic field associated with the disruption of the cross-tail current [after Baumjohann and Glaßmeier, 1984].

sector and later also in the pre- and postmidnight quadrants [e.g., Nagai, 1982]. The dipolarization of the magnetic field is accompanied by (dispersionless) injection of hot plasma into the near-Earth tail [e.g., McIlwain and Whipple, 1986].

In the deep tail plasma sheet, ISEE-3 observations have unambiguously identified tailward travelling closed magnetic structures, the so-called plasmoids [e.g., Hones et al., 1984; Scholer et al., 1984a], whose appearance is associated with substorms (cf. Figure 4). Evidence for the closedness of the structure comes from the observed B_z polarity change from northward to southward during the passage of the plasmoid as well as the detection of isotropic, i.e., trapped, hot electrons in its interior. The plasmoids have a typical length of 70–80 R_E [Scholer et al., 1984b]. They are ejected downtail with velocities of the order of 600 km/s.

The typical total kinetic, thermal, and magnetic energy contained in a plasmoid was estimated to be of the order of 10^{15} J [Scholer et al., 1984b]. This is about the same amount of energy as typically dissipated in the auroral ionosphere during the course of an average substorm [Baumjohann, 1986]. Since the average amount of substorm energy deposited in the ring current is also of the same order [Baumjohann and Kamide, 1984], one may say that the energy released in a substorm is roughly equipartitioned between the three different deposition regions, namely auroral ionosphere, ring current, and downtail solar wind.

4. The Near-Earth Neutral Line Model

In the most-accepted substorm model, the so-called near-Earth neutral line model, the unloading of tail energy is initiated by the formation of a near-Earth neutral line, where magnetic reconnection converts the stored magnetotail energy into particle heating. The most likely instability which leads to the onset of near-Earth reconnection is the ion tearing mode. Basically, the ion tearing mode goes unstable if the ions cannot preserve their magnetic moment when crossing the magnetic neutral sheet.

The near-Earth neutral line model is the only substorm model that explains the formation and release of the plasmoids observed by ISEE-3. As sketched in Figure 4, the plasmoids are created at substorm onset between the newly formed near-Earth and the presumably always existing distant neutral line. Due to the slingshot effect (magnetic tension) these plasmoids are then ejected downtail and engulf the ISEE-3 satellite at about 220 R_E downtail some 30 min after substorm onset.

Concerning the near-Earth tail and auroral ionosphere substorm features, the actual mechanism leading to the the formation of the substorm current wedge is still a matter of debate. But it may be explained by the near-Earth reconnection scenario if the third dimension is taken into account [see, e.g., Baumjohann, 1988]: three-dimensional computer simulations have shown that when one limits the azimuthal extent of the near-Earth neutral line, upward and downward field-aligned currents at the dusk- and dawnside edge are a natural consequence [Birn and Hones, 1981; Sato et al., 1983].

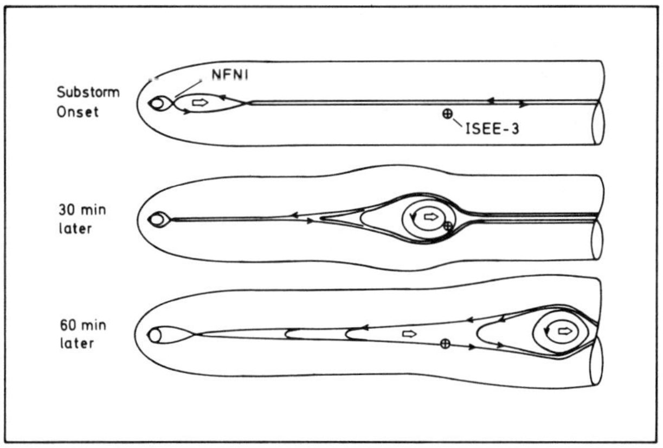

Fig. 4. Schematic diagram showing the tailward release of a plasmoid [after Hones et al., 1984].

However, observations at tail distances of 10–30 R_E revealed some problems with the classical near-Earth neutral line model described by, for example, Hones [1979]. Among others, Rostoker and Eastman [1987] pointed out that the classical near-Earth neutral line model did not concur with the observations in two key points. First, tailward ion flows are rarely observed in the near-Earth tail, in contrast to what should be expected from near-Earth neutral line formation between 10 and 20 R_E. Second, the classical model did not account for the existence of the plasma sheet boundary layer and the high-speed flows observed in this layer.

The first objection is supported by Figure 5, which is based on the IRM tail survey data set described by Baumjohann et al. [1988,1989] and Baumjohann and Paschmann [1989]. From all high speed ion flows with a velocity in excess of 300 km/s observed by the IRM satellite during the two 4-month periods it spent in the magnetotail at distances between 9 and 19 R_E (more than 9000 spin-averaged 4.5 s samples), the overwhelming majority was directed earthward. Less than 4% had a distinct tailward component. For bulk speeds in excess of 600 km/s, virtually no ion flow sample had a tailward component.

Since the ion bulk flow should be tailward directed at distances tailward of a neutral line, these observations and similar results by Cattell and Mozer [1984] indicate that the near-Earth neutral line is rarely, if ever, located inside 20 R_E, in contrast to what was proposed in the classical near-Earth neutral line model. However, if one assumes that the neutral line is formed rather at 30–40 R_E [Nishida et al., 1988], these observations do not pose any problem.

The second objection is also supported by data, even though not in its extreme position as voiced by Huang and Frank [1986] who argued that high speed flows can only be observed in the plasma sheet boundary layer. Figure 6 shows the occurrence rates of high speed flows in three different layers of the plasma sheet, again using the IRM tail survey data. From all spin-averaged ion flow samples taken in in the plasma sheet boundary layer (more than 73,000) about 6.6% exhibit velocities in excess of 300 km/s. This rate drops to 1.5% out of 148,000 samples in the outer central plasma sheet, but rises again, to slightly more than 4% of the 52,000 samples obtained in the neighborhood of the neutral sheet. Thus there is a 3:1:4 chance to detect ion flows with velocities greater 300 km/s when going from the neutral sheet to the boundary layer.

For even higher velocities, the ratio between the occurrence rates in the neutral sheet neighborhood and the plasma sheet boundary layer stays at about 1:1.5, but the chance to detect high speed flows in the outer central plasma sheet drops. For example, the relative chance to detect samples with $V_i > 600$ km/s in the three different layers is about 6:1:8. Thus high speed flows occur preferentially near the neutral sheet and in the boundary layer, but not so much in the remainder of the central plasma sheet.

But again, also the second objection is only valid against an overly simplified model. New simulations show that if realistic boundary conditions are taken into account, a plasma sheet boundary layer with earthward field-aligned ion flow is a natural result of reconnection at a distant or near-Earth neutral line. The formation of the boundary layer results from a diversion of (part

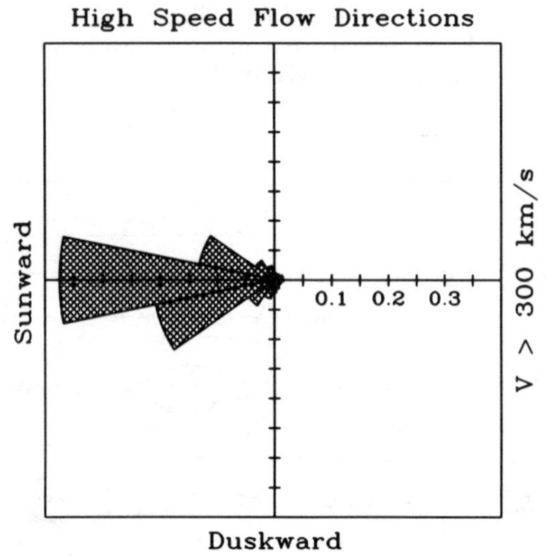

Fig. 5. Angular distribution of ion velocity vectors (in the X_{GSM}–Y_{GSM} plane) of more than 9000 high speed flow samples ($V_i > 300$ km/s) observed in the plasma sheet [after Baumjohann et al., 1988,1989].

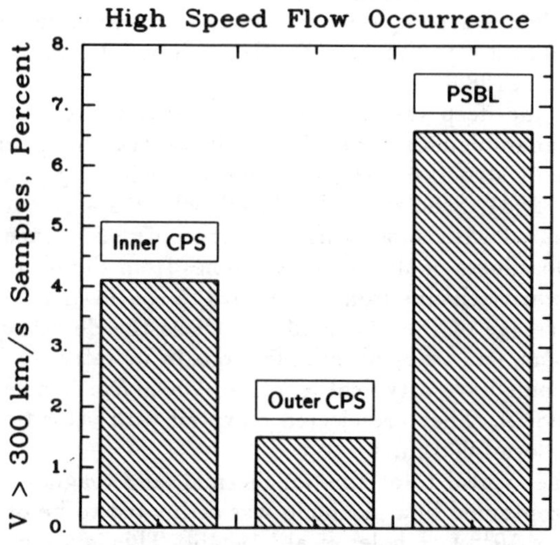

Fig. 6. Occurrence rates of high speed flows in the inner and outer central plasma sheet and the plasma sheet boundary layer (based on the tail survey data set of Baumjohann and Paschmann [1989]).

of) the earthward flow in the neutral sheet due to either one of the following three 'obstacles': ionospheric line-tying [Forbes, 1988], increasing magnetic field strength [Schindler and Birn, 1987], or high ion density in the central plasma sheet [Scholer, 1987] (the latter two may be closely related).

Hence, in the framework of an revised near-Earth neutral line model with a neutral line around 30–40 R_E and boundary layer formation, as proposed by Baumjohann [1988] and Lyons and Nishida [1988] and sketched in Figure 7, the aforementioned observations in the near and distant tail can be explained.

However, even the revised near-Earth neutral line model still does not incorporate a couple of important features. For example, the model does not yet explain the strong heating of the plasma sheet ions during substorms [Huang and Frank, 1986; Baumjohann et al., 1989] and does not include a mechanism to trigger substorm onsets by changes in the interplanetary medium [cf., Baumjohann, 1986]. Furthermore, the near-Earth neutral line model includes the substorm effects in the inner magnetosphere (injection and dipolarization) and, most important, the auroral ionosphere and its feedback to the tail processes only in a rather sketchy way. Some of these features are described in competing substorm models [e.g., Smith et al., 1986; Roux, 1985; Rothwell et al., 1988] Perhaps these models or parts thereof together with the neutral line model can serve as building blocks for a more comprehensive unified substorm model.

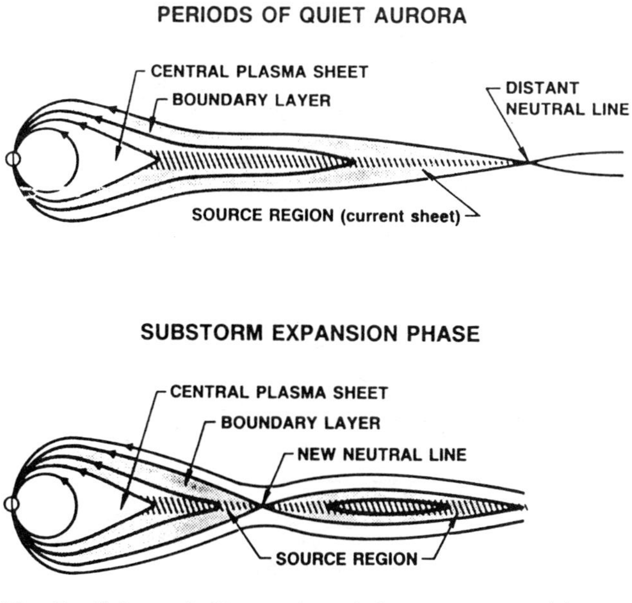

Fig. 7. Schematic illustration of the magnetotail (a) for quiet periods and (b) during the onset of a substorm expansion phase [adapted from Lyons and Nishida, 1988].

5. Flux Rope Structures in Substorms

Besides the global features of a magnetospheric substorm outlined above, this process includes also a number of smaller scale features. Among those are circular current tubes with small radii as well as flux rope-like structures most likely associated with magnetic reconnection.

The earliest evidence for the existence of circularly shaped tubes of intense upward field-aligned current came from the work of Baumjohann et al. [1981], Inhester et al. [1981] and Opgenoorth et al. [1983]. These authors combined ground magnetometer array measurements with radar measurements of the ionospheric electric field in order to model the electrodynamic structure of the auroral ionosphere in the vicinity of the westward traveling surge. As shown in Figure 8 for the case studied by Opgenoorth et al. [1983], the upward field-aligned current at the western edge of the active aurora region (in the head of the westward traveling surge) is fed by the substorm electrojet and thus constitutes the western part of the substorm current wedge. In all cases this current is concentrated in a circular tube and very intense.

The upward current tubes have typical diameters of 100 km, current densities of 5–10 $\mu A/m^2$ and move westward along with the surge with typical velocities of several kilometers per second. Since the diameters found are at the lower limit of the spatial resolution of the measurements used by these authors, the actual diameter may well be smaller and thus the current density higher. Perhaps the parameters of the low-altitude current tubes are similar to those of the even smaller current tubes with higher current densities found during substorms at geostationary orbit.

Robert et al. [1984] reported field-aligned current tubes observed during substorm onsets with typical diameters of 400 km and densities of 0.08 $\mu A/m^2$ which pass by the GEOS-2 spacecraft with azimuthal velocities of

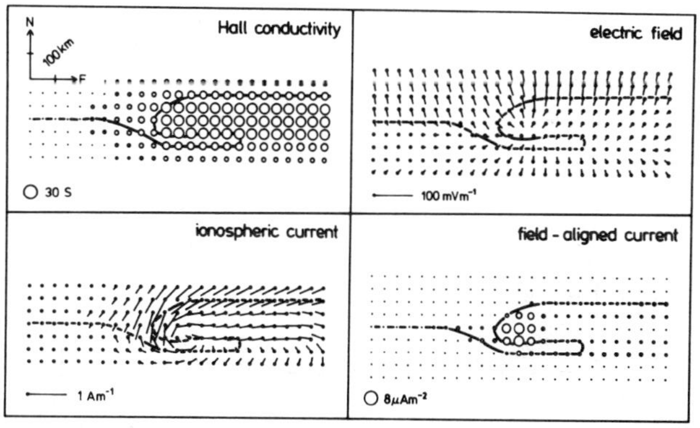

Fig. 8. Spatial distribution of conductance, electric field, and currents around a westward traveling surge [after Opgenoorth et al., 1983].

70 km/s (see Figure 9). Mapped into the auroral ionosphere, these current tubes have an average diameter of 20 km, drift with a velocity of about 4.5 km/s, and carry a current with a density of 40 μA/m^2. Note that these values for the current density extremely high and hardly ever observed or deduced for ionospheric heights.

There is no proof yet that the low-altitude and geostationary current tubes are the same animal, but because of the similarity of their features it seems likely. Anyhow, their internal structure should be similar, since the current densities are large enough to produce a substantial spiraling of the magnetic field and thus the current itself [Kaufmann and Larson, 1989]. Hence the internal structure of these upward field-aligned current tubes should resemble a flux rope. The maximum current densities observed at geostationary orbit come close to the Kruskal-Schwartzschild stability limit, which is defined as the maximum current that can flow along a field line before it becomes disrupted by its own helical motion. At geostationary orbit, this current density limit amounts to about 0.1 μA/m^2 [Robert et al., 1984].

While the internal structure of the current tubes remains unobserved at present, there have been a few observations of flux rope structures in the geomagnetic tail, each of them associated with substorm activity. The first such observation was reported by Sibeck et al. [1984]. As summarized in Figure 10, these authors recognized a particular magnetic field structure observed by the ISEE-3 satellite in the distant tail plasma sheet/lobe boundary which could be explained by assuming that the spacecraft was traversing a magnetic flux rope with a strong core field aligned with the magnetotail axis and and a helical field component at its borders. Scholer et al. [1985] inferred a minimal flux rope radius of about 3 R_E using gradients in the energetic proton fluxes.

Two years later, Elphic et al. [1986] reported the observation of two magnetic flux ropes in the vicinity of the near-Earth neutral sheet with similar scale sizes, i.e., 3–5 R_E. The near-Earth flux ropes also had a strong core field, but their axis was aligned perpendicular to the tail axis, along the dawn-dusk line. They moved rapidly across the spacecraft, with velocities of 200–600 km/s. Elphic et al. [1986] suggested that their flux rope structures were a natural consequence of a finite dawn-dusk B_y magnetic field component in the magnetic tearing mode islands. This suggestion is supported by the relaxation equilibrium calculations of Paranicas and Bhattacharjee [1989].

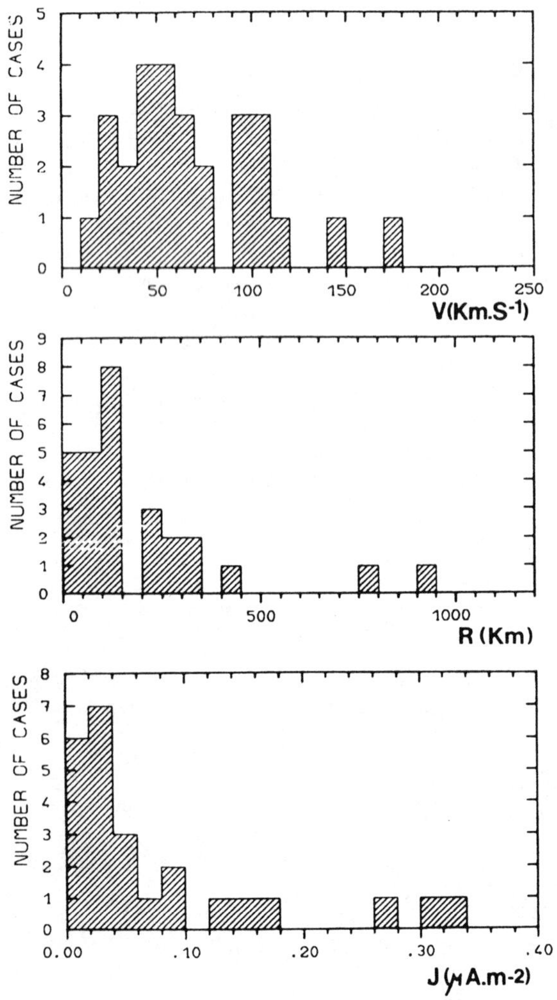

Fig. 9. Histograms of the geostationary orbit current tube characteristics [from Robert et al., 1984]. From top to bottom: velocity, radius, and current density.

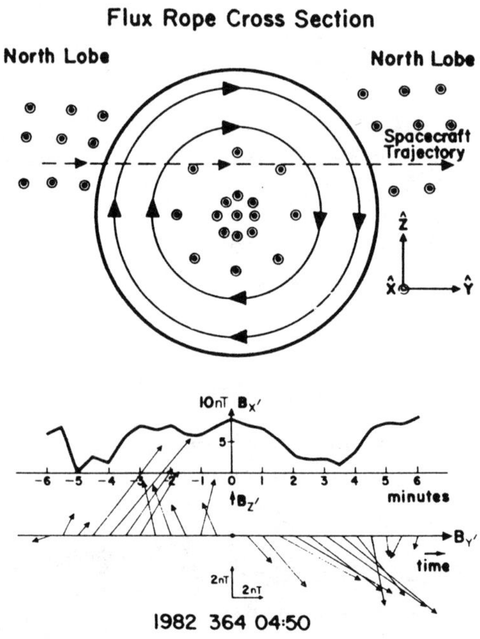

Fig. 10. Cross-section of a model flux rope embedded in the distant northern lobe and ISEE-3 magnetic field observations [from Sibeck et al., 1984].

Flux rope structures due to a finite B_y component do not only change the structure of the small-scale tearing mode islands, but also the topology of their larger-scale counterpart, the aforementioned plasmoids. Hughes and Sibeck [1987] were the first to point out that a net cross-tail magnetic field component breaks the symmetry of the two-dimensional plasmoid depicted in Figure 4 and that instead of closed loops helical field lines are produced. Their qualitative arguments were recently supported by the more quantitative calculations of Birn et al. [1989] and lead to the plasmoid release scenario depicted in Figure 11.

In its initial stage, a plasmoid is characterized by the formation of helical field lines which cross the neutral sheet typically more than once, but are still connected with the Earth. When reconnection proceeds to lobe field lines, the central plasmoid flux rope becomes enveloped by a sheath of open field lines that pull the plasmoid flux rope back toward the tail. During the gradual separation of the plasmoid helical field lines connected with the Earth are converted to open ones going tailward into interplanetary space. The actual magnetic field topology during the gradual separation of the plasmoid from the Earth is rather complicated, forming a layer of intermingled filamentary flux tubes of open, partially open and closed field lines inside the earlier mentioned sheath of open field lines.

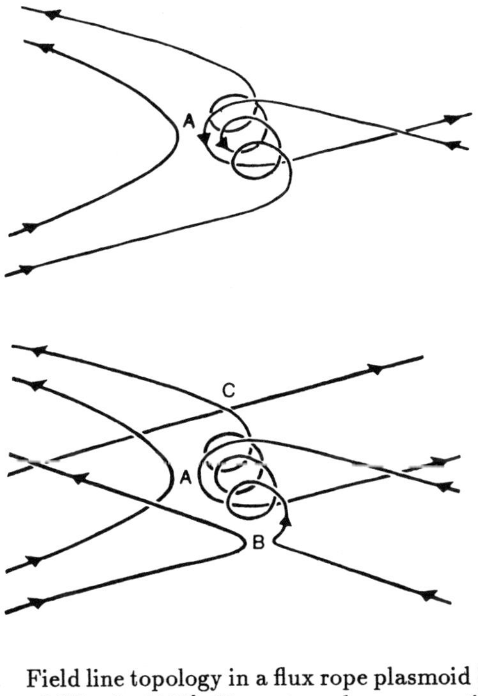

Fig. 11. Field line topology in a flux rope plasmoid [from Hughes and Sibeck, 1987]. Upper panel: reconnection of lobe field lines at A results in a new plasma sheet field line and an open field line wrapped around the closed field line flux rope. Lower panel: the flux rope plasmoid is finally released when flux rope field lines reconnect with lobe field lines near the tail flanks; this process has just happened at B and will soon happen at C.

The actual exploration of the flux rope plasmoid features is still in a very preliminary phase. For example, at present it is unclear how the substorm current wedge fits into the flux rope plasmoid picture. Fully three-dimensional simulations will be needed before the too simple symmetric plasmoid release scenario described in Section 4 can be converted into a working model of flux rope plasmoid release that can be compared with satellite observations. Due to the complicated field topology one might have to await multi-satellite missions like Cluster before an observational test of a three-dimensional model becomes possible. However, since during substorms a substantial part of the field-aligned current flows in the flux rope-like current tubes, studying these features seems essential to understand the substorm magnetosphere-ionosphere coupling, despite all possible observational difficulties.

References

Akasofu, S.-I., Energy coupling between the solar wind and the magnetosphere, *Space Sci. Rev.*, *28*, 121–190, 1981.

Baker, D. N., S.-I. Akasofu, W. Baumjohann, J. W. Bieber, D. H. Fairfield, E. W. Hones, Jr., B. Mauk, R. L. McPherron, and T. E. Moore, Substorms in the magnetosphere, in *Solar Terrestrial Physics — Present and Future*, ed. by D. M. Butler and K. Papadopoulos, chap. 8, NASA, Washington, 1984.

Baker, D. N., T. A. Fritz, R. L. McPherron, D. H. Fairfield, Y. Kamide, and W. Baumjohann, Magnetotail energy storage and release during the CDAW 6 substorm analysis intervals, *J. Geophys. Res.*, *90*, 1205–1216, 1985.

Baumjohann, W., Some recent progress in substorm studies, *J. Geomagn. Geoelec.*, *38*, 633–651, 1986.

Baumjohann, W., The plasma sheet boundary layer and magnetospheric substorms, *J. Geomagn. Geoelec.*, *40*, 157–175, 1988.

Baumjohann, W., and K.-H. Glaßmeier, The transient response mechanism and Pi2 pulsations at substorm onset: Review and outlook, *Planet. Space Sci.*, *32*, 1361–1370, 1984.

Baumjohann, W., and Y. Kamide, Hemispherical Joule heating and the AE indices, *J. Geophys. Res.*, *89*, 383–388, 1984.

Baumjohann, W., and G. Paschmann, Solar wind-magnetosphere coupling: Processes and observations, *Physica Scripta*, *T18*, 61–72, 1987.

Baumjohann, W., and G. Paschmann, Determination of the polytropic index in the plasma sheet, *Geophys. Res. Lett.*, *16*, 295–298, 1989.

Baumjohann, W., R. J. Pellinen, H. J. Opgenoorth, and E. Nielsen, Joint two-dimensional observations of ground magnetic and ionospheric electric fields associated with auroral zone currents: Current systems associated with local auroral break-ups, *Planet. Space Sci.*, *29*, 431–447, 1981.

Baumjohann, W., R. Nakamura, and G. Haerendel, Dayside equatorial-plane convection and IMF sector structure, *J. Geophys. Res.*, *91*, 4577–4560, 1986.

Baumjohann, W., G. Paschmann, N. Sckopke, C. A. Cattell, and C. W. Carlson, Average ion moments in the plasma sheet boundary layer, *J. Geophys. Res.*, *93*, 11,507–11,520, 1988.

Baumjohann, W., G. Paschmann, and C. A. Cattell, Average plasma properties in the central plasma sheet, *J. Geophys. Res.*, *94*, 6597–6606, 1989.

Birn, J., and E. W. Hones, Jr., Three-dimensional computer modeling of dynamic reconnection in the geomagnetic tail, *J. Geophys. Res.*, *86*, 6802–6808, 1981.

Birn, J., M. Hesse, and K. Schindler, Filamentary structure of a three-dimensional plasmoid, *J. Geophys. Res.*, *94*, 241–251, 1989.

Cattell, C. A., and F. S. Mozer, Substorm electric fields in the Earth's magnetotail, in *Magnetic Reconnection in Space and Laboratory Plasmas*, ed. by E. W. Hones, Jr., pp. 208–215, AGU, Washington, 1984.

Elphic, R. C., C. A. Cattell, K. Takahashi, S. J. Bame, and C. T. Russell, ISEE-1 and 2 observations of magnetic flux ropes in the magnetotail: FTE's in the plasma sheet? *Geophys. Res. Lett.*, *13*, 648–651, 1986.

Forbes, T. G., Magnetohydrodynamic boundary conditions for global models, in *Modeling Magnetospheric Plasma*, ed. by T. E. Moore and T. H Waite, pp. 319–328, AGU, Washington, 1988.

Hones, E. W., Jr., Transient phenomena in the magnetotail and their relation to substorms, *Space Sci. Rev.*, *23*, 393–410, 1979.

Hones, E. W., Jr., D. N. Baker, S. J. Bame, W. C. Feldman, J. T. Gosling, D. J. McComas, R. D. Zwickl, J. A. Slavin, E. J. Smith, and B. T. Tsurutani, Structure of the magnetotail at 220 R_E and its response to geomagnetic activity, *Geophys. Res. Lett.*, *11*, 5–7, 1984.

Huang, C. Y., and L. A. Frank, A statistical study of the central plasma sheet: Implications for substorm models, *Geophys. Res. Lett.*, *13*, 652–655, 1986.

Hughes, W. J., and D. G. Sibeck, On the 3-dimensional structure of plasmoids, *Geophys. Res. Lett.*, *14*, 636–639, 1987.

Inhester, B., W. Baumjohann, R. A. Greenwald, and E. Nielsen, Joint two-dimensional observations of ground magnetic and ionospheric electric fields associated with auroral zone currents: 3. Auroral zone currents during the passage of a westward travelling surge, *J. Geophys.*, *49*, 155–162, 1981.

Kaufmann, R. L., and D. J. Larson, Electric fields near auroral current systems, *J. Geophys. Res.*, *94*, in press, 1989.

Lyons, L. R., and A. Nishida, Description of substorms in the tail incorporating boundary layer and neutral line effects, *Geophys. Res. Lett.*, *15*, 1337-1340, 1988.

McIlwain, C. E., and E. C. Whipple, The dynamic behavior of plasmas observed near geosynchronous orbit, *IEEE Trans. Plasma Sci.*, *PS-14*, 874–890, 1986.

McPherron, R. L., Magnetospheric substorms, *Rev. Geophys. Space Phys.*, *17*, 657–681, 1979.

Nagai, T., Observed magnetic substorm signatures at synchronous altitude, *J. Geophys. Res.*, *87*, 4405–4417, 1982.

Nishida, A., S. J. Bame, D. N. Baker, G. Gloeckler, M. Scholer, E. J. Smith, T. Terasawa, and B. Tsurutani, Assessment of the boundary layer model of the magnetospheric substorm, *J. Geophys. Res.*, *93*, 5579–5588, 1988.

Opgenoorth, H. J., R. J. Pellinen, W. Baumjohann, E. Nielsen, G. Marklund, and L. Eliasson, Three-dimensional current flow and particle precipitation in a westward travelling surge (Observed during the Barium-Geos rocket experiment), *J. Geophys. Res.*, *88*, 3138–3152, 1983.

Paranicas, C., and A. Bhattacharjee, Relaxation of magnetotail plasmas with field-aligned currents, *J. Geophys. Res.*, *94*, 479–484, 1989.

Pellinen, R. J., W. Baumjohann, W. J. Heikkila, V. A. Sergeev, A. G. Yahnin, G. Marklund, and A. O. Melnikov, Event study on presubstorm phases and their relation on the energy coupling between solar wind and magnetosphere, *Planet. Space Sci.*, *30*, 371–388, 1982.

Robert, P., R. Gendrin, S. Perraut, and A. Roux, Geos 2 identification of rapidly moving current structures in the equatorial outer magnetosphere during substorms, *J. Geophys. Res.*, *89*, 819–840, 1984.

Rostoker, G., and T. E. Eastman, A boundary layer model for magnetospheric substorms, *J. Geophys. Res.*, *92*, 12,187–12,202, 1987.

Rostoker, G., S.-I. Akasofu, W. Baumjohann, Y. Kamide, and R. L. McPherron, The roles of direct input of energy from the solar wind and unloading of stored magnetotail energy in driving magnetospheric substorms, *Space Sci. Rev.*, *46*, 93–111, 1987.

Rothwell, P. L., L. P. Block, M. B. Silevitch, and C.-G. Fälthammar, A new model for substorm onsets: The pre-breakup and triggering regimes, *Geophys. Res. Lett.*, *15*, 1279–1282, 1988.

Roux, A., Generation of field-aligned current structures at substorm onset, in *Future Missions in Solar, Heliospheric & Space Plasma Physics*, ed. by E. Rolfe and B. Battrick, pp. 151–159, ESA, Noordwijk, 1985.

Sato, T., T. Hayashi, R. J. Walker, and M. Ashour-Abdalla, Neutral sheet current interuption and field-aligned current generation by three-dimensional driven reconnection, *Geophys. Res. Lett.*, *10*, 221–224, 1983.

Schindler, K., and J. Birn, On the generation of field-aligned plasma flow at the boundary of the plasma sheet, *J. Geophys. Res.*, *92*, 97–107, 1987.

Scholer, M., Earthward plasma flow during near-earth

magnetotail reconnection: Numerical simulations, *J. Geophys. Res.*, *92*, 12,425–12,431, 1987.
Scholer, M., G. Gloeckler, B. Klecker, F. M. Ipavich, D. Hovestadt, and E. J. Smith, Fast moving plasma structures in the distant magnetotail, *J. Geophys. Res.*, *89*, 6717–6727, 1984a.
Scholer, M., G. Gloeckler, D. Hovestadt, B. Klecker, and F. M. Ipavich, Characteristics of plasmoidlike structures in the distant magnetotail, *J. Geophys. Res.*, *89*, 8872–8876, 1984b.
Scholer, M., B. Klecker, D. Hovestadt, G. Gloeckler, F. M. Ipavich, and A. B. Galvin, Energetic particle characteristics of magnetotail flux ropes, *Geophys. Res. Lett.*, *12*, 191–194, 1985.
Sibeck, D. G., G. L. Siscoe, J. A. Slavin, E. J. Smith, S. J. Bame, and F. L. Scarf, Magnetotail flux ropes, *Geophys. Res. Lett.*, *11*, 1090–1093, 1984.
Smith, R. A., C. K. Goertz, and W. Grossmann, Thermal catastrophe in the plasma sheet boundary layer, *Geophys. Res. Lett.*, *13*, 1380–1383, 1986.

EVIDENCE FOR FLUX ROPES IN THE EARTH's MAGNETOTAIL

David G. Sibeck

The Johns Hopkins University Applied Physics Laboratory
Laurel, MD 20707

Abstract. Magnetic field reconnection is a fundamental process that occurs in the magnetotail during geomagnetic substorms. Some two-dimensional reconnection models predict the formation of a plasmoid, or closed loop of magnetic field lines, in the noon–midnight meridional plane at those times. When the three-dimensional magnetotail magnetic field is considered, it becomes clear that reconnection produces a flux rope with an axis transverse to the earth–sun line. Three signatures mark both two-dimensional plasmoids and three-dimensional flux ropes: (1) a bipolar magnetic field signature, (2) tailward flow of a hot plasma, and (3) convecting isotropic energetic particle distributions. Plasmoids and flux ropes may be distinguished by (4) the axial magnetic field that only flux ropes possess. All four signatures have been identified in near-earth ($X > -60\ R_E$), middle ($-60 > X > -120\ R_E$), and distant ($-120\ R_E > X > -240\ R_E$) magnetotail observations, but their interpretation is disputed. Thus, the existence of magnetotail flux ropes remains a controversial subject.

Introduction

The solar wind draws back geomagnetic field lines to form a magnetotail on the nightside of the earth. In the prevailing model [Dungey, 1961], this interaction is accomplished through merging of interplanetary and geomagnetic field lines at the dayside magnetopause. Following merging, the shocked solar wind flow in the magnetosheath drags the interconnected magnetic field lines antisunward and stretches them to lie nearly parallel to the earth–sun line on the nightside of the earth. These stretched magnetotail magnetic field lines sink slowly toward the magnetotail midplane.

In order that magnetic flux not be permanently lost from the dayside magnetosphere, there must be a return flow from the nightside to the dayside. This can be accomplished by reconnection of the stretched magnetic field lines at the magnetotail midplane. However, the site and nature of magnetotail reconnection remain controversial. In some models, reconnection occurs continuously along a line some 100 earth radii (R_E) from earth. In others it also occurs sporadically as little as 6.6 R_E from earth.

In one scenario, the brief appearance of a near-earth reconnection line results in the formation of a closed loop of magnetic field lines [Schindler and Ness, 1972; Russell, 1974; Schindler, 1974; Terasawa and Nishida, 1976], or plasmoid [Hones, 1976, 1977], between the near and distant reconnection lines. Because they are formed by reconnection, plasmoids should have distinctive signatures. First, reconnection fills them with an energized (hot) antisunward flowing plasma. Second, the closed magnetic field loops within plasmoids produce north–south magnetic field signatures when they pass by nearly stationary satellites. Finally, plasmoids contain convecting isotropic energetic particle distributions on closed magnetic field loops, rather than streaming distributions on open field lines.

Plasmoids are usually drawn as two-dimensional loops in the noon–midnight meridional plane. Clearly, they should be considered three-dimensional objects, both because they must have a finite cross-tail extent and because the magnetotail magnetic field has a nonzero cross-tail component. Several proposed three-dimensional models [Vasyliunas, 1976, 1980; Stern, 1979; Hones et al., 1982] did not gain the widespread attention they deserved. Recently, Hughes and Sibeck [1987], Schindler et al. [1988], Birn et al. [1989], Wright and Berger [1989], and Song and Lysak [1989] have considered the three-dimensional topology of plasmoids and have shown that it is rather complicated, with aspects resembling flux ropes. This paper constitutes an effort to reconcile previously reported observations with plasmoid/flux-rope model predictions.

Predictions

The top panel of Figure 1 (adopted from Hughes and Sibeck [1987]) shows a three-dimensional view of magnetotail reconnection from above the ecliptic plane. Because of the presence of a finite B_y component, north and south lobe

Geophysical Monograph 58

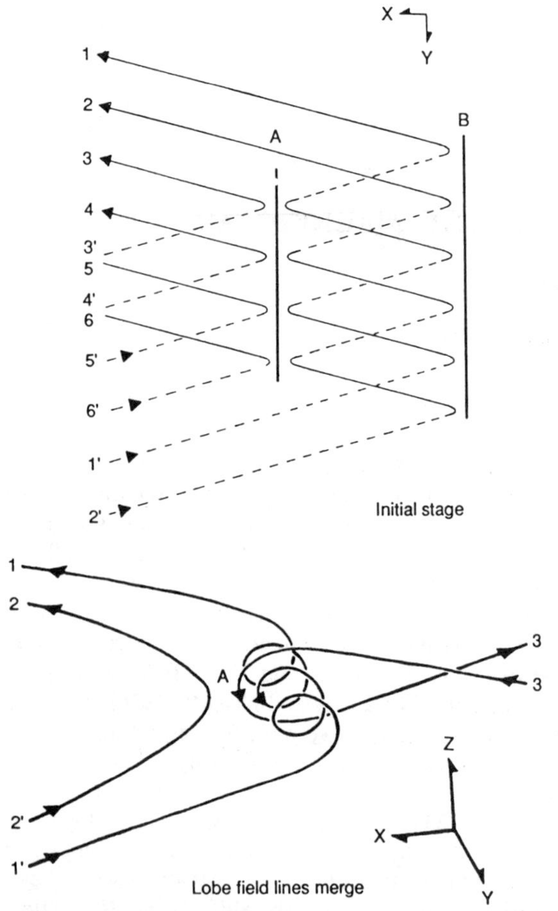

Fig. 1. The top panel shows a view of the magnetotail from above the ecliptic plane. The earth and sun lie to the left in the figure. Solid lines lie above the ecliptic plane, dashed lines below. The presence of a B_y component causes the field lines above and below the ecliptic plane to skew. Reconnection occurs at both near-earth (A) and distant (B) lines, resulting in the formation of a flux rope with an axis transverse to that of the magnetotail. Field lines within the flux rope (1–1′) remain connected to the earth. The lower panel of the figure shows a later stage in reconnection, when open lobe magnetic field lines have begun to reconnect at the near-earth line. Reconnection now produces pairs of field lines, one of which (2–2′) snaps back earthward, while the other (3–3′) loops around the flux rope and exerts an antisunward stress.

(and plasma sheet) magnetic field lines skew: they are not antiparallel. Reconnection occurs at both near-earth and distant lines, neither of which extends across the entire magnetotail. Initially, closed plasma sheet magnetic field lines reconnect at the near-earth line. The addition of the cross-tail (B_y) magnetic field component prevents the loops from closing in a single plane and results in the formation of a flux rope with spiraling magnetic fields that still have both ends on earth (1–1′). As the process continues, lobe field lines begin to reconnect at the near-earth line. As shown in the lower panel of Figure 1, when lobe field lines reconnect, the result is a closed field line with both ends on earth (2–2′) that snaps back earthward, and an open field line (3–3′) with both ends in the solar wind. The open lines loop around the closed lines of the flux rope and exert antisunward stresses. If reconnection proceeds more rapidly in the center of the magnetotail, the stress of the open field lines bows this portion of the rope tailward, as shown in Figure 2. This stress bends the flux rope axes, giving them significant components parallel to the magnetotail axis. Eventually one or both ends of the flux rope reconnect, the connection to earth is broken, and the flux rope is released to rapidly accelerate antisunward.

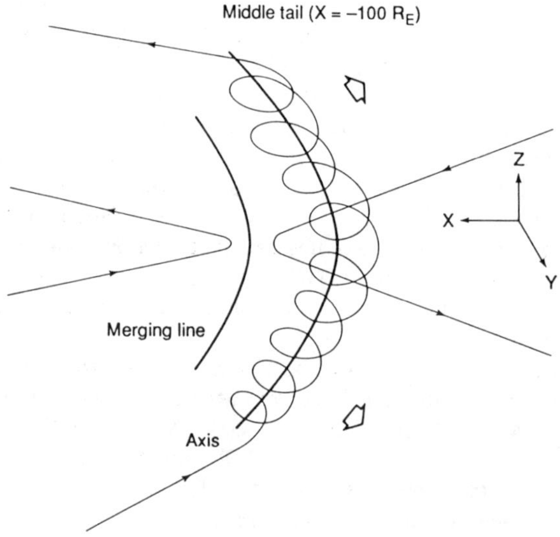

Fig. 2. Reconnection continues at the near-earth line. The stress of the open lines looped around the flux rope bows that structure. The axis of the rope is now curved. The rope is held in place by the balance of the earthward force on the closed interior magnetic field lines and the antisunward force of the lines looping around the rope. As reconnection continues, the antisunward force increases until such time as the flux rope magnetic field lines are disconnected from earth. The flux rope then accelerates rapidly antisunward.

A field-aligned current flowing along the flux tube axis is required to produce the spiraling field of the flux ropes. For the flux rope in Figure 2, the current would flow antisunward on the dawn side of the magnetotail, from dawn to dusk across the central portion of the magnetotail, and sunward on the dusk side of the magnetotail. Intense sunward and antisunward filamentary field-aligned currents are common features of numerical MHD simulations of reconnection in the near-earth magnetotail [Birn and Hones, 1981; Sato et al., 1984; Ogino et al., 1986].

The signatures of flux ropes should be very similar to those of plasmoids. Indeed, plasmoids form one limit of flux ropes, the limit in which the cross-tail magnetic field component is zero. Thus, nearly stationary satellites observe a bipolar north–south magnetic field signature accompanied by antisunward plasma flows when flux ropes move tailward. The same satellites observe streaming energetic particles on the open, outer magnetic field lines surrounding flux ropes but observe convecting isotropic energetic particle distributions on the closed field lines within the flux rope. Once one or both ends of the rope disconnect, the satellite may observe streaming particles throughout the rope.

There are, however, some features that distinguish flux ropes from plasmoids. The core magnetic field, at the center of flux ropes, points along their axes, while magnetic field strengths fall to zero at the center of plasmoids. Flux ropes may be held in place by the balance of sunward tension on the closed-core magnetic field lines within ropes and the antisunward tension of open lines enveloping the rope. Finally, the flux-rope model provides for the bowing of flux-rope axes in the magnetotail.

Review of Observations

We divide our search for the signatures of magnetotail flux ropes into a review of three magnetotail regions: near earth ($X > -60\ R_E$), middle ($-60 > X > -120\ R_E$), and distant ($-120\ R_E > X > -240\ R_E$).

Near-Earth Magnetotail

The evidence for plasmoids, not to mention flux ropes, in this region of the magnetotail is rather limited. Hones [1976, 1977, 1979a, b] and Nishida et al. [1981, 1983] have reported numerous instances of north–south magnetic field turnings accompanied by antisunward convective flow. The events last less than 10 min, and the structures they represent are some 10–20 R_E long. A strong cross-tail (B_y) magnetic field component is present during some events [Akasofu et al., 1978], suggesting that the preconditions for flux-rope formation are met. However, alternative explanations have been put forward for the obervations just noted. For example, Lui [1984] suggested wavy motion of magnetotail magnetic field lines as the cause of north–south turnings in the plasma sheet magnetic field, and Lui et al. [1978] suggested that field-aligned particle streaming could mimic the signatures of plasma convection.

Simultaneous observations of isotropic energetic particle distributions during such events would provide firmer evidence for the closed field line structures expected in plasmoids and flux ropes. Hones [1977, 1979a] interpreted IMP-8 plasma and magnetic field obervations from 1040–1048 and from 1248–1250 UT on April 18 and from 1827:30–1829:20 UT on October 8, 1974, as evidence for plasmoids. Particle and field observations for the events on April 18 are shown in Figure 3. The particle observations, discussed by Baker and Stone [1976, 1977], indicate small, nonzero, energetic ($E > 200$ keV) electron anisotropies during the events on April 18. Hones [1979a] stated that energetic electrons were nearly isotropic during the event on October 8, 1974, but it is not clear how small anisotropies must be for consistency with the plasmoid/flux-rope models. We note here that magnetic field observations reported by Hones [1977, 1979a] indicate the presence of a significant cross-magnetotail component during the events, requiring a three-dimensional (flux-rope?) model.

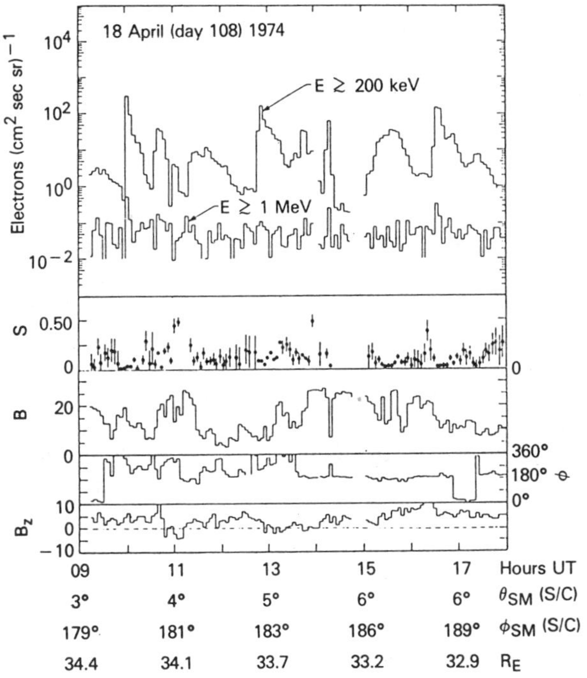

Fig. 3. IMP-8 energetic electron and magnetic field observations originally presented by Baker and Stone [1977]. The top panel shows IMP-8 energetic electron fluxes. The second panel shows the quantity S, a measure of the streaming anisotropy. S is 0 for isotropic distributions and 1 for purely streaming distributions. The remaining three panels show the magnetic field strength (B), the azimuthal magnetic field direction (ϕ = 0° is sunward, 90° is duskward), and the north/south component of the magnetic field, B_z. The magnetic field observations are in SM coordinates, and the location of IMP-8 is shown at the base of the plot.

Lui and Meng [1979] have reported energetic electron anisotropies for another plasmoid identified by Hones [1979b] in IMP-6 observations from 0940–0952 UT on September 26, 1973. Lui and Meng noted the presence of bidirectional and unidirectional antisunward streaming during this period but no evidence of isotropic distributions. While bidirectional streaming might be consistent with the closed magnetic field lines expected within plasmoids and flux ropes, tailward streaming surely is not. Lui and Meng [1979] noted the presence of a significant cross-tail magnetic field component during this event and remarked that it could not be adequately explained by existing two-dimensional models. Lui and Meng called this and similar events bubbles to distinguish them from plasmoids. They argued that bubbles have very-small-scale sizes and are present during geomagnetically quiet periods.

Evidence for plasmoids has recently been sought in ISEE 2 observations of the near-earth magnetotail. The top panels of Figure 4 show one possible example. Note the brief (20 s) bipolar north–south ($+/-$ B_z) magnetic field signature and antisunward ($-V_x$) flow near 1058:40 UT. The ratio of tailward to earthward streaming ions, shown in the lowermost panel, is near unity at this time, indicating a near-isotropic population until 1059 UT. Thus, McPherron and Manka [1985] have interpreted these observations as evidence for a plasmoid.

However, an alternative interpretation of the observations as a flux rope is equally attractive. Figure 4 shows that the event near 1058:40 UT was marked by a strong core magnetic field that pointed in the antisunward ($-x$) direction. The observations are consistent with an interpretation in which a rope, with its axis parallel to the earth–sun line, moved across the magnetotail, from dawn to dusk or vice versa. Such an interpretation would explain the strong core magnetic field and the bipolar north/south turning but would require the presence of a plasma velocity component in the cross-tail (y) direction. Figure 4 shows that this component was present and negative, i.e., dawnward.

Elphic et al. [1986] presented ISEE 1/2 plasma, magnetic-field, and electric-field observations from 1840–1930 UT on March 23, 1979. They interpreted two brief events (at 1911 and 1922 UT) as evidence for antisunward (1911 UT) and sunward (1922 UT) moving flux ropes with diameters of about 1.5-5 R_E and axes transverse to the earth–sun line. The rope at 1911 UT was marked by a bipolar north–south magnetic field and a strong sunward ($+x$) magnetic field. Elphic et al. [1986] interpreted the field variations during this event as evidence for the draping of north plasma sheet magnetic field lines over an east–west aligned flux tube, but the observations could also be interpreted as indicating a flux rope with an axis parallel to the earth–sun line. Figure 5 shows observations of the flux rope at 1922 UT. This event is more consistent with an explanation in terms of an east–west aligned flux rope than with one whose axis lies parallel to the earth–sun line, since this flux rope is marked by a bipolar north/south magnetic field signature and a core field that has its strongest component in the $-y$ direction, parallel to the simultaneously measured IMF. However, the earthward plasma flow during the 1922 UT event is puzzling; perhaps the tension of the field lines connecting the flux rope to earth exceeded any antisunward tension and the rope was accelerated earthward.

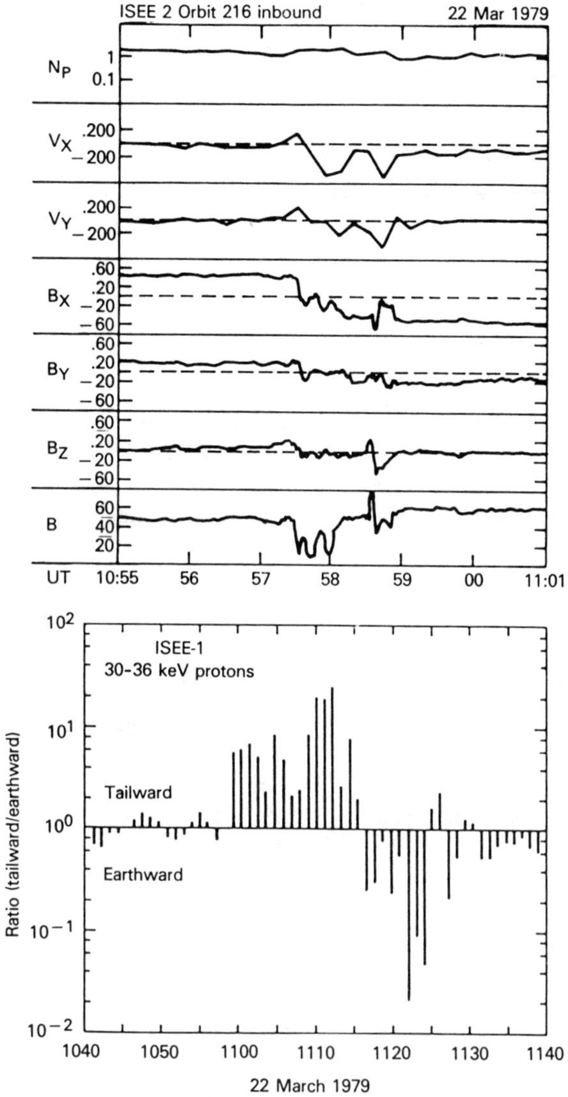

Fig. 4. The top panels show ISEE 2 plasma and magnetic field observations on March 22, 1979. This figure was adopted from Paschmann et al. [1985], who indicated that the plasma observations were in spacecraft (GSE) coordinates but did not state the coordinate system for the magnetic field observations. The lower panel depicts ISEE 1 energetic (30–36 keV) electron front-to-back ratios. This panel is taken from Ipavich et al. [1985].

Middle Magnetotail

Shull [1981] reported that magnetic bubbles (i.e., brief drops in magnetic field strength) at lunar orbit are attended by several-minute-long bursts of cold, fast plasma flow. They found no correlation of these bursts with geomagnetic activity. The relationship of such bubbles to plasmoids and/or flux ropes is not clear.

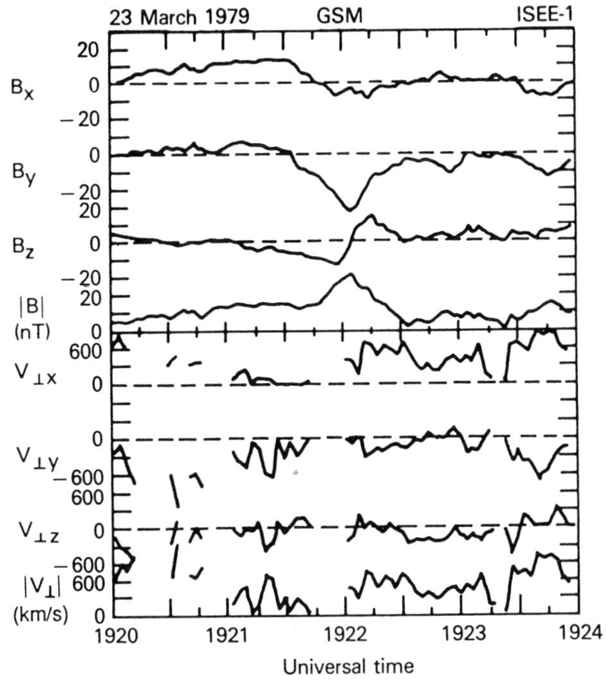

Fig. 5. ISEE-1 observations of a flux rope. This figure, adopted from Elphic et al. [1986], shows ISEE 1 magnetic field and calculated E × B flow velocities in GSM coordinates.

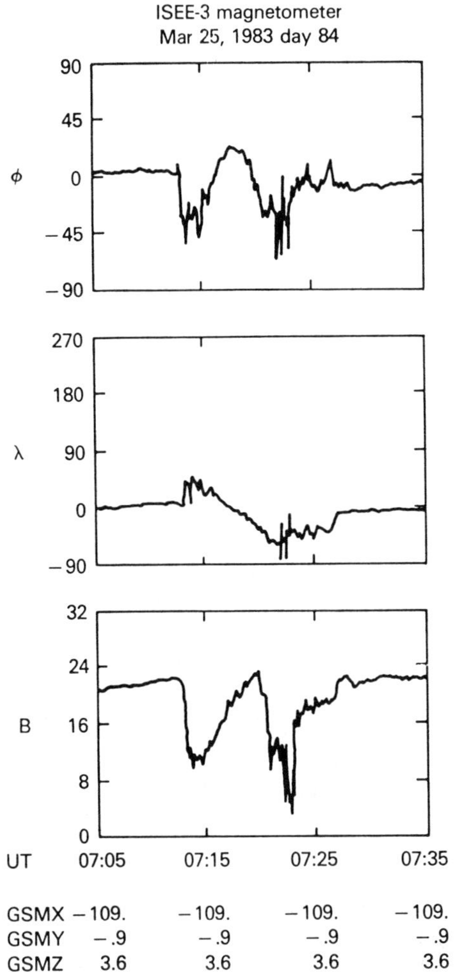

Fig. 6. Observations of a possible flux rope in the middle magnetotail. This figure shows ISEE 3 magnetometer observations in GSM coordinates of a bipolar event that occurred on March 25, 1983. Note the bipolar ϕ signature: ϕ turns first toward positive values, then negative, indicating a bipolar signature in the y-direction, i.e., dawn–dusk. The magnetic field strength reaches a relative maximum at the center of the event, near 0717 UT. The signature in λ is rather more complicated.

Flux ropes in this range of the magnetotail have had a rather checkered history. Siscoe et al. [1984] and Sibeck et al. [1984] noted three transient events (each lasting some 10 min) in ISEE 3 observations that were marked by bipolar magnetic field signatures, strong core magnetic fields pointing nearly along the earth-sun line, tailward plasma flow, bursts of energetic particles, and enhanced plasma-wave activity. The events on December 28 and 30, 1982, were marked by bipolar north–south signatures, but the event on March 25, 1983 (shown in Figure 6) was marked by a bipolar east-west signature. Sibeck et al. [1984] suggested that the east-west signature resulted from a spacecraft's north–south passage through the latter flux rope. Scholer et al. [1985] confirmed the presence of isotropic energetic electron distributions during those events and used energetic ion remote sensing techniques to demonstrate that the boundary orientations of the events on December 30 and March 25 were consistent with those expected for flux ropes. Kennel et al. [1986] presented in-situ ISEE 3 wave observations that indicate intensified broadband electrostatic noise during all the events and intense whistler mode noise during the March 25 event.

In contrast to the results of Scholer et al. [1985], Murphy [1987] used similar techniques for the remote sensing of energetic particles to demonstrate that boundaries of the event on December 30 were inconsistent with expectations for a flux rope. Murphy did show that the boundaries of the events on December 28 and March 25 were consistent with flux ropes, and he determined that the diameters of these ropes were on the order of 3.5–7.4 R_E. Figure 7 shows the rather circuitous route that ISEE 3 must have followed in order for the event on March 25 to be interpreted as a flux rope. Richardson et al. [1989] argued that the observations of the March 25 event, in which a south–north–south signature occurs, constituted an observation of the tail end of one plasmoid (south) and a full observation of the next (north–south). This explanation does not account for the presence of the strong bipolar B_y field within the structure nor the observations reported by Sibeck et al. [1984], which indicated an orientation parallel to the magnetotail axis. Fairfield et al. [1989] have also

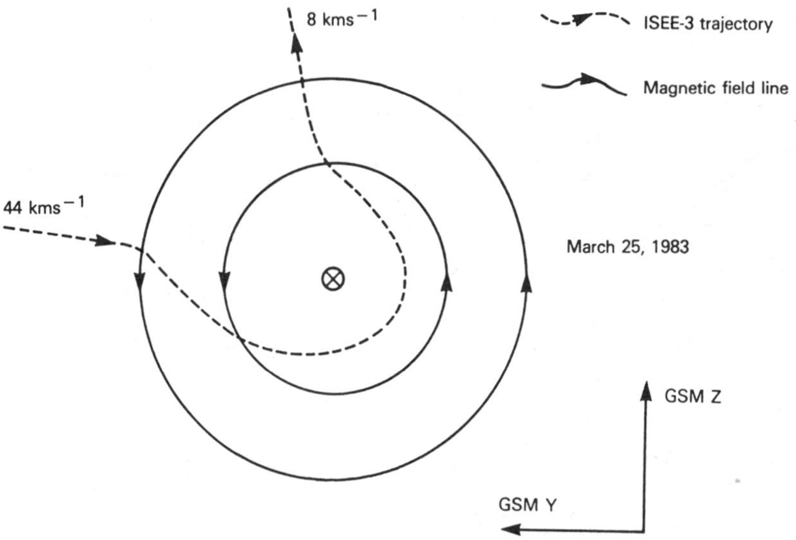

Fig. 7. Murphy [1987] proposed this convoluted relative trajectory between the ISEE 3 spacecraft and the flux rope on March 25, 1983 (Figure 6) in order to reconcile magnetic field observations with the energetic ion gradient anisotropies.

discussed the March 25 event and do not rule out the possibility that it is a flux rope.

Nishida et al. [1986] reported observations indicating the presence of stagnant plasmoids in the middle magnetotail. The magnetic field within those events points strongly southward, and streaming and isotropic energetic particle distributions are observed, but the events move antisunward rather slowly. They are observed for periods on the order of 10 min. As noted by Hughes and Sibeck [1987], the structures are consistent with observations of flux ropes still connected to earth but surrounded by open field lines that slowly pull the ropes antisunward.

Distant Magnetotail

The characteristics of plasmoids in the distant magnetotail are now well established: they are filled with a hot antisunward flowing plasma [Baker et al., 1984] and convecting isotropic energetic particle distributions [Scholer et al., 1984], and their passage is attended by a characteristic north then south magnetic field turning [Hones et al., 1984]. Plasmoids are observed for periods of about 20 min in the distant magnetotail. Figure 8 shows an example of a plasmoid in the distant magnetotail. There remains some controversy over whether distant-magnetotail plasmoids are associated with geomagnetic activity; Hones et al. [1984] and most other researchers believe that they are, although Tsurutani et al. [1987] found no evidence for such an association.

Hughes and Sibeck [1987] have noted the presence of a cross-tail magnetic field component within many plasmoids. Such a component demonstrates that they must be treated as three-dimensional objects and led to the proposal by Hughes and Sibeck that the simplest three-dimensional model consistent with the observations was a flux-rope model. Now it is becoming clear that many plasmoids have strong core fields that point nearly parallel to the magnetotail axis [Slavin et al., 1989]. Such strong core fields are difficult to reconcile with the weak core fields expected inside plasmoids but can be taken as evidence for flux ropes with axes nearly parallel to the magnetotail axis. Multipoint magnetic field and energetic particle anisotropy observations would be necessary to demonstrate conclusively, or rule out, the flux rope structure of distant magnetotail plasmoids.

Here it is of interest to consider the adiabatic expansion of near-earth plasmoids (or flux ropes) with cross-tail axes as they move into the regions of lower confining pressure in the distant magnetotail. For simplicity, we assume that near-earth reconnection has ceased before the expansion, the flux ropes have cylindrical shapes, the magnetic field within the ropes is uniform and axial, and the cross-tail lengths of the flux ropes do not vary with distance down the magnetotail. Under these assumptions, flux-rope temperatures (T) are related to their radii (R) by the equation $T \propto R^{2-2\gamma}$. Taking γ as 5/3, one finds (as expected) that the ropes cool as they expand.

Now, plasmoids have diameters of 10–20 R_E in the near-earth magnetotail [Hones, 1976, 1979a, b] and of 20–100 R_E in the distant magnetotail [Hones et al., 1984; Slavin et al., 1984], indicating that their radii increase by a factor of 2–5 and their cross-sectional areas by a factor of 4–25. Because of conservation of the flux passing through the ends of the tubes, one expects the axial (or cross-tail, B_y) component of the magnetic field within flux ropes to diminish by a factor of 4–25 from the near-earth to the distant magnetotails.

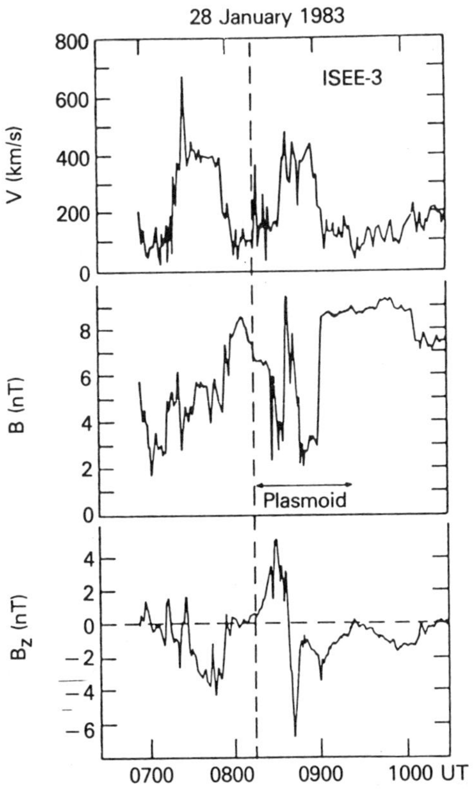

Fig. 8. Adopted from Baker et al. [1987], this figure shows ISEE 3 plasma and magnetic field observations of a plasmoid in the distant magnetotail on January 28, 1983. The event at 0845 UT demonstrates the strong antisunward flow and north/south magnetic field turning, which are often taken as evidence for plasmoids. Note the strong core magnetic field in this event, as indicated by the large total magnetic field strength at 0845 UT. This feature is indicative of a flux-rope structure.

Plasmoids begin their existence with B_y components typical of the near-earth magnetotail, i.e., about 50% of that measured simultaneously in the interplanetary magnetic field (IMF) [Lui, 1984]. As the quiet-time distant magnetotail plasma sheet is expected [Owen and Cowley, 1987] and observed [Murphy, 1987] to be quite thin ($\approx 1\ R_E$), most prolonged encounters with the distant magnetotail plasma sheet are associated with the passage of plasmoids. Thus, the survey of distant magnetotail plasma sheet B_y reported by Tsurutani et al. [1984] pertains primarily to plasmoids. They found that the cross-tail magnetic field component in the distant magnetotail was only 0.09–0.21 of that simultaneously measured in the IMF. This is a factor of 2.5–5 less than that observed nearer earth, consistent with our expectations.

For the observed increases in plasmoid dimensions from the near-earth to the distant magnetotail, one expects temperatures to fall by a factor of 2.5–21.5. Plasmoids begin their near-earth existence during reconnection events and contain plasma electrons at energies of 210–830 eV [Bieber et al., 1984]. In the distant magnetotail, plasmoid electron temperatures are in the range 40–400 eV [Richardson et al., 1987], two to five times cooler, consistent with our crude predictions. Gloeckler et al. [1984a, b] report a similar result on the basis of energetic ion observations. They find that temperatures within distant magnetotail plasmoids fall by factors of 3–10 from temperatures in the near-earth magnetotail.

Finally, Gloeckler et al. [1984b] use energetic ion observations to deduce that plasma flows within plasmoids have components toward the sides of the magnetotail, i.e., dawnward on the dawnside and duskward on the duskside. They conclude that these flows indicate the expansion of plasmoids with distance down the magnetotail. In the model of Hughes and Sibeck [1987], such flows could indicate the expected bowing of flux-rope structures: the portion of the flux rope nearest the center of the magnetotail moves antisunward faster than other portions of the flux rope, resulting in a component of the flow toward the flanks of the magnetotail, as illustrated by the large arrows in Figure 2.

Conclusions

In this review, we have examined observations of plasmoids and flux ropes throughout the explored length of the earth's magnetotail. We have found evidence suggesting the presence of flux ropes. Bubbles, plasmoids, or flux ropes have been reported on many occasions. Such features must be treated as three-dimensional objects, and the flux-rope model provides one way to organize the observations. However, we fully recognize that such an explanation is not unique. Only a comprehensive set of multipoint and multi-instrument observations could resolve the structure of small-scale magnetotail phenomenon.

Even without such observations, much work remains to be done on existing data sets. We do not know what fraction of substorms produces plasmoids and/or flux ropes. Nor have convincing plasmoid observations within 60 R_E of earth been presented. At a minimum, one wishes to see more observations of isotropic energetic electrons within plasmoids/flux ropes to confirm the closed-loop model. Only a few magnetotail flux ropes have been reported, and their interpretation as flux ropes is disputed.

As to theoretical questions, we do not know how plasmoids grow from their small sizes near earth to enormous blobs of plasma with dimensions comparable to those of the magnetotail. The axial magnetic field of the flux-rope model seems an effective way to hold plasmoids together as they expand and move antisunward. However, if flux ropes expand as they move antisunward, the strength of this axial field diminishes. At some point the flux rope must become unstable, buckle, and be destroyed. Little consideration has been given to the stability of magnetotail flux ropes or plasmoids.

Questions and Answers

Hollweg: In the flux-rope picture the isotropic electrons presumably come about due to mirroring, rather than being in a closed loop as stated by Haerendel. But, should there then be a loss cone? Do you see a loss cone?

Sibeck: The only spacecraft ever sent into the middle or distant magnetotail was ISEE-3. This spacecraft carried 2 energetic particle instruments, neither of which had sufficient resolution to resolve loss cone effects.

Lysak: Is there any observation or evidence for an ionospheric signature of a tethered flux-rope?

Sibeck: The edges of the flux-rope/plasmoid map down to the edges of the substorm current wedge. I do not know of any specific feature at the ground which could immediately be associated with the flux-rope in the magnetotail. Perhaps the signatures can be identified in the model of Ogino and Walker, which generates magnetotail flux-ropes and describes their ionospheric footprints.

Acknowledgments. The author thanks the referees for helpful suggestions. Work at APL was supported by NASA under Space and Naval Warfare Systems Command contract N00039-87-C-5301 of the Navy.

References

Akasofu, S.-I., A. T. Y. Lui, C. I. Meng, and M. Haurwitz, Need for a three-dimensional analysis of magnetic fields in the magnetotail during substorms, *Geophys. Res. Lett.*, *5*, 283–286, 1978.

Baker, D. N., and E. C. Stone, Energetic electron anisotropies in the magnetotail: Identification of open and closed field lines, *Geophys. Res. Lett.*, *3*, 557–560, 1976.

Baker, D. N., and E. C. Stone, Observations of energetic electrons ($E \gtrsim 200$ keV) in the earth's magnetotail plasma sheet and fireball observations, *J. Geophys. Res.*, *82*, 1532–1546, 1977.

Baker, D. N., S. J. Bame, R. D. Belian, W. C. Feldman, J. T. Gosling, P. R. Higbie, E. W. Hones, Jr., D. J. McComas, and R. D. Zwickl, Correlated dynamical changes in the near-earth and distant magnetotail regions: ISEE 3, *J. Geophys. Res.*, *89*, 3855–3864, 1984.

Baker, D. N., R. C. Anderson, R. D. Zwickl, and J. A. Slavin, Average plasma and magnetic field variations in the distant magnetotail associated with near-earth substorm effects, *J. Geophys. Res.*, *92*, 71–81, 1987.

Bieber, J. W., E. C. Stone, E. W. Hones, Jr., D. N. Baker, S. J. Bame, and R. P. Lepping, Microstructure of magnetic reconnection in earth's magnetotail, *J. Geophys. Res.*, *89*, 6705–6716, 1984.

Birn, J., and E. W. Hones, Jr., Three-dimensional computer modeling of dynamic reconnection in the geomagnetic tail, *J. Geophys. Res.*, *86*, 6802–6808, 1981.

Birn, J., M. Hesse, and K. Schindler, Filamentary structure of a three-dimensional plasmoid, *J. Geophys. Res.*, *94*, 241–251, 1989.

Dungey, J. W., Interplanetary magnetic field and the auroral zones, *Phys. Rev. Lett.*, *6*, 47–48, 1961.

Elphic, R. C., C. A. Cattell, K. Takahashi, S. J. Bame, and C. T. Russell, ISEE 1 and 2 observations of magnetic flux ropes in the magnetotail: FTE's in the plasma sheet?, *Geophys. Res. Lett.*, *13*, 648–651, 1986.

Fairfield, D. H., D. N. Baker, J. D. Craven, R. C. Elphic, J. F. Fennell, L. A. Frank, I. G. Richardson, H. J. Singer, J. A. Slavin, B. T. Tsurutani, and R. D. Zwickl, Substorms, plasmoids, flux ropes, and magnetotail flux loss on March 25, 1983: CDAW 8, *J. Geophys. Res.*, in press, 1989.

Gloeckler, G., M. Scholer, F. M. Ipavich, D. Hovestadt, B. Klecker, and A. B. Galvin, Abundances and spectra of suprathermal H^+, He^{++}, and heavy ions in a fast moving plasma structure (plasmoid) in the distant geotail, *Geophys. Res. Lett.*, *11*, 603–606, 1984a.

Gloeckler, G., F. M. Ipavich, D. Hovestadt, M. Scholer, A. B. Galvin, and B. Klecker, Characteristics of suprathermal H^+ and He^{++} in plasmoids in the distant magnetotail, *Geophys. Res. Lett.*, *11*, 1030–1033, 1984b.

Hones, E. W., Jr., The magnetotail: Its generation and dissipation, in *Physics of Solar Planetary Environments*, edited by D. J. Williams, AGU, Washington, D.C., pp. 558–571, 1976.

Hones, E. W., Jr., Substorm processes in the magnetotail: Comments on 'On hot tenuous plasma, fireballs, and boundary layers in the earth's magnetotail' by L. A. Frank, K. L. Ackerson, and R. P. Lepping, *J. Geophys. Res.*, *82*, 5633–5640, 1977.

Hones, E. W., Jr., Transient phenomena in the magnetotail and their relation to substorms, *Space Sci. Rev.*, *23*, 393–410, 1979a.

Hones, E. W., Jr., Plasma flow in the magnetotail and its implications for substorm theories, in *Dynamics of the Magnetosphere*, edited by S.-I. Akasofu, D. Reidel, Boston, pp. 545–562, 1979b.

Hones, E. W., Jr., J. Birn, S. J. Bame, G. Paschmann, and C. T. Russell, On the three-dimensional structure of the plasmoid created in the magnetotail at substorm onset, *Geophys. Res. Lett.*, *9*, 203–206, 1982.

Hones, E. W., Jr., D. N. Baker, S. J. Bame, W. C. Feldman, J. T. Gosling, D. J. McComas, R. D. Zwickl, J. A. Slavin, E. J. Smith, and B. T. Tsurutani, Structure of the magnetotail at 220 R_E and its response to geomagnetic activity, *Geophys. Res. Lett.*, *11*, 5–8, 1984.

Hughes, W. J., and D. G. Sibeck, On the 3-dimensional structure of plasmoids, *Geophys. Res. Lett.*, *14*, 636–639, 1987.

Ipavich, F. M., A. B. Galvin, M. Scholer, G. Gloeckler, D. Hovestadt, and B. Klecker, Suprathermal O^+ and H^+ ion behavior during the March 22, 1979 (CDAW 6) substorms, *J. Geophys. Res.*, *90*, 1263–1272, 1985.

Kennel, C. F., F. V. Coroniti, and F. L. Scarf, Plasma waves in magnetotail flux ropes, *J. Geophys. Res.*, *91*, 1424–1438, 1986.

Lui, A. T. Y., Characteristics of the cross-tail current in the earth's magnetotail, in *Magnetospheric Currents*, edited by T. A. Potemra, AGU, Washington, D.C., pp. 158–170, 1984.

Lui, A. T. Y., and C.-I. Meng, Relevance of southward magnetic fields in the neutral sheet to anisotropic distributions of energetic electrons and substorm activity, *J. Geophys. Res.*, *84*, 5817-5827, 1979.

Lui, A. T. Y., L. A. Frank, K. L. Ackerson, C.-I. Meng, and S.-I. Akasofu, Plasma flows and magnetic field vectors in the plasma sheet during substorms, *J. Geophys. Res.*, *83*, 3849-3858, 1978.

McPherron, R. L., and R. H. Manka, Dynamics of the 1054 UT March 22, 1979, substorm event: CDAW 6, *J. Geophys. Res.*, *90*, 1175-1190, 1985.

Murphy, N., The use of energetic ions to determine some aspects of the dynamics of the geomagnetic tail, PhD Thesis, Imperial College, University of London, October 1987.

Nishida, A., H. Hayakawa, and E. W. Hones, Jr., Observed signatures of reconnection in the magnetotail, *J. Geophys. Res.*, *86*, 1422-1436, 1981.

Nishida, A., Y. K. Tulanay, F. S. Mozer, C. A. Cattell, E. W. Hones, Jr., and J. Birn, Electric field evidence for tailward flow at substorm onset, *J. Geophys. Res.*, *88*, 9109–9113, 1983.

Nishida, A., M. Scholer, T. Terasawa, S. J. Bame, G. Gloeckler, E. J. Smith, and R. D. Zwickl, Quasi-stagnant plasmoid in the middle tail: A new preexpansion phase phenomenon, *J. Geophys. Res.*, *91*, 4245–4255, 1986.

Ogino, T., R. J. Walker, M. Ashour-Abdalla, and J. M. Dawson, An MHD simulation of the effects of the interplanetary magnetic field B_y component on the interaction of the solar wind with the earth's magnetosphere during southward interplanetary magnetic field, *J. Geophys. Res.*, *91*, 10029–10045, 1986.

Owen, C. J., and S. W. H. Cowley, Simple models of time-dependent reconnection in a collision-free plasma with an application to substorms in the geomagnetic tail, *Planet. Space Sci.*, *35*, 451–466, 1987.

Paschmann, G., N. Sckopke, and E. W. Hones, Jr., Magnetotail plasma observations during the 1054 UT substorm on March 22, 1979 (CDAW 6), *J. Geophys. Res.*, *90*, 1217–1229, 1985.

Richardson, I. G., S. W. H. Cowley, E. W. Hones, Jr., and S. J. Bame, Plasmoid-associated energetic ion bursts in the deep geomagnetic tail: Properties of plasmoids and the post-plasmoid plasma sheet, *J. Geophys. Res.*, *92*, 9997–10013, 1987.

Richardson, I. G., C. J. Owen, S. W. H. Cowley, T. R. Sanderson, M. Scholer, J. A. Slavin, and R. D. Zwickl, ISEE-3 observations during the CDAW-8 intervals: Case studies of the distant geomagnetic tail covering a wide range of geomagnetic activity, *J. Geophys. Res.*, in press, 1989.

Russell, C. T., The solar wind and magnetospheric dynamics, in *Correlated Interplanetary and Magnetospheric Observations*, edited by D. E. Page, D. Reidel, Boston, pp. 3–47, 1974.

Sato, T., R. J. Walker, and M. Ashour-Abdalla, Driven magnetic reconnection in three-dimensions: Energy conversion and field-aligned current generation, *J. Geophys. Res.*, *89*, 9761-9769, 1984.

Schindler, K., A theory of the substorm mechanism, *J. Geophys. Res.*, *79*, 2803–2810, 1974.

Schindler, K., and N. F. Ness, Internal structure of geomagnetic neutral sheet, *J. Geophys. Res.*, *77*, 91–100, 1972.

Schindler, K., M. Hesse, and J. Birn, General magnetic reconnection, parallel electric fields, and helicity, *J. Geophys. Res.*, *93*, 5547–5557, 1988.

Scholer, M., G. Gloeckler, B. Klecker, F. M. Ipavich, D. Hovestadt, and E. J. Smith, Fast moving plasma structures in the distant magnetotail, *J. Geophys. Res.*, *89*, 6717–6727, 1984.

Scholer, M., B. Klecker, D. Hovestadt, G. Gloeckler, F. M. Ipavich, and A. B. Galvin, Energetic particle characteristics of magnetotail flux ropes, *Geophys. Res. Lett.*, *12*, 191-194, 1985.

Shull, P., Jr., Fast plasma flows in the translunar magnetotail, *J. Geophys. Res.*, *86*, 4708–4714, 1981.

Sibeck, D. G., G. L. Siscoe, J. A. Slavin, E. J. Smith, S. J. Bame, and F. L. Scarf, Magnetotail flux ropes, *Geophys. Res. Lett.*, *11*, 1090–1093, 1984.

Siscoe, G. L., D. G. Sibeck, J. A. Slavin, E. J. Smith, B. T. Tsurutani, and D. E. Jones, ISEE 3 magnetic field observations in the magnetotail: Implications for reconnection, in *Magnetic Reconnection in Space and Laboratory Plasmas*, edited by E. W. Hones, Jr., AGU, Washington, D.C., pp. 240–248, 1984.

Slavin, J. A., E. J. Smith, B. T. Tsurutani, D. G. Sibeck, H. J. Singer, D. N. Baker, J. T. Gosling, E. W. Hones, and F. L. Scarf, Substorm associated traveling compression regions in the distant tail: ISEE-3 geotail observations, *Geophys. Res. Lett.*, *11*, 657–660, 1984.

Slavin, J. A., D. N. Baker, J. D. Craven, R. C. Elphic, D. H. Fairfield, L. A. Frank, A. B. Galvin, W. J. Hughes, R. H. Manka, D. B. Mitchell, I. G. Richardson, T. R. Sanderson, D. G. Sibeck, H. J. Singer, E. J. Smith, and R. D. Zwickl, ISEE 3 observations of plasmoid structure and dynamics: CDAW-8, *J. Geophys. Res.*, *93*, in press, 1989.

Song, Y., and R. L. Lysak, Evaluation of twist helicity of FTE flux tubes, *J. Geophys. Res.*, *94*, 5273-5281, 1989.

Stern, D. P., The role of O-type neutral lines in magnetic merging during substorms and solar flares, *J. Geophys. Res.*, *84*, 63–71, 1979.

Terasawa, T., and A. Nishida, Simultaneous observations of relativistic electron bursts and neutral-line signatures in the magnetotail, *Planet. Space Sci.*, *24*, 855–866, 1976.

Tsurutani, B. T., D. E. Jones, R. P. Lepping, E. J. Smith, and D. G. Sibeck, The relationship between the IMF B_y

and distant tail (150-238 R_e) lobe and plasmsheet B_y fields, *Geophys. Res. Lett.*, *11*, 1082–1085, 1984.

Tsurutani, B. T., M. E. Burton, E. J. Smith, and D. E. Jones, Statistical properties of magnetic field fluctuations in the distant plasmasheet, *Planet. Space Sci.*, *35*, 289–293, 1987.

Vasyliunas, V. M., An overview of magnetospheric dynamics, in *Magnetospheric Particles and Fields*, edited by B. M. McCormac, D. Reidel, Boston, pp. 99–110, 1976.

Vasyliunas, V. M., Upper limit on the electric field along a magnetic O line, *J. Geophys. Res.*, *85*, 4616–4620, 1980.

Wright, A. N., and M. A. Berger, The effect of reconnection upon the linkage and interior structure of magnetic flux ropes, *J. Geophys. Res.*, *94*, 1295–1302, 1989.

MAGNETIC ISLANDS IN THE NEAR GEOMAGNETIC TAIL AND ITS IMPLICATIONS FOR THE MECHANISM OF 1054 UT CDAW 6 SUBSTORM

N. Lin, R. J. Walker

Institute of Geophysics and Planetary Physics, University of California, Los Angeles, CA 90024

R. L. McPherron, and M. G. Kivelson

Institute of Geophysics and Planetary Physics and Department of Earth and Space Sciences, University of California, Los Angeles, CA 90024

Abstract. During the 1054 UT CDAW 6 substorm event two ISEE spacecraft observed dynamic changes in the magnetic field and in the flux of energetic particles in the near Earth ($\sim 15\ R_E$) plasma sheet. In the substorm growth phase, the magnetic field at both ISEE spacecraft became tail-like. Following expansion phase onset, two small scale magnetic islands were observed moving tailward at a velocity of ~ 580 km/s. The passage of these two magnetic islands was coincident with bursts of tailward streaming energetic particles. The length of the magnetic loops was estimated to have been ~ 2 to $3\ R_E$ while the height of the loops was less than $0.5\ R_E$. We suggest that the magnetic islands were produced by multi-point reconnection processes in the near tail plasma sheet which may have been associated with the formation of the near Earth neutral line and the subsequent formation of a large scale plasmoid. The near Earth neutral line retreated tailward later in the expansion phase as suggested by the reversal of the streaming of energetic particles.

Introduction

The CDAW 6 substorm reported here occurred on March 22, 1979 when the Earth's dipole axis was almost precisely orthogonal to the Earth-Sun line. This substorm has been intensively studied with measurements from well instrumented spacecraft and ground observatories [see the overview by McPherron and Manka, 1985, and references therein]. Although previous studies [e.g., Paschmann et al., 1985; Ipavich et al., 1985; Baker, 1984; Fritz et al., 1984] have revealed that the overall magnetic field and plasma signatures during the substorm event were reasonably consistent with the near-Earth neutral line model of substorms [Hones, 1979], details of the process in the near Earth magnetotail are still not clear. In this paper, we have examined in detail the magnetic field variation and the behavior of energetic particles in the near geomagnetic tail, using magnetometer data from ISEE-1 and 2 [Russell, 1978] and data from the ISEE-1 Medium Energy Particles Instrument (MEPI) which measured ions and electrons of ~ 24 keV and above in three dimensions [Williams et al., 1978]. We propose that a multi-point reconnection process occurred in the near tail plasma sheet at the expansion onset of the substorm. The reconnection process may lead to the formation of the near Earth neutral line.

Geophysical Monograph 58

Overview of the Substorm

The CDAW 6 1054 UT substorm was an isolated event following an interval of magnetic calm. As described in McPherron and Manka [1985], a large increase in solar wind velocity and an enhancement of the northward IMF, identified as a fast forward shock by Tsurutani et al. [1984], initiated a storm sudden commencement at ~ 0826 UT. At ~ 1010 UT, following a long period of northward B_z, the IMF turned southward and remained so for more than one hour. About ten minutes later (at ~ 1020 UT), the 57 station AE index began to increase. Although the IMF B_z turned southward at ~ 1010 UT [Baker et al., 1985] implying the rate of energy input into the magnetosphere from the solar wind increased, a sharp increase in AE values was not seen for more than 40 minutes when at about 1100 UT AE rose from slowly increasing, moderate values to more than 1000 nT. A northward fluctuation of the IMF, and possibly of the solar wind velocity, apparently triggered the onset of the expansion phase at 1054 UT. The timing was determined by McPherron and Manka [1985], using ground magnetometer measurements and an increase of electron fluxes at synchronous orbit. The field-aligned current system associated with the 1054 UT expansion onset was centered at about 0300 LT. After reaching a peak at ~ 1140 UT, the AE index decreased until 1300 UT when another substorm onset interrupted its recovery.

The ISEE spacecraft were inbound at ~ 0150 LT and at a distance of $\sim 15\ R_E$ behind the Earth. The scenario for the event derived from the near tail observations is illustrated schematically in Figure 1 and is presented here to motivate the observations discussed later. The southward turning of the IMF and a change in the solar wind velocity blew the axis of the magnetotail down over the two spacecraft which had initially been located $\sim 1\ R_E$ below the expected position of the neutral sheet (Figure 1a). The southward turning of the IMF increased the energy input to the magnetosphere. As the magnetic stress built up, the near Earth magnetotail became increasingly tail-like. The region of tail-like field extended from 6.6 R_E to beyond the ISEE location at $\sim 15\ R_E$ (Figure 1b).

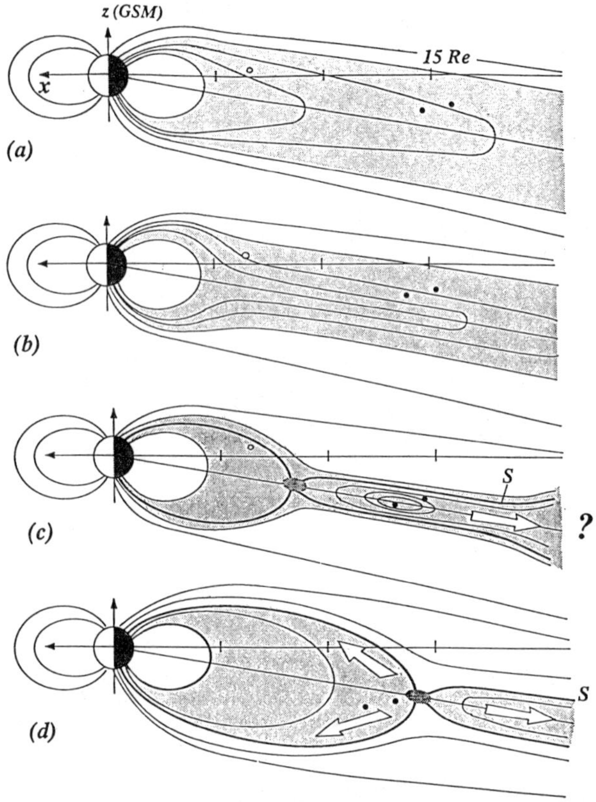

Fig. 1. Schematic diagrams illustrating the scenario of the CDAW 6 substorm. The plasma sheet region is shaded. The solid circles denote the ISEE spacecraft, and the open circle denote a geostationary spacecraft. (*a*) Southward turning of solar wind B_z and a change in the solar wind velocity blew the magnetotail southward. (*b*) During the growth phase, the near tail field became tail-like. (*c*) Reconnection occurred and expansion phase began. The thick line S represents the separatrix. The arrows indicate the streaming of particles. A question mark indicates a possible large scale plasmoid tailward of ISEE which we didn't observed. (*d*) The near-Earth neutral line retreated to the tailward side of the ISEE spacecraft (see text for detailed explanation).

The antiparallel field was conducive to the onset of reconnection, and it is plausible that a near Earth neutral line released a plasmoid at about 1054 UT when signatures of expansion onset were seen. The magnetic field at geostationary orbit became more dipolar at almost the same time as the onset of the expansion phase while the tail-like field lasted for about 15 minutes after the onset at the ISEE spacecraft. Figure 1c illustrates our interpretation of data obtained by the two ISEE spacecraft at about 1058 UT when observations suggest that a near-Earth neutral line had formed earthward of the spacecraft. The large scale plasmoid which we propose was released at $\sim$ 1054 UT has left the near-tail region. The two spacecraft (solid dots) were engulfed in the plasma sheet and were initially both above the neutral sheet and most of the tail current. A sudden upward movement of the plasma sheet at $\sim$ 1057 UT caused the neutral sheet to cross ISEE-2 leaving the two spacecraft in high field regions on opposite sides of the current sheet. The separatrix shown by a heavy line marked S in the diagram separates reconnected field lines and plasma sheet field lines which have not yet reconnected (as we shall discuss later, we note that it is not necessary that the separatrix be located at a constant height in the dawn-dusk direction). When the ISEE spacecraft crossed this separatrix (at $\sim$ 1058 UT for ISEE-1 and $\sim$ 1057:30 UT for ISEE-2) as a result of the upward motion of the plasma sheet, both spacecraft observed tailward plasma flow [Paschmann et al., 1985], and MEPI on ISEE-1 observed tailward streaming of energetic particles. Associated with a burst of enhanced fluxes of tailward streaming energetic particles, two bipolar signatures in B_z were observed. We interpret the two events as the signatures of two small scale magnetic islands which were produced successively by multiple reconnection in the plasma sheet. The multiple reconnection may have been associated with the formation of the near-Earth neutral line and the subsequent formation of a large scale plasmoid.

Part way through the expansion onset, the neutral line retreated tailward past the two ISEE spacecraft, and earthward streaming of particles was observed (Figure 1d). Near the beginning of the recovery phase the streaming of particles subsided and the particle distribution became isotropic again. In the next section we present the observations which led us to the interpretation sketched above.

Observations

Magnetic Data in Neutral Sheet Coordinates

Figure 2 shows the time variation of the magnetic field measured by ISEE-1 and -2. Since the magnetic data indicate that around the onset of the expansion phase the neutral sheet was not parallel to the Sun-Earth line, we transformed the coordinates into a neutral sheet coordinate system in which the new x axis is parallel to the tilted neutral sheet and positive sunward, while the y component of the assumed background magnetic field is near zero [McPherron et al., 1987]. This system is rotated 15° about z(GSM) and 10° about the new y axis as shown in Figure 3. All of the magnetic data presented in this paper have been plotted in the neutral sheet coordinate system.

Tail-like Magnetic Field Prior to the Expansion Onset

Figure 2a shows that in the early stage of the substorm (before $\sim$ 1025 UT), ISEE-1 was south of the neutral sheet and observed B_z and B_x components of comparable size,

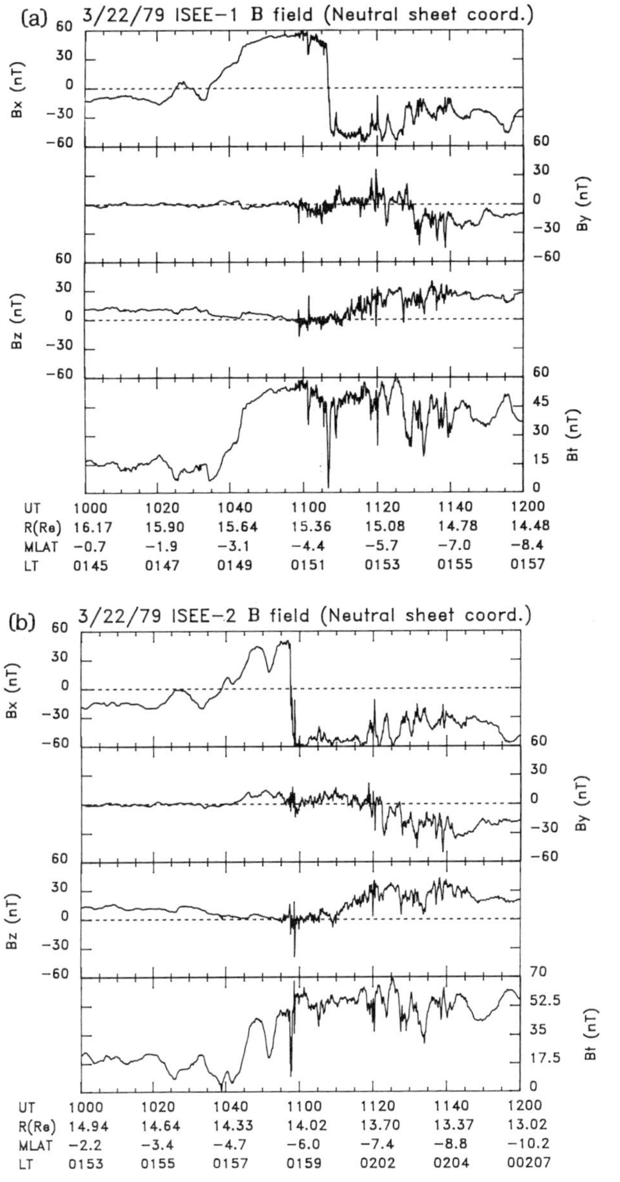

Fig. 2. (*a*) ISEE-1 magnetic data in the neutral sheet coordinates. B_t is the total magnetic field. The spacecraft geocentric radial distance in Earth radii (R_e), the magnetic latitude in degrees, and the local time are provided. (*b*) ISEE-2 magnetic data in the neutral sheet coordinate system.

implying a dipole-like field. Due to motion of the magnetotail, ISEE-1 crossed the neutral sheet three times between 1025 – 1035 UT, shown by the changes of the sign of B_x. After 1040 UT, the spacecraft was north of the neutral sheet and observed a large increase in B_x while B_z dropped to nearly zero, indicating a very tail-like magnetic field. ISEE-2 measured a similar magnetic variation (Figure 2b). The separation vector pointing from ISEE-2 to ISEE-1 near the onset time was about (−8800, 1200, 3700 km) in the tilted neutral sheet coordinates.

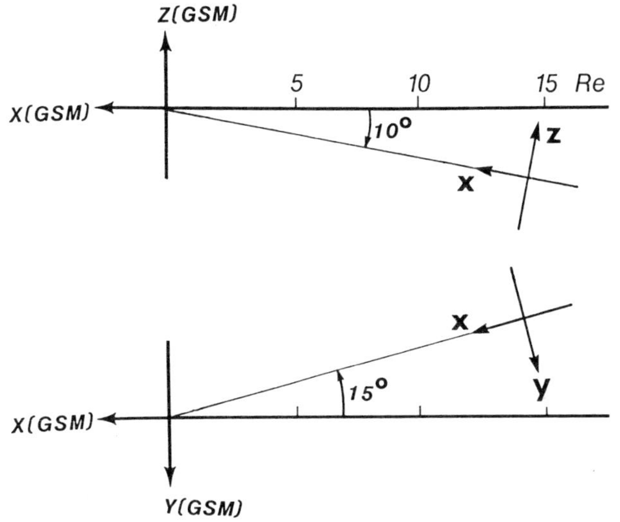

Fig. 3. Tilted neutral sheet coordinate system relative to GSM coordinates.

McPherron et al. [1987] used the observations in Figure 2 to model the current sheet thinning. They estimated that the current sheet had a thickness of ~ 5 R_E prior to the substorm and that it was less than 1 R_E thick at expansion onset.

The development of a tail-like field was also observed by the geostationary spacecraft GOES 3 which was near the same 0200 LT magnetic meridian as ISEE during the interval when the IMF was southward (1010 UT to 1110 UT) [McPherron and Manka, 1985; Barfield et al., 1985; Fritz et al., 1984]. These observations suggest that the magnetotail field between 6.6 R_E and 15 R_E was stretched to a tail-like configuration (as illustrated in Figure 1b). This stretching was associated with energy accumulation during the growth phase of the substorm before the expansion onset. It has been shown that following a southward turning of the IMF at ~ 1010 UT the energy input to the magnetosphere increased greatly [Baker et al., 1985; Tsurutani et al., 1985]. However, Baker et al. found that the energy dissipation in the ionosphere did not increase much until after the expansion onset. They concluded that the energy from the solar wind was stored in the magnetotail. Since the ISEE spacecraft never went into lobe regions during this period (the plasma density was always above ~ 0.1/cm^3 [Paschmann et al., 1985]), the large increase in the magnitude of the total magnetic field after 1040 UT is regarded as the signature both of increasing storage of magnetic energy in the tail and of spacecraft motion away from the central plasma sheet.

Expansion Onset and Streaming of Energetic Particles

As described in the Overview section, the onset of the expansion phase occurred at 1054 UT (when a large scale plasmoid is presumably released) and an upward motion of the magnetotail caused both ISEE spacecraft to cross the separatrix at ~ 1058 UT (Figure 1c). We have noticed that this separatrix is within the plasma sheet, since before and after the separatrix crossing both spacecraft remained in the plasma sheet. In the 2-D plasmoid picture, this separatrix separates lobe field lines and the (reconnected) plasma sheet field lines [Hones, 1979; and the 'outer separatrix' in Hones et al., 1984]. If the reconnection region is three dimensional, the separatrix may not extend infinitely in the dawn-dusk direction. We propose in Figure 4 a possible structure of the three dimensional separatrix in the dawn-dusk direction. Figure 4 illustrates schematically the cross section of the tail (at the location of the spacecraft) looking towards the Earth just before the spacecraft crossed the separatrix S, $\sim$ 4 minutes after the release of the plasmoid (the expansion onset). As shown in the figure, the separatrix is restricted in the dawn-dusk direction. Only around the center of the reconnection region does the separatrix separate plasma sheet field lines (reconnected field lines) from lobe field lines. At both edges of the region the separatrix S goes into the plasma sheet. Before the separatrix crossing, the two spacecraft (solid dots) were still located within the plasma sheet although the near-Earth neutral line had already formed. Once they crossed the separatrix, the spacecraft then observed streaming energetic particles as described below. Note that the magnetic field lines on either side of the post-plasmoid plasma sheet wrap over the separatrix. This produces a magnetic component in the dawn-dusk direction. High time resolution data plotted in Figure 8 indicates B_y perturbations consistent with this diagram. ISEE-1 above the neutral sheet observed a large negative B_y perturbation, while ISEE-2 below saw a smaller positive B_y perturbation.

The lower spacecraft ISEE-2 went further into the current sheet crossing the neutral sheet at 1057:30 UT, so that between $\sim$ 1058 UT and $\sim$ 1107 UT the two spacecraft were on the opposite sides of the neutral sheet. The magnetic data in Figure 2 show that the magnetic field during this period was still very tail-like and antiparallel at the two spacecraft locations. The field remained tail-like until $\sim$ 1110 UT when B_z began to increase rapidly to a level comparable to B_x, indicating that the magnetic field at the spacecraft location was again dipole-like.

After the spacecraft passed through the separatrix, field and particle signatures associated with the expansion onset were observed: the magnetic field exhibited larger fluctuations (at $\sim$ 1058 UT at ISEE-1 and $\sim$ 1057 UT at ISEE-2), and at the same time MEPI on ISEE-1 observed an abrupt and simultaneous increase in the ion fluxes in all energy channels. Figure 5 shows the ion fluxes averaged over 12 spin periods ($\sim$ 36 s). The ion fluxes increased abruptly at $\sim$ 1058 UT. This was followed by a period of rapid fluctuations, during which the ion intensities gradually increased until $\sim$ 1140 UT when the recovery phase of the substorm began on the ground. Note that the sharp decrease of fluxes at about 1125 UT was caused by a brief excursion of ISEE-1 into the southern lobe [Ipavich et al., 1985].

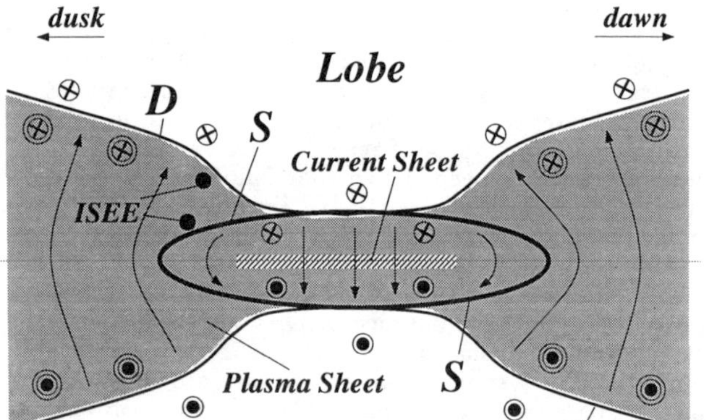

Fig. 4. The cross section of the tail at the location of the ISEE spacecraft looking towards the Earth just before the spacecraft crossed the separatrix S (see text). Lines D are the separatrices between plasma sheet field lines and lobe field lines. S and D coincide only near the center of the tail. Thin lines with arrows represent the direction of the magnetic field in the cross sectional plane. Solid dots (crosses) in single circles denote open magnetic field lines pointing out of (into) the paper, solid dots (crosses) in double circles denote closed field lines pointing out of (into) the paper. The extention of the thin current sheet in the dawn-dusk direction is not clear.

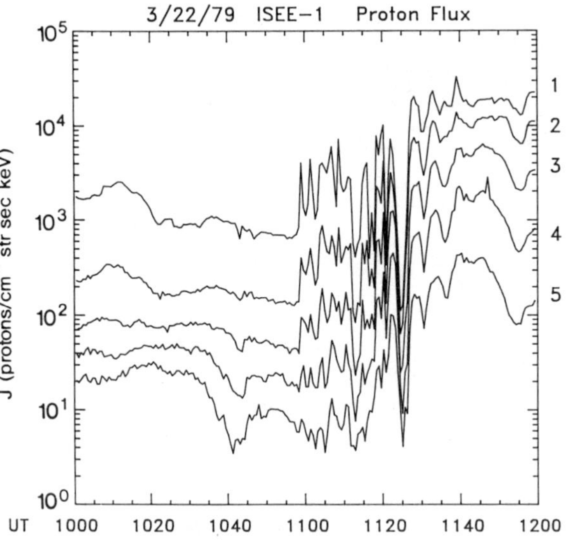

Fig. 5. 12-spin averaged proton fluxes. Only data of the first five energy channels are shown. Channel numbers are shown at the right side of the panel: (1) 24–44.5 keV, (2) 44.5–65.3 keV, (3) 65.3–95.5 keV, (4) 45.5–142 keV, (5) 142–210 keV.

The sudden increase in the energetic ion fluxes at 1058 UT consisted of predominantly tailward streaming ions. Figure 6 displays the ratio (r) of field aligned fluxes (with pitch angle $\leq 30°$ or $\geq 150°$) directed away from the Earth (tailward) to field aligned fluxes directed towards the Earth observed by ISEE-1. $r > 1$ corresponds to tailward fluxes, while $r < 1$ corresponds to earthward fluxes. Prior to the expansion onset, there was no field-aligned ion streaming. At 1058 UT, sudden strong tailward streaming appeared simultaneously in all ion energy channels. This streaming strongly suggests the formation of a neutral line earthward of the ISEE-1 location. The tailward streaming continued until $\sim$ 1118 UT when r dropped be-

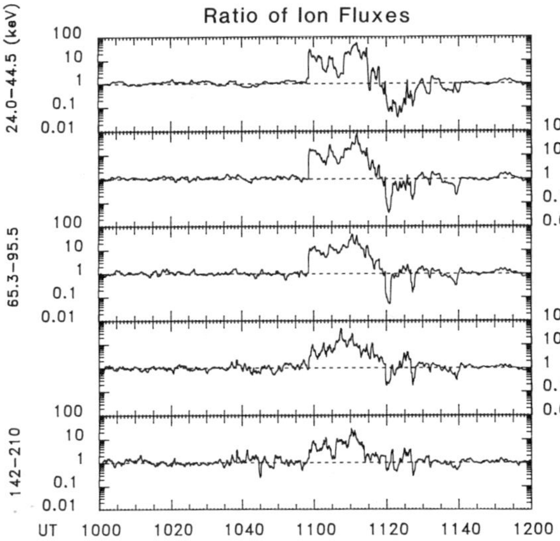

Fig. 6. The ratio of tailward fluxes to earthward fluxes for channels 1 to 5 of the ISEE-1 MEPE data.

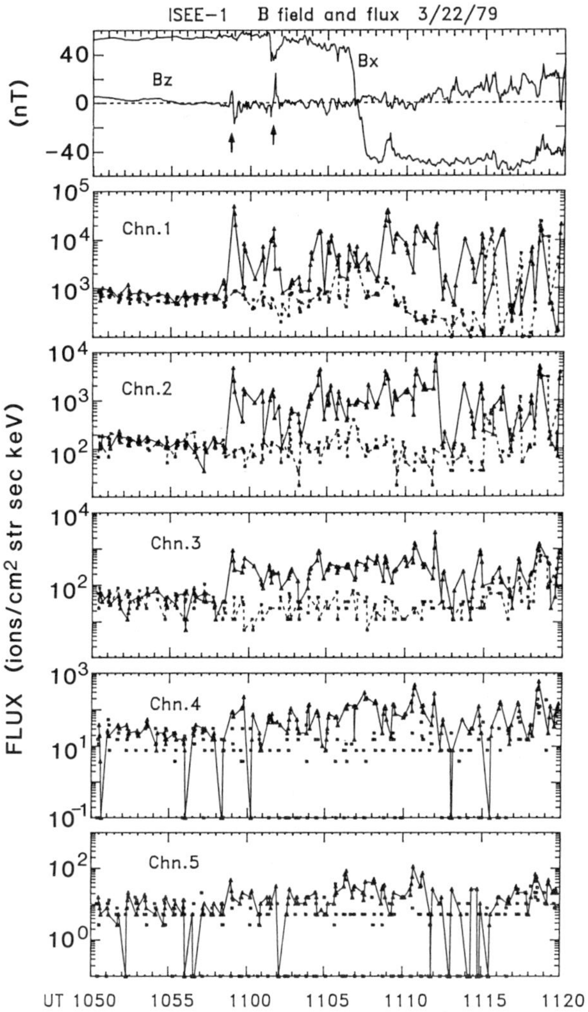

Fig. 7. The magnetic B_x and B_z components are plotted in the top panel. Panels 2 to 6: Tailward ion fluxes (triangles connected by solid lines) and earthward fluxes (squares, connected by dashed lines for channels 1 to 3) of channels 1 to 5 of the ISEE-1 MEPE data for the interval 1055–1120 UT.

low 1, corresponding to earthward streaming. The earthward streaming gradually decreased and the asymmetries of the field-aligned flux disappeared at $\sim$ 1140 UT. The sequence of tailward followed by earthward streaming was similar for all energy channels. Energetic electrons measured by MEPI also showed the same streaming variation as that of the ions [Lin et al., 1989]. We note also that the streaming pattern of the energetic particles is consistent with the variation of the direction of plasma bulk velocity reported by Paschmann et al. [1985].

Observations of Magnetic Islands

The start of tailward particle streaming at $\sim$ 1058 UT coincides with a bipolar signature in B_z (marked by the first arrow in Figure 7 which displays the magnetic B_x and B_z components for ISEE-1, and the tailward and earthward directed ion fluxes used for determining the flux ratio plotted in Figure 6). This signature was also seen by ISEE-2, which at this time was on the opposite side of the plasma sheet. There was another bipolar B_z in ISEE-1 data (the second arrow in the Figure 7) associated with another burst of tailward streaming ions. We believe that these two bipolar signatures were caused by the passage of magnetic island-like structures over ISEE.

Figures 8a and 8b present high resolution (0.25 s per sample) magnetic data for the interval 1058 to 1103 UT when the two loop-like structures were seen. The first bipolar signature in B_z was observed by both spacecraft. We propose that ISEE-1 just crossed the northern edge of the magnetic island since B_x was almost constant while B_z changed from +10 to -20 nT (Figure 8a). ISEE-2, penetrated deeper into the island structure since B_z first increased to $\sim$ 30 nT and then decreased to $\sim$ -50 nT (Figure 8b). The zero crossing of the first bipolar signature at ISEE-1 lagged behind that of ISEE-2 by about 15 seconds. Since ISEE-2 was $\sim$ 8800 km earthward of ISEE-1 on their inbound orbit, the tailward velocity of the loop structure was $\sim$ 580 km/s. This result is consistent with that from electric field measurements by Pedersen et al. [1985]. In the insert of their Figure 3, they showed that northward B_z and dawnward E_y preceded southward B_z and duskward E_y, which gave a tailward $\mathbf{E} \times \mathbf{B}$ drift velocity of $\sim$ 500 km/s. It took about 30 seconds for the spacecraft to pass the island structure (Figure 8), thus the length of the structure is estimated as $\sim 2 - 3\ R_E$.

The second loop was observed on ISEE-1 between 1101 UT and 1102 UT. This time the B_x component decreased significantly (by about 20 nT) while B_z showed another bipolar signature: first negative ($\sim$ -15 nT) then positive ($\sim$ +30 nT). However, ISEE-2 recorded no evidence whatsoever of this event. This implies that the second loop passed either north or dawnward of ISEE-2 (see Figure 4). In Figure 9 we present a schematic view of the two magnetic loop structures which provides a possible interpreta-

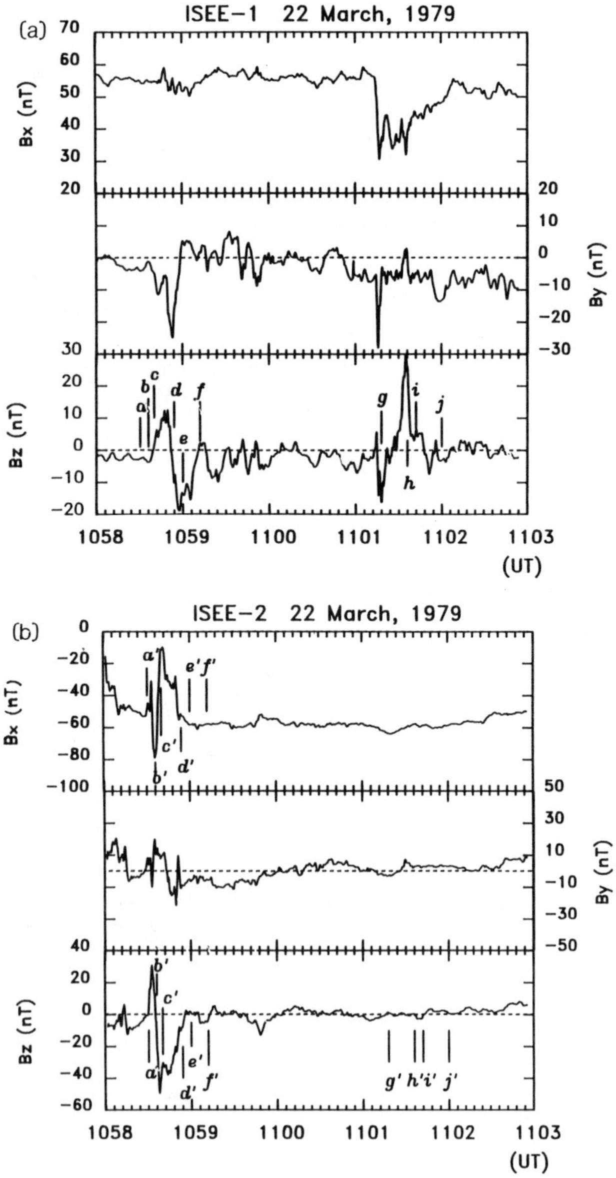

Fig. 8. (*a*) High resolution (0.25s/sample) magnetic data from ISEE-1 for the interval 1058 to 1103 UT. (*b*) Same as 7*a* but for ISEE-2. The labelled points correspond to points in a schematic interpretation illustrated in Figure 8.

tion of these observations assuming that the second loop passed primarily north of ISEE-2. The suggested paths of ISEE-1 and 2 with respect to the magnetic islands are plotted in the figure with dashed lines, and critical times are marked a to j, corresponding to the labels in Figure 8.

To better illustrate the magnetic evidence for the two magnetic islands we have plotted in Figure 10 magnetic vectors measured at ISEE-1 and ISEE-2 projected onto the $X - Z$ plane. Vectors measured on ISEE-1 are plotted for the interval 1058:30 to 1059 UT (Figure 10a) and 1101 to 1102 UT (Figure 10b). ISEE-2 vectors are plotted for intervals 15 seconds earlier than the ISEE-1 intervals (1058:15 to 1058:45 UT and 1100:45 to 1101:45 UT). In the diagrams, we assumed that static structures moved tailward past the spacecraft with a uniform velocity of 580 km/s. The vector pattern in each interval is consistent with the loop-like magnetic field lines which are overplotted schematically on the same diagrams. We note that there were finite B_y components in the two events, and will discuss them below.

Reversal of Streaming of Energetic Particles

We have noted in Figure 6 that midway through the expansion phase, at $\sim$ 1118 UT, streaming of energetic particles changed to the earthward direction as observed by ISEE-1 MEPI. The bulk velocity of plasma at the two ISEE spacecraft also shows a tailward to earthward reversal at about the same time [Paschmann et al., 1985]. As illustrated in Figure 1d, the neutral line region apparently retreated tailward past the two ISEE spacecraft, placing them on the earthward side of the near Earth X−line, thus the earthward streaming of particles was observed. The streaming of particles subsided near the beginning of the recovery phase and the flux of energetic particles increased to a level a few orders of magnitude higher than that before the expansion onset (Figure 5), which is evidence of conversion of magnetic energy accumulated during the growth phase into particle kinetic energy.

Discussion

We have presented observational data which support the neutral line model for the CDAW 6 event described in the Overview section. The near Earth reconnection model of substorms [Hones, 1979] has been taken as a framework for our interpretation, but we have provided a more detailed physical picture of what happened in the near magnetotail during the substorm.

During the growth phase of the substorm, the ISEE spacecraft, which were located at $\sim$ 15 R_E in the tail

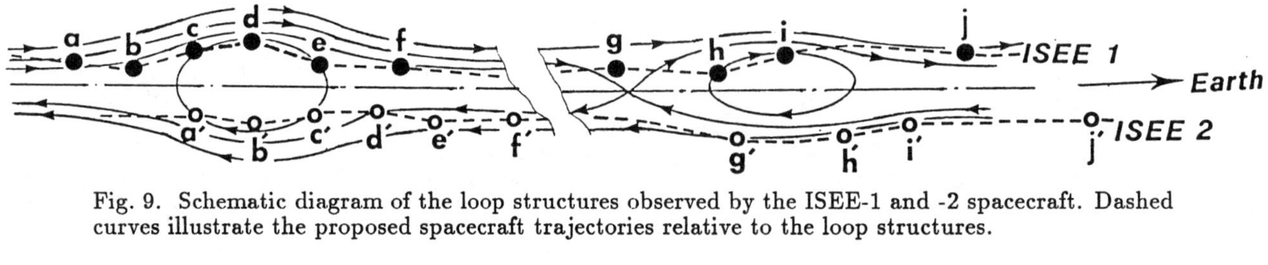

Fig. 9. Schematic diagram of the loop structures observed by the ISEE-1 and -2 spacecraft. Dashed curves illustrate the proposed spacecraft trajectories relative to the loop structures.

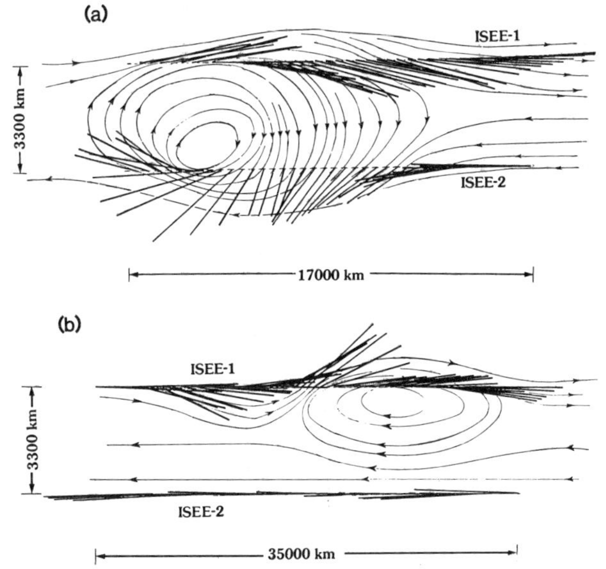

Fig. 10. Magnetic vectors (thick lines) measured at ISEE-1 and -2 projected onto the $X - Z$ plane to visualize the two loop structures: (a) the first loop observed between 1058:30 and 1059 UT by ISEE-1, and (b) the second loop observed between 1101 and 1102 UT by ISEE-1. Imaginary magnetic field lines (thin lines) are overplotted on each vector plot. Vectors are plotted every 0.75 seconds.

near the neutral sheet, observed that the magnetic field configuration changed from dipole-like into tail-like. The normal magnetic component, B_z, became very small after 1040 UT (Figure 2). Reconnection is very likely in such a tail-like magnetic field configuration. Schindler [1974] suggested that when the plasma sheet is sufficiently thin and/or the normal magnetic field component has been sufficiently reduced, the ion tearing mode will abruptly start and lead to the formation of multi $X-$lines and $O-$lines. He identified the breakup of ion tearing with the abrupt beginning of the substorm expansion phase. Recently, Lee [1988] studied time-dependent magnetic reconnection and suggested the possibility that multiple X line reconnection (MXR) might occur in the tail during substorms.

The two magnetic island structures observed by the ISEE spacecraft were probably produced by such multipoint reconnection. They moved tailward at a velocity of $\sim$ 500 to 600 km/s as indicated by magnetic and particle signatures. The second loop appeared two minutes later than the first. Thus, by the time the second loop was observed, the first one would have moved down the tail $\sim$ 10 R_E behind the spacecraft. Since we estimate that the near Earth neutral line was only $\sim$ 5 R_E earthward of the spacecraft, both loops could not have existed earthward of the ISEE spacecraft. Instead, the second loop was probably generated later than the first. The multipoint reconnection in this bottle-neck region of the tail is probably an intermittent process continuing for only a few minutes. The substorm onset was characterized by particle streaming parallel to the field. Simulations by Sato and Walker [1982] show that the tearing mode is excited much more violently in the presence of parallel plasma flow in the plasma sheet than in the case with no flow.

Recently, Richard et al. [1989] have studied the magnetic island coalescence instability for magnetotail reconnection. In the coalescence process small islands coalesce to form a large magnetic island which moves down the tail. They pointed out that reconnection in the tail could be a two step process: small scale island formation followed by the rapid coalescence of small islands into larger ones. There is no evidence that we were observing the two step process, but the occurrence of the small loop structures within 15 R_E may be evidence of small spatial scale reconnection in the near Earth magnetotail.

According to the near-Earth neutral line model, a localized X-line forms inside the plasma sheet sometime during the late growth phase. This neutral line slowly reconnects closed field lines threading the plasma sheet, creating a large bubble of closed magnetic loops centered about a point part way between the near-Earth $X-$line and the distant neutral line 100 R_E tailward. Expansion onset occurs when the $X-$ line severs the last closed field line connected to the distant neutral line. At this point the merging rate increases explosively as low density lobe plasma begins to flow into the $X-$ line. Subsequent reconnection of lobe magnetic field produces a sheath of interplanetary field lines around the back side of the plasmoid. Tension in these highly curved field lines causes them to contract accelerating the disconnected plasmoid away from the Earth. The $X-$ line remains at its initial location for sometime allowing continued reconnection of lobe magnetic field. As the plasmoid retreats it leaves behind a long, extremely thin current sheet threaded by weak southward magnetic field. This sheet persists until the neutral line moves away from the Earth filling the plasma sheet earthward of its location.

We propose that our observations of magnetic islands at $\sim$ 1058 and $\sim$ 1101 UT were made four minutes after the large scale plasmoid was released, and that the two spacecraft were on opposite sides of this thin current sheet. We believe that two, small magnetic bubbles were created by tearing of this current sheet, probably induced by its sudden upward motion. The data suggest that the $X-$ line remained earthward of the spacecraft until at least 1118 UT, and that no further islands were generated. The eventual fate of the islands that passed the ISEE spacecraft is unknown, but we speculate that they follow the main plasmoid down the tail, and may eventually catch up and merge with it through the coalescence instability [Richard et al., 1989].

As discussed earlier there were changes in the B_Y component associated with the first magnetic bubble to pass the ISEE 1 and 2 spacecraft. The sense of these perturbations, eastwards above the neutral sheet and westwards below, is consistent with our drawing of the separatrix closing across the neutral sheet as shown in Figure 4. However, the ISEE-2 signature for this event is not consistent with a flux rope, since deeper within the structure we would expect an even stronger B_y which was not observed. Most likely these events are small magnetic bubbles.

In summary, we propose that multi-point reconnection occurred in the near tail plasma sheet when the magnetic tail became very tail like and led to the expansion onset of the substorm under investigation. The tail-like field configuration was formed during the growth phase of the substorm. Our observations of magnetic field and energetic particles are consistent with the above conclusion.

We have considered the possibility of the phenomena we observed, such as the streaming pattern of energetic particles, being the results of the spacecraft traversing different regions, e.g. the central plasma sheet and the sheet/lobe boundary layer. But the onset of the particle streaming, its reversal and subsidence (for example in the period between 1055 UT and 1150 UT) were observed by two ISEE spacecraft (with a small time shift) while the two spacecraft were obviously travelling in different regions for some period (for example, from $\sim$ 1058 UT to $\sim$ 1107 UT, ISEE-2 was south of the neutral sheet while ISEE-1 was north of it). A similar streaming pattern observed by the two spacecraft located in the different regions is best interpreted as a time variation associated with a particle source (which is suggested as the near Earth reconnection region) rather than spatial effects (for example, the boundary layer effect). More detailed analysis of the data and discussion on the substorm are presented in Lin et al. [1989].

Acknowledgments. We are grateful to D. J. Williams for providing the MEPE data and C. T. Russell for magnetic data. The authors are grateful to D. G. Mitchell for helpful discussions on this event and on the data analysis problems associated with the use of the MEPE data. Work at UCLA has been supported by NSF grant ATM 83-18200 A01, JPL contract 955232, NASA NGL-05-007-004, and ONR N00014-85-K-0556.

References

Baker, D.N., Particle and field signatures of substorms in the near magnetotail, in *Magnetic Reconnection in Space and Laboratory Plasmas, Geophys. Monogr. Ser.*, Vol. 30, edited by E.W. Hones, Jr., p.193, AGU, Washington, D.C., 1984.

Baker, D.N., T.A. Fritz, R.L. McPherron, D.H. Fairfield, Y. Kamide, and W. Baumjohann, Magnetotail energy storage and release during the CDAW 6 substorm analysis intervals, *J. Geophys. Res.*, *90*, 1205, 1985.

Barfield, J.N., C.S. Lin, R.L. McPherron, Observations of magnetic field perturbations at GOES 2 and GOES 3 during the March 22, 1979, substorms: CDAW 6 analysis, *J. Geophys. Res.*, *90*, 1289, 1985.

Elphic, R.C., C.A. Cattell, K. Takahashi, S.J. Bame and C.T. Russell, ISEE-1 and 2 observations of magnetic flux ropes in the magnetotail: FTE's in the plasma sheet? *Geophys. Res. Lett.*, *13*, 648, 1986.

Fritz, T.A., D.N. Baker, R.L. McPherron, W. Lennartsson, Implications of the 1100 UT March 22, 1979 CDAW 6 substorm event for the role of magnetic reconnection in the geomagnetic tail, in *Magnetic Reconnection in Space and Laboratory Plasmas, Geophys. Monogr. Ser.*, Vol. 30, edited by E.W. Hones, Jr., p.203, AGU, Washington, D.C., 1984.

Hones, E.W., Jr., Plasma flow in the magnetotail and its implications for substorm theories, in *Dynamics of the Magnetosphere*, edited by S.I. Akasofu, p.545, D. Reidel, Hingham, Mass., 1979.

Hones, E.W., Jr., J. Birn, D.N. Baker, S.J. Bame, W.C. Feldman, D.J. McComas, R.D. Zwickl, J.A. Slavin, E.J. Smith, and B.T. Tsurutani, Detailed examination of a plasmoid in the distant magnetotail with ISEE 3, *Geophys. Res. Lett.*, *11*, 1046, 1984.

Lee, L.C., Toward a time-dependent magnetic reconnection model, *EOS, AGU, 69(49)*, p.1617, 1988.

Lin, N., R.L. McPherron, M.G. Kivelson, R.J. Walker, Multi-point reconnection in the near-Earth magnetotail: CDAW 6 Observations of energetic particles and magnetic field, submitted to JGR, 1989.

McPherron, R.L., and R.H. Manka, Dynamics of the 1054 UT March 22, 1979, substorm event: CDAW 6, *J. Geophys. Res.*, *90*, 1175, 1985.

McPherron, R.L., A. Nishida, C.T. Russell, Is near-Earth current sheet thinning the cause of auroral substorm onset? in *Quantitative Modeling of Magnetosphere-Ionosphere Coupling Processes*, edited by Y. Kamide and R.A. Wolf, p.252, Kyoto Sangyo University, Kyoto, Japan, 1987.

Ipavich, F.M., A.B. Galvin, M. Scholer, G. Gloeckler, D. Hovestadt, and B. Klecker, Suprathermal O^+ and H^+ ion behavior during the March 22, 1979 (CDAW 6), substorm, *J. Geophys. Res.*,, *90*, 1263, 1985.

Paschmann, G., N. Sckope, and E.W. Hones, Jr., Magnetotail plasma observations during the 1054 UT substorm on March 22, 1979 (CDAW 6), *J. Geophys. Res.*, *90*, 1217, 1985.

Pedersen, A., C.A. Cattell, C.-G. Falthammar, K. Knott, P.-A. Lindqvist, R.H. Manka, and F.S. Mozer, Electric fields in the plasma sheet and plasma sheet boundary layer, *J. Geophys. Res.*, *90*, 1231, 1985.

Richard, R.L., R.J. Walker, R.D. Sydora, and M. Ashour-Abdalla, the coalescence of magnetic flux ropes and reconnection in the magnetotail, *J. Geophys. Res.*, *94*, 2471, 1989.

Russell, C.T., The ISEE 1 and 2 fluxgate magnetometers, *IEEE Trans. Geosci. Electron., GE-16(3)*, 239, 1978.

Sato, T., and R.J. Walker, Magnetotail dynamics excited by the streaming tearing mode, *J. Geophys. Res.*, *87*, 7453, 1982.

Schindler, K., A theory of the substorm mechanism, *J. Geophys. Res.*, *79*, 2803, 1974.

Tsurutani, B.T., C.T. Russell, J.H. King, R.D. Zwickl, and R.P. Lin, A kinky heliospheric current sheet: Cause of CDAW-6 substorms, *Geophys. Res. Lett.*, *11*, 339, 1984.

Tsurutani, B.T., J.A. Slavin, Y. Kamide, R.D. Zwickl, J.H. King, and C.T. Russell, Coupling between the solar wind and the magnetosphere: CDAW 6, *J. Geophys. Res.*, *90*, 1191, 1985.

Williams, D.J., E. Keppler, T.A. Fritz, B. Wilken, and G. Wibberenz, The ISEE 1 and 2 Medium Energy Particles Experiment, *IEEE Trans. Geosci. Electron., GE-16(3)*, 270, 1978.

THE MAGNETIC TOPOLOGY OF THE PLASMOID FLUX ROPE IN A MHD-SIMULATION OF MAGNETOTAIL RECONNECTION

J. Birn and M. Hesse

University of California, Los Alamos National Laboratory, Los Alamos, NM 87545

Abstract. On the basis of a three-dimensional MHD simulation we discuss the magnetic topology of a plasmoid that forms by a localized reconnection process in a magnetotail configuration including a net dawn-dusk magnetic field component B_{yN}. As a consequence of $B_{yN} \neq 0$ the plasmoid assumes a helical flux rope structure rather than an isolated island or bubble structure. Initially all field lines of the plasmoid flux rope remain connected with the Earth, while at later times a gradually increasing amount of flux tubes becomes separated, connecting to either the distant boundary or to the flank boundaries. In this stage topologically different flux tubes become tangled and wrapped around each other, consistent with predictions on the basis of an ad-hoc plasmoid model.

Introduction

One of the major components of the common model of magnetospheric substorms, usually referred to as the near-Earth neutral line model, is the severance, and subsequent tailward ejection, of a part of the plasma sheet, called the plasmoid [e.g., Hones, 1977]. The simplest picture of the magnetic topology and of its changes during this process, guided by two-dimensional or symmetric models, is that the plasmoid represents a magnetic "bubble," an entity consisting of closed loops of field lines, formed from closed plasma sheet field lines through magnetic reconnection, and enclosed by a magnetic flux surface. It has been realized, however, that the picture where field lines reconnect with themselves to form closed loops is highly singular and requires a symmetry that will usually not exist [Hughes and Sibeck, 1987; Schindler et al, 1988; Birn et al, 1989]. The more general picture of a plasmoid involves helical field lines rather than closed loops, which must be connected with either the Earth or interplanetary space and thus form a flux rope of field lines, not necessarily topologically different from the usual plasma sheet, lobe, or open field lines.

In the present paper we will investigate the structure of such a plasmoid flux rope, found in a 3-D MHD simulation of magnetic reconnection in the Earth's magnetic tail. A companion paper [Hesse and Birn, 1989] addresses the properties of the electric field in this simulation, with the emphasis on the component parallel to the magnetic field and on possible acceleration effects. An extended discussion of the simulation is given by Birn and Hesse [1989]. An understanding of the particular structure of the plasmoid flux rope, developing in a self-consistent way, might also help to understand and interpret apparent flux rope structures in other space environments, such as the magnetopause (flux transfer events), the solar atmosphere, the solar wind, or other planetary environments.

Geophysical Monograph 58

Initial Configuration and Numerical Procedure

The simulation started from a magnetotail equilibrium configuration (Figure 1) that includes a net dawn-dusk magnetic field component, B_{yN}, which breaks the mirror symmetry around the center plane $z = 0$ of the plasma sheet. It was calculated using an integration procedure described by Birn [1987, 1989]. This method allows one to calculate an equilibrium configuration through basic integrations, if four functions of two variables each are specified, equivalent to the general boundary value problem. With suitable simplifications these functions can be represented by (1) the total pressure $\hat{p}(x,y)$, which is independent of z because of the stretched tail geometry, (2) the location of the neutral sheet $z_o(x,y)$, defined as the location, where B_x reverses sign, or where $|\mathbf{B}|$ assumes a minimum along a field line, (3) $L(y)$, the characteristic scale length in z (current sheet or plasma sheet half thickness) as a function of y at a chosen location in x, and (4) $B_N(p)$, a function that determines the magnitude of the net cross-tail field on different field lines; for two-dimensional equilibria $(\partial/\partial y = 0)$ B_N would be constant on field lines, for three-dimensional equilibria this is no longer the case, and $B_N(p)$ has to be prescribed at a chosen location in x. Here we have made the following choice

$$\begin{aligned} &\hat{p} = 10/(r-70) \qquad r = \sqrt{(80-x)^2 + y^2} \\ &z_o = 0 \\ &L = cosh(ky) \qquad k = 0.125 \\ &B_N = 0.033 \end{aligned} \tag{1}$$

where normalized quantities are used, based on the characteristic current sheet scale length L_c and the lobe magnetic field

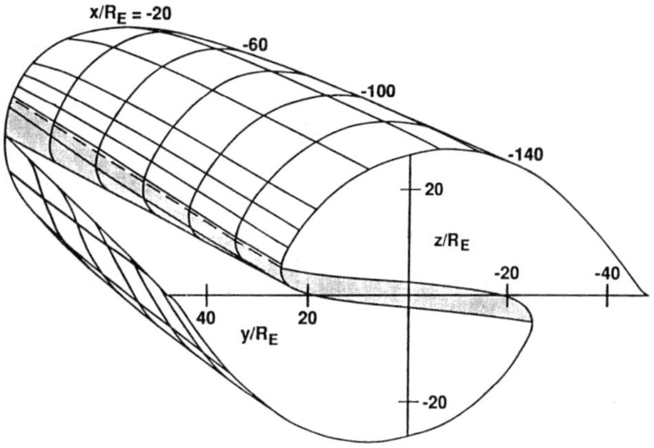

Fig. 1. Perspective view of the initial configuration as seen from the tail, showing a magnetic flux surface, originating from a circle $r = 20R_E$ at $x_{GSM} = -20R_E$. The shaded region consists of field lines with both ends inside this circle.

strength B_c, both taken at the center of the near-Earth boundary of the simulation box at $x = 0$, $y = 0$; L and B_N are also chosen at $x = 0$.

Characteristic features of the initial equilibrium, already included in earlier simulations [Birn and Hones, 1981; Birn, 1984], are gradients in the x direction, i.e., along the tail axis, associated with a field component, B_z, normal to the plasma/current sheet, and an increase of B_z, implied by the variation of the plasma sheet thickness, away from the center of the tail toward the dawn and dusk flanks ($\pm y$ directions).

The numerical code integrates the resistive MHD equations in a fully explicit way, using a standard leapfrog method described earlier [Birn and Hones, 1981]. No numerical smoothing or artificial diffusion was imposed, except when waves on the grid scale occurred, which gave rise to alternating gradients on more than two neighboring grid cells. In this case a local diffusion term was added to mass and momentum equations.

The simulation box was defined by $0 \geq x \geq -60$, $|y| \leq 10$, and $|z| \leq 10$. For a characteristic scale length $L_n = 2R_E$, this corresponds to a region in x_{GSM} between $-20R_E$ and $-140R_E$, if the total pressure is assumed to be proportional to $1/|x_{GSM}|$. Boundary conditions consisted of closed line tying boundaries at $x = 0$ and at $|y| = 10$ and $|z| = 10$, that is, assuming $\mathbf{v} = 0$, while ρ, p, and the tangential magnetic field $\mathbf{B_t}$ were assumed continuous with vanishing normal derivatives $\partial/\partial n = 0$, and a consistency condition $\nabla \cdot \mathbf{B} = 0$ was imposed on the normal magnetic field component. Line symmetry around the x axis was assumed, that is, symmetry for a rotation by 180°. This allowed us to restrict the simulation box to the region $z \geq 0$, with boundary conditions at $z = 0$ following from this symmetry. The distant boundary $x = -60$ was assumed open with vanishing normal derivatives of ρ, $\mathbf{v}$, and p. The tangential magnetic field components for outward flow were determined by a convective condition $d\mathbf{B_t}/dt = 0$, to ensure that topological features, such as neutral points $\mathbf{B} = \mathbf{0}$ could be transported across this boundary; for a simple von Neumann condition $\partial/\partial n = 0$ this is not the case. For inward flow, $\mathbf{B_t}$ was held fixed. A consistency condition $\nabla \cdot \mathbf{B} = 0$ was again imposed on the normal magnetic field component.

Field Evolution and Magnetic Topology

The dynamic evolution of the tail was initiated by imposing finite resistivity, which allowed an unstable tearing mode to develop from an initial diffusion. At the present time, when the actual dissipation mechanism that enables magnetic reconnection has not yet been clearly identified, a resistive term is the most convenient. For simplicity, and to avoid the prescription of the location and extent of the reconnection region, the resistivity η was chosen uniform, with a magnitude corresponding to a Lundquist number $S = \mu_o L_c V_c/\eta = 200$. Here L_c represents the scale length in the z direction, defined above, and V_c is a typical MHD wave speed, that is here, an Alfvén speed, defined by the characteristic lobe magnetic field strength and the plasma sheet density. The magnitude of the resistivity was chosen mostly for numerical reasons, to allow us to distinguish the Alfvén time scale, the resistive diffusion time scale, and the intermediate tearing growth time, within reasonable computing time. Nevertheless the growth time of the unstable modes of a few tens of Alfvén times, corresponding to a few minutes, is not far from observed time scales. The magnitude of the net cross-tail field component B_{yN} was chosen to be about 3% of the lobe field, consistent with average observed values [Fairfield, 1979]. For comparison, a symmetric case with $B_{yN} = 0$ was also studied.

Figure 2 shows the magnetic field evolution in the midnight meridian plane $y = 0$. The lines are trajectories of the vector field B_x, B_z only and not magnetic field lines or projections of field lines. Due to the smallness of B_y, and in particular of $\partial B_y/\partial y$, however, the shown configuration is very close to the case $B_{yN} = 0$, so that Figure 2 is almost identical with the corresponding figure in that case, where the lines shown would represent actual field lines. The field evolution exhibited by Figure 2 indicates the formation of a plasmoid and its tailward ejection found already in earlier simulations [e.g., Birn and Hones, 1981].

The southward turning of B_z in the equatorial plane, which marks the formation of a neutral line and of a plasmoid in the symmetric case $B_{yN} = 0$, occurs at about $t = 100$, where the time is normalized by a characteristic Alfvén crossing time L_c/V_c. The results demonstrated by Figure 2 confirm that the field evolution is not significantly altered by the presence of the net B_y, so that it seems reasonable to continue the terminology "plasmoid formation and motion," developed on the basis of the symmetric case.

The magnetic field topology, however, is altered by the presence of the net B_y, which does not drastically change in the reconnection region during the evolution [Birn and Hesse, 1989]. Instead of an isolated magnetic bubble we find indeed a magnetic flux rope, consisting of helical field lines, that are initially still connected with the Earth at both ends, but open this connection at later times (see Figures 3–5). Since plasmoid field lines are not topologically different from other field lines in a strict sense, we have used a pragmatic criterion to identify the plasmoid. Plasmoid field lines are defined as field lines that cross the neutral sheet, identical with the equatorial plane $z = 0$ in our case, more than once. This criterion picks out typical helical field lines. It is not a strict topological criterion, however, because it is based on the arbitrary choice of a certain plane. Figure 3 shows, in a perspective view, four

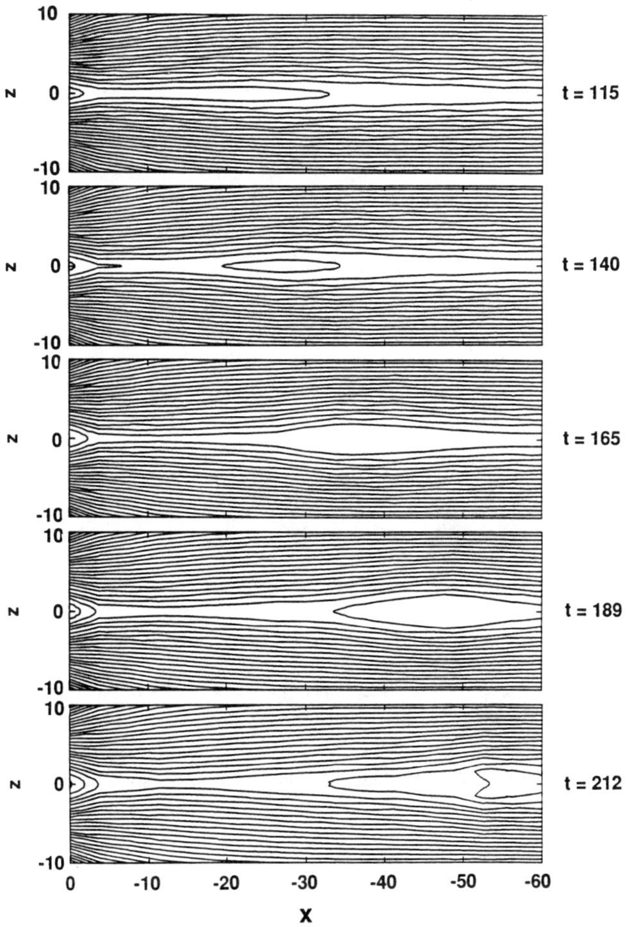

Fig. 2. Magnetic field evolution in the midnight meridian plane $y = 0$. The lines connect projections of field vectors in this plane and are not field lines or projections of field lines.

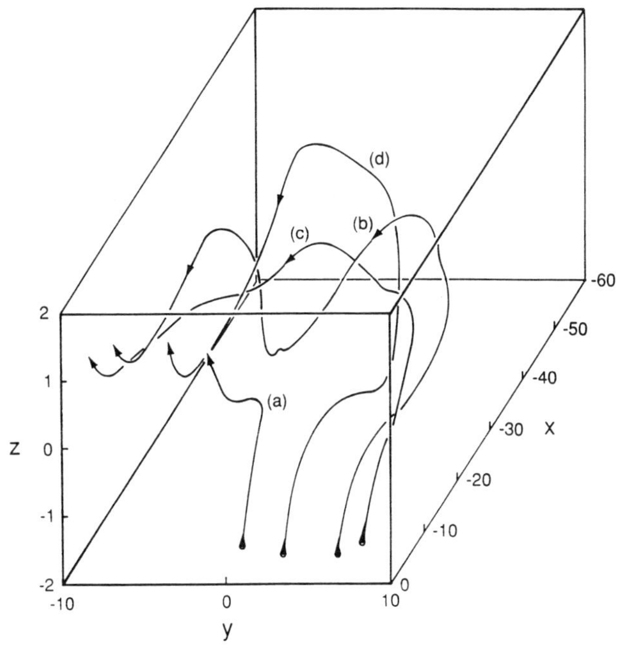

Fig. 3. Perspective view of four field lines as seen from the earthward side, for $t = 165$. Field lines labeled (a) and (d) represent ordinary plasma sheet field lines, crossing the equatorial plane just once; the field line labeled (b) represents a closed plasmoid field line, also connected with the near-earth boundary at both ends, but crossing the equatorial plane more than once; the field line labeled (c) near the center of the plasmoid flux rope, crosses the equatorial plane only once, due to its small pitch, but should also be considered as a plasmoid field line.

field lines at $t = 165$. One of these (labeled b) belongs to the plasmoid according to our criterion. One (labeled a) belongs to the plasma sheet earthward of the reconnection region, and another one (labeled d) belongs to the plasma sheet tailward from the plasmoid, which is not yet affected by the reconnection process. Both of these field lines cross the equatorial plane only once and can thereby be distinguished from the plasmoid field lines. The fourth field line (labeled c) crosses the equatorial plane also only once, due to its small pitch; it belongs therefore formally not to the plasmoid. Since it is located inside the region of plasmoid field lines with the major part of the plasmoid field lines winding around it, however, it should be considered as such a field line, too. This shows the problems with a criterion that is not based on the properties of the magnetic field only. Fortunately only very few field line are affected by this classification problem in the present case.

In addition to the plasmoid classification, we denote field lines as closed, if they have both ends at the near-Earth boundary, and as open, if they cross the distant boundary $x = -60$ or one of the boundaries at $|y| = 10$ or $|z| = 10$. It is quite obvious that this is not an invariant classification, because it depends on our choice of a simulation box, and that some of the field lines classified as open might in reality be closed or vice versa. We expect, however, that the qualitative results are not affected and that, for instance, a result that an increasing amount of initially closed magnetic flux tubes becomes connected with one of the flank boundaries indeed indicates that some such flux becomes connected with the magnetosheath. In the absence of physical boundaries in our system, this is the best we can do to get an impression of the topology change in the presence of real boundaries sufficiently close to the simulation boundaries.

A perspective view of the apparent plasmoid flux rope surface in our model is given by Figures 4 and 5 for two different times. These surfaces are defined as the boundaries between plasmoid field lines, crossing the equatorial plane more than once, and non-plasmoid field lines, that cross the equatorial plane either once, representing regular plasma sheet field lines (corresponding to types a or d in Figure 3), or not at all. Figures 4 and 5 were found by tracing plasmoid field lines close to this boundary. The flux rope represented in these figures is not only twisted, but also folded, so that field lines that are near the apparent surface in one region may become apparent interior field lines in another region. This is more obvious at the later time (Fig. 5), when the plasmoid consists not only of closed field lines, that are connected with the Earth at both ends, but also of open field lines connecting the earthward boundary $x = 0$ with the

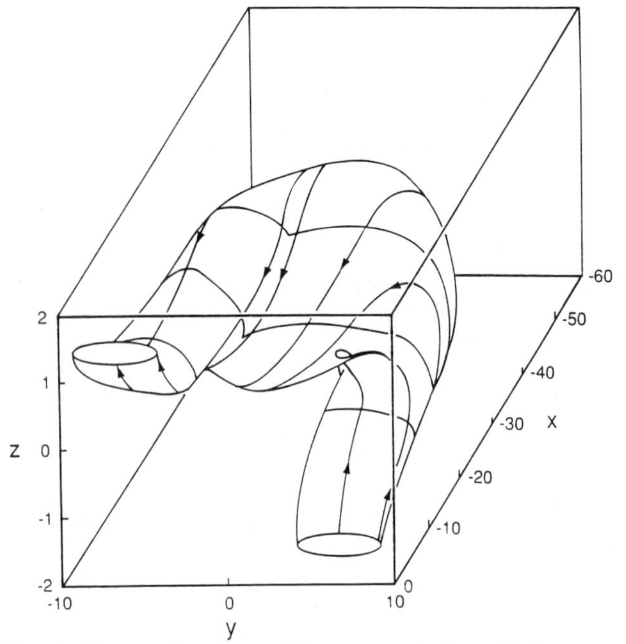

Fig. 4. Perspective view of the apparent plasmoid flux rope surface as seen from the earthward side, for $t = 165$.

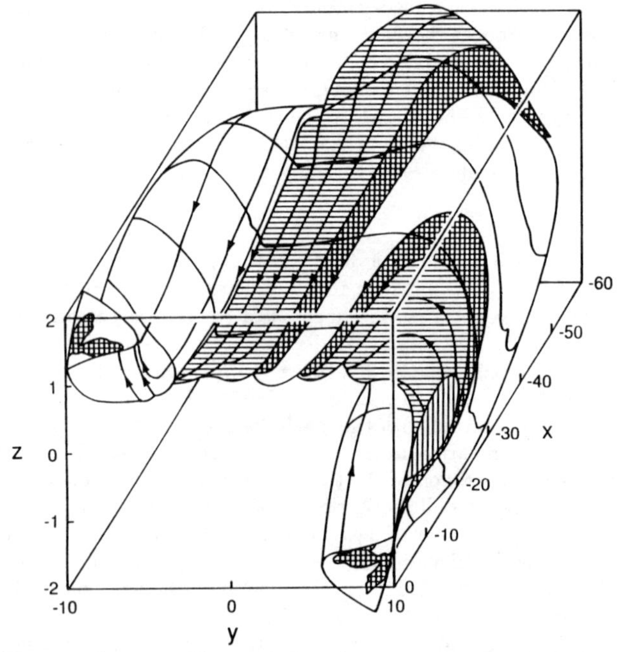

Fig. 5. Same as Figure 3, but for $t = 212$. Dark shading indicates open plasmoid field lines connecting the near-earth boundary $x = 0$ with the distant boundary $x = -60$; light shading indicates plasmoid field lines connected with one of the flank boundaries at $|y| = 10$.

distant boundary $x = -60$ (heavily shaded bundle) or with the flank boundaries $|y| = 10$ (lightly shaded bundles). A closed field line starting near the outer plasmoid surface on one side of the tail may end near the interior surface between open and closed field lines on the other side. Figure 5 shows how flux tubes with different connections get intertwined and wrapped around each other as predicted by Birn et al. [1989] on the basis of an ad hoc 3-D plasmoid model. Note that, despite the smallness of B_y, most field lines make only a few loops around the center of the plasmoid flux rope, as one can see from Figures 3–5. The reason is the large size of the plasmoid in the x direction, which leads to a considerable displacement of a field line in the y direction despite the smallness of B_y.

From Figure 5 it may look as if the amount of open plasmoid flux connected with the flank boundaries is comparable to, or even larger than the open plasmoid flux connected to the distant boundary. This, however, is misleading. The amount of flux connected to the flanks is in fact much smaller than the amount connected to the distant boundary, despite comparable surface areas at the boundaries, because the former one carries only the small B_y component across the boundary, while the distant flux carries the much larger B_x component across. This is the reason why only open plasmoid flux connected with the distant boundary (dark shading) is visible at the near-Earth boundary in Figure 5.

The detailed internal structure of the plasmoid and its evolution are further demonstrated by Figure 6, which shows the magnetic connections in several cross-sections $x = const$ of the tail at $t = 212$. The different types of field lines are indicated by different symbols. Plasmoid field lines are represented by the symbol #, if both ends are connected with the near-Earth boundary, and by a cross ×, if they connect the near-Earth boundary $x = 0$ with the distant boundary $x = -60$. Vertical bars represent regular plasma sheet field lines, crossing the equatorial plane just once, asterisks represent lobe field lines, which do not cross the equatorial plane, and horizontal bars represent field lines that are connected with one of the boundaries at $|y| = 10$. Note that no fully open field lines, connected at both ends with the distant boundary, are present, in contrast to what one might conclude from Figure 2. The full separation of the plasmoid thus apparently starts much later than suggested by Figure 2. Figure 6 demonstrates again the complexity of the field connections in the later stage, when the plasmoid separates from Earth. It is qualitatively similar to the structure of interwoven flux bundles of different topological type found earlier in an ad-hoc model of a plasmoid separating from Earth [Birn et al., 1989, Fig. 6].

For comparisons with observations of a plasmoid in the magnetotail or with other apparent flux rope structures in space, it is of particular interest to study the temporal variation of field components at a fixed location. This is done in Figures 7 and 8, which show the evolution of the three components of the magnetic field in the midnight meridian plane $y = 0$ for $z = 0$ and $z = 0.56$, respectively, and the locations $x = -30$ (solid lines) and $x = -45$ (dashed lines). The locations in x are tailward from the reconnection site, which is located near $x = -10$, at distances corresponding to about $40R_E$ and $70R_E$, if we assume a scale length $L_c = 2R_E$. Note that in our present simulation both locations are still within the initially closed plasma sheet region, where B_z is positive before the passage of the plasmoid. Both figures show the north-south signature of the passing plasmoid, becoming more pronounced with distance from the Earth. At $x = -30$, where the trailing part of the plasmoid can be observed longer due to its earlier passage,

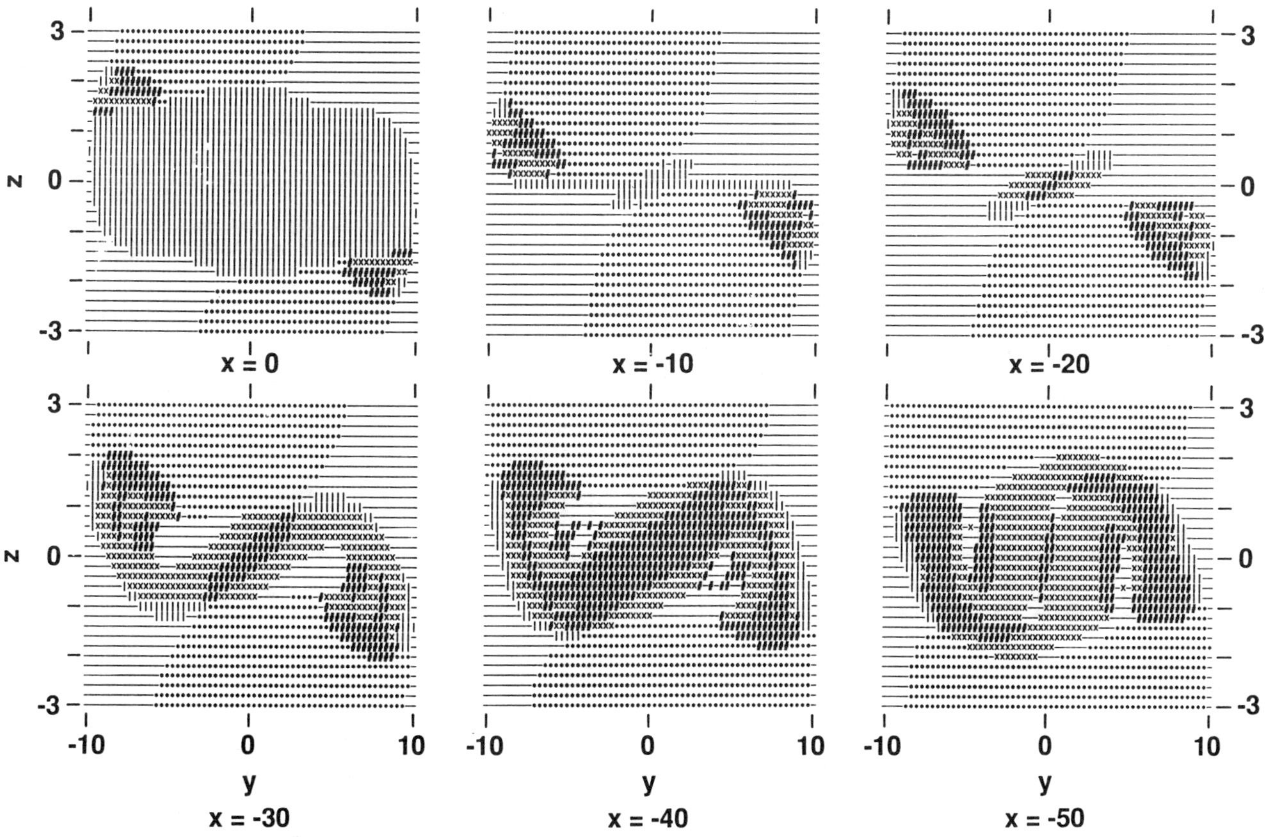

Fig. 6. Topological connections of field lines in different planes $x = const$ for $t = 212$. The symbols represent closed plasma sheet field lines (vertical bars), crossing the neutral sheet $z = 0$ just once; closed plasmoid field lines (symbol #), also with both ends at the near-earth boundary, but crossing the neutral sheet more than once; lobe field lines (asterisks), that do not cross the neutral sheet; lobe-like plasmoid field lines (symbol ×), also connecting the near-earth boundary $x = 0$ with the distant boundary $x = -60$, but with two or more neutral sheet crossings; and boundary field lines (horizontal bars) connected with one of the boundaries at $|y| = 10$.

the first set of maximum and minimum of B_z is followed by a second one during which B_z stays negative. This wavy pattern has already been observed in earlier simulations in a symmetric model without the net B_y [Birn, 1984]. In these simulations the second maximum of B_z could even get positive, so that the waves were associated with the presence of multiple neutral lines and multiple island structures. This waviness is apparently also a quite common feature of observed passages of plasmoids in the distant tail [e.g., Hones et al., 1984]. Figures 7 and 8 show that the waves in B_z are associated with waves in B_y, which exhibits maxima and minima with some phase difference to B_z. This indicates a helical distortion of field lines, associated with twisted flux tubes, visible in Figures 4 and 5, and the presence of field-aligned currents. The B_x component shows at all locations, except at $x = -45$, $z = 0.56$, a pronounced minimum, that coincides with a maximum of B_y. At this location B_x continues to decrease and even reverses sign. This signature, which would commonly be interpreted as a neutral sheet crossing, is related to the indented magnetic field structure in the rear end of the plasmoid at late times, visible in the last panel of Figure 2.

Summary

We have discussed the magnetic topology of a three-dimensional plasmoid, using results from a resistive MHD simulation of magnetic reconnection in the magnetotail. Due to the presence of a net dawn-dusk magnetic field component B_{yN} the plasmoid forms a flux rope with initially both ends at the near-earth boundary. As the plasmoid moves tailward and reconnection proceeds, an increasing amount of flux bundles opens to the tailward boundary and, to a smaller amount, to the flank boundaries. These open flux bundles are intertwined with the closed flux bundles, that are still fully connected with the earthward boundary. Despite the smallness of B_y the plasmoid field lines exhibit typically only a few loops, before they return to Earth or to the distant tail. The reason is the large size of the plasmoid in the x direction, i.e., along the tail, compared to its

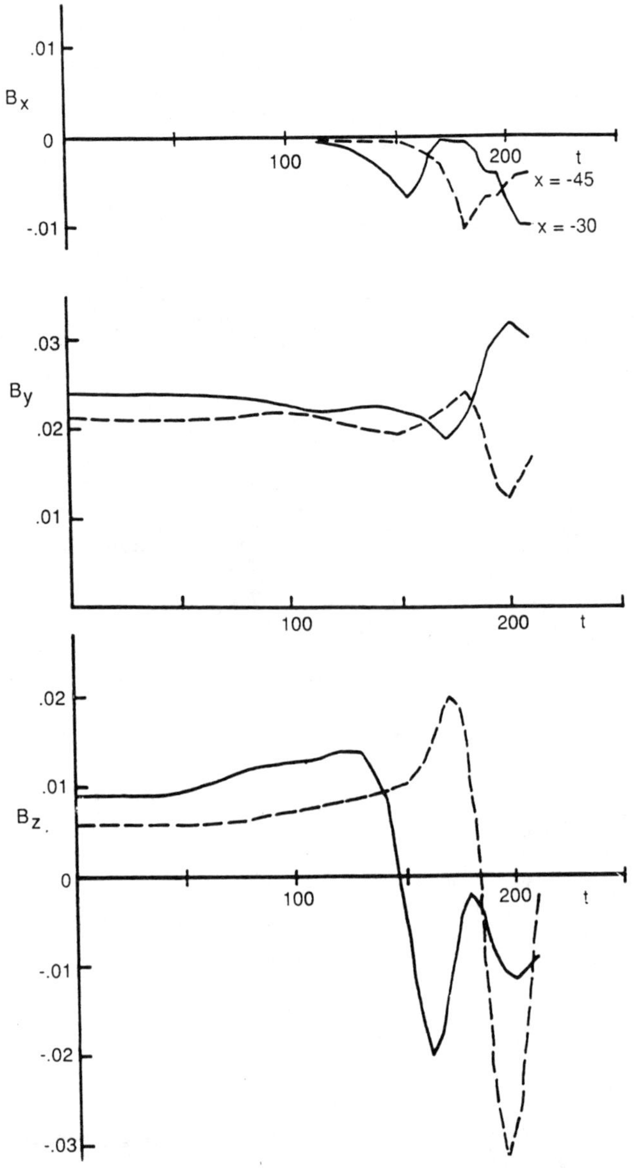

Fig. 7. Evolution of the three magnetic field components B_x, B_y, and B_z at fixed points with the coordinates $y = 0$, $z = 0$, and $x = -30$ (solid line), $x = -45$ (dashed line).

dawn-dusk extent. The plasmoid flux rope is not only twisted, but also folded, so that field lines that are located near its apparent surface in one region can become apparent interior field lines in another region.

Questions and Answers

Sonnerup: Am I correct in saying (a) that the dynamics you observe in your simulation are essentially uninfluenced by the presence of a nonzero

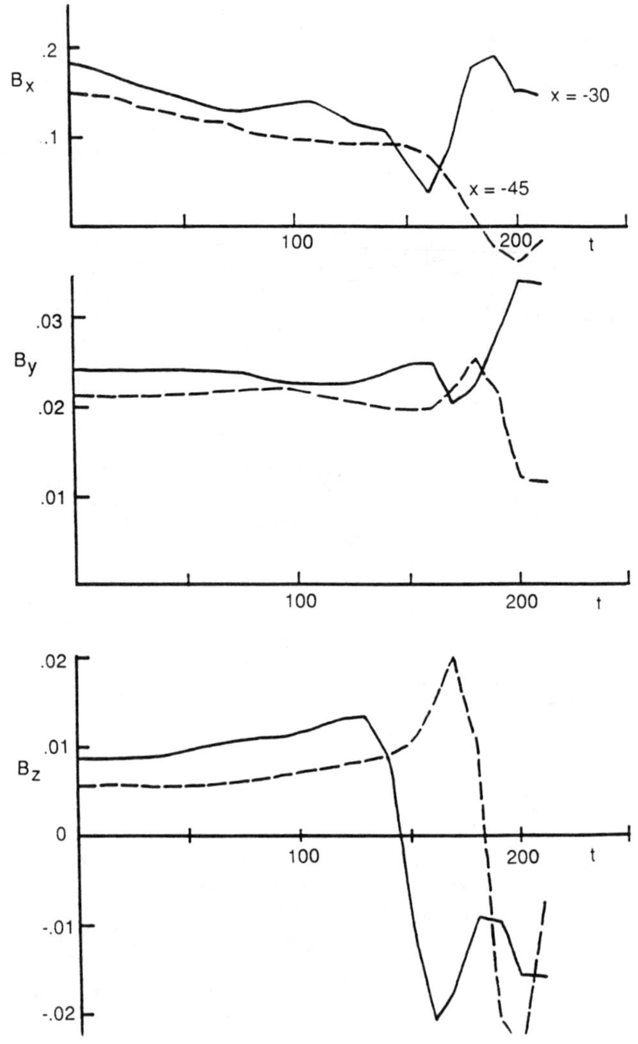

Fig. 8. Same as Figure 7, but for $z = 0.56$.

BY; (b) that those dynamics are strongly associated with, and controlled by the appearance of an X type null line in the BY = 0 simulations; (c) that it is this topological feature of the B field that has been eliminated in the new generalized definition of reconnection that you advocate. If the answer is yes, then a follow-up question is: Why not look for a definition of reconnection that incorporates this topological feature of the B field?

Birn: My answer to your first three questions is yes. The answer to your last question: A definition based on the vicinity of a symmetric case with the special topological features of an x-line and separatrices works well, when such a state is reasonably close and can be easily identified (this is why I continued to use the familiar terms "x-line" and so on, although with quotation marks). I would not know, however, what to do in a general case, further away from the symmetric one.

Hollweg: Solar folks get strong current sheets when flux ropes evolve dynamically. Do you get current sheet formation? Do you get formation of new current sheets which weren't there to start with?

Birn: Newly formed current sheets appear at the top and bottom of the plasmoid (at larger |z| values) and represent an outward squeezed and compressed region at the initially present crosstail current sheet. The region in the vicinity of the near-earth reconnection region ("x-line") is also considerably compressed compared to the initial cross-tail current sheet.

Sibeck: What is the cross-tail extent of a plasmoid in your model? What determines this extent?

Birn: The extent of the reconnection region, and thus of the plasmoid, in the y direction depends critically on the initial shape of the plasma sheet and the corresponding distribution of B_Z across the sail. In our earlier simulation, where B_Z had a more localized minimum region near $y=0$, the reconnection region occupied a smaller portion ($\sim$ 20-25%) of the tail width, whereas in the present case it extended over about 50% or more of the tail width.

Martens: As I understand it, when an $E_\parallel$ exists as in your results, electrons rather than protons are accelerated. Yet the observations show high energy protons. How do you reconcile this?

Birn: The parallel electric field in my model would accelerate both electrons and ions to about the same energy, but in opposite directions. The reason why I concentrated on electrons is that they are more likely adiabatic, so that their behavior and, in particular, the ejection regions of accelerated particles can be predicted by simply tracing field lines and not particle trajectories.

Forbes: What is the time-scale of the changes in the clumps in the plasmoid mapping in the near-earth region? Is it simply the Alfvén-scale time in the plasmoid region of the tail?

Birn: The "clumpy" regions of high field-aligned potential drop typically connect to both open and closed field line regions. The change of the location of these clumps does not directly reflect the change of filamentary connectivity, although the clumping is related to this structure. The speed of the motion in the z direction, toward higher latitude , is determined by the reconnection rate, i.e., corresponds to the inflow speed into the reconnection region, which is much smaller than the Alfvén speed. The speed of the motion in the y direction is also small, possibly also related to the reconnection rate.

Haerendel: In such numerical simulations, one has to assume a resistivity model whose appropriateness one can hardly assess. How good do you think is the derived parallel electric field?

Birn: The maximum parallel electric field in the diffusion region is close to the cross-tail field E_y in its vicinity, which is dominated by ideal MHD. I would therefore, expect that a different resistivity model will primarily affect the magnitude of $j_\parallel$ and the spatial concentration, that is, the size of the diffusion region, but to a lesser degree the maximum of $E_\parallel$. However, simulations with various resistivity models are needed to answer your question more satisfactorily.

Schindler: This is a comment regarding G. Haerendel's question on the physical relevance on assuming a resistivity. From particle simulations, one gets the impression that what counts is a suitable nonidealness in the sense of Ohm's law. The effect of the nonidealness (particle inertia in the case of the particle simulation) is restricted to a rather small region. The bulk of the field changes occur in the ideal domain. Therefore, dynamic models with different types of nonidealness tend to be quite similar. A final cautioning remark: The simulations I have in mind are still carried out with unrealistic ratios of electron to ion mass.

Acknowledgments. This work was supported by the U.S. Department of Energy through the Office for Basic Energy Sciences and by NASA under contract W-7405-ENG-36. M. H. gratefully acknowledges support from the Los Alamos National Laboratory Director's postdoctoral program.

References

Birn, J., Three-dimensional computer modeling of dynamic reconnection in the magnetotail, in *Magnetic Reconnection in Space and Laboratory Plasmas, Geophys. Monogr. Ser.*, vol. 30, edited by E. W. Hones, Jr., p. 264, AGU, Washington, D. C., 1984.

Birn, J., The distortion of the magnetotail equilibrium structure by a net cross-tail magnetic field, *J. Geophys. Res.*, in press 1989.

Birn, J., and M. Hesse, MHD simulations of magnetic reconnection in a skewed three-dimensional tail configuration, *J. Geophys. Res.*, submitted 1989.

Birn, J. and E. W. Hones, Jr., Three-dimensional computer modeling of dynamic reconnection in the geomagnetic tail, *J. Geophys. Res.*, *86*, 6802, 1981.

Birn, J., M. Hesse, and K. Schindler, Filamentary structure of a three-dimensional plasmoid, *J. Geophys. Res.*, *94*, 252, 1989.

Fairfield, D. H., On the average configuration of geomagnetic tail, *J. Geophys. Res.*, *84*, 1950, 1979.

Hesse, M., and J. Birn, Parallel electric fields in a simulation of magnetotail reconnection and plasmoid evolution, this issue.

Hones, E. W., Jr., Substorm processes in the magnetotail: Comments on 'On hot tenuous plasma fireballs and boundary layers in the Earth's magnetotail' by L. A. Frank, L. L. Ackerson, and R. P. Lepping, *J. Geophys. Res.*, *82*, 5633, 1977.

Hones, E. W., Jr., *et al.*, Structure of the magnetotail at 220 R_E and its response to geomagnetic activity, *Geophys. Res. Lett.*, *11*, 5, 1984.

Hughes, W. J., and D. G. Sibeck, On the 3-dimensional structure of plasmoids, *Geophys. Res. Lett.*, *14*, 636, 1987.

Schindler, K., M. Hesse, and J. Birn, General magnetic reconnection, parallel electric fields, and helicity, *J. Geophys. Res.*, *93*, 5547, 1988.

A 2½-DIMENSIONAL MAGNETIC FIELD MODEL OF PLASMOIDS

Mark B. Moldwin and W. J. Hughes

Center for Space Physics, Boston University, Boston, MA 02215

Abstract. The traditional two-dimensional picture of plasmoid formation predicts the creation of closed loops, field lines closed on themselves, which are called magnetic islands. Examination of plasmoid formation in three dimensions led Hughes and Sibeck [1987] to the conclusion that a flux rope is formed instead of a magnetic island. We use a 2 $\frac{1}{2}$-dimensional flux rope model to study the magnetic topology of plasmoids and examine the ability to distinguish between the two models using magnetometer data from a single satellite pass. Spacecraft data is simulated by sampling the magnetic field along a path through our model. We show that the principal axis directions are strongly dependent on the path of a satellite through the structure. We demonstrate that ISEE 3 magnetic field observations of plasmoids can be reproduced using a model of a flux rope with a significant axial component. It appears that principal axis analysis of magnetometer data of a single satellite pass is insufficient to differentiate between magnetic island and flux rope models, and can give misleading indications of the real axes of symmetry of the structure.

Introduction

The near-earth-neutral-line model of magnetospheric substorms predicts the formation of a separate plasma bubble, known as a plasmoid, in the earth's magnetotail at the time of substorm onset [Hones, 1977]. The plasmoid is subsequently ejected down the tail. The process of plasmoid formation and release consists of two stages. Initially the closed loops formed within the plasma sheet by magnetic reconnection are surrounded by closed field lines that have not been reconnected. The plasmoid is ejected down the tail when reconnection continues to lobe field lines that drape about the closed loops and pull it down the tail. This picture is two-dimensional and only incorporates the noon-midnight meridional plane. In this two-dimensional picture, the magnetic field within a plasmoid is closed on itself, a configuration we call a magnetic island. Hughes and Sibeck [1987] showed that the presence of a persistent and significant cross-tail magnetic field component, B_y, in the plasma sheet (as found by Akasofu et al. [1978]), alters the plasmoid formation process in such a way that helical field lines (flux ropes) are formed instead of closed loops. Therefore, a three-dimensional picture is needed to examine the magnetic structure of plasmoids. These three-dimensional plasmoids are still connected to the earth at their ends because of the finite extent of the reconnection region. If reconnection proceeds more rapidly at the center of the magnetotail open field lines wraped about the flux rope will bend or bow the structure anti-sunward. The plasmoid will be released when reconnection occurs at the ends of the structure.

These two pictures of plasmoid formation and evolution predict completely different topological structures and therefore might have different ramifications on the dynamics and configuration of the earth's magnetotail. Hence, the determination of the three-dimensional structure of plasmoids is important to understanding the configuration of the dynamic magnetotail. Unfortunately, the magnetic and plasma signatures for both closed loop and flux rope plasmoids are expected to be similar. They are: bi-polar magnetic field signatures, tailward flow of hot plasma, and isotropic energetic particle distributions [Sibeck, 1989]. Flux ropes should also have a significant axial field component. The existence of a significant axial magnetic field would lend support to the flux rope plasmoid picture. Magnetic signatures that have been interpreted as flux ropes have been observed in the deep tail by ISEE 3 and were thought to be related to plasmoids [Sibeck et al., 1984; Gloeckler et al., 1985]. Principal axis analysis (PAA) has been used in an attempt to determine the orientation of magnetic flux ropes and thus their axial field components in the Venus ionosphere [Elphic et al., 1980; Elphic and Russell, 1983] and in the geomagnetic tail [Sibeck et al., 1984; Elphic et al., 1986; Slavin et al., 1988]. In the case of the Venus flux ropes, which have strong core fields, Elphic et al. [1980] and Elphic and Russell [1983] showed that, assuming that the flux rope was cylindrically symmetric and that the spacecraft passed close to the axis of the flux rope, the maximum variance direction corresponds to the axis of the flux rope and the bi-polar trace is found in the intermediate direction. The flux ropes observed in the near-tail by Elphic et al. [1986] however, had a weaker core field so the axis of the flux rope was along the intermediate variance direction. The deep magnetotail flux ropes observed by Sibeck et al. [1984] were assumed to have the orientation of the axis parallel to the maximum variance direction. The physical dimensions of a plasmoid, approximately 100 R_E in length and a few times 10 R_E in height [Slavin et al., 1989], necessarily preclude the possibility for cylindrical symmetry. Nevertheless, several workers used PAA and the results of

Geophysical Monograph 58

Elphic and Russell's [1983] study of cylindrically and axially symmetric flux ropes to determine the orientation of plasmoids [Slavin et al., 1989; and Sibeck et al., 1984].

ISEE 3 magnetometer data of a typical plasmoid pass, that of March 22, 1983, is shown in Figure 1 in GSM cartesian coordinates. The goal of this paper is twofold; to determine if PAA of magnetic field observations from a single spacecraft pass, such as that shown in Figure 1, is sufficient to distinguish between non-cylindrically symmetric flux ropes and isolated magnetic islands, and to determine if the ISEE 3 magnetometer observations are consistent with a flux rope model. To accomplish these goals we developed a $2\frac{1}{2}$- dimensional magnetic field model of a flux rope. We were then able to run experiments that consisted of extracting the magnetic field along an arbitrary path through the structure and transforming the data into a principal axis coordinate frame. We show that the direction of the principal axis coordinates, determined from a satellite pass through a flux rope, are extremely dependent on the path of the satellite through the structure and, secondly, that the ISEE 3 magnetometer observations that have been interpreted as magnetic islands [Hones et al., 1984; Slavin et al., 1989] are also consistent with flux rope structures.

Model Description

We model the magnetic field topology of plasmoids with a $2\frac{1}{2}$-dimensional representation (i.e. $\frac{\partial}{\partial y}\equiv 0$, where we use the conventional GSM cartesian coordinates, x, y, and z) that can create the three-dimensional characteristics of magnetic flux rope plasmoids. Even though quantities in our model vary only in two dimensions, three dimensions are required to describe the magnetic topology, hence our use of the term $2\frac{1}{2}$-dimensional. The magnetic structure created in this model consists of three separate fields; a Harris neutral sheet [Harris, 1962] describing the basic tail structure, a circular closed loop structure centered at the neutral sheet, and an axial field which is perpendicular to the previous two fields. The axial field is determined to satisfy MHD equilibrium. The currents required to generate these fields are also calculated. These fields model a flux rope embedded in the neutral sheet of the earth's magnetotail, which results in a non-cylindrically symmetric structure.

This magnetic field model explicitly satisfies $\nabla \cdot \mathbf{B} = 0$ in the x-z plane and has $\frac{\partial B_y}{\partial y}\equiv 0$. The model is described by these equations:

$$B_x = B_{so}\tanh\left(\frac{z}{L_s}\right) + B_{po}\left(\frac{r^2}{L_o{}^2}\right)\exp\left(\frac{-r^2}{L_p{}^2}\right)\cos(\theta)$$

$$B_z = B_{po}\left(\frac{r^2}{L_o{}^2}\right)\exp\left(\frac{-r^2}{L_p{}^2}\right)\sin(\theta)$$

B_{so} is the value of the Harris neutral sheet field at infinity and L_s is its scale length. B_{po}, L_o and L_p are the characteristic field strength and scale lengths of the "plasmoid" field

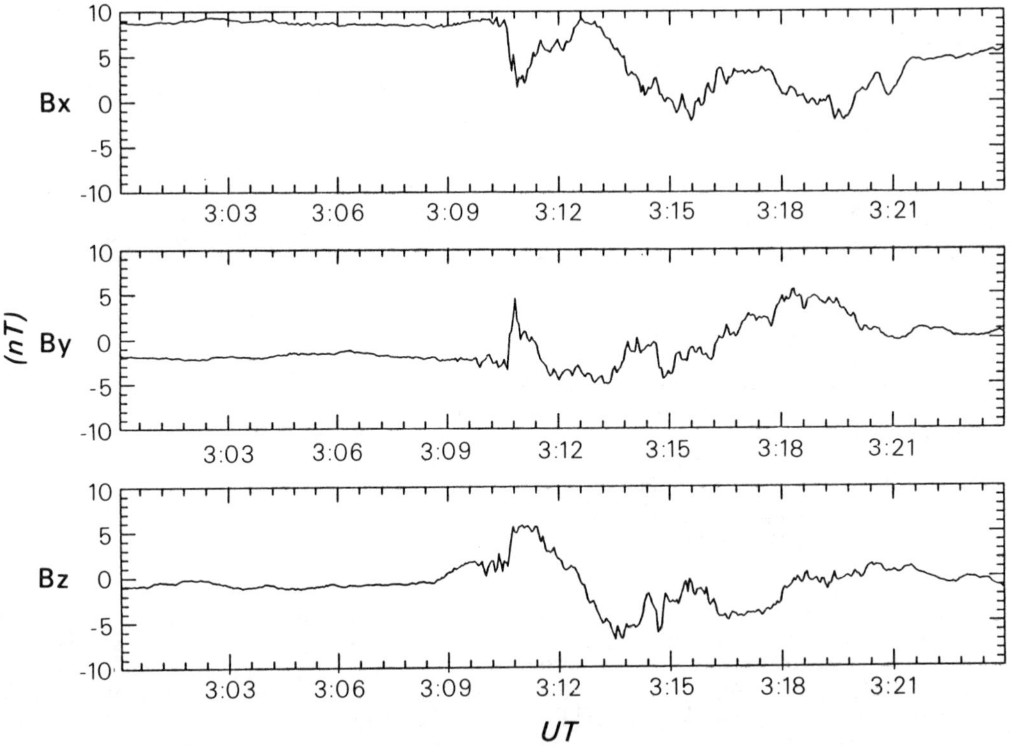

Fig. 1. The March 22, 1983 plasmoid event presented in GSM cartesian coordinates. ISEE 3 starts in the northern lobe and enters the plasmoid structure at about 03:09 UT and exits about 03:18 UT. The plasmoid signature consists of a a bi-polar variation in B_z and a strong core field. The core field for this event is predominatly in the B_x direction.

respectively. θ and r denote the usual azimuthal and radial cylindrical coordinates referenced to the axis of the flux rope which is in the y direction. The B_y magnetic component is calculated to satisfy MHD equilibrium or more specifically $(\mathbf{J} \times \mathbf{B})_y = 0$. This relationship requires the contours of the B_y field strength to be field-aligned with the magnetic field in the x-z plane. The method used to determine the B_y component is to specify the magnetic field strength in the y direction along the x-axis, then field lines are traced back to this axis from anywhere in the simulation box to determine the B_y component at that location. The analytic form of the boundary conditions imposed along the x-axis for the B_y field is:

$$B_y(x, y=0) = \frac{1}{2}\, B_{fo}(\cos(\tfrac{\pi x}{L_f})+1) + B_{yo}$$

where B_{yo} is the value of the y field in the lobes and B_{fo} is the value of the y field at the center of the plasmoid. The parameter L_f determines the size of the flux rope. The B_y field is set to B_{yo} outside the structure. Figure 2 shows field lines in this model projected onto the x-y plane, i.e., the plane perpendicular to the axis of the flux rope.

After a magnetic structure is calculated, different experiments can be run consisting of sampling the magnetic field along some arbitrary path through the structure (a satellite pass). The magnetic field time series so generated is cast into a PA coordinate system by determining the direction of minimum variance (see Sonnerup and Cahill [1967] for details). Currently, our model is constrained to have the flux rope axis strictly aligned parallel to the tail current. However, even with this restriction we can show that it is difficult to distingush between the closed loop and flux rope models.

Flux Rope Orientation

We used the model to test the relationship between the axis of a flux rope and the principal axi directions obtained from a single satellite pass through the structure. Elphic and Russell [1983] showed that the orientation of a cylindrically symmetric flux rope can be determined if the satellite passed close to the center of the structure. In the case of the flux ropes in the Venus ionosphere the maximum variance direction is parallel to the symmetry axis of the flux rope. The flux ropes observed in the near-earth tail by ISEE 1 and 2 [Elphic et al., 1986] had weaker core fields than the Venus flux ropes, therefore the symmetry axis of the structure was assumed to lie parallel to the intermediate variance direction. However, deep magnetotail flux rope plasmoids are not cylindrically symmetric and therefore the orientation of

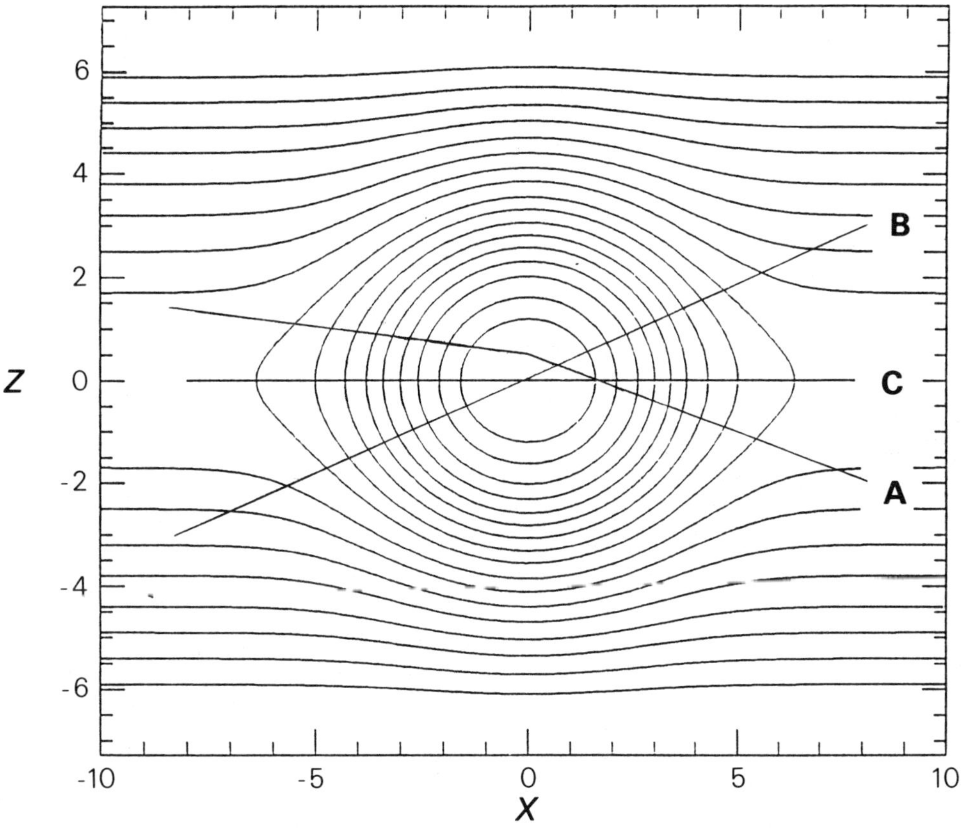

Fig. 2. The magnetic topology of a modelled plasmoid depicted in the noon-midnight meridional plane. The parameters used to create the structure are: B_{po}=15 nT, B_{so}=−4.5 nT, L_s=L_p=L_o=3. The flux rope magnetic field component (y direction) has a value of 10 nT in the center of the structure and drops to a value of 1 nT outside the closed loops. The lettered paths through the structure represent satellite trajectories.

TABLE 1. The principle axis coordinates of the three passes shown in figure 2 together with those derived from the ISEE data shown in figure 1. The table shows the average magnetic field strength, deviation and direction vectors.

Path	Axis	$\langle \mathbf{B} \rangle$	σ	direction vector		
	maximum variance	−1.12	3.81	(−.361	.063	.930)
A	intermediate variance	5.37	1.64	(.230	.973	0.02)
	minimum variance	3.23	.358	(.903	−.222	.366)
	maximum variance	0.07	3.90	(.714	0.02	−.698)
B	intermediate variance	4.00	3.40	(−.125	.999	0.02)
	minimum variance	−0.05	1.28	(.699	−.718	.715)
	maximum variance	4.40	3.37	(.000	.999	−.002)
C	intermediate variance	.001	3.18	(0.00	.002	.999)
	minimum variance	−.017	.005	(.999	−.001	0.00)
	maximum variance	−1.79	4.35	(−.128	.694	.708)
ISEE	intermediate variance	5.59	2.79	(.801	−.349	.487)
	minimum variance	1.40	1.01	(.585	.629	−.511)

the flux rope is more difficult to determine. The three paths labelled A, B, C through the modelled plasmoid shown in Figure 2 correspond to three possible satellite paths. The three trajectories were chosen to demonstrate that three completely different PAA coordinate frames can be generated by the same structure by just changing the path of the spacecraft. The paths were choosen to be as plausible and simple as possible and still give different maximum variance directions. Currently, the model is constrained to sample two straight line segments through the structure. This is the reason for the sudden change in motion in Path A. This change of direction can be explained by the waving and flapping of the distant magnetotail. The magnetic field was sampled at equal intervals along the three trajectories and presented in GSM cartesian coordinates. This data was then transformed into PA coordinates.

The parameters determined by casting the magnetic field variations along the three paths shown in Figure 2 into a PA coordinate frame are given in Table 1. Notice that the maximum variance direction lies predominately along a different axis for each pass through the same structure, all of which pass near the center of the structure. The intermediate and minimum variance directions are also changed from pass to pass. This ambiquity makes it difficult to determine the orientation of a flux rope and therefore, to differentiate between a closed magnetic loop and a flux rope on the basis of a single spacecraft trajectory.

Comparison to ISEE Observations

To determine if the ISEE 3 magnetometer observations are consistent with a flux rope picture, we compared ISEE observations with simulated observations derived from our model. We simulated real observations by first generating a time series plot obtained by sampling the magnetic field along a complete pass through a structure, that is a pass at the ends of which the only appreciable field is the Harris neutral sheet and the background uniform B_y component. We examined this plot for the signature of a plasmoid, the key criterion being a bi-polar trace in the GSM z-direction or the north-then-south turning of the magnetic field vector. We used this signature to determine the points of "entry" and "exit" from the plasmoid, a process similar to examining the actual ISEE 3 observations for plasmoid events. The subset of the pass so defined was then transformed into a PA coordinate frame. This data was then directly compared with similarly extracted and transformed ISEE 3 data. We did not search for the optimum fit to the data, as our purpose is simply to show that the data can be fit with a flux rope model. A number of different paths through the same model were found to give qualitatively comparable fits.

Figure 3 shows that the ISEE 3 magnetic field observations from March 22, 1983, which have been interpreted as a conventional plasmoid, can also be modelled with the three-dimensional structure illustrated in Figure 2. The figure shows the observational data (dots) and the modelled data from pass A (solid) both transformed into their own PA systems. The ISEE 3 data was band passed filtered to remove all short period oscillations (period $\leq$ 60 sec). The intermediate variance component of the model field is approximately in the y direction and is due to the strong core/axial field component of the flux rope. The maximum in its intensity occurs at the inflection point of the bi-polar trace seen in the maximum variance direction. The accuracy of fit is very good, lending support to the magnetic flux rope plasmoid picture. This is similar to the signatures for the near earth flux ropes seen by ISEE 1/2 [Elphic et al., 1986]. The principal axis coordinate vectors derived from both the ISEE data and the model data are given in Table 1. If the normal interpretation were given to the PA directions derived from the ISEE data, then the flux rope would have its symmetry axis approximately in the x direction but with finite y and z components. If the ISEE data is interpreted as a closed loop structure, the loops must be in the plane defined by the maximum and intermediate variance directions, which is tilted significantly out of the x-z plane. PAA cannot differentiate between these two pictures.

In modeling the ISEE data, we assumed the spacecraft had passed near the center of the structure, which severely constrained the model paramters, such that the main free parameter was the satellite path through the structure. Very different trajectories can give very similar principal

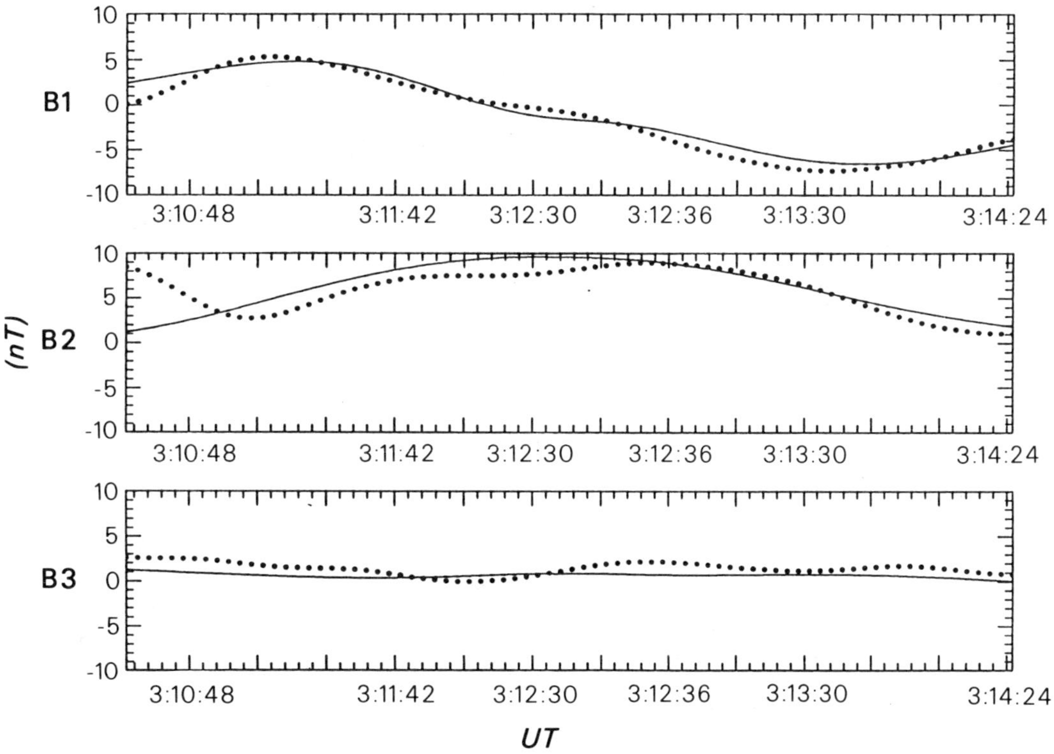

Fig. 3. A comparison of the ISEE 3 magnetic field observations from March 22, 1983 (dots), which have been interpreted as a conventional plasmoid, with a modelled fit (solid line). The model fit is obtained from Pass A in Figure 2. Both data are presented in their own PA coordinates. The accuracy of fit is very good, lending support to the magnetic flux rope plasmoid picture.

axis signatures, though the principal axis directions might be quite different. So the trajectory shown in Pass A in Fig. 1 is only one of possibly many trajectories that could reproduce the ISEE 3 data of March 22, 1983. Indeed, if we relax the condition that ISEE had passed close to the center of the structure, there are many other possible structures that could give the same PAA signatures. A closed loop structure with a off-axis pass would also give similar signatures. However, Figure 3 demonstrates that plasmoid events are consistent with a flux rope structure. The Febuary 6, 1983 plasmoid event (identified by Hones [1984] on the basis of the bipolar trace in the GSM θ component), was also successfully modelled as a flux rope though is not presented in this paper.

Conclusions

The existence of a persistent and significant magnetic field component in the cross-tail direction (B_y) in the earth's magnetotail, drastically changes the magnetic topological picture of plasmoids. The change in the plasmoid's configuration with a B_y component present was described qualitatively by Hughes and Sibeck [1984] and has been modelled by Birn et al. [1989]. They found instead of isolated closed magnetic loops, flux ropes are formed. The flux rope model developed in this paper reproduces the plasmoid magnetic signatures seen by ISEE 3 showing that the observations are also consistent with the flux rope picture. The strong dependence of the direction of the PA coordinates on spacecraft trajectory through the structure makes it very difficult to determine the orientation of that structure from the magnetic field data. Therefore, it is difficult to distinguish between closed loops and flux ropes from the ISEE 3 magnetic observations and care must be taken in interpreting the orientation of plasmoids by principal axis analysis.

The flexibility of our flux rope model allows us to examine other structures for flux rope signatures, such as travelling compression regions [Slavin et al. 1984] and flux transfer events. Plasma data can be used to help determine the magnetic topology of plasmoids as well as help constrain spacecraft trajectories, however, their use is beyond the scope of this study but will be included in future work.

Acknowledgements. We would like to thank Dr. J. A. Slavin for kindly providing us with the ISEE 3 magnetometer data and for his helpful comments and encouragement. We also thank Dr. R. Elphic for offering constructive and helpful criticism of this paper. This work was supported by NASA grant NAGW-1627.

References

Akasofu, S-I., A.T.Y. Lui, C.I. Meng, and M. Haurwitz, Need for a three-dimensional analysis of magnetic fields in the magnetotail during substorms, *Geophys. Res. Lett., 5,* 283, 1978.

Birn, J., M. Hesse, and K. Schindler, Filamentary structure of a three-dimensional plasmoid, *J. Geophys. Res. 94,* 241, 1989.

Elphic, R.C., C.T. Russell, J.A. Slavin, and L.H. Brace,Observations of the dayside ionopause and ionosphere of Venus, *J. Geophys. Res., 85,* 7679, 1980.

Elphic, R.C., and Russell, C.T.,Magnetic flux ropes in the Venus Ionosphere: Observations and models, *J. Geophys. Res. 88,* 58, 1983.

Elphic, R.C., C.A. Cattell, K. Takahashi, S.J. Bame and C.T. Russell, ISEE-1 and 2 observations of magnetic flux ropes in the magnetotail: FTE's in the plasma sheet?,*Geophys. Res. Lett., 13.* 648, 1986.

Gloecker, G., F.M. Klecker, D. Hovestadt, Energetic particle characteristics of magnetotail flux ropes, *Geophys. Res. Lett., 12,* 191, 1985.

Harris, E.G., On a plasma sheet seperating regions of oppositely directed magnetic fields, *Nuovo Cimento, 23,* 116, 1962.

Hones, E.W. Jr., Substorm processes in the magnetotail: Comments on 'On hot tenuous plasma, fireballs, and boundary layers in the earth's magnetotail' by L.A. Frank, K.L. Ackerson, and R.P. Lepping, *J. Geophys. Res., 82,* 5633, 1977.

Hones, E.W., Jr., D.N. Baker, S.J. Bame, W.C. Feldman, J.T. Gosling, D.J. McComas, R.D. Zwickl, J.A. Slavin, E.J Smith and B.T. Tsuratani, Structure of the magnetotail at 220 R_E and its response to geomagnetic activity, *Geophys. Res. Lett., 11,* 5, 1984.

Hughes, W.J., and D.G. Sibeck, On the 3-dimensional structure of plasmoids, *Geophys. Res. Lett., 14,* 636, 1987.

Sibeck, D.G., Evidence for flux ropes in the earth's magnetotail, presented at the AGU Chapman Conference on the Physics of Magnetic Flux Ropes, March 1989.

Slavin, J.A., E.J. Smith, B.T. Tsurutani, D.G. Sibeck, H.J. Singer, D.N. Baker, J.T. Gosling, E.W. Hones, Jr., and F.L. Scarf, Substorm associated travelling compression regions in the distant tail: ISEE-3 geotail observations, *Geophys. Res. Lett., 11,*, 657, 1984.

Slavin, J.A., D.N. Baker, J.D. Craven, R.C. Elphic, D.H. Fairfield, L.A. Frank, A.B. Galvin, W.J. Hughes, R.H. Manka, D.G. Mitchell, I.G. Richardsn, T.R. Sanderson, D.J. Sibeck, H.J. Singer, E.J. Smith, and R.D. Zwickl, CDAW-8 observations of plasmoid signatures in the geomagnetic tail: an assessment, preprint, 1988.

Sonnerup, B.U.O., and Cahill, L.J., Magnetopause structure and attitude from explorer 12 observations, *J. Geophys. Res., 72,* 171, 1967.

MAGNETIC FLUX ROPES IN 3-DIMENSIONAL MHD SIMULATIONS

Tatsuki Ogino[1], Raymond J. Walker[2] and Maha Ashour-Abdalla[2,3]

Abstract. We have used a three-dimensional time dependent global magnetohydrodynamic simulation of the interaction between the solar wind and the Earth's magnetosphere to model the generation of magnetic flux ropes at the magnetopause and in the magnetotail. When the Interplanetary Magnetic Field (IMF) has a large azimuthal component (B_Y) as well as a southward component (B_Z), strongly twisted and localized magnetic flux tubes similar to magnetic flux ropes appear at the subsolar magnetopause.

In the magnetotail plasmoids appear after the formation of a near-Earth magnetic neutral line. When the IMF has a finite B_Y component there is a large B_Y at the center of the plasmoid even in the noon-midnight meridian. The magnetic field lines have a helical structure connected from dawn to dusk. Near the edges of the plasmoid, the field lines which connect with the Earth, are bundled and form a structure similar to magnetic flux ropes along the X-direction near the tail magnetopause. The magnetic field increases while the plasma pressure decreases inside the flux rope. Eventually the flux rope disconnects from the Earth and propagates down the tail.

When a southward IMF was initially imposed throughout the magnetosphere, patchy magnetic flux ropes immediately appeared in the tail. These flux ropes are smaller than those associated with plasmoids and have a large B_Y component at their center.

1. Introduction

The interplanetary magnetic field (IMF) is thought to control much of the dynamics of the Earth's magnetosphere. When the IMF is southward, temporally and spatially limited magnetic flux tubes are often observed at the dayside magnetopause. These localized magnetic flux tubes are now called flux transfer events (FTE's) [Russell and Elphic, 1978; Elphic and Russell, 1979]. FTE's have a characteristic bipolar signature with a duration of 1 to 5 minutes in the component of the magnetic field normal to the magnetopause and frequently are associated with an increase in the magnetic field amplitude [Russell and Elphic, 1979; Rijnbeek et al., 1984]. The spatial scale size in the direction normal to the magnetopause is about 1 R_E while it is about 2 R_E in the tangential direction where R_E stands for the radius of the Earth [Saunders et al., 1984; Rijnbeek et al., 1984].

Russell and Elphic [1979] first realized that FTE's were the result of patchy and intermittent reconnection and noted that the observations were consistent with isolated elbow shaped flux tubes passing over the spacecraft. They suggested that FTE's are created near the subsolar magnetopause and are convected to higher latitudes. Lee and Fu [1985] presented an alternative model in which magnetic islands are formed by multiple x-line reconnection generated by the tearing mode in the magnetopause current layer. This causes twisted open flux tubes to form along the dayside magnetopause. The multiple X line configuration also was reproduced by 2-dimensional computer simulations [La Belle-Harmer et al., 1988; Shi et al., 1988]. Galeev et al. [1986] suggested the possibility of spontaneous patchy reconnection related to the growth and overlapping of magnetic islands. Recently Scholer [1988] and Southwood et al. [1988] have suggested that the reconnection associated with FTE's occurs at a single location but is time dependent. Sato et al.[1986] used a semi-global MHD model to demonstrate the creation of isolated magnetic flux tubes by repeated reconnection between the IMF and the geomagnetic field.

In one model of magnetospheric substorms, the stress imposed on the magnetosphere by reconnection at the dayside magnetopause, is released by the formation of a near-Earth magnetic neutral line. Much of the released energy is carried tailward as plasmoids [McPherron et al., 1973; Hones et al. 1984]. In this model of substorms, the plasmoid is thought of as fundamentally a two dimensional structure confined to the noon-midnight meridian. Recently Hughes and Sibeck [1987] considered the effect of the IMF B_Y component on the structure of plasmoids and proposed a new model in which the plasmoids are composed of helical magnetic field lines with the helix aligned along the Y-direction. In this model the edges of the helical field lines are initially connected with the Earth's ionosphere and this stops tailward motion of the plasmoid. As time elapses, the helical field lines disconnect from the Earth and the helical plasmoid begins to propagate down the tail.

Figure 1 shows cartoons giving the properties which have been inferred from observations of plasmoids and

[1] Research Institute of Atmospherics, Nagoya University, Toyokawa, Aichi 442, Japan

[2] Institute of Geophysics and Planetary Physics, University of California, Los Angeles, CA 90024

[3] Physics Department, University of California, Los Angeles, CA 90024

Geophysical Monograph 58

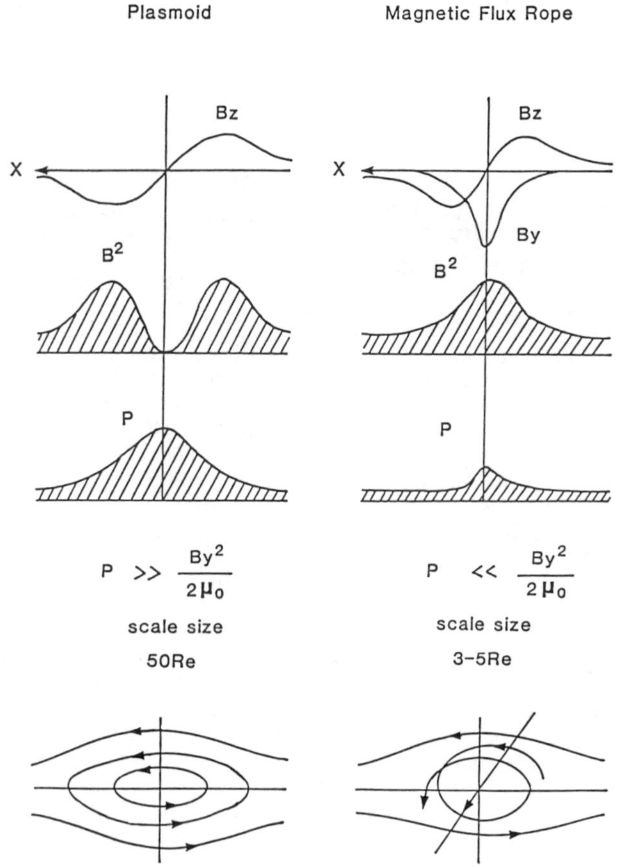

Fig. 1. Plasmoids and magnetic flux ropes in the magnetotail. In a plasmoid the B_Z component of the magnetic field has a bipolar signature and the hot plasma is confined to the center while for the magnetic flux rope a longitudinal magnetic field occurs in the center.

magnetic flux ropes in the magnetotail. The plasmoid is typically composed of closed loops of magnetic field lines and confines a hot plasma produced by tail magnetic reconnection. The typical scale size is about a third of the diameter of the tail magnetopause. On the other hand, the magnetic flux ropes which have been observed in the tail magnetopause [Sibeck et al., 1984; 1985] and in the central plasma sheet [Elphic et al., 1986], confine a longitudinal magnetic field instead of a hot plasma. In comparison with the plasmoids, the magnetic flux ropes have a smaller scale size and the maximum of the magnetic field is at the center. Both plasmoids and magnetic flux ropes have bipolar signatures in the B_Z component of the magnetic field. Sibeck et al. [1984; 1985] found evidence for magnetic flux ropes along the X-direction at the tail magnetopause and suggested that these magnetic flux ropes which contained magnetosheath like plasma may originate from FTE's at the dayside magnetopause. Elphic et al. [1986] found a magnetic flux rope aligned along the Y-direction in the central plasma sheet. This flux rope had a scale size of $3 \sim 5$ R_E in the X-Z plane and the helical field lines confined an enhanced longitudinal magnetic field directed along the Y-component.

In this study we have modeled magnetic reconnection at the dayside magnetopause and in the magnetotail by using a three-dimensional time-dependent global magnetohydrodynamic (MHD) simulation of the interaction between the solar wind and the Earth's magnetosphere. In particular we have modeled magnetic flux ropes at the magnetopause and in the plasma sheet during intervals when the IMF has both a southward and a dawnward component.

2. The Simulation Model

The simulation model will be reviewed briefly here since it has been described in detail elsewhere [Ogino, 1986; Ogino et al., 1989]. In order to study magnetic reconnection at the dayside magnetopause and in the tail, we used a high spatial resolution code with a long magnetotail.

This simulation code solves the MHD and Maxwell's equations as an initial value problem by using the modified two step Lax-Wendroff scheme. The normalized resistive MHD equations which we solved are written as follows:

$$\partial \rho/\partial t = -\nabla \cdot (\mathbf{v}\rho) + D\nabla^2 \rho$$

$$\partial \mathbf{v}/\partial t = -(\mathbf{v}\cdot\nabla)\mathbf{v} - (\nabla P)/\rho + (\mathbf{J}\times\mathbf{B})/\rho + \mathbf{g} + \Phi/\rho$$

$$\partial P/\partial t = -(\mathbf{v}\cdot\nabla)P - \gamma P \nabla\cdot\mathbf{v} + D_p \nabla^2 P$$

$$\partial \mathbf{B}/\partial t = \nabla \times (\mathbf{v}\times\mathbf{B}) + \eta \nabla^2 B$$

$$\mathbf{J} = \nabla \times (\mathbf{B} - \mathbf{B_d})$$

where ρ is the plasma density; $\mathbf{v}$ is the flow velocity; P the plasma pressure; $\mathbf{B}$ the magnetic field; $\mathbf{B_d}$ is the internal dipole field of the Earth; $\mathbf{J}$ the current density; $\mathbf{g}$ the gravity force; $\mathbf{\Phi} \equiv \mu \nabla^2 \mathbf{v}$ the viscosity; $\gamma = 5/3$ the ratio of specific heats; $\eta = \eta_o (T/T_o)^{-3/2}$ the resistivity; and T/T_o is the temperature normalized by its value in the ionosphere. The classical temperature dependence for the resistivity was chosen for simplicity. The units for distance , velocity, and time are the Earth's radius $R_E = 6.37 \times 10^6$ m, the Alfvén velocity at one Earth radius on the equator $v_A = 6.80 \times 10^6$m/s, and the Alfvén transit time $\tau_\alpha = R_E/v_A = 0.937$s. The numerical coefficients are $\eta_o = 0.005$, $\mu/\rho_{sw} = D = D_p = 0.002$, where ρ_{sw} is the solar wind density. The values of μ ,D and D_p were selected to be large enough to suppress start up fluctuations in the parameters but to be small enough that they had no significant influence on the global magnetospheric structure. The magnetic Reynolds number is $S = \tau_\eta/\tau_A = 100-2000$ where $\tau_\eta \equiv \Delta X^2/\eta$, and $\tau_A = \Delta X/v_A$. ΔX is the mesh size.

In the simulation, a uniform solar wind with $n_{sw} = 5/\mathrm{cm}^3$, $v_{sw} = 300$km/s and $T_{sw} = 2 \times 10^{5\circ}$ K flows into a simulation box from the upstream boundary at $x = x_1$ (Figure 2). The IMF is given by $B_{IMF} = (B_X, B_Y, B_Z) = B_{IMF}(0, cos\theta, sin\theta)$ where $B_{IMF} = 5$nT and $\theta = 210^\circ$ or 240°. Free boundary conditions, where the derivatives of all physical quantities are zero, were used at $X = X_o$, $Y = \pm Y_o$ and $Z = Z_o$. Mirror boundary conditions were used at $Z = 0$ and a simple fixed ionospheric boundary condition was used near the Earth [Ogino, 1986]. The MHD equations were solved on a $(N_X, N_Y, N_Z) = (96, 96, 48)$ point grid for the simulation of magnetic flux tubes at the day side magnetopause. The spatial mesh size for the higher resolution runs was $(\Delta X, \Delta Y, \Delta Z) = (0.5, 0.5, 0.5)\mathrm{R}_E$ and

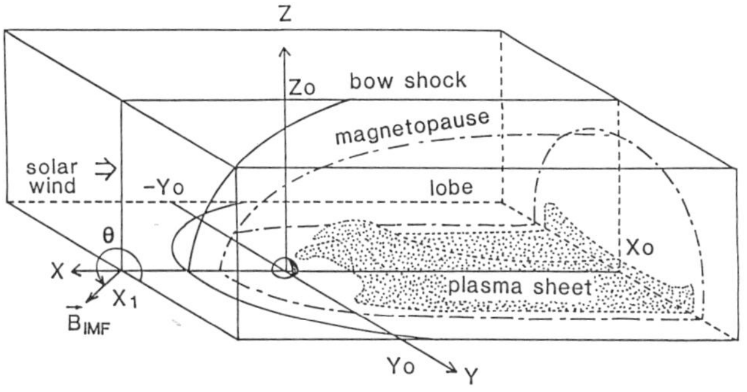

Fig. 2. Solar magnetospheric coordinate system used in the three-dimensional MHD simulation. The angle θ gives the direction of the interplanetary magnetic field.

the time step was selected as $\Delta t = 1.87$s in order to assure that the system was numerically stable. The physical domain of the calculation was given by $X_1 = -X_o = Y_o = Z_o = 24.25$ R_E. The parameters for the simulations of magnetotail dynamics were $(N_X, N_Y, N_Z) = (180, 120, 60)$, $(\Delta X, \Delta Y, \Delta Z) = (0.5, 0.5, 0.5)$ R_E, $\Delta t = 1.87$s, $X_1 = Y_o = Z_o = 30.25$ R_E and $X_o = -60.25 R_E$.

3. Simulation Results

Dayside Reconnection

In our earlier studies of dayside reconnection we found that two types of magnetic flux tubes formed at the magnetopause which depend on the orientation of the IMF [Ogino et al., 1989]. The dayside magnetic flux tubes occur only when the IMF has a southward component. When the IMF has a large B_Y component as well, a strongly twisted and localized magnetic flux tube similar to a magnetic flux rope forms at the magnetopause. Figures 3a and b and 4a and b show the time evolution of the magnetic configuration for a simulation in which the IMF had $B_X = 0$, B_Y = -4.3 nT and B_Z = -2.5 nT ($\theta = 210°$). These magnetic field lines were calculated by starting from the northern dayside polar cap. At the equator the symmetry boundary condition was used to complete the field lines. Figure 3a shows the configuration 24m after the IMF turned southward. At this time reconnection has already started at the locations where the IMF and the Earth's magnetic field are antiparallel. This region can be seen by noting the sharply bent field lines in the upper left and lower right parts of the figure. By $t = 28$m (Figure 3b) twisted field lines have started to form across the dayside magnetopause. This twisted flux tube forms on the last closed field lines at the magnetopause (Figures 3b, and 4a). By $t = 36$m (Figure 4b) the twisted flux tube has started to reconnect and twisted open flux tubes propagate across the dayside magnetosphere. This sequence of events has been cartooned in Figure 5. These flux tubes have many of the properties of flux transfer events [Ogino et al., 1989].

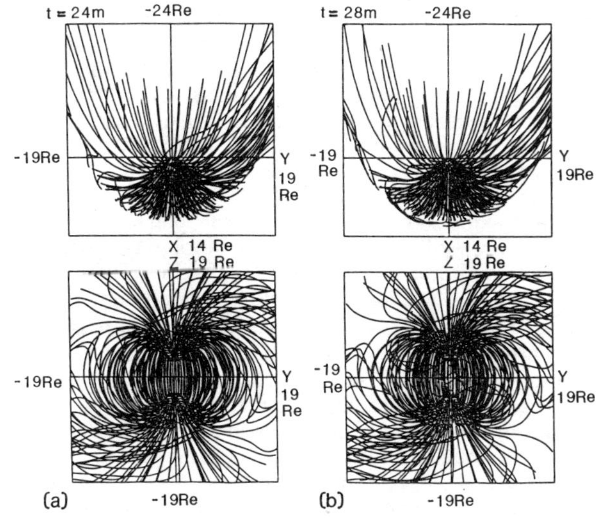

Fig. 3. Magnetic field lines with at least one end attached to the Earth viewed from the north pole (top) and the Sun (bottom). Figure 3a is a snapshot at $t = 24$m and Figure 3b is at $t = 28$m (from Walker and Ogino, 1989a).

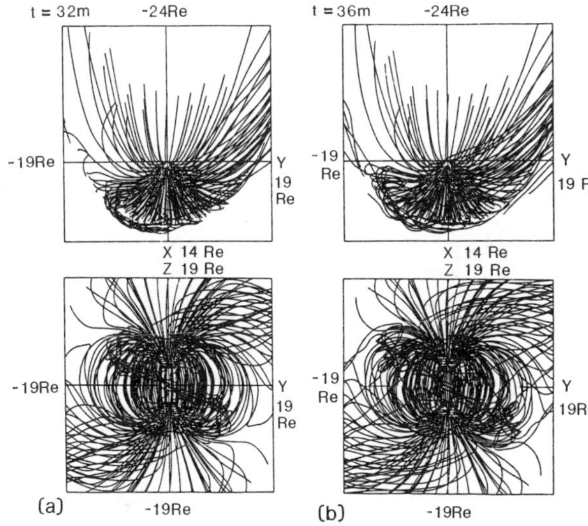

Fig. 4. The same as Figure 3 for $t = 32$m and $t = 36$m (from Walker and Ogino, 1989a).

The results in Figures 3 and 4 were obtained from a simulation with a spatial resolution of $(\Delta X, \Delta Y, \Delta Z) = (0.5, 0.5, 0.5)$ R_E. The twisted flux tube has dimensions of about 1.5 R_E in the X-direction and 3 R_E in the Z-direction. Although these flux tubes are near the minimum resolution of the numerical scheme, we believe they may be real [Ogino et al., 1989; Walker and Ogino, 1989a]. For a detailed analysis of the resolution of these events please see Walker and Ogino[1989a]. To further check these results we ran the same simulations with a higher resolution code with $(\Delta X, \Delta Y, \Delta Z) = (0.25, 0.4, 0.4)$ R_E and a $(N_X, N_Y, N_Z) = (192, 116, 56)$ point grid. This run reproduced the results in Figures 3 and 4 with almost the same dimensions for the twisted flux tubes so we believe that these flux tubes are real.

When the IMF B_Y component was small ($\theta = 240°$) twin open magnetic flux tubes were formed episodically on the dayside magnetopause near the subsolar point (see

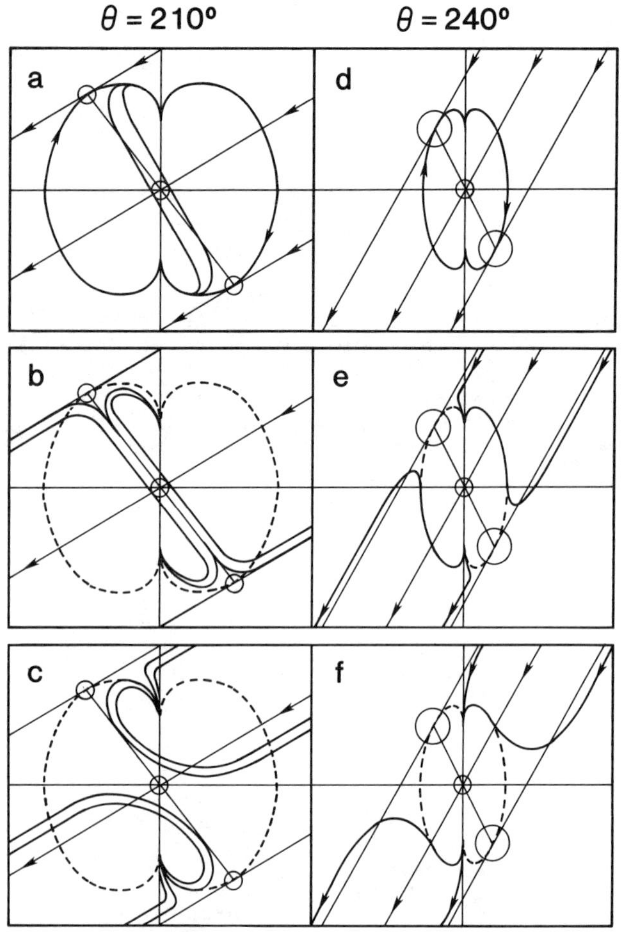

Fig. 5. A sketch of the time evolution of the isolated flux tube for $\theta = 210°$ (panels a, b, and c) and for $\theta = 240°$ (panels d, e, and f) (from Walker and Ogino, 1989a).

Figure 5 and Figures 10 and 11 of Walker and Ogino, [1989a]). These too were formed by anti-parallel reconnection and subsequently convected over the poles. However in this case the long twisted closed flux tube did not form first. These too resemble observed flux transfer events.

Magnetotail Reconnection

In our magnetotail studies we have varied two different parameters: the orientation of the IMF, θ, and the initial condition on the IMF, $\mathbf{B_{IMF}}(t = 0)$. As in the dayside studies, we considered two IMF orientations, $\theta = 210°$ and 240°. We also studied the effects of two initial conditions for the IMF. In one case the IMF was initially zero ($\mathbf{B_{IMF}}(t = 0) = 0$) and entered the simulation region from the upstream boundary after a magnetosphere had formed without an IMF. In the second approach, the IMF was imposed everywhere at $t = 0$ ($\mathbf{B_{IMF}}(t = 0) \neq 0$).

In Figure 6, we have plotted the profiles of plasma parameters on the Sun-Earth line in the Earth's magnetosphere at $t = 64$m for $\mathbf{B_{IMF}}(t = 0) = 0$ and $\theta = 210°$.

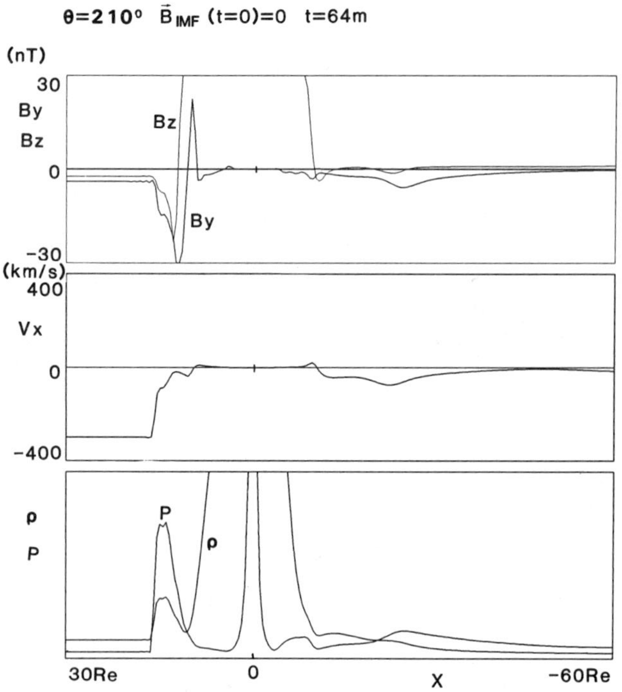

Fig. 6. Simulation parameters along the noon-midnight meridian for the case with $\mathbf{B_{IMF}}(t = 0) = 0$ at $t = 64$m. The top panel contains B_Y and B_Z, the middle panel contains the x-component of the velocity (V_x) while the bottom panel contains the density (ρ) and the pressure (P).

This snapshot was taken 64m after the southward and dawnward IMF began to flow into the simulation domain. Typical features of the Earth's magnetosphere such as the bow shock, magnetosheath, magnetopause and plasma sheet are clearly seen on the Sun-Earth line plot. As is shown in Figures 3 and 4 at the dayside magnetopause, a strongly twisted magnetic flux tube was formed due to magnetopause convection and antiparallel merging at $t = 32$m and after reconnection the separated flux tubes convected towards the northern and southern high latitude tail. The dayside reconnection is followed by reconnection in the near-Earth plasma sheet. In Figure 6 a plasmoid can be seen at $X = -25\mathrm{R}_E$ in the tail, where the plasma pressure, p increases, the plasma flows tailward ($v_X = -74$km/s), B_Z has a bipolar signature and B_Y has a minimum ($B_Y = -6$nT). The beta value, which is the ratio of the plasma pressure to the magnetic pressure, is $\beta = 2p/B_Y^2 = 9.0$ at the point where $B_Z = 0$. These are typical features for a plasmoid and the finite B_Y component at the center of the plasmoid implies a helical structure for the magnetic field lines.

In Figure 7 the magnetospheric configuration in the noon-midnight meridional and equatorial planes has been plotted in the left column while cross sectional patterns in the tail (Y-Z plane) have been plotted in the middle and right columns. The plasmoid can most readily be seen as the region of enhanced pressure in the tail. It extends approximately $30\mathrm{R}_E$, $24\mathrm{R}_E$, and $7R_E$ in the X, Y, and Z directions respectively. Note the plasmoid is associ-

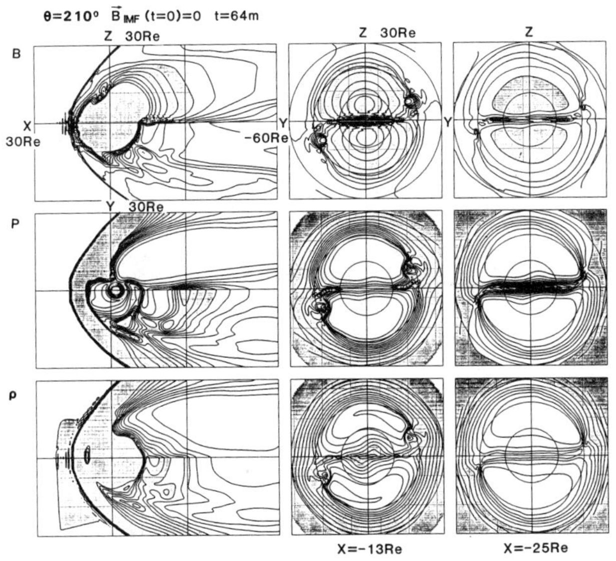

Fig. 7. Cross sectional plots of the magnetic field magnitude (B), the pressure (P) and the density (ρ). In the left column the top of each panel gives the values in the noon-midnight meridian while the bottom gives the values in the equatorial plane. The middle and right columns give the values in the Y-Z planes at $X = -13\mathrm{R}_E$ and $X = -25\mathrm{R}_E$ respectively. The values in this run are for $\theta = 210°$ and $\mathbf{B_{IMF}}(t = 0) = 0$ at $t = 64$m.

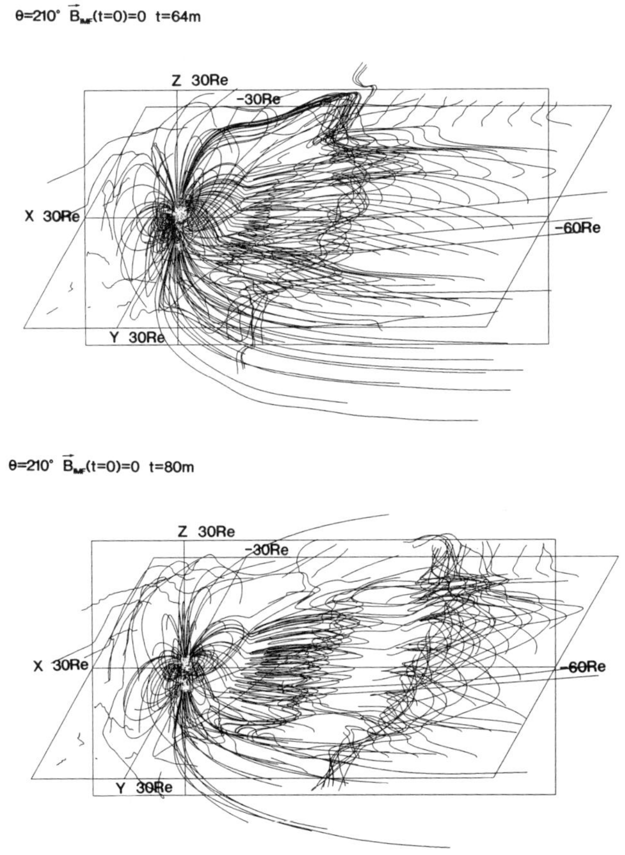

Fig. 8. The three dimensional magnetic field line configuration at $t = 64$m (a) and 80m (b) for $\theta = 210°$ and $\mathbf{B_{IMF}}(t = 0) = 0$.

ated with twin localized structures at the northern-dawn and southern-dusk magnetopauses in the tail cross section at $X = -13\mathrm{R}_E$. These structures are about $6\mathrm{R}_E$ in width and are found only on the cross sectional plot which is earthward of the plasmoid. In the localized structure, the magnetic field increases while the plasma pressure decreases in the center.

In Figure 8 we have plotted the three dimensional configuration of the magnetic field lines in the Earth's magnetosphere at $t = 64$m (Figure 8a) and $t = 80$m (Figure 8b) for $\theta = 210°$ and $\mathbf{B_{IMF}}(t = 0) = 0$. These field lines were calculated by starting at the equatorial plane and continuing the calculation to either the ionosphere or the simulation boundaries. Again the symmetry boundary condition was used in the southern hemisphere. The magnetic field lines within the plasmoid have a helical structure. The helical field lines at the edges of the plasmoid are bundled and are connected with the northern dawn and southern dusk ionosphere. Moreover, part of the helical field lines are connected with the IMF. The bundled magnetic field lines near the tail magnetopause also have a helical structure similar to magnetic flux ropes. Since they connect the edges of the plasmoid to the Earth's ionosphere, they have a substantial component in the X-direction. As time elapses, the tail reconnection proceeds up to the edges of the magnetosphere and all of the field lines at the edges of the flux rope become disconnected from the Earth and attached to the IMF. Then the helical plasmoid propagates down the tail ($t = 80$m).

In Figures 9 and 10 we show the profiles of the plasma parameters on the Sun-Earth line and the magnetospheric profiles at $t = 64$m for the initially imposed IMF ($\mathbf{B_{IMF}}(t = 0) \neq 0$) and $\theta = 210°$. A localized hump is seen in the plasma pressure at $X = -30\mathrm{R}_E$ in the tail. The scale size of this pressure increase is small in the X-direction ($\Delta X \sim 5\mathrm{R}_E$) and the B_Y component is strongly enhanced at the center of the bipolar signature in the B_Z component. In this case we find $\beta = 2p/B_Y^2 = 0.28$. These features are similar to the magnetic flux ropes in the last paragraph only the scale size is smaller in both the Y- and Z- directions. The three dimensional configuration of magnetic field lines for $\theta = 210°$ and $\mathbf{B_{IMF}}(t = 0) \neq 0$ at $t = 64$m (Figure 11) shows that several magnetic neutral lines have formed in the tail and that the magnetic field lines show a small scale wavy structure near the plasma sheet. Some of these magnetic field lines are open to the tail boundary at $X = -60\mathrm{R}_E$. Moreover, localized helical field lines similar to magnetic flux ropes appear near the noon-midnight meridian in the tail corresponding to the local magnetic field enhancement at $X = -30\mathrm{R}_E$ in Figure 9.

In Figures 12 and 13 the profiles of plasma parameters on the Sun-Earth line, and the magnetospheric profiles for $\theta = 240°$, $\mathbf{B}(t = 0) \neq 0$ and $t = 64$m have been plotted. The corresponding three dimensional magnetic field

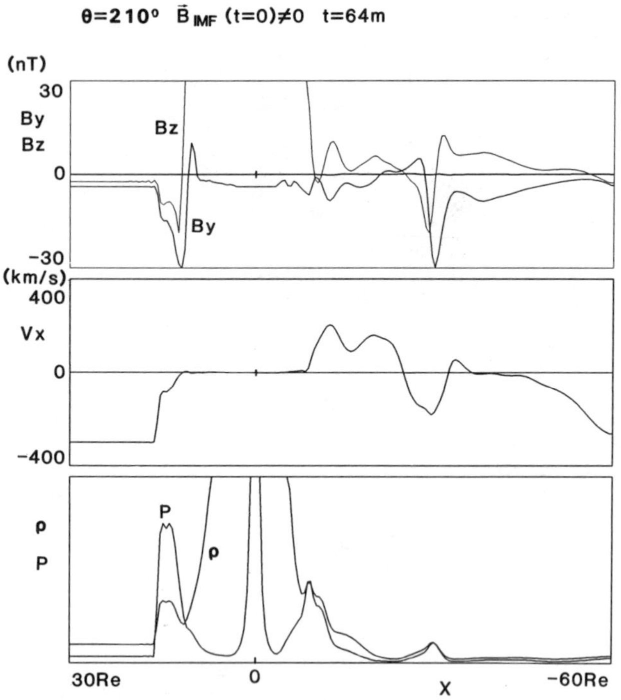

Fig. 9. The same as Figure 6 for the case with $\mathbf{B_{IMF}}(t = 0) \neq 0$.

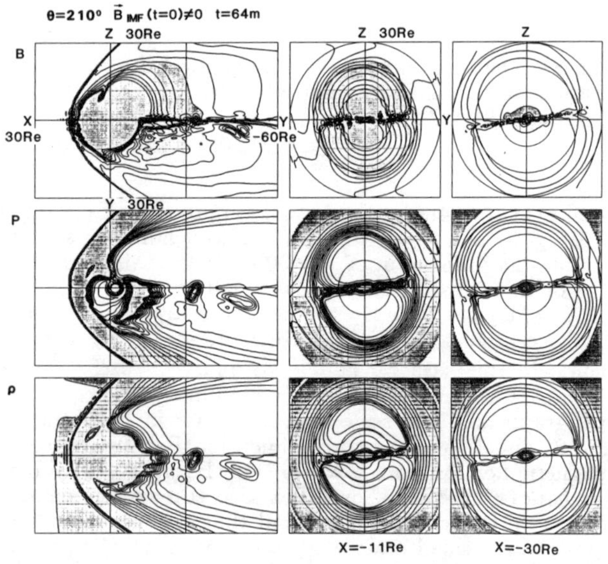

Fig. 10. The same as Figure 7 for the case with $\mathbf{B_{IMF}}(t = 0) \neq 0$. Note that this time the cross sections in the Y-Z plane are at $X = -11\mathrm{R}_E$ and $X = -30\mathrm{R}_E$.

lines are plotted in Figure 14. Two localized features in the magnetic field, which contain a bipolar signature in B_Z and an enhanced B_Y component, are clearly seen at $X = -30\mathrm{R}_E$ and -49R_E in the tail. For these magnetic

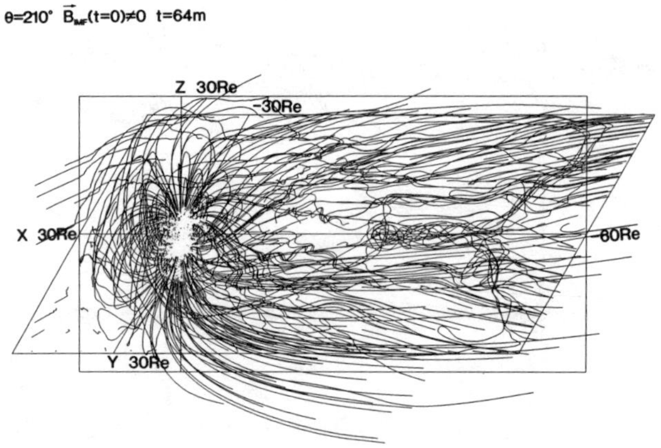

Fig. 11. The same as Figure 8a for the case with $\mathbf{B_{IMF}}(t = 0) \neq 0$.

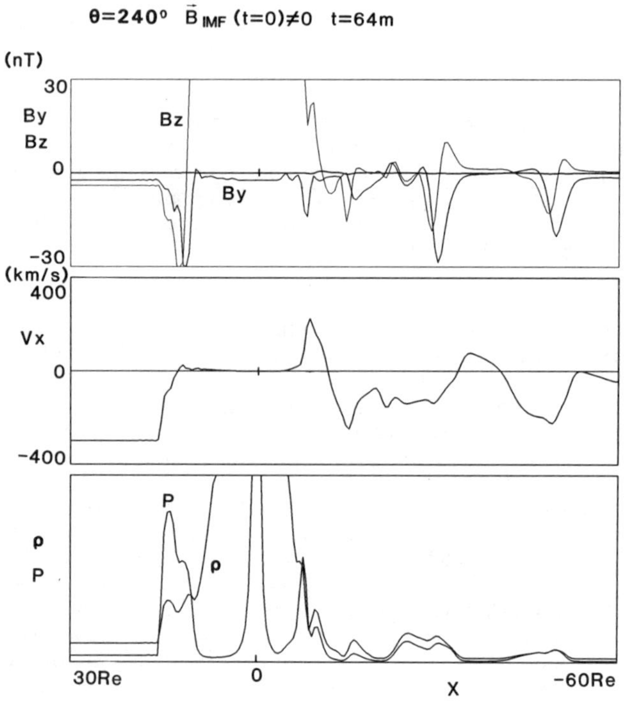

Fig. 12. The same as Figure 6 for the case with $\mathbf{B_{IMF}}(t = 0) \neq 0$ and $\theta = 240°$.

field features, the tailward plasma flow increases and the plasma pressure increases. The magnetic pressure of the B_Y component is greater than the plasma pressure at the center of these island like features ($\beta = 0.22 - 0.28$). On the equator, the localized islands have a scale size of about 7R_E in both the X- and Y- directions. The plasma sheet forms many small islands. This is caused by the combination of multiple reconnection in the tail, pinch effects in the regions where the cross-tail current is separated and the kink instability in the current channels along the Y-direction.

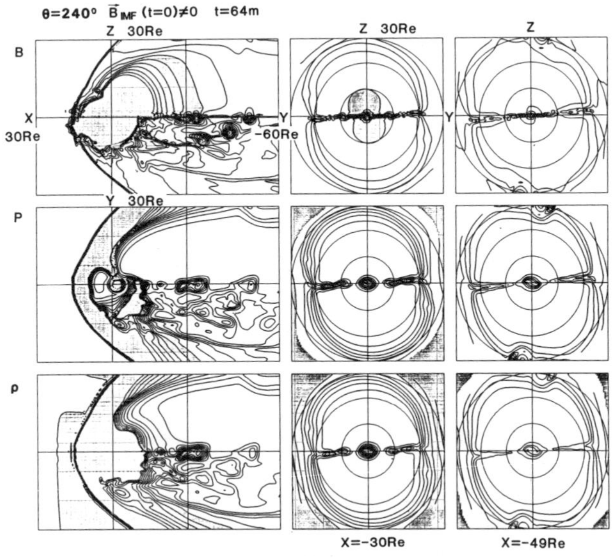

Fig. 13. The same as Figure 7 for the case with $\mathbf{B_{IMF}}(t = 0) \neq 0$ and $\theta = 240°$. Note that this time the cross sections in the Y-Z plane are at $X = -30\mathrm{R}_E$ and $X = -49\mathrm{R}_E$.

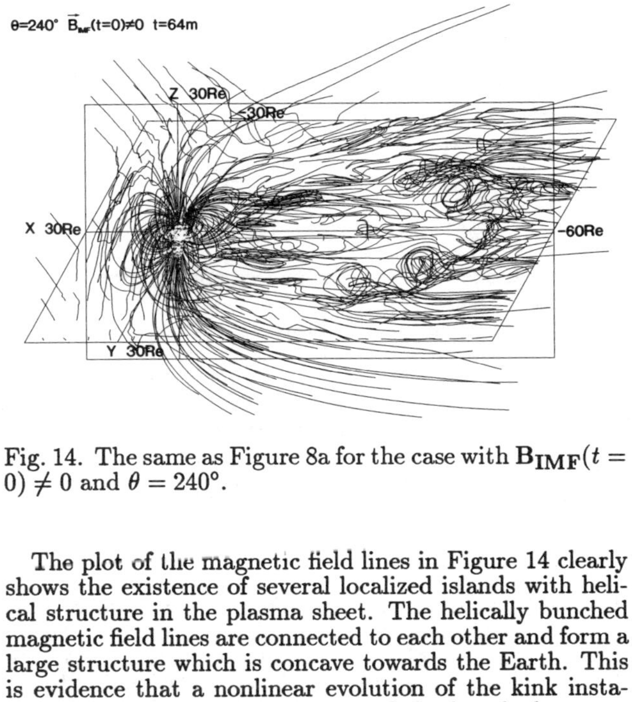

Fig. 14. The same as Figure 8a for the case with $\mathbf{B_{IMF}}(t = 0) \neq 0$ and $\theta = 240°$.

The plot of the magnetic field lines in Figure 14 clearly shows the existence of several localized islands with helical structure in the plasma sheet. The helically bunched magnetic field lines are connected to each other and form a large structure which is concave towards the Earth. This is evidence that a nonlinear evolution of the kink instability resulted in the occurrence of the bunched current channels in the Y-direction. The condition for the kink instability was satisfied when we estimated it from the longitudinal magnetic field B_Y and the longitudinal current, J_Y for the magnetic field islands in Figure 12. In this case the kink mode is unstable if the safety factor $g = L_Y J_Y / 2\pi B_Y \approx 0.07(L_Y/R_E) < 1$ where L_Y is a scale length in the Y- direction. The flux ropes in Figure 14 satisfy this condition for instability. These simulation results indicate that the magnetic flux ropes as well as the plasmoids might be a common feature in the plasma sheet if a southward IMF with finite B_Y can penetrate into the central plasma sheet.

4. Discussion and Conclusions

We have studied the formation and dynamics of magnetic flux tubes at the dayside magnetopause by using a time-dependent three-dimensional global MHD model with spatial resolution as high as $(\Delta X, \Delta Y, \Delta Z) = (0.5, 0.5. 0.5)\mathrm{R}_E$ and $(0.25, 0.4, 0.4)\mathrm{R}_E$. For southward IMF, we confirmed the results from lower resolution runs that two types of magnetic flux tubes were formed depending on the size of the IMF B_Y component. When B_Y was large, a strongly twisted and localized magnetic flux tube similar to magnetic flux ropes appeared in the subsolar region. This phenomenon originates from antiparallel merging at the high latitude magnetopause and the induced magnetopause convection [Ogino et al., 1989]. When the IMF B_Y component was small, twin flux tubes were found on the dayside magnetopause due to magnetic merging in the antiparallel region near the subsolar point.

We have drawn a cartoon showing the development and evolution of the twisted flux tube for the large B_Y case in panels a, b, and c of Figure 5. In the simulation the magnetic field lines near the magnetopause were distorted by viscous convection prior to the start of reconnection (see Figure 1 of Ogino et al. [1989]. In panel (a) magnetic reconnection has just started in the antiparallel field region where the IMF and the Earth's field are opposite (circled areas). The distortion of the field lines in the region between the two reconnection areas occurs because of a combination of the viscous convection and reconnection. As reconnection proceeds the distorted field lines just earthward of the reconnection region become twisted. The twisted flux tube is initially on closed field lines. That twisted flux tubes aren't found when the IMF is northward indicates that dayside reconnection is necessary for the generation of these twisted closed flux tubes. The reconnection enhances the distortion of the closed field lines by reducing the pressure in the reconnected areas.

In panel (b) the twisted flux tube has started to reconnect. The reconnection starts in both the northern and southern hemispheres when the twisted flux tube enters the antiparallel reconnection regions. When half of the flux tube has reconnected in both hemispheres the reconnection of the twisted flux tube stops as the newly reconnected flux tubes are convected tailward across the northern and southern magnetopauses (panel c).

The twisted flux tube in panel (c) is similar to that suggested by Lee and Fu [1985]. They suggested that component reconnection at multiple locations was responsible for the twisted structure. We have examined the flow and magnetic field pattern for evidence of component reconnection. Our examination indicates that while some component reconnection probably does occur, it is much less important for driving flows than is the antiparallel merging. In addition, the twisted flux tube initially forms on closed field lines a feature not found in the Lee and Fu model.

In this monograph, Kuznetsova and Zelinyi [1989] propose a stochastic percolation model for flux transfer events.

Their mechanism provides an alternative mechanism for the formation of the twisted bent closed flux tubes in Figure 4. They argue that some field lines start the percolation process but don't complete the formation of FTE's. When this happens the slowly convecting magnetosheath field lines will be bent similar to those in Figure 4.

The sequence of events which occurs when the IMF has a small B_y component are illustrated in panels (d),(e), and (f) of Figure 13. In this case, too, the dayside reconnection occurs in the antiparallel field region (panel d). Twin magnetic flux tubes form episodically at the magnetopause and convect over the poles (panels e and f). Our results for small B_Y are similar to those in the single neutral line calculation of Scholer [1988].

Using a semi-global simulation of a southward IMF interaction with the magnetosphere, Sato et al. [1986] proposed that twisted flux tubes could be formed by repeated reconnection between the IMF and field lines attached to the Earth (see their Figure 5). This does not appear to be a major effect in our results for small B_Y. (The small B_Y case is closest to the configuration studied by Sato et al.) There are two main differences between the global model and the semi-global calculation. First, there is no bow shock in the semi-global model and second the Sato et al. calculation did not include a B_Y component in the IMF. Both of these differences cause the configuration of the magnetic field lines in the magnetosheath to be different in the two cases. The almost straight IMF field lines and homogeneous sub-Alfvenic magnetosheath flow in the semi-global model caused the IMF field lines to meet the geomagnetic field for the first time at the subsolar point and then at higher latitudes. This lead to the repeated reconnection observed in the simulation. However in the global MHD model where we calculate the magnetosheath configuration self-consistently, we find that the effect of field line draping is to cause all parts of the IMF field lines to reach the boundary at almost the same time which probably does not allow the repeated reconnection to occur.

We also studied the corresponding magnetic reconnection in the magnetotail. In particular, we have modeled the structure of magnetic flux ropes and plasmoids during intervals when the IMF has both a southward and a dawnward component.

In Figure 15, we have sketched a simplified version of the magnetic field configuration in the tail for three initial IMF conditions. When the IMF was purely southward or had only a small B_Y, a symmetric plasmoid composed of closed loops of magnetic field lines formed in the noon-midnight meridian. This was due to the formation of a near-Earth neutral line (see Walker et al.,[1988] and Walker and Ogino, [1989b]). When the IMF had both a southward and a substantial dawnward component, a similar plasmoid was formed, however it had a finite B_Y component even in the noon-midnight meridian (panel b). The magnetic field lines had a helical structure connected from dawn to dusk and the B_Y component became larger than the other components near the center of the plasmoids. The helical field lines for the plasmoid were bundled at its edges where they connect to the ionosphere. Near the magnetopause the magnetic flux ropes were aligned along the X-direction. The width of the flux rope was about $6R_E$. The magnetic field increases while the plasma pressure decreases inside that part of the flux rope aligned with

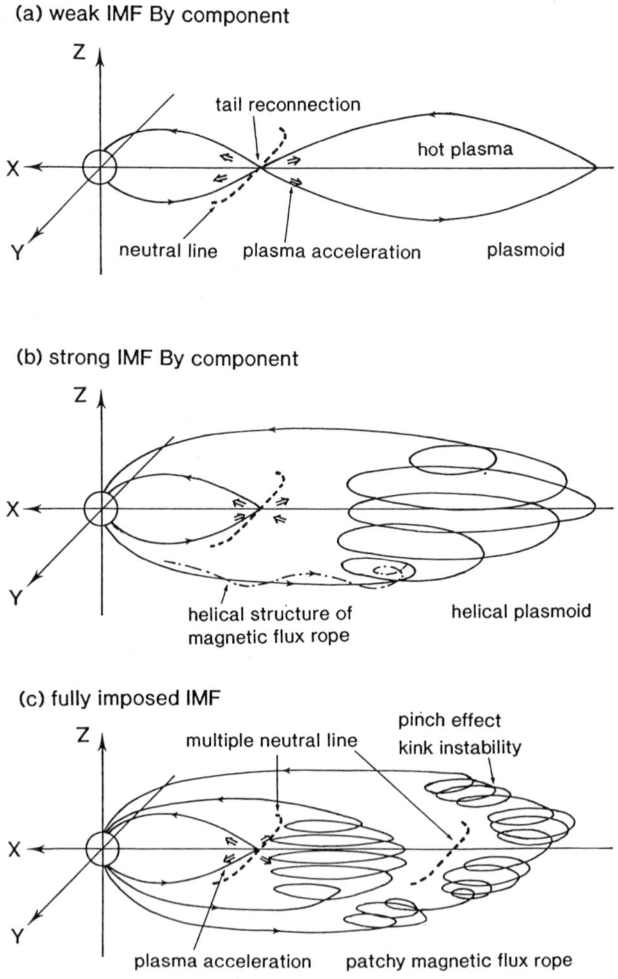

Fig. 15. A sketch of the magnetic configuration of plasmoids in the tail. In panel (a) the configuration for cases when the IMF has a southward component and only a small or no B_Y has been drawn. In panel (b) the configuration for the case when the IMF has a large B_Y in addition to a southward component is shown. In cases (a) and (b) the IMF initially entered the simulation box from the left edge. In panel (c) we have drawn the case which occurred when the southward and dawnward IMF initially was superimposed everywhere in the simulation box.

the X-axis. As time elapsed, the bundled and twisted field lines at the edge of the plasmoid became disconnected from the Earth and the plasmoid composed of helical field lines then propagated down the tail. The simulation results are consistent with the evolution of helical plasmoids proposed by Hughes and Sibeck [1987].

The magnetic configuration in the region of the plasmoid contains many topologically different types of field lines (Figures 8 and 15) (cf. Birn et al., 1989). At $t = 64$m (Figure 8a and 15b) most of the central core flux rope is connected at both ends to the Earth although at this time a few of the flux rope field lines have become connected to the IMF. The part of the central flux rope near midnight is overlaid with field lines both ends of which are attached to the IMF but closed field lines still drape the flux rope nearer to the magnetopause. By $t = 80$m (Figure 8b), the central flux rope has become completely attached to the IMF. Field lines with both ends attached to the IMF now drape the flux rope over almost its entire length.

When a uniform southward and dawnward IMF was imposed on the whole magnetosphere at the beginning of the simulation, multiple neutral lines occurred immediately in the thin plasma sheet (Figure 15, panel c). As a result, the cross tail current separated into several current channels in the X-direction. These were focused by the pinch effect. Next the kink instability worked in the current channels along the Y-direction and caused many small scale islands (patchy flux ropes) to form.

These global MHD simulation results indicate that magnetic flux ropes as well as plasmoids might be a common feature at the magnetopause and in the plasma sheet. However, we need further study to determine the amount of IMF B_Y component which can penetrate in the central plasma sheet and the rate at which it occurs.

Questions and Answers

Sibeck: Are flux ropes in your model force-free structures? What is their pitch?

Ogino: A force-free configuration is not formed at the early time when a helical flux rope is not well developed. However, a force-free configuration has been achieved after the helical flux-rope is completely formed. The pitch of helical field lines at the dayside magnetic flux tube is one to two turns from the dawn to dusk. This produces a total field-aligned current of 0.5MA. That of the tail magnetopause is also one to two turns from the edge of plasmoid to the ionosphere. In the helical plasmoids in the tail, the winding is the order of 10 times between the dawn and the dusk.

Schindler: Please explain whether your resistivity model depends explicitly on the spatial location. What is the relative importance of numerical resistivity?

Ogino: The resistivity in our model is $\eta = \eta_0 (T/T_0)^{-3/2}$ in the all calculation domain, where T/T_0 is the temperature normalized by its value in the ionosphere. In order to check our simulation, we changed the constant value of coefficient η_0 and also we used a constant resistivity in order to check the effect of $\nabla\eta$.

Birn: In your model, the B_y component in the plasmoid region is comparable to or even exceeds the applied solar wind B_y, whereas observations indicate that only 10% or less of the IMF B_y enters the tail. How do you think this affects your results?

Ogino: We used a resistivity model of $\eta = \eta_0 (T/T_0)^{-3/2}$ and η_0 is changed over a wide range of $0.1 \geqq \gamma_0 \geqq 0$. In the previous calculation, we used $\eta_0 = 0.002$, then B_y was less than 20% of the IMF B_y. We want to understand what percent of the IMF can impose into the center plasma sheet and the ratio can now be controlled in our model. If the IMF B_y enters the tail less, the pitch of helical plasmoids shrinks much more and the feature is similar to that of plasmoids for a pure southward IMF.

Zelenyi: Does the formation of closed bent flux tubes on the dayside (very evident in your pictures) depend on your assumptions about resistivity?

Ogino: The formation of closed bent flux tubes on the dayside magnetopause basically originates from magnetic merging at the high latitude magnetopause and the following magnetopause convection. As a result, the geomagnetic field lines in the throat region were strongly deformed toward the antiparallel field region. And, we found a similar flux tube for a constant resistivity. Therefore, we believe that the fundamental feature of the flux tube formation does not change so much by choosing a proper resistivity model.

Acknowledgements. This work was supported by a Grant-in-Aid for Science Research from the Ministry of Education, Science and Culture, by the NASA Solar Terrestrial Theory Program Grant NAGW-78 and by NASA Grant NGL-05-007-004. Computing support for our simulations was provided by the Computer Center of Nagoya University and by the Computer Center of the Institute of Space and Astronautical Science and by the San Diego Supercomputing Center.

References

Birn, J., M. Hesse and K. Schindler, Filamentary Structure of a Three-Dimensional Plasmoid, *J. Geophys. Res*, 94, 241, 1989.

Elphic, R. C., and C. T. Russell, ISEE-1 and -2 magnetometer observations of the magnetopause, in *Magnetospheric Boundary Layers*,43, ESA Scientific and Technical Publications, Noordwijk, The Netherlands, 1979.

Elphic, R. C., C. A. Cattell, K. Takahashi, S. J. Bame, and C. T. Russell, ISEE-1 and -2 observations of magnetic flux ropes in the magnetotail: FTE's in the plasma sheet?, *Geophys. Res. Lett.*, 13, 648, 1986.

Galeev, A. A., M. M. Kuznetsova, and L. M. Zelenyi, Magnetopause stability threshold for patchy reconnection, *Space Sci. Rev.*, 44, 1, 1986.

Hones, E. W., Jr., D. N. Baker, S.J. Bame, W. C. Feldman, J. T. Gosling, D. J. McComas, R. D. Zwickl, J. A. Slavin, E. J. Smith, and B. T. Tsurutani, Structure of the magnetotail at 220R_E and its response to geomagnetic activity, *Geophys. Res. Lett.*, 11, 5, 1984.

Hughes, W. J., and D. G. Sibeck, On the 3-dimensional structure of plasmoids, *Geophys Res. Lett.*, 14, 636, 1987.

Kuznetsova, M. M., and L. M. Zelenyi, The theory of FTE Stochastic Percolation Model, *The Physics of Magnetic Flux Ropes*, *Geophys. Monogr.*, in press, 1989.

La Belle-Harmer, A. L., Z. F. Fu, and L. C. Lee, A mechanism for patchy reconnection at the dayside magnetopause, *Geophys. Res. Lett.*, 15,152, 1988.

Lee, L. C., and Z. F. Fu, A theory of magnetic flux transfer at the Earth's dayside magnetopause, *Geophys. Res. Lett.*, 12, 105, 1985.

McPherron, R. L., C. T. Russell, and M. P. Aubry, Satellite studies of magnetospheric substorms on August 16, 1968, 9, Phenomenological model for substorms,*J. Geophys. Res.* , 78, 3131, 1973.

Ogino, T., A three dimensional MHD simulation of the interaction of the solar wind with the Earth's magnetosphere: The generation of field aligned currents,*J. Geophys. Res.*, 91, 6791, 1986.

Ogino, T., R. J. Walker, and M. Ashour-Abdalla, A magnetohydrodynamic simulation of the formation of magnetic flux tubes at the Earth's dayside magnetopause, *Geophys. Res. Lett.* , 16, 155, 1989.

Rijnbeek, R. P., S. W. H. Cowley, D. J. Southwood, and C. T. Russell, A survey of dayside flux transfer events observed by ISEE-1 and -2 magnetometers, *J. Geophys. Res.*, 89, 786, 1984.

Russell, C. T., and R. C. Elphic, Initial ISEE magnetometer results: Magnetopause observations, *Space Sci. Rev.*, 22, 681, 1978.

Russell, C. T., and R. C. Elphic, ISEE observations of flux transfer events at the dayside magnetopause, *Geophys. Res. Lett.* 6, 33,1979.

Sato, T., T. Shimada, M. Tanaka, T. Hayashi, and K. Watanabe, Formation of field twisting flux tubes on the magnetopause and solar wind particle entry into the magnetosphere, *Geophys. Res. Lett.*, 13, 801, 1986.

Saunders, M. A., C. T. Russell, and N. Sckopke, Flux transfer events: Scale size and interior structure, *Geophys. Res. Lett.*, 11, 131, 1984.

Scholer, M., Magnetic flux transfer at the magnetopause, *Geophys. Res. Lett.*, 15, 291, 1988.

Shi, Y., C. C. Wu, and L. C. Lee, A study of multiple X line reconnection at the dayside magnetopause, *Geophys. Res. Lett.*, 15, 295, 1988.

Sibeck, D. G., G. L. Siscoe, J. A. Slavin, E. J. Smith, S. J. Bame, and F. L. Scarf, Magnetotail flux ropes, *Geophys. Res. Lett.*, 11, 1090, 1984.

Sibeck, D. G., G. L. Siscoe, J. A. Slavin, E. J. Smith, B. T. Tsurutani, and S. J. Bame, Magnetic field properties of the distant magnetotail magnetopause and boundary layer, *J. Geophys. Res.*, 90, 9561, 1985.

Southwood, D. J., C. J. Farrugia, and M. A. Saunders, What are flux transfer events?, *Planet. Space Sci*, 92, 8613,1987.

Walker, R. J., and T. Ogino, The formation of isolated magnetic flux tubes on the dayside magnetopause, *Electromagnetic Coupling in the Polar Clefts and Caps*, Kluwer Academic Publishers, Dordrecht, The Netherlands, 11, 1989a.

Walker R. J., and T. Ogino, Global magnetohydrodynamic simulations of the magnetosphere, *IEEE Transactions on Plasma Science*, 17, 2, 135, 1989b.

Walker R. J., T. Ogino, and M. Ashour-Abdalla, A global magnetohydrodynamic model of magnetospheric substorms, *Physics of Space Plasmas*, Scientific Publishers, Cambridge, MA, SPI Conference Proceedings an Reprint Series, 7, 235,1988.

PARALLEL ELECTRIC FIELDS IN A SIMULATION OF MAGNETOTAIL RECONNECTION AND PLASMOID EVOLUTION

M. Hesse and J. Birn

University of California, Los Alamos National Laboratory, Los Alamos, NM 87545

Abstract. We investigate properties of the electric field component parallel to the magnetic field ($E_{\|}$) in a three-dimensional MHD simulation of plasmoid formation and evolution in the magnetotail in the presence of a net dawn-dusk magnetic field component. We emphasize particularly the spatial localization of $E_{\|}$, the concept of a diffusion zone and the role of $E_{\|}$ in accelerating electrons. We find a localization of the region of enhanced $E_{\|}$ in all space directions with a strong concentration in the z direction. We identify this region as the diffusion zone, which plays a crucial role in reconnection theory through the local break-down of magnetic flux conservation. The presence of B_y implies a north-south asymmetry of the injection of accelerated particles into the near-earth region, if the net B_y field is strong enough to force particles to follow field lines through the diffusion zone. We estimate that for a typical net B_y field this should affect the injection of electrons into the near-earth dawn region, so that precipitation into the northern (southern) hemisphere should dominate for duskward (dawnward) net B_y. In addition, we observe a spatial clottiness of the expected injection of adiabatic particles which could be related to the appearance of bright spots in auroras.

1. Introduction

A major characteristic of the magnetic reconnection process is the generation or existence of a localized region where the ideal MHD condition $\mathbf{E} + \mathbf{v} \times \mathbf{B} = 0$ breaks down, so that the global connection of plasma elements through magnetic field lines can change as these elements move through the system [Axford, 1984; Schindler et al., 1988; Hesse and Schindler, 1988]. In idealized configurations, and in particular if symmetries are present, such a region, commonly called the "diffusion zone", is associated with an X-type magnetic field topology [see, e.g., Vasyliunas, 1975] and with a magnetic null-line, or, more generally, a separator line defined by the intersection of separatrix surfaces between topologically distinct magnetic regions [e.g., Sonnerup et al., 1984]. In the general three-dimensional case, however, the presence of magnetic nulls, neutral lines or separators is not necessary [Schindler et al., 1988]. The remaining characteristic of the diffusion region is therefore the electric field property, which requires in particular the localized presence of an electric field component parallel to the magnetic field for global changes of the magnetic connection [Schindler et al., 1988; Hesse and Schindler, 1988]. In the idealized field geometries this parallel electric field in the diffusion region can be identified with an $E_{\|}$ along a separator or, if a magnetic neutral X-line exists, with an electric field along this X-line [Sonnerup et al., 1984].

In this paper we will investigate the properties of the diffusion zone, and in particular its spatial extent and the consequences for adiabatic particle acceleration, on the basis of a three-dimensional MHD simulation of magnetic reconnection in a geomagnetic tail configuration without magnetic nulls. An overview over the magnetic field evolution and the characteristic topology changes is given in this issue [Birn and Hesse, 1989a]. Further details are described elsewhere [Birn and Hesse, 1989b]. The evolution starts from a self-consistent tail equilibrium after Birn [1989], which includes a net dawn-dusk magnetic field component B_{yN}, that breaks the symmetry around the midnight meridian and the equatorial planes. Despite the strong influence of this field component on field line structures and topological connections, the evolution of the system remains similar to the symmetric case, showing plasmoid formation and ejection, starting at about 100 Alfvén times.

The calculations were performed with dimensionless quantities, normalized by the characteristic lobe magnetic field strength B_c, the characteristic scale length L_c for variations in the z direction, and the characteristic Alfvén speed V_c, based on the lobe magnetic field strength and the plasma sheet density, and by suitable combinations of these units, all defined at the near-Earth boundary of the simulation box $x = 0$. For illustration, we will use dimensional units $B_c = 40$nT, $L_c = 12000$km$\approx 2R_E$, and $V_o = 1000$km/s. This leads to a time unit $t_c = 12$s and an electric field unit $E_c = 20$mV/m.

2. Enhancements of Parallel Electric Fields

Using a general concept of magnetic reconnection, Schindler et al. [1988] and Hesse and Schindler [1988] found that in systems without zeros of $\mathbf{B}$, magnetic reconnection with global effects requires the presence of parallel electric fields in the diffusion zone with

$$\Phi = \int_{\text{fieldline}} E_{\|}\, ds \neq 0 \qquad (1)$$

where the integration extends along a field line through the diffusion zone, and s is the arc length.

Geophysical Monograph 58

In view of this theory the question arises, whether the simulations of plasmoid evolution show the presence of a diffusion zone with this property. In the simulation, the resistivity η was chosen to be constant. This might suggest that a possible reconnection region, defined by the spatial extent of $E_{\parallel}$ enhancements, were spread all over the simulation box rather that being localized. It is therefore of interest to investigate the spatial distribution of $E_{\parallel}$ during the simulation run.

We start our analysis with the investigation of the temporal variation of $E_{\parallel}$ during the simulation, looking first at the variation of $E_{\parallel}$ in the center of the plasma sheet, where the maximum $E_{\parallel}$ is found. Figure 1 shows the variation of $E_{\parallel}$ along the x axis of the simulation box at different times. There is an obvious concentration of $E_{\parallel}$ with a peak at about $x = -10$, which starts immediately and saturates even before the plasmoid has started to form. Note that the maximum values attained by $E_{\parallel}$ do not vary much with time. This implies that some other, possibly geometric, effect must be responsible for generating sufficiently high values of Φ for reconnection. We shall return to this problem in Section 3.

To demonstrate further the extent of the region of enhanced $E_{\parallel}$ Figure 2 shows as an example the contour lines of $E_{\parallel}$ in the equatorial plane (a) and in the midnight-meridional plane (b) at $t = 165$ (time units normalized to typical Alfvén crossing times). The shaded area in Figure 2a represents the region where $E_{\parallel}$ exceeds 50 % of its maximum value. The extent of this region in the y direction is approximately given by $\Delta y \approx 10$, about half of the width of the simulation box. An estimate of the extent in the z direction can be obtained from Figure 3, which shows the variation of $E_{\parallel}$ and of the cross-tail current density j_y (in dimensionless units) with z at $t = 165$. The figure shows the concentration of j_y due to the compression of

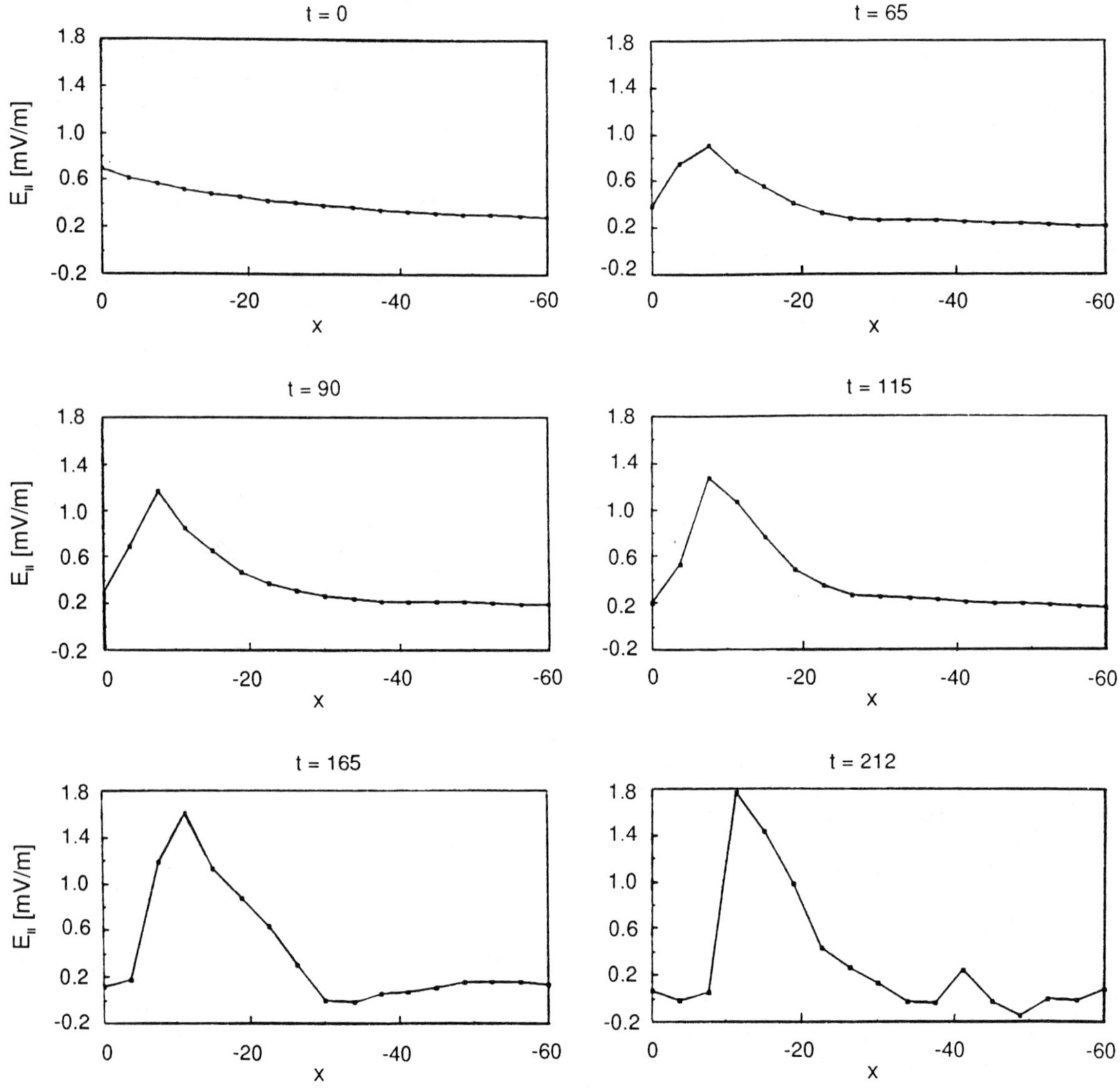

Fig. 1. Variation of the electric field component parallel to the magnetic field along the x axis at different times. The times are normalized to typical Alfvén crossing times and the electric field unit is 0.2mV/m. The plots show a strong localization of $E_{\parallel}$ around $x = -10$.

the plasma sheet in the reconnection region (Note that the z coordinate is normalized by the initial scale length of the current sheet), and the even stronger concentration of $E_{\|}$ in the center of the current sheet. We estimate the width of the region where $E_{\|}$ exceeds about 50% of its maximum value to $\Delta z \approx 0.1$, so

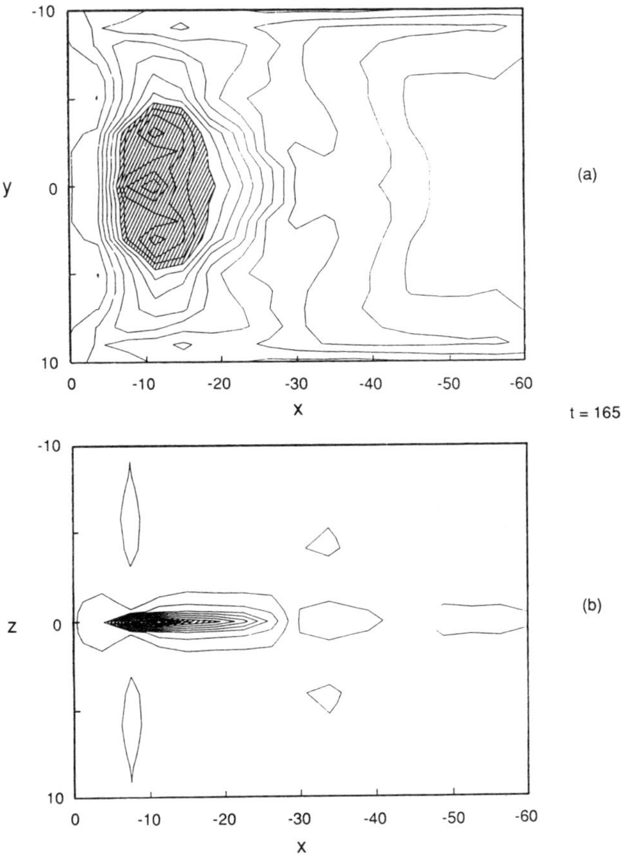

Fig. 2. Contour lines of the parallel electric field in the equatorial plane (a) and the midnight meridional plane (b) at $t = 165$ Alfvén times. The shaded areas indicate the region where $E_{\|}$ exceeds one half of its maximum value.

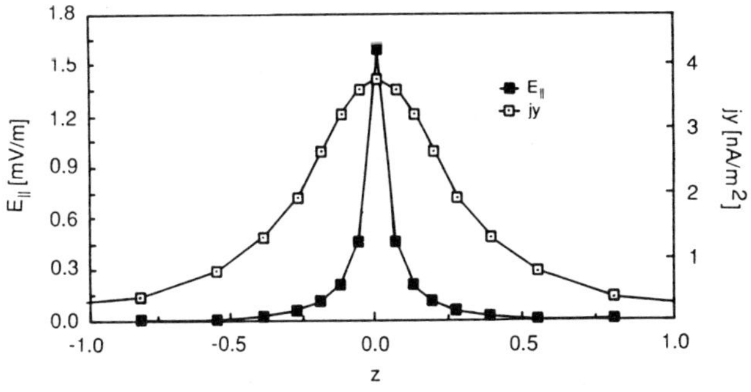

Fig. 3. Parallel electric field $E_{\|}$ and cross-tail current density j_y as functions of z for $y = 0$ and $x = -11.25$ at $t = 165$.

that the ratio of the volume of the enhanced $E_{\|}$ region to the entire volume is

$$\frac{\Delta x \Delta y \Delta z}{V} \approx 6 \times 10^{-4} \tag{2}$$

for this particular time.

The relative volumina of the region of enhanced $E_{\|}$ at times starting shortly before the plasmoid forms are plotted versus time in Figure 4. Obviously, the relative size of this region, which we identify as the diffusion region, stays rather small at all times considered. Note that the limitation in the x and y direction and the strong concentration in z take place although the resistivity η is constant.

The fact that the volume of the diffusion zone is nonzero, implies that not a single field line but an entire rope of finite magnetic flux passes through it and exhibits enhanced $E_{\|}$ with possible consequences for acceleration of particles along these field lines.

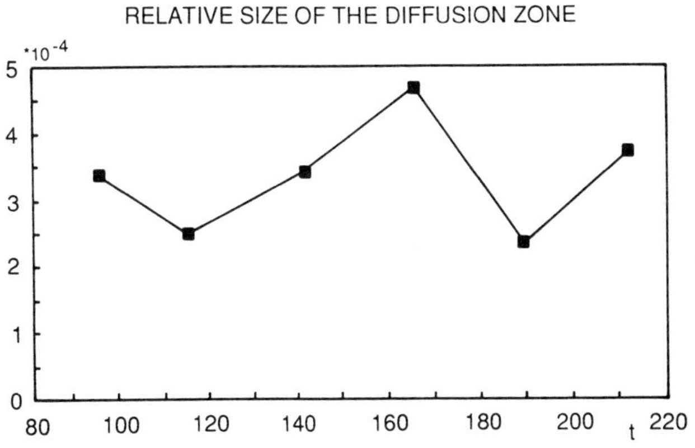

Fig. 4. Relative volume of the diffusion region (enhanced $E_{\|}$) as a function of time. Note that this volume is very small during the entire simulation run, which justifies the concept of a localized diffusion zone.

3. Particle Acceleration

The quantity relevant for the acceleration along a field line is the integrated potential difference Φ, given by (1). Thus we extend our analysis to an integration of $E_{\|}$ along the field lines to find the distribution and the maximum values of Φ during the simulation run. As an example, Figure 5a shows a typical spiralling field line, belonging to the plasmoid, at $t = 165$. Figure 5b shows the variation of Φ along that specific field line. Note that there is a strong variation close to $s = 0$, i.e., close to the center of the field line, located on the x axis at $x = -15$. The rather weak variation of Φ outside this central region is due to some field aligned currents at the plasma sheet boundary of the model, associated with parallel electric fields through the assumption of constant resistivity. The total voltage drop on this field line corresponds to about 14.4kV.

The temporal variation of the maximum values of Φ during the simulation run is plotted in Figure 6. It is obvious that the maximum values of Φ increase rapidly in a time interval, corresponding to less than three minutes, during the plasmoid formation phase. At later times, while flux is still added to the

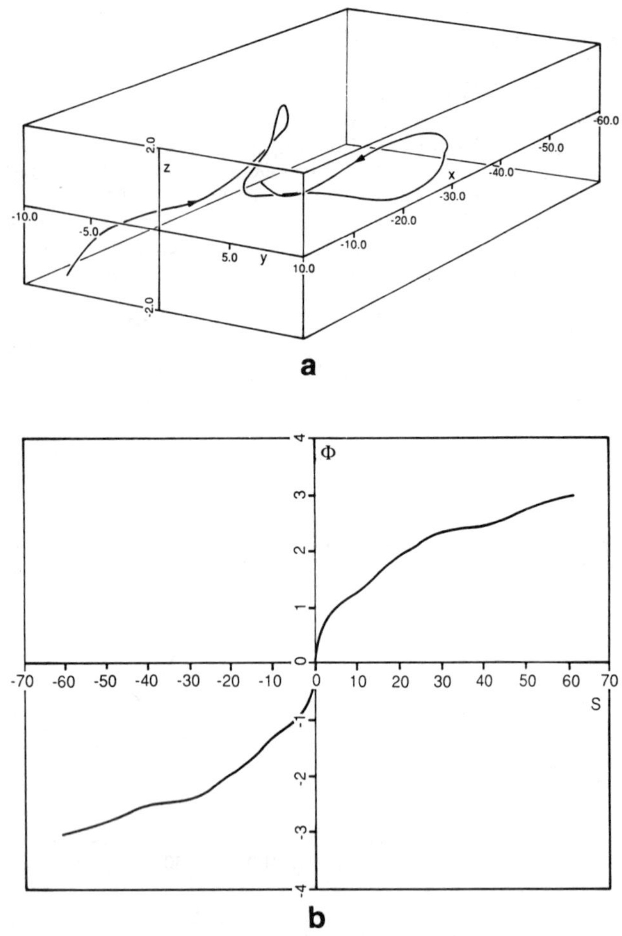

Fig. 5. (a) Typical spiralling plasmoid field line at time $t = 165$ Alfvén times. This field line was chosen to pass through the x axis at $x = -15$. As it is typical for plasmoid field lines at this stage of the evolution, it connects to the boundary closest to Earth with both ends. (b) The lower panel shows the variation of Φ with arc length s along the field line in (a). The location of $s = 0$ corresponds to the x axis at $x = -15$. Note that in the central region there is a steep increase in Φ due to the diffusion zone while the slow variation along the outer parts of the field line are due to some residual field aligned currents and the assumed constant resistivity.

plasmoid, the maximal Φ stays rather constant, near an average value of about 19kV. Note that the steep increase during the formation time occurs although the maximum of $E_{\parallel}$ does not vary much. This implies that the three-dimensional tearing mode changes the magnetic field geometry such that a set of field lines is picking up enough "field aligned potential drop" to allow reconnection to proceed. This property can also be seen from the form of the field line in Figure 5a. The part of that line of force closest to the x axis almost lies in the equatorial plane, due to the the fact that $\mathbf{B} \approx B_y e_y$ there such that it picks up a rather high value of $\int E_{\parallel}\, ds$.

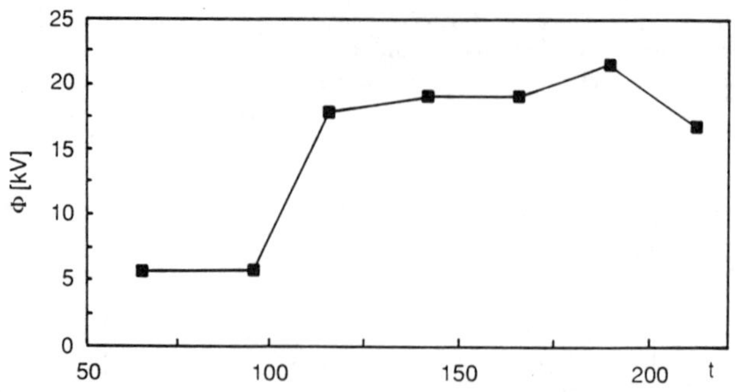

Fig. 6. Temporal variation of the maximum values of $\Phi = \int E_{\parallel} ds$. At early times, the graph shows a steep increase in the maximum values, indicating that the three-dimensional tearing mode changes the geometry of the system in order to accommodate sufficiently high values of Φ for magnetic reconnection. After the formation of the plasmoid, the highest values of Φ stay around 19kV due to the fact that magnetic flux is continuously added to the plasmoid structure.

4. Electron Injection Patterns

Apart from the very late times, all plasmoid field lines in the present simulation connect with both ends to the boundary closest to Earth, as shown in the example of Figure 5a. While it seems obvious that a concentration of $E_{\parallel}$ can act as a particle accelerator, up to energies of more than 10keV, it is not clear that particles can follow field lines after acceleration. Using the minimum magnetic field in the diffusion region of about 1nT, an electron energy of 15keV, and a plasmoid field line half length of $140R_E$ it is straightforward to estimate an electron Larmor radius to be less than $0.8R_E$, an $\mathbf{E} \times \mathbf{B}$ drift velocity to be at most of the order of 200km/s, and a typical travel time for an electron to the $x = 0$ plane of about 20s. This time is much smaller than the evolution time of the system.

Therefore, a considerable part of the accelerated electrons can follow field lines to the end of the simulation box closest to the Earth. In order to investigate the distribution of these accelerated electrons at this boundary, we map the concentrations of field aligned voltage drops along the magnetic field to the $x = 0$ plane. Figure 7 displays the contour lines resulting from this procedure for six different times of the simulation. The region of enhanced electron precipitation lies in the southern hemisphere for westward net B_y chosen in this simulation, and preferably on the dawn side. The regions bounded by the contour lines in the northern half-plane are not electron precipitation regions due to the direction of $E_{\parallel}$.

At t=90 Alfvén times the distribution of enhanced Φ is still rather broad (panel (a)) with a maximum Φ of about 5kV. At later times (panels (b)–(f)), one can see an increasing concentration of the regions of field lines carrying higher potential drops. Finally, at times greater than about 190 Alfvén times, only a very narrow clumpy region of the boundary plane is intersected by field lines with high potential drops. Assuming that these regions are actually regions of electron precipitation, we conclude that the temporal evolution of magnetotail reconnection may lead to an enhanced precipitation of energetic electrons in localized regions which decrease in spatial extent, resembling bright spot patterns as observed in the aurora.

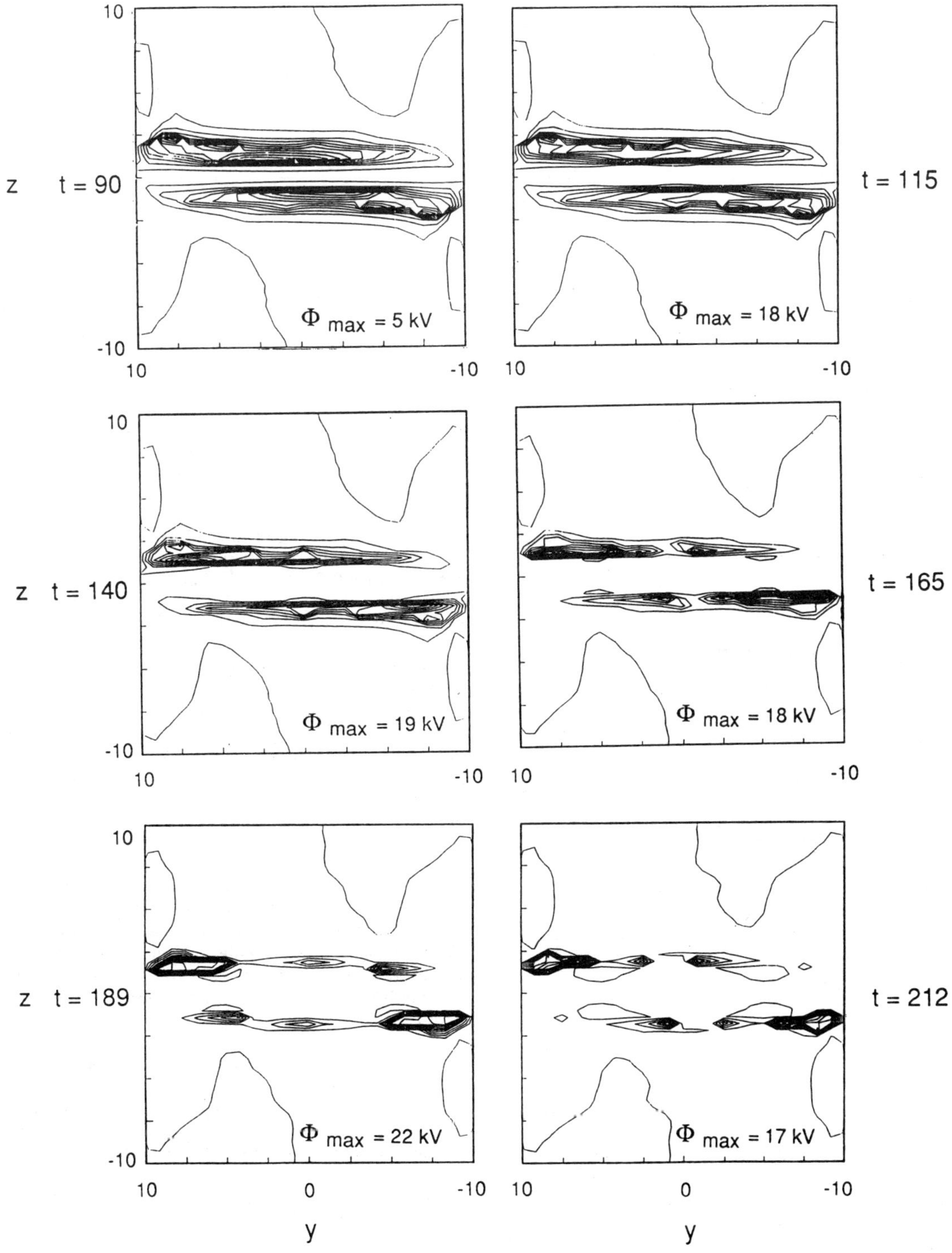

Fig. 7. Contours of constant $\Phi = \int E_{\parallel} ds$, integrated along field lines crossing the near-Earth boundary $x = 0$ of the simulation box, for positive net cross-tail magnetic field B_{yN} and different times indicated in the panels. The regions of dense contours indicate the regions of enhanced Φ at the boundary with the maximum values indicated in each panel. For the chosen sign of B_{yN} the parallel electric in the diffusion zone would accelerate electrons towards the southern hemisphere. Note that the plasmoid evolution causes a contraction of the region of enhanced Φ, from an initially rather broad diffuse pattern with rather low maximum values of Φ to a set of sharply bounded small regions with comparatively high maximum values of Φ.

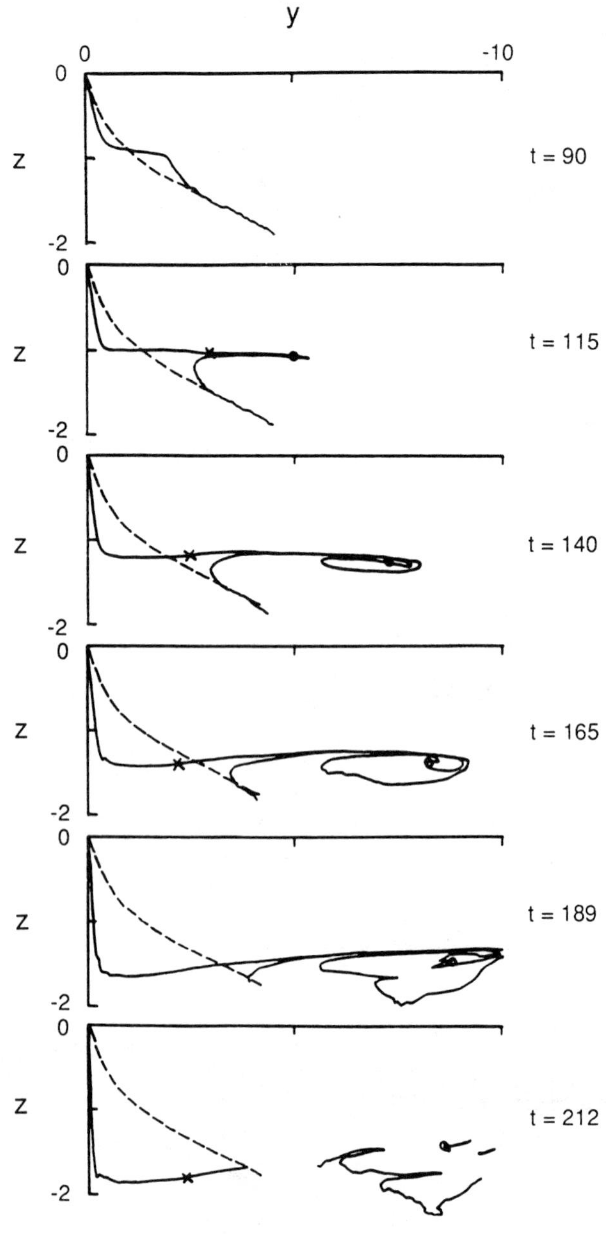

Fig. 8. Footpoints at the $x = 0$ plane of field lines intersecting the x axis, for different times indicated in the panels (solid lines) and for the initial equilibrium (dashed lines). The symbols X and O mark the points that are connected with the points $B_z = 0$ on the x axis, which in the absence of the net B_y would represent X- and O-points, respectively. These points indicate which regions are connected with the plasmoid. Note that the footpoints of such plasmoid field lines show a rather large separation in the y direction, as a consequence of reconnection. The folded parts of the footpoint lines coincide with the regions of high Φ in Fig. 7. The panels for the later times display an increased slope of the footpoint curves near $y = 0$ and an increase of the maximum $|z|$ value attained in this region, corresponding to a reduction of B_y and a dipolarization of the near-Earth field, associated with an expansion of the closed field line region to higher latitudes.

One way to relate the position of the enhanced parallel potential region to the dynamical changes of the magnetic field is to investigate the "footpoints" of the plasmoid field lines. By footpoints of a field line we mean the points where the field line intersects the $x = 0$ plane. Here we concentrate on those field lines that connect to the x axis. Their footpoints are displayed in Figure 8 for different times (solid lines) compared to the initial case (dashed lines). Only the curves in the regions $y \leq 0$, $z \leq 0$ are shown. The rest of the curves in the region $y > 0$, $z > 0$ follows from the assumed line symmetry, so that each footpoint at $y = -y_0$, $z = -z_0$ has a corresponding footpoint at $y = y_0$, $z = z_0$, connected through the same field line. Since our boundary conditions, discussed in the companion paper, assume $\mathbf{v} = 0$ at the boundary $x = 0$, the temporal change of the footpoint curves in Figure 8 demonstrates the consequences of reconnection, that is a change of the connection of plasma elements by field lines. The symbols X and O in Figure 8 mark the footpoints connected with points where B_z vanishes along the x axis, which in the absence of the net B_y would be identified as X- and O-type neutral points. This indicates that the most distorted parts of the footpoint curves are indeed connected to the plasmoid region. A comparison with Figure 7 reveals that these regions also closely match the main regions of enhanced parallel potential. The other enhancement regions in Figure 7 are apparently consisting of footpoints of different type field lines. For illustration, three different types of such field lines are plotted in Figure 9 for $t = 212$. The figure shows that high values of Φ exist not only on typical plasmoid field lines (solid line), but also on some open magnetic field lines (dashed line) and on closed field lines that cross the equatorial plane just once near the sides of the plasmoid flux rope (dotted line). High values of Φ on field lines of this kind can be interpreted as indications of magnetic reconnection on the flanks of the plasmoid.

The expected location of the precipitation region for adiabatic or nearly adiabatic particles relative to the equator depends on the sign of the net B_y (in the southern hemisphere for positive and in the northern for negative B_y). The electric field direction implies that the major part of the precipitation region of these electrons lies on the dawn side of the midnight meridional plane.

The ion behavior is impossible to predict without actually integrating particle orbits in the self-consistent fields. This is due to the fact that ions become nonadiabatic in the central plasma sheet region. Orbit calculations in ad-hoc field models [e.g., Lyons and Speiser, 1982] indicate that such particles should become injected from the neutral sheet into both hemispheres with about equal probability, so that no north-south asymmetry is to be expected. The same result should hold for electrons if the net B_y is so small that they also become non-adiabatic in the diffusion region.

While our present results suggest a north-south asymmetry of electron precipitation on the dawn side in connection with substorms, one must be aware that this property may be masked, or at least be superposed, by other influences on the electron precipitation pattern, such as electrostatic double layer structures [e.g., Block, 1984; Reiff et al., 1988]. Electrostatic structures above the ionosphere, however, usually exhibit potential drops of the order of only 5kV. Thus, our results indicate that there may be localized precipitation of electrons which have been accelerated my magnetic reconnection in the magnetotail.

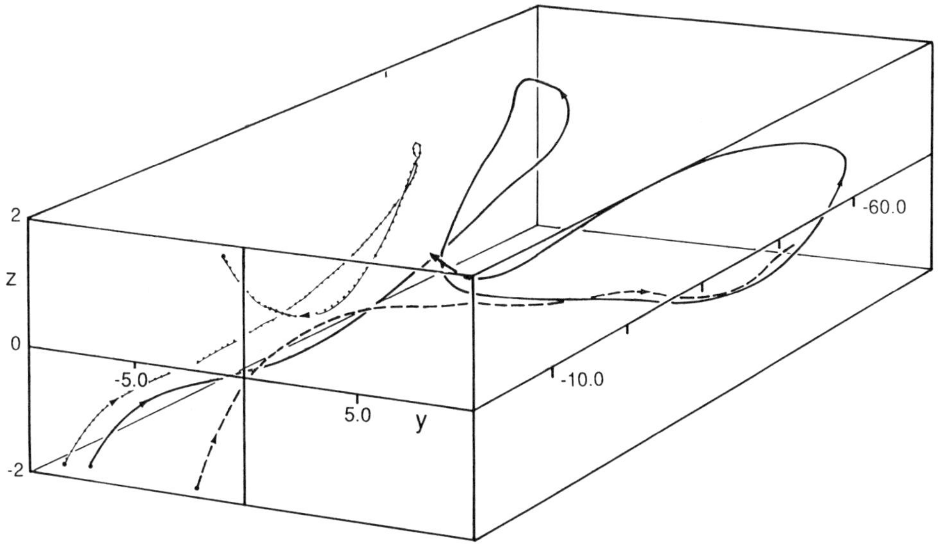

Fig. 9. Typical field lines ending in the regions of high Φ at $t = 212$. The majority of field lines passing through the diffusion zone are spiralling plasmoid field lines (solid line), but high values of Φ can also be found on open lobe-like field lines (dashed line) and on the flanks of the plasmoid, on distorted closed field lines (dotted line).

Acknowledgments. This work was supported by the U.S. Department of Energy through the Office of Basic Energy Sciences and by NASA. M. H. gratefully acknowledges support from the Los Alamos National Laboratory Director's postdoctoral program.

References

Axford, W. I., Magnetic Reconnection, in *Magnetic Reconnection in Space and Astrophysical Plasmas, Geophys. Monogr. Ser., vol. 30,* edited by E. W. Hones, Jr., p. 1, AGU, Washington D.C., 1984.

Birn, J., The distortion of the magnetotail equilibrium structure by a net cross-tail magnetic field, *J. Geophys. Res*, in press, 1989.

Birn, J., and M. Hesse, The magnetic topology of the plasmoid flux rope in a MHD simulation of magnetotail reconnection, this issue, 1989a.

Birn, J., and M. Hesse, MHD simulations of reconnection in a skewed three-dimensional tail configuration, submitted to *J. Geophys. Res.*, 1989b.

Birn, J., M. Hesse and K. Schindler, Filamentary structure of a three-dimensional plasmoid, *J. Geophys. Res., 94*, 241, 1989.

Block, L., Three-dimensional potential structure associated with Birkeland currents, in *Magnetospheric Currents, Geophys. Monogr. Ser., vol. 28,* edited by T. A. Potemra, p. 315, AGU, Washington D.C., 1984.

Hesse, M., and K. Schindler, A theoretical foundation of general magnetic reconnection, *J. Geophys. Res., 93*, 5559, 1988.

Lyons, L. R., and T. W. Spiser, Evidence for current sheet acceleration in the geomagnetic tail, *J. Geophys. Res., 88*, 2276, 1982.

Reiff, P. H., H. L. Collin, J. D. Craven, J. L. Burch, J. D. Winningham, E. G. Shelley, L. A. Frank, and M. A. Friedman, Determination of auroral electrostatic potentials using high- and low-altitude particle distributions, *J. Geophys. Res., 93*, 7441, 1988.

Schindler, K., M. Hesse and J. Birn, General magnetic reconnection, parallel electric fields and helicity, *J. Geophys. Res., 93*, 5547, 1988.

Sonnerup,B. U. Ö, et al., in *Solar Terrestrial Physics: Present and Future*, edited by D. M. Butler and K. Papadopoulos, *NASA Ref. Publ. 130*, Washington, D.C., 1984, ch. 1

Vasyliunas, V. M., Theoretical models of magnetic field line merging, 1, *Rev. Geophys. Space Phys., 13*, 303, 1975.